UNSATURATED SOILS FOR ASIA

PROCEEDINGS OF THE ASIAN CONFERENCE ON UNSATURATED SOILS
UNSAT-ASIA 2000/SINGAPORE/18 – 19 MAY 2000

Unsaturated Soils for Asia

Edited by

H. Rahardjo, D.G. Toll & E.C. Leong
Nanyang Technological University, Singapore

Taylor & Francis
Taylor & Francis Group

LONDON AND NEW YORK

Organised by the NTU-PWD Geotechnical Research Centre, Nanyang Technological University, Singapore, with the support of the Technical Committee on Unsaturated Soils (TC6) of the International Society of Soil Mechanics and Geotechnical Engineering.

Cover: The photograph shows a triaxial test set-up for unsaturated soil tests
at Nanyang Technological University.

The texts of the various papers in this volume were set individually by typists under the supervision of each of the authors concerned.

Authorization to photocopy items for internal or personal use, or the internal or personal use of specific clients, is granted by Taylor & Francis, provided that the base fee of US$ 1.50 per copy, plus US$ 0.10 per page is paid directly to Copyright Clearance Center, 222 Rosewood Drive, Danvers, MA 01923, USA. For those organizations that have been granted a photocopy license by CCC, a separate system of payment has been arranged. The fee code for users of the Transactional Reporting Service is: 90 5809 139 2/00 US$ 1.50 + US$ 0.10.

Published by Taylor & Francis
2 Park Square, Milton Park, Abingdon, Oxon, OX14 4RN
270 Madison Ave, New York NY 10016

Transferred to Digital Printing 2007

ISBN 90 5809 139 2
© 2000 Taylor & Francis

Publisher's Note
The publisher has gone to great lengths to ensure the quality of this reprint
but points out that some imperfections in the original may be apparent

Unsaturated Soils for Asia, Rahardjo, Toll & Leong (eds) © 2000 Taylor & Francis, ISBN 90 5809 139 2

Table of contents

3   *Numerical modelling*

3.1 Modelling of flow

3.2 Modelling of stresses and deformations

4 *Suction measurement techniques*

8   Strength and deformation

9 *Volume change*

9.1 Collapse

9.2 Shrinkage and swelling

10 *Engineering applications*

10.1 Foundations and roads

10.2 Slopes

Unsaturated Soils for Asia, Rahardjo, Toll & Leong (eds) © 2000 Taylor & Francis, ISBN 90 5809 139 2

Preface

The Asian Conference on Unsaturated Soils is organised by the NTU-PWD Geotechnical Research Centre, Nanyang Technological University, Singapore, with the support of the Technical Committee on Unsaturated Soils (TC6) of the International Society of Soil Mechanics and Geotechnical Engineering. This publication contains the papers presented at this conference. Although there have been several international and other regional conferences on unsaturated soils, this is the first Asian Conference on Unsaturated Soils. It is appropriate to organise this conference in Asia where abundant 'problematic soils' exist such as swelling soils, collapsible soils and residual soils. Common to these soils is their unsaturated nature and their problematic behaviour that arises when these soils come in contact with water. They are called problematic soils because their behaviour cannot be explained using conventional saturated soil mechanics. Unsaturated soil mechanics has developed to the stage where the theoretical frameworks, the required equipment and measuring techniques are readily available to be put into practice. Therefore, the theme 'From Theory to Practice' is adopted for this conference.

The First Asian Conference on Unsaturated Soils is timely in being organised when Asian countries are actively developing their infrastructures to cater for growth in their populations and economies. Many civil engineering developments involve unsaturated soils that require special consideration, analyses and measurements within the framework of unsaturated soil mechanics. The papers presented to this conference cover a wide range of topics from the fundamentals to practical experiences and from new equipment development to advances in numerical modelling. Flow, shear strength and volume change characteristics of unsaturated soils dominate the discussions of different problems such as heave, collapse, slope instability, foundation failures and contaminant transport. Suction and soil-water characteristics appear to play crucial roles in understanding and solving these problems that are closely related to water flow through unsaturated soils.

We hope that this publication helps to bridge the gap between the theory and practice of unsaturated soils.

H. Rahardjo, D.G. Toll and E.C. Leong
Editors

Unsaturated Soils for Asia, Rahardjo, Toll & Leong (eds) © 2000 Taylor & Francis, ISBN 90 5809 139 2

Organisation

ADVISORY COMMITTEE

Chairman:
Prof. D.G.Fredlund, Canada

Members:
Prof. S.N.Abduljauwad, Saudi Arabia
Prof. E.E.Alonso, Spain
Prof. T.Amirsoleymani, Iran
Prof. C.G.Bao, P.R.China
Prof. G.E.Blight, South Africa
Prof. S.Frydman, Israel

Prof. S.L.Houston, USA
Prof. D.Karube, Japan
Prof. S.K.Kim, Korea
Prof. Z.J.Shen, P.R.China
Assoc. Prof. C.I.Teh, Singapore
Prof. S.J.Wheeler, United Kingdom

TECHNICAL COMMITTEE

Chairman:
Assoc. Prof. H.Rahardjo   Nanyang Technological University, Singapore

Technical Secretary:
Dr D.G.Toll   Nanyang Technological University, Singapore

Members:
Prof. H.A.Faisal	University of Malaya, Malaysia
C.E.Ho	formerly Prescrete Engineering Pte Ltd, Singapore
Prof. R.Kitamura	Kagoshima University, Japan
Assoc. Prof. E.C.Leong	Nanyang Technological University, Singapore
Prof. H.D.Lin	National Taiwan University of Science and Technology, Taiwan
Assoc. Prof. C.W.W.Ng	Hong Kong University of Science & Technology, SAR, P.R.China
Dr T.Nishimura	Ashikaga Institute of Technology, Japan
Dr R.A.A.Soemitro	Institut Teknologi Sepuluh Nopember, Indonesia
Assoc. Prof. T.S.Tan	National University of Singapore
S.K.Tang	PWD Consultants Pte Ltd, Singapore
Assoc. Prof. K.Tateyama	Kyoto University, Japan
Dr D-Q.Yang	Moh and Associates (S) Pte Ltd, Singapore

SECRETARIAT

Er. J.S.Y.Tan   CI-Premier Pte Ltd, Singapore

1 Keynote lectures

Unsaturated Soils for Asia, Rahardjo, Toll & Leong (eds) © 2000 Taylor & Francis, ISBN 90 5809 139 2

Fundamentals of rockfill collapse

E. E. Alonso & L. A. Oldecop
Department of Geotechnical Engineering and Geosciences, Universitat Politecnica de Catalunya, Barcelona, Spain

ABSTRACT: Rockfill behaviour is known to be strongly influenced by water action. Collapse settlements typically occur in rockfill dams, in which the upstream shell is flooded during reservoir filling. Moreover, it seems that in some cases collapse deformations were related to heavy rainfall. With the aim of studying the behaviour of rockfill in non-saturated condition under variable water content, oedometer test were performed on a rockfill-type-material, a quartzitic slate, with control of relative humidity. The main conclusion of this work is that it is enough to impose a 100% relative humidity atmosphere within the rockfill voids in order to get the same collapse strain magnitude as obtained by flooding the specimen. Experimental results are interpreted and a basic mechanism of rockfill volumetric deformation, which is based on some fracture propagation mechanisms, is proposed.

1 INTRODUCTION

Rockfill shows some distinct features of behaviour due to their particular grain size distribution and grain shape. Time dependent deformations occur starting from early stages of construction and continue along very long periods. Settlements records of rockfill dams (Sowers et al 1965) showed that the deformation process of rockfill can last many decades. Another characteristic behaviour of rockfill is the development of collapse strains due to wetting. Upstream shells of rockfill dams show often collapse deformations upon flooding due to the reservoir filling. This behaviour was also observed in laboratory test programs in which the specimen was impounded under load (Sowers et al 1965, Marsal 1973, Nobari & Duncan 1972).

Collapse of rockfill has also been observed to be induced by wetting due to rainfall. Such a behaviour, which is still not completely understood, was reported for some ten rockfill structures over the world. Naylor et al (1997) reported a correlation between settlements of Beliche Dam (Portugal) and rainfall records in the site. A 360 mm rainfall caused very intense collapse deformations to occur in the 60 meters high Cogswell dam, when it was still under construction (Bauman 1960). In this concrete faced dam, the fill, made of quarried granitic gneiss, was placed by dumping with no compaction nor wetting. In contact with the rain water, vertical settlements suddenly increased up to 4% of the fill height and

the concrete facing was severely damaged due to bulging of the upstream slope in its lower half.

Figures 1 and 2 illustrate the post-construction behaviour of the 150 meters high Infiernillo dam build in Mexico in the 60's. It is a zoned rockfill dam with a thin central clay core. Shells where made of quarried diorite and conglomerates dumped dry and without compaction. The upstream shell collapse upon the reservoir filling, during 1964, may be recognised in the settlement and horizontal displacement records of surface marker M-10 (crest, El. 180). It is worth noting that after a nineteen months period with decreasing deformation rates (from January 1965 to July 1966), a sudden increase in settlement and downstream horizontal displacement rates of the surface markers M-10 and M-23 occurred in august 1966. This new deformation episode lasted until the end of 1968. In correspondence with the onset of such higher deformation rates, a rather intense rainfall occurred during the second semester of 1966 (as compared with the 1965 rainy season). This fact suggests that the rockfill of the downstream shell may be wetted significantly for the first time in the 1966 rainy season, and that the rain triggered the collapse. It could be expected that the wetting by rainfall is slower than the wetting due to reservoir impounding. Consistent with this suggestion, the deformation process during the downstream shell collapse lasted thirty months whereas the collapse of the upstream shell lasted only seven months. However both produced almost the same amount of crest settlements and horizontal displacement. The

3

rainy seasons of 1967 and 1968 are followed by further increments of the displacements rates, suggesting that the wetting of rockfill was not completed during 1966 and that the incorporation of further water quantities caused additional collapse strains. Marsal (1976) also suggested that the first flooding of the lower part of the downstream shell by the river water that, at the time, had a high level due to the spillway operation, may be the cause of this behaviour. However, the settlements records of cross-arm D-2 from July 1966 to March 1975 (Fig. 2), proves that the strains which occurred along that period affected the material from the foundation up to El. 120 which is clearly very much above the maximum river water level 68 reached in 1967 (see Fig. 1). Whereas it is reasonable to accept that the flooding of the lower part of the shell may contribute to the referred deformations, this effect alone does not justifies the straining of the upper part of the shell which was never flooded (from El. 68 to El. 120) and hence such deformations should be attributed to wetting by rainfall.

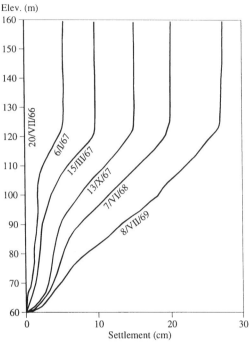

Figure 2. El Infiernillo Dam. Settlement record of cross-arm D-2 (see Fig. 1 for location) (data after Marsal, 1976)

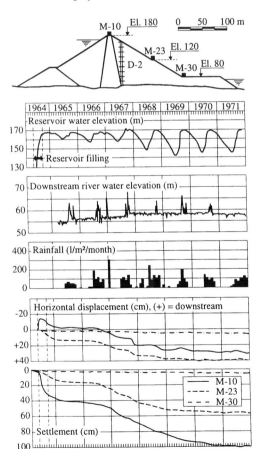

Figure 1. Post-construction behaviour of El Infiernillo Dam, Mexico (data after Marsal et al, 1976).

Soriano & Sánchez (1999) reported the post-construction behaviour of the Madrid-Sevilla (Spain) high-speed railway embankments, some of them exceeding 50 meters high and mostly constructed with slate and schist rockfill. Figure 3 shows a clear correlation, during 1996 and 1997, between rainfall and settlement rate recorded in one of those embankments. It should be noted that further rainfall at the end of 1997, of almost the same intensity as that occurred in the early 1996, does not cause any observable collapse behaviour. It can be concluded that previous rain events exerted a pre-consolidation effect in the fill.

Terzaghi (1960) suggested that a possible reason for the large deformations of rockfills could be the breakage of rock particles in the vicinity of highly stressed contacts and the subsequent rearrangement of the granular structure into a more stable position. This was confirmed by the large amount of rockfill experimental data produced during the 60s and 70s using large scale testing equipment (Sowers, 1965; Fumagalli, 1969; Marachi et al, 1969; Marsal, 1973; Penman et al, 1976; Veiga Pinto, 1983). Particle breakage was observed to occur in all of these tests. Breakage intensity depends on the applied stress level, particle shape and grading (angular particles and uniform grading cause more particle breakage), the rock strength, and the presence of water. In addition, a size effect was identified (Marachi et al,

1969) which causes that, other factors being constant, the larger the grain size, the more intense is the particle breakage because larger particles are statistically more prone to contain larger flaws.

Nobari and Duncan (1972) identified the factors having influence on the collapse of rockfill upon flooding, from the observation of the behaviour of crushed argillite in one-dimensional and triaxial compression tests. They found that the initial water content was the most important factor determining the amount of collapse upon flooding. The larger was the initial water content, the smaller was the collapse deformation. Sieve analysis carried out before and after flooding the sample, showed that particle crushing occurs not only due to stress increase but also during collapse due to flooding. This important observation suggests that the explanation for rockfill collapse should be sought in the influence of water in the rock fracture mechanisms.

With the aim of getting a better understanding of rockfill deformation mechanisms, a rockfill-type material was tested in one-dimensional compression applying a relative humidity (RH) control system. The test set up and the experimental results are presented in the following section. In the last part of the paper the experimental results are discussed and interpreted. A basic mechanism of rockfill volumetric deformation, which is based on a fracture propagation mechanism, is proposed. It is hoped that a better knowledge of rockfill deformation mechanisms will help to improve design and construction practices of rockfill structures, will improve our prediction capability and will also help in the interpretation of instrumentation records.

2 ONE-DIMENSIONAL COMPRESSION TESTS OF ROCKFILL

The previous review led to the idea that testing the rockfill under variable moisture conditions would provide a better insight in to the nature of water influence in the materials behaviour.

The experimental program included three classic oedometer tests with sample flooding at some particular confining stress and one oedometer test with RH control. The tested material is a slate obtained from the Pancrudo River outcrop (Aragón, Spain). The slate belongs to the Almunia formation, which has a Cambric origin. The outcrop is planned to serve as quarry for the construction of the shells for the Jiloca River Regulation Dam, a zoned earth and rockfill dam. Figure 4 shows the water retention curve for the Pancrudo slate rock.

A sample of crushed rock was obtained from the quarry with the aid of a digger. Recovered blocks having 200 to 400 mm maximum dimensions were broken down with the aid of a hammer. The material particle size was in the range of 0.4 to 40 mm with $D_{50} = 23$ mm and uniformity coefficient $C_u = 3$.

2.1 *Test procedure*

The oedometer test program was carried out in a Rowe type cell, 300 mm sample diameter, and an approximate sample height of 200 mm. Before specimen assembly, rockfill was allowed to reach equilibrium with the laboratory environment (approximately 50% RH and 22 °C) This resulted in a rather low initial water content (under 1% in weight). Specimens were compacted in four layers

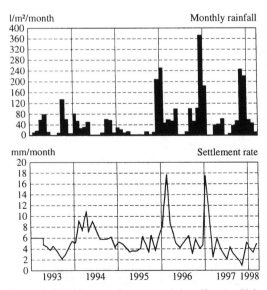

Figure 3. Rainfall and settlement records in a 40 meters high embankment from the Madrid-Sevilla high-speed railway built with slate and schist rockfill (after Soriano & Sánchez 1999).

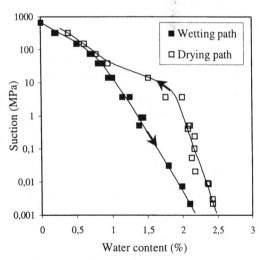

Figure 4. Retention curve for the Pancrudo slate rock.

5

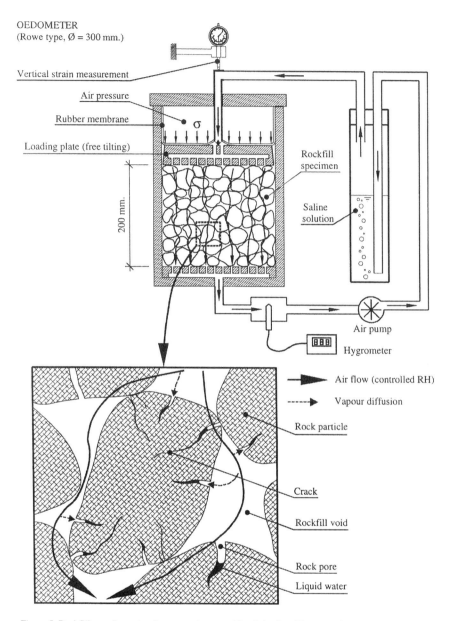

OEDOMETER
(Rowe type, Ø = 300 mm.)

Vertical strain measurement

Air pressure

Rubber membrane

σ

Loading plate (free tilting)

200 mm.

Rockfill
specimen

Saline
solution

Air pump

Hygrometer

Air flow (controlled RH)

Vapour diffusion

Rock particle

Crack

Rockfill void

Rock pore

Liquid water

Figure 5. Rockfill one-dimensional compression test with relative humidity control.

by means of a handy compaction hammer and applying a compaction energy in the range of 600 and 700 Joule/litre, which is slightly over the compaction energy used in the Standard Proctor Test (584.3 Joule/litre). All specimens tested reached a void ratio e = 0.55±0.03 after compaction.

In test 4, a RH control system was added to the oedometer. The aim of this system is to produce a gradual variation in rock moisture adding controlled quantities of water in a uniform manner across the specimen. In order to achieve this goal, water has to be transported in vapour state. The system consists in a closed-loop air circulation circuit as shown in Figure 5. The air, impelled by an electric pump, bubbles in a solution vessel and then flows across the specimen from the upper to the lower drainage plate. Saline solutions were used in order to impose a controlled RH to the airflow. An hygrometer inserted in the airflow coming out of the specimen allowed to monitor the progress of the test.

Rockfill can be considered as a double porosity material. Inter-particle spaces define a first set of large voids. Moreover, rock particles have their own natural porosity, and hence a second set of very much smaller voids can be identified within the rock particles. In the following, large inter-particle voids will be termed "rockfill voids" and the term "rock pores" will be used to refer to the small pores within the rock particles. In the later group, only open pores are considered (i.e. those connected with the rockfill voids).

As the test proceeds, the air flows through the rockfill voids, with a controlled RH value. During a wetting path, as in test 4, the RH imposed by the solution in the airflow is larger than the current RH within the rock pores. Water vapour is transported form the vessel to the rockfill voids by advection. Further transport of water vapour occurs from rockfill voids to rock pores by molecular diffusion due to the RH gradient created between rockfill voids and rock pores. Water vapour will condense within the rock pores wherever the width of the pore is smaller than approximately two times the equilibrium curvature radio of the gas-liquid interface (Fig. 5). The interface curvature depends on the surface tension of water and the difference between the gas pressure and the pressure in liquid water, i.e. the matric suction. Moreover, liquid water in rock pores may contain solutes and hence an osmotic component of suction can be defined. The sum of matric plus osmotic suction is usually termed the total suction, ψ. Total and matric suction would be equal in the case that rock pores contain only pure water.

The absorption of water by the rockfill specimen during the test was measured by recording the loss of weight of the solution vessel. Under constant imposed RH, the rate of water absorption decreased gradually until an equilibrium state (nil absorption rate) was reached. When the whole system (specimen and RH control device) becomes in thermodynamic equilibrium, the RH is equal at any point of the gas phase within the system and no further transport of water vapour can occur. In such an equilibrium state and under isothermal conditions, the RH in the gas phase is univocally related to total suction in the liquid water contained in the rock pores by the psychrometric relationship (Coussy 1995). Therefore RH and total suction, ψ, can be used as interchangeable parameters measuring the effect of water in the rockfill behaviour. In this work, total suction was used to plot the experimental results for the sole reason that it leads to simpler mathematical expressions for data fitting.

2.2 *Test results*

Tests were carried out in a load control mode, as in classic oedometer tests. Figure 6a shows the paths in

the vertical stress-total suction space, followed by the four oedometer tests carried out with the equipment described in the previous section. Figure 6b shows the plot of the measured vertical strain vs. the applied vertical stress. The load was applied by increments allowing the sample to deform a maximum of 24 hours before the application of the next load increment. Stress-strain data plotted in Figure 6b correspond to a constant time interval of 1000 minutes.

Test 1, 2 and 3 show the typical rockfill behaviour. A unique normal compression line (NCL) seems to exist for the dry (at initial water content) and for the saturated (flooded) states. Saturated rockfill is more compressible than dry rockfill and collapse strains develop if the specimen is flooded under load (test 3). The magnitude of collapse strains is such that, after flooding, the behaviour of the collapsed specimen reaches the NCL of the satu-

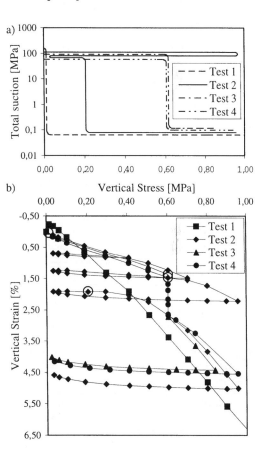

Figure 6. One-dimensional compression tests of rockfill. a) Test paths in the total suction-vertical stress space. b) Measured vertical strain vs. applied vertical stress. Large open circles indicate specimen flooding.

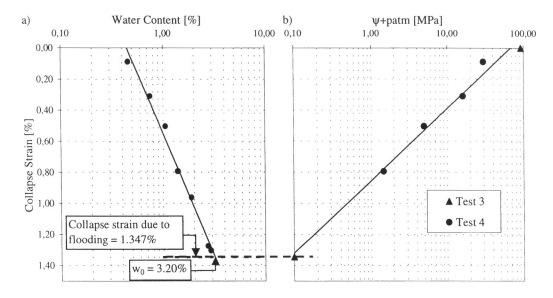

Figure 7. a) Collapse strain vs. water content. b) Collapse strain vs. total suction.

rated material. If after full collapse, specimens are further loaded, the material yields along the NCL corresponding to the saturated state. An elastic domain may also be identified from the unloading-reloading paths carried out in test 2.

However, some novel features were observed. If the rockfill is flooded under low load (test 1) or if it was strongly preconsolidated (test 2), no collapse strains occur during saturation. Instead, moderate swelling strains were measured, 0.27% and 0.016% vertical strain in test 1 and 2 respectively. Data from tests 1, 2 and 3 suggest that the collapse phenomenon occurs only if the applied vertical stress is higher than some threshold value. Beyond such a threshold stress, collapse strain increases linearly with the applied stress, at least in the stress range used in this test program (up to 1 MPa).

Test number 4 has an identical stress path as test number 3 but during the collapse process, the RH was controlled by means of the previously described set-up. The aim of this test was to induce collapse in a gradual manner as moisture increases. Figures 7a and 7b show plots of collapse strain vs. water content and collapse strain vs. total suction for the data recorded during test 4. It is worth noting that specimen 4 was never flooded and hence the "bulk degree of saturation" remained close to zero during the whole test (final water content at the end of collapse was 3.2%). By contrast, sample 3 was fully saturated during flooding and it reached a 19.5% water content. Nevertheless, the amount collapse strain, which occurred in both specimens, is almost the same.

Figure 8 shows the strain-log time records for specimen 3 for every load increment and the flood-

ing stage. A rapid strain increment (almost instantaneous) occurred after each load increment application followed by a time dependent deformation process. Strain vs. logarithm of time approached a linear relationship a certain time after load application (in most cases less than ten minutes). Its slope appears to increase with the stress level and for flooded specimens. A rapid increase in strain also occurred after sample flooding. This is the collapse deformation. However, collapse strains are not instantaneous since during flooding of the sample only a small amount of strain occurred while the major part of collapse occurred after completion of flooding. Time dependent strain seems to continue indefinitely although at a decreasing strain rate. Similar features can be observed in the results of oedometer tests on rockfill-type materials reported by Marsal (1973), Sowers (1965) and Nobari & Duncan (1972). As reported by Sowers (1965), after some stress increments a sudden increase in strain accompanied by acoustical emissions was observed. This occurred during the first minutes after load application, only for specimens in dry conditions and for an applied load in excess of 0.4 MPa. Unlike conventional granular soil behaviour, the rockfill time dependent strains are not negligible. Furthermore, the rockfill collapse deformation seems to involve only time dependent strains.

3 ROCKFILL DEFORMATION MECHANISM

The reported tests showed that both time dependent strains and collapse strains of rockfill are simultaneously controlled by stresses and water action. It seems likely that both behaviours are caused by a unique phenomenon acting at a micro-mechanical level within the rock particles. Test number 4, in which collapse was produced by rising gradually the relative humidity, showed that the water content at the end of the collapse path was a small fraction (~15%) of the water volume needed to fully saturate the rockfill, i.e. to fill with water both rock pores and rockfill voids. If the specimen water content at the end of the wetting path (w = 3.2%, see Fig. 7a) is referred to the water retention curve of the rock (Fig. 4), it appears that during test 4 the major part of the liquid water was essentially stored within the rock particles (i.e. in the rock pores) and that only a very small amount of liquid water was present in the rockfill voids. This fact was confirmed by visual inspection at the end of test 4. Hence, saturation of the rock particles themselves seems enough to produce the same collapse deformation as a full flooding of the rockfill specimen. This suggests once more that the mechanism causing the collapse behaviour occurs within the rock particles.

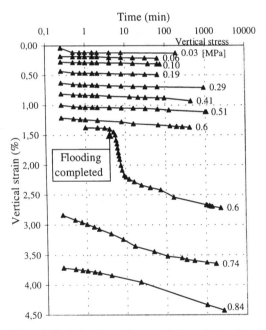

Figure 8. Vertical strain vs. time records for each stress increment for one-dimensional compression test 3.

From the above considerations, it seems reasonable to look for an explanation for the rockfill behaviour in some rock fracture phenomenon in which water plays a determinant role. Among the reported water-induced phenomena leading to rock weakening due to the influence of water (loss of cohesion

due to mineral superficial energy reduction, Vutukuri and Lama, 1978; suction reduction Vutukuri and Lama, 1978; expansion of clay minerals, De Alba and Sesana, 1978, Delgado et al, 1982) the subcritical crack propagation due to stress corrosion (Wiederhorn et al 1980, 1982, Freiman 1984, Atkinson and Meredith, 1987) offers comprehensive explanation for the rockfill behaviour.

3.1 Subcritical crack propagation

The theory of stress corrosion (Michalske and Freiman 1982, Atkinson and Meredith 1987) states that the highly stressed inter-atomic bonds at the tip of a crack are more vulnerable to the attack of a corrosive agent, such as water, than the unstressed material away from the tip. This is attributed to the reduction of the overlap of the atomic orbits due to the intense strain state present at the tip. The following reaction was proposed for the stress corrosion reaction of quartz in water environments (Freiman 1984):

$$H_2O + [Si-O-Si] \rightarrow \chi^{\ddagger} \rightarrow 2 [Si-OH] \qquad (2)$$

Water reacts with the crack tip material, i.e. the strained Si-O-Si bonds. This reaction yields the activated complex $\chi^{\ddagger}$ which has weaker bonds than the original material. Hence, bonds are broken under the applied loads. Reaction products are Si-OH groups on each crack surface. It can be expected that the crack propagation velocity, V, is controlled by the reaction rate. Hence, V depends on the applied load and the "availability" of the corroding agent. The presence of water in the testing environment is usually measured through the relative humidity, whereas the loads effect on the crack tip is conveniently represented by the stress intensity factor, K, as defined in linear elastic fracture mechanics (Broek 1987).

A typical relationship between V and K, as obtained from experiments performed in a constant RH environment are shown in Figure 9, and in qualitative terms in Figure 10. The fracture toughness, K_c indicated in Figure 10, is the critical value of K which onsets a catastrophic fracture propagation (quasi-instantaneous propagation).

The thermodynamic formulation of the reaction rate theory (Laidler 1987) applied to stress corrosion cracking (Wiederhorn et al 1980, Wiederhorn et al 1982, Freiman 1984) yields the following general expression for the crack propagation velocity with water as corrosive agent:

$$V = V_0 \, RH \, exp[-(E^{\ddagger} - b\,K)/RT] \qquad (3)$$

where V_0 is a proportionality constant, RH is the relative humidity in the testing environment, K is the stress intensity factor in the crack, T is the absolute temperature and R is the gas constant. The term ($E^{\ddagger}$ -

b K) can be interpreted as the standard Gibbs free energy of activation, ΔG_0, for the stress corrosion reaction per unit mole of the activated complex (also called "the energy barrier" of the reaction). $E^\ddagger$ involves the non-stress dependent terms of the energy barrier, i.e. the activation energy for the corrosion of the unstressed material in a water saturated environment, and b is the work-conjugated variable of K. It should be noted in eq. 3, that an increase in loads, measured as an increase in K, will cause a decrease in the activation energy and hence, an increase in V as observed in experiments (Fig. 9). V_0, $E^\ddagger$ and b become constants under fixed environmental conditions (constant T and RH) and hence they can be determined by experimental data fitting.

Equation 3 is implied in the bold part of curves in Figure 10. V_0, $E^\ddagger$ and b become constants under fixed environmental conditions (constant T and RH) and hence they can be determined by experimental data fitting. Eq. 3 was found to hold for a large body of experimental data, mostly obtained for glasses and ceramics (Wiederhorn et al 1980, Wiederhorn et al 1982 Freiman 1984), but also for rocks (Atkinson & Meredith 1987). It was observed in tests of soda lime-silica glass (Wiederhorn et al

1982, Freiman 1984) that proportionality between V and RH holds for RH ranging from 20% to 100%.

ΔG_0, can be expressed in terms of the chemical potentials of the reactants, μ_0^{rock} and μ_0^{water}, and the activated complex $\mu_0^\ddagger$, in their standard reference states (Wiederhorn et al 1982):

$$\Delta G_0 = E^\ddagger - b K = \mu_0^\ddagger - \mu_0^{rock} - \mu_0^{water} \quad (4)$$

Assuming that water vapour behaves as an ideal gas, its chemical potential is given by (Castellan, 1971):

$$\mu^{water} = \mu_0^{water} + R T \, Ln(p/p_0) \quad (5)$$

where p is the partial pressure of water vapour and μ_0^{water} is the chemical potential of pure water vapour at the reference state defined by pressure p_0 and temperature T. If the reference state is chosen as "pure water vapour at a given temperature T and pressure, p_0, equal to the saturation vapour pressure at temperature T", the quotient p/p_0 becomes equal to the relative humidity, RH. Hence, replacing eqs. 4 and 5 in 3 we get:

$$V = V_0 \exp[-(\mu_0^\ddagger - \mu_0^{rock} - \mu^{water})/R T] \quad (6)$$

In eq. 6, the term $(\mu_0^\ddagger - \mu_0^{rock})$ is the energy barrier for pure stress induced fracture, i.e. without the aid of the corrosive agent. For a given material and under constant K, eq. 6 shows that the crack propagation velocity is controlled uniquely by the chemical potential of water in contact with the crack tip. Hence, the use of RH in eq. 3 is, in fact, a practical way to measure the chemical potential of the corrosive agent.

Outside the region defined by the bold part of curves in Figure 2 different mechanisms, other than the reaction rate, control the crack propagation and therefore eq. 3 is not valid. It is usually assumed (although not yet experimentally confirmed for most rocks) that a stress corrosion limit, K_0, exists below which cracks do not propagate at all (Fig. 10).

On the other hand, a plateau in the V vs. K relationship is observed in some materials for high K values (Figs. 9-10). This is attributed to the fact that V becomes controlled by the diffusive transport rate of the corrosive agent towards the crack tip instead by the reaction rate. Moreover, if K is further increased approaching K_c, V increases rapidly and becomes independent of RH. It is assumed that in this region (curve marked as "dry environment" in Fig. 10), water is not able to reach the crack tip (since crack propagation is faster than water diffusion) and hence crack growth is due to pure thermal activation. This is confirmed by an almost equal V-K relationship (Fig. 9) obtained for soda-lime-silica glass tested in vacuum (Wiederhorn et al 1980). Therefore, possible V-K pairs are bounded by the stress corrosion curve corresponding to RH = 100% and by the "dry environment" curve.

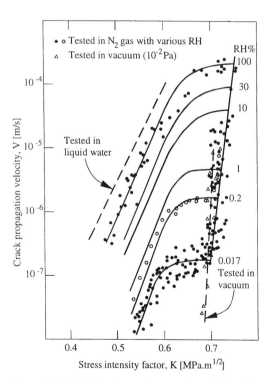

Figure 9. Subcritical crack growth curves for soda-lime-silica glass tested in nitrogen gas with various RH, in liquid water and in 10^{-2} Pa vacuum at room temperature. (data after Wiederhorn et al 1980).

In the case that crack growth occurs in a liquid environment, i.e. a solution of water and a non-aqueous solvent, it was observed that measured values of V are proportional to the RH of a gas in thermodynamic equilibrium with the solution (Freiman 1984). Thermodynamic equilibrium implies that both gaseous environment and liquid environment will have equal chemical potentials and hence, as could be derived from eq. 6, they would produce the same effect on the stress corrosion reaction, i.e., equal reaction rates and equal V (Wiederhorn et al 1982). This means that it does not matter if the corrosive agent (water) is in liquid or gas state, the stress corrosion reaction will proceed controlled by the RH measured in the environment surrounding the specimen, provided that thermodynamic equilibrium between the liquid and gas phases is ensured. This important conclusion establishes the theoretical basis for the testing procedure used in the oedometer tests with RH control previously presented.

3.2 *Conceptual model of rockfill volumetric deformation.*

Let us assume that we are able to compute the stress intensity factor K for every crack or flaw contained by the rock particles of a rockfill sample. The K value will depend on the geometry of the particle, the crack length and the applied macroscopic stresses. Then, each crack within the rockfill sample would occupy a defined position along the K axis of Figure 10. Cracks lying in region I ($K \leq K_0$) will not grow at all. Cracks lying in region II ($K_0 < K < K_c$) will grow at a velocity controlled by load and by the relative humidity. This crack growth implies in turn an increase in K for most crack configurations, since it usually increases with the increase in crack length (K may also decrease with the increase of crack length for 'negative geometries', Bazant and Planas, 1998). Eventually, when K values approach region III ($K \geq K_c$) an instantaneous particle breakage will occur. The associated rearrangement of the granular structure will lead to a macroscopic strain increment and a more stable configuration of contacts and contact loads.

By means of this mechanism, a conceptual explanation of the rockfill behaviour can be given. In a steady situation, all cracks will lie in region I and hence the rockfill does not deform. If a load increment is then applied, some cracks will move to region III producing an instantaneous deformation increment. Some other cracks will fall in region II and cause the delayed component of deformation. The number of source cracks lying in region II will decrease in time due the breakage process and this implies a reduction of strain rate with time.

If relative humidity increases at a given time, cracks in region II will increase their propagation velocity causing a sudden increase in the strain rate.

Additionally, it may be expected that the stress corrosion limit K_0 decreases as the RH increases (Atkinson 1984), as implied in Figure 10. This would cause a number of cracks to move from region I to region II leading to further time delayed breakages and hence additional strains. This is the nature of collapse: an increment of rockfill strain not related to an increment in load.

3.3 *Discussion*

The proposed conceptual model provides a unique micro-mechanism that explains both time dependent and collapse strains of rockfill. The existence of a threshold stress for the collapse phenomenon, as observed in the experimental results (Fig 6b), is also implied in the proposed model. If the applied macroscopic stress is not high enough to induce K values beyond the minimum K_0, which corresponds to RH = 100% (Fig. 10), no crack propagation will occur and hence collapse is not possible. Under the threshold stress value, strains are only due to the rearrangement of the granular structure by inter-particle slippage and rotation.

As can be derived form Figure 10, the proposed mechanism predicts that the collapse process is made up only from time-dependent strains. This conclusion was also derived from the experimental data presented in Figure 8.

It arises also from Figure 10 that full drying of rockfill (by testing in vacuum for example) would

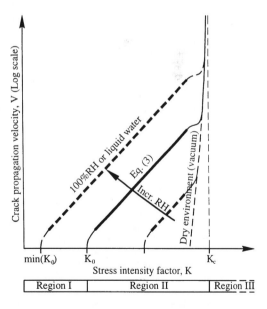

Figure 10. Schematic subcritical crack growth curves and conceptual model.

not fully eliminate time dependent strains. The "dry" V-K curve (Fig.10) determines a minimum crack propagation velocity which is not related to water action. This is an underlying deformation mechanism which provides a minimum time-dependent strain component regardless of the presence of the corrosive agent.

A physical interpretation was given to the nature of total suction in rockfill materials when the test procedure was discussed (Fig. 5): the sum of matric plus osmotic suction components in the liquid water contained in the rock pores and in thermodynamic equilibrium with the specimens gas phase. However, total suction has a more profound meaning if considered from a thermodynamical point of view: it can be considered as a measure of the chemical potential of water. Taking pure water at a given constant temperature and having a flat gas-liquid interphase as a reference state, the chemical potential of liquid water, μ_w, can be simply related to total suction by (Navarro, 1997):

$$\mu^{water} - \mu_0^{water} = -\nu\,\psi \qquad (7)$$

where ν is the molar volume of water. Under isothermal conditions μ^{water} and ψ become proportional because the molar volume ν is constant.

Considering that eq. 6 holds also for water in liquid state as corrosive agent (as previously discussed in 3.1) and considering eqs. 6 and 7 together with the proposed conceptual model, the fundamental role of total suction as a suitable variable for measuring the water influence on rockfill behaviour is explained. Due to the negative sign in eq. 7, a decrease in suction implies an increase in chemical potential of water, an increase in the crack propagation velocity (eq. 6) and therefore an increase in the macroscopic strain. In this way, a thermodynamic explanation is given to the experimentally obtained correlation between collapse strain for a given time interval and total suction (Fig. 7b). The collapse strain vs. water content correlation (Fig. 7a) is a consequence of the more fundamental collapse strain vs. total suction relationship and the rock retention curve (Fig. 4).

4 CONCLUSIONS

In the present paper, the occurrence of time dependent strains and collapse strains in rockfill is explained by means of a unique mechanism based on the subcritical crack growth phenomenon. The nature of water influence on rockfill behaviour has been analysed. Moreover, the proposed mechanism is able to explain a number of features of behaviour observed in the experimental program. Among them, the occurrence of collapse strains due to changes in water content in non-saturated states, a condition induced in the field by rainfall. Total suction (or RH)

was identified as a fundamental variable measuring the influence of water in rockfill behaviour.

An oedometer test equipment with relative humidity control was developed and a test program was carried out on a rockfill-type-material (quartzitic slate of Cambric origin). The main conclusion of this test program is that increasing the RH up to 100% within the rockfill specimen is enough to produce full collapse deformation (collapse under the same load when the rockfill is flooded). This experimental observation confirms that any situation leading to a change in moisture content in the rock pores is enough to cause collapse deformations and also increase the material compliance against further loading. This conclusion is consistent with the rain-induced settlements in rockfill structures.

5 ACKNOWLEDGEMENTS

Authors gratefully acknowledge the financial support given to the second author by the Universidad Nacional de San Juan (Argentina) and the Spanish Cooperation Agency (AECI), during the research work reported in this paper.

6 REFERENCES

Atkinson, B. K. (1984). Subcritical crack growth in geological materials. *J. Geophysical Research* **89**, No. B6, 4077-4114.

Atkinson, B. K. & Meredith, P. G. (1987). The theory of subcritical crack growth with applications to minerals and rocks. *Fracture Mechanics of Rock*, B. K. Atkinson, ed. London: Academic Press Inc., 111-166.

Bauman, P. (1960). Rockfill dams: Cogswell and San Gabriel Dams. *Transactions of the ASCE*, **125** part 2, 29-57

Bazant, Z. P. & Planas, J. (1998). *Fracture and Size Effect in Concrete and Other Quasibrittle Materials*. CRC Press.

Broek, D. (1986). *Elementary engineering fracture mechanics*. Dordrecht: Martinus Nijhoff Publishers.

Castellan, W. G. (1971). *Physical Chemistry*, 2nd ed, Addison-Wesley Publishing Co.

Coussy, O. (1995). *Mechanics of porous continua*. John Wiley & Sons Ltd. Chichester

De Alba, E. & Sesana, F. (1978). The influence of expansive minerals on basalt behaviour. *Proc. Int. Congress of Engineering Geology*, Madrid, 107-116.

Delgado Rodriguez, J., Veiga Pinto, A. & Maranha das Neves, E. (1982). *Rock index properties for prediction of rockfill behaviour*. Memória N°581, Laboratório Nacional de Engenharia Civil, Lisbon.

Freiman, S. W. (1984). Effects of chemical environments on slow crack growth in glasses and ceramics. *J. Geophysical Research* **89**, No. B6, 4072-4076.

Fumagalli, E. (1969). Tests on cohesionless materials for rockfill dams. *J. Soil Mech. Fdn. Engng,* ASCE, **95**, SM1, 313-330.

Laidler, K. J. (1987). *Chemical kinetics (3rd ed).* Harper Collins Publishers.

Marachi, N. D., Chan, C. K., Seed, H. B. & Duncan, J. M. (1969). *Strength and deformation characteristics of rockfill materials.* Department of Civil Engineering, Report No. TE-69-5, University of California.

Marsal, R. J. (1973). Mechanical properties of rockfill. *Embankment Dam Engineering. Casagrande Volume.* Hirschfeld, R. C. & Poulos, S. J., eds. John Wiley & Sons.

Marsal, R. J., Arellano, L. R., Guzmán, M. A. & Adame, H. (1976). El Infiernillo. *Behavior of Dams Built in Mexico.* Mexico: Instituto de Ingeniería, UNAM.

Michalske, T. A., & Freiman, S. W. (1982). A molecular interpretation of stress corrosion in silica. *Nature*, **295**, 511-512.

Navarro, V. (1997). *Modelo del comportamiento mecánico e hidráulico de suelos no saturados en condiciones no isotermas.* D. Thesis, Department of Geotechnical Engineering and Geosciences, Universitat Politecnica de Catalunya, Barcelona, Spain.

Naylor, D. J., Maranha, J. R., Maranha das Neves, E. & Veiga Pinto, A. A.(1997). A back-analysis of Beliche Dam. *Géotechnique* **48**, No. 2, 221-233.

Nobari, E. S. & Duncan, J. M. (1972). *Effect of reservoir filling on stresses and movements in earth and rockfill dams.* Department of Civil Engineering, Report No. TE-72-1, University of California.

Penman, A. D. M. & Charles, J. A. (1976). The quality and suitability of rockfill used in dam construction. *Dams and Embankments,* Practical Studies from the BRE London: The Construction Press, **6**, 72-85.

Soriano, A. & Sánchez, F. J. (1999) Settlements of railroad high embankments. *Proc. XII European Conf. on Soil Mech. and Geotech. Eng.*, Netherlands.

Sowers, G. F., Williams, R. C. & Wallace, T. S. (1965). Compressibility of broken rock and settlement of rockfills. *Proc. 6th ICSMFE,* **2**, Montreal, 561-565.

Terzaghi, K. (1960). Discussion on Salt Springs and Lower Bear River dams. *Trans. ASCE,* **125**, pt 2, 139-148.

Veiga Pinto, A. A. (1983). *Previsao do comportamento estrutural de barragens de enrocamento.* Laboratório Nacional de Engenharia Civil, Lisbon, PhD thesis.

Vutukuri, V. S. & Lama, R. D. (1978). *Handbook on Mechanical Properties of Rocks.* Claustahl: Trans Tech Publications.

Wiederhorn, S. M., Fuller, E. R. & Thomson, R. (1980). Micro-mechanisms of crack growth in ceramics and glasses in corrosive environments. *Met. Sci.*, **14**, 450-458.

Wiederhorn, S. M., Freiman, S. W., Fuller, E. R. & Simmons, C. J. (1982). Effects of water and other dielectrics on crack growth. *J. Mater. Sci.*, **17**, 3460.

Unsaturated Soils for Asia, Rahardjo, Toll & Leong (eds) © 2000 Taylor & Francis, ISBN 90 5809 139 2

Some thoughts and studies on the prediction of slope stability in expansive soils

C.G. Bao
Yangtze River Scientific Research Institute, Wuhan, People's Republic of China

C.W.W. Ng
Hong Kong University of Science and Technology, SAR, People's Republic of China

ABSTRACT: In addition to the Three Gorges Dam, the currently proposed South-to-North (S/N) Water Transfer Canal is likely to be one of China's major infrastructure projects in this new century. However, slope instability problems in unsaturated expansive soils along the middle-route of the Canal impose severe challenges to many academics and engineers. In this paper, a line of thoughts and some preliminary studies and reviews on the predictions of slope stability in expansive soils are discussed and presented. The studies and reviews include the effects of soil-water characteristic, rainfall infiltration, swelling potential and shear strength of the expansive soils. Moreover, some recent results from weathered soils are adopted for comparisons. The predictions of slope stability are intended to establish an early warning system for the S/N Water Transfer Canal project.

1 INTRODUCTION

1.1 *The expansive soil problem along the middle-route of the South-to-North Water Transfer project in China*

In this century, water resources are the most important requirement for the development of the north and west areas of China. It is well known that the volume of precipitation is disproportionately distributed throughout the country: deficient in the northern part and abundant in the southern part. For instance, the total annual discharge of the Yangtze River in the south, is 1000 billion m^3, about 20 times that of the Yellow River in the north. Transferring water from the south to the north is therefore an indispensable task for mitigating the existing crisis of water resources in northern China.

A water transfer project called the South-to-North (S/N) Water Transfer Canal was proposed for sometime. Three possible routes have been considered, namely the east, middle and west route. Each of these routes has its supplying regions and technical problems. Currently, the middle-route is considered to be the most favorable one for mitigating the existing crisis of water resources in northern China. The total length of the middle-route is approximately 1240 km, starting from Hubei and terminating at Beijing and Tianjing.

Along part of the middle-route, extending from the Hubei province to Beijing, about 180 km long canal will pass through the Nanyang expansive soil zone in the Henan province. The slope stability of the canal is one of the major engineering concerns for the construction of the canal. In the 1970's, a large approach canal was excavated with a depth of 40 m in this soil zone. During the excavation, 13 large-scale landslides occurred along the excavated canal. Some costly measures were taken to stabilize the landslides.

Some research work has been taken in recent years to study the slope stability in expansive soils (Liu, 1997; Bao et al, 1998). Not only is such a study meaningful to this project, but it provides an unprecedented opportunity for the development and application of unsaturated expansive soil theories (Fredlund, 1998).

Very often expansive soil is a problematic plastic unsaturated clay with distinct characteristics of swelling and shrinking when there is a change of moisture content. The Nanyang expansive soil is the quaternary miocene alluvial-pluvial clay with a brown-yellow or gray-white color. As given in Table 1, the Nanyang expansive soil has low to medium swelling pressure and shrinkage limit.

In this paper, a line of thoughts to develop an "early warning" system for the construction of the S/N Water Transfer Canal project is discussed.

Table 1. Basic indices for Nanyang expansive soil

Soil type	Colour	Water content (%)	Dry density (kN/m³)	Shrinkage limit (%)	Swelling pressure (kPa)	Swelling ratio (%)	Cation exchange ratio (ml/100g)
1	Gray-white	22.5 ~ 28.1 (24.7)*	14.8 ~ 16.3 (15.8)	11.4 ~ 13.7 (12.1)	56 ~ 780 (219)	4.8 ~ 26.4 (9.4)	45.63
2	Brown-yellow	21.7 ~ 27.0 (23.3)	15.3 ~ 16.8 (16.1)	8.1~ 12.0 (10.4)	31 ~ 200 (88)	1.0 ~ 11.6 (5.6)	33.51
3	Gray-brown	20.3 ~ 25.3 (23.0)	15.2 ~ 16.5 (16.0)	7.0 ~ 10.0 (9.3)	17 ~ 90 (31)	0.8 ~ 7.8 (3.2)	28.03

* Number in bracket is average value.

Some preliminary studies and reviews on the predictions of slope stability in expansive soils are also discussed and presented. The studies and reviews include the effects of soil-water characteristic, rainfall infiltration, swelling potential and shear strength of the expansive soils. Moreover, some recent results from weathered soils are adopted for comparisons. The predictions of slope stability are intended to assist in establishing the early warning system.

1.2 General descriptions of landslides in the expansive soil

The landslides in the Nanyang area may be broadly classified into two categories: sliding along the interface between different soil layers and failure along the pre-failure surface initiated, developed and propagated along some existing cracks. For the former, the mechanism of landslides is relatively simple to understand once the location of the interface can be identified. On the other hand, the failure mechanism of the latter is quite complicated. The development of the landslide is associated with the opening, propagation, enlargement of cracks in the expansive soil (Bao, 1998). Factors that control the formation, distribution and enlargement of the cracks may be classified as intrinsic and external conditions. The former refers to the so-called 'three characteristics' of expansive soils, namely, swelling and shrinkage, cracking and overconsolidation (e.g., k_0 lies between 1.5 ~ 3.0 for Nanyang expansive clay (Liu, 1997)). It is generally known that these three characteristics have significant influence on the strength and deformation of expansive soils. As for the external conditions, they refer to those factors leading to a variation in the distribution of moisture content or matric suction in the soil. Repeated wetting and drying cycles in the soil due to precipitation and evapotranspiration result in the repeated cycles of swelling and shrinking of soil throughout the year and hence lead to the initiation, propagation and enlargement of cracks, which in turn enhance physical weathering of the soil. As a result, the effect of the over-consolidation of the swelling soil is weakened (i.e., increase in void ratio) and the soil is softened progressively, i.e., the shear strength of the soil decreases with time. On the other hand, an increase in the water content due to rainfall infiltration will cause a reduction in matric suction in the soil and hence its shear strength. All the inter-related factors will potentially result in further deformation and even failure of a slope. However, the entire deterioration process of the soil may last a long period, say, several years. As significant variations of moisture content or matric suction associated with the climatic conditions are generally confined in the soil layer near the ground surface, many slope failures in the expansive soil are generally shallow.

1.3 Observations on a trial slope failure in the expansive soil

Figure 1 shows the measured deformations at two monitoring stations B1 and B2 along a cross-section of a 10m height trial excavated slope with a slope angle about 24° to the horizontal, in an expansive soil along the canal. Station B1 is located about 5m above B2 along the slope section. The measured deformation can be divided into four stages. The first stage deformation was caused by the excavation of the trial slope. The excavation was started in February and completed in May (Liu, 1997) and it resulted in a lateral movement approximately 35mm. Between May and middle of June, no significant movement or creep was recorded. On 12 June (Points C_1 & C_2), a heavy rainstorm occurred which marked the start of a raining season. This rainstorm appeared to initiate the movements at both stations (end of Stage 2). Subsequently, the slope continued to deform with an accelerated rate until a local failure occurred adjacent to Station B2 on 29 July (Stage 3). At Station B1, the recorded displacement was relatively small. The deformations recorded at both stations during this period (from C_2 to C_2*) appeared to be non-linear and ductile. In other words, if an early warning system and hence critical parameters can be identified and established for the S/N Water Transfer Canal project, damages caused by slope failures in this type of expansive soils would be minimized. At the last stage (Stage 4),

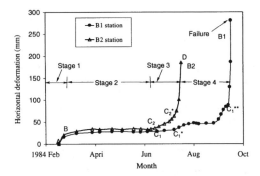

Figure 1. Measured deformation process of a trial canal slope

progressive deformations (from C_1^* to C_1^{**} had been recorded at Station B1 during the wet season between August and September and the slope eventually failed on 28 September.

1.4 *Prediction of slope failure in expansive soil*

As discussed in the previous section, the slope failure in the expansive soil was progressive in nature. Considering serious consequences of any major landslide, the concept of an 'early warning' system proposed in this paper will be of great significance. The proposed 'early warning' system has to be scientific based and is different from conventional predictions. The problem of the conventional predictions is that they generally focus on the determination of the stability of slopes and normally they do not identify key parameters or indices, which may affect the stability of the slopes. Preventative measure is therefore difficult to be carried out to minimize possible damages and loss of lives caused by any potential slope failure. For the proposed 'early warning' system, as illustrated in Figure 2, key parameters or indices such as moisture content will be identified by conducting rigorously analyses and then classified and monitored. Once a chosen parameter or index exceeds certain threshold, an early waning will be issued and preventative measures can be taken accordingly.

2 THE IDEA AND PATH OF DEVELOPING AN SIMPLE EARLY WARNING SYSTEM FOR EXPANSIVE SOIL SLOPES

2.1 *Air and water movements in unsaturated soils*

The mechanical and hydraulic properties of unsaturated soils are significantly influenced by the air and water phases in the soils (Fredlund & Rahardjo, 1993; Bao et al., 1998). When there is only a small amount of moisture in the soil, the air inside most of the pores is well-connected to the atmosphere. The

behaviour of the soil is mainly governed by the effective stress principle, despite the fact that very large suction may exist in a small number of pores. This unsaturated soil would just behave like a 'dry soil', which is not of interest in this paper.

On the other extreme, when water occupies most voids in the soil, the air will appear only in the form of occluded air bubbles suspended and movable in water. The soil can simply be regarded as a two-phase material containing a compressible fluid. Suction only plays a very limited role in the behaviour of the soil and also this case is not considered further in this paper.

For many civil engineering applications, only the partially-continuous air phase and the internally-continuous air phase are of great concerns when dealing with unsaturated soils (Bao et al., 1998). For the partially-continuous phase, air in the pores is still connected to the atmosphere. When the soil is compressed, the drained fluid is mainly air. However, for the internally-continuous phase, the drained fluid will be both air and water. Under this circumstance, the rate of consolidation will be slower than that of partially-continuous phase. This is because the rate of drainage of air is larger than that of water. This suggests that the mechanical behavior of these two phases is somewhat different from each other.

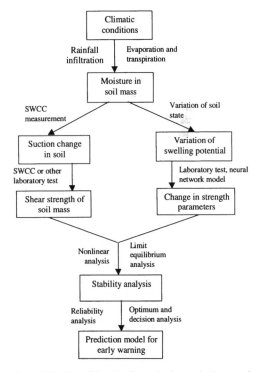

Figure 2. The line of thoughts for an "early warning" system for slope instability

17

2.2 A line of thoughts for the development of a simple early warning system

As discussed in the previous sections, the fundamental cause of landslides in expansive soils is the variation of moisture content in the soils. In turn, the variations of moisture content are mainly caused by changes in climatic conditions. Based on these, a line of thoughts for establishing a simple early warning system is developed and illustrated in Figure 2. The line of thoughts consists mainly of two chains. The left chain is for applications in general unsaturated soils whereas the right one with the assistant from the left chain, may be suitable for unsaturated expansive soils. From this figure, it implies that moisture (or degree of saturation) may be selected to be one of the key indices for the establishment of an early warning system.

2.3 Research topics for expansive soils

According to the flowchart shown in Figure 2, several research topics can be drawn for developing the simple early warning system and listed as follows:

(1) Studying soil-water characteristics for expansive soils;
(2) Investigation of the effects of rainfall infiltration and transient seepage on pore-water pressures;
(3) Development of the relationship between moisture content (or saturation degree, or volumetric water content), swelling potential and shear strength;
(4) Carrying out nonlinear mechanical analysis (Yi, 1995) and stress-deformation coupled analysis (Chen, 1998) for slope stability calculations;
(5) Establishment of some reliability and probabilistically based predictive models for providing early warning of slope instability in expansive soils.

Topics (1), (2) and (3) will be discussed in detail whereas topics (4) and (5) will not be covered in this paper.

3 STUDIES ON THE SOIL -WATER CHARACTERISTICS OF EXPANSIVE SOILS

3.1 Soil-water characteristic curve (SWCC)

A soil-water characteristic curve is commonly referred to as the relationship between matric suction and either water content or the degree of saturation. The curves are often plotted on a semi-logarithmic coordinate system. SWCC can usually be measured using a pressure plate apparatus in the laboratory. Figure 3 shows an idealized plot of SWCC. Two idealized characteristic points, A^* and B^*, can be identified. Points A^* and B^* correspond to the air-

entry value $((u_a-u_w)_b)$ and the residual water content (i.e., θ_r), respectively. For many soils, the portion between A^* and B^* may be approximated as a straight line. By comparing these two characteristic points with the characteristic points used for division of air phase patterns of unsaturated soils (Bao et al., 1998), it can be deduced that the portion between A^* and B^* were referred to be the partially-continuous and internally-continuous air phases.

3.2 Use of SWCC in unsaturated soil mechanics

SWCC is an important curve in the field of unsaturated soils. Not only hydraulic conductivity functions but also shear strength could be deduced empirically and mathematically from measured SWCCs. Vanapalli et al. (1996a, 1996b) proposed a formula for predicting the shear strength of unsaturated soils using a SWCC as follows:

$$\tau_f = c' + (\sigma_n - u_a)tg\varphi' + (u_a - u_w)\left(\frac{\theta - \theta_r}{\theta_s - \theta_r}\right)tg\varphi' \quad (1)$$

where, θ_s and θ_r are the saturated volumetric water content and the residual volumetric water content, respectively. If the saturated shear strength parameters (i.e., c and φ') are known, the shear strength of unsaturated soils can be estimated.

Based on Equation 1, Bao et al. (1998) idealized the portion between A^* and B^* of a SWCC to be a logarithmic linear form for practical applications. The equation can be simplified as,

$$\frac{\theta - \theta_r}{\theta_s - \theta_r} = m - n \cdot lg(u_a - u_w) \quad (2)$$

where m and n are the slope and the intercept of the linear part of SWCC with the water content axis, respectively.

Figure 4 shows the measured SWCC of a Nanyang expansive clay sample together with the best–fitted line using Fredlund & Xing's equation (Fredlund & Xing, 1994) and by using the simplified equation 2. It can be seen that the approximation

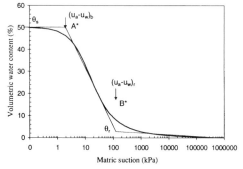

Figure 3. An idealized soil-water characteristic curve

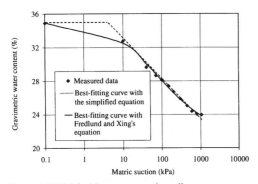

Figure 4. SWCC for Nanyang expansive soil

2. It can be seen that the approximation may be good enough for engineering purposes.

On the basis of Equations 1 and 2, a practical equation for estimating the shear strength of unsaturated soils is developed as follows:

$$\tau_f = c' + (\sigma_n - u_a) tg\varphi'$$
$$+ (u_a - u_w)[m - n \cdot \lg(u_a - u_w)] \cdot tg\varphi' \tag{3}$$

Thus, the prediction of shear strength for unsaturated soils probably becomes more convenient and practical.

3.3 Factors affecting SWCCs

The shape of a soil-water characteristic curve is influenced by many factors, which include soil type, initial water content during compaction, dry density, stress state, drying and wetting history, soil structure and many others.

3.3.1 Soil type

Different soils possess various particle and pore size distributions, soil plasticity, structure and fabric. Figure 5 shows some idealized SWCCs for three different soils. Generally speaking, the air-entry value of sand is relatively lower than that of silt, which in

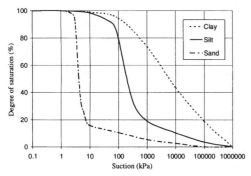

Figure 5. Idealized SWCCs for different types of soils

turn is lower than that of clay. The residual water content increases with an increase in plasticity. The desaturation rate is in proportion with the particle size. In other words, the larger the particle size, the faster the rate of drainage of water from the soil.

3.3.2 Initial water content and dry density

Figure 6 shows the measured SWCCs of two expansive clay specimens prepared at different initial water contents. It appears that the higher the initial water content (i.e. higher void ratio or larger pore size), the lower the air-entry value is. Similar results were obtained for some weathered soils in Hong Kong (Ng & Pang, 2000b). As the matric suction continues to increase, the water contents of the samples appear to approach each other.

The effects of initial dry density on SWCCs of two expansive soil samples are shown in Figure 7. The initial dry densities (ρ_d) for the samples are 1.51 g/cm^3 and 1.60 g/cm^3. It is clear that the initial dry density has a substantial effect on the air-entry value. The lower the dry density of the specimen, the smaller the air-entry value is. In addition, some studies on SWCCs were carried out by Chao et al. (1998) on a highly-overconsolidated expansive clay (Pierre shale) and a black plastic marine clay. Re-

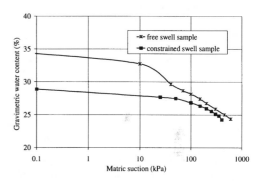

Figure 6. Influence of different initial water contents on SWCCs of Nanyang expansive soil

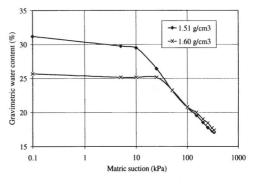

Figure 7. Influence of different dry densities on SWCCs of Nanyang expansive soil

19

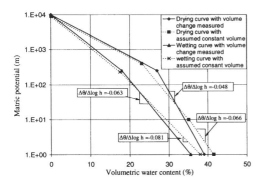

Figure 8. Comparison of SWCCs with and without volume change measurements for Pierre Shale (Chao et al, 1998)

sults from two of their tests during drying and wetting are shown in Figure 8. It was reported that the SWCCs with and without volume change measurements were quite different. The index of ($\Delta\theta/\Delta\log h$) differs by as much as 37.5% and 28.6% for the drying curve and the wetting curve, respectively. This implies that measurement of volume change is very important for expansive soils.

3.3.3 Influence of stress state
A soil-water characteristic curve of a soil is conventionally measured by means of a pressure plate extractor in which no vertical or confining stress is applied, whereas in practice the soil is usually subjected to some loading. To investigate the influence of stress state on the SWCC, Ng & Pang (1999, 2000a) developed a new stress-control pressure plate apparatus to measure the SWCCs under K_0 stress conditions. Some results are shown in Figure 9. It can be seen that the air-entry values of the specimens increase with the vertical load applied, and that the rate of reduction in volumetric water content decreases with the vertical load. The observed behaviour is attributed to the reduction in the pore sizes under load. The SWCCs subjected to the applied external stress are called stress dependent soil-water characteristic curves (SDSWCCs).

By using the same one-dimensional stress-control pressure plate apparatus and a newly developed isotropic stress-control apparatus (Ng et al., 2000a), three recompacted expansive soil specimens from Nanyang (ZY) were tested and some results are shown in Figure 10. These three specimens were compacted at the same initial water content and density but subjected to zero stress, one-dimensional and isotropic stress state of 50kPa after saturation. It is clear from the figure that the air-entry value is governed by the stress level. The higher the applied stress, the larger the air-entry value will be. Similar results on weathered soils are also reported by Ng and Pang (2000b). On the other hand, no significant

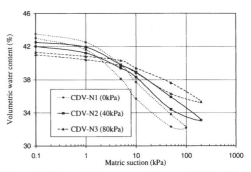

Figure 9. Effects of K_0 stress condition on SWCCs (Ng & Pang, 1999)

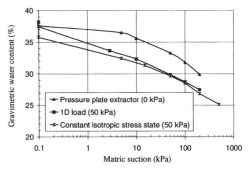

Figure 10. SWCCs of an expansive soil from Nanyang under different stress conditions

difference in the air-entry values can be seen between the one-dimensionally and isotropically loaded samples. The rate of desaturation of the three samples appear to be similar during the drying process for the suction range from 10 to 100 kPa. This observed behaviour is somewhat different from test results on a recompacted completely decomposed volcanic (CDV) soil. A decrease in the desaturation rate has been attributed to the applied stress (Ng & Pang, 2000b).

3.3.4 Influence of drying and wetting history
Very often unsaturated soils in the field are subjected to repeated cycles of drying and wetting throughout the year. Figure 11 shows the test results on a recompacted CDV sample subjected to three drying and wetting cycles (Ng et al., 2000b). As expected, a hysteretic loop appears between each drying and wetting paths. The size of the hysteretic loop is the largest in the first cycle but it becomes smaller with its shape remains almost constant during the subsequent drying and wetting cycles. In addition, the rate of desaturation is clearly higher in the first drying cycle than that of the subsequent drying cycles. Moreover, the rate of adsorption is faster during the first wetting and it becomes slower in the subsequent wetting cycles, especially between 5 kPa

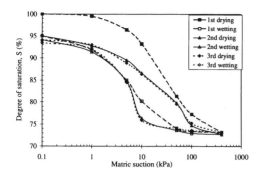

Figure 11. Influence of repeated cycles of drying and wetting on SWCC of CDV (Ng et al, 2000b)

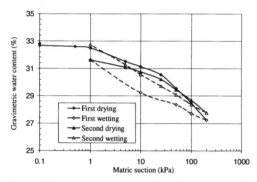

Figure 12. Influence of repeated cycles of drying and wetting on SWCC of expansive soil

and 40 kPa matric suction. Except the first wetting curve, the other two curves return approximately to their initial degree of saturation.

The observed relationship between the gravitimetric water content and matric suction during two drying and wetting cycles of a natural expansive soil sample is shown in Figure 12. As expected, hysteretic loops also exist. However, the size of the second hysteretic loop is substantially smaller than the first loop. For a given suction, there is a significant increase in water content between the first and the second wetting cycles, resulting in the reduction in size of the hysteretic loop. It is believed that the significant increase in water content during the second wetting path is attributed to the distinct swelling characteristics of the expansive soil.

4 STUDIES OF RAINFALL INFILTRATION

Variations in the water content of the soil in an unsaturated soil slope may be crucial to the slope stability. An increase in the water content will cause a reduction of matric suction (or negative pore-water pressure) in the soil or even destroy it completely. Moreover, it may lead to strong swelling in some

soils such as expansive soil and hence to reduce their shear strength. Rainfall infiltration is one of the major factors to cause variation of water content in soils. Therefore, a key issue for evaluating slope stability is to study transient seepage and to investigate the distributions and variations of pore-water pressure in the slope.

The physical processes of rainfall infiltration into unsaturated soils and the influence of infiltrated rainwater on matric suction and hence on unsaturated soil slope stability have been investigated by many researchers both in the laboratory (Fredlund & Rahardjo, 1993) and in the field (Liu & Yang, 1988; Lim et al., 1996; Zhang et al., 1997; Rahardjo et al., 1998). The common methodologies used include analytical and numerical analyses, simulated rainfall tests in the field and in-situ measurements. However, it is well-recognised that the measurement of soil suction is difficult both in the laboratory and on site and the accuracy of the measurement may not be desirable. Therefore, analytical and numerical methods can play an effective and complementary role to help academics and engineers to improve their understanding of the process and effects of rainwater infiltration on pore-water pressure distributions. The pore-water pressure or matric suction distribution depends not only on the air/moisture movement, but it is also governed by the deformation of soil due to an applied load or even ground temperature variations. Strictly speaking, a seepage-deformation-temperature coupled model should be constituted for simulating rainwater infiltration problems. However, the coupled model will be rather complicated and input parameters may be difficult to obtain. Therefore, at present, it may be reasonable to study uncoupled transient seepage using some analytical solutions and numerical approaches for investigating governing factors which affects rainwater infiltration and distribution of pore water pressures in unsaturated soils.

4.1 *Analytical solution for one-dimensional transient infiltration in homogeneous and layered soils*

For investigating transient seepage, an infinitely long soil slope with a slope angle, γ, to the horizontal is considered in Figure 13. The groundwater table is assumed to be parallel to the slope surface. For an isotropic soil, the two-dimensional (2D) differential equation governing water flow in unsaturated soil can be written as follows:

$$\frac{\partial}{\partial x}\left(k\ \frac{\partial \psi}{\partial x}\right)+\frac{\partial}{\partial z}\left(k\ \frac{\partial \psi}{\partial z}+k\right)=\frac{\partial \theta}{\partial t} \qquad (4)$$

For the long slope, it can be assumed that the equipotential lines of negative pore-water pressure above the main water table are also parallel to the slope surface. To obtain an analytical solution of the equation, the rotated coordinates, x* and z*, are used, as shown in Figure 13. Thus, the governing differential equation with respect to z* can be derived and simplified as follows:

$$\frac{\partial}{\partial z_*}\left(k\,\frac{\partial \psi}{\partial z_*}\right)+\frac{\partial k}{\partial z_*}\cos\gamma = \frac{\partial \theta}{\partial t} \qquad (5)$$

It can be seen that Equation 5 is identical to the 1D infiltration equation, except that the second term of it is multiplied by cos γ. To obtain the analytical solution, properly selected hydraulic characteristic funtions can be adopted to linearize the equation. Similar to Srivastava & Yeh (1991), exponential functional forms are assumed to represent the hydraulic conductivity function and SWCC,

$$\begin{cases} k = k_s e^{\alpha\psi} \\ \theta = \theta_r + \left(\theta_s - \theta_r\right)e^{\alpha\psi} \end{cases} \qquad (6)$$

where α is a coefficient representing the rate of reduction in the water content or hydraulic conductivity of the soil with increasing matric suction. The exponential assumption is generally accurate enough to represent the hydraulic properties for geotechnical engineering applications except that near full saturation (Bao et al., 1998).

With the two exponential functions, Equation 6 is transformed to the following linear equation,

$$\frac{\partial^2 k}{\partial z_*^2}+\alpha\cos\gamma\,\frac{\partial k}{\partial z_*} = \frac{\alpha\left(\theta_s - \theta_r\right)}{k_s}\frac{\partial k}{\partial t} \qquad (7)$$

By specifying some appropriate initial and boundary conditions, analytical solutions of Equation 6 can be obtained using the Laplace's transformation for two specific cases, (a) a single layered homogeneous isotropic soil slope, and (b) a two-layered homogeneous isotropic soil slope. Some details of the solutions are given by Zhan et al. (2000). The calculated negative pore-water pressure distributions for a 200cm thick (h), 45° (γ) slope with one and two soil layers are shown in Figures 14 and 15, respectively. An initial rainfall intensity (q_A) of 0.1 cm/s was adopted to calculate the steady-state conditions for a subsequent 24-hour rainfall of 2.225 cm/s (q_b). For the two soil layers (upper h_1 = lower h_2 = 100cm), two different saturated hydraulic conductivities, K_{s1}=2.5 cm/s and K_{s2}=10 cm/s, for the upper and lower soil layers, respectively. The movement of wetting band with time can be clearly seen from both the figures.

4.2 In-situ artificial rainfall infiltration tests

To investigate the physical process of rainfall infiltration and its influence on matric suction in unsaturated soil, artificial rainfall infiltration tests were

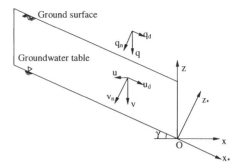

Figure 13. Coordinate systems and resolution of vectors

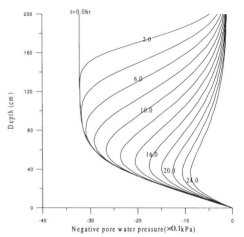

Figure 14 Pore-water pressure profiles at various time intervals α=0.1, h=200 cm, θ_s=0.45, θ_r=0.15, K_s=2.5 cm/s, q_A=0.1 cm/s, q_B=2.225cm/s, γ=45°

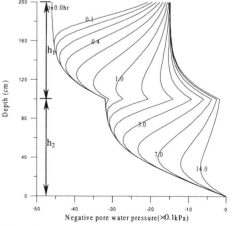

Figure. 15 Pore water pressure profiles at various time intervals α_1=α_2=0.1, h_1=h_2=100 cm, θ_{s1}=θ_{s2}=0.45, θ_{r1}=θ_{r2}=0.15, q_A=0.1 cm/s, q_B=2.225cm/s, γ=45°, K_{s1}=2.5 cm/s, K_{s2}=10 cm/s

22

Table 2. Details of rainfall infiltration simulations in the field

Stages	1		2	3
Duration (hour)	64		32	6.2
Time to start (time, date, month)	14:00, 5, 4	1:00, 6, 4	6:00, 8, 4	14:00, 9, 4
Elapsed time (hours)	11	64	96	102.2
Rainfall intensity (10^{-4}cm/s)	1.465	1.389	0.0	2.778

conducted on a horizontal bare ground surface with a test area of 1×1 m^2 by the Geotechnical Department of Yangtze River Scientific Research Institute (Zhang et al., 1997). The soil consists of 20 to 30 m thick completely-decomposed granite overlying granite weathered to different degrees. The saturated water coefficient of permeability of the completely-decomposed granite varies from 3.06×10^{-6} to 4.10×10^{-7} m/s. The entire in-situ simulation process was divided into three stages and details of the simulation stages are given in Table 2.

In the first stage, the intensity and duration of rainfall adopted were $1.389\sim1.465\times10^{-4}$ cm/s and 64 hours respectively. After a 32-hour cessation of rainfall, an intensive artificial rainfall (2.778×10^{-4} cm/s) was imposed for 6 hours and 12 minutes. Two axi-symmetric boreholes with respect to the center of the area were drilled for the installation of tensiometers. The distance of each hole from the center was 10 cm. In the first hole, 6 tensiometers were installed at 15cm interval along the depth from 10 cm below the ground surface. In the second hole, another 6 tensiometers were installed at the same interval along the depth from 25cm below the ground surface. Outside the artificial rainfall infiltration area, two additional boreholes were drilled at a distance of 10 cm from the 1m^2 rainfall infiltration boundary. In these two boreholes, 12 tensiometers were installed at the same 15cm interval along the depth of each hole. The former and the latter two holes were designed to investigate the influence of 1D rainfall infiltration and lateral flow of the infiltrated water on variations of pore water pressure, respectively. Due to the space constraint, only the measurements from the first two boreholes during the first stage (64 hours) of the rainfalls are reported in this paper. Figures 16 and 17 show the measured variations of negative pressure (suction) head and measured water content profiles along the depth with time, respectively. It can be seen that the initial suction of about 14 kPa was destroyed at the first 3 hours. This corresponds well with an increase in water content from about 15% to 27% (refer to Fig. 17). At the end of the rainfall (64 hours), the initial suctions from the ground surface to a depth of 80cm were completely destroyed due to rainwater infiltration. The measured reductions of suction with the duration of rainfall are consistent with the measured increase in water contents (compare Figs 16 & 17). However, the negative pore water pressures and water contents in the soil deeper than 80cm showed little changes during

the duration of the artificial rainfall. This might be attributed to an inclined crack existing at the depth of 78cm, draining the infiltrated rainwater from above. This observed phenomenon seems to imply that any randomly-distributed crack could be of great significance on the resulted distribution of water content and matric suction in the soil, due to the interruption of the physical process of infiltration. Realistic and proper numerical simulations of the in-situ infiltration tests would certainly help to improve the understanding of the observed phenomenon.

4.3 *Factors affecting the effects of rainfall infiltration*

It is recognised that there are many factors affecting the effects of rainfall infiltration on pore water pres-

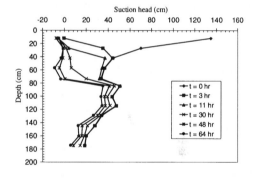

Figure 16. Measured variation of suction head profiles during rainfall (data from Zhang et al, 1997)

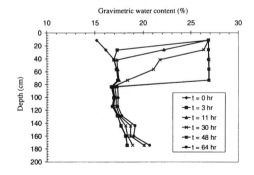

Figure 17. Measured variation of water content profiles during rainfall (data from Zhang et al, 1997)

23

sure distributions in the ground. Generally, the factors may be classified as two categories: (a) external conditions such as rainfall pattern, intensity and duration and (b) internal conditions such as hydraulic properties of soil such as SWCC and permeability and soil conditions such as initial density and water content, uniformity, homogeneity and distribution of cracks if any.

4.3.1 *Effects of rainfall patterns*

Ng et al. (1999 & 2000c) respectively carried out a 2D and 3D parametric study on the effects of rainfall patterns on the distribution of matric suction in an initially unsaturated cut slope.

In the 3D parametric study using the finite element method, three typical rainfall patterns in Hong Kong were considered (refer to Fig. 18), namely, advanced pattern (A), central pattern (C) and delayed pattern (D). Based on the computed results from simulations of a total amount of rainfall of 358 mm with different rainfall patterns within a 24-hour period (corresponding to 10-year return period) and a continuing simulation of 168 hours after the cessation of the rainfall; the following conclusions were drawn:
(a) Depending on the initial pore-water pressures in the ground, rainfall patterns can have different degrees of influence on the groundwater responses. When the initial pore water pressures in the ground are high (i.e. small suctions), the influence of different rainfall patterns on the total amount of rainfall infiltration is limited. This is because the high water permeability existed in the ground facilitates intensive infiltration. The variations of the actual infiltration rate with time follow closely with the rainfall patterns. By plotting changes in the ratio of surface flux to instantaneous water permeability (*q/k*) with time, the actual ground water responses can be easily interpreted.
(b) When the initial pore-water pressures in the ground are low (i.e. high suctions), the influence of different rainfall patterns on the ground water response is most significant. The advanced type A and the delayed type D lead to the highest and the lowest rate of infiltration and total amount of infiltration, respectively. This implies that the former pattern will be the most critical one for slope instability.

4.3.2 *Effects of stress-dependent SWCC (SDSWCC)*

In order to illustrate the influence of SDSWCCs (refer to Fig. 9) on pore water pressure distributions in an unsaturated completely decomposed volcanic (CDV) soil slope, 8.6m height inclined at 55° to the horizontal, two series of transient seepage analyses are conducted (Ng & Pang, 2000a). In the first series of analyses, all soil layers are assumed to have the same drying SWCCs and their corresponding water permeability functions. This series is a conventional approach. In the second series (unconventional) of analyses, different hydraulic properties are specified according to the soil depth (i.e. stress level) using the measured SDSWCCs under 40 kPa (CDV-N2) and 80 kPa (CDV-N3) applied net normal stresses and their corresponding permeability functions. The initial groundwater conditions for each series of transient seepage analyses are established by conducting two steady state analyses, during which a very small rainfall with an intensity of 0.001mm/day is applied on the top boundary surface. For subsequent transient analyses, two rainfall patterns with an average intensity of 394mm/day and 82mm/day are applied on the top boundary surface in both series of analyses to simulate a short and intensive 24-hour rainfall infiltration and a prolonged 7-day rainfall infiltration respectively. The computed pore water pressure distributions for each series are shown in Figure 19. It is clear that there is a substantial difference between the initial pore water pressure distributions computed using the conventional drying SWCCs and the unconventional wetting SWCC and SDSWCCs in the steady state analyses. The conventional analysis predicts a significantly higher soil suction profile than that computed by the unconventional analysis. This is because the soil in the former analysis, in comparison with the soil in the latter analysis, has a lower air-entry value and a faster rate of changing volumetric water content as values of soil suction increase and a higher water permeability function. In other words, the soil under the applied stress has a stronger capability to retain moisture for a given soil suction due to the presence of a smaller pore size distribution, as illustrated by a flatter soil water characteristic curve (Fig. 9). The computed results highlight the importance of considering stress effects on SWCCs.

During the short but highly intensive rainfall (24 hours), the pore-water pressure responses are similar in both the conventional and unconventional transient analyses. Only the soil suctions in the top 1 to 2m depth are destroyed irrespective of the magnitude of their initial values. A relatively shallow ad-

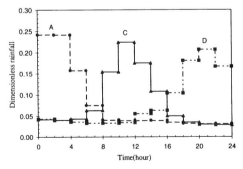

Figure 18. Selected typical rainfall patterns in Hong Kong

vancing "wetting front" is developed as most of the rainfall cannot infiltrate into the soil due to the relatively low water permeability to rainfall infiltration at high initial values of soil suction. On the contrary, the pore water pressure distributions predicted by the conventional and unconventional analyses are completely different during the 7-day low intensity prolonged rainfall. When the effects of stress state are considered, soil suctions at the mid-height of the slope are totally destroyed by the advancing "wetting front" to a depth of about 9m from the ground surface. This is because of the relatively low initial values of suction and the relatively high water permeability with respect to rainfall infiltration present in the soil. The less intensive but prolonged 7-day rainfall facilitates the advancement of "wetting front" into the soil to some great depths and cause significant reduction in soil suction which would have some devastating effects on slope stability. By using the limit equilibrium method, variations of factor of safety (FOS) with duration of rainfall for four almost equally-spaced circular trial surfaces (S1 to S4) along the cut slope are calculated and illustrated in Figure 20. S1 and S4 correspond to approximately 2m and 8m deep slip surfaces respectively. It can be seen that the limit equilibrium analyses, which adopted the pore-water pressures computed by using the conventional drying SWCCs, predicted substantially higher initial factors of safety of all four slip surfaces than those obtained from the analyses using the unconventional wetting SDSWCCs. This is attributed to the significant difference in the computed pore water pressure distributions (see Fig. 19) with and without considering the effects of the stress state and drying-wetting history. This implies that the traditional analyses using the conventional drying SWCCs may lead to unconservative designs.

4.3.3 *Effects of hydraulic parameters*
As discussed in Section 3.2 and Equation 6, an idealised SWCC may be characterised by three parameters, i.e., desaturation coefficient α, saturated and residual volumetric water contents, θ_s and θ_r, as illustrated in Figure 3. For analyzing transient seepage, saturated water permeability (K_s) and a permeability function are also needed. Zhan et al. (2000) used analytical solutions to investigate the influence of these four parameters during 1D rainfall infiltration in a 100cm (h) deep homogeneous unsaturated soil layer. Different initial (or antecedent) rainfall intensity (q_A) and subsequent rainfall infiltration rate (q_B) are considered.

4.3.4 *Effects of ground condition*
Very often natural ground conditions are not isotropic and also some impeding layers may exist. The effects of anisotropic water permeability on pore water pressure distributions and slope stability

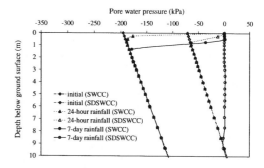

Figure 19. Pore water pressure distributions along the mid-height of the slope under various rainfall conditions

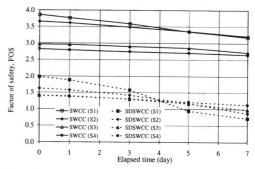

Figure 20. Variation of FOS with duration of rainfall

have been discussed and studied by a number of researchers (Fredlund & Rahardjo, 1993; Ng & Shi, 1998; Ng et al., 1999). Figure 21 shows the variation of FOS with ratio of horizontal (k_x) to vertical (k_y) water permeability for an unsaturated cut slope in Hong Kong. The results shown in the figure are based on numerical simulations of a 24-hour rainfall with 370mm rain depth. It can be seen that the ratio of k_x to k_y seems to play an important role in slope stability. This implies that a large anisotropic water permeability ratio (i.e., k_x/k_y) may induce adverse pore water pressure conditions in the ground and cause slope instability, as compared with isotropic conditions. Thus, it should be cautious to simplify any design analysis and calculation if the ground conditions are not known for sure.

The soil is rarely homogenous in many slopes. Very often a layer of relatively impermeable soil layer such as clay seam may exist in weathered slopes. This relatively impermeable soil layer may be referred to as an impeding layer. Ng et al (1999) carried out some parametric analyses to study the influence of an impeding layer on pore water pressure distributions and hence the slope stability. They assumed that the impeding layer lies parallel to a 55° cut slope in a weathered soil in Hong Kong at varying elevations during their parametric analyses. The

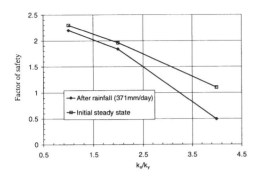

Figure 21. Influence of anisotropy permeability on FOS

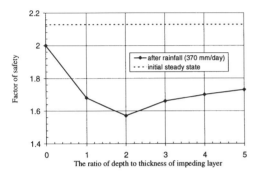

Figure 22. Influence of impeding layer on FOS

elevation of the impeding layer varies and an expression (d/t) is used to represent the elevation of the impeding layer for each parametric analysis, where d is the distance from the sloping ground surface to the bottom of impeding layer and t is the thickness of impeding layer, which is assumed to be 1m.

Figure 22 shows the variations of FOS with the (d/t) ratio. Initially the FOS decreases with an increase in the (d/t) ratio until a critical value, at which the FOS is the lowest, is reached. After reaching the critical ratio, the stability of the slope improves as the (d/t) value further increases. In this particular study, the critical d/t value is equal to 2. The existence of a critical (d/t) ratio may be explained by considering two extreme cases, d/t=1 and d/t=20 (say). At one extreme (i.e., d/t=1), this implies that the impeding layer is just located at the ground surface. Very little infiltration is expected and hence the stability of the slope is not likely to be affected by rainfall. At the other extreme (i.e., d/t=20), the presence of the impeding layer is not likely to affect the distributions of pore water pressures at shallow depths. The stability of the slope is also not likely to be affected by rainfall. Therefore, there exists an intermediate but critical d/t ratio at which the distributions of pore water pressures and hence the stability of the slope will be affected most.

It is generally recognised that the existence of cracks has a major impact on pore water pressure distributions, as illustrated in Figure 16 and the stability of slopes, particularly in expansive soils. Very often, well-developed crack network can be found in many expansive soils. However, it is a pity that relatively few research results can be found in the literature. This may be a good research and challenging topic for many researchers and engineers to take up in the future.

Although the review and study on the effects of rainfall infiltration discussed in the previous sections are in light of some non-expansive unsaturated soils, some of the general discussion points and conclusions are clearly also applicable to unsaturated expansive soils. Of course, problems for expansive soils would be more complex. This is because additional volume change will take place if there is an increase in water content in the soils, as illustrated in Figure 12. In other words, physical and mechanical properties of unsaturated expansive soils will change with any variation of moisture content. For example, an increase in water content will lead to a reduction in dry density and shear strength of soils, but an increase in water permeability. More discussion on unsaturated expansive soils is presented in the following section.

5 RELATIONSHIP OF SWELLING POTENTIAL, MOISTURE AND SHEAR STRENGTH

This section reviews the relationship between swelling potential, moisture content and shear strength of expansive soils, in particular for those along the middle-route of the S/N Water Transfer Canal project. The study of each component of the relationship and the relationship between them will, in fact, form the 'right chain' of the flow chart shown in Figure 2.

5.1 Predictive method for swelling potential

The term 'swelling potential' used here refers to either the ability of swelling capacity, or swelling pressure or swelling percentage of expansive soils. Many different soil properties are related to this potential including soil suction. The relation of the Atterberg limits, natural density and moisture, swelling potential and soil suction have studied and reported and some prediction models have been proposed in the literature (Komornik, 1980; Vijayvergiya, 1973).

It is generally known that swelling potential is influenced by many factors including both intrinsic and external factors. The intrinsic factors mainly consist of microstructure, mineralogy, physical states and various physio-chemical properties. The external factors refer to rates of evaporation and transpiration, humidity and temperature.

Behaviour of unsaturated expansive soils is very complex. It is extremely difficult to predict it accurately as the behaviour is governed by many factors. For the systematic study of the inter-relationships among these factors and their influence on the behaviour of expansive soils, the artificial neural network method should be an effective and useful tool for academics and engineers.

5.2 Brief descriptions of artificial neural network (ANN)

ANN is a network inter-connected by numerous neurons (or nodes), which resemble the biological neural cells in human beings. Its advantage lies in the fact that it can grasp the complicated non-linear relationships between inputs and outputs by learning from given samples. After learning and training, it can be used for predicting. In fact, the ANN method has already been applied to evaluate slope stability (Xu & Huang, 1994).

An ANN model should be established not only for processing a large amount of inter-related data, but it should also be trained for predicting. A popular network - Back Propagation (BP) is commonly used. It usually consists of three layers: input, hidden and output. They are feed forwardly and connected by interconnection weights of every layer. A schematic diagram of ANN is shown in Figure 23. In addition to the interconnection weights of every layer, there is also a highly non-linear function, sigmoid function in general, to reflect the relationship between input and output.

The learning process of network consists of two flow procedures: feedforward propagation and reverse propagation. The network is 'trained' using a known sample setting on the input layer, based on the feedforward propagation, calculating the output for every layer. If the error of output is beyond the tolerance, errors from the output layer are back propagated to the input layer by means of 'gradient descent method' in order to reduce the errors. By repeating iteratively such calculations until the square root deviation of actual output and desired output

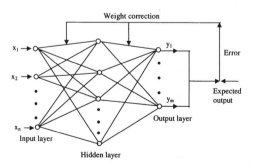

Figure 23. Schematic diagram of ANN

meets the given requirement, the 'trained network' can then be utilized for predicting.

5.3 Prediction of swelling potential for Nanyang expansive soils

5.3.1 Establishment of BP network
Li & Chen (1994) selected eight major governing factors to be the input nodes for establishing the network. These factors include natural moisture, natural density, degree of saturation and void ratio, liquid limit, plastic index, specific surface and cation exchange capacity (CEC). Test results from 16 natural samples were used for the network. Two output nodes, namely, swelling pressure and swelling capacity were predicted in terms of two BP networks. A three-layer BP model with a structure of 8×10×1 and the full connection between layers were selected in the analyses. Results from 12 out of the 16 samples were used for training network, the remaining 4 samples were reserved for verifying the output, named expected output. A computational program of BP network was specially developed. The network was then trained by using the 12 data sets. After learning and training, the trained network was used to predict the swelling potential and swelling pressure of the remaining 4 data sets. The comparisons between the predicted and the measured values and errors are illustrated in Table 3.

5.3.2 Determination of influential level of each factor
The influence level for each factor is summarized in Table 4. It is obtained by the summation of the absolute values of weights, corresponding to the neuron of every factor as shown in Table 4. It is found that factors, which have considerable influence, are the degree of saturation, liquid limit, specific surface and cation exchange capacity.

5.3.3 Comparisons of ANN and regression method
It is interesting to compare the predicted results between the ANN and a conventional regression method. Vijayvergiya & Ghazzaley (1973) proposed a method for predicting swelling pressure, P, and swelling ratio based on test results of 270 intact samples under $0.1t/ft^2$. He suggested an empirical formula according to the linear relation between values of swelling capacity or swelling pressure (in logarithm scale) and natural dry density, γ_d, liquid limit, LL, or water content, w. The formula for swelling pressure is shown as follows:

$$\log P = (0.4LL - w - 0.4)/12 \tag{8}$$

or

$$\log P = (\gamma_d + 0.65LL - 139.5)/19.5 \tag{9}$$

By making use of Equation 8 together with the re-

27

sults from the first 12 samples, the predicted values are listed in Table 5. It can be seen that the accuracy of predictions from the ANN model is much better than those from the regression method. ANN seems to be an appropriate and reliable tool for predicting behaviour of unsaturated expansive soils.

5.4 Prediction of shear strength for unsaturated expansive soils

Based on some direct shear tests on expansive soil specimens described by Liu (1997) in China, it is reported that the shear strength of expansive soils varies inversely with the moisture content in the field and laboratory direct shear tests.

Komornik et al (1980) described some test results on shear strength and swelling behavior of expansive clays. Relatively speaking, there is a lack of systematic study of shear strength and stress-strain behavior of unsaturated expansive soils under well-controlled suction conditions, together with reliable volume measurements.

It is generally accepted that the shear strength of unsaturated expansive soils is closely related to the moisture and swelling pressure. Some empirical equations were proposed to predict shear strength and swelling pressure by using water content with limited success. In recent years, the relationship of swelling pressure and shear strength has been studied by Lu et al. (1999). Alonso (1998) attempted to

Table 3. Comparison between measured and trained results of neural network

Sample No.	Swelling ratio			Swelling pressure		
	Measured (%)	Predicted (%)	Relative error (%)	Measured (kPa)	Predicted (kPa)	Relative error (%)
1	15.3	19.9	30.07	105	125	19.05
2	40.4	41.0	1.49	1190	1319	10.84
3	39.0	35.0	10.26	950	987	3.89
4	25.1	17.0	32.27	970	123	87.32
5	28.2	19.0	32.62	27	42	55.56
6	14.8	13.7	7.43	306	9	97.06
7	9.9	10.1	2.02	110	101	8.18
8	17.2	16.0	5.98	48	57	18.75
9	15.7	12.2	22.29	35	41	17.14
10	5.1	3.9	23.53	82	147	79.27
11	10.9	13.0	19.27	138	137	0.72
12	17.3	19.7	13.87	527	78	85.20

Table 4. Influence level on swelling properties

Parameter	Influence level (%)	Parameter	Influence level (%)
Natural water content	8.37	Liquid limit	12.94
Natural dry density	11.14	Plastic limit	11.97
Void ratio	10.95	Surface ratio	13.63
Degree of saturation	18.64	Cation exchange ratio	12.36

Table 5. Comparison between different predictive methods

Sample No.	Swelling pressure (kPa)						
	Measured (kPa)	BP network		Equation (3)		Equation (4)	
		Predicted (kPa)	Relative error (%)	Predicted (kPa)	Relative error (%)	Predicted (kPa)	Relative error (%)
13	14	17	21.43	117	735.71	39	178.57
14	87	102	17.24	345	296.55	110	26.44
15	1600	1258	21.38	284	82.25	438	72.62
16	38	35	7.89	70	84.21	48	26.32

develop a constitutive model for unsaturated expansive soils. However, the current understanding of shear behaviour and stress-strain relationship of unsaturated expansive soils is fairly limited. It is obviously that further work is needed to improve understanding and developing predictive tools.

6. DISCUSSION AND CONCLUSIONS

a) The proposed idea and path of developing an early warning system for unsaturated expansive slopes is not only of great significance for the construction of the South-to-North Water Transfer Canal but it will also help to improve understanding of unsaturated expansive soils.

b) There are two major "chains" in the development of the early warning system for the instability of expansive soil slopes. The first one is mainly concerned with pore-water pressure distribution and its variation associated with various intrinsic and external factors. The second one is to study the fundamental relationship between the state of expansive soils, swelling potential, stress-strain and shear strength of expansive soils. Based on the observed failure processes of a trial slope in unsaturated expansive soil, it is recognized that slope failure in an unsaturated expansive soil is likely to be ductile and hence the proposed early warning system would be easier to implement and be more effective.

c) The soil-water characteristic curve (SWCC) is important for research and application in the field of unsaturated soils. Hydraulic conductivity function and shear strength of unsaturated soils can be empirically predicted from SWCC with adequate accuracy for engineering practice. It is therefore important to investigate factors which will significantly affect SWCCs of unsaturated expansive soils. Based on the current reviews and studies, it is found that the most influential factors are soil type, initial and compaction water contents, initial dry density, stress state, swelling potential and drying and wetting cycles.

d) As it is very difficult to accurately and reliably measure soil suction in the field, analytical and numerical methods may be adopted as a complementary tool for calculating pore-water pressure distributions in unsaturated soil slopes. The prediction of pore-water pressure distributions is not only affected by intrinsic properties such as the shape of a SWCC, saturated permeability, permeability function and anisotropy, but it is also governed by external factors including rainfall pattern, intensity and duration, existence of cracks and impeding layer.

e) Conventionally, the influence of transient water flow, deformation and shear strength of soils on the stability of slopes has been investigated independently. As expansive soils posses the distinct swelling and shrinkage characteristics, a fluid flow-deformation-shear strength coupled model seems to be essential for accurate calculations. This is because any deformation and shear strength will be significantly affected by any volume change as a result of swelling and shrinkage of the soils.

f) The shear strength and swelling potential of an expansive soil mainly depend on the physical state of the soil such as moisture content and dry density. The complex relationship between the swelling potential and various physical parameters can be explored satisfactorily by the Artificial Neural Network (ANN). The study of the relationship of swelling potential and shear strength could be of significance for practical application and theoretical developments of expansive soils.

ACKNOWLEDGEMENTS

This research project is supported by research grants DAG99/00.EG23 & HKUST6046/98E provided by the Research Grants Council of the HKSAR. The authors would like to thank Messrs Zhan L.T. and Gong B.W. and Miss Wang Bin for conducting experiments and assisting in preparing figures presented in this paper. Their contributions are greatly acknowledged. Finally, the authors are also grateful to Professor Wilson Tang for his continual support to their research.

REFERENCE

Alonso, E. E. (1998). Modeling expansive soil behavior. Key-note lecture of 2nd Inter. Conf. on Unsat. Soils. Vol. 2, 37-70.

Bao, C. G. (1979). Air phase patterns of unsaturated compacted soils and pore pressure dissipation. Proc. of 3rd China Conference on Soil Mechanics and Foundation Engineering, Beijing (In Chinese).

Bao, C. G. (1998). On the analysis of canal slope stability for expansive soils. The Proc. of 2nd Symposium on Slope stability of expansive soils, Wuhan, 46-57 (In Chinese).

Bao, C. G. & Liu, T. H. (1988). Some properties of shear strength on Nanyang Expansive clay. Proc. of the International Conference on Engineering Problems of Regional Soils. International Acacdemic Publishers, 543-546.

Bao, C. G., Gong, B. W. & Zhan, L. T. (1998). Properties of unsatuarated soils and slope stability for expansive soils. Key-note lecture of 2nd Intern. Conf. on unsaturated soils. Proc. of 2nd International Conference on Unsaturated Soils. Vol. 2 71-98.

Barbour S. L. (1998). The soil-water characteristic curve: a historical perspective. Can. J. Geotech. J., Vol. 35, 873-894.

Chao, K. C., Durkee, D. B., Miller, D. J. & Nelson, J. D. (1998). Soil water characteristic curve for expansive soil. 13th Southeast Asia Geotechnical Conference, Taipei, 35-40.

Chen S. Y. (1998). Slope stability analysis considering the effects of infiltration and evaporation. The Proc. of 2nd Symposium on Slope Stability of Expansive Soils for South-to-North Canal Project, Wuhan, 58-62 (In Chinese).

Chen, Z. H. (1998) Nonlinear analysis on the interaction of water-air-deformation in expansive soil slopes. The Proc. of 2nd Symposium on Slope Stability of Expansive Soils for South-to-North Canal Project, Wuhan, 68-78 (In Chinese).

Fredlund, D. G. (1998). Some geotechnical engineering considerations related to the South-to-North Canal, China. The Proc. of 2nd Symposium on Slope Stability of Expansive Soils for South-to-North Canal Project, Wuhan, 32-45.

Fredlund, D. G. & Rahardjo, H. (1993). Soil mechanics for unsaturated soils. John Wiley & Sons, Inc., New York.

Fredlund, D. G. & Xing, A. (1994). Equations for the soil-water characteristic curve. Can. Geotech. J. 31: 521-532.

Komornik, A., Livneh, M. & Smucha, S (1980). Shear strength and swelling of clays under suctions. Proc. of the 4th International Conference on Expansive Soils. Vol. I, 206-226.

Lam, L., Fredlund, D. G. & Barbour, S. L. (1987). Transient seepage model for saturated-unsaturated soil systems: a geotechnical engineering approach. Can. Geotech.. J., Vol. 24, 565-580.

Li, Q. Y. & Chen, Z. Y. (1994). The study of predicting mechanical properties for swelling soil using artificial neural network technology. Yangtze River Scientific Research Institute Technical Report, No. 96-182 (In Chinese).

Lim, T. T., Rahardjo, H., Chang, M. F. & Fredlund, D. G. (1996). Effect of rainfall on matric suction in a residual soil slope. Can. Geotech. J. Vol. 33, 618-628.

Liu, T. H. (1997). The problem of expansive soils in engineering construction. Chinese Architecture and Building Press (In Chinese).

Liu, T. H. & Yang, X. K. (1988). Field monitoring on an expansive soils slope during failure. Proc. of the International Conference on Engineering Problems of Regional Soils, International Academic Pubilishers.

Lu, Z. J. (1999). Explorations on the suctional shear strength of unsaturated soils. Journal of Chinese Railway Science, Vol. 20, No. 2, 10-16 (in Chinese).

Lu, Z. J. & Wu, X. M. Sun, Y. Z. (1999). The role of swelling pressure in the shear strength theory of unsaturated soils. Chinese Journal of Geotechnical Engineering. Vol. 19, No. 5, 20-27 (In Chinese).

Miao, L. C., Zhong, X. C. & Yin, Z. Z. (1998). The relationship between strength and water content of expansive soil. Rock and Soil Mechanics, Wuhan, Vol. 20, No. 2, 71-75 (In Chinese).

Ng, C. W. W. & Pang, Y. W. (1998). Role of surface cover and impeding layer on slope stability in unsaturated soils. 13th Southeast Asian Geotechnical Conference, Taipei, 135-140.

Ng, C. W. W, Chen, S. Y. & Pang, Y. W. (1999). Parametric study of effects of rain infiltration of unsaturated slopes. Rock and Soil Mechanics, Vol. 20, No. 1, 1-14 (In Chinese).

Ng, C. W. W. & Pang, Y. W. (1999) Stress effects on soil-water characteristics and pore water pressure. 11th Asian Regional Conference on Soil Mechanics and Geotechnical Engineering, Korea, 371-374.

Ng, C. W. W. & Pang, Y. W. (2000a). Influence of stress state on soil-water characteristics and slope stability. J. Geotech. & Geoenviron. Engrg., ASCE. Vol. 126, No. 2. 157-166.

Ng, C. W. W. & Pang, Y. W. (2000b). Experiments investigations of soil-water characteristics of a volcanic soil. Provisionally accepted by Can. Geotech. J.

Ng, C.W.W. & Shi, Q (1998). A numerical investigation of the stability of unsaturated soil slopes subjected to transient seepage. Computers and Geotechnics, Vol. 22, No. 1, 1-28.

Ng, C. W. W., Wang, B., Gong, B. W. & Bao, C. G. (2000a). Preliminary study on soil-water characteristics of two expansive soils. Proc. of Asian Conference on Unsaturated Soils, Singapore.

Ng, C. W. W., Pang, Y. W. & Chung, S. S. (2000b) Influence of drying and wetting history on stability of unsaturated soil slopes. 8th International Symposium on Landslides, Cardiff, UK.

Ng, C. W. W., Tung, Y. K., Wang, B. & Liu, J. K. (2000c). 3D analysis of effect of rainfall patterns on pore water pressure in unsaturated slopes. Submitted to Geo Eng 2000 International Congress, Melbourne, Australia.

Rahardjo, H., Leong, E. C., Gasmo, J. M. and Deutscher, M. S. (1998). Rainfall-induced slope failures in Singapore: investigation and repairs. Proc. 13th Southeast Asian Conf. Taipei, Vol.. 1, 147-152.

Sun, H. W., Lam, W. H. Y., Ng, C. W. W, Tung, Y. K. & Liu, J. K. (1999). The Jul. 2 1997 Laiping road landslide. H. K. - hydrological characterization. Submitted to 8th International Symposium on Landslide, Cardiff, UK.

Sun, S. G., Chen, Z. H. & Huang, H. (1998). The method of stability analysis for expansive soils with the consideration of rainfall infiltration. The Proc. of 2nd Symposium on Slope Stability of Expansive Soils for South-to-North Canal Project, Wuhan, 63-67 (In Chinese).

Srivastava, R. & Yeh, T. C. J. (1991). Analytical solutions for one-dimensional, transient infiltration toward the water table in homogeneous and layered soils. Water Resour. Res., Vol. 27, No. 5, 753-762.

Van Genuchten (1980). A closed-form equation for predicting the hydraulic conductivity of unsaturated soils. Soil Science of America Journal, Vol. 44, No. 5, 892-898.

Vanapalli, S. K., Fredlund, D. G. & Pufahl, D. E. (1996 a). The relationship between the soil-water characteristic curve and the unsaturated shear strength of a compacted glacial till. Geotechnical Testing Journal, GTJODJ, Vol. 19, No. 3, 259-268.

Vanapalli, S. K., Fredlund, D. G., Pufahl, D. E. & Clifton, A. W. (1996 b). Model for the prediction of shear strength with respect to soil suction. Can. Geotech. J. 33:379-392.

Vijayvergiya, V. N. & Ghazzaley, O. I. (1973). Prediction of swelling potential for natural clay. Proc. of 3rd International Conference on Expansive Soils, Haifa, Vol. 1, 227-236.

Xu, Q. & Huang, R. Q. (1994). Artificial neural network methods for spatial Prediction of slope stability. 7th International IAEG Congress. Vol. VI, Balkema, 4725-4728.

Xu, Y. F. & Shi, C. L. (1997) The strength characteristics of expansive soils. Journal of Yangtze River Scientific Research Institute, Vol. 14, No. 1, 38-40 (In Chinese).

Yi, C. X. (1995). Nonlinear science and its application in Geology. Meteorological Press, Beijing (In Chinese).

Zhan, L. T., Ng, C. W. W., Wang, B. & Bao, C. G. (2000). Influence of some hydraulic parameters on pore pressure distribution in unsaturated soils. Proc. of Asian Conf. on Unsat. Soils, Singapore.

Zhang, J. F. (1997). 3D FEM modelling of saturated-nonsaturated and stable-nonstable seepage analysis. Journal of Yangtze River Scientific Research Institute, Vol. 14, No. 3 (In Chinese).

Zhang, J. F., Yang, J. Z. & Wang, F. Q. (1997). Rainfall infiltration test for high slope of shiplock in Three Gorge Dam. Yangtze River Scientific Research Institute, Technical Report, No. 97-264 (In Chinese).

Unsaturated Soils for Asia, Rahardjo, Toll & Leong (eds) © 2000 Taylor & Francis, ISBN 90 5809 139 2

Solar energy and evapotranspiration: A control of the unsaturation state in soils

G. E. Blight
University of the Witwatersrand, Johannesburg, South Africa

ABSTRACT: The depth of the water table in a soil profile, and the net flux of water through the soil surface, largely control the state of saturation in the soil and hence its engineering behaviour. In short, the state of saturation is dependent on the soil water balance, to which the major input is infiltration and the major output is evapotranspiration. Evapotranspiration, in turn, is driven by the solar energy reaching the soil surface.

Because the characteristics of solar energy are not always fully understood, it is virtually ignored as an energy source, particularly in geomechanics and waste management where its effects could be considered with great advantage. This paper will set out the principles of the availability of solar energy and its potential for drying soil and soil-like wastes. The effects of solar drying can be rationally considered when predicting the performance of soil strata and waste deposits.

NOTATION:

$\propto$	=	planetary albedo of earth
a	=	albedo at earth's surface
G	=	energy converted by heating earth's surface MJ/m^2
H	=	energy converted by heating air above surface MJ/m^2
L_e	=	energy converted to latent heat by evaporating water from surface MJ/m^2
n/D	=	ratio of actual to possible hours of sun
R_A	=	direct solar radiation at outer limit of atmosphere MJ/m^2
R_d	=	direct solar radiation at earth's surface MJ/m^2
R_n	=	net radiation at earth's surface MJ/m^2
R_s	=	scattered sky radiation MJ/m^2

1 INTRODUCTION

The depth of the water table in a soil profile and the net flux (infiltration minus evapotranspiration) through the soil surface largely control the state of saturation in the soil and hence its engineering behaviour. The net flux is a component of the water balance for the soil which can be written:

Water input to soil = water output + water stored in soil,

or Infiltration - evapotranspiration = recharge of water table + water stored in soil.

In general terms, if the soil water balance becomes positive, (i.e. in water surplus) the water table will rise and the soil will swell. If the water

balance becomes negative (i.e. in water deficit) the water table will fall and the soil will shrink. The engineering behaviour of the soil will change correspondingly. If the water balance is zero, water will be stored in the soil corresponding to its suction-water content relationship, with the suction at any point being equivalent to the height above the water table.

Evapotranspiration is a major term in the soil water balance and results from the conversion of solar energy reaching the soil surface. In arid and semi-arid regions of the earth, solar energy has a considerable capacity to evaporate water from the earth's surface, whether from free water surfaces, vegetated or bare soil or the surfaces of soil-like waste deposits. The evaporative capacity in humid regions is also not inconsiderable, but may be more than balanced by precipitation. Solar radiation nevertheless operates effectively in all climates and its contribution to overall water losses can be predicted and exploited with benefit.

This paper will consider the characteristics of radiant solar energy, how it reaches the earth's surface and how it is converted to effect evaporation and evapotranspiration.

2 THE THEORY OF EVAPORATION FROM A SOIL OR WASTE SURFACE

2.1 *Solar radiation*

The evaporative process is a large consumer of energy, which in the case of the surfaces of soils or waste deposits is freely supplied by the sun. If the amount of energy consumed by evaporation can be computed, the corresponding mass of water evaporated can be deduced.

The quantum of solar energy arriving at the earth's surface depends on:
- the solar constant
- the latitude of the location being considered and the time of year
- the influence of the atmosphere (dust and clouds)
- the albedo of the earth's surface
- the elevation of the location

The solar constant is the quantity of solar energy at normal incident angles to the atmosphere at the mean sun-earth distance. Some of this incident radiation is reflected or absorbed by the atmosphere and does not reach the earth's surface. Most of the radiation reaching the earth's surface is confined to short wavelengths, in the band 0.3 - 3 μm. On reaching the ground some of the short-wave radiation is also reflected by the surface.

Figure 1 (after Flohn 1969) shows diagrammatically how incoming solar radiation is split, with some being reflected or absorbed by the atmosphere, some being reflected by the ground surface, and some being absorbed at ground level. The diagram represents average annual conditions for the whole earth.

The overall radiation balance between space and the earth's surface may be described by the equations:

$$(R_d + R_s) = R_A(1 - \propto) \tag{1a}$$

$$R_n = (R_d + R_s)(1 - a) \tag{1b}$$

Here, R_A is the direct solar radiation received at the outer limit of the atmosphere and $\propto$ is the planetary albedo or reflectivity, the proportion of solar radiation reflected by the atmosphere. The planetary albedo $\propto$ varies with cloud cover and latitude, but in the earth's inhabited latitudes of 60°N to 45°S, lies in the range from 0.3 to 0.5 with 0.4 being a reasonable average value (Robinson 1966).

R_n is the net radiation, R_d is the amount of incident direct solar radiation, R_s is the amount of incident scattered sky radiation and a is the surface albedo or reflectivity of the ground surface.

The surface albedo a varies throughout the day and depends on the colour and texture of the earth's surface. Table 1, for example, gives average daily albedo values for various mineral tailings surfaces:

Table 1. Examples of surface albedo values

Light grey fly ash:	a = 0.22 when wet (darker)
	a = 0.33 when dry (lighter)
Dark brown tailings:	a = 0.06 when wet (darker)
	a = 0.14 when dry (lighter)
Yellow gold tailings:	a = 0.16 when wet (darker)
	a = 0.33 when wet (lighter)
Water over a light-coloured tailings surface:	a = 0.20

Figure 2 shows how the albedo varies throughout a typical day for four surfaces: shallow water over a light-coloured tailings surface, soil surfaces covered by long (0.5 m) grass and short (mown) grass and light grey fly ash. For solid surfaces, the albedo is greatest in the early morning and evening, when the sun's rays strike the surface at a low angle, and least at midday. For water the reverse applies presumably because of the influence of ripples on the surface.

Calculated direct solar radiations received at the outer limit of the atmosphere are given in Table 2 (taking the planetary albedo into account), as values of $(1 - \propto)R_A$.

Note that the annual value of R_A at the latitude of Stockholm (60°N) is 57% of that at the equator, while at the latitude of New York and Madrid (40°N) the annual value of R_A is 79% of that at the equator. Hence useful quanta of solar energy are available even at relatively high latitudes.

Apart from the planetary albedo $\propto$, the values of $(R_d + R_s)$ actually received at the earth's surface depend on cloud cover, elevation or altitude and the number of hours of sunshine. Examples of empirical relationships between $(1 - \propto)R_A$ and $(R_d + R_s)$ are as follows (Penman 1956):

For southern England (latitude 52°N):

$$R_d + R_s = (1 - \propto)R_A(0.18 + 0.55 \text{ n/D}) \tag{2a}$$

For Canberra, Australia (latitude 36°S)

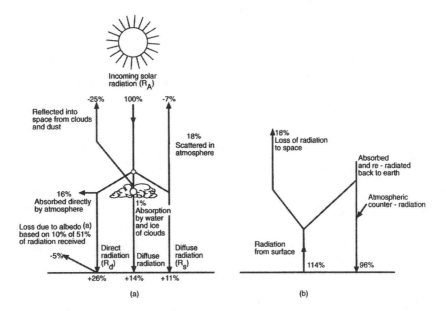

Figure 1. Annual radiation balance for earth; (a) short wave length solar radiation; (b) long wave length terrestrial re-radiation

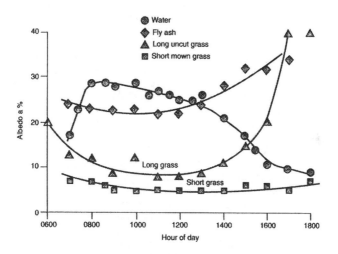

Figure 2. Variation of surface albedo during the day for a water surface and three land surfaces at 26°S latitude

$$R_d + R_s = (1 - \alpha)R_A(0.25 + 0.54 \, n/D) \qquad (2b)$$

Hence even on completely overclouded days ($n/D = 0$) about 20 to 25% of solar radiation reaches the earth's surface while on cloudless days 75 to 80% arrives (see also Fig. 1). On an average day in Stockholm, the value of $(1 - \alpha)R_A$ (from Table 2) will be 20 MJ/m² and if the day is cloudless, with an albedo of 0.3, the energy reaching the surface will (by equation (2a)) be about 10 MJ/m².

In Madrid, the corresponding energy would be about 14 MJ/m².

2.2 Surface energy balance

At the surface of the soil or a waste deposit, the net incident radiation R_n is converted in accordance with the surface energy balance.

$$R_n = G + H + L_e \qquad (3)$$

35

Table 2. Values of $(1 - \propto)R_A$, solar radiation at outer limit of earth's atmosphere

Latitude	January	February	March	April	May	June	July	August	September	October	November	December	Annual
60°N	112	274	550	863	1124	1235	1158	927	620	335	142	71	7411
40°N	464	630	860	1063	1207	1257	1221	1094	903	685	499	412	10294
20°N	819	932	1065	1148	1183	1189	1183	1151	1075	960	836	777	12317
0	1095	1129	1140	1100	1042	1008	1028	1064	1119	1124	1096	1076	13021
20°S	1259	1196	1080	925	789	728	763	882	1030	1158	1238	1269	12317
40°S	1295	1129	890	646	464	387	432	588	814	1060	1248	1340	10294

Units are $MJ/m^2/month$ (final column $MJ/m^2/year$)   After Angot, quoted by Wilson 1970

R_n can be measured directly and G can be estimated from changes in the temperature and temperature gradient below the surface together with the specific heat capacity for the soil or waste. H, the sensible heat may be calculated from temperature and relative humidity gradients above the surface. In principle, the calculation of the subdivision of R_n into G, H and L_e is simple, but in practice it is difficult because the equations tend to be ill-conditioned, especially when temperature and humidity gradients are small.

Data for energy balance calculations can be collected automatically by monitoring and recording the outputs of a radiometer, heat flux plates or thermocouples buried in the soil or waste and psychrometers, used to measure air temperature and relative humidity at two heights above the surface (500 mm and 1500 mm would be typical heights). Readings taken at two minute intervals, and averaged over longer periods (15 - 20 minutes) are recommended for best results. Hand-held portable instruments can also be used to measure the energy balance at significantly lower instrument cost and with much greater flexibility (but obviously with greater labour costs). The manual method is generally to be preferred if it is wanted to monitor several sites with less detail.

An 'A' class evaporation pan (or A-pan) should be monitored at every measurement site, for comparison to the quantity of evapotranspiration calculated from the energy balance. Wind speed is also measured, as advection of heat and water vapour due to gusts of wind may give rise to anomalies in the measured energy balance. (If the test site is located in the centre of a large uniform area, this effect is minimised). Precipitation should also be measured at each site.

Further details of applying the energy balance method for estimating evaporation and evapotranspiration have been given by Blight (1997) and Blight & Blight (1998). As mentioned above, the calcula-

tions can be difficult, and for this reason, the following simplification is useful:

Figure 3 shows a typical day's measurements on a landfill surface of incoming radiation $(R_d + R_s)$, reflected radiation a $(R_d + R_s)$ and the difference, $(1 - a)(R_d + R_s) = R_n$. The figure also illustrates the variation through the day of the albedo a and of $L_e = R_n - (G + H)$. Ordinates in Figure 3 are energy fluxes or rates in W/m^2 (1 Watt = 1 Joule/second). Areas under the curves represent energies in $J/m^2/day$. There is relatively little interchange of energy during the night, hence it is sufficient to take hand measurements only during daylight hours. The figure shows that $(G + H)$ makes up a relatively minor part of the net radiation R_n. Figure 4 shows the cumulative latent heat of evaporation (ΣL_e) plotted against the cumulative net radiation (ΣR_n) for the surfaces of three waste deposits in South Africa over a period of a year. These were measured on the soil-covered surfaces of two landfills and the bare ash surface of a power station fly ash dump. This plot shows that the latent heat of evaporation L_e makes up a relatively constant 75% to 85% of the net radiation R_n. Thus a preliminary estimate of L_e can be obtained by measuring R_n and taking a suitable percentage of it. The percentage probably varies with the type of surface, and may also vary with climate, but this knowledge does not appear to have been established as yet.

At this point, it is instructive to apply a simple check on the realism of the data given in Table 2. Annual A-pan evaporation in Johannesburg (latitude 26°S) amounts to 2200 mm or 2200 kg/m^2 of water. Multiplying this figure by the latent heat of evaporation of water (2.47 MJ/kg) gives 5430 $MJ/m^2/year$ as the corresponding value of energy converted, or an average of 14.9 $MJ/m^2/day$. If the value of R_A is taken as 12,000 $MJ/m^2/year$ and the percentage of sunshine n/D for Johannesburg as 0.7, then if

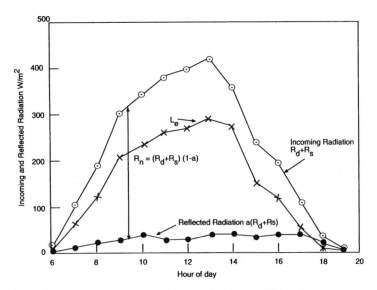

Figure 3. Variation of solar radiation components during a typical day at 26°S latitude

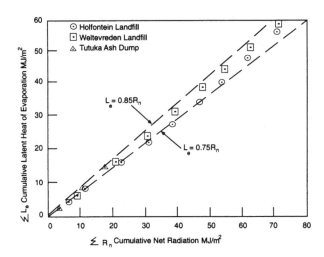

Figure 4. Relationship between cumulative net radiation ΣR_n and cumulative latent heat of evaporation ΣL_e for three waste deposits in South Africa (26°S latitude).

$(R_d + R_s) = (1 - \alpha)R_A(0.25 + 0.54 \, n/D)$ (equation 2b)

$(R_d + R_s) = 7540$ MJ/m²/year

Allowing for a surface albedo of 0.2, the value of R_n would be 6000 MJ/m²/year or 16.5 MJm²/day.

The energy converted into A-pan evaporation should be decreased by a small proportion to allow for the values of G and H that apply to the evaporation pan and the air above it. This decrease is about 10% for an evaporation pan (i.e. $L_e = 0.9 \, R_n$), which would give an average daily energy consumed by latent heat of evaporation of 0.9×16.5 MJ/m²/day $= 14.9$ MJ/m²/day. Thus according to this calculation, the figures in Table 2 are accurate.

As a further example, Figure 5 shows some detailed information on solar energy $(R_d + R_s)$ reaching the earth's surface at a landfill in Delaware, USA (which lies at 39°N) (Vasuki 1999). The annual value of $(R_d + R_s)$ is 2850 MJ/m²

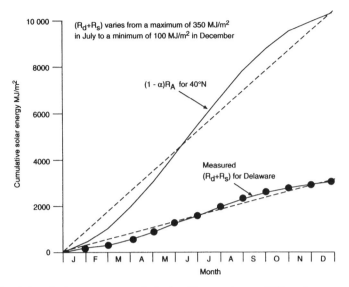

Figure 5. Annual variation of solar energy at outer limit of earth's atmosphere at 40°N , and energy actually received at earth's surface in Delaware, USA (39°N)

which, applying an albedo of 0.2 means that an R_n of approximately 2300 MJ/m^2 per year is available to cause evapotranspiration and surface heating. Dividing R_n by the latent heat of evaporation of water and adjusting the result by 10% gives a figure for annual evaporation of 840 mm of water. This is close to the observed annual figure for A-pan evaporation. Even though Delaware is generally considered to have a wet climate, the annual rainfall for the year of observation was only 850 mm and thus annual rainfall and evaporation from a free water surface were very similar. (It is interesting to note that the annual effect of cloud cover for Delaware could be represented by $(R_d + R_s)/(1 - \alpha)R_A = 2850/10{,}290 = 0.28$. Hence 72% of available solar energy did not reach the earth. Comparing this with the predictions of equations (2a) and (2b) indicates that n/D in Delaware has a low value.

2.3 Variation of evaporation or evapotranspiration through the year

Arid and semi-arid climates often have distinct wet and dry seasons, and rates of evaporation vary seasonally with incoming solar radiation. Referring to Figure 5, even in the nominally wet climate of Delaware, evapotranspiration varies seasonally from a high of 100 mm/month ($R_d + R_s = 350$ MJ/m^2/month) in July to a low of 30 mm/month ($R_d + R_s = 100$ MJ/m^2/month) in December.

Three conditions must be met in order for evaporation or evapotranspiration from a surface to proceed. There must be a: (1) source of energy (solar radiation) to provide the latent heat of evaporation; (2) vapour pressure (or relative humidity) gradient between the surface of the soil or waste and the atmosphere (with the atmosphere less humid than the surface); and (3) continuing supply of water from below the surface of the soil or waste to the surface where it can evaporate.

Satisfying the third condition depends on the suction or free energy in the near-surface pore water, as well as the permeability. Conduction of water to the surface declines both as the pore water suction increases, and as the permeability decreases (as it does with increasing suction). However, it should be noted that if the soil suction is, say 1 MN/m^2, this is equivalent to a free energy of the soil pore water of 10^6Nm per m^3 of water or 1 MJ/m^3. The latent heat of evaporation of water is 2.47 MJ/kg or 2470 MJ/m^3 of water. Hence if there is sufficient energy to evaporate the water, there will be more than sufficient energy eventually to draw it to the soil surface, making it available for evaporation.

As the suction at the surface increases, the tendency for lateral shrinkage also increases until the surface cracks. The presence of cracks in the surface increases the surface area from which evaporation can occur and this may partially compensate for the declining ability to conduct water to the surface as it dries out.

The seasonal variation of evaporation or evapotranspiration rates is illustrated in Figure 6 which compares cumulative A-pan evaporation with evapotranspiration determined from the energy balance for a soil surface covered by mown grass in Johannes-

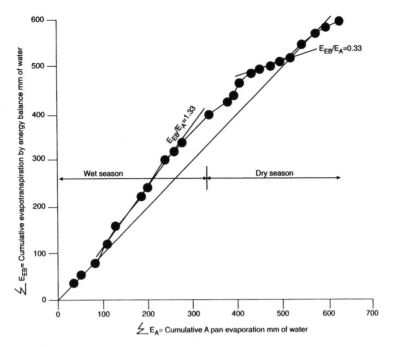

Figure 6. Variation of evapotranspiration (calculated from energy balance) with A-pan evaporation in wet and dry seasons in Johannesburg, South Africa (26°S latitude).

burg, South Africa. During the early part of the wet season (September) the evapotranspiration from the grassed surface equalled the A-pan evaporation rate. As the grass began to grow its rate of evapotranspiration increased to 133% of the A-pan evaporation rates. As the dry season began, water became less readily available in the soil (see condition (3) above) and the evapotranspiration rate declined until it was 33% of the A-pan rate, i.e. the evapotranspiration rate varied through the seasons by a factor of 4. In more arid climates evapotranspiration may almost cease during the dry season, through lack of ready availability of water in the soil.

2.4 Examples of evaporation predictions

The reasoning above can be applied to predicting evaporation from tailings dams or other waste deposits located or planned in various parts of the world. Tailings dams are a good example, as water will usually be freely available to be evaporated. Landfills and natural soil strata may be more difficult, as water may not be readily available to be evaporated at the end of a long dry season.

If, for the purpose of the example we take:

$$R_d + R_s = (1 - \alpha)R_A(0.25 + 0.5 \; n/D) \qquad (2c)$$

$$R_n = (1 - a)(R_d + R_s) \qquad (1b)$$

1. For a tailings dam in a cold semi-arid climate at latitude 50°N (e.g. Regina, Canada),

Take $a = 0.5$ and $L_e = 0.7 \; R_n$

Then, combining previous equations and numerical values:

$$L_e = 0.09(1 + 2 \; n/D)(1 - \alpha)R_A$$

n/D can be estimated from figures for average cloud cover for various periods of the day throughout the year. This information is usually available from local weather records.

With $n/D = 0.4$ and

$$(1 - \alpha)R_A = 8800 \; MJ/m^2/year \text{ (from Table 2)}$$

$$L_e = 1425 \; MJ/m^2/y$$

Dividing by the latent heat of evaporation of water (2.47 MJ/kg) gives annual evaporation of

$$E = 580 \; mm/y$$

The actual annual evapotranspiration for Regina is 500 to 600 mm/y (Barlishen 1998). Obviously, it was necessary to choose appropriate values for a and n/D to reach this result, but the values taken are reasonable.

2. For a similar tailings dam in a warm semi-arid climate (Johannesburg, South Africa) (see earlier example),

$$L_e = 0.8 \times 6000 = 4800 \text{ MJ/m}^2/\text{y}$$

which is equivalent to

$$E = 1940 \text{ mm/y}$$

Estimates such as these can be used, for example to prepare preliminary water balances for use in the design of tailings dams or other waste deposits.

3 CONCLUDING SUMMARY

The state of saturation in a soil and hence its engineering behaviour, is largely controlled by infiltration and available solar energy and evapotranspiration.

After describing the components of the solar energy balance at the earth's surface (Figs 1, 2, 3), it was shown that the radiant energy that is converted into latent heat of evaporation makes up a large proportion (75 to 85%) of the net solar radiation reaching the earth's surface during daylight hours (Fig. 4). The actual proportion depends on the availability of water at the surface and the rate at which water can be conducted to the surface under the evaporation gradient. The varying availability of water and also varying evapotranspiration by plant cover during the growing season, results in a seasonably variable rate of water loss from a soil or waste surface when compared with evaporation from a free water surface (Figs 5, 6).

It was shown that solar radiation can consciously be used as part of a designed process in geotechnical engineering and waste management. It is possible to predict average annual evaporation or evapotranspiration from a soil or waste surface, and to use that prediction as information for designing or predicting water interchanges in geotechnical or waste management processes.

REFERENCES

Barlishen, K. 1998. Personal communication with author.

Blight, G.E. 1997. Interactions between the atmosphere and the earth. *Geotechnique*, 42:715-766.

Blight, J.J. & G.E. Blight 1998. Using the radiation balance to measure evaporation losses from the surface of the soil. *2nd Int. Conf. on Unsaturated Soils, Beijing, China:* 327-332.

Flohn, H. 1969. *Climate and Weather*. London: Weidenfeld and Nicolson.

Penman, H.L. 1956. Vegetation and Hydrology, *Tech Com 53*, Commonwealth Agric Bur. UK.

Robinson, N. 1966. *Solar Radiation*. Amsterdam: Elsevier.

Vasuki, N.C. 1999. Personal communication with author.

Wilson, E.M. 1970. *Engineering Hydrology*. NY, USA: Gordon and Breach.

Unsaturated Soils for Asia, Rahardjo, Toll & Leong (eds) © 2000 Taylor & Francis, ISBN 90 5809 139 2

Unsaturated tropical residual soils and rainfall induced slope failures in Malaysia

H.A. Faisal
Department of Civil Engineering, University of Malaya, Kuala Lumpur, Malaysia

ABSTRACT: Residual soils derived from the in-place weathering of parent rocks represent the predominant type of earth materials in those parts of the world where slope failures are very prevalent. However, residual soils have received very little attention from the geotechnical community. Residual soils are invariably unsaturated and the stability of slopes in such soils is mainly governed evidently by the shear strength behaviour of unsaturated soils and the effect of rainwater infiltration. This paper summarizes some of the laboratory and field tests that had been carried out on unsaturated residual soils in Malaysia. Some of the catastrophic rainfall-induced slope failures in these soils are also described.

1 INTRODUCTION

Residual soils are found in many tropical areas where there is high temperature and rainfall. Slope failures in residual soils and weathered rock are common in tropical climates and often occur during periods of intense rainfall. The physical structure and engineering properties of these soils are unique and require special evaluation. Colluvium, which consists of gravity transported residual materials, are generally found on hill slopes. Since the problems of slope stability in residual soils and weathered rock, in practice, cannot be separated from problems related to colluvium mantle, it is also considered with residual soils (Deere and Patton, 1971).

The association of landslides with high intensity rainfalls are described by many research workers e.g. Brand (1985) and correlations between landslides and rainfall have been established. Such correlations may help us in understanding of landslide process. It has been observed that the residual soils are invariably unsaturated and the stability of slopes in such soils is mainly governed by the shear strength behavior of unsaturated soil and on the effect of rainwater infiltration. In the last two decades, there have been a number of significant advances in understanding the shear strength behavior of unsaturatcd soils (Fredlund et al, 1978; Fredlund et al., 1987; and Fredlund and Rahardjo, 1993).

From the point of view of analysis and design, residual materials are very difficult to deal with. Engineering problems in residual materials and colluvium are not usually amenable to the principles of soil mechanics or rock mechanics alone, but must be combined with the appropriate elements of geology, geomorphology and hydrology.

Landslides are endemic in the steep terrain of Southeast Asia, where high rainfall and deeply weathered rocks combine to form the main prereqiusites of slope instability. Lately there has been an increasing occurences of lanslides in Malaysia and some of them were catastrophic. The dire need for hillside development for road construction and residential area contributed to the increase in slope instability related problems.

2 TROPICAL RESIDUAL SOILS

2.1 Characteristics

The warm wet climate in tropical regions produces materials which are the products of in situ weathering of rocks and which are commonly referred to as 'residual' materials, the degree of weathering and the extent to which the original structure of the rock mass is destroyed varying with depth from the original ground surface. This process gives rise to weathering profiles which contain materials 'grades' from fresh rock to completely weathered materials, the latter usually being described by geotechnical engineers as 'soil'. For engineering purposes, it is difficult to separate the several component parts of the weathering profile, and the whole profile is therefore best treated as a single entity

The earth materials which comprise tropically weathered profiles are sometimes categorized simply as 'laterite', 'saprolite' and 'rock'. For the purposes of analysis and design, the engineering behaviour of both laterite and saprolite is usually considered to be governed by the principles of soil mechanics.

Residual materials have received very little attention from the geotechnical engineering community. This neglect, and the resulting paucity of relevant literature, stems largely from the fact that the very nature of residual material makes the application of soil mechanics principles problematical and renders the materials difficult to model for the purposes of engineering analysis and design. The main characteristics of residual materials which are responsible for this are :

(a) they are generally very heterogeneous, which makes them difficult to sample and test,

(b) they are nearly always unsaturated and ,

(c) they invariably have high permeabilities, which makes them subject to rapid changes in material properties because of external hydraulic influences.

As far as slope stability is concerned, the engineering behaviour of residual materials can be characterized very simply by: (a) the importance of the complete weathering profile, (b) the extreme difficulties of quantification, and (c) the dominant role of rainfall.

From the point of view of analysis and design, residual materials are very difficult to deal with. Geotechnical engineering in residual materials spans the narrowly separated fields of soil mechanics, rock mechanics and engineering geology, and the engineering geological approach is generally the most satisfactory one for these materials.

2.2 *Colluvium*

Large deposits of colluvium often exist in conjunction with residual materials, particularly as colluvial fans on footslopes of hillsides. Colluvium is material derived from the weathering of any parent rock which has been transported downhill by the agencies of gravity and water. It can range in general composition from a collection of matrixless boulders at one extreme to a fine slopewash material at the other. It poses many of the same general characteristics as residual material, particularly in the context of engineering behaviour. Because it is commonly found as slope cover over weathered rock profiles, it is sometimes difficult to distinguish between colluvium and the in situ material if only drillhole samples are available for examination. For geotechnical engineering purpose , colluvium can therefore be grouped with residual soil.

2.3 *Importance weathering profile*

The accurate logging of weathering profiles is fundamental to successful design and construction in residual profiles. These often contain a whole range of engineering materials from 'soil' to 'rock'. The weathering profile is therefore of great importance for the stability of slopes, because it usually controls :

(a) the potential failure surface, and therefore the 'mode' of failure for analysis and design, and

(b) the groundwater hydrology, and therefore the critical pore pressure distribution in the slope.

There is no universally accepted system for describing and classifying the component parts of a weathering profile. Classification in terms of weathering 'zones' and weathering 'grades' is essential for engineering design, and there have been several major attempts to provide a satisfactory description and classification system for engineering purposes. Any weathering description and classification system must be suitable for the particular geological conditions and engineering purpose to which it is applied. In Malaysia description and classification of weathering profile are commonly based on the BS 5930:1981 and the Geoguide 3,Geotechnical Engineering Office, Hong Kong.

There exists no engineering description and classification system for colluvium, although one is badly needed. Attempts have been made in many parts of the world to provide a framework for such a system, but these are entirely descriptive in character and require a great deal of further development.

3 SLOPE STABILITY ASSESSMENT IN TROPICAL RESIDUAL SOIL

The stability analysis of a slope is related directly to the prediction of the conditions under which the slope could fail. Apart from the application of sound judgement and experience alone, there are four basic methods available for the prediction of rain-induced failures in deeply weathered slopes. These are :

(a) correlations between slope failures and pattern of rainfall,

(b) terrain evaluation, mainly on the basis of geomorphological mapping,

(c) semi-empirical (or modified precedent) approach, which is based on an examination of the geomorphology and geology of stable and unstable slopes, and

(d) analytical methods, usually in the form of limit equilibrium analysis.

The first of these approaches can be considered to be directly related, since they apply to the stability of a land area in one particular location. The last three methods can be regarded as methods of analysis and design; all three have been used extensively in residual materials. The terrain evaluation and semi-empirical methods are closely related, in that both are based on an explicit assumption that the stability characteristics of a slope can be assessed on the basis of observations of the performance of others with similar characteristics.

Whereas the vast majority of cut slopes in residual materials were not designed on the basis of rigorous soil mechanics methods, the soil mechanics approach is being increasingly adopted even in this difficult material. This is certainly true in Malaysia, where design practice is governed largely by local design guidelines. The degree of safety of a slope can only be quantified on the basis of an analytical method, wheres methods (a), (b) and (c) provide no such quantification.

3.1 *Slope stability analysis*

Slope stability is usually based on the limit equilibrium approach, for which the equilibrium of a sliding mass is examined. The degree of stability is quantified in terms of a factor of safety which is most commonly defined as the ratio between the average shear resistance and the average shear stress along the most critical slip surface.

The distribution of shear stress along the critical slip surface is dependent on the loading and the method of analysis employed. The shear resistance along the slip surface is governed by the effective shear strengths of the materials of which the slope is composed and the normal effective stress distribution on the slip surface. The normal effective stress distribution is, in turn, a function of the pore pressure distribution at failure and the method of analysis. The five components of stability prediction are therefore :

(a) mode of failure,
(a) loading,
(b) method of analysis,
(c) shear strength, and
(d) pore pressure distribution at failure.

For a slope in residual material, the prediction of the pore pressure distribution is by far the most critical factor for slope stability. This is particularly so since most failures in residual slopes are caused by rainfall

The hydrological effects of rainfall on a permeable slope are depicted in Fig.1. Some of the

water runs off the slope and may cause surface erosion if there is inadequate surface protection. Because of the high soil permeability, however, the majority of the water infiltrates. This causes the water table to rise, or it may cause a perched water table to be formed at some less permeable boundary, usually dictated by the weathering profile. Above the water table, the degree of saturation of the soil increases, and the soil suction (i.e. negative pore pressure) therefore decreases.

To illustrate the effect of perched water table on the stability of slope Low et.al.(1999a) carried out numerical analyses by introducing 'impermeable' layers in the slope under consideration (Fig.2). The results of the analyses are shown in Figs.3,4&5. The magnitude of the effect depend on factors such as permeability ratio, angle of dipping number of impermeable stratum and intensity of rainfall.

Failures in residual cut slopes are thought to be caused mostly by the 'wetting-up' process by which the soil suction (and hence the soil strength) is decreased, but there is some evidence to suggest that transient rises in groundwater tables are responsible for some rain-induced landslides (Premchitt et al, 1985). Rain-induced slope stability failures thus occur as a direct result of pore pressure increases, and pore pressure distribution is therefore the variable of most concern.

The pore pressure distribution in a residual profile is dependent on the pattern of rainfall and the

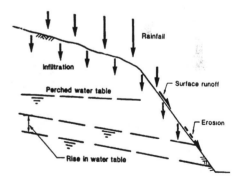

Fig.1 Effects of rainfall on a highly permeable slope

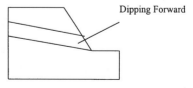

Figure 2 Typical profile showing the impermeable stratum

43

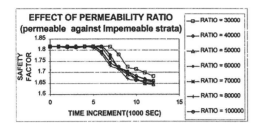

Fig.3 Effect of soil permeability ratio on slope safety factor (Permeability of 1x 10^{-8} m/s for impermeable stratum was used as datum)

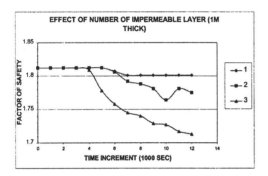

Fig. 4. Effect of number of impermeable strata on slope safety factor.

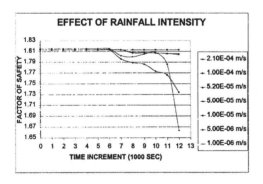

Fig.5 Effect of rainfall intensity on factor of safety.

hydrogeology of the slope. Pore pressure is therefore a variable which is independent of soil mechanics considerations, being imposed upon a slope by external influences. For this reason, it is extremely difficult to predict the appropriate pore pressures for slope design or stability assessment.

Some approximate methods exist for the prediction of pore pressure changes in slopes during rainfall infiltration. For analysis of slope stability in residual soils and colluvium, however, measured pore pressure data is much to be preferred to the

application of uncertain predictive methods. For such data to be meaningful, pore pressures must be monitored for a sufficiently long period of time In addition, an appropriate type of instrument must be used which can respond rapidly to pore pressure changes.

During most of the year, suctions exist in many residual soil slopes. These suctions, which can be of very high magnitudes (Fredlund, 1991), are reduced dramatically by the process of water infiltration during rainfall. There are almost certainly situations where even the heaviest rainfall does not completely destroy the soil suctions, and these continue to contribute to the stability of a slope. The extent to which suctions play significant part in slope stability in residual soil profiles depends on the pattern of rainfall and the infiltration characteristics of the material concerned.

Table 1 shows the assessment of the overall state –of-the-art of slope stability analysis made by Brand (1985) for slopes in residual soils. The determination of loads is straightforward. There are no reasons to believe that the available methods of analysis are anything but very good, as long as these are restricted to appropriate methods for the particular mechanism of failure. Brand (1985) said that our ability to select a correct mechanism of failure varies fairly widely depending on the geology and weathering profile for the slope. Because of the small amount of work that has been done to investigate the shear strength properties of residual earth materials, our knowledge of this component is only fair. However, by far the major difficulty with stability predictions in residual slopes is the poor state-of-the-art with respect to the prediction of pore pressure distribution at failure.

It is clear that the application of rigorous soil mechanics methods of slope stability analysis to residual soils is extremely difficult. It is prudent to adopt this approach only in conjunction with a thorough engineering geological assessment and with the liberal application of sound judgement. Analytical solutions alone cannot be relied upon.

4 LABORATORY TESTS ON UNSATURATED RESIDUAL SOILS IN MALAYSIA.

Residual granite rock soil and sedimentary rock soil occur extensively in Malaysia i.e. cover more than 80% of the land area. Yet, not much research work have been carried out on these materials. The situation is even worst in the case of unsaturated residual soil. Some investigation have been carried out since eight to nine years ago and it is still going on now. The interest in the behaviour of unsaturated

Table 1: Assessment of the state of the art of limit equilibrium stability analysis for slopes in residual profiles by Brand (1985)

COMPONENT	SUB-COMPONENT	KNOWLEDGE
MECHANISM	—	GOOD - POOR
SHEAR STRESSES	LOADING	V.GOOD - GOOD
	ANALYSIS METHODS	V.GOOD - GOOD
SHEAR STRENGTHS	STRENGTH PROPERTIES	FAIR - POOR
	PORE PRESSURES	POOR - V.POOR

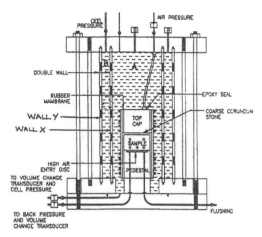

Fig.6 Schematic diagram of the double-walled cell

soil began to develop after the country was faced with a number of very serious slope stability problems. The coming sections summarise some of the laboratory studies that have been carried out on the materials.

4.1 Volume change

Some tests on compacted residual soil samples were carried out (Choong, 1998) in order to study :(a) the volume change behaviour of the soil subjected to an incrase in net mean stress at constant matric suction (b) the volume change behaviour of the soil subjected to reduction of matric suction at constant net mean stress (c) the effect of rate of loading of net mean stress on the volume change behaviour of soil, and the effect of stress history on the volume change. The tests were carried out using specially fabricated apparatus and a double-walled cell (Fig. 6) was used in the test as the structural volume change was measured using a high precision volume change indicator installed at the cell pressure line.

The following conclusions were arrived at based on the results of the tests:

(i) The loading rate of the net mean stress has a pronounced effect on the void ratio and degree of saturation but has insignificant effect on the water content of the soil subjected to constant applied matric suction. However, the loading rate of net mean stress has insignificant effect on the ,void ratio, water content and degree of saturation for soil not subjected to the applied matric suction

(ii) When the applied matric suction is increased at constant net mean stress condition, the void ratio of the soil decreases. However, when the matrix suction is reduced at constant net mean stress condition, the void ratio of the soil can either increase (swell) or decrease (collapse), depending

on the stress history of the soil and loading rate of net mean stress.

(iii) There may be a unique relationship (or uniqueness) between the void ratio, matric suction and net mean stress. The uniqueness in void ratio was normally observed in the study when there was a collapse (decrease in void ratio) due to the reduction in applied matric suction at constant net mean stress. However, the uniqueness in void ratio appears to be sensitive to stress history and loading rate.

(iv) The stress history may have a significant effect on the void ratio and degree of saturation of the soil. The increase in applied matric suction at constant net mean stress is found to have resulted in non-uniqueness in void ratio .

4.2 Shear strength tests.

Most of the unsaturated strength tests carried out were related to slope stability problems in Malaysia. The undistributed samples were prepared from the block samples taken from the slope being investigated.

4.2.1 Modified shear box tests

An ordinary shear box was modified to apply soil suction to the soil samples. Suction could be applied by controlling the pore air and pore water pressures. The direct shear box was placed in a special fabricated galvanized steel air chamber as shown in Fig.7. A 15 bar high air entry disc was placed at the lower block of the direct shear box. High air entry disc was used to separate soil samples with the water compartment underneath.

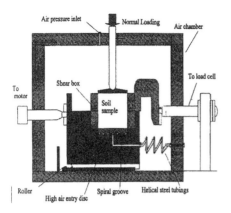

Fig.7 Modified direct shear apparatus

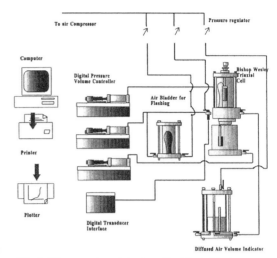

Fig.9 The experimental set up for triaxial tests

The total normal stress, σ, was applied vertically to the soil specimen through a loading ram as in the conventional shear box tests. The uplift pressure of the air in the air chamber on the loading ram was taken into account.   Undisturbed samples from a granitic residual soil cut slope were used in the tests.

Typical results obtained from the tests are shown in Fig.8 The effect of suction on the shear strength of the soil is clearly observed. One of the advantage of carrying out shear box test is that the time taken is shorter and easier to set up.

4.2.2 Triaxial tests.

The types of test that had been carried out on unsaturated residual soils were consolidated drained, consolidated undrained and constant water content test. The Bishop-Wesley triaxial cell set was modified to carry out the above test so that suction can be introduced into the specimen   and the experimental set up is shown in Fig.9

Different types of residual soil had been tested and some of them are presented below. Affendi and Faisal (1994a)carried out tests on samples taken from a granitic residual soil slope to study the effect of suction on the shear strength of the soil. Typical results of the tests are shown in Fig.10. The angle of shearing resistance,ϕ', and the angle indicating the rate of increase in shear strength with respect to suction, ϕ_b , are 26° and 17° respectively. Drained tests carried out by Hossain (1999) on more or less similar type of soil yielded values of ϕ' and ϕ_b equal to 26.5° and 17.2° respectively.

Saravanan et al. (1999) carried out tests on sedimentary residual soil samples taken from different weathering zones. Sample from Zone IV gave ϕ' value of 26° and maximum ϕ_b value of 21° . Sample from Zone III gave ϕ' value of 33° and maximum ϕ_b value of about 10°.

Hossain(1999) carried out consolidated undrained (CU) tests and constant moisture tests (CW) on unsaturated granitic residual soil samples and compared the effective strength parameters obtained with those from consolidated drained tests (CD). The values of ϕ_b obtained from CU and CW tests were 25.8° and 23.3° respectively. These are significantly different from ϕ_b value of 17.8° obtained from CD tests .

5 FIELD TESTS

As part of the research program on slope instability a number of field tests have been carried out which include field suction and rainfall measurement, soil moisture measurement and infiltration tests. Some of these tests are presented in the following sections.

Fig.8 Shear stress vs normal stress

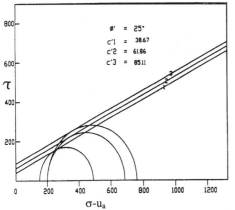

Fig.10a. Mohr circle plot for CD tests (Suctions: 50,100 & 200 kPa.)

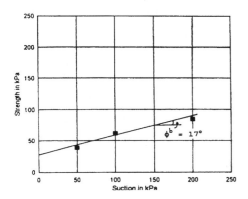

Fig.10b Increase of strength with suction.

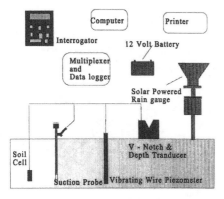

Fig.11 Schematic presentation of the field instrumentation

5.1 *Measurement of suction and rainfall*

The objective of the study is to determine the variation of field suction and rainfall. The measurement was done by carrying out field intsrumentation on each slope under consideration. Suctions were measured using tensiometers and the rainfall was measured using an automatic logging tipping bucket rain gauge. Fig.11 shows the schematic arrangement of the field instrumentation. An automatic data acquisition system which allowed continuous monitoring and supported by a solar powered set was used. Instrumentation detail at one of the tensiometer locations is shown in Fig.12.

5.1.1 *Granitic residual soil*

The site was an exposed profile of a cut slope along the Kuala Lumpur-Karak Highway. The weathering profiles were very distinctive and Fig.13 shows the lateral extension of the morphological horizons, within the weathering profiles and the gradings (Affendi & Faisal, 1994).

The total number of instruments and their locations are shown in Fig.13. The data from the 27 sensors were automatically recorded at a preset time interval and were periodically downloaded to a computer for further analysis.

Fig.14 shows the responses of the tensiometers located at berm 4. The suction reading for the shallower depth seems to be higher than for the deeper locations. Furthermore the response due to rainfall is less pronounced as the depth increases. There seems to be a limiting value for suction at various depths. The suction values level up to a certain value depending on the depth, after a long dry spell. The largest drop in suction is from a value of 89.2 kPa to 35.4 kPa.(depth 30.5 cm)

The drop in suction decreases as the depth

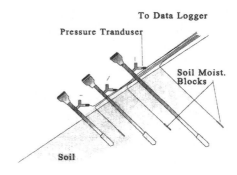

Fig.12 Instrumentation detail at one of the tensiometer location

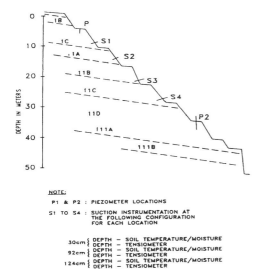

NOTE:

P1 & P2 : PIEZOMETER LOCATIONS

S1 TO S4 : SUCTION INSTRUMENTATION AT
THE FOLLOWING CONFIGURATION
FOR EACH LOCATION

30cm { DEPTH – SOIL TEMPERATURE/MOISTURE
      { DEPTH – TENSIOMETER
92cm { DEPTH – SOIL TEMPERATURE/MOISTURE
      { DEPTH – TENSIOMETER
124cm { DEPTH – SOIL TEMPERATURE/MOISTURE
       { DEPTH – TENSIOMETER

Fig.13 Cross-section of the cut slope.

increases. Similar behaviour was experienced in studies conducted on residual soils in Hong Kong.

Figs. 15, 16 & 17 show plot of suction against cumulative rainfall for various depths (30.5, 61.0 and 91.5 cm). The plots of suction variation at 30.5 cm depth for various berms show that the responses for the various grades of graniteare different.

The suction fluctuations for different berms at 30.5 cm depth are relatively large i.e. from below 10 kPa to above 85kPa while for depths 61.0 cm and 91.0 cm the fluctuations are smaller(i.e. between 25kPa and 65kPa

5.1.2 *Sedimentary residual soil*

The instrumentation was attempted to study the change of soil matric suction with the rainfall on a cut slope along the link road of The Kuala Lumpur International Airport (KLIA) Malaysia (Low et. Al, 1999b). The cut slope mainly consists of two types of weathered sedimentary residual soil, i.e.,

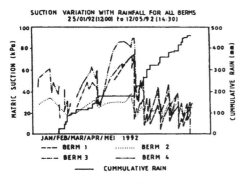

Fig.14 Suction variation for berm 4

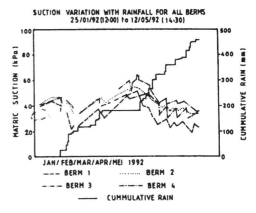

Fig.16 Suction variation – 61 cm deep

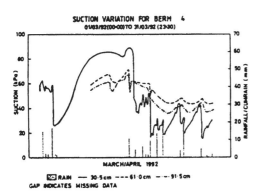

Fig.15 Suction variation – 30.5 cm deep

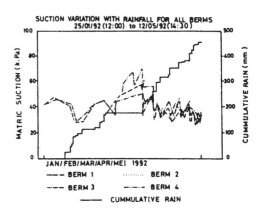

Fig.17 Suction variation – 91.5cm deep.

weathered sandstone and shale. These residual soils come in alternate bedding which is almost vertical. The weathered sandstone bed basically is the thicker bed and the study is concentrated in one of these beds. The soil consists of very fine sand and silt. 20 numbers of tensiometer and 20 numbers of moisture block and a rain gauge were installed on the slope to monitor the changes of matric suction with respect to rainfall. The tensiometers and moisture blocks were installed at different depths. At each berm, 4 numbers of tensiometers and moisture block were installed i.e., with depth of 0.5m, 1m, 3m and 3m.Fig.18 shows the instrumentation layout.

The normal coring tools could not be used because the soil is brittle and hard. A specially designed motorised auger was fabricated for the installation purpose (Fig.19)

Fig.20 shows a typical suction variation with rainfall (one month duration) for one of the berm at the study site. It clearly shows that as the depth increases, the matric suction reduces.

During the time interval of 14000 to 24000 minutes, there is no rainfall and all the four tensiometers are recording increments in matric suction. When the rain starts , matric suction does not reduce immediately. Due to the infiltration of rain water into the ground, after rainfall the matric suction continues to reduce slowly for all depths .

From Fig.20 , in the time interval of 0 to 13000 minutes, the 3.0m depth tensiometer gives very low suction values. This is mainly due to the water which has infiltrated during the earlier rainfall periods.

Fig. 21 shows one of the typical rainfall intensity patterns at the study site. The rainfall intensity at the site reaches as high as 1.13×10^{-4}m/s during the monitoring period.

Recently, infiltration study was performed by using rain simulator . Water was sprayed on the slope using sprinklers as shown schematically in Fig.22. A number of large water tanks were placed at some locations on the slope in order to have enough water supply for the study. In addition to the existing tensiometers, seven small tip tensiometers were installed at 75 mm below ground surface. Four

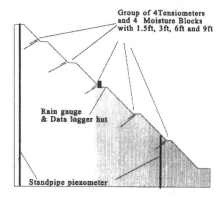

Fig.18 Instrumentation layout.

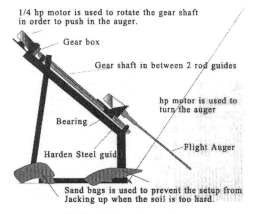

Fig.19 Specially fabricated augering machine

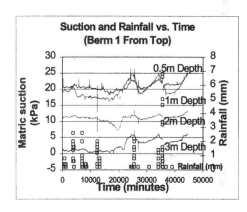

Fig.20. Typical suction variation with rainfall for one-month duration.

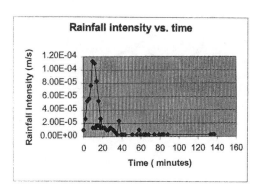

Fig.21 Typical rainfall intensity measured at the site

different surface conditions were studied i.e. grass + geotextile cover, geotextile only cover and bare slope. Fig.23 shows a typical variation of suction with time. In the figure shallow tensiometers are represented by 1 to 7 and existing tensiometers are denoted by A (0.5 m), B (1.0 m) and C (2.0 m). It is interesting to note that some of the shallow tensiometers indicate initial increase in suction as soon as the sprinkling begin. These were observed in every test carried out.

6.0 RAINFALL-INDUCED LANDSLIDES IN MALAYSIA

Lately there has been an increasing occurrences of landslide in Malaysia. Most of these landslides were on cut slopes or embankment along the road and highways. Some of these landslides occurred in the vicinity of high-rise apartment and residential area.

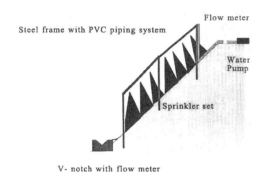

Fig.22 Infiltration test using rain simulator

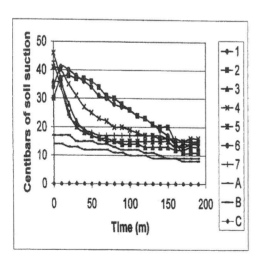

Fig.23 Variation of suction with time for bare slope

A few catastrophic landslides also occurred within the last five or six years. The landslides, which are still fresh in our memory, include the tragic Highland Tower landslide, Genting Sepah landslide and the Gua Tempurung landslide.

The dire need for hill slope development for road construction and residential area contributed toward more major landslides such as those along Ranau-Tamparulli Road (East Malaysia), East-West highway (West Malaysia) and a few residential areas in the island of Penang. Some of the major landslides are presented in the following sections.

6.1 *Highland Tower*

On Saturday afternoon of the 11th. December 1993, a 12-storey Block 1 Highland Towers at the Hillview Garden, Hulu Kelang, Selangor toppled and 49 people were buried in the rubble. Block 1 of the 3-block Highland Towers Condominium was built in 1979. It was located at the foot of a hill in Hulu Kelang, Selangor. The toppled Block 1 was founded on rail piles driven into weathered granitic formation. The earth materials here are of Grades V & VI sandy clay.The hill slope was terraced and protected by a series of rubble walls of differing heights and thickness. Altogether, there were 13 rubble walls built on the hillslope reaching some 100 meters above the foot of the slope

The collapse was caused by the landslide of the hillslope behind the condominium, involving some 50,000 cubic meters of earth. There had been a period of intense and prolonged rainfall in the area on the night and hours before the landslide . The rainfall records from nearby gauging stations showed that heavy rainfall started in early Nov. and this was maintained until the day the building collapsed. From 1st September the cumulative rainfall for the stations is between 800 mm and 900 mm. Aerial inspection suggested that there were four retrogressive slips immediately behind the collapsed structure.

6.2 .*Genting Sempah Landslide*

At about 5.30 p.m. on Friday 30 June 1995, a debris flow occurred on a slip road to Genting Highland. The debris flow resulted in 20 fatalities and 23 other people were injured. The area within the Genting Sempah vicinity is prone to landslide as evident by history of recurrences analysed from aerial photography records. The overburden soil of the area is relatively thin, overlying a metasedimentary rocks and rhyolitic rocks. As such, overburden soils had been continuously saturated from antecedent rainfall, especially in the latter part of June 1995 where the accumulation of rainfall through the months of June 1995 exceeded the normal long term average by more than 156%. The continuous 2 hour heavy rainfall on the evening of 30 June 1995, triggered major landslide into the stream of the developing debris flow.

6.3 *Paya Terubung Landslide*

On November 28th 1998, a massive rockslide occurred at Bukit Saujana, Paya Terubung.Tons of boulders and earth slided down on to a parking area of the adjacent Block 8 of the Sun Moon City Apartments, blocking Bukit Kukus Road, the only access to other apartments in that area. 17 vehicles parked along the roadside were buried by a huge boulder (about 14m in diameter) and landslide debris.

The slip occurred on a cut slope at an angle close to 60^0. The total height of the original slope was approximately 120m with huge boulders scattered on the surface of the slope covered by trees and shrubs.

The unique geological feature of this area is that there are hundreds of large surfaced boulders beyond the crest of the slope. A total of about 400 boulders ranging from two to ten meters in diameter were found scattering in the area. The area is underlain by medium to course-grained biotite-muscovite granite. From the daily rainfall record, the area was subjected to heavy rainfall few days before the slide. The rainfall data from various stations indicated that the month of November seemed to be one of the wettest month with the monthly rainfall intensity of up to 850 mm.

Observation from the soil investigation of the area suggested that the structure of the weathered granite get tighter as it gets dipper into the ground. The layer close to the surface will have higher permeability and higher infiltration rate. Subjected to intense rainfall, development of perched water table is highly favourable. The present of perched water table was confirmed based on the observation of water level in some boreholes.

6.4 *Bukit Antarabangsa Landslide*

Two major landslides occurred in May 1999 at Bukit Antarabangsa, about one kilo meter away from the tragic 1994 Highland Tower Landslide .The first landslide that occurred on 15th May 1999, is located about 100 meters to the west of the Wangsa Height condominium. The volume of the landslide was estimated to be 14,580 cubic meters. The slide had brought down about 70 meters stretch of Wangsa Road 3 leading to the JBA reservoir. The stretch of the road was part of the road embankment constructed on the fill ground.

Another landslide occurred the day before, May 14, 1999, at about 4:30 p.m, on a steep slope adjacent to the Athenaeum At The Peak condominium. The landslide scar measured about 31.0 meters wide and 140 meters long from the crown to the toe of the slide. The supersaturated slide debris of about 13,000 cubic meters piled up at the toe of the slope, occupying area of about 50 meters wide and 100 meters long.

Bukit Antarabangsa is underlain by light grey coarse-grained leucocrative granite. Granite boulders of various sizes were found scattered at the landslide scars. Generally, residual cover in the landslide area is relatively thick at the (about 15 to 30 m) and become relatively thinner on the mid-course of the slope, followed by exposed granite on the toe of the slope. Weathering of the medium to course-grained granite produced sandy clay to clayey sand residual soil.

There was an intense rainfall for a few hours before the first landslide occurred (4.30 p.m. on 14th May 1999). A total of 32.8 mm of rain was recorded between 3.00 p.m. to 4.00 p.m. on that day. Between 1.00 a.m. on 13 May to 4.00 p.m. on 14th May a total of 90.2 mm of rain was recorded with a peak hourly rainfall of 31.9 mm between 1.00 a.m. to 2.00 a.m. on 13th May 1999. The second landslide (5:30 a.m. on 15th May) occurred 13 hours later. The rainstorm on 14th May 1999 prior to the landslide was preceded by a heavy rainstorm on the 12th may 1999 with 80.5 mm of rain. A total of 308 mm of rain was recorded 13 days before the landslide.

7. CONCLUSIONS

Slope stability in residual soils is governed by the geologic environment and climatic conditions. The heavy rainfall in tropical areas is largely responsible for many slope failures that occur in steep terrain, rapid infiltration of water occurring because of the high permeability of residual materials and colluvium. The heterogeneity of the residual profiles and the random nature of the rainfall mean that the application of theoretical methods of stability analysis is problematical.

8 REFERENCES

Affendi .A & Faisal Haji Ali.1994a *Strength and suction characteristics related to slope stability.* Inter. Conf. On Landslides, Slope Stability and the Safety of Infra-Structures, Kuala Lumpur:1-8

Affendi. A. & Faisal Haji Ali. 1994b. *Field measurement of soil suction.*Proc. 13th Int. Conf. On Soil Mech. And Found. Engg. 3: 1013-1016

Brand, E. W .1985.*Predicting the performance of residual soil slopes.* Proceedings of 11th Internalional Confererenc on Soil Mechanics and Foundation Engineering, San Francisco, California. 5: 2541-2578.

Choong Foog Heng..1998.*Volume change behaviour of an unsaturated residual soil.*M.Eng Thesis, University of Malaya

Deere, D.U., and Patton, F.D. 1971.*Slope stability in residual soils:* Proceedings of the 4th Pan American Conferen. on Soil Mechanics and Foundation Engineering, Puerto Rico: 87-170.

Fredlund, D.G.. and Rahardjo, H. 1993 *Soil Mechanics for Unsaturated* Soils, John Wiley, New York.

Fredlund, D.G., et al. 1978 .*The shear strength of unsaturated soils,"* Canadian Geotechnical Journal, 15: 313.321.

Fredlund, D.G., et al. 1987 .*Non linearity of strength envelop for unsaturated soils.* Proceedings of thee 6th International Conference on Expansive Soils, New Delhi, India: 49.54.

Hossain M.K. 1999. *Shear strength characteristics of a decomposed granite residual soil in Malaysia.* Ph.D. Thesis. National University of Malaysia.

Low T.H. , Faisal Haji Ali, Saravanan M. & Phang K.S. 1999a. *Effects of perched water table on slope stability in unsaturated soil.* Proc. Int. Sympo. On Slope Stability Engg.1 : 393-397

Low T.H. , Faisal Haji Ali& Saravanan M.1999b. *Field suction variation with rainfall on cut slope in weathered sedimentary residual soil.* Proc. Int. Sympo. On Slope Stability Engg.1 : 399-403.

Premchitt,J., brand, E.W & Phillipson, H.B.1985. *Landslides caused by rapid grounwater changes.* Groundwater in Engineering Geology. Geological Society. Ed. J.C. Cripp et al: 87-94.

Saravanan M., Faisal Haji Ali, & Low T.H. 1999. *Determination of shear strength parameters of unsaturated sedimentary residual soil for slope stability analyses.* Proc. Int. Sympo. On Slope Stability Engg.2 : 687-691

Unsaturated Soils for Asia, Rahardjo, Toll & Leong (eds) © 2000 Taylor & Francis, ISBN 90 5809 139 2

Historical developments and milestones in unsaturated soil mechanics

D.G. Fredlund

Department of Civil Engineering, University of Saskatchewan, Saskatoon, Sask., Canada

ABSTRACT: The development of soil mechanics for unsaturated soils began about two to three decades after the commencement of soil mechanics for saturated soils. The basic principles related to the understanding of unsaturated soil mechanics were formulated mainly in the 1970's. There were two series of conferences that played a dominant role in our understanding of unsaturated soils. These were the series of conferences on expansive soils from 1965 to 1992, and the series of conferences on unsaturated soil mechanics from 1993 to the present.

The development of unsaturated soil mechanics has gone through several stages as it has moved towards implementation in standard geotechnical engineering practice. These stages are traced through an analysis of research papers published at a number of engineering conferences related to unsaturated soils.

Key words: unsaturated soils, expansive soils, history, research papers, conference papers.

1 INTRODUCTION

There is a rich heritage related to the emergence of unsaturated soil mechanics. It is the intent of this paper to highlight some of the historical aspects of this subject. It will not be possible to do justice to all the research that has taken place. As a result, comments and a brief analysis will be undertaken on a limited number of documents related to the subject of unsaturated soil behavior (i.e., mainly conference proceedings). The desire is to acknowledge and pay tribute to the many excellent studies on expansive and unsaturated soils that have been performed in numerous countries of the world.

Our historical heritage can be divided into seven decades. Each of the past five decades, in particular, has identified unique steps forward towards the emergence of unsaturated soil mechanics. An attempt will be made to identify these steps through a focus on the proceedings of key conferences and publication of several books. There are two series of conferences that have played a dominant role in communicating developments in our understanding of unsaturated soil behavior. These are the series of conferences on expansive soils from 1965 to 1992 and the series of conferences on unsaturated soils from 1993 to the present. There are also other one-time conferences that have also played an important role. An attempt has been made to analyze the contents of research papers at conferences in order to better understand the stages involved in the emergence of unsaturated soil mechanics.

In undertaking this synthesis, it is obvious that many valuable contributions will not be given proper credit. However, I trust that the synthesis will make mention of some of the key research publications and to a lesser extent, some of the many important researchers involved and the contribution that each has made. The focus is on research papers published at research conferences as opposed to those published in journals. A number of statistical indicators will be used to assist in the task.

2 HISTORICAL DEVELOPMENTS (1930'S TO 1950'S)

The first ISSMFE conference (International Society for Soil Mechanics and Foundation Engineering) in 1936 provided a forum for the establishment of principles and equations relevant to saturated soil mechanics. These principles and equations became pivotal throughout subsequent decades of research. This same conference was also a forum for numerous research papers on unsaturated soil behavior. Unfortunately, a parallel set of principles and equations did not immediately emerge for unsaturated soils. In subsequent years, a science and technology for unsaturated soils has been slow to develop. Not until the research at Imperial College in the late

1950's did the concepts for understanding unsaturated soils behavior begin to be established (Bishop, 1959). The research of Lytton (1967) in the United States did much to ensure that the understanding of unsaturated soil behavior was founded upon principles set forth in continuum mechanics.

One of the engineering problems observed in the 1930's related to unsaturated soils, involved the flow of water in the zone of negative pore-water pressure (i.e., capillary flow; Hogentogler and Barber, 1941). One of the problems that appeared to perplex civil engineers was that of the movement of water above the groundwater table. The term "capillarity" was adopted to describe the phenomenon of water flow upward from the static groundwater table. Terzaghi (1943) in his book, *Theoretical Soil Mechanics*, endorsed the concepts related to the capillary tube model. The importance of the air-water interface was emphasized with respect to its effect on soil behavior.

The historical review for the period up to the 1950's shows that most of the attention given to unsaturated soils was related to capillary flow. An attempt was made to use the capillary tube rise model to explain the observed phenomenon. Although this model was of some value, it had limitations that became increasingly obvious. In fact, attempts to heavily rely on the capillary tube rise model appear to be a significant factor in the slow development of unsaturated soil mechanics.

3 SUMMARY OF STATISTICS ON CONFERENCE PAPERS

The research literature is too extensive to review in detail. Therefore, it was decided that some of the primary publications would be selected for study and analysis. It was decided to focus primarily on the proceedings of conferences where unsaturated soils formed a significant theme of the conference. Most of the conferences were international in scope but some more regional conferences have also been selected. Mention will also be made of some of the books that have been written related to unsaturated soil behavior. A short review is provided on some of the conferences while a few statistics are simply given on other conferences. Unfortunately, space does not permit a thorough coverage of all the published information.

Table 1 provides a listing of the conference proceedings that have been reviewed. The table is divided into three sections; namely the period from the 1950's to 1964 (Table 1a), the period from 1965 to 1992 (Table 1b), the period from 1993 to the present (Table 1c). The period from 1965 to 1992 is the period when international conferences were held (approximately) every 4 years. Subsequently, this inter-national conference series was changed to focus more broadly on unsaturated soils. This change was at the request of the TC6 subcommittee of the ISS-MFE.

The chronology of the conferences related to unsaturated soil behavior is plotted on Figure 1. Also shown beside each conference is the number of research papers published on unsaturated soils. It is apparent that there has been a constantly growing interest in research related to unsaturated soils as shown in the plot of the number of research papers published during each decade (Fig. 2). The first research papers appeared in the late 1950's and have steadily increased until the present time.

3.1 *Categorization of Papers Based on the Soil Property Being Studied*

The proceedings listed in Table 1 contained more than 1000 research papers on unsaturated and expansive soils. All of the papers were analyzed with respect to three categorizations. The first categorization was with respect to the soil property being analyzed. Sub-categories were then selected that seemed to cover essentially all of the soil properties that had been researched. The sub-categories were: volume change, shear strength, soil suction measurements, classification of soils, permeability and chemical concentration properties. The category of "general" was also added since some of the papers did not fit the designated categories. It must be realized that the categorizations are approximate since it is difficult in some cases to put the research paper in only one category. The summary of the themes of the research papers is shown in Tables 2.

The cumulative number of research papers published on each of the soil properties for unsaturated soils is shown in Figure 3. A vertical line is shown between 1992 and 1993 since this is the approximate time when the scope of research was broadened to consider all aspects of unsaturated soils research. Volume change studies have formed the primary focus of research from the earliest of studies. At the same time, the volume change behavior remains the soil property about which the least is known from an engineering standpoint. Approximately 5 times as many papers have been published on volume change behavior as have been published on shear strength and permeability. And still, it is the shear strength and permeability (or hydraulic conductivity) related problems that have enjoyed the greatest success in implementation into engineering practice.

Figure 3 shows that there has been a significant increase in research related to the coefficient of permeability of unsaturated soils, in recent years. This area of research would appear to be largely driven by increased concern in the environmental and geoenvironmental areas. While less research has been

54

Table 1. List of conference proceedings reviewed and analyzed with respect to the development of unsaturated soil mechanics

Table 1a.) Early conference proceedings (1950's to 1964) emphasizing the behavior of unsaturated soils

1959 (October)	Conference on Theoretical and Practical Treatment of Expansive Soils, University of Colorado, Boulder, CO, U.S.A.
1960 (June)	Conference on Shear Strength of Cohesive Soils, ASCE, University of Colorado, Boulder, CO, U.S.A.
1961 (March)	Conference on Pore Pressures and Suction in Soils, Butterworths, London, England.
1963 (June)	Third Regional Conference for Africa on Soil Mechanics and Foundation Engineering, Salisbury, Rhodesia.

Table 1b.) Some of the conference proceedings (1965 to 1992) devoted primarily to the engineering behavior of expansive soils

1965 (September)	Engineering Effects of Moisture Changes in Soils, First International Research and Engineering Conference on Expansive Clay Soils, College Station, TX, U.S.A.
1965	Moisture Equilibria and Moisture Changes in Soils Beneath Covered Areas (A Symposium in Print) Butterworths, Australia, G, D. Aitchison, Editor.
1966 (July)	Symposium on Permeability and Capillarity of Soils, a Symposium presented at 69th Annual Meeting of ASTM, Atlantis City, NJ, U.S.A.
1969 (June)	Second International Research and Engineering Conference on Expansive Clay Soils, College Station, TX, U.S.A.
1971	Proceedings of the Fifth Regional Conference for Africa, Soil Mechanics and Foundation Engineering, Luanda, Angola.
1973 (August)	Proceedings of the Third International Conference on Expansive Soils, Haifa, Israel.
1980 (June)	Proceedings of the Fourth International Conference on Expansive Soils, Denver, CO, U.S.A.
1984 (May)	Proceedings of the Fifth International Conference on Expansive Soils, Adelaide, Australia.
1985 (February)	First International Conference on Geomechanics in Tropical Lateritic and Saprolitic Soils, TropicaLS '85, Brasilia, Brazil.
1987 (December)	Proceedings of the Sixth International Conference on Expansive Soils, New Delhi, India.
1988 (August)	Proceedings of the International Conference on Engineering Problems of Regional Soils, Beijing, China.
1992 (August)	Proceedings of the Seventh International Conference on Expansive Soils, Dallas, TX, U.S.A.

Table 1c.) Conference proceedings emphasizing the behavior of unsaturated soils from 1992 to the Present

1993 (July)	ASCE Geotechnical Special Publication, Unsaturated Soils Session, Geotechnical Special Publication, No. 39.
1994 (April)	Proceedings of the Second Brazilian Symposium on Unsaturated Soils, Recife, PE, Brazil
1995 (September)	Proceedings of the First International Conference on Unsaturated Soils, Paris, France.
1997 (April)	NSAT '97 Solos Não Saturados, Proceedings of the third Brazilian Symposium on Unsaturated Soils, Rio de Janeiro, Brazil.
1997 (July)	ASCE Geo-Logan Conference, Unsaturated Soil Engineering Practice, Geotechnical Special Publication, No. 68, Logan, Utah, U.S.A.
1998 (August)	Proceedings of the Second International Conference on Unsaturated Soils, Beijing, China.
2000 (May)	Asia 2000 Conference on Unsaturated Soils, Singapore.
2000 (August)	ASCE Geo-Denver Unsaturated Soils Specialty Session, Denver, CO, U.S.A.

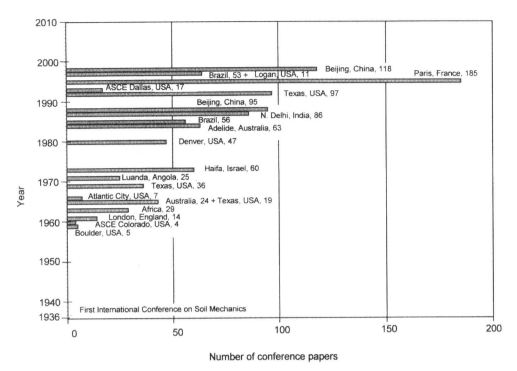

Figure 1. Chronology of the main conference with a theme related to unsaturated soil behavior.

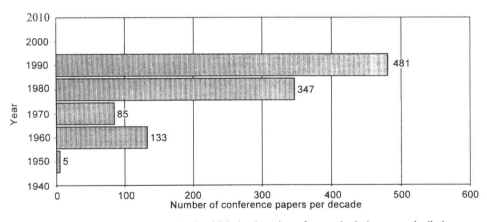

Figure 2. Number of research papers per decade published at the main conferences that had unsaturated soils themes.

undertaken on the shear strength of unsaturated soils, their behavior appears to have been easier to quantify. Research into the adsorption, diffusion and dispersion behavior of chemicals in unsaturated soils does not appear to have been extensive. However, probably the selection of research conferences is not the most appropriate for these topics.

The graph would also appear to indicate that although engineers have attempt to relate the behavior of unsaturated soils to soil suction, there has not been extensive research devoted to the measurement of soil suction. Fundamental studies on soil suction may have often been left to the soil science and soil physics disciplines. It would seem that more attention should be given to the measurement of soil suction and its components.

Figure 4 shows the soil property area distribution for all the papers published to-date. Here, volume

Table 2.   Summary of statistics on the research papers presented at selected conferences

Year	Soil property							Total
	Volume change	Shear strength	Permeability	Soil suction	Classification	Chemical	General	
1959 Boulder	5	0	0	0	0	0	0	5
1960 ASCE Colorado	0	4	0	0	0	0	0	4
1961 London	4	0	7	3	0	0	0	14
1963 Africa	16	5	3	2	3	0	0	29
1965 Texas	16	2	1	0	0	0	0	19
1965 Australia	5	0	9	6	4	0	0	24
1966 Atlantic City	0	0	7	0	0	0	0	7
1969 Texas	30	2	3	0	0	0	1	36
1971 Angola	17	5	0	0	3	0	0	25
1973 Haifa	43	1	5	5	4	2	0	60
1980 Denver	35	3	2	2	5	0	0	47
1984 Adelaid	51	2	0	3	7	0	0	63
1985 Brazil	23	7	5	1	15	3	2	56
1987 N. Delhi, India	72	5	0	1	7	0	1	86
1988 Beijing	62	19	4	1	3	5	1	95
1992 Texas	81	5	2	4	0	4	1	97
1993 ASCE Dallas	11	0	3	2	0	0	1	17
1995 Paris	82	21	49	18	3	5	7	185
1997 Brazil	17	12	8	4	9	2	1	53
1997 Logan	5	1	4	0	0	0	1	11
1998 Beijing	59	17	20	14	2	5	1	118
Total	634	111	132	66	65	26	17	1051

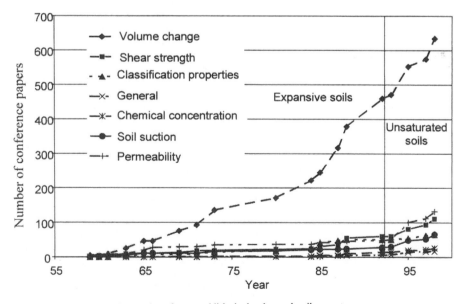

Figure 3.   The cummulative number of paper published related to each soil property

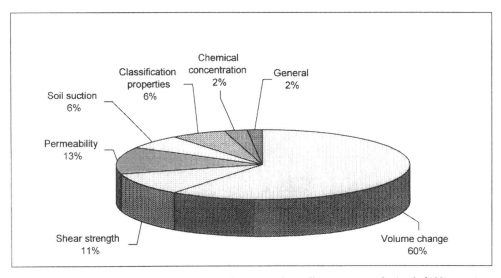

Figure 4. Percentage of research papers published relative to the various soil property categories (total of 1051 papers).

change studies account for 60% of the research while shear strength and permeability account for 11% and 13%, respectively. However, when the research papers are subdivided for the periods before and after 1992, the percentages are somewhat different. Volume change studies accounted for 77% of the papers before 1992 and have subsequently accounted for only 45% of the papers. On the other hand, shear strength papers accounted for only 5% of the papers prior to 1992 and presently account for 13%. Likewise, the number of publication related permeability has risen from 5% to 22% for similar time periods. The analysis of specific conferences show the same trends.

The analysis of the First and Second International Conferences on Unsaturated Soils are shown in Figure 5, indicating about 50% of the papers were on volume change. Figure 6 shows that the early international conferences on expansive soils were almost exclusively devoted to volume change studies. The Second International Conference on Expansive Soils had 83% of the papers on volume change but this reduced to72% by the Third International Conference on Expansive Soils. These results show a general increase in research in all aspects related to unsaturated soils behavior.

3.2 Categorization Based on Type of Research Being Undertaken

The second categorization of the research papers was based upon the type of research being undertaken; namely, whether the study was of an experimental or theoretical nature and whether the study was in the laboratory or the field. The four subcategories were theory, laboratory experiments, field

measurement and the application of formulations and theory to engineering problems. It is realized that all categories are not necessarily mutually exclusive and in some cases it was necessary to select the dominant aspect of the study. The results of the research paper analysis for categorization 2 and 3 are presented in Table 3. The analysis of categorization 3 will be discussed later.

Figure 7 shows a cumulative plot of the types of research studies being undertaken with time on unsaturated soils. Once again, an arbitrary dividing line is places between 1992 and 1993, denoting a change to a broader scope in unsaturated soils research. Throughout the period of research, the majority of studies were performed in the laboratory. On the other hand, the field measurements were the least common. The distributions shown are to be expected since certain types of research require greater research dollars to conduct. An analysis of all of the types of research studies undertaken is shown in Figure 8. The results indicate that about 20% of the studies were theoretical in nature while 25% of the studies were directed towards the application of the theory to practical problems. About 42% of the studies involved experimental measurements in the laboratory while 12% of the studies involved field measurements.

If the analysis separates the papers before and after 1992, it can be seen that there has been an increase in theoretical and laboratory studies. Theoretical studies accounted for 18% of the papers prior to 1992 and 24% of the papers after 1992. Laboratory studies accounted for 35% of the papers prior to 1992 and 50% of the papers after 1992. The field and application papers show a decrease during recent years. It would appear that the broadening of

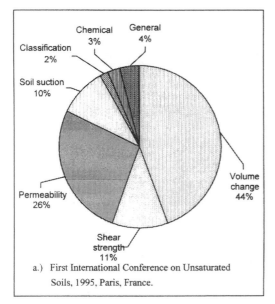
a.) First International Conference on Unsaturated
     Soils, 1995, Paris, France.

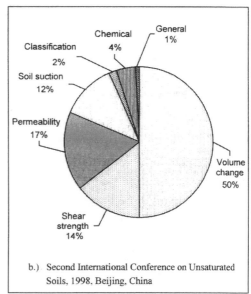
b.) Second International Conference on Unsaturated
     Soils, 1998, Beijing, China

Figure 5.   Percentage of research papers published relative to the various soil property categories, a.) First International Conference on Unsaturated Soils, 1995, Paris, France, and b.) Second International Conference on Unsaturated Soils, 1998, Beijing, China

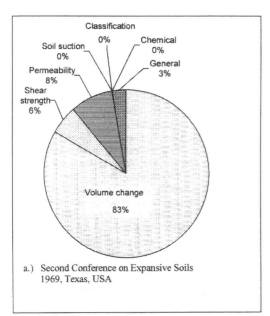
a.) Second Conference on Expansive Soils
     1969, Texas, USA

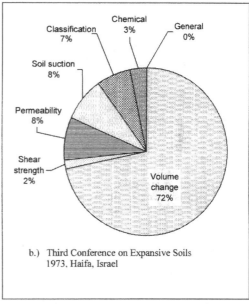
b.) Third Conference on Expansive Soils
     1973, Haifa, Israel

Figure 6.   Percentage of research papers published relative to the soil property categories, a.) Second Conference on Expansive Soils, 1969, Texas, USA, and b.) Third Conference on Expansive Soils, 1973, Haifa, Israel

the scope to embrace all types of unsaturated soils research resulted in a rejuvenation of theoretical and laboratory studies. Figure 9 shows the distributions for the Texas conference (1969) and the Paris conference (1995), illustrating the changes in emphasis with time. While these changes may be good, it also reveals that additional research is required in order to ensure that the research becomes implemented into engineering practice.

Table 3.   Summary of statistics on the research papers presented at selected conferences

Year	Type of research study					Region of the world						
	Theory	Lab	Field	Application	General	Africa	Asia	Australia	Europe	Middle East	North America	South America
1959 Boulder	2	0	0	3	0	0	0	0	0	0	5	0
1960 ASCE Colorado	0	4	0	0	0	0	0	0	4	0	0	0
1961 London	5	5	2	2	0	1	1	0	12	0	0	0
1963 Africa	2	9	5	12	1	28	0	1	0	0	0	0
1965 Texas	3	5	4	7	0	1	1	4	1	0	12	0
1965 Australia	2	7	6	9	0	8	2	8	4	2	0	0
1966 Atlantic City	0	4	2	1	0	1	0	1	0	0	5	0
1969 Texas	7	14	2	13	0	3	7	2	4	6	13	1
1971 Angola	5	9	1	10	0	14	1	0	5	0	3	2
1973 Haifa	14	25	3	15	3	7	8	14	7	12	12	0
1980 Denver	10	14	5	18	0	5	4	4	7	4	21	2
1984 Adelaide	10	16	22	15	0	7	5	16	7	7	21	0
1985 Brazil	4	29	7	15	1	5	9	3	9	4	3	23
1987 N. Delhi, India	18	33	6	29	0	5	44	1	15	7	13	1
1988 Beijing	16	45	9	24	1	2	74	1	10	2	5	1
1992 Texas	17	32	17	30	1	9	21	5	15	6	37	4
1993 ASCE Dallas	4	8	0	4	1	0	0	1	4	1	10	1
1995 Paris	43	93	12	34	3	6	35	18	64	14	28	20
1997 Brazil	13	25	9	6	0	0	0	1	0	0	8	44
1997 Logan	2	5	1	2	1	0	1	0	1	0	9	0
1998 Beijing	29	59	14	15	1	7	40	5	30	2	25	9
Total	206	441	127	264	13	109	253	85	199	67	230	108

3.3 Categorization Based on the Part of the World the Research is Being Conducted (Global Distribution)

The third categorization of the research papers was on the basis of the region of the world in which the research was conducted. The regions were subcategorized in terms of the the major land regions of the world. The numbers of papers from each region can be found in Table 3. It would appear that every region has problems related to unsaturated soils. However, it is also realized that the population base and the financial resources to address unsaturated and expansive soils problems, is not the same in every region.

The cumulative number of research papers published from each of the regions is shown in Figure 10. There are a number of observations that can be made from this graph. It can be observed that there has been a steady and gradual increase in research on unsaturated soils on all of the regions with time. Two regions have shown a dramatic increase in unsaturated soils research during the past two decades. For example, research in the mid-1980's dramatically increased in Asia. Later, in the early 1990's, there was a significant increase in research in South America (primarily in Brazil) in the area of unsaturated soils. However it should be noted that the research papers from the Brazil, 1992 conference were

not included in the study because only 2 papers were in English. In both of these cases, it would appear to be the broadening of the scope of unsaturated soils that precipitated studies on slope stability and other areas. The regions of Africa and the Middle East certainly have arid conditions and a need for unsaturated soil mechanics. There has been a slow but gradual increase in studies in these areas. At the same time, it must remembered that much of the early research on expansive soils was undertaken in South Africa and in Israel. The First International Conference on Expansive Soils came about as a result discussions between Professor Buchanan in United States and Dr. Aitchison in Australia. In fact, many regions of the world participated in the series of conferences on expansive soils.

Figure 11 shows the percentage distribution for all the research papers published from the various regions of the world. If the research papers are divided between those published before and after 1992, it can be observed that certain regions of the world have had a renewed interest in the subject. North America accounted for 30% of the papers prior to 1992 and 21% subsequent to 1992. The distribution of papers by region, published at the First and Second International Conferences on Unsaturated Soils is shown on Figure 12. The three largest contributors are Asia, Europe, and North America. It is important not to read too much into the regional

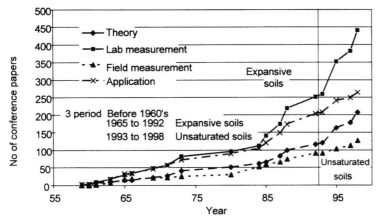

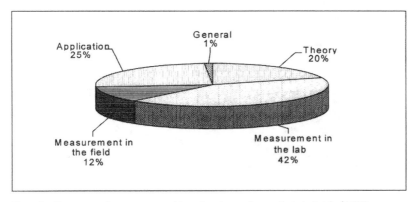

Figure 7. Number of papers in each year presented in various types of research study.

Figure 8. Percentage of papers presented in various types of research study (total of 1051).

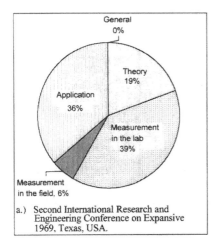

a.) Second International Research and Engineering Conference on Expansive 1969, Texas, USA.

b.) First International Conference on Unsaturated soils, 1995, Paris, France.

Figure 9. Percentage of papers presented in various aspects of soil property in the a.) Second International Research and Engineering Conference on Expansive Clay Soils, 1969, Texas, USA, and b.) First International Conference on Unsaturated Soils, 1995, Paris, France.

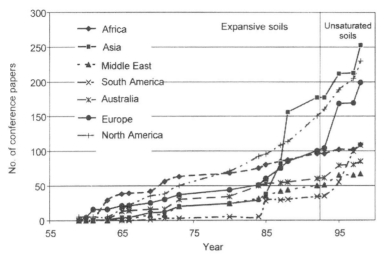

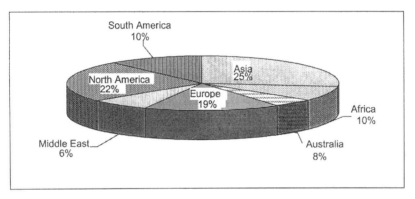

Figure 10 The cummulative number of paper published from various areas of the world.

Figure 11. Percentage of research paper published from various areas of the world (total of 1051 papers)

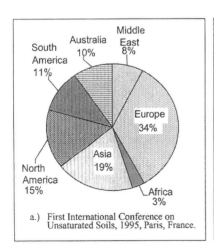

a.) First International Conference on Unsaturated Soils, 1995, Paris, France.

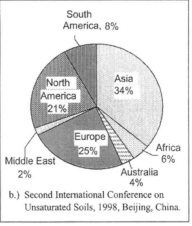

b.) Second International Conference on Unsaturated Soils, 1998, Beijing, China.

Figure 12 Percentage of research papers presented from various areas of the world in the a.) First International Conference on Unsaturated Soils, 1995, Paris, France, and b.) Second International Conference on Unsaturated Soils, 1998, Beijing, China.

analysis because there are many factors involved in the research paper distributions. For example, the commitment to unsaturated soils research may be very high in a particular area but the population or financial base may be low.

It is interesting that about 570 conference papers (i.e., 19 papers per year) were published up to 1991 and 481 papers (i.e., 69 papers per year) were published from 1992 to the present time. The figures show that there has been significant increase in research into the behavior of unsaturated soils.

3.4 Books Published on Unsaturated Soil Behavior and Analysis

There have been several books published that focus on the problems related to expansive and unsaturated soils. A list of books available is shown in Table 4. Three of the books deal specifically with expansive soils. The book published by Kassiff et al in 1969 contains extensive information on engineering in expansive soils areas. It is the authors impression that this book is an example of a valuable document that has not been given sufficient attention. Below is a short review of the contents of the book.

3.5 Short Review of Pavements on Expansive Clays by G. Kassiff, M. Livneh and G. Wiseman (1969)

Textbooks serve an important role in the development of unsaturated soil mechanics since they synthesize the "thinking" of many researchers into a somewhat a "cohesive" context. The text, "Pavements on Expansive Clays" focuses on expansive soils and pavements and appears to be the first book of its kind. The book does not address unsaturated soils in general but the experimental and experiential information presented is of great value as theories for unsaturated soil mechanics were later formalized. The authors had three objectives in mind when writing the book; namely:

1. To put forward those aspects of research into the behavior of unsaturated, expansive clays that are pertinent to the design of pavements.

2. To provide general background information that will enable the engineer with a basic knowledge of soil mechanics to follow the advances that are likely to be made in the coming years in the treatment of expansive clay subgrades.

3. To indicate design and construction recommendations that were believed to lead to successful pavement design on expansive clay subgrades.

The synthesis of material was undertaken primarily with the needs of the country of Israel in mind. However, research literature has been brought together in a logically organized manner from many countries of the world. The engineering basis for the subject is empirical and experimental.

The chapter on the *Methods of Classification and Identification of Swelling Clays* presents various procedures that have been used to identify expansive soils. The field measurements of pavement performance behavior are of great value. The final chapter uses the information on expansive soils in outlining pavement design methods. The treatment of expansive soils in pavement design is thorough and valuable for practicing geotechnical and transportation engineers. This book was written in 1969, and shows that the engineering difficulties associated with expansive were being systematically analyzed by a group of engineers in Israel. The initial intent was to produce a similar book devoted to structures on expansive clays. It is unfortunate that the second book was never completed.

4.0 UNSATURATED SOILS RESEARCH FROM THE LATE 1950'S TO THE 1960'S

Research into the volume change and shear strength of unsaturated soils commenced with new impetus in the late 1950's. Some of the researchers were Black and Croney (1957), Bishop et al., (1960), Aitchison (1967), and Williams (1957). The research resulted in the proposal of several so-called single-valued effective stress equations for unsaturated soils. During the next decade, reservations were expressed regarding the use of a single-valued effective stress equation. There was subsequently a slow change towards the acceptance of the use of two independent stress state variables (Coleman, 1962; Matyas and Radhakrishna, 1968; Fredlund and Morgenstern, 1977).

The first national conference on expansive soils was held in South Africa in 1957. Two years later, in 1959, a symposium on expansive soils was held at the Colorado School of Mines in Golden, Colorado. The proceedings from this symposium clearly described the nature of expansive soils problems and set the stage for an international series of conferences. A short review of the 1959 symposium held in Colorado is presented later.

A conference was held in London, England in 1960 where the focus was primarily of the measurement of negative pore-water pressure and its application in engineering practice. The proceedings provided a summary of valuable research information and a short review of this conference is later presented. The ASCE in United States held a Research Conference on Shear Strength of Cohesive Soils at the University of Colorado in Boulder, CO, in 1960. Much of the proceedings was devoted to better understanding the shear strength of saturated soils. However, there were a couple of important, and often referenced papers published in the proceedings. Once again, a short review of this conference is given below.

Table 4.   List of Books Published Related to the Behavior of Unsaturated Soils

1969	Pavements on Expansive Clays by G. Kassif, M. Livneh, and G. Wiseman, Jerusalem Academic press, Jerusalem, Israel.
1975	Foundations on Expansive Soils, Development in Geotechnical Engineering Vol. 12 by F. H. Chen, Elsevier Scientific Publishing Co., New York, N. Y.
1992	Expansive Soils Problems and Practice in Foundation and Pavement Engineering by John D. Nelson, Debora J Miller, John Wiley & Sons, Inc., New York, N. Y.
1993	Soil Mechanics for Unsaturated Soils by D. G. Fredlund, and H. Rahardjo, John Wiley & Sons, New York, N.Y.
1999	The Emergence of Unsaturated Soil Mechanics-Fredlund Volume by A.W. Clifton, G.W. Wilson and S. L. Barbour. National Research Council of Canada, Ottawa, Canada.

4.1 Short Review of Proceedings of the Colorado School of Mines conference on "Theoretical and Practical Treatment of Expansive Soils", October 1959.

The first soil mechanics conference organized by the Colorado School of Mines in U.S.A., was on the timely topic of "expansive soils". There were only 5 papers presented in 1 day to the conference; however, the conference proved to be of wide interest, drawing a good attendance. The presenters to the conference were selected so as to treat the topic with recognized experience and accomplishment.

Professor Means from Oklahoma State University, presented the first paper entitled "Building on Expansive Soils". He drew upon his extensive experience with the expansive soils problem in mid-west U.S.A. The paper clearly describes the interaction between climatic conditions and the expansive soils problem. There are numerous well-described examples given that illustrate the types of damage common to expanding clays. These are followed with a section on "methods of preventing damage". Material in this paper was later be incorporated into a soil mechanics book (i.e., Physical Properties of Soils by Means and Parcher) that dedicated a significant portion to engineering with expansive soils.

Professors Lambe and Whitman presented a paper on "The Role of Effective Stress in the Behavior of Expansive Soils"; a paper that has been often referenced. An attempt was made to direct engineers to think in terms of changes in effective stress. In other words, volume increases in the soil occur because of a decrease in effective stress. In the discussion following the paper, Dr. Hilf encouraged the search of the soil science literature "which provide a theory for negative pore-water pressures in unsaturated soils".

Measurements of vertical movement in expansive soils were presented by Professor Dawson, along with a suggested method of rating climatic conditions. Holtz presented information on the properties of expansive soils and the problems often encountered in practice. The main soils discussed are the clay shales encountered in Colorado and throughout the mid-west part of U.S.A. The plots of the percent swell, and the swelling pressure, under various density and water content conditions have been often referenced.

C. McDowell reported on the relevance of laboratory tests in the design of pavements and structures on expansive soils. A graph of the relationship between volumetric swell and load was developed. It was noticed that all soils produced curves of similar shape and it was suggested that "families" of curves could be produced. These curves defined the general character of results that could be anticipated from oedometer tests and provided a means of dealing with soils that extended over a wide region such as is encountered in the design of a highway.

In the discussions following the paper presentations, Prof. Dawson suggested that more attention should be given to the measurement of negative pore-water pressures. The discussions made reference to extensive research underway on expansive soils in England and South Africa. The double oedometer tests performed in South Africa by Jenning and Knight (1957) was suggested as a testing procedure to be considered.

In the closing remarks, Walker stated, "These are expansive soils, but they are much more expensive when you have to treat them after the fact, than before the fact. It was pointed out that each of the main presenters had close to 30 years experience with problems associated with expansive soils and there is reason for optimism. For a 1-day conference, these proceedings provided a wealth of material that has been often referenced.

4.2 Short Review of Pore Pressure and Suction in Soils, Conference held in March 1960 and the Proceedings were published in 1961

The conference, attended by over 160 people from 6 countries, was held in March, 1960 and the proceedings were published in 1961. In total, 17 papers were presented for discussion along with an opening address by Prof. Skempton on "Effective Stress in Soils, Concrete and Rocks". The conference focused on fundamental theories and behavior, with all but 3 papers dealing with unsaturated soils. The conference was also an attempt to synthesize efforts of the Institution of Civil Engineers in the international field.

The conference was most appropriately titled since the emphasis was on the measurement of negative pore-water pressures and soil suction. It had become obvious that matric suction (i.e., $u_a - u_w$ where u_a = pore-air pressure and u_w = pore-water pressure) was the primary stress variable to which unsaturated soil behavior must be related. There was a reinforcement of Bishop's (1959) effective stress equation where the soil property χ, was primarily related to the degree of saturation of the soil. Other similar forms for an effective stress equation for unsaturated soils were presented by Aitchison (Australia), Jennings (South Africa) and, Croney and Coleman (England). There was considerable discussion on the meaning of soil suction and the form that a single-valued equation should take but there was no question regarding the importance of soil suction in describing soil behavior.

A key set of experimental data was presented by Croney and Coleman. The data is the first complete soil-water characteristic curve published in geotechnical engineering. Later, it would become more common to interchange the variables on the abscissa and ordinate. The data for a heavy clay, also presents a shrinkage curve and the soil suction versus water content plot for a continuously disturbed soil. The data provides clear evidence of a soil suction value of 1,000,000 kPa at zero water content.

The proceedings of the Pore Pressure and Suction in Soils conference should be read by every student of unsaturated soil behavior. The laboratory and field test results have much to teach us about the importance of careful and meticulous data collection. Experimental testing information on high air entry disks continue to form the basis for most laboratory research on unsaturated soils.

4.3 Short Review of ASCE, Research Conference on Shear Strength of Cohesive Soils, University of Colorado, Boulder, CO, June, 1960

The purpose of the ASCE conference was to assemble, summarize and discuss present knowledge of factors governing the shear strength of cohesive soils. A portion of the conference was devoted to the shear strength of compacted cohesive soils. Approximately 400 engineers from 18 countries attended the conference. While the scope was narrow, the impact of the conference was substantial.

The reviewers of the research papers noted that there was controversy and uncertainty regarding negative pore-water pressures, and whether it was necessary to measure pore-air pressure. In general, Bishop's equation was endorsed but engineers were cautioned about assuming that χ is a given soil property dependent only on degree of saturation.

Seed and Hirschfeld (1960) noted that the shear strength of compacted, cohesive soils is a "subject that has probably undergone a more rapid development in the past ten years than any other aspect of strength". These early insights into the shear strength of compacted soils are likely the reason why shear strength became the first area to be developed in unsaturated soil mechanics. There was much discussion on the measurement of negative pore-water pressures that were felt to be inherently difficult to measure. At the same time, the papers by Bishop and Bjerrum (1960) (i.e., The Relevance of the Triaxial Test to the Solution of Stability Problems) and by Bishop, Alpan, Blight and Donald (1960) (i.e., Factors Controlling the Strength of Partly Saturated Cohesive Soils) provided a wealth of information on the laboratory testing of unsaturated soils. The "ingenious techniques and the use of very fine ceramic disks instead of the conventional coarse porous stones" opened the way for measuring very negative pore-water pressures.

The ASCE Boulder Conference of 1960 has gone down in history as an important conference on shear strength and this is equally true for unsaturated soils. Shear strength was addressed from a fundamental and classical soil mechanics standpoint. The "ground work" had been laid but it would take almost two more decades before the next significant steps forward would be taken in further understanding the shear strength of unsaturated soils.

5 UNSATURATED SOILS RESEARCH (PRIMARILY EXPANSIVE SOILS RESEARCH) FROM THE 1965 TO THE 1992

The First International Conference on Expansive Soils was held at Texas A & M University in 1965. Also delivered to the conference was a Symposium-in-Print edited by Dr. G. Aitchison of Australia. The proceedings of the conference, along with the Symposium-in-Print formed an informative, complimentary combination that would set the stage for subsequent conferences on expansive soils. The Symposium-in-Print started with a review of the theory related to soils with negative pore-water pressures, as viewed from the Soil Science and Soil Physics disciplines. This information has proved to be of value in forming an understanding of the differences between the way soil behavior was viewed in the earth sciences and in geotechnical engineering. With time, it has been found to be necessary to modify some of the theories of soil behavior to better fit the geotechnical engineering profession. The international series of conferences on expansive soils took place, approximately every four years over the period from 1965 to 1992 (Table. 1).

The series of conferences did much to bring attention to the lack of understanding associated with soils with negative pore-water pressures and in particular, expansive soils. Each of the conferences provided ample examples of buildings and other engineered structures that had cracked due to expansive soils. While these pictures and case histories were important, they did not appear to produce the development of a consistent theoretical framework for understanding the general behavior of unsaturated soils.

It is observed in the analysis of the papers presented to the conferences that the high percentage of research papers were devoted to volume change problems. In particular there was considerable emphasis on the formulation of mathematical models for the prediction of heave. These models were one-dimensional in character and were formulated to follow either the suction change stress path or the total stress change path. Even though many of the problems were two-dimensional in character, the models to predict heave remained one-dimensional. In general, there seemed to be a lack in fully understanding the theory associated with volume change behavior. Even with the largest number of research papers having been presented on the volume change problem, it would still appear that even today, it is the volume change behavior that still requires the most research. It can quite clearly be stated that the volume change problem is the most difficult aspect of unsaturated soil behavior to fully understand and use in geotechnical engineering practice.

There were several other conferences during the period from 1965 to 1992 that provided further understanding of the expansive soils problem as well as extending the scope of problems to include problems related to residual soils, collapsing soils, compacted soils and soils in arid regions. The conference on tropical Lateritic and Saprolitic soils in Brasilia in 1985, brought attention to the engineering problems unique to residual soils. However, there was a common aspect that placed the behavior of residual soils into the same category as expansive soils; namely, the pore-water pressures were negative.

A conference on regional problematic soils was held in Beijing in 1988. The proceedings contained considerable information on the behavior of collapsible soils as well as compacted, residual and expansive soils. In fact, four of the five main problematic soils discussed at the conference were soils with negative pore-water pressures.

6 UNSATURATED SOILS RESEARCH FROM 1993 TO THE PRESENT

In the early 1990's, there was general concensus that the scope of the expansive soils conferences should be expanded to more accurately reflect problems encountered in geotechnical engineering practice. Under the leadership of the President of the ISSMFE,

Dr. Morgenstern, it was moved that the TC6 sub-committee on expansive soils should be renamed to the sub-committee on unsaturated soils. This change in emphasis was reflected in the cross-section of research papers submitted to the 7th International Conference on Expansive Soils held in Dallas, Texas in 1992.

The change in name from expansive soil to unsaturated soils rekindled an interest in seeing a soil mechanics developed for unsaturated soils that was parallel in context to that enjoyed within saturated soil mechanics. The First International Conference on Unsaturated Soils was held in Paris, France in 1995. The interest in the broader scope of topics in unsaturated soils was obvious from the response to the call for papers. The conference drew a record number of participants from all over the world. Similarly, the Second International Conference on Unsaturated Soils in Beijing, China in 1998 attracted widespread interest. The Third International Conference on Unsaturated Soils is presently being planned for Brazil in 2002.

In addition to the series of international conferences, a number of national conferences have been organized. Table 1 lists other unsaturated soils conferences that have been organized, such as those in the United States and Brazil. There presently appears to be considerable interest in the general area of unsaturated soils behavior. It appears that there are new opportunities on the horizon for the implementation of unsaturated soils theories in standard geotechnical engineering practice.

It is of value to review the steps that are required in moving from a science that is studied in the laboratory to its implementation in engineering practice. Such a review will provide an understanding of what has been accomplished to-date and what aspects still require further research.

7 STAGES IN MOVING TOWARDS THE IMPLEMENTATION OF UNSATURATED SOIL MECHANICS

The progress in developing a science basis for unsaturated soil mechanics can be viewed in terms of a series of stages leading towards implementation. The stages leading towards implementation are listed in Figure 13 (Fredlund, 1999). Research studies over the past 6 decades has been directed at all stages leading towards an appropriate technology for implementation.

7.1 The State Variable Stage

The state variable stage is the most basic and fundamental level at which a science for unsaturated soil behavior can be initiated. The most important state variables for an unsaturated soil are the stress state variables; namely the net normal stress, $[(\sigma - u_a)$ where σ = total stress and u_a = pore-air pressure], and the matric suction, $[(u_a - u_w)$ where u_w = pore-water pressure)]. These state variables have become widely accepted and illustrate the need to separate the effects of total stress and pore-water pressure when the pore-water pressures are negative. There is also a smooth transition between the saturated and unsaturated states. The matrix form of the stress tensors for an unsaturated soil are shown in Equation (1).

$$\begin{bmatrix} (\sigma_x - u_a) & \tau_{yx} & \tau_{zx} \\ \tau_{xy} & (\sigma_y - u_a) & \tau_{zy} \\ \tau_{xz} & \tau_{yz} & (\sigma_z - u_a) \end{bmatrix} \tag{1a}$$

$$\begin{bmatrix} (u_a - u_w) & 0 & 0 \\ 0 & (u_a - u_w) & 0 \\ 0 & 0 & (u_a - u_w) \end{bmatrix} \tag{1b}$$

7.2 The Constitutive Stage

The constitutive stage becomes the point at which empirical, semi-empirical and possibly theoretical relationships between state variables are proposed and verified. The verification of proposed constitutive relations must be conducted for a wide range of soils in order to ensure uniqueness, and subsequently confidence on the part of the practicing engineer. Extensive research studies on constitutive relationships for unsaturated soils were made in the 1970's but earlier and later developments have also contributed to our understanding of unsaturated soil behavior.

7.3 The Formulation Stage

The formulation stage involves combining the constitutive behavior of a material with the conservation laws of physics applied to an elemental volume. The result is generally a partial differential equation that describes a designated process for an element of the continuum.

7.4 The Solution Stage

The solution stage involves solving specific examples representative of a class of problems. At the solution stage, the partial differential equations are converted to a numerical solution that becomes known as a software package.

7.5 The Design Stage

The is a gradual increase in engineering confidence as research progresses from the formulation stage, to the solution stage and on to the design stage. The de-sign stage focuses on the primary unknowns that must be quantified from a practical engineering standpoint. The design stage generally involves a quantification of geometric and soil property variables that become part of an engineering design. The computer has an important role in the design of earth systems (Fredlund, 1996) and has changed the way in which geotechnical designs are conducted and design generally takes the form of a parametric type study.

7.6 The Verification and Monitoring Stage

There is need to "observe" the behavior of any infrastructure during construction and subsequent to construction in order to provide feedback to the designer. Only through field monitoring and feedback can confidence be firmly established in the design procedures. The "observational method" as defined by Peck (1969) goes even beyond the verification of design and is considered to be a part of the design process.

7.7 The Implementation Stage

The implementation stage may not be realized in engineering practice even when the theoretical formulations and related design procedures have been fully studied and verified. Implementation is the final stage in bringing an engineering science into routine engineering practice. Other factors that need to be addressed at the implementation level are: i.) the cost of undertaking any special site investigations, soil testing and engineering analyses, ii.) the human resistance to change, and iii.) the political, regulatory and litigation factors that may be involved.

The slowness in the implementation of unsaturated soil mechanics appears to be related to the cost of soil testing for the quantification of soil properties. The old soil mechanics paradigm involving the direct measurement of soil properties becomes extremely costly when measuring unsaturated soil property functions. However, there are a number of other procedures that provide a new paradigm for evaluating unsaturated soil property functions. New procedures have been proposed, and are necessary for the assessment of unsaturated soil property functions (Fredlund, 1999).

The field of unsaturated soil mechanics has developed through a series of stages and today engineers are faced with the challenge of implementing a new science as part of geotechnical engineering. This may prove to be the greatest challenge; one on which further research is still required. At the same time it is the challenge that can open the door to unique technological applications and financial rewards.

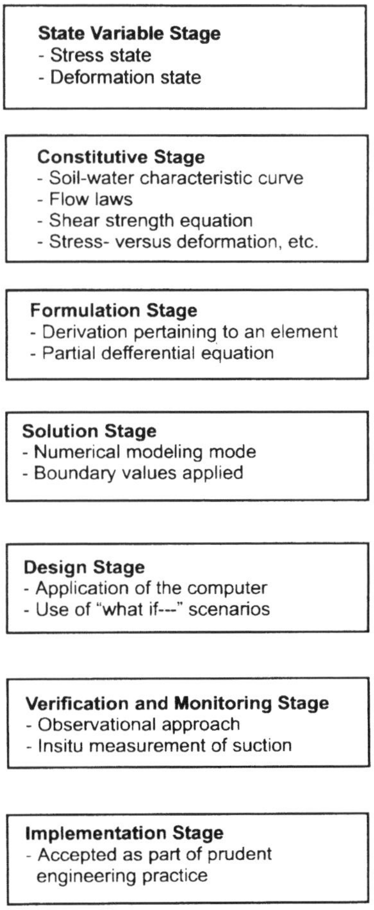

```
┌─────────────────────────────────────┐
│  State Variable Stage               │
│  - Stress state                     │
│  - Deformation state                │
└─────────────────────────────────────┘

┌─────────────────────────────────────┐
│  Constitutive Stage                 │
│  - Soil-water characteristic curve  │
│  - Flow laws                        │
│  - Shear strength equation          │
│  - Stress- versus deformation, etc. │
└─────────────────────────────────────┘

┌─────────────────────────────────────┐
│  Formulation Stage                  │
│  - Derivation pertaining to an element │
│  - Partial defferential equation    │
└─────────────────────────────────────┘

┌─────────────────────────────────────┐
│  Solution Stage                     │
│  - Numerical modeling mode          │
│  - Boundary values applied          │
└─────────────────────────────────────┘

┌─────────────────────────────────────┐
│  Design Stage                       │
│  - Application of the computer      │
│  - Use of "what if---" scenarios    │
└─────────────────────────────────────┘

┌─────────────────────────────────────┐
│  Verification and Monitoring Stage  │
│  - Observational approach           │
│  - Insitu measurement of suction    │
└─────────────────────────────────────┘

┌─────────────────────────────────────┐
│  Implementation Stage               │
│  - Accepted as part of prudent      │
│    engineering practice             │
└─────────────────────────────────────┘
```

Figure 13. Primary stages leading towards the successful implementation of unsaturated soil mechanics.

REFERENCES

Aitchison, G. D. 1967. Separate roles of site investigation, quantification of soil properties, and selection of operational environment in the determination of foundation design on expansive soils. *Proc. 3rd Conf. Soil Mech. Found. Eng.* Haifa, Israel **3**: 72 – 77.

Bishop, A. W. 1959. The principle of effective stress. Lecture delivered in Oslo, Norway, in 1955; *published in* Teknisk Ukeblad, **106** (39): 859 – 863.

Bishop, A. W. & L. Bjerrum 1960. The relevance of the triaxial test to the solution of stability problems. *ASCE Res. Conf. on Shear Strength of Cohesive Soils,* Boulder, Colorado: 437 - 501.

Bishop, A. W., I. Alphan, G. E. Blight & I. B. Donald 1960. Factors controlling the shear strength of partly saturated cohesive soils. *ASCE Res. Conf. Shear Strength of Cohesive Soils,* Boulder, Colorado: 503 – 532.

Black W. P. M. & D. Corney 1957. Pore water pressure and moisture content studies under experimental pavements. *Proc. 4th Int. Conf. in Soil Mech. Found. Eng.,* **2**: 94 –103.

Coleman, J. D. 1962. Stress/strain relations for partly saturated soils. *Geotechnique* (Correspondence) **12**(4): 348 – 350.

Fredlund, D. G. 1996. Microcomputers and saturated-unsaturated continuum modelling in geotechnical engineering. Symp. *Computers in Geotechnical Engineering, INFOGEO '96,* Sao Paulo, Brazil, Aug. 28-30, **2**: 29-50.

Fredlund, D. G. 1999. The implementation of unsaturated soil mechanics into geotechnical engineering. The 1999 R. M. Hardy Lecture, *52 Canadian Geotechnical Conf.,* Regina, Saskatchewan, Oct. 25 – 27.

Fredlund, D. G. & N. R. Morgenstern 1977. Stress state variables for unsaturated soils. *Journal of the Geotechnical Engineering Division, ASCE,* **103**(GT5): 447-466.

Hogentogler C. A. & E. S. Barber 1941. Discussion in soil water phenomena. *Proc. Hwy. Research Board,* **21**: 452 – 465.

Jennings, J. E. & K. Knight 1957. The prediction of total heave from the double oedometer test. *Proc. Symp. Expansive Clays* (South Africa Institute of Civil Eng. Johansperg, 7(9):13 – 19.

Kassiff, G., M. Linveh. & G. Wiseman 1969. *Pavement on expansive clays.* Israel: Jerusalem Academic Press.

Lytton, R. L. 1967. Isothermal water movement in clay soils. *Ph. D. Dissertation, University of Texas, Austin, TX.*

Means, R. E. & J. V. Parcher 1963. *Physical properties of soils.* Ohio: Charles E. Merrill Books Inc.

Mtyas E. L. & H. S. Radhakrishna 1968. Volume change characteristics of partially saturated soils, *Geotechnique* **18**(4): 432 – 448.

Peck, R. B. 1969. Advantages and limitations of the observational method in applied soil mechanics. *Geotechnique* **19**(2): 171-187.

Seed H. B. & R. C. Hirschfeld 1960. Shear strength of compacted, cohesive soils. *Final report ASCE Res. Conf. on Shear Strength of Cohesive Soils,* Boulder, Colorado: 1141 - 1150.

Terzaghi, K. 1943. *Theoretical soil mechanics.* New York: Wiley.

Williams, A. A. B. 1957. Studies of shear strength and bearing capacity of some partially saturated sands. *Proc. 4th Conf. Soil Mech. Found. Eng.* London, England **3**: 453 – 456.

Unsaturated Soils for Asia, Rahardjo, Toll & Leong (eds) © 2000 Taylor & Francis, ISBN 90 5809 139 2

Geotechnical aspects of groundwater recharge in arid regions

S. L. Houston

Department of Civil and Environmental Engineering, Arizona State University, Tempe, Ariz., USA

ABSTRACT: The use of ground surface infiltration basins for the replenishment of groundwater supplies is common practice in arid regions. Both relatively clean water and water of impaired quality, such as treated wastewater, are used for this purpose. The water is infiltrated through an unsaturated soil region to an underlying unconfined aquifer. The recharged water is subsequently recovered by pumping and reused. As water demands have grown, the recycling of wastewater has increased. When treated wastewater is used as the recharge source the process of infiltration through the saturated and unsaturated soil zones is referred to as Soil Aquifer Treatment (SAT) because additional cleanup of the wastewater occurs as it travels through the soil. Although the recycled wastewater is commonly used for nonpotable purposes such as landscape irrigation and power plant cooling, interest in reclaimed water as a potable supply is great. Several SAT sites are under study as a part of research conducted through the Center for Sustainable Water Supply at Arizona State University. The geotechnical considerations for SAT include primarily the evaluation hydraulic loading capacity of the site, maintenance of hydraulic loading capacity during basin operation, and study of the impact of soil type on water treatment.

1 INTRODUCTION

Urbanization in arid regions is on the rise. In the United States the fastest growing cities are located in the arid and semi-arid regions of the southwest. Cities such as Las Vegas, Nevada, and Phoenix, Arizona, are experiencing rapid expansion, and along with this expansion comes a major concern for the maintenance of a sustainable clean water supply. Past over pumping of aquifers has led to problems such as groundwater level decline, subsidence, and earth fissures. Groundwater recharge projects are considered necessary for meeting present and future demands on limited water resources. Continued growth in arid regions will require the development of a sustainable water supply, which will no doubt include improvement of our ability to store and utilize reclaimed wastewater.

Groundwater recharge using either relatively clean water or treated wastewater involves the infiltration of surplus or impaired quality water from the ground surface to underground water storage aquifers. Water is captured and stored underground in times of surplus for use in times of deficit. Underground storage avoids some of the disadvantages of conventional surface water reservoirs, such as evaporation, potential exposure to contaminants, and large capital investment become apparent.

Particularly in arid, water-limited regions treated wastewater has gained acceptance as a water supply, and as an augmentation to the more conventional surface and groundwater supplies. Infiltration land treatment systems, wherein the groundwater is recharged with pretreated wastewater effluent have been developed. This process is referred to as soil aquifer treatment (SAT). SAT operations and extensive research studies have been conducted for over 20 years in Los Angeles County, California, where no measurable impact on groundwater quality or human health has been found (Nellor, et al., 1984). Bouwer (1996) has reported that the costs of soil aquifer treatment systems have been shown to be less than 40% of the costs of equivalent above ground treatment.

A variety of treatment processes are operating as the wastewater flows vertically downward through the unsaturated soil of the vadose zone to the underlying aquifer. These physical and chemical processes include filtration, adsorption, ion exchange, precipitation, and biological degradation. These processes can be effective in removal of nitrogen, phosphorus, biochemical oxygen demand (BOD), suspended solids, organics, and trace metals. In general, the removal of nitrogen and organics is a renewable process as biodegradation is involved. Certain mechanisms such as adsorption and other physical process, however, are not generally considered to be sustainable.

Surface infiltration basins are supplied water for several days, followed by several days of basin drying. The cyclic wetting and drying of the basin is necessary to improve infiltration rates and to control aerobic/anoxic conditions in the soil. During wetting, the water percolates through an unsaturated soil region to the groundwater table for storage in an unconfined aquifer system, as depicted in Figure 1.

Infiltration rates are most rapid during the early stages of the wetting cycle when negative pore water pressures are relatively high. Infiltration rates gradually decline toward a rate corresponding to the saturated hydraulic conductivity and unit gradient. However, over time the surface of the basin will clog as a result of sedimentation of fines and organic matter and growth of algae, often resulting in a sustantial decrease in overall hydraulic loading capacity of the basin.

Geotechnical aspects of groundwater recharge using surface infiltration basins have been examined as a part of a recent SAT feasibility study conducted in the Phoenix, Arizona, metropolitan area. Field observations from this and other studies indicate that the hydraulics of water recharge basins are affected by many factors, including soil type, subsurface nonhomogeneity, the formation of a surface clogging layer, application times for wetting/drying cycles, water quality, and climatic conditions. While the formation of a surface clogging layer or presence of a clay soil may enhance certain aspects of water treatment, such fine-grained layers control the hydraulic loading capacity. Clearly maintenance of a reasonable hydraulic loading capacity is of primary importance to successful SAT operations. Thus, a balanced approach to surface clogging layer and soil type selection for obtaining appropriate levels of water quality as well as rate of recharge must be achieved. The studies described herein address many aspects of the hydraulics of infiltration for SAT, based on a combination of laboratory and field test results. The studies include evaluation of the influence of soil type, surface clogging material, basin water depth, and wetting/drying cycle times. The overall research team for the SAT and recharge studies is highly multidisciplinary, including geotechnical, water resources, and environmental engineers and microbiologists. Summaries of the team findings to date will be presented briefly for completeness.

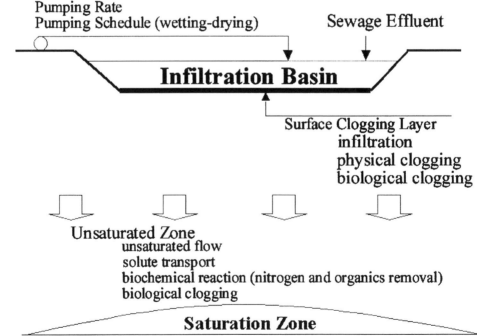

Process Related to Operation of the SAT System

Figure 1. Schematic of a Soil Aquifer Treatment Recharge Basin

2 OVERVIEW OF HYDRAULIC LOADING CAPACITY CONSIDERATIONS

Geotechnical aspects of the use of surface basins for groundwater recharge are primarily related to the ability of the surficial clogging material and the underlying soil profile to transmit water to the groundwater at an economically acceptable rate. It has been discovered through years of field observations that the hydraulics of wastewater infiltration basins are controlled to a large extent by the formation of a low-conductivity clogging layer on and within the upper few millimeters of the surface soils (Bouwer, 1996). Even with fresh water recharge basins, the formation of a surface clogging layer may occur due to algae growth.

Clogging problems were recognized and studied early in the development of artificial recharge facilities. Allison (1947) identified the effects of microbial clogging by comparing long term infiltration rates from tests on similar specimens, which were conducted under both "sterile" and microbially-active conditions. Others, including Behnke (1969), have attempted to develop empirical models that quantify changes in hydraulic conductivity resulting from suspended matter in the water. As recently as about 1985, basin operators were advocating the use of deeper ponds as a means of overcoming basin clogging (Oaksford, 1985). However, Bouwer and Rice (1989) speculated that this loose surface clogging material is compressible under seepage forces exerted by the infiltrating water, resulting in decreased hydraulic conductivity with increased pond water depth.

The characteristics of the native soils in the profile between the infiltration basin at the surface and the unconfined aquifer at depth are of great importance in establishing the hydraulic capacity of the site. Although the laboratory-determined saturated hydraulic conductivity of the surface soils can be a meaningful indicator of the suitability of a candidate site for SAT/recharge in terms of hydraulics, it is actually the overall profile conductivity for unsaturated flow that controls hydraulic capacity. Field rates of infiltration should be expected to be different than laboratory column rates for numerous reasons. These reasons are: (1) decreased degree of saturation in the field compared to the laboratory, (2) some horizontal movement of water, (3) field non-homogeneities and preferential flow paths, and (4) algae growth and formation of the surface clogging layer at the bottom of the field basins.

The presence of clay lenses of significant lateral extent can have adverse impact on the suitability of a site for cost-effective SAT or recharge. When wastewater is used, regulatory issues may restrict methods that can be utilized to route flow through any natural clay lenses, further limiting the possible

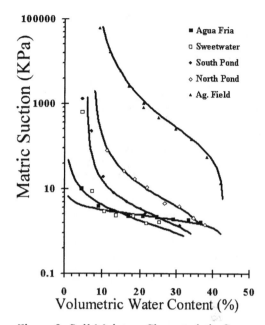

Figure 2. Soil Moisture Characteristic Curves

locations of recharge basins on the basis of soil profile considerations.

In general, infiltration rates for secondary effluent are lower than those observed for clean water recharge. Although the characteristics of the original soil are very important, significant reduction in hydraulic conductivity of the soil can occur during SAT due to soil clogging. Surface clogging is greater when wastewater is the source water.

3 SOILS STUDIED

The five soils included in this study represent near-surface soils from recharge locations in the state of Arizona. The soils were selected to cover a wide range in soil type, and consist of poorly-graded sand, SP (Agua Fria); poorly-graded, silty sand, SP-SM (Sweetwater); silty sands, SM (North and South Pond); and low-plasticity clay, CL (Agricultural Field). Soil tests included gradation by sieve analysis and hydrometer, Atterberg limits, specific gravity, in-situ density and water content. The soil cation exchange capacity and total organic content were also measured. A general description of these soils is given in Table 1.

The soil moisture characteristic curve data for the test soils are shown in Figure 2. The data shown are from matric filter paper soil suction measurements, using filter papers buried in the soil (Houston, et al, 1994).

4 DESCRIPTION OF WASTEWATER COLUMN INFILTRATION STUDIES

4.1 *Column operation*

Information on the wastewater effluents used in the column studies is provided in Table 2. Although the size, location, and character of the infiltrating water were varied, the five soil types described in the preceding section were used in all column effluent infiltration tests. These column tests resulted in data and specimens for use in more detailed studies of the surface clogging layer. In addition, the effluent column studies were used to study the effect of soil type on treatment efficiency.

The laboratory infiltration columns consisted of vertically oriented 1 meter long, 8.6 cm inside-diameter clear acrylic columns containing the test soils recompacted to field density and water content. The columns were fitted with tensiometer ports, water sampling ports, and, in some cases, ports for taking direct, small diameter drive-tube, water content specimens.

The laboratory wastewater infiltration columns that were operated by researchers at the University of Arizona used source water of clarified secondary waste effluent from the Roger Road Waste Water Treatment Plant in Tucson. Constant head conditions, were maintained during the infiltration tests. The waste effluent columns were operated in a stable indoor temperature environment in two-week cycles consisting of seven days of flooding followed by seven days of drying. Artificial light sources were used to stimulate algae growth in the ponded effluent and on the soil surfaces, consistent with field conditions. The soil columns were operated for a period of about 18 months. A schematic of the column experimental setup is provided in Figure 3. The larger pilot scale columns were operated by Arizona State University Researchers at the Phoenix 91st Avenue Waste Water Treatment Plant in Phoenix, Arizona, primarily for water quality studies. The columns were made of 2.4-m long, 30 cm inside-diameter stainless steel pipe. The soils were compacted in a series of 10-cm lifts to in-situ density and water content. Thirty-six sampling ports were installed in the columns to measure soil suction and to take water quality samples. The tops of the columns were fitted with overflow weirs to maintain pond depths of 17, 33, or 45 cm. Overflow from the weirs was collected in effluent reservoirs and recycled to the columns by peristaltic pumps. The boundary condition at the base of the columns was maintained at 100 cm water head, negative (suction). The base negative pore fluid pressure boundary condition was used to more closely simulate field wetting conditions, which are expected to result in a degree of saturation of around 60 to 80%, depending on the soil type and length of wetting period. The soil water characteristic data was used to obtain reasonable base soil suction conditions. As with the smaller scale laboratory tests, the surfaces of the columns were exposed to sunlight to match field prototype conditions as closely as possible. The columns were wrapped with cooling lines and foam rubber insulation to maintain typical sub-surface soil temperatures during the hot summer months.

The columns were operated for approximately 15 months using wetting/drying cycles consistent with field SAT operations. Algae and suspended solids surface clogging layers formed on all of the silt and clay columns. However, a distinct clogging surface did not form on the larger 91st avenue sand columns, as it was not possible to sustain a pond at the column surface

4.2 *Sub-sampling of column surface materials*

At the end of the 18-month operation of the larger diameter pilot scale waste effluent infiltration studies, soil specimens were collected by pushing 30 cm-long, 7 cm inside diameter Shelby tubes into the column from the surface. These specimens were used for various purposes, including the clogging layer studies described in the subsequent section. The smaller diameter (8.6 cm) laboratory columns were sampled for the clogging layer consolidation and hydraulic conductivity tests by extruding the material directly into stainless steel conventional consolidometer rings, 2 cm in height.

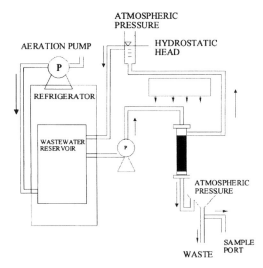

Figure 3. Schematic of Laboratory Soil Aquifer Treatment Column Studies

5 SUMMARY OF EFFECT OF SOIL TYPE ON SAT AND REMOVAL SUSTAINABILITY

Kopchenski, et al (1996) have described the results of water quality studies from the Phoenix 91st Avenue column studies. Recent findings from on-going research of several field SAT operations provide additional data that can be used to assess the effect of soil type of quality of recharged wastewater (Fox, et al, 1999). It is clear that soil type does have some impact on the rate and degree of water quality improvement that is realized during SAT of wastewater effluent. Important aspects of the soil are the gradation, degree of saturation achieved during the wetting cycle, and rate of aeration (air permeability) during the drying cycle. The ability to remove ammonia under ponded conditions (wetting cycle) is dependent on physical/sorptive removal, which is greater for the finer-grained soils with higher specific surface areas and cation exchange capacity. Finer-grained soils also tend to enhance removal by decay due to longer retention times. Ponded conditions during the wetting cycles of SAT lead to anoxic conditions and negligible nitrification during wetting. Therefore, nitrification is dependent on aeration of the soil during drying, which is enhanced by high air permeability of coarser grained soils. So, it is clear that with respect to water quality there are many tradeoffs for selection of any given soil type over another.

When the practical issues of system hydraulics are added and consideration of both hydraulic loading rates and treatment are taken together, studies to date indicate that the more pervious silty sands (SP/SM) appear to be optimal for SAT. However, other soil types may also be used successfully for SAT. When groundwater recharge is the primary objective, and treatment of wastewater is not an issue, highly pervious soil profiles are clearly the most desirable.

An overview of the SAT removal efficiencies and sustainability based on the collection of column and field studies to date has been made (Fox, et al, 1999). (1) The majority of the nitrogen removal appears to occur in early stages of SAT. (2) Removal of organics appears to be primarily biological, and therefore sustainable. (3) Removal of organics will occur under saturated and unsaturated flow conditions. Unsaturated flow conditions promote higher removal efficiency, however. (4) The removal efficiency for organics seems to depend more on the effluent quality than on soil type. Effluent quality depends on the characteristics of the source drinking water, including soluble microbial products. (5) The majority of data on pathogens has been qualitative, and definite conclusions on the fate of pathogens for SAT have not been reached. However, it does appear that maintenance of an unsaturated flow region

is useful in pathogen removal/deactivation, although it may not be essential.

Certain factors have added to the difficulty of data interpretation, particularly in the case of the field studies. These difficulties include limited sample volumes and difficulty of sample collection from the vadose zone (even during wetting cycles) and complex flow patterns in the subsurface. Additionally, mixing of reclaimed water with other intentional and unintentional recharged waters makes interpretation of long term water quality changes more difficult.

Recent research has included the development of detection methods for assessment of the presence of reclaimed water, both qualitatively and quantitatively. Improved methods of detection are important due to the particular concern that observed persistent organics of wastewater origin might pose a problem with the formation of harmful disinfection by products upon chlorination. Therefore separation of water from wastewater origin is considered important.

6 IMPACT OF CLOGGING LAYER ON INFILTRATION RATES

Long-term cumulative hydraulic loading for the small-scale waste effluent infiltration studies on the Agua Fria sand and the Sweetwater sandy loam are shown in Figure 4. The observed decreased rate of accumulation of hydraulic loading is indicative of the formation a soil clogging and decreased hydrau-

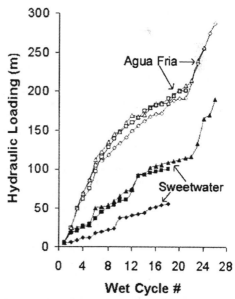

Figure 4. Cumulative Hydraulic Loading Over 25 Wetting Cycles

lic conductivity of the column. At approximately Cycle No. 23, surface scarification of the clogging layer on the top of the soil columns produced a dramatic increase in infiltration rate. The infiltration rates (during the wetting cycle) for a typical soil columns are shown in Figure 5. The saturated conductivity for this soil was determined to be approximately 0.20 m/day (Table 1), which corresponds closely to the infiltration rates observed in the early operation times. There is a gradual degradation of infiltration rate over the reported 400-day operation of the column, consistent with the observed formation of a surface clogging layer.

For the larger scale 91st Avenue columns similar patterns in infiltration rates were observed. The initial rate of infiltration, immediately following a drying cycle, is greatest, and the rate gradually declines through the wetting cycle as the soil becomes more saturated and the surface clogging layer re-establishes. This pattern can be observed in the infiltration rates observed over several cycles of wetting and drying on one of the North Pond Silt (SM/SP) columns that are shown in Figure 6.

7 DESCRIPTION OF CLOGGING LAYER STUDIES

The clogging layer is made up of algae originating in either the soil or the effluent, organic and inorganic suspended solids filtered from the effluent, dust blown into the pond from surrounding areas, microbes from the soil and effluent, and salts precipitated from solution. As these components accumulate on the bottom of the basin to form the clogging layer, they begin to restrict infiltration.Due to its soft compressible nature, the surface layer is susceptible to compression by the seepage forces exerted by the infiltrating water.

When clay lenses are not present, the surface clogging layer commonly has the lowest hydraulic conductivity in the recharge profile, restricting the rate of infiltration. This phenomenon is particularly prevalent when a large component of the clogging layer is comprised of filamentous algae. It has been found that the amount of algae can be controlled to some extent through use of shortened wetting cycles (e.g. 3 days rather than the typical 7 to 14 days of wetting) during the cyclic wetting/drying of the basins. However, surface clogging is almost impossible to avoid entirely.

Surface clogging is an extremely complex process. It is therefore desirable to develop a simplified picture of the clogging layer. The clogging layer can be defined as that zone of material over which a sharp drop in hydraulic head occurs as water infiltrates into a soil profile. It appears, therefore, that the behavior of the clogging layer should be mod-

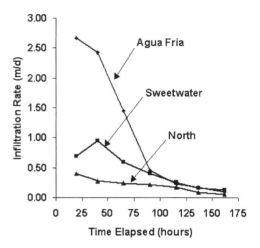

Figure 5. Degradation of Infiltration Rate Over the Course of One Cycle

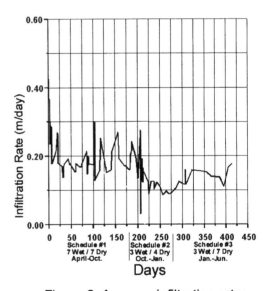

Figure 6. Average infiltration rates for wet/dry cycle applications to North Pond Soil

eled as two distinct parts. The conductivity of the upper layer, which consists of material that accumulated above the original ground surface, is governed by consolidation under seepage forces. The lower layer, consisting of native soil augmented by entrained organic and inorganic solids, has hydraulic conductivity controlled by the loss of preferred flow paths resulting from clogging of the coarse pore space in the native soil.

For purposes of modeling flow through the clogging layer, infiltration is assumed to occur vertically

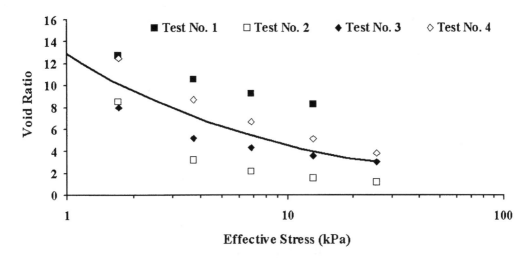

Figure 7. Consolidation Test Results on Mixed Culture Algae

downward at a unit hydraulic gradient. Where unsaturated fine-grained soil and/or significant nonhomogeneities are encountered directly beneath the basin surface, the soil suction profile may deviate from this simplified representation. However, field observations indicate that at least within the first few centimeters below the clogging surface that soils become essentially saturated.

As particulate material initially accumulates on the bottom of the basin to form the clogging layer, it exists in a loose, compressible state. This loose material is subject to compression due to the action of the seepage forces exerted by the infiltrating water. As a result, there is a reduction in the hydraulic conductivity of the clogging layer and in the infiltration rate that might occur when pond depth is increased. Increased pond depth also leads to increased hydraulic gradient across the clogging layer, which would tend to increase the infiltration rate if all other factors remained the same. The net effect is a result of the relative importance of these two phenomena, and an increase in pond depth may either raise or lower the infiltration rate, or the effects may compensate, resulting in an unchanged rate of infiltration.

Because the clogging layer thickness is not well defined, the expected order of magnitude of the effective stresses induced by seepage forces was estimated for a range of clogging layer thickness (up to 16 cm) and water depths (15 to 60 cm). The hydraulic gradient was computed by neglecting any soil suction at the base of the clogging layer and assuming a uniform-property clogging layer thickness (Houston, et al., 1999). The seepage force per unit volume was computed as the hydraulic gradient multiplied by the unit weight of water. While the ef-

fective stress at the top of the clogging layer equals zero, the maximum effective stress at the base of the clogging layer equals the thickness of the layer multiplied times the seepage force per unit volume. Although this procedure gives a reasonable estimate of the maximum effective stress at the base of the layer, it cannot be used to provide the actual stress distribution within the clogging layer.

The maximum stress occurs at the base of the clogging layer. The stress range of 1 to 10 kPa covers the range of interest for these studies for expected clogging layer thicknesses and typical water depths of basins or 0.5 to 2 meters. The effective stress is more strongly dependent on water depth than clogging layer thickness for the cases considered.

Characteristics of the clogging layer and some of its individual components were studied using a combination of consolidation and hydraulic conductivity testing. The tests were conducted on samples of clogging material obtained from the long term wastewater column experiments and also separately on the soil specimens and specimens of mixed culture algae. The specimens were tested under saturated one-dimensional flow conditions. All tests were performed in conventional consolidometer rings. The consolidation tests were periodically interrupted for the independent measurements of hydraulic conductivity.

The relationship between effective stress and void ratio for the specimens of mixed culture algae are shown in Figure 7. The corresponding hydraulic conductivity data are shown in Figure 8. Although there is some significant scatter in the data, largely related to difficulties in reproducing similar specimens of algae, it is clear that loose algae can

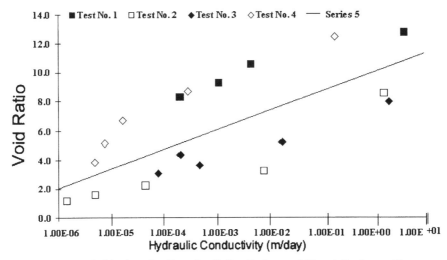

Figure 8. Hydraulic Conductivity Data on Mixed Culture Algae

compress under effective stress levels consistent with the expected seepage forces. Further, the hydraulic conductivity values for the compressed algae are significantly less than the conductivity values for the SP and SP/SM soils. The hydraulic conductivity of the algae specimens approached that of the CL at an effective stress of only 3 to 4 kPa.

The specimens collected from the surface layers of the effluent columns were also subjected to consolidation and hydraulic conductivity testing. With respect to SAT and recharge, the hydraulic conductivity profile through the upper region of the soil profile is of great importance. Therefore 2 cm thick consecutive specimens were collected from the columns beginning at the column surface and continuing to a depth of approximately 12 cm. The hydraulic conductivity data for the specimens collected from a column packed with the Sweetwater soil and permeated for 18 months with tertiary effluent are shown in Figure 9.This test result is typical for the clogging layer specimens from the silt and sand columns. In general, the lowest hydraulic conductivity was observed in the top 2 cm specimen taken from the effluent columns.

For all tests conducted, the conductivity of the upper specimen was substantially less than that of the specimens from other depths in the soil profile, all of which were found to be of approximately equal conductivity. The lower portion of the clogging layer, created by the entrapment of organic and inorganic particulate matter is not particularly susceptible to compression and reduction of conductivity under seepage forces, probably because the soil matrix itself is relatively stiff within the resultant stress range.

The total organic content profile through the effluent columns at the end of 18months of operation is provided in Table 3. For the Sweetwater silt of Figure 9, the upper 2 cm specimen was determined to have a total organic content of 1.5 percent, and specimens from 2 to 10 cm depth had organic contents ranging from 0.77 to 0.66. Compared with the background amount of organic matter in the Sweetwater soil, which is about 0.7 percent by weight, the increased organic content provides evidence of an accumulation of organic matter at the soil surface. The addition of a thin, primarily organic layer (on the order of a few mm) was commonly observed visually on essentially all of the soil column surfaces. This thin organic layer would be expected to make marked difference in hydraulic conductivity, consistent with the laboratory observations.

8 EFFECT OF WATER DEPTH

The large-scale, 0.3m diameter columns operated at the Phoenix 91st Avenue site were used to study the effect of depth of pond water on infiltration rates. This study was conducted at the end of the 15 month long effluent infiltration experiments. As discussed previously, these columns were operated under cycles of wetting and drying to simulate actual SAT operation. By neglecting the initial period of rapid infiltration following a drying cycle, an average rate of infiltration was established for a relatively shallow water depth. On the seventh day of a twelve-day wetting cycle, the overflow weir was adjusted to allow the water level to increase to a second, greater depth. Then, based on data from the remaining five days of the wetting cycle, an average rate of infiltration was determined for the second depth.

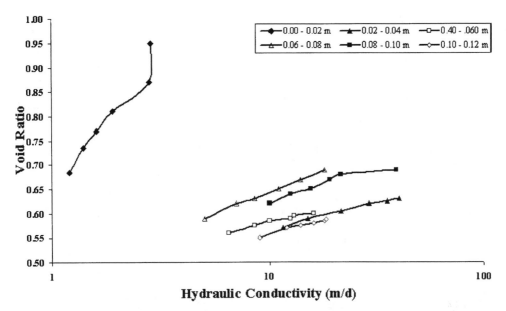

**Figure 9. Test Results on University of Arizona
Tertiary Effluent Column Sweetwater Soil**

Details of the analyses performed on these depth of pond studies are provided by Houston, et al., 1999. In summary, it was found that the depth of the ponded water did affect the hydraulic conductivity of the surface material and that increased depth of water resulted in reduced hydraulic conductivity. This result is consistent with the findings of the consolidation and hydraulic conductivity testing on the specimens of clogging layer material collected from the effluent columns. There were several experimental tests for which an increase in pond depth resulted in a reduction in infiltration rate. In other columns, however, increasing the depth of the ponded water did result in some increase in actual rate of infiltration, but the rate of increase was typically substantially less than would have been anticipated if the conductivity of the clogging layer surface had remained constant. The competing effects of increased hydraulic gradient from increased pond depth and decreased hydraulic conductivity of clogging layer material resulting from increased seepage gradients make the direction of change in actual rate of infiltration difficult to predict. Nevertheless it is clear that the conductivity of the surface layer is reduced substantially as depth of pond is increased.

9 INFLUENCE OF CLOGGING AND WATER DEPTH ON BASIN MAINTENANCE AND OPERATION

In addition to the laboratory testing of the clog-

ging material, an ecological model of the infiltration basins was developed to define the interaction between microorganisms such and algae, suspended solids from the effluent, and the nutrient rich water (Duryea and Houston, 1996). This model was used to aid in the quantification of the amount and nature of the material that accumulates on the basin bottom to form the clogging layer. As a result of all of these studies, a series of design and operation recommendations have been developed that seek to reduce clogging and thereby maximize infiltration for a given application. While periodic drying of infiltration basins helps to restore the infiltration rate for subsequent wetting cycles by desiccating the clogging layer and drying the underlying soil, there is ample evidence of a long-term degradation of rates in spite of these efforts.

In such instances, additional maintenance steps must be taken. The most common method is to scarify the soil surface using a disk harrow. Given the surficial nature of the typical clogging phenomena, this should be highly effective in restoring infiltration. Less frequently, operators perform a "deep rip" of the basin bottom using other equipment. However, there may not be enough added benefit from such treatment, compared to that realized by simple scarification, to justify its use. In fact, the deep rip could exacerbate clogging in some cases by forcing the particulate matter deeper into the soil.

From an operational perspective, the depth of water was found to have a marked effect on surface clogging and, since the water depth of the recharge basins affects the geometry and layout of the basins, it influences the design process as well. The effect of basin depth on clogging is strongly dependent on specific site and climatic conditions. In an environment that is rich in algae and/or suspended solids, increasing the depth of the water in the basin can decrease infiltration rates by seepage forced induced compression of the clogging material. Under other conditions, increasing the pond depth will result in an increased rate of infiltration. Water depth is also significant in that it influences the average hydraulic retention time of the basin, which in turn impacts the growth of algae. Longer retention times allow for more algae growth in the basin and therefore increase the potential for clogging. Therefore, relatively shallow infiltration basins (e.g. less than 0.5 m) and relatively short retention times (e.g. 3 days or less) are recommended for SAT to reduce surface clogging.

10 UNSATURATED FLOW ASPECTS OF SAT AND GROUNDWATER RECHARGE

10.1 *Overview*

The study of unsaturated flow through soils has become an important and active topic in environmental and geotechnical engineering, being relevant to such diverse topics as contaminant transport in the vadose zone and collapse and expansion of foundation soils upon wetting. Infiltration associated with soil aquifer treatment and groundwater recharge systems is a subset of the complex problem of unsaturated flow. Infiltration is the term applied to the process of water entry into unsaturated soils. Unsaturated flow through soils is generally difficult to model quantitatively due to the high nonlinearity of the process, associated with the fact that the flow properties and the gradient for flow are a function of the changing soil water content (soil suction). In spite of the difficulties, engineers and scientists have developed numerous computer codes for estimating unsaturated flow. These codes utilize a constitutive model that is a form of Darcy's law, and require information on the variation of the unsaturated hydraulic conductivity and the soil and the variation of the soil suction with soil water content.

The focus of the unsaturated flow studies herein is on groundwater recharge and SAT. One very important question relates to the degree of saturation achieved during surface infiltration of effluent because the degree of saturation of the soil has an influence on the efficiency of treatment. Also adequate modeling of the SAT process, including the tracking of biochemical changes in the recharged water, requires as a first step accurate modeling of the infiltration process. Clean water infiltration studies, both in the lab and in the field, provided the

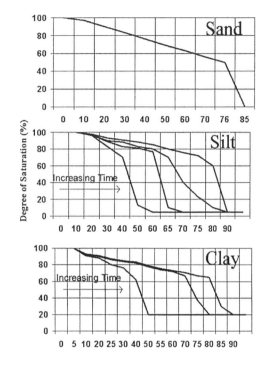

Figure 10. Profiles of Degree of Saturation

data for verification of unsaturated flow computer models. As a part of the model calibration unsaturated flow properties were back calculated, using inverse methodology from the one-dimensional column tests.

10.2 *Clean water column infiltration studies*

Figure 10 shows the degree of saturation profiles for clean water infiltration tests on silt, sand, and clay soils. These curves show that the degree of saturation behind the wetting front is well below 100% for all three soils. The degree of saturation drops below 100% fairly quickly after leaving the ponding source. Although the infiltration tests shown in Figure 10 represent transient flow conditions, similar results have been found even for long term infiltration. True steady state conditions are not achieved with recharge operations given the cyclic nature of the wetting.

Results from numerous infiltration columns show that the wetting profiles can be divided nominally into four zones, as noted by Bodman and Colman (1943). A thin zone at the pond base exists in which the degree of saturation is approximately 100%.

Although the spacing of instrumentation and sampling ports did not allow precise measurement of the thickness of this zone, it was found to be less than 4 cm in all cases studied, with some evidence that the thickness of the zone increases somewhat (although not dramatically) with time of wetting. Below the very thin saturated zone is a transition zone of very small thickness (also less than 4 cm) in which the degree of saturation decreases rapidly. The third zone consists of a substantially thick region of relatively constant degree of saturation, with a minor gradual decrease in degree of saturation with depth. The thickness of this zone increases with increased time of wetting, and the degree of saturation in this region tends to increase slightly with increased time of ponding.

10.3 Clean Water Recharge Field Studies

Two small-scale and two large-scale field infiltration basins were operated at potential recharge locations in the vicinity of Phoenix, Arizona. These basins used local groundwater as the source for infiltration. One purpose of the field studies was to assess differences between laboratory and field infiltration response, and to investigate potential perching layers at the field locations. Difference between lab column and field infiltration tests result primarily from heterogeneity in the field and some lateral flow that occurs for field profiles. Sample disturbance and geometrical effects can also contribute to the differences. The field tests provide the most direct measure of the hydraulic performance of a given soil profile.

The small-scale field ponding studies consisted of two tests, one with a 100 cm diameter pond and the other with a 50 cm diameter pond. A pond depth of 20 cm was used in both small-scale field tests. Moisture contents were monitored using a capacitance probe at the centers and edges of the ponds, as well as 50 cm radically outward from the pond edges. These capacitance probe-determined moisture contents were supplemented and confirmed by several direct water content measurements. Inflow infiltration rates were obtained by measuring water depth inside the reservoir used to supply the ponds.

The results from these small-scale field percolation tests, conducted in the near-surface silty sands (North and South Pond Soils), are consistent with the lab-column studies with respect to the degree of wetting achieved during initial infiltration into dry soils. Typical results show an average degree of saturation in the wetted region under the center of the pond of about 70 percent.

The large-scale field infiltration tests consisted of approximately 1/3-acre ponds located at two candidate recharge sites in the Salt River Valley (Greeley and Hansen, 1994). The South Pond location soils consisted of approximately 2 m of silty sand underlain by approximately 5 m of poorly-graded sand and 9 m of gravel and sand mixtures; the groundwater table was located approximately 50 m below the ground surface. The North Pond location had soils consisting of approximately 3 m of silty sand, underlain by sand, gravel and cobbles extending to approximately 10 m depth. A 3.5 m thick layer of clayey sand and gravel (SC and GC) existed beneath the sand/gravel/cobble layer. Beneath the clayey layer, soils consisted primarily of sandy gravel, with some clay. Neutron logging access tubes were installed around and adjacent to the basins to a depth of 9 m. Piezometers were installed at each basin to a depth below the groundwater table of approximately 30 m. These piezometers were used for monitoring fluctuations of the groundwater table and neutron logging to determine moisture content variations. Initial moisture content profiles were obtained from samples taken every 3 m along the entire piezometer length during installation. A flow meter was installed, and the depth of the basin was determined on regular intervals to track surface infiltration rates.

Wetting cycles lasted about one month, followed by a 2 to 3 week drying cycle. Based on the percolation test results, the average long term water infiltration rate was found to be approximately 0.75 m/day at each location. This remarkable agreement in field infiltration rates for these two different locations is believed to be largely coincidental. However, growth of algae was observed in both ponds that could have created a somewhat restricted rate of surface infiltration at both pond locations.

Neutron log data for these two tests indicate that there were significant differences in the manner in which water entered the profiles at the two sites. There is evidence, for example, that there was perching of water and lateral movement along the clayey soil layer at approximately 7 m depth at the South Pond location, as indicated by a dramatic change in neutron probe reading at this depth as well as by an apparent lack of rise in the groundwater table beneath the South Pond. At the North Pond location, however, the neutron log showed a very substantial rise in the groundwater table, and no spikes in the neutron probe readings to indicate any perched conditions. At the North Pond it appears as if the water infiltrated essentially vertically downward to the groundwater table.

A few direct degree of saturation/water content measurements were obtained at the end of a wetting cycle and in the center of the ponds. These samples were taken just as the water disappeared into the ground surface; small areas of standing water were still in the lower-lying areas of the basin bottom during this sampling operation. The direct water content measurements showed that the average degree of saturation beneath the center of the ponds, within the upper meter, was only about 60 percent, even after one month of wetting. This data indicates that, even with long-term field infiltration, the degree of saturation of the soil profile between the basin and the groundwater table is likely to remain well below 100 percent, except in areas of perching. However, the soil suction associated with the more or less steady state infiltration conditions is likely to be quite low, leading to a gradient of near unity for long-term infiltration conditions.

10.4 *Numerical Simulations of Unsaturated Flow*

In view of some of the parameter uncertainties for application of Darcy's law to infiltration problems, it is suggested that the Darcy's law model be firmly calibrated with field or laboratory infiltration tests. In most cases it would be desirable to use a field ponding test to calibrate a numerical code that is based on Darcy's law and which uses a governing differential equation based on mass balance, for example, Richards' equation. Following a common procedure, saturated conductivity would be measured for the soils as well as suction versus water content points on the soil moisture characteristic curves. Then a curve fit could be used to extrapolate the soil water characteristic curve, for example, using the process described by van Genuchten, 1980. Finally a function that related unsaturated conductivity to soil suction or soil water content is adopted.

In the SAT/recharge studies computer code MSTS (White and Nichols, 1993) was used to simulate the column and field infiltration tests. The quality of the simulation was found to be highly dependent upon the specific soil moisture characteristic curve and the unsaturated hydraulic conductivity relationship, as would be expected. Simulations were first conducted using the laboratory-determined soil moisture characteristic curves shown previously in Figure 2. Simulations were also conducted using flow parameters back calculated from the lab column test results. The back-calculated unsaturated hydraulic conductivity relationships and soil moisture characteristic curves were determined using computer code FLOFIT for inverse computations (performed iteratively using parameter corrections based on nonlinear least squares problem solutions) developed by Kool and Parker (1988). The van Genutchen/Mualem formulations were used for the inverse computations:

$$\Theta_e = \frac{\theta - \theta_r}{\theta_s - \theta_r} = \left[\frac{1}{1 + (\alpha h)^n} \right]^m \quad (1)$$

$$K_r = \frac{K}{K_s} = \Theta_e^{\frac{1}{2}} \left[1 - (1 - \Theta_e^{\frac{1}{m}})^m \right]^2 \quad (2)$$

where Θ_e is effective water content; θ is volumetric

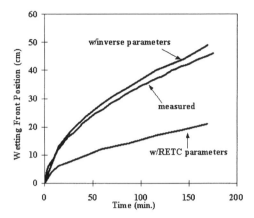

Figure 11. Wetting Front Progression for North Pond Soil Column

water content; θ_s and θ_r are saturated and residual water content, respectively; h is pressure head (suction positive); the parameters α, m and n are often obtained using curve-fitting techniques (note: typically m=1-1/n); K_s is saturated hydraulic conductivity, which can be measured directly in the laboratory or field. The saturated hydraulic conductivity and the saturated volumetric water content were measured independently and were not varied during the inverse computation. The residual water content was assumed to be the air-dried volumetric water content, as suggested in the studies of Dane and Hruska (1983).

Typical results of numerical simulations for a one-dimensional column experiment are shown in Figure11. It was possible to match the rate of advancement of the wetting front using back-calculated parameters, as shown in Figure 11. Agreement between measured and simulated flow for lab column experiments would be expected to be quite good as the column experiments were used to back analyze unsaturated flow parameters. Although the predictions of the rate of progression of the wetting front were excellent, the degree of saturation was again over-predicted by the computer model.

Simulations of field ponding tests also led to over-estimate of the degree of saturation. Numerical simulation of infiltration tended to predict an average degree of saturation of essentially 100 % within the wetted bulb, compared to the 60 to 85 % observed experimentally in the laboratory. The inverse-method determined unsaturated flow parameters and corresponding soil characteristics curves could be used quite successfully, however, for predicting the rate of advancement of the wetting front or ground surface infiltration rates. Others have reported numerical simulation results for infiltration into dry soils wherein the degree of saturation computed is essentially 100% behind the wetting front in

Table 1. Soil Characteristics

Soils	USCS	K_{SAT} m/day	CEC (meq/100g)	TOC (% by wt)	PI
Agua Fria	SP	72	2.4	0.32	NP
Sweet water	SP-SM	16	5.7	0.70	NP
South Pond	SM	1.1	6.9	0.97	NP
North Pond	SM	0.2	7.1	1.14	NP
Ag. Field	CL	0.03	22.3	3.03	12

spite of the fact that observed ground surface infiltration rates have been well-matched by the simulator (Touma and Vauclin, 1986, Kool and van Genuchten, 1991).

11 SUMMARY AND CONCLUSIONS

The desire to replenish groundwater supplies and to store water in underground aquifers for future use has led to great interest in groundwater recharge operations, particularly in rapidly growing arid regions. Sustainable development in arid water limited areas will almost certainly require augmentation of the more traditional surface and groundwater supplies with reclaimed water. For this reason, groundwater recharge with treated waste effluent is currently being conducted in many arid regions and is under extensive study.

Reclaimed water is normally considered for nonpotable water use such as landscape irrigation. However, municipalities and other government agencies express interest in use of reclaimed water as an indirect potable water supply. Recharge of groundwater supplies with treated waste effluent together with the use of soil aquifer treatment (SAT) to improve water quality represents one method for maintaining adequate water supplies in arid regions. Studies to date show that SAT is a feasible treatment process and that treated wastewater represents a viable source for replenishment of groundwater supplies.

The typical recharge or SAT operation involves the percolation of source water through an unsaturated soil region to an underlying unconfined aquifer. Therefore it is important to understand the movement of the recharged water through the unsaturated soil and to understand the role of unsaturated soils in the treatment when the source for the recharged water is waste effluent.

Although there are advantages and disadvantages to various soil types in terms of their appropriateness for soil aquifer treatment (SAT) applications, when all issues of wastewater treatment and system hydraulics are considered, studies to date indicate that the more pervious silty sands, SP/SM, are optimal for SAT. In the case of clean water recharge operations the coarse high conductivity soils are optimal.

ACKNOWLEDGMENTS

This work was supported in part by the cities of Phoenix, Mesa, Tempe, Glendale, and Youngtown, Arizona, the Salt River Project, Tucson Water, the USDA Water Conservation Laboratory, the American Water Works Association Research Foundation (AWWARF), the Water Environment Research Foundation (WERF), and the National Water Resources Institute (NWRI). The contributions of graduate students and faculty co-investigators at Arizona State University, University of Arizona, Stanford University, and University of Colorado are gratefully acknowledged

Table 2. Water Quality Characteristics of Effluent

Effluent Type		NH_3-N (mg/L)	Nox – N (mg/L)	Total – N (mg/L)	TOC/DOC* (mg/L)
Phoenix 91st Avenue WWTP Effluent	Dechlorinated Dentrified	1.0-3.0	1.0-6.0	4-10	8-10
	Chlorinated Dentrified	1.0-3.0	1.0-6.0	4-10	8-10
	High Rate Activated Sludge (Chlorinated)	10-30	0-8.0	10-40	13-15
Tucson Roger Road WWTP Effluent	Secondary Trickling Filter	15-25	0-2.0	15-30	15-25
	Tertiary-pressure Filter	15-25	0-2.0	15-30	10-15
	Primary	20-35	0-1.0	25-40	40-50

* DOC is a measure of residual organic carbon following filtration using a 0.45 micrometer filter

Table 3. Samples of Total Organic Content and Hydraulic Conductivity Data from Subsamples of Silt Soil Effluent Infiltration Tests After 18 Months of Operation

Column	TOC (%)	Ksat (cm/sec)	Void Ratio	Depth (cm)
UA1-Sweetwater	1.86	2.48 E-03	0.7895	0 to 2
UA1-Sweetwater	1.06	6.57 E-03	0.8048	2 to 4
UA2-Sweetwater	1.48	3.07 E-03	0.9447	0 to 2
UA2-Sweetwater	0.77	4.64 E-02	0.6272	2 to 4
UA2-Sweetwater	0.72	2.09 E-02	0.6871	4 to 6
UA2-Sweetwater	0.72	2.09 E-02	0.6871	6 to 8
UA2-Sweetwater	0.66	4.4 E-02	0.6882	8 to 10
UA2-Sweetwater	0.62	2.05 E-02	0.5975	10 to 12
ASU1-South Pond	2.34	2.59 E-03	0.9247	0 to 2
ASU1-South Pond	0.88	6.40 E-03	0.7152	2 to 4
ASU1-South Pond	1.06	1.14 E-02	0.8047	4 to 6
ASU1-South Pond	0.83	8.36 E-03	0.8047	6 to 8

REFERENCES

Allison, L.E. 1947. Effect of microorganisms on permeability of soil under prolonged submergence. Soil Science, 63:439-450.

Asano, Takashi, ed. 1985. *Artificial recharge of groundwater.* Boston, MA: Butterworth Publishers.

Behnke, Jerold J. 1969. Clogging in surface spreading operations for artificial groundwater recharge. Water Resources Research, 5 (August) 4: 870-876.

Bodman, G. B. and Colman, E. A. (1943) "Moisture and energy conditions furing downward entry of water into soils," Soil Science of America Proceedings, Vol. 8, pp 116-122.

Bouwer, Herman, and R.C. Rice. 1989. Effect of water depth in groundwater recharge basins on infiltration. *Journal of Irrigation and Drainage Engineering*, ASCE, 115(4):556-567).

Bouwer, Herman. 1996. New developments in groundwater recharge and soil aquifer treatment of wastewater for reuse. In *Water Reuse 96: Proceedings of the Conference held in San Diego, California 25-28 February 1996.* Denver, CO: American Water Works Association.

D'Angelo, Salvatore. 1996. AWWA and WEF prepares guidelines for using reclaimed water to augment potable water resources. In *Water Reuse 96: Proceedings of the Conference held in San Diego, California 25-28 February 1996.* Denver, CO: American Water Works Association.

Dane, J. H. and Hruska, S. 1983. "In situ determination of soil hydraulic properties during drainage." Soil Sci. Soc. Am J., 47, 619-624.

Duryea, Peter D., and Sandra L. Houston. 1996. Hydrological and mechanical properties of clogging layers. In *Water Reuse 96: Proceedings of the Conference held in San Diego, California 25-28 February 1996.* Denver, CO: American Water Works Association, 1031 - 1062.

Fox, P., Nellor, M., Arnold, B., Lansey, K., Bassett, R., Gerba, C., Yanko, W., Rincon, M., Reinhard, M., Houston, S.,

Westerhoff, P., and Drewes, J. 1999. Progress Report on Soil Aquifer Treatment for Sustainable Water Reuse, Report to American Water Works Association Research Foundation, July.

Greeley and Hansen Engineers. 1994. City of Phoenix 91st Avenue Wastewater Treatment Plant reclaimed water study phase II element B pilot percolation basin studies. Phoenix, AZ: Greeley and Hansen Engineers.

Houston, Sandra L., William N. Houston and A.M. Wagner. 1994 Laboratory filter paper suction measurements. *Geotechnical Testing Journal*, ASTM, 17(2):185-194.

Houston, S., Duryea, P., and Hong, R. 1999. Infiltration considerations for groundwater recharge with waste effluent. Journ of Irrigation and Drainage Engineering, ASCE, Sept./Oct., Vol. 125, No. 5, pp 264 – 272.

Kool, J.B., and van Genuchten, M. Th. (1991). "Computer Program - HYDRUS." Riverside, CA: U.S. Salinity laboratory.

Kool, J. and Parker, J., 1988. Environmental Systems and Technologies. Computer Program - FLOFIT A program for estimating soil hydraulic and transport properties from unsaturated flow and tracer experiments. Blacksburg, VA.

Kopchinski, T., Fox, P. Almadi, B., and Berner, M. 1996. The effects of soil type and effluent pre-treatment on soil aquifer treatment. Water Science Tech. Vol. 34, No. 11, pp 235 – 242.

Mays, Larry W., et al. In Press. *Soil treatability pilot studies to design and model soil aquifer treatment systems.* By Arizona State University, University of Arizona and University of Colorado at Boulder. Denver, CO: American Water Works Association Research Foundation (AWWARF) No. 90731.

Oaksford, Edward T. 1985. Artificial recharge: methods, hydraulics, and monitoring. In Artificial recharge of groundwater. Takashi Asano, ed. Boston, MA: Butterworth Publishers.

Touma, J., and Vauclin, M. 1986. "Experimental and numerical analysis of two-phase infiltration into a partially saturated soil." Trans, in Porous Media, 1, 27-55.

Treweek, Gordon P. 1985. Pretreatment processes for groundwater recharge. In *Artificial recharge of groundwater.* Takashi Asano, ed. Boston, MA: Butterworth Publishers.

van Genuchten, M. Th. 1987. Computer Program - RETC Analysis of soil hydraulic properties. Riverside, CA: USDA-ARS U.S. Salinity Laboratory.

van Genuchten, M.Th. 1980. A closed form equation for predicting the hydraulic conductivity of unsaturated soils. *Soil Science Society of America Journal*, 44:982-998.

White, M.D., and W.E. Nichols. 1993. Computer Program - MSTS. Richland, WA: Pacific Northwest Laboratory.

2 Fundamentals and theoretical developments

Unsaturated Soils for Asia, Rahardjo, Toll & Leong (eds) © 2000 Taylor & Francis, ISBN 90 5809 139 2

Prediction of cone-index from the state parameter of unsaturated soils

M.Z. Abedin & M.A. Rashid
Department of Farm Structure, Bangladesh Agricultural University, Mymensingh, Bangladesh

ABSTRACT: This paper elucidates a technique for prediction of cone-penetrometer index of unsaturated soil. The solution is based on identifying the state parameter establishing the linkage between the cone surface stress and pore volume within the critical state space of soil. The state parameter associated with soil type and moisture content is a two-parametric linear function that provides the basis for correlation analysis. The cone penetrometer performance under the laboratory and field situations require the establishment of the basic Mohr-Coulomb failure criteria and the critical state parameters of the soil over a range of moisture contents. The experimental evidences show a systematic change in state parameter of c-φ soils over a range of moisture contents. The state parameter coefficients estimated as a function of moisture content could be used a ready tool for converting the penetrometer readings to bulk density or pore-space estimates. The predictive performance of the cone-index model was found most encouraging.

1 INTRODUCTION

There are several recent developments, which have helped to advance our understanding of the complex processes involved in soil-machine inter-actions. The most significant impact on the discipline of Soil-Machine Mechanics are (a) the characteristics method (Sokolovski 1960) and (b) the Cambridge Critical State concept (Roscoe et al. 1958). The former is purely a theoretical approach of higher Mathematics and the later is based on the advanced Plasticity Theory. A drawback of the Sokolovski's solution is that it deals exclusively with 2-D stress fields and this requirement is some-what restrictive to 3-D soil failure induced by the cone penetrometers. The solution applies to rigid-plastic materials failing according to the Mohr-Coulomb failure criterion. It takes no account of the intermediate principal stress σ_2, nor the volume-change phenomena. The latter factor is obviously compatible with the Critical State model that does specify the condition $\sigma_2 = \sigma_3$ and allows for all changes in pore space during loading.

One of the exciting developments for sand is the state parameter concept that measures the soil behaviour correlating directly to the cone penetration test (CPT) explored by Been and his associates (Been et al. 1985, 1986 & 1987). According to their proposition, the normalised tip-resistance is a unique function of state parameter ψ which has a wide applicability as the normalising parameter for the constitutive modeling of soil behaviour (Abedin 1995, Abedin et al.1997).

Cone index is linked directly with the soil strength and indirectly with the soil bulk density. This study attempts to develop a constitutive model for prediction of cone-index removing the current empiricism associated with interpretations.

2 THEORETICAL FORMULATION

The soil behaviour characterised by the state parameter ψ combines the influences of specific volume and stresses while the fabric parameter characterises the arrangement of soil grains (Been & Jefferies 1985). The mean normal stress, p is enough to measure ψ as the deviator stress q reflects directly the soil fabric parameter. The state parameter embodies a combination of specific volume, ambient stress and orientation relative to the critical state line (CSL) of soil failures. A given

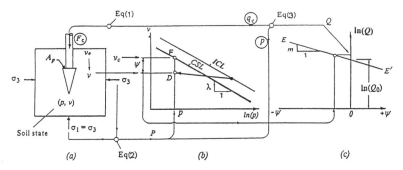

Figure 1. Flows of calibrating the state parameter coefficients m and Q_0

soil of initial specific volume, v_0 is compressed isotropically and then allowed to swell to v at D (Fig.1b). The soil at D has the ambient state p, v and the specific volume on the CSL (at F) corresponding to p is v_c. By definition, $\psi = v - v_c$. Intrinsically, ψ incorporates the information of p, current v and how far away is the CSL. The accuracy and practicability of the concept greatly depend on the ability to measure ψ in-situ. The cone index q_c is the penetration resistance per unit base area of the cone and is expressed by

$$q_c = F_c / A_p \qquad (1)$$

where F_c = force applied to penetrate the cone and A_p = projected base area of cone normal to its axis.

The value of q_c is normalised relative to the ambient stress p by two convenient reference stress levels (Wroth, 1986). The first possibility is based on the octahedral stress $\sigma_{oct} = (\sigma_1 + 2\sigma_3)/3$ at the isotropic condition $\sigma_1 = \sigma_3$ that leads the ambient stress to

$$p = \sigma_{oct} = \sigma_3 \qquad (2)$$

and the normalised cone index

$$Q = (q_c - p)/p \qquad (3)$$

The second possibility is $p = \sigma_{v0}$, where σ_{v0} is the vertical geo-static stress on the horizontal planes at the cone level. The present investigation is based on eq(3) with p defined by eq(2) for establishing a connection between ψ and q_c. There exists a close correlation between Q and ψ of sand (Been et al. 1985, 1986 & 1987). Incrementally, $\delta Q/Q = - m \, \delta \psi$ or in linearity takes on the form as

$$\ln(Q) = \ln(Q_0) - m\psi \qquad (4)$$

where Q_0 = normalised cone index of soil shearing

at constant volume. The slope m and the intercept $\ln(Q_0)$ of the regression line EE' (Fig.1c) depend mainly on the soil type and m is sensitive to p (Sladen 1989). The coefficient m for sands bears a linear relationship to the gradient λ of the CSL (Been et al. 1986 & 1987). The use of ψ in the interpretation of penetrometer data for saturated to unsaturated soils ranging from clay to sand needs information about λ, Γ, m, Q_0 and q_c with reference to p at penetration depth z.

A schematic diagram of the steps taken in calibrating m and Q_0 from σ_1, σ_3 together with measured q_c is shown in Fig.1. Referring to Fig.1a & b, the in-situ v (point D) and p leads to an estimate of ψ from known values of λ and Γ of the CSL as $v_c = \Gamma - \lambda \ln(p)$. The values of Q, estimated from p and q_c, and ψ for a range values of p and v plotted on log-linear scale gives the straight line EE' (Fig.1c) from which the values of m and Q_0 can be derived at any degree of soil saturation.

The stages in interpretation of q_c at field level are outlined in Fig.2. The value of p is to be estimated from earth pressure theory, $p = \sigma_v + 2\sigma_H$ the equivalent form of which is:

$$p = \gamma z (1 + 2K_0) \qquad (5)$$

where $\sigma_v = \gamma z$ and $K_0 = \sigma_H / \sigma_v = (1 - \sin\varphi)$ and φ is the soil friction angle obtained from triaxial tests.

The field value of q_c is calculated using p measured from eq(5) and taking F_c at depth z. The normalized cone index Q then can be calculated by eq(3) from measured field values of p (eqn.5) and q_c. The value of ψ associated with Q, is extrapolated from Fig.2b which on transferring together with p to the v-$\ln(p)$ plot (Fig.2c) leads to measure the required v of the soil. This allows to convert the field q_c to a measure of pore space status and hence the bulk density of field soil.

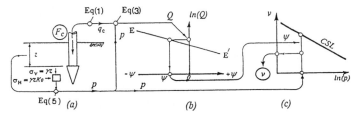

Figure 2. Conversion of cone index to soil specific volume

3 MATERIALS & METHODS

The experimental soil was a sandy clay loam (sand = 65.2, silt = 14.5 & clay = 20.3 percents) having the cone penetrometer plastic limit, CPPL= 18.5%, liquid limit, LL = 33% and specific gravity, G = 2.67. The experiment was conducted on five moisture levels w (7, 11, 15, 18 & 22 % by weight basis) to keep the degree of saturation S_r within 80% at 250 kPa confining pressure.

The sample preparation for the triaxial calibration chamber test was similar to that of triaxial tests (Head 1986). The diameter and height of the sample were 100 and 185mm respectively. The dry density γ_d was maintained to be 1.2 Mg/m^3. A known weight of soil was put into the mould in several steps (10 nos.) each of equal amount and was compacted evenly after each addition. A number of trials were needed to prepare a uniform sample of $\gamma_d = 1.2$ Mg/m^3.

The technique described by Abedin (1995) is followed for placing the sample into the calibration chamber. The sample was compressed isotropically at a confining pressure 250 kPa and then allowed to swell by reducing the cell pressure to 25, 50, 100, 150, 200 & 250 kPa and the respective penetration was carried out in each time. The penetration force F_c, penetration depth z and volume change in the sample were continuously recorded. The tests were repeated on three replications at individual w.

Parkin and Lunne (1982) found that for loose soil (relative density D_r < 30%), chamber size effects are not significant, when the chamber-to-cone diameter ratio exceeds 20. In practice the larger the sample size, the more difficult it is to prepare samples of uniform density. Hence, it is impractical to conceive apparatus that would give a value of D_r sufficiently favourable to eliminate such boundary effects (Bellotti 1985). The size of the calibration chamber could be selected around 15 to 20 times the cone diameter to make an acceptable compromise between end effects and experimental convenience (Abedin 1995). For the present investigation, a modified triaxial cell is used as a calibration chamber that can accommodate a maximum of 100 mm diameter soil sample with 10 mm free space for the water jacket around the sample. This, rather smaller sample, allows to prepare a sample of uniform density. As an experimental convenience, the 10 mm diameter cone is selected though it is not entirely satisfactory.

A field situation was created by processing the soil in the bin. The soil was produced with an expected accuracy for $\gamma_d \cong 1.2$ Mg/m^3. Standard cones (20 mm diameter and 30^0 smooth cone) were used in the tests. The penetrometer was advanced up to 300 mm. Five penetrations were carried out to obtain an average load. The tests were conducted at different moisture contents and dry densities. The soil-bin was kept covered between the test runs for moisture conservation.

4 RESULTS & DISCUSSION

The cone-index q_c calculated from F_c at a particular w was normalized by p (= σ_3). The normalized cone index Q and state parameter ψ were plotted on the ln(Q)-ψ plane. The slope m of the regression line and its intercept ln(Q_0) at $\psi = 0$ at different moisture content are shown in Fig.3. Notice that m and ln(Q_0) at different w changes in a systematic fashion.

In order to establish the state-parameter coefficients m and Q_0 precisely as a function of w, interpolation within the w-range and extrapolation near saturation (about 27.0%) was required. This was achieved by fitting a curve through the experimental data points. Nature of the curve fitted to a polynomial minimax is taken as

$$P(x) = a_0 + a_1 x + a_2 x^2 + a_3 x^3 + .. + a_{r+1} x^r \qquad (6)$$

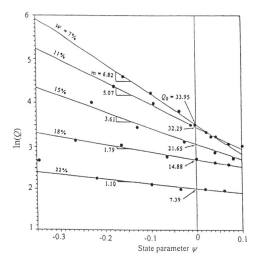

Figure 3. ln(Q) - ψ relationship at different moisture w

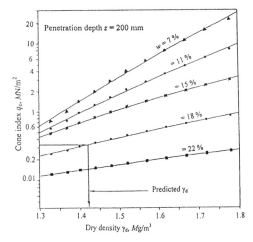

Figure 4. Cone index q_c vs γ_d at various soil moisture w

Table 1 Polynomial coefficients for Q_0 and m

Coefficients	
for Q_0	for m
$A_0 = -17.22621$	$B_0 = 6.65840$
$A_1 = 15.06277$	$B_1 = 0.37832$
$A_2 = -1.43390$	$B_2 = -0.06779$
$A_3 = 0.04889$	$B_3 = 0.00212$

such that $2|P(x_i) - y_i|$ is minimum for a given set of x_i, y_i. This routine uses in the exchange algorithm to compute rth order polynomial. This analysis accounts up to 3rd order polynomial coefficients

A_0, A_1, A_2 & A_3 for Q_0 and B_0, B_1, B_2 & B_3 for m have been computed and are given in Table 1.

The coefficients m and Q_0 characterised as a function of w takes the forms as

$$Q_0 = -17.22621 + 15.06277w - 1.43390w^2 + 0.04889w^3 \quad (7)$$
$$m = 6.65840 + 0.37832\,w - 0.06779w^2 + 0.00212\,w^3 \quad (8)$$

Once m and Q_0 are established in terms of w then it is possible to interpret the field q_c as follows:
$$q_c = Q\,p + p \quad (9)$$

where $Q = Q_0 \exp(-m\psi)$.

A computer software was developed to predict q_c against γ_d at different w with Q_0 and m obtained from eqs.(7) & (8). A typical output given in Fig.4 shows the predicted values of q_c at a depth of 200 mm for γ_d ranging from 1.33 to 1.78 Mg/m^3 with w from 7.0 to 22.0%.

A typical output of converted q_c obtained from the cone penetrometer-runs in the soil-bin against z is shown in Fig.5. Once Fig.5 is prepared for a field soil, then it is a simple matter to predict γ_d at any moisture content w of interest using Fig.4 at a particular depth of penetration. The sequential steps to predict γ_d are as follows:

(a) Find q_c corresponding to depth z from Fig.5.
(b) With q_c, find γ_d at particular w from Fig.4.

A typical output of the predicted γ_d at 17.5% moisture content is shown in Fig.6. It is to observe from Fig.6 that the predicted γ_d lies well above the in. situ estimated γ_d. The actual γ_d was obtained from core-sampler data carried out on the surface layer of the soil in the bin. Although this procedure appears to over-predict γ_d ($\leq 10\%$), the deviation reflects higher in-situ densities in deeper soil layers

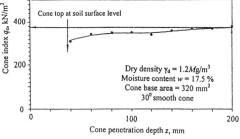

Figure 5. Estimation of q_c at different z

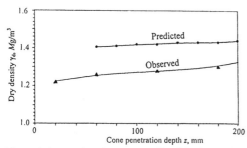

Figure 6. Comparison between predicted & observed γ_d

of bin. On this basis, the performance would appear to be in the acceptable range.

The interpretation of data obtained from in-situ test is difficult and in most cases, it is both incomplete and imprecise. Wroth (1984 & 1986) emphasised the difficulties in interpreting the in-situ testing. The mode of these difficulties is due to the complex behaviour of soils together with the lack of control and choice of boundary conditions in any field test. The calibration chamber investigation may help to allay some of these uncertainties.

5 CONCLUSIONS

The penetrometer readings are readily convertible to bulk density or pore-space estimates provided the chart for cone-index against the penetration depth for a given field soil is known beforehand. This technique over-predicts the bulk density to a maximum limit of 10%. The translation of cone index into a more useful evaluation of in-situ bulk density and pore space is most encouraging. The actual field trials may give a high degree of confidence in extending the state parameter concept to unsaturated soils. This might be a massive undertaking even for a single soil. However, it opens a possible route to develop some short cuts to make the technique more manageable. The published data are limited and confined to dry sand only. Hence, the performance may deviate in a clay soil.

REFERENCES

Abedin, M.Z.1995. The characterization of unsaturated soil behaviour from penetrometer performance and the ctritical state concept. *Ph.D. Thesis*, The Univ. of Newcastle-upon-Tyne, UK.

Abedin, M.Z., D.R.P. Hettiaratchi & M.A. Rashid 1997. Determination of cone index of agricultural soil from state parameter concept. *Proc. 14th Int. Conf., ISTRO*, Pulawy, Poland, 2A: 11-14.

Been, K. & M.G. Jefferies 1985. A State parameter of Sands. *Geotechnique*, 35(2): 99-112.

Been, K., J.H.A. Crooks, D.E. Becker & M.G. Jefferies 1986. The cone Penetration in sands: part-I, state parameter interpretation. *Geotechnique*, 36(2) : 239-249.

Been, K., M.G. Jefferies, J.H.A. Crooks & L. Rothenburg 1987. The cone Penetration in sands: part-II, general inference of state. *Geotechnique*, 37(3): 285-299.

Bellotti, R. 1985. Laboratory validation of in-situ tests. *In: Geotechnical Engineering,Italy* (Associazione Geotenica Italiana, Published for ISSMFE Golden Jubilee), pp. 251-270.

Head, K.H. 1986. Manual of Soil Laboratory Testing. Pentech Press, London: Plymouth, Vol. 2 & 3.

Parkin, A.M. & T. Lunne 1982. Boundary effects in the laboratory calibration of a cone penetrometer for sand. *Proc. 2nd Euro.Symp. on Penetration Testing*, Amsterdam, 2: 761-768.

Roscoe, K.H., A.N. Schofield & C.P. Wroth 1958. On yielding of soils. *Geotechnique*, 8: 22-53.

Sladen, J.A. 1989. Problems with interpretation of sand state from cone penetration test. *Geotechnique*, 39: 223-332.

Sokolovski, V.V. 1960. Statics of Soil Media. *Butterworths Scientific Publications*, London.

Wroth, C.P. 1984. The interpretation of in-situ soil tests. *Geotechnique*, 34(4): 449-489.

Wroth, C.P. 1986. Field testing: Interpretation of cone penetration test. *Geological Society, Engg. Geology Special Publi.*, Site Investigation Practice, 2: 17-19.

Unsaturated Soils for Asia, Rahardjo, Toll & Leong (eds) © 2000 Taylor & Francis, ISBN 90 5809 139 2

Physical definition of residual water content in unsaturated soils

T. Aoda
Faculty of Agriculture, Niigata University, Japan

ABSTRACT: The residual water content is one of the most important parameters to express the soil water characteristics and hydraulic conductivity in unsaturated soils. However we do not have a clear and rigorous definition of residual water content from the physical viewpoint. Therefore even fundamental knowledge of the capillary height of liquid water in soil is still involved. In this study, we focused on the pressure change of glass beads within various water levels. Experimental results show that water breaks up into discontinuous patches (pendular ring) at much higher water content than we assumed. Though pendular rings have continuity to film water at the solid surface, pendular rings are not able to transmit water pressure. Consequently water content, which consists of pendular ring and film water, could be thought as the minimum limit of Darcy's law with respect to liquid phase water movement.

1 INTRODUCTION

Numerical models for water flow in unsaturated soils are generally based on a second order, non-linear partial differential equation commonly referred to as Richards' equation (Bear & Bachmat 1991). The motion of water is incorporated in Richards' equation through Darcy's Law. This law was originally derived for saturated soils. Its application to unsaturated flow hinges on the assumption that the water phase is continuous. Hence the assumption leave some phenomena unclear e.g. preferential flow, salinization and root water uptake.

This focused on the soil water flow in liquid phase under gravity, without osmotic and thermal gradient. The pressure of the water phase in a column filled with glass beads is measured.

2 MATERIAL AND METHOD

Glass beads with diameter of 24 mm are placed in a zigzag state in a transparent acrylic case in a constant temperature laboratory. The top of the acrylic case is covered by a plastic film to reduce vapor diffusion. Deaired and desalted water was used for the experiment. The experiment focused only on the liquid phase water movement under gravity, without osmotic and thermal gradient. The water pressure is measured using a pressure

transducer through an injection needle with inside diameter of 0.2 mm. The time-dependent pressure measurement of discontinuous water patches (pendular ring) was made under the condition of various water levels, after the acrylic case was fully saturated. Simultaneously, we photographed the radii of menisci's curvature of pendular ring by digital camera. Figure 1 shows a schematic diagram of experimental beads column and the pinpoint pressure measurements.

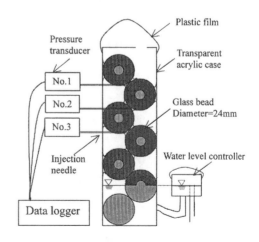

Figure 1. Schematic diagram of experimental apparatus named pinpoint pressure measurement

3 RESULT AND DISCUSSION

3.1 *Pressure Measurement*

Figure 2 shows the water pressure and water level with time. Lowering the water level, the water surrounding the glass beads break up at some critical height from the water. Then the pressure in the discontinuous water phase shows almost constant value and is independent of the water table. Raising the water table saturates the soil pore and the water pressure again follows the water level.

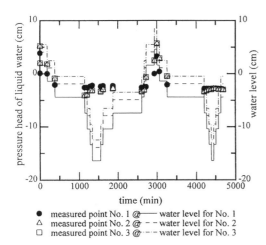

Figure 2. Water pressure and water level with time

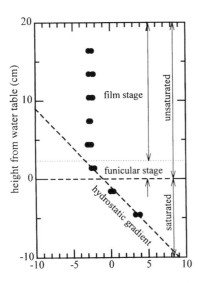

Figure 3. Relationship between pressure head and water level at measurement point No.1

Figure 3 shows change of pressure head at measurement point No.1. Below the water table, water pressure strictly depends on water level and the hydrostatic gradient. In unsaturated condition, pressure positively relates to water level until the critical height i.e. 3 cm. At this water saturation, liquid water connects to the water table and indicates funicular stage. After the liquid water breaks up, pressure is constant and is independent of water level. At this water content, which consists of pendular ring and film water (film stage), no pressure could be transmitted from one ring to another ring.

Considering water continuity, water has to respond to a variation in water level, even in the non-equilibrium state. We understand that in a film stage, water could move with phase transitions e.g. vapor phase to liquid phase. In practice, water content of the film stage is the lower limit of Darcy's law for liquid water flow in unsaturated soil. The critical pressure of the pendular ring is a function of the diameter of the solid particle.

Figure 4 shows the pendular ring at the highest measurement point at 1875 minutes. The principal radii of menisci's curvature are not dependent on water level, same as the measured pressure. At this pendular ring, pressure is not −7.4 cm as in hydrostatic pressure, but is −2.84 cm by measurement.

The pressure of the pendular ring, which defines menisci's radii, is almost constant, and is independent of the water level. Though film water and ring water exists continuously in the film stage, no pressure can be transmitted from one ring to the next though the water phase.

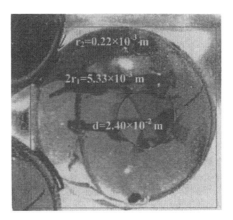

Figure 4. Pendular ring at point No.1 at 1875 minutes

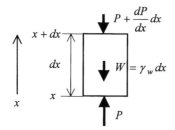

Figure 5. Force balance of liquid water element. Where x = vertical coordinate, positive upward; P = water pressure; W = mass of element; γ_w = unit weight of water. The cross section = 1.

3.2 Theoretical analysis

At gas-liquid interface, Laplace equation is

$$P = P_0 + \alpha\left(\frac{1}{r_1} + \frac{1}{r_2}\right) \qquad P = P_0 + \left(\frac{1}{r_1} + \frac{1}{r_2}\right). \qquad (1)$$

Where P_0 = atmospheric air pressure; α = the surface tension; r_1 and r_2 are principal radii of interface curvature (positive in convex shape and negative in concave). P_0 = 101325 Pa and α = 72.75×10^{-3} N m^{-1} in 20 °C in standard condition.

Taking into account the transmission of pressure of water path in unsaturated porous media (Fig.5), force upward and downward balances in this element, and we get

$$P + \gamma_w dx = P + \frac{\partial P}{\partial x} dx \qquad (2)$$

Where x = vertical coordinate, positively upward; P = water pressure; W = mass of the element; γ_w = unit weight of water.

Integrating Equation 2 from the water surface $x = 0$ to height $x = X$, one is able to predict the pressure at height X

$$\int_0^x \frac{dp}{dx} dx = \int_0^x -\gamma_w dx \qquad (3)$$

Reducing Equation 3, one obtain the hydrostatic pressure,

$$P = \gamma_w X \qquad (4)$$

Equation 4 is usable only under the condition of continuity of the P function within the limits of $0 \le x \le X$. Limit of x indicates capillarity of liquid water in the funicular stage. From Equation 1, pressure of ring water in gage is –3.06 cm, measured value is –2.84 cm and is –7.4 cm from Equation 4.

Though in many studies on unsaturated soils, Equation 1 and Equation 4 are taken as equivalent (Iwata, 1961). However in practice only Equation 1 is able to represent the pressure of ring water.

If $X < x$, water is not able to transmit pressure in the field of gravity without osmotic and thermal gradient. Under vacuum condition, we are able to predict minimum value of cross section in water path from Equation 1, i.e. 2.88×10^{-6} m. Diameter of water molecule is around 3×10^{-10} m, water is able to transmit the pressure in a set of 10,000 molecules. Therefore film water on soil surface is thinner than the 10,000 layers, and is impossible to transmit pressure.

Though in film stage average water pressure indicates the unit hydraulic gradient, no water could flow in the liquid phase. Considering Darcy's law, it might be rigorous that hydraulic conductivity of the film stage is zero (Aoda et al., 1992). Water diffuses from one ring to another ring in the vapor phase. Hence one should consider not hydraulic conductivity but vapor diffusivity in the film stage.

4 CONCLUSION

Lowering the water level within the column, the water become discontinuous rings, and its pressure remains constant and independent of the water level. In this film stage, water movement is only in the vapor phase. By focusing only on the soil water flow in liquid phase under gravity, it could be possible to understand that residual water content is made up of ring water and film water.

Vapor phase water transport plays an important role in water movement in unsaturated soils (Nakano, 1991). Vapor transport should be separated from the liquid water movement for further understanding of water dynamics in unsaturated soils, especially in coarse grained soils.

ACKNOWLEDGEMENT

We express our thanks the partly support from Arid Land Research Center, Tottori University, Japan, for this study.

REFERENCES

Aoda, T. Yoshida, S. Nakano. T. & Yamada, A. 1992. Fundamental studies on rain infiltration into sandy soil with consideration given to hysteresis. Transactions on Jap. Soc. of Irri. Drai. Rec. Eng. 157: 35-44.
Bear, J. & Bachmat, Y. 1991. Introduction to modeling of transport phenomena in porous media. Dordrecht: Kluwer Academic.
Iwata S. 1961. Energy concept on soil water. Jap. J. Soil Nutrition 32(11): 572-580.
Nakano, M. 1991. Material dynamics in soils. Tokyo: University of Tokyo press.

Unsaturated Soils for Asia, Rahardjo, Toll & Leong (eds) © 2000 Taylor & Francis, ISBN 90 5809 139 2

The behaviour of unsaturated porous media in the light of a micromechanical approach

X. Chateau
L.M.S.G.C. (UMR 113 CNRS-LCPC), Champs sur Marne, France

L. Dormieux
E.N.P.C.-CERMMO, Marne la Vallée, France

ABSTRACT: The macroscopic mechanical behaviour of unsaturated porous media under isothermal conditions is studied within the framework of upscaling techniques. First the main features of the homogenization method are recalled. Then, the macroscopic state equation including the capillary effects is derived. Finally, a morphological model is used in order to clarify the link between the sorbtion-desorbtion hysteretic phenomena and the macroscopic mechanical behaviour.

1 INTRODUCTION

Even if many contributions have been devoted to the constitutive behaviour of partially saturated porous, several questions of paramount importance are still debated (Coussy 1995), (Gens 1995) or (Schrefler & Gawin 1996). For instance, there is no consensus about the validity of an effective stress formulation and, if appropriate, about the form of the effective stress principle to adopt. The reasons why theses questions, among others, remain to receive satisfactory answers can be easily understood considering the difficulties encountered to modelize problems involving unsaturated porous media; according to the macroscopic point of view developed in the works cited above, the choice of the state variables and the form of macroscopic constitutive laws are to be postulated at the macroscopic scale and besides the model has to be validated with respect with experiments. The couplings between many interacting phases (at least 3: the solid and two fluids), the experimental evidence that most of the unsaturated porous media have a non linear behaviour and the lack of experimental results to quantitatively validate the models are three of the main difficulties encountered.

Because of its ability to incorporate at the macroscopic level the physics of the microscopic level as well as the morphological aspects, at least partly, the micro-macro methods may be a helpful modelling tool. In (Auriault 1987) and (Auriault et al. 1989), the homogenization theory of periodic structures is used to obtained the macroscopic description of the behaviour of a deformable porous medium saturated by two immiscible fluids from the description of the pore scale. In spite of the strong hypotheses concerning the

fluid-fluid interface (which is supposed to not intersect the surface of the skeleton) and the periodicity of the structure of the medium, this is, to our knowledge, the first theoretical justification of the generalized Darcy flow and of the generalization of Biot's equation for an unsaturated porous medium with the help of an homogenization theory.

Here, the mechanical behaviour of unsaturated porous media under isothermal condition is studied within the framework of change of scale methods. No attention is paid to the description of the flow of the fluids. Performing such a change of scale method requires to establish the relationship between the total macroscopic stress tensor Σ and the microscopic stress fields in the different phases of the porous medium. In the case of a 3 phase medium (solid, liquid, gas), the result obtained by (Chateau & Dormieux 1995), and (Chateau & Dormieux 1998) is recalled in the following section. Then, the link between hysteretic phenomena which can occur in cyclic sorbtion-desorbtion experiments, the reversibility of the macroscopic behaviour and the evolutions of the system at the microscopic scale is studied. Thereafter, two morphological models are proposed.

2 STATICS OF UNSATURATED POROUS MEDIA

The Representative Elementary Volume Ω is made up of a solid phase, a liquid phase and a gaseous phase which respectively occupy the domains Ω^s, Ω^ℓ and Ω^g. $\omega^{\alpha\beta}$ denotes the interface between the $\alpha-$ and $\beta-$phases. At the microscopic scale, the internal forces are described by a Cauchy stress tensor field σ in the domains Ω^s, Ω^g and Ω^ℓ. The Cauchy stress in the fluid phases is defined by the fluid pressure p^α ($\alpha = \ell, g$), that is $\boldsymbol{\sigma} = -p^\alpha \boldsymbol{\delta}$, where $\boldsymbol{\delta}$ denotes the unit tensor of second order. The capillary effects introduce internal forces of the membrane type located in the interfaces between phases. The latters are represented by a tensor field of surface tension $\gamma^{\alpha\beta} \boldsymbol{\delta}_{T_\omega}$ in the surface $\omega^{\alpha\beta}$, where $\boldsymbol{\delta}_{T_\omega}$ denotes the unit tensor to the tangent plane to surface ω and $\gamma^{\alpha\beta}$ the surface tension in the $\alpha\beta$ interface, with $(\alpha\beta) = (s\ell)$, (sg) and (ℓg). As usual within the framework of homogenization approach to the behaviour of the porous media skeleton (Auriault & Sanchez Palencia 1977) and (Auriault 1987), this internal forces must comply with no body force equilibrium conditions. Thus, to be statically admissible, the microscopic stress field has to comply with the momentum balance equation $\mathrm{div}\,\boldsymbol{\sigma} = 0$. In particular, this implies that p^ℓ and p^g are uniform in Ω^g and Ω^ℓ respectively. Taking the capillary effects into account, the classical condition of continuity of the stress vector at the interfaces $\omega^{s\alpha}$ is replaced by the following conditions:

$$(\forall\, \underline{x} \in \omega^{s\alpha})(\alpha = g, \ell)$$
$$\boldsymbol{\sigma} \cdot \underline{n} = -p^\alpha \underline{n} - \gamma^{s\alpha}(\boldsymbol{\delta}_{T_\omega} : \boldsymbol{b})\underline{n} \tag{1}$$

where $\underline{n}$ denotes the outer unit normal vector to Ω^s and $\boldsymbol{b}$ is the tensor of curvature. $\boldsymbol{\delta}_{T_\omega} : \boldsymbol{b}$ thus represents the mean curvature of $\omega^{s\alpha}$. The corresponding condition at the interface $\omega^{g\ell}$ is classically referred to as Laplace law:

$$(\forall\, \underline{x} \in \omega^{\ell g})$$
$$\gamma^{\ell g} \boldsymbol{\delta}_{T_\omega} : \boldsymbol{b} = p^\ell - p^g = -p^c \tag{2}$$

p^c will be referred to as the capillary pressure. Hill's lemma provides the link between the macroscopic stress tensor $\boldsymbol{\Sigma}$ and the microscopic stress field (Chateau & al. (1995)):

$$\boldsymbol{\Sigma} = (1 - f) < \boldsymbol{\sigma} >_s - f^\ell p^\ell \boldsymbol{\delta} - f^g p^g \boldsymbol{\delta}$$
$$+ \frac{1}{|\Omega|} \int_\omega \gamma\,(\underline{x})\, \boldsymbol{\delta}_{T_\omega} dS \tag{3}$$

where f denotes the porosity, $< \cdot >_\alpha$ the volume average in Ω^α and f^α the volume fraction of the α-phase.

The last term of (3) represents the mechanical contribution of the surface tensions to the macroscopic Cauchy stress. This corresponds to the fourth phase's effect introduced by (Fredlund & Morgenstern 1977) and (Houlsby 1997).

In the sequel, we examine the situation of an homogeneous linear elastic solid matrix. The microscopic state equation for the solid phase thus reads

$$(\forall\, \underline{x} \in \Omega^s) \begin{cases} \boldsymbol{\sigma}(\underline{x}) &= \ \mathbb{C}^s : \boldsymbol{\varepsilon}(\underline{x}) \\ \boldsymbol{\varepsilon}(\underline{x}) &= \ \mathbb{S}^s : \boldsymbol{\sigma}(\underline{x}) \end{cases} \tag{4}$$

where $\mathbb{C}^s$ denotes the elastic moduli and $\mathbb{S}^s$, the elastic compliances.

3 MORPHOLOGICAL ASPECTS

The aim of a micro-macro approach is to take into account the effect of morphological aspects at the microscopic scale into the macroscopic behaviour. This requires to define a mechanical problem on the representative elementary volume.

As far as unsaturated porous media are concerned, the "stability" of the evolution of the representative elementary volume, and more specificaly of the liquid-gas interface is a question of paramount importance. The concept of "stability" corresponds to the continuity of the location of the liquid-gas interface with respect to time. When this "stability" is achieved, the micro-macro approach performed by (Chateau & Dormieux 1998) proves the existence of a macroscopic thermodynamic potential depending solely on the macroscopic strain tensor $\boldsymbol{E}$ and on the fluid mass increases m^ℓ and m^g. In this case, the state equations take the form:

$$d\boldsymbol{\Sigma} \ = \ \mathbb{C}^0 : d\boldsymbol{E} - \boldsymbol{B}^\alpha dp^\alpha \tag{5}$$

$$dp^\alpha \ = \ M_{\alpha\beta}\left(-\boldsymbol{B}^\beta : d\boldsymbol{E} + \frac{dm^\beta}{\rho^\beta}\right) \tag{6}$$

where $\mathbb{C}^0$ is the tensor of elastic drained moduli, whereas $M_{\alpha\beta}$ and $\boldsymbol{B}^\alpha$ are the generalized Biot's moduli and coefficients respectively. This result confirms the macroscopic thermodynamic approach proposed by (Coussy 1995).

For a given macroscopic stress $\boldsymbol{\Sigma}$, equations (5) and (6) define a one to one relation between the fluid pressures, the macroscopic strain of the skeleton and the fluid mass increases. This is clearly incompatible with hysteretic phenomena which can occur in cyclic sorbtion-desorbtion experiments. This macroscopic hysteresis corresponds to "unstable" evolutions of the liquid-gas interface at the microscopic scale, as described for instance by (Dullien 1992). In fact, it is believed that the validity of the stability assumption depends on the morphology of the fluid phases. As an example, stability is likely in situation of low liquid saturation, in which Ω^ℓ consists in separated liquid menisci. In this case, the morphologies of domains Ω^ℓ and Ω^g are of differents types.

In the sequel, we restrict ourselves to the case where there is no morphological difference between Ω^ℓ and Ω^g. This hypothesis proves to be natural as long as the porous space is made up of pores of similar shape, respectively filled by liquid or by gas. In such a case, it is clear that "unstable" evolutions and, hence, hysteretic phenomena are expected.

96

4 SIMPLIFIED MODEL

Let us first assume that the surface tension $\gamma(\underline{x})$ can be neglected in equations (1) and (3). In such an approach the surface tension and the curvature of liquid-gas menisci are only taken into account through the difference between fluid and gas pressures. Therefore, equation (3) writes:

$$\Sigma = (1-f) < \boldsymbol{\sigma} >_{\Omega^s} + f^\ell p^c \boldsymbol{\delta} \qquad (7)$$

For the sake of simplicity, it is assumed that the boundary of the elementary representative volume $\partial\Omega$ is a subset of the boundary of the solid domain $\partial\Omega^s$.

Then, the solid matrix Ω^s is subjected to a loading characterized by the capillary pressure p^c and by the tensor $\boldsymbol{E}$ which represents the macroscopic strain tensor of Ω. $\boldsymbol{E}$ defines the classical Hashin condition on the displacement at the boundary $\partial\Omega$ whereas a traction equal to p^c is applied to the solid at $\omega^{s\ell}$:

$$\begin{cases} (\forall \, \underline{x} \in \omega^{s\ell}) & \boldsymbol{\sigma} \cdot \underline{n} = p^c \underline{n} \\\\ (\forall \, \underline{x} \in \omega^{sg}) & \boldsymbol{\sigma} \cdot \underline{n} = 0 \qquad (8) \\\\ (\forall \, \underline{x} \in \partial\Omega) & \underline{\xi} = \boldsymbol{E} \cdot \underline{x} \end{cases}$$

where $\underline{\xi}$ denotes the displacement vector of the solid particles in Ω^s.

The linearity of the elastic boundary value problem defined on the solid part by equations (8) and (4) implies that the strain tensor $\boldsymbol{\varepsilon}(\underline{x})$ in the solid matrix is a linear function of both loading parameters $\boldsymbol{E}$ and p^c:

$$\boldsymbol{\varepsilon}(\underline{x}) = \mathbb{A}(\underline{x}) : \boldsymbol{E} + \boldsymbol{\alpha}(\underline{x}) p^c \qquad (9)$$

where $\mathbb{A}(\underline{x})$ and $\boldsymbol{\alpha}(\underline{x})$ are the two concentration tensors respectively associated with the loading parameters $\boldsymbol{E}$ and p^c.

Then, introducing (9) into (7) yields an expression of the macroscopic stress Σ as a function of $\boldsymbol{E}$ and p^c:

$$\Sigma = \mathbb{C}^{hom} : \boldsymbol{E} + \boldsymbol{P} p^c \qquad (10)$$

where:

$$\mathbb{C}^{hom} = (1-f)\mathbb{C}^s : \mathbb{A}^s \qquad (11)$$

with:

$$\mathbb{A}^s = <\mathbb{A}>_{\Omega^s} \qquad (12)$$

$$\boldsymbol{P} = f^\ell \boldsymbol{\delta} + (1-f)\mathbb{C}^s :< \boldsymbol{\alpha} >_{\Omega^s} \qquad (13)$$

An homogenization method aims at estimating the tensors $\mathbb{C}^{hom}$ and $\boldsymbol{P}$. $\mathbb{C}^{hom}$, identical to $\mathbb{C}^0$ (see equation (5)) can be interpreted as the tensor of drained elasticity moduli. It described the overall elasticity of a porous medium constitued of a linear elastic solid matrix and of an empty porous space. As far as this tensor is concerned, the derivation of estimates (or bounds) is classical (see (Zaoui 1997)). In order to assess the debatted question of the validity of an "effective-stress" approach for unsaturated porous media, the sequel is devoted to the estimation of tensor $\boldsymbol{P}$.

An alternative approach to the problem defined by (8) and (4) consists in considering the representative elementary volume as an hererogeneous medium with a different constitutive law for each of the 3 phases.

$$\boldsymbol{\sigma} = \mathbb{C}^\alpha : \boldsymbol{\varepsilon} + \boldsymbol{P}^\alpha p^c \qquad (14)$$

with $(\alpha = s, \ell, g)$ and $\boldsymbol{P}^\ell = \boldsymbol{\delta}$, $\mathbb{C}^\ell = \mathbb{C}^g \to 0$ and $\boldsymbol{P}^s = \boldsymbol{P}^g \to 0$.

Contrary to the point of view defined by equation (8), the capillary pressure is no more a loading parameter for the boundary conditions defined over the solid matrix but appears in equation (14) as an pre-stress. Within this framework, Levin's theorem provides the relationship between the macroscopic pre-stress $\boldsymbol{P} p^c$ in (10), and the microscopic ones in (14).

$$\boldsymbol{P} =< {}^t\mathbb{A} : \boldsymbol{P}^\alpha > \qquad (15)$$

where $\mathbb{A}(\underline{x})$, introduced in (9), denotes the strain concentration tensor at point $\underline{x}$ associated with the solution for the representative elementary volume with empty porosity submitted to the macroscopic strain $\boldsymbol{E}$ according to Hashin boundary conditions.

In the case of no morphological difference between the domains Ω^g and Ω^ℓ under consideration, it is natural to write:

$$\mathbb{A}^g = \mathbb{A}^\ell \qquad (16)$$

where $\mathbb{A}^\alpha =< \mathbb{A} >_{\Omega^\alpha}$.

Using the relation $f^\alpha \mathbb{A}^\alpha = \mathbb{I}$, where $\mathbb{I}$ is the unit fourth-order tensor, we obtain:

$$\mathbb{A}^g = \mathbb{A}^\ell = \frac{1}{f}\left(\mathbb{I} - (1-f)\mathbb{A}^s\right) \qquad (17)$$

Equation (11) allows to substitute $\mathbb{C}^{hom}$ for $\mathbb{A}^s$:

$$\mathbb{A}^g = \mathbb{A}^\ell = \frac{1}{f}\left(\mathbb{I} - \mathbb{S}^s : \mathbb{C}^{hom}\right) \qquad (18)$$

Thus, the equation (15) takes the form:

$$\boldsymbol{P} = S_r\left(\mathbb{I} - \mathbb{C}^{hom} : \mathbb{S}^s\right) : \boldsymbol{\delta} \qquad (19)$$

where S_r denotes the liquid saturation equal to $f^\ell/(f^\ell + f^g)$. In this approach, the tensor $\boldsymbol{P}$ proves to be a linear function of the liquid saturation.

If the gas pressure is taken into account, the macroscopic behaviour of the skeleton writes:

$$\Sigma + (p^g - S_r p^c)\boldsymbol{B} = \mathbb{C}^{hom} : \boldsymbol{E} \qquad (20)$$

where $\boldsymbol{B} = \boldsymbol{P}/S_r$ is defined by (19) and:

$$\Sigma' = \Sigma + (p^g - S_r p^c)\boldsymbol{B} \qquad (21)$$

defines the relevant effective stress expression in the situation described by the hypothesis done above.

Putting $S_r = 1$ into (20), our estimate coincides with the result obtained by (Auriault & Sanchez Palencia 1977) for saturated periodic media. In the case of incompressible solid matrix ($\mathbb{S}^s : \boldsymbol{\delta} \to 0$), one obtains $\boldsymbol{B} = \boldsymbol{\delta}$, which is the form proposed by (Schrefler 1990). It also corresponds to the classical Bishop effective stress with $\chi = S_r$. The approach performed here shows that this form can be justified if surface tension effect at the gas-solid interface as well as the surface tension contribution to the Cauchy stress tensor can be neglected. Equations (19) and (20) generalize Schrefler's approach to the case of a compressible anisotropic solid matrix.

5 SURFACE TENSION EFFECT

For the sake of simplicity, it is assumed in the sequel that the liquid perfectly wets the solid and that the surface tension in the solid-liquid interface is naught. Then we have $\gamma^{\ell g} = \gamma^{sg} = \gamma$ and $\gamma^{s\ell} = 0$.

In order to introduce the capillary effects in a more rigorous way than in section 4, the capillary terms in (1) and (3) which have been neglected in the simplified model must be taken into account. This requires to specify the geometry of the interfaces between phases. Let us consider the case where the porous space can be satisfactory described by N subsets of spheres, each subset described by the radius R_β and the volume fraction f_β ($\beta = 1, \dots, N$). Moreover it is assumed that the pores belonging to a subset β are all filled by liquid or all filled by gas. The distribution of liquid and gaseous phases in the pore network is described by the set I_g, such as:

$\beta \in I_g$: the pores of the subset β are filled by gas.

$\beta \notin I_g$: the pores of the subset β are filled by liquid.

The porous medium is then described at the microscopic scale by the couple (p^c, I_g). Although the evolutions of these two quantities are linked, there is generally no one to one relation between p^c and the set I_g. Hence, the set I_g is considered, like p^c, as a known variable for the problem under consideration. No attempt is made in this section to relate the capillary curve (which is the macroscopic counterpart for the imbibition-desorption of the porous space by the liquid phase) to the evolutions of the set I_g.

Taking now into account mechanical effects of surface tension, the second equation of (1) writes:

$$(\forall \beta \in I_g) \qquad \boldsymbol{\sigma} \cdot \underline{n} = p^\beta \underline{n} \qquad (22)$$
$$(\forall \underline{x} \in \partial\Omega^{s\beta})$$

with, since the solid-gas interface ∂S_β is a sphere of radius R_β:

$$p^\beta = \frac{2\gamma}{R_\beta} \qquad (23)$$

In comparison with section 4, the boundary value problem on Ω^s is now defined by the first and the third equations of (8), the second one being replaced by (22).

The contribution of the solid-gas interface ∂S^β to the macroscopic stress writes:

$$\int_{\partial S^\beta} \gamma \boldsymbol{\delta}_{T_\omega(\underline{x})} dS \quad = \quad \frac{2\gamma}{R_\beta} \frac{4}{3} \pi R_\beta^3 \boldsymbol{\delta} \qquad (24)$$

$$= \quad p^\beta |S^\beta| \boldsymbol{\delta}$$

where $|S^\beta|$ denotes the volume of a sphere belonging to the class β.

Equations (22) and (24) indicate that it is possible to take into account the mechanical effect of surface tension at the interface $\partial\Omega^{s\beta}$ by considering that these pores are filled by a fluid at pressure p^β. Indeed, the loading applied by such a fictitious fluid on the solid matrix is defined by (22), while equation (24) shows that the contribution of the surface tension to the macroscopic stress tensor Σ is identical to that of a negative pressure $-p^\beta$ in a volume $|S^\beta|$.

Equation (9) must be remplaced by:

$$\varepsilon(\underline{x}) = \mathbb{A}(\underline{x}) : \boldsymbol{E} + \boldsymbol{\alpha}(\underline{x})p^c \qquad (25)$$

$$+ \sum_{\beta \in I_g} \boldsymbol{\alpha}^\beta(\underline{x})p^\beta$$

where $\boldsymbol{\alpha}^\beta(\underline{x})$ are additionnal concentration tensors respectively associated with the loading parameters p^β, $\beta \in I_g$.

Then, introducing (25) in (3) yields an expression of the macroscopic stress Σ as a function of $\boldsymbol{E}$ and p^β:

$$\Sigma = \mathbb{C}^{hom} : \boldsymbol{E} + \sum_{\beta=1}^{N} \boldsymbol{P}^\beta p^\beta \qquad (26)$$

with $p^\beta = p^c$ for $\beta \notin I_g$, $p^\beta = 2\gamma/R_\beta$ for $\beta \in I_g$. $\mathbb{C}^{hom}$ is still defined by (11).

As in (17), it is possible to write in a formal manner the behaviour of phase β in the form:

$$\boldsymbol{\sigma} = \mathbb{C}^\beta : \boldsymbol{\varepsilon} + p^\beta \boldsymbol{\delta} \qquad (27)$$

For the sake of simplicity, (27) is used for the solid phase with $p^\beta = p^s = 0$ when $\beta = s$.

According to Levin's theorem (15), the macroscopic pre-stress $\sum_{\beta=1}^N \boldsymbol{P}^\beta p^\beta$ of (26) is related to the microscopic pre-stresses $p^\beta \boldsymbol{\delta}$ of (27) by:

$$\sum_{\beta=1}^N \boldsymbol{P}^\beta p^\beta = \sum_{\beta=1}^N f^\beta p^\beta {}^t\mathbb{A}^\beta : \boldsymbol{\delta} \qquad (28)$$

with:

$$\mathbb{A}^\beta = <\mathbb{A}>_{\Omega^\beta} \qquad (29)$$

Observing that there exists no morphological difference between the domains Ω^β, the same kind of analysis as in section 4 yields

$$(\forall \beta) \qquad \mathbb{A}^\beta = \frac{1}{f}\left(\mathbb{I} - \mathbb{S}^s : \mathbb{C}^{hom}\right) \qquad (30)$$

Introducing (30) into (28), it can be concluded that $\boldsymbol{P}^\beta$ depends on phase β solely through the ratio f^β/f:

$$\boldsymbol{P}^\beta = \frac{f^\beta}{f}\left(\mathbb{I} - \mathbb{C}^{hom} : \mathbb{S}^s\right) : \boldsymbol{\delta} \qquad (31)$$

Finally, if the gas pressure p^g is taken into account, the macroscopic behaviour writes:

$$\begin{aligned}
\boldsymbol{\Sigma} &= \mathbb{C}^{hom} : \boldsymbol{E} + \boldsymbol{\Sigma}(I_g) \\
&+ \left(S_r p^c - p^g\right)\left(\mathbb{I} - \mathbb{C}^{hom} : \mathbb{S}^s\right) : \boldsymbol{\delta}
\end{aligned}$$

with:

$$\boldsymbol{\Sigma}(I_g) = \sum_{\beta \in I_g} \frac{f^\beta p^\beta}{f}\left(\mathbb{I} - \mathbb{C}^{hom} : \mathbb{S}^s\right) : \boldsymbol{\delta} \quad (32)$$

$\boldsymbol{\Sigma}(I_g)$ takes the surface tension into account. It depends on the geometry of the domain Ω^g.

6 PORE NETWORK

Although the evolutions of I_g and p^c are linked, I_g is not, in general, a one-to-one function of p^c. We now study a simpler situation in which the pore network is made up of spherical pores of decreasing radius R_β, related to one another by capillary necks of decreasing radius r_β as indicated on Figure 1.

In fact, this model defines a one to one relation between S_r and I_g. For an incremental variation dS_r of the current saturation ratio S_r, a single β-subset is being filled with liquid or with gas $(dS_r = \pm f^\beta/f)$. It is then possible to relate the variation of $\boldsymbol{\Sigma}(I_g)$ to dS_r

$$d\boldsymbol{\Sigma}(I_g) = -\boldsymbol{B}dS_r p^\beta \qquad (33)$$

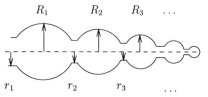

Figure 1: *Morphological model*

The hysteresis of the capillary curve is classically explained by the concepts of access radius and pore radius (Dullien 1992). The capillary pressure at which a gas-filled pore is being filled with liquid (imbibition) depends on the pore radius R_β whereas the capillary pressure at which a liquid-filled pore is being filled with gas (drainage) is controlled by the neck radius r_β (Figure 2).

Imbibition

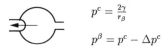

$$p^c = \frac{2\gamma}{R_\beta}$$
$$p^\beta = p^c$$

Drainage

$$p^c = \frac{2\gamma}{r_\beta}$$
$$p^\beta = p^c - \Delta p^c$$

Figure 2: *Sorbtion-desorbtion mechanism*

Taking (33) and the sorbtion-desorbtion mechanism described above into account, the differentiation of (32) yields the differential form of the macroscopic state equation:

If $dS_r > 0$:
$$p^\beta = p^c \quad \rightarrow \quad \delta\overline{\boldsymbol{\Sigma}} = \mathbb{C}^{hom} : d\boldsymbol{E}$$

If $dS_r < 0$: $\qquad\qquad\qquad\qquad\qquad (34)$
$$p^\beta \neq p^c \quad \rightarrow \quad \delta\overline{\boldsymbol{\Sigma}} - \boldsymbol{B}\Delta p^c(S_r)dS_r = \mathbb{C}^{hom} : d\boldsymbol{E}$$

with

$$\delta\overline{\boldsymbol{\Sigma}} = d\boldsymbol{\Sigma} + (dp^g - S_r dp^c)\,\boldsymbol{B} \qquad (35)$$

and where $\Delta p^c(S_r) = p^c(S_r, dS_r < 0) - p^c(S_r, dS_r > 0)$ denotes the value of the hysteresis in the sorbtion desorbtion curve (Figure 3).

According to the present micromechanical approach, we note that the state equation

$$\delta\overline{\boldsymbol{\Sigma}} = \mathbb{C}^{hom} : d\boldsymbol{E} \qquad (36)$$

proposed by (Coussy 1995) requires that the hysteresis in the capillary curve be negligible.

Consider a loading in which the capillary pressure is subjected to a cycle, whereas the macroscopic

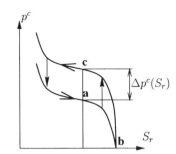

Figure 3: *Sorbtion-desorbtion curve*

stress Σ and the gas pressure p^g are kept constant. Starting from point **a** on the imbibition curve (Figure 3), the sample is first totally saturated (point **b**) and then dried such as the saturation reaches its initial value (point **c** on the drainage curve). The integration of equation (34) over this loading cycle yields an irreversible strain:

$$\Delta \boldsymbol{E} = \mathbb{C}^{hom-1} : \boldsymbol{B} \Delta p^c(S_r) S_r \qquad (37)$$

$\Delta \boldsymbol{E}$ appears as the mechanical counterpart of the hysteresis in the capillary pressure curve. Though the matrix behaves elastically, an inelastic behaviour is observed at the macroscopic scale. The macroscopic constitutive law when the porous space is saturated with liquid, obtained by putting $S_r = 1$ and $I_g = \emptyset$ in equation (32), writes $\Sigma + p^\ell \boldsymbol{B} = \mathbb{C}^{hom} : \boldsymbol{E}$. The comparison of this equation and (37) shows that the mechanical effect of the capillary hysteresis is the same as the one due to a variation $\Delta p = \Delta p^c(S_r) S_r$ of the pore pressure applied to a saturated sample at constant macroscopic stress.

Within the framework of the pore network of Figure 1, it is emphasized that the capillary pressure curves and the drained mechanical properties allow to characterize the mechanical behaviour in unsaturated conditions.

7 CONCLUSIONS

In a multiphase medium, the relationship (3) between the macroscopic total stress and the internal stresses at the microscopic level in a representative elementary volume must include a term taking into account the surface tensions in the interfaces between the different phases. Accordingly, in the macroscopic state equation (32), the effect of surface tension appears through a correcting term to be added to the classical Bishop "effective stress" with $\chi = S_r$ (Bishop, 1959). The hysteresis in the capillary pressure curve is incompatible with an (even non linear) poroelastic modelling. The interest of the micro-macro approach is to relate the capillary hysteresis to the inelastic nature of the macroscopic mechanical behaviour. Refering to a particular morphology, it has been shown that it is possible to link the macroscopic mechanical properties to the capillary pressure curve.

REFERENCES

Auriault, J. L. (1987). Nonsatured deformable porous media : quasistatics. *Transport in Porous media 2*, 45–64.

Auriault, J. L., Lebaigue, O., & Bonnet, G. (1989). Dynamics of two immiscible fluids flowing through deformable porous media. *Transport in Porous media 4*, 105–128.

Auriault, J. L. & Sanchez Palencia, E. (1977). Etude du comportement macroscopique d'un milieu poreux saturé déformable. *Journal de Mécanique 16*(4), 575–603.

Chateau, X. & Dormieux, L. (1995). Homogenization of a non-saturated porous medium: Hill's lemma and applications (in french with abridged english version). *C. R. Acad. Sci., t. 320, Série II b*, 627–634.

Chateau, X. & Dormieux, L. (1998). Comportement élastique d'un milieu poreux non saturé par homogénéisation. In C. Petit, G. Pijaudier-Cabot, & J. M. Reynouard (Eds.), *Ouvrages, Géomatériaux et Interactions - Modélisations multi-échelles*, Paris, pp. 337–352. Hermes.

Coussy, O. (1995). *Mechanics of porous continua*. New York: John Wiley.

Dullien, F. A. L. (1992). *Porous media - Fluid transport and pore structure*. London: Academic Press.

Fredlund, D. G. & Morgenstern, N. R. (1977). Stress state variables for unsaturated soils. *J. Geotech. Eng. Div. ASCE 103*(GT5), 447–466.

Gens, A. (1995). Constitutive laws. In A. Gens, P. Jouanna, & B. A. Schrefler (Eds.), *Modern issues in non-saturated soils*, Wien New York, pp. 129–158. Springer-verlag.

Houlsby, G. T. (1997). The work input to an unsaturated granular material. *Géotechnique 47*(1), 193–196.

Schrefler, B. A. (1990). Recent advances in numerical modelling of geomaterials, key note lecture. In *III EPMESC CONT., Meccanica*, Macau.

Schrefler, B. A. & Gawin, D. (1996). The effective stress principle: incremental or finite form. *Int. j. numer. anal. methods geom. 20*, 785–814.

Zaoui, A. (1997). Structural morphology and constitutive behaviour of microheterogeneous materials. In P. Suquet (Ed.), *Continuum micromechanics*, Wien New York, pp. 291–347. Springer.

Unsaturated Soils for Asia, Rahardjo, Toll & Leong (eds) © 2000 Taylor & Francis, ISBN 90 5809 139 2

Applicability of a general effective stress concept to unsaturated soils

F.Geiser
Soil Mechanics Laboratory, EPFL, Lausanne, Switzerland

ABSTRACT: Currently most experimental results on unsaturated soils are interpreted using the widely independent state variable approach. However it is believed that a single stress approach could lead to significant simplifications for practical purposes. In this paper the applicability of the effective stress based approach proposed by Khalili & Khabbaz (1998) is examined for the determination of the shear strength. The relationship is applied to unsaturated triaxial tests on remoulded silt, using both peak strength and critical state strength values. Other results from the literature are also analysed to confirm the applicability of the effective stress approach to predict shear strength of unsaturated soils.

1 CONTEXT

Generally the independent state variables approach is used to determine the shear strength of unsaturated soils (e.g. Fredlund et al. 1978). The total stress in excess of the pore-air pressure and the matric suction are considered separately and the shear strength is evaluated as:

$$\tau = c' + (\sigma - u_a)\tan\phi' + (u_a - u_w)\tan\phi^b \quad (1)$$

where σ is the total stress, u_a the pore-air pressure, u_w the pore-water pressure, c' the effective saturated cohesion, ϕ' the effective saturated angle of friction and ϕ^b the angle of friction with respect to the matric suction $(u_a - u_w)$.

However it is believed that an effective stress approach could lead to significant simplifications for practical purposes. It would reduce the model parameters and would eliminate the need for shear strength tests in an unsaturated state. In addition it would enable saturated and unsaturated states to be considered simultaneously in current geotechnical problems, as the transition from saturated to unsaturated state is straightforward.

Jennings & Burland (1962) have shown that the volume change and shear strength could not be related to a single effective stress, since the collapse phenomenon observed during wetting could not be explained with a single stress. Their arguments were subsequently widely used in numerous papers. However recently some authors have shown that with a complete elastoplastic model, it is possible to predict within an effective stress approach the main features of unsaturated soils, including the collapse (e.g.

Kogho et al. 1993, Modaressi & Abou-Bekr 1994, Bolzon et al. 1996).

2 EFFECTIVE STRESS CONCEPT

The effective stress is defined after Bishop (1959) as:

$$\sigma' = (\sigma - u_a) + \chi(u_a - u_w) \quad (2)$$

where χ is the effective stress parameter, which has a value of 1 for saturated soils and 0 for dry soils. The shear strength τ can then be evaluated simply as:

$$\tau = c' + \sigma'\tan\phi' \quad (3)$$

which is identical to the saturated relation and c' and ϕ' are the saturated shear parameters.

In the Bishop relationship (Eq. 2), χ is an empirical parameter representing the proportion of soil suction that contributes to the effective stress. Several attempts have been made to define this parameter χ in the past. Many authors have correlated χ to the degree of saturation, with no real success. The more recent approaches consider χ as a function of the suction (Kogho et al. 1993, Modaressi & Abou-Bekr 1994, Khalili & Khabbaz 1998). In these approaches, the importance of determining the air entry suction is always outlined.

The relationship established by Khalili & Khabbaz is tested in this paper. They propose a unique relationship between the effective stress parameter χ and the ratio of suction over the air entry value, based on the shear strength data of 14 soils reported in the literature. χ is expressed as:

$$\chi = \left[\frac{\left(u_a - u_w\right)}{\left(u_a - u_w\right)_e} \right]^{-0.55} \qquad (4)$$

where $\left(u_a - u_w\right)_e$ is the air entry suction.

3 APPLICABILITY OF THE SINGLE STRESS APPROACH TO A REMOULDED SILT

3.1 Test program and experimental equipments

An extensive experimental program has been performed on silt at the Swiss Federal Institute of Technology at Lausanne (Laloui et al. 1995, 1997, Geiser 1999). The studied soil is a washing silt from the region of Sion (Switzerland). Its characteristics are given in Table 1.

Table 1: Characteristics of the Sion silt

$w_L(\%)$	$w_P(\%)$	I_P	$\%<2\mu m$	$\%>60\mu m$	$\gamma_s(kN/m^3)$
25	17	8	8	20	27.4

The sample preparation is explained in Laloui et al. (1995). Standard experimental apparatus were adapted to run unsaturated tests with the air pressure technique. Air pressure is imposed at the top of the sample, while the pore-water pressure is measured or imposed at the bottom, where a special high air entry value ceramic is located. The suction s is then obtained as the excess of pore-air pressure u_a to pore-water pressure u_w.

The air entry value was obtained using the pressure plate method. The observed air entry suction s_e is between 50 and 80 kPa. A constant value of 65 kPa is used in the following.

3.2 Saturated critical state

This paper deals only with the shear behaviour of unsaturated soils. In order to discuss the predictions obtained with a single stress approach for unsaturated soil, the saturated results are first shown as reference.

Figure 1 shows critical state values of the deviatoric stress q plotted against the effective mean stress p' for around 30 triaxial shear tests performed at different stress levels. The critical deviatoric stress q is defined as:

$$q = Mp' \qquad (5)$$

The critical state slope M is equal to 1.3, which corresponds to a critical friction angle $\phi_{CSL} = 32.5°$. The grey zone corresponds to estimation of the critical friction angle of $\pm 1°$, which is accurate enough for most practical purposes. To enable comparison on the predictions of the shear strength with either the independent stress or the single effective stress

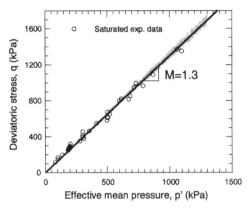

Figure 1. Saturated critical state (Sion silt)

approach, the same grey area is plotted on all following figures.

3.3 Peak resistance

Most existing experimental data come from suction controlled shear box and describe the peak value evolution with the suction rather than the critical state. Consequently the peak resistance evolution is first analysed for the Sion silt.

Several normally consolidated triaxial shear tests were conduced at different suction levels. The points representing the peak strength versus the corresponding net mean pressure p-u_a at different suction levels are plotted on Figure 2 for the entire unsaturated tests. The saturated case is represented by the critical state line (CSL) as the peak strength and the critical state are assumed to be the same in normally consolidated tests. Lines on the figure represent the peak strengths points determined at various suction levels.

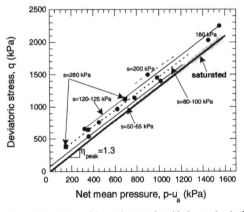

Figure 2. Evolution of the peak strengths with the suction in the $(p$-$u_a)$-q plane (Sion silt)

102

The present data on the Sion silt show that the apparent cohesion increases with the suction, while the friction angle remains almost. The increase of the apparent cohesion with the suction has been observed on several soils in the literature (e.g. Wheeler & Sivakumar 1995, Cui & Delage 1996) in this net mean pressure representation. It can be observed that for the tests conducted at higher suction levels (s=280 kPa, $S_r \approx 25$ %), the deviatoric stress starts to decrease. This can be explained as for highly unsaturated sample, a clear brittle failure was observed before the samples could reach the peak value.

The same peak strengths are plotted on Figure 3 in an effective stress approach (see Eq. 2 & 4). It can be observed that this time most points are closer to the critical state line defined in Equation 5. However some points are slightly above the line, suggesting an increase of the peak value with suction, except for the case where a clear early brittle failure was observed. This increase in the peak strength can be explained in terms of the increase in the preconsolidation pressure with suction and the overconsolidated response of the soil.

3.4 Critical state

Suction may affect the peak strength, however the critical state should be uniquely defined if the effective stress approach is correct.

In an independent stress approach the critical state line evolves similarly to what was observed for the peak strengths: a clear increase of the apparent cohesion with the suction can be observed for the Sion silt (Geiser 1999).

Figure 4 shows critical state values of the deviatoric stress q plotted against the effective mean stress p' for triaxial shear tests performed at different suction levels. They are less experimental points than on the previous figures, as the critical state was not al-

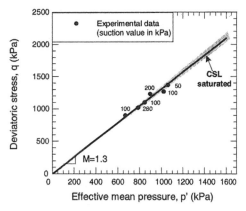

Figure 4. Evolution of the critical state with the suction in the p'-q plane (Sion silt).

ways clearly reached for all the tests. A unique critical state line (CSL) can be observed for this soil in the range of the suctions tested.

3.5 Influence of the parameter choice in the Khalili & Khabbaz relationship

Figure 5 shows for the Sion silt the influence of the determination of the air entry suction s_e on the prediction of the shear strength. Experimentally it was established that s_e was between 50 and 80 kPa for the Sion silt. It can be observed that in this range of suction the effect of a change in the air entry evaluation has a small effect. The maximal error for those tests would lead to an overestimation around 9 % of the deviatoric stress, an error which is still acceptable. However this illustrates that it may be possible to determine experimentally the air entry value with sufficient degree of accuracy for all practical purposes using for example the pressure plate or the filter paper technique.

In Equation 4, the exponent was determined statistically as equal to 0.55 on the basis of numerous experimental data. After Khalili & Khabbaz, the observed coefficient ranges from 0.4 to 0.65. Figure 6 shows the influence of this range on the predictions. The maximal observed error for the Sion silt is around 8 % of the deviatoric stress, which is once again acceptable.

4 CRITICAL STATE EVOLUTION FOR OTHER SOILS

To confirm these observations, results from the literature were also analysed with this effective stress approach. While numerous researches focus on the shear strength evolution with the suction, only a few

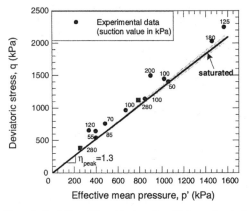

Figure 3. Evolution of the peak strengths with the suction in the p'-q plane.

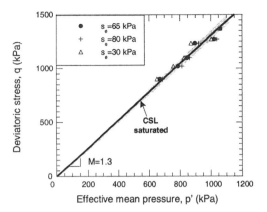

Figure 5. Influence of the determination of the air entry suction s_e on the prediction of the shear strength.

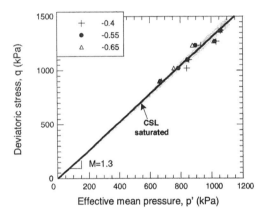

Figure 6. Influence of the coefficient in Equation 4 (s_e=65 kPa).

have examined the critical state evolution. The three analysed data were: a kaolin (Wheeler & Sivakumar, 1995), the Trois-Rivières silt (Maâtouk et al. 1995) and the Jossigny silt (Cui & Delage 1996).

4.1 Kaolin

In 1995 Wheeler & Sivakumar published results on an unsaturated compacted speswhite kaolin. The critical state hyperline were defined in a net mean stress versus deviatoric stress plane. Both the critical state slope and the cohesion were observed to be function of the suction.

The same results are represented on Figure 7 with an effective stress approach. As the air entry suction has not been experimentally established for this soil, it has to be estimated. On the base of the published consolidation results on the same soil, a change in compressibility is observed at a suction level of 100 kPa, suggesting that the air entry value must be less than 100 kPa (Sivakumar 1993). Therefore the air

entry value of 85 kPa backcalculated by Khalili and Khabbaz (1998) is adopted in this analysis.

As can be noted, most of the points plotted on Figure 7 are close to the saturated CSL, confirming the applicability of an effective stress approach.

4.2 Jossigny silt

Cui & Delage presented several triaxial results on the Jossigny silt in 1996. Vicol (1990) has studied the hydric behaviour of this soil. An air entry value around 85 kPa was observed and is chosen in the following.

Figure 8 shows critical state values of the deviatoric stress q plotted against the effective mean stress p' for triaxial shear tests performed at different suction levels. For some tests a clear fall in deviatoric stress was observed after the peak suggesting a possible brittle failure and that the critical state line may not have been reached; those points are outlined by hollow symbols. As there are only two results on saturated soil at low pressure level, both the CSL fitting best to those saturated tests and to the whole tests are represented. The experimental results deviate more from the CSL than for the previous soils. However, considering the range of the tested suctions, the predicted deviatoric stresses with the effective stress approach are still satisfactory.

4.3 Trois-Rivières silt

Finally results obtained on the Trois-Rivières silt are analysed (Maâtouk et al. 1995). Their critical state envelope at various suction in an independent stress approach show an increase in apparent cohesion with suction, while the critical state slope decreases.

Once again the air entry suction was not determined in a laboratory. A value of s_e=10 kPa was chosen on the basis of the shear strength tests. It is not possible to independently confirm this value on the basis of the existing results as no test has been conducted on this soil for suctions under 80 kPa. However the isotropic consolidation tests show that the air entry value must be less than 80 kPa. Figure 9 shows that all the experimental points are close to the CSL.

5 CONCLUSIONS

The applicability of the single effective stress relationship proposed by Khalili & Khabbaz (1998) has been tested. The quantitative prediction of the shear strength on an unsaturated remoulded silt is good especially for critical state. Additionally a unique critical state line can be observed for saturated and unsaturated state. The observations are confirmed for three other soils.

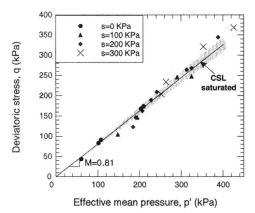

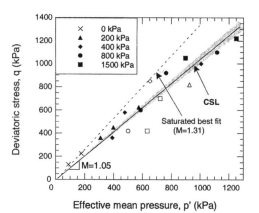

Figure 7. Evolution of the critical state with the suction in the p'-q plane (kaolin).

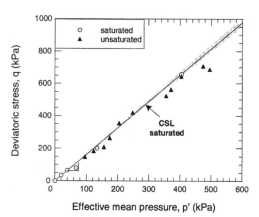

Figure 8. Evolution of the critical state with the suction in the p'-q plane (Jossigny silt).

Figure 9. Evolution of the critical state with the suction in the p'-q plane (Trois-Rivières silt).

The results show that a single stress approach is applicable for the prediction of unsaturated soil shear strength. To be complete and definitively prove the validity of this effective stress approach, it has yet to be applied for the prediction of the volumetric behaviour.

This approach, if proven to be valid, is of great interest for practical purposes, as it offers significant simplifications for constitutive modelling of unsaturated soils.

REFERENCES

Bishop, A.W. 1959. The principle of effective stress. *Tecknisk Ukelbad.* 39:859-863.

Bolzon, G., B.A. Schrefler & O. Zienkiewicz 1996. Elastoplastic soil constitutive laws generalised to partially saturated states. *Géotechnique* 46(2): 279-289.

Cui, Y.J. & P. Delage 1996. Yielding and plastic behaviour of an unsaturated compacted silt. Géotechnique 46(2): 291-311.

Fredlund, D.G, N.R. Morgenstern & R.A. Widger 1978. The shear strength of unsaturated soils. *Can. Geotech. J.* 15(3): 313-321.

Geiser, F. 1999. *Comportement mécanique d'un limon non saturé: étude expérimentale et modélisation constitutive.* PhD Thesis, Swiss Federal Institute of Technology, Lausanne.

Jennings, J.E.B. & J.B. Burland 1962. Limitations to the use of effective stresses in partly saturated soils. *Géotechnique* 12(2): 125-144.

Khalili, N. & M.H. Khabbaz 1998. A unique relationship for χ for the determination of the shear strength of unsaturated soils. *Géotechnique* 48(5): 681-687.

Kogho, Y., M. Nakano & T. Myazaci 1993. Theoretical aspects of constitutive modelling for unsaturated soils. *Soils and Foundations* 33(4): 49-63.

Laloui, L., L. Vulliet & G. Gruaz 1995. Influence de la succion sur le comportement mécanique d'un limon sableux. . In Alonso & Delage (eds), *Proc. Firs Int. Conf. on Unsaturated Soils, Paris: 133-138.* Rotterdam: Balkema.

Laloui, L., F. Geiser, L. Vulliet, X.L. Li, R. Charlier. & A. Bolle 1997. Characterization of the mechanical behaviour of an unsaturated sandy silt. *Proc. 14th Conf. on Soil Mech. and Found. Eng., Hambourg:* 347-350. Rotterdam: Balkema.

Maâtouk, A., S. Leroueil & P. La Rochelle 1995. Yielding and critical state of collapsible unsaturated silty soil. *Géotechnique* 45(3): 465-477.

Modaressi, A. & N. Abou-Bekr 1994. A unified approach to model the behavior of saturated and unsaturated soils. In Siriwardane (ed), *Proc. 8th Int. Conf. on Computer Methods and Advances in Geomech., Morgantown:1507-1513.* Rotterdam: Balkema.

Vicol, T. 1990. *Comportement hydraulique et mécanique d'un limon non saturé: application à la modélisation.* PhD Thesis, Ecole Nationale des Ponts et Chaussées, Paris.

Wheeler, S. J. & V. Sivakumar 1995. An elasto-plastic critical state framework for unsaturated soil. *Géotechnique* 45(1): 5-53.

Unsaturated Soils for Asia, Rahardjo, Toll & Leong (eds) © 2000 Taylor & Francis, ISBN 90 5809 139 2

A neural network framework for unsaturated soils

G. Habibagahi, S. Katebi & A. Johari
Department of Civil Engineering, Shiraz University, Iran

ABSTRACT: In this paper, a neural network approach is used to describe the mechanical behavior of unsaturated soils. The network has a sequential architecture, that is, a multi-layer perceptron network with feedback capability. The input layer consists of ten neurons. Four of the input neurons, namely, initial water content, dry density, degree of saturation and suction represent initial soil conditions. The remaining six neurons, namely, axial strain, volumetric strain, deviatoric strain, net mean pressures with respect to air and water pressure, and change in suction are continuously updated for each increment of axial strain based on outputs from the previous increment. The output layer consists of three neurons representing values of deviatoric stress, volumetric strain, and change in suction at the end of each increment. Next, a database was developed from triaxial test results available in literature. The database was used to train and test the network. Neural network simulations were compared with experimental results. The comparison indicates the good performance of the proposed network for predicting mechanical behavior of unsaturated soils.

1 INTRODUCTION

Limitations in describing mechanical behavior of unsaturated soils based on a single effective stress equation, similar to the one proposed by Bishop & Donald (1961), has led to different approaches for modeling the observed behavior of these soils. Consequently, stress-strain behavior of unsaturated soils has been topic of numerous investigations in recent years. Among important contributions are the works of, Matyas & Radhakrishua (1968), Fredlund and Morgenstern (1977), Toll (1990) and Alonso et al (1990). Fredlund and Morgenstern (1977) proposed the idea of a "state surface " defined in terms of independent state variables such as $(\sigma - U_a)$ and $(U_a - U_w)$. Using these state variables, suitable state surfaces were defined for the stress-strain-strength behavior of unsaturated soils. Alonso et al (1990) and Toll (1990) proposed a critical state framework by considering the effect of total stress and suction separately. Though successful in presenting the critical state parameters in terms of degree of saturation or soil suction, the models are as yet incapable of dealing with many aspects of unsaturated soil behavior. Some of these aspects include dependence of collapse potential on confining stress and hysteresis effects involved in wetting and drying of the soils.

Neural networks (NN) have received considerable attention in recent years with wide range of applications in civil engineering. Some examples are the works of Goh (1994), Goh (1995) and Habibagahi (1998). Neural networks have also been employed for constitutive modeling. Ellis et al (1995) employed NN to model stress-strain behavior of sands based on results obtained from triaxial compression tests. Ghaboussi and Sidarta (1998) proposed an adaptive neural network for constitutive modeling. They tested the proposed network on results derived from drained compression tests prepared from Sacramento River sand. More recently, Penumadu and zhao (1999) used NN to model drained behavior of sand and gravel.

In this paper, a multi-layer perceptron network with feedback is used to model the stress-strain behavior of unsaturated soils. Test results from triaxial tests on unsaturated soils available in literature were used and the network was trained and tested using the database thus developed. Details of the network and the procedure involved are described in the following sections.

2 NEURAL NETWORK

In multi-layer perceptron network, artificial neurons are arranged in layers with full connections of each neuron to all neurons of the next layer. This type of network consists of an input layer, a hidden

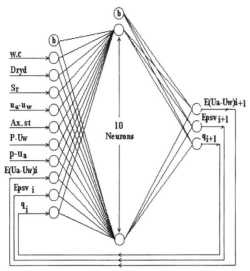

Figure 1. Architecture of neural network used.

layer (or hidden layers), and an output layer. The neurons in the input layer represent the number of input variables considered, while the output neurons identify the desired outputs. Each neuron in the network has an activation function, usually expressed by a sigmoid function. Weights are assigned to all the connections inside the network and the goal is to find the optimum values of these weights such that the error measure of the network is minimized. The error measure is expressed in terms of the mean sum squared of the errors (MSSE) expressed by

$$\text{MSSE} = \frac{1}{NK} \sum_{p=1}^{N} \sum_{i=1}^{K} (t_i^p - o_i^p)^2$$

where t_i and o_i are the target and network values for the pattern p and output node i, respectively. N and K are the number of patterns (data sets) in the database and the number of output neurons, respectively. A multi-layer perceptron neural network is basically identified by its architecture, the way the neurons are arranged inside the network, and a training rule. A database is also required for training and testing the network. For the network considered in the present study, these components are further described in the subsequent sections.

2.1 Database

Triaxial test results on Lateritic gravel reported by Toll (1988) were used to develop the necessary database. Table 1 indicates the basic properties of the soil. Results from twenty-three unsaturated specimens prepared using static and dynamic compression were available. However, for the sake of consistency, only nineteen specimens prepared with static compression were considered in this investigation. Table 2 presents the specimens and their initial conditions. Specimens MGU17 through MGU20 were prepared by a different procedure (dynamic compression) and hence, they were not included in Table 2. Triaxial test results reported on these specimens were then digitized to obtain the necessary database. For digitization, an incremental value of axial strain equal to 0.08% was adopted. The database thus developed had a total of 1643 patterns.

2.2 Architecture

As indicated in Figure 1, a three-layer network with feedback capability was used in this study. Ellis et al (1995) and Penumadu and Zhao (1999) had successfully employed this type of network to model stress-strain behavior of sand and gravel.

The input layer had ten neurons. The first four input neurons, namely, soil water content, dry density, degree of saturation and suction pressure represent the initial condition of the specimens before shearing. The other six neurons, namely, axial strain, change in suction pressure, mean effective stress with respect to pore air and water pressures, volumetric strain, and deviatoric stress are input variables that are to be updated incrementally during the training and testing of the network based on the outputs from the previous increment of training or testing.

The output layer had three neurons representing deviatoric stress, volumetric strain and change in suction at the end of current increment. To find the optimum number of hidden neurons, they were increased from a minimum of 5 neurons and checking the error measure of the network. This resulted in a total number of 10 neurons for the hidden layer. Among the input variables that are updated incrementally, the deviatoric stress (q), volumetric strain, and suction change were updated explicitly by feeding back the corresponding incremental values from the output layer. However, the other two variable, namely, the effective mean pressure with re-

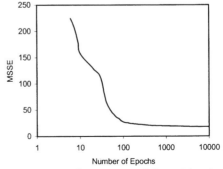

Fig. 2. Variation of error measure during training phase

Table 1. Basic properties of the specimens (From Toll (1988))

Liquid limit	56-66%	Clay fraction	8-9%
Plastic limit	27-31%	Specific gravity	3.2
Plasticity index	29-35%		

				Grading					
Sieve size(mm)	25	9.5	5.0	2.5	1.0	0.63	0.40	0.30	0.15	.063
%Passing	99	96	67	50	31	26	21	19	17	15

Table 2. Initial condition of soil specimens (From Toll (1988))

Sample	water content %	Dry density Mg/m^3	Degree of saturation %	Initial suction kPa	Ua kPa	σ_3 kPa	Neural Network Pattern
MGU1	19.6	1.442	51.4	384	496	552	Train
MGU2	25.5	1.632	84.9	4.0	247	302	Train
MGU3	20.8	1.531	61.1	149	297	350	Train
MGU4	21.4	1.551	64.4	22	227	300	Train
MGU5	20.7	1.646	70.2	105	300	353	Train
MGU6	21.0	1.489	58.5	256	445	500	Train
MGU7	17.0	1.474	46.5	450	450	500	Train
MGU8	21.1	1.587	66.4	186	304	352	Train
MGU9	18.9	1.625	62.4	407	400	451	Test
MGU10	25.1	1.508	71.6	11	300	350	Train
MGU11	24.9	1.506	70.8	26	299	350	Train
MGU12	21.9	1.716	81.0	161	300	350	Test
MGU13	20.0	1.674	70.2	547	500	550	Train
MGU14	26.0	1.706	95.0	5	299	350	Train
MGU15	25.0	1.702	90.9	12	300	399	Train
MGU16	25.9	1.709	95.0	0.0	302	503	Train
MGU21	26.3	1.498	74.1	2.0	300	350	Test
MGU22	24.3	1.708	89.0	78	450	473	Train
MGU23	25.8	1.705	94.2	54	299	324	Train

spect to air and water pressures were updated implicitly from the outputs considering the fact that the total stress path has a constant slope during triaxial compression tests (dp/dq=1/3) and also taking into account the fact that the pore air pressure was kept constant during each test.

2.3 Training and Testing

The network was trained using a generalized delta rule algorithm (back-propagation of error) proposed by Rumelhart et al (1986). The learning rate was varied during training from a value of 0.1 to 0.35. In order to avoid trapping in local minima a momentum factor was also introduced.

From the 19 triaxial tests used in this study, sixteen tests were used for training and the remaining three tests were used for testing the network. Consequently, from a total of 1643 patterns generated, 1418 patterns were used as training patterns. The remaining 225 patterns were used for testing the network. Table 2 indicates the specimens that were used for training and testing the network.

The network error measure decreased rapidly with the number of epochs of training. Fig. 2 indicates the variation of error measure during training. Each epoch is equal to one cycle of presentation of all the training patterns.

3 RESULTS

After training, the network was used to simulate the stress-strain, volumetric strain and change in suction of unsaturated soils during triaxial compression test. The network was used to simulate the behavior of all the 19 triaxial tests. Figure 3 shows stress-strain and volume change behavior as well as change in suction predicted for a typical specimen used in training of the network (MGU23). From this figure it may be concluded that the network has a good potential for predicting the behavior of unsaturated soils with reasonable accuracy. Similarly, Figure 4 presents the prediction of network for a typical test specimen (MGU9). From this figure it may be concluded that the network is also capable of simulating new test results, though not as good as previous case. The two specimens used to indicate the performance of the network are "typical" among the network predictions for "training" and "testing" specimens. These results indicate the powerful feature of the neural networks to learn and to predict the mechanical behavior of unsaturated soils without making any assumption or simplifications.

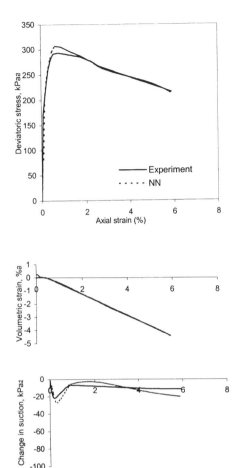

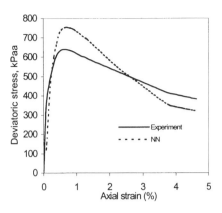

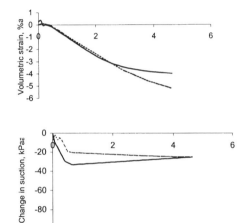

Figure 3. Typical simulation of training data (MGU 23)

Figure 4. Typical simulation of testing data (MGU9)

4 CONCLUSION

A neural network with feedback from the outputs was used to simulate the mechanical behavior of unsaturated soils subjected to the conventional triaxial stress path. It was shown that the neural network is capable of simulating complex behavior of unsaturated soils without making any assumption and/or simplifications. The network could predict the behavior of unsaturated soil specimens reasonably well both for specimens used in training as well as test specimens that the network had not been exposed to before. However, the authors believe that the present network may not be an optimum network for such a complex problem. Currently, more research is pursued to improve the performance of the network by changing the input –output combinations as well as trying other probable networks and architectures. Once trained, the network may also be used as a tool to assess the influence of various parameters and to get better understanding of the mechanical behavior of unsaturated soils. This can be achieved by carrying out a parametric study on the trained network.

5 ACKNOWLEDGEMENTS

The authors wish to express their gratitude to the Research Council of Shiraz University for providing financial support under research project No. 77-EN-1070-624.

6 REFERENCES

Alonso, E.E, Gens, A. and Josa, A. 1990. A constitutive model for partially saturated soils. *Geotechnique*, 40(3), 405-430.
Bishop, A.W. and Donald, I.B. 1961. The experimental study of partly saturated soil in the triaxial apparatus. *Proc 5th Int. conf. Soil Mech. Fdn Engineering*, 1, 13-21.

Ellis, G.W., Yao, C., Zhao, R. and Penumadu, D. 1995. Stress-strain modeling of sands using artificial neural networks. *J. of Geotech. Engineering*, ASCE, 121(5), 429-435.

Fredlund, D.G. and Morgenstern, N.R. 1977. Stress- strain variables for unsaturated soils. *J. Geotech. Engineering Div.* ASCE, 103, GTS, 447-446.

Ghaboussi, J. and Sidarta, D.E. 1998. New nested adaptive neural networks (NANN) for constitutive modeling. *Computers and Geotechnics*, 22(1), 29-52.

Goh, A.T.C. 1994. Seismic liquefaction potential assessed by neural network. *J. of Geotech. Eng.*, ASCE, 120(9), 1467-1480.

Goh, A.T.C. 1995. Modeling soil correlations using neural networks. *J. of computing in civil Eng.*, ASCE, 9(4), 275-278.

Habibagahi, G. 1998. Reservoir induced earthquakes analyzed via radial basis function networks. *Soil Dynamics and Earthquake Eng.*, 17(1), 53-56.

Matyas, E.L. and Radhakrishna, H.S. 1968. Volume change characteristics of partially saturated soils. *Geotechnique*, 18(4), 432-448.

Penumadu, D. and Zhao, R. 1999. Triaxial compression behavior of sand and gravel using artificial neural networks (ANN). *Computers and Geotechnics*, 24, 207-230.

Rumelhart, D.E., Hinton G.E. and Williams, R.J. 1986. Learning internal representation by error propagation. In *Parallel Distributed Processing*, 1, Cambridge: MIT Press.

Toll, D.G. 1988. The behavior of unsaturated compacted naturally occurring gravel. *Ph.D. thesis*, University of London.

Toll, D.G. 1990. A framework for unsaturated soil behavior. *Geotechnique*, 40(1), 31-44.

111

Unsaturated Soils for Asia, Rahardjo, Toll & Leong (eds) © 2000 Taylor & Francis, ISBN 90 5809 139 2

Study of the influence of adhesion force on deformation and strength of unsaturated soil by DEM analysis

S. Kato
Department of Civil Engineering, Kobe University, Japan

S. Yamamoto
Obayasi Construction Company Limited, Japan

S. Nonami
Oyotisitsu Company Limited, Japan

ABSTRACT:In this study, using the distinct element method analysis, simulations of the biaxial compression test for two-dimensional granular material is carried out. In this analysis, the problems encountered in the triaxial compression for unsaturated soil are canceled. The influence of meniscus water in unsaturated soil is reproduced by introducing constant intergranular adhesive force that acts perpendicular to the tangential plane at contact point. The influence of the intergranular adhesive force on the stress-strain relation and the strength constants are examined. From the analytical results, it is found out that the internal friction angle and the cohesion of the assembly of the disk particle increases with the inrergranular adhesive force.

1 INTRODUCTION

In the case of the triaxial test for unsaturated soil, it is not possible to determine the volume change of the specimen based on the displacement, because pore air exists inside the specimen. The double cell method (Bishop and Henkel, 1957) is often used to obtain the volume change. However, in the double cell method, the unobstructed capacity of the inner cell varies with the lateral pressure changes. Accordingly, a correction to the measured change of the unobstructed capacity for the lateral pressure change is required in order to determine the volume change of the specimen. The correcting values have to be obtained before the test, and the test results are arranged by applying them. It is considered that to obtain test results with reproducibility, a degree of skill is needed, and even then the specimen installation has effect on the test results obtained.

In the triaxial compression test for unsaturated soil, the axis translation technique is often used to apply suction to a specimen. The capacity of the suction applied is limited by the air entry value (AEV) of the ceramic disk used. It is possible to use a ceramic disk of about AEV=1500kPa in practice, but a ceramic disk of about AEV=500kPa is mainly used, since the coefficient of the permeability of the ceramic disk decreases with the increase of AEV. Consequently, as a testing condition, the suction from 0 to about 500kPa is often applied to the specimen.

The intergranular adhesive force acting between particles does not correspond to the suction value applied. It is known that the relationship between

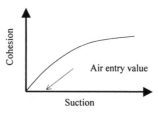

Figure 1 Relationship between cohesion and suction

suction and the cohesion obtained from the triaxial compression test result is approximated by the hyperbola as shown in figure 1 (Fredlund et al.,1995) . As shown in this figure, the cohesion increases proportionally with suction, until the suction reaches the AEV. When the suction exceeds the AEV, the cohesion approaches some constant value with increase in the suction. Generally, 100kPa or less represents the bounds for the value of the AEV in an unsaturated soil. Under a suction of about 500kPa, the rate of increase of the cohesion considerably decreases. This non-linearity means that the intergranular adhesive force acting between particles does not correspond to the suction value applied. Consequently, it is difficult to reproduce the condition in which large intergranular adhesive force acts, with the triaxial compression test apparatus. Accordingly, there exists a limitation in clarifying the effect of the intergranular adhesive force by the triaxial compression test using the axis translation technique.

Void ratio influences the strength and the rigidity of the soil. Accordingly, it is desirable that a soil is tested under the condition in which only the suction

Table 1 Parameters and material properties of the sample

	Between particles	Between particle and platen
Normal spring constant (N/m/m)	0.9×10^{10}	1.8×10^{10}
Tangential spring constant (N/m/m)	3.8×10^{8}	6.0×10^{8}
Normal damping constant (N/m/m)	7.9×10^{5}	1.1×10^{5}
Tangential damping constant (N/m/m)	1.4×10^{4}	2.0×10^{4}
Friction coefficient (deg.)	16	10
Material density (kg/m³)	2700	

differs and the void ratio is the same, for clarifying the effects of suction. However, considerable skill is required to adjust such condition, because void ratio also changes when suction varies.

The triaxial compression test for unsaturated soil using the axis translation technique has the problems mentioned above. Consequently, the triaxial test for unsaturated soil itself is difficult to carry out, and the test results are fewer than those for saturated soil. There are some aspects for which the effect of suction has not been clarified experimentally.

In this study, a biaxial compression test for two-dimensional granular material is simulated by distinct element method analysis. In this analytical method, the volume change of the specimen is clear, and the size of the intergranular adhesive force can also be optionally set. It seems to remove many of the problems in triaxial compression test for unsaturated soil mentioned above.

Based on the analytical results, the effects of the intergranular adhesive force on deformation and strength characteristics of an assembly of disk particles are examined, and they are compared with the tendencies observed in the triaxial compression test results for unsaturated soil.

2 ANALYSIS METHOD AND ANALYSIS CONDITION

2.1 Outline of DEM analysis with intergranular adhesion force

The distinct element method (DEM) is one of the discontinuous corpora analysis method proposed by Cundall(1971).

Unlike continuum analysis of the finite element method or the boundary element method, it is suitable for analyzing the dynamic behavior of the granular material. In the DEM analysis, simple dynamic models (they are generally the Voigt model and Coulomb's friction rule) are introduced at contact points and contacting surfaces between the disk particles with the assumption that the particles are rigid. An independent equation of motion at every element is solved forwardly in the time domain, and interactions between particles and deformation of particle aggregates are traced. This method has the merit that necessary output data such as stress, strain, and rotation angle of some disk particles, and the setting of boundary conditions are easy.

In this study, the analysis is carried out using the granular material distinct element analysis system (GRADIA) programmed by Yamamoto(1995).

The effect of meniscus water observed in unsaturated soil is expressed by introducing an intergranular adhesive force described below.

The intergranular force acting between contacting particle i and j is denoted as P_{ij}. The x-direction component, y-direction component, and moment component of the resultant force of the intergranular adhesive force at time t are given by the following equations considering that the direction of P_{ij} is normal for the line direction of the particle tangent plane.

$$[F_{x_i}^P]_t = \sum_j P_{ij} \cos \alpha_{ij} \qquad (1)$$

$$[F_{y_i}^P]_t = \sum_j P_{ij} \sin \alpha_{ij} \qquad (2)$$

$$[M_i^P]_t = 0 \qquad (3)$$

Where, $[F_{x_i}^P]_t, [F_{y_i}^P]_t, [M_i^P]_t$ are the x-direction component, y-direction component and moment component of the resultant force of the intergranular adhesive force and α_{ij} is the angle between the normal direction of the contact plane and the x-axis.

It is possible to introduce an intergranular adhesive force by deducting these components from each component of the total intergranular force except for the intergranular adhesive force. By integrating these equations of motion for each particle by the Euler method with respect to time, the analysis was carried out. Consequently, the solution becomes stabilized conditionally on the integral time increment.

Cundall(1971) proposed the following equation for the integral time increment.

$$\Delta t < \Delta t_c = 2\sqrt{\frac{m}{k}} \qquad (4)$$

Where, k is mass of the disk particle, and m is the spring constant.

It has been found experientially that sufficient stability and accuracy can be ensured in the quasistatic problem at about $1/10 \Delta t_c$, though equation (4) is deduced from the equation of motion for a single-degree-of-freedom system. But taking this equation as a standard, the integral time increment must be decided by trial and error. In this study, $\Delta t=1/10\Delta t_c$ $=5\times10^{-5}$ sec. is used.

The parameters and the material properties necessary for the analysis are listed in Table 1.

Yamamoto(1995) carried out the simulation, in which disk particles are arranged in a similar position to a biaxial compression test carried out for aluminum bars, using the material property shown in Table 1, and confirmed that both macroscopic internal frictional angles were about 22~24 degrees and agreed with each other.

2.2 Analytical condition

As an analytical model, a specimen in whom two kinds of particles of 5mm and 9mm diameters were placed randomly at the proportion of mixing weight ratio of 3:2, was used. These were 6372 5mm particles and 1269 9mm diameter particles. The sum of them was 7641.The shape of the specimen was a rectangle of about 582mm height and about 438mm width as shown in figure 2. The four edges of the rectangle were surrounded by the rigid line elements that correspond to the loading platens. Both of the side rigid line elements were controlled in order that the lateral stress was constant, and their vertical movements were restricted. The bottom rigid line element was also fixed. When the shear process of the biaxial compression test was simulated, the upper rigid line element was moved down perpendicularly at a constant rate of 10cm/s.

In this study, firstly, without the intergranular adhesive force, the specimen was compressed isotropically under a confining pressure of 49 kPa. Then, the desired intergranular adhesive force was introduced between particle contact points. Afterwards, the shear process of the biaxial compression test was simulated under a confining pressure of 49 kPa, or the specimen was compressed isotropically to σ_3 =490 kPa, and the shear process was simulated under confining pressure of σ_3 =490 kPa. In addition, a isotropic compression test to σ_3=980kPa compressive stress was simulated with an intergranular adhesive force of P=0.0N, 9.8N and 49N in order to observe the effect of the intergranular adhesive force on the compression process.

3 ANALYTICAL RESULTS AND DISCUSSIONS

3.1 Effect of intergranular adhesive force on compression characteristic

In the isotropic compression process of an unsaturated soil under constant suction, the void ratio change during the compression decreases with an increase of the suction. This phenomenon means that the stiffness of the soil skeleton increases with increase of the suction, and that the compression index decreases. Taking void ratio at a compression pressure σ_c =49kPa as a standard, figure 3 shows plots

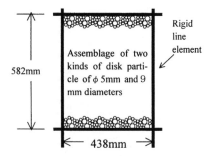

582mm — Assemblage of two kinds of disk particle of ϕ 5mm and 9 mm diameters

Rigid line element

438mm

Figure 2 Schematic diagram of the specimen analyzed

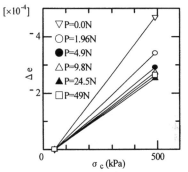

$[\times10^{-4}]$

$\triangledown$ P=0.0N
$\bigcirc$ P=1.96N
$\bullet$ P=4.9N
$\triangle$ P=9.8N
$\blacktriangle$ P=24.5N
$\square$ P=49N

Δ e

σ_c (kPa)

Figure 3 Void ratio change, Δ e to σ_c =490kPa

of the void ratio change, Δ e to σ_c =490kPa for every intergranular adhesive force P. It is found that the void ratio change decreases when the intergranular adhesive force increases. This result is similar to that observed in the isotropic compression test for unsaturated soil under constant suction, because the larger intergranular adhesive force condition corresponds to the higher suction condition.

Karube et al. (1986) evaluated the effect of the suction as a suction stress, and suggested that the plastic volumetric strain in the isotropic compression is determined by the following equation.

$$\varepsilon_v^p = \frac{\lambda-\kappa}{1+e_0}\ln\frac{p_{net}+p_s}{p_{net_0}+p_s} \tag{5}$$

Where, ε_v^p is the plastic volumetric strain, λ,κ are the compression and swelling indexes, e_0 is the referential void ratio, p_{net}, p_{net_0} are the net stress and the referential net stress and p_s is the suction stress.

Kato (1997) proposed the next equation in which the converted normal net stress is taken into account.

$$\varepsilon_v^p = \frac{\lambda-\kappa}{1+e_0}\ln\frac{p_{net}+\sigma_0}{p_{net_0}+\sigma_0} \tag{6}$$

Where, σ_0=ccot ϕ is the converted normal net stress and c, ϕ are the cohesion and the internal frictional angle that are derived from the triaxial compression test result under constant suction.

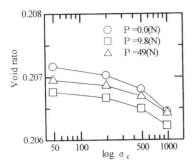

Figure 4 Void ratio arranged with log σ_c

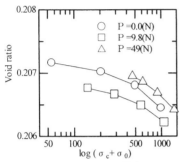

Figure 5 Void ratio arranged with log($\sigma_c + \sigma_0$)

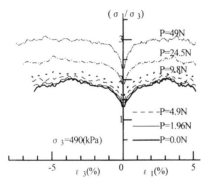

Figure 6 $\sigma_1/\sigma_3 \sim \varepsilon_1 \sim \varepsilon_3$ for the case of σ_3=49kPa

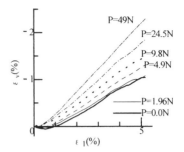

Figure 7 $\varepsilon_1 \sim \varepsilon_3$ relation for the case of σ_3=49kPa

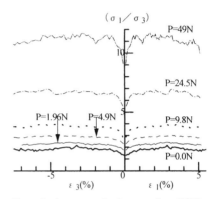

Figure 8 $\sigma_1/\sigma_3 \sim \varepsilon_1 \sim \varepsilon_3$ for the case of σ_3=490kPa

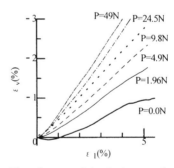

Figure 9 $\varepsilon_1 \sim \varepsilon_3$ relation for the case of σ_3=490kPa

Features of equations (5) and (6) are that the plastic volumetric strain depend on the constant compression index, λ and the constant swelling index, κ regardless of the suction condition. The later condi-

Table 2 Comparison of compression indexes

Intergranular adhesive force (N)	Compression index by log σ_c	Compression index by log($\sigma_c + \sigma_0$)
0.0	12.021×10^{-4}	12.134×10^{-4}
9.8	8.963×10^{-4}	10.192×10^{-4}
49	8.950×10^{-4}	14.303×10^{-4}

tion has recently achieved general acceptance in some constitutive models, (for example, Alonso et al. (1990)) and its validity has been confirmed. In the present analytical results, we examined whether the constant compression index condition is observed in spite of the intergranular adhesive force.

Figures 4 and 5 show plots of void ratio with log σ_c and log($\sigma_c + \sigma_0$) respectively for the cases of P=0.0, 9.8 and 49N. Table 2 shows the comparison of the compression indexes based on the void ratio change from σ_c = 490kPa to 980kPa for different intergranular adhesive forces. From the result shown in Table 2, when the compression indexes in each

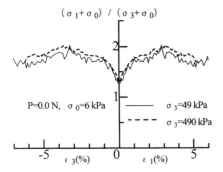

Figure 10 $(\sigma_1+\sigma_0)/(\sigma_3+\sigma_0)\sim\varepsilon_1\sim\varepsilon_3$ relation for the case of P=0.0N (σ_0=6kPa)

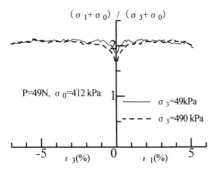

Figure 11 $(\sigma_1+\sigma_0)/(\sigma_3+\sigma_0)\sim\varepsilon_1\sim\varepsilon_3$ relation for the case of P=49N (σ_0=412kPa)

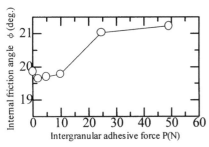

Figures 12 Relationship between the intergranular adhesive force P and the internal friction angle, ϕ

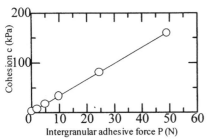

Figures 13 Relationship between the intergranular adhesive force P and the adhesion, c

case are compared, it is found out that the difference between the compression indexes in the arrangement by log($\sigma_c+\sigma_0$) is smaller than that by logσ_c. This result corresponds to the expectation from equations (5) and (6).

3.2 Stress and strain relation in shear process under constant intergranular adhesive force

Figures 6 and 7 show $\sigma_1/\sigma_3\sim\varepsilon_1\sim\varepsilon_3$ relation and $\varepsilon_1\sim\varepsilon_3$ relation for the case of σ_3=49kPa. Figures 8 and 9 show a similar relation for the case of σ_3=490kPa. It is found that the $\sigma_1/\sigma_3\sim\varepsilon_1\sim\varepsilon_3$ relation shifts at the top, and the shear strength increases when the intergranular adhesive force increases. The characteristics observed in figures 6 and 8 show that the principal stress ratio increases in the initial stage of the shear process when the intergranular adhesive force is higher, and that the change of principal stress ratio decreases after this. From the $\varepsilon_1\sim\varepsilon_3$ relations shown in figures 7 and 9, it is found that, when there is no intergranular adhesive force, the volume of the specimen changed from compression to expansion during the shear process, and that, in other cases, expansive volumetric strain occurs with an increase in the intergranular adhesive force. These results correspond to the tendency observed in direct shear tests for sand under constant suction (Shimada,1998). In his results, the volumetric expansion of the specimen during the shear process increased when the suction increased.

Kato(1997) conducted triaxial compression tests under constant suctions using compacted cohesive clay specimens, and showed that the $(\sigma_1+\sigma_0)/(\sigma_3+\sigma_0)\sim\varepsilon_1\sim\varepsilon_3$ relations under constant converted normal net stresses agree in spite of the size of the confining pressure.

Figure 10 shows the $(\sigma_1+\sigma_0)/(\sigma_3+\sigma_0)\sim\varepsilon_1\sim\varepsilon_3$ relation for the case of P=0.0N (σ_0=6kPa). Figure 11 shows a similar relation for the case of P=49N (σ_0=412kPa). For the case of P=0.0N, the relation shows a tendency which agrees for the whole area of the shear process. For the case of P=49N, it is found that the difference appears in the initial stage of shear, and that it has come to the upper part, when confining pressure is smaller. When the cases of σ_3=49kPa and 490kPa (shown in figures 10 and 11) are compared, since the void ratios before the shear process are almost equal, it can be considered that there is no effect of void ratio. Consequently, it is supposed that the difference observed in stress-strain relations shown in these figures exist as an effect of the intergranular adhesive force.

3.3 Strength constant and dilatancy characteristics

Figures 12 and 13 show the relationships between the intergranular adhesive force, P and the internal friction angle, ϕ and the adhesion, c. From figure

12, it is found that the internal frictional angle, ϕ increases with the increase in the intergranular adhesive force. From figure 13, it is found out that the adhesion, c is proportional to the increase in the intergranular adhesive force.

In the research based on triaxial compression test results on unsaturated soil, there are some interpretations for the relationship between suction and strength. According to the proposal by Fredlund et al. (1995), the internal frictional angle is a constant value in spite of the suction. On the other hand, Karube et al. (1986) showed triaxial test results for unsaturated soil, in which the internal frictional angle slightly increases with the increase in suction. From the present analytical result shown in figure 12, it is found out that the internal frictional angle tends to increase with the increase of the intergranular adhesive force. This result corresponds to the test results shown by Karube et al.

4 CONCLUSIONS

In this study, isotropic compression tests and biaxial compression tests for two-dimensional granular material were simulated using a DEM program, which introduces the intergranular adhesive force between the disk particles. The effects of the intergranular adhesive force on deformation and strength was examined. From the results and discussions, conclusions are summarized as follows. These conclusions almost agree with the tendencies that are observed in triaxial compression test results of unsaturated soils under constant suction.

1) The stiffness of the soil skeleton increases, when the intergranular adhesive force increases. However, it is observed that the compression indexes approach a constant value, when the results are arranged by taking the converted normal net stress into account as a confining pressure.

2) The tendencies observed between the principal stress ratios, the principal strain and the volumetric strain in the shear process differed when the intergranular adhesive forces increased. Under the same confining pressure, when the intergranular adhesive force become bigger, the increase of the principal stress ratio for principal strain suddenly became more, and more expansive volumetric strain was observed.

3) The internal frictional angle increased with the increase in the intergranular adhesive force. This result corresponds to the test results shown by Karube et al(1986). The adhesion increased in proportion to the intergranular adhesive force.

REFERENCES

Alonso, E.E., Gens, A. and Josa, A. 1990, A Constitutive Model for Partially Saturated Soils, *Geotechnique*, Vol.40, No. 3, pp.405-430,.

Bishop, A.W. & Henkel, D.J. 1957, The measurement of soil properties in triaxail test, Edward Arnold Ltd.

Cundall, P.A. 1971, A computer model for simulation progressive, large scale movement in blocky rock system, Symp. ISRM, Nancy, France, Proc.2, pp.129-136.

Fredlund, D.G., Vanapalli, S.K., Xing, A. and Pufahl, D.E. 1995, Predicting the shear strength function for unsaturated soils using the soil-water characteristic curve, *Proc. of 1st Int. Conf. on Unsaturated Soils*, Vol.1, pp.43-46.

Karube, D., Kato,S. & Katsuyama, J. 1986, Effective stress and soil constants in unsaturated kaolin, Journal of Japanese society of civil engineering, No.370/ III -5, pp.179-188. (In Japanese)

Kato, S. 1997, Effects of soil water distribution on behavior of unsaturated soil and its evaluating method, PhD thesis, Kobe University. (In Japanese)

Shimada, K. 1998, Effect of matric suction on shear characteristics of unsaturated fraser river sand, *Jour. of Faculty Environmental Science and Technology*, Okayama Univ., Vol.3, No.1, pp.127-134.

Yamamoto, S. 1995, Fundamental study for behavior of granular material by DEM analysis, PhD thesis, Nagoya Institute of Technology. (In Japanese)

Unsaturated Soils for Asia, Rahardjo, Toll & Leong (eds) © 2000 Taylor & Francis, ISBN 90 5809 139 2

Application of the effective stress principle to volume change in unsaturated soils

N. Khalili

School of Civil and Environmental Engineering, The University of New South Wales, Sydney, N.S.W., Australia

ABSTRACT: The application of the effective stress concept to volume change in unsaturated soils is investigated. Several laboratory test data as well as data from the literature have been analysed. Extremely good agreement is obtained between the measured and predicted values in all cases. It is shown that quantitative predictions of deformation in unsaturated soils can be made using the effective stress concept.

1 INTRODUCTION

The application of the effective stress concept to unsaturated soils has been a controversial issue, particularly in relation to volume change associated with collapse observed during wetting. Arguments have also been made, from a microscopic point of view, against combining total stress with pore air and pore water pressures, and thus the use of effective stress in unsaturated soils (Burland 1965). Although not examined rigorously, these arguments have been widely accepted in the literature and several researchers have proposed the use of two independent variables for characterisation of the behaviour in unsaturated soils (e.g. see Matyas & Radhakrishna 1967, Fredlund & Morgenstern 1977, Alonso et al 1990, Wheeler & Sivakumar 1995). A major difficulty in this approach is that it requires the determination of two sets of material parameters, for each of the stress variables, which in fact may not be independent.

Concerns have also been raised as to the uniqueness of the effective stress parameter, χ, with the effective stress, σ_{ij}, defined as (Bishop 1959),

$$\sigma'_{ij} = (\sigma_{ij} - u_a \delta_{ij}) + \chi(u_a - u_w)\delta_{ij} \quad (1)$$

in which σ_{ij} is the total stress tensor, u_w is the pore water pressure, u_a is the pore air pressure and δ_{ij} is the Kronecker's delta; $\sigma_{ij} - u_a \delta_{ij}$ is referred to as the net stress, and $u_a - u_w$ is the matric suction. However, as pointed out by Khalili & Khabbaz (1998a) most of these concerns have been related to finding a relationship between χ and a volumetric parameter such as the degree of saturation, S_r. In fact, they showed that by plotting the values of χ obtained from the literature against a more appropriate parameter such as the ratio of matric suction over the air entry value[1] (suction ratio) a unique relationship may be obtained for most soils, Figure 1.

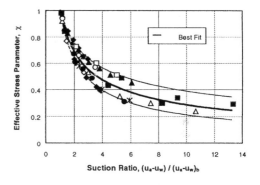

Figure 1. χ versus suction ratio, after Khalili and Khabbaz (1996, 1998a).

The best-fit to the data in Figure 1 is expressed as,

$$\chi = \begin{cases} \left[\dfrac{(u_a - u_w)}{(u_a - u_w)_b} \right]^{-0.55} & if \quad (u_a - u_w) \geq (u_a - u_w)_b \\[2mm] 1 & if \quad (u_a - u_w) \leq (u_a - u_w)_b \end{cases} \quad (2)$$

in which $(u_a - u_w)_b$ is the air entry value.

---

[1] The air entry value corresponds to the matric suction above which air recedes into the soil pores.

To examine the validity of the relationship in (2), Khalili & Khabbaz (1998a) conducted 17 suction controlled shear strength tests on two laboratory compacted soils: (1) a compacted kaolin, and (2) a compacted sand (75%) and kaolin (25%) mixture. Extremely good agreement was obtained in all the cases between the calculated and measured results. Further experimental evidence as to the validity of relationship (2) was provided in Khalili & Khabbaz (1998b).

The main objective of this paper is to examine the validity of relationship (2) for volumetric change in unsaturated soils. Both data from the literature and data from a programme of laboratory testing are used for this purpose. Wetting as well as drying loading paths are considered.

2 EFFECTIVE STRESS AND VOLUME CHANGE

According to the effective stress principle, the volumetric change in unsaturated soils may be described as,

$$\frac{\Delta V}{V} = c\Delta[(p - u_a) + \chi(u_a - u_w)] \qquad (3)$$

in which $p = \sigma_{ii}/3$ is the mean total stress, $\Delta V/V$ is the volumetric strain, and c is the drained compressibility of the soil matrix. If the applied stress is kept constant, the volumetric strain can be written as,

$$\frac{\Delta V}{V} = c\Delta[\chi(u_a - u_w)] \qquad (4)$$

Equations (3) and (4) can be used predictively if the value of c is determined based on the current effective stress and the elasto-plasticity of the soil matrix. For the elastic response, the coefficient of compressibility for re-loading/unloading must be used, whereas for the elasto-plastic response the coefficient of compressibility associated with first-time loading must be utilised. Other important factors to be considered are:

1) The process of desaturation commences only at suction values greater that the air entry value. At suction values below the air entry value the soil will be in a saturated state and the mechanics of saturated soils will apply.
2) Suction has a dual effect on the behaviour of unsaturated soils. It affects both the effective stress, and the preconsolidation pressure or the yield stress of the soil skeleton. Much of the debate in the literature as to the applicability of the effective stress concept to unsaturated soils has been due to this dual effect. Collapse, for instance, is a direct consequence of a shift in the preconsolidation pressure with suction.

3) Upon desaturation, the soil response enters the elastic region (e.g. see Fleureau et al 1993). (Mathematically, this is equivalent to the rate of increase in the preconsolidation pressure with suction being greater than the rate of increase in the effective stress with suction.) A direct consequence of this is that the elastic coefficient of compressibility must be used in predicting volume change in non-collapsing unsaturated soils subject to drying and/or wetting at constant net stress. This fundamental aspect in the behaviour of unsaturated soils has not been taken into account in the previous analyses of the volume change in unsaturated soils, leading to erroneous conclusions as to the applicability and/or uniqueness of the effective parameter in unsaturated soils.

3 TEST PROGRAMME

To investigate the application of the effective stress principle to volume change in unsaturated soils, a series of desaturation tests were conducted on two laboratory prepared soils: a pure kaolin, and a silty soil from the Glenmore Park area in West Sydney. Prior to testing, each soil was carefully mixed and wetted with sufficient distilled water using a spray gun to a high water content (close to liquid limit). The soils were then placed in a sealed plastic bag and allowed to cure for at least 2 days, to ensure uniform distribution of water throughout the soil and moisture equilibrium. The index properties of the soils are given in Table 1. The soil water characteristics of the samples are given in Figure 2.

Table 1. Index properties of soils used in volume change experiments.

Property	Soil	
	Kaolin	Silt
Liquid Limit, LL (%)	63	21
Plastic Limit , PL (%)	30	14
Specific Gravity	2.61	2.78
Dry Density (g/cm³)	1.237	1.967
Finer than 75 μm (%)	-	54
Finer than 10 μm (%)	100	-
Finer than 3 μm (5%)	70	-
USCS Symbol	CH	CL-ML
Air Entry Value (kPa)	85	15

All volume change tests were conducted in a modified oedometer, capable of independent measurement and control of pore air, pore water pressures, total deformation of the sample and water volume change within the sample. The pore air pressure was controlled through a coarse porous stone placed at the top of the sample, and the water pressure was controlled through a 15 bar saturated porous stone placed at the bottom of the sample. The suction within the sample was controlled using the axis translation technique, which is a common technique to prevent cavitation in the laboratory testing of unsaturated soils.

Following setting up in the equipment, the kaolin sample was consolidated to 600 kPa, in load increments to normal stresses 50, 100, 200, 400 and 600 kPa. The sample was then subjected to a series of loading and unloading cycles in the range of 100 kPa to 600 kPa, in order to obtain the saturated elastic response. The Glenmore Park silt was consolidated to 400 kPa and was cycled in the range of 50 kPa to 400 kPa. Once the elastic response was obtained, each sample was unloaded to the target net stress, and then subjected to increasing values of suction. The target net stress for kaolin was 100 kPa and for the Glenmore Park silt was 50 kPa.

The results of the tests along with a comparison of the predicted and calculated volumetric changes are given in Figure 3. The air entry values shown on Figure 3 were obtained using the pressure plate technique and the soil water characteristic curves shown on Figure 2. The predicted volumetric change values were calculated using the effective stress concept and the relationship given in (4). The effective stress parameter χ was obtained using the relationship given in Equation (2). The drained compressibility, c, was obtained using the saturated elastic response of the soil, shown as the dashed line.

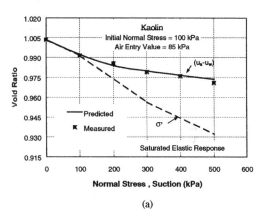

(a)

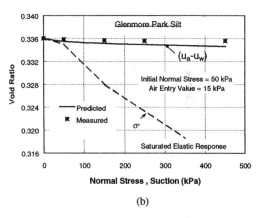

(b)

Figure 3. Comparison between predicted and measured volume change test results: a) kaolin, b) Glenmore Park silt.

As can be observed from Figure 3, a good agreement is obtained between the calculated and the measured values, implying that a quantitative prediction of deformation in unsaturated soils can be made using the effective stress concept.

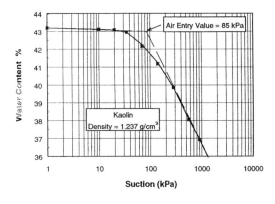

(a)

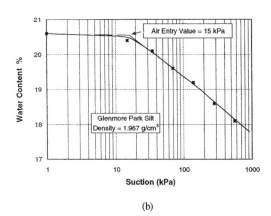

(b)

Figure 2. Soil water characteristic curve: a) kaolin, b) Glenmore Park silt.

4 DATA FROM LITERATURE

To further examine the validity of the effective stress concept to volume change in unsaturated soils, the data from Fleureau et al (1993) and Blight (1965) were analysed. Both data sets provide the saturated elastic response of the medium, required for a predictive analysis of volume change in unsaturated soils.

4.1 Data from Fleureau et al (1993)

Fleureau et al (1993) conducted a series of drying and wetting tests on 11 slurried clayey soils. The results on Jossigny loam, White clay (kaolin), and Montmorillonite (Ca-montmorillonie) are presented on Figures 4 to 6. The index properties of the soils, as reported by Fleureau et al, are given in Table 2.

Table 2. Index properties of Jossigny loam, White clay (kaolin) and Montmorillonite (Ca-montmorillonite), after Fleureau et al (1993).

Property	Soil*		
	1	2	3
Liquid Limit, LL (%)	37	61	170
Plastic Limit , PL (%)	18-21	31	60
Finer than 80 μm (%)	80	100	100
Finer than 2 μm (%)	28	85	40
D_{60} (μm)	22	0.5	11
D_{10} (μm)	<0.1	<0.1	<0.1

* Soil 1= Jossigny loam, 2= White clay, 3= Montmorillonite

The results show a normally consolidated isotropic compression with increasing suction until the point of air entry in all three samples except for the White clay which appears overconsolidated up to a suction value of 30 kPa. Beyond the point of air entry, the samples enter the elastic region, marked by a drastic reduction in the volume change with suction. Upon wetting (i.e. unloading), the samples undergo elastic dilatancy until the suction attains the air entry value. The results show little or no hysteresis during the elastic dilatancy. At suctions below the air entry value, wetting results in stress paths increasingly entering the saturated elastic region, yielding a greater level of elastic dilatancy than in the unsaturated region, in accordance with the effective stress concept.

The effective stress predictions of the volume change in the unsaturated region of the soils (i.e. after the point of air entry) are shown on Figures 4 to 6 by the solid lines. The predictions are based on an extension of the saturated elastic response of the soils shown by the dashed lines, and values of the air entry obtained from the soil water characteristic curves reported in Fleureau et al (1993) for each soil. The results show a good agreement between the predicted and the experimental data in all three cases.

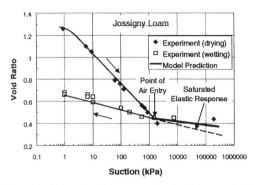

Figure 4. Comparison between predicted and measured volume change results (Jossigny loam), after Fleureau et al (1993).

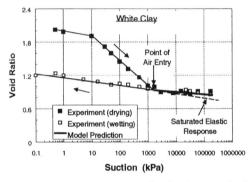

Figure 5. Comparison between predicted and measured volume change results (White clay), after Fleureau et al (1993).

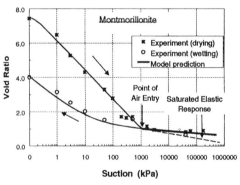

Figure 6. Comparison between predicted and measured volume change results (Montmorillonite), after Fleureau et al (1993).

4.2 Data from Blight (1965)

Blight (1965) conducted several volume change tests on an expansive soil from Vereeniging in the Transvall, South Africa, with the index properties LL=55%, PI=35%. The results from two swelling tests conducted at net stresses 14 kPa and 70 kPa are presented in Figures 7 and 8.

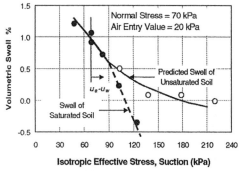

Figure 7. Comparison between predicted and measured volume change results (net stress=70 kPa), after Blight (1965).

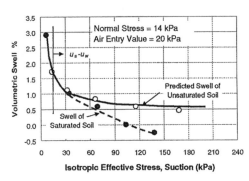

Figure 8. Comparison between predicted and measured volume change results (net stress=14 kPa), after Blight (1965).

Also, included in Figures 7 and 8 are the predicted response of the unsaturated soil, represented by the solid lines. Once again the predictions are based on the effective stress relationship (1), the saturated elastic response of the soil shown by the dashed lines, and the air entry values defined as the point of separation between the saturated and unsaturated volumetric responses.

Good agreements are obtained in both cases.

5 PLASTIC COLLAPSE

The phenomenon of collapse upon wetting in unsaturated soils has been extensively used in the past as an indication of the breakdown of the effective stress concept in unsaturated soils. However, as has been shown in recent years by several investigators, the process can be readily explained in terms of effective stresses by invoking an appropriate plasticity model (Kohgo et al 1993, Modaressi & Abou-Bekr 1994, Bolzon et al 1996, Khalili & Khabbaz 1996, Khalili & Loret 1999, Loret & Khalili 2000). Indeed, collapse upon wetting has no bearing as to the applicability or otherwise of the effective stress concept to unsaturated soils, and is a direct consequence of a

shift in the preconsolidation pressure with suction.

To explain collapse in unsaturated soils, consider consolidation tests at constant suction on two identical soil samples, Figure 9. Sample A is consolidated in a saturated state, whereas Sample B is consolidated at a suction greater that the air entry value. The soils are initially overconsolidated so that paths C_1-C_2 and C_1-D_2 occur elastically. Yielding occurs at points C_2 and D_2 for samples A and B, respectively.

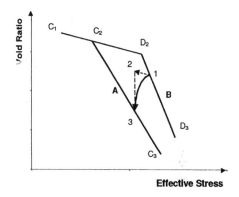

Figure 9. Description of collapse in unsaturated soils

To induce collapse, Sample B is flooded at Point 1, reducing the suction within the sample to zero, and resulting in a tendency for elastic dilatancy from Point 1 to 2. However, as suction is reduced, the state boundary retreats from D_2-D_3 to C_2-C_3, rendering the stress state at Point 2 unattainable. Thus, the sample undergoes plastic volumetric contraction at constant effective stress, to collapse onto the new state boundary at Point 3. During the entire process the effective stress within the sample is given by Equation 1.

6 CONCLUSIONS

The validity and the application of the relationship proposed by Khalili and Khabbaz (1996, 1998) to volume change in unsaturated has been examined through a new set of laboratory test data, and data from the literature. It is shown that quantitative predictions of deformation in unsaturated soils can be made using the effective stress concept. Extremely good agreement has been obtained between the measured and predicted values in all cases examined.

ACKNOWLEDGEMENT

The financial support of the Australian Research Council (ARC) is gratefully acknowledged.

REFERENCES

Alonso, E.E., A. Gens & A. Josa 1990. A constitutive model for unsaturated soils. *Geotechnique*. 40: 405-430.

Bishop, A.W. 1959. The principle of effective stress. *Tecknish Ukebla*. 106: 859-863.

Blight, G.E. 1965. A study of effective stress for volume change. *Proc. Symp. Moisture Equilibria and Moisture Changes in the Soils Beneath Covered Areas*. 259-269. Butterworths. Sydney. Australia.

Burland, J.B. 1965. Some acpects of the mechanical behavior of partly saturated soils. *Proc. Symp. Moisture Equilibria and Moisture Changes in the Soils Beneath Covered Areas*. 270-278. Butterworths. Sydney. Australia.

Bolzon, G., A. Schrefler & O.C. Zienkiewicz 1996. Elasto-plastic soil constitutive laws generalised to partially saturated states. *Goetechnique*, 46: 279-289.

Fleureau, J.M., S. Kheirbek-Saoud, R. Soemitro & S. Taibi 1993. Behaviour of clayey soils on drying-wetting paths. *Can. Geotech. J.*, 30: 287-296.

Fredlund, D.G. & N.R. Morgenstern 1977. Stress state variables for unsaturated soils. *Soil Mechanics and Foundation Division, ASCE*. 103: 447-466.

Khalili N. & M.H. Khabbaz 1996. The effective stress oncept in unsaturated soils. *UNICIV Report No. R-360*. ISBN 85841 327 2 : 25p.

Khalili N. & M.H. Khabbaz 1998a. A unique relationship for χ for the determination of the shear strength of unsaturated Soils. *Geotechnique*. 48: 681-688.

Khalili, N. & M.H. Khabbaz 1998b. An effective stress based approach for shear strength determination of unsaturated soils. Proc. 2nd Int. Conf. on Unsaturated Soils, UNSAT'98. 27-30 August. Beijing. China. 1: 84-89.

Khalili, N. & B. Loret 1999. An elasto-plastic model for the behaviour of unsaturated soils. *Proc. 4th Asia Pacific Conf. on Computational Mechanics, APCOM'99*. 15-17 December, Singapore.

Loret, B. & N. Khalili. 2000. A three phase model for unsaturated soils. In press. *Int. J. Num. Analy. Methods in Geomechanics*.

Kohgo, Y., M. Nakano & T. Miyazaki 1993. Theoretical aspects of constitutive modelling for unsaturated soils, *Journal of Soil Mechanics and Foundation Engineering, SMFE Jap. Soc.* 33: 49-63.

Modaressi, A. & N. Abou-Bekr 1994. A unified approach to model the behavior of saturated and unsaturated soils. In Siriwardane (ed), *Proc. 8th, Int. Conf. On Computer Methods and advances in Geomech., Morgantown*: 1507-1513. Roterdam: Balkema.

Matyas, E.L. & H.S. Radhakrishna 1967. Volume change characteristics of partially saturated soils. *Geotechnique*. 18: 432-448.

Wheeler S.J. & V. Sivakumar 1995. An elasto-plastic critical state framework for unsaturated soils. *Geotechnique*. 45: 35-53.

Unsaturated Soils for Asia, Rahardjo, Toll & Leong (eds) © 2000 Taylor & Francis, ISBN 90 5809 139 2

CT discrimination of fabric change of unsaturated compacted loess during compression process

X.J.Li, Z.Wang & J.Z.Yin
Guangdong Provincial Highway Construction Company, Guangzhou, People's Republic of China

ABSTRACT: The triaxial compression tests of compacted loess samples with different lateral pressures are conducted with a new set of modified triaxial compression apparatus which made the soil sample can be scanned with CT machine at the same time during compression process. The different damage process of compacted loess sample is directly observed for the first time with CT images and CT numbers. The production mechanisms of the soil micro-crack in the CT images during different process are analyzed.

1 GENERAL INSTRUCTIONS

The constitution relationship presented nowadays for unsaturated soil are mainly concentrated on describing the influence to strength and strain caused by the change of suction. Recently, the study on soil fabric is more and more valued. Loess is a kind of typical unsaturated structural soil. Many researchers agree that the influence of soil fabric should be considered in the constitution relationship to describe the collapsible behavior of loess. One of the prerequisites of this kind of constitution relationship is the observation and quantitatively analysis of soil fabric during compression process.

In fact, scanning electron micrograph (SEM) was used in most experiments on soil fabric observation. The results obtained from SEM only show the fabric change after elastic deformation recovery. On the another hand, when the little SEM sample is made, the disturbance to soil fabric is very difficult to be omitted.

In order to overcome those shortcomings, the computer tomography (CT) technique, which is widely used in medical area, is utilized in our experiments.

The triaxial compression tests of compacted loess samples with different lateral pressure are conducted with a new set of modified triaxial compression apparatus which made the soil sample can be scanned with CT machines at the same time during compression process. The different damage process of compacted loess sample is directly observed for the first time with CT images and CT numbers. The production mechanisms of the soil micro-crack in the CT images during different damage process are

simply analyzed.

With a set of newly developed software, the CT images are processed. A new method and a quantitative index are also presented to analyze the change of soil fabric during process on the basis of CT images and stereoligical technique.

2 EXPERIMENT FACILITIES

All the tests referred in the paper were performed with a new set of modified triaxial compression apparatus at State Key Laboratory of Frozen Soil Engineering, LIGG CAS. The resolving power of CT machine is $0.0081mm^2$. In the tests, computer storage technique is adopted so that the CT images can be transferred directly from CT machine to personal computers. With this technique, the loss of image message is avoided and accuracy of image analysis is assured. One of the differences between modified compression apparatus and conventional triaxial compression apparatus is that the cavity of modified triaxial compression apparatus is made of special metal material. Another difference is that the major part of the modified triaxial compression apparatus is fixed in the CT bed. So it can move with the movement of CT bed. Those measures are taken to suit CT scan.

3 PREPARATION OF SPECIMENS

The laboratory specimens were compacted in a 61.8mm diameter, 125mm high mold. Each specimen was compacted in five layers of approximately equal height to 100% of the laboratory maximum dry density at optimum moisture content. This mold size provided a specimen with minimum disturbance during sample preparation. It also provided a sample with a height-to-diameter ratio of at least 2.

Disturbed soil specimens were obtained from a in-service subgrade in Xi'an. Specimens were air dried until the loess became friable, and then were pulverized with a mallet.

After measuring the moisture content of the processed and air-dried soil, a calculated amount of water was added to obtain an optimum moisture content, 14%. Subsequently, the soil was sealed in a plastic bag and stored in a humidity room for two days to allow the soil moisture to equilibrate the weight of the compacted soil. The compacted specimens were trimmed and measured to determine the weight of compacted soil. Subsequently, the specimens were carefully extruded using a hydraulic jack. The extruded specimens were wrapped in a plastic bag and stored in a humidity room for two days before being tested.

After storing the compacted samples for two days in a humidity room, a rutted membrane was applied to the sample. Then the sample was stuck with special glue on the loading piston to avoid eccentric phenomenon during compression process. Fig.1 shows test results of sample without sticking to the piston. To each specimen, three scanning positions are chose and the intermission among every scanning position is 30mm (Fig.2). The interfaces produced when the specimen was compacted are carefully avoided. The undrained compression tests were performed with different lateral pressures. After each loading, when the vertical deformation of soil sample is less than 0.01mm, the soil sample is scanned and the stress and strain are recorded manually.

Fig .1 Images of eccentric phenomenon

4 TEST RESULTS AND ANALYSIS

Fig.3 is a plot of test results showing the relationship between stress and strain of soil sample during triaxial compression test with different lateral pressure. The letters around the plot represent different CT scan stages. The corresponding CT

Fig 2   Scan positions

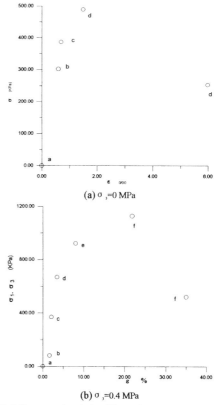

(a) $\sigma_3 = 0$ MPa

(b) $\sigma_3 = 0.4$ MPa

Fig 3 Stress-strain relationships and corresponding scan stages

images and means as well as variance of CT numbers are observed in Fig.4-8 and Table 1.

In each image, the white parts represent high-density areas and the black parts represent low-density areas. In Table 1, the means of CT numbers in location I are larger than the ones in location II and the means of CT numbers in location II are larger than those in location III. This means that for single soil specimen local densities in different position are not uniform. From the test results we obtained that the soil densities in location I are the largest and the densities in the location III are the

126

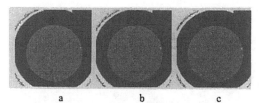

a   b   c

Fig.4 Images of I position from stage a-c (σ_3=0Mpa))

a   b   c

Fig.5 Images of II position from stage a-c (σ_3=0Mpa)

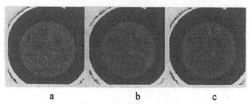

a   b   c

Fig.6 Images of III position from stage a-c (σ_3=0Mpa)

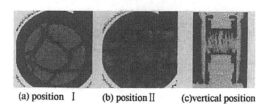

(a) position I   (b) position II   (c)vertical position

Fig.7 Images of stage d (σ_3=0Mpa)

Table 1 Mean and variance of CT numbers with different scan Table1 Means and variances of CT numbers with different positions and stages (σ_3=0Mpa)

stage	item	I	II	III
a	mean	1294.6	1257.4	1221.7
	variance	40.90	51.51	83.82
b	mean	1295.1	1257.0	1220.9
	variance	40.37	52.04	85.47
c	mean	1294.4	1253.2	1216.8
	variance	40.27	52.58	88.12
d	mean	987.1		760.3
	variance	572.68		663.33

obviously observed. During whole compression process we do not observe lineal shear band in plane. From Fig.9(g) and Fig.10(g) we can deduce that the with the specimens we prepared there is a group of shear bands ,which looks like cone around the inner part of soil sample in our experiment.

5 APPLICATION OF STEREOLOGICAL METHOD

Current methods of estimating volume fractions from measurement on a two dimensional section rely on an equivalence between the volume fraction and the intercepted area, line, on point fractions (Hilliard, 1968). The lineal analysis is used in the current study. In this analysis, the porosity of the soil is defined as the ratio of the average length of interceptions of voids on a test line placed at random in a spherical domain to the total length of the line (Hilliard, 1968). The above technique can be extended to determine the parameters of a directional porosity function from microscopic observations (Kanatani, 1984; Muhunthan, 1997). It has been shown that the directional porosity function can be expressed by (Kanatani, 1984; Muhunthan, 1997):

$$n(l) = n_m \left(1 + F_{ij} l_i l_j \right) \tag{1}$$

where n_m is the average porosity , i and j is a permutation of 1,2,3, the components of unit vector l are given by

$$l_1 = \sin\theta \sin\phi, l_2 = \cos\theta, l_3 = \sin\theta \cos\phi \tag{2}$$

and F_{ij} is the second order porosity tensor.

F_{ij}, the void fabric tensor falls into the general class of second rank symmetric tensors. Consequently, it is possible to define the three invariants of this tensor:

$$(J_f)_1 = F_{ii} = 0 \tag{3}$$

smallest. The same trend is also observed in other samples. We believe that those differences are due to the way with which the specimens are prepared. Fig.4-Fig.11 show gradual deformation of soil samples with gradual increase of vertical loading with different lateral pressures. In Fig.9(f), we find that high density-areas are mixed with low- density areas and the shapes of low-density areas and high-density areas are regular and symmetrical. With the increase of plastic region, some short, radically well-distributed cracks, which are located in the edge of soil sample, are observed.(Fig.9(g)) We can still observe in Fig.9(g) that in the middle of soil sample the soil density is high and around this high density area the soil density is lower and around this lower density area is an area which is full of short cracks. In Fig.10(g), besides similar short, well-distributed short cracks, an circular continuous cracks is

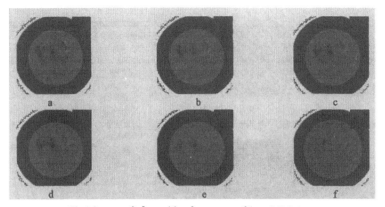

Fig 8 Images of Ⅰ position from stage a-f (σ₃=0.4Mpa)

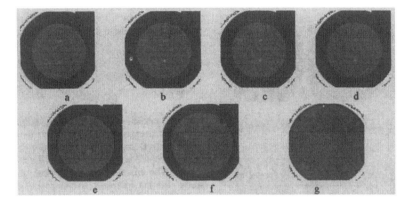

Fig 9 Images of Ⅱ position from stage a-g (σ₃=0.4Mpa)

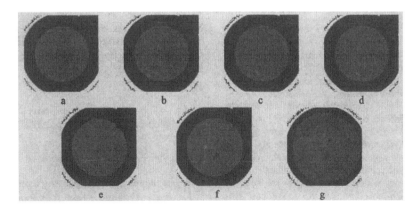

Fig 10 Images of Ⅲ position from stage a-g (σ₃=0.4Mpa)

a   d   g

Fig 11 Images of vertical section of different stages

$$(J_f)_2 = \frac{1}{2}(F_{ig}F_{gi}) \qquad (4)$$

$$(J_f)_3 = \frac{1}{3}(F_{ig}F_{jk}F_{ki}) \qquad (5)$$

The two dimensional form of $(J_f)_2$ is similar to the vector magnitude and is termed void fabric index. The subscript 2 is omitted for clarity and J_f will be used to denote the void fabric index in the subsequent analysis.

For two-dimensional observations, J_f reduces to

$$J_f = F_{11}^2 + F_{12}^2 \qquad (6)$$

The experimental procedure to determine the parameters of description has been detailed elsewhere (Kanatani, 1984, Muhuathai, 1997). With a set of newly developed software, the CT images are processed and fabric indexes are calculated, (Fig.12).

To single soil specimen the value of J_f are different in every different position. The bigger the

value of J_f is, the stronger anisotropy soil sample has.

When lateral pressure is zero, the inherent anisotropy in location I and location II is strong. With the increase of deformation, the value of J_f is decrease and at last tends to be relatively constant. However even when the soil sample is destroyed, the value of J_f is not tends to be zero. When lateral pressure is 0.4MPa, the inherent anisotropy in location I and III are stronger. After loading, the changes of J_f show that firstly the induced anisotropy to be strong. Then with the increase of deformation, the anisotropy gradually vary to almost isotropy.

6 CONCLUSION

From the test results, the following conclusion are drawn:

1. CT machine can be used in geotechnical area and shortcomings of SEM can be overcome.

2. The densities in single sample are not uniform. With the increase of deformation, high-density parts are mixed with low-density parts and the shapes of them are regular and symmetrical.

3. In our experiment, besides short, radically well distributed short cracks, circular continuous crack or circular interface of high density-area with low-density area is observed.

4. With the increase of deformation, the inherent anisotropy gradually changes to isotropy and the values of J_f, defined as fabric index, have same corresponding change.

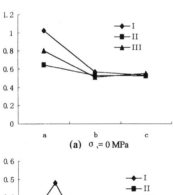

(a) $\sigma_3 = 0$ MPa

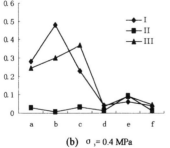

(b) $\sigma_3 = 0.4$ MPa

Fig.12 Change of J_f with different stage

REFERENCES

Hilliard, J.E., 1968. Measurement of volume involume, *Quantitative Microscopy*, DeHOff, R. T.,and Rhines, F.N.,(Eds),McGraw-Hill, NY.

Kanatani, K., 1984a. Distribution of directional data and fabric tensors, *Int. Jour. of Eng. Sci.*, Vol.22,No.2:149-164

Muhunthan, B.,& J.L.Chameau, 1997. Void fabric tensor and ultimate state surface of soils, *Journal of Geotechnical and Geoenvironmental Engineering*. Vol.123,No.2:173-180

Unsaturated Soils for Asia, Rahardjo, Toll & Leong (eds) © 2000 Taylor & Francis, ISBN 90 5809 139 2

Generalized pore pressure model considering damage for unsaturated soil

X. H. Luo

Department of Civil Engineering, Wuhan Urban Construction Institute, People's Republic of China

ABSTRACT: Most of unsaturated soil has notable structure. Them shear failure is closely relevant to structural damage. In this paper, from the idea of generalized pore pressure (Shen Z J 1997), elastic-plastic constitutive equation considering damage for unsaturated soil is created by the definition of damage variable on the base of the datum of soil test. A compositive model for between matric suction and axial strain is proposed. Finally, a slope of unsaturated soil, as an example, is taken to calculate, which shows application of suggested method.

1 GENERAL INSTRUCTIONS

Soil mechanics for unsaturated soil has been formed integratively theoretical system and become an important branch of soil mechanics because of the research of many scholars. As for expression of stress state for unsaturated soil, using the effective stress $(\sigma - u_a)$ and the matric suction $(u_a - u_w)$ as the stress state variables has been generally accepted and constructed shear strength theory of unsaturated soil. Relevant Mohr-Coulomb failure envelope is a curved surface in 3D composed of effective stress, matric suction and shear stress (in general, it is assumed as a plane in application). It is common knowledge that, based on the classical theory of soil mechanics, not only deformation but also shear failure of soil mass is rearrangement of soil structure owing to the relative displacement among soil grains. Something of its characteristic is that stress-strain relationship shows nonlinear strain softening or strain hardening (but Bishop's test indicated any soil mass all have the feature of stain softening and the difference only to varying degrees). For example, the elastic modulus (not only tangent modulus but also secant modulus) would gradually decrease with the increase of the stress. The results of consolidated test also show that the expansion index is strikingly smaller than compression index. When the consolidation pressure is lower than preconsolidation pressure P_c, the compressibility of soil is expressed by expansion index C_s. But when the consolidation pressure is larger than preconsolidation pressure, which the structure of soil is damaged and the consolidation curve becomes to be steep, its compressibility is described by compression index C_c. That unconfined compression strength of disturbed soil is evidently lower than that of undisturbed soil yet sufficiently indicates that the strength and deformation of soil have begun on essential change from the structural damage to the failure. As regards the conception of failure in soil mechanics, it means ultimate limit equilibrium, but on the base of the viewpoint damage mechanics, the failure reaching limit equilibrium is an ultimate expression of the damage change of material property. So the decreasing procedure of mechanic parameters could be thought to be a gradually accumulative phenomenon of soil structure damage. When this accumulation attains limit, it is the failure that stress state reaches limit equilibrium. That is the same as the reason of slide, which interior sliding surface of the slope is expanded, elongated and finally formed seriatim shear zone.

Since Introduction to Continuum Damage Mechanics (Kachanov 1986) was published, the damage mechanics has formed a more completely theoretical system. It has two important branches. First is Microdamage Mechanics. It is a method of mechanic mean, based on the analysis of deformation and development of the typical damage element in material interior, for example mini-fissure, that the relationship between micro-damage parameters and the procedure of material deformation and damage. Second is continuum damage mechanics. It is said that using the continuous medium mechanics and the continuous medium thermomechanics research that the damage exert the influence on macro-mechanics property, procedure of structural damage and law of damage evolution. Its analytical method is to create constitutional model and damage evolution model.

There are coupled analysis (Shun H, Zhao X H, 1999) and non-coupled analysis. If using non-coupled analysis, constitutional model is built by continuous medium mechanics, the establishment of damage evolution model is based on the results of test. In this paper, damage analytical model for un-saturated soil is to use non-coupled analytic method.

2 DESCRIPTION OF UNSATURATED SOIL DAMAGE

On the based of equivalent strain principle (Lemactre 1971), the strain of damaged material under the action of stress is equivalence to that of undamaged material under the action of real effective stress that exists in damaged material (in this paper, effective stress defined by damage mechanics is titled as equivalent stress differentiating effective stress defined by soil mechanics). So the definition damage variables $\omega_m(m=1,1,3)$ are follow

$$\omega_m = 1 - \frac{\Delta_{ijkl}}{D_{ijkl}} \quad (1)$$

The stress-strain relationship of damaged soil is

$$\varepsilon_{ij} = \frac{\Sigma_{ij}}{D_{ijkl}} = \frac{\sigma_{ij}}{D_{ijkl}(1-\omega_m)} = \frac{\sigma_{ij}}{\Delta_{ijkl}} \quad (2)$$

where ε_{ij} = real strain tensor, Σ_{ij} = equivalent stress tensor, σ_{ij} = real stress tensor, Δ_{ijkl} = damage elastic-plastic tensor, D_{ijkl} =undamaged elastic-plastic tensor. Applying the definition of eq.(1), Figure.2 is the relationship between damage variables and deviator stress of UD triaxial consolidated test of silty soil in Figure.1 (Bight 1967). This relationship could be fitted as follow

$$\omega = \rho[(\sigma_1 - \sigma_3) - (\sigma_1 - \sigma_3)_d] \quad (3)$$

where ρ = slope of straight line, deviator stress on horizontal coordinate axis for every straight line is defined as damage threshold stress $(\sigma_1-\sigma_3)_d$. The relationship between damage threshold stress and matric suction could be simulated with exponential

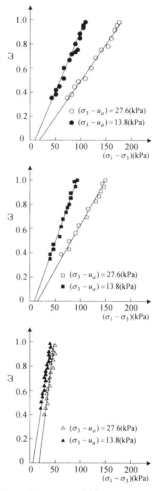

Figure.2 Damage variables versus stress

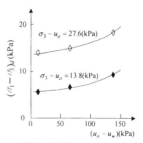

Figure.3 Damage threshold stress versus matric suction

function as the same as Figure.3.

$$(\sigma_1 - \sigma_3)_d = \eta \exp[\lambda(u_a - u_w)] \quad (4)$$

where η and λ are parameters of test.

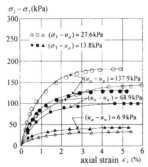

Fig.1 Stress versus strain for UD triaxial consolidation text of silty soil

3 GENERALIZED SUCTION AND GENERALI-ZED EFFECTIVE STRESS

Frendlund (1978) proposed shear strength formula with two variables of stress state for unsaturated soil

$$\tau_f = c' + (\sigma - u_a)_f \tan\varphi' + (u_a - u_w)_f \tan\varphi^b \quad (5)$$

where c' = effective cohesion when $(u_a - u_w)=0$. $(\sigma - u_a)_f$ and $(u_a - u_w)_f$ are effective normal stress and matric suction on the shear surface, respectively. φ = angle of effective internal friction, φ^b = increasing ratio of shear strength with the increase of matric suction.

Being similar to Bishop's method, total cohesion could be shown as $c = c' + (u_a - u_w)_f \tan\varphi^b$ on condition that matric suction and effective normal stress are all equal to zero. At the same time, effective normal stress is defined as $\sigma' = (\sigma - u_a)_f$. So Frendlund's stress strength formula could be rewritten as

$$\tau_f = c + \sigma' \tan\varphi' \quad (6)$$

Defining $\zeta = c \, \mathrm{ctg}\varphi'$, so eq.(6) could be rewritten as

$$\tau_f = \tilde{\sigma} \tan\varphi' \quad (7)$$

where $\tilde{\sigma} = \sigma' + \zeta$, which ζ could be titled as generalized suction (Shen Z J 1997). It is composed of two sections. First part is $c \, \mathrm{ctg}\varphi'$, which is bonding force among soil grain and is determined by real cohesion c' because of the cementation of soil. Second part is $(u_a - u_w)_f \tan\varphi^b \mathrm{ctg}\varphi'$. Its mechanism indicates the capillary suction among grain, which is determined by matric suction $(u_a - u_w)_f$ of shear failure. From this analysis, shear strength of un-saturated soil could be thought to be the sum of effective frictional force $\sigma' \tan\varphi'$, which is generated by the external load, and generalized frictional force $\zeta \tan\varphi'$, which belongs to the action of gene-ralized suction among grain. Therefore, $\tilde{\sigma}$ could be called generalized effective stress.

In the light of Terzaghi's principle of effective stress, defining p as the pore pressure, effective force could be shown as

$$\sigma' = \sigma - p \quad (8)$$

so generalized effective stress could be rewritten as

$$\tilde{\sigma} = \sigma - p + \zeta \quad (9)$$

Because the shear failure of soil could be considered to be a procedure of the formation and the development of shear zone, at the same time, initial state of soil structure has an striking effect on relative displacement among soil grains in that procedure, so initial generalized suction ζ_0 is defined thinking for intrinsic initial state of unsaturated soil. On the other hand, the shear failure could also be thought to be damage procedure owing to change of soil structure, so defining generalized pore pressure $\tilde{p}$ would express an influence of generalized suction on effective stress among soil grains.

$$\tilde{p} = \zeta_0 - \zeta \quad (10)$$

Characteristic described by $\tilde{p}$ is that the decrease of generalized suction results to the increase of generalized pore pressure from the initial damage of soil structure to the residual stress state. Therefore, eq.(9) could be rewritten again as

$$\tilde{\sigma} = \sigma - p + \zeta_0 - \tilde{p} \quad (11)$$

since initial generalized suction ζ_0 is a constant and is only relevant to initial soil structure, so equilibrium equation of unsaturated soil on the base of the description of generalized effective stress is follow

$$\sigma_{ij,j} - \delta p_{,j} - \delta\tilde{p}_{,j} = f_i \quad (12)$$

where δ = Kronecker's delta. Constitutive equation of effective stress field is showed as follow

$$\sigma_{ij} - \delta p_j - \delta\tilde{p}_j = \Delta_{ijkl}\varepsilon_{kl} \quad (13)$$

If geometrical equation and constitutive equation are applied in equilibrium equation being conditional upon total drainage $\Delta p=0$, incremental analyzable equation of FEM is follow

$$[K]\{\Delta u\} + [C]\{\Delta\tilde{p}\} = \{\Delta F\} \quad (14)$$

where $[K]$ = effective stiffness matrix. $\{\Delta u\}$ = the displacement increment. $[C]$ = coupled matrix. $\{\Delta F\}$ = vector of load.

4 ELASTIC-PLASTIC CONSTITUTIVE RELA-TIONSHIP THINKING DAMAGE

From plastic theory, orientation of plastic strain increment is only determined by flow rule. Because the shear failure is thought to be a procedure of structural damage, plastic potential g could is defined as $g(\sigma_{ij}, \kappa, \omega_m)=0$ $(i,j,m=1,2,3)$, which κ is equivalent plastic strain. So the relationship between plastic strain increment $d\varepsilon_{ij}^p$ and stress state on any a point M of stress space could be expressed as

$$d\varepsilon_{ij}^p = d\lambda \frac{\partial g}{\partial \sigma_{ij}} \quad (15)$$

where $d\lambda$ is plastic factor. As for non-correlative flow rule, definition of yield function is follow

$$f(\sigma_{ij}, \kappa, \omega_m) = 0 \quad (16)$$

Compatibility equation is

$$\frac{\partial f}{\partial \sigma_{ij}} d\sigma_{ij} = -\frac{\partial f}{\partial \kappa} d\kappa - \frac{\partial f}{\partial \omega_m} d\omega_m \quad (17)$$

Increment of equivalent plastic strain is indicated as

$$d\kappa = (d\varepsilon_{ij}^p d\varepsilon_{ij}^p)^{1/2} = d\lambda \left(\frac{\partial g}{\partial \sigma_{ij}} \frac{\partial g}{\partial \sigma_{ij}}\right)^{1/2} \quad (18)$$

Thinking for couple of elasticity and plasticity, strain increment is written as

$$d\varepsilon_{ij} = d\varepsilon_{ij}^R + d\varepsilon_{ij}^I = d\varepsilon_{ij}^R + d\varepsilon_{ij}^p + d\varepsilon_{ij}^c \qquad (19)$$

where $d\varepsilon_{ij}^R$ = reversible strain, $d\varepsilon_{ij}^p$ = plastic strain and $d\varepsilon_{ij}^c$ = elastic-plastic coupled strain. $d\varepsilon_{ij}^R$ is

$$d\varepsilon_{ij}^R = C_{ijkl} d\sigma_{kl} \qquad (20)$$

where C_{ijkl} = flexible tensor of damaged soil, which is function of damage variables. The strain increment of elastic-plastic coupling is generated with the change of flexible tensor of damaged soil owing to the change of damage variables

$$d\varepsilon_{ij}^c = dC_{ijkl}\sigma_{kl} \qquad (21)$$

where $dC_{ijkl} = \dfrac{\partial C_{ijkl}}{\partial \omega_m} d\omega_m$. Since

$$d\varepsilon_{ij}^R + d\varepsilon_{ij}^c = C_{ijkl} d\sigma_{kl} + dC_{ijkl}\sigma_{kl} = d\varepsilon_{ij}^e \qquad (22)$$

so the sum of reversible strain and elastic-plastic coupled strain could be called generalized elastic strain. Eq.(19) could be rewritten as

$$d\varepsilon_{ij} = d\varepsilon_{ij}^e + d\varepsilon_{ij}^p = d\varepsilon_{ij}^e + d\lambda \frac{\partial g}{\partial \sigma_{ij}} \qquad (23)$$

On the base of eq.(17), there is

$$\frac{\partial f}{\partial \sigma_{ij}} \Delta_{ijkl} d\varepsilon_{kl} = \frac{\partial f}{\partial \sigma_{ij}} \Delta_{ijkl} d\varepsilon_{kl}^e + d\lambda \frac{\partial f}{\partial \sigma_{ij}} \Delta_{ijkl} \frac{\partial g}{\partial \sigma_{kl}} \qquad (24)$$

since

$$d\sigma_{ij} = \Delta_{ijkl} d\varepsilon_{kl}^e + \frac{\partial \Delta_{ijkl}}{\partial \omega_m} d\omega_m \varepsilon_{kl}^e \qquad (25)$$

so

$$\frac{\partial f}{\partial \sigma_{ij}} \Delta_{ijkl} d\varepsilon_{kl} = \frac{\partial f}{\partial \sigma_{ij}} d\sigma_{ij} - \frac{\partial f}{\partial \sigma_{ij}} \frac{\partial \Delta_{ijkl}}{\partial \omega_m} d\omega_m \varepsilon_{kl}^e$$
$$+ d\lambda \frac{\partial f}{\partial \sigma_{ij}} \frac{\partial \Delta_{ijkl}}{\partial \omega_m} \frac{\partial g}{\partial \sigma_{kl}} \qquad (26)$$

Eq.(26) would be used into compatibility equation and defining

$$A_1 = \frac{\partial f}{\partial \kappa}\left(\frac{\partial g}{\partial \sigma_{ij}}\frac{\partial g}{\partial \sigma_{ij}}\right)^{1/2} \qquad A_2 = \frac{\partial f}{\partial \sigma_{ij}} \Delta_{ijkl} \frac{\partial g}{\partial \sigma_{kl}}$$

plastic factor would be attained as follow

$$d\lambda = \frac{\dfrac{\partial f}{\partial \sigma_{ij}} \Delta_{ijkl} d\varepsilon_{kl} + \left(\dfrac{\partial f}{\partial \omega_m} + \dfrac{\partial f}{\partial \sigma_{ij}}\dfrac{\partial \Delta_{ijkl}}{\partial \omega_m}\varepsilon_{kl}^e\right)d\omega_m}{A_2 - A_1} \qquad (27)$$

If eq.(23) and eq.(27) would be applied in eq.(25), so elastic-plastic coupling constitutive model thinking damage of soil structure is attained. The form is

$$d\sigma_{ij} = \left[\Delta_{ijkl} - \frac{A_2 \Delta_{ijkl}}{A_2 - A_1}\right]d\varepsilon_{kl}$$
$$+ \frac{\left(A_1 \dfrac{\partial \Delta_{ijkl}}{\partial \omega_m}\varepsilon_{kl} - \Delta_{ijkl}\dfrac{\partial f}{\partial \omega_m}\dfrac{\partial g}{\partial \sigma_{kl}}\right)d\omega_m}{A_2 - A_1} \qquad (28)$$

where A_1 is parameter to indicate strain hardening or strain softening of damaged soil. When $A_1<0$, it shows strain hardening. When $A_1>0$, it means strain softening. When $f = g$, $\omega_m = \omega$ and $\Delta_{ijkl} = D_{ijkl}(1-\omega)$ are used into eq.(28), so elastic-plastic constitutive equation thinking isotropic damage in condition of the correlation would be written as follow

$$d\sigma_{ij} = \left[D_{ijkl}(1-\omega) - \frac{A_3 D_{ijkl}(1-\omega)}{A_3 - A_1}\right]d\varepsilon_{kl}$$
$$+ \frac{\left(A_1 \dfrac{\sigma_{ij}}{1-\omega} - D_{ijkl}(1-\omega)\dfrac{\partial f}{\partial \omega}\dfrac{\partial f}{\partial \sigma_{kl}}\right)d\omega}{A_3 - A_1} \qquad (29)$$

$$A_3 = \frac{\partial f}{\partial \sigma_{ij}} D_{ijkl}(1-\omega)\frac{\partial f}{\partial \sigma_{kl}}$$

Since the yield surface would be changed owing to influence of structural damage of soil, moreover, from thinking application, this paper suggests that the yield function in condition of damage use Drucker-Prager criterion. Using analogism, the yield function is follow

$$f(\sigma_{ij}, \kappa, \omega) = \alpha I + \sqrt{J} - K \qquad (30)$$

where $I = I_1/(1-\omega)$, I_1 = first effective stress invariant. $J = J_2/(1-\omega)^2$, J_2 = second effective stress invariant of deviatoric stress. α and K are coefficient of test, which are follow

$$\alpha = \sqrt{3}\big/3\sqrt{3 + \sin^2 \varphi'}$$

$$K = \sqrt{3}\left[c' + (u_a - u_w)\tan\varphi^b\right]\cos\varphi'\big/3\sqrt{3 + \sin^2 \varphi'}$$

5 EXPRESSION OF GENERALIZED PORE PRESSURE

The generalized pore pressure defined by eq.(10) is under the influence of matric suction and is determined by real effective cohesion. If φ^b is defined as a constant, increment of shear strength owing to the action of generalized suction could be showed as

$$\Delta\tau_f = \Delta(u_a - u_w)_f \tan\varphi^b \operatorname{ctg}\varphi' \qquad (31)$$

so increment of $\Delta\tilde{p}$ could be written as

$$\Delta\tilde{p} = -\Delta\zeta = -\Delta(u_a - u_w)_f \tan\varphi^b \qquad (32)$$

From eq.(32) and Figure.4 (Frendlund 1993), it illus-trate that the generalized pore pressure could

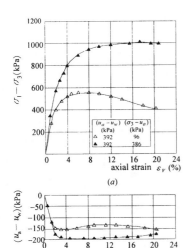

(a)

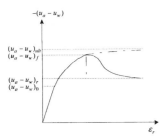

(b)

Figure.4 Triaxial text of Dhanauri 'clay
(a) stress versus strain
(b) matric suction versus strain

$-(u_a - u_w)$

$(u_a - u_w)_{ult}$
$(u_a - u_w)_f$

$(u_a - u_w)_r$
$(u_a - u_w)_0$

ε_r

Figure.5 Fitting sketch of matric
suction versus strain

not be dissipated but be gradually accumulated in procedure of structural damage. The mutual relationship between matric suction and axial strain could be fitted by hyperbolic model or compositive model being composed of straight line and hyperbola. In addition, this relationship is analogous to the specialty of strain softening to some extent. As a consequence, thinking universal law of fitting method, calculated fitting curve of mutual relationship between matric suction and axial strain is assumed as compositive model of straight line and hyperbola (Figure.5). If $(u_a - u_w)_0$ is supposed as proportional limit of matric suction, $(u_a - u_w)_f$ as failure matric suc-tion, $(u_a - u_w)_r$ as residual matric suction, so calculating formula of fitting curve could be written as follow

$$(u_a - u_w) = H(u_a - u_w)_0 + \frac{\varepsilon_r - H\varepsilon_0}{\chi + H\beta(\varepsilon_r - H\varepsilon_0)} \quad (33)$$

where ε_0 = axial strain when $(u_a - u_w) = (u_a - u_w)_0$. χ and β are hyperbolic parameters. H = Heaviside's function, when $(u_a - u_w) \geq (u_a - u_w)_0$, $H = 1$, or else H

=0. So eq.(14) could be rewritten with iterative calculated method

$$[K]\{\Delta u\} = \{\Delta F\} - [C]\{\Delta \tilde{p}(\Delta u)\} \quad (34)$$

6 SLOPE STABILITY ANALYSIS OF UNSATURATED SOIL

A slope of unsaturated soil (Figure.6), as an example, is taken to calculate with proposed calculating model of this paper. The height of slope is 6m. Excavating procedure of the slope is divided into two steps. The depth of every excavating step is 3m. Calculated parameters are in Table.1.

Table.1 Chief calculating parameters for unsaturated soil slope

Calculating parameters	Value
Elastic modulus	180 MPa
Poisson's ratio	0.4
Effective cohesion	18.6 kPa
Angle of effective internal friction	$26°$
Proportional limit of matric suction	200kPa
Strain of proportional matric suction	0.03
Unit weight	20kN/m^3

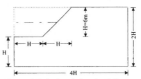

Figure.6 Calculated sketch of slope

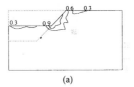

(a)

unit: kPa

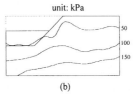

(b)

Figure.7 Calculated results of 1st
excavating step
(a) Contour of damage variable ω
(b) Contour of major principal stress

The chief results for the excavation of two steps, respectively, are shown in Figure.7 and Figure.8. From contour of damage variable and major principal stress, they could be explained that chief distributing law of soil damage as compared with that of major principal stress is in common. If the ratio of limit

135

bearing capacity and needed capacity is defined as safety factor for stability analysis (Song E X 1997. Zou J Z, et al. 1995), the curve of safety factor (F_s) versus displacement (d) of point at the top of slope is shown in Figure.9. It illustrates that the safety factor of excavated slope in first step would gradually incline to a constant, but in second step it is strikingly reduced.

7 CONCLUSIONS

1. On the base of principle of continuum damage mechanics, damage variable for unsaturated soil is defined. The relationship both damage variable and matric suction could be fitted by exponential function.
2. The shear strength of unsaturated soil is considered as the sum of effective frictional force and generalized frictional force.
3. The curve of matric suction versus axial strain could be simulated by compositive model of straight line and hyperbola, which can be conveniently used into iterative increment equation of FEM.
4. Elastic-plastic damage constitutive law is attained if damage variable is assumed as internal variable of plastic potential function.
5. The results of stability analysis for unsaturated soil slope indicate that the distributive law of damage variable and major principal stress is in common.

REFERENCES

Bight G. B., 1967. Effective stress evaluation for unsaturated soils. ASCE J. Soil Mech. Found. Eng. Div., 93(SM2):125-148.
Frendlund D.G., Rahardjo H., 1993. Soil mechanics for unsaturated soil. New York: John Wiley & Sons.
Lemaitre J., 1971. Evaluation of dissipation and damage in metals submitted to dynamic loading. Proceedings of ICM-1.
Kachanov L M. 1986. Introduction to Continuum Damage Mechanics. Martinus Nijhoff Publishers, Dordrecht, The Netherlands.
Shen Z J, 1997. Generalized pore pressure model for strain softening materials. Chinese J. Geot. Eng., 19(3):14-21
Song E X, 1997. Finite element analysis of safety factor for soil structure. Chinese J. Geot. Eng., 19(2):1-7.
Sun H, Zhao X H, 1999. An analysis of elastoplastic anisotropic damage on soft soil. Chinese Rock and soil Mech., 20(3):7-12
Zou J Z, Williams D J, Xiong W L, 1995. Search for critical slip surface based on finite element method. Canadian Geot. J., 32(1):233-246.

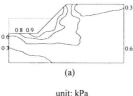

(a)

unit: kPa

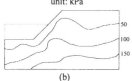

(b)

Figure.8 Calculating results of 2nd excavating step
(a) Contour of damage variable ω
(b) Contour of major principal stress

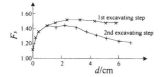

Figure.9 Safety factor versus displacement at the top of slope

Unsaturated Soils for Asia, Rahardjo, Toll & Leong (eds) © 2000 Taylor & Francis, ISBN 90 5809 139 2

Constitutive models for triaxial consolidation and shearing of unsaturated soil

M.A. Rashid & M.Z. Abedin

Department of Farm Structure, Bangladesh Agricultural University, Mymensingh, Bangladesh

ABSTRACT: The consolidation and shearing parameters of soil largely depend on their constitutive models. The widely accepted v-$\ln(p)$ model of soil consolidation is unable to describe the specific volume v at the mean normal stress $p = 0$ and its linear parameters vary with the soil water content w and stress p. The normalized $\ln(v)$-p/p_n model successfully describes v at any p as a function of w. In fact, the parameters of dry soil at $p = 0$ would be the basic characteristics for the basis on soil fabrics. The soil states are precisely described by the p,v,S_r instead of p,v,w for continuous changes of soil saturation S_r with consolidation pressure. For the soil states p, v, S_r, the normalized model extends to the $\ln(v)$-p/p_n-S_r form that rolls on a surface generated in its self-contained orthogonal axes. Accordingly, for the soil states p, v, S_r, the model of shearing stresses takes on the M-p/p_n-S_r form. The linear parameters of the extended models of consolidation and triaxial shearing being independence of the stress and moisture status are the basic characteristics of soil fabrics.

1 INTRODUCTION

The natural soil consists mainly of solid particles and pore space e. The pore space may be occupied by air or water. The air and water occupy their respective voids e_a and e_w with pressures u_a and u_w. The variable relevant to volume change is e_a as e_w changes very little with consolidation pressure. The critical state theory relates the shear strength with e in the form of a constitutive model. The basic concepts of critical state mechanics originally developed from the plasticity theory of saturated soils are equally applicable to unsaturated soils (Toll 1990, Wheeler & Sivakumar 1995, Abedin 1995). The effective stress concept as used in the saturated soil leads to difficulty when used for unsaturated soils but is valid in terms of total stresses (Kirby 1989, 1991).

The standard triaxial test used to investigate the volumetric deformation of soil loaded by a combination of two stress-components p and q is carried out dividing the test into the consolidation and shearing parts. In the consolidation part, the soil sample of loose state is loaded isotropically by the all-round cell pressure ($p = \sigma_3$) through a repeated unloading-reloading process until a pre-determined consolidation pressure p_n is achieved. The values of v corresponding to a set of p stresses form a compression line and a family of unloading-reloading lines in the $p{:}v$ plane with zero excess pore pressure. In the shearing part, the soil is sheared by increasing the ram pressure, keeping the cell pressure at or below p_n. The shear tests conducted at different ambient stresses measure the p, q and v values at soil failure. The linear parameters of the compression lines are the characteristics of loaded soil.

The volumetric deformation with consolidation depends on the soil states at loading time. The soil behaviour under compression is characterized by the state and fabric parameters. The state parameter ψ combines the influences of p and v and the fabric parameter characterizes the soil grain arrangements (Mitchell 1976). The mean normal stress, p is enough for description of ψ, because q directly reflects the soil fabric parameter (Been & Jefferies 1985). The linkages of p and q with the octahedral stresses under triaxial loading condition, $\sigma_2 = \sigma_3$ are as follows.

$$p = \sigma_{oct} = \frac{1}{3}(\sigma_1 + 2\sigma_3) \qquad (1a)$$

$$q = \frac{1}{\sqrt{2}}\tau_{oct} = \frac{1}{3}(\sigma_1 - \sigma_3) \qquad (1b)$$

Here p is the mean normal stress and q is the mean deviator stress. The difference of p and q is noticed always constant with respect to the cell pressure σ_3.

The fabric or soil grain arrangement is largely affected by its moisture status commonly described by w. The specific gravity G together with w produces the pore water void e_w, a constant irrespective of soil volume deformations. The soil moisture status is precisely described by $S_r = e_w/e$ as it always alters with changes of v. Attempts are, therefore, made to constitute the consolidation and shearing models taking p, v, S_r as the soil states instead of p,v,w.

2 SOIL CONSOLIDATION MODEL

The Cambridge model of soil consolidation given by v-ln(p) has been constituted on a linear basis, paying little attention to the mathematical principles of data-scale transformation. Moreover, the model is applicable to $p \geq 1$ kPa thus limiting its use. The Bailey and Johnson's (1989) model with v in log-scale describe v at any p (Petersen 1993) is too complicated for a large number of derived parameters. The three-parametric model of Harris (1993) takes on the awkward v-ln(σ_1) relationship at $\sigma_1 > \sigma_3$.

Mathematically, the stress p for possible value of zero can never be transformed to the log-scale. Moreover, a model comprising of dissimilar variables (e.g. p pressure, v ratio) always produces biased values of linear parameters. In general, the pre-determined pressure p_n be constant for all tests conducted for a specified purpose. Normalizing p by p_n and taking in arithmetic scale for its possible value of zero and v in log-scale for its minimum value of unity, the soil consolidation model takes on ln(v)-p/p_n relationship. The different compression lines, namely the normal consolidation line (NCL), critical state line (CSL), tension cut off line (TCL) and elastic reloading line (ERL) plotted on the ln(v): p/p_n plane are shown in Figure 1. The differential equation of a line on this plane is:

$$\delta v / v = -\lambda \delta(p/p_n) \qquad (2)$$

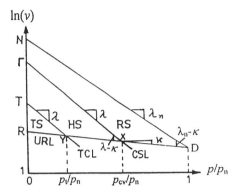

Figure 1. Compression lines on the ln(v):p/p_n plane

The ln(v)-p/p_n model describes v at any level of p as a function of w. It is, therefore, essential to constitute a model that will produce the soil parameters based on soil fabric only.

The volumetric deformation is governed by the soil fabric and moisture status described precisely by S_r. The ln(v)-p/p_n model for soil states p,v,S_r extends to ln(v)-p/p_n-S_r relationship, which at linearity takes on the form:

$$\ln(v) = \ln(v_0) - \lambda(p/p_n) - \mu S_r \qquad (3)$$

where, v_0 = specific volume of soil at $S_r=0$ and $p=0$, λ = soil parameter relating to p, and μ = soil parameter relating to S_r. All these are soil fabric parameters for independence on p and S_r.

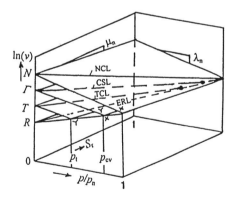

Figure 2. Typical plots of compression lines & planes

Typical compression planes on the ln(v): p/p_n : S_r space are depicted in Figure 2. All these compression lines roll over their self-contained surfaces with changes in p and S_r. The equations of orthogonal compression lines, NCL, CSL, TCL and ERL are as follows.

NCL : $\ln(v_n) = \ln(N) - \lambda_n (p_i/p_n) - \mu_n S_{ri}$ (4a)

CSL : $\ln(v_c) = \ln(\Gamma) - \lambda(p_c/p_n) - \mu S_{rc}$ (4b)

TCL : $\ln(v_t) = \ln(T) - \lambda(p_t/p_n) - \mu S_{rt}$ (4c)

ERL : $\ln(v) = \ln(R) - \kappa (p/p_n) - \omega S_r$ (4d)

where, N, Γ, T, R = specific volumes of soil at $S_r = 0$ and $p = 0$, $\lambda_n, \lambda, \kappa$ = soil parameters relating to p and μ_n, μ, ω = soil parameters relating to S_r with respect to the NCL, CSL or TCL and ERL. TCL is taken parallel to the CSL.

Notice from Figure 2 that the ERP (elastic reloading plane) intersects both the CSP and TCP. The intersection of ERP and CSP along XX line represents soil shearing at the ambient state satisfying the conditions $v = v_c$, $S_r = S_{rc} = S_{rv}$ (say) and $p = p_c = p_{cv}$ (say). Substituting these values into Equations 4(b) and 4(d) and then solving for the stress ratio,

$$p_{cv}/p_n = [\ln(\Gamma/R) - (\mu - \omega) S_{rv}] / (\lambda - \kappa)$$ (5)

Similarly, the intersection of ERP and TCP along the YY line represents the soil-state at zero tension having $v = v_t$, $S_r = S_{rt}$, $p = p_t$. Substituting these values in Equations 4(c) and 4(d) and then solving

$$p_t/p_n = [\ln(T/R) - (\mu - \omega) S_{rt}] / (\lambda - \kappa)$$ (6)

Dividing this by Equation 5, we get,

$$\frac{p_t}{p_{cv}} = \frac{\ln(T/R) - (\mu - \omega)S_{rt}}{\ln(\Gamma/R) - (\mu - \omega)S_{rv}}$$ (7)

Equations 5 through 7 without S_r reduce to those obtained from the $\ln(v)$- p/p_n model.

The state parameter ψ that embodies p, q and v relative to the CSL as a dimensionless factor is actually the difference in volumes measured from the ERL to CSL at the same stress level. The state parameter of a soil having ambient states p, S_r and failure state p_c, S_{rc} is described in Figure 3. By trigonometry, from ΔOAB, we get,

$$\psi = (\lambda - \kappa) (p - p_{cv}) / p_n$$ (8)

This shows that ψ is negative, positive and zero with respect to the stress conditions $p < p_{cv}$, $p > p_{cv}$

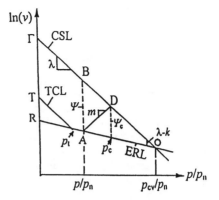

Figure 3. State parameters of soil

and $p = p_{cv}$. A negative ψ associates with soil dilation in the super-critical zone and a positive ψ with soil contraction in the sub-critical zone. The condition $\psi = 0$ indicates soil failing at the ambient state or constant volume. The nature of ψ thus depends on the position of p relative to p_{cv} which is consistent with the postulation of Been et al. (1986 & 1987). From the trigonometry of ΔABD

at A: $\psi = (m + \lambda)(p - p_c)/p_n$ (9)

at p_c: $\psi_c = (m + \kappa)(p - p_c)/p_n$ (10)

Dividing ψ by ψ_c and then solving for m, we get,

$$m = \frac{\lambda \psi_c - \kappa \psi}{\psi - \psi_c}$$ (11)

The slope m, in fact, is the coefficient of soil dilatancy and $\tan^{-1}(m)$ is the angle of dilation or contraction with respect to the positive and negative values of m.

3 SOIL SHEARING MODEL

The stress variables relevant to the Mohr-Coulomb failure condition of unsaturated soils are $(\sigma - u_a)$, $(u_a - u_w)$ and τ (Fredlund & Morgenstern 1977). A change in e or e_a has negligible influence on u_w as the pore space of topsoil has a free path to the atmosphere. This reduces the stress variables to σ, τ and u_w (suction). The stress variables of dry soil reduce to σ and τ and their magnitudes depend solely on the soil fabrics. The Mohr-Coulomb equation for soil stresses at failure is

$$(\sigma_1 - \sigma_3) = (\sigma_1 + \sigma_3) \sin\varphi + 2C\cos\varphi$$

which in terms of p and q is

$$q = q_0 + Mp \qquad (12)$$

where, $M = \dfrac{2\sin\varphi}{3-\sin\varphi}$ and $q_0 = MC\cot\varphi$

Toll (1990) proposed a similar expression for triaxial stresses of soil. Equation 12 produces an intercept q_0 on the ordinate of $q{:}p$ plane when soil cohesion C involves in shearing. The stress circle satisfying Equation 12 as depicted in Figure 4 shows that the circle has the diameter of $3q$ and its rear diametric point locates at a distance $\sigma_3 = p - q$ from the intersection of axes.

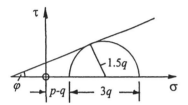

Figure 4. Mohr's circle of triaxial stress

Actually, the critical stress ratio M is an indicator of the mechanical behaviour of soil under shearing and its value varies with fabric and ambient states of the soil. Strictly speaking, the dry-soil under zero confining stress ($p = 0$) will contain a unique mechanical behaviour that will depend only on soil fabric. The linear expression for $M = q/p$ in terms of p/p_n, S_r takes on the form:

$$M = M_0 - \eta(p/p_n) - \beta\, S_r \qquad (13)$$

where, M = critical stress ratio of soil failure, M_0 = critical stress ratio at $p = 0$ and $S_r = 0$, η = fabric parameter for p and β = fabric parameter for S_r. The soil parameters M_0, η and β are free from the effects of p and S_r. Equation 13 represents as orthogonal line on the $M{:}p/p_n{:}S_r$ space lying on the surface as shown in Figure 5.

The boundary surfaces outside which the combination of p, q and v is impossible, can be described by plotting q and p as shown in Figure 6(a). The geometry of plot and derived parameters remain unaltered even at normalization of p and q by the pre-consolidation pressure p_n.

The ranges in stress coordinates furnishing the different state surfaces bounded by the TCL and

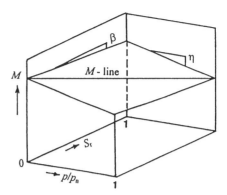

Figure 5. Typical plot for M - line

CSL are $(0,0)$ to (p_t, q_t) for the tension surface (TS), (p_t, q_t) to (p_{cv}, q_{cv}) for the Hvorslev surface (HS) and (p_{cv}, q_{cv}) to $(p_n, 0)$ for the Roscoe surface (RS). The HS cuts an intercept q_0 in the q axis when extended. From Figure 6(a), the intercept,

$$q_0 = (1-h)\,p_t = (M-h)\,p_{cv}$$

or, $p_t/p_{cv} = (M-h)/(1-h) \qquad (14)$

For dry soil, Equation 7 reduces to

$$p_t/p_{cv} = \ln(T/R)/\ln(\Gamma/R) \qquad (15)$$

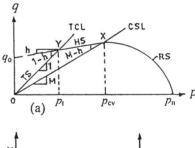

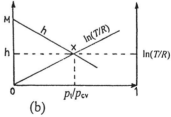

Figure 6. Representation of soil stress : (a) p-q plane, and (b) interpolation of h and p_t/p_{cv}

Table 1. Comparison in values of soil parameters derived from two-parametric consolidation models.

M.C.	Model : v - $\ln(p)$						Model : $\ln(v)$ - p/p_n					
w, %	NCL		CSL		ERL		NCL		CSL		ERL	
	N	λ_n	Γ	λ	R	κ	N	λ_n	Γ	λ	R	κ
7	2.66	0.136	2.34	0.093	1.90	0.004	2.22	0.184	1.94	0.028	1.88	0.007
11	2.78	0.173	2.42	0.133	1.85	0.011	2.25	0.251	1.95	0.142	1.80	0.012
15	2.80	0.171	2.60	0.166	1.88	0.014	2.24	0.229	2.02	0.181	1.84	0.022
18	2.90	0.213	2.58	0.172	1.84	0.024	2.22	0.267	1.99	0.203	1.75	0.032
22	2.92	0.213	2.76	0.195	1.84	0.020	2.22	0.258	2.09	0.224	1.78	0.026

Table 2. Soil parameters derived from extended models of consolidation and shearing.

Items	Sp. vol. at $p = 0$, $S_r = 0$	Regression for p/p_n	Regression for S_r	Determination R^2	Direct contribution, P^2	
					for p/p_n	for S_r
NCL	$N = 2.292$	$\lambda_n = 0.218$	$\mu_n = 0.093$	0.963	**0.779**	0.044
CSL	$\Gamma = 2.117$	$\lambda = 0.108$	$\mu = 0.170$	0.721	**0.418**	**0.233**
TCL	$T = 2.029$	do	do			
ERL	$R = 1.921$	$k = 0.016$	$\omega = 0.122$	0.733	0.031	**0.690**
Ratio M	$M_0 = 0.586$	$\eta = 0.097$	$\beta = 0.500$	0.934	0.079	**0.834**

Separating h and $\ln(T/R)$, we get

$$h = \frac{M - p_t/p_{cv}}{1 - p_t/p_{cv}} \quad \text{and} \quad \ln(T/R) = \ln(\Gamma/R)\frac{p_t}{p_{cv}}$$

The plottings of h and $\ln(T/R)$ vs p_t/p_{cv} given in Figure 6(b) show that h and $\ln(T/R)$ intersect each other at X that locates their values at the optimum p_t/p_{cv}. By simple inspection, we get,

$$h = \ln(T/R) \tag{16}$$

Combining Equations 14 to 16, and solving for h, we get,

$$h = \frac{1}{2}\left[F - \sqrt{F^2 - 4M\ln(\Gamma/R)}\right] \tag{17}$$

in which F = 1 + $\ln(\Gamma/R)$. The slope h is thus a dependent parameter of the soil.

4 RESULTS & DISCUSSIONS

The predictability of consolidation and shearing models are tested on a sandy-clay loam soil (sand = 65.2, silt = 14.5 and clay = 20.3 percents) having plastic limit, PL = 18.5%, liquid limit, LL = 33% and specific gravity, G =2.67. The experiment was conducted at five moisture levels (7, 11, 15, 18 and 22 % by weight basis) to keep S_r within 80%

at the pressure, p_n = 250 kPa. The correlation between p and v following the v-$\ln(p)$ and $\ln(v)$-p/p_n relationships are measured for the soil at each moisture content w. The comparative values of linear parameters given in Table 1 show that the magnitudes of N and Γ in the v-$\ln(p)$ model vary with w. These variations significantly reduce in case of the $\ln(v)$-p/p_n model. These reductions are mainly due to transformation of variables and dimensionless form of the model. The soil parameters calculated from the normalized model are more realistic as it agrees closely with the actual values. The extended models further refine the linear parameters making them independent of the stress and moisture status. The soil parameters derived from 36 data sets are presented in Table 2.

The coefficients R^2 and P^2 respectively quantify the overall and direct contribution of p/p_n and S_r on v and M. For higher P^2 value, the mean normal stress p in normal consolidation, and the soil saturation S_r in elastic swelling governs the soil behaviour which at the critical states depends on both the stress and moisture status. The critical stress ratio M is mostly governed by the S_r as its direct contribution $P^2 = 0.834$ is very large when compared to that of p/p_n. For dry soil, the critical stress ratio $M_0 = 0.586$.

The stress ratio M decreases with gradual increase in initial soil moisture w and is linearly

141

related to w (Figure 7) below the plastic limit (Abedin 1995). The values of M obtained from the q-p model and plotted against w as given in Figure 7 show that at dry state, $M_0 = 0.589$ which is equal to that obtained from the M - p/p_n - S_r model. It is further noted from Figure 7 that the use of normalized water content $\theta = w/PL$ instead of w yields the same M_0.

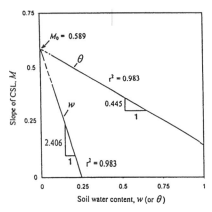

Figure 7. Relationship between M and w or θ

5 CONCLUSIONS

The critical state concept of saturated soil is equally applicable to unsaturated soils in terms of total stresses and the standard triaxial test is sufficient for evaluating the soil parameters. The consolidation model of p/p_n-$\ln(v)$-S_r relationship and the shearing model of M-p/p_n-S_r form are elegant concepts for constitutive modeling of soil behaviour under triaxial loading. The techniques and analyses introduced here are applicable to all soils and open a new avenue of research on critical state soil mechanics modeling to the actual field operations.

REFERENCES

Abedin, M.Z. 1995. The characterization of unsaturated soil behaviour from penetrometer performance and the critical state concept. *Ph.D. Thesis*. The University of Newcastle-upon Tyne, UK.

Bailey, A.C. & C.E. Johnson 1989. A soil compaction model for cylindrical stress-states. *Trans. ASAE.*, 32(3): 822-825.

Been, K. & M.G. Jefferies 1985. A State parameter of Sands. *Geotechnique*, 35 (2): 99-112.

Been, K., J.H.A Crooks, D.E. Becker & M.G. Jefferies 1986. The cone penetration in sands: part I, state parameter interpretation. *Geotechnique*, 36 (2): 239-249.

Been, K., M.G. Jefferies, J.H.A. Crooks & L. Rothenburg 1987. The cone penetration in sands: part II, general inference of state. *Geotechnique*, 37 (3): 285-299.

Fredlund D.G. & N. R. Morgenstern 1977. Stress- state variables for unsaturated soils. *J. Geotech. Eng. Div. ASCE,* 103 (GT 5): 447- 466.

Harris H.D. 1993. A critical state interpolation of soil compaction by anisotropic consolidation. *J. Agric. Engng. Res.*, 55: 265-276.

Kirby J. M. 1989. Measurement of the yield surfaces and critical state of some unsaturated agricultural soils. *J. Soil Sc.*, 40: 167-182.

Kirby J.M. 1991. Critical state soil mechanics parameters and their variation for vertisols in Eastern Australia. *J. Soil Sc.*, 42: 487-499.

Mitchell J.K.1976. Fundamentals of Soil Behaviour. *John Wiley and sons Inc.*, NY.

Petersen C.T. 1993. The variation of critical state parameters with water content for two agricultural soils. *J. Soil Sc.* , 44: 397-410.

Toll, D.G. 1990. A framework for unsaturated soil behaviour. *Geotechnique*, 40: 31-44.

Wheeler, S.J. & V. Sivakumar 1995. An elasto-plastic critical state framework for unsaturated soils. *Geotechnique*, 45 (1): 35-53.

Unsaturated Soils for Asia, Rahardjo, Toll & Leong (eds) © 2000 Taylor & Francis, ISBN 90 5809 139 2

Damage function and a masonry model for Loess

Z.J. Shen
Nanjing Hydraulic Research Institute, People's Republic of China

Z.Q. Hu
Xian University of Technology, People's Republic of China

ABSTRACT: Natural soils usually have their peculiar structure and their behavior cannot be properly modeled by traditional elasto-plastic theory. In this paper a so-called masonry model is proposed in which a damage function is defined and used in parallel with the yield function to predict the deformation of natural loess during loading and wetting.

1 INTRODUCTION

By introducing suction as a new variable, the Cambridge critical state model has been generalized to model the behaviour of unsaturated soils. According to Alonso et al.(1990), a so-called LC (Loading and collapse) locus is assumed for predicting the plastic behaviour of unsaturated soils due to an increase of load and a decrease of suction. This kind of elasto-plastic model seems satisfactory for many compacted soils, but trouble will be encoun-tered when one deals with some natural soils with open macro pores such as loess which usually has a peak deformation at a definite pressure when wetted under different vertical pressures (Figure 1).

Many natural soils have their structure and exhibit quite different behaviour before and after the damage of the structure. A new kind of constitutive model which takes account of the damage of the soil structure during loading and other mechanical effects has been proposed by the first author as a structural model (Shen, 1998). In the following sections, such a model will be developed for loess.

2 DAMAGE FUNCTION

In classic plasticity theory, the yield function is the key variable in characterizing the evolution of plastic deformation during loading. Usually the yield function is expressed as

$$F(\{\sigma\},h) = f(\{\sigma\}) - p(h) = 0 \qquad (1)$$

where h is a hardening parameter. Damage function

Supported by Chinese National Science Foundation (No.19772019)

is the generalization of yield function to describe the evolution of the damage of soil structure. In accordance with the two variable approach, the damage function must be a function of both net stress and suction, and its general formulation can be expressed as

$$G(\{\sigma^*\},s,d) = g(\{\sigma^*\},s) - q(d) = 0 \qquad (2)$$

where $\{\sigma^*\} = \{\sigma\} - u_a\{\delta\}$ is net stress,
$\{\delta\} = \{1\ 1\ 1\ 0\ 0\ 0\}^T$, $s = u_a - u_w$ is suction, d is the damage parameter. q can be regarded as an equivalent stress which causes the structure to be damaged. If an elliptic function for net stress is used as in the Cambridge model, the following expression for g can be suggested

$$g = \sigma_m^* (1 + \frac{\sigma_s^{*2}}{M^2 \sigma_m^{*2}}) \frac{1}{1 + \alpha(s/p_a)^m} \qquad (3)$$

where $\sigma_m^* = (\sigma_1^* + \sigma_2^* + \sigma_3^*)/3$,
$\sigma_s^* = [(\sigma_1^* - \sigma_2^*)^2 + (\sigma_2^* - \sigma_3^*)^2 + (\sigma_3^* - \sigma_1^*)^2]^{1/2}/\sqrt{2}.$

M, m and α are 3 constants, and p_a is atmospheric pressure. However for better fitting the available

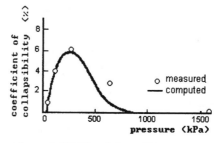

Figure 1 Wetting deformation

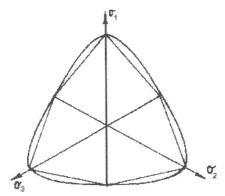

Figure 2 Yield locus

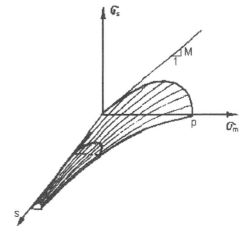

Figure 3 Damage surface

experimental data both in the π plane and in the meridian plane,the following new expression will be used

$$g = \frac{\overset{\bullet}{\sigma}_m}{1-(\overset{\bullet}{\eta}/M)^n}\frac{1}{1+\alpha\,(s/p_a)^m} \quad (4)$$

where

$$\overset{\bullet}{\eta} = \frac{1}{\sqrt{2}}[(\frac{\overset{\bullet}{\sigma}_1-\overset{\bullet}{\sigma}_2}{\overset{\bullet}{\sigma}_1+\overset{\bullet}{\sigma}_2})^2+(\frac{\overset{\bullet}{\sigma}_2-\overset{\bullet}{\sigma}_3}{\overset{\bullet}{\sigma}_2+\overset{\bullet}{\sigma}_3})^2+(\frac{\overset{\bullet}{\sigma}_3-\overset{\bullet}{\sigma}_1}{\overset{\bullet}{\sigma}_3+\overset{\bullet}{\sigma}_1})^2]^{1/2}$$

The curve for $\overset{\bullet}{\eta}$ *=constant is shown in Figure 2 (Shen, 1989). A damage surface in σ_m— σ_s— s space for n =1.2 and m=1 is shown in Figure 3. Equation 4 explains that either increase of σ_m* and η * or decrease of s can cause the damage of the soil structure.

The damage parameter d is a measure of damage. A common definition of d is damage ratio, i.e. the ratio of volume occupied by the damaged part to the total volume of the examined element. However, this ratio is unable to be measured, and in the following another definition will be given.

3 A MASONRY MODEL

3.1 *Basic assumption*

Many natural soils exhibit essentially elastic behaviour when loaded at low stress levels. Therefore we can regard the natural soil sample as an elastic body intersected by randomly distributed joints, i.e. like a masonry. When the load reaches a threshold value the weakest joints will break down first making the sample like an assembly of several lumps. When the load increases further, the lumps are crushed again and their size get smaller and smaller. Finally, when all the soil aggregates have been destroyed, a sample with dispersed particles is obtained. Figure 4 demonstrates the process of shear band formation of a masonry sample.

Accordingly, the behaviour of soil after the threshold of loading has been exceeded can be characterized by an assembly of stone-like lumps

vulnerable to crushing. An elastoplastic model can be used for it, but an additional term of plastic strain due to the crushing of stones must be added. Therefore, when the associated flow law is adopted the stress-strain relationship in the incremental form can be likely written as

$$\{\Delta\varepsilon\} = [C]\{\Delta\overset{\bullet}{\sigma}\}+A_p\{\frac{\partial f}{\partial\overset{\bullet}{\sigma}}\}\Delta f+A_d\{\frac{\partial g}{\partial\overset{\bullet}{\sigma}}\}\Delta g \quad (5)$$

where $[C]$ is elastic compliance matrix, A_p is plastic coefficient due to yielding, A_d is the same coefficient due to damage.

3.2 *Plastic deformation due to yielding*

For the stone-like assembly a yield function similar to that used in Equation 4 will be used

$$f = \frac{\overset{\bullet}{\sigma}_m}{1-(\overset{\bullet}{\eta}/M)^n} \quad (6)$$

Adopting the plastic volumetric strain as a hardening parameter, a hardening rule similar to that of the Cambridge model can be expressed as

$$p = p_0\,exp(\frac{1+e_0}{\lambda-\kappa}\varepsilon_v) \quad (7)$$

where p_0 is a reference pressure when ε_v =0, λ and κ is the slope of e-ln$p$ curve for virgin compression and rebound respectively. The incremental form of this equation will be

$$\Delta\varepsilon_v^p = \frac{\lambda-\kappa}{1+e_0}\frac{\Delta p}{p} \quad (8)$$

In the case of isotropic compression, Δf= Δp, $\{\partial f/\partial\sigma^*\} = \partial f/\partial\overset{\bullet}{\sigma}_m$
Comparison of Equation 8 with the second term of Equation 5 yields the following expression for the plastic coefficient A_p

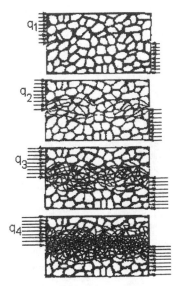

Figure 4 Shearing of a masonry sample

$$A_p = \frac{\lambda - \kappa}{1+e_0} \frac{1}{p \partial f / \partial \sigma_m^{\cdot}} \qquad (9)$$

3.3 Plastic deformation due to damage

For determinating plastic coefficient A_d we must choose the damage parameter d first. As mentioned above the damage ratio commonly used in the classical damage mechanics will be abandoned here. A reasonable way is taking the size of stone-like lumps as the damage parameter which reduces progressively in the process of damage. But this quantity also cannot be measured. In the following we shall define the parameter d as

$$d = \frac{e_0 - e}{e_0 - e_s} \qquad (10)$$

where e is current void ratio, e_0 is its initial value, and e_s is the stable void ratio obtained from compression test of a saturated reconstituted sample of the same soil (Shen, 1996, Figure 5). This definition gives a complete damage (d=1) when $e=e_s$, and no damage at all (d=0) when $e=e_0$.

Now a damage law will be postulated as follows

$$q = q_0 + a \frac{d}{1-d} \qquad (11)$$

where a is the initial slope of $q \sim d$ curve. This equation satisfying the following condition: $q=q_0$, d =0 and $q \to \infty$, d=1. From Equation 10, Δd = $(-\Delta e + d \Delta e_s)/(e_0 - e_s)$, and accounting for $\Delta e_s = -\lambda$ $\Delta p/p$, $d=(e-e_s)/a(e_0-e_s)$) and $\Delta \varepsilon_v = -\Delta e/(1+e_0)$, then for isotropic compression $\Delta p = \Delta q = \Delta g$, and $\{\partial g/\partial \sigma^{\cdot}\} = \partial g/\partial \sigma_m^{\cdot}$

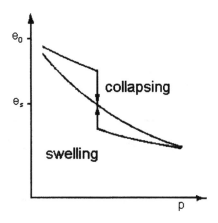

Figure 5 Stable void ratio

the following expression for A_d is obtained from Equation 5

$$A_d = \frac{(e-e_s)/a + (d \lambda_1 - \lambda)/p}{(1+e_0)\dfrac{\partial g}{\partial \sigma_m^{\cdot}}} \qquad (12)$$

3.4 Final expression

After inserting $\quad \Delta f = \{\dfrac{\partial f}{\partial \sigma^{\cdot}}\}^T \{\Delta \sigma^{\cdot}\} \quad$ and

$$\Delta g = \{\frac{\partial g}{\partial \sigma^{\cdot}}\}^T \{\Delta \sigma^{\cdot}\} + \frac{\partial g}{\partial s} \Delta s$$

the final expression of stress-strain relationship is as follows

$$\{\Delta \varepsilon\} = ([C] + A_p[C]_f + A_d[C]_g)\{\Delta \sigma^{\cdot}\}$$
$$\{+A_d[C]_g \frac{\partial g}{\partial s}\}\{\delta\}\Delta s \qquad (13)$$

where

$$[C]_f = \{\frac{\partial f}{\partial \sigma^{\cdot}}\}^T\{\frac{\partial f}{\partial \sigma^{\cdot}}\}, \quad [C]_g = \{\frac{\partial g}{\partial \sigma^{\cdot}}\}^T\{\frac{\partial g}{\partial \sigma^{\cdot}}\}$$

and

$$\frac{\partial g}{\partial s} = \frac{\sigma_m^{\cdot}}{1 - (\eta^*/M)^n} \frac{a\, m(s/p_a)^{m-1}}{p_a[1 + a\, (s/p_a)^m]^2} \qquad (14)$$

4 PARAMETER STUDY

In the above-mentioned Equations, 6 parameters M , λ , κ , a , q_0 and a are encountered. Among them M and λ must vary with the damage parameter, because with the decrease of the size of the stone-like lumps the shear strength of the sample decreases and its compressibility increases. However the rebound coefficient κ will be taken as a constant for simplicity.

M is closely related with the internal friction angle and can be calculated as follows

145

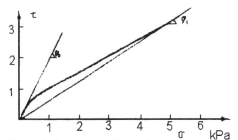

Figure 6 Shear strength envelop

$$M = \sqrt[n]{1+n} \sin \phi \qquad (15)$$

If ϕ_0 and ϕ_1 is the friction angle for undamaged (d=0) and completely damaged (d=1) sample respectively, the following linear interpolation formula will be used to get ϕ at an intermediate value of d

$$\phi = \phi_0 - d(\phi_0 - \phi_1) \qquad (16)$$

ϕ_0 and ϕ_1 can be obtained from the result of triaxial test of natural samples under low and high confining pressures, (see Figure 6). A similar interpolation formula will be used for parameter λ

$$\lambda = \lambda_0 + d(\lambda_1 - \lambda_0) \qquad (17)$$

λ_0 and λ_1 can be determined from the slope of e-ln$p$ curve at small and high stress level respectively as shown in Figure 7. From this figure we can obtain another pair of parameters p_s and p_u for saturated and unsaturated sample respectively, from which the q_0 in Equation 11 and α in Equation 4 can be deduced under assumption that $q \approx 1.2p$ and m=2. Finally, the value a in Equation 11 is obtained by trial and error to obtain a best fit to the experimental data.

5 COMPUTED EXEMPLES

A series of test with artificially prepared loess samples having similar structure with natural loess has been tested in order to determine the model constants and to verify the model behaviour. The obtained constants are: e_0=1.0, s_0=160kPa, w_0=17.7%, λ_0=0.015, λ_1=0.09, sin ϕ_0=0.90, sin ϕ_1=0.52, q_0=30kPa, a=0.4, α =1.4. In addition $\kappa = \lambda_0$ is assumed and $Ei=q_0/\lambda_0$ and v=0.33 are used for the elastic behaviour of the samples before the threshold value q_0 has been exceeded. The computed results are also shown in Figure 1 and Figure 7.

6 CONCLUSIONS

1. Natural loess has a peculiar structure with open

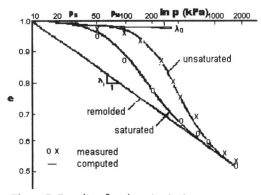

Figure 7 Results of oedometer test

macropores, and its deformation behaviour during loading and wetting cannot be properly modeled by simple extension of existing soil plasticity and introducing only an additional term to take account of suction.

2. The masonry model proposed in this paper regards the soil sample as an assembly of lumps connected with weak bonds which can be destroyed during loading and wetting

3. According to this model the deformation of a sample consists of 3 parts:① elastic deformation of lumps, ② plastic deformation due to sliding between lumps and ③ irrecoverable deformation due to crushing of lumps which can be described by a damage function.

4. The damage function can be defined similar to the yield function, but an extra term with suction as an independent variable must be added to reflect the action of wetting.

5. The preliminary verification shows that the new model might offer an effective way in modeling many natural soils.

REFERENCES

Alonso E.E., Gens A.& Josa A.1990. A constitutive model for partially consolidated soils,*Geotechnique*, 40(3):405-439

Shen Z.J.1998. Advances in numerical modeling of deformation behaviors of unsaturated soils, State -of-art report,*2nd Int. Conf. Unsaturated Soils*, Beijing, II: 180-193

Shen Z.J. 1989. A stress-strain model for sands under complex loading,*Advances in Constitutive Laws for Engineering Materials*, Edited by: Fan J. and Murakami S. I:303-308

Shen Z.J. 1996. Generalized suction and unified deformation theory of unsaturated soils,*Chinese J. of Geot. Eng.*18(2):1-8

Unsaturated Soils for Asia, Rahardjo, Toll & Leong (eds) © 2000 Taylor & Francis, ISBN 90 5809 139 2

Change of shear strength due to surface tension and matric suction of pore water in unsaturated sandy soils

K. Shimada
Faculty of Agriculture, Tokyo University of Agriculture and Technology, Japan

H. Fujii, S. Nishimura & T. Nishiyama
Faculty of Environmental Science and Technology, Okayama University, Japan

T. Morii
Faculty of Agriculture, Niigata University, Japan

ABSTRACT : This paper discusses the change of the normal stress due to the surface tension and the matric suction of the pore water in unsaturated assemblies of spheres, and also discusses the change of the shear strength of sandy soils with the matric suction. Theoretically derived are the relationships between the change of the normal stress and the matric suction in unsaturated assemblies of regularly/randomly packed equal-sized/non equal-sized spheres. Shear tests with poorly- and well-graded sandy soils were also carried out. Experimental results show the same tendency of the relationships between the change of the normal stress and the matric suction as that from theoretical considerations.

1 INTRODUCTION

Water held among soil particles forms capillary menisci in an unsaturated soil. The curved air-water interface produces the negative pressure in the pore water and then generates interparticle attractive forces among the soil particles. Fisher(1926) has shown a formula giving the interparticle force at the contact point of two equal-sized spheres. Although the forces act in all directions, the forces acting normal to a shear plane are essential because they increase the normal stress, then increase the shear strength of the sphere assembly. Shimada et al. (1993) have presented the change of the normal stress due to the pore water in unsaturated assemblies of regularly and randomly packed equal-sized spheres.

This paper first reviews the theoretical consideration for unsaturated assemblies of regularly and randomly packed equal-sized spheres. Secondly, it extends the theory to assemblies of non equal-sized spheres. Finally, it compares the theoretical results with experimental results of sandy soils.

2 INTERPARTICLE ATTRACTIVE FORCE AT A SINGLE CONTACT POINT

Figure 1 shows two equal-sized spheres placed vertically and holding water between them. When we assume that the contact angle between water surface and solid surface is zero, the radii of curvature of the menisci in its two principal directions, R_1 and R_2 can be expressed in the next equation (1) with the water retention angle θ from geometrical consideration (Fisher, 1926),

$$R_1 = R (\sec\theta - 1), \quad R_2 = R (1 + \tan\theta - \sec\theta) \quad (1)$$

where R is the radius of the spheres. Then, the expression for the matric suction (S_u) can be,

$$S_u = T (1/R_1 - 1/R_2) \quad (2)$$

where T is the surface tension of pore water.

The tension exerted by the air-water interface also contributes to the interparticle attractive force. The tension (F_t) is equal to the circumference of the interface multiplied by the surface tension of water. Then, the total interparticle force (F_1) at a single contact point is the sum of the force due to the matric suction (F_s) and the force due to the surface tension (F_t) as follows:

$$F_1 = F_s + F_t = \pi R_2^2 S_u + 2\pi R_2 T = 2\pi RT / \{1 + \tan(\theta/2)\} \quad (3)$$

The arrangement in Figure 1 occupies the square whose area is $4R^2$ in the plane. This area is called 'Area of unit (A_1)'. Then, we can calculate the increase of the normal stress ($\Delta\sigma$) between the spheres as follows:

$$\Delta\sigma = F_1 / (4R^2) = \pi T / [2R \{ 1 + \tan(\theta/2)\}] \quad (4)$$

When the water retention angle (θ) is given, $\Delta\sigma$ can be calculated with Eq.(4) and the matric suction (S_u) can be calculated with Eqs.(1) and (2). Then, we can obtain the relationship between the matric suction and the increase of the normal stress. The bold line in Figure 2 shows the change of $\Delta\sigma$ with S_u for the case of $R=5$ μm. Shown are also the component due to the matric suction ($\Delta\sigma_s$) and that due to the surface tension ($\Delta\sigma_t$). Since the fringes of menisci meet each other when θ becomes 45 degrees, the calculation is carried out up to $\theta=45$ degrees. $\Delta\sigma_s$ increases with S_u, but $\Delta\sigma_t$ decreases. Then, the summation of them ($\Delta\sigma$) does not show the great change with S_u. When θ becomes greater than 45 degrees, the fusion of the menisci occurs. Then, we cannot apply Eq.(4) for that range. However, since we know that $\Delta\sigma$ is equal to zero when $S_u = 0$, we can assume that $\Delta\sigma$ will greatly decrease to zero with the decrease of S_u.

3 INCREASE OF NORMAL STRESS DUE TO PORE WATER

3.1 in assembly of regularly packed equal-sized spheres

Five different types of packing of equal-sized spheres (Mitchell, 1993) are listed and their properties are shown in Table 1. The meanings of each property are explained in the bottom of the table.

The change of the normal stress ($\Delta\sigma$) can be expressed as the summation of the vertical components of F_1 of all contact points on the shear plane:

$$\Delta\sigma = F_1 K' Z'/A_1 = \beta F_1 / R^2 \qquad (5)$$

where A_1 is the area of unit and is constant for each type of packing.

Figure 3 shows the calculation results for each type of packing of spheres whose radius is $R = 5\,\mu m$. As mentioned previously, Eq.(5) is valid provided that the water retention angle (θ) is smaller than θ_c. Then, the lines for the lower part of the matric suction are not presented in the figure. As the form of Eq.(5) is the same as Eq.(4), $\Delta\sigma$ does not show the great change with S_u.

3.2 in assembly of randomly packed equal-sized spheres

Since it is difficult to estimate the area of unit in an assembly of randomly packed spheres, we calculate the change of the normal stress from the number of spheres in a unit area of the assembly. The calculation procedure is shown in Figure 4. The left column is for the calculation of the interparticle force (F_1) at a single

contact point. In the central column, we are trying to calculate the number of contact points per unit area (N_c) from the number of the spheres per unit cross-sectional area (N_p) and the coordination number (K). The change of the normal stress ($\Delta\sigma$) can be calculated F_1 multiplied by N_c. However, as each F_1 directs arbitrary directions, $\Delta\sigma$ can be then expressed with the correction factor Z :

$$\Delta\sigma = F_1 N_c Z \qquad (6)$$

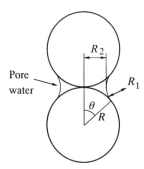

Figure 1. Pore water held between two equal-sized spheres

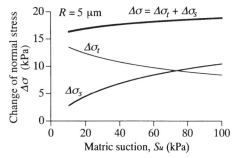

Figure 2. Change of normal stress due to matric suction ($\Delta\sigma_s$) and surface tension of water ($\Delta\sigma_t$)

Table 1. Proprties of regular packings of equal-sized spheres

Type of packing	A_1	K'	Z'	β	Void ratio (e)	θ_c (deg.)
Simple cubic	$4R^2$	1	1.0	1/4	0.910	45
Cubical-tetrahedral	$4R^2$	2	$\sqrt{3}/2$	$\sqrt{3}/4$	0.654	30
Tetragonal-sphenoidal	$2\sqrt{3}R^2$	2	$\sqrt{3}/2$	1/2	0.432	30
Pyramidal packing	$4R^2$	4	$1/\sqrt{2}$	$1/\sqrt{2}$	0.350	45
Tetrahedral	$2\sqrt{3}R^2$	3	$\sqrt{2}/\sqrt{3}$	$1/\sqrt{2}$	0.350	30

A_1 : area of unit.
K' : the number of the contact points where F_1 has the vertical component.
Z' : the directional cosine of F_1 to vertical axis.
β : the factor in the next equation : $\Delta\sigma = F_1 K' Z' / A_1 = \beta F_1 / R^2$.
θ_c : the water retention angle when a meniscus fuses with the next one.

148

Next, we try to obtain the correction factor Z. Figure 5 shows the contact points on the shear plane. The line segments between the particles represent the interparticle force (F_1) and bold one shows F_1 related to the change of the normal stress on the shear plane. L denotes the particles whose lower parts are intersected by the imaginary shear plane and U *vice versa*. The number of U and L particles can be equal provided that the spheres are randomly packed.

As shown with the bold line segments in Figure 5, only the contact points located on the upper hemisphere of the U particles are related to the increase of the normal stress for the U particles. When we project all the contact points on the upper hemisphere of U particles to a single U particle, the contact points spread uniformly on the surface of the upper hemisphere of the U particle. The number of contacts $(N_c)_U$ can be as follows:

$$(N_c)_U = 1/2 \ N_p/2 \ K/2 = N_p K/8 \tag{7}$$

where N_p is the number of particles in unit area and can be calculated as Figure 4, K is the coordination number in random packing.

Figure 6 shows the interparticle force (F_1) acting on the surface of the upper hemisphere of the U particle. Since $\Delta(N_c)_U$ (the number of contact points on the shaded area) is proportional to its surface area, it can be expressed by the next equation,

$$\Delta(N_c)_U = (N_c)_U \sin\alpha \, d\alpha. \tag{8}$$

The summation of the vertical components of F_1 acting on the shaded area $(F_v)_i$ is

$$(F_v)_i = F_1 \cos\alpha \ \Delta(N_c)_U = F_1 (N_c)_U \cos\alpha \sin\alpha \, d\alpha \tag{9}$$

Making the summation of $(F_v)_i$ over the upper hemisphere of the U particle, i.e., integrating Eq.(9) from zero to $\pi/2$ with respect to α, we can obtain the summation of vertical components of F_1 acting over the upper hemisphere of the U particle, i.e., the increase of the normal stress on the shear plane caused by the U particles $(\Delta\sigma)_U$, as follows:

$$(\Delta\sigma)_U = F_1 (N_c)_U \int_0^{\pi/2} \cos\alpha \sin\alpha \, d\alpha = 1/2 \, F_1 (N_c)_U \tag{10}$$

We can also estimate the increase of the normal stress caused by the L particles $(\Delta\sigma)_L$, and it can be the same as Eq.(10). Then the total increase $(\Delta\sigma)$ can be as follows:

$$\Delta\sigma = F_1 (N_c)_U Z = F_1 N_c /4 \tag{11}$$

Then, the correction factor Z can be 1/4.

When we employ the relationship between the coordination number (K) and the void ratio (e), proposed by Rumpf (1962),

$$K e / (1+e) = \pi, \tag{12}$$

the number of contact points (N_c) in unit cross-sectional area can be,

$$N_c = N_p K/2 = 3/(4eR^2). \tag{13}$$

Substituting Eq.(12) and Eq.(13) into Eq.(11), we can get the next equation.

$$\Delta\sigma = 3/(16e) \ F_1/R^2 \tag{14}$$

The calculation results for randomly packed $R = 5 \, \mu m$ spheres with the range of void ratio from 0.35 to 0.91 are shown in Figure 3.

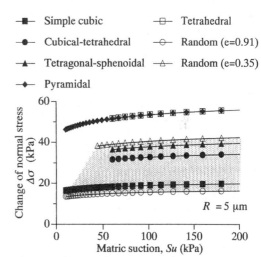

- ■— Simple cubic
- □— Tetrahedral
- ●— Cubical-tetrahedral
- ○— Random (e=0.91)
- ▲— Tetragonal-sphenoidal
- △— Random (e=0.35)
- ◆— Pyramidal

Figure 3. Change of normal stress in regularly packed assemblies of equal-sized spheres

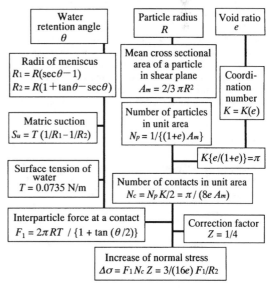

Figure 4. Flow chart

Eq.(14) is valid until the menisci fuse each other, that is, $\theta < \theta_c$. However, it is difficult to estimate θ_c in random packing of equal-sized particles. Then, we assume that menisci fuse each other when the surface area where the pore water contacts with the particle becomes equal to the total surface area of the particle. According to this assumption, θ_c can be calculated from the next equation,

$$\theta_c = \cos^{-1}(1 - 2/K) \quad (15)$$

The results presented in Figure 3 are calculated within the range of $\theta < \theta_c$. Since the form of the Eq.(14) is similar to the Eq.(5) for a regularly packed assembly, the relationships between the matric suction and the change of the normal stress shows the same tendency as that of the regularly packed assembly.

3.3 in assembly of randomly packed non equal-sized spheres

No analytical solution is presented for the Laplace-Young equation of the matric suction between two particles of the different radii (Lian et al., 1993). We here assume that an assembly of non equal-sized spheres consists of many subassemblies of equal-sized spheres. We then make the following assumptions for the cubic assembly of $1\ cm^3$.

(1) The cubic assembly is composed of j^3 subassemblies which consist of the equal-sized spheres. Figure 7 shows the shear plane intersecting j^2 subassemblies.
(2) The effect of interaction among the subassemblies is ignored.
(3) The void ratio of subassemblies is equal to that of the assembly of $1\ cm^3$.
(4) The change of the normal stress varies linearly with the matric suction after the menisci fuse each other. This assumption is illustrated in Figure 8.

Figure 9 shows the S_u-$\Delta\sigma$ relations for randomly packed subassemblies of four different radii spheres. If an assembly is composed of subassemblies of different radii spheres, its S_u-$\Delta\sigma$ relation can be the summation of the each S_u-$\Delta\sigma$ relation as presented in Figure 9. Then, the total interparticle forces along the shear plane in Figure 7 can be the summation of that in each subassembly $(\Sigma(F_{1v})_k)$ which can be,

$$\Sigma(F_{1v})_k = F_1 (N_c)_k Z \quad (16)$$

where $(N_c)_k$ is the number of contact points in the k-th subassembly and it can be obtained from Eq.(13) as follows:

$$(N_c)_k = 1/j^2 \{ 3/(4eR^2) \} \quad (17)$$

According to the 3rd assumption and Eq.(12), the number of contact points in each subassembly is equal to each other in subassemblies.

Let the number of the subassemblies of the specific radius spheres along the shear plane be n, the total

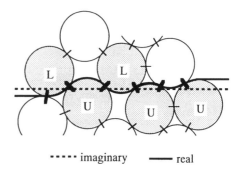

- - - - imaginary ——— real

Figure 5. Contact points on shear plane

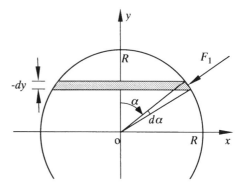

Figure 6. F_1 acting on the surface of U particle

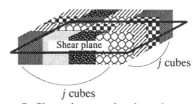

Figure 7. Shear plane passing through an assembly of subassemblies of randomly packed equal-sized spheres

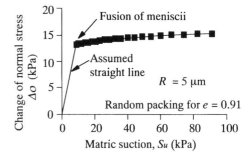

Figure 8. S_u-$\Delta\sigma$ relation for assembly of randomly packed equal-sized spheres

interparticle force along the shear plane becomes as follows:

$$n \Sigma (F_{1v})_k = n F_1 (N_c)_k Z \qquad (18)$$

where n depends on the particle size distribution.

Figure 10 shows particle size distributions of three imaginary assemblies, A, B & C. Assembly A for equal-sized sphere packing of $R = 4.5 \mu m$, Assembly B for equal-sized sphere packing of $R = 1.2 \mu m$ and Assembly C for non equal-sized sphere packing.

The assembly C consists of spheres of ten kinds of radii, and each content is 10%. Then n for Assembly C is equal to $0.1 j^2$. Substituting Eqs.(3), (17) and n into Eq.(18), we can obtain the result of Eq.(18) as follows:

$$n \Sigma (F_{1v})_j = [0.1 j^2] \cdot [2 \pi R T / \{ 1 + \tan (\theta / 2) \}] \cdot [1/j^2 \{ 3/(4eR^2) \}] \cdot 1/4 \qquad (19)$$

Calculating Eq.(19) for each radii, we can get the change of the normal stress ($\Delta \sigma$) with the matric suction. Figure 11 shows the calculation result of Eq.(19) for Assemblies A, B and C. Although the change of $\Delta \sigma$ of Assemblies A and B with the matric suction is very small, that of Assembly C increases gradually with the matric suction in the range of the low matric suction, and then it becomes small in the higher matric suction.

4 DISCUSSION WITH EXPERIMENTAL RESULTS OF SANDY SOILS

It is not necessary in geotechnical engineering to evaluate the interparticle force itself. Essential is the change of the shear strength due to the change of the matric suction.

When soils follow the Coulomb failure criterion, we can estimate the change of the shear strength due to the change of the normal stress ($\Delta \sigma$) as follows :

$$\tau_f = c + (\sigma + \Delta \sigma) \tan\phi, \quad \Delta \tau_f = \Delta \sigma \tan\phi \qquad (20)$$

where $\Delta \tau_f$ is the change of the shear strength, c the cohesion and ϕ the internal friction angle.

The change of the shear strength parameters, c and ϕ, with the change of the matric suction has been under discussion (Shimada et al., 1999). However, ϕ is supposed not to change with the change of the matric suction because friction between soil particles does not change under the existence of some quantity of pore water. Then, we can conclude that the internal friction angle is essentially constant with the change of the matric suction.

When ϕ is constant in Eq.(20), the change of $\Delta \sigma$ affects linearly the change of the shear strength. Then, we can compare the tendency of the change of the normal stress and that of the shear strength. We are trying to compare the change of the shear strength ($\Delta \tau_f$) with the change of $\Delta \sigma$ presented in the previous chapter.

4.1 *Materials and test procedure*

Fraser River sand and Masa96 are used in this experiment. Fraser River sand is a poorly-graded sandy soil (its uniformity coefficient (U_c) is 2.2), and Masa96 a well-graded sandy soil ($U_c = 54.8$). Figure 12 shows the particle size distribution curves of the two soils.

Shear tests for Fraser River sand (Shimada, 1998) were carried out with NGI type simple shear apparatus, which used the wire reinforced membrane. When the horizontal displacement exceeded 4 mm, where the shear strain was about 25 %, no further increment of the shear stress was applied and the shear stress at this point is referred to the shear strength in this paper. Direct shear

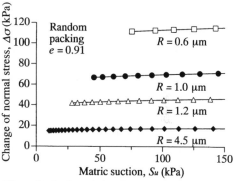

Figure 9. Su-$\Delta\sigma$ relation for assembly of random packing of different radius spheres

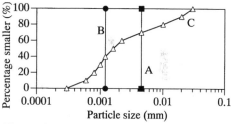

Figure 10. Assumed particle size distribution curves

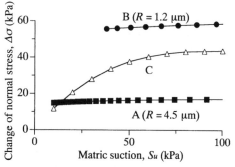

Figure 11. Su-$\Delta\sigma$ relation for assemblies of randomly packed equal-sized (A&B) and non equal-sized spheres (C)

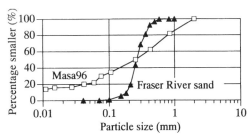

Figure 12. Particle size distribution curves of Fraser River sand and Masa96

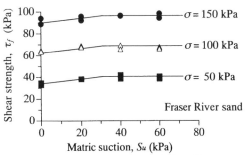

Figure 13. S_u and shear strength relation of Fraser River sand

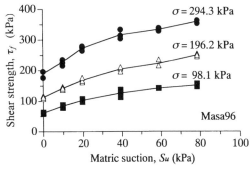

Figure 14. S_u and shear strength relation of Masa96

box tests were carried out for Masa96 (Shimada et al., 1999). Both tests were stress-controlled and the matric suction was kept constant during shear with the suction method.

4.2 Discussion with test results

Figure 13 summarizes test results of Fraser River sand with the relationships between the matric suction (S_u) and the shear strength (τ_f). Bilinear relationships are intentionally drawn for each different normal stresses (σ). The symbols representing the test results fairly fit to the bilinear relations. It is clearly shown that the change of τ_f is negligibly small in the region of S_u values greater

than 20 kPa.

Fraser River sand has a low value of U_c and is a poorly -graded sand as shown in Figure 12. Since the particle size distribution of this sand is similar to that of the assembly of equal-sized spheres assumed in the theoretical consideration for the estimation of the change of the normal stress, the sand has shown the same tendency as that predicted from the theoretical consideration.

Figure 14 shows the test result of Masa96 with the relation of S_u and τ_f. As Masa96 has a high value of U_c and is a well-graded sandy soil, its tendency of the shear strength can be predicted from the theory for an assembly of non equal-sized spheres. That is, the shear strength increases gradually with the matric suction, and this tendency is shown in Figure 14.

5 CONCLUSIONS

Theoretical considerations and experimental results on the increase of the normal stress in unsaturated assemblies of spheres and on the shear strength of sandy soils have given the following conclusions.

(1)The normal stress in the unsaturated assemblies of regularly and randomly packed equal-sized spheres does not show the great change with the change of the matric suction.

(2)The normal stress in the unsaturated assembly of randomly packed non equal-sized spheres changes gradually with the change of the matric suction.

(3)Experimental results of poor-graded and well-graded sandy soils show the same tendency of the relationship between the change of the normal stress and the matric suction as that for the unsaturated assemblies of equal and non equal-sized spheres.

REFERENCES

Fisher, R.A. 1926. On the capillary forces in an ideal soil. *Journal of Agricultural Science* 16:492-505.

Lian, G., Thornton, C. & Adams, M.J. 1993. Effect of liquid bridge forces on aggromerate collisions. *Powders & Grains 93* (*Proc. of 2nd. Int. Conf. on Micromechanics of Granular Media*):59-64.

Mitchell, J.K. 1993. *Fundamentals of Soil Behavior* (2nd Ed.). 136-139.New York : John Wiley and Sons.

Rumpf, H. 1962. The strength of granules and agglomerates. *Agglomeration* edited by W.A.Knepper. 379-418 : Interscience.

Shimada, K., Fujii, H., & Nishimura, S. 1993. Increase of shear strength due to surface tension and suction of pore water in granular material. *Proc. of Symp. on Mechanics of Granular Materials*, JSGE. 17-20. (in Japanese)

Shimada, K. 1998. Effect of Matric Suction on Shear Characteristics of Unsaturated Fraser River Sand. *J. of the Faculty of Environmental Science and Technology, Okayama University* 3:127-134.

Shimada, K., Fujii, H., Nishimura, S. & Morii, T. 1999. Change of Shear Strength of Unsaturated Decomposed Granite Soils with Matric Suction. *Trans. of Japanese Society of Irrigation, Drainage and Reclamation Engineering* 201:99-104.

Unsaturated Soils for Asia, Rahardjo, Toll & Leong (eds) © 2000 Taylor & Francis, ISBN 90 5809 139 2

Three-dimensional elasto-plastic model for unsaturated soils

D.A. Sun & H. Matsuoka
Department of Civil Engineering, Nagoya Institute of Technology, Japan

Y.P.Yao
Department of Civil Engineering, Beijing University of Aeronautics and Astronautics, People's Republic of China

W. Ichihara
Network Business Division, NTT Communications, Tokyo, Japan

ABSTRACT: A three-dimensional elasto-plastic model for unsaturated soil is presented, using a transformed stress tensor based on the Extended SMP criterion. The influences of suction on the mechanical behavior of unsaturated soil are taken into account in the model by using a parameter named as suction stress $\sigma_0(s)$, which is added to net stress to form an effective stress, and associating a yield stress and model parameters with suction through $\sigma_0(s)$. The prediction of the proposed model shows good agreement with the measured behavior of unsaturated soil in various stress conditions under the conditions of constant suction and reduction in suction, which may induce collapse.

1 INTRODUCTION

Several elasto-plastic models for unsaturated soils have been developed by extending the critical state theory (e. g. Alonso et al. 1990, Kohgo et al. 1993, Wheeler & Sivakumar 1995, Karube et al. 1997). As well known, however, the mechanical behavior of saturated or unsaturated soil in three-dimensional (3D) stresses except for triaxial compression stress can not be predicted accurately by the critical state model (e. g. Matsuoka et al. 1999). The existing elasto-plastic models for unsaturated soils are essentially applicable to predict the mechanical behavior of unsaturated soil only in triaxial compression stress, and nobody has concisely predicted the stress-strain behavior (including collapse) of unsaturated soil in three-dimensional stress excluding isotropic and K_0-consolidation stress using their models. In addition, it is hard to say that the existing elasto-plastic models can be easily applied to engineering practice because there are many material parameters and the determination of these material parameters needs complicated tests on saturated and unsaturated soils, although the models have their own characteristics to predict the behavior of unsaturated soils. This study is aimed at developing a three-dimensional elasto-plastic model for unsaturated soil in which the model parameters are selected as few as possible and are determined as easily as possible for practical use.

The following symbols are used for stress and pressure.

σ_{ij} : total stress

s : suction($=u_a$-u_w; u_a : pore air pressure, u_w : pore water pressure)

σ_{ij} : net stress($= \sigma_{ij}$-$u_a \delta_{ij}$; δ_{ij} : Kronecker's delta)

p : mean net stress($= \sigma_{ii}/3$)

q : deviatoric stress($= \sqrt{3(\sigma_{ij} - p\delta_{ij})(\sigma_{ij} - p\delta_{ij})/2}$)

2 A TRANSFORMED STRESS BASED ON EXTENDED SMP CRITERION

The criterion of the Extended Spatially Mobilized Plane (Extended SMP) has been successfully applied to some frictional and cohesive materials such as cemented soils (Matsuoka & Sun 1995). As the shear yield or failure criterion for frictional and cohesive materials, the Extended SMP criterion can be expressed as

$$\hat{I}_1 \hat{I}_2 / \hat{I}_3 = \text{const.} \tag{1}$$

where $\hat{I}_1$, $\hat{I}_2$ and $\hat{I}_3$ are the first, second and third inviariants of the translated stress tensor, which is defined by

$$\hat{\sigma}_{ij} = \sigma_{ij} + \sigma_0 \delta_{ij} \tag{2}$$

in which σ_0 is called a bonding stress($\sigma_0 = c \cdot \cot\phi$; c=cohesion, ϕ=angle of internal friction). It is considered that the Extended SMP criterion is more reasonable than the Extended Mises criterion with cohesion ($q/(p + \sigma_0)$ =const.), which is adopted as the shear yield and failure criteria in the Cam-clay model improved for unsaturated soil. In order to introduce the Extended SMP criterion to the existing elasto-plastic model verified only in triaxial compression stress, a following transformed stress $\tilde{\sigma}_{ij}$ based on the Extended SMP criterion has been proposed (Sun et al. 1998). The transformed stress $\tilde{\sigma}_{ij}$ is deduced from what makes the Extended SMP criterion become a cone with the axis being the space diagonal line in the transformed principal stress($\tilde{\sigma}_i$) space, that is, what makes the Extended SMP curve become a circle in the π-plane of the transformed principal stress($\tilde{\sigma}_i$) space(Fig. 1).

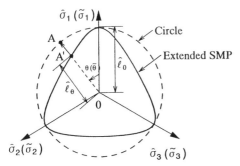

Figure 1. Extended SMP criterion in π-planes of ordinary and transformed principal stress spaces

$$\tilde{\sigma}_{ij} = \hat{p}\delta_{ij} + \frac{\hat{\ell}_0}{\ell_\theta}(\hat{\sigma}_{ij} - \hat{p}\delta_{ij}) \qquad (3)$$

where

$$\hat{p} = \hat{\sigma}_{ii}/3 \qquad (4)$$

$$\left.\begin{array}{l} \hat{\ell}_0 = \dfrac{2\sqrt{6}\hat{I}_1}{3\sqrt{(\hat{I}_1\hat{I}_2 - \hat{I}_3)/(\hat{I}_1\hat{I}_2 - 9\hat{I}_3)} - 1} \\[4mm] \hat{\ell}_\theta = \sqrt{(\hat{\sigma}_{ij} - \hat{p}\delta_{ij})(\hat{\sigma}_{ij} - \hat{p}\delta_{ij})} \end{array}\right\} \qquad (5)$$

$\tilde{\sigma}_{ij}$ can be used to take place of σ_{ij} for extending the existing elasto-plastic model using only p and q as stress parameters to a reasonable 3D elasto-plastic model for frictional and cohesive materials, because this extension means that the Extended SMP criterion is introduced as the shear yield and failure criteria in the model.

3 FORMURATION OF MODEL FOR UNSATURATED SOILS

3.1 *Strength of unsaturated soil*

An extension of the general Mohr-Coulomb failure criterion to include partial saturation has been proposed by Fredlund et al. (1978). Because this extension is acceptable only in triaxial compression stress, more reasonable failure criterion for unsaturated soil should be the Extended SMP failure criterion in general stresses, that is

$$\tilde{q} = M(s)\tilde{p} = M(s)(p + \sigma_0(s)) \qquad (6)$$

where $\sigma_0(s)$ is called suction stress related to suction s, and is taken place of σ_0 in the Extended SMP criterion for applying it to unsaturated soil; M(s) is the slope of $p \sim q$ relation in triaxial compression stress at failure, and is assumed to be linear with regard to $\sigma_0(s)$, i.e.

$$M(s) = M(0) + M_s\sigma_0(s) \qquad (7)$$

M(0) is the value of M(s) for s=0, M_s is a material parameter. $\tilde{p}$ and $\tilde{q}$ are the invariants of the transformed stress tensor $\tilde{\sigma}_{ij}$, i. e.

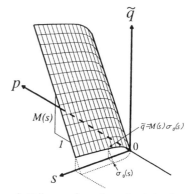

Figure 2. Failure surface of unsaturated soil under three-dimensional stress

$$\left.\begin{array}{l} \tilde{p} = \tilde{\sigma}_{ii}/3 \\[2mm] \tilde{q} = \sqrt{3(\tilde{\sigma}_{ij} - \tilde{p}\delta_{ij})(\tilde{\sigma}_{ij} - \tilde{p}\delta_{ij})/2} \end{array}\right\} \qquad (8)$$

Eq. 6 has been verified by the results of true triaxial tests on unsaturated soil under constant suction (Matsuoka et al. 1998). When suction changes, Eq. 6 can be drawn schematically as Fig. 2.

3.2 *Formulation of model in isotropic stress state*

Figure 3 shows the compression curves for saturated and unsaturated soils along the virgin isotropic loading in the e~ln $\hat{p}$ plane. The normal compression curves for saturated soil and for unsaturated soil with a constant suction s are

$$e = e_i(0) - \lambda(0)\ln\frac{p_y^*}{p_i} \qquad (9)$$

$$e = e_i(s) - \lambda(s)\ln\frac{p_y + \sigma_0(s)}{p_i} = e_i(s) - \lambda(s)\ln\frac{\hat{p}_y}{p_i} \qquad (10)$$

where $\sigma_0(s)$ is a suction stress, p_i is a reference net stress, and p_y^* and p_y are the yield stresses on the normal compression curves for saturated and unsaturated soils in the isotropic stress. In Fig. 3, we assume that there exits a normal compression line between void ratio (e) and mean effective stress ($\hat{p}_y$) for a given unsaturated soil and constant suction, which is essentially the same as the test results or proposal of some investigators (e.g. Karube et al. 1989, Kohgo et al.1993) and different from Alonso et al. (1990)'s proposal. We do not adopt the linear e~ln p relationship but adopt the linear e~ln $\hat{p}$ relationship because $\hat{p}$ is considered to be the mean effective stress for unsaturated soils. Eq. 10 can also be rewritten by using the coordinate of point N as

$$e = e_n - \lambda(s)\ln\frac{p_y + \sigma_0(s)}{p_n} \qquad (11)$$

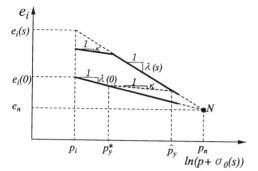

Figure 3. Relationship between compression curves for saturated and unsaturated soils

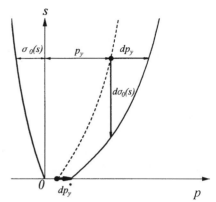

Figure 4. Relationships among dp_y, dp_y^* and $d\sigma_0(s)$ in isotropic stress

where p_n and e_n are coordinates of the point N(Sun et al. 1998). When p_y or/and $\sigma_0(s)$ change in the normal compression region, the increment of void ratio (e) can be obtained by differentiating Eq. 11.

$$de = \frac{\partial e}{\partial p_y}dp_y + \frac{\partial e}{\partial \sigma_0(s)}d\sigma_0(s)$$

$$= \frac{-\lambda(s)}{p_y + \sigma_0(s)}dp_y + (\lambda_s \ln\frac{p_n}{p_y + \sigma_0(s)} - \frac{\lambda(s)}{p_y + \sigma_0(s)})d\sigma_0(s)$$

(12)

in which $\lambda(s)$ is assumed to be linear with regard to $\sigma_0(s)$, i.e.

$$\lambda_s = \frac{d\lambda(s)}{d\sigma_0(s)} = \frac{\lambda(s) - \lambda(0)}{\sigma_0(s)}$$

(13)

so,

$$\lambda(s) = \lambda(0) + \lambda_s\sigma_0(s)$$

(14)

From the same plastic volumetric strain of saturated and unsaturated soils(Fig.3), the following equation can be obtained with the assumption that the swell indexes κ of saturated and unsaturated soils are the same.

$$(\lambda(s) - \kappa)\ln\frac{p_n}{p_y + \sigma_0(s)} = (\lambda(0) - \kappa)\ln\frac{p_n}{p_y^*}$$

(15)

Substituting Eq. 14 into Eq. 15 and solving for p_y gives

$$p_y = p_n(p_y^*/p_n)^{\frac{\lambda(0) - \kappa}{\lambda(0) - \kappa + \lambda_s\sigma_0(s)}} - \sigma_0(s)$$

(16)

Figure 4 show the relationship among dp_y, $d\sigma_0(s)$ and dp_y^*, which can be obtained by differentiating Eq. 16 as follows:

$$dp_y = \frac{\partial p_y}{\partial p_y^*}dp_y^* + \frac{\partial p_y}{\partial \sigma_0(s)}d\sigma_0(s)$$

(17)

It can be seen from Fig. 4 that the increase in the plastic volumetric strain of unsaturated soil is induced by the increase in the yield stress or decrease in suction, while the increase in the plastic volumetric strain of saturated soil is induced only by the increase in the yield stress.

3.3 Formulation of model in general stress

In order to describe the stress path dependency of the stress-dilatancy relation, the plastic strain increment is divided into two components (Nakai & Matsuoka 1986). They are the plastic strain increment $d\varepsilon_{ij}^{p(AF)}$, which obeys an associated flow rule in the $\tilde{\sigma}_{ij}$-space, and the plastic strain increment $d\varepsilon_{ij}^{p(IC)}$, which is caused by the increase in mean effective principal stress, i. e.

$$d\varepsilon_{ij}^p = d\varepsilon_{ij}^{p(IC)} + d\varepsilon_{ij}^{p(AF)}$$

(18)

1) Determination of $d\varepsilon_{ij}^{p(IC)}$

Following Nakai & Matsuoka (1986)'s proposal for saturated clay, the plastic strain increment $d\varepsilon_{ij}^{p(IC)}$ for unsaturated soil is assumed as (Sun et al. 1998)

$$d\varepsilon_{ij}^{p(IC)} = \frac{\delta_{ij}}{3}\frac{M(s) - \tilde{q}/\tilde{p}}{M(s)}\frac{-de^p}{1 + e_i(s)}$$

(19)

where

$$de^p = \frac{\kappa - \lambda(s)}{p + \sigma_0(s)}\langle dp \rangle + (\frac{\lambda(s) - \kappa}{p + \sigma_0(s)} - \lambda_s\ln\frac{p_n}{p + \sigma_0(s)})\langle -d\sigma_0(s)\rangle$$

(20)

< > is the Macauley symbol.

2) Determination of $d\varepsilon_{ij}^{p(AF)}$

The yield function for unsaturated soils can be obtained from the original Cam-clay model by using $\tilde{\sigma}_{ij}$ instead of σ_{ij} and considering the effect of cohesion as follows

$$g = f = \ln\tilde{p} + \frac{\tilde{q}}{M(s)\tilde{p}} - \ln(p_y + \sigma_0(s)) = 0$$

(21)

Figure 5 shows the geometrical shape of Eq. 21 when suction is a given value for unsaturated soil or zero for saturated soil. When suction changes, the yield surface

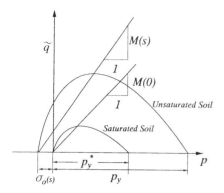

Figure 5. Yield curves under constant suction

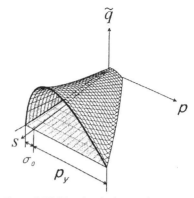

Figure 6. Yield surface in (p, $\tilde{q}$, s) space

can be drawn as Fig.6 by allowing for Eq. 16 and the relation between $\sigma_0(s)$ and s.

Because the plastic strain increment $d\varepsilon_{ij}^{p(AF)}$ obeys an associated flow rule in the $\tilde{\sigma}_{ij}$-space, so

$$d\varepsilon_{ij}^{p(AF)} = \Lambda \frac{\partial f}{\partial \tilde{\sigma}_{ij}} \qquad (22)$$

where the proportionality constant Λ can be determined from the consistency condition. Eq. 21 can be rewritten as $f = f(\tilde{p}, \tilde{q}, p_y, \sigma_0) = f(\sigma_{ij}, p_y, \sigma_0) = 0$, so we have

$$df = \frac{\partial f}{\partial \sigma_{ij}} d\sigma_{ij} + \frac{\partial f}{\partial p_y} dp_y + \frac{\partial f}{\partial \sigma_0(s)} d\sigma_0(s) = 0 \qquad (23)$$

Substituting Eq. 17 into Eq. 23 and arranging it give

$$df = \frac{\partial f}{\partial \sigma_{ij}} d\sigma_{ij} + \frac{\partial f}{\partial p_y} \frac{\partial p_y}{\partial p_y^*} dp_y^* \\ + \left(\frac{\partial f}{\partial p_y} \frac{\partial p_y}{\partial \sigma_0(s)} + \frac{\partial f}{\partial \sigma_0(s)} \right) d\sigma_0(s) = 0 \qquad (24)$$

in which, the isotropic yielding stress p_y^* for saturated soil is related to the volumetric strain ε_v^p, as the same as that used in the Cam-clay model. Because the plastic volumetric strain ε_v^p is hardening parameter in our model, $d\varepsilon_v^p$ of saturated soil induced by dp_y^* is the same as $d\varepsilon_v^p$ of unsaturated soil induced by dp_y and/or $d\sigma_0(s)$. Allowing for Eqs 9, 18 and 22, we have

$$dp_y^* = \frac{1 + e_i(0)}{\lambda(0) - \kappa} p_y^* d\varepsilon_v^p = \frac{1 + e_i(0)}{\lambda(0) - \kappa} p_y^* (d\varepsilon_v^{p(IC)} + d\varepsilon_v^{p(AF)}) \\ = \frac{1 + e_i(0)}{\lambda(0) - \kappa} p_y^* (d\varepsilon_v^{p(IC)} + \Lambda \frac{\partial f}{\partial \tilde{\sigma}_{ij}} \delta_{ij}) \qquad (25)$$

Substituting Eq. 25 into Eq. 24 can obtain Λ. When $\sigma_0(s) = 0$ and $d\varepsilon_v^{p(IC)} = 0$, the yield function (Eq. 21) and the proportionality constant Λ for unsaturated soil become those of the original Cam-clay model revised by the SMP criterion (Matsuoka et al. 1999). Therefore, the proposed elasto-plastic model for unsaturated soil includes the original Cam-clay model for saturated soil as a special case.

4 COMPARISON OF MODEL PREDICTION WITH EXPERIMENTAL RESULTS

4.1 Model parameters and their determination

In order to calculate the plastic strain, it is necessary to determine the material parameters $\lambda(0)$, λ_s, κ, $M(0)$, M_s and $\sigma_0(s)$, and the initial state values p_i, $e_i(0)$ and $e_i(s)$. These model parameters are determined from the results of isotropic consolidation tests with loading-unloading-reloading process and subsequent triaxial compression tests on saturated and unsaturated soils under constant suction and constant p or constant confining net stress. In details, $\lambda(0)$, λ_s, κ, p_i, $e_i(0)$ and $e_i(s)$ are determined from the results of isotropic consolidation tests on saturated and unsaturated soils with loading-unloading process, as shown in Fig. 3; $M(0)$ and M_s are determined from the envelope lines of Mohr's circles at failure measured by a triaxial compression test on saturated soil and two triaxial compression tests on unsaturated soil under constant suction at different confining net stress, using Eq. 7; $\sigma_0(s)$ is the distance between the origin and the intersection point of the σ-axis and the envelope line of measured Mohr's circles at failure (Matsuoka & Sun 1995).

The elastic component is calculated from Hooke's law, while Poisson's ratio is assumed to be zero, and the elastic modulus is calculated from that in the same way as the Cam-clay model, i. e.

$$E = \frac{3\hat{p}(1 + e_i(s))}{\kappa} \qquad (26)$$

4.2 Model prediction versus experimental results

The triaxial compression and extension tests on unsaturated compacted kaolin have been conducted by authors (1999). The tests include collapse induced by decreasing imposed suction in the isotropic and shear states. The values of the relevant model parameters used in prediction are summarized in Table 1.

156

Table 1. Material parameters and initial state values

Material parameter							Initial state value	
$M(0)$	Ms	$\lambda(0)$	λs	κ	$\sigma_0(s)$	p_i	s	$e_i(s)$
1.05	0.625	0.08	0.10	0.03	0.4	1.0	1.5	1.27
							0	1.11

Notes: unit of $\sigma_0(s)$, p_i and s : kgf/cm^2(98kPa);
unit of M_s and λ_s: cm^2/kgf(1/98kPa)

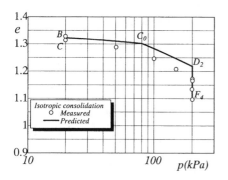

Figure 7. Model prediction for variation of void ratio (e) with p along isotropic stress path

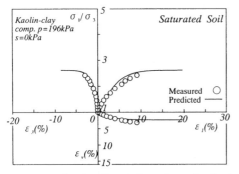

Figure 8. Predicted and experimental result of triaxial compression test on saturated soil p=196kPa and s=0kPa

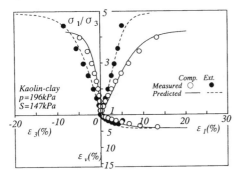

Figure 9. Predicted and experimental result of triaxial compression and extension tests on unsaturated soil under p=196kPa and s=147kPa

Figure 7 shows the comparison between the measured and predicted results of unsaturated compacted kaolin under the conditions of increasing suction (s=95kPa(initial suction of the compacted sample)→147kPa) and constant mean net stress (B→C), increasing mean net stress and constant suction (C→C$_0$→D$_1$), and constant mean net stress and decreasing suction (D$_1$→F$_4$; s=147kPa→0kPa) in the isotropic stress. Unsaturated soil behaves elastically in the stress path B→C→C$_0$, i. e. Hooke's law is adopted while the elastic modulus is calculated from κ as expressed in Eq. 26 regardless of the suction increase (B→ C) or the mean net stress increase (C→C$_0$).

Figure 8 shows the comparison between the measured and predicted results of triaxial compression test on saturated compacted kaolin under constant mean effective stress. In Fig. 8 and the following figures, σ_1/σ_3 denotes the major-minor principal stress ratio, and ε_1, ε_3 and ε_v denote the major, minor principal strains and volumetric strain, respectively. Figure 9 shows the comparison between the measured and predicted results of triaxial compression and extension tests on unsaturated compacted kaolin under constant mean net stress (p=196kPa) and constant suction (s=147kPa). It can be seen from Figs 8 and 9 that the model can predict accurately the mechanical behavior of saturated and unsaturated soils under constant suction in triaxial compression and extension stresses.

Figures 10 and 11 show the comparison between the measured and predicted results of triaxial compression and extension tests on unsaturated compacted kaolin under constant mean net stress, in which stress paths include suction reduction (s=147kPa→0kPa) at different stress ratios during shearing. It can be seen from the measured results shown in Figs 10 and 11 that when the shear strain (i. e. $\varepsilon_1 - \varepsilon_3$) increment induced by collapse at high stress ratio (σ_1/σ_3) (E$_2$→F$_2$ in Fig.10 and H$_2$→I$_2$ in Fig.11) is greater than that at low stress ratio (E$_1$→F$_1$ in Fig.10 and H$_1$→I$_1$ in Fig.11), while the volumetric stain increment is almost the same at the two tested stress ratio. This phenomenon is explained that the volumetric stain increment induced by collapse from unstable unsaturated soil to stable saturated soil is mainly dependent of mean stress, while the shear strain increment is mainly dependent of the ratio of shear stress, at which suction decreases, to shear strength of saturated soil. The model describes well this behavior, i.e. the shear strain increment induced by decreasing suction at high stress ratio (E$_2$→F$_2$ in Fig.10 and H$_2$→I$_2$ in Fig.11) is greater than that at low stress ratio (E$_1$→F$_1$ in Fig.10 and H$_1$→I$_1$ in Fig.11), while the volumetric stain increment is almost the same at the two tested stress ratio. It is interesting to know that the model can predict well the whole stress-strain relation of unsaturated soil during the process of constant suction (D$_2$ →E$_1$ and D$_2$→E$_2$ in Fig.10 and D$_2$→H$_1$ and D$_2$→H$_2$ in Fig.11), the suction reduction (E$_1$→F$_1$ and E$_2$→F$_2$ in Fig.10 and H$_1$→I$_1$ and H$_2$→I$_2$ in Fig.11) and zero suction (F$_1$→G$_1$ and F$_2$→G$_2$ in Fig.10 and I$_1$→J$_1$ and I$_2$→J$_2$ in Fig.11) in triaxial compression and triaxial extension.

It can be concluded from Figs 7~11 that the proposed model predicts well the stress-strain behavior including

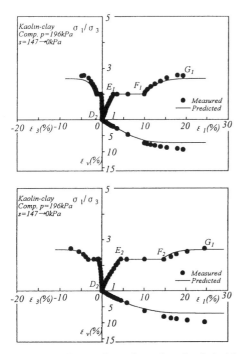

Figure 10. Predicted and experimental results of triaxial compression tests under p=196kPa and decreasing suction (s=147kPa→0kPa) during shearing

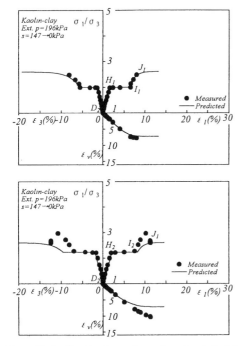

Figure 11. Predicted and experimental results of triaxial extension tests under p=196kPa and decreasing suction (s=147kPa→0kPa) during shearing

collapse of unsaturated soil under the isotropic or general stress state, using the compression and swelling indexes and strength parameters of saturated and unsaturated soils.

5 REFERENCES

Alonso F. E., Gens A. & A. Josa 1990. A constitutive model for partially saturated soils, *Geotechnique*, 40(3): 405-430.

Fredlund D. G., Morgenstern N. R. & R. S. Widger 1978. The shear strength of unsaturated soils, *Canadian Geotechnical Journal*, 15(3): 313-321.

Karube D., Honda M., Kato S. & K.Tsurugasaki 1997: the relationship between shearing characteristics and the composition of pore-water in unsaturated soil, *Proc. of JSCE*, No.575/III-40: 49-58 (in Japanese).

Karube D., Katsyama J., Nishiumi K. & N. Niwa 1989. Yield function of unsaturated soil under triaxial compression, *Proc. JSCE*, No.406/III-11: 205-212 (in Japanese).

Kohgo Y., Nakano M. & T. Miyazaki 1993. Theoretical aspects of constitutive modelling for unsaturated soils, *Soils and Foundations*, 33(4):49-63

Matsuoka H. & D. A. Sun 1995. Extension of spatially mobilized plane (SMP) to frictional and cohesive materials and its application to cemented sands, *Soils and Foundations*, 35(4): 63-72.

Matsuoka H., Sun D. A., Ando M., Kogane A. & N. Fukuzawa 1998. Deformation and strength of unsaturated soil by true triaxial tests, *Proc. 2nd Int. Conference on Unsaturated Soils*, 1:410-415.

Matsuoka, H., Yao, Y. P. & D. A. Sun 1999. The Cam-clay models revised by the SMP criterion, *Soils and Foundations*, 39(1): 81-95.

Nakai T. & H. Matsuoka 1986. A generalized elastoplastic constitutive model for clay in three-dimensional stresses, *Soils and Foundations*, 26(3): 81-98.

Sun D. A., Matsuoka H., & Y. Takiwaki 1998. A three-dimensional elasto-plastic model for unsaturated soil based on "Extended SMP", *Proc. 2nd Int. Conference on Unsaturated Soils*, 1: 521-526.

Sun D. A., Matsuoka H., Yao Y. P. & W. Ichihara 1999. An elastoplastic model for unsaturated soil and its verification, *Proc. 34th Japan National Conference on Geotechnical Engineering*, 753-756 (in Japanese).

Wheeler S. J. & V. Sivakumar 1995. An elasto-plastic critical state framework for unsaturated soil, *Geotechnique*, 45(1): 35-53.

Unsaturated Soils for Asia, Rahardjo, Toll & Leong (eds) © 2000 Taylor & Francis, ISBN 90 5809 139 2

Reconsideration of suction controlled techniques by thermodynamics theory of water in soil

H.Suzuki

Department of Civil Engineering, The University of Tokushima, Japan

ABSTRACT: Soils in general consist of three phases: soil solid, water and air. The water in a soil has very complicated phenomena, that are related to the repetitions of freezing, liquefying and evaporating. Therefore, it is necessary to consider soil suction, a potential energy of the water in the soil, from the viewpoint of thermodynamics theory. In this paper, we discuss the thermodynamic consistencies of the suction controlled techniques, namely, the axis translation technique, the suction method, the osmotic suction and the vapor pressure controlled methods, which are commonly used in laboratory tests of unsaturated soil, before considering the concept of soil suction itself.

1 INTRODUCTION

Study of thermodynamics of water in soil has been performed since the 1900's in the field of soil science. Schofield (1935) proposed the concept of pF, namely, the logarithm of the specific Gibbs free energy and established the energy concept of water in soil. Edlefsen & Anderson (1943) emphasized the use of the chemical potential in his book entitled "thermodynamics of soil moisture". After that, some researches related to soil moisture movement were carried out using thermodynamics of water in soil. The International Society of Soil Science defined the total potential of water in 1963. Iwata (1972a, 1972b) proposed a more general energy concept of water in soil, which includes the definition of the total potential. In geotechnical engineering, Aitchison (1965) organized the international symposium on the definition of soil suction considering thermodynamics, then soil suction was expressed as a function of vapor pressure. This concept of soil suction was introduced into geotechnical engineering researches, for example, Fredlund and Rahardjo (1988).

To examine the mechanical properties of unsaturated soils, the axis translation technique, suction method, osmotic suction and vapor pressure methods are used as a suction controlled technique. The axis translation technique is especially widely introduced into mechanical soil test apparatuses such as suction controlled axis conversion test device., but the technique was only justified by Bocking and Fredlund (1980) who did not consider using thermodynamics theory of water in soil. In the present study, we make clear the meaning of suction controlled technique using the thermodynamics theory established by Iwata et al.(1995), and try to confirm the usefulness of the experimental results of unsaturated soils.

2 . THERMODYNAMICS OF WATER IN SOIL

2.1 *Necessity of thermodynamic consideration*

Gotechnical engineers must recognize that water in soil is a aggregate of water molecules with extensive properties, for almost solid condition strongly combining with the soil particles and also affected by the effects of electric charges and dissolution ions, needless to say, water in soil can be partly treated as the liquid condition. Considering unsaturated ground, water in the soil usually has a lower pressure than atmospheric pressure when subjected to surface tension, which is the interface energy between water and air. On the other hand, the water has higher pressure caused by the surface energy of the solid near the soil particle. As mentioned above, conditions of water in soil are very complicated and pressure depend on its location. However, in thermodynamic consideration, we can use the law that the total free energy of water in soil is minimized under the equilibrium condition and the chemical potential has a constant value regardless of the phase and the location. A water molecule can have various pressure and phase conditions due to various potential components under the condition of constant chemical potential. Therefore, it is very useful to understand macroscopically the behavior of water in soil which is a mixture of aggregate with various states of molecules of water. We can make thermodynamic consideration using an index expressing the potential of the various states of water in soil from ice to vapor by introducing the chemical potential into geomechanics.

2.2 Formulation of chemical potential

The chemical potential, which is a thermodynamic quantity, can be described according to the result of research established by Iwata et al. (1995).

$$d\mu_w = -\underline{S}_w dT + \underline{v}_w dp + \frac{v_w}{4\pi}\left(\frac{1}{\underline{\varepsilon}} - 1\right)dD$$

$$+ \left(\frac{\partial\psi}{\partial z}\right)dz + gdh + \sum_{j=1}^{k-1}\left(\frac{\partial\mu_w}{\partial C_j}\right)dC_j$$

$$+ \frac{\partial}{\partial r}\left(\frac{2\sigma}{r}\underline{v}_w\right)dr \qquad (1)$$

where, μ_w is the chemical potential of water, $\underline{S}_w$ is the partial entropy of water and T is the temperature. $\underline{v}_w$ is the partial volume of water, $\underline{\varepsilon}$ is the partial dielectric constant and D is the electric displacement. ψ is the potential energy of intermolecular interaction between a clay particle and water molecules of unit mass, C_j is the concentration of jth component and k is the number of components, σ is the surface tension of water, p is the total sum of eternal and internal forces, p_{ex} and p_{in}, respectively.

2.3 Configurations of water in soil

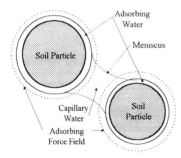

Fig.1. Schematic diagram of configuration of water in soil

Figure 1 shows a schematic diagram of the configuration of water in soil with reference to Iwata et al.(1995). The field of strong interaction between water molecule and soil particle is made by a force of hydrogen combination and van der Waals forces, which are shown by the dotted line as covering soil particle. The water molecules in this domain are strongly adsorbed by forming a film to cover the soil particle. On the other hand, water encircled by a concave surface near a contact point of two soil particles is retained by surface tension occurring between water and air. The capillary force due to the contact angle occurs between water and soil particle. Namely, water retained by soil is partly combined by the strong adsorbing forces at the surface of the soil particle (retaining water at surface), and partly by capillary forces at the contact points of soil particles (retaining water in void). Such effects from soil particles confine water in the soil, and its free energy is lower than that of pure water. In an equilibrium state, chemical potential must be constant at any place. If chemical potential depends on a location, it is not in equilibrium because water in soil moves from a high chemical potential to a low chemical potential place.

3. CONSIDERATIONS OF SUCTION CONTROLLED TECHNIQUES BY THERMODYNAMICS THEORY WATER IN SOIL

The outline of the thermodynamics of water in soil can make clear the meanings of the suction controlled techniques.

3.1 The axis translation technique

Hilf (1956) proposed this technique to measure negative pore water pressure in a ground. Fredlund and Rahardjo (1995) explained it in detail from the historical and experimental viewpoints. In general, this technique is used under condition of constant temperature, humidity and concentration of water in soil. Therefore, the chemical potential can be expressed by neglecting the first and sixth terms of Equation.1 as shown below:

$$d\mu_w = \underline{v}_w dp + \frac{v_w}{4\pi}\left(\frac{1}{\underline{\varepsilon}} - 1\right)dD + \left(\frac{\partial\psi}{\partial z}\right)dz$$

$$+ gdh + \frac{\partial}{\partial r}\left(\frac{2\sigma}{r}\underline{v}_w\right)dr \qquad (2)$$

Figure 2 shows a schematic diagram of the axis translation technique applied to triaxial compression test when soil suction is only controlled by u_a and

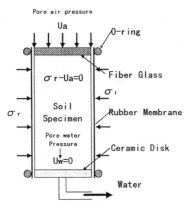

Fig.2 Schematic diagram of the axis translation technique for triaxial compression test

σ r without shear stress. Pore air pressure is usually controlled under the condition that pore water pressure is equal to atmospheric pressure. As shown in Figure 2, confining pressure σ r is equal to pore air pressure u_a, therefore, the first term in Equation 2 can be neglected. Since u_a is isotropically applied to the soil particles and water in a specimen, the gravitational work expressed the fourth term is not changed and the surface tension of water expressed by the fifth term is also the same state. The potential energy of the intermolecular interaction between a soil particle and water molecules of unit mass ϕ strongly depends on the size of soil particle, but the materials are the same, therefore, the effects of ϕ and the electric field expressed in Equation 2 do not need to be considered. We can conclude that the axis translation technique is useful to geotechnical engineering by theory of thermodynamics of water in soil.

3.2 *Suction method*

In this method, suction is applied to the soil specimen by negative pore water pressure under atmospheric pressure. The suction method make use of the

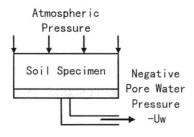

Fig.3 Schematic diagram of suction method

that water in soil has a constant value of chemical potential at each place in equilibrium, therefore, it is in consistency with the thermodynamic theory of water in soil.

3.3 *Osmotic suction controlled method*

This method was introduced into oedometer tests (for example, Mukhtar 1995, Dineen and Burland 1995). Soil suction is exactly controlled by the difference of osmotic pressure due to the concentration between water in soil specimen and the saturated salt solution without changing external and internal pressures. Therefore, we must only consider the sixth term in Equation 1.

$$[d\mu_w]_{P,T,\cdots} = \sum \left(\frac{\partial \mu_w}{\partial C_j}\right)_{P,T,\cdots,C_k} dC_j \qquad (3)$$

This equation assuming an ideal solution,. can be written as,

$$[d\mu_w]_{P,T,\cdots} = d\left[\frac{RT}{M}\ln x_w\right] = d\left[\frac{RT}{M}\ln\left(1-\sum x_j\right)\right] \quad (4)$$

where, x_w and x_j are mole fractions of water and solute j, respectively. The decrement of chemical potential due to the solutes, μ_0, is

$$\mu_0 = \int_1^{x_w} d\left[\frac{RT}{M}\ln x_w\right] = \frac{RT}{M}\ln\left(1-\sum x_j\right) \qquad (5)$$

If $\sum x_j \ll 1$, Eq.(5) becomes

$$\mu_0 = -\frac{RT}{M}\sum x_j = -\frac{RT}{1000}\sum n_j = \Pi \underline{v}_w \qquad (6)$$

where, n_j is the mole concentration of solute j and Π is the osmotic pressure. As shown in Equation 6, the chemical potential μ_0 is equivalent to the osmotic potential. Therefore, it is concluded that the osmotic suction controlled method is consistent with the thermodynamic theory.

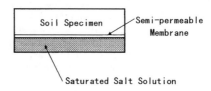

Fig, 4 Schematic layout of osmotic suction controlled method

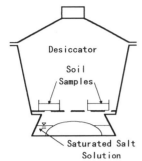

Fig.5 Testing using vapor pressure controlled method

3.4 *Vapor pressure controlled method*

This method measures the amount of water absorption in soil under condition of constant vapor pressure. The moisture state of soil samples is due to the vapor pressure caused by the saturated salt solution whose concentration is already known and usually set up in a desiccator in a constant temperature room.

The surface tension effects of retaining water in soil can be expressed by the seventh term in Equation 1, which is the following equation:

$$\mu_c = \int_{\infty}^{r} \frac{\partial}{\partial r}\left(\frac{2\sigma}{r}\underline{v}_w\right)dr = \frac{2\sigma}{r}\underline{v}_w \qquad (7)$$

μ_c causes the condensation of water vapor from soil air when the chemical potential of water vapor relative to that of saturated vapor is larger than μ_c. The chemical potential of vapor μ_v is given by the following equation:

$$\mu_v = \frac{RT}{M}\ln\frac{p}{p_0} \qquad (8)$$

where, p_0 is the saturated vapor pressure at T. Equation 8 expresses the vapor pressure controlled method itself, therefore, this is also consistent with the thermodynamic theory

4. CONCLUSIONS

We have examined the validity of the suction controlled methods, which are usually used for investigating the mechanical properties of unsaturated soils from the viewpoint of thermodynamics theory. Consequently, these methods are fully explained by the theory, because our discussions were performed under the very ideal condition concerned with the philosophy of the suction controlled methods. However, we cannot refer to the accuracy of the methods because of complicated problems such as the measuring system of chemical potential and the unifor-

mity of suction in soil specimen and so on. Although we have relied on the thermodynamic theory established by Iwata et al.(1995) to understand the phenomena of water in soil, but we need to propose a developed theory for solving our geotechnical problems. Next it is our objective to consider the definition of soil suction by the thermodynamics theory of water in soil which is the basis of soil mechanics, especially, unsaturated soils.

REFERENCE

D. G. Aitchison 1965. Engineering concepts of moisture equilibria and moisture changes in soils. Statement of the Review Panel, Ed., published in *Moisture Changes in Soils Beneath Covered* pp.7-21.
Areas, A Symp.-in-print Australia, Butterworths,
K. A. Bocking and D. G. Fredlund 1980. Limitation of the Axis Translation Technique. in *Proc. 4th Int. Conf. Expansive Soils*, Denver, CO, pp117-135.
K. Dineen and J. B. Burland 1995 A new approach to osmotically controlled oedometer testing, *1st Int. Conf. of Unsaturated Soils*, pp.459-465.
Edlefsen, N. E., and B. C. Anderson 1943. Thermodynamics of soil moisture. *Hilgardia* 15, pp.31-298.
D. G. Fredlund and H. Raharjo 1988. State-of-development in the measurement of soil suction, in *Proc. Int. Conf. Eng. Problems on Regional Soils*, Beijing, China, pp.582-588.
D. G. Fredlund and H. Rahardjo 1995 *Soil mechanics for unsaturated soil*, Jhon Wiley & Sons INC, pp.64-106.
J. W. Hilf 1956. An investigation of pore-water pressure in compacted cohesive soils, Ph. D. dissertation. Tech. Memo. No.654, U. S. *Dep. of the Interior, Bureau of Reclamation, Design and Construction* Div, 654pp
International Society of Soil Science 1963. *Soil Physics Terminology, Bull.* 23, p.7.
S. Iwata 1972a. Thermodynamics of soil water. Vol.1, *Soil Science*, No.113, pp.162-166.
S. Iwata 1972b. On the definition of soil water potential as proposed by the ISSS in 1963, *Soil Science*, No.114, pp.88-92.
S. Iwata, T. Tabuchi and B. P. Warkentin 1995. *Soil-water interactions*. Marcel Dekker, Second Edition, pp.1-38.
M. A. L. Mukhtar 1995 Coupling phenomena of hydraulic and mechanical stress in low porosity clays, *Int. Symp. of Compression and Consolidation of Clayey Soils* Vol.1, pp.15-20.
Schofield, R. K. 1935. The pF of the water in soil. *Trans. 3rd Int. Cong. Soil Science* Vol.2, pp.37-48.

Unsaturated Soils for Asia, Rahardjo, Toll & Leong (eds) © 2000 Taylor & Francis, ISBN 90 5809 139 2

Understanding of some behaviour of unsaturated soil by theoretical study on capillary action

K. Tateyama & Y. Fukui
Department of Civil Engineering Systems, Kyoto University, Japan

ABSTRACT: The effect of capillary water on the behavior of partially saturated soil is simulated through numerical calculation. In the calculation, soil particles are replaced by uniform spheres and water is supposed to be distributed around the surface of particles as adsorbed water and at the contact points between spheres as capillary water. The inter-particle force induced by the surface tension of the capillary water is calculated to discuss the relation of the physical properties of partially saturated soil such as particle size, water content and void ratio to the mechanical properties of suction and strength. The experiments are carried out to study the results of the calculation. In the experiments, the suction and strength are measured for the partially saturated sample of glass beads and their results are compared with the results of the calculation. We try to understand the mechanical properties characteristic of the partially saturated soil through the discussion.

1 INTRODUCTION

In partially saturated soil, water is retained around soil particles and at contact points of soil as shown in Figure 1. We call the former water as adsorbed water and the latter one as capillary water in this paper. The shape of the capillary water is a center-pinched cylinder to form a meniscus, and inter-particle force is generated between the soil particles due to the surface tension of the capillary water acting on the meniscus. The mechanical behavior of partially saturated soil is affected by the inter-particle force. In this paper, numerical simulation is carried out to simulate the effect of the capillary water on the mechanical behavior of partially saturated soil.

Through the discussion, we try to understand the mechanical properties characteristic of the partially saturated soil, such as suction and strength.

2 NUMERICAL SIMULATION

2.1 *Model for simulation*

Figure 2 shows the model for soil particles and water retention employed in this research. The soil particles are replaced by uniform spheres and some part of water is consumed to cover the surface of the spheres as adsorbed water and the rest will be distributed as capillary water at the contact points between spheres.

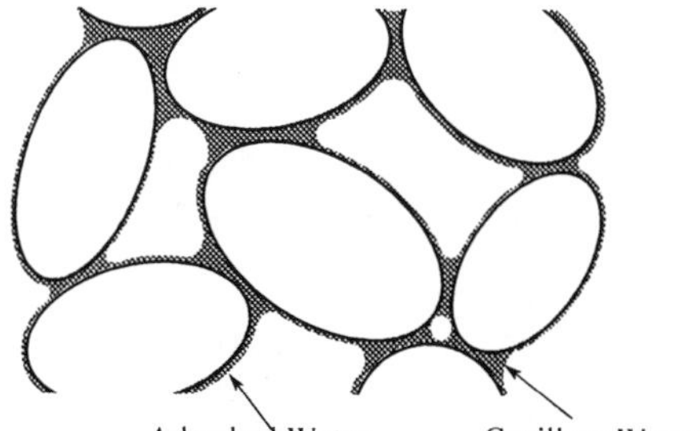

Figure 1   Water in partially saturated soil

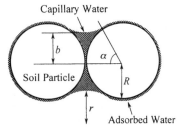

Figure 2   Model of partially saturated soil

The capillary water creates the meniscus on its surface and then generates the suction S_U and the inter-particle force H due to the surface tension of the capillary water, which are expressed respectively by equation (1) and (2),

$$S_U = T\left(\frac{1}{r} - \frac{1}{b}\right) \tag{1}$$

$$H = \pi b^2 T\left(\frac{1}{r} - \frac{1}{b}\right) + 2\pi b T \tag{2}$$

where
S_U : suction in the sphere sample (N/m^2)
H : inter-particle force between particles (N)
R : radius of the sphere (m)
b : radius of water cylinder at its center (m)
r : radius of meniscus (m)
α : angle of water retention (degree)
T : surface tension of water (N/m).

The surface tension of water is 0.0735N/m when its temperature is 15 degrees C. The contact angle between the meniscus and the surface of the particle is assumed to be zero at their boundary for ease. b and r can be expressed with R and α from the geometrical consideration on the meniscus by equation (3) and (4).

$$b = R(\tan\alpha + 1 - \sec\alpha) \tag{3}$$

$$r = R(\sec\alpha - 1) \tag{4}$$

There are many number of inter-particle contact points in the soil and the total of the capillary action induced at those contact points will affect the mechanical behavior of the partially saturated soil. In the calculation, we try to simulate the effect of the physical properties of the sphere sample such as particle size, water content and void ratio on the mechanical properties of suction and strength.

2.2 Flow-chart for the simulation

Figure 3 shows a flow chart for the simulation. In the simulation, the input data are radius R (m) and density ρ_S (kg/m^3) of the sphere, water content w (%) and void ratio e of the sphere sample. The volume of water containing in the sample can be obtained from the water content w and the void ratio e. This water is distributed to the surface of the particles as the adsorbed water and to the contact points among the sphere sample as the capillary water. The suction can be obtained by equation (1) after the angle of water retention is determined from the volume of the capillary water distributed to each contact point. The inter-particle force between spheres can also be determined by equation (2). The internal stress induced by the capillary water can be obtained by summing up all the inter-particle forces acting on unit area in the sphere sample.

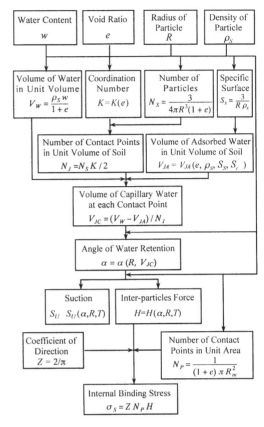

Figure 3   Flow chart for simulation

The detail process of the flow chart is as follows.

1) *Volume of water in unit volume of soil V_W*

The volume of water contained in a unit volume of the sphere sample V_W can be determined from the water content w and the void ratio e by equation (5).

$$V_W = \frac{\rho_S}{1+e}\frac{w}{100} \tag{5}$$

where ρ_S is density of the sphere particle (kg/m^3).

2) *Volume of adsorbed water V_{JA}*

The water in the sample is distributed to the surface of the particles as the adsorbed water and to the inter-particle contact points as the capillary water. The volume of the adsorbed water is assumed to depend on the water content and the particle size. We employ the specific surface of the sample S_S (m^2/kg) defined by the total surface area of the particles containing in its unit mass to represent the particle size.

When soil particles are uniform spheres such as those employed in this simulation, the specific surface of the sample S_S can be expressed with the radius R and density ρ_S of the particle by equation (6).

$$S_S = \frac{3}{R \, \rho_s} \qquad (6)$$

Here, M is defined as the mass of the water (kg), adsorbed on the surface of particles in saturated soil whose total mass is 1 kg under dry condition. In this paper, we assume that the value of M is determined from the specific surface S_S by equation (7), referring to the research by Rode (1955).

$$M = \sqrt{S_s} \times 10^{-2} \qquad (7)$$

In the unsaturated soil, we assume that the volume of the adsorbed water in unit volume of the sample is proportional to the degree of saturation of the sample S_r (%). Under this assumption, the volume of the adsorbed water in unit volume of the sample V_{JA} (m^3) is expressed by equation (8),

$$V_{JA} = \frac{\rho_s}{1 + e} \frac{M}{\rho_w} \frac{S_r}{100} \qquad (8)$$

where ρ_W is density of water (kg/m^3).

The water except the adsorbed water is distributed to the inter-particle contact points as the capillary water. The total volume of this water can be obtained by subtracting the volume of the adsorbed water V_{JA} from the total volume of the water contained in unit volume of the sample V_W. In this simulation, it is assumed that the capillary water is distributed equally to all the contact points in the sample. Therefore, the volume of the capillary water distributed to each contact point can be determined by the total volume of the capillary water divided by the total number of the contact points containing in the sample. The total number of the contact points in unit volume of the sample can be determined by multiplying the number of sphere particles and a half of the coordination number K.

3) Coordination Number K

The coordination number can be defined by the number of the contact points that each particle has, and depends on the packing conditions of spheres, that is, the void ratio e.

Graton et al. (1935) discussed the relationship between the void ratio and the coordination number when uniform spheres are packed. He supposed four types of packing patterns of spheres shown in Table 1 and studied the relationship between the void ratio and the coordination number for the packing patterns.

Figure 4 shows the relation of the void ratio to the coordination number, obtained from Table 1. Their relation can be approximated by equation (9).

$$K = 5.363 - 6.045 \ln e \qquad (9)$$

4) Number of particles in unit volume N_S

Number of particles contained in unit volume of the sample N_S is obtained form the radius of the particle R and void ratio e by equation (10).

$$N_S = \frac{3}{4 \pi R^3 (1 + e)} \qquad (10)$$

5) Total number of the contact points N_J

Total number of the contact points N_J in unit volume of the sample is obtained by multiplying the contact points N_S and the coordination number K,

Table 1 Packing patterns of particle and void ratio

Simple Cubic	Orthorhombic	Wedge Shaped Tetrahedrom	Hexagonal Close
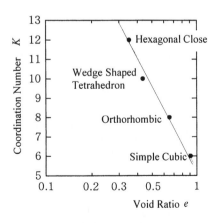			
$e = 0.908$	$e = 0.653$	$e = 0.433$	$e = 0.351$
$K = 6$	$K = 8$	$K = 10$	$K = 12$

Figure 4   Coordination number and void ratio

165

although the coordinate number must be divided by 2 because any contact point between particles is common for both particles each other.

$$N_J = \frac{N_S K}{2} \tag{11}$$

6) Volume of water for each contact point V_{JC}

The volume of water distributed to each contact point is determined by the total volume of the capillary water divided by the total number of the contact points in unit volume of the sample, as shown by equation (12).

$$V_{JC} = \frac{V_W - V_{JA}}{N_J} \tag{12}$$

7) Angle of water retention α

Angle of water retention α can be determined from geometrical study on the shape of the capillary water. Equation (13) gives the relationship between α and V_{JC}. In the simulation, the value of α should be determined to satisfy the equation (13).

$$V_{JC} = 2\pi \frac{R^3 (1 - \cos\alpha)^2}{\cos^2\alpha}$$
$$\left\{ 2 - \cos^2\alpha + \tan\alpha \left(\alpha - \frac{1}{2}\sin 2\alpha - \frac{\pi}{2} \right) \right\} \tag{13}$$

8) Inter-particle force H and suction S_U

Inter-particle force between particles H is given by the equation (2) and the suction S_U is obtained by equation (1).

9) Internal binding stress σ_S

Internal binding stress σ_S is defined by the stress induced by the capillary action on unit area in the sphere sample. It can be calculated by summing up the inter-particle forces on unit area in the sphere sample.

Figure 5 shows the contact condition among sphere particles. The thick line is a line linking the contact points among the spheres. This contact line is replaced by the thin straight line A-A in the simulation. The number of contact points on unit area N_P is given by equation (14),

$$N_P = \frac{1}{(1 + e)\, \pi R_m^2} \tag{14}$$

where R_m is the average radius of cross sections which appear on the line A-A, and is expressed by equation (15).

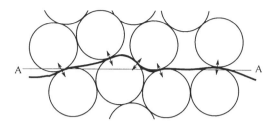

Figure 5   Contact conditions in sphere sample

$$R_m = \frac{\pi}{4} R \tag{15}$$

The direction of contact points varies randomly in the sample as shown in Figure 5. The stress component for the specific direction can be obtained by introducing the direction coefficient z expressed by equation (16).

$$z = \frac{2}{\pi} \tag{16}$$

The internal binding stress σ_S is finally determined by equation (17).

$$\sigma_S = z H N_P \tag{17}$$

The suction S_U and the internal stress σ_S can be calculated by the calculation process shown in Figure 3 from the input data of water content w, void ratio e, radius R and density ρ_S of the particle.

3 SUCTION DUE TO CAPILLARY ACTION

3.1 Measurement of suction

Experiments have been carried out to study the reliability of the numerical simulation. In the experiments, we measured the suction, which was induced in the partially saturated sphere sample. The glass beads were employed as the sphere sample because it is almost perfect sphere and has an uniform size. The glass beads were packed in the acrylic cylinder whose diameter and height were respectively 10cm and 30cm after the water content of glass beads had been controlled. The water content was controlled by adding the water for seven steps.

A tensiometer was set in the packed glass beads to measure the suction of the sample. The diameter of glass beads was 0.07mm and the density of a bead was 2.53 x 10^3 kg/m^3. Figure 6 shows the experimental apparatus.

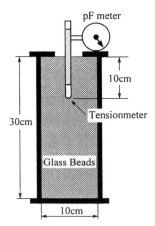

Figure 6   Measurement of suction in glass beads

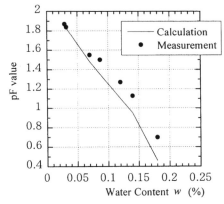

Figure 7   Results of experiments and calculation
for suction in partially saturated glass beads

3.2 *Calculation of suction and comparison with experimental results*

Suction in partially saturated glass beads was calculated by the simulation mentioned in the previous section. The parameters employed in the calculation were set to express those in the experiments. The calculated suction was transported to *pF* value by converting it to the water height in logarithm.

Figure 7 shows the results of experiments and calculation, where the *pF* values are plotted against water content.

It is clear in this figure that the calculated results agree with the experimental ones, although a little difference is recognized at high water content. The assumption in the estimation of the adsorbed water volume expressed by equation (8) is thought to be one of the reasons for the difference.

4   STRENGTH DUE TO CAPILLARY ACTION

4.1   *Measurement of strength*

In partially saturated soils, the additional strength will be induced by capillary water. The additional strength can be evaluated by the flow chart of Figure 3. We carried out some experiments to measure the additional strength due to capillary action and to compare their results with the results of calculation.

In the experiments, the partially saturated glass beads, which were the same ones as those used in the experiments for suction, were packed in a soil bin and a slip failure was induced in the glass beads sample. Figure 8 shows the experimental apparatus. One side of the soil bin is movable, and the slip failure was induced by putting the sheets of paper on the surface of the sample for surface load. The weight of surface load can be controlled by changing the number of the sheets. In the experiments, two ways of packing conditions of loose and dense, and four steps of water contents of glass beads (2.5, 4.1, 5.9, and 6.3 %) were employed. The paper weight, which induces the slip failure in the sample, was measured for those conditions.

4.2   *Calculation of strength due to capillary action and comparison with experimental results*

The paper weight, which induced the slip failure in the experiments, was predicted by the calculation. In the calculation, the slip plane was supposed to be a flat plane and the paper weight q (N/m^2) was calculated by equation (12), (13) and (14),

$$c = \sigma_S \tan \phi \tag{18}$$

$$\tan 2\theta = \frac{1}{\tan \phi} \tag{19}$$

$$q = \frac{2\,c}{\tan \theta} - \frac{1}{2}\,h\,\gamma \tag{20}$$

where

c :   cohesion due to the capillary action (N/m^2)
ϕ :   internal friction angle of the sample (degree)
γ :   unit weight of the sample (N/m^3)
θ :   slip angle from the side wall (degree)
h :   height of the sample (m)

The cohesion c due to the capillary action is thought to be caused by the friction between particles, which is induced by the internal binding stress σ_S. This means that the internal stress σ_S due to the capillary action binds soil particles each other and it creates the friction between particles to restrict the movement of the particles. In this paper, we suppose that this process induces the cohesion in

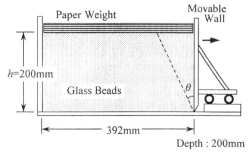

Figure 8   Experimental apparatus for strength

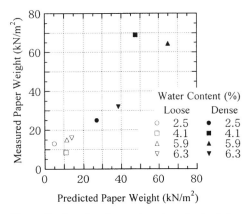

Figure 9   Results of experiments and calculations
for strength in partially saturated glass beads

of the drying or water supplying process.   The second one is the introduction of the variety in size and shape of soil particles to the simulation.

REFERENCES

Graton and Fraser (1935). J. Geology, 43, pp. 785.

Rode A.A. (1955). Soil Science (Pochvovedeniye), Transl. by Yamazaki,F. in Japanese., Yokendo (1969)

the partially saturated soil, and then the cohesion was defined by equation (18).

Figure 9 shows the results of experiments and calculations, where the paper weight at failure observed in the experiments are plotted against the predicted ones.   It is clear from this figure that the calculated results agree with the experimental ones although there is a scatter in some cases.

5   CONCLUSION

The effect of capillary water on the suction and strength of the partially saturated sphere sample is simulated through numerical simulation, and its reliability is examined by some model experiments. The result of the study made it clear that the simulation can express the behavior of the partially saturated soil although there are some factors to be improved such as the estimation of the adsorbed water.

The following factors should be studied more strictly to express the behavior of real soil: The first one is the estimation of the adsorbed water, including the effect of the difference in controlling method of water content, especially the difference

3   Numerical modelling

3.1 Modelling of flow

Unsaturated Soils for Asia, Rahardjo, Toll & Leong (eds) © 2000 Taylor & Francis, ISBN 90 5809 139 2

Analysis of non-Darcy flow in unsaturated soils modeled by finite element method

W. Gasaluck

Department of Civil Engineering, Khon Kaen University, Thailand

ABSTRACT: The flow equations obtained from the falling-head permeability tests conducted on unsaturated soils compacted in cubical molds show non-linear characteristic. It is very complicated to solve such type of flow in earth dam using the finite element method. The simple technique called fictitious coefficient of permeability based on Darcy's law was applied to the computer program. The program can also solve the non-Darcy flow in anisotropic non-homogeneous porous media.

1 INTRODUCTION

Seepage problems in soil mechanics can be solved by various methods. Simple methods such as Dupuit method, Casagrande method, etc. are practical when the problems are not complicated. The finite element method, a very powerful numerical technique whose rise and development was promoted by the rapid development of computers, was introduced into geotechnical engineering field in 1960. It became the favorite method for applying to numerous problems of flow through porous media, groundwater flow, flow with a phreatic surface, etc. It can handle any shape of boundary and any combination of boundary condition, non-homogeneous and anisotropic media (Bear 1979).

There are many published techniques for finite element analysis of seepage problems, for example, variable mesh technique (Taylor & Brown 1967; Finn 1967; Desai 1972), constant mesh technique (Bathe & Khoshgoftaar 1979). In the mentioned publications, the validity of Darcy's law is usually assumed in the majority of practical problems because the basic differential equations of flow cannot be solved mathematically, except with the use of a constant coefficient of permeability (Kovacs 1981). Volker (1969) presented the analysis of non-Darcy flow by finite element method but his work was for high velocity flow and the equations being used were complicated and quite different from those for Darcy flow. Neuman (1973) used an interactive Galerkin type finite element method to solve the equation of transient seepage in saturated-unsaturated porous media. His equations were complicated as well.

In 1990, Kazda introduced the so-called fictitious

coefficient of permeability technique based on Darcy's law for use a finite element computer program. The program is able to solve even non-Darcy flow problems. In other words, the complicated differential equations of non-Darcy flow can be eliminated using this technique.

2 VALIDITY OF DARCY'S LAW

Several studies have been made to investigate the range over which Darcy's law is valid. Scheidegger (1974) discussed several reasons that flow through very small openings may not follow Darcy's law in soft clay. Some investigators (Swartzendruber 1962, Miller & Low 1963) have claimed that in clayey soils, low hydraulic gradient may cause no flow or only low flow rate that is less than proportional to the gradient. Ping (1963) and Gill (1977) also reported on the threshold gradient. Basak (1977) introduced the probable velocity-hydraulic gradient relationship including non-Darcy flow in low hydraulic gradient zone. In contrast, Miller et al. (1969) concluded that there was no threshold gradient in their experiments. Seepage consolidation may cause apparent deviation from Darcy's law (Pane et al. 1983). For falling-head permeability test, a long time required to measure low flow rates which might cause the problems related to evaporation and temperature change (Remy 1973).

There is a suggestion that all of the mentioned researches were done on special apparatus and the moisture condition was controlled to be saturated. In such tests, air eliminating process or otherwise back pressure is applied to saturate soils sample and it is needed for maintaining the saturated state throughout the test (Dun & Mitchell 1984). But in the field, water flow through soils naturally as the conventional falling-head permeability test. Consequently, the saturated state does not occur in real earth structure and Darcy's law cannot be adopted in the analysis.

The objective of this paper is to present the finite element computer program for analyzing non-Darcy's flow problem but the differential equations employed in the program are based on Darcy's law.

3 FICTITIOUS COEFFICIENT OF PERMEABILIT

The governing partial differential equations for anisotropic flow in the xy plane in an earth dam are as follow (Segerlind 1984).

$$k_x \frac{\partial^2 \phi}{\partial x^2} + k_y \frac{\partial^2 \phi}{\partial y^2} + Q = 0 \tag{1}$$

where k_x, k_y = horizontal and vertical coefficient of permeability respectively; ϕ = potential; Q = recharge, pumping is a negative Q.

The minimization of a functional related to Equation 1 give the following matrices:

$$[K]\{\Phi\} = \{F\} \tag{2}$$

$$[K] = \sum_{e=1}^{E} [k^{(e)}] \tag{3}$$

$$[k^{(e)}] = \int_{v^{(e)}} [B^{(e)}]^T [D^{(e)}][B^{(e)}] \tag{4}$$

$$[B^{(e)}] = \begin{bmatrix} \dfrac{\partial N_i^{(e)}}{\partial x} & \dfrac{\partial N_j^{(e)}}{\partial x} & \dfrac{\partial N_k^{(e)}}{\partial x} \\ \dfrac{\partial N_i^{(e)}}{\partial y} & \dfrac{\partial N_j^{(e)}}{\partial y} & \dfrac{\partial N_n^{(e)}}{\partial y} \end{bmatrix} \tag{5}$$

$$[D^{(e)}] = \begin{bmatrix} k_x & 0 \\ 0 & k_y \end{bmatrix} \tag{6}$$

$$\{F\} = -\sum_{e=1}^{E} \{f^{(e)}\} \tag{7}$$

$$\{f^{(e)}\} = \int_{v^{(e)}} [N^{(e)}]^T Q \, dV \tag{8}$$

$$[N^{(e)}] = [N_i^{(e)} \ N_j^{(e)} \ N_k^{(e)}] \tag{9}$$

where ϕ is the potential at any node of the model under consideration; E is the number of elements; $N_i^{(e)}$ $N_j^{(e)}$ and $N_k^{(e)}$ are shape functions, one for each node. The two-dimensional simplex element is the triangle shown in Figure 1. This element has

straight sides and three nodes, one at each corner. The nodal values of the scalar quantity ϕ are denoted by Φ_i, Φ_j and Φ_k. The coordinate pairs of the three nodes are (X_i, Y_i), (X_j, Y_j) and (X_k, Y_k).

The element equation has three shape functions as shown in the following equations:

$$\phi = N_i \Phi_i + N_j \Phi_j + N_k \Phi_k \tag{10}$$

$$N_i = \frac{1}{2A}[a_i + b_i x + c_i y] \tag{11}$$

$$N_j = \frac{1}{2A}[a_j + b_j x + c_j y] \tag{12}$$

$$N_k = \frac{1}{2A}[a_k + b_k x + c_k y] \tag{13}$$

where :

$$a_i = X_j Y_k - X_k Y_j \; ; \; b_i = Y_j - Y_k \; ; \; c_i = X_k - X_j$$

$$a_j = X_k Y_i - X_i Y_k \; ; \; b_j = Y_k - Y_i \; ; \; c_j = X_i - X_k$$

$$a_k = X_i Y_j - X_j Y_i \; ; \; b_k = Y_i - Y_j \; ; \; c_k = X_j - X_i$$

$$2A = \begin{vmatrix} 1 & X_i & Y_i \\ 1 & X_j & Y_j \\ 1 & X_k & Y_k \end{vmatrix}$$

Evaluating the matrix product yields

$$[k^{(e)}] = \frac{k_x}{4A} \begin{bmatrix} b_i b_i & b_i b_j & b_i b_k \\ b_i b_j & b_j b_j & b_j b_k \\ b_i b_k & b_j b_k & b_k b_k \end{bmatrix} + \frac{k_y}{4A} \begin{bmatrix} c_i c_i & c_i c_j & c_i c_k \\ c_i c_j & c_j c_j & c_j c_k \\ c_i c_k & c_j c_k & c_k c_k \end{bmatrix} \tag{14}$$

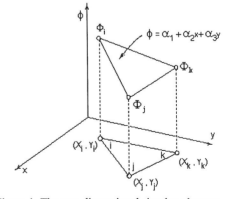

Figure 1. The two-dimensional simplex element

$$\{f^{(e)}\} = \frac{QA}{3}\begin{Bmatrix}1\\1\\1\end{Bmatrix} \qquad (15)$$

The boundary conditions associated with the phreatic line in an earth dam can be expressed as (Bathe & Khoshgoftaar 1979):

$$\phi = y \qquad (16)$$

$$\frac{\partial \phi}{\partial n} = 0 \qquad (17)$$

Where y = elevation at the point under consideration; n = the normal to the surface.

For saving CPU time, $\partial \phi / \partial n \leq 0.01$ is employed in this paper. Since ordinary earth dam does not have any recharge and pumping, Q is zero.

The equations explained above require the coefficient of permeability. Non-Darcy flow then cannot be analyzed directly by the use of those equations since the discharge velocity is a function of the hydraulic gradient. Kazda (1990) introduced the so-called technique of fictitious coefficient of permeability that is helpful to make Darcy flow differential equations usable equations for non-Darcy analysis. The simple equation employed is

$$k^{(e)} = \frac{v^{(e)}}{i^{(e)}} \qquad (18)$$

Where $k^{(e)}$ and $v^{(e)}$ are the fictitious coefficient of permeability and the discharge velocity in an element, respectively, with the hydraulic gradient of $i^{(e)}$. $k^{(e)}$ is not the actual coefficient of permeability that is a slope of the straight relationship between discharge velocity and hydraulic gradient in accordance with Darcy's law. But it is a ratio of discharge velocity to hydraulic gradient for each element as schematically shown in Figure 2. By substituting $k^{(e)}$ into the differential equations a further quasilinear state is obtained.

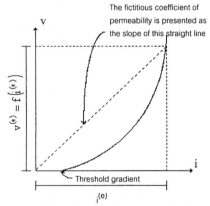

Figure 2. The schematic diagram of fictitious coefficient of permeability.

According to Equation 18, $k^{(e)}$ of the elements in the earth dam is different from each other since it varies depending upon the hydraulic gradient and discharge velocity which are not constant. In other words, $k^{(e)}$, $v^{(e)}$ and $i^{(e)}$ vary depending upon each other. Therefore the algorithm shown in Figure 3 is applied. First all elements are specified to have the same k value. The first loop is the quasilinear state analyzed by the finite element computer program based on Darcy's law (Finn 1967). Equation 16 and 17 are employed for getting the actual phreatic lines. The second loop is the fictitious coefficient of permeability technique.

4 CALCULATION SAMPLE

The earth dam model as shown in Figure 4 was considered. It was composed of two soil types overlying the impervious layer. The model was subdivided into 1712 nodes and 1574 quadrilateral elements as shown in Figure 5. Each element was subdivided again into four triangular elements by the program.

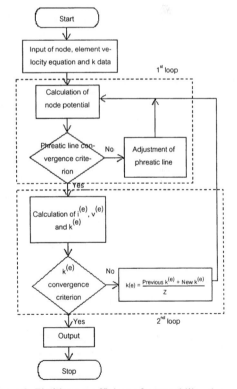

Figure 3. Fictitious coefficient of permeability algorithm.

173

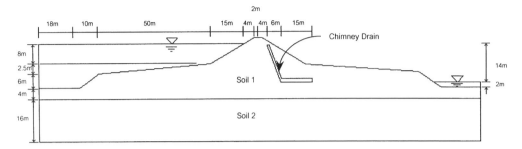

Figure 4. The model taken into consideration

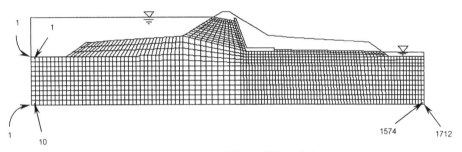

Figure 5. The model subdivided into 1712 nodes and 1574 quadrilateral elements.

The discharge velocity equations of the soils were obtained from the falling-head permeability tests conducted on the soils compacted in cubical molds (Gasaluck et al. 1995). The unsaturated samples gave non-linear relationship between discharge velocity and hydraulic gradient as shown in the following equations.

$$v_{1x} = 0.0381i^{-0.14} - 0.627 + 1.035i + 0.052i^2 \quad (19)$$

$$v_{1y} = -1.21 \times 10^{-50}i^{-52} - 0.0015 + 0.02i + 0.00039i^2 \quad (20)$$

$$v_{2x} = 0.026i^{-0.32} - 0.131 + 1.785i + 0.062i^2 \quad (21)$$

$$v_{2x} = 0.0026i^{-1.03} - 0.114 + 0.737i + 0.026i^2 \quad (22)$$

Where v_{1x} and v_{1y} (cm/day) = soil1 horizontal and vertical discharge velocity respectively: v_{2x} and v_{2y} (cm/day) = soil2 horizontal and vertical discharge velocity respectively.

The threshold gradients are 0.5506, 0.1250, 0.0273 and 0.1239 for Equation 19, 20, 21 and 22 respectively.

The best $k^{(e)}$ convergence criterion is that the last $k^{(e)}$ value of all elements does not change from the previous one, in other words, the $k^{(e)}$ becomes constant for each element which could not be achieved. One possible method is to let the program execute for many rounds and report the change of $k^{(e)}$ in

Table 1. The difference specified as the convergence criterion.

Difference	Soil 1		Soil 2	
Mode	k_x	k_y	k_x	k_y
Maximum	1.93×10^{-3}	1.23×10^{-4}	2.23×10^{-4}	6.97×10^{-4}
Average	1.45×10^{-4}	9.31×10^{-6}	4.88×10^{-5}	4.67×10^{-5}

every elements. The difference between the k(e) value computed in iteration 62 and 63 is shown in Table1. The unit of k is cm/day.

According to the author's experience, the $k^{(e)}$ changes are great in the first 5 to 10 iterations.

The results are shown as equipotential lines in Figure 6. The equipotential lines do not deflect at the interface between soil1 and soil2. The reason is that the hydraulic gradient in this area is very small and makes $k^{(e)}$ of both soil layers become zero or near zero. In other words both soil layers have approximately the same $k^{(e)}$.

5. CONCLUSIONS

The difficulty of solving the flow equation of unsaturated soils can be avoided with the help of the fictitious coefficient of permeability technique. The technique enables a simple computer program based on Darcy's law to solve the seepage problems of

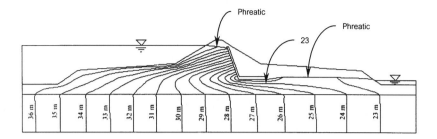

Figure 6.   Phreatic Lines and Equipotential Lines in Non-Homogeneous Earth Dam Found by FEM under Anisotropic Flow Condition.

various conditions, even non-linear flow in anisotropic non-homogeneous earth dam. The simple program could be easily improved by only changing the k parameter to be an array and increasing the if clause for checking the $k^{(e)}$ convergence.

REFFERENCES

Basak, P. 1977. Non-Darcy flow and its implication to seepage problems. *J.Irrigation and Drainage Div.* IR4: 459-473.

Bathe, K.J. & Khoshgoftaar, M.R. 1979. Finite element free surface seepage analysis without mesh iteration. *Int.J.Numerical and Analytical Methods in Geomechanics* Vol.3: 13-22.

Bear, J. 1979. *Hydraulics of groundwater.* Israel: McGraw-Hill.

Desai, C.S. 1972. Seepage analysis of earth banks under drawdown. *J.Soil Mechanics. and Foundation Div.* SM11: 1143-1162.

Dun, R.J. & Mitchell, J.K. 1984. Fluid conductivity testing of fine grained soils. *J.Geotechnical Eng.* 110: 1648-1665.

Fin, L.W.D. 1967. Finite element analysis of seepage through dams. *J.Soil Mechanics. and Foundation Div.* SM6: 41-48.

Gasaluck, W. et al. 1995. On the non-linearity and anisotropy of seepage flow. *Tranc.JSIDRE* No.178: 1-10.

Gill, M.A. 1977. Analysis of permeameter data to obtain parameters of non-Darcy's flow. *J.Hydrology* 32: 165-173.

Kazda, L. 1990. *Finite element techniques in groundwater flow studies.* Czechoslovakia: Elsevier.

Kovacs, G. 1981. *Seepage hydraulic.* Hungary: Elsevier.

Miller, R.J. & Low, P.F. 1963. Threshold gradient for water in clay systems. *Proc. Soil Sci. Soc. Am.* 27: 605-609.

Mill, R.J. et al. 1969. The absence of threshold gradients in clay-water systems. *Proc. Soil Sci. Soc. Am.* 33: 183-187.

Neuman, S.P. 1973. Saturated-unsaturated seepage by finite elements. *J. Hydraulic Div.* HY12: 2233-2250.

Pane, V. et al. 1983. Effects of consolidation on permeability measurements for soft clay. *Geotechnique* 33: 67-72.

Ping, L.S. 1963. Measuring extremely low flow velocity of water in clays. *J. Soil Sci.* 95: 410-413.

Remy, J.P. 1973. The measurement of small permeabilities in the laboratory. *Geotechnique* 23: 454-458.

Scheidegger, A.E. 1974. *The physics of flow through porous media 3rd ed.* UK: U. of Toronto Press.

Segerlind, L.J. 1984. *Applied finite element analysis 2nd ed.* USA: John Wiley & Sons.

Swartzendruber, D. 1962. Non-Darcy flow behavior in liquid-saturated porous media. *J. Geophysical Research* 67: 5205-5213.

Talor, R.L. & Brown, C.B. 1967. Darcy flow solutions with a free surface. *J. Hydraulic Div.* HY2: 25-33.

Volker, R.E. 1969. Non-linear flow in porous media by finite elements. *J. Hydraulic Div.* HY6: 2093-2114.

Unsaturated Soils for Asia, Rahardjo, Toll & Leong (eds) © 2000 Taylor & Francis, ISBN 90 5809 139 2

A technology for unsaturated zone modelling under variable weather conditions

M. I. Gogolev & P. Delaney
Waterloo Hydrogeologic Incorporation, Waterloo, Ont., Canada

ABSTRACT: The WHI UnSat Suite Plus software package integrates a tool for generating daily weather predictions with several popular models for unsaturated zone simulation. A simulation technology, which combines the advantages of different models of the package has been developed for effective modelling of unsaturated flow and transport under variable weather conditions. The use of technology is illustrated with the simulation experiment for two cities with different types of climate. It is found that the accuracy of the transport simulation may be very sensitive to the level of averaging of the flow at the upper boundary of the unsaturated profile. The differences in concentration caused by differences of inflow averaging depend on climate being smaller in tropical climate (Malacca). However, even in tropical climate, the concentration may differ up to 5.8 times when monthly and annual levels of averaging are compared.

1 INRODUCTION

While practicing soil scientists and engineers recognize the importance of modelling physical and chemical processes within the unsaturated zone, a lack of practical tools and technologies restricts their abilities to properly address these important problems. Most of the popular unsaturated zone models were developed as DOS applications and cannot utilize advantages of the WINDOWS computer environment. In addition, the majority of these models rely upon simplified steady-state boundary conditions at the soil surface (constant head or constant flux) which strongly impacts the accuracy, reliability and validity of the simulation results.

In an effort to address the problems and difficulty associated with obtaining reliable and realistic boundary data for these models, Waterloo Hydrogeologic, Inc. (WHI) has developed a software package WHI UnSat Suite Plus that integrates a new tool for generating seasonal weather predictions with several popular models for simulating water flow and contaminant transport through the unsaturated zone.

A new software technology, which combines the advantages of different infiltration and percolation models, has been developed for effective modelling of unsaturated flow and transport under variable weather conditions.

The purpose of this paper is twofold: (1) to pres-ent the modelling technology and its major components and (2) to illustrate the use of the technology by examining the same unsaturated zone profile at different time scales under two different patterns of weather conditions.

2 MODELLING TECHNOLOGY

2.1 Component Models

Two of the four models included into the WHI Un-Sat Suite Plus are used in the described technology. These models were selected after analysis and comparative testing of their performance. The models are:

HELP (Hydrology Evaluation of Landfill Performance) (Schroeder et al 1994) - a versatile model for predicting landfill hydrologic processes and testing the effectiveness of landfill designs.

VS2DT (Variably Saturated 2D flow and Transport) (Healey 1990) - a finite difference numerical model for determining the water movement as well as the fate of agricultural chemicals, landfill leachate, and accidental chemical spills in the unsaturated zone.

As the name indicates, the HELP model is commonly used to simulate landfill hydrology. However, HELP uses numerical solution techniques that account for the effects of precipitation, surface storage, snowmelt, runoff, infiltration, evapotranspiration, vegetative growth, soil moisture storage, lateral subsurface drainage, unsaturated vertical drainage, and leakage through soil. This exhaustive list of soil hydrologic processes makes HELP an ideal tool for determining more realistic predictions of upper moisture flux boundary conditions for unsaturated zone models. HELP cannot be used alone for the assessment of contaminant concentration because it does not simulate unsaturated transport. This problem may be effectively solved by VS2DT, which is able to simulate a wide list of chemical processes but is very limited in capabilities to present upper boundary condition of the profile. The combination of advantages of both HELP and VS2DT bears a lot of potential for unsaturated zone simulations.

2.2 Global Weather Generator

A reliable source of weather data is a factor of major importance for obtaining accurate values of water fluxes at soil boundary. HELP was equipped by its developer Dr. P. Schroeder with the Weather Generator – an effective USDA program that produces statistically reliable sets of daily values of precipitation, air temperature and solar radiation for periods up to 100 years (Richardson 1984).

However, the original DOS version of the Weather Generator program included a database with parameters for generating synthetic weather data for only 139 U.S. cities. We have developed a global database that includes more than 3000 meteorological stations and a GIS feature for locating the nearest stations. As a source of raw weather data the NOAA database, which contains 14 years (1977-1991) of observed daily precipitation and temperature data for nearly 10,000 stations across the world, was used. It took a great deal of research and programming to decode the database files and develop filters for the records with missing data. Finally, the valid records were mathematically processed and coefficients for weather generation determined for more than 3000 locations.

To verify the reliability of the newly obtained coefficients, two series of tests were performed. In the first series of tests the generated daily precipitation and temperature were compared to the observed sets for six US cities for which Dr. Clarence Richardson, the developer of the original Weather Generator database, performed his comparison in 1984. Results of the tests proved that errors between statistics for generated and observed data were in the range or less than when Dr. Richardson performed his tests using another set of observed data and coefficients.

Table 1. Some statistics obtained in comparison of generated and observed data for total monthly precipitation.

City	Mean Absolute Error, in	Mean Relative Error, %
Beijing, China	-0.03	4.44
Tokyo, Japan	0.08	3.85

Table 2. Some statistics obtained in comparison of generated and observed data for monthly air temperature.

City	Monthly Statistic	Mean Absolute Error, F°	Mean Relative Error, %
Beijing, China	Mean daily temperature	0.02	0.09
	Mean daily maximum	0.25	0.09
	Mean daily minimum	0.58	1.52
Tokyo, Japan	Mean daily temperature	0.06	0.1
	Mean daily maximum	1	1.57
	Mean daily minimum	-0.25	-0.68

During the second series of tests, synthetic weather data generated, using the new weather coefficient databases, were compared with observed weather data for twelve sites around the world. The analysis revealed that the new weather databases could be successfully used to synthesize reliable weather data. Mean monthly temperature was consistently accurate within 3 degrees Fahrenheit. Mean monthly precipitation was consistently accurate within 0.3 inches. Test results for Beijing, China and Tokyo, Japan (the two cities from Asia in the global test set) are presented below in Tables 1, 2.

The analysis shows that the developed database could be used to synthesize reliable weather data. The new Global Weather Generator database contains several hundred stations for each continent. It contains more than 400 stations for the new countries of Asian part of the former USSR and 609 stations for the remaining Asian countries. National databases include 322 stations for China, 57 stations for Japan, 63 stations for India, 45 Stations for Korea, 34 stations for Thailand and 21 stations for Turkey.

2.3 The technology

The models HELP, VS2DT and the Weather Generator used for simulation of unsaturated flow and transport under variable weather conditions were developed and thoroughly tested by the major environment protection organizations (US EPA and USDA). Each of them has certain advanced features that will increase accuracy and reliability of simulation results if they are combined. The role of the technology is to combine advantages of these models, and to integrate them the most efficient way.

At the first stage of this technology, the upper part of unsaturated zone (specific top layer or the root zone) is simulated with the HELP model and the Weather Generator to provide the high accuracy in assessment of the values of water fluxes at the bottom of the upper layer. At the second stage, the VS2DT model is used to simulate the unsaturated flow and transport of a specific chemical in the major part of the soil profile using the results of the HELP simulation as an upper boundary condition.

In an efford to make unsaturated zone simulations easy to perform and analyze, the following interface features have been developed:
- graphical tools that allows effective editing of the profile structure and layer parameters,
- database tools for easy editing and managing soil and chemical properties,
- the GIS searcher for Weather Generator that allows the quick finding nearest weather station,
- A special interface that allows the transfer of percolation data assessed with HELP into the VS2DT upper flow boundary input module.

The user has a choice to specify at which level of generalization – daily, monthly or annual - the moisture flow through the upper soil boundary has to be specified.

3 SIMULATION EXPERIMENT

To illustrate the application of the technology we performed a study that has a certain methodological interest. In the computational experiment we attempted to assess the effect of the inflow averaging on the contaminant transport. The neighboring question - the effect of averaging of weather conditions on infiltration into deep (up to 700 meters) unsaturated zone - has been studied by S.A. Stothoff (1997). This study revealed that weather averaging over a month produced underpredictions of infiltration as large as a factor of 5. This fact lets expecting that inflow averaging may has an impact on the results of the contaminant transport simula-

tion. The current study is to demonstrate quantitatively this impact.

To study how this effect might appear in different weather conditions, the experiment was performed for two different locations: Buffalo, USA and Malacca, Malaysia.

3.1 Conditions of the experiment

The conditions of the numeric experiment were based on a practical case for which WHI UnSat Suite was previously applied. Some innovative companies have developed a process for recycling powerplant bottom ash, the by-product of coal combustion. The processed ash product is used as construction fill to a depth of 2 feet. The ash contains chlorine, which can potentially leach from the ash and contaminate groundwater. Lysimeter experiments showed that, under natural weather conditions, concentration of chlorine in the water drained from the bottom of the ash layer may have the following pattern (Barnes & Roethel 1999):

Table 3. Concentration of chlorine in the drainage from the bottom of the ash layer.

Time, days	Chlorine concentration, mg/l
0-30	20000
30-90	7000
90-210	1400
210-365	200

The natural unsaturated zone profile (under the ash layer) is formed by a 20 feet sandy layer.

3.2 Stage 1: Simulation of the upper boundary condition for water with HELP

At the first stage, HELP was used to assess the values of water percoiating through the 2 feet ash layer. The compacted aggregated ash has the following parameters: saturated hydraulic conductivity of 0.8 m/day, total porosity of 0.57 and field capacity of 0.29. The initial conditions for soil profiles were near steady-state as provided by the standard HELP procedure. The length of simulation was one year.

The following results have been obtained for Buffalo, USA (Figs 1, 2).

As it might be seen from the Figures 1, 2, the precipitation at Buffalo (annual total is 1081 mm) is distributed quiet unevenly during the year which, in combination with temperature pattern (accumulation of precipitation in the forms of snow and ice), cause the absence of percolation at the beginning of the year. The peak percolation 70.5 mm at day 105

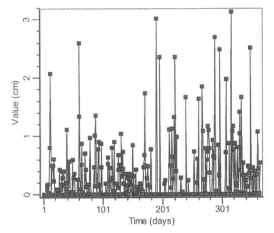

Figure 1. Generated daily precipitation for Buffalo, USA.

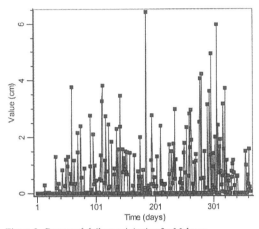

Figure 3. Generated daily precipitation for Malacca, Malaysia.

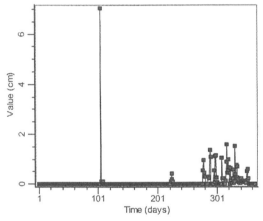

Figure 2. Simulated daily percolation through the ash layer for Buffalo, USA.

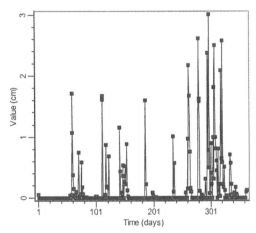

Figure 4. Simulated daily percolation for Malacca, Malaysia.

is caused by the melting of accumulated snow a few days before. The total percolation through the ash layer for Buffalo is 322mm.

Results of the HELP simulation for Malacca are presented in Figures 3, 4.

Results for Malacca, where the total annual precipitation is 1967 mm, show relatively more even percolation through the ash layer. The total percolation through the ash layer for Malacca is 589 mm.

To allow the comparison of different scales of averaging, percolation values for both locations were saved being averaged at daily, monthly and annually levels.

3.3 Stage 2: Simulation of chlorine transport with VS2DT

At the second stage, VS2DT was used to simulate the unsaturated flow and transport of chlorine in the zone below the bottom of the ash layer. The natural unsaturated zone profile (under the ash layer) is formed by a 20 feet sandy layer with saturated hydraulic conductivity of 2 m/day, porosity of 0.36, and residual moisture content of 0.03. The unsaturated flow was simulated using van Genuchten's approximation (Healey 1990) with parameters -α' equal to 20 cm and β' equal to 5.0. The contaminant transport was simulated using the dispersivity value of 40 cm and molecular diffusion of 0.5 cm2/day.

Calculations for both locations were performed using the pattern of chlorine concentration at the upper boundary presented in Table 3. Because the concentration of leachate depends on the pattern of water inflow, it may not be absolutely correct using data typical for Northern USA for Malacca. How-

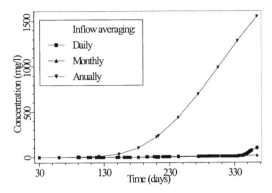

Figure 5. Chlorine concentration at depth 603 cm for different scales of inflow averaging for Buffalo, USA.

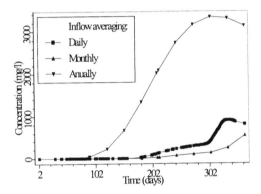

Figure 6. Chlorine concentration at depth 603 cm for different scales of inflow averaging for Malacca, Malaysia.

ever, use of the available data from other region is for range assessment is commonly done in engineering practice. Consequently, we have the right to use this approach in our study, which has the methodological character.

The results of the VS2DT simulation are presented in Figures 5, 6 and Tables 4, 5.

3.4 Discussion

The difference between simulation results obtained for different levels of averaging is increasing with depth and time both for Buffalo and Malacca. Not only the values of concentration but the shape of the breakthrough curve differs for different ways of inflow averaging (Fig. 6). Figures 5, 6 present chlorine concentration close to groundwater level where the maximum difference between simulation results occurs. As it might be seen, the difference in concentration for monthly and annual averaging of inflow may be as much as a factor of 86 for Buffalo and as a factor of 5.8 for Malacca. However, at the points closer to the soil surface significant differ-

Table 4. Chlorine concentration (mg/l) at different depth for different scales of inflow averaging at the end of the year for Buffalo, USA.

Depth, cm	Inflow averaging		
	Daily	Monthly	Annually
53	210	229	364
209	268	383	1338
603	103	18	1550

Table 5. Chlorine concentration (mg/l) at different depth for different scales of inflow averaging at the end of the year for Malacca, Malaysia.

Depth, cm	Inflow averaging		
	Daily	Monthly	Annually
53	206	210	229
209	255	257	459
603	810	539	3130

ence may still be observed (Tabs 4, 5). For depths 53 and 209 cm this difference is especially high when daily and monthly averaging are compared with annual averaging of inflow. The difference may be as much as a factor of 5 for Buffalo and as much as factor of 1.8 for Malacca (depth 209 cm, daily averaging in both cases). The overpredictions of the chlorine concentration for the annual averaging of inflow is explained by the fact the even inflow rate in this case brings a lot of chlorine into the profile during the first several months of simulation when the concentration of inflow is extremely high. The inflow with daily and monthly averaging, contrary, is very small (Malacca) or even equal to zero for Buffalo.

In general, the difference in simulated concentrations for Buffalo is much higher than that for Malacca. This phenomenon is explained by peculiarities of the climate at two locations. Percolation through the ash layer at Malacca is more stable, freezing does not interrupt it, and the rates of percolation are nearly two times higher.

4 CONCLUSIONS

The accuracy of transport simulation may be very sensitive to the level of averaging of the flow at the upper boundary of the unsaturated profile.

The differences in concentration caused by differences of inflow averaging depend on the climate being smaller in tropical climate (Malacca). However, even in tropical climate, the concentration may differ up to 5.8 times when monthly and annual levels of averaging are compared.

To obtain reliable results of contaminant transport simulation, daily variation of inflow has to be simulated. The efficient way to reach this goal is the application of the unsaturated zone modelling technology based on the functions of the WHI UnSat Suite Plus software package.

The Global Weather Generator included into the modelling suite provides a fast and efficient way for assessing daily weather characteristics for around 3000 locations globally.

REFERENCES

Barnes, J. & F. Roethel 1999, pers. comm.
Healey, R.W. 1990. Simulation of solute transport in variably saturated porous media with supplemental information on modifications to the US Geological Survey computer program VS2D. Water resource investigations report 90-4025. Denver, Colorado.
Richardson, C.W. & D.A. Wright 1984. WGEN: a model for generating daily weather variables. US Department of Agriculture, Agricultural Research Service, ARS – 8.
Schroeder, P.R. et al. 1994. The hydrologic evaluation of landfill performance (HELP) model, EPA report EPA/600/R-94/168b.
Stothoff, S.A. 1997, Sensitivity of long-term bare soil infiltration simulations to hydraulic properties in an arid environment. Water Resources Research, vol. 33, NO.4: 547-558.

Unsaturated Soils for Asia, Rahardjo, Toll & Leong (eds) © 2000 Taylor & Francis, ISBN 90 5809 139 2

Volume averaged parameters for two-phase flow

Marc Hoffmann
Department of Environmental Sciences, Wageningen University, Netherlands

During the last years the theory of volume averaging became widely used among theoretical research on the transport parameters for porous media flow. Practical applications of this theory appeared slowly in the literature because of the complex closure problem. By using idealized geometrical porous media descriptions, the closure problem can be solved approximately. One of the first applications of this concept is the work of (du Plessis & Masliyah 1988) for single phase flow. The volume averaging technique is used to model viscous flow including hydrodynamic drag effects in porous media. In this presentation further developments in the application of volume averaging for two-phase flows are described. The main emphasis lies on slow gas-liquid flows which occur in natural soils.

The presence of two mobile fluid phases (e.g. water and air) introduces extra terms in the volume averaging process. We describe the solution of the closure problem with approximate analytical equations. The final volume averaged equations have a non-local structure, but are suited for numerical solution of flow problems in porous media.

LIST OF SYMBOLS

$\mathcal{A}$: area in pore (corner) occupied by fluid $[m^2]$

A_p : pore area normal to flow $[m^2]$

C : factor depending on x

d : microscopic characteristic length (RUC) $[m]$

f : Fanning friction factor for fully developed flow $[-]$

$f(\theta, \delta)$: function describing geometric components $[-]$

g : gravity acceleration $[m/s^2]$

$\mathbf{I}$: vector integral expression

k : number of corners in channel $[-]$

l_g : flow length inside RUC $[m]$

$\mathcal{L}$: length of interface $[m]$

n : fraction of porosity occupied by liquid or gas $[-]$

p : pressure $[kg/ms^2]$

P_p : perimeter of pore normal to flow $[m]$

q : volumetric flow rate $[m^3/s]$

Q : specific discharge over RUC (summed over individual corners) $[m^3/s]$

r : radius of curvature of gas-liquid interface $[m]$

r_{crit} : critical radius of curvature of gas-liquid interface during drainage $[m]$

r_i : critical radius of curvature of gas-liquid interface during imbibition $[m]$

Re : Reynolds number, $\rho v_p D_h/\mu$, $[-]$

$\mathcal{S}$: surface $[m^2]$

v : velocity $[m/s]$

V_p : volume of pores in averaging volume $[m^3]$

V_0 : volume of averaging volume $[m^3]$, d^3

β : resistance factor $[-]$

δ : angle of corner rad

ϵ : porosity $[m^3/m^3]$, V_p/V_0

γ : surface (interfacial) tension at interface $[kg/s^2]$

ρ : density $[kg/m^3]$

μ : dynamic viscosity $[kg/ms]$

θ : contact angle liquid - solid rad

ν : normal vector on $\mathcal{S}$ $[-]$

subscripts:

c : mean inside corner

l : liquid

n : volumetric intrinsic phase average

p : mean inside pore

s : solid

symbols:

$\langle ( ) \rangle$: average of quantity

$( )$: deviation from mean

$\|$: streamwise pores

$\perp$: transverse pores

1 INTRODUCTION

The description of the flow of multiple phases through a porous medium is important in many areas of science and engineering. Examples are unsaturated zone hydrology, petrophysics and catalytic flow reactors in chemical engineering.

The equations describing the movement of fluids through a porous medium are usually based on Darcy type equations for slow flow. These equations require the description of parameters like conductivity and the water-content vs. pressure relationship. In practice these parameters are estimated using empirical relationships or through experiments on specific porous media. This paper uses the method of *Volume Averaging* (Bachmat & Bear 1986) to describe the behaviour of a wetting liquid inside a porous medium with the presence of a non-wetting liquid. The volume averaged equations have a different form than the Darcy type equations and are used to directly estimate porous media properties.

Volume averaging has been used in the last decennia, mostly for single-phase flow in porous media, although examples of multiphase flow are described in (Bear & Bensabat 1989). Volume averaging is a complex mathematical technique and despite the enormous theoretical developments, up to now the practical applications of volume averaging have been limited due to the so called closure problem. Different possibilities exist to solve this closure problem: numerical methods (Quintard & Whitaker 1990), approximations based on empirical equations (Whitaker 1980) and analytical solutions for prototype unit cells (du Plessis & Masliyah 1988; du Plessis & Diedericks 1997). The approach taken in this paper is based on the closure scheme developed in (du Plessis & Masliyah 1988). This scheme is augmented with additional boundary conditions which are introduced due to fluid-fluid interfaces. This has become possible due to the recent work of (Zhou et al. 1997), who derived an analytical approximation for the resistance factor to flow in corners.

The paper starts with a description of the volume averaging process and introduces the volume averaged equations. From these new equations the description of the volume averaged parameters for two-phase flow are derived.

2 THE BASIC INTERSTITIAL EQUATIONS

As a basis we start with the steady-state Navier-Stokes equations inside the pores of a porous medium. These equations are then averaged over a Representative Elementary Volume (REV). To make this possible the REV is modelled as a typical configuration of the pore space, called a Representative Unit Cell (RUC). For each of the fluid phases, e.g. gas and liquid, these equations are:

$$\nabla \cdot (\rho v v) = \rho g + \nabla p - \mu \nabla^2 v \qquad (1)$$

They are augmented by the mass conservation equations:

$$\nabla \cdot v = 0 \qquad (2)$$

with boundary conditions:

$$v_g = 0 \quad \text{at gas-solid interface} \quad (3)$$

$$v_l = 0 \quad \text{at liquid-solid interface} \quad (4)$$

$$v_g = v_l \quad \text{at gas-liquid interface} \quad (5)$$

$$(p_g - p_l)|_x = \gamma_{gl} C|_x \quad \text{in longitudinal di-} \quad (6) \\ \text{rection of capillary}$$

The last boundary condition describes the surface forces acting on the fluids. It is assumed that the only surface force is surface tension, which is modeled by the Young-Laplace equation. If the fluid contents vary slowly in the longitudinal direction of the capillary (see region C of figure 3), the pressure of the non-wetting liquid is assumed constant and taken as reference pressure, the pressure in the wetting phase becomes:

$$p = -\frac{\gamma}{r} \qquad (7)$$

Only the equations describing the liquid movement will be retained.

3 VOLUME AVERAGING FOR VARIABLE FLUID VOLUME FRACTIONS

In the literature several examples exist of the use of volume averaging inside a porous medium with constant volume fractions of fluids inside an averaging volume. Averaging with variable fluid fractions is not common and introduces extra terms in the averaged equations which are similar to the case when the porosity varies. The volume averaged form of equation (1) for the liquid is:

$$\rho_l \nabla \cdot (\epsilon n_l v_{ln} v_{ln}) = \epsilon n_l \rho_l g + \epsilon n_l \nabla p_{ln}$$

$$+ \mu_l \nabla^2 (\epsilon n_l \, v_{ln}) - \rho_l \nabla \cdot (\epsilon n_l \langle \dot{v}_l \dot{v}_l \rangle_n)$$

$$+ \frac{1}{V_0} \int_S \left(\nu \dot{p}_l + \mu_l \frac{\partial}{\partial \nu}(v_l) \right) dS \qquad (8)$$

The surface integral terms have two contributions: from the gas-liquid and from the liquid-solid interface. Rearranging equation (8), dropping the terms containing momentum advection and convective terms and writing terms which generate momentum left and dissipate momentum right gives the following steady-state equation:

$$\epsilon n_l \rho_l g + \epsilon n_l \nabla p_{fn} = -\frac{1}{V_0} \int_S \left(\nu \dot{p}_l + \mu_l \frac{\partial}{\partial \nu}(v_l) \right) dS \qquad (9)$$

4 MODELLING OF THE VOLUME AVERAGED EQUATIONS

In order to find a simple approximation or analytical solution to the closure problem the integral terms in equation (9) have to be rewritten. To make this possible the RUC concept of (du Plessis & Masliyah 1988) is used. Their RUC concept is changed to accommodate the specifics of gas-liquid flows. Flow channels are approximated by triangular channels. This is necessary in order to accurately model the saturation vs. capillary pressure relationship of porous materials.

4.1 Geometrical properties of the RUC

The triangular pores are modelled as composed of right-angled triangles. These fit into corners of the RUC (figure 1).

The void (pore) volume of the RUC is given by:

$$V_p = \frac{3}{2}abd - \frac{1}{3}(a^2b + ab^2) \qquad (10)$$

As a first approximation the flow length inside the RUC is computed as:

$$l_g = 3d - \frac{2}{3}(a+b) \qquad (11)$$

The geometry of flow inside a corner is shown in figure 2. The geometrical properties $\mathcal{L}_{gl}$, $\mathcal{L}_{sl}$ and $\mathcal{A}$ can be expressed in terms of r, δ and θ. r is directly linked to the capillary pressure by equation (7).

4.2 Capillary pressure relationship

The relation among pressure difference between the fluids and saturation is modelled similar to (Mason & Morrow 1991, eq. 21) using the MSP method. The major difference is that also the longitudinal changes in the fluid gas interface have to be approximated and

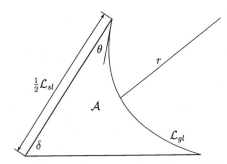

Figure 2: *Geometry of interface between liquid and gas inside a corner, the direction of flow is perpendicular to the paper.*

that different contact angles are taken into account. The derived relationship does not take into account the effects of surface adhesion forces, but could be modified accordingly.

If a non-wetting fluid enters a pore filled with a wetting fluid, an interface between the two fluids is formed. This interface is called a MTM (Main Terminal Meniscus). As soon as the two principal radii of the meniscus have the same magnitude, the non-wetting fluid enters the capillary. This radius is called the critical radius for drainage, r_{crit}. During imbibition, when a wetting fluid fills a capillary a different critical radius corresponds to complete filling of the capillary with the wetting fluid. This radius r_i corresponds to the case when the liquid from the separate corners starts to touch each other. r_{crit} is always smaller than r_i. Both of these critical radii depend on the contact angle. As a simple example the capillary pressure relations for zero contact angle are used in the following.

$$r_{crit} = \frac{P_p + \sqrt{4\pi A_p}}{\frac{P_p^2}{2A_p} - 2\pi} \qquad (12)$$

If there are saturation gradients, the average saturation inside the pores is calculated using the saturation profile given by (Dong & Chatzis 1995):

$$n_l = \frac{3f(\theta,\delta)l_g\gamma^2}{5V_p} \frac{p_l^{-5} - p_0^{-5}}{p_l^{-3} - p_0^{-3}} \qquad (13)$$

4.3 Flow resistance relationships

In order to find an approximate analytical solution for the closure problem, the resistance to flow in the RUC channel needs to be specified. In case the channel is filled with one fluid only, standard solutions from (Shah & London 1978) are available. For two-phase flow the solution procedure from (Zhou et al. 1997) is followed.

These relationships are valid for laminar flow. The

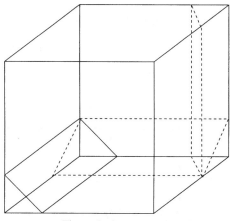

Figure 1: *Geometry of RUC.*

assumption of laminar flow is generally valid in the regime of slow two-phase flow. The flow resistances are written in terms of dimensionless resistance term β and the geometrical properties of the flow itself for the two-phase flow.

The derivation of the resistance factors for two-phase flow is mainly based on the work of (Zhou et al. 1997) and (Mogensen & Stenby 1998). In contrast to single-phase flow, where $f\mathrm{Re}$ is given for the whole capillary, in two-phase flow the resistance factor is given for each corner in the capillary separately. This resistance factor is based on the average velocity of the fluid inside a corner of a capillary. This average velocity is defined as:

$$v_c = \frac{1}{\mathcal{A}} \int_{\mathcal{A}} v_l \, d\mathcal{A} \qquad (14)$$

with inside the corner for the volumetric flow rate:

$$q_c = \frac{\mathcal{A} r^2}{\beta \mu_l} \nabla p \qquad (15)$$

r depends on the capillary pressure and β is defined as in (Mogensen & Stenby 1998).

4.4   Modelling of the surface terms

In order to solve the closure problem, the integral term in equation (9) needs to be expressed in terms of the resistance factors and geometric terms. The surface integral in equation (9) can be split in the parts containing the liquid-solid interface and the liquid-gas interface. These two terms give the stress components. The tangential stresses at the gas-liquid interface are assumed to be zero, but because of the capillary effects, normal stresses do exist.

$$\mathbf{I}_\mu = \frac{1}{V_0} \int_{\mathcal{S}_{sl}} \left(\mu_l \frac{\partial}{\partial \nu}(v_l) + \nu \dot{p}_l \right) d\mathcal{S}$$

$$+ \frac{1}{V_0} \int_{\mathcal{S}_{gl}} \left(\mu_l \frac{\partial}{\partial \nu}(v_l) + \nu \dot{p}_l \right) d\mathcal{S} \qquad (16)$$

To evaluate the above integral expression, the following assumptions are made:

1. The geometry is given by the RUC shown in figure 1.

2. The mean pore velocity v_p is directed axially along that specific pore section.

3. Pore sections (and thus also corners) are oriented perpendicularly and transversally with respect to the local average velocity.

4. Shear stresses which occur along transverse sur-

faces will not contribute to the streamwise component of $\mathbf{I}_\mu$. They will contribute to the pressure deviation term. Because the transverse pore sections have two times the length of the streamwise pore section, they contribute with a factor of two.

$$\mathbf{I}_\mu = \mathbf{I}_\mu(\mathcal{S}_\parallel) + \mathbf{I}_\mu(\mathcal{S}_\perp) = \frac{1}{V_0} \int_{\mathcal{S}_\parallel + 2\mathcal{S}_\perp} \mu_l \frac{\partial v_l}{\partial \nu} \, d\mathcal{S}$$

$$(17)$$

5. The shear stress on the liquid is a function of the resistance factor β. The mass flow is constant in each capillary and the two terms which generate momentum for dissipation are the gravity term and the pressure term. For a single corner the resistance is:

$$\tau \equiv \mu_l \frac{\partial v_l}{\partial \nu} \qquad (18)$$

6. The part of $\mathcal{S}_\parallel$ contributing to momentum dissipation is: $\mathcal{S}_{sl}$

7. Normal stresses at the interface between the liquid and the gas are given by the Young-Laplace equation.

8. The form of the interface in longitudinal direction is modelled as in (Dong & Chatzis 1995). This implies that the volume averaged equation is valid inside region C (figure 3). Inside region A a the saturated case is recovered.

The integral term in equation (17) is rewritten using the conditions stated above. As before, the equations are developed for a single corner and have to be assembled for the specific form of the capillary. Using equation (15, 17, 18) and some algebraic manipulation:

$$\frac{1}{V_0} \int_{\mathcal{S}_{sl}} \tau \, d\mathcal{S} = \frac{-3\beta \epsilon \mu_l n_l q_p (p_l - p_0)}{f(\theta, \delta)(p_l^{-3} - p_0^{-3}) \gamma^4} \qquad (19)$$

5   COMBINATION OF THE VOLUME AVERAGED EQUATION

If all volume averaged terms are used and the ex-

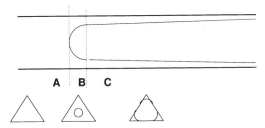

Figure 3: *Regions where solution is valid (A, B); A: saturated, B: region of MTM [$r \geq r_{crit}$], C: two-phase region [$r \leq r_{crit}$].*

pression for $\mathbf{I}_\mu$ is substituted in equation (9) and rearranged:

$$Q = -\left[\sum_{j=1}^{k} \frac{f_j(\theta,\delta)}{\beta_j}\right] \frac{\gamma^4(p_l^{-3} - p_0^{-3})}{3\mu_l(p_0 - p_l)}$$

$$\times (\nabla p_{fn} + \rho_l g) \qquad (20)$$

The summation term inside the square brackets is a purely geometric term, which is constant for a specific RUC chosen. Equation (20) contains no specific reference to the length scales of the RUC anymore. The length scale terms are implicitly contained in the boundary conditions p_0 and p_l. These boundary conditions together with the gradient term depend on the flow itself. If this equation is rewritten in terms of the apparent (or "Darcy") velocity, the similarity to the single-phase volume averaged equation becomes apparent:

$$v_d = -\left[\sum_{j=1}^{k} \frac{f_j(\theta,\delta)}{\beta_j}\right] \frac{d\epsilon\gamma^4(p_l^{-3} - p_0^{-3})}{3n_l V_p \mu_l(p_0 - p_l)}$$

$$\times (\nabla p_{fn} + \rho_l g) \qquad (21)$$

The main difference is that the saturation $n_l V_p$ directly enters the equation and that nonlinear contributions from the interaction between liquid pressure and saturation are present. Explicit reference to porous media structure parameters such as RUC length d and pore volume V_p can be measured directly in homogeneous media as shown by (du Plessis & Masliyah 1988).

The difference with respect to the classical Darcy type equation is the presence of the non-linear term containing p_0 and p_l. The first part of equation (20) on the right hand side can be interpreted as a conductivity function. The main difference to a traditional conductivity-pressure relationship, e.g. (van Genuchten 1980), which is defined in terms of a local pressure, is that no local pressures are used, but that boundary conditions p_0 and p_l directly enter the equations.

6 EFFECTIVE PARAMETERS FOR NUMERICAL MODELLING UNSATURATED FLOW

When numerically solving the equations describing the movement of a fluid through a porous medium, usually these equations are discretized and an approximation like finite differences or finite elements is used. The conductivity term needs to be defined for a discrete volume. Usually this is done by averaging nodal values of the conductivity. This averaging introduces approximation errors in the discrete solution. Equation (21) is an *"effective"* equation in terms of its parameters and the RUC scale. Because of these properties no averaging is necessary in the discretiza-

tion step and it is suited to discretization in standard edge centered finite difference codes or finite volume codes. The non-local structure is due to the incooperation of boundary conditions.

A straightforward 1-D discretization for horizontal flow is:

$$v_{di} = \left[\sum_{j=1}^{k} \frac{f_j(\theta,\delta)}{\beta_j}\right] \frac{d\epsilon\gamma^4}{3V_p\mu_l} \frac{(p_{i+1/2}^{-3} - p_{i-1/2}^{-3})}{n_{li}\Delta x} \quad (22)$$

with i the node number. This is a "simple" non-linear equation and together with the saturation profile given in (Dong & Chatzis 1995) the pressure in node i can be calculated exactly.

The first part of equation (21):

$$\left[\sum_{j=1}^{k} \frac{f_j(\theta,\delta)}{\beta_j}\right] \frac{d\epsilon\gamma^4}{3V_p\mu_l} \qquad (23)$$

contains only porous media and fluid dependent properties. These are usually temporally constant and the porous medium properties are allowed to change in space. These space dependent parameters can be upscaled by standard real space renormalization procedures (Hoffmann 1998).

Due to the volume averaged nature of equations (20, 21) the boundary conditions need to be specified in volume averaged form. Earlier research (Saffman 1971) showed that at low porosity or equivalent at low water contents standard microscopic boundary conditions can be used with small approximation errors. Further research is needed to specify these boundary conditions at large porosities and fluid contents.

7 EXAMPLE

As a simple example consider the behaviour of the discrete, simplified form of equation (20) for horizontal flow:

$$Q_i = K\frac{p_{i-1/2}^3 - p_{i+1/2}^3}{p_{i+1/2}^3 p_{i-1/2}^3} \qquad (24)$$

with K denoting all constant fluid and porous media properties. The behaviour of this equation subject to

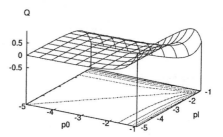

Figure 4: *Example solution of eq. (24), scaled.*

changing boundary conditions p_0 and p_l is shown in figure 4. As can be seen from the figure, the scaled solution shows that at low negative pressures (implying high water content, see eq. (7)) relative high volumetric flows occur. Clearly an averaged pressure can not model the complete behaviour of this solution.

8 CONCLUSIONS

The volume averaged form of the transport equation for two-phase flow has a different structure than the single-phase equation. The most significant is the introduction of boundary conditions in the equation itself. Due to its simple structure it can be solved more efficiently than Richards equation after discretization. This is mainly due to the fact that no interpolation and averaging of pressure heads or velocities is necessary. Parameters for this equation follow directly from structural data of the porous matrix and fluid dependent properties. The non-linear contribution from the boundary conditions together with the parameters directly replaces the traditional conductivity-pressure relationship. A point of further research is the specification of external boundary conditions. Because the volume averaged nature of the equations, also the boundary conditions need to be volume averaged.

REFERENCES

Bachmat, Y. & Bear, J. (1986). Macroscopic Modelling of Transport Phenomena in Porous Media. 1: The Continuum Approach. *Transport in Porous Media 1*(3), 213 – 240.

Bear, J. & Bensabat, J. (1989, October). Advective Fluxes in Multiphase Porous Media Under Nonisothermal Conditions. *Transport in Porous Media 4*(5), 423 – 448.

Dong, M. & Chatzis, I. (1995, June). The Imhibition and Flow of a Wetting Liquid along the Corners of a Square Capillary Tube. *Journal of Colloid and Interface Science 172*(2), 278 – 288.

du Plessis, J. P. & Diedericks, G. P. J. (1997). Pore-Scale Modelling of Interstitial Transport Phenomena. In P. du Plessis (Ed.), *Fluid Transport in Porous Media*, Volume 13 of *Advances in Fluid Mechanics*, Chapter 2, pp. 61 – 104. Computational Mechanics Publications.

du Plessis, J. P. & Masliyah, J. H. (1988, April). Mathematical Modelling of Flow Through Consolidated Isotropic Porous Media. *Transport in Porous Media 3*(2), 145 – 161.

Hoffmann, M. (1998). Fast real space renormalization for two-phase porous media flow. In *Proc. Int. Symp. on Computer Methods for Engineering in Porous Media, Giens, France, sept. 28 - oct. 2 1998*.

Mason, G. & Morrow, N. R. (1991, January). Capillary behaviour of a Perfectly Wetting Liquid in Irregular Triangular Tubes. *Journal of Colloid and Interface Science 141*(1), 262 – 274.

Mogensen, K. & Stenby, E. H. (1998, September). A Dynamic Two-Phase Pore-Scale Model of Imhibition. *Transport in Porous Media 32*(3), 299 – 327.

Quintard, M. & Whitaker, S. (1990, October). Two-Phase Flow in Heterogeneous Porous Media II: Numerical Experiments for Flow Perpendicular to a Stratified System. *Transport in Porous Media 5*(5), 429 – 472.

Saffman (1971, June). On the Boundary Condition at the Surface of a Porous Medium. *Studies in Applied Mathematics L*(2), 93 – 101.

Shah, R. K. & London, A. L. (1978). *Laminar Flow Forced Convection in Ducts – A Source Book for Compact Heat Exchanger Analytical Data*, Volume Supplement 1 of *Advances in Heat Transfer*. New York: Academic Press.

van Genuchten, M. T. (1980). A Closed-form Equation for Predicting the Hydraulic Conductivity of Unsaturated Soils. *Soil Sci. Soc. Amer. J. 44*, 892 – 898.

Whitaker, S. (1980). Heat and Mass Transfer in Granular Porous Media. In A. S. Mujumdar (Ed.), *Advances in Drying*, Volume 1, Chapter 2, pp. 23 – 61. Washington: Hemisphere.

Zhou, D., Blunt, M., & Orr, Jr., F. M. (1997, March). Hydrocarbon Drainage along Corners of Noncircular Capillaries. *Journal of Colloid and Interface Science 187*(1), 11 – 21.

Unsaturated Soils for Asia, Rahardjo, Toll & Leong (eds) © 2000 Taylor & Francis, ISBN 90 5809 139 2

Application of FE consolidation analysis method to fill-type dams

Y. Kohgo, I. Asano & H. Tagashira
National Research Institute of Agricultural Engineering, Tsukuba, Japan

ABSTRACT: The purpose of this paper is to apply a method of consolidation analysis to fill-type dams to estimate the movement behavior of the dams during construction and reservoir filling. The consolidation analysis employs the force equilibrium equations for soil and the Richards' mass conservation equation of pore water as the governing equations and was formulated using FEM. An elastoplastic model for unsaturated soils is used to express the stress-strain behavior of the fill materials. The numerical results were compared with measurements. The analysis results were consistent with the movements observed. The method can continuously analyze the behavior of dams from construction to reservoir filling periods.

1 INTRODUCTION

Fill-type dams are constructed by compacting unsaturated soils. After the construction, the dams are saturated with water from the reservoir. The behavior of fill-type dams during reservoir filling is very complex and hard to estimate. Nobari & Duncan (1972) pointed out that four separate effects of a rise in water level behind a typical zoned dam might be present. These are (1) downstream and downward movements due to the water-load applied on the core, (2) upstream and downward movements due to the water-load applied on the upstream foundation, (3) upward movements within the upstream shell zone and downstream rotation of the dam due to the buoyant uplift forces in this shell, and (4) downward movements within the upstream shell zone and upstream rotation of the dam due to saturation collapse of this shell. Moreover, deformations due to pore water pressures within the core and the changes of material properties due to wetting should occur in the dam. The behavior cannot be analyzed until the consolidation analysis methods with elastoplastic models for unsaturated soils are introduced.

The purposes of this paper are to apply a consolidation analysis method coupled with an elastoplastic model for unsaturated soils to fill-type dams, and to discuss the behavior of the dams during construction and reservoir filling.

2 CONSOLIDATION ANALYSIS METHOD

At first, suction s and effective suction s^* are defined as follows:

$$s = u_a - u_w \tag{1}$$

$$s^* = \langle s - s_e \rangle \tag{2}$$

where u_a = pore air pressure; u_w = pore water pressure; s_e = air entry suction; and the brackets $< >$ denote the operation $<z>=0$ at $z < 0$ and $<z>=z$ at $z \geq 0$.

The consolidation analysis method includes the governing equations and an elastoplastic model for unsaturated soils. They will be described in the following sections.

2.1 Generalized elastoplastic model

Kohgo et al. (1993a) clarified that the mechanical properties of unsaturated soils were controlled not only by suction but also by saturation conditions (or soil water retention conditions). The saturation conditions are divided into three, namely insular saturation, pendular saturation and fuzzy saturation conditions and it was found to be necessary to consider two suction effects to represent the mechanical properties of unsaturated soils. The two suction effects are (a) an increase in suction increases effective stresses and (b) an increase in suction enhances yield stress and affects the resistance to plastic deformations.

The suction effect (a) may be expressed by the following empirical effective stress equations:

$$\sigma' = \sigma - u_{eq} \tag{3}$$

$$u_{eq} = u_a - s \qquad (s \leq s_e) \tag{4}$$

$$u_{eq} = u_a - \left(s_e + \frac{a_e}{s*+a_e} s* \right) \quad (s > s_e) \tag{5}$$

where u_{eq} = equivalent pore pressure; σ' = effective stress; σ = total stress; a_e = a material parameter. The equations are derived from the relationship between suction and shear strength values at the critical state.

The suction effect (b) may be formulated by evaluating the state surface, which defines elastoplastic volume change behavior in unsaturated soils (Kohgo, 1987). The state surface can be plotted in the space with the axes; effective mean stress p', effective suction $s*$, and void ratio e. The figures of state surfaces depend on the types of soils. Here, the postulated state surface can be expressed as follows:

$$e = -\lambda * \log p' + \Gamma * \tag{6}$$

$$\lambda * = \frac{\lambda}{1+y} \tag{7}$$

$$\Gamma * = \frac{\Gamma + e_0^0 y}{1+y} \tag{8}$$

$$y = \left(\frac{s*}{a_s} \right)^{n_s} \tag{9}$$

where $\lambda = \lambda *$ in saturation; $\Gamma = \Gamma *$ in saturation; $\lambda *$ = slopes of e - $\log p'$ curves in the elastoplastic range; $\Gamma *$ = void ratios of e - $\log p'$ curves in the elastoplastic range at p'= unit; a_s, n_s, e_0^0 = material parameters. More details can be found in Kohgo et al. (1993a).

An elastoplastic model with the two suction effects described above is employed as a stress - strain relationship. The model is schematically shown in Figure 1. In the following, compression stresses are treated as negative. The suction effect (a) may be automatically taken into account by formulating the elastoplastic model in terms of effective stress invariants I_1, J_2 and θ (the Lode angle). The suction effect (b) may be evaluated as follows. If the elastoplastic model has plastic volumetric strain ε_v^p as a

hardening parameter, the yield stress I_c in the model (see Figure 2) can be calculated using Eqs. (11)~(13). Once I_c is identified, the parameters I_0, $K*$, a, and $b*$ can be calculated using Eqs. (14), (15) and yield surfaces may be determined as Eqs. (17), (18). Besides, $\alpha*_{cs}$, $\alpha*$ and R are material parameters and $\alpha*_{cs}$ and $\alpha*$ can be calculated using Eq. (16). A subloading surface model is also introduced into this model to express the smooth transition from elasticity to plasticity. More details can be found in Kohgo et al. (1993b).

2.2 Governing equations and FE formulation

In many geotechnical problems, it seems to be reasonable that pore air is always connected with the atmosphere and the permeability for airflow is sufficiently high. Pore air pressure is therefore always the same as the atmospheric pressure. The field equations for consolidation problems of unsaturated soils may be presented as the following two weak form equations:

$$\int_V \left(\sigma'_{ij,j} + \delta_{ij} u_{eq,j} + \gamma F_i \right) \cdot \delta u_i^* dV = 0 \tag{20}$$

$$\int_V \left(q_{i,i} - \dot{a}_{ii} + S_r \dot{\varepsilon}_{ii} \right) \cdot \delta u_w^* dV = 0 \tag{21}$$

where σ'_{ij} = components of effective stress tensor; δ_{ij} = Kronecker delta; γ = unit weight of the soil; F_i = components of the body force vector; q_i = components of the relative displacement velocity vector of water with respect to the soil skeleton; $\dot{a}_{ii}$ = change of storage of water due to change of degree of saturation; $\dot{\varepsilon}_{ii}$ = volumetric strain of the soil skeleton; δu_i^* = virtual displacement vector; δu_w^* = virtual pore water pressure. Subscripts after a comma denote spatial differentiation. Repeated indices indicate summation and a superposed dot denotes differentiation with respect to time. Equation (20) is the equilibrium equation of the soil and Equation (21) is the Richards' mass conservation equation of the pore water.

The governing equations are introduced to the discrete system using the finite element method and the solution procedure is based on the modified Newton-Raphson method. More details can be found in Kohgo (1995).

2.3 Consideration of soil properties

Nonlinear properties of elasticity, permeability and soil water retention need to be considered in this consolidation analysis method. Elastic moduli (shear and bulk moduli) are assumed to be functions of p' and $\sqrt{J_2}$ as follows:

$$K = \frac{-2.3(1+e_0)}{\kappa} + K_i \tag{22}$$

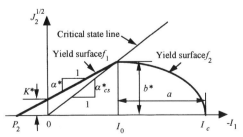

Figure 1. Yield surfaces of the elastoplastic model (Kohgo et al. 1993b).

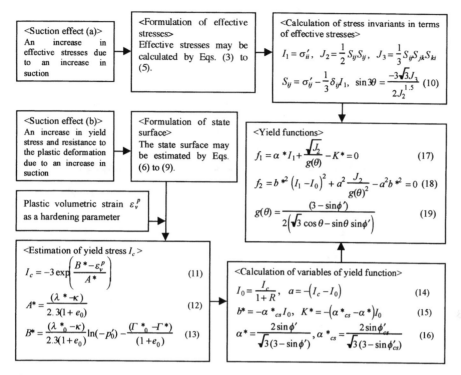

Figure 2. Suction effects and the elastoplastic model.

$$G = G_i + \gamma_J \sqrt{J_2} - \gamma_p p'. \qquad (23)$$

Where K = bulk modulus; G = shear modulus; e_0= initial void ratio; κ = slope of $e - \log p'$ curve at unloading; K_i, G_i, γ_J and γ_p = material parameters.

The permeability property is postulated to be a function of e and S_r (Kohgo, 1995).

Tangential model is used to express the soil water retention curve. The model satisfies the continuity of tangential slope c of the soil water retention curve. More details can be found in Kohgo (1995).

3 CONSOLIDATION ANALYSES

The method has been already applied to analyze the behavior of a virtual earth dam (height H =50 m and width B =305 m) as shown in Figure 3 (Kohgo, 1992 & 1997). Some interesting results were obtained. The most interesting one is movement of the up-stream crest. Figure 4 shows horizontal displace-ments with time at the upstream crest. The dam showed an upstream deformation after commence-ment of seepage for some time and then moved back downstream. This behavior is consistent with typical dam behavior during reservoir filling. It could not be analyzed until the consolidation analysis method

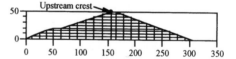

Figure 3. A virtual earth dam and finite mesh.

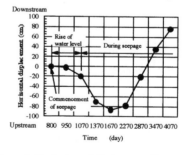

Figure 4. The horizontal displacement history at the upstream crest during reservoir filling.

taking into account the saturated collapse was adopted.

In this paper, the method is applied to analyze a rock fill dam named Oroville dam during construc-tion and reservoir filling. The parameters are shown

191

Table 1. Material prameters for analysis of Oroville dam.

	Elasticity					Subloading		Initial condition			
	κ	K_i (kPa)	G_i (kPa)	γ_j	γ_p	h (kPa)	a_h	S_{r0}	s_0	e_0	γ (kN/m³)
Impervious Core	0.015	15700	13900	0	54.9	2×10^6	2	0.778	9.4	0.3	23.5
Transition	0.005	10000	51400	0	35.4	2×10^6	2	0.05	15	0.252	23.5
Shell	0.005	10000	59700	0	39.8	2×10^6	2	0.05	15	0.252	23.5

	Plasticity					Effective stress		State surface				
	ϕ'	$\phi'cs$	R	I_c (kPa)	p_0' (kPa)	a_r (kPa)	s_r (kPa)	λ	Γ	e^0_0	a_s	n_s
Impervious Core	21	25	1	-2000	-398	0	0	0.07	0.482	0	1×10^{10}	1
Transition	33.8	38.8	1	-2E+06	-127000	0	0	0.014	0.309	0.282	27.6	1.28
Shell	33.8	38.8	1	-2E+06	-127000	0	0	0.014	0.309	0.282	27.6	1.28

	Permeability			Soil water retention							
	k_s (m/d)	m_p	n_p	h_s	S_{rs}	S_{rm}	S_{rf}	s_m (kPa)	s_f (kPa)	c_m (kPa⁻¹)	c_f (kPa⁻¹)
Impervious Core	3×10^{-3}	3	3	0.02	0.91	0.5	0.1	17	60	3.75×10^{-2}	1.25×10^{-3}
Transition	2.4×10^2	3	3	0.02	0.91	0.5	0.1	2.6	9	2.5×10^{-1}	8.33×10^{-3}
Shell	3×10^2	3	3	0.02	0.91	0.5	0.1	2.6	9	2.5×10^{-1}	8.33×10^{-3}

in Table 1. The dimensions of the rock fill dam are $H=235$ m and $B=1098$ m and it has an inclined impervious core as shown in Figure 5 (a). The finite element mesh and the zoning of the dam are shown in Figure 5. The embankment was completed in 671 days and the speed of construction was constant at 0.35 m/d. Pore water pressures in shell and transition zones were assumed to maintain the initial values during construction. Then, the consolidation behavior of the impervious core was only considered. As boundary conditions, the base was rigid and impervious. The elements used here were isoparametric for displacements and super-parametric for pore water pressures. Reduced integration (2x2 Gauss point integration) was also adopted. The reservoir was filled in 590 days at a constant rate and the full water level was 228 m. The results of this analysis are shown in Figures. 6~ 13. Figure 6 shows comparisons of the measured and calculated settlements during construction. As would be expected, the values of settlement in the impervious core were larger than those induced in the transition and shell zones. On the whole, the agreement between the measured and calculated settlements was quite good.

Figure 7 shows comparisons of the measured and calculated horizontal displacements during construction. Though the calculated horizontal displacements at measuring point No. 3 are larger than those observed, the agreement between the measured and calculated displacements is quite good on the whole.

Figure 8 shows the deformations of the dam. From Figure 8 (a), at the completion of construction, settlements concentrated in the impervious core. Horizontal displacements were smaller than the settlements. The upstream part of the dam moved so as to lean against the downstream shell zone. From Figure 8 (b), it was found that the dam moved more downstream and the tendency was especially re-

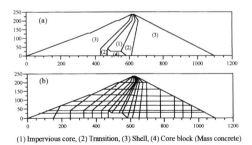

Figure 5. A section of Oroville dam and finite element mesh used in this analysis.

(1) Impervious core, (2) Transition, (3) Shell, (4) Core block (Mass concrete)

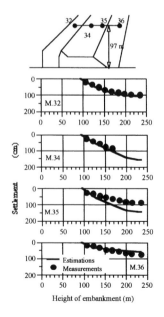

Figure 6. Settlements at $H=97$m during construction.

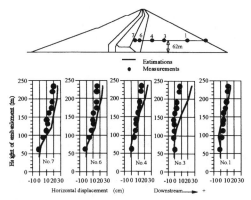

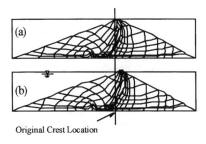

Figure 7. Horizontal displacements at $H = 62$m during construction.

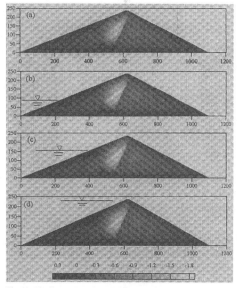

Original Crest Location

Figure 8. Deformation results (a) Just after construction, (b) High water level.

markable near the crest.

Figure 9 shows contours of horizontal displacements of the embankment during reservoir filling. In this figure, the downstream movements are expressed as positive values. The parts which existed more upstream than the boundary line between the core and the downstream transition zone moved upstream and the more downstream parts moved downstream. As the water level rose, the embankment moved downstream.

Contours of the calculated values of settlements during reservoir filling are shown in Figure 10. The maximum value of settlements appears at mid height of the core. The maximum settlement was about 1.8m. The upstream parts of the embankment moved downward as the water level rose. The downstream parts had little influence of reservoir.

Figure 11 shows contours of pore water pressure during reservoir filling. The maximum value of pore water pressures in the core induced by self-weight of the core material was about 150 kPa and the value still remained small in comparison with the size of the dam. It arises from the fact that the initial degree of saturation of the core material ($S_{r0}=77.8\%$) is not so high. As the water level rises, the pore water

pressures accumulated during construction may rapidly disappear and steady state seepage appears in the core.

Figure 12 shows vertical stress distributions in the dam. From Figure 12 (a), at the completion of construction, the arching action induced by the concentration of vertical stresses at the boundaries between the impervious core and the transition zones can be clearly seen. However, at full water level, the

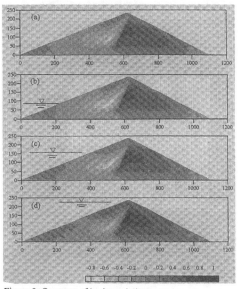

Figure 9. Contours of horizontal displacement.

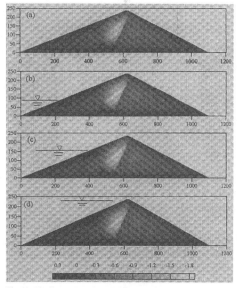

Figure 10. Contours of settlements.

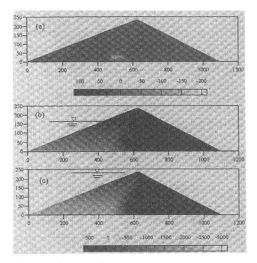

Figure 11. Contours of pore water pressures.

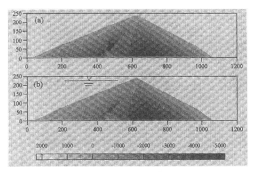

Figure 12. Vertical stress distribution (a) Just after construction, (b) High water level.

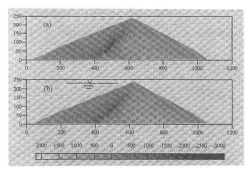

Figure 13. Shear stress distribution (a) Just after construction, (b) High water level.

arching action at the upstream boundary is not so remarkable. It still remains at the downstream boundary (see Figure 12 (b)).

Figure 13 shows shear stress distributions in the dam. The shear stresses concentrated at the bound-

ary between the impervious core and the upstream transition zone. The maximum absolute value of shear stress at completion of construction was higher than that at high water level.

4 CONCLUSIONS

A finite element method of consolidation analysis was applied to a fill-type dam (Oroville dam) to analyze its behavior during construction and reservoir filling. The numerical results were consistent with the field movements observed. The method may continuously analyze the behavior of dams from construction to reservoir filling.

REFERENCES:

Kohgo, Y. 1987. The interpretations and analyses of mechanical behaviour of partly saturated soils using elastoplastic models. *Proc. Symposium Unsaturated Soils*: 69-78. Osaka, Japan: J.S.S.M.F.E. (in Japanese).
Kohgo, Y. 1992. Deformation analysis of fill-type dams during reservoir filling. *Proc. 4th Int. Symposium Numer. Models Geomech* NUMOG IV 2: 777-787: Balkema.
Kohgo, Y. 1995. A consolidation analysis method for unsaturated soils coupled with an elastoplastic model. *Proc. 1st Int. Conf. on Unsaturated Soils* 2: 1085-1093: Balkema.
Kohgo, Y. 1997. Method of analysis of saturation collapse. *JIRCAS J.* 4: 1-24.
Kohgo, Y., Nakano, M. & Miyazaki, T. 1993a. Theoretical aspects of constitutive modelling for unsaturated soils. *Soils and Foundations* 33(4): 49-63.
Kohgo, Y., Nakano, M. & Miyazaki, T. 1993b. Verification of the generalized elastoplastic model for unsaturated soils. *Soils and Foundations* 33(4): 64-73
Nobari, E. S. & Duncan, J. M. 1972. Effect of reservoir filling on stresses and movements in earth and rockfill dams. *Report No. TE-72-1*: University of California, Berkeley.

Unsaturated Soils for Asia, Rahardjo, Toll & Leong (eds) © 2000 Taylor & Francis, ISBN 90 5809 139 2

Finite element modelling of geomechanics and geoenvironmental engineering for subsurface systems

N.Abd. Rahman
Faculty of Civil Engineering, Universiti Teknologi Malaysia, Johor Bahru, Malaysia

ABSTRACT: This paper describes numerical simulation study of geomechanics and groundwater contamination. The problem of coupled multiphase flow, heat and contaminant transport in a subsurface system is important in environmental engineering. For this purpose, the main objective of this study was to develop a finite element model that describes the flow of multiphase fluids, together with coupled contaminant transport in a deforming porous medium for both isothermal and nonisothermal problems. The multiphase flow model of Brooks and Corey, which express the dependence of saturation and relative permeability on capillary pressure, is presented to simulate subsurface flow and pollutant transport in the system. Subsurface flow is simulated through numerical solution of fluid mass balance equation, where the subsurface system may be either saturated, or party or completely unsaturated. The governing partial differential equation, in term of soil displacement, fluid pressure, energy balance and concentrations are fully coupled and behave non linearly but can be solved by the finite element method. Numerical implementation of the formulation is discussed and examples related to subsurface problems are used to demonstrate the model.

1 INTRODUCTION

The simulation of groundwater contamination in subsurface systems by nonaqueous phase liquids, such as petroleum hydrocarbons and immiscible industrial chemicals, requires a solution of the multiphase flow equations for deforming porous media. The spill and leakage of organic chemical into the global environment have resulted in widespread contamination of subsurface systems. Many of these pollutants are slightly water- soluble and highly volatile fluids (NAPLs). In the unsaturated zone, residual NAPL, as well as NAPL dissolved in the water phase, may also volatilize into the soil gas phase.

The remaining nonaguaeous phase liquids may persist for long periods of time, slowly dissolving into the groundwater and moving in the water phase via advection and dispersion processes. Several numerical models have been developed in order to study the movement of organic pollutants in subsurface systems and a few of the reported simulators consider interphase mass transfer of one or more NAPL components to the gas and water phase but assume that the NAPL is immobilized in

the unsaturated zone (Baehr and Corápcioglu 1987 and Abd. Rahman and Lewis 1997).

Recently, a nonisothermal model has been developed by Falta et al. (1992), Adenekan et al. (1993) and Panday et al. (1995) for examining nonaqueous phase liquid contamination and remedial scenarios (i.e. stream injection). Fully coupled models for water flow, airflow and heat flow in deformable porous media have been studied by Schrefler et al. (1995) and Thomas and He (1995) which do not consider interphase mass transfer or other limitation (i.e. fluid density is assumed to be function of pressure, temperature and contaminant) in the simulation model for pollutant transport behavior.

The problem of coupled multiphase flow (namely gas, water and nonaqueous phase liquids), heat and contaminant transport in a subsurface system is important in environmental engineering. At present, there exists a critical need for the physical observation of coupled transport phenomena in porous media. These observations are required both as an aid to our understanding of coupled process and to evaluated current predictive methods for such phenomena. For this purpose, the main objective of this study was to develop a finite element model that describes the flow of multiphase fluids, together with coupled contaminant transport, in a deforming porous medium for both isothermal and nonisothermal problems. The flow of all multiphase fluids, as well as the water gas phase transport, are included.

The governing equations which describe the displacement of soil, multiphase fluid pressures, temperature and pollutant transport are coupled and the resulting non- linear partial differential equation are solved by the finite element method. Non- linear saturation and relative permeability functions are incorporated into a partially water saturated porous media. The physical model and model description of the study based on the reference by Lewis et. al. (1998), Abd. Rahman and Lewis (1997, 1999) and Abd. Rahman (1998).

2   MATHEMATICAL MODEL

Contaminant transport models for miscible components cannot describe the migration of an immiscible contaminant. The flow of an immiscible contaminant is controlled by its own flow potential, which depends on pressure, gravity and surface forces and is not necessarily similar to the groundwater flow potential. In order to describe mathematically the flow of immiscible fluids through a porous medium, it is necessary to determine functional expressions that best define the relationship between the hydraulic properties of the porous medium, i.e. saturation, relative permeability and capillary pressure.

A mathematical model of the physical processes occuring in multiphase flow with heat and pollutant transport and soil deformation is given in this section, to present the mathematical model as a formulation suitable for numerical solution by the finite element method. The principles of the conservation of mass and energy are employed to derive the governing differential equation for fluid (water, gas and NAPL), heat and pollutant transport in terms of seven primary variables; i.e. U, P_w, P_g, P_w, T, C_w and C_g.

These conservation equations are implicitly linked through the dependency of the fluid density, fluid viscosity and matrix porosity upon the fluid pressure, fluid temperature and contamination mass fraction. In a multiphase, water, NAPL and gas in a porous medium, depends primarily on the gravitational force and the capillary pressures between the multiphase fluids.

The seven, fully coupled, partial differential equation i.e the multiphase fluid flow equations and the equilibrium equation are an extension of the study of the model developed by Lewis and Schrefler (1998), Schrefler (1995), Abd. Rahman and Lewis et al. (1999). Nonisothermal deforming porous medium are modelled by utilizing the equilibrium equation together with the governing equation of heat, fluid flow and pollutant transport through such a porous medium.

1) Equilibrium equation

$$\int_\Omega \delta\varepsilon^T D_T \frac{\partial\varepsilon}{\partial t} d\Omega - \int_\Omega \delta\varepsilon^T m \frac{\partial\overline{p}}{\partial t} d\Omega +$$

$$\int_\Omega \varepsilon^{T\delta} D_T m \frac{\partial\overline{p}1}{\partial t 3K_s} d\Omega - \int_\Omega \delta\varepsilon^T D_T C d\Omega -$$

$$\int_\Omega \delta\varepsilon^T D_T m \frac{\partial T\beta_s}{\partial t 3} d\Omega - \int_\Omega \delta\varepsilon^T D_T \frac{\partial\varepsilon_o}{\partial t} d\Omega -$$

$$\int_\Omega \delta u^T \frac{\partial b}{\partial t} d\Omega - \int_\Gamma \delta u^T \frac{\partial s}{\partial t} d\Gamma = 0 \qquad (1)$$

2) Multiphase flow equation

$$-\nabla^T [\frac{Kk_{r\alpha\rho\alpha}}{\mu_\alpha B_\alpha} \nabla(P_\alpha + \rho_\alpha gh)] +$$

$$\rho_\alpha \frac{S_\alpha}{B_\alpha}[(m^T - \frac{m^T D_T}{3K_s})\frac{\partial\varepsilon}{\partial t} +$$

$$\frac{m^T D_T C}{3K_s} + (\frac{(1-\phi)}{K_s} - \frac{m^T D_T m}{(3K_s)^2})\frac{\partial\overline{p}}{\partial t} +$$

$$\{-(1-\phi)\beta_s + \frac{m^T D_T m}{3K_s}\frac{\beta_s}{3}\}\frac{\partial T}{\partial t}]\times$$

$$\phi\frac{\partial}{\partial t}(\frac{\rho_\alpha S_\alpha}{B_\alpha}) + P_\alpha Q_\alpha + \Gamma_\alpha = 0 \qquad (2)$$

3) Pollutant transport equation

$$\frac{\partial}{\partial t}(\phi S_\theta C_\theta) + \nabla.(V_\theta C_\theta) -$$

$$\nabla.(\phi S_\theta D_{ij\theta}\nabla C_\theta) + \overline{C}_\theta Q_\theta + \Gamma_\theta = 0 \qquad (3)$$

4) Energy transport equation

$$\frac{\partial}{\partial t}[(1-\phi)\rho_s C_{ps} + \phi\rho_w S_w c_{pw} + \phi\rho_g S_g c_{pg}$$

$$+ \phi\rho_n S_n c_{pn}]T + \rho_w S_w c_{pw} V_w.\nabla T +$$

$$\rho_g S_g c_{pg} V_g.\nabla T + \rho_n S_n c_{pn} V_n.\nabla T = \nabla.(\lambda_T \nabla T)$$

$$+ (1-\phi)\rho_s Q_s + \phi\rho_w S_w Q_w +$$

$$\phi\rho_g S_g Q_g + \phi\rho_n S_n Q_n] \qquad (4)$$

where, α represents the water(w), gas(g) and NAPL (n) phase and Γ_α is the interphase mass transfer. Detailed notation can be found in Lewis. et. al (1998) and Abd. Rahman (1998).

3 FINITE ELEMENT MODEL AND DISCRETIZATION

Due to complexity of the theoretical formulation in the mathematical model, numerical approximation methods are required to achieve the simultaneous solution of the governing differential equations. The finite element method, based on Galerkin's weighted residual approach (Abd. Rahman, 1998), is used for discretization of the governing equations and is implemented as the numerical solution method because analytical solutions are incapable of dealing with the highly complex and non- linear governing equations. The approach adopted in this study employs two- dimensional, nine noded isoparametric element. Fluid pressures, temperatures, concentrations and displacements are taken as the primary unknown variables.

3.1 Numerical Formulation

The procedure used is based on Galerkin's weighted residual method, which implements shape functions to approximate the known variables. In the simulations carried out , there are seven degrees of freedom viz, u, P_g, P_w, Pn, Cw, C_g and T which are approximated as

$$x = N^T X \qquad (5)$$

where N^T is the transpose of the shape function and X is the nodal value of the variable. In this case nine- noded elements and the associated shape of the equations in the mathematical model, may now be expressed in terms of the nodal displacements , u, v, nodal fluid pressures, i.e. P_n, P_w, and P_g, nodal temperature , T and the nodal concentrations. i.e. Cw and Cg using the Galerkin's method. Discretization of the governing equations is obtained by applying Galerkin's procedure of weighted residuals. Terms involving second spatial derivatives are transformed by means of Gauss's theorem. The field variables are then approximated in space as is usual in finite element techniques, and expressed in terms of their nodal variables.

The analysis of a model deals with a seven degree of freedom field problem. Written in a matrix form, the discretized forms of mathematical equations are written in the following form:

$$A \bar{x} + Bx = F \qquad (6)$$

Where x = [u, P_w, P_g, P_n, T, C_w, C_g] and the matrices A, B and F are obtained by inspection . Equation (6) is the final form of the governing equations where displacements, fluid pressures, temperature and concentrations are the primary unknowns. This equation forms a fully coupled nonsymmetrical and highly non-linear system of ordinary differential equations in time and is solved by an implicit scheme with the time weighting parameter set to one. In an implicit scheme, since the entire non-linear coefficient is dependent on the unknowns, iterative procedures are usually performed within each time step to obtain the final solution.

The coupling terms are evaluated for each element using this numerical technique and then assembled into the global matrix. The format in the elemental matrices and the global matrix differs in the ordering of unknown variables. Thus it is necessary to transform the coupling terms from the elemental matrices to the global matrix, when assembling the global matrix. Once the global matrix has been fully assembled, the partial differential equations have been transformed into first order ordinary differential equations. A linear variation of the unknown variables in time is assumed to approximate the first order time derivatives. The generalised mid- point family of methods is employed to discretize the time derivatives, which yields the following recurrence scheme,

$$x^{n+1} = \frac{1}{[A + \theta \Delta t B]} \quad \text{x}$$

$$([A - (1-\theta)\Delta t B]x^n + \Delta t [\theta F^{n+1} + (1-\theta)F^n]) \qquad (7)$$

For coupled heat, fluid flow and pollutant transport in deforming porous media, the discretization is accomplished through a fully implicit finite difference scheme. A linear variation of the unknown variables in time is assumed to be a good approximation for the first order time derivative. The generalized mid-point family of methods is employed to discretize the time derivatives. Because of the non-linearity involved, a solution scheme of the fixed-point type is used within every time step. The convergence criteria implemented are based on a maximum change in the unknown variables between successive time step.

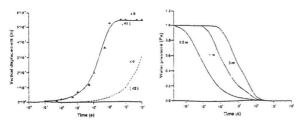

Figure 1. (a) Surface settlement and (b) water pressures vs. time for the one-dimensional saturated problem

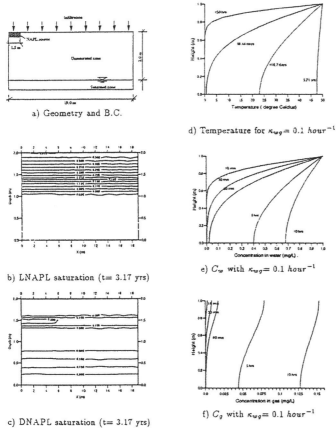

a) Geometry and B.C.

b) LNAPL saturation (t= 3.17 yrs)

c) DNAPL saturation (t= 3.17 yrs)

d) Temperature for $\kappa_{wg} = 0.1 \; hour^{-1}$

e) C_w with $\kappa_{wg} = 0.1 \; hour^{-1}$

f) C_g with $\kappa_{wg} = 0.1 \; hour^{-1}$

Figure 2. Pollutant Transport Examples

4 NUMERICAL EXAMPLES

Several analyses have been carried out in order to study the present model for isothermal and heat transfer affected to multiphase fluids flow and pollutant transport in deforming porous media. For this purpose a one dimensional, fully saturated, consolidation problem is solved and compared with a previous solution by Lewis and Schrefler (1987). A column of linear elastic material is 7 m height and 2 m wide and the pore pressure is equal to zero at the top surface. The top surface is the only drained boundary whilst the rest are assumed to be closed. The vertical displacement vs. time of the top surface is shown in Figure 1 (a) for the case of isothermal consolidation, and good agreement is observed when compared with the results of Lewis and Schrefler (1987). Obviously, the soil permeability $K_2 = 1 \times 10^{-11} m^2$ has a lower rate of settlement when compared with $K_1 = 4 \times 10^{-9} m^2$. The pore pressure vs. time at several depths shown in Figure 1 (b).

The first example demonstrates the impact of an

undected leak of non-aqueous phase liquid in a porous medium for an isothermal case. This problem was designed to study the migration of NAPL, from a continuous source at the soil surface. Two cases are considered: one in which the contaminant phase has a density greater than water, and one which the contaminant has a density less than that of water. A schematic diagram showing the geometry of the simulated cross section and boundary condition is presented in Figure 2 (a). The system is subjected to water infiltration at a rate of 100 kg/yr/m^2. The NAPL source, with a release rate of 900 kg/yr, is located at the top left hand edge. The assumption is that the pressure in the gas phase are negligible and leads to two partial differential equations (Pw,Pn). The left and right-hand boundaries of the domain were treated as impermeable, and the bottom boundary was assigned constant-pressure conditions with Pw = 101325 Pa and Pn = 101325 Pa. The simulation domain in Figure 2 (a) is 2 metre thick and 19.0 metre in length.

The domain was discretized using a 2-D rectangular grid consisting of 40 elements and 187 nodes. As was noted, two base cases were simulated. For an immiscible fluid with a density greater than water, gravity effect could be expected to be dominant. As a consequence, an anticipated downward migration of the contaminant in both the unsaturated zone and below the water table, would take place. For a contaminating fluid less dense than water, the contaminant would be expected to pool in zone near the water table. These conditions were observed in the result of simulation in Figure 2 (b, c).

The second example is for a soil column of linear elastic material, 1 metre in height, simulates the transport of a small amount of pollutant introduced at the top of the soil column with an initial water saturation of 0.445 and with a surface temperature jump of 50 °C above the reference temperature, T_{ref} of the column. The initial pore water pressure was 420 kN/m^2 and the boundary pore water pressure was instantaneously changed to a value of -280 kN/m^2 at the surface. Figure 2 (d) shows the temperature distribution versus height for different times and it can be seen that a longer time is required to achieve the surface temperature boundary condition value throughout the soil column. Figure 2(e,f) show the corresponding concentration profile in the water and gas phase with interphase mass transfer values equal to 0.1 /hour. The results indicate that the mass transfer coefficients have a significant influence on the immobilized NAPL.

5 CONCLUSIONS

The investigation of the formulation of fully coupled multiphase flow (namely, water, gas and NAPL), heat flow and pollutant transport in deforming porous media is the main concern of this study. The mathematical formulation and numerical model have been tested for accuracy, stability and the ability to simulate different condition where u, P_g, P_w, Pn, Cw, C_g and T are the primary unknowns in the example.

Result of the multiphase flow example show the behaviour of NAPL infiltration in the model where the dominant gravity effect in a DNAPL situation is a downward migration of the pollutant and for LNAPL a horizontal spreading and flotation on the water table. For immobilized NAPL, the results show the usefulness of the present model in deforming porous media, including the mass transfer process and density effect for a nonisothermal model of pollutant transport.

REFERENCES

Abd. Rahman, N and R. W. Lewis 1997. Finite element simulation of isothermal multhiphase flow and pollutant transport in deforming porous media, In John Chilton et al., *Proc. of the XXVII IAH Congress on groundwater in the urban environment, Cardiff*, UK, 149-154.

Abd. Rahman, N. and R. W. Lewis 1997. Numerical modelling of multiphase immiscible flow in deforming porous media for subsurface system, In R. N. Yong and H. R Thomas (eds.), *Proc. of the conference on Geonvironmental Engineering- Contaminated ground: fate of pollutants and remediation, Cardiff*, UK, 254-263.

Abd. Rahman, N. and R. W. Lewis 1999. Finite element modelling of multiphase flow and pollutant transport in deforming porous media for subsurface system, *Journal of Computers and Geotechnics*, vol.24, 41-63.

Abd. Rahman, N. 1998. Finite element analysis of multiphase flow heat flow and pollutant transport in deforming porous media for subsurface system, *Ph.D . Thesis* , Uni. of Wales, Swansea, U.K.

Adenekan, A.E., T. W. Patzek and K. Pruess 1993. Modelling of multiphase transport of multicomponent organics contaminants and heat in the subsurface: Numerical model formulation, *Water Resour. Res.*, 29 (11), 3727-3740.

Baehr, A. L. and M. Y. Corapcioglu 1987. A compositional multiphase flow and pollutant transport in deforming porous media, In John Chilton et al., *Proc. Of the XXVII IAH Congress on groundwater in the urban environment, Nottingham*, UK, 149-154.

Falta, R.W., K. Pruess, I. Javandel and P. A. Witherspoon 1992. Numerical modelling of steam injection for the removal of nonaqueous phase liquids from the subsurface, 1, numerical formulation, *Water Resour. Res.*, 28 (2), 443-449.

Lewis, R. W. and B. A. Schrefler 1987. *The finite element method in the deformation and consolidation of porous media*, John Wiley & Sons, Chichester.

Lewis, R.W. and B. A. Schrefler 1998. *The finite element method in the static and dynamics deformation and consolidation of porous media*, John Wiley & Sons, Chichester.

Lewis, R.W., J. T. Cross, D. T. Gethin, R. S. Ransing, M. T. Manzari and N. A. Rahman 1998. Recent development in heat transfer and porous media, *Proc. of the ICMF '98-3rd Iternational Conference on advance Computational Methods in Heat Transfer, Cracow*, Poland.

Lewis R.W., B. A. Schrefler and N. Abd. Rahman 1998. A finite element analysis of multiphase nimmiscible flow in deforming porous media for subsurface system, *Comm. Numerical Methods Eng*. Vol 1492, 135-149.

Panday, S., P. A. Forsyth, Yu-Shu Wu and P. S. Huyakorn 1995. Considerations for robust compositional simulations of subsurface nonaqueous phase liquid contamination and remediation, *Water Resour. Res.,* 31 (5), 1273-1289.

Schrefler, B.A. 1995. F.E. in environmental engineering; coupled thermo-hydro-mechanical processes in porous media including pollutant transport, *Achieve of Computational Methods in Engineering*, 2, 3, 1-54.

Schrefler, B.A., X. Zhan and L. Simoni 1995. A coupled model for waterflow, airflow and heat flow in deformable porous media, *Int. J. Numer. Methods Heat Fluid Flow*, 5, 531-547.

Thomas, H. R. and Y. He 1995. Analysis of coupled heat, moisture and air transfer in a deformable unsaturated soil, *Geotechnique*, 45, 4 677 – 689.

Unsaturated Soils for Asia, Rahardjo, Toll & Leong (eds) © 2000 Taylor & Francis, ISBN 90 5809 139 2

General partial differential equation solvers for saturated-unsaturated seepage

N.T.M.Thieu, D.G.Fredlund & V.Q.Hung
Department of Civil Engineering, University of Saskatchewan, Saskatoon, Sask., Canada

ABSTRACT: This paper presents the application of a general purpose partial differential equation solver, called PDEase2D and the associated soil property functions to the analysis of saturated-unsaturated seepage problem. Examples of steady state and transient seepage through an earth dam are used to illustrate the use of PDEase2D in solving saturated-unsaturated seepage problems.

1 INTRODUCTION

Seepage analyses form an important and basic part of geotechnical engineering. The pore-pressure distribution in a soil are necessary to describe the flow processes and are also necessary to predict the volume change and shear strength changes associated with a change in stress state. Seepage through an unsaturated soil is mathematically characterized by a partial differential equation that is non-linear and soil properties that can be highly non-linear. As a result, the modeling of saturated-unsaturated soil systems has proven to be a challenge. There is need for numerical software packages that ensure convergence when solving seepage problems involving saturated-unsaturated soil systems (Fredlund, 1996).

This paper illustrates the use of a general-purpose partial differential equation solver, called PDEase2D, to solve seepage problems involving saturated-unsaturated soil systems. Several forms of permeability functions and water storage functions are used when solving the seepage problem. These forms are demonstrated through two example problems, one involving steady state seepage, and another involving transient seepage through unsaturated soil systems.

2 BACKGROUND

The governing partial differential equation for seepage through a heterogeneous, anisotropic, saturated-unsaturated soil can be derived by satisfying conservation of mass for a representative elemental volume, assuming that flow follows Darcy's law. If it is assumed that the total stress remains constant during a transient process and that pore-air pressure is at-mospheric, the differential equation can be written as follows for the two-dimensional transient case:

$$\frac{\partial}{\partial x}\left(k_x \frac{\partial h}{\partial x}\right) + \frac{\partial}{\partial y}\left(k_y \frac{\partial h}{\partial y}\right) = m_2^w \gamma_w \frac{\partial h}{\partial t} \qquad (1)$$

where h = total head (i.e., pore-water pressure head plus elevation head); k_x and k_y = coefficient of permeability of the soil in the x- and y-direction, respectively; γ_w = the unit weight of water (i.e., 9.81 kN/m^3); and m_2^w = the slope of the soil-water characteristic curve (i.e., water storage).

For steady state seepage, only the coefficient of permeability is required because the time dependent term in eq. (1) disappears and the storage function drops out. The coefficient of permeability must be considered as a function of the stress state for an unsaturated soil. The coefficient of permeability can be written as follows for an unsaturated soil:

$$k_w = func[k_s, (\sigma - u_a), (u_a - u_w)] \qquad (2)$$

where k_s = saturated coefficient of permeability; $(\sigma - u_a)$ = net normal stress; and $(u_a - u_w)$ = matric suction. However, the coefficient of permeability of an unsaturated soil is predominantly a function of the matric suction.

Many permeability equations have been suggested in the literature and the following equations are used in this paper: Gardner (1958) equation; van Genuchten-Burdine (1980) equation; van Genuchten-Mualem (1980) equation; and Fredlund and Xing (1994) equation. In addition, linear interpolation between experimental data points is also used in this paper.

In addition to the non-linear coefficient of permeability function, the water storage function is also required for solving saturated-unsaturated transient seepage problems. The water storage indicates the amount of water taken or released by the soil as a result of a change in the pore-water pressure and it is the slope of the soil-water characteristic curve. Therefore, the water storage function is obtained by differentiating the soil-water characteristic curve with respect to the matric suction. Several equations have been proposed to describe the soil-water characteristic curve. These equations involve finding best-fit parameters, which produces a curve that fits the measured data. Any of the proposed equations can be used to describe soil-water characteristic curve, but only the van Genuchten (1980) equation and the Fredlund and Xing (1994) equation are used in this paper. The analysis of several soil types showed that an extended form of an error function equation can also approximate the water storage function. Therefore, an extended error function is also used in this paper. The solution of a transient seepage problem is obtained and compared with respect to van Genuchten (1980) equation, Fredlund and Xing (1994) equation, and extended error function.

3 PARTIAL DIFFERENTIAL EQUATION SOLVERS

In the last two decades, the development and application of the computer to solving complex problems has been extensive. Computer programs make use of numerical methods such as finite element technique. These computer programs have become necessary tools for solving engineering problems. An unsaturated soil problem involves the soil properties that are highly non-linear, such as coefficient of permeability and water storage functions. The partial differential equations to be solved become highly non-linear and require the input from persons specially trained in the area of mathematics. This has given rise to the use of general partial differential equation solvers that are designed to solve equations from many areas of engineering.

A finite element computer program, PDEase2D, marketed by Macsyma Corp, is one of the first general-purpose partial differential equation solvers to be marketed. Another similar software package called FlexPDE, is marketed by PDE Solution Ltd. While PDEase2D can analyze only two-dimensional problem, FlexPDE is extended to solve problems in three-dimensions. These software packages have several special features that are of interest to geotechnical engineers and have the potential to interface with many other software packages such as a computing program, MathCad, SoilVision (1998), graphics programs, AutoCAD and a database program, ACCESS.

The user of the general partial differential equation solver must specify the governing partial differ-

ential equation to be solved. The material properties can be described in the tabular form or as a mathematical equation. The boundary conditions can be specified as a dependent variable type (i.e., "head " type boundary condition) or a derivative of a dependent variable type (i.e., "flux" type boundary condition).

4 EXAMPLE PROBLEMS INVOLVING SEEPAGE THROUGH A SATURATED-UNSATURATED SOIL SYSTEM

A large number of example problems as well as a parametric study have been performed using the PDEase2D (Thieu, 1999). However, as part of this paper, only two example problems, the first associated with steady state seepage and the second associated with transient state seepage, will be presented. The soil material is assumed to be silt and isotropic with respect to the coefficient of permeability. Experimental data showing the coefficient of permeability versus matric suction, and the volumetric water content versus matric suction are obtained from Ho (1979). The saturated coefficient of permeability is 2.5 x 10^{-7} m/sec and the saturated volumetric water content is 0.381.

4.1 *Steady state seepage example*

This example is presented to illustrate the different forms of input the coefficient of permeability when analyzing steady-state seepage through an isotropic earth dam with a horizontal drain. The permeability functions used are the Gardner (1958) equation, van Genuchten-Mualem (1980) equation, van Genuchten-Burdine (1980) equation, Fredlund and Xing (1994) equation and a series of data points. The parametric study was done to find the best-fit values for the permeability functions using MathCad program. The permeability functions and their values of fitted parameters are shown in Table 1. Specified permeability functions together with the experimental data used to analyze the problem are presented graphically in Fig. 1.

The geometry, boundary conditions and finite element mesh used in running PDEase2D for steady-state seepage example are shown in Fig. 2. A maximum error of 0.1% was specified. The results of the total head distribution are presented in Fig. 3. The pore-water pressure distribution and flow vectors under steady state seepage are shown in Fig. 4 for different forms of data input. The position of the total head contours and phreatic lines are the same for all permeability functions.

The coefficient of permeability function does not need to be specified precisely when computing the distribution of pore-water pressure. Therefore, an approximate permeability function is adequate for

Table 1. Permeability functions and fitted parameters for steady-state seepage analyses

	Permeability functions	Fitted parameters
Gardner (1958)	$k_w = \dfrac{k_s}{1 + a\psi^n}$	$a = 1.969 \times 10^{-10}$, $n = 6.912$
van Genuchten-Burdine (1980)	$k_w = \dfrac{k_s}{\left[1 + (a\psi)^n\right]^{\left(1 - \frac{2}{n}\right)}}$	$a = 4.127 \times 10^{-2}$, $n = 9.401$ $m = 0.787$
van Genuchten-Mualem (1980)	$k_w = \dfrac{k_s}{\left[1 + (a\psi)^n\right]^{\left(1 - \frac{1}{n}\right)}}$	$a = 4.049 \times 10^{-2}$, $n = 8.852$ $m = 0.887$
Fredlund and Xing (1994)	$k_w = k_s \dfrac{\displaystyle\int_{ln(\psi)}^{b} \dfrac{\theta(e^y) - \theta(\psi)}{e^y} \theta'(e^y)\,dy}{\displaystyle\int_{ln(\psi_{aev})}^{b} \dfrac{\theta(e^y) - \theta_s}{e^y} \theta'(e^y)\,dy}$	$b = ln(1,000,000)$, $y =$ dummy variable of integration representing the logarithm of suction, $\psi_{aev} =$ soil suction at air entry value

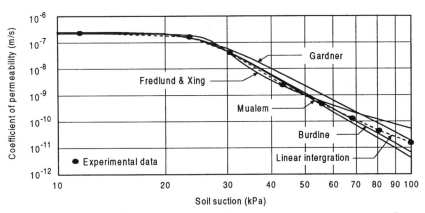

Figure 1. Specified permeability functions for the steady state seepage example

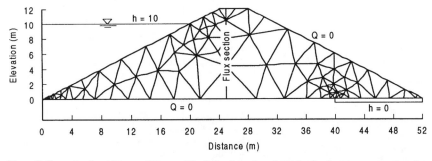

Figure 2. Geometry, boundary conditions and the finite element mesh for the steady state seepage example

analysis purposes. However, the quantity of flux through a section is slightly different, as shown in Table 2. The differences shown in Table 2 are not significant from an engineering standpoint and the results would indicate that any one of several possible permeability functions would yield satisfactory results.

4.2 Transient state seepage example

The second example is presented to show the different forms that can be used to input the coefficient of water storage when analyzing transient seepage through an isotropic earth dam with a horizontal drain. The base of the dam is selected as the datum.

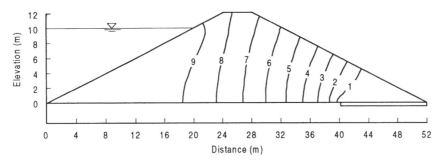

Figure 3. Total head distribution for steady state seepage example

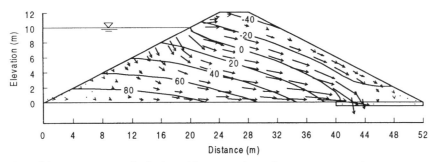

Figure 4. Pore-water pressure distribution and flow vectors for steady state seepage example

Table 2. Value of flux quantities using various permeability functions

Functions	Quantity of flux ($\times 10^{-7}$ m^3/s)	Deviation (%)
Gardner	7.449	0.8
van Genuchten-Burdine	7.432	0.5
van Genuchten-Mualem	7.430	0.5
Fredlund & Xing	7.376	0.2
Linear interpolation	7.272	1.6
Average	7.392	0

Initially, the dam is at steady-state conditions with the reservoir water level of 4 m above the datum. At a time assumed to be equal to zero, the water level in the reservoir is instantaneously raised to a level of 10 m above the datum.

Gardner (1958) equation is used to describe the permeability function. The water storage function is obtained by differentiating the soil-water characteristic curve. The soil-water characteristic curve is described using the van Genuchten (1980) equation, the extended error function, the Fredlund and Xing (1994) equation and the step-wise values. The parametric study was done to find out the best-fit parameters for the water storage function using Math-Cad program. The software, SoilVision (1998) can also be used for this purpose. The derivatives of the soil-water characteristic curve with respect to matric suction and their best-fit values are shown in Table 3 and Fig. 5.

The geometry and boundary conditions for this example are shown in Fig. 6. A steady-state condition with water level of 4 m is first performed in order to get the distribution of initial pore-water pressure head. A maximum error limit of 0.5% was specified. The number of elements and nodes used in running this process varies with time from 181 to 1006, and from 428 to 2215, respectively. The finite element mesh at time equal to 15 hours is presented in Fig. 7.

The pore-water pressure distributions and the phreatic lines for elapsed time equal to 15 hours corresponding to different forms of inputting the data are shown in Fig. 8. The results indicate that the water storage functions defined are quite sensitive to the transient state seepage predictions at early times. The phreatic line obtained using various water storage functions would be closer at latter time steps, and approaches the same location at steady state conditions. The solution of the problem at steady state condition was presented previously in Figs. 4 and 5.

5 CONCLUSION

General partial differential equation solvers, such as PDEase2D, written with a focus on the mathematical aspects of the problem, show great potential for solving saturated-unsaturated seepage problems.

Table 3. Water storage functions and fitted parameters for transient seepage analyses

Water storage functions		Fitted parameters
van Genuchten (1980)	$$\frac{d\theta}{d\psi} = -\frac{\theta_s}{(1+(a\psi)^n)^m}(a\psi)^n\frac{mn}{\psi(1+(a\psi)^n)}$$	$a = 4.391 \times 10^{-2}$, $n = 46.754$, $m = 0.036$
Extended error function	$$\frac{d\theta}{d\psi} = Ce^{a(|\psi-\omega|)^n}$$	$a = -0.183$, $n = 1.073$, $c = 0.028$, $\omega = 28.783$
Fredlund & Xing (1994)	$$\frac{d\theta}{d\psi} = -\theta_s\left[\frac{1}{\ln\left[e+\left(\frac{\psi}{a}\right)^n\right]}\right]^m \frac{m}{\ln\left[e+\left(\frac{\psi}{a}\right)^n\right]}\left(\frac{\psi}{a}\right)^n\frac{n}{\psi\left[e+\left(\frac{\psi}{a}\right)^n\right]}$$	$a = 26.127$, $n = 14.030$, $m = 0.622$

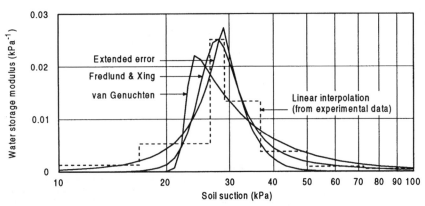

Figure 5. Specified water storage functions for transient seepage example

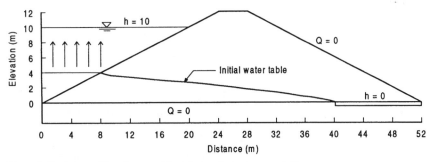

Figure 6. Geometry and boundary conditions for transient seepage example

These programs are particularly well-suited for solving unsaturated soils problems because of the attention given to: i) ensuring convergence when solving non-linear equations; ii) allowing material properties to be input in a variety of forms; and iii) allowing material properties to be non-linear in character.

It is possible to use a variety of formats for the input of soil property functions the PDEase2D. The formats for data input can vary from being a series of data points to a closed-form mathematical equation. In addition, MathCad and SoilVision software can be used in conjunction with PDEase2D to compute acceptable mathematical functions for unsaturated soil properties.

In general, it has been concluded, based upon the results of this study, that the philosophical approach behind this study is particularly valuable to the field

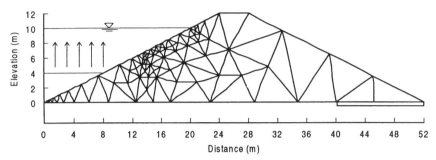

Figure. 7 Finite element mesh at an elapsed time, T = 15 hr, transient seepage example

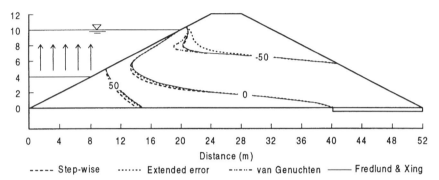

----- Step-wise   ······· Extended error   –··–··– van Genuchten   ——— Fredlund & Xing

Figure 8. Comparison of the location of pore-water pressure contours for various water storage functions, after an elapsed time, T = 15 hr, transient seepage example

of unsaturated soil mechanics. The philosophical approach infers that optimum use should be made of science and math specialties that overlap with the needs in geotechnical engineering. Unsaturated soil analyses are complex and demanding. Therefore, the research work associated with related disciplines in mathematics and computing should be adopted wherever possible. This approach should prevent the ongoing "re-discovery" of numerical software solutions.

REFERENCE

FlexPDE Manual, Version 2.12. 1999. PDE Solution Inc. Fremont, CA, USA.

Fredlund, D.G., and Rahardjo, H. 1993. Soil Mechanics for Unsaturated Soil. John Wiley & Sons, New York, 560 p.

Fredlund, D.G., and Xing, A. 1994. Equations for the Soil-Water Characteristic Curve. Canadian Geotechnical Journal, **31**(3): 521-532.

Fredlund, D.G., Xing, A., and Huang, S. 1994. Predicting the Permeability Function for Unsaturated Soils using the Soil-Water Characteristic Curve. Canadian Geotechnical Journal, **31**(3): 533-546.

Fredlund, D.G. 1996. Microcomputers and Saturated-unsaturated Continuum Modeling in Geotechnical Engineering. Symposium on Computers in Geotechnical Engineering, Sao Paulo, Brazil, pp. 29-50.

Fredlund, M.D. 1997. Application of PDEase Finite Element Analysis Software to Geotechnical Engineering, Class Project, Texas A&M College Station, Texas.

Fredlund, M.D. 1998. SoilVision Software, User's Manual, 1st Edition, SoilVision Systems Ltd., Saskatoon, SK, Canada.

Freeze, R.A. 1971a. Three-dimensional, Transient, Saturated-unsaturated Flow in a Groundwater Basin. Water Resources Research, **7**(2): 247-366.

Freeze, R.A. 1971b. Influence of the Unsaturated Flow Domain on Seepage through Earth Dam. Water Resources Research, **7**(4): 929-940.

Gardner, W.R. 1958. Some Steady State Solutions of the Unsaturated Moisture Flow Equation with Application to Evaporation from a Water Table. Soil Science, **85**: 228-232

Ho, P.G. 1979. The Prediction of Hydraulic Conductivity from Soil Moisture Suction Relationship. B.Sc. Thesis, University of Saskatchewan, Saskatoon, SK, Canada.

Lam, L., Fredlund, D.G., and Barbour, S.L. 1988. Transient Seepage Model for Saturated-Unsaturated Systems: A Geotechnical Engineering Approach. Canadian Geotechnical Journal, **24**(4): 565-580.

Leong, E.C., and Rahardjo, H. 1997. Review of Soil-Water Characteristic Curve Equations. Journal of Geotechnical and Geoenvironmental Engineering, pp. 1106-1117.

Leong, E.C., and Rahardjo, H. 1997. Permeability Functions for Unsaturated Soils. Journal of Geotechnical and Geoenvironmental Engineering, pp. 1118-1126.

PDEase2D 3.0 Reference Manual, 3rd Edition. 1996. Macsyma Inc. Arlington, MA, 02174 USA.

Thieu, N.T.M. 1999. Solution of Saturated-Unsaturated Seepage Problems Using a General Partial Differential Equation Solver. M.Sc. Thesis, University of Saskatchewan, Saskatoon, SK, Canada.

van Genuchten, M.T. 1980. A Closed-form Equation for Predicting the Hydraulic Conductivity of Unsaturated Soils. Soil Science Society of America Journal, **44**: 892-898.

Unsaturated Soils for Asia, Rahardjo, Toll & Leong (eds) © 2000 Taylor & Francis, ISBN 90 5809 139 2

Numerical modelling of soil water plant interaction

M.K.Trivedi, K.S.Hariprasad, A.Gairola & D.Kashyap
Department of Civil Engineering, University of Roorkee, India

ABSTRACT: The mathematical modelling of transient soil water flow in the presence of plant roots has been undertaken by several authors. In most of the models, the water uptake by roots is represented by a volumetric sink term, which is added to the continuity equation for soil water flow in the unsaturated zone. The sink term in the model needs detailed information about the root system such as root radius, root length, root density etc. Experimental evaluation of root system is both time consuming and costly. Therefore, to overcome this, the present paper deals with the solution of transient two dimensional axi -symmetric radial flow Richards equation where the volumetric sink term is computed by assigning known suction head at the root-soil interface as a boundary condition. The governing Richards equation is solved numerically by using an Iterative Alternating Direction Implicit explicit (IADIE) finite difference scheme.

1 INTRODUCTION

The availability of water for plant roots is an important topic which has been explored by a number of investigators, e.g. Gardner, (1960,1964), Whisler et al., (1968), Molz and Remson, (1970, 1971,1973), Belmans et al., (1979), de Jong and Cameron , (1979), McCoy et al., (1984). The research has followed basically in two different approaches: Microscopic approach and Macroscopic approach. In microscopic approach radial moisture flow towards a single root was analysed (Philip,1957; Gardner, 1960 and Molz et al.,1968). In macroscopic approach (Gardner, (1964), Whisler et al. (1968), and Feddes et al., (1974)) the attention is focussed on the removal of moisture from a differential volume of a soil as a whole without considering the effect of individual roots. In such an approach, the removal of moisture from roots is represented by a volumetric sink term. The sink term in these models requires detailed information of root characteristics. Molz et al.(1970) and Nimah et al.(1973a) have incorporated the sink term while modelling the water uptake by plants with some assumptions. However, these approximations may not accurately describe the root water uptake. Feddes et al. (1978) and Hoogland et al. (1981) have given more simpler and practical functions for the estimation of sink term. The numerical solution of

Richard's equation with sink term requires definition of an additional constituent parameter in the form of sink term. The sink term is described by a geometrical shape which is a function of soil water content at the field capacity, permanent wilting point, change in moisture storage and root characteristics.

To overcome this prior knowledge of sink term for the solution of Richard's equation, a suction head at the root-soil interface as a boundary condition is assigned because it is easier to know the suction head at the root-soil interface rather than the experimental evaluation of root characteristics and prior knowledge of the variation of soil moisture content in the root zone. The model requires only the information of depth of root length and the root radius.

The objective of this study is to develop a mathematical model for the water uptake by plant and to use it for some field data. Here, uniform root growth is considered

2 MODEL DEVLOPMENT

To model flow towards root in unsaturated zone, suction head based two dimensional axi-symmetric nonlinear parabolic differential Richard's equation is preferred which is given by:

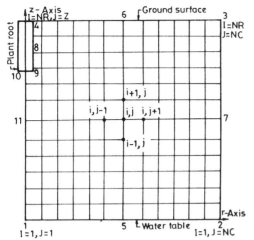

Fig. 1 Finite Difference Mesh & Boundaries

$$C\frac{\partial h}{\partial t} = \frac{1}{r}\frac{\partial}{\partial r}\left[rK_r\frac{\partial(-h)}{\partial r}\right] + \frac{\partial}{\partial z}\left[K_z\frac{\partial(-h+z)}{\partial z}\right] \quad (1)$$

Where, r is the distance outward from the main plant axis, z is vertical coordinate in the upward direction, h is suction head, K_r and K_z are unsaturated hydraulic conductivities in the r and z direction respectively, C ($=d\theta/dh$) is specific moisture capacity and t is the time.

Referring to Figure 1, the initial and boundary conditions are as follows:

Initial condition

(a) $\theta(z,r,0) = \theta_0(z)$,   t=0, 0≤z≤L, 0<r<R

Where L is the depth of the domain in the z-direction and R is the width of the domain in r direction.

Boundary conditions

(b) Boundary 1-2 (Lower boundary)

Water table is considered as the lower boundary conditon i.e.,

h = 0, t>0, z=0, 0<r<R

(c) Boundary 2-3(Far end constant head boundary)

Here Dirichlet type boundary condition is assigned:

h = HH, t>0, 0<z<L, r = R

where, HH is the initial suction head.

(d) Boundary 3-4(Upper boundary)

Infiltration due to either rainfall or applied irrigation is considered as the upper boundary condition, which is as follows:

$$q = k_z\left(\frac{\partial h}{\partial z} - 1\right), \; t>0, \; z=L, \; 0<r<R$$

(e) Boundary 4-9(Root-Soil interface)

constant suction head at root-soil interface is assigned as follows:

h(z,r) =H_root ;   0≤z≤root length at all times.

(f) Boundary 1-9(Water divide)

Boundary 1-9 is the vertical axis extending from the centre of the bottom centre of the root to the ground water table. As the flow is axis symmetric, this can be considered as a water divide and hence no flow across the boundary. Neuman type boundary condition is assigned with q=0 as follows:

$$\frac{\partial h}{\partial r} = 0.0 \quad \text{at } (z,0); \; 0 \le z \le C \text{ at all times.}$$

Where C is the depth between bottom of the root to the ground water.

3 NUMERICAL SOLUTION

Equation (1) in terms of finite difference, subject to the initial and boundary conditions (a)-(f), is solved by the Iterative Alternating Direction Implicit Explicit (IADIE) method (Remson et al.1971; Ahmed et al.1990,1991) proposed by Peacemen and Rachford(1955), where the flow domain is discretised by a finite number of nodal points. The time domain is also discretised by a finite number of discrete time steps. The soil characteristic data (θ, K_r, K_z and C) for the unsaturated zone are generated within the model, in accordance with the following equation:

3.1 K – Characteristic

The following relations, proposed by Brooks and Corey(1964) are adopted:

$$K(r,\theta) = K_{sat}\left(\frac{\theta - \theta_r}{\phi - \theta_r}\right)^\lambda \; ; \theta \ge \theta_r \quad (2)$$

$$K(r,\theta) = 0.0 \ ; \ \theta < \theta_r \tag{3}$$

$$K(z,\theta) = K_{sat}\left(\frac{\theta - \theta_r}{\phi - \theta_r}\right)^{\lambda} \ ; \ \theta \geq \theta_r \tag{4}$$

$$K(z,\theta) = 0.0 \ ; \ \theta < \theta_r \tag{5}$$

Where, θ_r is residual moisture content; ϕ is porosity; K_{sat} is hydraulic conductivity at saturation; θ is volumetric moisture content and λ is pore size index.

3.2 θ - Characteristic

The following relation (Rao et al. 1986) is adopted:

$$\theta = \exp\left[\frac{\ln(\phi - 2\theta_r)}{h_b}h\right] + \theta_r \ ; \quad h \geq h_b \tag{6}$$

$$\theta = \phi - \frac{\theta_r}{h_b}h \quad\quad ; \ 0 \leq h \leq h_b \tag{7}$$

Where, h is matric suction head; h_b is air entry value (bubbling pressure) and other variables are defined earlier.

The solution obtained in the form of suction head at different times at all nodes of the domain by IADIE method is again solved by Picard's iteration scheme.

3.3 Convergence criteria

At the end of a discrete time step the difference of h values of successive iteration at each of the finite difference nodes is computed. The IADIE iteration are continued until the biggest of these differences is less than a pre-stipulated convergence factor ε. Thus, for IADIE as well as Picard's solution, the convergence criteria is as follows:

$$\text{Max}[\text{Abs}(h_l - h_{l-1})] < \varepsilon \tag{8}$$

Where h_l= the suction head at a node after l^{th} IADIE iteration, h_{l-1}=the suction head at the node after $(1-1)^{th}$ iteration.

4 MODEL APPLICATION

The data for the initial soil moisture content at the time of crop sowing and soil physical properties like porosity, saturated hydraulic conductivity etc. are observed at the demonstration farm of Water Resources Development Training Centre, University of Roorkee, Roorkee (India) during Rabi(winter) 1990-1991.

The initial soil moisture at the time of sowing are given below:

Soil depth(cm): 00-30 30-60 60-90 90-120 120-150

Soil moisture : 0.2451 0.2552 0.2650 0.2751 0.2848 (cc/cc)

Soil properties (physical): Bulk density =1.36 g/cc; Particle density = 2.53g/cc; Porosity (%) = 0.462; Saturated hydraulic conductivity = 0.5517mm/hr; Field capacity =21.00(g/g %); Permanent wilting point = 13.50g/g%; Bubbling pressure (estimated) = 655.62mm ; Pore size index (λ) (estimated) = 4.0

Plant properties:

The root radius and its length are observed as 16mm and 200mm after the initial crop stage (26 days). 30mm irrigation on 22nd day and 15mm irrigation on 24th day of sowing is applied.

The suction head at the root-soil interface which depends upon the atmospheric demand and type of the crop, is taken as 0.2 bar for the wheat crop (initial stage) from available published data.

The proposed numerical model simulates the above observed data in the form of moisture content and suction head distribution at the root zone depth with respect to time. Figure 2 shows the contour of moisture content at the root zone depth with respect to time. Figure 3 shows the variation of amount of water uptake by plant with respect to time and compared with field estimated data.

5 CONCLUSIONS

The numerically simulated moisture content distributions at the root zone depth with respect to time would help in the estimation of the amount of water uptake by the plant if free flow movement of water is considered from root surface to the top of the plant leaves to the atmosphere which leads to the computation of the transpiration rate. This approach is easier and numerically suitable for the estimation of transpiration rate.

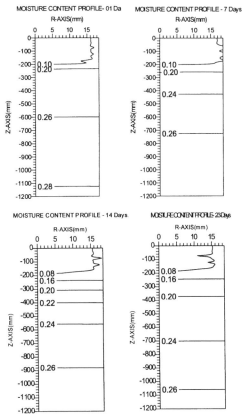

Figure 2 Contours of moisture content on 1,7,14 and 23 days.

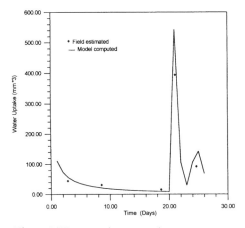

Figure 3 Water uptake versus time.

REFERENCES

Ahmed, S., Kashyap, D., and Mathur, B.S.(1991), "Numerical modelling of two dimensional transient flow to ditches." *J. Irrig. And Drain. Engineering , ASCE*,117(6), 839-851

Belmans, c., Feyen,J. and Hillel, D., 1979. An attempt at experimental validation of macroscopic scale models of soil moisture extraction by roots. *Soil Sci.*,127:174-186

Feddes, R.A., Breseler, E. and Neuman, S.P., 1974. Field test of a modified numerical model for uptake by root systems. *Water Resources Research*, 10(6):1199-1206.

Feddes, R.A., Kowalik,P.J. and Zardny, H., 1978. Simulation of field water use and crop yield. PUDOC, *Wageningen, Simulation Monographs*, 189pp.

Hoogland, J., Belmans, C. and Feddes, R.A.,1981. Root water uptake model depending on soil water pressure head and maximum extraction rate. *Acta Hortic.*,119:123-136.

Gardner, W.R.,1960. Dynamic aspects of water availability to plants. *Soil Sci.* 89:63-73.

Gardner,W.R.,1964. Relation of root distribution to water uptake and availability. *Agron. J.* 56, 41-48.

Rao., K. M. M., Kashyap, D., and Satish Chandra. (1986). "Hydrological response of unsaturated zone upto water table" *Ph.D. thesis, Univ. of Roorkee, Roorkee, India.*

Molz, F. J., Remson, I., Fungaroli, A.A.and Darke, R.L. 1968. Soil moisture availability for transpiration. *Water Resources Rec.* 4:1161-1169

Molz, F. J. and Remson, I.1970.Extraction term models of soil moisture use by transpiring plants. *Water Resour. Res.* 6(5):1346-1356.

Nimah, M.N. and Hanks, R.J., 1973. Model for estimating soil water, plant and atmosphere interrelations. I. Description and sensitivity. *Soil Sci. Soc. Am. Proc.*, 37:522-527.

Peaceman, D.W., and Rachford, H.H.(1955), "The numerical solution of parabolic and elliptic differential equations." *J. Soc. Ind. Appl. Math.*, 3,24-41.

Philip,J.R. 1957. The theory of infiltration:1. *Soil Sci*, 83: 345-357.

Remson, I., Hornberger, G. M., and Molz, F. D. (1971), Numerical methods in subsurface hydrology, *John Wiley and Sons, New York, N. Y.*

Whisler, F.D., A. Klute, and R. J. Millington.1968. Analysis of steady state evapotranspiration from a soil column. *Soil Sci. Soc. Am. Proc.* 32:167-174.

Unsaturated Soils for Asia, Rahardjo, Toll & Leong (eds) © 2000 Taylor & Francis, ISBN 90 5809 139 2

Water flow in unsaturated soil slopes with horizontal drains

D.Q.Yang & S.K.Kong
Moh and Associates (S) Pte Limited, Singapore

H.Rahardjo & E.C.Leong
NTU-PWD Geotechnical Research Centre, Nanyang Technological University, Singapore

ABSTRACT: There are numerous slope failures caused by periods of heavy and intense rainfall in Singapore, Malaysia, China and other countries in the world. During rainfall, not only will the water content in the unsaturated vadose zone increase, but the ground water table in the saturated vadose zone may also rise as rain water infiltrates into the slopes. One of the methods to lower the ground water table and to maintain the matric suction in the slopes is by horizontal drains. To study the flow mechanism of water in the slopes with horizontal drains and to examine the effectiveness of the horizontal drains, a three-dimensional finite element computer programme for seepage analysis in unsaturated soils has been developed. The horizontal drains in the slopes are modelled as internal flux boundary conditions. The flow mechanism of water in the slopes with horizontal drains and the effectiveness of the drainage are discussed in detail using an idealised model slope in this paper.

1 INTRODUCTION

There have been a number of reported slope failures caused by periods of heavy and intense rainfall in the regional countries and other countries in the world. During rainfall, not only will the water content in the unsaturated vadose zone increase, but the ground water table in the saturated vadose zone may also rise through rainfall infiltration into the slopes. The increase in water content of the unsaturated vadose zone will reduce the soil matric suction, and consequently lead to the decrease of the shear strength of the soil. A rising ground water table in the slope may also reduce the shear strength of the soil. Therefore the lowering of the ground water table and the maintenance of the matric suction in the slope are important aspects in slope stabilization work. One of the methods to reduce the level of the groundwater table and to maintain the matric suction in the slopes is through the installation of horizontal drains. The beneficial effects of horizontal drains have been empirically well known, but it seems to be very difficult to evaluate quantitatively the effectiveness of the horizontal drains.

To study the flow mechanism of water in the slopes with horizontal drains and to examine the effectiveness of the horizontal drains, a three-dimensional finite element computer programme for seepage analysis in unsaturated soils has been developed. The horizontal drains in slopes are modelled as internal flux boundary conditions. The computer code has been verified by comparing the numerical results of a selected case and the corresponding theoretical solution. The flow mechanism of water in the slopes with horizontal drains and the effectiveness of the horizontal drains will be discussed in detail using an idealised slope in this paper.

2 THEORY BACKGROUND

2.1 Governing equations for water flow

The mass flux of isothermal pore-liquid water can be expressed using a generalized form of Darcy's law as shown below:

$$\vec{q}_w = -\frac{k_w}{g}\nabla(u_w + \rho_w g z) \tag{1}$$

where: $\vec{q}_w$ is the mass flux vector of the pore-liquid water; k_w is the unsaturated permeability matrix of the liquid water phase; g is the gravitational constant; u_w is the pore-water pressure; z is the vertical space coordinate (upward positive); ρ_w is the density of water. $\underline{k}_w$ is a function of the porosity and the liquid water content or the suction of the soil. The unsaturated permeability can be formulated as

$$\underline{k}_w(u_a - u_w, n) = \underline{k}^r_{ws}(n) k^r_{rw}(u_a - u_w) \qquad (2)$$

where: (u_a-u_w) is the matric suction of the soil; n is the porosity of the soil; $\underline{k}^r_{ws}$ is the saturated permeability matrix related to the porosity of the soil at room temperature; k^r_{rw} is the relative unsaturated permeability associated with the suction of the soil at room temperature.

Applying the principle of mass conservation to the pore-liquid water yields

$$\frac{\partial}{\partial t}(\rho_w n S) = -\nabla \cdot \left(\overrightarrow{q}_w \right) + f^w \qquad (3)$$

where: n is the porosity of soil; f^w is the evapotranspiration mass rate per unit soil volume due to the uptake of water by evaporation and transpiration or external supply rate of the moisture mass per unit volume of soil.

2.2 Material Parameters

The saturated permeability matrix for the pore-liquid water at room temperature $\underline{k}^r_{ws}(n)$ and the unsaturated relative permeability for the pore-liquid water at room temperature $k^r_{rw}(u_a - u_w)$ or $k^r_{rw}(S)$ can be obtained from laboratory tests. The $\underline{k}^r_{ws}(n)$ can be assumed to be constant in many cases. Many empirical expressions of $k^r_{rw}(u_a - u_w)$ or $k^r_{rw}(S)$ are available (Alonso et al., 1987).

Beside the permeabilities for the pore-liquid water, the degree of saturation is another important soil parameter. The degree of saturation is related to matric suction. The state surface with respect to the degree of saturation can be determined from experiments. The following relationship between volumetric water content and matric suction has been proposed by Fredlund & Xing (1994):

$$\theta_w = C(u_a - u_w) \frac{\theta_{ws}}{\left\langle \ln\left\{ e + \left[\frac{(u_a - u_w)}{a} \right]^n \right\} \right\rangle^m}$$

(4a)

where:

$$C(u_a - u_w) = 1 - \frac{\ln\left[1 + \frac{(u_a - u_w)}{(u_a - u_w)_r} \right]}{\ln\left[1 + \frac{1000000}{(u_a - u_w)_r} \right]} \qquad (4b)$$

is a correction function; $(u_a - u_w)_r$ (kPa) is the matric suction corresponding to the residual volumetric water content θ_r; a, n and m are three material parameters; θ_w and θ_{ws} are the volumetric water content (i.e., Sn) and the saturated volumetric water content (i.e., n), respectively.

3 NUMERICAL METHOD

In this paper, the Galerkin finite element method is used to discretise Equation 3, and the formulation $\int_{t_k}^{t_k + \Delta t} ( ) \, dt = \Delta t \left[(1 - \alpha)( )_{t_k} + \alpha( )_{t_k + \Delta t} \right]$ is applied to discretise the time domain, where: Δt is the time step; $( )_{t_k}$ and $( )_{t_k + \Delta t}$ are the values of the function at time t_k and $t_k + \Delta t$, respectively; α is an integral parameter. If the increment of the pore-liquid water pressure Δu_w is accepted as the independent variable, the finite element formula for solving water flow in unsaturated soils can be expressed as

$$\sum_{j=1}^{N} [\underline{k}_{ij}]^{t_k + 0.5 \Delta t} \{\Delta V_j\} = \{\Delta F_i\} \quad (i = 1, 2, 3, \cdots, N) \ (5)$$

where: $\{\Delta V_j\} = [\Delta u_{wj}]^T$ is the column matrix of the increments of the nodal variables at node j within time interval Δt; $\{\Delta F_i\} = [\Delta f^1_i]^T$ is the column matrix of the increments of the generalized equivalent nodal water mass at node i within time interval Δt; the elements of the sub-matrix $\underline{k}_{ij}$ can be derived using the finite element technique; N is the total number of nodes; $( )^{t_k + 0.5\Delta t}$ is the values of the function at time $t_k + \Delta t / 2$. Based on the formulae described above, a three-dimensional finite element programme for seepage analysis in unsaturated soils has been developed. Two kinds of spatial isoparametric elements, i.e. eight-noded and six-noded spatial isoparametric elements are incorporated in this programme. Within this programme, solution of the resulting set of global matrices is achieved using a nonsymmetric solution algorithm. An iterative technique is used to solve the nonlinear equations established above and convergence of solution is assumed to be achieved when the difference between successive iterations falls is less than a specified tolerance.

4 FLUX BOUNDARY CONDITIONS

The flux boundary conditions for modelling water flow in slopes with horizontal drains is a fundamental and important issue that has not been studied well so far. The questions that may be asked are: (1) what should be the infiltration flux into slopes during rainfall infiltration?, and (2) what should be the criterion for discharge of horizontal drains? Basically there are two types of flux boundary conditions as discussed in the following sections.

4.1 Internal conditions along horizontal drains

For convenience of analysis, the computer code treats the drain as a line of nodal points without consideration of the diameter of the drain. At each nodal point "i", the following criteria for discharge of the horizontal drains are set up:

$$\begin{cases} (u_{wi})_{l+1} = 0 \, , & \text{if } (u_{wi})_l > 0 \\ (u_{wi})_{l+1} < 0 \, (\text{unknown}), & \text{if } (Q_{wi})_l \leq 0 \end{cases} \quad (6)$$

where Q_{wi} (Positive denotes discharge of water and negative denotes absorption of water) is the discharge of water at the nodal point "i"; u_{wi} is the pore-liquid water pressure at the nodal point "i"; l represents numerical iterative level number.

4.2 External conditions on slope surface

Some work published so far indicated that the mass flux on external slope surface should be equal to the saturated coefficient of permeability if the rainfall rate is greater than the saturated coefficient of permeability. This assumption appears to be inappropriate. In fact, the actual normal mass flux on slope surface should be equal to the input of normal rainfall rate, if the pore-liquid water pressure, u_w, on the slope surface is negative. On the other hand, if the pore-liquid water pressure, u_w, on the slope surface becomes positive, then it should be set to

zero. This means that no excess pore-liquid water pressure is allowed to build up on the slope surface.

$$\begin{cases} \vec{q}_w \cdot \vec{n} = q_w, & \text{if } u_w < 0 \\ u_w = 0, & \text{if } u_w > 0 \end{cases} \quad (7)$$

where $\vec{q}_w \cdot \vec{n}$ is the normal mass flux on slope surface; q_w is the rainfall rate.

5 VERIFICATION OF NUMERICAL MODEL

Water movement in unsaturated soils is a complex natural phenomenon with highly nonlinear characteristics. Theoretical analytical solutions of water movement in unsaturated soils seem to be impossible, except for a few special cases. The following special case shows an example of horizontal infiltration in an unsaturated soil. A horizontal unsaturated cylindrical soil specimen of one metre length is assumed to be homogeneous with an initial volumetric water content of 10 % and porosity of 0.287 under an initial matric suction of 602 kPa. A constant head of water pressure of -205 kPa or a constant volumetric water content of 26.7 % is applied to the specimen on the left boundary. This problem is one-dimensional with one variable, that is the pore-water pressure, u_w. The deformation of the soil structure and the movement of the pore-air are assumed to be negligible. Temperature is kept constant at room temperature. The unsaturated coefficient of permeability, k_w, of the soil is given as follows:

$$k_w = \frac{k_{ws}}{1 + A_1 P_c^{C_1}} \quad \text{(m/s)} \quad (8)$$

where: P_c is the normalised suction [i.e., (u_a-u_w) (Pa)/1000(Pa)], the saturated coefficient of permeability k_{ws} is 9.44×10^{-5} m/s and the constants, A_1 is 1.696×10^{-11} and C_1 is 4.74. The relationship of the degree of saturation and the matric suction is expressed as:

$$S = \frac{1 - S_r}{1 + A_0 P_c^{B_0}} + S_r \quad (9)$$

where: S_r is the residual degree of saturation (i.e., 26.13 %) and the constants, A_0 is 7.3434×10^{-11} and B_0 is 3.96. Figure 1 shows the comparison between the numerical and analytical volumetric water

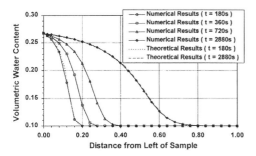

Figure 1. Comparison between numerical and theoretical volumetrical water content distribution.

content distribution along the length of the specimen. The numerical solution is obtained using the programme developed and the analytical results are from Philip & Knight (1974). Figure 1 demonstrates the good agreement between the numerical and the analytical results.

6 WATER FLOW IN UNSATURATED SOIL SLOPE WITH HORIZONTAL DRAINS

In order to reveal the flow mechanism of water in a slope with horizontal drains and the effectiveness of the horizontal drains, an idealised model slope with a gradient of 1V:1.5H and height of 20m as shown in Figure 2 is used for the numerical study. The slope is assumed be homogeneous. Three horizontal drains with length of 20m are installed in the slope at three different heights, i.e. 2m, 8m and 14m from the toe of the slope. The longitudinal spacing of horizontal drains is assumed to be 9m. Due to the regular pattern of horizontal drain installations, only a half section (i.e., 4.5m length) of the slope with three horizontal drains is analysed using the developed three-dimensional finite element computer programme. The idealised slope was discretised into 1077 elements and 1588 nodal points. Ten internal nodal points were assigned at equi-distances of 2m apart for each of the horizontal drains.

The initial pore-liquid water pressure in the slope is obtained by assuming that the ground water table on the right boundary is maintained at 2m below the ground surface. For the present study, the duration and rate of rainfall are assumed to be 45 days and 1.157×10^{-6} m/s, respectively.

The problem is three-dimensional with one variable, that is the pore-liquid water pressure, u_w. The deformation of the soil structure and the

movement of the pore-air are assumed to be negligible. Temperature is assumed to be constant. The unsaturated coefficient of permeability of the soil is given in Equation 10.

$$k_w = \begin{cases} k_{ws}; & P_c \le P_{cb} \\ k_{ws} \exp \left\{ \begin{array}{l} c_0 + c_1 \times \ln(P_c) + c_2 \times \\ [\ln(P_c)]^2 + c_3 \times [\ln(P_c)]^3 \end{array} \right\}; & P_c > P_{cb} \end{cases}$$

(10)

where: P_{cb} is the normalised bubbling suction of the soil (i.e., 1.00); k_{ws} is the saturated permeability of the soil (i.e., 5.0×10^{-7} m/s); c_0, c_1, c_2 and c_3 are four (4) coefficients given as 0, -1.5269, -0.0454 and 0.001, respectively.

The relationship of the degree, of saturation and the matric suction is expressed in Equation 4, where: θ_{ws} is saturated volumetric water content (i.e., 0.589); a, n and m are three coefficients given as 26.92 kPa, 0.3183 and 0.4024, respectively. On the basis of the computed results, the following three main aspects, i.e. (1) effect of horizontal drains on ground water table, (2) discharge of horizontal drains, and (3) infiltration rates into slope are discussed in detail in the following sections.

6.1 Effect of horizontal drains on ground water table

The movement of the ground water table (drop or rise) may have a significant influence on slope stability. Figure 3 presents the distribution of pore-liquid water pressures in the slope without horizontal drains before and after a prolonged rainfall of 45 days. From Figure 3, the rise of the ground water table (i.e., zero pore-liquid water pressure line) after the prolonged rainfall (45 days) appears to be very significant.

Comparison of the pore-liquid water pressures in the slope with and without horizontal drains after a prolonged rainfall of 45 days is shown in Figure 4. Figure 5 illustrates the distribution of the pore-liquid

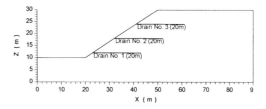

Figure 2. Idealised slope for numerical study.

water pressures in the slope with horizontal drains after a prolonged rainfall of 45 days along different slope sections (i.e., Y = 0.0m and 4.5m). The significant drop of ground water table in the slope with horizontal drains indicates that ground water drainage using horizontal drains can be effective as part of landslide control works. The maximum drop of the ground water table and the extent of the influenced zone are dependent on factors such as length of horizontal drains, spacing of horizontal drains and number of horizontal drains.

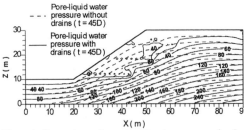

Figure 3. Comparison of pore-liquid water pressures in slope without horizontal drains before and after a prolonged rainfall (45 days).

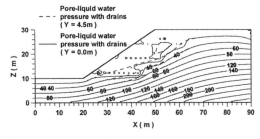

Figure 4. Comparison of pore-liquid water pressures in slope with and without horizontal drains after prolonged rainfall (45 days).

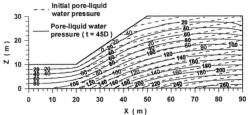

Figure 5. Comparison of pore-liquid water pressures in slope with horizontal drains after prolonged rainfall (45 days) along different sections.

6.2 Discharge of horizontal drains

The discharge of the horizontal drains is an important index for design of ground water drainage using horizontal drains as it will be useful in the selection of diameter size and number of horizontal drains. Figures 6 and 7 are plots of normalised discharge and discharge rate of horizontal drains versus time, respectively. The discharge and discharge rate as shown in Figures 6 and 7 are normalised by the spacing of the horizontal drains. On the basis of the discharge and discharge rate of drains shown in Figures 6 and 7, it can be summarised as: (1) the lower drains play a more important role in lowering the ground water table, (2) the upper drains discharge infiltrated water from the slope in order to prevent the ground water table from rising and to maintain the matric suction in the slope, and (3) the size of the upper level drains can be smaller as compared with the lower levels due to the lesser discharge from the upper level drains.

6.3 Infiltration rates into slope

The normal infiltration rates for slopes with and without drains are presented in Figures 8 and 9, respectively. Figures 8 and 9 demonstrate the

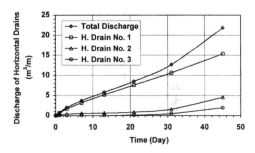

Figure 6. Plots of normalised discharge of horizontal drains versus time.

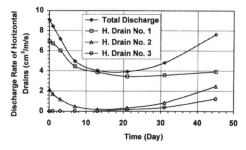

Figure 7. Plots of normalised discharge rate of horizontal drains versus time.

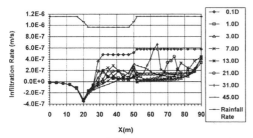

Figure 8. Plots of normal infiltration rate along slope surface at different time without horizontal drains.

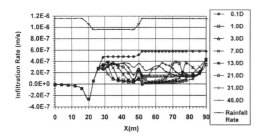

Figure 9. Plots of normal infiltration rate along slope surface at different time with horizontal drains.

following phenomena: (1) the computed normal infiltration rate along the surface for the idealised slope is much smaller than the saturated coefficient of permeability, (2) the computed normal infiltration rate decreases with time during the early stage, but increases with time during the later stages while the surface soil becomes saturated, and (3) the normal infiltration rate into the slope is related to factors such as slope angle, soil suction profile, soil properties etc.

7 CONCLUDING REMARKS

In order to study the flow mechanism of water in slopes with horizontal drains and to examine the effectiveness of the horizontal drains, a three-dimensional finite element computer programme has been developed to account for internal flux boundary conditions along horizontal drains in the slopes. Through numerical modelling of an idealised slope, three main aspects, i.e. effect of horizontal drains on ground water table, discharge of horizontal drains and infiltration rates into slope have been investigated.

On the basis of the results of the above study, the following conclusions can be made:

- The maximum drop in ground water table and the extent of the influenced zone are dependent on factors such as length of horizontal drains, spacing of horizontal drains and number of horizontal drains;
- The lower level drains play a more important role in lowering ground water table in slopes;
- The upper level drains discharge infiltrated water from the slope in order to prevent the ground water table from rising and to maintain the matric suction in the slope;
- The diameter of the upper level drains can be smaller than the lower level drains as the discharge is less from the upper level drains;
- The computed normal infiltration rate decreases with time during the early stage, but increases with time during the later stages when the surface soil becomes saturated;
- The normal infiltration rate into a slope is related to factors such as slope angle, soil suction profile, soil properties etc.

REFERENCES

Alonso, E.E., Gens, A. and Hight, D.W. 1987. Special problem soils, general report. *Proc. 9th Eur. Conf. Soil Mech.*, Dublin. 3: 1087-1146.

Fredlund, D.G. and Xing, A. 1994. Equations for the soil-water characteristic curve. *Canadian Geotechnical Journal.* 15(3): 313-321.

Philip, J.R. and Knight, J.H. 1974. On solving the unsaturated flow equation: 3. New quasi-analytical technique. *Soil Science.* 117: 1-13.

Unsaturated Soils for Asia, Rahardjo, Toll & Leong (eds) © 2000 Taylor & Francis, ISBN 90 5809 139 2

Governing equations of unsaturated seepage taking the effects of air flow and volume change into consideration

H. Zhang & S. Y. Chen
Institute of Rock and Soil Mechanics, The Chinese Academy of Science, Wuhan, People's Republic of China

ABSTRACT: Unsaturated seepage is one of the most important components in unsaturated soil mechanics. Due to the existence of air phase, seepage in unsaturated soil is more complex than that in saturated soil. In this paper, the general governing equations of unsaturated seepage with the effects of air flow and volume change are derived. Some simplified cases are given and some problems in numerical solution are discussed.

1 INTRODUCTION

Unsaturated seepage is one of the most important components in unsaturated soil mechanics. In the research field of unsaturated soil mechanics, matric suction has been taken to be one of the two independent stress state variables. This concept has been widely accepted by most researchers. In many problems, the solution of stress state variables is closely associated with the analysis of unsaturated seepage. Due to the existence of air phase in unsaturated soil, the analysis of fluid flow in unsaturated soil is more complex than that in saturated soil. Air phase not only takes part in the flow process, but also interferes with water phase during unsaturated seepage. For the cases of transient flow, moisture transport in unsaturated soil will lead to the change in the degree of saturation, and it will result in a significant change in permeability. In cohesive soil, especially in expansive soil, it will also cause the changes in porosity and pore size distribution as well as permeability. As a result, the effects of two-phase water-air flow and volume change should be considered in a rigorous unsaturated seepage theory. For non-isothermal flow condition, the effect of temperature gradient should also be considered.

Multiphase flow theory in porous media was first developed and then further investigated in the petroleum engineering. Bear (1972) gave an introduction about the theory in detail. The coupling of two-phase flow and soil deformation were investigated in consolidation problem of unsaturated soil by Dakshanamurthy (1984) and Fredlund et al. (1986). This kind of works were also done by the others but using more complex constitutive model for soil deformation in recent years.

However, in ordinary analysis for the other problems concerning with transient unsaturated flow, such as in the soil slope stability analysis, the effects of air phase flow and volume change are usually neglected and a simplified single-phase flow model is adopted. The governing equation of the model is the well-known Richands' equation. Freeze et al. (1979) considered that the approximation of single-phase flow in unsaturated flow was accurate enough for most practical problem, while in some unsaturated flow problems, two-phase water-air flow must be considered, such as infiltration analysis under strong rainfall. Sun et al. (1995) compared the different analyzing results between single-phase flow and two-phase flow in the investigation of soil slope stability during rainfall. He found that using two-phase flow analysis method, the water seepage toward the depth of slope became slow and the shallow region of slope was fast saturated thus causing the shallow slope failure. While using single-phase flow analysis method, water infiltrated easily into the depth of slope, thus leading to a higher safety factor than that in two-phase flow analysis. Consequently, the result of two-phase flow analysis well reflected the practical condition.

According to the above analysis, whether and in what condition the simplified analysis model will cause significant deviation are still worth making a systematic study. The authors are now working on it. In this paper, a group of general governing equation of unsaturated seepage with the effects of two-phase water-air flow and volume change considered is derived. And some simplified equations in the special conditions are given. Some problems in numerical solution are also discussed in the end of this paper.

2 TWO-PHASE FLOW THEORY WITH VOLUME CHANGE

Water phase flow and air phase flow in unsaturated soils are both governed by Darcy's law, the same as in the general single-phase flow theory.

$$\boldsymbol{v}_w = -\boldsymbol{k}_w \cdot \nabla h_w \tag{1}$$

$$\boldsymbol{v}_a = -\boldsymbol{k}_a \cdot \nabla h_a \tag{2}$$

where $\boldsymbol{v}_w$ and $\boldsymbol{v}_a$ are the average velocity vector of water flow and air flow respectively; $\boldsymbol{k}_w$ and $\boldsymbol{k}_a$ are the permeability tensor of water and air respectively; h_w is the hydraulic head of water and h_a is the pressure head of air. Hubber (1940) (see Bear 1972) extended the concept of piezometric head to compressible fluid. In general, h_w and h_a can be formulated as:

$$h_w = z + \int_{u_{w0}}^{u_w} \frac{du_w}{\rho_w g} \tag{3}$$

$$h_a = \int_{u_{a0}}^{u_a} \frac{du_a}{\rho_a g} \tag{4}$$

then

$$\nabla h_w = \nabla z + \frac{1}{\rho_w g} \nabla u_w \tag{5}$$

$$\nabla h_w = \frac{1}{\rho_w g} \nabla u_w \tag{6}$$

where, u_w and u_a are the pressure of water and air phase; ρ_w and ρ_a are the density of water and air respectively; u_{w0} and u_{a0} are the pressure of water and air in a datum state respectively; z is the elevation and g is the acceleration of gravity.

The equation of mass conservation for the fluid in porous media is

$$\frac{\partial(\rho_\alpha n S_\alpha)}{\partial t} + \nabla \cdot (\rho_\alpha n S_\alpha \boldsymbol{v}_\alpha) = 0 \tag{7}$$

where ρ_α is the density, α is water or air and $\boldsymbol{v}_\alpha$ is the average velocity vector of α phase. It is not difficult to obtain the governing equations of water phase flow and air phase flow. Using u_w and u_a as the variables, the equations can be expressed as

$$\frac{\partial(\rho_w g n S_w)}{\partial t} - \frac{\partial}{\partial x}\left(k_{wx}\frac{\partial u_w}{\partial x}\right) - \frac{\partial}{\partial y}\left(k_{wy}\frac{\partial u_w}{\partial y}\right)$$
$$- \frac{\partial}{\partial z}\left(k_{wz}\frac{\partial u_w}{\partial z}\right) - \frac{\partial(\rho_w g k_{wz})}{\partial z} = 0 \tag{8}$$

$$\frac{\partial(\rho_a g n S_a)}{\partial t} - \frac{\partial}{\partial x}\left(k_{ax}\frac{\partial u_a}{\partial x}\right) - \frac{\partial}{\partial y}\left(k_{ay}\frac{\partial u_a}{\partial y}\right)$$
$$- \frac{\partial}{\partial z}\left(k_{az}\frac{\partial u_a}{\partial z}\right) = 0 \tag{9}$$

where S_w and S_a are the degree of saturation of water phase and air phase; k_{wz}, k_{wy}, k_{wz} are the water coefficient of permeability in x, y, z direction respectively and k_{ax}, k_{ay}, k_{az} are the air coefficient of permeability in x, y, z direction respectively.

There are 7 unknown variables: ρ_w, ρ_a, n, S_w, S_a, u_w, u_a in equations of (8),(9). In order to solve the governing equations, there are still 5 equations needed.

The relationship of the degree of saturation between water phase and air phase is

$$S_w + S_a = 1 \tag{10}$$

Due to the soil-water characteristic curve, the relation between matric suction and the degree of saturation of water phase is

$$S_w = f(u_a - u_w) \tag{11}$$

Denote the partial derivation of S_w by f , i.e.

$$\frac{\partial S_w}{\partial(u_a - u_w)} = f' \tag{12}$$

The compressibility of water phase can be expressed as

$$\frac{d\rho_w}{\rho_w} = \beta du_w \tag{13}$$

where β is the coefficient of compressibility of water.

Assuming air phase is ideal gas, in isothermal condition, we have

$$\rho_a = \frac{\rho_{a0} u_a}{u_{a0}} \tag{14}$$

where ρ_{a0} and u_{a0} are air density and pressure in a datum state respectively.

The volume change behavior of a unsaturated soil can be expressed as

$$dn = C_t d(\sigma_{mean} - u_a) + C_m d(u_a - u_w) \tag{15}$$

where C_t and C_m are the coefficients of compressibility associated with net normal stress and matric suction respectively.

Assuming the mean normal stress σ_{mean} is constant during volume change, using the chain rule of partial differentiation to extend the first term of equations (8) and (9) and substituting equations (10), (11), (13), (14), (15) into (8) and (9), the general governing equations of the two-phase flow of water and air through a compressible porous media are obtained.

$$C\frac{\partial u_w}{\partial t} + D\frac{\partial u_a}{\partial t} - \frac{\partial}{\partial x}\left(k_{wx}\frac{\partial u_w}{\partial x}\right) - \frac{\partial}{\partial y}\left(k_{wy}\frac{\partial u_w}{\partial y}\right)$$

$$-\frac{\partial}{\partial z}\left(k_{wz}\frac{\partial u_w}{\partial z}\right)-\frac{\partial(\rho_w gk_{wz})}{\partial z}=0 \quad (16)$$

$$E\frac{\partial u_w}{\partial t}+F\frac{\partial u_a}{\partial t}-\frac{\partial}{\partial x}\left(k_{ax}\frac{\partial u_a}{\partial x}\right)-\frac{\partial}{\partial y}\left(k_{ay}\frac{\partial u_a}{\partial y}\right)$$

$$-\frac{\partial}{\partial z}\left(k_{az}\frac{\partial u_a}{\partial z}\right)=0 \quad (17)$$

where

$$C=\left(nS_w\frac{\partial \rho_w}{\partial u_w}-n\rho_w f'-S_w\rho_w C_m\right)g \quad (18)$$

$$D=\left(n\rho_w f'+S_w\rho_w(C_m-C_t)\right)g \quad (19)$$

$$E=\left(n\rho_a f'-(1-S_w)\rho_a C_m\right)g \quad (20)$$

$$F=\left(n(1-S_w)\frac{\rho_{a0}}{u_{a0}}-n\rho_a f'+(1-S_w)\rho_a(C_m-C_t)\right)g \quad (21)$$

3 SIMPLIFIED GOVERNING EQUATIONS

If some effect factors are negligible, the governing equations (16) and (17) can be simplified into different cases which discussed as follows.

3.1 *Governing equation of two-phase water-air flow without volume change*

If effect of volume change is negligible, the coefficient of compressibility C_t and C_m are equal to zero. Then,

$$C_1\frac{\partial u_w}{\partial t}+D_1\frac{\partial u_a}{\partial t}-\frac{\partial}{\partial x}\left(k_{wx}\frac{\partial u_w}{\partial x}\right)-\frac{\partial}{\partial y}\left(k_{wy}\frac{\partial u_w}{\partial y}\right)$$

$$-\frac{\partial}{\partial z}\left(k_{wz}\frac{\partial u_w}{\partial z}\right)-\frac{\partial(\rho_w gk_{wz})}{\partial z}=0 \quad (22)$$

$$E_1\frac{\partial u_w}{\partial t}+F_1\frac{\partial u_a}{\partial t}-\frac{\partial}{\partial x}\left(k_{ax}\frac{\partial u_a}{\partial x}\right)-\frac{\partial}{\partial y}\left(k_{ay}\frac{\partial u_a}{\partial y}\right)$$

$$-\frac{\partial}{\partial z}\left(k_{az}\frac{\partial u_a}{\partial z}\right)=0 \quad (23)$$

where

$$C_1=\left(S_w\frac{\partial \rho_w}{\partial u_w}-\rho_w f'\right)ng \quad (24)$$

$$D_1=n\rho_w f'g \quad (25)$$

$$E_1=n\rho_a f'g \quad (26)$$

$$F_1=\left((1-S_w)\frac{\rho_{a0}}{u_{a0}}-\rho_a f'\right)ng \quad (27)$$

equations (22) and (23) are the general governing equations of two-phase immiscible and compressible water-air flow. More complete analysis should also consider the effect of air dissolving in water. But it will not be considered in the present study If the compressibility of water phase is neglected further, then $\frac{\partial \rho_w}{\partial u_w}=0$, $C_1=-D_1$.

3.2 *Governing equation of single-phase flow with volume change*

If the effect of the air phase is neglected, air phase pressure is usually assumed to be constant. If an atmospheric pressure is defined as datum zero pressure, then pore water pressure in unsaturated soil is negative. According to the above assumption, the governing equation (17) vanishes naturally and the term $\frac{\partial u_a}{\partial t}$ is equal to zero. Thus, the simplified governing equation is:

$$C\frac{\partial u_w}{\partial t}-\frac{\partial}{\partial x}\left(k_{wx}\frac{\partial u_w}{\partial x}\right)-\frac{\partial}{\partial y}\left(k_{wy}\frac{\partial u_w}{\partial y}\right)$$

$$-\frac{\partial}{\partial z}\left(k_{wz}\frac{\partial u_w}{\partial z}\right)-\frac{\partial(\rho_w gk_{wz})}{\partial z}=0 \quad (28)$$

3.3 *Governing equation of single-phase flow without volume change*

On the base of the case discussed above, the effect of volume change can be neglected further. That is to say, C_t and C_m are assumed to be zero and the porosity n is considered to be constant. Thus the governing equation is

$$C_1\frac{\partial u_w}{\partial t}-\frac{\partial}{\partial x}\left(k_{wx}\frac{\partial u_w}{\partial x}\right)-\frac{\partial}{\partial y}\left(k_{wy}\frac{\partial u_w}{\partial y}\right)$$

$$-\frac{\partial}{\partial z}\left(k_{wz}\frac{\partial u_w}{\partial z}\right)-\frac{\partial(\rho_w gk_{wz})}{\partial z}=0 \quad (29)$$

If water phase is considered to be incompressible, then $\frac{\partial \rho_w}{\partial u_w}$ term vanishes. Thus the (29) can be rewritten as

$$\rho_w gnf'\frac{\partial u_w}{\partial t}-\frac{\partial}{\partial x}\left(k_{wx}\frac{\partial u_w}{\partial x}\right)-\frac{\partial}{\partial y}\left(k_{wy}\frac{\partial u_w}{\partial y}\right)$$

$$-\frac{\partial}{\partial z}\left(k_{wz}\frac{\partial u_w}{\partial z}\right)-\frac{\partial\left(\rho_w g k_{wz}\right)}{\partial z}=0 \qquad (30)$$

Equation (30) is the well-known Richard's equation in the form of water phase pressure.

4 SOME PROBLEMS IN NUMERICAL SOLUTION

Governing equations (8), (9) as well as their simplified equations can be solved by the finite element method. The detailed FEM solutions are not included in this paper. Only some problems encountered in numerical solution are discussed briefly.

4.1 *The coefficients of volume change: C_t and C_m*

The behavior of unsaturated soil has complex nonlinearity. General speaking, the coefficients or functions of physical property in the constitutive equation of unsaturated soil are highly non-linear. C_t in loading ($d(\sigma_{mean}-u_a)>0$) or unloading ($d(\sigma_{mean}-u_a)<0$) condition and C_m in drying ($d(u_a-u_w)>0$) or wetting ($d(u_a-u_w)<0$) condition both have different values. Furthermore, these coefficients are also related to the values of stress state variables. That is to say, C_t and C_m change with the stress path and the stress state. While during the research of the effect of air phase flow in unsaturated seepage problem, in order to simplify the problem, C_t and C_m can be assumed to be constant.

4.2 *Permeability functions k_w and k_a*

Permeability functions of water phase and air phase are one of the key factors in unsaturated seepage. At present, empirical equation or soil-water characteristic curve is usually used to obtain the permeability functions k_w and k_a indirectly. In the analysis of seepage without considering volume change, k_w and k_a are usually assumed to be a single-valued function of the degree of saturation or matric suction. While in the analysis of seepage with considering volume change, k_w and k_a are the function of the degree of saturation and porosity. How to choose the suitable permeability function in analysis is important.

4.3 *Complex boundary conditions.*

In transient unsaturated flow, most boundary conditions are flux changing boundary conditions. Furthermore, rain infiltration and moisture evaporation are both involving very complex flux boundary conditions. At present, some experience has been accumulated in rain infiltration analysis. But there are not enough knowledge on the boundary conditions for evaporation and circulation of drying and wetting under the effect of climate. Boundary conditions must be reasonable and representative in analysis.

4.4 *Solution of the equations*

Equations (16), (17), are coupling equations of water phase and air phase pressure. The coefficients in (16) and (17) : C, D, E, F are all the function of pressure. Thus, it is very difficult to solve this highly non-linear coupling governing equations by the finite element method. A reasonable discretization of the non-linear terms and a good uncoupling method will lead to good convergence and highly accurate solution. Otherwise, it will be unstable or dispersed in solution. The efficient solution method is worth further researching.

5 CONCLUDING REMARKS

Unsaturated flow is a very complex physical process. Two-phase water-air flow model with considering the effect of volume change will be more reasonable to describe it. However, the complex model includes more independent variables, so that the numerical solution is more difficult. Simplified single-phase flow includes less independent variable and the numerical solution is less difficult. Rich experience has been accumulated in single-phase flow numerical solution. But in some conditions, the results of single-phase flow model deviate from the practical condition. Therefore, it is necessary to further research on coupling flow model in unsaturated soil mechanics. In this paper, the governing equations of isothermal transient unsaturated flow are discussed. Further research results will be presented in succession.

6 REFERENCES

Bear, J. 1972. Dynamics of fluids in porous media. American Elsevier Publishing Company Inc.

Dakshanamurthy, V., Fredlund, D. G., & Rahardjo, H. 1984. Coupled three –dimensional consolidation theory of unsaturated porous media. 5th Int. Conf. on Expansive Soil, Adelaide, South Australia:99-103.

Fredlund, D. G. & Rahardjo, H. 1993. Soil mechanics for unsaturated soils. New York : Wiley

Freeze, R. A. & Cherry, J. A. 1979. Groundwater. Englewood Cliffs, NJ : Prentice-Hall

Sun, Y., Nishigaki, M. & Kohno, I. 1995. A study on stability analysis of shallow layer slope due to raining permeation. Proceedings of the 1st Int. Conf. on unsaturated soils:1135-1141. Rotterdam: Balkema.

3   Numerical modelling

3.2 Modelling of stresses and deformations

Unsaturated Soils for Asia, Rahardjo, Toll & Leong (eds) © 2000 Taylor & Francis, ISBN 90 5809 139 2

Using soil diffusion to design slab-on-ground foundations on expansive soils

B. M. El-Garhy & A. A. Youssef
Menoufia University, Shebin El-Koom, Egypt

W. K. Wray
Ohio University, Athens, Ohio, USA

ABSTRACT: The development of a three dimensional soil-structure interaction model based on the finite element method for calculating the structural design parameters (i.e. moments, shears, and deflections) in a slab-on-ground foundation (stiffened or constant thickness) resting on expansive soil is described. The model requires only a statement of the initial soil suction conditions in the supporting expansive soil mass and the changes in the boundary conditions to predict the response of the slab-on-ground foundation to these boundary condition changes. The model can estimate the distorted mound shape by considering soil suction distribution changes in the supporting soil mass and the associated soil volume changes (shrink/heave) with respect to time. The model can consider various combinations of conditions that cause soil moisture and soil volume changes such as climate, moisture barriers, trees, and adjacent ponded water. Examples of the model analysis compared favorably with the observed behavior of experimental slab-on-ground foundations.

Expansive or swelling soils are classified as clay soils which undergo volumetric changes (shrink/heave) in response to changes in soil water content. As the soil suction decreases (the soil water content increases) the ground surface typically moves upward and as the soil suction increases (the soil water content decreases) the ground surface typically recedes. These cyclic movements (shrink/swell) in the supporting expansive soils create design and performance problems, particularly for slab-on-ground foundation.

When a slab-on-ground foundation is constructed over expansive soils, the direct effect of evaporation and precipitation on the soil mass beneath the slab-on-ground foundation is prevented, but the climatic conditions still affect the surrounding soil and cause non-uniform suction or moisture distribution in the supporting soil mass beneath the slab-on-ground foundation. Consequently, differential soil movements (shrink/heave) occur between the edge and the interior of the slab-on-ground foundation. It has been found (U. S. Army Corps of Engineers, 1983) that differential rather than total movement of expansive soils are generally responsible for major structural damage to structures constructed over expansive soils.

The performance of slab-on-ground foundations resting on expansive soil is largely dependent on moisture distribution in the supporting expansive soil. Therefore, if an adequately accurate moisture distribution in the supporting soil can be predicted, the volume changes (shrink/heave) in the soil and the behavior of the slab-on-ground foundation resting on it can be reasonably estimated. Moisture changes in expansive soils may be caused by several sources such as climate, trees, and ponded water, in addition to evaporation and water inflow.

In the United States, the use of slab-on-ground foundations (stiffened or constant thickness) has been favored because of its relative ease of construction, economy, and satisfactory performance (Lytton & Woodburn, 1973). In Egypt, Aboulid & Reyad (1985) reported that the conventional design methods and construction of foundations which were commonly used for other types of soils had failed to overcome the problems of expansive soils.

1 LIMITATIONS IN THE EXISTING DESIGN METHODS

The available practical design procedures for slab-on-ground foundations (stiffened or constant thickness) resting on expansive soil suffer from some limitations on which the method described herein should improve. These limitations have been discussed by Wray (1978), Holland et al. (1980), and Mitchell (1988).

The major shortcoming is that they assume the interaction to be between the slab-on-ground foundation and an already distorted mound shape and they require an estimation of the initial distorted mound shape. The initial distorted mound shape is one of the important variables affecting analysis of slab-on-ground foundations resting on expansive soils in all of the design methods (Mitchell 1988). The initial distorted mound shape is defined by the edge moisture variation distance, e_m, the maximum differential heave, y_m, and/or the mound exponent, m. The reliable determination of edge moisture variation distance is considered to be a major problem. Also, estimation of the initial mound shape has the disadvantage of not considering the moisture movement through the supporting expansive soils and the resulting volume change when the soil mass is subjected to localized different edge effects that commonly cause moisture changes such as climate, trees, lawn irrigation, and ponded water.

2 MODELING THE PROBLEM

The analysis method developed in the study being described herein considers two problems that are modeled and solved independently: (1) three-dimensional moisture movement through the unsaturated expansive soils and the associated volume changes, and (2) analysis of structural interaction between the slab-on-ground foundation and the movements in the supporting expansive soils. Actually, the analysis of each problem influences the other, since moisture movements in the supporting expansive soils depend on the vertical stress beneath the slab-on-ground foundation. Structural analysis of slab-on-ground foundations depends on the movements in the supporting expansive soils or the results of the first problem. For computational simplicity, each problem is solved separately.

In this paper, a moisture diffusion and volume change model called SUCH was developed to determine the distribution of soil moisture suction through the supporting unsaturated expansive soil mass with respect to time and the resulting volume changes when the soil mass is subjected to different types of edge effects that commonly cause moisture changes, such as climate, trees, adjacent ponded water, and moisture barriers. To complete the framework, a soil structure-interaction model called SLAB97 was developed by coupling between the moisture diffusion and volume change model and the structural model to analyze the interaction between the slab-on-ground foundation and the vertical movements occurring in the supporting expansive soils.

2.1 Moisture Flow Model

It has been suggested (Mitchell 1979) that the movement of moisture through unsaturated expansive soils can be adequately represented by a transient diffusion equation in terms of soil suction expressed in pF units, e.g. Eq. (1). The transient suction diffusion equation can be rapidly solved in three dimensions using a desktop computer.

$$\frac{\partial^2 u}{\partial x^2} + \frac{\partial^2 u}{\partial y^2} + \frac{\partial^2 u}{\partial z^2} + \frac{f(x,y,z,t)}{p} = \frac{1}{\alpha}\frac{\partial u}{\partial t} \quad (1)$$

where

u = soil suction expressed as a pF (kPa = 0.1×10^{pF})
α = diffusion coefficient (cm^2/sec)
p = unsaturated permeability (cm/sec)
t = time (sec)
x, y, z = space coordinate
$f(x,y,z,t)$ = internal source of moisture

The diffusion coefficient, α, can be measured in the laboratory (Mitchell 1979) or calculated from empirical equations (McKeen & Johnson 1990; Bratton 1991; Lytton 1994). The different methods used to determine the value of diffusion coefficient, α, for expansive soils are presented and discussed by El-Garhy (1999).

2.2 Numerical Method of Analysis

The soil mass is descretized and represented as a grid of points as shown in Figure 1. The finite difference equation representing the change of soil suction with time at each nodal point takes the form:

$$u_{i,j,k}^{t+1} = R\left[\begin{array}{l} R_x^2\left(u_{i+1,j,k}^t + u_{i-1,j,k}^t\right) + R_y^2\left(u_{i,j+1,k}^t + u_{i,j-1,k}^t\right) + \\ \left(u_{i,j,k+1}^t + u_{i,j,k-1}^t\right) \end{array}\right]$$
$$+\left[1 - 2R\left(1 + R_x^2 + R_y^2\right)\right]u_{i,j,k}^t \quad (2)$$

where

$$R_x = \frac{\Delta z}{\Delta x}, \quad R_y = \frac{\Delta z}{\Delta y}, \text{ and } R = \frac{\alpha\Delta t}{\Delta z^2}$$

For convergence and stability, the ratio R must be:

$$R \le \frac{1}{2\left(1 + R_x^2 + R_y^2\right)} \text{ or } \Delta t \le \frac{\Delta z^2}{2\alpha\left(1 + R_x^2 + R_y^2\right)} \quad (3)$$

where

Δt = time interval
Δx = division length in x-direction
Δy = division length in y-direction
Δz = division length in z-direction
α = diffusion coefficient (cm^2/sec)

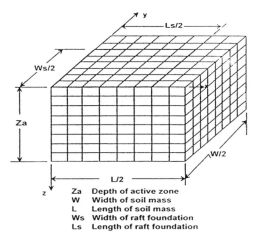

Za   Depth of active zone
W   Width of soil mass
L   Length of soil mass
Ws   Width of raft foundation
Ls   Length of raft foundation

Figure 1. Idealization of Raft Foundation and Supporting Soil Mass for Moisture Movement and Soil-Structural Interaction Analysis

$u_{i,j,k}^{t}$ = soil suction at time step, t, at grid point (i,j,k)

$u_{i,j,k}^{t+1}$ = soil suction at time step, $t+1$, at grid point (i,j,k)

In a mathematical sense, the type of transient suction diffusion equation Eq. (1) is called a parabolic partial differential equation. Two sets of information must be known to solve this equation: (1) the initial conditions, i.e. specify the initial value of suction at each node in the soil mass at the initial time, and (2) boundary conditions, i.e. specify the values of suction on the boundaries of the soil mass at each time step. If the values of suction, u^{t}, at time step t are known at each point in the soil mass, the values of suction, u^{t+1}, at time step $t+1$ can then be calculated by Eq. (2). This forward difference method was used successfully by Richards (1965) to solve the two-dimensional problem of moisture movement through expansive soils. The value of suction u^{t+1} at time step $t+1$ at each point depends on the values of suction u^{t} at time step t at that point and the six surrounding points at the previous time step.

2.3 Volume Change Model

Total and differential vertical volume change are important parameters in the selection and design of foundations resting on expansive soils. After determining the suction distribution throughout the unsaturated expansive soil mass with respect to time, the vertical volume change can be calculated by one of the volume change prediction methods which is based on the concept of soil suction. In the present work, Wray's model (1997) was used, Eq. (4), to calculate the vertical volume change at each nodal point in the soil mass.

$$\Delta H_{i,j,k} = \Delta z \, \gamma_{h_{i,j,k}} \left[\Delta pF_{i,j,k} - \Delta pP_{i,j,k} \right] \qquad (4)$$

where

$\Delta H_{i,j,k}$ = incremental change in elevation (shrink or heave) at grid point (i,j,k) over the vertical distance Δz

Δz = division length in z-direction over which shrink or heave occur

$\gamma_{h_{i,j,k}}$ = suction compression index at grid point (i,j,k)

$\Delta pF_{i,j,k}$ = change of soil suction expressed as a pF (kPa = 0.098×10^{pF}) at grid point (i,j,k)

$\Delta pP_{i,j,k}$ = change of soil overburden over vertical distance Δz at grid point (i,j,k)

The surface vertical volume change (shrink/heave) of each nodal point at the top of the soil mass and beneath the slab-on-ground foundations can be calculated by summation of the vertical movements of the nodal points on the vertical line passing through that point extending from the top to the bottom of the soil mass in the active zone. The maximum and minimum expected surface movements (shrink/heave) beneath the slab-on-ground foundation can be determined by comparing the magnitudes of the expected surface movements of nodal points at the top of the soil mass. Differential movement at the top of the soil mass can then be calculated by finding the difference between the expected surface movement at each nodal point and the nodal point with the minimum surface movement.

The different methods used to calculate the value of suction compression index, γ_{h}, are discussed by Wray (1998) and El-Garhy (1999). McKeen's chart and McKeen's equations are the most frequently recommended methods in the literature to estimate the value of γ_{h}, (i.e. McKeen 1980; Lytton 1994; and Wray 1997, 1998).

2.4 Boundary Conditions

In a mathematical sense, two types of boundary conditions can be considered. These are: (1) a prescribed value of suction on the boundary of the soil mass, and (2) a prescribed value of suction gradient perpendicular to the boundary of the soil mass. These well-known types are called Dirichlet and Neumann, respectively, by Champra & Canale (1998).

All of the engineering boundary conditions can be expressed as a boundary value or a boundary gradient. Figure 2 shows a set of typical common problems that cause a change in soil moisture in the sup-

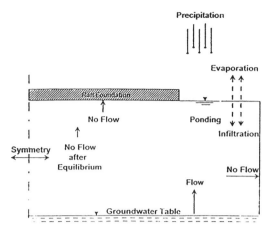

Figure 2. Some Types of Boundary Conditions

porting expansive soils and the various types of boundary conditions.

2.5 *Soil-Structure Interaction Model*

A FORTRAN computer program SUCH utilizes the finite difference technique to solve the transient suction diffusion equation in three dimensions to determine the distribution of soil suction through the unsaturated expansive soil mass with respect to time and the resulting volume changes beneath a slab-on-ground foundation when the soil mass is subjected to one or more of seven different edge effects that commonly cause moisture changes in the supporting expansive soils. These edge effects are: (1) climate only (i.e. no tree, no adjacent ponded water, and no barriers), (2) vertical moisture barrier around the perimeter of the foundation, (3) horizontal moisture barrier around the perimeter of the foundation, (4) combined vertical and horizontal moisture barriers around the perimeter of the foundation, (5) tree roots adjacent to the edge of the foundation, (6) trees and vertical moisture barrier at the edge of the foundation, and (7) ponded water adjacent to the edge of the foundation (no barriers).

To complete the framework, a soil-structure interaction model called SLAB97 was developed by coupling between the moisture diffusion and volume change model SUCH and the finite element method to analyze a slab-on-ground foundation (stiffened or constant thickness) resting on expansive soil. The program SLAB97 was based on the classical theory of thin plates which considers the plane before bending to remain plane after bending. The footing is assumed homogeneous, isotropic, and linearly elastic and supported on an elastic half space. The general method for formulating the soil stiffness matrix follows the procedures developed by Cheung & Zienkiewicz (1965).

SLAB97 is able to calculate the values of moments, shears, and deflections at each node in the finite element mesh, as well as the maximum and minimum values of moments, shears, and deflections, which are needed for design. The model being described is able to analyze slab-on-ground foundations when the supporting expansive soil mass is subjected to one of the seven different edge effects that commonly cause moisture changes in the supporting expansive soils and which were considered in the moisture diffusion and volume change model SUCH.

3 COMPARISONS WITH FIELD MEASUREMENTS

SLAB97 was tested by comparing the predicted and observed displacements of actual slab-on-ground foundations resting on expansive soils. Holland et al. (1980) reported the results of observations of a lightly loaded stiffened slab-on-ground foundation constructed over a highly expansive soil at the end of a dry season (April 1976) at Sunshine, Melbourne, Australia.

The slab-on-ground was 7.4 m (34.3 ft) square, and stiffened by 35 cm wide × 30 cm (13.8 x 11.8 in.) deep stiffening beams with reinforcement of 0.17% top and bottom, the spacing between the stiffening beams was 3.7 m (12.1 ft) in each direction. The slab was 10 cm (3.9 in.) thick reinforced with Australian F72 reinforcement. The slab-on-ground was loaded by 7.1 kN/m (486.5 lb$_f$/ft) on the perimeter and 4.7 kN/m (322.1 lb$_f$/ft) on the center, giving a uniformly distributed load of 8 kPa (167 lb$_f$/ft^2) over the entire slab area.

The soil profile at the site consisted of very stiff gray-brown and gray clays from the surface to at least 5 m (16.4 ft). Figure 3 shows the measured wet and dry suction profiles at the site. The active zone depth and the equilibrium soil suction are taken to be 2.0 m (6.6 ft) and 4.28 pF (1905 kPa), respectively, from Figure 3. Typical contour plots of the slab-on-ground movements after two years from the site construction is presented in Figure 4.

Unfortunately, no measurements of the diffusion coefficient and suction compression index were made. However, a differential soil movement of 6.3 cm (2.48 in.) was measured, as a result of a difference in wet and dry suction profiles shown in Figure 3. These measurements are used by the model SUCH to determine the in-situ diffusion coefficient and suction compression index of the soil. Therefore, numerous computer runs using different values of diffusion coefficient were made to predict the exact suction distribution profile and differential soil movements as measured in the field. Through these

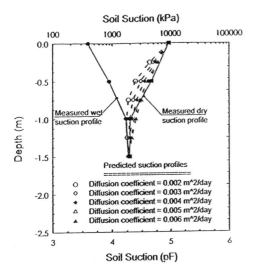

Figure 3. Comparison of Predicted and Measured Soil Suction Profiles at Sunshine Site, Melbourne, Australia

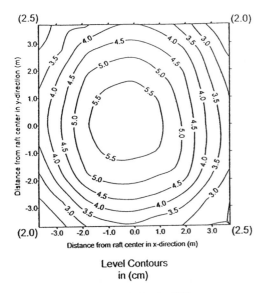

Level Contours
in (cm)

Observed, March, 1978

Figure 4. Observed Distortion Pattern After Two Years (After Holland et al. 1980)

trial computer runs, the measured wet suction profile is assumed as the initial suction profile and the predicted dry suction profiles are compared to the measured dry suction profile as shown in Figure 3. The values of diffusion coefficient and suction compression index of 6.9×10^{-4} cm^2 / sec and 0.07, respectively, seem to be the best match for the expansive soil at this site.

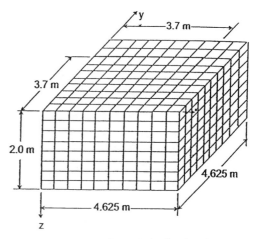

Figure 5. Idealization of Experimental Raft and Supporting Soil Mass for Moisture Flow and Soil-Structural Interaction Analysis

The idealization of the slab-on-ground foundation and the supporting soil mass is shown in Figure 5. Because loads and soil suction changes are assumed to be symmetric, only the upper right quarter of the soil-slab system is considered in the analysis. The dry suction profile shown in Figure 3 is assumed to be the initial soil suction profile within the supporting soil mass. The surface boundary suction is considered to change as a sinusoidal form with frequency equal to 1 cycle/year (Mitchell 1980).

A contour plot of the predicted slab-on-ground deformation pattern after two years is presented in Figure 6. This pattern compared reasonably well with the observed distortion pattern of the slab-on-ground shown in Figure 4. The predicted slab-on-ground surface displacements along the x-axis compared favorably with the observed values as shown in Figure 7.

4 CONCLUSIONS

A three dimensional time-dependent soil-structure interaction model called SLAB97 based on the finite element method has been developed for calculating the structural design parameters in a slab-on-ground foundation resting on expansive soils. Structural design parameter results from SLAB97 from the example were not reported in this paper because of paper length considerations. SLAB97 overcomes the major shortcoming present in all current existing design procedures, namely the estimation of the distorted mound shape and directly employing the distorted shape of the supporting soil to perform a structural analysis of the slab foundation. It is capa-

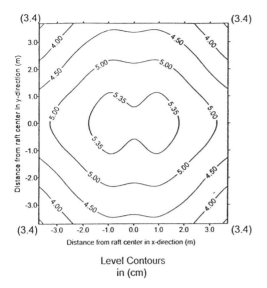

Level Contours
in (cm)

Predicted, March, 1978

Figure 6. Predicted Distortion Pattern After Two Years

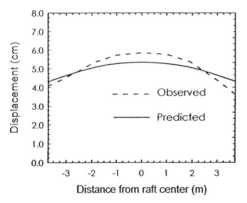

Figure 7. Comparison of Predicted and Observed Raft Displacements After Two Years

ble of estimating the distorted mound shape by considering the changes of soil suction distribution throughout the supporting expansive soil mass and the associated volume change with respect to time under a set of different edge conditions that commonly cause moisture changes (e.g. climate, trees, and adjacent ponded water). However, it does require the likely final boundary conditions to be estimated which may result in the most severe soil suction distribution through the soil mass under the slab-on-ground foundation during the lifetime of the structure being predicted. The analysis method reported in this study appears to be an improvement on other design methods presently available for the design of slab-on-ground foundations resting on expansive soil. The results of the analysis agreed well with the observed behavior of an experimental slab-on-ground foundation.

REFERENCES

Aboulid, A. F. & Reyad, M. M. 1985. Study of the Behavior of Isolated Footings on Expansive Soils. *Proc. 11th Int. Conf. on Soil Mechanics and Foundation Engineering* 2: 965-967.

Bratton, W. L. 1991. Parameters for Predicting Shrink/Heave Beneath Slab-on-Ground Foundations Over Expansive Clays. *Ph.D. Dissertation, Department of Civil Engineering, Texas Tech University.* Lubbock, Texas.

Champra, S. C. & Canale, R. P. 1998. Numerical Methods for Engineers with Programming and Software Applications. 3rd Edition. WCB/ McGraw-Hill, Inc.

Cheung, Y. K. & Zienkiewicz, O. C. 1965. Plates and Tanks on Elastic Foundations-An Application of Finite Element Method 1: 451-461.

El-Garhy, B. M. 1999. Soil Suction and Analysis of Slab-on-ground Foundation Resting on Expansive Soils. *Ph.D. Dissertation, Civil Engineering Department, Menoufia University.* Shebin El-Koon, Menoufia, Egypt.

Fredlund, D. G. 1997. An Introduction to Unsaturated Soil Mechanics. *Unsaturated Soil Engineering Practice.* Geotechnical Special Publication No. 68, ASCE: 1-37.

Holland, J. E., Pitt, W. G., Lawrance, C. E. & Cimino, D. J. 1980. The Behavior and Design of Housing Slabs on Expansive Clays. *Proc. 4th Intl. Conf. on Expansive Soils, Colorado* 1: 448-468.

Lytton, R. L. & Woodburn, J. A. 1973. Design and Performance of Mat Foundations on Expansive Clay. *Proc. 3rd Int. Conf. on Expansive Soils, Haifa* 1: 301-307.

Lytton, R. L. 1994. Prediction of Movement in Expansive Clay. *Vertical and Horizontal Deformations of Foundations and Embankments*, Geotechnical Special Publication No. 40, A. T. Yeung and G. Y. Felio, Editors, ASCE, 2: 1827-1845.

McKeen, R. G. 1980. Field Studies of Airport Pavements on Expansive Clay. *Proc. 4th Int. Conf. on Expansive Soils* 1: 242-261. Denver, Colorado.

McKeen, R. G. & Johnson, L. D. 1990. Climate-Controlled Soil Design Parameters for Mat Foundations. *Journal of Geotechnical Engineering, ASCE* 116, 7: 1073-1094.

Mitchell, P. W. 1979. The Structural Analysis of Footings on Expansive Soil. *Research Report No. 1, Kenneth W. G. Smith and Associates.* Adelaide South Australia.

Mitchell, P. W. 1980. The Structural Analysis of Footings on Expansive Soil. *Proc. 4th Int. Conf. on Expansive Soils* 1: 438-447. Denver, Colorado.

Mitchell, P. W. 1988. The Design of Footings on Expansive Soils. *Engineering Problems of Regional Soils, Proc. Int. Conf.*: 127-135. Beijing, China.

Richards, B. G. 1965. An Analysis of Subgrade Conditions at the Horsham Experimental Road Site Using Two-Dimensional Diffusion Equation on a High-Speed Digital Computer. *Moisture Equilibria and Moisture Changes in Soils Beneath Covered Area*: 243-258. Butterworths, Australia.

U.S. Army Corps of Engineers. 1983. Foundations in Expansive Soils, *Technical Manual No. 5-818-7.*

Wray, W. K. 1978. Development of a Design Procedure for Residential and Light Commercial Slabs-on-Ground Constructed Over Expansive Soils. *Ph.D. Dissertation.* Texas A & M University, College Station, Texas.

Wray, W. K. 1997. Using Soil Suction to Estimate Differential Soil Shrink or Heave. *Unsaturated Soil Engineering Practice, Geotechnical Special Publication No. 68, ASCE*: 66-87.

Wray, W.K 1998. Mass Transfer in Unsaturated Soils: A Review of Theory and Practices. *Proc., 2ⁿᵈ Int. Conf. on Unsaturated Soils* 2: 99-155. Beijing.

NOTATION

The following symbols are used in this paper:

$f(x, y, z, t)$ = internal source of moisture

p = unsaturated permeability (cm / sec)

$R = \dfrac{\alpha \Delta t}{\Delta z^2}$

$R_x = \dfrac{\Delta z}{\Delta x}$

$R_y = \dfrac{\Delta z}{\Delta y}$

t = time (sec)

u = soil suction expressed as a pF (kPa = 0.098×10^{pF})

$u_{i,j,k}^{t}$ = soil suction at time step, t, at grid point (i, j, k)

$u_{i,j,k}^{t+1}$ = soil suction at time step, $t + 1$, at grid point (i, j, k)

x, y, z = space coordinate

$\gamma_{h_{i,j,k}}$ = suction compression index at grid point (i, j, k)

α = diffusion coefficient (cm^2 / sec)

$\Delta H_{i,j,k}$ = incremental change in elevation (shrink or heave) at grid point (i, j, k) over the vertical distance

Δz = change of soil suction expressed as a pF (kPa =) at grid point (i, j, k) = change of soil overburden over vertical distance at grid point (i, j, k)

Δt = time interval

Δx = division length in x-direction

Δy = division length in y-direction

Δz = division length in z-direction

Unsaturated Soils for Asia, Rahardjo, Toll & Leong (eds) © 2000 Taylor & Francis, ISBN 90 5809 139 2

Volume change predictions in expansive soils using a two-dimensional finite element method

V.Q. Hung & D.G. Fredlund

Department of Civil Engineering, University of Saskatchewan, Saskatoon, Sask., Canada

ABSTRACT: The solution of a two-dimensional volume change problem associated with an unsaturated, expansive soil is proposed in a manner consistent with the theory of unsaturated soil behavior. The elastic modulus functions must be computed from conventional laboratory test results. Procedures for the calculation of elastic modulus functions associated with an unsaturated soil are presented. Predictions of two-dimensional heave are performed in an uncoupled manner through the use of a general-purpose, partial differential equation solver, called PDEase2D. An example problem involving volume change due to changes in matric suction is used to illustrate the proposed method of volume change prediction and the use of the PDEase2D computer program.

1 INTRODUCTION

Volume change in an unsaturated soil is caused by changes in either or both the stress state variables, net normal stress and matric suction. The volume change and water flow processes are dependent processes, requiring a coupled analysis. The soil properties associated with an unsaturated soil are dependent on stress state variables, and the governing partial differential equations become highly non-linear. Therefore, a coupled solution is difficult to obtain. An approximate solution can be obtained by performing the analysis in an uncoupled manner, where the continuity and equilibrium equations are solved independently.

This paper presents an uncoupled solution of volume change problems associated with an expansive soil under plane strain loading conditions. The governing partial differential equations for seepage and stress-deformation are based on the general theory of unsaturated soil behavior. The air phase is assumed to be continuous and at atmospheric pressure. The net normal stress is therefore equal to net total stress and the matric suction is equal to the absolute value of the pore-water pressure. The soil is assumed to be isotropic, non-linear and elastic. It is also assumed that the elastic modulus with respect to net normal stress, E, obtained from the net normal stress plane, and the elastic modulus with respect to matric suction, H, obtained from matric suction plane can be applied to the entire constitutive surface.

Theory associated with saturated-unsaturated seepage is presented in Thieu et al. (2000), and is not repeated in this paper. Solutions of non-linear partial differential equations for both saturated-unsaturated seepage and stress-deformation are obtained using PDEase2D computer program.

2 VOLUME CHANGE THEORY

Two independent stress state variables are needed to describe the volume change behavior of an unsaturated soil (Fredlund and Morgenstern, 1977). The two stress state variables are net normal stress, $(\sigma - u_a)$, and matric suction, $(u_a - u_w)$, where σ is total normal stress, u_a is pore-air pressure and u_w is pore-water pressure.

Assuming the soil behaves in an incrementally isotropic, linear elastic material, the soil structure constitutive relations can be written as follows:

$$d\varepsilon_{ij} = \frac{1+\mu}{E} d(\sigma_{ij} - u_a) - \frac{\mu}{E} d(\sigma_{kk} - 3u_a)\delta_{ij} + \frac{d(u_a - u_w)}{H}\delta_{ij} \quad (1)$$

where ε_{ij} = components of the strain tensor for the soil structure, δ_{ij} = the Kronecker delta, μ = Poisson's ratio, E = modulus of elasticity for the soil structure with respect to a change in net normal stress, and H = modulus of elasticity for the soil structure with respect to a change in matric suction.

The total volumetric deformation of an unsaturated soil element, $d\varepsilon_v$, can be written as the sum of the normal strains:

$$d\varepsilon_v = \frac{dV_v}{V_0} = d\varepsilon_x + d\varepsilon_y + d\varepsilon_z \quad (2)$$

where: $d\varepsilon_x$, $d\varepsilon_y$, $d\varepsilon_z$ = normal strain components in x, y, and z-direction, respectively.

2.1 Calculation of elastic moduli from volume change indices.

Fredlund and Rahardjo (1993) presented the relationship between the volumetric strain change, $d\varepsilon_v$, and changes in stress state variables in an elasticity form for various loading conditions. These constitutive equations are based on Eqs. (1) and (2), and can be written in a compressibility form as follow:

$$d\varepsilon_v = m_1^s d(\sigma - u_a) + m_2^s d(u_a - u_w) \quad (3)$$

where: m_1^s = coefficient of volume change with respect to a change in net normal stress, and m_2^s = coefficient of volume change with respect to a change in matric suction.

The relationships between the coefficients of volume change and the elastic moduli are presented in Table 1 for general, three-dimensional loading, and plane strain loading conditions.

The constitutive relationship for the soil structure of an unsaturated soil can be presented graphically as a three-dimensional surface (Fig. 1). Coefficients of volume change corresponding to the unloading surface can be subscripted with an "s" to represent the word swelling (i.e., m_{1s}^s and m_{2s}^s).

The constitutive surface can also be obtained when void ratio, e, is plotted with respect to the logarithms of the stress state variables (Fig. 2). The logarithmic plots are essentially linear over a relatively large stress range on the extreme planes (i.e., the $\{log(\sigma - u_a) \approx 0\}$ plane and $\{log(u_a - u_w) \approx 0\}$ plane) (Fredlund and Rahardjo, 1993). The volume change indices for the unloading surface are subscripted with an "s" as C_{ts} and C_{ms}.

Using a mathematical conversion between a semi-logarithmic scale and arithmetic scale, the coefficients of volume change can be written in term of

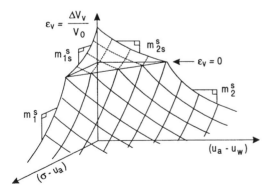

Figure 1. Three-dimensional constitutive surface for soil structure of an unsaturated soil (Fredlund and Rahardjo, 1993)

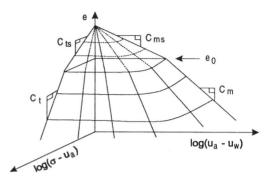

Figure 2. Semi-logarithmic plot of void ratio versus stress state variables (Fredlund and Rahardjo, 1993)

the volume change indices:

$$m_1^s = \frac{0.434 C_t}{1 + e_0} \frac{1}{(\sigma - u_a)_{ave}} \quad (4)$$

$$m_2^s = \frac{0.434 C_t}{1 + e_0} \frac{1}{(u_a - u_w)_{ave}} \quad (5)$$

where: $(\sigma - u_a)_{ave}$ = average of the initial and final net normal stress for an increment, $(u_a - u_w)_{ave}$ = average of the initial and final matric suction for an increment.

The elastic moduli, E and H, can be calculated from the volume change indices, initial void ratio and Poisson's ratio by substituting Eqs. (4) and (5) into conversion relationships shown in Table 1.

$$E = n_t (\sigma - u_a)_{ave} \quad (6)$$

$$H = n_m (u_a - u_w)_{ave} \quad (7)$$

where: n_t = coefficient that relates net normal stress with elastic modulus E, and n_m = coefficient that relate matric suction with elastic modulus H.

Table 2 presents the n_t and n_m coefficients for general, three-dimensional loading, and plane strain loading conditions. Figure 3 illustrates the relationship between elastic modulus, E, and net normal stress for various value of swelling index, when the initial void ratio is equal to 1.0 and Poisson's ratio equal to 0.35. The relationship between elastic modulus, H, and matric suction, is plotted for various values of swelling index in Fig. 4.

2.2 Stress-deformation formulation

Equations of equilibrium for the soil structure of an unsaturated soil are:

$$\sigma_{ij,j} + b_i = 0 \quad (8)$$

232

Table 1 Coefficient of volume change for some loading conditions (From Fredlund and Rahardjo, 1993)

Loading	m_1^s	First stress state variable	m_2^s	Second stress state variable
Three-dimensional (General)	$3\dfrac{(1-2\mu)}{E}$	$d(\sigma_{mean}\text{-}u_a)$	$\dfrac{3}{H}$	$d(u_a\text{-}u_w)$
Plane strain (Two-dimensional)	$\dfrac{2(1+\mu)(1-2\mu)}{E}$	$d(\sigma_{ave}\text{-}u_a)$	$\dfrac{2(1+\mu)}{H}$	$d(u_a\text{-}u_w)$

Table 2 Coefficients n_t and n_m for some loading conditions

Loading	n_t	First stress state variable	n_m	Second stress state variable
Three-dimensional (General)	$\dfrac{6.908(1-2\mu)(1+e_0)}{C_t}$	$(\sigma_{mean}\text{-}u_a)_{ave}$	$\dfrac{6.908(1+e_0)}{C_m}$	$(u_a\text{-}u_w)_{ave}$
Plane strain (Two-dimensional)	$\dfrac{4.605(1+\mu)(1-2\mu)(1+e_0)}{C_t}$	$(\sigma_{ave}\text{-}u_a)_{ave}$	$\dfrac{4.605(1+\mu)(1+e_0)}{C_m}$	$(u_a\text{-}u_w)_{ave}$

Note: $\sigma_{mean} = (\sigma_x+\sigma_y+\sigma_z)/3$; $\sigma_{ave} = (\sigma_x+\sigma_y)/2$

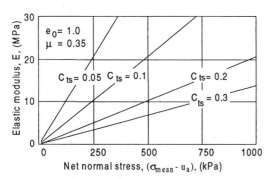

Figure 3. Relationship between elastic modulus, E, and net normal stress for various values of swelling index

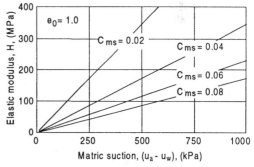

Figure 4. Relationship between elastic modulus, H, and matric suction for various values of swelling index

where σ_{ij} = components of the net total stress tensor, and b_i = components of the body force vector.

The partial differential equations for the soil structure can be derived from constitutive equations (Eq. (1)) and the force equilibrium equations (Eq. (8)). The partial differential equations in term of displacements in x and y-direction (i.e., u and v) for plane strain loading ($d\varepsilon_z = 0$) are as follows:

Equations 9 and 10 are non-linear since the elastic moduli are functions of the stress state variables. An incremental procedure can be used to solve these equations (Desai and Christian, 1977). The incremental procedure for solution is illustrated in Fig. 5. The total load (or suction change) is divided into increments and one load (or suction) incremental step is applied at a time. During each incremental step,

$$\frac{\partial}{\partial x}\left\{c\left[(1-\mu)\frac{\partial u}{\partial x}+\mu\frac{\partial v}{\partial y}-\frac{(1+\mu)}{H}(u_a-u_w)\right]\right\}+\frac{\partial}{\partial y}\left\{G\left(\frac{\partial v}{\partial x}+\frac{\partial u}{\partial y}\right)\right\}=0 \tag{9}$$

$$\frac{\partial}{\partial x}\left\{G\left(\frac{\partial v}{\partial x}+\frac{\partial u}{\partial y}\right)\right\}+\frac{\partial}{\partial y}\left\{c\left[\mu\frac{\partial u}{\partial x}+(1-\mu)\frac{\partial v}{\partial y}-\frac{(1+\mu)}{H}(u_a-u_w)\right]\right\}+\rho g=0 \tag{10}$$

where $c=\dfrac{E}{(1-2\mu)(1+\mu)}$; $G=\dfrac{E}{2(1+\mu)}$; ρ = density of the soil, g = acceleration due to gravity

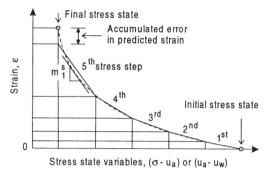

Figure 5. Incremental procedure for stress-deformation analysis

the elastic moduli are assumed to be unchanged. New moduli are selected at the beginning of the new incremental step based on the stress level, and the displacements from each incremental step are accumulated to give to total displacements. The computer program, PDEase2D, was used to perform the stress-deformation analysis.

3 EXAMPLE PROBLEM AND COMPUTER RESULTS

A typical volume change example was analyzed to demonstrate the application of the two-dimensional volume change prediction method (Hung, 2000). The example considers a 5 m thick layer of expansive clay soil (Fig. 6). The coefficient of permeability of the soil was described using Gardner's equation (1958) with a saturated coefficient of permeability equal to 1.16×10^{-9} m/s, a parameter a equal to 0.001 and a parameter n equal to 2. The initial void ratio of the soil is equal to 1.0, and the swelling index with respect to matric suction, C_{ms} is equal to 0.07.

The initial matric suction was taken to be constant throughout the depth and equivalent to 700 kPa. It was then assumed that a leaking water line produced zero pore-water pressure under the cover. The soil is then watered at the surface with an infiltration rate of 1.16×10^{-10} m/s, The water table is 15 m below the ground surface.

Deformations in the soil profile due to changes in matric suction from initial to final state (i.e., steady state) are predicted.

The elastic modulus function H with respect to matric suction for the soil with a given initial void ratio, swelling index, and an assumed Poisson's ratio equal to 0.3 can be calculated from Eq. 7 and Table 2 and written as follows:

$$H = 171.05(u_a - u_w)_{ave} \qquad (11)$$

A seepage analysis was performed to predict final matric suction conditions. Boundary conditions and the finite element mesh for the seepage analysis is shown in Fig. 7. Zero flow is specified at the left and the right boundaries. A zero total head is specified at the flexible cover and –15 m total head is specified at the lower boundary. A boundary flow value of 1.16×10^{-10} m/s is specified along the uncovered surface. The maximum error of 0.1% is specified for the problem. The finite element mesh satisfying the specified error has 89 six-node-triangular elements and 206 nodes.

The matric suction distribution in the soil at equilibrium under specified boundary condition is shown in Fig. 8. The suction change below the cover is more than that below the exposed area, and therefore more heaves would be expected at the cover area.

Figure 9 shows the finite element mesh and the boundary conditions specified for the stress-deformation analysis. At the left and right sides of the domain, the soil is free to move in the vertical direction and is fixed in horizontal direction. The lower boundary is fixed in both directions.

The problem is analyzed using twenty-five equal suction increments. The size of stress step varies from 23 kPa to 27 kPa. Maximum error specified for stress-deformation analysis is 1%. The finite element mesh varies with step numbers, from 25 elements and 78 nodes for the first step, to 55 elements and 134 nodes for the last step.

The cumulative heave at surface is shown in Fig. 10 for various stress steps. The total heave at ground

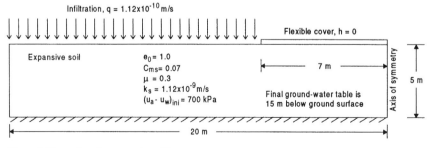

Figure 6. Illustration of an example problem

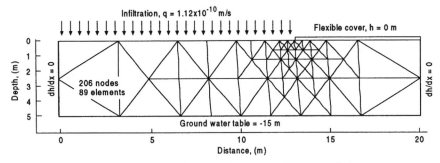

Figure 7. Finite element mesh and boundary condition for unsaturated seepage analysis

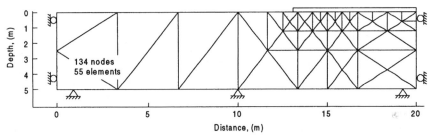

Figure 8. Final matric suction profile (steady-state conditions)

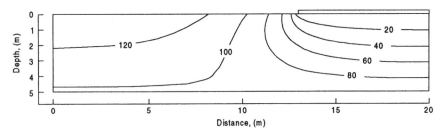

Figure 9. Finite element mesh and boundary conditions for stress-deformation analysis

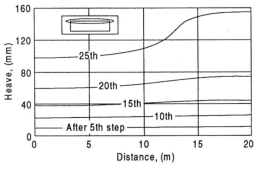

Figure 10. Cumulative heave at surface at various stress steps

stress steps, because the value of elastic modulus get lower as matric suction is reduced. Contours of total heaves are presented in Fig. 12 and the distribution of the deformation vectors is shown in Fig. 13.

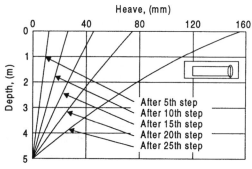

Figure 11. Cumulative heave below the cover (at x = 20 m)

surface under the cover is 155 mm, while it is only about 100 mm at the exposed area. The differential heave is 55 mm. Figure 11 presents the cumulative heave below the cover. Figures 10 and 11 show that most of the displacements take place in the later

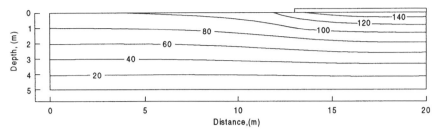

Figure 12. Distribution of total heave contours (mm)

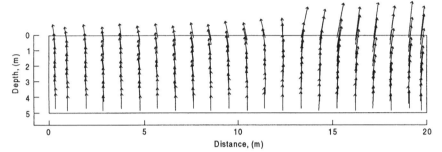

Figure 13. Distribution of heave vectors

4 CONCLUSION

The volume change index with respect to net normal stress in the net normal stress plane and the initial void ratio are required for the calculation of the elastic modulus function, E. The volume change index with respect to matric suction in the matric suction plane and the initial void ratio are required for the calculation of elastic modulus function, H. A value for Poisson's ratio must be assumed.

Two-dimensional volume change of an expansive soil has been analyzed in an uncoupled manner using partial differential equation solvers. The matric suction conditions in soils can be predicted by performing a saturated/unsaturated seepage analysis. Stress-deformation analysis then can be performed to predict the deformations.

General-purpose partial differential equation solvers, such as PDEase2D, appear to be a powerful tool for solving seepage and volume change problems associated with an unsaturated soil.

REFERENCES

Desai, C.S., and Christian, J.T. 1977. Numerical Methods in Geotechnical Engineering. McGraw-Hill, New York, 783 p.

Fredlund, D.G., and Rahardjo, H. 1993. Soil Mechanics for Unsaturated Soil. John Wiley & Sons, New York, 560 p.

Fredlund, D.G., and Morgenstern, N.R. 1977. Stress State Variables for Unsaturated Soils. Journal of the Geotechnical Engineering Division, Proceedings, American Society of Civil Engineering (GT5), **103**: 447-466.

Fredlund, D.G., and Morgenstern, N.R. 1976. Constitutive Relations for Volume Change in Unsaturated Soils. Canadian Geotechnical Journal, **13**(3): 261-276.

Fredlund, D.G. 1987. The Prediction and Performance of Structures on Expansive Soils. Proceedings, International Symposium on Prediction and Performance in Geotechnical Engineering, Calgary, Canada, pp. 51-60.

Ho, D.Y.F., Fredlund, D.G., and Rahardjo, H. 1992. Volume Change Indices During Loading and Unloading of an Unsaturated Soil. Canadian Geotechnical Journal, **29**(2): 195-207.

Hung, V.Q. 2000. Finite Element Method for the Prediction of Volume Change in Expansive Soils. M.Sc. Thesis, University of Saskatchewan, Saskatoon, SK, Canada.

Hwang, C.T., Morgenstern, N.R., and Murray, D.W. 1971. On Solutions of Plane Strain Consolidation Problems by Finite Element Methods. Canadian Geotechnical Journal, Vol. 8, pp. 109-118.

Lambe, T.W., and Whitman, R.V. 1969. Soil Mechanics. John Wiley & Sons, New York, 553 p.

Lewis, R.W. 1991. Coupling versus Uncoupling in Soil Consolidation. Inter. J. Numer. Anal. Meth. Geomech., Vol.15, pp. 533-548.

PDEase2D 3.0 Reference Manual, 3rd Edition. 1996. Macsyma Inc. Arlington, MA, 02174 USA.

Pereira, J.H.F., and Fredlund, D.G. 1977. Constitutive Modelling of a Metastable-structured Compacted Soil. Proceddings, International Symposium on Recent Developments in Soil and Pavement Mechanics, Rio de Janeiro, Brazil, pp. 317-326.

Thieu, N.T.M., Fredlund, D.G, and Hung, V.Q. 2000. General Partial Differential Equation Solvers for Saturated-Unsaturated Seepage. Proceeding of Asian Conference on Unsaturated Soils, Singapore.

Unsaturated Soils for Asia, Rahardjo, Toll & Leong (eds) © 2000 Taylor & Francis, ISBN 90 5809 139 2

A numerical model of ground moisture variations with active vegetation

K.G. Mills
School of Engineering, University of South Australia, Adelaide, S.A., Australia

ABSTRACT: Soil profiles of reactive clays will exhibit shrink/swell cycles when subject to seasonal variation in moisture. For foundations on such soils, it is usual to assume some normal range of wetness in design, but this can be influenced by site conditions such as the presence of trees or infiltration trenches. This paper presents a numerical model which may give some insights into the effect of such features. It mimics the moisture movements in reactive clays and includes the effects of infiltration and moisture extraction by plant roots and surface evaporation. Matric and solute suctions have been tracked separately since they influence clay swell, moisture movement and plant usage differently. The model was implemented in 2D FLAC, with FISH subroutines.

1 INTRODUCTION

Most commonly the designer's philosophy when confronted with a moisture reactive site, is to design a foundation system to withstand "normal" seasonal variations in moisture conditions, and to insulate the foundation soil from any extreme changes. However, this restricts the choices the building owner has for such things as the provision of surface and subsurface drainage, the location of gardens and trees, and grading and paving around the periphery of the building. Vegetation, moisture barriers and soakage pits alter conditions from the "design" values.

To aid the development of rational methods for design that include such factors, the numerical study in this paper attempts to model the infiltration and evaporation processes in the vicinity of a house foundation, including in particular, the uptake of moisture by active tree roots. The simulation reproduces the seasonal variations in conditions, and may be "run" over several years. The work is one part of an on-going program of investigation into reactive soils at the University of South Australia.

2 REVIEW

In present Australian design approaches for light structures on reactive soils, (Australian Standard AS 2870 - 1990, *Residential Slabs and Footings*), the range of soil moisture states to be used in design, is given a prescribed value prudently based on regional observation. Site management practices are set out to minimise moisture variations and limit site movements. Generally, this means restricting tree planting, providing perimeter aprons, and promoting good site drainage. There are growing pressures to provide design procedures that can account for the influence of trees, and possibly on-site water retention through soakage trenches. (Argue 1995) Local practice in South Australia is to provide guidance on tree species selection, and to increase the range of soil moisture suction used in calculating site movement, but to exclude any infiltration devices from the site.

In general terms, roots exacerbate the seasonal drying under the foundations, a region normally isolated from evaporative loss (Biddle 1983, Richards et al 1983). The return to "normal" conditions after tree removal may take decades (Cheney 1988), and to establish the full impact of a tree on the foundation moisture regime may likewise take years. Experimental studies could be of long duration and expensive or impracticable. Therefore a numerical study was commenced to give insights into the processes and guide a later stage of field investigation.

Previous investigators modelling the response of a foundation to seasonal conditions, have used a formulation with suction as a boundary condition. (Richards 1967, Li.1996) and have treated soil suction as a single variable. If vegetation is to be included, then it is to be established just what the surface suctions are, and they cannot be validly taken as a boundary condition. Furthermore moisture movement is probably related more directly to matric suction than total suction, (Hillel 1980a and Fredlund & Rahardjo 1993 Ch 5). Contrarily, vegetative uptake is by osmosis, and total suction is the variable that will regulate water extraction, (Hillel 1980b). An extra consideration is that when a tree's roots extend beneath a covered area, the roots will extract moisture from the soil by osmosis, but selectively leave soil salts behind. If the soil is recharged by soil water, (not rainfall since the region is covered), then there may be a prolonged accumulation of salt beneath the foundation. This will lead to soil shrinkage, but it could also mean that the current reactivity testing methods are not necessarily appropriate as these tests generally preserve the salt content, (Fredlund & Rahardjo 1993, Richards 1984)

Evidently for a simulation to be realistic, the modelling of the site processes needs well-founded mechanistic representations. The method described herein, which draws on simplified theory, rather than experiment, captures many of the major interactions identified in the field problem.

3 MODEL FORMULATION

3.1 Implementation of the model

A first step in formulating the model was to identify all the processes that co-exist and interact in the foundation system. Numerical representations were then set up for each of these processes, linkages established and initial conditions defined. A time stepping procedure was devised to advance the simulation in time. The model has been implemented using the geotechnical code FLAC as a base vehicle (ITASCA 1993). Only a two dimensional analysis was attempted.

3.2 Implementation of the model

The processes included in the model are:-

- Stress deformation response in the soil,
- Moisture redistribution in the foundation with time,
- Soil reactivity and stiffness adjustment as soil water conditions change
- Vegetative uptake from the soil water
- Infiltration and evaporation at the exposed surface.
- Salt movement within the soil

The first two equate to the consolidation problem of conventional soil mechanics. Richards(1980) demonstrates that such problems may be treated in-

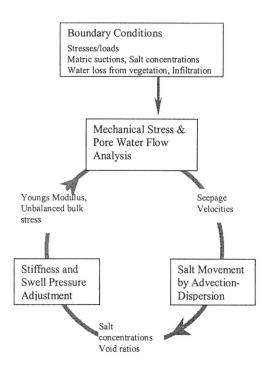

Figure 1 Linkages in the Composite Model

crementally, allowing first seepage flow for a time increment, then stress readjustment. He reproduced the results of a coupled model (Mitchell et al 1973). De-coupling also allows for the introduction of other interactions.

The soil response used is the quasi-theoretic model of Mills (1998 and 2000). It is responsive to solutes in the soil water and clay mineral type. Vegetative uptake was estimated using a form of the Penman equation, for evapotranspiration. Infiltration is related to rainfall data. Salt redistribution follows the methods of Bear & Verruijt (1987).

Figure 1 shows the inter-linked processes that comprise the simulation. A cycle is repeated for each incremental advance in time. The boxes are sub-models in the complete process, the information carried between them is indicated on the circular path line joining the sub-models.

The coding necessary was implemented on a PC. The hydro-mechanical (consolidation and seepage) response, could be included directly using the provisions within FLAC itself. Other processes required for the complete model needed special code modules to be developed.

The adjustment procedure for each time step is perhaps akin to using the chain rule for differentiation, $dw = \frac{\partial w}{\partial x} dx + \frac{\partial w}{\partial y} dy + \dots$; cross-linking is not done sequentially, but the consolidation and the salt concentration changes are done independently for the time step. The mechanical consolidation process was modelled for a semi-time step, the salt diffusion (using the resultant velocities for flow) was then performed for the full time step, and the second semi-step for mechanical consolidation followed with updated boundary values and soil moduli. This completes one full time step.

3.3 Sub-model description

3.3.1 Soil reactivity and stiffness
A general model for soil stiffness and reactivity was devised by Mills (2000) using experimental results for pure clays, and the work of Bolt (1956) on osmotic swell pressures. It is assumed that the soil is pore space saturated, and matric suction is below the air entry value. Under these conditions, matric suction contributes to total stress in accord with the usual saturated effective stress principles and FLAC can reproduce its action as a normal but negative pore pressure. To allow for soil reactivity, the propensity for swell is represented as an increment of bulk stress. Thus when there is a change to solute conditions as a result of salt movement, the clay soil, if notionally restrained, develops a swell pressure. In FLAC, at each adjustment cycle, the bulk stress can be changed directly to reflect this unbalance. In subsequent time stepping, the stress is redistributed until equilibrium is re-attained.

3.3.2 Water and Salt flow
Taken separately, water flow would be calculated with Darcy's law, and salt flux as the sum of advection and diffusion. Irreversible thermodynamics predicts a coupling between the two fluxes (Mitchell et al 1973) and this would mean they may not be considered independently. Stated differently, and noting the soil reactivity model already implies clay acts as an osmotic membrane, it would be expected that for salt flux, advection would be inhibited, and that any osmotic pressure gradient would contribute to water flow. This coupling needs careful consideration. The coupled flow of water and salt J^V and J^S were formulated (Mills 1998) as

$$J^v = q = - k \, \nabla (H + \frac{X}{\rho g} \Pi) \qquad (1)$$

$$J^s = C_S (1 - X) k \, \nabla(-(H + \frac{X}{\rho g} \Pi))$$
$$+ D (1 - X^2 \frac{C_S R T k}{\rho g D}) \, \nabla(-C_S) \qquad (2)$$

where J^v and J^s are the water and salt fluxes, H is the hydraulic potential, and $\Pi = - C_S R T$ the osmotic

or solute potential. Cs is the salt concentration, R the gas constant, T the temperature. J^s will be in units compatible with those of C_S. D is the coefficient of diffusion. Inspection of these equations shows that an osmotic gradient contributes to water flow through some efficiency factor X, and the advective component of salt flux is diminished by $(1-X)$.

Pore spaces in clays are known to be of two types, (larger) inter domain pores, and (smaller) inter-layer pore in the space between platelets of clay. (Alonso et al 1990) The osmotic effects for a clay, as used in the reactivity theory, arise between pore water in these two different pore types. Provided therefore the hydraulic flow in the clay is preferentially through the larger voids, which do not exclude salt, little cross-coupling of salt and water fluxes would be expected. Mitchell et al (1973)in experiments on osmotic effects in clays, drew the conclusion that osmosis as a driver of fluid flow would only be of any consequence with extreme reactivity as in bentonite. Most field clays certainly will not comprise bentonite, and in addition will generally contain even larger defects such as cracks, which carry unimpeded water flow.

In short, the coupling is strongly mitigated by the clay structure and may be neglected. The osmotic effect remains in the swell pressure adjustments from clay reactivity. These generate matric suction changes and ultimately cause fluid flow during the resulting consolidation.

The water and salt flux equations then simplify to:-

$$J^v = q = - k \, \nabla (H) \qquad (3)$$
$$J^s = C_S \, J^v + D \, \nabla(-C_S) \qquad (4)$$

The fluid flow relationship need not be addressed further, since it is available within FLAC, except to mention the difficulty of assigning a value to permeability.

Seminal works on contaminant transport in soils contain the salt flux equation in some form. It is both practicable and conventional to treat the dispersion produced during advection due to velocity variations, as additive to molecular diffusion. D is then a lumped coefficient representing the dispersion-diffusion processes that must be assigned from experience. (Domenico & Schwartz 1990)

Bear & Verruijt .(1987) present a finite element formulation for the solution of Equation with the time advance using finite differences. This was adapted using FISH, for the present study, with some modification. Salt flux does not leave the system at an evaporating boundary, but may do so for lateral percolation. Extra element forms were constructed to allow for this difference.

One informative way to consider the finite element equations is to regard each one as a continuity condition for a region surrounding each node. The terms then have a physical meaning. Using this approach and considering only the advective terms, Bear and Verruijt's expressions and the equivalent ones with accumulation at the node are easily formulated.

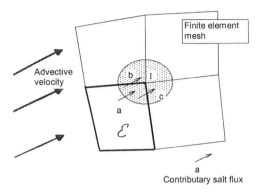

Figure 2   Salt Fluxes ascribed to a node

In Bear's form, the fluxes labelled in Figure 2, a, b and c are ascribed to node I when assembling the contributions from element $\mathcal{E}$, to the global equations. Generally, b and c are cancelled by the contributions from neighbouring elements. If node I is on a boundary and salt is leaving the system, this is correct. If salt is not crossing the boundary even though there is a water flux, salt fluxes b and c are zero, and are not included in the alternative formulation.

3.3.3   Vegetative Usage

Roots draw this water from a large volume of soil which they permeate intensively, eg Hillel (1980b) notes a single plant may have hundreds of kilometers of roots. This permits the mathematical modelling to represent the root extraction as a distributed sink, where the roots are effectively "smeared" over the soil zone. The rate of usage depends mainly on incident radiation on the leaf canopy, with plants evaporating water from their stomata, or leave openings. The Penman equation relates maximum potential plant usage ET_m to radiation R_n , (Hillel 1980b)

$$ET_m \approx \frac{R_n}{1 + \mathrm{fn}(r_s/r_a)} \qquad (6)$$

where r_s is the stomatal resistance, and r_a the atmospheric resistance. The actual usage is less than the potential value, the plant increasing its stomatal resistance to flow for dry conditions in the atmosphere or soil. (Landsberg & McMurtie 1984). Equation 7 uses a modified form of the Penman equation that substitutes pan evaporation for radiation and reduces

potential usage under environmental stress, (Mills 1998)

$$ET_m \approx \frac{E_p}{1 + \frac{1}{4}\,r_s\Big/r_a} \qquad (7)$$

$$\frac{r_s}{4\,r_a} = \frac{r_o}{4\,r_a} \frac{1 + \alpha_1/E_p}{\left\{\left(E_{p\,\max}\Big/E_p\right)^n - 1\right\}\left\{1 - \frac{\Psi_{soil}}{\Psi_{wilting}}\right\}^n} \qquad (8)$$

Soil moisture deficit Ψ_{soil} is the total moisture potential, pF units being used in this section. Estimates of α_1 and the limiting values for the environment variables are needed, as is the ratio $\dfrac{r_o}{4\,r_a}$. This last term compares the stomatal resistance at low stress to the equivalent atmospheric resistance for vapour movement away from the leaf surface. This ratio has been estimated as about one and one half for low growing crops and about an order of magnitude greater for taller vegetation (Stewart 1984). Hence the combined term $\dfrac{r_o}{4\,r_a}$ is expected to be about 0.4 to 4 in practice, the limits applying for low growing crops and isolated trees respectively.

3.3.4   Infiltration and Surface evaporation

A surficial infiltration zone was assumed to suffer evaporation as per the potential evaporation for roots, as well as becoming saturated during rainfall. In FLAC it was necessary to specify an infiltration flow during wetting to produce zero matric suction, rather than using zero pore pressure conditions. This flow was calculated at each time cycle to bring the pore pressure back to zero. It was limited to available rainfall if precipitation could not satisfy soil moisture demand.

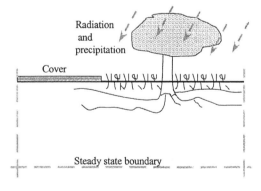

Figure   3   Schematic   of   Field   Problem   Simulation.

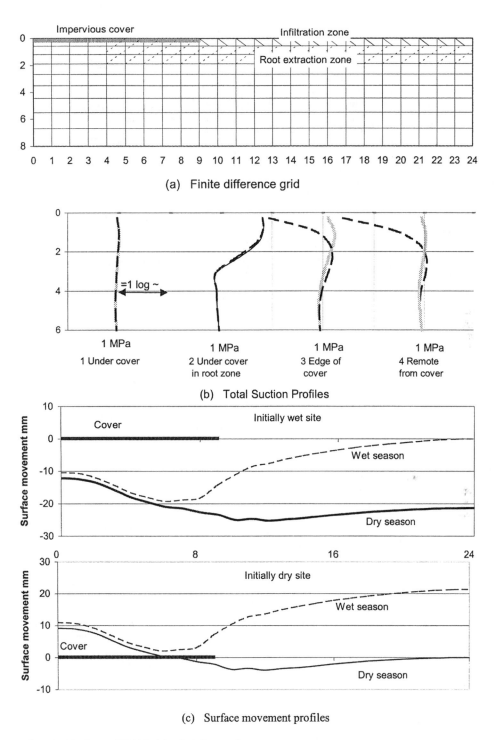

(a) Finite difference grid

(b) Total Suction Profiles

(c) Surface movement profiles

Figure 4 Simulation of Field Problem for influence of tree roots.

4 DEMONSTRATION OF THE COMPOSITE MODEL

A simulation of field conditions (Figure 3) demonstrates the model. Parameters used were reasonable for Adelaide, South Australia. Mean annual rainfall is 600mm and pan evaporation 1800mm. Winter is the wet season. Boundary conditions imposed were a constant soil moisture state at 8m depth, with surface infiltration and evaporation over the exposed portion of the ground surface. Tree roots were assumed to extend beneath the covered area and evapotranspiration was smeared over the root zone. The program was run for a simulated 20 years, and key variables were tracked, the most interesting being the accumulation of salt leading to an increase in total suction, beneath the covered area.

Figure 4 has some details for the demonstration. Provided roots could have an area of influence as assumed, then a zone of high solute suction and soil shrinkage will exist under the cover. The soil near the edge of the cover is not so strongly affected as it benefits from rainfall recharge. (Figure 4 (b)).

5 CONCLUSIONS

1. A two dimensional numerical model for moisture variation over time in clay soils in the presence of active vegetation, has been established using FLAC as a base vehicle. Salt is treated as distinct phase, and solute and matric suctions considered separately. Infiltration from rainfall is applied by reducing matric suction to zero over a surface zone. Moisture extraction by vegetation is simulated by a smeared sink over the root zone. The vegetative demand is linked to available soil moisture.

2. A demonstration of the model has shown the possible salt accumulation in a root zone below a covered area.

3. The model and the theory involved in its formulation can be adapted to field and laboratory situations such as: -

Estimation of the effects of tree roots on of shallow foundations, when subsidence is induced by evapo-transpiration,

Specification of testing regimes for reactivity in soils subject to salt increase,

Prediction of the ability of recharge wells to reverse drying shrinkage in foundation soils beneath distressed buildings.

It is planned to relate this study to some field work on treed sites. Ultimately it is hoped the program will contribute to the basic understanding of complex field processes, and extend available design technologies.

REFERENCES

Alonso, F E, Gens, A, & Josa, A, (1990), A constitutive model for partially saturated soil, *Geotechnique*, 40, N0. 3, 405-430

AS 2870 - 1990, (1990), *Residential Slabs and Footings*, Standards Australia, Sydney.

Argue, J R, (1995). Stormwater management in Australian residential developments : towards a common practice,. *Proc. I.E. Aust. 2nd International Symposium on Urban Stormwater Management,* Melbourne. ISBN (BOOK) 0858 25628 2.

Bear, J, & Verruijt, A, (1987), *Modelling Groundwater Flow and Pollution,* D Reidel Publishing Co. Dordrecht 1987

Biddle, P G, (1983), Patterns of soil drying and moisture deficit in the vicinity of trees on clay soils, *Geotechnique* 33, pp107-126.

Bolt, G H, (1956), Physico-Chemical Analysis of the Compressibility of Pure Clays, *Geotechnique,* Vol. 6 No. 1, p86-93

Cheney, J E, (1988), 25 years' heave of a building on clay due to tree removal, *Ground Engineering,* 1988, 21, No. 5, pp13-27.

Domenico P A, & Schwartz F W, (1990), *Physical and Chemical Hydrogeology,* John Wiley and Sons, New York.

Hillel, Daniel, (1980 a), *Fundamentals of Soil Physics,* Academic Press, New York

Hillel, Daniel, (1980 b), *Applications of Soil Physics,* Academic Press, New York

ITASCA, (1993) *FLAC Fast Lagrangian Analysis of Continua Version 3.2, User's Manual,* Itasca Consulting Group, Minneaplois

Fredlund, D G, & Rahardjo, H, (1993), *Soil Mechanics for Unsaturated Soils,* John Wiley and Sons, New York.

Landsberg, J J, & McMurtie, R, (1984) Water use by Isolated Trees, *Evapotranspiration from Plant Communities,* ed Sharma M L, p223-242, Developments in Agriculture and Managed Forest Ecology, Elsevier, Amsterdam

Li, J, (1996), *Analysis and Modelling of Performance of Footings on Expansive Soils,* Ph.D. Dissertation, University. of South Australia.

Mills, K G, (1998), *The Response of Reactive Clay Soils to Wetting in the Presence of Active Vegetation,* Ph.D. Dissertation, Queensland University. of Technology.

Mills, K G, (2000), Clay stiffness and reactivity derived from clay mineralogy *Proc Asian Conf on Unsaturated Soils, May 2000,* Singapore, Balkema, Rotterdam.

Mitchell, J K, Greenberg, J A, & Witherspoon P A, (1973) Chemico-Osmotic Effects in Fine Grained Soils, *Proc. ASCE, J. Soil Mechanics and Foundation Division,* Vol. 99, N[O.] SM4, April, 1973.

Richards, B G, (1967), A Mathematical Model for Moisture Flow in Horsham Clay, *Civil Eng Transactions,* Inst of Engrs, Aust CE22(3), p252-261.

Richards, B G, Peter, P, & Emerson W W, (1983), The effects of Vegetation on the swelling and Shrinking of Soils in Australia, *Geotechnique,* Vol. 33(2), p127-139

Richards, B G, Peter, P, & Martin, R, (1984), The Determination of Volume Change Properties in Expansive Soils, *Forth International Conference on Expansive Soils,* Adelaide, 21 23 May 1984.

Unsaturated Soils for Asia, Rahardjo, Toll & Leong (eds) © 2000 Taylor & Francis, ISBN 90 5809 139 2

A stress-strain relationship to solve the singularity resulting from coupling the yield surfaces of Barcelona model

K. Nesnas
Department of Civil Engineering, Heriot Watt University, Edinburgh, UK

I.C. Pyrah
Napier University, Edinburgh, UK

ABSTRACT: The Barcelona model is a constitutive model for unsaturated soil using the deviatoric stress q, the net mean stress p and the suction as stress state variable. The model has two yield surfaces in the (s, $\overline{p}$) plane: the loading collapse yield curve and the suction increase yield curve. The coupling of the yield surfaces results in a corner where the vector of incremental plastic strain is not unique. A common procedure used to avoid the numerical problems caused by this singularity is to smooth the yield surface in the vicinity of the corner. This procedure is cumbersome when the yield surface is three dimensional in ($\overline{p}$, q, s) space. Another alternative is proposed which assumes that the plastic strain at corners is a combination of the plastic strain resulting from the movement of the two yield surfaces, forming the corner, by introducing two plastic scalar multipliers. In this paper the proposed stress-strain relationship which consistently takes into account the coupling of the loading collapse yield curve and the suction increase yield curve is derived in the framework of classical plasticity.

1 INTRODUCTION

Constitutive modelling of unsaturated soils requires the use of two state variables; the net stress $\overline{\sigma} = \sigma - u_a$ and the suction $s = u_a - u_w$ (Fredlund & al 1977). Thus, researchers developed elastoplastic constitutive models for unsaturated soil which include these two stress variables. These models are characterized by the loading collapse yield surface (LC) introduced to model the phenomenon of collapse observed when soaking a highly confined unsaturated soil sample in a triaxial test (Alonso et al. 1990 and Wheeler & Sivakumar 1995).

The LC yield surface is expressed mathematically in terms of net stress, suction and plastic volumetric strain ε_v^p, and for the purpose of the development presented below as:

$$F = F(\overline{\sigma}, s, \varepsilon_v^p) = 0 \qquad (1)$$

To this yield surface is associated a plastic potential expressed as:

$$Q = Q(\overline{\sigma}, s, \varepsilon_v^p) = 0 \qquad (2)$$

Alonso et al. (1990), also introduced a suction increase yield surface (SI) to model the occurrence of irreversible volumetric strain when the suction is increased beyond a maximum value s_o. It can be expressed mathematically as:

$$f = f(s, \varepsilon_v^p) \qquad (3)$$

It is observed that Equation (3) is chosen to be independent of net stress $\overline{\sigma}$ because there is no experimental evidence on the shape of the SI yield surface. Alonso et al (1990) represented it as a straight line parallel to the net mean stress axis ($\overline{p}$), (Figure 1).

Also, plastic potential can be associated to the SI yield curve thus:

$$q = q(s, \varepsilon_v^p) \qquad (4)$$

In the following, the only case considered is when the two yield surfaces are both activated by a stress path such as illustrated in Figure 1. First, the stress-strain relationship coupling the LC and SI yield surfaces is developed, then an implicit backward Euler method (Zienkiewicz & Taylor 1991 and Crisfield 1991), for the integration of the stress strain relation is presented.

2 STRESS-STRAIN RELATION WITH TWO ACTIVE YIELD SURFACES

The SI yield surface and the LC yield surface are both activated only if the stress path, starting for example from a stress state inside the yield surface, has reached the stress state at corner D, Figure 1. The plastic strain when the two yield surfaces SI and LC are both activated is given as:

$$\{\delta\varepsilon^{P}\} = \delta\lambda\left\{\frac{\partial Q}{\partial\overline{\sigma}}\right\} + \delta\lambda_1\frac{\partial q}{\partial s}\{m\} \tag{5}$$

where $\delta\lambda$ and $\delta\lambda_1$ are the positive plastic multipliers corresponding respectively to the LC yield surface and the SI yield surface, and m is the unit vector defined as: $\{m\}^{T} = \{1\ \ 1\ \ 1\ \ 0\ \ 0\ \ 0\}$

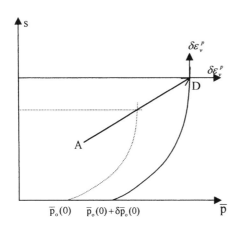

Figure 1. Activation of the suction increase and the loading collapse yield curves

The total strain is computed as:

$$\{\delta\varepsilon\} = [D_e^{-1}]\{\delta\overline{\sigma}\} + \frac{1}{3}\{m\}\beta(s)\delta s + \lambda\left\{\frac{\partial Q}{\partial\overline{\sigma}}\right\} + \lambda_1\frac{\partial q}{\partial s}\{m\} \tag{6}$$

where $\beta(s)$ is the soil elastic compressibility with respect to suction and defined for the Barcelona model as:

$$\beta(s) = \frac{\kappa_s}{v(s + P_{at})} \tag{7}$$

where v is the specific volume, κ_s is the slope of the loading-unloading line in (s, v) space and P_{at} is the atmospheric pressure.

Assuming isotropic hardening, the consistency condition is satisfied by the following two equations:

$$\left\{\frac{\partial F}{\partial\sigma}\right\}^{T}\{\delta\overline{\sigma}\} + \frac{\partial F}{\partial s}\delta s + \frac{\partial F}{\partial\varepsilon_v^P}\delta\lambda\frac{\partial Q}{\partial p} = 0 \tag{8}$$

$$\frac{\partial f}{\partial s}\delta s + \frac{\partial f}{\partial\varepsilon_v^P}\delta\lambda_1\frac{\partial q}{\partial s} = 0 \tag{9}$$

Equation (6) is multiplied by the factor $\{\partial F/\partial\overline{\sigma}\}^{T}D_e$:

$$\left\{\frac{\partial F}{\partial\sigma}\right\}^{T}[D_e]\{\delta\varepsilon\} = \left\{\frac{\partial F}{\partial\overline{\sigma}}\right\}^{T}\{\delta\overline{\sigma}\} + \left\{\frac{\partial F}{\partial\overline{\sigma}}\right\}^{T}[D_e]\{m\}\times$$
$$\frac{\kappa_s}{3v(s + P_{at})}\delta s + \left\{\frac{\partial F}{\partial\overline{\sigma}}\right\}^{T}[D_e]\left\{\frac{\partial Q}{\partial\overline{\sigma}}\right\}\delta\lambda + \tag{10}$$
$$\left\{\frac{\partial F}{\partial\overline{\sigma}}\right\}^{T}[D_e]\frac{\partial q}{\partial s}\{m\}\delta\lambda_1$$

Inserting for the term $\{\partial F/\partial\overline{\sigma}\}^{T}\{\delta\overline{\sigma}\}$ obtained from equation (8) as:

$$\left\{\frac{\partial F}{\partial\overline{\sigma}}\right\}^{T}\{\delta\overline{\sigma}\} = -\frac{\partial F}{\partial s}\delta s - \delta\lambda\frac{\partial F}{\partial\varepsilon_v^P}\frac{\partial Q}{\partial p} \tag{11}$$

gives:

$$\left\{\frac{\partial F}{\partial\overline{\sigma}}\right\}^{T}[D_e]\{\delta\varepsilon\} = -\frac{\partial F}{\partial s}\delta s - \delta\lambda\frac{\partial F}{\partial\varepsilon_v^P}\frac{\partial Q}{\partial p} -$$
$$\delta\lambda_1\frac{\partial F}{\partial\varepsilon_v^P}\frac{\partial q}{\partial s} + \left\{\frac{\partial F}{\partial\overline{\sigma}}\right\}^{T}[D_e]\{m\}\frac{\kappa_s}{3v(s + P_{at})}\delta s + \tag{12}$$
$$\left\{\frac{\partial F}{\partial\overline{\sigma}}\right\}^{T}[D_e]\delta\lambda\left\{\frac{\partial Q}{\partial\overline{\sigma}}\right\} - \left\{\frac{\partial F}{\partial\overline{\sigma}}\right\}^{T}[D_e]\delta\lambda_1\frac{\partial q}{\partial s}\{m\}$$

The $\delta\lambda_1\partial q/\partial s$ in Equation (12) is obtained from Equation (9) as:

$$\delta\lambda_1\frac{\partial q}{\partial s} = -\frac{\dfrac{\partial f}{\partial s}}{\dfrac{\partial f}{\partial\varepsilon_v^P}}\delta s \tag{13}$$

Experimental evidence suggests that the SI yield curve should be coupled to the LC yield curve in such a way that when the SI yield curve translates, it induces translation of the LC yield curve. This may be done by expressing that plastic volumetric strain is:

$$\delta\varepsilon_v^p = \delta\lambda\frac{\partial Q}{\partial \bar{p}} = \delta\lambda_1\frac{\partial q}{\partial s} \tag{14}$$

Thus $\delta\lambda_1\partial q/\partial s$ is re-expressed as:

$$\delta\lambda_1\frac{\partial q}{\partial s} = -\frac{1}{2}\frac{\dfrac{\partial f}{\partial s}}{\dfrac{\partial f}{\partial \varepsilon_v^p}}\delta s + \frac{1}{2}\delta\lambda\frac{\partial Q}{\partial \bar{p}} \tag{15}$$

Inserting Equation (15) into Equation (12) gives:

$$\left\{\frac{\partial F}{\partial \bar{\sigma}}\right\}^T [D_e]\{\delta\varepsilon\} = -\frac{\partial F}{\partial s}\delta s - \delta\lambda\frac{\partial F}{\partial \varepsilon_v^p}\frac{\partial Q}{\partial \bar{p}} +$$
$$\left\{\frac{\partial F}{\partial \bar{\sigma}}\right\}^T [D_e]\{m\}\frac{\kappa_s}{3v(s+P_{at})}\delta s + \delta\lambda\left\{\frac{\partial F}{\partial \bar{\sigma}}\right\}^T [D_e]\left\{\frac{\partial Q}{\partial \bar{\sigma}}\right\} -$$
$$\frac{1}{2}\left\{\frac{\partial F}{\partial \bar{\sigma}}\right\}^T [D_e]\{m\}\frac{\dfrac{\partial f}{\partial s}}{\dfrac{\partial f}{\partial \varepsilon_v^p}}\delta s + \delta\lambda\frac{1}{2}\left\{\frac{\partial F}{\partial \bar{\sigma}}\right\}^T [D_e]\{m\}\frac{\partial Q}{\partial \bar{p}} \tag{16}$$

Solving for $\delta\lambda$:

$$\delta\lambda = \frac{1}{H}\left\{\frac{\partial F}{\partial \bar{\sigma}}\right\}^T [D_e]\{\delta\varepsilon\} + \frac{1}{H}\frac{\partial F}{\partial s}\delta s - \frac{1}{H}\left\{\frac{\partial F}{\partial \bar{\sigma}}\right\}^T [D_e]\{m\}$$
$$\times\frac{\kappa_s}{3v(s+P_{at})}\delta s + \frac{1}{H}\frac{1}{2}\left\{\frac{\partial F}{\partial \bar{\sigma}}\right\}^T [D_e]\{m\}\frac{\dfrac{\partial f}{\partial s}}{\dfrac{\partial f}{\partial \varepsilon_v^p}}\delta s \tag{17}$$

where,

$$H = -\frac{\partial F}{\partial \varepsilon_v^p}\frac{\partial Q}{\partial \bar{p}} + \left\{\frac{\partial F}{\partial \bar{\sigma}}\right\}^T [D_e]\left(\left\{\frac{\partial Q}{\partial \bar{\sigma}}\right\} + \{m\}\frac{1}{2}\frac{\partial Q}{\partial \bar{p}}\right)$$

Inserting Equation (17) into Equation (15) gives:

$$\delta\lambda_1\frac{\partial q}{\partial s} = -\frac{1}{2}\frac{\dfrac{\partial f}{\partial s}}{\dfrac{\partial f}{\partial \varepsilon_v^p}}\delta s - \frac{1}{H}\frac{1}{2}\frac{\partial Q}{\partial \bar{p}}\left\{\frac{\partial F}{\partial \bar{\sigma}}\right\}^T [D_e]\{\delta\varepsilon\} +$$
$$\frac{1}{H}\frac{1}{2}\frac{\partial Q}{\partial \bar{p}}\frac{\partial F}{\partial s}\delta s - \frac{1}{H}\frac{1}{2}\frac{\partial Q}{\partial \bar{p}}\left\{\frac{\partial F}{\partial \bar{\sigma}}\right\}^T [D_e]$$
$$\times\{m\}\frac{\kappa_s}{3v(s+P_{at})}\delta s + \frac{1}{H}\frac{1}{4}\frac{\partial Q}{\partial \bar{p}}\left\{\frac{\partial F}{\partial \bar{\sigma}}\right\}^T [D_e]\{m\}\frac{\dfrac{\partial f}{\partial s}}{\dfrac{\partial f}{\partial \varepsilon_v^p}}\delta s \tag{18}$$

Inserting (17) and (18) into Equation (6) rearranged as:

$$\{\delta\bar{\sigma}\} = [D_e]\{\delta\varepsilon\} - [D_e]\{m\}\frac{\kappa_s}{3v(s+P_{at})}\delta s -$$
$$[D_e]\left\{\frac{\partial Q}{\partial \bar{\sigma}}\right\}\delta\lambda - \{D_e\}\frac{\partial q}{\partial s}\{m\}\delta\lambda_1$$

gives, with some additional algebra:

$$\{\delta\bar{\sigma}\} = [D_{ep}]\left[\{\delta\varepsilon\} - \left(\{m\}\frac{\kappa_s}{3v(s+P_{at})} + \{m\}\frac{\dfrac{\partial f}{\partial s}}{\dfrac{\partial f}{\partial \varepsilon_v^p}} +\right.\right.$$
$$\left.\left.\frac{\left\{\dfrac{\partial Q}{\partial \bar{\sigma}}\right\}\dfrac{\partial F}{\partial s}}{-\dfrac{\partial F}{\partial \varepsilon_v^p}\dfrac{\partial Q}{\partial \bar{p}}} + \frac{\{m\}\dfrac{\partial Q}{\partial \bar{p}}}{-2\dfrac{\partial F}{\partial \varepsilon_v^p}\dfrac{\partial Q}{\partial \bar{p}}}\left(\frac{\partial F}{\partial s} - \frac{\dfrac{\partial f}{\partial s}\dfrac{\partial F}{\partial \varepsilon_v^p}}{\dfrac{\partial f}{\partial \varepsilon_v^p}}\right)\right)\delta s\right] \tag{19}$$

where,

$$[D_{ep}] = [D_e] - \frac{1}{H}\{D_e\}\left(\left\{\frac{\partial Q}{\partial \bar{\sigma}}\right\} + \{m\}\frac{1}{2}\frac{\partial Q}{\partial \bar{p}}\right)\left\{\frac{\partial F}{\partial \bar{\sigma}}\right\}^T$$

$$H = -\frac{\partial F}{\partial \varepsilon_v^p}\frac{\partial Q}{\partial \bar{p}} + \left\{\frac{\partial F}{\partial \bar{\sigma}}\right\}^T [D_e]\left(\left\{\frac{\partial Q}{\partial \bar{\sigma}}\right\} + \{m\}\frac{1}{2}\frac{\partial Q}{\partial \bar{p}}\right)$$

The last term in Equation (19) results from coupling the SI and the LC yield curves.

3 INTEGRATION OF THE STRESS-STRAIN RELATION

Integration of Equation (19) over the increment between the old state and the new state (illustrated in Figure 2) designated by Δ is written as:

$$\{\Delta\overline{\sigma}\} = [D_e]\{\Delta\varepsilon\} - \int_\Delta [D_e]\{\delta\varepsilon^P\} \tag{20}$$

The integral in Equation (20) is approximated as:

$$\Delta\varepsilon^P = \Delta\lambda\left[(1-\theta)\frac{\partial Q}{\partial\overline{\sigma}}\Big|_B + \theta\frac{\partial Q}{\partial\overline{\sigma}}\Big|_D\right] + \Delta\lambda_1\left[(1-\theta)\frac{\partial f}{\partial s}\Big|_B + \theta\frac{\partial f}{\partial s}\Big|_D\right] \tag{21}$$

For $\theta = 0$, Equation (21) reduces to the forward Euler method, which is equivalent to using Equation (19) with all the terms calculated at the old state. This method will produce a drift from the yield surface due to error accumulation, thus a relaxation procedure needs to be devised to minimize it to a suitable tolerance. For $\theta = 1$, Equation (19) reduces to the backward Euler method, which is a stable procedure (Zienkiewicz & Taylor 1991)

For $\theta = 1$, the combination of Equations (20) and (21) gives:

$$\{\overline{\sigma}_D\} = \{\overline{\sigma}_A\} + [D_e]\{\Delta\varepsilon\} - \Delta\lambda\left\{[D_e]\frac{\partial F}{\partial\overline{\sigma}}\right\}\Big|_D - \Delta\lambda_1[D_e]\frac{\partial f}{\partial s}\Big|_D\{m\} \tag{22}$$

Normally Equation (22) will not give the true stress state at D, therefore a further correction is obtained by minimizing the residual which can be expressed as:

$$\Re = \{\overline{\sigma}_D\} - \left(\{\overline{\sigma}_B\} - \Delta\lambda[D_e]\frac{\partial F}{\partial\overline{\sigma}}\Big|_D - \Delta\lambda_1[D_e]\frac{\partial f}{\partial s}\Big|_D\{m\}\right) \tag{23}$$

where, $\{\overline{\sigma}_B\} = \{\overline{\sigma}_A\} + [D_e]\{\Delta\varepsilon\}$

A first order expansion of Equation (23) gives:

$$\Re^{k+1} = \Re^k + \{d\overline{\sigma}\}\Big|_{k+1} + [D_e]\left\{\frac{\partial F}{\partial\overline{\sigma}}\right\}\Big|_k d(\Delta\lambda)_{k+1} +$$

$$\Delta\lambda[D_e]\left[\frac{\partial^2 F}{\partial\overline{\sigma}^2}\right]\Big|_k \{d\overline{\sigma}\} + \Delta\lambda[D_e]\frac{\partial^2 F}{\partial\overline{\sigma}\partial s}\Big|_k \{m\}ds +$$

$$[D_e]\frac{\partial f}{\partial s}\Big|_k \{m\}d(\Delta\lambda_1)_{k+1} + \Delta\lambda_1[D_e]\frac{\partial^2 f}{\partial s^2}\Big|_k \{m\}ds = 0 \tag{24}$$

where k is the iteration number.

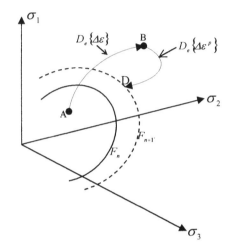

Figure 2. Visualization of Equation (20)

In the following development, the iteration number indicator is omitted for convenience. From Equation (24), the stress correction $\{d\overline{\sigma}\}$ is obtained as:

$$\{d\overline{\sigma}\} = -[\Xi]^{-1}\left(\Re^k + \Delta\lambda_1[D_e]\frac{\partial^2 f}{\partial s^2}\{m\}ds + \Delta\lambda[D_e]\frac{\partial^2 F}{\partial\overline{\sigma}\partial s}\right.$$

$$\left.\times\{m\}ds + d(\Delta\lambda)[D_e]\left\{\frac{\partial F}{\partial\overline{\sigma}}\right\} + d(\Delta\lambda_1)[D_e]\frac{\partial f}{\partial s}\{m\}\right) \tag{25}$$

where, $\Xi = I_d + \Delta\lambda[D_e]\frac{\partial^2 F}{\partial\overline{\sigma}^2}$

The stress correction, $\{d\overline{\sigma}\}$, is completely defined when determining the quantities $d(\Delta\lambda)$ and $d(\Delta\lambda_1)$ by making the new corrected state satisfy the yield functions F and f. This is expressed in the vicinity of $\{d\overline{\sigma}\}$ by the following two equations:

$$F^{k+1} = F^k + \left\{\frac{\partial F}{\partial \bar{\sigma}}\right\}^T d\bar{\sigma} + \frac{\partial F}{\partial s} ds +$$
$$\frac{\partial F}{\partial \varepsilon_v^p} d(\Delta\lambda) \frac{\partial Q}{\partial p} = 0 \tag{26}$$

$$f^{k+1} = f^k + \frac{\partial f}{\partial s} ds + \frac{\partial f}{\partial \varepsilon_v^p} d(\Delta\lambda_1) \frac{\partial f}{\partial s} = 0 \tag{27}$$

Inserting Equation (25) into Equation (26) and rearranging gives:

$$F^{k+1} = F^k - \left\{\frac{\partial F}{\partial \bar{\sigma}}\right\}^T [\Xi]^{-1} \Re^k + d(\Delta\lambda)\left(-\left\{\frac{\partial F}{\partial \bar{\sigma}}\right\}^T \times \right.$$
$$[\Xi]^{-1}[D_e]\left\{\frac{\partial F}{\partial \bar{\sigma}}\right\} + \frac{\partial F}{\partial \varepsilon_v^p} \frac{\partial Q}{\partial p}\right) + \left(-\Delta\lambda_1 [D_e]\left\{\frac{\partial F}{\partial \bar{\sigma}}\right\}^T [\Xi]^{-1} \times$$
$$[D_e]\frac{\partial^2 f}{\partial s^2}\{m\} - \Delta\lambda\left\{\frac{\partial F}{\partial \bar{\sigma}}\right\}^T [\Xi]^{-1}[D_e]\frac{\partial^2 F}{\partial \bar{\sigma}\partial s}\{m\} + \frac{\partial F}{\partial s}\right) ds -$$
$$d(\Delta\lambda_1)\left\{\frac{\partial F}{\partial \bar{\sigma}}\right\}^T [\Xi]^{-1}[D_e]\frac{\partial f}{\partial s}\{m\} = 0 \tag{28}$$

From Equation (27), $d(\Delta\lambda_1)$ is obtained as:

$$d(\Delta\lambda_1) = \frac{f^k + \frac{\partial f}{\partial s} ds}{\frac{\partial f}{\partial \varepsilon_v^p}\frac{\partial f}{\partial s}} \tag{29}$$

Inserting for $d(\Delta\lambda_1)$ in Equation (28) and rearranging, $d(\Delta\lambda)$ is obtained as:

$$d(\Delta\lambda) = \frac{F^k}{\Theta} - \frac{1}{\Theta}\left\{\frac{\partial F}{\partial \bar{\sigma}}\right\}^T [\Xi]^{-1}\Re^k + \frac{1}{\Theta}\left(-\Delta\lambda_1\left\{\frac{\partial F}{\partial \bar{\sigma}}\right\}^T [\Xi]^{-1}\right.$$
$$\times [D_e]\frac{\partial^2 f}{\partial s^2}\{m\} - \Delta\lambda\left\{\frac{\partial F}{\partial \bar{\sigma}}\right\}^T [\Xi]^{-1}[D_e]\frac{\partial^2 F}{\partial \bar{\sigma}\partial s}\{m\} +$$
$$\frac{\partial F}{\partial s} + \frac{1}{\Theta}\frac{\frac{\partial f}{\partial s} ds}{\frac{\partial f}{\partial \varepsilon_v^p}\frac{\partial f}{\partial s}}\left\{\frac{\partial F}{\partial \bar{\sigma}}\right\}^T [\Xi]^{-1}[D_e]\frac{\partial f}{\partial s}\{m\}\right) ds +$$
$$\frac{1}{\Theta}\frac{f^k}{\frac{\partial f}{\partial \varepsilon_v^p}\frac{\partial f}{\partial s}}\left\{\frac{\partial F}{\partial \bar{\sigma}}\right\}^T [\Xi]^{-1}[D_e]\frac{\partial f}{\partial s}\{m\} \tag{30}$$

where $\Theta = \left(\left\{\frac{\partial F}{\partial \bar{\sigma}}\right\}^T [\Xi]^{-1}[D_e]\left\{\frac{\partial F}{\partial \bar{\sigma}}\right\} - \frac{\partial F}{\partial \varepsilon_v^p}\frac{\partial Q}{\partial p}\right)$

Equations (25), (29) and (30) define, respectively, the corrections that should be done on the stress $\bar{\sigma}$, the plastic multipliers and $\Delta\lambda_1$ and $\Delta\lambda$.

Now two cases need to be distinguished. The first case is related to the drained behaviour of the unsaturated soil element. This simplifies considerably Equations (25), (29) and (30), because there is no need to do corrections on the suction (i.e. $ds = 0$ in Equations (25), (29) and (30)) which is either constant, or varied externally for the particular geotechnical problem at hand. Thus, the corrections $d\bar{\sigma}$, $d(\Delta\lambda_1)$ and $d(\Delta\lambda)$ for the drained case are written as:

$$\{d\bar{\sigma}\} = -[\Xi]^{-1}\left(d(\Delta\lambda)[D_e]\left\{\frac{\partial F}{\partial \bar{\sigma}}\right\} + d(\Delta\lambda_1)[D_e]\frac{\partial f}{\partial s}\{m\}\right)$$
$$\tag{31}$$

$$d(\Delta\lambda_1) = \frac{f^k}{\frac{\partial f}{\partial \varepsilon_v^p}\frac{\partial f}{\partial s}} \tag{32}$$

$$d(\Delta\lambda) = \frac{F^k}{\Theta} - \frac{1}{\Theta}\left\{\frac{\partial F}{\partial \bar{\sigma}}\right\}^T [\Xi]^{-1}\Re^k +$$
$$\frac{1}{\Theta}\frac{f^k}{\frac{\partial f}{\partial \varepsilon_v^p}\frac{\partial f}{\partial s}}\left\{\frac{\partial F}{\partial \bar{\sigma}}\right\}^T [\Xi]^{-1}[D_e]\frac{\partial f}{\partial s}\{m\} \tag{33}$$

The second case concerns the undrained behaviour for which the suction is no longer defined as for the drained behaviour, but changes with stress $\bar{\sigma}$ in a manner to maintain the water content or, equivalently the specific water volume constant. Nesnas et al. (1998) obtained the change in suction in terms of change in net stress $\bar{\sigma}$, which is written here in a condensed form as:

$$\Delta s = -\{\aleph\}\{\Delta\bar{\sigma}\} \tag{34}$$

where the term $\{\aleph\}$ is a line vector, with components which are a function of the net stress $\bar{\sigma}$ and the suction s.

The suction correction ds (in Equations 25, 29 and 30) should satisfy the undrained constraint defined by Equation (34). To do so, the first order of the following residual should equate to zero:

$$r = s_D - s_A + \{\aleph\}(\{\bar{\sigma}_D\} - \{\sigma_A\})$$ (35)

Thus,

$$r = ds - \left(\frac{\partial\{\aleph\}}{\partial s}ds + \{d\bar{\sigma}\}^T[\Omega]^T\right)\{\bar{\sigma}_A\} +$$
$$\{\aleph\}\{d\bar{\sigma}\} + \left(\frac{\partial\{\aleph\}}{\partial s}ds + \{d\bar{\sigma}\}^T[\Omega]^T\right)\bar{\sigma}_D$$

(36)

where, $\Omega = \partial\{\aleph\}/\partial\bar{\sigma}$ and $\bar{\sigma}_c$ is calculated at the previous iteration as: $\{\bar{\sigma}_D\} = \{\bar{\sigma}_A\} + [D_e]\{\Delta\varepsilon\} + \{d\bar{\sigma}\}^k$, with $d\bar{\sigma}^k$ the stress correction for the previous iteration.

Rearranging Equation (36) and neglecting the infinitely small, the suction correction ds is obtained as:

$$ds = \frac{\{\aleph\}\{d\bar{\sigma}\} + \{d\bar{\sigma}\}^T[\Omega]^T[D_e]\{\Delta\varepsilon\}}{\left(1 + \frac{\partial\{\aleph\}}{\partial s}\{m\}\right)}$$ (37)

The iteration procedure is summarised in the following steps:

Step 1: Obtain an initial guess of the new stress $\bar{\sigma}_D$ using Equation (22), where all the terms are calculated at B (i.e. the stress state violating the consistency conditions).

Step 2: Calculate the residual $\Re$ from Equation (23)

Step 3: If residual is lower or equal to a fixed tolerance, $\|\Re\| \leq Tol$, then stop iteration. If not go to step 4

Step 4: Calculate the correction $d\bar{\sigma}$, ds, $d(\Delta\lambda)$ and $d(\Delta\lambda_1)$. If drained problem use Equations 31, 32 and 33. If undrained problem, use Equations 29, 30, 37 and 25.

NB: All terms in equations should be calculated using results of the previous iteration.

Step 5: Correct variables as:

$$\{\bar{\sigma}\}_D^{k+1} = \{\bar{\sigma}\}_D^k + d\bar{\sigma}$$
$$s_D^{k+1} = s_D^k + ds$$

$$\Delta\lambda^{k+1} = \Delta\lambda^k + d(\Delta\lambda)$$
$$\Delta\lambda_1^{k+1} = \Delta\lambda_1^k + d(\Delta\lambda_1)$$

Step 6: keep going to Step 2 until $\|\Re\| \leq Tol$

4 CONCLUSIONS

In this paper, a stress-strain relationship suitable for solving the singularity resulting from coupling the two yield surfaces of the Barcelona model is proposed. A detailed algorithm to integrate the stress-strain relationship for finite element implementation is also proposed for the drained case and the undrained case. The case where one of the yield surfaces is activated without reaching the singularity point is not presented in this paper, but it is easily derived. It must be noted that if the algorithm is used in conjunction with a quadratic convergence of the global solution, a consistent tangent matrix including second order terms must be obtained instead of Equation (19).

REFERENCES

Alonso E.E., Gens A. & A. Josa 1990. A constitutive model for partially saturated soils. *Geotechnique*. 40: 405-430.
Crisfield M.A. 1991. *Non-linear finite element analysis of solids and structures*. London: McGraw-Hill.
Fredlund, D.G. and N.R. Morgenstern 1977. Stress state variables for saturated soils. *Proc. ASCE*. 103(5): 313-312.
Nesnas K., Pyrah I.C. and S.J. Wheeler 1998. A methodology to simulate the drained and undrained behaviour of unsaturated soils. *Second International Conference on Unsaturated Soils*: 503-508. Beijing: Balkema.
Wheeler SJ and V. Sivakumar 1995. An elastoplastic critical state framework for unsaturated soil. *Geotechnique*. 45: 35-53.
Zienkiewicz,O.C. & R.L. Taylor 1991. *The finite element method*. London: McGraw-Hill.

Unsaturated Soils for Asia, Rahardjo, Toll & Leong (eds) © 2000 Taylor & Francis, ISBN 90 5809 139 2

Undrained conditions of unsaturated soil using an integral form of the constraints and full Newton-Raphson procedure

K. Nesnas
Department of Civil Engineering, Heriot Watt University, Edinburgh, UK

I.C. Pyrah
Napier University, Edinburgh, UK

ABSTRACT: The undrained condition is the initial stage of a consolidation problem for unsaturated soil and for the particular case of saturated soil. The undrained behaviour is simulated in a coupled formulation using a small time step; several iterations are necessary due to convergence difficulties. A simple procedure, which does not introduce the time factor but is compatible with the formulation of a consolidation problem, is to formulate the undrained problem using the integral form of the constraints. The undrained constraints derived from physical consideration and a general relationship in terms of net stresses and pore pressures are added to the equation governing the equilibrium of an element of unsaturated soil through an integral form of the constraints. The linearisation process results in set of highly non-linear algebraic equations with net stresses, pore water pressures and pore air pressures as unknowns. The algorithm solving these equations using a full Newton procedure is presented for use in a finite element program.

1 INTRODUCTION

Numerical modelling has already been carried out (Nesnas, Pyrah & Wheeler 1998, Nesnas & Pyrah, unpubl.) to simulate the undrained analysis of unsaturated soils. The numerical model is a displacement formulation where the undrained constraint is introduced into the equilibrium equations resulting in additional terms which necessitate use of an iterative solution technique; sufficient iterations were performed to ensure that both the equilibrium and undrained conditions are satisfied.

The numerical development presented in this paper to simulate the undrained condition of unsaturated soils is a formulation which involves the three unknowns: the displacement, pore water pressure and air pressure formulation. This makes the development suitable to be used as an initial stage in conjunction with the consolidation of unsaturated soils.

The general procedure followed in deriving the mathematical model is based on establishing first variational relationships for degree of saturation (defined as the ratio of the volume of water V_w and the void volume V_V) and porosity (defined as the ratio of the void volume V_V and the total volume V) in terms of changes in net stresses, water pressure and air pressure. Then, the weak form of these relationships is introduced through an integral form in conjunction with the principal of virtual work to obtain the governing equations.

2 RELATIONHIPS FOR DEGREE OF SATURATION AND POROSITY

In this section the relationships which link the change in porosity and degree of saturation to change in water pressure and air pressure are derived.

An element of soil, assumed to be a three-phase material with air, water and solid particles, when subjected to an increment of load results, in a volume change δv written as:

$$\delta v = \delta v_a + \delta v_w + \delta v_s \tag{1}$$

where, δv_a, δv_w and δv_s are, respectively, change in volume of air, water and solid.

Assuming that the solid particles are rigid then the change in volume of the unsaturated soil element is:

$$\delta v = \delta v_a + \delta v_w \tag{2}$$

If S_r is the degree of saturation, n is the porosity and v is the volume of the soil element, then the volume of the water phase is:

$$v_w = nS_r v \tag{3}$$

The increment of volume of the water phase can be linked to its excess in water pressure as:

$$\delta v_w = v_w C_w \delta u_w \tag{4}$$

where, C_w is the compressibility of the water phase.

Inserting Equation (3) into Equation (4) gives:

$$\delta v_w = nS_r v C_w \delta u_w \tag{5}$$

The volume of the air phase in the soil element is:

$$v_a = n(1 - S_r(1 - H))v \tag{6}$$

The differential of Equation (6) is expressed as:

$$\delta v_a = (1 - S_r(1 - H))v\delta n - nv(1 - H)\delta S_r \\ + n(1 - S_r(1 - H))\delta v \tag{7}$$

where δn, δS_r and δv are, respectively, change in porosity, change in degree of saturation and change in volume; and H is Henry's constant.

Inserting for Equations (6) and (7) into (2) gives:

$$\delta v = \frac{v(1 - S_r(1 - H))}{(1 - n(1 - S_r(1 - H)))}\delta n - \frac{vn(1 - H)}{(1 - n(1 - S_r(1 - H)))}\delta S_r \\ + \frac{vnS_r C_w}{(1 - n(1 - S_r(1 - H)))}\delta u_w \tag{8}$$

The effect of air pressure is introduced in Equation (6) by considering the law of perfect gases, which states that the product of gas volume and its pressure is constant, thus:

$$(1 - S_r(1 - H))nvu_a = constant \tag{9}$$

The differential of Equation (9) gives:

$$-u_a nv(1 - H)\delta S_r + (1 - S_r(1 - H))u_a v\delta n \\ (1 - S_r(1 - H))u_a n\delta v + (1 - S_r(1 - H))nv\delta u_a = 0 \tag{10}$$

Rearranging Equation (10) gives:

$$\delta v = -\frac{v}{n}\delta n + \frac{v(1 - H)}{(1 - S_r(1 - H))}\delta S_r - \frac{v}{u_a}\delta u_a \tag{11}$$

Identifying Equation (11) with Equation (8) and rearranging gives:

$$\frac{\delta n}{n} - \frac{(1 - H)}{1 - S_r(1 - H)}\delta S_r = \\ \frac{1 - n(1 - S_r(1 - H))}{u_a}\delta u_a - nS_r C_w \delta u_w \tag{12}$$

Introducing now, the undrained condition through the variation of the specific volume (defined as $v_w = 1 + S_r e = 1 + S_r n/1 - n$) which should be zero:

$$\frac{S_r}{1 - n}\delta n + n\delta S_r = 0 \tag{13}$$

Solving for Equation (12) and (13) we obtain the expression for δn and S_r as:

$$\delta n = -\frac{n(1 - n)(1 - S_r(1 - H))(1 - n(1 - S_r(1 - H)))}{((1 - S_r(1 - H))(1 - n) + S_r(1 - H))u_a}\delta u_a - \\ \frac{S_r n^2 C_w(1 - n)(1 - S_r(1 - H))}{(1 - S_r(1 - H))(1 - n) + S_r(1 - H)}\delta u_w \tag{14}$$

$$\delta S_r = \frac{S_r(1 - S_r(1 - H))(1 - n(1 - S_r(1 - H)))}{((1 - S_r(1 - H))(1 - n) + S_r(1 - H))u_a}\delta u_a + \\ \frac{S_r^2 n C_w(1 - S_r(1 - H))}{(1 - S_r(1 - H))(1 - n) + S_r(1 - H)}\delta u_w \tag{15}$$

Constitutive relations for porosity n and degree of saturation S_r are generally expressed in terms of mean net stress, deviatoric stress q, suction s and the preconsolidation pressure $\bar{p}_o$:

$$n = n(\bar{p}, q, s, \bar{p}_o) \tag{16}$$

$$S_r = S_r(\bar{p}, q, s, \bar{p}_o) \tag{17}$$

Noting that $\delta s = \delta u_a - \delta u_w$ the differential form of Equations (16) and (17) gives:

$$\delta n = \left[\left(\frac{\partial n}{\partial \bar{p}} + \frac{\partial n}{\partial \bar{p}_o}\frac{\partial \bar{p}_o}{\partial \bar{p}}\right)\left\{\frac{\partial \bar{p}}{\partial \bar{\sigma}}\right\} + \right.$$
$$\left.\left(\frac{\partial n}{\partial q} + \frac{\partial n}{\partial \bar{p}_o}\frac{\partial \bar{p}_o}{\partial q}\right)\left\{\frac{\partial q}{\partial \bar{\sigma}}\right\}\right]\{\delta\bar{\sigma}\} \qquad (18)$$
$$+ \left(\frac{\partial n}{\partial s} + \frac{\partial n}{\partial \bar{p}_o}\frac{\partial \bar{p}_o}{\partial s}\right)\delta u_w - \left(\frac{\partial n}{\partial s} + \frac{\partial n}{\partial \bar{p}_o}\frac{\partial \bar{p}_o}{\partial s}\right)\delta u_a$$

$$\delta S_r = \left[\left(\frac{\partial S_r}{\partial \bar{p}} + \frac{\partial S_r}{\partial \bar{p}_o}\frac{\partial \bar{p}_o}{\partial \bar{p}}\right)\left\{\frac{\partial \bar{p}}{\partial \bar{\sigma}}\right\} + \right.$$
$$\left.\left(\frac{\partial S_r}{\partial q} + \frac{\partial S_r}{\partial \bar{p}_o}\frac{\partial \bar{p}_o}{\partial q}\right)\left\{\frac{\partial q}{\partial \bar{\sigma}}\right\}\right]\{\delta\bar{\sigma}\} \qquad (19)$$
$$+ \left(\frac{\partial S_r}{\partial s} + \frac{\partial S_r}{\partial \bar{p}_o}\frac{\partial \bar{p}_o}{\partial s}\right)\delta u_w - \left(\frac{\partial S_r}{\partial s} + \frac{\partial S_r}{\partial \bar{p}_o}\frac{\partial \bar{p}_o}{\partial s}\right)\delta u_a$$

Combination of Equation (14) and (15) respectively with Equation (18) and (19) gives:

$$\{A_1\}\{\delta\bar{\sigma}\} + A_2\delta u_w + A_3\delta u_a = 0 \qquad (20)$$

$$\{B_1\}\{\delta\bar{\sigma}\} + B_2\delta u_w + B_3\delta u_a = 0 \qquad (21)$$

where

$$A_1 = \left(\frac{\partial n}{\partial \bar{p}} + \frac{\partial n}{\partial \bar{p}_o}\frac{\partial \bar{p}_o}{\partial \bar{p}}\right)\left\{\frac{\partial \bar{p}}{\partial \bar{\sigma}}\right\} + \left(\frac{\partial n}{\partial q} + \frac{\partial n}{\partial \bar{p}_o}\frac{\partial \bar{p}_o}{\partial q}\right)\left\{\frac{\partial q}{\partial \bar{\sigma}}\right\}$$

$$A_2 = \left(\frac{\partial n}{\partial s} + \frac{\partial n}{\partial \bar{p}_o}\frac{\partial \bar{p}_o}{\partial s}\right) + \frac{S_r n^2 C_w (1-n)(1-S_r(1-H))}{(1-S_r(1-H))(1-n) + S_r(1-H)}$$

$$A_3 = -\left(\frac{\partial n}{\partial s} + \frac{\partial n}{\partial \bar{p}_o}\frac{\partial \bar{p}_o}{\partial s}\right) - \frac{n(1-n)(1-S_r(1-H))(1-n(1-S_r(1-H)))}{((1-S_r(1-H))(1-n) + S_r(1-H))u_a}$$

$$B_1 = \left(\frac{\partial S_r}{\partial \bar{p}} + \frac{\partial S_r}{\partial \bar{p}_o}\frac{\partial \bar{p}_o}{\partial \bar{p}}\right)\left\{\frac{\partial \bar{p}}{\partial \bar{\sigma}}\right\} + \left(\frac{\partial S_r}{\partial q} + \frac{\partial S_r}{\partial \bar{p}_o}\frac{\partial \bar{p}_o}{\partial q}\right)\left\{\frac{\partial q}{\partial \bar{\sigma}}\right\}$$

$$B_2 = \frac{\partial S_r}{\partial s} + \frac{\partial S_r}{\partial \bar{p}_o}\frac{\partial \bar{p}_o}{\partial s} + \frac{S_r^2 n C_w (1-S_r(1-H))}{(1-S_r(1-H))(1-n) + S_r(1-H)}$$

$$B_3 = -\frac{\partial S_r}{\partial s} + \frac{\partial S_r}{\partial \bar{p}_o}\frac{\partial \bar{p}_o}{\partial s} - \frac{S_r(1-S_r(1-H))(1-n(1-S_r(1-H)))}{((1-S_r(1-H))(1-n) + S_r(1-H))u_a}$$

3 GOVERNING EQUATIONS

The equilibrium of an element of unsaturated soil of volume V, which is assumed to be a continuous medium, subjected to a body force {b} and an external pressure {τ} (which is a force per unit area) is expressed by the equation of virtual work (Nesnas et al. 1998) as:

$$\int_V \{\delta^*\varepsilon\}^T [D_{ep}]\{\delta\varepsilon\}dV = \int_V \{\delta^*\varepsilon\}^T [D_{ep}]\{\delta\varepsilon_o\}dV -$$
$$\int_V \{\delta^*\varepsilon\}^T \{m\}\delta u_a dV + \int_V \{\delta^*d\}^T \{\delta b\}dV + \int_A \{\delta^*d\}^T \{\delta\tau\}dA \qquad (22)$$

where, $\{\delta^*\varepsilon\}$ is the virtual strain vector and $\{\delta^*d\}$ is the virtual nodal displacement vector, $[D_{ep}]$ is the tangent elastoplastic stiffness matrix, $\{\delta\varepsilon\}$ is the vector of total strain increments for the soil skeleton and $\{\delta\varepsilon_o\}$ is a vector of initial strain increments caused directly by a change of suction, defined as:

$$\{\delta\varepsilon_o\} = C(\bar{\sigma}, s)(\delta u_a - \delta u_w) \qquad (23)$$

In addition, the integral forms of Equations (20) and (21) expressing weak forms of the undrained constraints are introduced as:

$$\int_V \delta^*u(\{A_1\}\{\delta\bar{\sigma}\} + A_2\delta u_w + A_3\delta u_a)dV = 0 \qquad (24)$$

$$\int_V \delta^*u(\{B_1\}\{\delta\bar{\sigma}\} + B_2\delta u_w + B_3\delta u_a)dV = 0 \qquad (25)$$

where, δ^*u is a virtual pore pressure.

In the finite element discretisation, the increment of displacement vector {δd} and the increment of strain vector {δε} are expressed in terms of the nodal displacement vector {δa_d} as:

$$\{\delta d\} = [N]\{\delta a_d\} \qquad (26)$$

where [N] is the shape function matrix, and

$$\{\delta\varepsilon\} = [B]\{\delta a_d\} \qquad (27)$$

where [B] is the matrix linking strain to displacements.

Also, the finite element discretisation of the pore pressure assuming the same shape function $[N_p]$ for air and water is written as:

$$\{\delta u_w\} = [N_p]\{\delta a_w\} \tag{28}$$

$$\{\delta u_a\} = [N_p]\{\delta a_a\} \tag{29}$$

where $\{\delta a_w\}$ is the nodal water pressure and $\{\delta a_a\}$ is the air water pressure.

The discretisation of Equation (22) with (23), and Equation (24) and (25) results in:

$$\int_V [B]^T [D_{ep}][B]dV\{\delta a_d\} + \int_V [B]^T [D_{ep}][C][N_p]dV\{\delta a_w\}$$

$$+ \left(\int_V [B]^T \{m\}[N_p]dV - \int_V [B]^T [D_{ep}][C][N_p]dV \right)\{\delta a_a\}$$

$$= \int_A [N]^T dA\{\delta\tau\} + \int_V [N]^T dV\{\delta b\} \tag{30}$$

$$\int_V [N_p]^T \{A_1\}[D_{ep}][B]dV\{\delta a_d\} + \int_V [N_p]^T A_2[N_p]dV\{\delta a_w\}$$

$$+ \int_V [N_p]^T A_3[N_p]dV\{\delta a_a\} = 0 \tag{31}$$

$$\int_V [N_p]^T \{B_1\}[D_{ep}][B]dV\{\delta a_d\} + \int_V [N_p]^T B_2[N_p]dV\{\delta a_w\}$$

$$+ \int_V [N_p]^T B_3[N_p]dV\{\delta a_a\} = 0 \tag{32}$$

In the following, the solution of the governing equations based on a full Newton-Raphson procedure is presented.

4 SOLUTION ALGORITHM

The governing equations defined by (30), (31) and (32) form a set of non-linear algebraic equations which can be written in the following condensed form as:

$$\{R(U)\} = [H]\{U\} - \{F_{ext}\} \tag{33}$$

where $[H]$ is the global matrix, $\{\delta U\}$ is the global vector of unknown displacement, pore water pressure and air pressure and $\{F_{ext}\}$ is the forcing vector. The term $\{R(U)\}$ is the residual vector. Its variation should satisfy, at each iteration k, the following:

$$\{R(U^{k+1})\} = \{R(U^k)\} + \left.\frac{\partial R(U)}{\partial U}\right|_k \Delta(U)^{k+1} = 0 \tag{34}$$

Thus, Equation (34) becomes:

$$\left.\frac{\partial R(U)}{\partial(U)}\right|_k \Delta U^{K+1} = -\{R(U^k)\} \tag{35}$$

A solution for the problem defined by Equation (33) for the current increment i+1 is obtained as:

$$U^{i+1} = U^i + \sum \Delta U^k \tag{36}$$

when the iterative solution satisfies:

$$\frac{\|\Delta U^{k+1} - \Delta U^k\|}{\|\Delta U^k\|} \leq Tol \tag{37}$$

The first derivative of the residual $\{R(U)\}$ in Equation (35) is hardly possible to calculate explicitly when complex elastoplastic constitutive models are involved. It is possible to get around this by performing numerical perturbations of the degrees of freedom. Let η be the numerical perturbation of the value of the variable U_m corresponding to the degree of freedom m which results in a change of $U_m \pm \eta$. This perturbation yields the residuals expressed as $R(\{U\} + \eta\{e\})$ and $R(\{U\} - \eta\{e\})$, where $\{e\}$ is a column vector with all components equal to zero except the component corresponding to the perturbed degree of freedom which equals to 1. These residuals are used to obtain a central difference approximation of the derivative in Equation (35) as:

$$\frac{\partial R(U)}{\partial U} = \frac{R(U + \eta e) - R(U - \eta e)}{2\eta} \tag{38}$$

In recapitulation the full Newton-Raphson procedure defined by Equations (35) and (38) consists of:

1. Solving the set of algebraic equations:

$$\left[\frac{\partial R}{\partial (U)}\left(U^k\right)\right]\Delta U^{K+1} = -\left\{R\left(U^k\right)\right\}$$

with

$$\left[\frac{\partial R}{\partial (U)}\left(U^k\right)\right] = \left[\frac{R\left(U^k + \eta e\right) - R\left(U^k + \eta e\right)}{2\eta}\right]$$

The matrix $\partial R/\partial U$ is assembled as:

$$\frac{\partial R}{\partial U} = \begin{bmatrix} \dfrac{R_1\left(U_1 + \eta, U_2, \ldots U_n\right) - R_1\left(U_1 - \eta, U_2, \ldots U_n\right)}{2\eta} & \cdots & \dfrac{R_1\left(U_1, U_2, \ldots U_n + \eta\right) - R_1\left(U_1, U_2, \ldots U_n - \eta\right)}{2\eta} \\ \dfrac{R_2\left(U_1 + \eta, U_2, \ldots U_n\right) - R_2\left(U_1 - \eta, U_2, \ldots U_n\right)}{2\eta} & \cdots & \dfrac{R_2\left(U_1, U_2, \ldots U_n + \eta\right) - R_2\left(U_1, U_2, \ldots U_n - \eta\right)}{2\eta} \\ \vdots & \vdots & \vdots \\ \dfrac{R_m\left(U_1 + \eta, U_2, \ldots U_n\right) - R_m\left(U_1 - \eta, U_2, \ldots U_n\right)}{2\eta} & & \dfrac{R_m\left(U_1, U_2, \ldots U_n + \eta\right) - R_m\left(U_1, U_2, \ldots U_n - \eta\right)}{2\eta} \end{bmatrix}$$

2. A solution is obtained as: $U^k = U^{k-1} + \Delta U^k$

3. If Equation (35) is satisfied then the solution of Equation (31) is: $U = U^k$

5 CONCLUSIONS

In this paper, the numerical development for analysis of the undrained behaviour of unsaturated soil based on integral forms of the undrained constraints and a full Newton-Raphson procedure is presented.

In the full Newton-Raphson procedure the derivative of the residual is calculated by numerical perturbation through a central difference. The choice of the perturbation η, is computer dependent and an initial test to chose a suitable value should be performed in order to avoid rounding errors if η is too small, and large truncation errors if it is too large.

REFERENCES

Nesnas K., Pyrah I.C. and S.J. Wheeler 1998. A methodology to simulate the drained and undrained behaviour of unsaturated soils. *Second International Conference on Unsaturated Soils*: 503-508. Beijing: Balkema.

Nesnas, K. and Pyrah, I.C. Finite element implementation and verification of triaxial drained and undrained conditions for unsaturated soils. (Submitted to *the journal of computer and geotechnics*).

Unsaturated Soils for Asia, Rahardjo, Toll & Leong (eds) © 2000 Taylor & Francis, ISBN 90 5809 139 2

Performances of HiSS $\delta_{1\text{-unsat}}$ model for soils under saturated and unsaturated conditions

A. Rifa'i, L. Laloui, L. Vulliet & F. Geiser
Soil Mechanics Laboratory, EPFL, Lausanne, Switzerland

ABSTRACT: HiSS-$\delta_{1\text{-unsat}}$ model is a general elasto-plastic constitutive model for saturated and unsaturated soils including undrained paths and damage effects. This model uses saturated effective stress ($\sigma' = \sigma - u_w$) and matric suction ($s = u_a - u_w$) as two independent stress state variables. In this paper, numerical simulations are presented, making use of experimental results obtained by Maâtouk (1993) and Sivakumar (1993) on unsaturated silt and kaolin. These two materials have been prepared and desaturated by different procedures. After a brief review of the experimental results, the determination of the constitutive parameters of the model is explained. All constitutive parameters could be obtained from the existing isotropic and shear tests. The comparison between numerical simulation and experimental data shows the capacity of the HiSS-$\delta_{1\text{-unsat}}$ model.

1 INTRODUCTION

HiSS $\delta_{1\text{-unsat}}$ is an elasto-plastic constitutive model dedicated to the modelling of saturated and unsaturated soils and including undrained paths and damage effects (Geiser *et al.* 1997a and 1999). It was developed at LMS-EPFL on the base of the HiSS family of models proposed by Desai (1994). It is formulated within the framework of hardening plasticity using saturated effective stress ($\sigma' = \sigma - u_w$) and matric suction ($s = u_a - u_w$) as the two independent sets of stress variables. The detailed description of the model is given in Geiser (1999).

2 MATHEMATICAL FORMULATION

The mathematical formulation of the model is based on a decomposition of the strain increment into an elastic and a plastic part:

$$\dot{\varepsilon}_{ij} = \dot{\varepsilon}_{ij}^e + \dot{\varepsilon}_{ij}^p \tag{1}$$

2.1 Elastic component

The elastic increment $\dot{\varepsilon}_{ij}^e$ is composed of a mechanical and a hydric strain increment:

$$\dot{\varepsilon}_{ij}^e = \dot{\varepsilon}_{ij}^{e^m} + \frac{1}{3}\dot{\varepsilon}_v^{e^h}\delta_{ij} = \mathbf{D}_{ijkl}^{e-1}\dot{\sigma}_{kl}' + \frac{1}{3}A^{-1}\dot{s}\delta_{ij} \tag{2}$$

where $\dot{\varepsilon}_{ij}^{e^m}$ is the elastic mechanical strain increment induced by the variation of the effective stress σ', $\frac{1}{3}\dot{\varepsilon}_v^{e^h}$ is the reversible hydric strain increment, $\mathbf{D}^e$ is the classical elastic tensor and A is a proportionality coefficient which describes the hydric behaviour.

2.2 Plastic component

The plastic strain increment is also deduced from mechanical and hydric considerations.

The model has two yield surfaces: F_1 represents the plastic limit in the effective stress plane and F_2 gives the yield limit in the suction against mean effective stress plane.

The mechanical yield surface F_1 is expressed as:

$$F_1 \equiv \frac{J_{2D}}{p_a^2} - \left[-\alpha\left(\frac{J_1' + R}{p_a}\right)^n + \gamma\left(\frac{J_1' + R}{p_a}\right)^2 \right]F_s \tag{3}$$

where $F_s = \left[1 - \beta.\left(\sqrt{27}/2.J_{3D} \cdot J_{2D}^{-3/2}\right)\right]^{-0.5}$

J_1' is the first invariant of the effective stress tensor; J_{2D} is the second invariant of the deviatoric stress tensor; J_{3D} is the third invariant of the deviatoric stress tensor; p_a is the atmospheric pressure; R is a bonding stress parameter which is assumed as zero in this paper; γ and β are ultimate parameters and n is the phase change parameter expressed as:

$$n = \cfrac{2}{1-\left(\cfrac{J_{2D}}{J_1^{'2}}\right)\cfrac{1}{F_s\gamma}} \qquad (4)$$

at zero volume change

α is the hardening parameter related to the volumetric plastic deformation. A hyperbolic equation is proposed by Desai & Wathugala (1987):

$$\alpha = \frac{a_1}{\xi^{\eta_1}} \qquad (5)$$

where a_1 and η_1 are parameters giving the hardening evolution and ξ is the trajectory of total plastic strains given by:

$$\xi = \int (d\varepsilon_{ij}^p d\varepsilon_{ij}^p)^{1/2} \qquad (6)$$

In Equation 5, a_1 is a function of the suction such that F_1 increases when the saturation decreases.

The hydric yield surface, F_2 is expressed as:

$$F_2 \equiv -\left[-\alpha_h\left(\frac{3s}{p_a}\right)^n + \gamma\left(\frac{3s}{p_a}\right)^2\right]F_s \qquad (7)$$

where α_h is the hardening parameter related to the hydric plastic deformation:

$$\alpha_h = \frac{a_3}{\xi^{\eta_1}} \qquad (8)$$

where a_3 is a hydric parameter giving the hardening evolution.

The model is associated for the hydric path (F_2) and non-associated for the mechanical path (F_1) which adds two non-associative parameters κ_0, κ_∞, instead of α in Equation 3.

For the generation of the plastic deformation on the hydric path, the following condition is imposed:

$$\begin{cases} F_2 < 0 \text{ when } s \geq s_e \\ F_2 \equiv \text{Equation (7) otherwise} \end{cases} \qquad (9)$$

in which s_e is the air entry suction.

This condition is motivated by the experimental observation that on the drying path, the deformation is elasto-plastic till the suction reaches the air entry value. After that the behaviour is mainly reversible.

2.3 Hydro-mechanical coupling

The hydro-mechanical coupling is introduced in the model by using a three-phase mixture theory (Hutter et al. 1999). In particular, the compressibility of the fluid (β_f) is assumed to be a function of the degree of saturation (S_r) as suggested by Bishop and Eldin (1950):

$$\beta_f = (1 - S_r + HS_r)u_{a0}/u_a^2 \qquad (10)$$

where H is Henry's constant, generally assumed to be 0.02; u_a is the average air pressure in the sample and u_{ao} is the initial air pressure.

The constitutive relation of Seker (1983) is used to evaluate the degree of saturation as a function of the suction:

$$S_r = \cfrac{1}{1+\left(\cfrac{\log(10*s)}{\psi_0}\right)^{1/\psi_1}} \qquad (11)$$

where ψ_0 and ψ_1 are material parameters and s is expressed in kPa.

2.4 Constitutive parameters of the $\delta_{1\text{-}unsat}$ model

The model parameters for saturated and unsaturated conditions are given in Tables 1 and 2.

To determine the model parameters, ideally three triaxial saturated tests (E, ν, γ, β, n, R, κ_0, κ_∞), one isotropic triaxial saturated test (a_1, η_1), one drying unsaturated test with measurement of the void ratio (A, a_3, s_e), three isotropic triaxial unsaturated tests (a_2) and three triaxial unsaturated tests (r) are used.

3 REVIEW OF THE USED EXPERIMENTAL RESULT

To evaluate the performance of the HiSS $\delta_{1\text{-}unsat}$ model, experimental results for a silt (Maâtouk 1993, Maâtouk et al. 1995) and for a kaolin (Sivakumar 1993, Wheeler & Sivakumar 1995) are used. These two materials have been prepared and desaturated by different procedures (Table 3).

Table 1. Parameters of the $\delta_{1\text{-}unsat}$ model for saturated state.

Symbol	Parameter	Equation
E, ν	Elastic parameters.	Eq. (2) inside $\mathbf{D^e}$.
γ, β	Ultimate parameters.	Eq. (3).
a_1, η_1	Growth parameters.	Eq. (5).
n	Phase change parameter.	Eq. (4).
R	Bonding stress parameter.	Eq. (3).
κ_0, κ_∞	Non-associative parameter.	Eq. (3).

Table 2. Parameters of the $\delta_{1\text{-}unsat}$ model for unsaturated state.

Symbol	Parameter	Equation
A	Elastic proportionality coefficient.	Eq. (2).
s_e	Air entry suction.	
a_2	Parameter for the evolution of a_1 parameter when suction $s>s_e$.	Eq. (14).
a_3	Hydric hardening parameter.	Eq. (8).

256

Table 3. Description of materials.

	Maâtouk (1993)	Sivakumar (1993)
Material	Silty soil from Trois-Rivières, Québec. Saturated and loose in its natural condition.	Kaolin.
Properties	Based on BNQ 2501-025 method: 66 % silt, 18 % clay and 16 % sand, w_L= 22 %; w_P= 15 % and the soil is not significantly expansive.	Dry density, γ_d= 1.2 g/cm^3; water content, w= 25 %; initial void ratio, e= 1.21 and S_r= 54 % at the maximum dry density.
Dimension of specimen	Height = 71 mm and diameter = 38.1 mm.	Height 100 mm and diameter = 50 mm.
Preparation technique	Individual forming by slight tamping in a mould with 200 g dry soil and soil was at a water content of about 8.5%. Slight static compaction. Initial conditions were thus established under u_w= 0 kPa, u_a= 150 kPa, σ_3= 155 kPa and σ_1= 160 kPa.	Compaction in 9 layers with 9 blows for each layer by static compaction.
Used method to impose the suction	Imposing a positive u_a, while keeping u_w equal to the atmospheric pressure.	Imposing a positive u_a, while keeping a constant u_w

4 DETERMINATION OF THE CONSTITUTIVE PARAMETERS

The determination of the soil parameters for saturated and unsaturated states is explained in this section; parameters are summarized in Table 4.

4.1 Elastic parameters in saturated state: E, ν

The elastic parameters are usually determined from unloading-reloading slopes of shear tests. Unfortunately, there is no unloading test in the two experimental works used in this paper. Consequently the parameters have been evaluated from the initial slope of ε_1-q and ε_1-ε_v curves for a shear test (at axial strain, ε_1= 0.2%). As the Young's modulus E varies with the stress level, an average value was determined.

4.2 Ultimate parameters: γ, β

The ultimate parameters can be evaluated from the conventional angle of friction (ϕ) as:

$$\gamma = \frac{\left(\dfrac{2}{\sqrt{3}} \dfrac{\sin \phi}{(3 - \sin \phi)} \right)^2}{F_s} \quad (12)$$

$$\beta = \frac{1 - P^{-4}}{1 + P^{-4}} \quad \text{where} \quad P = \frac{\left(\dfrac{2}{\sqrt{3}} \dfrac{\sin \phi}{(3 - \sin \phi)} \right)}{\left(\dfrac{2}{\sqrt{3}} \dfrac{\sin \phi}{(3 + \sin \phi)} \right)} \quad (13)$$

4.3 Growth parameters: a_1, η_1

These parameters are determined on the base of one isotropic saturated test.

4.4 Parameter a_2

This parameter describes the evolution of parameter a_1 when s≥s_e and it is equal to zero if s<s_e. Its determination is based on isotropic test at different constant suction levels.

$$a_2 = \frac{\ln\left(\dfrac{a_1 - 0.1\, a_{1_0}}{0.9\, a_{1_0}} \right)}{(s_e - s)} \quad \text{for } s \geq s_e \quad (14)$$

where a_{1_0} is the value of parameter a_1 at s equal to zero.

4.5 Parameter s_e

This parameter should be determined on the base of a drying test. It corresponds to the suction above which air recedes into the soil pores. As we do not have any information on s_e for the used soils, the estimations of air entry values were based on the soil properties (for the silt) and on an analysis made by Khalili & Khabbaz (1998) (for the kaolin).

4.6 Parameter a_3

The determination of a_3 is similar to the determination of the parameter a_1, but based on the hydric drying test at constant mechanical pressure.

Many authors showed that the hydric slope and the saturated mechanical slope are identical (Zerhouni 1991 for example). So, this parameter can estimated as $a_3 \equiv a_1$ if there is not an hydric drying test available, as it is the case with the tested soils.

4.7 The calibration paths

The used experimental results for the determination of the soil parameters of silt are two saturated undrained shear tests at σ_3=96 and 300 kPa, four isotropic tests with suction s=0, 80, 400 and 600 kPa, and three unsaturated shear tests with s=150 kPa & σ_3= 180 kPa; s= 400 kPa & σ_3= 450 kPa; and s=600 kPa & σ_3= 725 kPa. For the kaolin two saturated shear tests at σ_3=75 and 150 kPa, four isotropic tests with suction s=0, 100, 200 and 300 kPa and three unsaturated shear tests with s=100 kPa & σ_3= 250 kPa; s= 200 kPa & σ_3= 350 kPa; and s=300 kPa & σ_3= 450 kPa are used. The obtained soil parameters are given in Table 4.

Table 4. Summary of $\delta_{1\text{-unsat}}$ model parameters.

Parameter / Soil	Silt	Kaolin
E [MPa]	100	40
ν	0.3	0.3
γ	0.058	0.0304
β	0.702	0.58
n	3.1	3.0
a_1	$3.33\ 10^{-5}$	$8.45\ 10^{-4}$
η_1	2.5	0.627
κ_0	1.0	0.2
κ_∞	0.7	0.41
R	0	0
A [MPa]	100	40
s_e [kPa]	80	86
a_2	0.0062	0.0052
a_3	$3.33\ 10^{-5}$	$8.45\ 10^{-4}$

5 NUMERICAL SIMULATIONS

Numerical simulations of the experimental results on silt and kaolin are done in saturated and unsaturated states.

5.1 Numerical simulations of the silt

Figure 1 shows the back prediction of the saturated consolidated undrained (CU) tests. The numerical simulation of this test is good in the deviatoric stress path. For the pore pressure generation, the model shows a rapid increase of the pore pressure, then a decrease and finally a relatively constant value. The prediction of the pore pressure is not as good. This difference between the experimental results and the numerical one is also observed in the p'-q plane.

The simulation (back prediction) of the isotropic consolidation at different initial suction values shows that the model is able to reproduce the main aspect of the behaviour such as the value of the plastic yield point and the curve slope (Figure 2).

The simulation (back prediction) of the three triaxial tests at different confining pressures and suction (Figure 3) shows that the model is able to reproduce the evolution of the shear resistance. The volumetric behaviour is qualitatively good but quantitatively the model underestimate the volume variation.

5.2 Numerical simulation of the kaolin

The numerical simulation (back prediction) of the fully saturated drained triaxial test is given in Figure 4. The isotropic consolidation test at four suction values are simulated in Figure 5. These tests were used in the determination of the model parameters. The back prediction shows the good ability of the

model to reproduce the suction effect on the isotropic behaviour. For the deviatoric path, the back prediction of the three unsaturated triaxial tests show that the model gives a good reproduction of the tendency of the behaviour of the unsaturated soils (Figure 6).

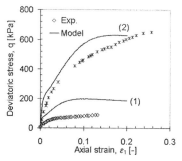

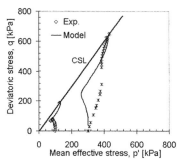

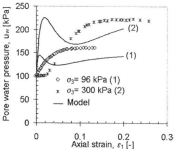

Figure 1. Back prediction of saturated undrained test on silt.

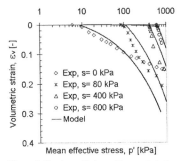

Figure 2. Back prediction of isotropic behaviour of silt.

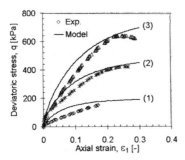

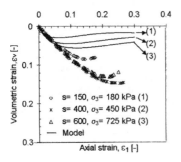

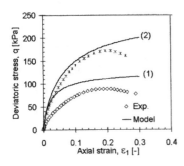

Figure 3. Back prediction of unsaturated drained test on silt.

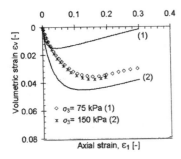

Figure 4. Back prediction of saturated drained test on kaolin.

Figures 7 and 8 show the prediction of constant mean effective stress at different constant suction values.

For the two states, saturated (Fig. 7) and unsaturated (Fig. 8), the model predictions are very close to

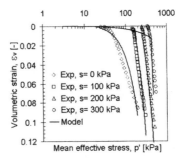

Figure 5. Back prediction of isotropic behaviour of kaolin.

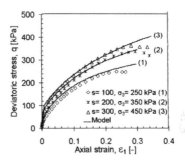

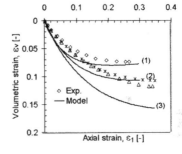

Figure 6. Back prediction of unsaturated drained test on kaolin.

the experimental results in the deviatoric plane (q-ε_1). The prediction of the volumetric variation shows that the model overestimates the dilatancy aspect of the behaviour in the saturated case. For the unsaturated case, the prediction of the volumetric deformation is better. From the predicted results given in Figure 8, one can appreciate the ability of the model to predict the behaviour of the soils in unsaturated conditions.

6 CONCLUSIONS

The HiSS $\delta_{1\text{-unsat}}$ model was developed and used successfully by Geiser *et al.* (1997a, 1997b and 1999) based on original tests performed on a remoulded silt. Here, existing experimental results on compacted silt and kaolin were used to evaluate the

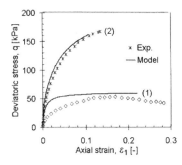

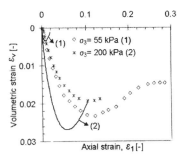

Figure 7. Prediction of saturated constant mean effective stress test on kaolin.

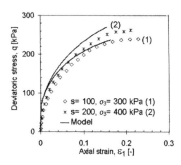

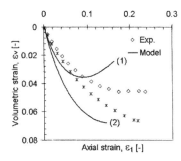

Figure 8. Prediction of unsaturated constant mean effective stress on kaolin.

performance of the model. The capability of the model is demonstrated, and shows particularly good performance in predicting the shear strength of the soil in saturated and unsaturated states. However, the predicted volumetric variation is less satisfactorily, but qualitatively conform to the experimental observations.

REFERENCES

Bishop, A.W. & Eldin,G., 1950. Undrained triaxial tests on saturated sands and their significance in the general theory of shear strength, *Géotechnique* 2 (1): 13-32.

Desai, C.S. & Wathugala, G.W. 1987. Hierarchical and unified models for solids and discontinuities (joints/interfaces), *Implementation of constitutive laws for engineering materials*, Second international conference and short course on constitutive laws for engineering materials: 31-124, Arizona.

Desai, C.S. 1994, Hierarchical single surface and the disturbed state constitutive models with emphasis on geotechnical applications, *Geotechnical engineering: Emerging trends in design and practice*, Saxena Ed.: 115-154.

Geiser F., Laloui L., Vulliet L., Desai C.S., 1997 (a). Disturbed state concept for partially saturated soils. *Numerical Models in Geomechanics, NUMOG VI*:129-133, Montréal.

Geiser, F., Laloui, L. & Vulliet, L. 1997 (b). Constitutive modelling of unsaturated sandy silt, *Proc. Computer and Advances in Geomechanics, Wuhan*: 899-904.

Geiser F., Laloui L., Vulliet L., 1999, Unsaturated soil modelling with special emphasis on undrained conditions. *Numerical Models in Geomechanics, NUMOG VII*: 9-14, Graz.

Geiser, F., 1999. *Comportement mécanique d'un limon non saturé: étude expérimentale et modélisation constitutive*, Thèse n° 1942, EPFL, Lausanne.

Hutter, K., Laloui, L. & Vulliet, L. 1999. Thermodynamically based mixture models of saturated and unsaturated soils, *Journal of mechanics of cohesive-frictional materials*, 4 (2): 295-338.

Khalili,N. and Khabbaz, M.H. 1998, A unique relationship for χ for the determination of the shear strength of unsaturated soils, *Géotechnique*, 48 (5): 681-687.

Maâtouk, A. 1993. *Application des concepts d'état limite et d'état critique à un sol partiellement saturé effondrable*, Doctoral thesis, University of Laval.

Maâtouk, A., S. Leroueil & P. La Rochelle 1995. Yielding and critical state of collapsible unsaturated silty soil. *Géotechnique* 45(3): 465-477.

Seker, 1983, *Etude de la déformation d'un massif de sol non saturé*, Thèse n° 492, EPFL, Lausanne

Sivakumar, V. 1993. *A critical state framework for unsaturated soil*, Ph.D. thesis, University of Sheffield.

Wheeler, S. J. & V. Sivakumar 1995. An elasto-plastic critical state framework for unsaturated soil. *Géotechnique* 45(1): 5-53.

Zerhouni, M.I. 1991. *Rôle de la pression interstitielle négative dans le comportement des sols- Application aux routes*. Thèse, Ecole Centrale, Paris.

4 Suction measurement techniques

Unsaturated Soils for Asia, Rahardjo, Toll & Leong (eds) © 2000 Taylor & Francis, ISBN 90 5809 139 2

A new matric suction calibration curve

R. Bulut, S.-W. Park & R. L. Lytton
Department of Civil Engineering, Texas A&M University, College Station, Tex., USA

ABSTRACT: This paper reports on constructing a calibration curve for matric suction measurements by constructing a calibration curve using both pressure plate and pressure membrane devices. The pressure plate apparatus can measure matric suction values up to 150 kPa. However, with the pressure membrane device matric suction values can be extended up to 10,000 kPa. In developing the filter paper calibration curve, the capabilities, pitfalls, and limitations of the method are also discussed.

1 INTRODUCTION

The filter paper method is a soil suction measurement technique. Soil suction is one of the most important parameters describing the moisture condition of unsaturated soils. The measurement of soil suction is crucial for applying the theories of the engineering behavior of unsaturated soils. The filter paper method is an inexpensive and relatively simple laboratory test method. It is also the only known method that covers the full range of suction.

With the filter paper method, both total and matric suction can be measured. If the filter paper is allowed to absorb water through vapor flow (non-contact method), then only total suction is measured. On the other hand, if the filter paper is allowed to acquire water through fluid flow (contact method), then only matric suction is measured.

With a reliable soil suction measurement technique, the initial and final soil suction profiles can be obtained at convenient depth intervals. The change in suction with seasonal moisture movement is valuable information for many engineering applications.

2 HISTORICAL BACKGROUND

The use of paper for estimating the water potential in the United States was first reported by Gardner (1937). Gardner (1937) used an ash free quantitative filter paper (Scheleicher & Schuell No. 589 White Ribbon). The filter papers were calibrated using sulfuric acid solutions for higher water potentials and a centrifugal force method for lower water potentials. William & Sedgley (1965) used a new type of filter paper (Whatman No. 50) and the calibration was established for the filter paper water content versus water potential (suction) by using only the pressure membrane. Both Gardner (1937) and William & Sedgley (1965) used the filter paper method for agricultural applications.

McQueen & Miller (1968) improved the filter paper method much further in a research supported by the U. S. Geological Survey (USGS). The calibration curve was established using the Schleicher & Schuell No. 589 White Ribbon filter papers with the use of saturated salt solutions, pressure plates, pressure membranes, and soil columns at equilibrium above the groundwater table. Al-Khafaf & Hanks (1974) followed a slightly different procedure and technique for the calibration of Schleicher & Schuell No. 589 White Ribbon filter paper by using salt solutions, thermocouple psychrometers, pressure plates, and soil columns at equilibrium above the groundwater table.

Swarbrick (1992) also established a calibration curve using Whatman No. 42 filter papers and a combination of common suction measurement devices such as thermocouple psychrometers, pressure plates and membranes. All the calibration curves constructed from Gardner (1937) to Swarbrick (1992) appear to have been established as a single curve by using different filter papers, a combination of different soil suction measurement devices, and different calibrating testing procedures.

Houston et al. (1994) developed two different calibration curves; one for total suction and one for matric suction measurements using Fisher quantitative coarse filter papers. For the total suction calibration curve, saturated salt solutions and for the matric suction calibration curve tensiometers and pressure membranes were used. Houston et al. (1994) reported that the total and matric suction calibration curves were not compatible. This simply implies that two different calibration curves, one for matric and one for total suction, need to be used in soil suction measurements.

3 SOIL SUCTION

The basic research regarding the effect of the pore fluid on the soil was first initiated by soil scientists and agronomists towards the end of the 19th century. The importance of the soil suction concept was first recognized by civil engineers in the 1950s. Initially, civil engineers adopted soil suction terminology from soil science and worked on the behavior of unsaturated soils using this terminology. More recently, engineering researchers have adopted these definitions to engineering applications.

Soil suction is an energy quantity based on the principles of thermodynamics and is simply defined as the capability of a soil to retain water. In engineering practice, soil suction consists of matric and osmotic suction components. The sum of the matric and osmotic suctions is called total suction. Matric suction arises from the capillary forces, soil texture, and adsorption forces of clay particles. Osmotic suction results from salts present in the soil structure.

3.1 Principles of matric suction measurements

Matric suction is measured if filter paper makes contact with the soil because dissolved salts present in the soil water will move with it. Osmotic suction which results from the salt concentrations in the soil will not impact the matric suction measurements. Thus, only matric suction is measured when the filter paper acquires water through fluid flow. It is very important that the contact between filter paper and soil is established intimately enough for the transfer of soil water only through fluid flow not through vapor flow.

Actually, in practice, it is very difficult (or may be even impossible) to keep a full contact between filter paper and soil, but the degree of contact may be increased by simply burying the filter paper into the soil. Also, it should be kept in mind that the movement of soil water to filter paper is a thermodynamic process based on suction differences between the soil and filter paper; therefore, the system will eventually come to equilibrium even in a partial contact between the soil and filter paper, but in an expended equilibration period.

For the measurement of matric suction, filter papers are simply buried into a soil specimen (in a very good contact manner) in closed containers. Then, the sealed containers are put in relatively constant temperature environment for equilibrium. After equilibrium is established between the filter paper and soil, the filter paper water content is measured as quickly and accurately as possible. The matric suction value of the soil is found from a pre-established calibration curve. The ASTM standard has a procedure for the measurement of matric suction using filter paper (ASTM D 5298 1994).

3.2 Principles of total suction measurements

If no contact is provided between the filter paper and soil, the dissolved salts will not leave the soil water. Therefore, total suction, which is the sum of matric and osmotic suction, is measured because both matric and osmotic suction components exist in the soil. Since there is no contact between the filter paper and soil when measuring the total suction, the filter paper absorbs water through vapor flow.

If a filter paper is placed in a sealed container (in a non-contact manner) with a soil specimen in a constant temperature environment, at the end of the equilibration time (e.g. when transfer of the water vapor between the soil and filter paper has stopped) the total suction value of the filter paper and soil will be the same. Then, the water content of the filter paper is used to get the total suction value of the soil from a calibration curve.

4 MATRIC SUCTION CALIBRATION

Tensiometers, pressure plates, and pressure membranes are commonly used matric suction measuring devices. The calibration curves for filter paper have been constructed using a combination of soil suction measuring devices (e.g. a combination of both total and matric suction devices).

For matric suction calibration, a contact path is provided between the filter paper and the measuring device in order not to have an osmotic suction component of total suction. In other words, if transfer of soil water is allowed only through fluid flow, dissolved salts will move with soil water, and the measuring device will not detect an osmotic suction component. As a result, matric suction measuring devices usually operate by imposing suction on a given specimen which can be a soil specimen or filter paper. A typical schematic drawing of a pressure plate or pressure membrane device is depicted in Figure 1.

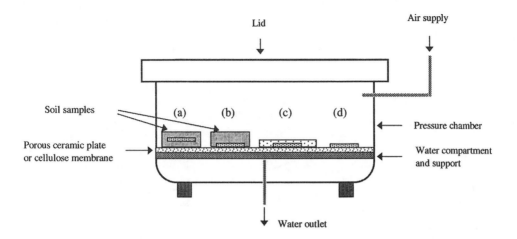

(a) One filter paper between two larger size protective filter papers embedded into the soil sample.

(b) One filter paper makes contact with the porous plate or membrane and covered on top with a larger size protective filter paper in the soil sample.

(c) One filter paper makes contact with the porous plate or membrane and covered on top with two larger size protective filter papers.

(d) One filter paper on the porous plate or membrane.

Figure 1. Schematic drawing of a pressure plate or membrane device.

4.1 *Pressure plate and pressure membrane apparatus*

The pressure plate and pressure membrane devices and methods were developed in soil science field to study the water uptake and retention of soils. The soil water characteristic curve which is obtained by plotting various applied pressures (matric suctions) against the water contents of soil specimens has wide areas of applications in soil science and agronomy. The main components of the pressure plate and membrane apparatus are a pressure chamber, a porous ceramic plate or cellulose membrane, and an air compressor. The main difference between the pressure plate and pressure membrane devices is that the former uses ceramic porous disk that can be used for pressures up to 150 kPa and the later uses cellulose membranes with which pressures can be extended up to 10,000 kPa. The ceramic disks are rigid enough to carry the soil specimens on them, but a support is provided for the highly flexible membrane. A typical schematic drawing of a pressure plate or pressure membrane apparatus is shown in Figure 1.

4.2 *Filter paper calibration with pressure plate and pressure membrane devices*

For matric suction calibration or soil suction measurements, the filter paper or the soil is put into the suction measuring device in a manner that ensures good contact with the porous plate or cellulose membrane. In this process, main concern is that an intimate contact needs to be established between the water inside the filter paper and the air entry disk so that transfer of the water is allowed only through continues water films, not by any other means. For this reason, a calibration testing program was undertaken to investigate the degree of contact between the filter paper, soil, and porous disks as shown in Figure 1.

This paper describes the calibration of Schleicher & Schuell No. 589-WH filter papers with the pressure plate and membrane devices. For the calibration of filter papers with pressure plate and membrane devices, a combination of procedures suggested by Houston et al. (1994), Lee (1991), and McQueen & Miller (1968) was adopted. Figure 2 depicts the calibration curve developed in this study.

The filter paper water contents and matric suction values obtained from the pressure plate and membrane were plotted by a single curve as shown in Figure 2. Equation 1, below, represents the best-fit curve that relates the water content values of Schleicher & Schuell No. 589-WH filter papers versus matric suction.

$$h_m = 1.1247w^{-0.8473} \tag{1}$$

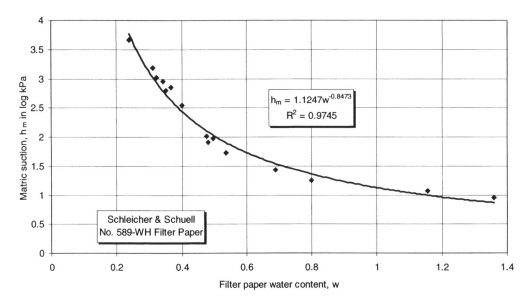

Figure 2. Matric suction calibration curve.

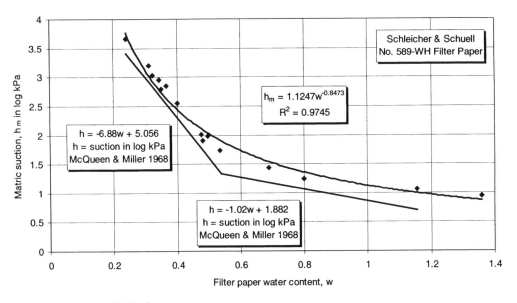

Figure 3. Comparison of calibration curves.

where h_m = matric suction in log kPa, w = filter paper water content in decimals.

Each data point on Figure 2 is at least an average of three tests and in each test data is average of at least four filter papers. The pressure membrane was adopted for suctions above 150 kPa and the plates were used for lower suctions. With the membrane suctions as high as 4570 kPa were obtained.

4.3 Calibration testing procedure

Prior to each test, the porous disk or membrane was saturated with distilled water and then sealed within the pressure chamber along with the soil specimens containing the filter papers or filter papers alone which rest on the surface of the saturated porous plate or cellulose membrane (Figure 1).

Three different soils (e.g. fine clay, sandy silt, and pure sand) were used in the calibration process of the filter papers. The results of the filter paper water contents when used in the soils will be discussed in the next section. To hold the soil specimens in place with filter papers inside, small plastic rings were used. The soil samples were saturated with distilled water for at least a day prior to their use. After putting the soil specimens with the filter papers on the saturated disks, plenty of distilled water was added on the plate to a level that all the filter papers were under water. The contamination of the filter papers was avoided by using larger size protective filter papers. With the influence of the applied pressure, the water inside the soil specimen and the ceramic disk or membrane was expelled out and collected in a graduated cylinder until a suction equilibrium between the soil or filter paper and the applied air pressure has established. At equilibrium, the soil suction of the specimen equals the applied air pressure. At the end of the equilibrating period, the filter paper water contents were measured as quickly and accurately as possible. The suggested equilibrating times for matric suction measurements using the pressure plates usually change between 3 to 5 days. The equilibration periods used for this study changed between 3, 5, and 7 days depending on testing set up. For instance, when filter papers were embedded in the soil it was 5 or 7 days, but when only filter papers were used it was 3 days.

5 DISCUSSION OF RESULTS

The main reasons for using soils in the filter paper calibration process are

1. To use the fine structure of the soil as a tool in creating an intimate contact between the plate or membrane and filter paper,

2. To avoid the moisture gains into the filter paper after releasing the air pressure from the pressure chamber.

However, very high filter paper water contents were obtained when all the three soils were used as in the set up (a) (Figure 1). Fortunately, the filter paper water contents were all comparable as obtained from the set ups (b), (c), and (d) (Figure 1). The results from (b) were slightly wetter than (c) and the results from (d) were slightly drier than (c). In obtaining the calibration curve (Figure 2), the filter papers from the set up arrangements (b), (c), and (d) were used.

The calibration curve obtained in this study is plotted along with the calibration curve established by McQueen & Miller (1968) in Figure 3 above. Both curves were obtained using the same brand filter papers (Schleicher & Schuell No. 589 type filter papers). The differences between the two are most probably due to overtime changes in the pore struc-

ture of filter papers, the type of curve fitting tech niques techniques used, and most importantly at high suction values McQueen & Miller (1968) used salt solutions in the calibration testing, which is a wetting process; however, the pressure membrane results in a drying curve.

6 CONCLUSION

A calibration curve was obtained for Schleicher & Schuell No. 589-WH filter papers using both the pressure plate and pressure membrane devices. Very high filter paper water content measurements were obtained when a filter paper was sandwiched between two larger size protective filter papers in the soil samples (Figure 1). There was almost no change in the filter paper moisture content values even after 7 seven days of equilibration times at very high applied air pressures. Therefore, care must be taken when embedding the filter papers into the soil for the filter paper calibration using the pressure plates or measurements of soil matric suction using the filter paper method. The use of different types of soils did not have any impact in the filter paper calibration testing as expected.

ACKNOWLEDGMENT

Authors would like to thank to Prof. Kirk Brown and Mr. Jim Thomas for their generous help in providing their laboratory and equipment to us.

REFERENCES

Al-Khafaf, S. & Hanks, R. J. 1974. Evaluation of the filter paper method for estimating soil water potential. *Soil Science* 117(4): 194-199.

ASTM D 5298 1994. Standard test method for measurement of soil potential (suction) using filter paper. *American Society of Testing and Materials.* 1998 Annual Book of ASTM Standards: 156-161.

Gardner, R. 1937. A method of measuring the capillary tension of soil moisture over a wide moisture range. *Soil Science* 43(4): 277-283.

Houston, S. L., Houston, W. N., & Wagner, A. M. 1994. Laboratory filter paper suction measurements. *Geotechnical Testing Journal* (GTJODJ) 17(2): 185-194.

Lee, H. C. 1991. *An evaluation of instruments to measure soil moisture condition.* M.Sc. Thesis, Texas Tech University, Lubbock, Texas, USA.

McQueen, I. S. & Miller, R. F. 1968. Calibration and evaluation of a wide-range gravimetric method for measuring moisture stress. *Soil Science* 106(3): 225-231.

Swarbrick, G. E. 1995. Measurement of soil suction using the filter paper method. *Proceedings, 1st International Conference on Unsaturated Soils, Paris*: 653-658.

William, O. B. & Sedgley, R. H. 1965. A simplified filter paper method for determining the 15-atmosphere percentage of soils. *Australian Journal of Experimental Agriculture and Animal Husbandry* 5(18): 201-202.

Unsaturated Soils for Asia, Rahardjo, Toll & Leong (eds) © 2000 Taylor & Francis, ISBN 90 5809 139 2

Comparison of total suction values from psychrometer and filter paper methods

R. Bulut, S.-W. Park & R. L. Lytton
Department of Civil Engineering, Texas A&M University, College Station, Tex., USA

ABSTRACT: This paper reports on total suction measurements using the transistor psychrometer and the filter paper method that were carried out on three different types of soils. The soil samples were compacted at different moisture contents well above their optimum values and representative samples were taken from each soil sample to conduct the experiment. By comparing the two methods, the capabilities, pitfalls, and limitations of the calibration and measurement procedures for both methods are discussed.

1 INTRODUCTION

Soil suction is one of the most important parameters for applying the theories of engineering behavior of unsaturated soils. Suction controls different properties of unsaturated soils such as strength, stiffness, and hydraulic conductivity. In order to predict a soil's behavior, the suction in the soil must be measured. Both the transistor psychrometer and the filter paper methods are soil suction measurement techniques. With the filter paper method, both total and matric suction measurements are possible. However, the transistor psychrometer can only be used for total suction measurements. Therefore, only a total suction comparison is made between the transistor psychrometer and the filter paper method.

1.1 *Transistor psychrometer*

The transistor psychrometer is an instrument for the measurement of the relative humidity of the air within a confined space. The transistor psychrometers, therefore, can be used to measure the total suction of a soil by measuring the relative humidity in the air in proximity of a soil. Psychrometers function on the basis of temperature differences between a dry bulb and wet bulb. The relative humidity is related to the difference in the temperatures between these two bulbs.

In the case of the transistor psychrometer, the evaporating wet bulb is wetted by placing a drop of water into a small ring in all the psychrometer probes. These probes are inserted into a thermally insulated container, which ensures that the probes

remain at a near constant temperature during the period of the test. The output from the probes is in terms of millivolts, which can be monitored with a millivoltmeter or a millivolt recorder. These millivolt values are related to the relative humidity through a calibration procedure, by which a correlation is established between the millivolts obtained from the probes and total suction. The calibration of each probe is performed by using different concentrations of salt solutions that refer to an appropriate suction range. Sodium Chloride is the common salt for that purpose. A more detailed information for the measurement of soil suction using the transistor psychrometer can be found in the transistor psychrometer user manual by Soil Mechanics Instrumentation, Adelaide, Australia.

1.2 *Filter paper method*

The filter paper method is a soil suction measurement technique. It is an indirect laboratory test method. The filter paper method is the only method which covers the full range of suction measurement. Furthermore, both total and matric suction measurements are possible with the method. Basically, the filter paper comes into equilibrium with the soil either through vapor flow or liquid flow. At equilibrium, the suction value of the filter paper and the soil will be the same. If the filter paper is allowed to absorb water through vapor flow (non-contact method), then only total suction is measured. However, if the filter paper is allowed to absorb water through fluid flow (contact method), then only matric suction is measured.

For the measurement of soil suction, filter papers are simply placed above or in a soil specimen (in a non-contact or contact manner) in sealed containers. Then, the whole set up is put in a constant temperature environment. Temperature fluctuations need to be kept as minimal as possible during the equilibrium period. At the end of the equilibration time, the filter paper water content is measured as quickly and accurately as possible. The equilibration period suggested by researchers for total suction measurements is about one week. More detailed information on the measurement of soil suction using filter paper method can be found in ASTM D 5298 Standard Test Method for Measurement of Soil Potential (Suction) Using Filter Paper (1994).

2 CALIBRATION

Both the psychrometer and the filter paper methods are indirect techniques of measuring total suction. Both methods need calibration. The transistor psychrometer is calibrated using salt solutions and the filter papers are usually calibrated using a combination of salt solutions, pressure plates and membranes, and tensiometers.

2.1 Transistor psychrometer

The transistor psychrometer was calibrated using Sodium Chloride (NaCl) salt solutions following the manufacturer's (Soil Mechanics Instrumentation, Adelaide, Australia) procedure. Sodium Chloride solutions were prepared between 2 to 4 log(kPa) osmotic suction ranges. The psychrometer probes were stabilized overnight with 1 log(kPa) solution and each probe was calibrated with 2, 2.5, 3, 3.5, and 4 log(kPa) solutions. The same sodium chloride concentration was applied three times to each probe for the calibration process. In other words, three millivolt measurements were taken at one suction value for each probe by applying the same concentration three times in three runs of tests.

During the calibration as well as the soil suction measurements, the psychrometer was kept in a temperature-controlled room in which temperature fluctuations were kept below ±1°C. One hour of equilibration time was adopted for the calibration process. A calibration curve was obtained for each probe by plotting total suction in log(kPa) versus millivolt outputs from the psychrometer on a semi-log paper.

2.2 Filter paper method

The calibration curves relating soil suction to water content of filter papers have been established using filter papers, salt solutions, pressure plates and membranes, and tensiometers. The salt solutions are usually used for the high suction range and the pressure plates and tensiometers are employed for low suction range. Until recently, a single calibration curve has been constructed using both total and matric suction measuring devices (Fredlund & Rahardjo 1993). However, Houston et al. (1994) established two calibration curves, one for total suction and other for matric suction measurements using the Fisher quantitative coarse filter papers.

In this study, the calibration relationship developed by McQueen & Miller (1968) for the Schleicher & Schuell No. 589-WH type filter papers was adopted. This relationship is also provided in the ASTM D 5298 (1994) standard on the filter paper method of measuring soil suction. The provided curve consists of two equations for a wide range of filter paper water contents. The part of the curve that gives suction range between about 1.5 log(kPa) and about 5 log(kPa) was used in comparison. The relationship is given in Equation 1.

$$h_t = 5.056 - 0.0688w \tag{1}$$

where h_t = log suction in kPa, w = filter paper water content in percent. Equation 1 is valid for the filter paper water contents equal or lower than 54.16%.

3 SOIL TOTAL SUCTION MEASUREMENTS

The total suction measurements on three different soils have been performed using the transistor psychrometer and the filter paper method. The soil samples were compacted at different moisture contents and left in sealed plastic bags in a relatively constant temperature room for about one week. Then, representative samples were taken from the same soil to conduct the psychrometer and the filter paper measurements. Both the set up for the filter paper measurements and the loading of the psychrometer probes were initiated at the same time and conditions with the same soil.

3.1 Soil samples

Three different soils were obtained from three different locations and some amount of these soils, which were already processed and sieved through US No.40 sieve for an ongoing project, were used for this research. The physical properties of the three soils adopted in this study with low, medium, and high plasticity indexes are summarized in Table 1.

Table 1. Physical properties of the three soils.

Soil	Liquid Limit (LL)	Plastic Limit (PL)	Plasticity Index (PI)	Optimum Moisture Content (%)
High PI	57	16	41	15
Medium PI	40	14	26	13.5
Low PI	24	12	12	10.5

3.2 Transistor psychrometer suction measurements

In this study, the 12-probe transistor psychrometer was used for the suction measurements. After stabilizing the probes overnight with 1 log(kPa) solution, 11 probes of the 12-probe psychrometer (one of the probes currently not functioning) were loaded with one of the soil samples and the probes were inserted into the thermally insulated container for one hour of equilibration time. The psychrometer was kept in a relatively constant temperature room during the testing process.

At the end of the equilibration time the output millivolt values of the each probe were recorded and with the corresponding calibration curve the total suction values of the soil samples were obtained. The total suction results are summarized in Tables 2-4. In these tables under the psychrometer column are the suction values from each of the 11 probes from the same soil sample. The average suction values and the standard deviations of the soil samples from the 11 probes are also tabulated under the corresponding columns.

3.3 Filter paper method suction measurements

In this study, the Schleicher & Schuell No. 589-WH type filter papers were used for the total suction measurements. The procedures suggested by the ASTM D5298 (1994) standard, (Lee 1991), and (Bulut 1996) on the measurement of total suction using the filter paper method were followed closely.

For the measurement of total suction, filter papers were simply placed above a soil specimen (in a non-contact manner) in sealed containers. Then, the whole set up was put in a constant temperature environment. Temperature fluctuations were kept as minimal as possible during the equilibrium period, which was one week for this study. At the end of the equilibration time, the filter paper water content was measured as quickly and accurately as possible. Finally, the total suction value of the soil sample was found from a calibration curve that was constructed for that particular filter paper.

The calibration relationship developed by McQueen & Miller (1968) for the Schleicher & Schuell No. 589-WH filter paper was adopted in this study, which is given in Equation 1. The total suction results obtained from the filter paper test are

Table 2. Total suction results of high and medium PI soils.

High PI Soil ($w^* = 23.5\%$)		Medium PI Soil ($w^* = 22.1\%$)	
Psychrometer (log kPa)	Filter Paper (log kPa)	Psychromter (log kPa)	Filter Paper (log kPa)
2.54	2.74	2.82	2.49
2.84	2.61	2.78	2.41
2.69	2.56	2.57	2.67
2.74	2.86	2.67	2.48
2.44	2.78	2.49	2.38
2.56	2.54	2.47	2.41
2.86	2.52	2.50	2.48
2.37	2.57	2.50	2.39
2.50	2.73	2.50	-
2.57	2.74	2.32	-
2.47	-	1.75	-
Average	Average	Average	Average
2.60	2.66	2.49	2.46
Std. Dev.[**]	Std. Dev.[**]	Std. Dev.[**]	Std. Dev.[**]
0.16	0.12	0.29	0.09

[*]w = Soil water content.
[**]Std. Dev. = Standard Deviation calculated with n-1 method.

Table 3. Total suction results of medium PI soil.

Medium PI Soil ($w^* = 14.3\%$)		Medium PI Soil ($w^* = 16.7\%$)	
Psychrometer (log kPa)	Filter Paper (log kPa)	Psychromter (log kPa)	Filter Paper (log kPa)
3.00	3.05	2.57	2.59
2.78	3.07	2.74	2.61
2.75	3.03	2.67	2.70
2.74	3.03	2.80	2.87
2.79	3.01	2.53	2.44
2.91	2.94	2.57	2.51
2.83	2.73	2.73	2.72
2.80	2.83	2.69	2.70
2.72	-	2.47	-
2.76	-	2.50	-
2.74	-	2.60	-
Average	Average	Average	Average
2.80	2.96	2.62	2.64
Std. Dev.[**]	Std. Dev.[**]	Std. Dev.[**]	Std. Dev.[**]
0.08	0.12	0.11	0.13

[*]w = Soil water content.
[**]Std. Dev. = Standard Deviation calculated with n-1 method.

summarized in Tables 2-4. In these tables under the filter paper column are the suction values from each filter paper from the same soil sample. The average suction values and the standard deviations of all the filter papers are also tabulated under the corresponding columns.

Filter papers should not make direct contact with the soil sample in total suction measurements using the filter paper method. For this purpose some kind of a non-corrosive support, which prevents filter papers from making contact with the soil, were placed on the soil sample in sealed containers. The support

Table 4. Total suction results of low PI soil.

Low PI Soil (w* = 11.4%)		Low PI Soil (w* = 13.4%)	
Psychrometer (log kPa)	Filter Paper (log kPa)	Psychromter (log kPa)	Filter Paper (log kPa)
1.81	2.51	1.55	1.91
2.52	2.42	2.24	1.83
2.41	2.68	1.76	1.90
1.97	2.52	0.76	2.03
2.15	2.62	2.10	1.97
2.39	2.64	1.97	2.16
2.46	2.44	1.90	2.46
2.51	2.28	2.15	2.43
1.89	-	2.05	-
2.06	-	1.58	-
2.29	-	1.75	-
Average	Average	Average	Average
2.22	2.51	1.80	2.09
Std. Dev.**	Std. Dev.**	Std. Dev.**	Std. Dev.**
0.26	0.13	0.41	0.24

*w = Soil water content.
**Std. Dev. = Standard Deviation calculated with n-1 method.

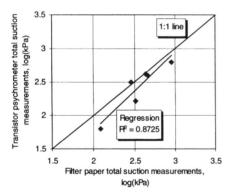

Figure 1. Comparison of data from both methods.

mentioned above should be of such a form that makes minimum contact with filter papers in order to prevent the condensed moisture that can move into filter papers through the support.

The filter paper suction results in Tables 2-4 contain both top and bottom filter papers. The average suction values from Tables 2-4 are plotted on the line of equality in Figure 1 to see how the suction values from both methods are scattered.

4 DISCUSSION OF RESULTS

This paper reports on comparison of total suction measurements using the transistor psychrometer and the filter paper methods. Total suction measurements were performed on three different soils at different moisture contents well above their optimum

values. Suction results from all of the probes and from each filter paper are depicted in Tables 2-4.

Both the psychrometer and filter papers become very sensitive to changes in total suction values at high moisture contents. As is mentioned in the transistor psychrometer user manual it becomes very difficult to measure reliable relative humidity values below about 2 log(kPa) suction. This is because when the relative humidity approaches about 99.9% total suction becomes very sensitive to very minor moisture changes. This sensitivity was also observed during the calibration process; scattered suction values were obtained from the probes when using the 2 log(kPa) sodium chloride salt solution. However, the results become more consistent with decrease in the relative humidities below about 99%. This trend of the sensitivity can also be observed from Tables 3-4 and Figure 1. For the same soils, from Table 3 or 4, the psychrometer gives more scattered results with the increase in the soil moisture content.

The psychrometer and the filter paper method work in a similar way; both predict the relative humidity thus total suction in a closed system. For this reason, filter papers are also very sensitive when dealing with low suction measurements. This sensitivity point starts at about below 1.5 log(kPa) suction for the Schleicher & Schuell No. 589-WH type filter papers. Houston et al. (1994) also pointed out this sensitivity in the development of a total suction calibration curve for the Fisher quantitative coarse filter papers. For instance, it can be seen in Table 4 that when the low PI soil water content increases from 11.4% to 13.4% the suction values become more scattered. This can also be concluded from the standard deviation results of the same low PI soil.

5 CONCLUSION

This paper reported on total suction measurements using the transistor psychrometer and the filter paper method that were conducted on three different soils that were compacted well above their optimum moisture contents. The total suction measurements indicate that suction becomes very sensitive to very small changes in the relative humidity at high moisture contents. When measuring total suction using the psychrometer and the filter paper methods at high water contents, care must be taken about the sensitivity of suction at high water contents at low suctions. Although both methods are sensitive to suction changes at high moisture contents, it is seen from the standard deviation results (Table 2-4) that the filter paper method gives more consistent results. Both methods give close results at low relative humidities.

REFERENCES

ASTM D 5298 1994. Standard test method for measurement of soil potential (suction) using filter paper. *American Society of Testing and Materials.* 1998 Annual Book of ASTM Standards: 156-161.

Bulut, R. 1996. *A re-evaluation of the filter paper method of measuring soil suction.* M.Sc. Thesis, Texas Tech University, Lubbock, Texas, USA.

Fredlund, D. G. & Rahardjo, H. 1993. *Soil mechanics for unsaturated soils.* New York: John Wiley & Sons, Inc.

Houston, S. L., Houston, W. N., & Wagner, A. M. 1994. Laboratory filter paper suction measurements. *Geotechnical Testing Journal* (GTJODJ) 17(2): 185-194.

Lee, H. C. 1991. *An evaluation of instruments to measure soil moisture condition.* M.Sc. Thesis, Texas Tech University, Lubbock, Texas, USA.

McQueen, I. S. & Miller, R. F. 1968. Calibration and evaluation of a wide-range gravimetric method for measuring moisture stress. *Soil Science* 106(3): 225-231.

Unsaturated Soils for Asia, Rahardjo, Toll & Leong (eds) © 2000 Taylor & Francis, ISBN 90 5809 139 2

Use of a new thermal conductivity sensor for laboratory suction measurement

D.G. Fredlund, F. Shuai & M. Feng
Department of Civil Engineering, University of Saskatchewan, Saskatoon, Sask., Canada

ABSTRACT: The thermal conductivity matric suction sensor has been proven to hold great promise for the *in situ* and laboratory measurement of soil suction. Some difficulties have been associated with the use of early versions of the thermal conductivity sensors. An improved thermal conductivity sensor has been developed at the University of Saskatchewan, Saskatoon, Canada to overcome these difficulties. The new sensor has some design aspects that are superior to that of previous sensors. These design aspects make the new sensor a more suitable device for both laboratory and *in situ* soil suction measurements.

1 INTRODUCTION

The measurement of soil suction is essential to the study of unsaturated soil behavior. Volume change, shear strength and seepage analyses all require an understanding of the matric suction in the soil. One of the more common methods used for continuous suction monitoring is the use of thermal conductivity matric suction sensors. This method uses a measurement of the thermal conductivity of a standard ceramic tip. The thermal conductivity of the ceramic tip is correlated with the matric suction of the ceramic which is in equilibrium with the surrounding soil. This method has been proven to be a promising device for the *in situ* and laboratory measurement of soil suction (Fredlund and Rahardjo, 1993).

The attractiveness of the thermal conductivity soil suction sensor lies primarily in its ability to produce a reasonably reliable measurement of soil suction over a relatively wide range of suctions over a long period of time. The measurements are essentially unaffected by the salt content of the soil (Lee and Fredlund, 1984 and Fredlund and Wong, 1989). The thermal conductivity sensors are also versatile and able to be connected to a data acquisition system for continuous and remote monitoring.

Some of the past difficulties associated with previous thermal conductivity sensors are the low strength of the ceramic, poor durability of the ceramic and electronics, and the low accuracy over portions of the suction range. The influences of factors, such as the soil temperature and the hysteresis properties of the ceramic on the suction measurement were not previously taken into account.

An improved thermal conductivity matric suction sensor was developed at the University of Saskatchewan, Canada, for the purpose of measuring soil suctions over a wide range of values. This paper describes the use of the new thermal conductivity sensors to measure matric suction in the laboratory.

2 GENERAL CHARACTERISTICS OF THE NEW THERMAL CONDUCTIVITY SOIL SUCTION SENSOR

The new thermal conductivity soil suction sensor (Fig. 1) has some characteristics that make it superior to previous sensors. Those characteristics make the new soil suction sensor more suitable for both laboratory and *in situ* soil suction measurement.

A new ceramic with a high porosity (i.e., more than 60%) and a wide range of pore sizes (i.e., ranging from 0.05 mm to less than 0.0001 mm) is used as the ceramic tip for the new sensor. This ceramic tip can be used to measure suctions from approximately 5 kPa to 1500 kPa. In addition, there is a near-linear relationship between water content and logarithm of soil suction between 5 kPa and 500 kPa.

The electronic properties of the new sensor have also been improved by using superior integrated circuitry for the temperature sensing device. Advanced signal conditioning (i.e., amplification, isolation and filtering) was used to further increase the resolution and accuracy of the new sensor. A comparison of heating curves measured before and after improvement is presented in Fig. 2 and the main technical

Figure 1 The thermal conductivity soil suction sensor developed at the University of Saskatchewan, Saskatoon, Canada.

Table 1. Technical specification.

Measurement parameters	Soil suction Soil temperature
Measurement range	Soil suction 5 to 1500 kPa Temperature -40 °C to 110 °C
Accuracy	± 5% for suction measurement ± 0.5 °C for temperature measurement
Resolution	0.33 mV
Soil types	Suitable for all soil types
Protection	Suitable for long-term burial
Temperature	0 to 40 °C for suction measurement (no damage when used in frozen soils, but suction readings do not represent soil suction)
Power supply	12V ~ 15V DC, 250 mA
Size	Diameter: 28 mm, Length: 38 mm
Cable length	Standard: 8m, Maximum: 100m

specifications for the new sensor are listed in Table 1.

3 CALIBRATION OF THE NEW SENSOR

The accuracy of the soil suction measurements obtained when using the thermal conductivity suction sensor is dependent upon their calibration. The calibration of the new sensor was performed by applying a range of matric suction values to the sensor. The sensor was embedded in a soil that was placed in a Tempe cell (Fig. 3). The soil in the Tempe cell provided continuity between the water phase in the porous ceramic tip and the water in the high air entry

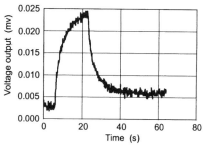

A. Measured heating curve before improvement

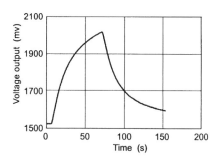

B. Measured heating curve after improvement

Figure 2 The heating curves measured before and after improvement.

disk. A matric suction was applied by increasing the air pressure in the Tempe cell while maintaining the water pressure below the high air entry disk at at-

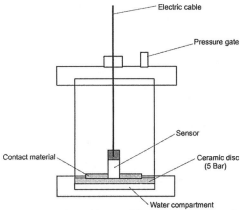

Figure 3 The physical layout of the test apparatus for calibration of sensors.

where a = parameter designating the output voltage under saturated conditions; c = parameter designating the output voltage under dry conditions; d = parameter designating the slope of the calibration curve, and b = parameter related to the inflection point on the calibration curve.

A typical calibration curve for a sensor is shown in Fig 4 along with its parameters. Since there are four parameters in Eq. 1, only five calibration points are required to establish the calibration curve. As a result, the calibration process is simplified and the time required for calibration is significantly reduced. Equation 1 facilitates the calculation of the soil suction from the output voltage of the sensor and increases the accuracy of the suction measurement.

mospheric conditions. The change in voltage output from the sensor was monitored periodically until suction equilibrium was achieved. The above procedure was repeated for various applied suctions ranging from 0 to 400 kPa so that a calibration curve could be obtained.

The calibration curve obtained from the calibration process is non-linear. In order to use the calibration curve to calculate the soil suction from the output voltage of the sensor, it is important to have a reasonably accurate characterization of the calibration curve. The following equation is proposed to fit the relationship between output voltage, ΔV, and soil suction, ψ.

$$\psi = \left[\frac{b \cdot (\Delta V - a)}{c - \Delta V} \right]^d \qquad (1)$$

4 FACTORS INFLUENCING THE SOIL SUCTION MEASUREMENT

Some factors may influence the suction measurement. This factors are soil temperature, freeze-thaw cycles and hysteresis of the ceramic.

4.1 Soil Temperature

The influence of soil temperature on the suction measurement is shown in Fig. 5. This influence is attributed to the change in thermal conductivity of the water with ambient temperature. In order to eliminate the influence of ambient temperature, a temperature correction was developed and can be expressed as follows,

$$\Delta V_{23\,^\circ C} = \frac{0.0014t + 0.561}{0.593} \Delta V_t \qquad (2)$$

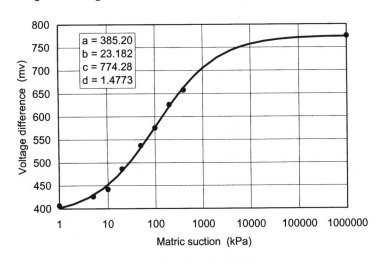

Figure 4 A typical calibration curve for a thermal conductivity sensor.

277

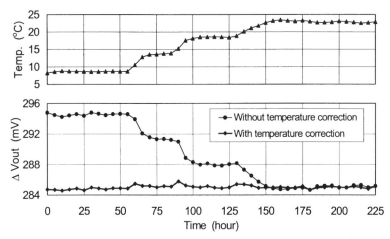

Figure 5. The influence of the ambient temperature and temperature correction technique verification (suction = 10 kPa).

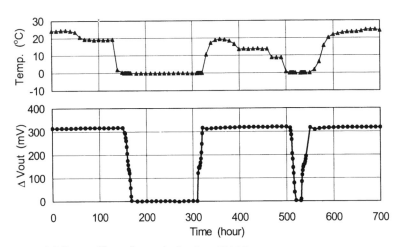

Figure 6 Influence of freeze-thaw cycles (suction = 50 kPa).

where t = the soil temperature; $\Delta V_{23°C}$ = the output voltage at 23°C; and ΔV_t = the output voltage at temperature, t.

Figure 5 shows the output voltage with or without temperature correction for one thermal conductivity sensor. The soil temperature during the measurements is also shown in the figure. It can be seen that the temperature influence is significantly accounted for by using the proposed temperature correction technique.

4.2 Freeze-Thaw Cycles

Experiments in the laboratory have shown that there are difficulties associated with measuring negative pore-water pressures in freezing soils when using thermal conductivity soil suction sensors (Fredlund etc., 1991). Distinct drops in the output voltage have been observed during the freezing and thawing processes (Fig. 6). These drops are attributed to the effect of the latent heat of fusion on thermal conductivity. As soil freezes, the proportions of unfrozen and frozen water changed and since ice and water have different thermal conductivities, the output voltages are difficult to interpret and to convert to soil suctions. However, the freeze-thaw cycles do not appear to influence the soil suction measurements after thawing. In the other words, the quality and the calibration properties do not appear to be affected by the freeze-thaw cycles (Fig. 6).

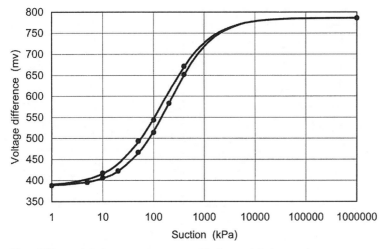

Figure 7 Hysteresis in the sensor response caused by hysteresis in the ceramic.

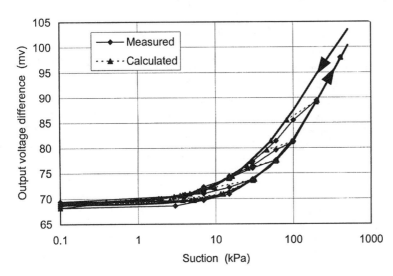

Figure 8. A comparison between the applied suction and the predicted suction using the hysteresis model.

4.3 *The Hysteresis of the Ceramic*

The water content versus matric suction curves for any porous material during wetting and drying are generally not the same. The hysteresis in the soil-water characteristics of the ceramic may cause hysteresis in the sensor response upon wetting and drying (Fig. 7) The research conducted by Feng (1999) indicated that the maximum possible relative error in the suction measurement caused by hysteresis is about 30%. However, the hysteresis loop for the new ceramic tip is stable and reproducible. In the other words, the wetting and drying curves for a ceramic

tip do not change with time.

In order to eliminate the influence of the hysteresis associated with the ceramic, a hysteresis model was proposed (Feng, 1999). This model can be used to modify data obtained in engineering practice according to the wetting or drying history of the sensor. As a result, it is possible to obtain greater accuracy in the measurement of soil suction. A comparison between actual soil suction changes and the predicted soil suction changes using the hysteresis model are shown in Fig. 8. Good agreement was observed between the predicted soil suction changes and actual soil suction changes.

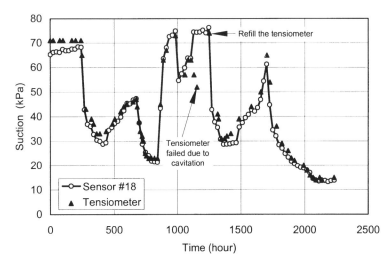

Figure 9   Soil suction readings from thermal conductivity sensor installed in a Indian Head till soil specimen.

5   LABORATORY SOIL SUCTION MEASUREMENTS

The new thermal conductivity sensor was installed into a soil specimen in the laboratory to measure soil thermal conductivity sensor. The test result is presented in Fig. 9. Good agreement was observed between the soil suction measured using the thermal conductivity sensor and those measured using the tensiometer (prior to cavitation of the water in the tensiometer).

6   CONCLUSIONS

An improved thermal conductivity soil suction sensor was developed to eliminate the difficulties encountered with early versions of the device. The new thermal conductivity sensor has been found to be quite sensitive and accurate in measuring soil suctions in the range from 5 to 1500 kPa. The strength and durability of the new sensor have also been significantly improved.

A laboratory calibration procedure has been set up. A calibration equation has been proposed to facilitate in the calculation of soil suction from the output voltage. The overall result is an increase in the accuracy of soil suction measurements.

Soil temperature and hysteresis in the ceramic have negative influences on soil suction measurements. A temperature correction and a hysteresis model have been proposed to eliminate these influences and to increase the accuracy of the soil suction measurements.

REFERENCES

Fredlund, D.G., Gan, J. K-M., and Rahardjo, H. 1991. Measuring negative pore-water pressures in a freezing environment. *Transportation Research Record* 1307: 291-299.

Fredlund, D.G. and Rahardjo, H. 1993. *Soil Mechanics for Unsaturated Soils*. New York, NY, John Wiley & Sons.

Fredlund, D.G. and Wong, D.K.H. 1989. Calibration of thermal conductivity sensors for measuring soil suction. *ASTM Geotechnical Testing Journal* 12(3): 188-194.

Lee, R.K.C. and Fredlund, D.G. 1984. Measurement of soil suction using the MCS 6000 gauge. *Proc. 5th Int. Conf. Expansive Soils*. Adelaide, Australia. 50-54.

Unsaturated Soils for Asia, Rahardjo, Toll & Leong (eds) © 2000 Taylor & Francis, ISBN 90 5809 139 2

A comparison of in-situ soil suction measurements

B.A. Harrison
P.D. Naidoo and Associates, Johannesburg, South Africa

G.E. Blight
University of the Witwatersrand, Johannesburg, South Africa

ABSTRACT: A number of soil suction measuring devices were employed to measure total and matrix suction *in-situ*. Two sites were selected, each underlain by different subsoil and groundwater conditions. Measurements of suction were taken over a period of two years, from which it was established that none of the instruments yielded comparable values of absolute suction or suction change. The instruments did, however, record suctions that follow seasonal wet and dry cycles. The paper presents possible reasons for the disparity in measuments for the devices.

1 INTRODUCTION

The role of soil suction in the prediction of unsaturated soil behavior has been well documented. Over the past decades numerous instruments have been developed to measure matrix and solute suction. However, most of these instruments are amenable for carrying out suction measurements in the laboratory with only a few suitable for operation *in-situ*.

Depending on the suction range of interest, these instruments employ either direct or indirect methods of measurement. Whereas the latter require appropriate calibration prior to use, the former do not. Thus, if inappropriate calibration methods are adopted, inconsistent measurements result, particularly when making comparisons of the various measuring techniques.

Whilst comparative suction measurements employing a range of instruments and techniques have been reported in the literature (Lee & Wray - 1992, Guan & Fredlund - 1997) all of these comparisons have been carried out on soil samples prepared and tested under controlled conditions in the laboratory. Few reports have been published comparing the results of *in-situ* soil suction measurements employing different instruments.

This paper presents the results of *in-situ* soil suction measurements using various devices over a period of two years. Two sites were selected, each underlain by different soil types and groundwater conditions.

2 DESCRIPTION OF THE SITES

2.1 Climate

Two sites, referred to as Site A and Site B, were selected to carry out the soil suction measurements *in-situ*. Both sites are in South Africa and enjoy warm temperate conditions with summer rain. The area is in a water deficient region where annual evaporation exceeds precipitation. Some climatic statistics are as follows:

Annual average rainfall: 713mm
Annual A pan evaporation: 2306mm
Annual average temperature: 16°C
Mean of warmest month: 22°C
Mean of coldest month: 10°C

2.2 Subsoils

The subsoil underlying Site A comprises a deep residual andesite in the form of a slightly clayey silt, which classifies as MH according to the Unified Soil Classification (USC) system. The regional water table is at some 15 to 20m below the ground surface in this area.

Site B is underlain by transported colluvium comprising clayey sand, which classifies as SC according to the USC system. A ferruginised (ferricrete) layer is present beneath the colluvium at a depth of approximately 1m below the ground surface. The regional water table in the area is fairly deep. However, the ferricrete horizon forms a relatively impermeable layer on top of which a perched water table develops in the wet summer seasons.

3 INSTRUMENTATION

The instruments used to measure the soil suction *in-situ* at the two sites included thermocouple psychrometer probes, gypsum resistance blocks and fibreglass moisture cells, all of which employ indirect methods to measure soil suction. In addition to these instruments, direct suction measurements were carried out at Site B employing a jet-fill tensiometer and Imperial College (IC) tensiometer.

Ridley & Wray (1996) have described the manner in which these instruments measure soil suction.

3.1 Calibration

Calibration of the psychrometer probes was carried out by suspending them above salt solutions of varying concentrations, and hence known suctions, and recording the corresponding voltage readout.

Both the gypsum blocks and fibreglass cells were calibrated in the laboratory by burying the probes in compacted specimens of the soil encountered at each of the sites. The soil, container and probes were periodically weighed and the corresponding outputs from the readout devices recorded. This process was continued as the soil was permitted to dry evenly through slow evaporation to the atmosphere. Thereafter the soil was gradually wetted and readings repeated.

Upon completing the measurements, oven drying the soil enabled its water content to be determined at each of the instrument readings. From separately measured soil-water characteristic curves determined in the pressure plate apparatus (suction vs water content relationship) for each soil, a calibration relationship was derived between soil suction and readout units for each device and soil type.

The gypsum resistance blocks were also calibrated directly in the pressure plate apparatus. In this instance the block was inserted in a cylinder which was in turn filled with a slurried specimen of soil. The arrangement was placed onto the saturated ceramic and the apparatus assembled. A specially designed lid enabled the leads from the block, situated in the chamber, to be connected to the recording meter situated outside of it. Air pressures were applied and meter readings taken when moisture equilibrium had been achieved. A hysteresis curve was established by incrementally reducing the air pressure.

4 *IN-SITU* SUCTION MEASUREMENTS

Each psychrometer, gypsum block and fibre cell was inserted into a 100mm diameter hand augered hole drilled to a depth of 600mm below the ground surface. The soil removed from the hole was lightly tamped around each of the sensors and the wire electrical leads were sealed in a jar buried just below the surface.

The jet fill tensiometer was inserted into a pre-drilled hole of the same diameter as the tensiometer probe. The hole was prepared just prior to inserting the instrument. The IC tensiometer was bedded into the base of a hole augered near to the sensors, which was also drilled just prior to taking a reading.

Periodic readings from the buried sensors were taken over a period of two years and, based on the calibration data, converted to suction pressures. These *in-situ* suction readings are presented in Figures 1(a) and (b) for sites A and B respectively.

Neither the jet fill nor the IC tensiometers were able to function at site A, due to the high suctions encountered in the area, however, some readings were possible with the jet fill tensiometer at site B when a perched water table developed in the wet summer seasons. Cavitation occured in the IC tensiometer and no meaningful measurements were possible with this instrument for the soil encountered at this site.

Suctions recorded with the jet fill tensiometer are illustrated in Figure 2. Readings were only possible when a perched water table developed on the underlying ferricrete layer, and the instrument was unable to record soil suctions above 75kPa.

Superimposed on the figures 1 and 2 is the mean monthly rainfall for the region.

5 DISCUSSION OF RESULTS

It is evident from Figures 1(a) & (b) that none of the devices measure comparable soil suctions. Whilst it can be argued that since the psychrometer measures total suction and the others matrix, the psychrometer should record higher suctions than the other instruments, which in most instances it does.

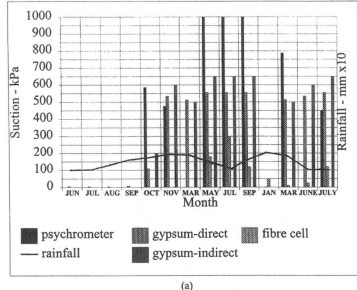

(a)
Site A

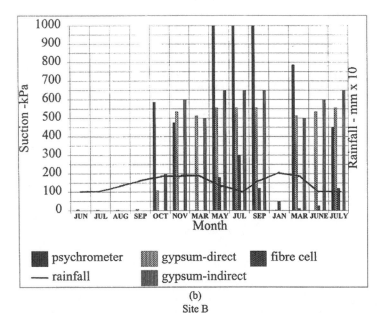

(b)
Site B

Figure 1: *In-situ* suction readings at Sites A & B

However, total suctions should be consistently greater than matrix suctions, which is not evident from the data.

Matrix suction measurements are similarly not in agreement for each of the two indirect and direct instruments and different calibration techniques. Possible reasons for this include hysteresis and moisture continuity, discussed below.

5.1 Hysteresis

Unsaturated soils undergo substantial hysteresis in suction during wetting and drying, as do the porous sensors of the measuring instruments. Thus, depending on whether the soil is wetting or drying, its suction for the same water content will be

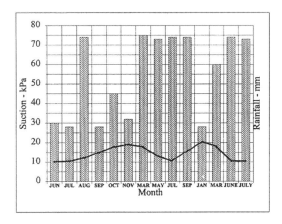

Figure 2: Suction readings using the jet fill tensiometer at Site B

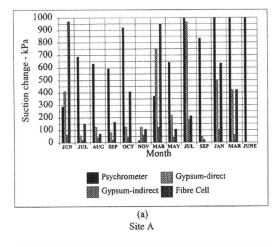

(a)
Site A

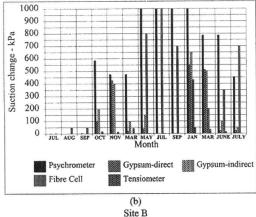

(b)
Site B

Figure 3(a) & (b): Monthly suction changes

different. Similarly, the readout from the porous measuring sensor will also record a different value when it is wetting to that when it is drying. Due to the large time (monthly) intervals over which the suction readings were taken, it was not possible to establish whether the soil was in the process of wetting up or drying out at the time of taking the readings. An attempt was nevertheless made to account for the hysteresis effects of the instruments by assuming that they wet up in the summer rainfall months and dry out as the dryer winter approaches.

5.2 Moisture Continuity

The devices employed to indirectly measure matrix suction utilise a porous sensor, its purpose being to communicate, via the pores, with the moisture in the soil. For this to occur moisture must be transferred to the soil from the sensor or from the sensor to the soil. If moisture continuity is not maintained between the measuring device and the soil, then the instrument will not record the matrix suction of the soil.

As far as possible, the suction-water content curve of a porous sensor should be similar to that of the soil, and with similar hysteresis characteristics. However, the limited range of sensors commercially available makes this difficult to achieve.

It may be argued that in many instances absolute suction is less relevant to the prediction of unsaturated soil behavior than the moduli of suction change. Month-to-month suction changes within the soil at each of the two sites have been plotted in Figures 3(a) & (b). Again it is evident from the figure that none of the devices recorded similar changes in suction over the two year period.

What is perhaps noteworthy, and the very least to be expected, is that suction measurements do change in sympathy with the rainfall patterns, with relatively high and low suctions being recorded in the dry and wet months respectively.

6 CONCLUSIONS

Three indirect and one direct *in-situ* soil suction measuring devices have been employed to record, over a period of two years, the soil suction at two sites underlain by two soil types with different groundwater conditions. None of the instruments recorded comparable suctions nor were suction changes inferred from the various instruments comparable. Suction was, however, seen to change in response to the wet and dry seasons.

There are many reasons for the lack of agreement between the various devices, some of which have been discussed earlier, while others have yet to be identified.

The study shows that after close to a century of efforts to measure soil suction *in-situ*, we still do not seem to have a consistent, simple and reliable method of doing this. The various devices that are available indicate the direction of suction change, but we still lack a method for reliably measuring suction magnitude.

7 REFERENCES

Guan, Y. & Fredlund, D. G. (1997) *Use of the tensile strength of water for the direct measurement of high soil suction.* Can. Geot. J. Vol. 34(4), pp. 604-614.

Ridley, A. M. & Wray, W. K. (1996) *Suction measurement: A review of current theory and practices* Proc. 1st Int. Conf. on Unast. Soils. Paris, France. pp 1293-1322.

Unsaturated Soils for Asia, Rahardjo, Toll & Leong (eds) © 2000 Taylor & Francis, ISBN 90 5809 139 2

Comparative measurements with sand box, pressure membrane extractor and pressure plate extractor

E. Imre – *Geotechnical Research Group for the Hungarian Academy of Sciences, Budapest, Hungary*

K. Rajkai – *Research Institute for Soil Science and Agricultural Chemistry of the HAS, Budapest, Hungary*

Z. Czap & T. Firgi – *Department of Geotechnics, Budapest University of Technology and Economics, Hungary*

G. Telekes & L. Aradi – *Ybl Miklós School of Engineering, St. István University, Budapest, Hungary*

ABSTRACT: Two soil-water characteristic curve (SWCC) measuring methods (in the 0 to 500 kPa suction range) are compared in an ongoing research. The first method comprises the sand box (in the low suction range) in conjunction with the pressure membrane extractor (for higer suction values). The second method uses a pressure plate extractor where the water content is determined in such a way that the mass of the sample and the equipment are measured together. Initial results indicate that the air mass is not negligible for high suction values using the pressure plate extractor method. The application of a correction table is suggested since the non-monotonous air pressure increase may destroy the water continuity at the sample-plate contact. It is also found that a constant error may occur in the pressure membrane extractor method for plastic soils.

1 INTRODUCTION

The soil-water characteristic curve (SWCC) is a relation between the water content (gravimetric w or volumetric v) and the stress state variable called suction ($\psi = u_a - u_w$) of the soil under the condition where the other stress state variable, net normal stress, is zero. This constitutive relation is used in transient seepage analyses and, for the calculation of some soil property functions.

Various procedures are known for the experimental determination of the SWCC. A point on this curve is generally "measured" in such a way that the equilibrium water content of the soil sample is determined at a prescribed suction value. The water content determination is generally made by weighing the wet and the dry sample. No standard exists to embrace every type of soils and methods (e.g. ASTM D 2325 is for a pressure plate extractor up to 100 kPa of suction).

The equipment used for suction control can be categorised into 3 groups: (i) The air pressure is atmospheric, the water pressure is decreased (e.g. simple mechanical methods below the cavitation pressure, $\psi < 100$ kPa); (ii) The water pressure is controlled and the air pressure is increased (axis-translation technique); and (iii) The relative humidity is controlled (e.g. chemical method for high suctions $\psi > 2000$ kPa).

A semi-permeable membrane is necessary for the first two cases that is closed for air and open for water. For the sand box method, this is a saturated sand layer; for the pressure plate method, this is a high air entry ceramic disk, and for the pressure membrane method, this is a cellulose membrane.

The aim of the common research is to compare two methods known as the method of the Research Institute for Soil Science and Agricultural Chemistry of the Hungarian Academy of Sciences (SSI) and the method of the Budapest University of Technology and Economics, Department of Geotechnics (TU).

The first method comprises the sand box method (in the low suction range) in conjunction with the pressure membrane extractor method (for higher suction values). Each measurement is made on a separate soil core.

The second method uses the pressure plate extractor manufactured at the University of Saskatchewan. The water content is determined on the same sample in such a way that the mass of the sample and the equipment are measured together.

Initial results indicate: (i) the sample-plate contact may be destroyed by the non-monotonous air pressure increase in the pressure plate method, and (ii) there is a possible constant error in the pressure membrane extractor method for plastic soils.

2 METHODS

2.1 The SSI method

The SSI method (Rajkai, 1993) comprises the sand box in the low suction range ($\psi < 50$ kPa) and the pressure membrane extractor for the higher suction values (Figs 1 and 2).

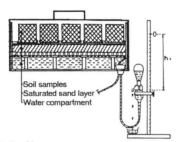

Figure 1. Sand box.

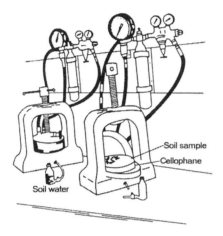

Figure 2. The pressure membrane extractor.

The saturated sand layer plays the role of the semi-permeable membrane in the sand box method. The suction is applied by the decrease of the pore water pressure [i.e. lowering the outlet level of the water compartment (Fig. 1)].

The pressure membrane extractor is 20 to 30 cm in diameter. The cellophane membrane is situated at the bottom of the extractor. The air pressure is increased inside the extractor only. Fredlund & Rahardjo (1996) suggested a counter-pressure below the membrane. The soil specimen is 1 cm in height and 4 cm in diameter (Fig. 2).

The water content was determined at the following suction values: 1 kPa, 10 kPa, 20 kPa, 50 kPa, 250 kPa, 500 kPa. Each measurement was made on a separate soil core.

The duration of one suction stage was at least three days, the mass was checked every second day afterwards.

2.2 The TU method

The Laboratory of the Geotechnical Department, Budapest University of Technology and Economy used the pressure plate extractor manufactured at the University of Saskatchewan, Department of Civil Engineering.

Photo 1. Parts of the pressure plate extractor.

The cell is a transparent extractor apt for the testing of a single soil sample. The main parts are shown in Photo 1. These are: (i) pedestal with the high air entry disk, (ii) the cylinder (h=9.5 cm, d=8.3 cm), (iii) the cup with the air valve, (iv) the spring and (v) the screws.

Photo 2. Spiral on the bottom of the pedestal.

The sample container is kept in contact with the disk by a spring. The air pressure is regulated through the upper valve.

The fine ceramic high air entry disk an air entry value of 500 kPa air. The pore water pressure compartment under the disk is spiral shaped so that the diffused air could be flushed (Photo 2).

The operation can be seen in Photos 3 and 4. The cell can be disconnected from the one air and two water lines and the whole equipment can be weighed. The equilibrium is reached when the mass change is less than 0.05 g within 30 minutes.

The water content of the soil sample was determined at the following pressure values: 1 kPa, 5 kPa, 10 kPa, 20 kPa, 30 kPa, 50 kPa, 250 kPa, and 400 kPa.

Photo 3. Equipment during operation.

Photo 4. Mass measurement.

3 SOILS

The main physical properties of the tested soils are shown in Tables 1 to 3. The grain size distribution of sandy soil 1 is shown in Figure 3.

Soil 1 is an eolian sand, soil 2 is a highly OC Quaternary Pannon clay, soils 3 is a lightly OC Pleistocen Szeged clays. The geological features of these soils are described in Ronai (1968), Imre (1995).

Table 1. Some soil physical parameters of the sand.

Sample	d_{60} (mm)	d_{10} (mm)	U (-)	void ratio (-)
1.	0,3	0,13	2,5	0,66

Table 2. Some soil physical parameters of the plastic soils

Sample	w_L (%)	I_P (%)	e (-)	w (%)
2.	35,7	17,6	0,720	27,18
3.	53,6	29,1	0,641	23,04

Table 3. Sampling conditions.

Sample	Intactness	Saturation condition
1.	disturbed	air dry
2.	undisturbed	below groundwater level
3.	undisturbed	below groundwater level

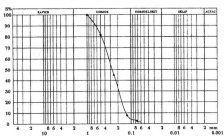

Figure 3. Grain size distribution for soil 1.

4 RESULTS, DISCUSSION

The SSI test results are shown in Table 4, the TU test results for soils 1 to 3 are shown in Table 5. Results are compared in Figures 4 to 6.

It is known, that the graph of a SWCC generally can be approximated by three straight lines with two breaking points on a semi-logarithmic plot.

Results indicate good agreement of the two methods in the case of the sand (soil 1) and a characteristic difference for the plastic soils (soils 2 and 3).

In the latter case the third straight-line portion was not determined. The slope of the first and, in particular the second part seems to be significantly different (with the same trend) for the two methods.

Table 4. Results of SSI.

Sample	1	2	3.
ψ (kPa)			W (%)
1	20	29.4	25,25
	20.625	29.53	24,94
10	6.25	28.67	24,14
	5.25	28.7	23,83
20	2.44	28.27	23,83
	2.875	28.2	23,64
50	1.75	26.2	23,33
	2.5	26.53	23,09
250	1.25	23.53	22,35
	2.125	25.4	22,16
500	0.94	22.3	21,91
	1.8125	24.87	21,73

Table 5 Results of TUB.

sample	1.	2.	3.
ψ (kPa)	w (%)	w (%)	w (%)
1	22,03	27,00	25,26
5	20,27	-	-
10	18,29	26,96	24,63
20	8,57	26,38	24,42
30	3,23	26,38	24,21
50	2,68	24,68	23,58
250	2,07	18,22	21,91
400	1,58	16,40	21,28

Table 6. Correction for air mass (measured).

air pressure (kPa)	air mass (g) for sample volume of 18,8 cm³	air mass (g) for sample volume of 37,6 cm³
5	0,00	0,00
20	0,05	0,05
50	0,10	0,10
250	0,70	0,75
400	1,20	1,30

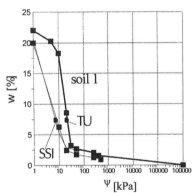

Figure 4. SWCCs for soil 1.

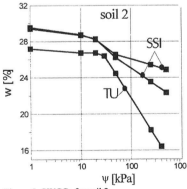

Figure 5. SWCCs for soil 2.

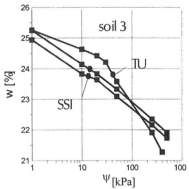

Figure 6. SWCCs for soil 3.

The equalization time for the sand box and the pressure membrane extractor method was at least 72 hours in each case.

The equalization time for the pressure plate extractor method varied between 3 to 15 hours for high water contents and, 14 to 72 hours for lower water contents.

The difference is partly attributed to the applied measurement regime and partly to the fact that the water permeability of the membrane is less than the plate.

The necessity of the correction for the air pressure mass was investigated for the pressure plate extractor method where the mass of the sample and the equipment are measured together.

It was found that the air mass is not negligible in case of high air pressures (Table 6). It was also found that that if the air pressure is decreased before the measurement then the water continuity of the sample-plate contact is generally destroyed. Therefore, the non-monotonous air pressure increase is not suggested for the pressure plate extractor method.

5 CONCLUSION

1. The advantages of the pressure plate methods are as follows: (i) the equalization time is much less than that for the pressure membrane extractor method, and (ii) only one soil core is tested throughout the test.
2. In case of the applied pressure plate extractor method (i) correction is necessary for the air mass, (ii) the air pressure increase is suggested to be monotonous otherwise the water continuity at the soil-plate contact may be destroyed.
3. The ongoing research aims at the clarification of the deviation in the two methods for plastic soils.

ACKNOWLEDGEMENT

This research is part of the research project of the

Hungarian National Research Fund (OTKA No T020446). The help of Professor D.G. Fredlund is greatly acknowledged.

REFERENCES

Fredlund D.G. – Rahardjo H.(1993) „Soil mechanics for un-saturated soils" John Wiley & Sons, New York, 560p.

Imre, E. (1995). Statistical evaluation of simple rheological CPT data. *Proc. of XI. ECSMFE, Copenhagen,* Vol. 1. 155-161.

Rajkai K.(1993) „A talajok vízgazdálkodási tulajdonságainak vizsgálati módszerei" Búzás I.(ed): Talaj- és agrokémiai vizsgálati módszerkönyv 1. INDA4321 Kiadó, Budapest. 115-160.

Rónai, A. (1968). The Quaternary of the Hungarian Basin. Guide to Excursion 41c. Hungarian Academy of Sciences, Budapest

Unsaturated Soils for Asia, Rahardjo, Toll & Leong (eds) © 2000 Taylor & Francis, ISBN 90 5809 139 2

Suction calibration curve of filter paper made in China

G.Jiang, Z.Wang, G.Tan & J.Y.Qiu
Wuhan University of Hydraulic and Electric Engineering, People's Republic of China

ABSTRACT: The filter paper method for measuring the suction in soil is described in this thesis. The homemade quantitative filter papers are calibrated with pressure plate apparatus, and the suction calibration curves during drying and drying-wetting are obtained. The characteristics of the curves are compared with classical filter papers. Some conclusions are summarized.

1 INTRODUCTION

Suction is a important parameter for engineering properties of unsaturated soil, one of variable for dominating unsaturated soil mechanical character. The coefficient of permeability, shear strength and deformation characteristic of unsaturated soil are related with soil suction. So, correctly predicting suction change in soil is the premise for unsaturated soil mechanics.

Gardner (1937) first proposed a method in which a commonly available filter paper is used to determine the soil suction. Subsequent researchers have further developed and endorsed the method. In authors' opinion, the filter paper method is a simple, low cost and easily piggybacked on site investigation for way of determining soil suction.

2 EXISTED METHOD

The filter paper method is based upon the water content equilibrium between soil and filter paper. The filter paper may be regarded as a sensor. Water could transfer within soil sample and filter paper either by capillarity effect with filter paper in direct contact with soil, or by vapor absorption with filter paper in the atmosphere directly above the soil. Under equilibrium, soil suction is equal to filter paper suction. The total suction can be divided into two components, matric and osmotic suction. Osmotic suction arises from the salt content in the pore fluid. Osmotic potentials arise from variations in salt content from one point to another. Matric suction is usually ascribed to capillarity and texture

and is therefore strongly related to geometrical factors such as pore size and shape. A simplified definition of matric suction is the affinity a soil has for water in the absence of any salt content gradients in the water. The filter paper is the only known technique for obtaining suction measurements that covers the full range of interest (near zero to perhaps 100mPa).

2.1 Calibration curve

Fawcett & Collis-George had described the first calibration curve with Whatman's No.42 filter paper. They used paper from eight different batches, treated with 0.005% Hgcl solution to avoid fungal and bacterial growth. Hamblin(1981), who repeated the study with both treated and untreated paper, reported that the treatment didn't affect the calibration curve. McQueen & Miller (1968) have developed similar correlations for Schleicher & Schuell No.589 White Ribbon filter paper. They pointed out that the paper placed in close contact with the soil measure matric potential, whereas when the paper in supported indirectly contact with soil in a closed chamber both matric and osmotic potentials are measured. It is common practice to measure either the total suction or the matric suction, but many researchers believe that it is more important for many applications in unsaturated soil mechanics to have matric suction values rather than total suction values. So, most of their attentions are focused on filter matric suction.

McQueen indicated that whether the filter paper is wetting or drying, it is no difference about the filter calibration curve, and needn't consider the filter paper hysteresis. But Chanlder, Swarbrick refers that filter paper exhibit hysteresis, their calibration and use should take this into account. The usual method is to always use the filter paper in a wetting or drying cycle, and to calibrate the paper accordingly. In general, the filter paper calibration curves could be classified with total suction calibration curves and matric suction curves.

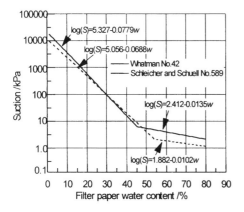

Figure 1. Classical filter paper calibration curve (Fredlund & Rahardjo, 1993)

2.2 Total suction

The total suction calibration curve of filter paper could be obtained through placing the filter paper in the headspace above the liquid in such a way that transfer of water to the filter occurs only by vapor absorption. Under isothermal conditions the relative humidity in the headspace above water in a sealed container approach a fixed, reproducible value. In the case of distilled water the equilibrium relative humidity is essentially 100%. If the water has salt in it, the equilibrium relative humidity may be significantly less than 100% depending on the type and concentration of salt solution. It is presumed that the salt resides in the liquid water and that the vapor in the headspace is pure water vapor when only water and salt make up the solution. Under these conditions, the relative humidity can be uniquely related to the total suction. Houston et al (1994) developed the total suction calibration curve of Fisher quantitative coarse filter paper with the salt solution method.

2.3 Matric suction

The matric suction calibration curve was determined with pressure membrane apparatus and tensiometers in the low suction range. The pressure membrane data were obtained by placing filter paper directly on the ceramic membrane and bringing the system to equilibrium at a set pressure. In general, the pressure

membrane apparatus data represent a drying curve. Tensiometers is used to obtain data for the case of wetting the filter paper. Dry filter paper, along with the tensiometer were placed in the soil, under equilibrium, the filter paper water content was determined.

Since Fawcett & Collis-George (1967) determined the first calibration curve of Whatman's No.42 filter paper, more reports about filter paper calibration curve were discussed by Hamblim (1981), MeQueen & Miller (1968), Al-Khafaf & Hanks (1974), Houston et al (1994). Calibration curve of another filter paper, namely Schleicher & Schuell No.589 White Ribbon filter paper, is similar to the Whatman's No.42. Classical two curves calibrated by several authors are generalized in Fig.1.

When the calibration curve of filter paper is determined, soil suction could be calibrated with the filter paper technique. It had been used by Mckeen(1980) for studies of airport pavement subgrades, and by Ching & Fredlund (1984) for the examination of swelling clay profiles.

2.4 Suction measurement with filter paper technique

Using the filter paper method, total or matric suction is determined for soils with filter paper. The filter paper must be non-ash quantitative filter. The same batch paper can have the same calibration curves. When measuring total suction a spacer is placed on top of the sample and then a filter paper placed upon it to prevent contact between the two. For matric suction measurement the filter is placed in direct contact with the sample, either underneath or weighted down on top of the sample.

Some implements are required with the filter paper technique, a sealed container, a electronic balance (0.0001g) and thermostat, infra-red oven, tweezers and so on. Different authors have used different time periods for equalization of the filter paper with the suction of the soil sample. Usually 7 days are allowed, but at least 5 days are required. The most important point of the technique is that drying and wetting from the oven-dry state both occur quickly in the laboratory atmosphere. Consequently, it is recommended that weighing be done within 30s.

3 CALIBRATION

Several filter papers had been calibrated, such as Whatman's No.42 filter paper and Schleicher & Schuell No.589 White Ribbon filter paper. The following filter paper, ShuangQuan No.202 quantitative filter paper is made in China, its calibration data was determined.

Filter paper is ShuangQuan quantitative filter made in China, diameter is 7cm, filter rate is medium, type is 202, ash is 0.35×10^{-4} g/per.piece.

One of calibration apparatus is No.1600 pressure plate extractor, range 0-1500kPa. Another is No.1250 volumetric pressure plate extractor and hysteresis attachment, range 0-200kPa.

Calibration experiment during drying was made with No.1600 pressure plate extractor, 8 piece of paper were placed directly contact upon membrane. Water seeping from filter paper was measured with 0.05ml burette. Experiment of the cycle from drying to wetting was taken with No.1250 volumetric pressure Plate extractor and hysteresis attachments. 20 piece of laminated paper were placed directly upon membrane, because the diameter of compression chamber had only 12cm. The water seeping from filter paper was measured with 0.1ml burette.

4 RESULTS AND DISCUSSION

The experiment results are shown in Figure 2 and Figure 3. Compared with calibration curves of Whatman's No.42 and Schleicher & Schuell No.589 White Ribbon filter paper, some discussion are taken as follows.

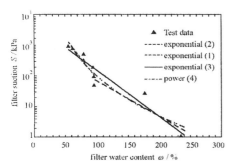

Figure 2. 1500kPa filter paper calibration curves

Curves of exponential function and power function fitted with the data of 1500kPa during drying are shown in Figure 2. Sectionalized exponential function curves fitted with Whatman's No.42 filter paper and Schleicher & Schuell No.589 White Ribbon filter paper are shown in Figure 1. The same analysis result is shown in Table 1. The data of 1500kPa during drying are reliable from the result with these filter paper curves. The coherence with analysis of all data is better than sectionalized analysis.

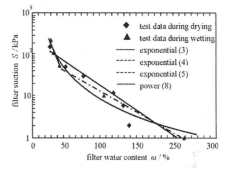

Figure 3. 200kPa filter drying-wetting calibration curves

Table 1. 1500kPa drying calibration curve of filter paper

Function	Range	Relative coefficient
Log(S)=2.989-0.0113 ω	0<S<50	0.79
Log(S)=1.556-0.025 ω	50<S<1000	0.83
Log(S)=3.663-0.0147 ω	0<S<1000	0.91
S=(4E+10) . $\omega^{-4.302}$	0<S<1000	0.90

The data of filter paper 200kPa during drying-wetting are shown in Figure 3. The data of during drying and the data of during wetting are interfluent, the hysteresis effect is not apparent. The fitted curves of filter paper 200kPa during drying-wetting

Table 2. 200kPa drying-wetting calibration curve of filter paper

		Function		Range	Relative coefficient
	Upside	Log(S)=2.425-0.011 ω	(1)	20<S<200	0.93
Drying	Underside	Log(S)=1.657-0.0064 ω	(2)	0<S<20	0.72
	Full	Log(S)=2.286-0.0097 ω	(3)	0<S<200	0.91
	Upside	Log(S)=3.165-0.037 ω	(4)	50<S<200	0.97
Wetting	Underside	Log(S)=1.989-0.0082 ω	(5)	0<S<50	0.99
	Full	Log(S)=2.316-0.0098 ω	(6)	0<S<200	0.96
Drying-wetting		Log(S)=2.261-0.0095 ω	(7)	0<S<200	0.93
Drying-wetting		S=1.58469$\times 10^5 \times \omega^{-2.1008}$	(8)	0<S<200	0.95

are shown in Table 2 in accordance with sectionalized and diverse process. The filter paper calibration curves of during drying (3), during wetting (6), and during drying-wetting are almost having no difference. ShuangQuan No.202 quantitative filter paper has no observable difference between the drying and wetting calibration curve and is regarded with no hysteresis phenomenon.

The calibration curves of 1500kPa paper and 200kPa have some variance. At the same pressure, water of 1500kPa is large than 200kPa. The filter paper placement way is not same because of differential diameter of calibration apparatus. Further more, saturated degree of membrane, stable time at pressure, the humidity and temperature of laboratory all invisible affect the result. The further study experiments need to be made.

The calibration curves of ShuangQuan No.202 quantitative filter paper, Whatman's No.42 filter paper and Schleicher & Schuell No.589 White Ribbon filter paper have good consistency.

5 CONCLUSION

ShuangQuan No.202 quantitative filter paper is calibrated with pressure plate apparatus, 1500kPa suction calibration curve during drying, and 200kPa calibration curve during drying-wetting are obtained. No observable hysteresis phenomenon is detected between drying and wetting.

REFERENCE

Fredlund, D.G. & Rahardjo, H. 1993. Soil mechanics for unsaturated soils. John wiley and sons. , New York.

Fawcett, R.G. & Collis-George, N 1967. A filter paper method for determining the moisture characteristics of soil. *Australian J. exp. agriculture and animal husbandry* 7: 162-167.

McQueen, I.S. & Miller, R.F. 1968. Calibration and evaluation of a wide range method of measuring moisture stress. *Journal of Soil Science* 106(3): 225-331.

Hamblin, A.P. 1981. Filter-paper method for routine measurement of field water potential. *Journal of Hydrology* 53: 355-360.

Chandler, R.J. & Gutierrez, C.I. 1986. The filter paper method of suction measurement. *Geotechnique* 36: 265-268.

Houston.S.L., Houston.W.N, Wayner, A.M. 1994. Laboratory filter paper suction measurement. *Geotechnical Testing Journal* 17(2):185-194.

Swarbrick, G.E. 1995. Measurement of soil suction using the filter paper method. In E.E. Alonso & P. Delage(eds), *Unsaturated soils, Proceeding of the first international conference on unsaturated soils*, Paris, 6-8 September 1995. Rotterdam: Balkema.

Gourley, C.S. & Schreiner, H.D. 1995. Field measurement of soil suction. In E.E. Alonso & P. Delage(eds), *Unsaturated soils, Proceeding of the first international conference on unsaturated soils*, Paris, 6-8 September 1995. Rotterdam: Balkema.

Unsaturated Soils for Asia, Rahardjo, Toll & Leong (eds) © 2000 Taylor & Francis, ISBN 90 5809 139 2

Investigation of suction generation in apparatus employing osmotic methods

E.E.Slatter, C.A.Jungnickel, D.W.Smith & M.A.Allman
The University of Newcastle, Callaghan, N.S.W., Australia

ABSTRACT: The osmotic oedometer is a device that enables one-dimensional consolidation testing of soils under conditions of controlled pore water suction. The device develops suction in the sample pore water using a solution with a high osmotic suction, typically polyethlyene glycol (PEG), separated from the sample by a semi-permeable membrane.

It has been found that the suction generated in soil in the apparatus is not equivalent to the osmotic suction of the PEG solution used. The mechanism of the phenomenon is described. This paper investigates this feature of osmotic systems using the results of psychrometer testing, osmotic oedometer calibrations and osmotic pressure cell testing. Comparisons are made between the osmotic suctions exerted by PEG solutions of different molecular weights and in systems using two different membrane types, in an attempt to explain the discrepancy between solution and soil suction.

1 INTRODUCTION

Suction controlled laboratory testing is an essential component of the investigation of unsaturated soil behaviour and theory. There are currently several techniques available for suction controlled testing; these techniques include the axis translation method, pressure membrane apparatus, vapour pressure methods and osmotic methods. Analysis of the advantages and disadvantages of each of these methods (Slatter & Allman, 1998) has shown that osmotic methods are a viable alternative to the more traditional techniques for suction controlled testing.

1.1 *Principle of Osmotic Methods*

Figure 1 shows a diagram of a system of two compartments separated by a semipermeable membrane. Compartment 1 contains pure solvent and compartment 2 contains a solution of the same solvent. The semi permeable membrane separating the two compartments is permeable to the solvent but not the solute. In a system such as this the solvent will spontaneously flow across the membrane. This flow is called osmosis. If the water level in the columns off both compartments is initially equal the solvent will flow across the membrane from compartment 1 to compartment 2. This will cause the pressure in compartment 1 to drop and the pressure in compartment 2 to rise, and this will result in a difference in water level between the two columns. Flow

across the membrane will continue until the osmotic force driving the flow is balanced by the hydrostatic pressure difference between the two compartments. The pressure difference required to produce zero flow between the two compartments is called the osmotic pressure of the solution in compartment 2. The water in compartment 1 has a lower free energy when the water level between the two compartments differs. This decrease in free energy is commonly referred to as the osmotic suction of the solution in compartment 2.

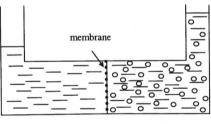

Figure 1. Diagram of osmotic system. Fluid flows from compartment 1 to compartment 2.

In thermodynamics terms the osmotic pressure of the system can be related to chemical potential and solute concentration using equation (1).

$$\mu_{2w} = \mu_{1w}^0 + RT \ln a_{2w} + \int_{p_1^0}^{p_1^0 + \pi} \overline{V}_{2w} dp \qquad (1)$$

Where μ_{2w}= chemical potential of the water in compartment 2, μ^0_{1w}= standard state for the pure water in compartment 1, R= universal gas constant, T= absolute temperature. a_{2w}= activity of component $2w = p_{2w}/p^0_{1w}$, p_1=the vapour pressure for the solution, p^0_1= normal vapour pressure for the pure liquid, π = osmotic pressure, $\overline{V}_{2w}$=partial molal volume of the water in compartment 2. Assuming equilibrium $\mu_{2w}=\mu^0_{1w}$. Integrating Equation 1, assuming $\overline{V}_{2w}$ is constant, and rearranging gives Equation 2.

$$\pi = -\frac{RT \ln a_{2w}}{\overline{V}_{2w}} \qquad (2)$$

Equation 2 shows that the osmotic pressure is related to the concentration of the solute and the vapour pressure of the solution.

A number of theories on the mechanism of osmosis have been proposed in the last century. Much of the uncertainty in these theories is due to lack of knowledge about the way in which semi permeable membranes function (Thain, 1967).

Osmotic methods of soil suction control and measurement use a solution with a high molecular weight, typically polyethylene glycol (PEG), separated from the soil sample by a semi permeable membrane. The membrane is chosen with a molecular weight cut-off lower than mean molecular weight of the osmotic species in solution. The solute molecules are then unable to pass across the membrane barrier. Water, with a much smaller molecular weight and size, is able to move freely across the membrane.

1.2 The Problem

A variety of soil suction control devices employing the osmotic method have been developed in the last thirty years. It has become apparent that the pore water suctions induced in soil samples by these apparatus are less than the osmotic suctions of the PEG solutions used. When reporting on the development of an osmotic oedometer Dineen and Burland (1995) noted the existence of a discrepancy between the osmotic suction of PEG solutions and the suctions measured in the apparatus, during the calibration of the suction control system. Dineen (1997) presented a calibration plot for the Imperial College tensiometer that also exhibited this behaviour and results from the calibration of the osmotic tensiometer of Peck and Rabbidge (1969) showed a similar phenomenon. Figure 2 shows a plot of the data reported for these calibrations. Also plotted is a calibration of PEG 20 000 solutions presented by Williams and Shaykewich (1969). Solution suctions were estimated using a thermocouple psychrometer. This method did not involve the use of membranes. It can be seen in Figure 2 that the calibration curves of the osmotic apparatus are all significantly lower

than the suctions found by Williams and Shaykewich. This indicates that some effect exists within apparatus, that reduces the suctions generated by the osmotic solution.

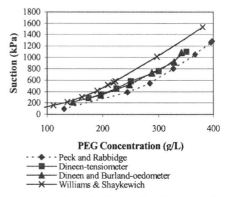

Figure 2. Comparison of calibration data for a range of apparatus from the literature.

2 EXPERIMENTAL

2.1 Apparatus

The experimental program was designed to enable investigation of the mechanism behind the reduction of osmotic suction in osmotic apparatus. To this end, three apparatus were used and these are described below.

2.1.1 Osmotic Oedometer

An osmotic oedometer is a device that enables soil to be one-dimensionally consolidated under conditions of controlled vertical stress and pore water suction.

The apparatus incorporates features of earlier devices (Delage et al 1992, Kassif & Ben Shalom 1971, Dineen & Burland, 1995) and, as a new feature includes lateral strain gauges. Figure 3 shows the oedometer cell, used by the authors, into which soil is placed and then loaded via a mass hanger resting on the top cap. Vertical deformation is measured by an LVDT on the mass hanger and lateral stress can be measured by a strain gauge on the cell wall. The suction in the sample pore water is developed using a PEG solution with an average molecular weight of 20 000, separated from the sample by a semi-permeable membrane. The PEG solution is pumped across the base of the sample. The soil water is drawn out of the soil by the osmotic suction of the solution, until the suction of the pore water is equal to the osmotic suction of the solution.

There is no direct means of measuring the suction of the soil while it is in the apparatus. Suction measurements are made at the end of tests by placing a subsample of soil in a transistor psychrometer.

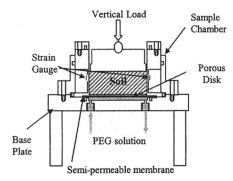

Figure 3. Schematic diagram of osmotic oedometer.

2.1.2 *Osmotic Pressure Cell*
The osmotic pressure cell is shown schematically in Figure 4. This device enables the direct measurement of the osmotic pressure generated across a membrane, using a pressure transducer.

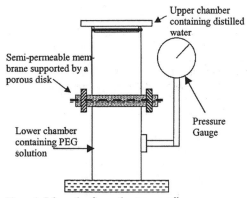

Figure 4. Schematic of osmotic pressure cell

PEG solution is placed in the lower, stainless steel chamber. A semi permeable membrane is then placed on top of the lower chamber and clamped down by the upper, perspex chamber. A porous disk supports the membrane. The upper chamber is filled with distilled water. A magnetic stirrer is used to agitate the solution in the lower chamber to ensure even concentration of solution throughout. The system is left to equilibrate. A pressure transducer in the wall of the lower chamber monitors the increase in pressure as water is drawn from the upper chamber. The test is ended when pressure readings reach equilibrium.

2.1.3 *Transistor Psychrometer*
An eight channel transistor psychrometer was used to measure the total suction of both soil and solution

samples. The technique involves placing a sample in a purpose-designed chamber and allowing it to equilibrate with the psychrometer probe. The relative humidity of the vapour above the sample is measured by a voltage difference between wet and dry transistors within the probe head. The relative humidity can then be related back to the chemical potential of the moisture in the sample. This device has an air space separating the sample and the measuring probe, and so was used to obtain suction measurements without the presence of a semipermeable membrane.

2.2 *Experimental Program*

2.2.1 *Osmotic Oedometer Tests*
A series of slurry tests were performed in the osmotic oedometer to calibrate it for PEG concentration vs soil suction. Slurry tests were performed by placing slurry in the sample chamber of the oedometer. A PEG solution of known concentration was then circulated through the apparatus and the soil water was allowed to equilibrate with the PEG solution. The soil sample was then removed from the apparatus and the suction of the sample was determined using the transistor psychrometer. This method was repeated for a range of PEG solution concentrations.

2.2.2 *Osmotic Pressure Cell Tests*
The osmotic pressure cell was used to determine the osmotic suction across a membrane for a range of PEG solution concentrations. Tests were run for PEG solutions with an average molecular weight of 20 000 and 35 000.

The osmotic pressure cell was also used to measure the osmotic pressures generated across two different membrane types. A cellulose acetate (CA) membrane and a polyether sulphone (PES) membrane were calibrated. The calibrations were performed using PEG solutions with a molecular weight of 20 000 at a range of concentrations.

2.2.3 *Psychrometer Tests*
The transistor psychrometer was used to determine the osmotic suction for a range of PEG solution concentrations. Calibrations were performed on PEG solutions with an average molecular weight of 20 000 and 35 000.

3 RESULTS

3.1 *Osmotic Oedometer Tests*

Figure 5 shows a plot of the results from the calibration of the osmotic oedometer. The results from the psychrometric measurement (Williams and

Shaykewich, 1969) of PEG suction are plotted for comparison. The calibration of the osmotic pressure cell (with CA membrane and PEG 20 000) is also plotted. It can be seen that the suctions generated across the membrane in the oedometer and the pressure cell are lower than the values from the psychrometer. This is consistent with the results from Peck and Rabbidge (1969) and Dineen (1997) for the calibration of other osmotic apparatus, plotted in Figure 2.

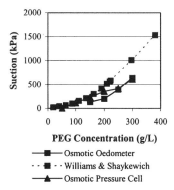

Figure 5. Suctions generated in soil in the osmotic oedometer compared with PEG solution suctions from Williams & Shaykewich.

3.2 *Osmotic Pressure Cell Tests*

The results of the calibration of PEG concentration vs suction for PEG 20 000 and PEG 35 000 are plotted in Figure 6. Again, the results from Williams and Shaykewich are plotted for comparison. This plot shows that both solutions produce very similar osmotic suctions across the membrane in the osmotic pressure cell.

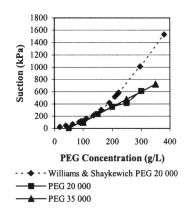

Figure 6. Results of pressure cell tests on PEG 20 000 solutions and PEG 35 000 solutions.

Figure 7 shows a plot of the calibration of the two membrane types in the osmotic pressure cell.

The plot shows that the solutions acting across the PES membrane produced higher osmotic suctions than across the cellulose acetate membrane but these suctions were still less than the osmotic suctions predicted by Williams and Shaykewich (1969)

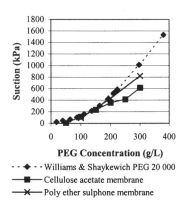

Figure 7. Results of pressure cell tests using PEG 20 000 at various concentrations with cellulose acetate membrane and poly ether sulphone membrane.

3.3 *Psychrometer Tests*

Figure 8 shows a plot of the suction values estimated with the psychrometer for PEG solutions with a molecular weight of 20 000 and 35 000 for a range of concentrations. This plot shows that the PEG concentration vs suction curve is the same, within the bounds of experimental error, for both of the molecular weights tested.

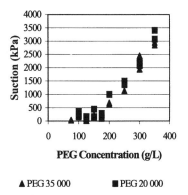

Figure 8. Psychrometer test results for PEG 35 000 solutions and PEG 20 000 solutions.

4 ANALYSIS

Data from the literature (Dineen 1997, Peck & Rabbidge 1967, Dineen & Burland 1995), and the results presented here, has shown that osmotic suctions generated by PEG solutions are consistently lower for systems involving membranes, than for systems without membranes.

This appears to indicate that the reduction of osmotic suction is due to some effect arising from the presence of the membrane. Possible mechanisms for this phenomenon include;

- Leakage of PEG molecules through the membrane
- Errors in the measurement of suctions, and
- Physical interaction of the membrane with the PEG molecules.

Each of these possible mechanisms will be discussed in the following sections.

4.1 Membrane Leakage

Leakage of PEG molecules through the membrane would lower the PEG concentration difference that acts to create the suction across the membrane. This would then lower the apparent osmotic suction of the PEG solution. This mechanism was investigated by comparing osmotic suctions generated by solutions of PEG with molecular weights of 20 000 and 35 000. The molecular weight of the solution is defined as the mean molecular weight of molecules in the solution. Within the solution there will be a normal distribution of molecular weights present, around this mean value. The molecular weight cut-off of the semi permeable membrane is defined as the mean molecular weight of molecules retained by the membrane after a standard time period.

If the membrane leakage was the dominant mechanism in osmotic suction reduction, fewer PEG molecules in the PEG 35 000 solution would be able to pass across the membrane than in the PEG 20 000 solution. This would show in the results as higher osmotic suction measurements for PEG 35 000 solutions in systems including membranes. The results, in Figure 6, show that leakage of molecules through the membrane is not the dominant mechanism of osmotic pressure reduction, as the solutions of the two different molecular weights produced similar suctions.

4.2 Measurement Errors

The possibility of errors in the measurement of suctions is reduced due to the variety of apparatus used and the consistency of the results presented, both in this paper and in the literature.

The results obtained from transistor psychrometers lack a high degree of precision but the consistency of results, from the literature and tests performed by the authors, appears to confirm their accuracy.

The osmotic pressure cell used in this investigation gives measurements with a high degree of both accuracy and precision.

4.3 Membrane Interaction

It appears that the most likely cause of the reduction of osmotic suctions generated across semi permeable membranes in osmotic apparatus involves some physical or chemical interaction between the semi permeable membrane and the PEG molecules in solution. For the effect to lead to a lower suction measurement across the membrane at equilibrium the interaction must raise the chemical potential of the solution. This results in lower osmotic suction values.

A change in the chemical potential of a solution is equivalent to a change in the Gibbs Free energy and this can be expressed as shown in Equation 3.

$$\Delta G = \Delta H - T\Delta S \tag{3}$$

Where ΔH = change in the enthalpy, T = the absolute temperature, and ΔS = change in entropy.

The entropy of water molecules rises as the solute is added. This is called the entropy of mixing. In addition a decrease in the entropy of the solvent occurs as the solvent molecules hydrate the solute molecules and are thus bound in a more ordered state than in the pure solvent, and heat is given out as the water molecules bind to the PEG. The combined effect of these three processes is an overall decrease in the free energy of any solution in which the solute dissolves spontaneously, as is the case for PEG and water.

It is proposed that the likely mechanism for the reduction in osmotic suction seen in systems incorporating semi permeable membranes is some degree of structuring of the PEG molecules in solution. This would result from the PEG molecules binding together, decreasing the entropy of mixing and so raising the free energy.

This structuring would effect all PEG solutions to some extent. For a solution with a free surface, in a system enclosed from the atmosphere the proposed effect is depicted in Figure 9a. The darker shading indicates a higher degree of structuring of the PEG molecules. The figure shows how the structuring extends some distance from the walls of the chamber, this distance of influence depending on the surface material.

 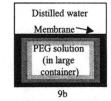

Figures 9a and b. Diagrams of extent of PEG structuring in solutions with a free surface and membrane boundary in a large container. A smaller solution container would result in a greater proportion of the solution being more structured.

The effect of the addition of an upper confining surface to this system, as occurs in a system bounded by a membrane, is shown in Figure9b. It is apparent from this diagram that the chemical potential of the PEG solution confined on all sides would be significantly higher than for the same solution with a free upper surface.

Figures 10a and b are qualitative plots of the expected Gibbs free energy for systems of pure solvent, shown as G_{pure}, for a solution with a free surface as the upper boundary G_{soln} (Figure 10a), and for a solution with a membrane forming the upper boundary G_{memb} (Figure 10b). The y-axis represents vertical displacement through the container. The zero position is at the bottom of the solution container and the upper surface of the solution is indicated as either a free surface or membrane boundary. The dashed lines indicate the relative contribution to ΔG made by ΔH and $T\Delta S$, for each system. The plots show that free energy is constant throughout the solutions. The relative positions of the free energy plots indicate that maximum free energy occurs for the pure solvent and the solution with the free surface has the lowest free energy.

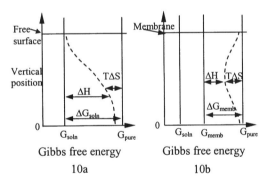

Figure 10a and b. Qualitative plots of free energy through the system for pure solvent, solution with free surface and membrane bound solution.

The degree to which the structuring of PEG molecules occurs would depend on a number of factors including the size of the container of PEG and the chemical nature of the confining surfaces. It can be extrapolated from Figures 9a & b that structuring at the walls of a larger container would influence a smaller proportion of the PEG solution. It would be expected that the osmotic suction of this system would be relatively higher than for the same solution in a smaller container. It would also be reasonable to expect that the chemical nature of the container walls would influence the degree to which PEG structuring occurs. For example the cellulose acetate membrane is highly hydrophilic, with many acetate groups acting as binding sites for water and

for the glycerol groups on the PEG molecules. Glass and plastic containers are much less hydrophilic than the cellulose acetate membranes and as such could be expected to initiate significantly lower degrees of PEG structuring. This would then explain the significant decrease in osmotic pressure seen on the addition of the membrane surface.

Measurements involving hydrophobic membranes, and containers of various sizes could be used to investigate these predictions and test the hypothesis.

5 CONCLUSIONS

It has been shown that membranes significantly decrease the osmotic suction generated by PEG solutions across the membrane. It has been proposed that structuring of the PEG molecules occurs at confining surfaces and results in a rise in the chemical potential of the solution and a corresponding decrease in the osmotic suction of the solution.

The results of the experimental investigation presented here appear to support the hypothesis. Further experiments are proposed.

REFERENCES

Delage, P., Vicol, T. & Suraj de Silva, G.P.R. 1993. Suction Controlled testing of non-saturated soils with an osmotic consolidometer. In Proc. 7th Intern. Conf. on Expansive Soils, Dallas.

Dineen, K. & Burland, J.B. 1995. A new approach to osmotically controlled oedometer testing. In Alonso and P. Delage (eds.) Proceedings 1st Int. Conf. on Unsaturated Soils, Paris. Balkema.

Dineen, K. 1997. The Influence of Soil Suction on Compressibility and Swelling, PhD Thesis. Imperial College.

Hiemenz, P.C. & Rajagopalan R. 1997. Principles of Colloid and Surface Chemistry: Marcell Dekker.

Kassif, G. & Ben Shalom, A. 1971. Experimental Relationship between Swell Pressure and Suction. *Geotechnique* 2: 245 - 255.

Peck, A.J. and Rabbidge, R.M. 1969. Design and performance of an osmotic tensiometer for measuring capillary potential. *Proc. from the Soil Soc of America Journal* 33: 196-202

Slatter, E.E. & Allman, M.A. 1998. Suction Controlled Testing of Unsaturated Soils. In N. Vitharana & R. Colman (eds), Proc. 8th Australia New Zealand Conf. On Geomechanics, Hobart: Australian Geomechanics Society.

Thain, J.F. 1967. Principles of Osmotic Phenomena: Royal Institute of Chemistry.

Williams, J. & Shaykewich C.F. 1969. An Evaluation of PEG 6000 and PEG 20 000 in the Osmotic Control of Soil Water Matric Potential. *Canadian Journal of Soil Science* 49:397 – 401.

Unsaturated Soils for Asia, Rahardjo, Toll & Leong (eds) © 2000 Taylor & Francis, ISBN 90 5809 139 2

A study of the efficiency of semi-permeable membranes in controlling soil matrix suction using the osmotic technique

A. Tarantino & L. Mongiovɪ
Dipartimento di Ingegneria Meccanica e Strutturale, Università degli Studi di Trento, Italy

ABSTRACT: This paper presents results of experimental investigations carried out to improve the performance of osmotically controlled apparatuses. A rigid porous interface was placed behind the semi-permeable membrane to reduce errors in the vertical strain measurement. It was verified that water exchange between the specimen and the osmotic solution was not slowed down by this porous element. The performance of three different semi-permeable membranes was then evaluated. Experimental results indicated that all membranes experienced chemical breakdown as suction exceeded a threshold value. This value was found to depend on the type of membrane. Furthermore, two membranes were used in the osmotic system, one on the top of the other. It was observed that the threshold suction could be displaced to higher values when using two membranes.

1 INTRODUCTION

The mechanical behaviour of unsaturated soils is mostly investigated using two experimental techniques: axis translation technique and osmotic technique.

Experimental evidence of the validity of the axis translation technique has been provided by Fredlund and Morgenstern (1977) for high degrees of saturation and more recently by Tarantino et al. (in press) for the range of continuous air phase. Limitations of the axis translation technique lie in the relatively short range of suctions that can be applied to the soil specimen in conventional oedometer and triaxial tests. Also, the interpretation of experimental results is not straightforward when testing samples at high degrees of saturation, where the air phase is in the form of cavities within the pore water.

To investigate the behaviour of clays and silty clays the osmotic technique seems indeed more suitable. Higher values of suction may be applied to the soil and the transition from saturation to the state where air and water are both continuous in the interstices can be investigated. In addition, the air pressure around the sample remains atmospheric as in real conditions.

The osmotic technique has been successfully used to investigate the unsaturated soil behaviour (Delage et al. 1987; Cui & Delage 1996). Most recently, improvements have been made by extending the suction range of this method to 10 MPa (Delage et al. 1998).

In the osmotic technique, soil suction is controlled by a semi-permeable membrane which is in contact with the soil specimen and allows water exchange between the soil and a solution circulating on the other side of the membrane. Details of this technique can be found elsewhere (Cui 1993).

Limitations of the osmotic technique lie in the need for direct measuring the negative pore water pressure. Dineen & Burland (1995) showed that pore water pressure cannot be estimated from the concentration of the osmotic solution. As a result, one or more tensiometers should be incorporated in the osmotically controlled apparatuses.

Furthermore, when using this technique, an accurate measurement of vertical strains is often difficult to achieve. The drainage system at the bases of the specimen generally includes a metal or plastic mesh in contact with the semi-permeable membrane. The specimen tends to penetrate into the mesh when loaded and this causes error in the measurement of vertical strain.

Another drawback of this method was found while performing osmotic consolidation tests at the University of Trento. It was observed that semi-permeable membranes may undergo a chemical breakdown as a threshold suction is exceeded.

Moving from the above considerations, some investigations were carried out in an attempt to improve the performance of osmotically controlled apparatuses. In the first stage, in order to reduce errors in the measurement of vertical strains, a rigid porous interface was placed in contact with the semi-permeable membrane on the side of the osmotic solution. This scheme is advantageous as it can be easily implemented in oedometer and triaxial apparatuses. The design of a new oedometer base incorporating this interface is detailed in the paper.

As a second step in the experimental study, two standard cellulose membranes of different brands and one reinforced cellulose membrane were tested. The study was carried out to determine which membrane exhibits the highest threshold suction. The results of these investigations are also presented and discussed in the paper. As a final step, it was verified whether the use of two membranes, one on the top of the other, could displace the threshold suction to a higher value, thus improving the performance of the osmotic system.

2 APPARATUSES

An osmotic consolidometer was first developed at the University of Trento in an attempt to reconstitute unsaturated samples (Tarantino, 1998). The base and the cap of the consolidometer were equipped with semi-permeable membranes to allow soil water to be drained by an osmotic solution. Samples were initially consolidated from slurry to a given vertical total stress and suctions were applied through the osmotic solution to desaturate the samples. Only vertical settlements were measured and soil suction was merely estimated from the osmotic solution concentration.

Then a 100 mm diameter osmotically controlled oedometer was purchased from the Imperial College of London (Dineen & Burland 1995). Using this apparatus matrix suction was directly measured by the tensiometer introduced and then modified by Ridley & Burland (1993; 1996) and moisture transfer determined by monitoring the weight of the osmotic solution. Underdrainage from beneath the semipermeable membrane was ensured by a woven nylon mesh resting on the oedometer base.

Some components of the Imperial College oedometer were subsequently modified at the University of Trento to improve measurement accuracy. The cap was redesigned to allow two tensiometers to be installed (Figure 1). In addition, the annular space between the tensiometer and the hole in the cap was sealed with an O-ring. This made it possible to prevent water evaporation from the soil surface where suction is measured. An elastomer membrane was therefore placed over the annular gap between the cap and the inner oedometer ring to prevent soil water evaporation. The membrane attachment was designed to allow this gap to be minimised.

The main shortcoming in the original apparatus lay in the woven mesh below the semipermable membrane. Although the mesh was found to be effective in uniformly diffusing the solution, the specimen penetrated into the mesh when loaded during testing. This caused errors in measurement of vertical settlements which were difficult to calibrate.

A different design was therefore pursued. Rigid semipermable membranes seem to be unavailable on the market. To prevent penetration into the mesh a rigid porous interface seems thus inevitable. If positioned above the semi-permeable membrane, such interface should be a high air entry value ceramic. As a main disadvantage, water would remain under tension inside the ceramic and hence subject to cavitation. Alternatively the interface can be placed below the semi-permeable membrane. As a first attempt a 100 mm diameter bronze porous disc of the type used in conventional triaxial apparatuses was inserted between the semi-permeable membrane and the mesh. Two oedometer tests were therefore carried out with and without the porous disc. It was found that the disc did not slow down in any way the water exchange rate between the sample and the osmotic solution (Tarantino & Mongiovi 1999). Moving from this satisfactory result, the base of the oedometer was modified to settle the 3 mm thick bronze porous disc over the woven mesh.

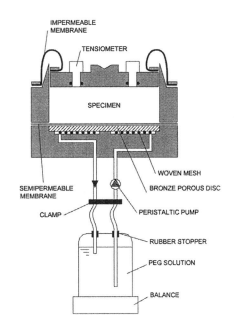

Figure 1. Schematic layout of the apparatus

It is worth noticing that if the disc remained on the woven mesh, a significant error in the measurement of vertical settlements would occur as a result of the mesh deformability. In order to overcome this problem, the oedometer base was machined to avoid contact between the disc and the mesh. As shown in the figure, the disc is held by a 5 mm wide lateral support and a 10 mm diameter central support. The woven mesh fitted accurately the space between the base and the porous disc.

To measure the change in soil water content the osmotic solution was placed in a closed bottle instead of a beaker covered with silicon oil. This was expected to minimise water losses due to evaporation and allow more accurate measurement. The bottle had two openings where two rubber stopper had been pushed in. The tubes of the peristaltic pump were therefore forced through a hole drilled into each stoppers so as to ensure air-tightness. The tubes were then firmly secured at mid-length using a clamp to avoid swaying which could affect the weight measurement. It was verified that the flexible tubes did not prevent the free movement of the balance plate thus allowing an accurate measurement of the change in the solution mass.

3 EARLIER INVESTIGATIONS

The osmotic consolidometer was employed at the onset of a major research programme on unsaturated soil stress variables in an attempt to desaturate kaolin samples previously consolidated from slurry. Pure polyethylene glycol (PEG) of 20000 molecular weight and one sheet of standard Viskase cellulose membranes of 14000 molecular weight cut off (MWCO) were used for the osmotic system. At that stage, data presented by Dineen and Burland (1995) had been used to obtain the calibration curve of PEG concentration against matrix suction. This calibration was later found reliable for the Viskase membranes used in the study.

Some osmotic consolidation tests were performed applying matrix suctions up to 1500 kPa through the PEG solution. Despite these high suction values, samples always had matrix suctions not greater than about 800 kPa after removal from the consolidometer. The settlement versus time curves were then checked. An osmotic consolidation curve is reported in Figure 2 for a test where a suction of 500 kPa and subsequently of 1000 kPa were applied to the sample. It can be noticed that vertical settlement increases almost suddenly about one day after the application of a 1000 kPa suction. Such increase would be associated to a membrane failure and the subsequent ingress of water into the sample. As there was not evidence of mechanical rupture of the membrane after removal from the consolidometer, it was inferred that this breakdown was chemical.

In order to gain a better understanding of the behaviour of semi-permeable membranes, the osmotically controlled oedometer was set up. This apparatus made it possible to monitor the response of the semi-permeable membranes by measuring the changes in both pore water pressure and water mass of the specimens.

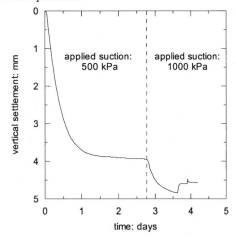

Figure 2. Osmotic consolidation under constant vertical stress

4 MATERIALS

Three semi-permeable membranes were selected for this study, all having a MWCO equal to 14000. There were one Spectrum standard cellulose membrane (No. 4), one standard Viskase cellulose membrane, and one Spectrum reinforced cellulose membrane (No. 5). The latter membrane has a layer of paper within the regenerated cellulose sheet so as to improve its mechanical properties. This membrane was selected to verify whether chemical breakdown was induced by the compression of the membrane during testing. It is worth recalling that semi-permeable membranes are manufactured for dialysis and are certainly not designed to retain their chemical properties if squeezed.

The other two membranes had standard characteristics and were chosen because less expensive and more permeable than the reinforced one. All tests were performed using pure polyethylene glycol (PEG) of 20000 molecular weight.

5 EXPERIMENTAL PROCEDURE

Prior to use, the semi-permeable membranes were first soaked in water for 30 minutes, thoroughly rinsed and stored again in water where penicillin was added. According to the manufacturer's recommendations, deionized water was uses during this process. Great care was taken to not allow the membrane to dry out during pre-treatment and while fixing it into the oedemeter cell.

The bronze porous disc was first saturated in boiling water and then assembled underwater in the oedometer base to avoid desaturation. Only after clamping the semi-permeable membrane on top of the porous disc, the oedometer base was removed from the water basin where it was assembled. The glass bottle, already connected to the oedometer base, was positioned at the lower point of the plumbing system. At this stage the bottle was only filled with water.

The water was let circulate behind the membrane for one hour using the peristaltic pump. Since water pressure in the plumbing system was less than atmospheric, possible leakage could be detected by ingress of air into the system.

The soil sample was then cut using the cutting edge oedometer ring and set in place into the base. A vertical pressure of 20 kPa was applied to ensure contact between the soil sample and the top cap. The tensiometers were installed through the cap after placing a saturated kaolin paste on the tensiometer porous stone.

The bottle containing water was finally replaced with the bottle containing a PEG solution of given concentration. Some penicillin was added to the solution to prevent bacterial growth on the membrane. After the pump started to circulate the solution, changes in pore water pressure and water content were continuously monitored.

6 TEST RESULTS

In six tests equilibrium conditions were attained after circulating the PEG solution. The pore water pressures recorded in these tests were used to calibrate the PEG concentration against matrix suction. The calibration curve obtained for the Spectrum membranes is given in Figure 3 together with the calibration curve obtained by Dineen and Burland (1995).

These two calibration curves differ from each other and only the datum relative to the Viskase membrane seems to be fitted by the Dineen and Burland's calibration curve. Such disagreement is however not surprising. The polyethylene glycol and the semi-permeable membranes used in the two experimental programme are likely to be of different brands and, as a result, the osmotic system responds differently. These results would confirm that PEG calibration curves should be determined independently for each type of PEG and semi-permeable membrane, as also pointed out by Dineen and Burland.

In the remaining tests equilibrium conditions were not attained. In Figure 4 it is shown the performance of a single sheet of Spectrum No. 4 and No. 5 membranes in response to applied suctions of 915 and 1380 kPa. After circulating the solution, the pore water pressure initially decreased and, accordingly, the weight of the bottle containing the osmotic solution increased.

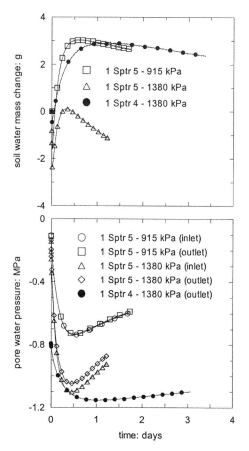

Figure 4. Response of a single semi-permeable membrane sheet.

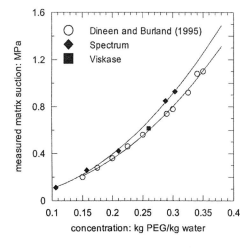

Figure 3. PEG calibration.

However, pore water pressures reached a minimum and then started to increase linearly with time. Similarly, the weight of the osmotic solution started to decrease indicating that water had entered the sample. This confirm that once a threshold suction is exceeded the PEG molecules are no longer retained by the membrane and slowly pass into the specimen. The concentration gradient decreases and, as a result, the flux of water is reversed.

This threshold suction does not correspond to the minimum pressure recorded by the tensiometers. Water pressure can attain lesser values during a transient period. This is made clear by the different response of the Spectrum No. 5 membrane when imposing suctions of 915 kPa and 1380 kPa (Figure 4).

For the case where a suction of 915 kPa was applied, the pore water pressure reached a minimum of about -730 kPa prior to increasing toward zero. Clearly, the threshold suction for this membrane is not greater than 730 kPa.

Nevertheless the pore water pressure attained a value of about -1100 kPa when the suction of 1380 kPa was applied. In this case the osmotic solution was still capable to remove some water from the specimen even though the threshold suction had been exceeded. The different pressures recorded by the two tensiometer confirm that that suction was not in equilibrium throughout the specimen. As it would be expected, pore water pressure is more negative at the inlet, where the solution had the highest concentration. It is worth noticing that pore water pressure decreased more rapidly for the case where the greater suction was applied. The rate of decay of pore water pressure can thus be taken as an indicator of the difference between the threshold suction and that imposed through the osmotic solution.

In order to make it possible the comparison between Spectrum No. 4 and Spectrum No. 5 the same suction of 1380 kPa was thus imposed through the osmotic solution (Figure 4). It was observed that Spectrum No. 5, despite its improved mechanical properties, exhibited the lowest threshold suction. The pore water pressures recorded for this membrane were in fact higher and increased more rapidly toward zero after the minimum value was attained. It might be therefore inferred that the compression of the membrane is not the cause of the chemical breakdown.

Because of its poor performance the Spectrum No. 5 membranes were no longer used in the further experiments.

Another investigation was therefore carried out, using only Spectrum No. 4 and Viskase semipermeable membranes. In this case two membranes one on top of the other were used. The results obtained for the same imposed suction of 1380 kPa are shown in Figure 5. For comparison the results of the test performed on a single sheet of Spectrum No. 4 are also reported in the figure.

The pore water pressure in the test on the Spectrum No. 4 could not be recorded continuously as cavitation occurred in the suction probe after 1 day. The probe was then taken out of the oedemeter and reinstalled after 6 days. In the figure the part of the curve corresponding to the missing measures is indicated in dot line. It was drawn accordingly to corresponding curve for the change in soil water mass.

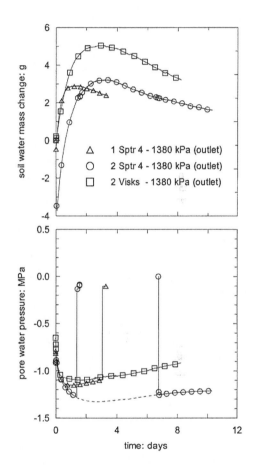

Figure 5. Response of a double sheet of semi-permeable membranes.

It can be noticed that the use of two membranes Spectrum No. 4 lowered to about 200 kPa the minimum pressure recorded by the tensiometers. Also the rate of pressure decay after breakdown appears to be to be less significant.

The displacement of the threshold suction to higher values is likely to apply also to the Viskase membranes. When using of one single membrane, the minimum pore water pressure induced in the sample was of about 800 kPa, according to the results of the osmotic consolidation tests. Then the use of two sheets of Viskase membranes also have increased the threshold suction.

One possible explanation for this improvement is that chemical breakdown is controlled by the concentration gradient across the membrane. When using two membranes one on top of the other the overall thickness increases and, as a result, the concentration gradient decreases.

From Figure 5 it can also be observed that Spectrum No. 4 has a threshold suction higher than Viskase. On the basis of these results the latter membrane was discarded and the two sheets of Spectrum No. 4 finally adopted for the osmotically controlled oedometer tests.

7 CONCLUSIONS

The osmotic technique seems more suitable than the axis translation technique for testing unsaturated samples at high degree of saturation. In order to improve the performance of the osmotically controlled apparatuses some investigations were carried out at the University of Trento.

In order to improve accuracy of settlement measurements a rigid porous disc was tentatively placed at the backside of the semi-permable membrane. Since the porous disc was found to not slow in any way the water exchange rate, a new oedometer base was designed to incorporate the porous disc.

An experimental study was therefore carried out to evaluate the performance of three different semi-permeable membranes for use in the osmotic technique: two standard cellulose membranes of different brands and one reinforced cellulose membrane. Experimental results indicated that all membranes experienced breakdown as the osmotic pressure of the PEG solution exceeded a threshold value. Beyond this value, solute molecules were no longer retained by the semi-permeable membrane and passed into the soil sample, with a reduction of the concentration gradient and resulting decay of soil suction. This threshold value was found to depend on the type of membrane, ranging from less than 700 kPa for the reinforced cellulose membrane to a value likely close to 1000 kPa for one of the two standard cellulose membrane.

It was also observed that the maximum suction which may be applied to the soil specimen increased to about 200 kPa when using two membranes one on the top of the other. This experimental result suggests that a decrease in the the concentration gradient across the membrane displaces the threshold suction to a higher value.

ACKNOLEGEMENTS

The authors are grateful to Prof. Pierre Delage and Dr. Yu Jun Cui for their support and their precious advise. They are also thankful to Dr. Kieran Dineen as this research started from some of his suggestions and remarks.

REFERENCES

Cui, Y.J. 1993. Etude du comportement d'un limon compacté non saturé et sa modelisation dans un cadre elasto-plastique. *PhD Thesis*, Ecole Nationale des Ponts et Chaussées (CERMES), Paris, France.

Cui, Y.J. & Delage, P. 1996. Yielding and plastic behaviour of an unsaturated silt. *Géotechnique* 46 (2): 291-311

Delage, P., Suraji De Silva, G.P.R. & De Laure, E. 1987. Un nouvel appareil triaxial pour les sols non-saturés. *Proc. 9th ECSMFE*, Dublin, 1: 25-28.

Delage, P., Howat, M.D. & Cui, Y.J 1998. The relationship between suction and swelling properties in a heavily compactes unsaturated clay. *Engineering Geology* 50: 31-48.

Dineen, K. & Burland, J.B. 1995. A new approach to osmotically controlled oedometer testing. In E.E. Alonso & P. Delage (eds.), *Unsaturated Soils, Proc. 1st Int. Conf. on Unsaturated Soils*, Paris, 2: 459-465. Rotterdam: Balkema.

Ridley, A.M. & Burland, J.B. 1993. A new instrument for the measurement of soil moisture suction. *Géotechnique* 43(2): 321-324.

Ridley, A.M. & Burland, J.B. 1996. A pore pressure probe for the *in situ* measurement of soil suction. In Craig (ed.), *Advances in site investigation practice*: 510-520. London: Thomas Telford.

Tarantino, A. 1998. Le variabili di stato tensionale per i terreni non saturi. *PhD Thesis*, Politecnico di Torino, Italy.

Tarantino, A. & Mongiovì, L. 1999. Impiego della tecnica osmotica per il controllo della pressione negativa dell'acqua interstiziale. In *Atti del XX Convegno Italiano di Geotecnica*, Parma, Italy: 295-300. Pàtron Editore: Bologna.

Tarantino, A., Mongiovì, L. & Bosco, G. (in press). An experimental investigation on the isotropic stress variables for unsaturated soils. *Géotechnique*.

Unsaturated Soils for Asia, Rahardjo, Toll & Leong (eds) © 2000 Taylor & Francis, ISBN 90 5809 139 2

Response of the IC tensiometer with respect to cavitation

A. Tarantino, G. Bosco & L. Mongiovì
Dipartimento di Ingegneria Meccanica e Strutturale, Università degli Studi di Trento, Italy

ABSTRACT: In recent years tensiometers for direct measurement of negative pore water pressure have been developed at Imperial College and later on at the University of Saskatchewan. The major drawback of these instrument is water cavitation which may occur before pressure equalisation. A better understanding of the mechanisms that control cavitation inside the tensiometer may help optimise their design and define adequate experimental procedure. This paper presents results of several suction measurements carried out with three IC tensiometers. In particular the history of tension breakdown of each tensiometer was recorded. It was thus possible to recognise some factors that affect the maximum sustainable tension as well as measurement duration. Finally, a possible mechanism of cavitation in the tensiometer has been outlined on the basis of some data recorded before and after cavitation.

1 INTRODUCTION

In recent years, several studies have been undertaken to develop reliable procedures and simple equipment for soil suction measurement in the laboratory as well as in the field. Among the newly presented instruments the tensiometer developed at the Imperial College (Ridley & Burland, 1993; 1996) and more recently the one constructed at the University of Saskatchewan (Guan & Fredlund, 1997) seem to provide an affordable, easy to use and reliable mean for soil suction measurement.

Unfortunately the major drawback of these instruments is the possibility of water cavitation inside the tensiometer. Water cavitation may occur before pore pressures equalisation is reached thus causing test interruption and instrument resaturation. A knowledge of the conditions leading to cavitation is therefore essential to optimise instrument design and to determine adequate experimental procedure to allow long lasting measurements.

Several models have been presented in the literature to explain the occurrence of cavitation in water subjected to tensile stresses. The most widely accepted is the one proposed by Harvey et al. (1944). They assume that cavitation originates from undissolved gas nuclei existing in the interstices in the container's wall rather than from free cavities in the liquid. A free spherical gas nucleus is usually unstable and tends to go into solution. On the contrary, a gas nucleus in a cavity in the wall of the container may remain undissolved even under a high water

pressure. When pressure is decreased to negative values, these nuclei can expand and eventually trigger cavitation. This process is controlled by gas diffusion across the gas-liquid boundaries and creeping of the gas-liquid-solid junction determined by the advancing and receding of contact angles.

Water tensile stress has been measured by physicists using glass and steel Berthelot tubes. This tube is initially almost completely filled with water while the remaining volume is occupied by a mixture of air and water vapour. The tube us then heated to expand the enclosed liquid and force the air into solution. On subsequent cooling the liquid adheres to the wall of the tube and becomes subjected to gradually increasing tension until it breaks down at the onset of cavitation. It was shown that pre-pressurisation is needed before water tension is produced in the tube. High positive pressures cause in fact the majority of gas nuclei to dissolve and enable the liquid to sustain tension. Moreover, it was found that few heating-cooling cycles were necessary before tension could be generated (Chapman et al. 1975) and that the breaking tension increased with repeated heating-cooling cycles (Jones et al. 1981).

The pre-pressurisation of water has been considered essential to saturate the tensiometer both at Imperial College and at the University of Saskatchewan. Some differences however exist between the two pre-pressurisation procedures. Guan & Fredlund apply many pre-pressurisation cycles that begin with a vacuum of -85 kPa followed by a positive pressure up to 12000 kPa. In fact it has been suggested that tension breakdown is affected primarily by the number of cycles and the magnitude of the positive pressure applied (Guan et al. 1998).

Conversely, Ridley & Burland (1999) assert that the pre-pressurisation procedure is less important while the initial saturation of the porous stone plays a more important role. Provided that the suction probe is thoroughly saturated, the stress required to cause breakdown would be uniquely related to the air entry value of the ceramic. Differently from Guan & Fredlund, they saturate the tensiometer by applying a constant pressure of 4000 kPa that is maintained for at least 24 hours.

At the University of Trento three Imperial College tensiometers have been used over the past four years to measure matrix suction. Measurements were carried out during oedometer tests, during null-tests and also on air dried kaolin samples to define appropriate experimental procedures and to evaluate measurement reliability. Selecting among the experimental data collected so far, this paper presents the results of measurements where cavitation occurred. In particular the history of the tension breakdown of each tensiometer was recorded. Based on the obtained results, a possible mechanism of cavitation in the tensiometer is then suggested.

2 EQUIPMENT

The tensiometer used in the experimental programme is manufactured at the Imperial College and has been previously described by Ridley & Burland (1996). It carries a 1.5 MPa air entry value porous ceramic tip and has a water reservoir of reduced size.

A saturation pressure chamber, also manufactured at Imperial College, was used to re-saturate the suction probes after cavitation. The chamber is equipped with a screw piston for manual pressurisation. Although cheap and safe, this system shows a disadvantage. It requires continuous adjusting of the screw piston to maintain the water pressure in the range required to saturate the porous stone. In fact, once a pressure is applied air cavities are forced into solution causing expansion of water in the chamber and the pressure to drop. Such drop was found to be significant when the system was left overnight and believed to be a critical drawback of the experimental procedure.

To ensure a constant 4 MPa water pressure during the 24 hours pressurisation period, a different pressurisation system was constructed. An air/water bladder cell was connected to the pressure chamber as shown in Figure 1. The upper compartment was filled with nitrogen so as to increase gas pressure up to 4 MPa. Nitrogen was used because of the availability of high pressure nitrogen cylinders in the laboratory. Water volume changes induced by air dissolution were thus compensated by the expansion of the nitrogen in the cell. Because of the high compressibility of nitrogen, the pressure in the upper compartment, and hence the water pressure, could be maintained constant over a length of time.

This pressurisation device was preferred to an automatic controlled water pump to avoid overpressure in case of failure of the control procedure. Water has a low compressibility and even a small volume change can cause a sudden pressure variation. Clearly all precautions were taken to make the high pressure system safe.

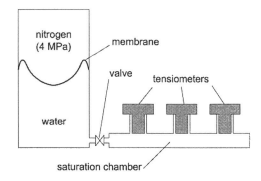

Figure 1. Pressurisation system

Special attention was also given to the energy supply system and the data acquisition system. The tensiomter has a low sensitivity, approximately 0.7 µV/ kPa. A resolution of 1-2 µV was therefore required for the electronic system. An AC energy supplier was specifically designed at the Physics Laboratory of the University of Trento and it was coupled with a 16-bit A/D board. This instrumentation permitted a resolution of about 1 µV.

3 EXPERIMENTAL PROCEDURE

Calibration for the tensiometer was first derived for positive pressures and then extrapolated into the negative range. The accuracy was found to be better than 1.5 kPa. The calibration procedure has been described elsewhere (Tarantino, 1998).

When stored, the tensiometers were always kept in the saturation chamber. After removal from the chamber, they were first placed in free water until pressure equalisation was attained. Thereafter the porous stone was wiped and left to dry out until a water pressure of approximately -1000 kPa was attained. The tensiometer was then immediately replaced in free water to check whether water pressure regained the same initial value. This was taken as an indictor of saturation of the porous stone. If pressure did not return to zero, the tensiometer was not considered ready for measurement. In this case it was placed on a dry sample in order to trigger cavitation and thereafter resaturated for at least 24 hours.

Tensiometers ready for measurement were put in contact with the kaolin samples via a saturated kaolin paste. The sample and the tensiometer were thus placed in a plastic bag to prevent soil water evaporation. A string was used to tighten the plastic bag around the electric wires of the tensiometer.

4 WATER TENSILE STRESS MEASUREMENT

The three tensiometers that have been used throughout the experimental programme described herein were stored inside the saturation chamber continuously for 1 year since not in use in this period. For simplicity they will be later referred to as pr2, pr3, pr4. The air/water bladder cell ensured a constant pressure of 4 MPa and there was no need to refill the system or restore the nitrogen pressure. After 1 year saturation the three tensiometers were calibrated while in the saturation chamber and then retrieved to carry out suction measurements.

Table 1 lists the tension breakdowns of pr2 during the 5 month experimental programme.

Table 1. Tension breakdown history for tensiometer *pr2*

Test No.	Tension breakdown	Measurement duration
	kPa	min
1	-655	1
8	-2260	490
13	-2584	885
17	-2403	200

It can be noticed that after removing the tensiometer from the saturation chamber, cavitation occurred after just 1 min at a very low tensile stress. Such a result is certainly surprising as the suction probe was expected to be fully saturated after the one year period of pressurisation. However it can be also observed that after the first cavitation and subsequent pressurisation the probe has been capable of measuring very high suctions, much higher than the nominal air entry value of the porous stone. These data are remarkable as direct measurement of suctions greater than 2 MPa have not been reported in the literature. In particular during test 13 it was also possible to attain equalisation as reported by Tarantino & Mongiovi (1999). It would thus appear that cycles of cavitation and subsequent pressurisation produce an increase in the maximum sustainable tension. This experimental procedure was indeed suggested by Steve Ackerley (pers. comm.) as it had been earlier observed at Imperial College that repeated cycles of cavitation improved the performance of the probes.

The importance of these cycles is better emphasised by the cavitation history of pr3 and pr4. Their porous stones accidentally dried out during shipping and were resaturated at the University of Trento following the procedure described by Ridley & Burland (1999). The vacuum chamber was obtained by adapting a triaxial cell chamber. It is possible that the large volume of the cell did not permit to reduce the air pressure inside the chamber as much as necessary. As a result the porous stone could not be fully saturated. This would explain the different performance of pr3 and pr4 (Tables 2 and 3) when compared to probe pr2.

Table 2. Tension breakdown history for tensiometer *pr3*

Test No.	Tension breakdown	Measurement duration
	kPa	min
1	-1306	10
2	-1270	14
3	-1299	8
4	-948	2
5	-1225	47
6	-1238	475
7	-1230	162
8	-1886	74
9	-1258	396
10	-1236	220
11	-1876	80
14	-2137	113
17	-2049	100
21	-1275	2000
23	-1240	2049

Table 3. Tension breakdown history for tensiometer *pr4*

Test No.	Tension breakdown	Measurement duration
	kPa	min
1	-1400	12
2	-1268	64
3	-1500	380
5	-1215	22
6	-1190	10
7	-1255	1270
8	-1995	145
10	-1275	5600
11	-1870	70
14	-2040	100
17	-2110	60

Tests were carried out mainly using two kaolin sample having suction of approximately 1300 kPa and 2600 kPa respectively. When the former sample was used, cavitation always occurred at water tensions in the range 1200-1300 kPa. However measurement duration increased progressively with the number of cavitation.

For the case where the 2600 kPa suction sample was used, tension breakdown also increased gradually. Tensile stresses greater than 2000 kPa could be measured after some cycles of cavitation and pressurisation.

5 DATA INTERPRETATION

The results obtained so far differ from those reported by Guan & Fredlund (1997) who recognise the pre-pressurisation pressure as an important factor affecting the maximum sustainable tension.

The experimental data illustrated above show that a mere pressurisation is not sufficient to dissolve cavitation nuclei. This is proved by the almost instantaneous cavitation experienced by pr2 after 1 year of uninterrupted pressurisation at 4 MPa.

However, still holding the same pre-pressurisation pressure, it was possible to record water tensions greater than 2 MPa after repeated cycles of cavitation. Thus the number of cavitations and subsequent pressurisation seems to be one of the factors affecting the maximum sustainable tension. This is in agreement with the results obtained by physicists using the Berthelot tube. In the same manner as water tension was found to increase with repeated heating-cooling cycles it can be assumed that cycles of pressurisation and cavitation increase the tension breakdown.

Accordingly a mechanism of cavitation inside the tensiometer is proposed. Small gas nuclei exist in the tensiometer interstices which are not dissolved by the high pressures applied because of the curvature of the air-water meniscus. If the tensiometer is left for a long period under high pressure, not only these small gas nuclei remain stable but air may diffuse towards those nuclei where the pressure is lower. These are likely to trigger cavitation when water pressure is reduced to negative values. Such mechanism would explain why cavitation occurred after just 1 min in pr2 despite 1 year of pre-pressurisation.

On the other hand, when tension breakdown occurs, a large air cavity forms inside the tensiometer. In this large cavity the absolute pressure is almost zero and, as a result, any surrounding air is driven towards the cavity. Also the small gas nuclei in the interstices can be drawn into it with the result that the number of potentially cavitation nuclei is reduced. On the subsequent pressurisation, the large cavity may be easily forced into solution as opposed to the small gas nuclei in the interstices which remained stable under high water pressures.

In other words, the gas nuclei that remain undissolved upon pressurisation might be eliminated through cavitation followed by pressurisation. Accordingly, tension breakdown and measurement duration increase.

Another point which is worth highlighting is that cavitation would take place in the ceramic porous stone and not in the water reservoir of the tensiome-

ter. Cavitation nuclei are likely to be entrapped in the small pores of the ceramic rather than in the crevices of the water reservoir walls. Besides small air cavities would remain inside the porous stone even though water was under tension.

This assumption is supported by an interesting result obtained while carrying out measurement n. 11 using tensiometer pr 2. After measuring a water tensile stress of 2530 kPa for 315 min, this tensiometer was removed from the sample and placed in free water. As shown in Figure 2 water pressure increased very slowly to zero and 10 minutes passed before complete equalisation was reached. Commonly 3 minutes were sufficient for pressure equalisation. This time lag was probably caused by the growth of the small gas nuclei existing inside the porous stone. The tensiometer had experienced for the first time that value of water tension and this caused the small nuclei in the ceramic to enlarge. As a result, the permeability of the porous stone decreased thus delaying equalisation.

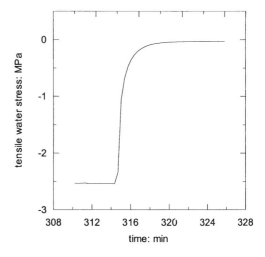

Figure 2. Water pressure equalisation after expansion of gas nuclei in the porous stone

Another anomalous result was often recorded while increasing the water pressure after cavitation occurred. A typical result is shown in Figure 3 where the pressurisation of pr3 when it had cavitated after 113 min at a pressure of - 100 kPa is reported. The tensiometer was left on the sample for about 200 min and during this time the water pressure slowly increased to -95 kPa. After installing the tensiometer in the saturation chamber the valve connecting the chamber with the air-water bladder cell was opened to increase water pressure. It can be noted that the pressure recorded by the tensiometer did not increased instantaneously but dropped of about 35 kPa prior to rising towards 4000 kPa. The drop in pressure would indicate that a tensile stress was transmitted to the measurement diaphragm and hence water still completely filled the reservoir. This would confirm that cavitation had occurred inside the porous stone.

The reason of this pressure drop is not clear. However, it is possible that the positive pressure applied on the porous stone produces a change in the configuration of the menisci inside the ceramic and a temporary drop in pressure.

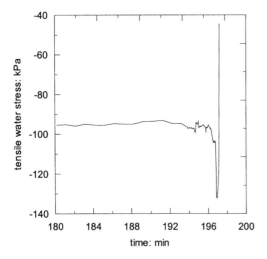

Figure 3. Response of the tensiometer upon pressurisation.

6 CONCLUSIONS

Tensiometers for direct measurement of matrix suction have been developed at the Imperial College and later at the University of Saskatchewan. These instruments are well suited for laboratory and in situ soil suction measurements. Their major drawback is water cavitation which may occur before pore pressures equalise. A better understanding of the cavitation mechanisms may therefore help to optimise instrument design and define adequate experimental procedure.

The tension breakdown history of three IC tensiometers has been presented in the paper. Based on the experimental data presented herein it can be inferred that a mere pre-pressurisation of the water in the probe is not sufficient to dissolve potential cavitation nuclei within the probe. Instead, the experiments indicated that cycles of cavitation and subsequent pressurisation can produce a significant increase in the maximum sustainable tension. In

particular water tension greater than 2 MPa was recorded by all three tensiometers used in this programme. Cycles of repeated cavitation also increased the measurement duration.

To interpret the response of the tensiometers it has been assumed that small gas nuclei are difficult to dissolve by applying a pre-pressurisation pressure. In fact in this case dissolution is prevented by the cavity air-water interface, which is convex on the air side and carries the high water pressure.

On the contrary the large cavities which form inside the tensiometer after cavitation are capable to suck out the air from the small cavitation nuclei. Air from nuclei is driven into the larger cavities by their nearly zero absolute pressure. As opposed to the small gas nuclei, the large cavities can be easily forced into solution by the subsequent pressurisation.

Finally, it would appear that small air cavities remain in the porous stone even if water is under tension. Cavitation would be triggered by the expansion of these nuclei and therefore would take place in the porous stone.

ACKNOLEGEMENTS

The authors are grateful to Dr. Andrew Ridley for his suggestions and advice. They are also in debt to Mr. Steve Ackerley for the help in solving many of the experimental problems encountered.

REFERENCES

Ackerley, S. 1997. Personal communication.
Chapman, P.J., Richards, B.E. & Trevena, D.H. 1975. Monitoring the growth of tension in a liquid in a Berthelot tube. *J. of Physics E: Scientific Instruments* 8: 731-735
Guan, Y. & Fredlund 1997. Use of tensile strength of water for the direct measurement of high soil suction. *Canadian Geotechnical Journal* 34: 604-614.
Guan, Y., Fredlund, D.G. & Gan, J.K.M. 1998. Behavior of water subjected to high tensile stresses. *Proc. 2nd Int. Conf. on Unsaturated Soils*, Beijing, China, 356-361. International Academic Publishers: Beijing.
Harvey, E.N., Barnes, D.K., McElroy, W.D., Whiteley, A.H. Pease, D.C. & Cooper, K.W. 1944. Bubble formations in animal, 1- Physical Factor. *J. Cellular and Comparative Physiology*, 24(1): 1-22.
Jones, W.M., Overton, G.D.N. & Trevena, D.H. 1981. Tensile strength experiments with water using a new type of Berthelot tube. *J. of Physics D: Applied Physics* 14: 1283-1291.
Marinho, F.A.M. & Chandler, R.J. 1995.Cavitation and direct measurement of soil suction. In E.E. Alonso & P. Delage (eds.), *Unsaturated Soils, Proc. 1st Int. Conf. on Unsaturated Soils*, Paris, 2: 623-630. Rotterdam: Balkema.
Ridley, A.M. & Burland, J.B. 1993. A new instrument for the measurement of soil moisture suction. *Géotechnique* 43(2): 321-324.
Ridley, A.M. & Burland, J.B. 1996. A pore pressure probe for the *in situ* measurement of soil suction. In Craig (ed.), *Ad-

vances in site investigation practice: 510-520. London: Thomas Telford.

Ridley, A.M. & Burland, J.B. 1999. Use of tensile strength of water for the direct measurement of high soil suction: Discussion. *Canadian Geotechnical Journal* 36: 178-180.

Tarantino, A. & Mongiovì, L. 1999. Misura diretta della pressione negativa dell'acqua interstiziale in terreni non saturi e saturi: accuratezza e problemi sperimentali. In *Atti del XX Convegno Italiano di Geotecnica*, Parma, Italy: 295-300. Bologna: Pàtron Editore.

Tarantino, A. 1998. *Le variabili di stato tensionale per i terreni non saturi*. Doctoral Thesis, Politecnico di Torino, Italy.

5 Soil-water characteristics

Unsaturated Soils for Asia, Rahardjo, Toll & Leong (eds) © 2000 Taylor & Francis, ISBN 90 5809 139 2

Mineralogy and microfabric of unsaturated residual soil

K.K.Aung, H.Rahardjo, D.G.Toll & E.C.Leong
NTU-PWD Geotechnical Research Centre, Nanyang Technological University, Singapore

ABSTRACT: Microstructure characteristics are essential factors for identifying the degree of weathering in residual soil. This study encompasses the determination of mineral composition, microfabric examination and micro porosity measurement. Scanning Electron Microscope (SEM) examination and mercury porosimetry measurement were conducted to analyze the microfabric characteristics of the Bukit Timah granitic residual soil from Singapore. The results are presented and discussed in this paper in the light of unsaturated soil mechanics principles. Depending on the degree of weathering and parent rock types, the mineral contents of residual soil vary with depth. The range and distribution of micropores in a residual soil from different depths have also been investigated in order to obtain a better understanding of the unsaturated nature of a residual soil. It has been found that there is a close relationship between pore size distribution and matric suction. Soil-water characteristic curves (SWCC) for the residual soil have been analyzed with respect to the micropore size distribution of the soil.

1 INTRODUCTION

Residual soils are formed by the weathering of different parent rocks. Most of the residual soils around the world are normally found in tropical regions and are usually unsaturated in nature. The degree of weathering varies with depth from fresh rock to a completely weathered condition and finally residual soils are formed. This varied degree of weathering makes the residual soils different in terms of mineralogy and soil fabric from their parent rocks. During the weathering process, parent rock minerals are decomposed and soil structures are disintegrated. In this paper, the mineralogy and soil microfabric of residual soils from the Bukit Timah granitic formation in Singapore are described.

2 BACKGROUND

The mineralogy of residual soils vary in content as well as in proportion around the world. Although the mineralogy of residual soil is derived from the parent rock minerals, its final mineral content can vary according to the degree of weathering. The granitic residual soils in Japan vary in colour from light grey to red depending on the quantity of limonite present (Onodera, 1976). Their clay fraction varies from 2 to 26% depending on the relative quantities of feldspar, mica, and quartz in the parent rock. However, grani-

tic residual soils in Ghana were formed from a deep profile of sand containing feldspar crystals (Ruddock, 1967). The residual soils in Hong Kong are mostly boulderly and core stones. Their mineralogy consists of kaolinite, quartz, gibbsite and some ferroxide minerals (Irfan, 1996).

Mineralogy of residual soil varies with depth according to the weathering (Irfan, 1996). Topography also controls the minerals forming in a residual soil even under the same bedrock formation and climatic conditions. It controls the drainage condition and leaching of soluble minerals in the residual soil. Van der Merwe (1965) studied the development of clay minerals from three topographic locations where basic igneous rock types were present. Samples from the high slope with good runoff contained great amounts of kaolinite and vermiculite clay minerals. A flatter site showed chlorite, vermiculite, kaolinite and montmorillonite mineral contents. In the flat site with impeded drainage, montmorillonite was found to be the dominant clay mineral. Fitzpatrick and Le Roux (1977) found that the depth of weathering in hillsides increased down the slope. Kaolinite and halloysite were found to be the predominant clay minerals at the top of the slope whereas montmorillonite was found to dominate at the bottom of the slope.

Residual soil from the Bukit Timah granitic formation covers nearly one third of Singapore's land area. The Bukit Timah granitic formation was formed during the Lower to Middle Triassic periods (Pitts, 1984). The formation includes acidic rocks of granite, adamellite and granodiorite. It has a gradual variation in the distribution of quartz, feldspar, sodic plagioclase and their ferromagnesium materials. Under the hot and humid tropical weathering, the Bukit Timah granite is rapidly weathered and residual soils are formed. The uppermost layer at the Bukit Timah granitic formation is completely transformed into residual soil. The average thickness of residual soil layer ranges from 10 to 35m (Poh et al., 1985).

Mineralogy of the Bukit Timah granite consists mainly of quartz (30%), feldspar (60 to 65%) and Biotite and Hornblende (5-10%). Orthoclase feldspar forms approximately 28% and the rest (35%) is plagioclase (Hutchison, 1964). The granite is usually light grey in colour.

Soil microfabric has been studied since the early development of soil mechanics (Terzaghi, 1925, Goldschmidt, 1926 and Casagrande, 1932). These early concepts were primarily concerned with the interaction of individual clay platelets or group of clay platelets. During the 1950s, Schofield and Samson (1954) and Lambe (1953, 1958) made new developments in soil microfabric based on the double layer theory and the behavior of clay platelets in dilute colloidal solutions. Collins and McGown (1974) conducted a scanning electron microscope (SEM) study on the form and function of microfabric features in a variety of natural soils. After one decade, Collins (1985) introduced a new characterization scheme for the study of microfabric study of residual soils in the tropical region.

Study of soil microfabric includes the examination of the spatial arrangement and distribution of soil particles as well as the associated pore spaces. Micropores between individual soil particles are related to the weathering conditions of the residual soils and also affect their engineering characteristics. The mercury intrusion test is usually used for the determination of micropore size distribution (Diamond, 1970).

3 METHODOLOGY

In this study, SEM examinations were performed to analyse the microfabric of residual soil from the Bukit Timah granitic formation. X-ray diffraction (XRD) tests were conducted to determine the mineral contents of this residual soil. Mercury intrusion tests were executed to find out the micropore size distribution. Samples from a soil slope in Yishun (Northern part of Singapore) were used for these tests. Soil samples at different depths (0 to 20m) from a single borehole were used to analyse the variation of mineralogy and soil microfabric at different degrees of weathering.

The freeze drying method was used to prepare the specimens for SEM examination to minimize changes of soil microfabric due to shrinkage. Magnification of 500X, 1000X and 5000X were used for SEM images. Air-dried samples were used for XRD tests and mercury intrusion tests. Reliability method based on intensity of peak value (I), and basal spacing (d) was used to determine the most probable minerals from the XRD tests. The basal spacing, d can be calculated from the specific angle, 2θ, using Bragg's equation (Mitchell, 1976).

Micropore size distribution is one of the factors in considering the soil microfabric. Mercury intrusion tests were used to determine the micropore size distribution of residual soil from the Bukit Timah granitic formation at different depths. Washburn's equation was used to transform the applied mercury pressure to mean micropore diameter (Diamond, 1970).

4 RESULTS AND DISCUSSION

From visual inspection of the soil specimens, two different degrees of weathering, Grade V and Grade VI (Little, 1969) are found to be present. The upper portion of the soil layer down to 10m depth is reddish brown to yellowish brown in colour. The deeper residual soil is yellowish to whitish in colour. The natural water content is between 20 to 60%. Liquid limits range from 40 to 60 while plastic limits vary in the range of 20 to 40. Specific gravity varies from 2.55 to 2.78 and generally decreases with depth. Grain size distribution shows that fine content is 60 to 70% in the upper layer soil and less than 50% in the lower layer soil. A certain amount of sand size particles are found to be present below the depth of 13~14m.

XRD results indicate that kaolinite is the most common clay mineral throughout the soil profile. In the upper portion of the soil layer (4~5m depth), kaolinite-montmorillonite are present in high proportions. A certain amount of iron oxide and hydroxide compounds are found at this depth. A small amount of quartz is observed in the upper portion. At 8.5~9.5m depth, kaolinite remains the most abundant mineral. Kaolinite-montmorillonite is also found to be the second most abundant mineral. A certain amount of halloysite is found at this depth. A small amount of illite is also present. At the greater depth, 20.5~21.5m, kaolinite is still the most common clay mineral. However, the montmorillonite content decreases considerably. Halloysite is still observed at this depth in a small amount. Quantitative measurement of mineral contents has not yet been defined.

The XRD results are confirmed by the SEM examination. Almost all of the SEM at different depths show that platy shape kaolinite clay minerals are

Figure 1. SEM image for residual soil from the Bukit Timah granitic formation at 8.5~9.5m depth (1000X magnification).

Figure 2: SEM images for the residual soil from the Bukit Timah granitic formation at 20.5~21.5m depth (1000X magnification).

Figure 3. SEM image for residual soil from the Bukit Timah granitic formation at 8.5~9.5m depth (5000X magnification).

Figure 4. SEM image for residual soil from the Bukit Timah granitic formation at 20.5~21.5m depth (5000X magnification).

present. However, no evidence of tubular shape halloysite minerals is found in the SEM images. The results of XRD for residual soil at a nearby borehole do not indicate the content of halloysite (University of East London, 1999).

SEM images of magnification 1000X at 8.5~9.5m and 16~17m are shown in Figure 1 and 2, respectively. These figures reveal that soil particles at 8.5~9.5 m are well bonded and those at 20.5~21.5m are weakly bonded.

According to the characterization introduced by Collins in 1985, the elementary level at 8.5~9.5m depth is dominated by clay size clusters whereas the fine particles at 20.5~21.5m have clothed contact in granular form. At the assemblage level, soil microfabric at 8.5~9.5m depth shows that some granular particles are covered with kaolinite and montmorillonite clay matrix. Soil particles at 20.5~21.5m are

mostly granular matrix with few examples of bonding at inter-particle level. Platy shape kaolinite particles are found to exist separately in the soil microfabric at this depth.

As can be seen from Figure 3, a 5000x magnification of SEM image shows surfaces of soil particle at the depth of 8.5~9.5m are weathered and highly porous. However, the surfaces of the particle at 20.5~21.5m depth are smooth and there is no sign of intra-elementary pores (Fig. 4).

The qualitative analysis of micropore size distribution can be explained by the mercury intrusion test results. The relation between cumulative pore volume and mean diameter of the micropore at different depths of residual soil profile from the Bukit Timah granitic formation is shown in Figure 5. This figure indicates that the majority of micropore size ranges from 0.1 to 6μm. The cumulative micropore volume is found to decrease with depth. Flattening of the curves for soil samples at 13~14m and 20.5~21.5m shows that fewer micropores with sizes less than 0.1μm exist in the deeper soil layers. The trend for the two curves at 5.5~6.5m and 8.5~9.5m reveals that various micropore sizes less than 0.1μm still exist in the soil sample at the shallower depths. This result indicates that the cumulative micropore volume decreases with depth and the higher the degree of weathering, the greater the range of sizes of micropores.

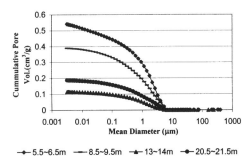

Figure 5. Variation of cumulative micropore volume versus mean micropore diameter at different depths of residual soil from the Bukit Timah granitic formation.

A soil-water characteristic curve (SWCC) is the relation between the matric suction and the water content of the soil. The SWCC can be used to calculate the coefficient of permeability for an unsaturated soil under air and water phase conditions and to predict the shear strength properties of an unsaturated soil (Fredlund et al. 1994; Vanapalli et al., 1996).

The result of micropore size distribution reflects the behavior of the soil-water characteristic curve of residual soil at higher matric suction as the water phase retreats into finer micropores during high matric suction. The initial loss of water content at lower matric suction is controlled by macropores.

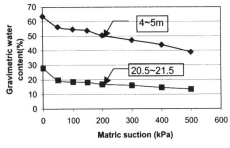

Figure 6. Soil water characteristic curves for the residual soil samples from the Bukit Timah granitic formation at different depths.

The SWCCs of the residual soil from Bukit Timah granitic soil at two different depths are shown in Figure 6. As can be seen from Figure 6, the SWCC from 20.5~21.5m is much flatter than that at 4~5m due to the more restricted range of micropore size present. As the matric suction becomes higher (50kPa and above), the slope of the SWCC depends on the micropore size distribution of the soil. The larger the range of micropore sizes is, the steeper the slope of the SWCC curve at higher matric suctions.

5 CONCLUSIONS

The initial stage of a study of the mineralogy and microfabric of a residual soil from the Bukit Timah granitic formation has been introduced. It has been observed from the XRD results that kaolinite is the most common clay mineral throughout the soil profile from 0~21m depth. Montmorillonite is also present and reduces in proportion as the depth increases. Halloysite is found at 8.5~9.5m although SEM images do not indicate the tubular form of the halloysite clay mineral.

Mercury intrusion test results on the residual soil indicate that micropore size distribution changes with depth, probably due to a reduction in the degree of weathering. The soil-water characteristic curve is flatter at greater depths due to more restricted range of micropore sizes. This micropore size distribution can be related to the slope of soil-water characteristic curve at high matric suctions.

ACKNOWLEDGEMENTS

This study was funded by the National Science and Technology Board of Singapore (NSTB 17/6/16: "Rainfall-induced slope failures"). The first author acknowledges the research scholarship provided by Nanyang Technological University.

REFERENCES

Casagrande, A. 1932. The structure of clay and its importance in foundation engineering. *Journal of Bostom Society of Civil Engineering*. 19: 168-208

Collins, K.1985. Towards characterization of tropical microsoil structure. *First international conference on Geomechanics in tropical lateritic and saprolitic soils, Brasilia*. 1: 85-96

Collins, K. & McGown, A. 1974. The form and function of microfabric features in a variety of natural soils. *Geotechnique*. Vol. 24, No.2: 223- 254

Diamond, S .1970. Pore size distribution in clays. *Clays and clay minerals*. 18: 7-23

Fitzpatrick,R.W & Le Roux, J. 1977. Mineralogy and chemistry of a Transvaal black clay topo sequence. *Jour. Soil Sci.* 28: 165-179

Fredlund, D. G., Xing, A. & Huang, S. 1994. Predicting the Permeability Function for Unsaturated Soils using the Soil Water Characteristic Curve. *Canadian Geotechnical Journal.* 31: 533- 546

Goldshmidt, V.M. 1926. Undersokelser over lersedimenter. *Nordisk jordbrugsforskning.* 47: 434-445

University of East London 1999. *Report on the soil Mineralogy of Samples Using X-Ray Diffraction Analysis.* Unpublished report to Geotechnical Research Centre, Nanyang Technological University, University of East London.

Hutchison, C.A. 1964. A Gabbro- Granodiolite Association in Singapore Island, *Quarterly Journal of Geological Society*, London. 120: 283- 297

Irfan, T.Y. 1996. Mineralogy, fabric properties and classification of weathered granites in Hong Kong, *Quarterly Journal of Engineering Geology.* 29: 5-35

Lambe, T.W. 1953. The structure of inorganic soils. Proc. Am. Soc. Engrs 79, publ. No. 315

Lambe, T.W. 1958. The structure of compacted clay, *Proc. Am. Soc. Engrs.* 84, SM2: 1-34

Little, A. L. (1969), The Engineering Classification of Residual Tropical Soils. 7th *Int. Conf. on Soil Mechanic and Foundation Engineering*, Mexico. 1: 1- 10

Mitchell, J. K. (1976). Fundamentals of Soil Behaviour. *John Wiley and Sons Inc.* New York. 437

Onodera, T. 1976. Shear strength of undisturbed sample of decomposed granite soil. *Soil and Found.* 16(1): 17-26

Pitts, J . 1984. A Survey of Engineering Geology in Singapore, *Geotechnical Engineering.* 15: 1-20

Poh, K.B., Chuah, H.L & Tan, S.B. 1985. Residual granite soil of Singapore. *Eight Southeast Asian Geotechnical conference.* 3-1 to 3-9

Ruddock, E.C. 1967. Residual soils of the Kumasi district in Ghana. *Geotechnique.* 17(4): 359-377

Schofield,R.K. & Samson,H.R. 1954. Flocculation of kaolinite due to the attraction of oppositely charged crystal faces. *Faraday Society*, London. Discussion, 18: 135-145

Terzaghi, K. 1925. *Erdbaumechanik auf Bodenphysikalischer Grundlage Deuticke*, Vienna.

Vanapalli, S. K., Fredlund, D.G., Pufahl, D.E. & Clifton, A.W. 1996. Model for the prediction of shear strength with respect to soil suction, *Canadian Geotechnical Journal.* 379- 392

van der Merwe, D.H. 1965. The soils and the engineering properties of an area between Pretoria north and Brits, Transvaal. *DSc thesis.* University of Pretoria.

Unsaturated Soils for Asia, Rahardjo, Toll & Leong (eds) © 2000 Taylor & Francis, ISBN 90 5809 139 2

The use of indicator tests to estimate the drying leg of the soil-water characteristic curve

B.A. Harrison
P.D. Naidoo and Associates, Johannesburg, South Africa

G.E. Blight
University of the Witwatersrand, Johannesburg, South Africa

ABSTRACT: Wetting and drying soil-water characteristic curves were determined for undisturbed soil samples, of both transported and residual origin. Indicator tests were performed on the samples and relationships between them and the corresponding characteristic curve investigated. It was found that for fine grained soils of transported origin, the drying legs of the characteristic curves are related to the indicator tests. No relationship is evident between the indicators and characteristic curves for residual soils or for the wetting leg of the curve. From this information a preliminary estimate of moisture content changes during drying, associated with suction changes, can be assessed, and likely changes of volume, shear strength, permeability and water storage in fine grained undisturbed unsaturated transported soils subsequently estimated.

1 INTRODUCTION

The soil-water characteristic curve presents a constitutive relationship between water content and suction. It provides both conceptual aid to the understanding of unsaturated soil behaviour, and is employed for quantifying unsaturated shear strength, estimating soil volume changes, as a field moisture capacity curve and as an aid to the derivation of permeability functions. However, for a given soil, the water content suction relationship is not unique but exhibits hysteresis, and depends on whether the soil is drying or wetting.

This paper, an extension of an earlier paper (Harrison & Blight, 1999), illustrates how both the wetting and drying legs of soil-water characteristic curves were derived in the pressure plate apparatus for undisturbed transported and residual soils. Factors influencing the form and shape of the curve are discussed, and its properties in relation to simple indicator tests examined.

2 DESCRIPTION OF SOILS

Undisturbed soil specimens were prepared from block samples extracted from trial holes excavated with a backacter excavator. In total 21 samples were prepared, 11 of which were of transported origin and the remaining 10 residual. A clearly defined pebble marker horizon demarcated the separation between the overlying transported and underlying residual soils. The transported soils comprised alluvium, gulleywash, aeolian and hillwash whilst the residual soils were derived from the *in-situ* decomposition of igneous rocks including andesite, dolerite, syenite, granite, norite and basalt.

Laboratory indicator tests carried out on the samples included sieve and hydrometer analyses, Atterberg limits and bar linear shrinkage tests.

3 DETERMINATION OF THE SOIL-WATER CHARACTERISTIC CURVE

The soil-water characteristic curve for each soil specimen was determined in a pressure plate apparatus employing the axis translation technique.

Each specimen was prepared by trimming the undisturbed block sample to fit into a 20mm diameter and 15mm deep copper ring. After trimming, each ring was immersed in a container of water and left to soak for at least 2 days. The soil filled rings were then placed onto the saturated ceramic of the pressure chamber and a predetermined air pressure (u_a) applied.

After moisture equilibrium had been achieved each specimen was removed from the chamber and its mass accurately determined on an analytical balance.

The sample and ring were returned to the chamber and the procedure repeated for the next increment in pressure. Air pressures were applied in 10 and 50kPa increments and typically ranged from 10 to 900kPa on the drying leg.

After the final maximum de-saturation pressure of 900kPa had been reached, the soil specimens were subjected to decreasing chamber pressures, in increments of 200kPa, to induce re-wetting. During the wetting phase the specimens were weighed every third day to establish a suitable equilibrium period under a given chamber pressure. After some three weeks in the chamber at constant pressure, little notable change in their wet mass was observed.

After equilibrating at the final pressure, and determining the wet soil mass, each specimen was oven dried and the soil water content at each pressure increment, or suction, determined. This enabled a soil-water characteristic curve to be derived for both the drying and wetting states. Three of these curves are presented in Figure 1. The form or shape of these curves is typical of the 21 specimens tested.

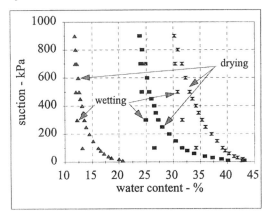

Figure 1: Soil-water characteristic curves

4 A DESCRIPTION OF THE CHARACTERISTIC CURVE

A simplistic description of the forces operative within the matrix of a homogeneous soil mass of both reducing (drying) and increasing (wetting) water contents, as reflected by the characteristic curve, follows.

The Drying Curve

The flat portion of the curve at low suctions reflects the outflow of water from the larger soil pores, which are less able to retain water against the applied suction. Capillary tension effects are thus the primary forces responsible for the release of water at low suctions. The water phase within the soil is continuous over this pressure range until air enters the system and the air entry suction is reached.

On approaching the air entry suction, progressively smaller pores are emptied and, where present, the hydrated water layer enveloping secondary clay minerals reduces. The transition curve thus reflects matrix suction associated with capillary tension and adsorption. Soil water continuity progressively diminishes over this range as air filled channels begin to form within the soil voids.

At high suctions water retention is considered to be primarily due to adsorptive forces, with capillary tension operative in micro voids that still retain capillary water. The soil matrix contains water in isolated pockets, which continuously reduce in volume as the suction increases.

The Wetting Curve

As water is introduced to a relatively dry soil, it is taken into the voids after first being adsorbed by the clay minerals that may be present. The clay imbibes water prior to the voids being filled since it attracts water with greater tenacity than that of the voids ie. the adsorbtive forces exceed those of capillary tension. As additional water becomes available to the soil, the adsoption forces of the clay are attenuated and the voids gradually fill.

The path of this wetting curve does not follow that of its drying counterpart and, as illustrated in Figure1, for a given water content the suction for the drying curve is greater than for the wetting. There are numerous reasons for this, some of which are briefly discussed below.

- The soil structure, and hence void sizes, change as a result of shrinkage (drying) or swelling (wetting).
- Air is entrapped during wetting, which results in the formation of occluded air bubbles and hence a reduction in water content.
- The contact angle of the menisci is greater for a wetting process than a drying one and hence the radius of curvature is greater and the suction correspondingly lower.
- The irregular size and shape of the soil voids affect the manner and ease with which water enters and leaves the soil.

An inspection of Figure 1 indicates that at high suctions the wetting curve undergoes a substantial reduction in suction as a result of a nominal increase in water content. It seems unlikely that the factors causing the hysteresis effects, alluded to above, will result in the very steep wetting curve obtained for all of the soil specimens tested. An explanation for the initial steep wetting curve is outlined below and is believed to reflect limitations of the pressure plate apparatus.

Soil water continuity is absent in soils exhibiting high suctions since it is contained in the soil voids as isolated pockets. Thus, the water within a soil specimen of high suction resting on the ceramic of the pressure plate apparatus may not be in direct contact with the water in the ceramic. Moreso, since at the high chamber pressures water will not be present at the surface of the ceramic, as it will have been driven into the recesses of its pores. These pores are illustrated in Figure 2, which is a micrograph of the ceramic used in the pressure plate apparatus.

Water is thus only able to pass from the ceramic to the soil in the vapour phase during wetting at high suctions. As the chamber air pressure is gradually reduced so water progressively migrates to the surface of the ceramic, from the underlying reservoir, until a critical chamber pressure is reached when contact between the water in the ceramic and water in the soil takes place. The test results suggest this pressure to be of the order of 300kPa.

Figure 2: Micrograph of ceramic showing coarse surface pores

5 INDICATOR TESTS AND THE CHARACTERISTIC CURVE

It appears that certain soil properties play a significant role in determining the water content–suction relationship. These properties include heterogeneity, distribution of void size, and the abundance and nature of the clay minerals present.

With the exception of heterogeneity, it can be argued that index tests are also related to these properties.

For example, the void size distribution is largely dictated by the particle size distribution, as determined in the sieve and hydrometer tests. The hydrometer test also quantifies the proportion of clay sizes ($<2\mu m$) present, whilst the Atterberg limits provide an indication as to their activity. Based on these observations a relationship, albeit tenuous, should exist between the index tests and some characteristics of the soil-moisture curve.

Burland (1990), Marinho & Chandler (1993), and Ridley & Peres-Romero (1998) have presented the drying portion of the characteristic curves of reconstituted compacted soils as a semi-logarithmic plot. They found a linear relationship between log suction and water content at suctions above about 100kPa. Similar relationships were established for the soils tested in this work, which are illustrated in Figure 3 for suctions ranging from 10 to 900kPa (1 to 90m of negative water head).

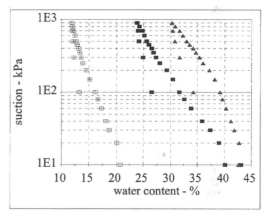

Figure 3: Log-suction vs water content

The gradient of the curve for suctions greater than 100kPa has been termed the Suction Capacity (C – in logkPa/%). Plots of C against liquid limit for reconstituted samples have been presented by the aforementioned authors, from which a substantially direct relationship was derived. A similar plot is presented in Figure 4, for all of the undisturbed soil samples tested in this work.

On initial inspection it appears that only a poor correlation exists. If, however, the residual soils are removed from the plot, with one or two exceptions, fairly strong interdependence for the transported soils is evident.

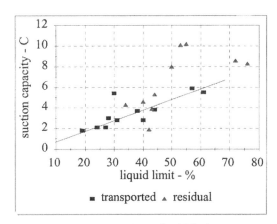

Figure 4: Suction capacity vs liquid limit

The suction capacity derived for each of the transported soils, above 100kPa suction, has been plotted against the corresponding plasticity index, linear shrinkage, clay fraction and grading modulus. The plots are presented in Figures 5 a) to d), from where it can be seen that satisfactory correlations exist. No correlation was exists for the wetting curve.

As shown in Figures 6 a) to d), the water content at a suction of 100kPa also correlates well with the corresponding plasticity index, linear shrinkage, clay fraction and grading modulus. Hence a value of any one of these familiar index parameters can be used, via Figures 5 and 6, to define the suction capacity and the water content at a suction of 100kPa, and thus the complete drying leg of the soil water characteristic curve for suctions from 100kPa upwards.

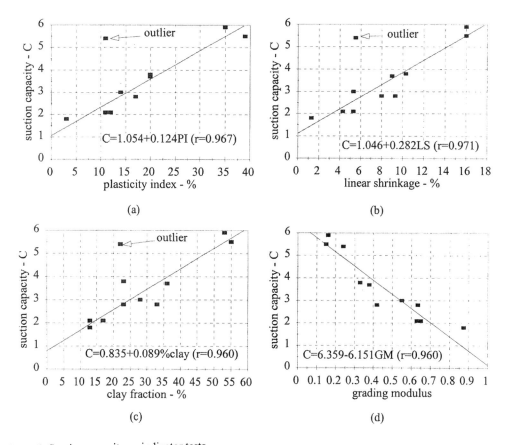

Figure 5: Suction capacity vs indicator tests

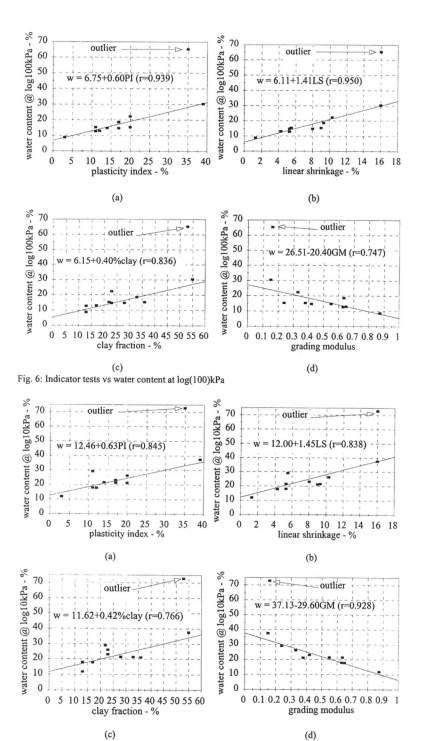

Fig. 6: Indicator tests vs water content at log(100)kPa

Fig. 7: Indicator tests vs water content at log(10)kPa

Figure 3 illustrates that the semi-logarithmic plots change slope markedly at suctions less than 100kPa. However, as shown in figures 7 a) to d), the water content at 10kPa suction correlates very well with the grading modulus but less so with the other soil indicators. Thus, employing the information from Figures 6 and 7, or 7 d) in particular, the lower leg of the soil-water characteristic curve can be established and hence the entire curve for drying.

It is of interest to speculate as to why the index tests correlate with the water contents at suctions of 10 and 100kPa. In the latter case better correlation is achieved with the Atterberg limits and linear shrinkage than with the other indicator tests, whilst moisture contents at 10kPa suction correlate well with the grading modulus but less so with the Atterberg limits and linear shrinkage.

This corroborates the conceptual model described earlier, where void size distribution, inferred from the grading modulus, substantially influences the development of capillary tension at low suctions, whilst clay mineralogy, as reflected by the Atterberg limits and linear shrinkage, plays a more predominant role in the retention of water, through adsorption, at the higher suctions.

6 CONCLUSIONS

Undisturbed samples of fine grained, homogeneous transported soils subjected to drying were found to exhibit reasonable correlation between indicator tests, suction capacity and water content at 10 and 100kPa suction. Similar relationships with residual soils, on the other hand, were not attainable due to their inherent saprolitic fabric and inherent non-homogeneity, both natural features established during their formation. No correlation was evident for properties of the wetting curve and indicator tests.

Employing the relationships presented in this paper, a preliminary estimate of moisture content changes associated with suction changes, during drying, can be established for transported soils. This may be useful for determining the magnitude of volume change, shear strength, permeability and water storage in undisturbed, fine grained, unsaturated transported soils.

Determining the wetting leg of the characteristic curve in the pressure plate apparatus is not recommended at suctions above about 300kPa. This is believed to be due to little, if any, communication taking place between the water in the ceramic and water in the soil at suctions above this value. The true wetting curve may be much closer to the drying curve.

7 REFERENCES

Burland, J.B. 1990. On the compressibility and shear strength of natural clays. *Geotechnique* 40(3): 329-378.

Harrison, B.A. & Blight, G.E. 1999. The determination of soil-water characteristic curves from indicator tests. *Geotechniques for Developing Africa*. 12th Reg. Conf. for Africa on Soil Mech. & Geot. Eng. Durban, South Africa. 325-329. Rotterdam Balkema.

Marinho, F.A.M. & R.J. Chandler, 1993. Aspects of the behavior of clays on drying. *ASCE*, Geotech. Spec. Pub. No. 39: 77-90.

Ridley, A.M. & J. Perez-Romero 1998. Suction-water content relationships for a range of compacted soils. *2nd Int. Conf. On Unsat. Soils*, Beijing: 114-118. Rotterdam: Balkema.

Unsaturated Soils for Asia, Rahardjo, Toll & Leong (eds) © 2000 Taylor & Francis, ISBN 90 5809 139 2

The model of water retention curve considering effects of void ratio

K. Kawai & S. Kato
Department of Civil Engineering, Kobe University, Japan

D. Karube
Graduate School of Science and Technology, Kobe University, Japan

ABSTRACT: Water retention curves derived from some initial conditions based on experimental results and some factors affecting the water retention curves are examined. On the drying process of the water retention curve, degree of saturation, water content and void ratio tend to converge to a specific value with increase in suction. And, the value can be regarded as a constant in terms of the water content. The suction value at which the desaturation begins on the drying process from the saturated state is called "Air Entry Value" and depends on the void ratio. The relationship between the air entry value and void ratio can be described as a unique "Air Entry Value Line". On the wetting process, it is found that the shape of the water retention curve depends on the degree of saturation. Likewise, the water entry value on the wetting process can be expressed as a function of void ratio as the "Water Entry Value Line". Thus, in this paper, a new model for the water retention curve adopting the concept of "Virgin Drying Line" presented by Toll is proposed. It is shown that the model can predict the hysteresis loop in the water retention curves when the initial states (suction, void ratio, water content) of specimen are provided.

1 INTRODUCTION

Unsaturated soil contains both water and air in its void. Accordingly, the pressure difference between pore air and pore water yields the difference of surface activities of water and air. This pressure difference is called "suction" and characterizes the behavior of the unsaturated soil. It is known that decrease of suction causes collapse phenomenon which cannot be explained by Terzaghi's effective stress theory for saturated soils. Moreover, the hydraulic gradient caused by the suction affects the infiltration characteristics of unsaturated soil media. Usually,

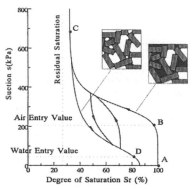

Figure 1 Water retention curve

the water retention curve expressed as the relation between suction and water content is adopted when suction is determined in the seepage analysis for unsaturated soils. However, the water retention curve expressed in terms of water content cannot be uniquely specified. It depends on state variables such as void ratio and draws different curves in the drying process and the wetting process. In this paper, this hysteresis response is discussed, and the factors affecting the water retention curves are examined. Finally, a new model, which can quantitatively predict the hysteresis response, is proposed to express the water retention relation.

2 BRIEF REVIEW OF STUDIES ON WATER RETENTION CHARACTERITICS

The water retention curve is an expression to quantify the retained water in a unit volume of soil mass in terms of the volumetric water ratio or the degree of saturation, it gives the relation with the suction as shown in Figure 1. The suction is defined as,

$$s = u_a - u_w \qquad (1)$$

in which s is suction, u_a is the pore air pressure and u_w is the pore water pressure. Desaturation in the soil increases suction and absorption decreases suction. The degree of saturation hardly changes until a

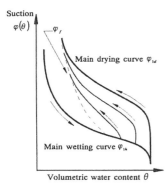

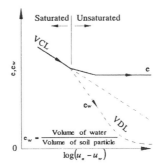

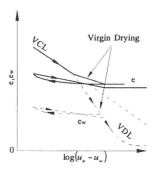

Figure 2. Water retention curve
presented by Vachaud et al. (1971)

(a) Drying from saturated state   (b) Effect of suction hysteresis
Figure 3. Water retention model presented by Toll (1995)

certain suction value (from point A to point B in Figure 1), when desaturation begins from the fully saturated state (point A in Figure 1). After the suction value exceeds point B the degree of saturation decreases at a fairly quick rate. The suction value at point B is called "air entry value". There after it reaches a certain residual value (point C in Figure 1). Contrary to it, when resaturation begins from the point C, the wetting process curve does not correspond to that from the drying process. The degree of saturation reaches a certain suction value (point D in Figure 1), which is called "water entry value". Studies on such hysteresis response of the water retention curve are introduced below.

2.1 Study by Vachaud et al.

Vachaud et al. (1971) carried out seepage tests for unsaturated soil specimens with several initial conditions and finally expressed the water retention characteristics as shown in Figure 2. They found that every water retention response in the drying process and in the wetting process could be expressed by two unique curves. They were named "Main wetting curve" and "Main drying curve", respectively. They also found that the water content in on arbitrary unsaturated state is plotted inside the region bounded by these two curves and the water retention path in the desaturation process converges into the main drying curve φ_f under high suction state. However, since sand specimens were used in the experiment, the effect of volumetric change was not taken into account in their discussion.

2.2 Study of Toll

Toll (1995) expressed the water content in terms of equivalent void ratio defined as,

$$e_w = V_w/V_s = S_r' \cdot e = G_s \cdot w' \qquad (2)$$

in which e_w is equivalent void ratio, V_w and V_s is volume of water and soil respectively, $S_r' = S_r/100$,

S_r is degree of saturation, e is void ratio, G_s is specific gravity of soil, $w' = w/100$ and w is water content. Figure 3 shows his model to describe the water retention characteristic using the equivalent void ratio. The Virgin Drying Line (VDL) expresses the change of equivalent void ratio with suction for the normally consolidated soil, while the Virgin Consolidated Line (VCL) shows the change of void ratio of the soil. As shown in the figure, when these two curves agree, the soil remains in the saturated state. Every state of unsaturated soil is plotted under VDL in the saturated state. He interpreted that the equivalent void ratio changed slightly until the suction reached the largest value of suction that the soil had experienced in the past. And once the path reaches the VDL, it goes along the VDL. However, the applicability of this model to over-consolidated clay is not clear.

3 EFFECTS OF INITIAL CONDITIONS AND SUCTION HYSTERESIS ON WATER RETENTION CURVE

We examined factors affecting the shape of water retention curve as follows:

3.1 Experimental procedure

Two series of experiments were carried out for specimens of silty clay whose material properties are listed in Table.1.
The purpose of one series of experiments is to examine the effect of the initial conditions on the shape of the water retention curve. We prepared unsaturated specimens by static compaction of powder clay. The initial conditions of the specimens are shown in Table 2. Each specimen was set in the oedometer ap-

Table 1. Material properties of clay used

Gs	w$_P$	w$_L$	I$_P$
2.70	29.6%	43.0%	13.4

Table 2. Initial condition

	Void ratio e	Water content w (%)	Degree of saturation Sr (%)
a	2.11	23.2	29.7
b	1.62	30.2	50.5
c	1.59	17.5	29.7
d	1.20	23.1	52.0
e	0.99	30.4	82.9

Table 3. Suction hysteresis

	Confining pressure p (kPa)	Suction s (kPa)
A	19.6	0→490→0
B	19.6	0→392→49→245
C	19.6	0→294→49→245

paratus modified for unsaturated soil, and suction was applied by means of the pressure plate method. The purpose of another series of experiments was to investigate the effect of suction hysteresis on water retention characteristics. Slurry clay specimens were pre-consolidated under the saturated condition. Each specimen was set in the triaxial apparatus modified for unsaturated soils and the suction was applied under the conditions shown in Table 3 by means of the pressure plate method.

3.2 *Experimental results and discussion*

Figures 4 (a), (b), (c) show degree of saturation, water content and void ratio of specimen with suction respectively. As for degree of saturation, the dependency on initial void ratio appears. Degrees of saturation converge at a certain value in the drying process except for the specimen "e" whose void ratio is relatively large. It is found that specimen with smaller void ratios converge at higher degree of saturation in the lower suction region. As for water content, the drying curves converge at a unique value even if the initial water contents are different. This convergence value seems to correspond to the adsorbed water quantity. When the adsorbed water on the surface of soil particle crystallized by chemical combination, it can never be separated even if suction is applied as an external force. The ratio of the adsorbed water to water content is constant since the adsorbed water quantity is proportional to the particle surface, if the particle size distribution is the same. Therefore, the water contents converge to a unique value. As for the void ratio, the volume of the specimen hardly changes during the drying process. Collapse occurs in the wetting process, when the void ratio of the specimen is large. If the suction value at which collapse occurs is regarded as the water entry value, the volume of the specimen hardly

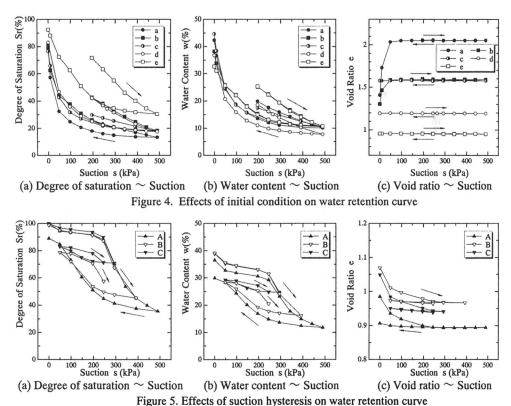

(a) Degree of saturation ～ Suction   (b) Water content ～ Suction   (c) Void ratio ～ Suction

Figure 4. Effects of initial condition on water retention curve

(a) Degree of saturation ～ Suction   (b) Water content ～ Suction   (c) Void ratio ～ Suction

Figure 5. Effects of suction hysteresis on water retention curve

331

changes in the high suction region and it is possible to ignore volume change during seepage.

Figures 5 (a), (b), (c) shows change of the states of the specimen applied suction hysteresis. As for void ratio, unlike the drying process from the unsaturated state, the specimen was greatly compressed in the initial stage of drying process. Simultaneously, the water content changes are large. Therefore, the degree of saturation remains at about 100%.

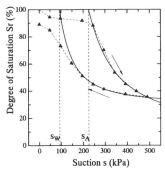

Figure 6. Application of Brooks and Corey model

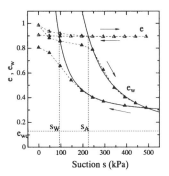

Figure 7. Application for Toll's model

In this suction region, the specimen has not desaturated yet, and the drainage agrees with the volume change. In this state, suction acts merely as negative pore-water pressure, and the Terzaghi's effective stress theory is applicable. Since the initial states of these three specimens are almost the same, it can be considered that the three tests results are wetting curves from the same drying curve. Even if the suction is reduced, the degree of saturation hardly recovers for specimens on the drying curve from the saturation state. This phenomenon is more remarkable, as the degree of saturation for the wetting process is higher. However, three specimens whose void ratio is almost equal converged into one wetting curve. Therefore, the degree of saturation recovers more quickly in the low suction region, when the degree of saturation of the specimen is higher. On the redrying process, it is not possible to judge these results and it is considered that it may converge into the initial drying curve.

3.3 Applying empirical formula for experimental results

The experimental results are quantified using existing empirical formula. The model proposed by Van Genuchten (19980) is popular for approximating the water retention curve. However, it is difficult to obtain the optimum parameters, since trial and error is necessary for the approximation. In this paper, the model proposed by Brooks and Corey (1966) is adopted, since the form of the equation is simple and the parameters has physical meanings. The model is shown below:

$$S_e = \frac{S_r - S_{ra}}{S_{rf} - S_{ra}} = \left(\frac{s_b}{s}\right)^{\lambda} \qquad (3)$$

Here, S_e is the effective degree of saturation, S_{ra} is S_r at $s \to \infty$, S_{rf} is S_r at $s \to 0$. In this paper, $S_{rf} = 100$ is used, and λ, s_b are the shape parameters. The value of s_b is the maximum suction value when the degree of saturation is 100%, and corre-

sponds to the air entry value in the drying process and water entry value in the wetting process. The value of λ affects the curvature of the water retention curve. When the value of λ is large, the gradient of the curve changes more rapidly. The curve fitting for a typical test result is shown in Figure 6. The value of s_A and s_W in the figure shows air entry value and water entry value respectively.

Equation 3 shows water retention curve in the high suction region rather than the suction region near the air entry value or the water entry value. Since void ratio hardly changes in this suction region, Equation 3 can be transformed into the following equation using the equivalent void ratio proposed by Toll (1995):

$$S_e = \frac{e_w - e_{wa}}{e - e_{wa}} = \left(\frac{s_b}{s}\right)^{\lambda} \qquad (4)$$

Here, e_{wa} is the equivalent void ratio at $s \to \infty$. Figure 7 shows as example of applying Equation 4 for a test. Figure 8 and Figure 9 show relations of air entry value and of water entry value with void ratio. These figures show that both air entry value and water entry value are unique function of void ratio. Though the value of λ in the drying process is constant, dependency of the value of λ in the wetting process on degree of saturation is found. Figure 10 shows relation of λ and S_{ra} ($= 100 \times e_{wa}/e$).

4 NEW MODEL OF WATER RETENTION CURVE CONSIDERING EFFECT OF VOID RATIO

4.1 Drying process

Figure 11 shows the drying curve model from the saturated state. The "air entry value line" shown in the figure is the same curve shown in Figure 8. The dashed lines show the change of void ratio, and the solid line show the change of the equivalent void ratio.

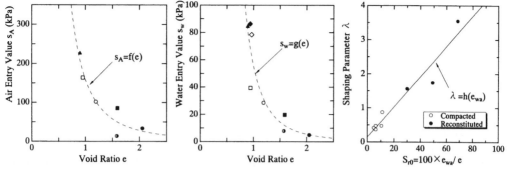

Figure 8. Relation of the air entry value and void ratio

Figure 9. Relation of the water entry value and void ratio

Figure 10. Relation of λ and Sr_0

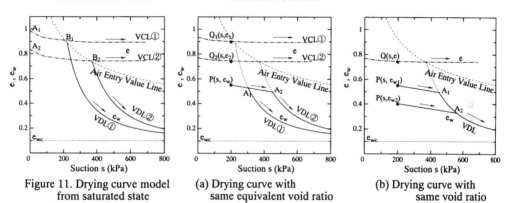

Figure 11. Drying curve model from saturated state

(a) Drying curve with same equivalent void ratio

(b) Drying curve with same void ratio

Figure 12. Drying curve model from unsaturated state

We assume that the suction region, smaller than air entry value, is in the saturated state in which suction can be regarded as negative pore-water pressure. Consequently, the change of void ratio can be calculated by applying the consolidation theory for saturated soil. In this suction region, void ratio and equivalent void ratio are equivalent and their results are shown as $A_1 \rightarrow B_1$.

The specimen desaturate, when the suction reaches the air entry value (point B_1). As for equivalent void ratio, the equation of Brooks and Corey is available for the higher suction region than air entry value. The equivalent void ratio converges into unique value, since it is shown as volumetric water content. This convergence value is volumetric adsorption water content e_{wc}. It is possible to obtain the equivalent void ratio curve, if the value of λ which shows the curvature is assumed constant. Toll (1995) has named this drying curve, which begins from saturation state, as "Virgin Drying Line (VDL)", and postulated that "VDL" is unique. However, in the proposed model, drying curve (VDL②) of saturated specimen with smaller void ratio (point A_2) crosses the air entry value line at a different point (point B_2). The change in void ratio due to the applied suction in the suction region, higher than the air entry value,

can be calculated with the theory of Karube (1994) using the following equations:

$$\frac{\partial e}{\partial p'} = \frac{p_m}{p' + p_m} \qquad (4)$$

$$p' = p + p_b \qquad (5)$$

$$p_b = \frac{S_{rb}}{100 - S_{r0}} \times s = \frac{S_r - S_{rd}}{100 - S_{rd}} \times s \qquad (6)$$

$$p_m = \frac{S_{rm}}{100 - S_{r0}} \times s = \frac{(100 - S_r)(S_{rd} - S_{r0})}{(100 - S_{r0})(100 - S_{rd})} \times s \qquad (7)$$

Here, p_b is bulk stress and calculated using volumetric ratio of bulk water S_{rb} and p_m is meniscus stress and calculated using volumetric ratio of meniscus water S_{rm}. The value of S_{rb} and S_{rm} is calculated using S_{rd}, degree of saturation on "Driest Curve".

The drying curve model from unsaturated state is shown in Figure 12. Figure 12(a) show the drying curves when the equivalent void ratio is the same and when the void ratio is different. It can be assumed that unsaturated specimens experienced suction hysteresis. Since the "VDL" depends on void ratio of the specimen, the drying curve of the specimen with a smaller void ratio become as if the specimen had experienced a larger suction hystere-

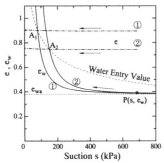

(a) Wetting curve with same equivalent void ratio

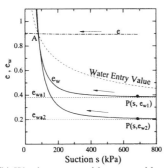

(b) Wetting curve with same void ratio

Figure 13. Wetting curve model

sis. Though the equivalent void ratio hardly changes until the suction value reaches the VDL, the equivalent void ratio goes along the VDL after the suction value passes the VDL. Figure 12(b) shows drying curve when the void ratio is same and the equivalent void ratio is different. In this case, they have the same VDL. However, it is difficult to predict the change of the equivalent void ratio in the suction region under the VDL quantitatively.

4.2 Wetting Process

Figure 13 shows the wetting curve model. The wetting curve has its own convergence value and its own curvature. Accordingly, it is necessary to obtain convergence value of the equivalent void ratio, which satisfies the following equation in order to obtain the wetting curve of the specimen given the void ratios, equivalent void ratio and suction:

$$\frac{e_{w1} - e_{wa}}{e_1 - e_{wa}} = \left(\frac{s_b(e_1)}{s_1}\right)^{\lambda(e_{wa}/e_1)} \tag{6}$$

Here, $s_b(e)$ is a function of e and $\lambda(e_{wa}/e)$ is a function of e_{wa}/e. It must be calculated numerically, since this equation cannot be developed using e_{wa}.

Figure 13(a) shows the wetting curve model when the equivalent void ratio is same and the void ratio is different. In this case, the convergence value of the equivalent void ratio is different, and the wetting curves have different water entry value and different curvature.

Figure 13(b) shows wetting curve model when the equivalent void ratio is different and the void ratio is same. In this case, the wetting curves have the same water entry value and different curvature. Specimens A, B, C shown in Table 3 belong to this case.

5 CONCLUSIONS

In this paper, the factor affecting the shape of the water retention curve was examined. It was found that the shape of the water retention curve depends on the void ratio, and that its effect appears in the air entry value and the water entry value. Both the air entry value and the water entry value become larger when the void ratio of the specimen is smaller. And, it was found that the curvature of the water retention curve in the wetting process depends on the degree of saturation of the specimen.

A new model adopting the concept of "Virgin Drying Line" proposed by Toll considering the effects of void ratio was proposed. This model introduces the air entry value line and the water entry value line, and can express the effect of the void ratio. It is shown that the model can predict the hysteresis loop in the water retention curves when the initial states (suction, void ratio, water content) of the specimen are provided. However, further examination is necessary to predict the behavior of the specimen in the suction region under the air entry value or the water entry value.

REFERENCES

Vachaud. G. and Thony. J. L. 1971. Hysteresis during infiltration and redistribution in a soil column at different initial water contents. *Water Resources Research*. Vol.7. No.1 : 111-127.

D. G. Toll. 1995. A conceptual model for the drying and wetting of soil. *Proc. 1st Int. Conf. on Unsaturated Soils.* Vol.2 : 805-810

Van Genuchten. 1980. A closed-form equation for predicting hydraulic of unsaturated soils. *Soil. Sci. Soc. Amer.* Vol.44 : 892-898

R. H. Brooks and A. T. Corey. 1966. Properties of porous media affecting fluid flow. *Proc. ASCE92.* IR(92) : 61-88

D. Karube. And S. Kato. 1994. An ideal unsaturated soil and the Bishop's soil. *Proc. 13th Int. Conf. SMFE.* 1 : 43-46

Unsaturated Soils for Asia, Rahardjo, Toll & Leong (eds) © 2000 Taylor & Francis, ISBN 90 5809 139 2

A new test procedure to measure the soil-water characteristic curves using a small-scale centrifuge

R. M. Khanzode & D. G. Fredlund
Department of Civil Engineering, University of Saskatchewan, Saskatoon, Sask., Canada

S. K. Vanapalli
Royal Military College of Canada, Kingston, Ont., Canada

ABSTRACT: The soil-water characteristic curve is conventionally measured using a pressure plate apparatus or a Tempe cell. Considerably long periods of time are required to measure the soil-water characteristic curves using the conventional equipment. A new test procedure is proposed, using a small-scale medical centrifuge to measure the soil-water characteristic curves for compacted, fine-grained soil specimens. Soil specimen holders were designed for the small-scale centrifuge. The soil-water characteristic curves of statically compacted specimens for three different fine-grained soils with varying percentages of clay were measured using the centrifuge for a suction range between 0 to 500 kPa. There is good comparison between the soil-water characteristic curves measured using the small-scale centrifuge and the conventional laboratory equipment. The results of this study are encouraging as soil-water characteristic curves can be measured in a shorter period of time, resulting in considerable savings.

1 INTRODUCTION

The engineering behavior of an unsaturated soil can be interpreted in terms of two stress state variables namely; net normal stress, $(\sigma_n - u_a)$, and matric suction, $(u_a - u_w)$, using experimental test results (Fredlund and Rahardjo 1993). Experimental techniques to determine the unsaturated soil properties are however costly and time consuming. In the last five years, several simple procedures have been proposed in the literature to predict the engineering behavior of unsaturated soils using the soil-water characteristic curve and the saturated soil properties. The procedures to predict the engineering behavior of unsaturated soils are simple and economical, and therefore useful to practicing geotechnical and geo-environmental engineers.

The soil-water characteristic curve defines the relationship between the soil suction and soil gravimetric water content, w, or volumetric water content, θ_w, or the degree of saturation, S. Soil-water characteristic curves are commonly measured in the laboratory for a suction range between 0 to 1,000 kPa using conventional equipment. This suction range is of interest to the geotechnical and geo-environmental engineers. Typically, six to eight data points are measured such that the important features of the soil-water characteristic curve (i.e., the air-

entry value and the residual state conditions) can be determined from the measured data.

Conventional equipment used for the measurement of the soil-water characteristic curve includes the pressure plate or the Tempe cells. These apparatuses are reliable for measuring the soil-water characteristic curve behavior of both coarse and fine-grained soils, but considerable time is required. More details of these equipment and testing procedures are available in Fredlund and Rahardjo (1993).

The time period required for the measurement of the soil-water characteristic curve, using conventional testing procedures for soils such as sand or silt is between 6 to 8 days (for obtaining 6 to 8 data points). In other words, approximately one day is required to obtain one data point for relatively coarse-grained soils. Longer periods of time are required to measure the soil-water characteristic curves for fine-grained soils such as tills and clays. A time period of approximately 5 to 6 days is required for specimens to equilibrate under each value of soil suction. Typically, 4 to 6 weeks of time is required to obtain the soil-water characteristic curve for a fine-grained soil with a suction range of 0 to 1,000 kPa (i.e., for 6 to 8 data points).

Soil-water characteristic curves were measured for some coarse-grained and fine-grained soils using the centrifuge technique (Gardner 1937, Russell and Richards 1938, Croney et al. 1952, Skibinsky 1996). However, centrifuge techniques are not conventionally used for the measurement of soil-water characteristic curves. Limited studies have been undertaken to measure the soil-water characteristics of fine-grained, compacted soil specimens using centrifuge techniques.

The centrifuge principle and technique for measuring the soil-water characteristic curve are provided in this paper. The design details of soil specimen holder are also described. Test result comparisons between the measured soil-water characteristic curve using conventional equipment and the proposed centrifuge technique for three different, fine-grained soils with varying percentages of clay is also presented and discussed.

2 PRINCIPLE OF THE CENTRIFUGE TECHNIQUE

A high gravity field is applied to the soil specimen supported on a saturated, porous ceramic column using the centrifuge. The base of the ceramic stone has a water table that is at atmospheric pressure conditions. The water content profile in soil specimen after attaining equilibrium conditions is similar to water draining under field conditions to a groundwater table where gravity is several times that on earth.

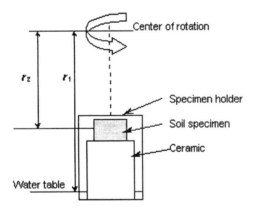

Figure 1. Suction measurement principle of the centrifuge.

Figure 1 demonstrates the suction measurement principle of the centrifuge method detailed earlier. The suction in the soil specimen in a centrifuge can be calculated using Eq. [1] proposed by Gardner (1937).

$$\psi = \frac{\rho \omega^2}{2} \left(r^2_2 - r^2_1 \right) \qquad [1]$$

where :

ψ = suction in the soil specimen
r_1 = radial distance to the free water surface
r_2 = radial distance to the midpoint of the soil specimen
ω = angular velocity
ρ = density of the pore fluid

Equation 1 defines a non-linear relationship between suction and centrifugal radius. The soil suction, ψ, becomes a function of the difference of the squares of the centrifugal radii, r_1 and r_2 keeping the density, ρ, and the angular velocity, ω, constant. The distance from the centre of rotation to the free water surface, r_1, is fixed and is a constant.

Different values of suction can be induced in the soil specimen by varying the radial distance to the midpoint of the soil specimen, r_2. This can be achieved by using ceramic cylinders of different heights. Higher values of suction can also be induced in soil specimens by increasing the test speed (i.e., angular velocity, ω).

3 SMALL-SCALE MEDICAL CENTRIFUGE

A J6-HC small-scale medical centrifuge with JS-4.2 rotor assembly with an operable radius of 254 mm was used in the research program. The JS- 4.2 rotor assembly of the centrifuge consists of six swinging type buckets (Fig. 2). The buckets in the centrifuge are able to rotate at angular velocities varying from 300 to 4200 rpm. The maximum suction that can be induced in the specimen at 4200-rpm using this centrifuge is equal 2800 kPa. The swinging type buckets of the small-scale centrifuge, assumes horizontal position when the centrifuge is spinning. All the six buckets can be used simultaneously with six specimen holders for testing. The mass in all the specimen holders, however, should be the same to avoid rotary imbalance.

Specially designed soil specimen holders are required to accommodate the specimens in the centrifuge buckets. Six data points of water content versus suction can be obtained in a single test run using a J6-HC small-scale medical centrifuge at one particular angular velocity, ω. Identical soil specimens have to be placed at different heights in the six specimen holders such that the six data points result in different water contents and suction values. The water content in the specimen can be measured from mass-volume relationships and the suction in the specimen can be estimated using Eq. [1].

Figure 2. J6-HC centrifuge with six swinging type buckets of the JS-4.2 rotor assembly.

In the present study, only two swinging buckets were used during one test spin with two specimen holders. The remainder four buckets of the centrifuge were left empty without any specimen holders.

4 SOIL SPECIMEN HOLDERS

Two aluminum soil specimen holders were specially designed for use in the J6-HC centrifuge to hold 10 to 15 mm thick soil specimens at different heights. Figure 3 shows a typical aluminum soil specimen holder used in the study. The soil specimen holder consists of five individual outer rings (inner diameter of 75mm and 15 mm thick), a drainage plate with a free water surface reservoir to accommodate a ceramic cylinder. A reservoir cup serves as a collection area for water extracted from the soil specimens at the base of the holder.

A porous cylinder was designed to act as a filter to prevent the movement of soil from the specimen to the drainage plate. This plate facilitates drainage into the reservoir cup through eight evenly spaced drainage ports drilled horizontally through sides of the plate. The horizontal overflow ports are connected to vertically drilled drainage holes to allow the removal of water from the soil specimen that flows down from the drainage plate into the reservoir cup.

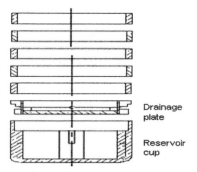

Figure 3. Aluminum soil specimen holder (from Skibinsky 1996).

4.1 Ceramic cylinders

The ceramic cylinders used in the drainage plate were made up of 60% Kaolinite and 40% Aluminum Oxide. The porosity of the ceramic cylinders was equivalent to 45%. Ceramic cylinders of different heights in combination with different test speeds can be used to apply different suctions in the soil specimens.

In the present study, four ceramic cylinders with different heights of 15mm, 30mm, 45mm and 60 mm were used to keep the soil specimen at four different distances from the centre of rotation. This would enable the application of different suctions in the soil specimens at one constant test speed.

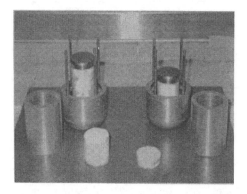

Figure 4. Saturated soil specimens on top of the saturated ceramic cylinders in the drainage plate.

The saturated ceramic cylinders are held in the drainage plate as shown in Figure 4 at the start of the test. The ceramic cylinder bottom rests in the free water surface reservoir in the drainage plate. The ceramic cylinder remains saturated during the entire period of spinning. The statically compacted, saturated soil specimens are placed on top of the saturated cylinders such that there is a direct hydraulic connection between the pore-water in the soil specimens and the free water surface reservoir at the base of the ceramic cylinder.

Water from the saturated soil specimen escapes into the bottom reservoir through the ceramic cylinder and reaches equilibrium conditions when the specimen is centrifuged. The suction in the soil specimen will be equivalent to the applied centrifugal force after reaching equilibrium conditions.

Ceramic cylinders of two different heights were used in one test run to position the soil specimens at two different distances from the centre of rotation of the centrifuge. The soil specimens were subjected to two different centrifugal forces and different values of suction were induced in two identical soil specimens placed in the soil specimen holders, subjected to the same speed.

Different cylinder heights and increasing test speeds of 300, 500, 1000, 1500, 2000 and 2500 rpm were used in the present study. The soil-water characteristics were measured for a suction range from 0 kPa to 600 kPa. Table 1 shows the calculated suction values at the midpoint of the soil specimens using Eq. [1] for different distances from the centre of rotation at different test speeds.

Table 1. Suction associated with different test speeds and different ceramic cylinders

Test Speed in rpm	Suction in the soil specimen (kPa)			
	15 mm cylinder	30 mm cylinder	45 mm cylinder	60 mm cylinder
300	6.04	8.38	10.51	12.41
500	16.69	23.18	29.06	34.32
1000	67.11	93.26	116.8	138.8
1500	151.1	210.0	263.1	310.8
2000	268.7	373.3	467.7	552.5
2500	420.0	583.6	731.1	863.6

5 TEST PROGRAM

Three different fine-grained soils, namely, the Processed silt (w_L = 24 %, I_p = 0, and Clay = 7%, G_s = 2.7), Indian Head till (w_L = 35.5%, I_p = 17%, and Clay = 30%, G_s = 2.73) and Regina Clay (w_L = 75.5% and I_p = 21%, and Clay = 70%, G_s = 2.75) were used for testing in the centrifuge. All the three soils were first air-dried and then pulverized. Precalculated amounts of water content was added to the soil and stored in polythene bags in a humidity-controlled room for 24 hours to attain uniform water content.

The Processed silt specimens were statically compacted at an initial water content, w, of 22% and a dry density, ρ_d, of 1.57 Mg/m^3. For Indian Head till specimens, representing the wet of optimum conditions, soil specimens were statically compacted at an initial water content of 19.2% and ρ_d of 1.77 Mg/m^3. The Regina clay specimens were statically compacted at an initial water content of 38% and ρ_d of 1.30 Mg/m^3. All the specimens were statically compacted in steel rings of 50mm diameter and 15mm height. More details of soil properties and specimen preparation are available in Khanzode (1999).

6 TEST PROCEDURE

Ceramic cylinders of two different heights (i.e., 30mm and 60 mm) and the statically compacted soil specimens were saturated at the start of the test by submerging in a water bath for 24 hours. The centrifuge was started and allowed to run at 300 rpm for half an hour to adjust the temperature of the rotating chamber. All the tests were conducted at a constant temperature of 20° C. The ceramic cylinders and the drainage plates were then placed in respective specimen holders (Figure 4). The bottom end of the ceramic cylinder was placed such that it just dips into the reference free water reservoir in the drainage plate.

The masses of the saturated soil specimens were determined and the soil specimens were placed on the top of the ceramic cylinders. The top of ceramic cylinders was wetted before placing the specimens. A filter paper was placed between the saturated soil specimens and the ceramic cylinders to prevent loss of any soil from the soil specimen. The soil specimens were covered on top with an aluminum foil to prevent moisture loss by evaporation. Hollow aluminum spacer cylinders were then placed around the ceramic cylinders and the soil specimens.

The spacer cylinders were required as the soil specimens and the ceramic cylinders used had a diameter of 50 mm and the outer aluminum spacer rings of the holder had an inner diameter of 75 mm. The aluminum spacer rings were pushed down the side bolts around the aluminum spacer cylinders and tightened with nuts on the top.

The mass in the specimen holders was weighed before subjecting to spinning. An additional mass was placed in the reservoir cup of one of the specimen holders to balance the rotor. It was ensured that the difference in the masses between both the specimen holders was less than 0.5gms. The soil specimen holders were then placed in the centrifuge buckets as shown in Figure 5 before subjecting for centrifugation.

Figure 5. Soil specimen holders in the centrifuge buckets ready for centrifugation.

Table 2. Centrifugation time at different testing speeds

#	Test speed in rpm	Time of rotation in hrs		
		Silt	Till	Clay
1	300	2	2	4
2	500	2	2	6
3	1000	2	4	8
4	1500	2	4	8
5	2000	2	6	10
6	2500	2	6	12

The specimens were centrifuged initially at a speed of 300 rpm until equilibrium conditions were attained. Two hours of rotation time was found to be sufficient to attain equilibrium conditions for soil specimens tested with a thickness of 15mm for silty soils. However, it was found that 2 hrs of centrifugation time was not sufficient to attain equilibrium conditions for the specimens of Indian Head till and Regina clay. Both these soils had higher percentage of fines in comparison to Processed silt. The time of centrifugation was increased in steps for these specimens to achieve equilibrium conditions. Table 2 summarizes the testing speeds along with the equilibration times used for all soils tested.

The centrifuge was stopped after attaining equilibrium conditions at each speed tested and the masses of the soil specimens were determined. After the 2500 rpm run, the soil specimens were kept in an oven for water content determination. The water content values for the earlier test speeds were then back-calculated.

7 EXPERIMENTAL RESULTS & DISCUSSIONS

Figure 6 shows the comparison between the soil-water characteristic curves for the Processed silt, Indian Head till and Regina clay measured using Tempe cell, Pressure plate and the centrifuge method.

The soil-water characteristic curves were measured using Tempe cell for the Processed silt specimens compacted at an initial water content of 23% and dry density, ρ_d, of 1.68 Mg/m³ in two weeks time (Wright 1999). The time period required for measuring the soil-water characteristic curve for Processed silt specimens compacted at an initial water content of 22% and dry density, ρ_d, of 1.57

Mg/m³ using the centrifuge method was only 12 hours. The small differences in the soil-water characteristics for the Processed silt specimens may be associated with the differences in the dry densities and initial water contents at which the silt specimens were prepared.

A time period of 24 hours was required for Indian Head till specimens compacted at 19.2% (wet of optimum) initial water contents to obtain the soil-water characteristic curves using the centrifuge. However, the time required for identical Indian Head till specimens for the same suction range using the Tempe cell was 6 weeks. There is a good correlation between the soil-water characteristic curves obtained by both the two methods for 19.3% (wet of optimum) specimens.

Figure 6 also shows the comparison between the soil-water characteristic curves for the Regina clay measured using pressure plate and the centrifuge method. The soil-water characteristic curve using the centrifuge for Regina clay specimens were measured in 36 hours. A time period of almost 16 weeks was required to obtain the soil-water characteristic curve using pressure plate apparatus (Shuai 1996). The differences in soil-water characteristics can be associated mainly due to the variations in initial water content conditions.

Table 3 summarizes the time required by all the three types of soils to obtain the centrifuge as well as the Tempe cell and pressure plate soil-water characteristic curves.

8 SUMMARY & CONCLUSIONS

The small-scale centrifuge can be used to obtain multiple water content versus suction data points of the soil-water characteristic curve. The time period for measuring the soil-water characteristic curves for compacted, fine-grained soils reduces considerably when using the centrifuge method. There is a good comparison between the experimental results obtained by using conventional procedures and the centrifuge for the three soils tested. The results of

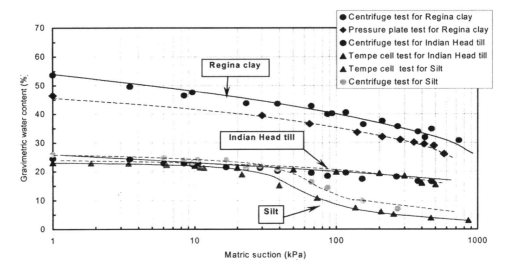

Figure 6. Comparison between the soil-water characteristic curves measured using the conventional procedure and the proposed new procedure using the small-scale medical centrifuge.

Table 3. Time periods to obtain the soil-water characteristic curve using centrifuge and conventional testing methods

Test Method	Silt	Indian Head till	Regina Clay
Centrifuge (time in days)	0.5	1	2
Tempe cell (time in days)	14	42	112

this study are encouraging to use the proposed centrifuge method for the determination of the soil-water characteristic curves for compacted, fine-grained soils.

9 REFERENCES

Croney, D., Coleman, J.D., and Bridge, P.M. 1952. The suction of moisture held in soil and other porous materials. *Road Research Technical Paper,* No. 24. London.

Fredlund, D.G. and Rahardjo, H. 1993. Soil Mechanics for Unsaturated soils. Inc., New York, NY, John Wiley and Sons

Gardner, R. A. 1937. The method of measuring the capillary tension of soil moisture over a wide moisture range, *Soil Science.* **43**: 277-283.

Khanzode, R.M. 1999. An alternative centrifuge method to obtain the soil-water characteristic curves for fine-grained soils. M.Sc. thesis, University of Saskatchewan. In progress.

Russell, M.B., and Richards L.A. 1938. The determination of soil moisture energy relations by centrifugation. *Soil Sci. Soc. America.* Proceedings, **3**: 65-69.

Skibinsky, D.N. 1996. A centrifuge method to obtain the soil-water characteristic curve. M.Sc. thesis, University of Saskatchewan.

Shuai, F. 1996. Simulation of swelling pressure measurements on expansive soils. Ph.D. thesis, University of Saskatchewan.

Wright. A. 1999. Shear strength of unsaturated soils in residual conditions. Undergraduate thesis. Civil Engineering Department, University of Saskatchewan.

Unsaturated Soils for Asia, Rahardjo, Toll & Leong (eds) © 2000 Taylor & Francis, ISBN 90 5809 139 2

A simple method of determining the soil-water charactistic curve indirectly

L.-W. Kong & L.-R. Tan
LRSM Laboratory, Institute of Rock and Soil Mechanics, Chinese Academy of Sciences, Wuhan, People's Republic of China

ABSTRACT: Based on the results of mercury intrusion porosimeter (MIP) test and shrinkage test of soil, the relationship between the water content and pore volume, pore size distribution of the tested samples during the drying process is obtained. A simple method of determining the soil-water characteristic curve indirectly is presented by MIP test for freeze-drying specimen and shrinkage test under the condition of the initial water content, in which the capillary model is employed. Several typical empirical equations for the soil-water characteristic curve can be derived by the given general formula in this paper. The validity of the proposed general formula is also preliminarily proved by the laboratory test.

1 INTRODUCTION

The soil-water characteristic curve is used as the basis for calculation and prediction of other unsaturated soil parameters. The matric suction is proved to be one of the two stresses state variables controlling the behavior for an unsaturated soil, which plays very key roles in soil mechanics for unsaturated soils (Fredlund, D.G & Rahardjo, H.1993). Measurement to suction correctly and prediction its change with environmental condition is the presupposition of applying the soil mechanics for unsaturated soils to solve the practical engineering problems. Thus, determining the soil-water characteristic curve is very useful in academic and application.

In theoretical analysis and practical works in soil mechanics for unsaturated soils, it is usually necessary to convert water content into matric suction through the soil-water characteristic curve. In order to convert and analyses conveniently, many empirical equations were established, but, in practice, it is always encountered difficulties to determine the parameters of empirical equations that can be only used in special soil types and areas. Recently, based on the assumption that the shape of the soil-water characteristic curve is dependent upon the pore-size distribution of the soil, Fredlund et al proposed a general equation for the soil-water characteristic curve, which provides a good fit for sand, silt and clay soils over the entire suction range from 0 to 10^6kPa (Fredlund,D.G.&Xing,AnQing.1994). However, some difficulties still exist to determine the four parameters for the proposed equation in different area and soil type previously without measurement to suction. The present devices for measuring suction such as ceramic plate extractor, thermal conductivity sensors have some limitations and disadvantages, moreover, some test methods are not complete mature. Therefore, some researchers try to search the relationship between the matric suction and results of other ordinary geotechnical tests so that the soil-water characteristic curve can be determined indirectly. In general, it is not well accepted to determine the soil-water characteristic curve by the results of ordinary geotechnical tests and further study should be done (Fredlund, M.D et al, 1998,Lu Zhaojin et al, 1992). On the other hand, some researchers obtained the soil-water characteristic curve through the pore-size distribution and capillary model (Ragab, R. & Feyen, J.1982, Prapaharan, S. et al, 1985, Olson, K.R.1985), in which it is not considered the fact that the volume of soil will decrease as the water content is reduced or increase or swell during wetting.

Since the MIP is a standard commercial item, and the freeze drying equipment is both simple and inexpensive, the measurement of pore size distribution may be practical for large geotechnical laboratories. The pore size distribution concept is somewhat new to geotechnical engineering even though several researchers have successfully used MIP in the past to investigate soil properties (Griffiths & Joshi 1991, Kong, L.W et al, 1995). In this paper, according to the results of shrinkage test and MIP test, the soil-water characteristic curve is determined. We wish that the soil-water characteristic curve could be indirectly determined through the combination of macro-test and micro-test method.

2 GENERAL FORMULA FOR THE SOIL-WATER CHARACTERISTIC CURVE DETERMINED INDIRECTLY

Matric suction is mainly shown in the form of capillary force in a certain range, which reflects the action of the capillary force to the soil water. In the viewpoint of microstructure, soil particle itself is mainly consisted of silicate mineral and the primary bond forms its strength, whereas the secondary bond and hydrogen bond form the attractive force of interparticle and soil particle to water molecular. The research result of physical chemistry shows that the range of influence for surface force (capillary bond), var der Waals force, hydrogen bond and covalent bond are $0.0001\sim10\mu m$, $0.3nm\sim0.1\mu m$, $0.2\sim0.3nm$ and $0.1\sim0.2nm$, respectively (Hong Yukang, 1998). Thus, when the pore radius in the range of $0.1\sim10\mu m$, it can be supposed that the matric suction is mainly caused by the capillary force. Based on the capillary model, the corresponding matric suction are in the range of 30kPa~3.0MPa, which can basically satisfy the demand of soil mechanics. We can assume that the soil suction results from the capillary force of a certain circular pore diameter in the soil-water system for an unsaturated soil.

According to the previous hypothesis, the soil-water characteristic curve can be indirectly expressed as a relationship between the maintained water content and pore equivalent diameter or pore volume filled with water under the different suction.

$$u_a - u_w = \frac{4\sigma \cos(\alpha)}{d} \qquad (1)$$

in which u_a-u_w is matric suction, u_a is pore-air pressure, u_w is pore-water pressure, σ is surface tension of water, d is diameter of capillary tube, α is contact angle, equation (1) can be simplified into equation(2) because the α is usually zero:

$$u_a - u_w = \frac{4\sigma}{d} \qquad (2)$$

From the soil-water characteristic curve, it is seen that a soil is under a saturated state when the matric suction is zero. If a little suction is exerted on the soil, in the initial stage, no water is drained off. As the suction increases and exceeds a critical value $(u_a-u_w)_b$, the water in large pores of the soil begins to decrease. As the suction increases further, some pores can not resist against the exerted pressure to keep up original water content, resulting in the reduction of water content further. In term of equation (2), we can convert the matric suction coordinate axis into equivalent pore diameter coordinate axis. When the suction is equal to $(u_a-u_w)_i$, the corresponding capillary water in equivalent pore diameter more than or equal to d_i will be drained off and the rest of the pore diameter less than d_i capillary can maintain water. The corresponding water content is

w_i, as the matric suction $(u_a-u_w)_i$ increases to $(u_a-u_w)_j$. The corresponding equivalent pore diameter is d_j, and the capillary water in equivalent pore diameter more than d_j will be drained away. Only the capillary water of the pore diameter less than d_j can be maintained, a corresponding water content is w_j. That is to say, when the matric suction changes from $(u_a-u_w)_i$ to $(u_a-u_w)_j$, the part of capillary water filled with the pore diameter range of d_i to d_j will be drained off, the corresponding change of water content is w_i-w_j, and the change of void ratio is e_i-e_j due to the soil shrinking with the reduction of water content.

According to the conversion relation between physical parameters of soil mass, the total pore volume per gram of soil particle can be got as the water content is w:

$$V_t = \frac{e}{G_s \rho_w} \qquad (3)$$

where G_s is specific gravity of soil particle, ρ_w is density of water.

In order to determine the pore size distribution of soil with water content w, the rapid freeze-drying technique is used to prepare soil sample and the MIP device is employed. The soil total pore volume per gram of soil particle in the measurable range is V_{mt}, the measurable pore volume of the pore diameter more than d_r (d_r is the corresponding pore diameter of the residual water content) is expressed as V_m. At the same time, the total pore volume of pore diameter less than d_r is expressed as V_r (V_r including the measureless fine pore volume), and the V_s is defined as the measureless very fine pore volume (pore volume of pore diameter less than the lower limit in the measurable range), then:

$$V_r = V_t - V_m \qquad (4a)$$

$$V_s = V_t - V_m \qquad (4b)$$

From the shrinkage test, we can get the void ratio versus water content relation curve. The void ratio varies with water content linearly in the range of shrinkage limit to initial water content. It can be expressed as the following form equation (5):

$$e_i = e_0 - k(w_0 - w_i) \qquad (5)$$

in which e_0 is the initial void ratio, w_0 is the initial water content; k is the slope of the line.

In term of equation (3) and equation (5), it is not difficult to obtain the reduced pore volume resulted from the drying shrinkage process due to surface tension, as the soil water content decreases from w_0 to w_i,

$$V_{t0} - V_{ti} = \frac{k(w_0 - w_i)}{G_s \rho_w} \qquad (6)$$

Because the d_r and d_s are usually very small, we can suppose that the V_r and V_s do not change with the water content in the range of w_0 to w_i. From equation (4) and equation (6), the following can be obtained,

$$V_{mi} = V_{m0} - \frac{k}{G_s \rho_w}(w_0 - w_i) \qquad (7)$$

Similarly, we can suppose that the cumulative pore volume filled with water for pore diameter less than d_i is $V(d_i)$. The measurable pore volume of pore diameter less than d_i and more than d_r is $V_m(d)$, then:

$$V(d_i) = V_m(d_i) + V_r \qquad (8a)$$
$$w = 100 \ \rho_w V(d_i) \qquad (8b)$$

The $V(d_i)$ can be obtained from the measured pore size distribution and the equation (3)and equation (4).

From the above derivation process, we can get the general formula of the soil-water characteristic curve by the results of shrinkage test and MIP test.

$$\left.\begin{array}{l} (u_a - u_w)_i = \dfrac{4\sigma}{d_i} \\[2mm] w_i = 100 \ \rho_w V(d_i) \\[2mm] S_i = \dfrac{100 \ V(d_i)}{V_{mi} + V_r} \\[2mm] \theta_i = S_i \dfrac{e_i}{1 + e_i} \end{array}\right\} \qquad (9)$$

in which S_i is the degree of saturation at the water content w_i, θ_i is the volumetric water content

From the equation (9), it is shown that the soil-water characteristic curve can be determined through the soil physical parameters at the initial water content, pore size distribution regularity and the result of shrinkage test.

3 VERIFICATION OF THE GENERAL FORMULA USEING EMPIRICAL EQUATION

Some empirical equations for the soil-water characteristic curve are proposed by several researchers (Fredlund, D.G.,1994, Bruce, R.R & Luxmoore, R.J. 1986). We use the following several typical empirical equation to verify the presented general formula of determining the soil-water characteristic curve indirectly.

$$\psi(\theta) = a_1 e^{(-b_1\theta)} \qquad (10)$$

$$\psi(\theta) = a_2 \theta^{-b_2} \qquad (11)$$

$$\frac{\theta - \theta_r}{\theta_s - \theta_r} = \left(\frac{\psi_b}{\psi}\right)^{\lambda} \qquad (12)$$

$$\frac{\theta - \theta_r}{\theta_s - \theta_r} = e^{-\frac{\psi - a_4}{b_4}} \qquad (13)$$

$$\frac{\theta - \theta_r}{\theta_s - \theta_r} = \frac{1}{1 + a_5 \psi^{b_5}} \qquad (14)$$

in which $\psi = u_a - u_w$ is matric suction, ψ_b is air entry matric suction, θ is volumetric water content, θ_r is residual volumetric water content, θ_s is saturated volumetric water content.

For discussion conveniently, we use the equations (15) and (16) to connect the pore volume with volumetric water content.

$$S_e = \frac{V(d) - V_r}{V_i - V_r} \qquad (15)$$

$$S_e = \frac{\theta - \theta_r}{\theta_s - \theta_r} \qquad (16)$$

After the measured pore size distribution curve is modified by the result of shrinkage test, the θ versus d relation and the S_e versus d relation can be derived from equation (9) and equation (15), respectively. It can be seen that the empirical equation (10)~(14) are suited to the soil-water characteristic curve if the following relations exist, respectively.

$$\theta = A_1 + B_1 \ln d \qquad (17)$$
$$\ln \theta = A_2 + B_2 \ln d \qquad (18)$$
$$\ln S_e = A_3 + B_3 \ln d \qquad (19)$$
$$\ln S_e = A_4 + B_4 \frac{1}{d} \qquad (20)$$
$$\ln \frac{1 - S_e}{S_e} = A_5 + B_5 \ln d \qquad (21)$$

From the equations (10)~(14) and equations (17)~(21), it is shown that the each empirical equations mentioned above has limitations, which are mainly controlled by the regularity of pore-size distribution and shrinkage property. In other words, the empirical equations are valid under the condition that the pore volume or volumetric water content versus equivalent pore diameter exists special function, which is also proved in some reviewed literatures. Thus, the validity and suited area of the presented method of determined the soil-water characteristic curve indirectly is more than the empirical equations mentioned above.

4 VERIFICATION OF THE GENERAL FORMULA THROUGH LABORATORY TEST

In order to verify the validity of the general formula presented in this paper, the expansive soil from Yichcng, Hubei province, China is studied in laboratory. Table 1 is the physical properties for the expansive soil. Fig.1 and Fig. 2 are the void ratio and degree of saturation varies with water content obtained by the shrinkage test for intact expansive soil, respectively. From Table 1, Fig.1 and Fig. 2, it can be seen that the swelling and shrinkage properties are very significant, the degree of saturation can be relatively very high when the water content is close to the shrinkage limit.

The expansive soil is prepared by freeze-drying; air-drying and oven-drying .The pore-size distribution curve of freeze-drying intact expansive soil by MIP is shown in Fig.3. In contract the result of MIP test for drying and oven-drying samples are also plotted in Fig.3. The data measurement is used by PoreSizer 9310,Micromeritics Instrument Corporation, USA, and the measurable pore diameter range from 360 to 0.006μm. Table 2 is the comparison of pore volume of specimens dried in the three different methods.

From Fig.3 & Table 2, it can be seen that the specimens dried in the air and 105^0C do nor have pore volume of diameter more than 0.1um as compared with the specimen dewatered by freeze drying, whereas the pore volume of pore diameter less than 0.006μm is almost equal. It indicates that the V_r and V_s as constants in the previous derivation is correct.

Table 1 Physical properties of the expansive soil

Property	Value	Remark
Specific gravity	2.77	
Liquid limit	81.7 (%)	
Plastic limit	34.7(%)	
Plasticity index	47.0	
Free swell	103.0(%)	
Shrinkage ratio	12.1(%)	Vertical
	11.7(%)	Horizontal
Grain size distribution	64.5(%)	<10μm
	50.0(%)	<5μm
	25.0(%)	<2μm

Table 2 Comparison of pore volume of specimens dried in the three different methods.

Item	Freeze drying	Air drying	Oven drying
Dry density (g/cm³)	1.2237	2.0220	2.0438
Void ratio	1.2636	0.3699	0.3553
Total pore volume (mm³/g)	456.2	133.5	128.3
Measured pore volume (mm³/g)	399.7	85.8	68.1
Measureless pore volume (mm³/g)	56.5	47.7	60.2

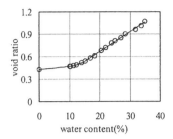

Fig.1 Void ratio versus water content for expansive soil

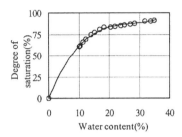

Fig.2 Degree of saturation versus water content

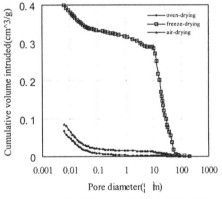

Fig.3 The pore size distribution curve

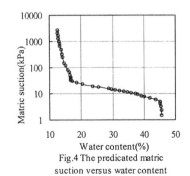

Fig.4 The predicated matric suction versus water content

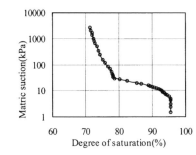

Fig.5 The predicated matric suction
versus degree of saturation

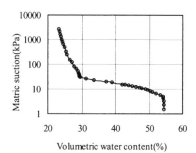

Fig.6 The predicated matric suction
versus volumetric water content

It is noted that the physical parameters of the specimens for shrinkage test and for MIP test are not quite uniform, which may be caused the two reasons that one is the different precision of measurement for shrinkage test and MIP test, another is the soil not completely homogeneous. However, in order to verify the validity of the general formula preliminarily through the shape and basic tendency of the soil-water characteristic curve, the shrinkage test result is still used when the soil-water characteristic curve are expressed in degree of saturation or volumetric water content coordinate axis.

According to Fig. 3, Table 2 and equation (9), the predicted soil-water characteristic curve are shown in Fig. 4~ Fig. 6.It can be seen that the shape and change tendency of predicted soil-water characteristic curve is basically reasonable compared to the Fig. 2. Maybe, the studied soil is very special; more laboratory test should be performed to verify the general formula further.

5 CONCLUSIONS

The soil-water characteristic curve can be indirectly determined by the results of MIP test (specimens dewatered by rapid freeze-drying) and shrinkage test of specimens with initial water content, in which the capillary model is employed. The several well-known empirical equations for soil-water characteristic curve can be derived by the presented general formula in this paper, which can be only used in special condition. The validity of the proposed general formula is preliminarily verified by the laboratory test. This procedure may provide a new approach to the problem.

6 ACKNOWLEDGMENTS

The National Natural Science Foundation Commission of China Supports the Project (Grant no: 19772068)

REFERENCES

Bruce, R.R & Luxmoore, R.J.1986.Water Retention: Field Method (p679 in Method. Soil Analysis; part 1), Second Edition, Madison, Wisconsin, USA

Fredlund, D.G & Rahardjo, H.1993. Soil Mechanics for Unsaturated Soils. John Wiley & Sons.Inc. New York.

Fredlund, D.G. & Xing AnQing.1994. Equations for the Soil-Water Characteristic Curve. Can. Geotech. J.31: 521-532

Fredlund, M.D., Wilson, G.W. & Frelund, D.G. 1998. Estimation of Hydraulic Properties of an Unsaturated Soil Using a Knowledge-Based System. Proc. 2nd Inter. Conf. On Unsaturated Soil. Beijing, China.Vol. 1:479-483.

Griffiths, F.J., & Joshi, R.C.1991.Change in Pore Size Distribution Owing to Secondary Consolidation of Clays. Can. Geotech. J.28: 20~24

Hong Yukang. 1998.Soil Behavior and Soil Mechanics. People Transportation Press: 23

Kong, L.W, Luo, H.X & Tan, L.R.1995.Fractal Study on Pore Space Distribution of Red Clay in China.Proc.10th ARC-SMFE, Beijing, China.Vol.1: 139~142

Lu Zhaojin, Zhang Huiming, Chen Jianhua & Feng Man.1992. Shear Strength and Swelling Pressure of Unsaturated Soils. Chinese Journal of Geotechnical Engineering. 3: 1-8

Olson, K.R. 1985. Characterization of Pore Size Distribution within Soils by Mercury Intrusion and Water –release Methods. Soil Science, 139(5): 400~404

Prapaharan, S. Altschaeffl, F. & Dempsey, B.J.1985.Moisture Curve of Compacted Clay: Mercury Intrusion Method. J. Geotechnical. Engineering. ASCE, 9: 1139~1143

Ragab, R. Feyen, J. & Hillel, D.1982. Effect of the Method for Determining Pore Size Distribution on Predication of the Hydraulic Conductivity Function and of Infiltration. Soil Science, 2: 141~145

Unsaturated Soils for Asia, Rahardjo, Toll & Leong (eds) © 2000 Taylor & Francis, ISBN 90 5809 139 2

Preliminary study on soil-water characteristics of two expansive soils

C.W.W.Ng & B.Wang
Hong Kong University of Science and Technology, SAR, People's Republic of China

B.W.Gong & C.G.Bao
Yangtze River Scientific Research Institute, Wuhan, People's Republic of China

ABSTRACT: Some preliminary results of laboratory tests on soil-water characteristics for two expansive soils from mainland China are presented in this paper. The test equipment adopted include conventional pressure plate extractors, newly developed 1D and isotropic stress control apparatuses. Effects of initial water content, soil structure, drying and wetting cycles and stress state on the soil-water characteristic curves of the expansive soils are addressed and compared with results from a weathered volcanic soil. Stress effects are more significant on SWCCs of the expansive soils than the weathered soil. Due to the distinct swelling characteristics, the volumetric water contents along the wetting path of the second are substantially higher than those of the first wetting cycle.

1 INTRODUCTION

According to Nelson & Miller (1992), expansive soil is a term for any soil or rock that has a potential for shrinking or swelling due to any change of water content. The shear strength of expansive soils decrease sharply and the volume increases greatly with increasing water content. Due to these special characteristics, damages caused by this type of soils have been reported in the many countries worldwide such as United States and China (Steinberg, 1998).

China has a wide distribution of expansive soils over the whole territory, especially in the southern and southwestern areas. There will be a 180km long excavated canal made up of the expansive soil along the Middle Route of the huge 'South-to-North' Water Transfer Project, which aims at alleviating the problems caused by disproportionate water distribution throughout China (Liu, 1997). Thus, the slope stability problem becomes the major concern of the design and construction of the canal in these areas.

Research work on expansive soils is currently carried out worldwide (Fredlund et al., 1995; Alonso, 1998; Bao et al., 1998; Bao & Ng, 2000).

Soil-water characteristic curve (SWCC) is an important hydraulic property for unsatuated soils. It is defined as the relationship between water content and matric suction (Williams, 1982). Some equations were established by many researchers to predict SWCCs (Brooks & Corey, 1964; Van Genuchten, 1980, Fredlund & Xing, 1994).

To directly measure shear strength and permeability for unsaturated soils is not only expensive but also time-consuming, therefore, SWCC is widely used to predict other useful properties of unsaturated soils (Van Genuchten, 1980; Fredlund et al., 1994; Fredlund et al., 1995; Vanapalli et al., 1996). Furthermore, the study on stress effects on SWCCs for unsaturated soils has attracted much attention lately (Vanapalli et al., 1999; Ng & Pang, 1999, 2000).

However, there is relatively little literature on SWCCs for expansive soils. In this paper, some preliminary laboratory results are presented on SWCCs for some unsaturated expansive soils from Nanyang, China. Moreover, a newly modified triaxial apparatus was used to investigate the influence of isotropic stress state (ISO) on SWCCs of an expansive soil (Wang, 2000). The result will be compared with those of zero and 1-dimensional (1D) stresses.

The study for the SWCCs of these soil samples is indispensable for seepage analysis for slopes made up of these problematic soils and believed to be helpful to the further study of slope stability analysis. Hopefully, some of the results are applicable to better design and subsequently to the maintenance of the South-to-North canal.

2 TESTING PROGRAM AND PREPARATION PROCEDURES

In this study three different types of apparatus were used, namely, volumetric pressure plate/membrane extractor of 200, 500 and 1500 kPa, with zero stress state (manufactured at Soil Moisture, CA, US), 1D volumetric pressure plate extractor (Ng & Pang, 1999, 2000) and modified isotropic triaxial apparatus (Wang, 2000).

The soil samples, namely, LZ and ZY were taken from Liangzhuang, Henan Province and Zaoyang, Hubei Province, central China, respectively. LZ and ZY samples both belong to the category of Nanyang expansive soil, the quaternary miocene alluvial-pluvial clay with gray-brown and brown-yellow color, respectively (Liu, 1997). According to USCS, LZ and ZY can be classified as clay with low (CL) and intermediate (CI) plasticity, respectively. LZ contains 17-32% of illite, 15-17% of montmorillo-nite and 8-10% of kaolinte, and ZY is made up of 31-35% of illite, 16-22% of montmorillonite and 8% of kaolinite (Liu, 1997). The respective plastic indices for LZ and ZY are 28.5 and 35.5 %.

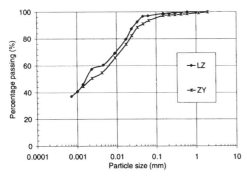

Figure 1. Particle size distributions for LZ and ZY

The particle size distributions for these two types of soils are shown in Figure 1. LZ soil is generally finer than ZY, whereas the PI of LZ is lower than that of ZY. This might be due to the difference in mineral components of the two soils.

2.1 Sample preparation

Two types of soil samples were used in this study, namely, intact and recompacted samples. They were prepared in different ways.

Each of intact soil specimens was cut by a stainless steel ring from a block sample obtained from the field. The trimmed specimens should be in a good contact with the inner side of the ring.

To prepare recompacted specimens, pieces of natural soils were firstly dried in an oven at a temperature of 40°C to 50°C for 24 hours and grounded with a pestle. Before the soil was wetted and com-pacted, it was passed through a 2mm sieve. The soil particles passing through the sieve were carefully mixed with a certain amount of de-aerated water so that the soil had a particular moisture content. The sample should be mixed thoroughly by a spatula un-til no big soil clods or dry soil particles can be seen. The soil-water mixture was left for 24 hours in an airtight container to equalize the moisture in the soil.

Afterwards, a certain amount of wet soil was weighed and statically compacted into a stainless steel ring to make a recompacted sample with spe-cific water content and a specific dry density.

2.2 Saturation

The saturation processes for intact and recompacted samples were the same. Swelling after compaction should be completely constrained. Each specimen was sandwiched between two coarse porous stones along with two pieces of filter paper. All the speci-mens and porous stones were fixed with two steel plates by two clamping rods. Provided that all coarse porous stones and steel plates were rigid, all the samples could be assumed to have no swell during this procedure.

They were then placed into a transparent plastics desiccator in which a vacuum of about 90 kPa is ap-plied. 6 hours later, de-aerated water was added to the desiccator under vacuum until all the specimens are immersed with water. The applied vacuum was maintained for 12 hours. Then the specimens while under water were left open to the atmosphere for 6 more hours. The process of saturation was thus completed. From calculation, it was found that the degree of saturation could be up to nearly 100%. All samples used in this study correspond to the state be-fore saturation, i.e. either the natural state or right after compaction are listed in Table 1.

2.3 Determination of soil-water characteristics

After saturation, all the samples are not constrained any longer. If the sample is tested under zero stress, it is under a condition of free swell. For those tested under certain stress state, swell is partially con-strained, if the stress applied is not higher than the swelling pressure (Table 1). The details in test proc-ess of pressure plate extractors are not presented in detail here.

3 RESULTS AND DISCUSSION

All the test results shown in this paper are drying curves unless stated otherwise.

Table 1 Some details of test samples

Sample identity	Initial dry density ($\times 10^3$ kg/m^3)	Initial water content (%)	Initial void ratio	Initial degree of saturation (%)	Swelling pressure (kPa)	Testing apparatus and testing condition
LZ-N1	1.56	26.4	0.74	97	50	Pressure plate extractor (0-1500 kPa)
LZ-N2						Pressure plate extractor (0-1500 kPa)
LZ-N3						Pressure plate extractor (0-500 kPa)
LZ-R1	1.59	25.7	0.71	98.3		Pressure membrane extractor (0-1500 kPa)
ZY-N1	1.37	29.3	0.97	0.81	70	Pressure plate extractor (0-500 kPa)
ZY-N2	1.39	30.3	0.96	93.0		Pressure plate extractor (0-200 kPa)
ZY-R1						Pressure plate extractor (0-500 kPa)
ZY-R2						Pressure plate extractor (0-200 kPa)
ZY-R3	1.37	30.3	0.99	83.6		1-dimensional pressure plate extractor (0-200 kPa) with 50 kPa external vertical load
ZY-R4						Triaxial apparatus (0-500 kPa) with 50 kPa isotropic stress state

*All data were calculated from initial values. N and R represent natural and recompacted sample, respectively.

3.1 Comparison of intact and recompacted samples

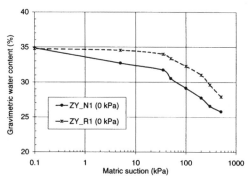

Figure 2. Comparison of SWCCs of intact and recompacted ZY samples

These two samples were tested under zero stress. It can be seen from Figure 2, two samples, ZY-N1 and ZY-R1, possess the same initial water content of about 35%, which implies they have the same initial void ratio. The intact sample starts to desaturate once the matric suction increases, whereas the desaturation rate for the recompacted sample is low at low suction range. It could be attributed to that there are some natural fissures and large pores inside the intact sample, whereas the recompacted sample no longer has these large drainage paths after recompaction. The desaturation rates of these two samples are close to each other when the matric suction is higher than 100 kPa, since the two curves become parallel.

A similar trend can be observed from Figure 3, except that the difference between LZ-N1 and LZ-R1 is smaller. The initial water content is a little lower than that in Figure 2. This may be due to the overall particle sizes are smaller for LZ sample than

for ZY (Fig. 1), which makes the LZ intact sample has a lower initial water content as well as a smaller desaturation rate, compared with ZY intact sample. On the other hand, it may be postulated that finer particles would make it easier to compact a more uniform sample. That might be able to explain why the two curves in Figure 3 are closer than those in Figure 2.

According to Ng & Pang (2000), there are differences in SWCCs for natural and recompacted residual soils. The air entry value of natural samples is greater than that of recompacted ones and the desaturation rate for both categories are very close in the high suction range, which agrees with what observed in this study on expansive soils.

It can be concluded that a recompacted expansive soil sample has a higher water retaining ability than an intact sample, even though they are with a same dry density. The pore size distribution in recompacted samples might be generally smaller than in natural samples. Moreover, it is possible that the pores in intact samples are better interconnected.

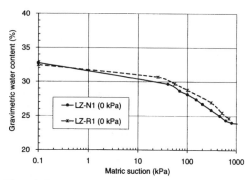

Figure 3. Comparison of SWCCs of intact and recompacted LZ samples

3.2 *Effects of different initial water contents*

For two recompacted specimens compacted in different water content, it is believed that their microstructure will be different and in turn they have different SWCCs (Vanapalli et al, 1999). It might be also valid that if two intact samples have different pore size distributions, their SWCCs would also be different. This is verified by the results shown in Figure 4.

Figure 4 shows two curves, representing two specimens of intact LZ sample, LZ-N2 and LZ-N3, with a same initial water content and dry density, moreover, they must have the same structure and fabric in their natural state. However, after two different saturation processes, free swell (LZ-N2) and constrained swell (LZ-N3) during saturation, these two specimens no longer have the same dry density and void ratio. LZ-N2 must possess a higher void ratio and a more open structure, compared with LZ-N3, so that its initial water content is higher. It can also be seen that LZ-N2 has a lower air entry value of 30 kPa, while the air entry value of LZ-N3 is much higher, about 150 kPa.

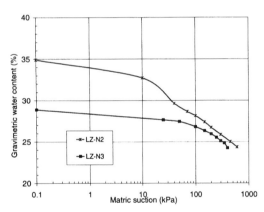

Figure 4. Comparison of SWCCs of intact LZ samples at different initial water contents

In the low suction range, from10 to 100 kPa, the desatuation rate of LZ-N3 is lower by that of the free swell sample, which means that there might be more interconnected pores corresponding to this low suction range in the free swell sample. However, as the suction increases to higher than 100 kPa, these two curves tend to converge. That means the different saturation methods applied on these two samples, LZ-N2 and LZ-N3, result in different macrostructures but have less effect on the microstructures of the soils. Free swell soils normally have a relatively low air entry value and a higher water retention curve.

3.3 *Hysteresis of expansive soils*

The hysteresis is an interesting phenomenon in the investigation on SWCCs. Figure 5 shows the SWCC with three cycles of a typical residual soil in Hong Kong (Ng et al, 2000). The drying and wetting loop of the first cycle is larger in size than those of later cycles. And except for the first cycle, there is nearly no difference between the second and the third cycles. This is probably due to a significant change in soil structure in the first drying and wetting cycle. After that, the structure might become more stable, such that the other cycles are nearly overlapped.

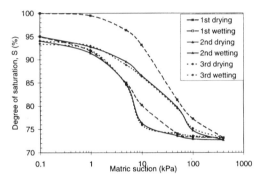

Figure 5. Influence of repeated cycles of drying and wetting on SWCC of a residual soil (Ng et al, 2000)

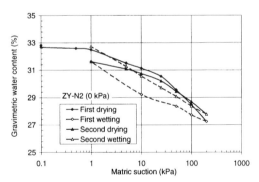

Figure 6. Influence of repeated cycles on SWCC of an intact ZY sample

Some of observations from Ng & Pang (2000) can also be seen from Figure 6, in which two loops of ZY-N2 are shown. The first hysteric loop is much larger than the second one. It should be particularly noted that the second wetting curve is not below the first wetting, which is different from the SWCC in Figure 5. It could be possible. Because the soil sample is under zero stress, it must freely swell subjected to wetting and then have a larger void ratio. Moreover, as drying and wetting cycles going on, more pores become interconnected and the adsorption path becomes smoother and the adsorptive force might also be greater due to its expansive minerals, e.g. smectite. Therefore, it is possible the second wetting curve can even rise higher than the first one, which might be a distinct characteristic of expansive soils.

3.4 Stress effects on SWCC of expansive soils

The results presented in Figure 7 show the stress effects on SWCCs of a CDV in Hong Kong (Ng & Pang, 2000). This series of tests were conducted in the modified 1D pressure plate extractor. K_0 condition is assumed in that study. In the low suction range, the higher the stress is, the lower the water retention curve is. However, when the suction is high, the greater the stress, the higher the water retention curve. This might be resulted from a smaller void ratio in low suction, whereas the smaller pore size distribution resulted from the higher stress state leads to a higher water retention curve in the high suction range.

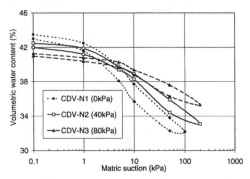

Figure 7. Effect of K_0 stress conditions on SWCCs of a residual soil (Ng & Pang, 1999)

In Figure 8, three curve represent three samples under three different stress conditions during the drying and wetting process, but they are prepared in the same way and with same properties before test. ZY-R2 was conducted in a volumetric pressure plate extractor with zero stress. ZY-R3 and ZY-R4 were tested in the 1D pressure plate extractor with 50 kPa vertical load and the triaxial apparatus with 50 kPa isotropic stress, respectively.

The initial water content is about 37.5% for all three samples, however, there is a reduction in water content for either ZY-R3 or ZY-R4 due to the consolidation procedure. For ZY-R3 the reduction is not very significant, while the change in ZY-R4 is quite large, it decreased to 35.8 % when the consolidation is completed. Once the consolidation is assumed finished, the matric suction is then applied onto the sample. The stress effects can be easily seen from this figure. It can be noticed that ZY-R2 has a clear

air entry value, which can not be seen for the two samples under nonzero stress. It is seen that the isotropic stress state results in a lower water retention curve. The desaturation rates in the suction from 10 to 100 kPa are very close for all three samples, which is contradictory from the results reported by Ng & Pang (2000), in which the higher the stress applied, the lower the desaturation rate was in the high suction range. Further studies will be presented in detail in the future.

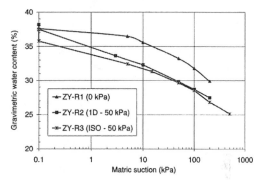

Figure 8. Influence of different stress conditions on SWCCs of recompacted ZY samples

4 PRELIMINARY CONCLUSIONS

From the laboratory test results presented, some conclusions can be drawn as follows:
1) For expansive soils, an intact sample has a lower air entry value than that of a recompacted sample, but the desaturation rates are approximately the same for suction higher than about 30 kPa. For a given dry density, recompacted samples posses a higher water retention curve than that of the intact samples.
2) Different saturation methods result in different soil structures for natural expansive soil samples. The free swell sample has a lower air-entry value and a higher desaturation rate than those of the constrained one at low suctions. The difference in the desaturation rates tends to diminish as the matric suction increases.
3) Hysteresis is most significant for the first wetting and drying cycle. Due to the presence of expansive minerals, the water retaining ability increases and the volumetric water content become higher at the second than that at the first wetting path.
4) It is found that stress effects on SWCCs of the expansive soils are important. Due to the stress applied, air entry value decreases almost to zero but the rate of desaturation is nearly the same as that of the conventional SWCCs for suction ranging from 10 to 100 kPa.

ACKNOWLEDGEMENTS

This research project is financially supported by research grant F-HK99/00.EG05 of the France/Hong Kong Joint Research Scheme and also by research grants HKUST6046/98E & DAG99/00.EG23, provided by the Research Grants Council of the Hong Kong Special Administrative Region. Finally, the authors are also grateful to Professor Wilson Tang for his continual support to their research.

REFERENCES

Alonso, E. E. (1998). Modeling Expansive Soil Behavior. Proc. of The 2nd International Conference of Unsaturated Soils, Beijing, China. Vol. 2, 37-70.

Bao, C. G., Gong, B. W.& Zhan, L. T. (1998) Properties of Unsaturated Soils and Slope Stability for Expansive Soils. Proc. of The 2nd International Conference on Unsaturated Soils, Beijing, China. Vol. 2, 71-98.

Bao, C.G. & Ng, C.W.W. (2000). Keynote paper: Some thoughts and studies on the prediction of slope stability in expansive soils. *Asian Conference on Unsaturated Soils (Unsat-Asia 2000), May, Singapore.*

Chao, K. C., Burkee, D. B., Miller, D. J. & Nelson, J. D. (1998). Soil water characteristic curve for expansive soils. Proc. of Thirteenth Southeast Asian Geotechnical Conference, Taipei, Taiwan, 35-40.

Fredlund, D. G. & Rahardjo, H. (1993). Soil Mechanics for Unsaturated Soil Mechanics. Wiley Inter Science. New York.

Fredlund, D. G. & Xing, A. (1994). Equations for the soil-water characteristic curve. Can. Geotech. J. 31: 521-532.

Fredlund, D. G., Clifton, A. W., Barbour, S. L., Huang, S. K., Wang, Q., Ke, Z. J. & Fan, Q. Y. (1995). Matric suction and deformation monitoring at an expansive soil site in southern China. Proc. of the1st International Conference on Unsaturated Soils, Paris. Vol. 2, 855-861.

Fredlund, D. G., Xing, A. & Huang, S. (1994). Predicting the permeability function for unsaturated soils using the soil-water characteristic curve. Can. Geotech. J., 31: 553-546.

Fredlund, D. G., Xing, A., Fredlund, M. D. & Barbour, S. L. (1995). The relationship of the unsaturated soil shear strength to the soil-water characteristic curve. Can. Geotech. J. 32: 440-448.

Fredlund, D.G. & Rahardjo, H. (1993). Soil Mechanics for Unsaturated Soil Mechanics. Wiley Inter Science, New York.

Liu, T. H. (1997). The problem of expansive soils in engineering construction. Chinese Architecture and Building Press, 4-5 (In Chinese).

Nelson, J. D. & Miller, D. J. (1992). Expansive soils - Problems and Practice in Foundation and Pavement Engineering. John Wiley & Sons, Inc.

Ng, C. W. W. & Pang, Y. W. (1999) Stress effects on soil-water characteristics and pore water pressure in unsaturated soil slopes. 11th Asian Regional Conference on Soil Mechanics and Geotechnical Engineering, Korea, Vol. 1, 371-374.

Ng, C. W. W. & Pang, Y. W. (2000) Influence of stress states on soil-water characteristics and slope stability. Journal of Geotechnical and Geo-environmental Engineering, ASCE, Vol. 126, No. 2. 157-166.

Ng, C. W. W., Pang, Y. W. & Chung, S. S. (2000) Influence of drying and wetting history on stability of unsaturated soil slopes. Accepted by 8th International Symposium on Landslide, Cardiff, UK.

Steinberg, M. (1998). Geomembranes and The Control of Expansive Soils in Construction. McGraw-Hill, New York, NY.

Van Genuchten, M. T. (1980). A closed-form Equation for predicting the hydraulic conductivity of unsaturated soils. Soil Sci. Soc. Am. J. 44: 892-898.

Vanapalli, S. K., Fredlund, D. G. & Pufahl, D. E. (1996). The relationship between the soil-water characteristic curve and the unsaturated shear strength of a compacted glacial till. Geotechnical Testing Journal, GTJODJ, Vol. 19, No. 3, 259-268.

Vanapalli, S. K., Fredlund, D. G. & Pufahl, D. E. (1999). The influence of soil structure and stress history on the soil-water characteristic of a compacted till. Geotechnique 49(2), 143-159.

Wang, B. (2000). Stress effects of SWCC on slope stability in expansive soils. M. Phil. Thesis, Dept. of Civil Engineering, HKUST (in preparation).

Williams, P. J. (1982). The surface of the earth, an introduction to geotechnical science. Longman Inc., New York.

Unsaturated Soils for Asia, Rahardjo, Toll & Leong (eds) © 2000 Taylor & Francis, ISBN 90 5809 139 2

Engineering properties of unsaturated soils (volume change and permeability)

K. Premalatha

Division of Soil Mechanics and Foundation Engineering, Anna University, Chennai, India

ABSTRACT: The water content of unsaturated expansive soil is a function of the suction pressure of the soil. This relationship between the water content in the soil and the suction can be expressed in a plot of volumetric water content versus suction known as the soil-water characteristic curve. More established uses of this soil-water characteristic curve are to predict the volume-change characteristics, derivation of permeability function and shear strength function of expansive clays. In this study, an attempt is made to study the soil-water characteristic curves of certain expansive clays of Chennai and hence to evaluate the volume change characteristics and permeability functions which may be used to evaluate the percentage swell and permeability of the soil at different degrees of saturation at a given density.

1. INTRODUCTION

Lightly located structures founded on desiccated unsaturated soils commonly suffer severe distress subsequent to their construction. Changes in the environment around the structure result in changes in the negative pore-water pressure (suction) thereby producing volume changes in the soil.

Generally an unsaturated soil is considered to be a three-phase system. However the independent properties and continuous bonding surfaces of the air-water boundary require its consideration as a fourth phase (Davies and redial, 1963). Several effective stress equation have been proposed for unsaturated soil. All equations have included a soil property in the description of the stress variables controlling the soil behaviour.

The volume change of an unsaturated soil can be converted into perdition of swell or shrinkage by substituting the change in the stress state variables of an unsaturated soil. The water content in an unsaturated soil is a function of the suction present in the soil. This relationship between the water content in the soil and the suction can be expressed in a plot of volumetric water content versus suction, known as the soil-water characteristic curve.

Fredlund and Morgenstern (1977) concluded that net normal stresses are the stress-state variables of an unsaturated soil. The water content in an unsaturated soil is a function of the suction present in the soil. This relationship between the water content in the soil and the suction can be expressed in a plot of volumetric water content versus suction, known as the soil-water characteristic curve.

Soil suction is believed to be the key factor in controlling moisture movements in portly saturated clay. The relationship between soil suction, moisture content and the volume changes suggest a possible analytical method for estimating heave or settlement. The co-efficient of volume change with respect to matric suction is given by the slope of the soil-water characteristic curves. Also these curves are used to derive the permeability and shear strength functions of both saturated and unsaturated soils.

Table 1 Index properties of collected soils samples :

Properties	Thiruneermalai	Anna Nagar	K.K. Nagar	Korattur	Thiruvanmiyur
Sand (%)	4	4	7	9	10
Silt (%)	25	42	37	39	37
Clay (%)	71	56	49	52	53
Liquid Limit (%)	78	70	57	60	64
Plastic Limit (%)	34	31	26	25	27
Plasticity Index (%)	44	39	31	35	37
Shrinkage Limit (%)	11.7	8.9	12	11	10
I.S Classification	CH	CH	CH	CH	CH

2. COLLECTION OF SOIL SAMPLES

To study the soil-water characteristic of expansive clays in Chennai city, disturbed and undisturbed soil samples from five different locations of Chennai are collected. They are Thangam colony of Anna Nagar, K.K. Nagar, Thiruvanmiyur, Korattur, and Thiruneermalai. Detailed laboratory test were done on the collected disturbed samples to classify the soil as per unified soil classification system. The Index properties of the soil of the five different locations are given in Table 1.

3.SOIL-WATER CHARACTERISTIC CURVE

The soil-water characteristic curve of a soil can be obtained using the pressure plate-extractor device in the laboratory. Using the axis-translation technique, air pressure above atmospheric pressure is applied to the soil specimen while the water pressure is kept at a lower value that is usually atmospheric. This is made possible through the use of a high air entry disk that separates the air phase from the water phase. The difference between the air and the water pressure is known as matric suction. The water content of the soil specimen at various matirc suction can therefore be determined and the soil-water characteristic curve is drawn. The obtained soil water characteristic curve of the different soil are shown in Figure 1 and 2.

3.1 Percent Swell and heave evaluation

The percent swell values from suction studies were obtained using Brakely (1980) empirical equation. The heave is also calculated using wisemann's (1984) equation

SW = Percent swell
Z = Active depth

While computing the heave the Restrained factor value of 0.5 was used in relation of the degree of saturation in the soil (wisemen, 1987). The calculated values of percent swell and heave are presented in Table 2.
To compare these values the percent swell and heave are also evaluated by conducting the expanded loaded test on the different soil samples collected from these locations at various depth. To calculate heave from the expanded loaded method the U.S. army corps Engineers method is used.

$$\text{Percent swell, SW} = \frac{PI - 10}{10} \log S/P \quad \text{Where}$$

PI = Plasticity Index (%)
S = Soil suction (Kpa)
P = Foundation and over burden pressure.

Heave = R.F X SW X Z/100

Where R.F. = Restrained factor

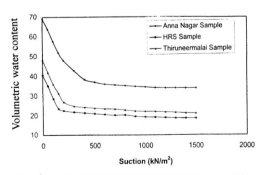

Fig. 1. **VOLUMETRIC WATER CONTENT Vs SUCTION**

354

Table 2 Percent swell and heave

SlNo	Description	%Swell		Heave (mm)		
		*	**	*	**	***
1	K.K Nagar Water content 21%	3.35	3.35	33.6	58.8	50.0
2.	Korattur water content 20%	4.75	4.87	47.8	83.3	60.3
3	Thiurvanmiyur water content 20%	5.2	5.46	52.1	93.5	70.5
4	Anna Nagar Water Content 20%	6.5	6.83	65.1	117.2	90.0
5	Thiruneermalai water content 30%	2.7	-	40	41	37.5
	* from Expanded loaded method ** from soil water characteristic curve *** observed data in field/ lab					

3.2 Evaluation of Permeability function

Many Geotechnical and geoenvironmental problems involve water flow through unsaturated soils. Coefficient of permeability in an unsaturated soil is a function of pore water pressure and water content. Direct measurement of permeability in the laboratory can be time consuming, especially for low water content conditions. Indirect measurements of permeability are commonly performed by establishing permeability functions through the use of relationship between water content and pore water pressure i.e, for different soils at their respective density were obtained using the equation suggested by Averjanov (1950) and are shown in Figure 2. For each soil if the saturated

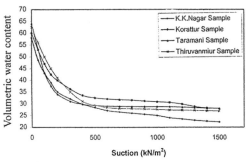

Fig. 2.   VOLUMETRIC WATER CONTENT Vs SUCTION

permeability is known, from the graph the permeability at different degree of saturation at a given density can be calculated.

4. CONCLUSION

The Percent swell and heave that are obtained from the soil water characteristic are compared in the observed data of field and data obtained from other laboratory studies. From the data it is concluded the percent swell and heave evaluation from the soil water characteristic curves are comparable with the observed data in the field/laboratory. The permeability function $k_r(\Psi)$ is obtained as a function of volumetric water content (θ) for the different expansive clays.

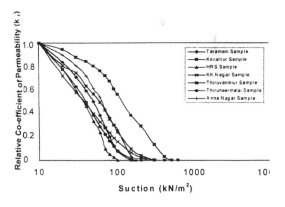

Fig. 3.   RELATIVE COEFFICIENT OF PERMEABILITY Vs SUCTION

REFERENCES

1.   Fredlund and Zing, D.G. and Shangyan Huang., (1994), "Predicting the permeability

function for unsaturated souls using the soil – water characteristic curve", Canadian Geotechnical journal, vol. 31, pp. 533-546

2. Long E.C. and Rhardjo, H. (1997), "Permeability functions for unsaturated soils", Journal of Geotechnical and Geoenvironmental Engineering vol. 123, pp. 1118-1126.

3. Sujatha, B. (1993), "performance of heave of an expansive soil using unsaturated theory", A thesis submitted for the Degree of Master of Engineering, Anna University,Chennai – 25.

4. Vasumathi, G.(1998), "Predication of heave of expansive clays", A Thesis submitted for the Degree of Master of Engineering, Anna University,Chennai –25.

5. Sanjay Kumar R. (1999) "A Study on Volume change characteristics of expansive clays", A Thesis submitted for the Degree of master of Engineering, Anna University, Chennai-25

Unsaturated Soils for Asia, Rahardjo, Toll & Leong (eds) © 2000 Taylor & Francis, ISBN 90 5809 139 2

Soil-water characteristic curves of peaty soils

J.L.Tan, E.C.Leong & H.Rahardjo
NTU-PWD Geotechnical Research Centre, Nanyang Technological University, Singapore

ABSTRACT: In Singapore, peaty soils are found in several locations at shallow depths between 3m and 5m. Peaty soils, having low strength and high compressibility, have always been regarded as a poor geomaterial. In construction where peaty soils are encountered, they are generally replaced with more competent soil. One important characteristic of peaty soils is its high water retention ability, a key to understanding the poor engineering properties of peaty soils. The water retention ability of the peaty soils can be studied using principles of unsaturated soil mechanics. The soil-water characteristic curves of peaty soils from two different sites were obtained in the laboratory using the pressure plate apparatus and the salt solution method. A study of the soil-water characteristic curves of the peaty soils and how they can be related to its engineering properties using principles of unsaturated soil mechanics are presented in the paper.

1 INTRODUCTION

The distinction between saturated soil mechanics and unsaturated soil mechanics lies in the state of stress of the water phase within the soil system. Soils with positive pore-water pressures and a degree of saturation S of 100% are regarded as saturated soils while unsaturated soils are those which contain air bubbles and have negative pore-water pressures. The relationship between the moisture retained within the soil mass and its suction can be clearly illustrated by its moisture retention curve. The moisture retention curve is more commonly known as soil-water characteristic curve (SWCC) in geotechnical engineering. Plots of gravimetric water content w or volumetric water content θ_w or the degree of saturation S versus matric suction ψ are but different variations of the SWCC.

According to Fredlund & Rahardjo (1993a), there exists a unique relationship between the properties of unsaturated soil and its SWCC. It is possible to derive approximate estimates of the unsaturated coefficient of permeability k_w and shear strength τ_{ff} of the soil based on its SWCC together with some of its conventional saturated soil properties like effective angle of shearing resistance ϕ', effective cohesion c', and saturated coefficient of permeability k_s. The aim of this paper is to study the SWCCs of two Singapore peaty soils and how they can be related to their engineering properties using principles of unsaturated soil mechanics.

2 MATERIALS & PROCEDURES

2.1 Basic soil properties

The peaty soils were collected from two different sites on the main island of Singapore. These soils will be referred to as Bedok (BK) soil and Jalan Boon Lay (JBL) soil in this paper.

The basic index properties of the peaty soils were determined in accordance with ASTM (1997) as summarised in Table 1. Based on the Unified Soil Classification System, the peaty soils are classified as organic clays of high plasticity. The JBL soil has higher organic and fibre contents compared with the BK soil which may account for its higher natural water content, higher void ratio, lower bulk unit weight and lower specific gravity. Although the JBL soil has higher liquid and plastic limits compared with the BK soil, the plasticity indices of both soils are the same.

2.2 Experimental procedures

The pressure plate test and the salt solution method were used to obtain the SWCCs of the soils at suctions ranging from 50 kPa to 1000 kPa and 3000 kPa to 10000 kPa, respectively. In most cases, the water content of the soil can generally be assumed to be equivalent when subjected to matric suction or total suction greater than 1500 kPa (Fredlund & Xing,

Table 1. Basic index properties of peaty soils.

Geotechnical Properties	BK	JBL
Organic Content (%)	13	31
Fibre Content (%)	1.0	14.7
Natural Water Content, w_n (%)	95	165
LL (%)	115	199
PL (%)	49	133
PI (%)	66	66
Bulk Unit Wt, γ (kN/m^3)	14.1	12.3
Specific Gravity, G_s	2.40	2.10
Initial Void Ratio, e_o	2.27	3.58
Grain Size Distribution		
a) Sand (%)	1.5	0.8
b) Silt (%)	14.3	31.0
c) Clay (%)	84.2	68.2

1994). Saturated soil specimens of known initial volumes were placed in the pressure plate apparatus and these soil specimens were sheared using the laboratory miniature vane shear apparatus on reaching equilibrium at the desired matric suction values. The salt solution method was used to apply higher suctions to the soil specimens. Filter papers were placed in the desiccators to confirm the suction values achieved by the salt solutions.

2.2.1 Pressure plate and vane shear test

The initial volume of the soil specimens in the pressure plate apparatus is governed by the requirements of the laboratory miniature vane shear test D4648-87 (ASTM, 1997) using a vane blade of diameter 12.7 mm and height 12.7 mm. In accordance to the recommendations of ASTM-D4648-87, the soil specimen should have a diameter of at least 5 times the blade's diameter and a minimum height of at least 3 times the blade's height. Hence, the minimum size of the soil specimen should be 63.5 mm in diameter and 38 mm in height. The soil specimens were carved out from block samples into PVC rings, 72 mm internal diameter and 40 mm height, with the aid of a sharp edge ring cutter.

Eight specimens were placed in the pressure plate apparatus and were weighed daily. The matric suctions applied to the soil specimens were 50 kPa, 100 kPa, 200 kPa, 400 kPa, 600 kPa, 800 kPa and 1000 kPa. The soil specimens were sheared upon reaching equilibrium at the desired matric suction value. Dimensions of the soil specimen, namely the height and diameter, were also taken before each pressure increment and vane shear test. The water content of the specimen was determined after conducting the vane shear test. Using the water content and shear strength obtained at the respective matric suction values, plots of the SWCC and shear strength versus matric suction of the peaty soils were obtained.

2.2.2 Salt solution method

Three desiccators were partially filled with salt solutions of known concentrations to provide total suction values of 3000 kPa, 5000 kPa and 10000 kPa. Soil specimens of 5 g to 8 g were placed in 5 cc porcelain crucibles. These crucibles were then placed in the desiccators. The crucibles were weighed weekly and upon reaching equilibrium the soil specimens and the filter paper were oven dried to determine the water content and the total suction in the respective desiccator.

3 RESULTS

3.1 Soil water characteristics curve (SWCC)

The BK and JBL SWCC data are shown in Figures 1 and 2, respectively. It is convenient if the SWCC can be described using an equation. A number of SWCC equations have been proposed. A comprehensive review of the SWCC equations can be found in Leong and Rahardjo (1997). In this paper, the SWCCs are presented in the form of normalised volumetric water content Θ versus suction ψ. The experimental SWCCs obtained for both soils are S-shaped and are typical of clayey soils.

The suitability of three SWCC equations is evaluated; they are the Fredlund & Xing (1994), van Genutchen (1980) and Weiss et al. (1998) equations.

The three-parameter normalised form of the Fredlund & Xing (1994) equation is:

$$\Theta = 1/\{\ln[\exp + (\psi/a)]^n\}^m \tag{1}$$

where $\Theta = \theta_w/\theta_s$; θ_s = saturated volumetric water; exp = natural base of logarithms; ψ = suction pressure (kPa); and a, n & m = curve fitting parameters.

The three-parameter normalised form of the van Genutchen (1980) equation is:

$$\Theta = 1/\{1 + (\alpha\psi)^n\}^m \tag{2}$$

where $\Theta = (\theta_w - \theta_r)/(\theta_s - \theta_r)$; α = curve fitting parameter; and θ_r = residual volumetric water.

The two parameter normalised form of the Weiss et al. (1998) equation is:

$$\Theta = 1/\exp\{[\ln(\theta_s/\theta_{wilt})]*[(\log \psi + 1)/4.2]^k\} \tag{3}$$

where $\Theta = \theta_w/\theta_s$; θ_{wilt} = wilting point water content at 1.554 MPa.

The curves produced by the Fredlund & Xing and the van Genuchten equations are sigmodal in nature (Leong and Rahardjo, 1997). The Weiss et al. equation was developed specifically for peaty soils.

The results of the curve fitting exercise are shown in Figures 1 and 2. The goodness of fit of the equations to the SWCC data was examined using the sum of squared residual (SSR) value. The SSR value is given by the squared difference between the normalised volumetric water content from the experiment and that of the equation (Leong & Rahardjo,

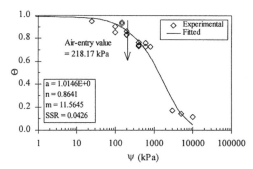

(a) - Fredlund & Xing (1994) equation [Eq. 1]

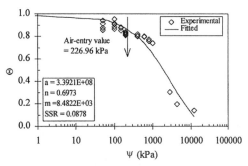

(a) - Fredlund & Xing (1994) equation [Eq. 1]

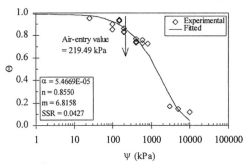

(b) - van Genuchten equation (1980) [Eq. 2]

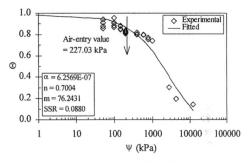

(b) - van Genuchten (1980) equation [Eq. 2]

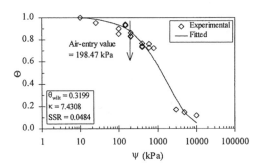

(c) - Weiss et al. equation (1998) [Eq. 3]

Figure 1. Summary of curve fits for BK peaty soil.

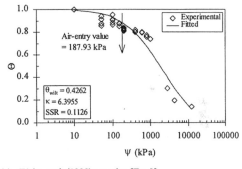

(c) - Weiss et al. (1998) equation [Eq. 3]

Figure 2. Summary of curve fits for JBL peaty soil.

1997). The smaller the SSR value, the better is the fit. The values of the SSR of the equations are summarised in Table 2.

The BK SWCC generally has lower SSR values than the JBL SWCC. Nevertheless, all the three equations appear to give a good fit to the data of both soils. The SSRs of the Fredlund & Xing (1994) and van Genuchten (1980) equations are almost the same for both soils.

Air-entry values of the fitted SWCCs were obtained via the estimation method recommended by Vanapalli et al. (1998). The air-entry values obtained from the curves using Equations 1 and 2 are almost identical, whereas the air-entry values obtained from Equation 3 are always smaller compared to the other two equations (Figs 1 and 2). It is perhaps surprising that both soils have almost the same air-entry value.

3.2 Permeability

The unsaturated coefficients of permeability k_w, of the soil specimens were derived using the program ACUPIM (Automated Computation of Unsaturated Permeability from Indirect Measurements) developed by Leong (1995). The statistical model adopted

Table 2. SSRs of SWCC curve fitting exercise.

Site	Equation 1	Equation 2	Equation 3
BK	0.0426	0.0427	0.0484
JBL	0.0878	0.0880	0.1126

in ACUPIM is the model first proposed by Childs and Collis-George (1950) and subsequently modified by Mashall (1958) and Kunze et al. (1968). Further information can be found in Fredlund & Rahardjo (1993b).

The k_s and the SWCCs generated, using Equation 1, for each soil were input into the program. The k_s of the BK and JBL peaty soils are 4.277E-08 m/s and 4.228E-09 m/s, respectively. The permeability functions $k_w(\psi)$ for both soils are shown in Figure 3. The difference in the k_w values of the BK and JBL peaty soils seems to diminish as suction values increases.

3.3 *Shear strength*

The laboratory vane shear strength of the BK and JBL peaty soils at different ψ values are shown in Figures 4 and 5, respectively. Fredlund et al. (1995) and Vanapalli et al. (1996) had proposed that the shear strength of an unsaturated soil can be predicted using the following equation:

$$\tau_{ff} = [c' + (\sigma_n - u_a) \tan \phi'] + (u_a - u_w)[(\Theta^\kappa)(\tan \phi')] \quad (4)$$

where τ_{ff} = shear strength of an unsaturated soil; c' = effective cohesion of the soil; ϕ' = effective angle of shearing resistance for a saturated soil; $(\sigma_n - u_a)$ = net normal stress; $(u_a - u_w)$ = matric suction; σ_n = normal stress; u_a = pore-air pressure; u_w = pore-water pressure; Θ = normalised volumetric water content; and κ = a fitting parameter.

The first part of Equation 4, $[c' + (\sigma_n - u_a) \tan \phi']$, is the saturated shear strength of the soil, where $u_a = u_w$, and the second part is the contribution of shear strength due to soil suction. Therefore, the saturated shear strength parameters, c' and ϕ', and the normalised SWCC of the specimens are required. Although c' and ϕ' of the peaty soils have yet to be determined, $[c' + (\sigma_n - u_a) \tan \phi']$ can be obtained from the vane shear test conducted at zero matric suction on fully saturated specimens. In this paper, $(u_a - u_w)$ is taken to be equivalent to ψ. As no experimental value of ϕ' is available, ϕ' and κ in Equation 4 were varied to give the best fit to the experimental data. The fitted ϕ' and κ values for the BK and JBL peaty soils are summarised in Table 3.

It is evident from the curves of shear strength prediction using SWCC (Figs 4 and 5) that a better fit is obtained for the JBL peaty soil. The fitted shear strength curves superimposed onto one another despite using three different SWCC equations.

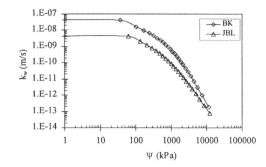

Figure 3. Permeability function $k_w(\psi)$ for peaty soils using ACUPIM.

Table 3. Summary of ϕ' and κ values for peaty soils.

Site	Equation 1 ϕ', κ	Equation 2 ϕ', κ	Equation 3 ϕ', κ
BK	27.2°, 2.8031	27.2°, 2.8148	26.2°, 2.6039
JBL	26.4°, 4.3190	26.4°, 4.3197	24.3°, 3.7971

4 DISCUSSION

The average air-entry value of the BK and JBL peaty soils is approximately 200 kPa. Despite having different liquid limits and plastic limits, both soils have identical plasticity index and fines content which may explain why they have almost identical air-entry value.

The Fredlund & Xing (1994), van Genuchten (1980) and the Weiss et al. (1998) equations fit the experimental SWCCs well. The SSR values for the SWCC curve fitting exercise varies within a narrow range of 0.0058 and 0.0248 for the BK and JBL peaty soils, respectively. However, based on the SSR values, a better fit was obtained for the BK soil. It is observed that the air-entry values and the SSR values obtained via Equation 1 and 2 are almost identical. This may be attributed to the fact that the pore-size distribution function used in Equation 1 is similar to Equation 2.

The permeability of an unsaturated soil is a function of the matric suction. Although the JBL soil has a higher void ratio than the BK soil, its saturated permeability is the lower of the two. Hence, the unsaturated permeability of the BK soil is always higher than that of the JBL soil at each corresponding matric suction value. The predicted permeability function indicates that k_w changes by almost five orders of magnitude as suction increases from 0 kPa to 10000 kPa. Since direct measurements of the unsaturated permeability of the peaty soils have not been conducted, the goodness of fit of the predicted k_w to the experimental k_w is not known.

The shear strength of both peaty soils increases linearly with respect to matric suction up to a matric suction value close to their respective air-entry values. The κ value for the BK and JBL peaty soil obtained from the SWCC Equations 1 to 3 ranges from 2.6039 to 2.8148 and 3.7971 to 4.3197, respectively. Fredlund et al. (1995) observed that the value of κ generally increases with the plasticity of the soil and can be greater than 1. Although both the BK and JBL peaty soils have the same plasticity index, the κ value of the BK peaty soil is lower than that of the JBL peaty soil. The vane shear strength of the BK peaty soil increases linearly up to a matric suction close to its air-entry value, peaks at a matric suction of about 400 kPa and then starts to reduce. Unlike the BK peaty soil, no distinct peak was observed in the vane shear strength of the JBL peaty soil.

The value of ϕ' ranges from $26.2°$ to $27.2°$ and $24.3°$ to $26.4°$ for the BK and JBL peaty soils, respectively. The value of ϕ' obtained is quite consistent as it varies over a range of only $2°$. The predicted shear strength curves obtained using different SWCC equations are so close that they virtually superimpose onto one another (Figs. 4 and 5). Since little variations are noted in the SSRs in Table 2, it can be deduced that the equations provide equally good fit to the experimental SWCCs and hence resulting in the almost identical predicted shear strength curves.

The effective angle of shearing resistance ϕ' decreases with increasing void ratios and natural water contents (Holtz & Kovacs, 1981). This may explain why the JBL peaty soil has a lower ϕ' value than the BK peaty soil as the former has the higher void ratio and natural water content of the two.

The Singapore peaty soils used in this study is found at a shallow depth, above the upper marine clay layer. The vane shear strengths of the saturated BK and JBL peaty soils at zero matric suction are 7.12 kPa and 22.64 kPa, respectively. This range of undrained shear strength falls within the undrained shear strength range of 5 kPa to 50 kPa reported by Tan (1983) for the Singapore upper and lower marine clay layers. Hence, it is postulated that both the peaty soils and marine clays may have similar ϕ' values. Tan (1983) reported an average ϕ' value of $21.5°$ and Ahmad & Peaker (1977) reported ϕ' values in the range of $12°$ to $22°$ for the Singapore upper and lower marine clays. The ϕ' values obtained from Equation 4 are slightly above the reported values of ϕ'. Therefore, it is recommended that further experimental studies are to be conducted to verify the ϕ' value of the peaty soils.

5 CONCLUSION

SWCCs for two Singapore peaty soils were obtained using the pressure plate apparatus and the salt solu-

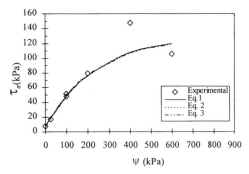

Figure 4. Vane shear strength of the BK peaty soil.

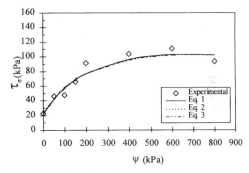

Figure 5. Vane shear strength curves of the JBL peaty soil.

tion method. The SWCCs showed that the air-entry values of the two peaty soils are about 200 kPa. The narrow range of SSRs obtained via the curve fitting exercise indicates that the Fredlund & Xing (1994), van Genuchten (1980) and Weiss et al (1998) equations are capable of providing a good fit to the SWCCs of the peaty soils. Permeability functions $k_w(\psi)$ for the peaty soils were obtained using the saturated coefficient of permeability k_s and the SWCC via a statistical model. The relationship of shear strength with matric suction using the SWCC was examined. The κ values of the peaty soils obtained through the shear strength curve fitting method are greater than one, in the range of 2.6 to 4.3. The predicted shear strength curves obtained from SWCC equations, Equations 1 to 3, are almost identical as the equations provide a similar goodness of fit to the experimental SWCCs. Although the ϕ' values obtained varies over a narrow range of about $2°$, further research is recommended before any conclusions are drawn upon the results obtained in this paper as the ϕ' values appear to be higher than that of the soft Singapore marine clays.

ACKNOWLEDGMENTS

The study is funded by a research project grant, RG 9/98, from the Ministry of Education, Singapore. The authors wish to thank Mr John Wei, Mr James Kerr and Mr Chua Kok Eng from the Housing Development Board of Singapore (HDB) for their assistance in the acquisition of the peaty soils. The first author acknowledges the research scholarship provided by the Nanyang Technological University.

REFERENCES

Ahmad, S.A., & Peaker, K.R. (1977). Geotechnical properties of soft marine Singapore clay. *International Symposium on Soft Clays,* Bangkok, Thailand, 3 -14.

ASTM (1997). *Annual book of ASTM standards.* Vol. 4.08, Soils and Rock (I): D420 – D4914, 207 – 216.

Childs, E.C., & Collis-George, G.N. (1950). The permeability of porous materials. *Proceedings of the Royal Society of London.* Series A, 201, 392 - 405.

Cox, J.B. (1970). The distribution and formation of recent sediments in South East Asia. *Proceedings of 2nd South East Asian Conference on soil Engineering,* Singapore, 29 -47.

Fredlund, D.G., & Rahardjo, H. (1993a). An overview of unsaturated soil behaviour. *Proceedings of ASCE Specialty Series on Unsaturated Soil Properties.* Dallas, Texas, October.

Fredlund, D.G., & Rahardjo, H. (1993b). *Soil mechanics for unsaturated soils.* New York: Wiley & Sons, Inc.

Fredlund, D.G., & Xing, A. (1994). Equations for the soil-water characteristic curve. *Canadian Geotechnical Journal,* 31, 521 - 532.

Fredlund, D.G., Xing, A, Fredlund, M.D., & Barbour, S.L. (1995). The relationship of the unsaturated shear strength to the soil-water characteristic curve. *Canadian Geotechnical Journal,* 32, 440 - 448.

Holtz, R.D., & Kovacs, W.D. (1981). *An introduction to geotechnical engineering.* New Jersey: Prentice-Hall, Inc.

Kunze, R.J., Uehara, G., & Graham, K. (1968) Factors important in the calculation of hydraulic conductivity. *Proceedings of Soil Science Society of America,* 32, 760 - 765.

Leong, E.C. (1995). *ACUPIM - Automated computation of unsaturated permeability for indirect measurements.* NTU-PWD Geotechnical Research Centre, Nanyang Technological University, School of Civil & Structural Engineering, Singapore.

Leong, E.C., & Rahardjo, H. (1997). Review of soil-water characteristic curve equations. *Journal of Geotechnical and Geoenvironmental Engineering, 123, 12, 1106 - 1117.*

Marshall, T.J. (1958). A relation between permeability and size distribution of pores. *Journal of Soil Science,* 9, 1 - 8.

Tan, S.L. (1983). Geotechincal properties and laboratory testing of soft soils in Singapore. *International Seminar on Construction Problems in Soft Soils,* Singapore, TSL1 - TSL42.

van Genuchten, M. Th. (1980). A closed-form equation for predicting the hydraulic conductivity of unsaturated soils. *Soil Science Society of American Journal,* 44, 892 - 898.

Vanapalli, S.K., Fredlund, D.G., Pufahl, D.E., & Clifton, A.W. (1996). Model for the prediction of shear strength with respect to soil suction. *Canadian Geotechnical Journal,* 33, 379 - 392.

Vanapalli, S.K., Sillers, W.S., & Fredlund, M.D. (1998). The meaning and relevance of residual state to unsaturated soils. *Proceedings of 51st Canadian Geotechnical Conference,* Edmonton, Alberta, October.

Weiss, R., Alm, J., Laiho, R., & Laine, J. (1998). Modeling moisture retention in peat soils. *Soils Science Society of American Journal,* 62, 305 - 313.

6 Permeability and flow

Unsaturated Soils for Asia, Rahardjo, Toll & Leong (eds) © 2000 Taylor & Francis, ISBN 90 5809 139 2

A triaxial permeameter for unsaturated soils

S.S.Agus, E.C.Leong & H.Rahardjo
NTU-PWD Geotechnical Research Centre, Nanyang Technological University, Singapore

ABSTRACT: Many geotechnical and geoenvironmental engineering problems involve fluid movement through soil. The rate of water flow through soil is characterised by its water coefficient of permeability, while its air coefficient of permeability controls the rate of air flow through soil. Both coefficients of permeability are functions of the degree of saturation and the void ratio. The degree of saturation of the soil is related to matric suction in the soil. As matric suction increases, the degree of saturation of a soil decreases and consequently, the water coefficient of permeability of the soil decreases while the air coefficient of permeability increases. The magnitude of the coefficients of permeability may change by several orders as the degree of saturation decreases from 100% to near 0%. The direct measurement of the water and air coefficients of permeability therefore poses a challenging problem especially at high matric suction values. This paper presents the development of a triaxial permeameter to measure the coefficients of permeability with respect to the water and air phases for unsaturated soils. The performance of the triaxial permeameter was evaluated using a Singapore residual soil whereby the water and air coefficients of permeability are measured following the drying and wetting paths. Variables that may affect the accuracy of the permeability measurements such as magnitude of the hydraulic gradient, ambient temperature and volume change of the specimen are also discussed in the paper.

1 INTRODUCTION

Fluid movement through soil is an important problem in geotechnical and geoenviromental engineering. Water flow through the unsaturated zone in an earth dam, for instance, is one of the problems where unsaturated water coefficient of permeability of the soil is involved. Concern about migration of contaminated gas through soils is an example where the determination of the air coefficient of permeability is needed.

In saturated soils, the coefficient of permeability with respect to water phase is a function of the void ratio of the soil only. However, in unsaturated soils, the water and air coefficients of permeability are functions of both the void ratio, e, and the degree of saturation, S, of the soil. As matric suction of the soil increases, S decreases and accordingly water coefficient of permeability decreases while air coefficient of permeability increases.

The unsaturated water and air coefficients of permeability can be determined directly in the laboratory or indirectly using macroscopic and statistical models. Although more costly and more time consuming, the direct measurement of permeabilities gives more reliable results. Generally, the direct methods of measurement can be classified into two main categories: steady-state and unsteady-state measurements. The tests can be performed using either constant-head or variable-head. In the case of unsaturated soils, steady-state measurement of permeability using constant-head is commonly adopted.

A number of researchers have measured the water and air coefficients of permeability of unsaturated soils using a triaxial setup (Barden & Pavlakis, 1971, Fleureau & Taibi, 1994, Huang et al., 1998). A summary of the capabilities of the triaxial set-ups used by past researchers is listed in Table 1.

In this study, a triaxial permeameter was developed to measure both the water and air coefficients of permeability of unsaturated soils. In the measurement of the water coefficient of permeability, high hydraulic gradients were used to reduce the test duration especially at high matric suction values, up to 300 kPa. Total volume change of the specimen as matric suction changes was monitored using a digital water pressure/volume controller.

Table 1. Summary of triaxial permeameter systems used by other researchers

References	Range of k_w measured (m/s)	Range of k_a measured (m/s)	Suction range (kPa)	Measurement of volume change of specimen	Soil tested
Barden and Pavlakis (1971)	10^{-12}-10^{-10}	10^{-8}-10^{-3}	0-95	Not measured	Compacted soils
Fleureau and Taibi (1994)	up to 10^{-8}	up to 10^{-8}	0-80	Not measured	Undisturbed clay, loam and sand-kaolin mixture
Huang et al. (1998)	10^{-11}-10^{-8}	Not measured	0-90	Using non-contacting transducers	Silty sand

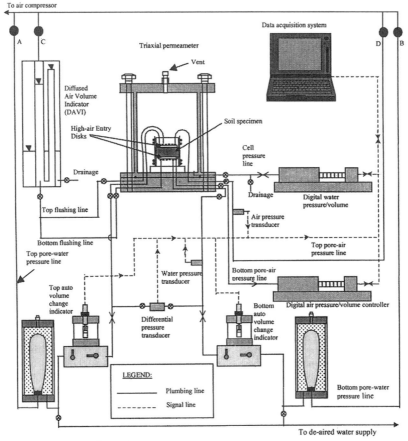

Figure 1. Schematic drawing of triaxial permeameter system

2 DESCRIPTION OF TRIAXIAL PERMEAMETER

The triaxial permeameter system (Fig. 1) consists mainly of a triaxial permeameter cell, pressure lines, flow and specimen volume-change measurement devices.

The triaxial permeameter cell is modified from a standard triaxial cell. The cylindrical perspex cell wall was replaced with a stainless steel cell wall in order to prevent expansion of the cell under high cell pressures. The top cap and the pedestal, both of 71 mm diameter, were specially designed to have a grooved water reservoir and a recess for high-air entry ceramic disks. Two 5-bar high-air entry ceramic disks with average coefficient of permeability of

4.45x10^{-10} m/s were sealed into the top cap and pedestal using Epoxy glue. Grooves of 2mm wide the and 1mm deep were cut onto the faces of the ceramic disks in contact with the soil specimen to provide free flow of air into the soil specimen.

The top and bottom water pressures were applied through air-water interfaces that transmit water through volume change indicators (Fig. 1). The water pressures were monitored using pore-water pressure and differential pressure transducers connected to the bottom line and the bottom-top connection line, respectively. The pore-air pressures were applied and regulated through both top and bottom air pressure lines.

Permeability measurements for unsaturated soils usually take a long time and therefore automation of the measurement becomes a necessity. Two auto volume-change measurement devices (Fig. 1), were installed to measure the volumetric water flow during the water permeability measurements. The devices are connected to two air-water interfaces that provide the water pressure for water permeability measurement.

A digital air pressure/volume controller was used to measure the volumetric air flow during the air permeability measurements. Adams et al. (1996) had used a digital air pressure/volume controller to measure the air volume change in triaxial tests under isothermal conditions (±2°C). It was found that the precompression of the air and minimisation of the volume of the confined air in the digital air pressure/volume controller can effectively decrease the error in the measurement. By taking the above precautions, the digital air pressure/volume controller can be employed for measuring the air flow rate through the soil specimen. In addition, the duration of the test is typically short so that the readings were not substantially affected by fluctuations of the ambient temperature.

The volume change measurements of the soil specimen were obtained via a digital water pressure/volume controller which also provides the confining pressure in the cell. At a certain confining pressure, the controller adjusts itself to maintain a constant pressure whenever the soil specimen expands or compresses during the test.

3 SOIL PROPERTIES

A residual soil from the Bukit Timah granitic formation of Singapore was used in this study. The basic soil properties of the sample are listed in Table 2.

Table 2. Basic Properties of the Soil Sample

Soil property	
Saturated water content (%)	37.3
Bulk unit weight (kN/m^3)	19.0
Specific gravity	2.6
Initial void ratio, e_0	0.974
Saturated permeability, k_s (m/s)	6.65x10^{-8}
Liquid limit	53.7
Plastic limit	23.2
Plasticity index	20.5
Sand (%)	43
Silt (%)	45
Clay (%)	12
USCS Classification	MH

4 TEST PROCEDURES

The test procedures used in this study comprise of three main stages: (1) specimen preparation and saturation, (2) consolidation, and (3) water and air permeability measurements. The test was performed with the soil specimen undergoing a drying-wetting cycle.

The residual soil sample was cut and trimmed to the required size (i.e. 71 mm in diameter and 30 mm in height) and then subsequently saturated. After saturation, the specimen was consolidated at a confining pressure of 50 kPa to the desired matric suction value. Consolidation was deemed to be completed when there was negligible water measurement either into or out of the soil specimen.

After consolidation, the water coefficient of permeability was determined first. A hydraulic gradient (from 27 at 50 kPa to as high as 81 at 300 kPa of matric suctions) was applied to the soil specimen and the water inflow and outflow were monitored until a steady-state flow condition was established, that is when the inflow rate was approximately equal to the outflow rate (Fig. 2).

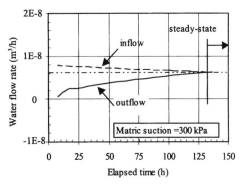

Figure 2. Steady-state flow condition

The water coefficient of permeability may be different for different applied hydraulic gradients. Fox (1996) noted that the magnitude of the water coefficient of permeability of a clay slurry at saturation decreased by an average factor of 2 as the hydraulic gradient was increased from 8.8 to 17. It was also reported that for compressible soils, excessive hydraulic gradients applied result in lower apparent saturated coefficients of permeability being measured. Therefore, tests were performed to investigate the effect of high hydraulic gradients on the water coefficient of permeability of an unsaturated soil at different matric suction values. It was found that the water coefficient of permeability does not vary significantly for the residual soil tested, especially at high matric suction values (Fig. 3).

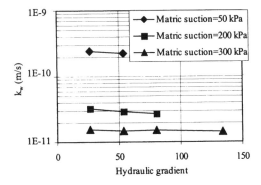

Figure 3. Variation of water coefficient of permeability with hydraulic gradient

5 ANALYSIS AND RESULT

5.1 Determination of Water and Air Coefficients of Permeability

The water coefficient of permeability was calculated using Darcy's law. As the coefficients of permeability of the ceramic disks may affect the determination of the water coefficient of permeability, the impedance of the ceramic disks were taken into account. The water coefficient of permeability of the soil specimen, k_w, was determined by treating the disk-soil-disk arrangement as a three-layered system. Hence:

$$k_w = \frac{H_s}{\left(\frac{H}{k} - \left(\frac{H_t}{k_t} + \frac{H_b}{k_b}\right)\right)} \quad (1)$$

where: H_s = height of the soil specimen; H_t = thickness of the top ceramic disk; H_b = thickness of the

bottom ceramic disk; $H = H_t + H_s + H_b$, k_t = coefficient of permeability of the top disk; k_b = coefficient of permeability of the bottom disk and k = coefficient of permeability of the three-layered system determined as $k = Q_w/i_wAt$, with Q_w = water flow reading, i_w = applied hydraulic gradient, A = cross-sectional area of the soil specimen and t = elapsed time.

Since the viscosity of water changes with temperature, a correction factor with respect to temperature fluctuations was applied to the water flow reading and a corrected water coefficient of permeability at 20°C was determined.

Darcy's law is assumed to be valid for the calculation of the air coefficient of permeability, k_a (Barden & Pavlakis, 1971, Fredlund & Rahardjo, 1993, Grant & Groenvelt, 1993, Eischens & Swanson, 1996). As the air pressures were applied directly to the soil specimen, the system was therefore considered as a single-layered system. Therefore k_a is given as:

$$k_a = \frac{Q_a}{i_a A t} \quad (2)$$

where: Q_a = volumetric air flow reading and i_a = applied pneumatic gradient.

Compared with the determination of the water coefficient of permeability, which takes several days to reach the steady state condition, the determination of the air coefficient of permeability requires a much shorter time (Fig. 4).

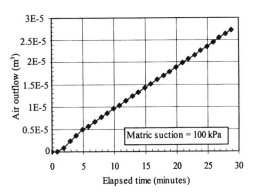

Figure 4. Air outflow versus elapsed time

As air is compressible, both pressure and temperature affect its volumetric flow measurement. The measured air flow rates were corrected to standard temperature and pressure (STP). The STP condition is usually taken as 20°C (293 K) and atmospheric pressure (101.3 kPa). Therefore the corrected volumetric air flow, Q_{corr} at STP condition is (Eischens & Swanson, 1996):

$$Q_{corr} = \frac{Q_a \left(\frac{P_t + P_b}{2} + P_B \right) T_s}{P_s \qquad T} \qquad (3)$$

where: P_t = air pressure at top of the soil specimen; P_b = air pressure at bottom of the soil specimen, P_B = test barometric pressure; P_s = pressure at STP (101.3 kPa); T_s = temperature at STP in Kelvin (293 K) and T = test temperature in Kelvin.

The water and air coefficients of permeability of the residual soil at different matric suction values are presented in Figure 5.

5.2 Determination of Total Volume Change of Soil Specimen

The volume change of the soil specimen during consolidation was determined using the digital water pressure/volume controller. However, the measurements were affected by the fluctuations of the ambient temperature approximately ±1.3°C (Fig. 6). A dummy test using a steel specimen was conducted to determine the coefficient of thermal expansion of the system, α.

expansion of the system, α, is known. The value of α will be different from the coefficient of thermal expansion of water, α_w, as it includes the thermal expansion of the cell, valves, tubing and the digital water pressure/volume controller. The correction is determined by first fitting a Fourier series to the temperature measurements:

$$T(t) = a_0 + \sum_{m=1}^{n} (a_m \sin mt + b_m \cos mt) \qquad (4)$$

The volume change, ΔV_c due to temperature fluctuations can be computed as:

$$\Delta V_c = \alpha [T(t) - T_{ref}] V_0 \qquad (5)$$

For the dummy test, the measured volume change, ΔV_m is caused by temperature fluctuations. Therefore, the value of α can be determined by matching ΔV_m to ΔV_c.

It was found that α is 11.5×10^{-5} /°C which is about half the value of the coefficient of thermal expansion of water, α_w, that is 21×10^{-5} /°C (Kaye & Laby, 1973). By applying the correction, not all the errors were completely eliminated. There is still an error of -0.05 ± 0.17 cm^3 as shown in Fig. 6.

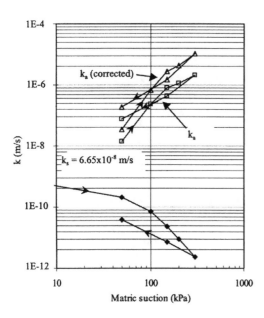

Figure 5. Measured water and air coefficient of permeability for a Singapore residual soil

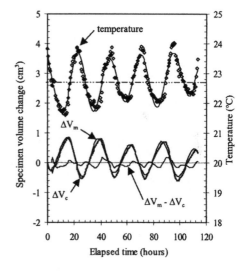

Figure 6. Fluctuations in temperature and specimen total volume change

Figure 6 shows a correspondence between temperature fluctuations and volume fluctuations in the dummy test. If the volume of water in the confining system, V_0 is known, a correction to the volume change measurements due to temperature fluctuations can be calculated if the coefficient of thermal

Using the above method of correcting volume measurement due to temperature fluctuation, the void ratio changes of the soil specimen, Δe can be determined (Fig. 7). Each point in the plot represents the change in void ratio at the end of the determination of the water coefficient of permeability at respective matric suction values.

369

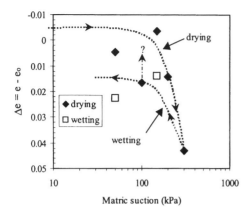

Figure 7. Change in void ratio versus matric suction values

6 DISCUSSION AND CONCLUSIONS

The triaxial permeameter system is capable of measuring the water and air coefficients of permeability of residual soils at matric suction values up to 300 kPa (Fig. 5). The system can measure the water coefficient of permeability as low as 10^{-12} m/s and even lower if the matric suction is increased up to the air entry value of the ceramic disk that is 500 kPa. The highest water coefficient of permeability which can be determined in this permeameter is 4.45×10^{-10} m/s, which corresponds to the coefficient of permeability of the ceramic disks. The system can also be used to measure the air coefficient of permeability ranging from 10^{-8} to 10^{-5} m/s.

Test duration for determining the air coefficient of permeability is much shorter than the water coefficient of permeability. Hence, it is worthwhile to convert air coefficient of permeability into water coefficient of permeability by relating degree of saturation with the water and air coefficients of permeability of soil as mentioned by Brooks & Corey (1964). However, further verifications are still required.

Since the steady-state flow condition for the determination of the water coefficient of permeability is established after a very long period of time, the need of using a short and large diameter soil specimen becomes a necessity. However, the determination of the air coefficient of permeability requires a longer soil specimen as recommended by Eischens and Swanson (1996). Indeed, verifications are needed to assess the most suitable and representative size of soil specimen to be used.

The ambient temperature fluctuations cause errors in the specimen volume change measurement. The error is due to the large volume of the confined water in the permeameter cell. By accounting for the ambient temperature fluctuations the system is able to measure volume change to an accuracy of ±0.17

cm^3. The accuracy may not be sufficient for very stiff soils where volume change is small. The volume change measurement accuracy can be improved by reducing the volume of the triaxial permeameter cell.

ACKNOWLEDGMENTS

This research is funded by grants from National Science and Technology Board, grant no. NSTB 17/6/16, and the Ministry of Education, grant no. ARC 12/96. The first author acknowledges the research scholarship provided by Nanyang Tecnological University.

REFERENCES

Adams, B.A., Wulfsohn, D., & Fredlund, D.G. (1996). Air volume change measurement in unsaturated soil testing using a digital pressure-volume controller. *Geotechnical Testing Journal*. GTJODJ. 19, 1: 12-21.

Barden, L. & Pavlakis, G. (1971). Air and water permeability of compacted unsaturated cohesive soil. *Journal of Soil Science*. 22: 302-317.

Brooks, R.H. & Corey, A.T. (1964). *Hydraulic properties of porous media*. Hydrology Paper. No. 3. Colorado State University. Fort Collins. Col.

Eischens, G. & Swanson, A. (1996). Proposed standard test method for measurement of pneumatic permeability of partially saturated porous materials by flowing air. *Geotechnical Testing Journal*. GTJODJ. 19, 2: 232-238.

Fleureau, J.M. & Taibi, S. (1994) New apparatus for the measurement of water-air permeabilities. *Proceeding of the Geoenvironmental Conference*. Edmonton.

Fox, P.J. (1996). Analysis of hydraulic gradient effects for laboratory hydraulic conductivity testing. *Geotechnical Testing Journal*. GTJODJ. 19, 2: 181-190.

Fredlund, D.G. & Rahardjo, H. (1993) *Soil Mechanics for Unsaturated Soils*. John Wiley and Sons.

Grant, C.D. & Groenvelt, P.H. (1993). Air Permeability. *Soil Sampling and Methods of Analysis*. M.R. Carter (ed.). Lewis.

Huang, S., Fredlund, D.G., & Barbour, S.L. (1998). Measurement of the coefficient of permeability for a deformable unsaturated soil using a triaxial permeameter. *Canadian Geotechnical Journal*, 35: 411-435.

Kaye, G.W.C. & Laby, T.H. (1973). *Tables of physical and chemical constant and some mathematical functions*. 14th edition. Longman.

Unsaturated Soils for Asia, Rahardjo, Toll & Leong (eds) © 2000 Taylor & Francis, ISBN 90 5809 139 2

An explicit relationship for the average capillary suction at the wetting front of soils

M.Y. Fard

Department of Civil Engineering, University of Tabriz, Iran

ABSTRACT: In infiltration problems and unsaturated flow studies, knowledge of capillary suction is essential. Capillary suction measurement at any point in the soil can be carried out using different instruments in laboratory or field and all need particular accuracy and are not simple methods. To calculate the hydraulic gradient between two ends of wet region of the soil column in infiltration problems, capillary suction at the wetting front is required. In the present paper, the author has developed an explicit relationship for capillary suction at wetting front using Mein and Larson (1971) definition for average suction and the inverse method of Zayani et al (1991) and Fard (1996) for unsaturated flow problems. The method requires only a laboratory infiltration test for the material under consideration to obtain the inverse method parameters.

1. INTRODUCTION

In the inverse method of determination of soil hydraulic properties, the matric suction 'S', hydraulic conductivity 'k' and diffusivity 'D' are defined as functions of water content as follows (Zayani et al, 1991 and Fard, 1996):

$$S = p(\theta - \theta_i)^{-m}$$
$$k = k_0(\theta - \theta_i)^{v}$$
$$D = D_0(\theta - \theta_i)^{n} \qquad (1)$$

in which the parameters p, k_o, v, m, D_o, and n are constants obtained by a horizontal infiltration test in which the cumulative infiltration and wetting front position are measured with time. In these equations θ_i is the initial volume water content which is a constant and θ is the variable water content with time at any point in the horizontal infiltration test.

To calculate the hydraulic gradient between two ends of wet region of the soil column in infiltration problems, the capillary suction at the wetting front is required.

For downward vertical infiltration, it is assumed that the wet region is nearly saturated. Mein and Larson (1971) defined the average capillary suction

at wetting front from S-k relationship in which 'S' is the capillary suction and 'k' is the hydraulic conductivity functions (equation 2).

2. THE GOVERNING EQUATIONS

The Mein & Larson (1971) equation for average suction at wetting front is stated as follows:

$$S_{av} = \frac{\int_{k(\theta_i)}^{k(\theta_s)} S\,dk}{\int_{k(\theta_i)}^{k(\theta_s)} dk} \qquad (2)$$

For the case of a saturated wet region the integrals will be evaluated between θ_i and θ_s [initial and saturated volume water contents respectively].

If the average hydraulic conductivity of the wetted region is assumed as saturated hydraulic conductivity, it is more convenient to use the relative conductivity, k_r instead of the absolute conductivity 'k' as $k_r(\theta) = k(\theta)/k(\theta_s)$.

Substituting this relationship in equation 1, the following will be obtained:

$$S_{av} = \frac{\int_{k_r(\theta_i)}^{1} S \, dk_r(\theta)}{1 - k_r(\theta_i)}$$

(3)

as $k(\theta_i)$ is nearly zero, then:

$$S_{av} = \int_{0}^{1} S \, dk_r$$

(4)

In the case of the absorption of water from a water table, the wet region is unlikely to be saturated. The average degree of saturation for this case can be obtained by a simple absorption test from a water table in the laboratory for the similar compaction rate of the material. For simplicity, in this case $k(\theta_{av})$ or simply k_{av} is considered instead of $k(\theta_s)$ at wet region. To predict the water content at any point in the field an indirect method of measuring the water content such as electrical resistivity method of McInnes, 1972 or Fard and Eslami, 1999 can be used.

By substituting for the capillary suction and the hydraulic conductivity from the inverse method, as stated in equations 5 and 6, an explicit relationship is obtained for S_{av} by a direct integration.

$$S = p(\theta - \theta_i)^{-m}$$

(5)

$$k = k_0 (\theta - \theta_i)^{v}$$

(6)

$$k_r = \frac{k}{k_{av}} = \frac{k_0}{k_{av}} (\theta - \theta_i)^{v}$$

(7)

$$dk_r = \frac{v k_0}{k_{av}} (\theta - \theta_i)^{v-1} d\theta$$

(8)

$$S_{av} = \int_{0}^{1} S \, dk_r = \int_{\theta_i}^{\theta_{av}} p(\theta - \theta_i)^{-m} \cdot \frac{v k_0}{k_{av}} (\theta - \theta_i)^{v-1} =$$

(9)

$$\frac{p v k_0}{k_{av}} \int_{\theta_i}^{\theta_{av}} (\theta - \theta_i)^{v-m-1} d\theta$$

(10)

$$S_{av} = \frac{p v k_0}{k_{av}(v - m)} (\theta_{av} - \theta_i)^{v-m}$$

(11)

The Equation 11 is an explicit relationship to find the S_{av} from the inverse method parameters.

3. RESULTS

In the special case of saturated assumption of the saturated wet region, Equation 11 will have the following form:

$$S_{av} = \frac{p k_0 v}{k_s(v - m)} (\theta_s - \theta_i)^{v-m}$$

(12)

in which 'k_s' and 'θ_s' are the hydraulic conductivity and the water content for the material in saturated condition respectively. All parameters such as p, k_0, v, m, and k_s are determined for the material using the inverse method. (Zayani, 1991 and Fard, 1996).

The results of the inverse method and the corresponding S_{av} values for the following two cases (a and b) for the ASTM upper bound sub base material are given in Table 1 as follows:
a) Fully saturated assumption, for the wet region,
b) Fifty percent saturated assumption for the wet region.

Table 1- The results of the inverse method and the corresponding S_{av} values for two cases (a) and (b), Fard(1996).

$p=0.091$
$k_0=1.188$
$v=4.17$
$m=2.58$
$\theta_s =0.20$
$\theta_i =0.039$

Case (a): $k_{av}=k_s=0.00055$ and $\theta_{av}=\theta_s=0.20$, these data result in $S_{av}=27.85$ cm.
Case (b): $k_{av} =0.0000096$ and $\theta_{av}= 0.10$, these data result in $S_{av}=344$ cm.
(For any degree of saturation we have: $\theta_1/\theta_2=s_1/s_2$ in which s_1 and s_2 are different degrees of saturation and θ_1 and θ_2 are corresponding water contents).

4. CONCLUSIONS

There are a number of instruments to measure soil-water capillary suction in the field or laboratory, but all need particular accuracy and are not simple methods. By combination of the inverse method theory and a simple horizontal infiltration test one can achieve an explicit relationship for capillary suction rather than other hydraulic characteristics as hydraulic conductivity and diffusivity. The method presented here is applicable for any material (soil). What we need, is performing a simple infiltration test to determine the inverse method parameters. To use the results in the field, we need to measure the water content at any point under consideration by an indirect method such as using built in electrodes at different depths and to know the degree of saturation in order to use the explicit relationship obtained.

REFERENCES:

Fard M. Y. Z., 1996; Infiltration of water in road shoulders. PhD thesis, University of Birmingham.

Fard M. Y. Z., and S. Eslami 1999; Four electrode method for indirect measurement of soil moisture, Bull. 21, pp. 35 – 43 Faculty of Engineering, University of Tabriz.

McInnes D. B. 1972; Moisture measurements in pavement materials using electrical resistivity methods. ARRB Vol. 4 No. 10, pp. 69-78.

Mein R. G. and C. L. Larson 1971; Modelling of the infiltration component of the rainfall-runoff process, Bull. 43, pp. 21-29, Water Resour. Res. Centre, Univ. of Minn., Minneapolis.

Zayani K., J. Tarhouni, G. Vachaud and M. Kutilek 1991; An inverse method for estimating soil core water characteristics. J. of Hydrology. Vol. 122, pp. 1-13.

Unsaturated Soils for Asia, Rahardjo, Toll & Leong (eds) © 2000 Taylor & Francis, ISBN 90 5809 139 2

A method for determining hydraulic conductivity and diffusivity of unsaturated soils

M.Y. Fard

Department of Civil Engineering, University of Tabriz, Iran

ABSTRACT:The study of moisture flow in soils in unsaturated condition is based on the so-called Richard's second order partial differential equation in which the coefficients of the equation are: hydraulic conductivity and soil water diffusivity. The main difficulty of solving the equation is that these coefficients are unknown functions of water content. There are a few methods of measuring these parameters in laboratory which need special apparatus and accurate instruments which are not available everywhere. In the present paper a new method of determining the hydraulic conductivity and diffusivity of soils has been developed using the indirect measurement of soil moisture called electrical resistivity method and cumulative infiltration in a vertical infiltration test at time intervals which can be carried out in any soil laboratory. The results are compared by Zayani's (1991) and Fard's (1996) inverse method of determining soil water hydraulic characteristics and there is a very good agreement.

1. INTRODUCTION

The governing equation of unsaturated flow is as follows (Child, 1969, Hillel, 1971, Hanks and Ashcroft, 1980, Marshall and Holmes, 1988, and Koorevaar et al, 1991):

$$\partial\theta / \partial t = \nabla K \nabla H \qquad (1)$$

For horizontal infiltration and vertical infiltration (Philip, 1957), the equation is written in diffusivity form as:

$$\partial\theta / \partial t = \partial/\partial x(K\partial H / \partial x) = \partial/\partial x(K\partial H_m/\partial x) \qquad (2)$$
$$\partial\theta / \partial t = \partial/\partial z(K\partial H / \partial z) = \partial/\partial z(K\partial H_m/\partial z) + \partial K /\partial z \qquad (3)$$

in which :
x = horizontal direction coordinate,
z = vertical direction coordinate,
t = time,
K = hydraulic conductivity,
H = total suction potential,
H_m = matric suction,
θ = volumetric water content,
∇ = vectorial differentiation of hydraulic head in the direction of flow.

There are different methods of determining and measuring the hydraulic conductivity and diffusivity which some of them are summarised as follows:

Ahuja et al, 1988, suggested a method for determining hydraulic characteristics by a laboratory method using a tensiometer data. Clothier and White, 1981, have shown a field method to measure sorptivity and soil-water diffusivity. Meyer and Warrick, 1990, have shown an analytical expression of soil water diffusivity from horizontal infiltration experiments which is a special case of study. Mualem, 1976, has worked on a model to predict the hydraulic conductivity of unsaturated soils. Zayani et al, 1991, and Fard, 1996 have worked on an inverse method of determination of soil characteristics.

Using the indirect method of soil moisture measurement of Fard and Eslami, 1999, in combination with the inverse method just mentioned makes it possible to find a simple laboratory method for determining the hydraulic conductivity of unsaturated soils which is presented in this paper.

2. THE PROCEDURE AND THE GOVERNING EQUATIONS

Infiltration of water in a soil sample under a ponding water of depth (h) is studied (Fig. 1). During infiltration, water passes through the soil and drains out into a container. the method is explained by the following steps.

I) The moisture contents are measured at different depths at time intervals ΔT, by a number of moisture meters (Resistivity Meters), installed in the soil (at equal spaces of ΔZ) during compaction of the layers in the model). They have previously calibrated for the material at desired compaction for different moisture contents (McInnes, 1972 and Fard, 1996).

II) The suction of the material is measured in a separate test using a tensiometer (or pressure plate apparatus) for different moistures at the same compaction conditions. In other words a calibration table of suction against water content should already be prepared (Hillel, 1971).

III) The total hydraulic potential (H) for any point at any depth is determined:

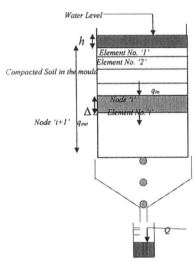

Fig. 1- Schematic of the laboratory model for measurement of hydraulic conductivity

$$H = H_p + H_m + H_z \qquad (4)$$

where:
H_p = Pressure potential which assumed equal to zero for shallow ponding and for unsaturated condition,
H_m = Matric potential which is known by measuring the moisture content of the point under consideration and using calibration table (No. II above), and

H_z = Gravitational potential which can be determined by knowing the position of the point and is equal to 'Z' if the soil water interface is considered as datum line.

IV) At any time for any point or any element with thickness ΔZ between two respective points dH/dZ is determined by finite differences as:

$$\Delta H / \Delta Z = (H_{i+1} - H_i)/(Z_{i+1} - Z_i) \qquad (5)$$

in fact dH/dZ can be determined by derivation rules for any one point if the function of H with respect to Z is available. Here, because of discrete measurements, $\Delta H / \Delta Z$ is defined by finite differences instead of dH/dZ. Equation 5 determines the dH/dZ value for the element 'i' or in other words for the midpoint of the element.

V) Darcy's Law may be used for any element as:

$$q = kdH/dZ$$
$$k = q/(dH/dZ) = q \, \Delta H / \Delta Z \qquad (6)$$

where:
q = Flux or amount of water passing through a unit area per unit time with dimension of (LT^{-1}).
k = Hydraulic conductivity with dimension of (LT^{-1}).
dH/dZ = Potential gradient in the direction of flow.
Darcy's Law states that the flux is directly related to potential gradient and the relationship coefficient is the hydraulic conductivity 'k'.

VI) Value of 'k' is calculated by Darcy's Law at any known time for any element by the knowledge of flux 'q' which is explained below.

3. DETERMINATION OF 'q'

VII) Assume that 'h' is the depth of ponding water as shown in Fig. 1 and 'Q' is the volume of outflow from soil sample. The water contents $\theta_1, \theta_2, \theta_3, ... \theta_n$ for each element are measured at time 'T'. The same parameters as h', $\theta'_1, \theta'_2, \theta'_3, ... \theta'_n$ and Q' can be measured at time $T' = T + \Delta T$. Theta primes (θ') are water contents at different points at any time.
Now the flux inflow and outflow for any element may be calculated.
The element water content is defined as average of two water contents at two boundary points of the element.

VIII) The amount of water held in any element can be calculated by the basic definitions as follows:

$$\theta_1 = V_{w1}/V_T = V_{w1}/(A. \Delta Z) \qquad (7)$$
$$\theta'_1 = V_{w2}/(A. \Delta Z) \qquad (8)$$
$$\theta'_1 - \theta_1 = \Delta \theta = (V_{w2} - V_{w1})/(A. \Delta Z) \qquad (9)$$
$$V_{w2} - V_{w1} = A. \Delta \theta. \Delta Z \qquad (10)$$

where:

V_{w1} = Volume of water in the element at time 'T'.

V_{w2} = Volume of water in the element at time 'T+ΔT'.

V_T = Total volume of the element.

By definition:

$$q = (V_{w2} - V_{w1})/(A.\Delta T) = \Delta\theta.\Delta Z/\Delta T = q_{held} \quad (11)$$

IX) At any time interval the flux (q) is calculated as described below. The $q_{i\,in}$ and $q_{i\,out}$ denote the inflow and outflow flux for element 'i' respectively. $q_{i\,held}$ denotes the volume of water held in the element 'i' per unit area per unit time.

4. THE CONTINUITY EQUATION

From the initial definition of flux, $q_{i\,in}$ and $q_{i\,out}$ can be described as follows:

$q_{1\,in} = (h-h')/\Delta T = q_{1\,in} \qquad \Rightarrow$

$q_{1\,held} = (\theta'_1 - \theta_1)\Delta Z\Delta/T \qquad$ (Because of continuity)

$q_{1\,out} = q_{1\,in} - q_{1\,held} = q_{2\,in} \qquad \Rightarrow$

$q_{2\,held} = (\theta'_2 - \theta_2)\,\Delta Z/\Delta T$

$q_{2\,out} = q_{2\,in} - q_{2\,held} = q_{3\,in} \qquad \Rightarrow$

$q_{3\,held} = (\theta'_3 - \theta_3)\,\Delta Z/\Delta T$

$\qquad\qquad ... \qquad\qquad\qquad \Rightarrow$

$q_{n-1\,out} = q_{n-1\,in} - q_{n-1\,held} = q_{n\,in} \qquad \Rightarrow$

$q_{n\,held} = (\theta'_n - \theta_n)\,\Delta Z/\Delta T$

$q_{n\,out} = q_{n\,in} - q_{n\,held} = (Q-Q')/(A.\Delta T) \qquad (12)$

By summing up all equations of the first column and cancelling the similar parameters from the two sides of the equations, the continuity equation is obtained as follows:

$$(h-h')/\Delta T - \Sigma_{i=1}^{n} q_{i\,held} = (Q-Q')/(A.\Delta T) \quad (13)$$

$$\Sigma_{i=1}^{n} q_{i\,held} = (h-h')/\Delta T - (Q-Q')/(A.\Delta T) \quad (14)$$

$$\Delta Z/\Delta T)\Sigma_{i=1}^{n}(\theta'_n - \theta_n) = (h-h')/\Delta T - (Q-Q')/(A.\Delta T) \quad (15)$$

This is the equation for conservation of water.

This relationship is valid if there is no evaporation. If there is any evaporation as 'E' per unit time the relationship will be as follows:

$$\Delta Z/\Delta T)(\Sigma_{i=1}^{n}\theta' - \Sigma_{i=1}^{n}\theta) = (h-h')/\Delta T - (Q-Q')/(A.\Delta T) - E \quad (16)$$

- If there is no outflow, then $(Q-Q') = 0$
- If there is no sharp change in water content for any element at the time interval ΔT, it means that the wetting front has not yet reached that

element. By comparing water content values at two different times, the position of wetting front at any time can be predicted.

X) As described in IX, $q_{i\,in}$ and $q_{i\,out}$ are obtained for any element at any time starting from the first element to the last one respectively. By using $q_{i\,in}$ and $q_{i\,out}$, 'k' is determined. The dH/dZ will be evaluated for two respective points of the element.

$$q_{i\,in} = q_{(i-1)\,out} = q_{(i-1)\,in} - q_{(i-1)\,held} \quad (17)$$

$$q_{(i-1)\,held} = (\theta'_{i-1} - \theta_{i-1})\,\Delta Z/\Delta T \quad (18)$$

5. CALCULATING THE DIFFUSIVITY

XI) By having 'k' and suction calibration against water content, diffusivity 'D' for any element corresponding to a known water content can be determined as follows:

$$D = kdH/d\theta = k\Delta H/\Delta\theta = k(H_2-H_1)/(\theta_2-\theta_1) = q\Delta Z/\Delta\theta \quad (19)$$

in which:

ΔH = Difference between total hydraulic potentials at two ends of an element at any known time.

ΔZ = Thickness of element, and

$\Delta\theta$ = Difference between water contents of two ends of the element at the same time.

XII) As described above, using moisture meters and suction meters placed in the material during compaction it is possible to characterise the curves of H_m, k, and D, against water content.

6. VALIDATION OF THE METHOD

By performing vertical infiltration tests on a larger scale under a pond of water and using moisture meters at different depths of the compacted soil, the unsaturated hydraulic conductivity for the soil has been calculated using the above mentioned method and the results are illustrated in Fig. 2. As can be noticed, the results for the unsaturated hydraulic conductivity function for different water contents using the inverse method (Zayani et al, 1991 and Fard, 1996) is shown for the same material as well. The comparison of the results for unsaturated hydraulic conductivity from the two different approaches shows some disagreement in the results. By taking into account that hydraulic conductivity is a very sensitive function of water content and the material is not a homogenous material, this type of agreement seems very good.

For comparison of these two methods of measurement, the method recommended by Altman and Bland, 1983, has been used This method of comparison is a simple yet powerful graphical and statistical technique for the comparison of methods of measurements which deal with the fundamental issues which are: i) is there any relative bias between the methods; and ii) what additional variability is there? In this method of comparison, the difference of measurements are shown against the mean of measurements.

Relative bias is any systematic difference between the methods such as one consistently giving a higher value than the other. The additional variability is any random differences possibly arising from causes such as random behaviour of components of the measuring system or interaction between measuring devices and external factors. As can be noted from Fig. 3, there is not relative bias between the methods because the data points are around the bisecting line of the first quarter of the co-ordinate axes. Therefore the unsaturated hydraulic conductivity of soils can be measured by the method suggested here by the author.

7. CONCLUSIONS

Using tensiometers and moisture meters in laboratory and performing a vertical infiltration test under a pond of water, a practical method to obtain the hydraulic conductivity and diffusivity of unsaturated soils for different water contents is introduced in this paper. By comparison of the results of two different methods, the present method has been validated. It can be used for any material.

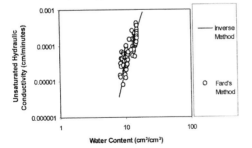

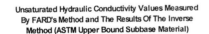

Fig. 2- Comparison of the results of unsaturated hydraulic conductivity obtained from the author's method and the inverse method for the upper bound of the ASTM subbase material

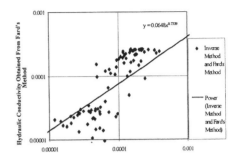

Fig. 3- Comparison of hydraulic conductivity for 90 data values for ASTM upper bound subbase material for similar water contents obtained from two methods; a) the inverse method, b) The author's method.

What should be noted, is that the field condition such as compaction rate must be produced in the laboratory to achieve good results for the hydraulic conductivity and diffusivity.

REFERENCES:

AHUJA, L. R., J. D. ROSS, R. R BRUCE,. and D. K. CASSEL, 1988; Determining unsaturated hydraulic conductivity from tensiometer data alone. Soil Sc. Soc. Am. J., Vol. 52, pp. 27-34.

ALTMAN, D. G. And J. M. Bland, 1983; Measurements in medicine: the analysis of method comparison studies. The statistician, Vol. 32, pp. 307-317.

CHILDS E. C. 1969; An introduction to the physical basis of soil water phenomena. Published by Wiley, Chichester.

CLOTHIER B. E. and I. WHITE 1981; Measurement of sorptivity and soil water diffusivity in the field. J. Soil Sc. Soc. Am. Vol. 45, pp. 241-245.

FARD M. Y. Z., 1996; Infiltration of water in road shoulders. PhD thesis, University of Birmingham.

FARD M. Y. Z., and S. ESLAMI, 1999; Four electrode method for indirect measurement of soil moisture, Bull. 21, pp. 35 - 42 .Faculty of Engineering, University of Tabriz.

HANKS R. J. and G. L. ASHCROFT 1980; Applied soil physics. Berlin, New York, Springer, Verlag.

HILLEL D. 1971; Soil and water physical principles and processes. Academic press New York and London.

KOOREVAAR P, G. MENELIK and C. DIRKSEN C. 1991; Elements of soil physics. Elsevier, 228p.

MARSHALL T. J. and J. W. HOLMES 1988; Soil Physics, Cambridge University press.

MCINNES D. B. 1972 ; Moisture measurements in pavement materials using electrical resistivity methods. ARRB Vol. 4 No. 10, pp. 69-78. Dec. 1972.

MEYER J. J. and A. W. WARRICK 1990; Analytical expression for soil water diffusivity derived from horizontal infiltration experiments. Soil Science Society of America, Vol. 54, pp. 1547-1552.

MUALEM 1976; A new model for predicting the hydraulic conductivity of unsaturated porous media. Water Resour. Res. 12, pp. 513-522.

PHILIP J. R. 1957; The theory of Infiltration 1, the infiltration equation and its solution. Journal of Soil Sc. Vol. 83, pp. 345-357.

ZAYANI K., J. TARHOUNI, G. VACHAUD and M. KUTILEK 1991; An inverse method for estimating soil core water characteristics. J. of Hydrology. Vol. 122, pp. 1-13.

Unsaturated Soils for Asia, Rahardjo, Toll & Leong (eds) © 2000 Taylor & Francis, ISBN 90 5809 139 2

A new laboratory method for the measurement of unsaturated coefficients of permeability of soils

J.K.-M. Gan
M.D. Haug and Associates Limited, Saskatoon, Sask., Canada

D.G. Fredlund
Department of Civil Engineering, University of Saskatchewan, Saskatoon, Sask., Canada

ABSTRACT: A new laboratory method was recently developed at the University of Saskatchewan, Saskatoon, for the direct measurement of the unsaturated coefficients of permeability. The method was verified in a testing program involving residual and saprolitic soils from Hong Kong. The method makes use of the pressure plate concept for the direct control of the pore-air and pore-water pressures in the soil specimen. The test procedure is similar to the "inflow equal outflow" method. The coefficients of permeability measured ranged from 10^{-6} m/s to 10^{-10} m/s. The choice of porous disks had to have the right relationship between the air entry value and the saturated coefficient of permeability. Experimental results are presented. The experimental results were compared with predicted permeability functions obtained from soil-water characteristic curves along with the saturated coefficients of permeability.

1 INTRODUCTION

The permeability function is often required in the numerical modelling of water and solute transport in unsaturated soil problems. Although techniques are available for the prediction of the permeability function from soil-water characteristic curves together with the saturated coefficient of permeability, measurement of the permeability function is required for confirmation of the correct function.

The coefficient of permeability can decrease by several orders of magnitude as the degree of saturation of the soil decreases (or as the suction of the soil is increased). The unsaturated coefficients of permeability corresponding to changes in soil suction, are difficult to measure because their variation can be over several orders of magnitude. The variation in the coefficient of permeability over several orders of magnitude makes it difficult to design a single apparatus that can be used for testing all soils. In addition, when dealing with residual and saprolitic soils, the highly heterogeneous nature of the soils adds further complications to the laboratory measurement.

2 NEW DIRECT LABORATORY METHOD

A new methodology was developed to provide a direct measurement of unsaturated coefficients of permeability with special attention given to the nature of residual and saprolitic soils from Hong Kong. The methodology is likely suitable for most soils.

The method makes use of the pressure plate concept for the direct control of the pore-air and pore-water pressures in the soil specimen. The pore-water pressure was controlled at both ends of the soil specimen with the assistance of high (or low) air entry porous disks. The air pressure was controlled through an air inlet.

It was not possible to install small tensiometer tips at two points in the soil specimen for the measurement of pore-water pressures due to the coarse and fragile nature of the soils. Therefore, it became important to ensure that the coefficients of permeability of the soil were always lower than the adjacent air entry porous disks.

The set up in Figure 1 was used for determining the coefficient of permeability at suction values ranging from 10 kPa to 100 kPa. The air entry values of the soils being tested were low, generally less than 10 kPa. The soils would be unsaturated and their coefficients of permeability were anticipated to be low. Consequently, 100 kPa high air entry ceramic disks were used. The 100 kPa high air entry ceramic had a saturated coefficient of permeability of approximately 1×10^{-7} m/s.

The set up in Figure 2 was used for measuring the coefficient of permeability at suction values less than 1.5 kPa. At low suctions, the soil has its highest coefficient of permeability and highly permeable porous disks were required. On the basis of hydraulic conductivity and air entry tests conducted on steel, brass and corundum porous disks, porous brass disks were selected. The brass porous disks have an air

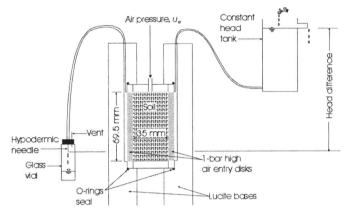

Figure 1 - Permeameter set up for matric suctions greater than 10 kPa

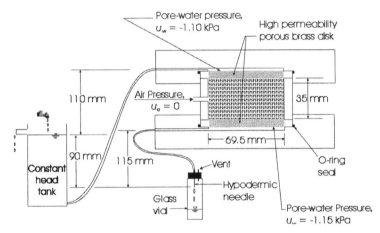

Figure 2 - Permeameter set up for low matric suctions of less than 1.5 kPa.

entry value of about 1.5 kPa and a coefficient of permeability of 4.5×10^{-5} m/s.

3 TEST PROCEDURE

The soil specimen was trimmed into the transparent Lucite specimen ring holder. The specimen ring holder has a height of 35mm and a diameter of 69 mm.

The soil specimen in the permeameter was saturated prior to performing the permeability tests. Saturation of the soil specimen was achieved by maintaining a positive head of water at both ends of specimen. The air inlet was open to the atmosphere so water could rise into the air inlet, driving out air from the specimen.

In tests conducted using the set up in Figure 1, an air pressure was applied to the air inlet. The water head at one end of the specimen was lowered in order to maintain a head gradient across the specimen.

The difference between the applied air pressure and the average water pressure gave the average matric suction in the soil specimen when equilibrium was attained.

In tests conducted using the set up in Figure 2, the air pressure was maintained at atmospheric condition by leaving the air inlet to the permeameter open to the atmosphere. Negative water pressure was applied to both ends of the specimen. The head at the outflow end was maintained at a slightly lower value in order to maintain a positive head gradient across the specimen for flow to occur.

Initially, water would flow from the soil specimen in response to both the applied air pressure and the total head gradient across the specimen. Once a matric suction corresponding to the applied air pressure and the applied water pressures was established, water would flow only under the influence of the total head difference across the specimen. When the flow rate from the specimen became constant with time, the inflow to the specimen was assumed to be

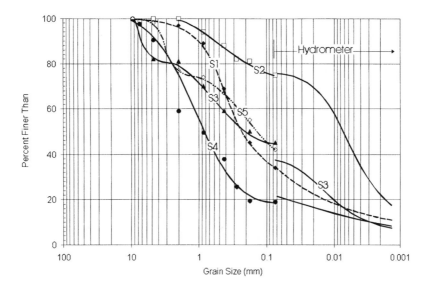

Figure 3 - Grain size distribution curves for the five saprolitic soils.

equal to the outflow from the specimen. The coefficient of permeability corresponding to the applied matric suction was then computed from the steady outflow rate.

Tests were also conducted under positive porewater pressures (i.e., negative matric suctions and the specimens were saturated).

4 SOILS TESTED

Measurements of the permeability function were conducted on five saprolitic soils from Hong Kong. Only one specimen of each soil was tested. The soils were highly heterogeneous and specimen preparation was difficult. Grain size distribution curves for the five saprolitic soils are presented in Figure 3. The grain size curves were each obtained using a separate specimen from the same block sample as the specimen used for the permeability test. Color description and specific gravity values for the five saprolitic soils are presented in Table 1.

In an earlier research program (Gan and Fredlund, 1998) saturated permeability tests and soil-water characteristic tests were conducted on the above five saprolitic soils. These tests are not described in this paper but the results are presented. The saturated permeability tests were conducted using a double ring permeameter under constant head conditions. The saturated coefficients of permeability for the five saprolitic soils are presented in Table 2. The soil-water characteristic curves are presented in Figure 4. A total of 13 soil-water characteristic curves were obtained.

Table 1. Color description and specific gravity values

Soil Identificaton	Description	Specific gravity
Soil S1	Light whitish yellow spotted light brown	2.61
Soil S2	Reddish brown spotted with white	2.67
Soil S3	Light whitish brown striped with black	2.66
Soil S4	Greyish brown mottled with light yellow	2.63
Soil S5	Light greyish brown mottled orangish yellow	2.58

Table 2 – Coefficients of saturated permeability.

Soil	This test program		Previous test program (Gan and Fredlund, 1998)	
	k_{sat} (m/s)	gradient, i	k_{sat} (m/s)	gradient, i
S1	2.84E-09	10.6	2.05E-05	2.2
	6.26E-08	3.3	2.66E-06	4.4
S2	7.66E-09	15.8	2.37E-05	3.4
			3.36E-05	4.3
S3	1.11E-08	15.7	8.23E-06	1.1
	1.20E-06	1.43	5.83E-06	1.5
S4	2.29E-06	1.43	3.92E-05	4.3
	8.45E-09	15.7	2.23E-05	2.1
S5	1.16E-08	1.43	8.08E-06	1.2
			8.60E-05	11.5

5 RESULTS AND DISCUSSION

The coefficients of permeability as a function of matric suction for the five saprolitic soils are presented in Figure 5. The highest coefficient of permeability measured was 2.29×10^{-6} m/s (i.e., k_{sat} for S4) which is more than one order of magnitude lower than the saturated coefficient of permeability of the porous brass disk. The highest coefficient of permeability measured using setup No. 1 was 10^{-8} m/s (i.e., S5 and S3 at a matric suction of 10 kPa) which was more than one order of magnitude lower than the saturated coefficient of permeability of the ceramic disks. Hence, the ceramic disks and the porous brass disks should not have presented any significant impedance effects in any of the tests. The lowest coefficient of permeability measured was

about 10^{-10} m/s (i.e., S2 at a matric suction of 100 kPa). The coefficients of permeability measured span a range of 4 orders of magnitude.

In general, the shapes of the coefficient of permeability functions (Figure 5) loosely resemble the shapes of the soil-water characteristic curves (Figure 4).

The permeability functions were also predicted for each of the five saprolitic soils from the soil-water characteristic curves and the corresponding

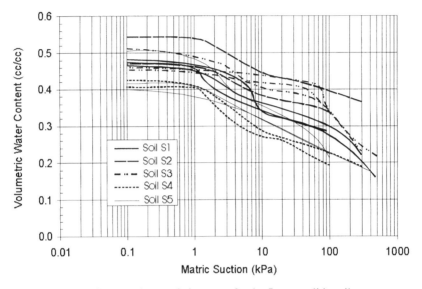

Figure 4 – Soil-water characteristic curves for the five saprolitic soils.

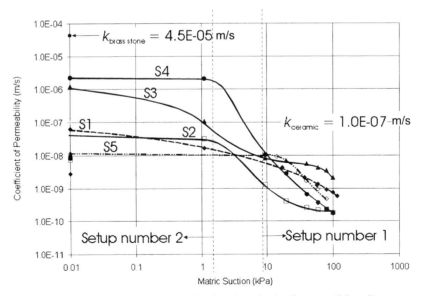

Figure 5 – Experimental permeability functions for the five saprolitic soils

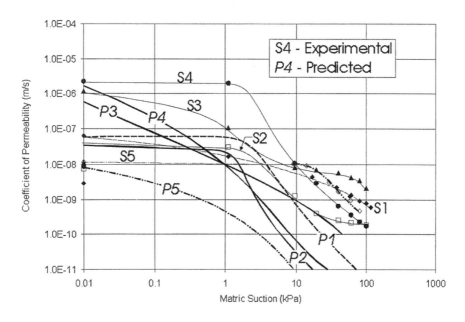

Figure 6 – Comparison of predicted permeability functions with experimental values.

saturated coefficient of permeability. The saturated coefficients of permeability obtained using setup No. 2 in this test program were used in the predictions. The coefficients of saturated permeability from Gan and Fredlund (1998) which were obtained using a double ring permeameter under constant head conditions were significantly higher than the values obtained using setup No. 2 (see values presented in Table 2). The high values from Gan and Fredlund (1998) were due in large part to significant side-wall leakage. Trimming of heterogeneous saprolitic soils was difficult and the resulting specimens often had rough and jagged sides. As a result, the specimens do not fit snugly in the permeameter rings. Side-wall leakage can be largely eliminated by the application of a small matric suction less than the air entry value of soil, in conducting the permeability test using the setup in Figure 2.

The theory for the prediction of the permeability functions has been presented by Fredlund, Xing and Huang (1994). The soil-water characteristic curve is first best-fit using a non-linear regression analysis. Then the procedure involves integrating along the soil-water characteristic curve, starting from saturated conditions. The end result is a series of computed points that can be connected to form the predicted permeability function. The theory has been programmed into the computer software package called SoilVision (Fredlund, 1996),

The predicted permeability functions for the five saprolitic soils are presented in Figure 6, together with the experimental data for comparison. The pre-

dicted permeability functions show considerable variation from the experimental permeability functions. Possible explanations for the deviations would include the followings:
1.) the saprolitic soils from Hong Kong are highly heterogeneous.
2.) the highly heterogeneous soils have a variation in the soil-water characteristic curves (Figure 4).

6 CONCLUSIONS

One of the main difficulties encountered in the measurement of the permeability function over a wide range of matric suctions is the availability of porous disks with the right combination of air entry value and coefficient of permeability. Two conditions related to the porous disks must always be satisfied:
1.) the ceramic disk must remain saturated, and
2.) the coefficient of saturated permeability of the porous disks should be about one order of magnitude greater than the coefficient of permeability of the soil being tested.

The experimental permeability functions have the characteristic "S" shape similar to the soil-water characteristic curves. The experimental results thus appear to be reasonable.

The predicted permeability functions deviate from the experimentally measured functions probably as a result of the highly heterogeneous nature of the soils.

ACKNOWLEDGEMENTS

The authors would like to acknowledge the financial support provided by the Geotechnical Engineering Office, Government of Hong Kong. This study allowed for an opportunity to design a new permeability testing apparatus that could be used on a residual soil that is difficult to test for unsaturated coefficient of permeability.

REFERENCES

Fredlund, M.D. 1996. *SoilVision User's Guide, a Knowledge-Based System for Geotechnical Engineering*, Version, 1.20, SoilVision Systems Ltd., Saskatoon, Sask., Canada.

Fredlund, D.G., Xing, A. and Huang, S.Y. 1994. Predicting the permeability function for unsaturated soils using the soil-water characteristic curve. *Canadian Geotechnical Journal, Vol. 31, No. 3, pp.*521-532.

Gan, J.K.-M. and Fredlund, D.G. 1998. Study of the application of soil-water characteristic curves and permeability functions to slope stability, Final report on Part 1 – Permeability and soil-water characteristic curve tests, and the computations of permeability functions, *Report for Geotechnical Engineering Office, Civil Engineering Department, Hong Kong, Oct. 16.*

Unsaturated Soils for Asia, Rahardjo, Toll & Leong (eds) © 2000 Taylor & Francis, ISBN 90 5809 139 2

Water content and soil suction in the capillary zone

W. N. Houston & S. L. Houston
Arizona State University, Tempe, Ariz., USA

D. Al-Samahiji
University of Bahrain

ABSTRACT: Soil Columns were constructed for capillary rise measurements for 16 soils from Bahrain and 3 soils from the Phoenix, Arizona, metropolitan area. These included 5 sands that were non-plastic, 13 sandy silts and clays of low to moderate plasticity, and one clay of high plasticity. The capillary rise measurements showed that, for all soils, the initial rate of rise was high, then the rate dropped dramatically and slowly decayed with time. However, the rate of rise was still easily measurable after 100 days or more. Assuming the directly measured values of degree of saturation are reliable, the test results show that the degree of saturation rather quickly drops below 90% and eventually may drop to 40 or 50% with increasing distances above the groundwater table. Soil water characteristic curves were measured for each test soil. The data indicate that the soil suction is much higher than the value calculated by y times γ_w, where y is the distance above the groundwater table.

1. INTRODUCTION

It is well established that the mechanical properties of soil are water content dependent. Although soil properties such as shear strength and compressibility are normally the most inferior when the soil is wettest, it is also typically true that significant changes in water content, particularly from dry to wet, are the greatest source of engineering problems. Urban development results in changes in both surface and subsurface water regimes. Infrastructure construction, such as the building of highways tends to modify surficial and groundwater flows. Construction associated with urbanization normally results in local or regional groundwater table rise and increases in water content of the near-surface soil, especially in arid climate regions. Wetting of initially dry soils may be due to downward infiltration from a surface source, rising groundwater table, or capillary rise.

Whatever the source of wetting, it is important for geotechnical engineers to be able to anticipate these changes in water content and the associated changes in mechanical properties of the soil. Fredlund et al. (1995) and Fredlund and Rahardjo (1993) have established that unsaturated soil behavior is best characterized in terms of two stress state variables: the net normal stress, $\sigma_n - u_a$, and the soil suction, $u_a - u_w$. Further, the soil water characteristic (SWCC) links the water content to the soil suction.

The vadose zone is very important for many engineering applications. In fact, proper characteriza-tion of this zone is often the key to the development of satisfactory solutions to engineering problems. Although capillary rise and the capillary zone above the groundwater table is only part of the total picture of water content change in the unsaturated materials, it is an important aspect. This paper is directed toward adding to our knowledge base of water contents and soil suctions in the capillary rise zone.

2. COMMON ASSUMPTIONS RELATIVE TO THE CAPILLARY REGION

Upward moisture movement by capillary rise is a potential problem long recognized by engineers. The deterioration of mechanical properties of wetted soils, capillary rise through concrete floors and slabs, and the exacerbation of frost heave problems (Padilla, et al., 1997) are all very familiar issues for soil foundation engineers. The most commonly applied solution is the placement of open-graded gravel beneath the slab or somewhere in the soil profile as a capillary break (Day, 1996). Unfortunately, this solution is not as effective as is widely assumed. Although moisture movement may be slowed by the capillary break, it continues to occur via vapor transfer and the material above the gravel does get wetted.

The first step in assessing the magnitude of this problem is to estimate the probable future position of the groundwater table. The second step is to make an estimate of the height of capillary rise, at least semi-quantitatively. As examination of most any text on soil mechanics will show, it is commonly assumed that capillary rise results in a zone of capillary saturation above the groundwater table (Sowers, 1979). Even though the water in this zone is in tension it is assumed that the degree of saturation is 100%. It is further assumed that the soil suction develops at the rate given by y times γ_w, where y is the distance above the groundwater table and γ_w is the unit weight of water, as indicated graphically in Figure 1. Some authors project a fairly rapid drop-off in the degree of saturation, S, above the zone of capillary saturation. Others simply make the simplifying assumption that the wetted material above the zone of capillary saturation is "dry". The capillary rise measurements by the authors that form the basis for this paper will be presented in the next section. These measurement results are qualitatively consistent with the commonly employed assumptions cited above, but differ significantly quantitatively. We do not suggest that these commonly employed assumptions relative to the capillary zone are inappropriate for all applications. For example, if the objective is to evaluate the total vertical normal stress and the effective stress at points beneath the groundwater table, then these assumptions are typically adequate. If, however, the focus of attention is on the degree of saturation and the soil suction in the capillary zone, then the differences between these common assumptions and the actual degree of saturation (soil suction) could be important.

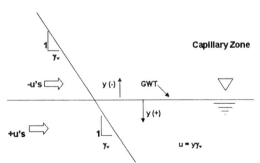

Fig. 1- Schematic of Water Pressure Variation Above and Below GWT

3. CAPILLARY RISE MEASUREMENTS

3.1 *Materials Used for Measurements*

Most of the soils tested were transported from the state of Bahrain in the Persian Gulf to Arizona State University by the second author of this paper. These soils consisted primarily of clayey, silty, and fine-sand materials. The percent finer than the no. 200 sieve ranged from 39 to 84 and averaged about 65%. Values of PI ranged from 0 to 16 and averaged about 8 percent. Unified soil classification symbols included SP, SM, ML, CL, and MH.

In addition to the 16 Bahrainian soils tested, three soils local to Phoenix, Arizona, were tested. The first of these Phoenix soils was a clean, SP sand derived from the Salt River, with a PI of zero. The second was a silt referred to as Price Club Silt, classified as ML with 54% passing the no. 200 sieve and a PI of 4. The third soil, Fountain Hills Clay, classified as a CH, had 94% finer than the no. 200 sieve, and a PI of 41.

3.2 *Test Specimen Preparation*

The Bahrainian soils had to be heated to 225°F, due to import regulations. This caused clumping and aggregation, of course, and required break-up with a mortar and pestle. The index properties cited above were measured after this processing. All capillary rise measurements were made by compacting each soil in a 7.5 cm inside diameter clear plastic tube 90 cm high, using 5 cm lifts. Placement dry densities ranged from 1.1 to 1.7 Mg/m3 and compaction water contents from 0.2 to 15%. The plastic tubes had ports on the side at 7.5 cm vertical spacing, for obtaining water content samples or measurements of suction. For this test series water content samples were taken and SWCCs were used to infer soil suction. A filter material was placed at the base of the soil column and a pot of water was used to maintain constant groundwater table level at about 1 cm above the bottom of the soil column. Plastic wrap was placed at the top of the soil column so as to allow the passage of air, but severely retard evaporation.

3.3 *Results*

For each soil column the height of capillary rise was recorded versus time, until the water reached the top of the column or until the rate of change was very small. The results for the Phoenix sand, silt, and clay are shown in Figure 2. The relative positions of the curves are as expected. However, it is remarkable that, even after 150,000 minutes (104 days), the curves did not truly flatten out. All three soils show a steep initial rate, followed by a substantial slow down, and then a rate that decreases very gradually with time.

The results for the 16 Bahranian soils are summarized in Figure 3. These soils were divided into two groups: non-plastic and low-moderate plasticity, with the curves for the non-plastic soils falling in the shaded region and the curves for the low-moderate plasticity soils generally falling in the unshaded region indicated. The curves from Figure 2 are consistent with those for the Bahrainian soils of Figure 3 in the following respect. The clean sand is non-plastic and its curve falls in the shaded region of Figure 3. The silt is of low plasticity and it falls in the low-moderate region of figure 3. The clay is of high plasticity and its curve falls on or just above the upper bound of the low-moderate plasticity soils of Figure 3.

4 COMPARISON OF COLUMN RESULTS WITH SWCCS AND DISCUSSION OF RESULTS

The SWCC was determined for each test soil using the filter paper method. The SWCC was represented as degree of saturation versus the log of matric suction, with suction expressed in cm of water. Results for two of the Bahrain soils are presented in Figures 4 and 5.

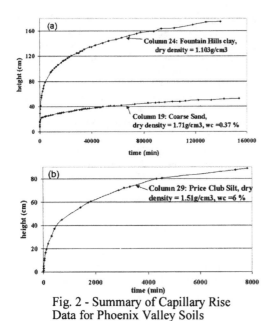

Fig. 2 - Summary of Capillary Rise Data for Phoenix Valley Soils

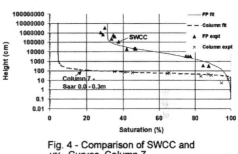

Fig. 4 - Comparison of SWCC and yγw Curves, Column 7

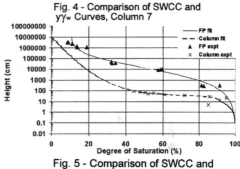

Fig. 5 - Comparison of SWCC and yγw Curves, Column 8

The triangles are measured data points for the SWCC determined by the filter paper method. The crossed data points represent directly measured degrees of saturation from the column experiment. If the straight line extrapolation depicted in Figure 1 were used to get soil suction values in the capillary zone, then soil suction would be given by y times γ_w, if units of stress were to be obtained for soil suction. However, if the desired units were in height of water column then the soil suction would simply be "y", the height above the groundwater table. This second option was chosen for presentation in Figures 4 and 5. Thus the crossed data points, the column experiments, correspond to the height in centimeters above the groundwater table for various points along the column at the end of the capillary rise measurements. At these various points a sample was taken and degree of saturation measured. If the extrapola-

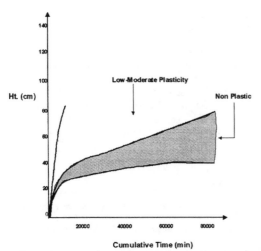

Fig. 3 - Summary of Capillary Rise Data for Bahrain Soils

tion depicted in Figure 1 were to give valid values of soil suction, then these values of height, corresponding to the crossed data points, would be soil suction values in centimeters of water.

There are at least two interesting observations coming from Figures 4 and 5. First, the column experiment data show that the degree of saturation drops below 90% rather quickly, within 10 to 20 cm above the groundwater table, and continue to diminish down to 40 or 50% within about 1 m of the groundwater table. Secondly, at any degree of saturation below about 90% the soil suction inferred from the SWCC is much, much higher than the value depicted in Figure 1 (corresponding to the crossed data points) – perhaps an order of magnitude more in the 70 to 90% saturation range and maybe two orders of magnitude in the 40 to 50% saturation range.

In searching for an explanation for the large differences between the two sets of curves of Figures 4 and 5, the first thought that comes to mind is "Are the SWCCs themselves reliable?" In all cases the experimental data were fitted with a curve using a fitting algorithm from Soil Vision (Fredlund, et al., 1998). Obviously some experimental error is present in these curves, but the experimental error seems far too small to account for the large differences between the curves. Another possible explanation that was considered relates to the boundary condition at the top of the columns. The plastic covering used was intended to allow escape of air, so that pore air pressure would not build up during capillary rise, and at the same time severely retard evaporation at the top. If evaporation had actually been quite significant then a moderately high flux rate at the top of the columns would help to explain the somewhat surprisingly high suctions in the capillary rise zone. However, upon reconsideration of the boundary conditions at the top of the columns, it was decided that no significant rates of evaporation were possible.

Nevertheless, flow conditions were transient, even at the end of the tests. This conclusion is inescapable because of the large upward gradient in potential required by the soil suction values derived from the SWCC for points in the capillary rise zone. In addition, the easily measurable rates of capillary rise observed at the ends of the tests (see Figure 2) show that flow conditions were transient, even after more than 100 days.

For the laboratory conditions used in this study the upper and lower boundary conditions were quite fixed and constant. Yet, after more than 100 days large upward gradients existed in the columns and capillary rise continued. By contrast, in most prototype situations: (a) the groundwater table may move up and down with time, (b) uncovered ground surfaces experience changes in temperature and relative humidity with time, and (c) even areas covered by slabs experience changes in temperature and variable edge conditions. These observations, together with the data reported in this study, cause one to ask "How often do we encounter truly static conditions in the field, wherein water is not moving through the soil in any direction?"

5. SUMMARY AND CONCLUSIONS

Capillary rise measurements were performed for 19 soils ranging from clean sands (non-plastic) to silts and clays of low to moderate plasticity and included one clay of high plasticity. The results of this study indicate that the degree of saturation drops to less than 90% rather quickly with distance, typically within 10 to 20 cm of the groundwater table, and continues to drop to as low as 50% within a meter of the groundwater table. Although these degrees of saturation may well rise somewhat with additional time when more nearly static conditions are reached, the time required to reach static conditions may be very large in typical cases. In fact, the results of this study indicate that truly static conditions may be very elusive in the field, where boundary conditions tend to vary with time.

To the extent that the SWCCs and degrees of saturation measured for these soils are valid and accurate, it appears that the soil suction is typically much, much larger than the value calculated by y times γ_w, where y is the distance above the groundwater table.

ACKNOWLEDGEMENTS

The support provided by the Fulbright Program of the Council of International Exchange of Scholars for the Visiting Professor ship of Dr. Al-Samahiji at Arizona State University is gratefully acknowledged. The support provided by the University of Bahrain is also gratefully acknowledged. A part of this work is based upon research supported by the National Science Foundation under grant no. CMS9612073. Technical support in the form of laboratory testing by Bret Howie is also acknowledged.

REFERENCES

Day, R. 1996. Moisture Penetration of Concrete Floor Slabs, Basement Walls, and Flat Ceilings. *Practice Periodical on Structural Design and Construction*, Vol. 1, No. 4, pp. 104 – 107.

Fredlund, D.G., and H. Rahardjo. 1993. *Soil Mechanics for Unsaturated Soils*. Wiley.

Fredlund, D.G., Vanapalli, S.K., Xi, A., and D.E. Pufahl. 1995. Predicting the Shear Strength Function for Unsaturated Soils Using the Soil-Water Characteristic Curve. *Unsatu-*

rated Soils, Proc. 1st International Conf. on Unsaturated Soils, Paris, Vol. 1, pp. 63 - 69.

Fredlund, M.D., Wilson, G.W., & Fredlund, D.G., 1998. Estimation of Hydraulic Properties of an Unsaturated Soil Using a Knowledge-Based System, *Proc 2nd International Conference on Unsaturated Soils, Beijing,* August 27-30.

Padilla, F., Villeneuve, J., and J. Stein. 1997. Simulation and Analysis of Frost Heaving in Subsoil and Granular Fills of Roads. *Cold Regions Science and Technology*, Vol. 25, No. 2, pp. 89 – 99.

Sowers, G.E. 1979. *Introductory Soil Mechanics and Foundations: Geotechnical Engineering,* 4th edition, Macmillan, New York, NY.

Unsaturated Soils for Asia, Rahardjo, Toll & Leong (eds) © 2000 Taylor & Francis, ISBN 90 5809 139 2

Examination of the inverse methods for hydraulic properties of unsaturated sandy soil

T. Ishida
Department of Civil and Environmental Engineering, Toyo University, Saitama, Japan

H. Taninaka
Suncoh Consultants Company Limited, Tokyo, Japan

Y. Nakagawa
Techno Sol Company Limited, Tokyo, Japan

ABSTRACT: To obtain the hydraulic properties of unsaturated soils, it is necessary to analyze the saturated-unsaturated seepage flow, carrying out laboratory and field experiments. However, the time required for measurement is long and it requires a number of laboring jobs. To overcome the problems, this study is to compare the different independent methods, such as the pressure plate method, the apparatus of Richards and the Instaneous profile method using FDR with tensiometers, and Multi-step outflow and Evaporation method approach for estimating the soil hydraulic properties using inverse modeling. This paper discusses the influence of each method on the hydraulic properties and the modeling

1 INTRODUCTION

Rain infiltration often causes failure of earth structures such as river embankments. Rain infiltration and resulting flooding are problems related to unsteady seepage flow involving saturated zones and unsaturated zones. Elucidating these problems is of considerable significance to the safety evaluation of embankments.

Saturated-unsaturated seepage analysis using the finite element method can be applied to this kind of seepage flow, but the seepage flow analysis requires the determination of the infiltration properties of soil (saturated permeability k_s; volumetric water content θ~relative hydraulic conductivity k_r; θ~negative pressure head ψ). However, experimental determination of moisture characteristic curves, particularly unsaturated permeability (k_r~θ), is too complex and time-consuming for practical use only by carrying out laboratory and field experiments. At present, therefore, function models are used predominantly to estimate soil properties, and the model proposed by (van Genuchten, 1980; Mualem, 1976), which has proved highly versatile, is now used widely (Akai *et al.*, 1977). However, It is difficult that rain infiltration test result of the full-scale model, such as embankment or retaining wall structure, is accurately simulated using the soil hydraulic properties required in the laboratory test. (Yamamura *et al.*, 1995; Ishida *et al.*, 1998).

Recently, it is noticed that the measurement method of soil hydraulic conductivity from dielectric constant of soil such as TDR method and FDR method are frequency utilized. However, these methods have problems of their complex automatic measuring system in terms of calibration and its application.

In this comparative study, for the purpose of examination from the results of the measurement methods of unsaturated hydraulic properties carried out using various methods with probable soils which are available for construction work view-points. The various methods of laboratory experiments were used to determine the hydraulic properties by independent methods, such as the pressure plate method, Richards type apparatus, Instaneous profile method using FDR with tensiometers, and Multi-step outflow method and Evaporation method approach for estimating hydraulic properties using inverse modeling.

The objective of this study is to compare the results of independent methods and approaches for estimating the soil hydraulic properties using inverse methods and to determine whether the application of these modeling techniques is possible.

2 INDEPENDENT METHODS AND GROUND MATERIAL

2.1 *Richards type apparatus*

The steady-state method is the most basic and reliable for measurement of unsaturated hydraulic conductivity of various soils. But the measurement requires a lot of time and labor and is not so easy to obtain reliable and accurate values that are needed.

The volumetric water content in the steady con-

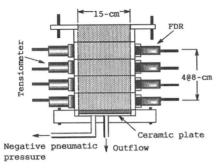

Figure 1. Cross-section of the Instaneous profile method using FDR (Frequency-domain reflectometry) with tensiometers

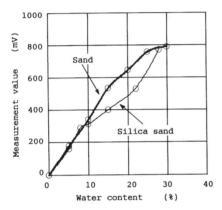

Figure 2. Calibration curves of FDR

dition at each flow rate must be beforehand obtained in order to acquire the relationship between unsaturated hydraulic conductivity and volumetric water content. The adjustment of degree of saturation and water content is very difficult for this method. In this experiment a volume of core of 314 cm^3 (10-cm diameter and 4-cm long soil column) was used. The tension cups of 1.5-cm length and 0.6-cmϕ were horizontally inserted into drill holes in the soil specimen at 1 and 3-cm from the specimen surface.

2.2 Instaneous profile method using FDR

In this study, Instaneous profile method using FDR (Frequency-domain reflectometry) and tensiometers obtained unsaturated hydraulic conductivity and moisture characteristic curve. The test apparatus is shown in Figure 1. Five acrylic columns (15-cm

Table 1. Soil properties

Index properties	Sand	silica sand
Specific gravity	2.63	2.63
National water content (%)	0.3	0.1
Uniformity cefficient Uc	2.09	5.24
Coefficient of curvature Uc'	1.08	1.92
Fine-grained soil content (%)	4.0	9.7

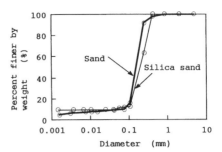

Figure 3. Particle size distribution curves

diameter and 8-cm long soil column) were piled up. FDR and tensiometer were horizontally inserted in this apparatus at 8cm interval to measure matric head and volumetric water content.

The one-dimensional seepage flow was drained from the bottom, after saturation state was confirmed. The times required for the measurement from the start of the drainage of sandy soil and cohesive soil were 1 to 2 hours and 2 to 3 days, respectively. Figure 2 shows calibrations relation between the sandy soil moisture and the readings of FDR. The obtained data from this method were arranged based on this experimental data.

2.3 pF tests

By controlling the pore water pressure in the specimen to be equal with atmospheric pressure, the pressure plate method dehydrated soil specimens, until the air pressure is balanced. A volume of the specimen is 100 cm^3. It was placed on a ceramic plate. The air entry value of ceramic plate was 0.1 MPa which operating pneumatic pressure head can range from 20 cm to 1000 cm and when more over used the ceramic plate of 0.3 MPa. It was pressurized through the nitrogen gas in the equipment.

Moreover, the sand column method was used for pneumatic pressures head below 32 cm, and the centrifuge method was employed for measurements at pressures ranging from 1200 to 20,000 cm. Three methods covered different ranges, in order to determine the moisture characteristic curve.

2.4 Ground material and experimental results of independent methods

Table 1 shows the soil test results of the ground materials used in this comparative study and Figure 3 shows particle size distribution curves. From each result fitted curves were drawn according to the Mualem-van Genuchten model. These parameters were known as MVG parameters. Authors used to obtain the soil hydraulic properties determined in the numerical model using the method of optimization of the MVG parameters, as described by Takeshita

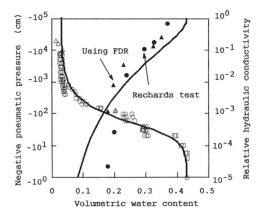

(a) Sand sample (dry density, 14.7 kN/m³)

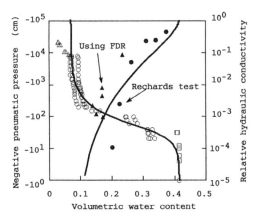

(b) Silica sand sample (dry density, 15.1 kN/m³)

Figure 4. Soil hydraulic properties from independent methods

& Kono (1993). Figure 4 for example, is measured retention data and fitted MVG curves and the parameters of these curves are listed in Table 2. Based on these parameters comparison of results of two inverse methods is given below.

3 INVERSE METHODS

3.1 *Multi-step outflow method*

Initial applications of the inverse model is derived from one-step outflow by (Kool *et al.*, 1985; van Dam *et al.*, 1992). It is dominated by the Richards' equation for unsaturated flow.

$$C \frac{\partial h}{\partial t} = \frac{\partial}{\partial z}\left[K(h)\left(\frac{\partial h}{\partial z}+1\right)\right]$$ (1)

where, C is the water capacity, h is the soil matri potential, t is denotes time, z is the vertical coordinate (positive upward) and $K(h)$ are the unsaturated

Table 2. Results of MVG parameters from independent methods

Sample	Drydensity $\gamma d(kN/m^3)$	MVG parameters $\alpha(cm^{-1})$	n	θr	θs	ks $(10^{-2}cm/s)$
	15.7	0.015	2.153	0.02	0.392	2.83
Sand	15.2	0.022	1.867	0.02	0.411	3.73
	14.7	0.032	1.824	0.03	0.430	4.09
	15.7	0.019	1.893	0.03	0.391	1.01
Silica	15.4	0.025	1.811	0.05	0.403	1.23
sand	15.1	0.026	1.934	0.07	0.414	1.64

hydraulic conductivity.

Boundary and initial conditions for the Multi-step outflow experiment are:

$$\begin{array}{llll} h = h(0,z) & t = 0 & 0 \le z \le L \\ \partial h/\partial z = -1 & t > 0 & z = L \quad (2) \\ h = h(t,0) - h_a & t > 0 & z = 0 \end{array}$$

where, $h(0,z)$ is the initial soil matric head distribution, L is the height of the soil core with the stainless steel filter, $h_a(t)$ is the applied pneumatic pressure.

Equations (1) and (2) were solved with the Galerkin finite element model, and the soil hydraulic properties employed in the numerical model are as follows:

$$S_e = (\theta - \theta_r)/(\theta_s - \theta_r)$$
$$h(S_e) = \left[\left(S_e^{-1/m} - 1\right)^{1/n}\right]/\alpha$$ (3)
$$K(S_e) = K_s S_e^{0.5}\left[1 - \left(1 - S_e^{1/m}\right)^m\right]^2$$

where S_e is the normalized volumetric water content, α and n are empirical parameters, θ_r and θ_s are the residual and saturated hydraulic conductivity, and $m = 1 - 1/n$.

To estimate the parameters, the objective function $O(\mathbf{b})$ has to be solved by the following equation. This $O(\mathbf{b})$ is the weighted least squares problem.

$$O(\mathbf{b}) = \sum_{i=1}^{N1}\left\{w_i\left[Q_0(t_i) - Q_c(t_i,\mathbf{b})\right]\right\}^2$$
$$+ \sum_{i=1}^{N2}\left\{W_1 v_i\left[\theta_0(h_i) - \theta_c(\mathbf{b},h_i)\right]\right\}^2$$ (4)

where, w_1 and v_1 are weighting factors, N1 and N2 are the number of observations of $Q_0(t_i)$ and $\theta(h_i)$ and W_1 is

$$W_1 = \left[\sum_{i=1}^{N1}Q_0(t_i)\bigg/N1\right]\bigg/\left[\sum_{i=1}^{N2}\theta_0(h_i)\bigg/N2\right]$$ (5)

The weight factors are used to account for differences in accuracy of measurements and in correlation associated with residuals. When $O(\mathbf{b})$ contained outflow data only, the outflow measurements could be assumed to have equal accuracy, which resulted in the ordinary least squares criterion:

395

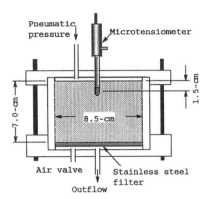

Figure 5. Cross-section of the Multi-step outflow method

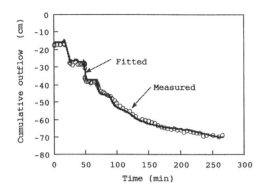

Figure 6. Relation between measured and fitted cumulative outflow from Multi-step outflow method

$$O(\mathbf{b}) = \sum_{i=1}^{N1} \left[Q_0(t_i) - Q_c(t_i, b) \right]^2 \qquad (6)$$

The calculation of algorithm was carried out by the Levenberg-Marquardt method in order to minimize the difference between matric potential and the amount of outflow in the time step by the experiment data

In the experiment, a volume of core of 400 cm³ (8.5-cm diameter and 7-cm long soil column) was used. The bottom of soil column was a high-flow stainless steel filter (k_s=0.01078 cm/min) of 2-mm thick. A micro-tensiometer was installed close to the center of the column at 1.5-cm below the soil surface (see Figure 5). The soil sample was saturated from the bottom and subsequently equilibrated to an initial soil water pressure head of around -15 cm at the soil surface. The soil surface were applied the pneumatic pressure of 10,20 and 30 cm in consecutive steps at 40 minutes, and ranging from 40, 50, 60, 80 and 100 cm at 1 hour intervals. Time intervals for measuring of the tensiometer and cumulative outflow were 15 seconds. When the measurements finished, the soil core sample was removed from the column and the water content was determined by oven drying for 1 day at 110°C. The saturated permeability k_s of each sample was determined by using the constant head permeability test (see in Table 2).

The soil hydraulic parameters were estimated by numerical inversion of the observed cumulative outflow data and the measured water content at the applied pneumatic pressure. It was calculated the effect of the number of optimized parameters by repeating the optimizations with two (α and n), three (α, n, and θ_r), three (α, n, and θ_s) and four (α, n, θ_r and θ_s) fitting parameters on all samples.

The optimizations were repeated referring to data of independent methods, until it obtained the near-equilibrium outflow data. For example, Figure 6 compares the measured with optimized data in relation between cumulative outflow and elapsed times. This sand sample has 15.7 kN/m³ dry density. The

final fit had a correlation of 0.99942.

3.2 Evaporation method

The results of experiment were analyzed using the modified Wind method as illustrated by Wendroth et al. (1993). They described the water retention curve with a fitting polynomial, as the following equation.

$$\theta = \theta_r + \frac{\theta_s - \theta_r}{\left(1 + |\alpha h|^n\right)^m} \qquad (7)$$

Parameters in the equation (1) were considered equal to the empirical fitting MVG parameters. Hydraulic parameters were obtained by simultaneous fitting of the resulting data using the RETC code of nonlinear least-squares optimization program by van Genuchten.

To setup for experiment a mold (10-cm diameter, 12.7-cm long, 1000 cm³ volume) was used. The soil sample was placed on a glass filter of high permeability. Two micro-tensiometers, part of the ceramic cup was 6.3-mmϕ and 15-mm long, were horizontally inserted in the soil sample at 2-cm and 4.5-cm below the soil surface. A bulb for drying (i.e., 100W) exists 20-cm above the soil surface (see Fig-

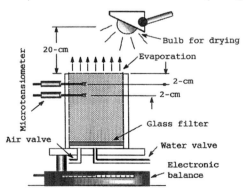

Figure 7. Cross-section of Evaporation method

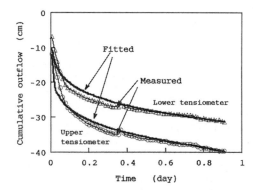

Figure 8. Relationship between measured and fitted cumulative outflow from Evaporation method

ure 7). The operating temperature was 37 °C zone near the soil surface. This equipment was placed on an electronic balance for measuring water volume according to the evaporation function. The soil sample was saturated from the bottom. After saturation, readings of two micro-tensiometers and weight of the total equipment were recorded at 10 minute intervals. When the measurement finished, the soil sample was removed from the mold and the water content was measured.

At first, the experimental data were analyzed using the modified Wind's method. The soil hydraulic parameters were obtained by simultaneous fitting using the RETC code program. Figure 8 shows an example of relationship between the readings of cumulative outflow and elapsed times using the measured data and estimation for the sand sample of the same dry density as example of Multi-step method in Figure 6. The final fitting of this case had the correlation at 0.9968.

4 COMPARISON OF RESULTS

From the Multi-step outflow method, the relationship between cumulative flow rate and elapsed time shown in Figure 6, the values of $\alpha=0.025$ $n=6.76$, $\theta_r=0.01$ and $\theta_s=0.395$ were obtained, when the MVG parameter was estimated. This case used optimizations with four (α, n, θ_r and θ_s) fitting parameters. The MVG parameters of Evaporation method were $\alpha=0.009$ $n=5.22$, $\theta_r=0.07$ and $\theta_s=0.392$ in the case of Figure 8 using the sample of equal dry density condition as in Figure 6. This fixed parameter was θ_s. The final values of optimized parameters for each method are shown in Table 3. Estimated range of the α values is from 0.025 to 0.027 (cm^{-1}) in the Multi-step outflow method without sample D. It corresponds to what were obtained by the independent methods data in Table 2. However, the α values estimated from Evaporation method are smaller than 0.017 (cm^{-1}), therefore cap-

Table 3. Lists final values of optimized parameters

Method	Sample No.	MVG parameters				γd (kN/m³)
		α(cm⁻¹)	n	θr	θs	
	A	0.025	6.76	0.010	0.395	15.7
	Sand B	0.025	6.00	0.020*	0.433	15.2
Multi step	C	0.027	5.00	0.030*	0.430*	14.7
	Silica D	0.060	1.98	0.030	0.390	15.7
	Sand E	0.027	3.00	0.041	0.420	15.4
	F	0.026	5.54	0.020	0.402	15.1
	G	0.009	5.22	0.070	0.392*	15.7
	Sand H	0.006	4.18	0.020	0.410*	15.2
Evaporation	I	0.008	4.53	0.030*	0.430*	14.7
	Silica J	0.006	4.38	0.030*	0.391*	15.7
	Sand K	0.017	5.51	0.050*	0.403*	15.4
	L	0.010	6.70	0.070*	0.414*	15.1

* fixed parameters

illary fringe rose because the parameter α is said to be nearly equal to the reciprocal of the critical capillary pressure head.

The n values in optimizing are bigger than the result of independent methods in Table 2. Figures 9

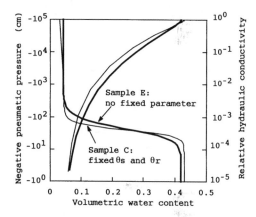

Figure 9. Estimated soil hydraulic properties from Multi-step outflow method

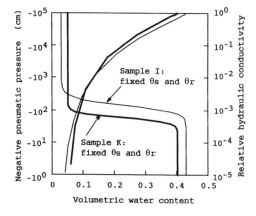

Figure 10. Estimated soil hydraulic properties from Evaporation method

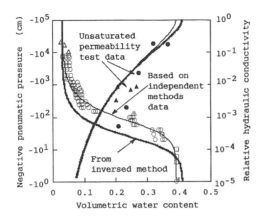

Figure 11. Comparison of independent methods data and inverse method data (Sample D: Silica sand, $\gamma_d = 1.57$kN/m³)

and 10 show for examples the estimated hydraulic property curves for both Multi-step outflow method and Evaporation method. Sample C and I correspond to Figure 4a. However, the shapes of the moisture characteristic curves are different from the relationship between values of α and n, therefore inverse methods overestimated relative hydraulic conductivity curves compared with independent methods data according to the MVG model (see Figure 4).

Being the closest values compared to results of independent methods and inverse methods, showed in sample D of Multi-step outflow method. These retention data of the relation between volumetric water content and relative hydraulic conductivity, and the relation between volumetric water content and negative pressure head, are depicted in Figure 11 with the fitted curves according to the MVG model. Inverse technique approach is located under the curve of the independent methods from obtained retention data, because the α value is small. While, curves of relative hydraulic conductivity are corresponded. Moreover, in this case, θ_s values were different so this inverse analysis did not used to estimate the fixed parameter in the optimization.

5 CONCLUSIONS

Multi-step outflow method and Evaporation method in the laboratory tests and combined with the numerical modeling and nonlinear parameters were used to estimate the hydraulic properties of unsaturated sandy soils. The objective of this study is to investigate the possibility of these techniques. In order to accomplish our purpose, obtained that data from both independent methods and inverse methods were compared. From this, authors can get the following conclusions.

The optimizations were necessary changed fitting parameters in order to agree the measured and opti-

mized cumulative out flow curves for soil samples. In other words, it is not possible to limit fixed parameters even in the case of the experiment for similar samples.

In initial optimization, the value of θ_r tends to become 0. The parameters θ_s and k_s should depend on the measured data because they easily obtained during the experiments. The α values from Multi-step outflow method are shown to be nearly equal to the results of independent methods compared to Evaporation method. It seems necessary that the impact given to the sample by the experiment because Multi-step outflow method is more abounding, such as pressures applied of subsequently in consecutive steps, than Evaporation method on this.

The n values range from 4 to 7 in optimizing were bigger than the result of independent methods, therefore relative hydraulic conductivity curves are overestimated.

Multi-step outflow method is an effective method, not only simplify the set of test equipment, but also measurement time can be shortened. In the future, the applicability of inverse methods will be examined using the cohesive soil.

REFERENCES

Akai, K., Ohnishi, Y. & Nishigaki, M. 1977. Finite element analysis of saturated-unsaturated seepage in soil, *Proc. of the Japan society of civil engineers*, 264: 87-96 (in Japanese).

Ishida,T., Nakagawa, Y. & Tatsui, T. 1998. Analysis of drainage system for rain infiltration of retaining wall structures, *3rd. Intern. Conference on Hydroscience and Engineering*, Abstract:144, paper on CD-ROM, Cottbus/Berlin Germany.

Kool, J. B., Parker, J. C. & van Genuchten, M. Th. 1985. ON-ESTEP: A nonlinear parameter estimation program for evaluating soil hydraulic properties from one-step outflow experiments: *Bulletin* 85-3, Virginia Agr. Exp. Station

Mualem, Y. 1976. A new model for predicting the hydraulic conductivity of unsaturated porous media, *Water Resour. Res.*, 12(3): 513-522.

Takeshita, Y. & Kono, I. 1993. A method to predict hydraulic properties for unsaturated soil and its application to observed data, *Ground Engineering* , 11(1): 95-113 (in Japanese).

van Dam, J. C., Stricker, J. M. N. & Droogers, P. 1992. Inverse method for determining soil hydraulic functions from one-step outflow experiment, *Soil Sci. Soc. Am. J.*, 56: 1042-1050.

van Genuchten, M. Th. 1980. A closed-form equation for predicting the hydraulic conductivity of unsaturated soils, *Soil Sci. Soc. Am. J.* 44: 892-898.

Wendroth, O., Ehlers, W., Hopmans, J. W., Kage, H. Halbertsma, J. & Wösten, J. H. M. 1993. Reevaluation of the evaporation method for determining hydraulic functions in unsaturated soils, *Soil Sci. Soc. Am. J.*, 57: 1436-1443.

Yamamura, K. Zhu, W. & Ishida, T. 1995. Soil Properties on Rain Infiltration into Embankment, *10th Asian Regional Conference on Soil Mechanics and Foundation Engineering*, ISSMFE: 361-364, Beijing/China.

Unsaturated Soils for Asia, Rahardjo, Toll & Leong (eds) © 2000 Taylor & Francis, ISBN 90 5809 139 2

Grain size and void diameter distributions of sands

K. Kamiya & T. Uno
Department of Civil Engineering, Gifu University, Japan

ABSTRACT: The purpose of this study is to understand the intricate structure of soil void space and to evaluate the hydraulic properties of unsaturated soil. In this paper, the relationships between the void diameter distributions produced by the "air intrusion method" and by the usual "moisture characteristic curve method" and grain size distribution are examined for sandy soils. It is shown that the void diameter distribution by the moisture characteristic curve method tends to be parallel to the mass-based grain size distribution, and that by the air intrusion method is similar to the number-based grain size distribution. The void diameter in both methods is estimated using the proposed β_r-value and particle size. The results provided an understanding of the void structure using the relationship between void diameter and grain size distributions.

1 INTRODUCTION

Knowledge of the soil void structure is necessary to evaluate the hydraulic properties of unsaturated soil. The method for predicting unsaturated hydraulic conductivity from void diameter distribution estimated in the "moisture characteristic curve method" is popular. However, the meaning of void diameter distribution is not clear.

The "air intrusion method" to measure void diameter distribution was suggested by Uno et al. (1995). The method is based on the technique developed by Zagar (1956). The shape of void diameter distribution of sand by the air intrusion method becomes more uniform than that by the moisture characteristic curve method (Uno et al., 1998). The discrepancy between both distributions was caused by the void structure in void space as a tree of knot (Uno et al., 1999).

In this paper, the relationships between void diameter distributions by both methods and particle mass-based and number-based grain size distributions are examined for sandy soil. The purpose of this study is to understand the intricate structure of soil void on the evaluation of void diameter distribution.

2 VOID DIAMETER DISTRIBUTIONS

2.1 *Air intrusion method*

For estimation of the void diameter distribution by the air intrusion method, the relationship between

head of air pressure, h_a (cm), and air flow rate, Q_a (cm³/s), is required. The relationship is obtained when the air intrudes into a saturated soil sample forcing the water out of the voids and passing through the sample (Uno et al., 1995). This relationship is shown as curve B in Figure 1.

By using curve B and the capillary tube model, h_a is related to void diameter, d_e (mm), as follows.

$$d_e = \frac{4\sigma}{\rho_w g h_a} \cdot 10^4 \tag{1}$$

where σ = surface tension of water (N/m); ρ_w = water density (g/cm³); and g = gravitational acceleration (cm/s²). The percentage of cumulative void volume, V_b (%), is defined as:

$$V_b = \left(1 - \frac{n_b}{n}\right) \cdot 10^2 \tag{2}$$

where n_b = cumulative porosity and is estimated from the air permeability (Kamiya et al., 1996); and n = porosity.

The void diameter distribution by the air intrusion method is denoted as the relationship between d_e and V_b.

2.2 *Moisture characteristic curve method*

In the moisture characteristic curve method, it is assumed that the moisture characteristic curve of the soil shown as the relationship between volumetric water content, θ, and head of suction, h_p (cm), in Figure 2, is equivalent to the relationship between

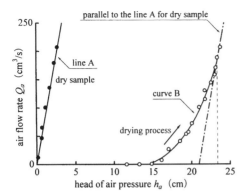

Figure 1. Relationship between head of air pressure and air flow rate for Nagara river sand (porosity: $n = 0.410$).

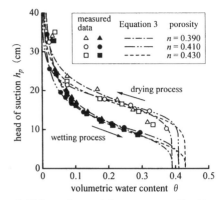

Figure 2. Moisture characteristic curve measured by the suction plate method for Nagara river sand.

the water content and height of rise of water in the capillary model. In Figure 2, the curves are fitted using Equation 3 (van-Genuchten, 1980).

$$\frac{\theta - \theta_r}{\theta_s - \theta_r} = \frac{1}{\left[1 + \left(\alpha h_p\right)^{n^*}\right]^{1-1/n^*}} \qquad (3)$$

where θ_r = residual volumetric water content; θ_s = saturated volumetric water content; and α, n^* = empirical parameters. In this paper, it is assumed that $\theta_r = 0$ and $\theta_s = n$ for sandy soils.

As a result, the moisture characteristic curve in Figure 2 should be rotated counterclockwise 90°, and h_p is converted into void diameter, d_m (mm), and θ is converted into the percentage of cumulative void volume, V_r (%), as follows (Uno et al., 1998):

$$d_m = \frac{4\sigma}{\rho_w g h_p} \cdot 10^4 \qquad (4)$$

$$V_r = \frac{\theta}{n} \cdot 10^2 \qquad (5)$$

The void diameter distribution by the moisture characteristic curve method can be expressed as the relationship between d_m and V_r.

In the above conversion, only the drying curve in Figure 2 is adopted owing to the resemblance of the drying process to the air intrusion method.

2.3 Comparison between two methods

Figure 3 shows the examples of distributions of sand void diameter by the two methods for two distributions of grain size described later. The lines in the distributions by the moisture characteristic curve method represent the following relationship between d_m and V_r given by substituting Equations 4 and 5 into Equation 3 on the assumption that $\theta_r = 0$ and $\theta_s = n$.

$$V_r = \frac{10^2}{\left[1 + \left(\alpha \dfrac{4\sigma}{\rho_w g d_m} \cdot 10^4\right)^{n^*}\right]^{1-1/n^*}} \qquad (6)$$

In Figure 3, the shape of the void diameter distributions by the air intrusion method becomes more uniform than the distributions by the moisture characteristic curve method and their distributions are very different from each other.

It was recognized that the void diameter, d_e, by the air intrusion method is in agreement with the diameter, d_m, by the moisture characteristic curve method (Uno et al., 1998). Therefore, the discrepancy between both distributions is caused by the difference between the percentage of cumulative void volume, V_b, and V_r. It is considered that the void volume estimated by the air intrusion method corresponds to only that of the air flow channel. On the contrary, the one by the moisture characteristic curve method corresponds to the volume of water evacuated from the three-dimensional void space. Furthermore, the adaptability of the air intrusion method was verified by the measurement of void diameter distribution of an artificial capillary tube sample as the capillary model itself (Kamiya et al., 1996).

The discrepancy of the distribution by the moisture characteristic curve method from that by the air intrusion method can be regarded as a discrepancy from the one-dimensional capillary model. In part of the voids as illustrated in Figure 4, only the inshaded part of the void is estimated by the air intrusion method because the air is stagnant in the shaded part of the void. As a result, it was considered that the discrepancy between the distribution by the air intrusion method and that by the moisture characteristic curve method is affected by the void structure in the void space as a tree of knot in Figure 4 (Uno et al., 1999).

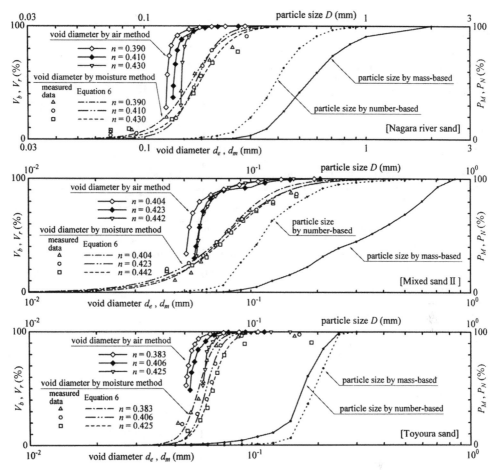

Figure 3. Void diameter distributions by two methods, with grain size distributions by mass-based and by number-based, for Nagara river sand, Mixed sand II and Toyoura sand.

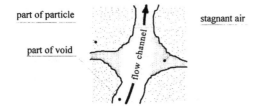

Figure 4. A part of void estimated by air method

3 RELATION OF GRAIN SIZE DISTRIBUTION TO VOID DIAMETER DISTRIBUTION

3.1 *Shape of distributions*

New particle number-based grain size distributions (Fukuda et al., 1997) can be drawn with the usual particle mass-based distributions as shown in Figure 3. The number-based distribution is denoted as the relationship between particle size, D (mm), and the

percent finer by number, P_N (%), and the mass-based one as the relationship between D and the percent finer by mass, P_M (%). The number-based distribution is converted from the mass-based distribution by calculating the number using the mass and intermediate size with the range of sieve/particle size. The frequency of small size of particle is higher on the number-based distribution than the mass-based one. Subsequently, the number-based distributions become more uniform than the mass-based one.

In Figure 3, it is recognized that the distribution of void diameter by the moisture characteristic curve method is parallel to the mass-based distribution of grain size, and that by the air intrusion method has a strong resemblance to the number-based distribution. It seems that the discrepancy between the distributions of void diameter by the air intrusion method and by the moisture characteristic curve method is similar to the gap between the number-based and the mass-based distributions of grain size.

3.2 Void diameter by moisture characteristic curve method and particle size

Owing to the above result, the distribution obtained when particle sizes, D, in the mass-based distribution is respectively converted to void diameters, d_m, by the following relationship, is equivalent to the void diameter distribution by the moisture characteristic curve method as described by Haverkamp et al. (1986).

$$d_m = \beta_r D \qquad (7)$$

where β_r = empirical parameter ($\beta_r < 1$). Substituting Equation 7 into Equation 6 with $V_r = P_M$ gives the following relationship between D and P_M:

$$P_M = \frac{10^2}{\left[1 + \left(\alpha \dfrac{4\sigma}{\rho_w g \beta_r D} \cdot 10^4\right)^{n^*}\right]^{1-1/n^*}} \qquad (8)$$

The mass-based grain distribution can also be expressed by Equation 8.

The value of β_r is estimated by fitting the mass-based distribution with Equation 8 using the values of α and n^* in Equation 6 shown such as Figure 3. Figure 5 shows the relationship between β_r-value and void ratio, e. The values of β_r range from 0.2 to 0.4; for sand. Furthermore, it is recognized that the β_r-value for each sample tends to increase with increasing void ratio.

3.3 Estimation of β_r-value

According to the theoretical relationship between particle size and void diameter described by Carman (1937), the β_r-value can be expressed as:

$$\beta_r = \frac{4}{K_s} e \qquad (9)$$

where K_s = shape factor of particles and is evaluated from the specific surface area of particles (Uno et al., 1993). The value of K_s is 6 for spherical particles and is greater than 6 for angular particles.

The relationships between shape factor, K_s, and the value of β_r/e calculated from the measured data in Figure 5 are compared with the theoretical relationship given by Equation 9 and shown as the solid line in Figure 6. The experimental relationship agrees with the theoretical one for all samples except for Mixed sand I and Mixed sand II.

Figure 7 shows the relationship between uniformity coefficient, U_c, of mass-based grain distribution and the value of $\beta_r/(4e/K_s)$ calculated by using the data in Figure 6. It is clear that the values of $\beta_r/(4e/K_s)$, that is, β_r-values decrease with increasing values of U_c.

On the other hand, the solid line in Figure 7 represents the relationship between D_c (mm) from

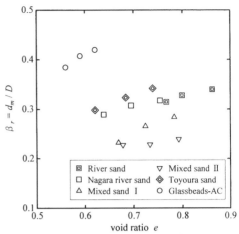

Figure 5. Relationship between void ratio and β_r-value.

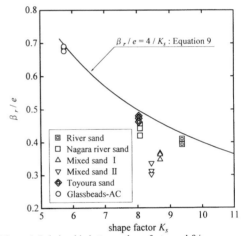

Figure 6. Relationship between shape factor and β_r/e.

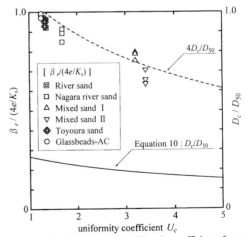

Figure 7. Relationship between uniformity coefficient of mass-based grain size distribution and $\beta_r/(4e/K_s)$, with relationship between uniformity coefficient and D_c/D_{50}.

402

the number-based grain distribution, 50% diameter of particle, D_{50} (mm), from the mass-based grain distribution and U_c (Fukuda et al., 1997) given by:

$$\frac{D_c}{D_{50}} = \frac{0.3}{\exp\left[0.5(0.484 + 0.420 \cdot \ln U_c)^2\right]} \quad (10)$$

The value of D_c/D_{50} decreases with increasing U_c values and this is similar to the tendency for $\beta_r/(4e/K_s)$.

The values of four times D_c/D_{50} calculated by Equation 10 are shown as the broken line in Figure 7, and it is found that the values are in close agreement with the values of $\beta_r/(4e/K_s)$. Therefore, β_r can be expressed by the following equation that is slightly different from the theoretical Equation 9:

$$\beta_r = \frac{4}{K_s} e \cdot 4 \frac{D_c}{D_{50}} \quad (11)$$

The value of β_r is estimated from the shape factor of the particles, K_s, the void ratio, e, and the relationship between particle sizes from number-based grain distribution and from mass-based grain distribution, which is expressed using the uniformity coefficient, U_c, as given in Equation 10.

Substituting Equation 11 into Equation 7 and substituting Equation 10 into the result, we obtain:

$$d_m = \beta_r D = \frac{16}{K_s} e D_c \cdot \frac{D}{D_{50}}$$

$$= \frac{4.8e}{K_s \exp\left[0.5(0.484 + 0.420 \cdot \ln U_c)^2\right]} D \quad (12)$$

As a result, the void diameter, d_m, by the moisture characteristic curve method is influenced by K_s, e and particle size, D_c, from the number-based grain distribution because both D and D_{50} are particle sizes from the mass-based grain distribution.

3.4 Void diameter by air intrusion method and particle size

It is expected that the void diameter, d_e, by the air intrusion method can be expressed as an equation like Equation 12, because d_e by the air intrusion method is equivalent to d_m by the moisture characteristic curve method.

Replacing d_m by the mean diameter of void, d_e^* (mm), from the air intrusion method and substituting D_{50} for D in Equation 12, the diameter, d_e^*, is calculated. The relationship between the calculated d_e^* and the measured d_e^* is shown in Figure 8. The values of the measured d_e^* are smaller than the calculated values shown by the broken line in Figure 8. Consequently, void diameter, d_e^*, by the air intrusion method is estimated from β_r-value and particle size as follows:

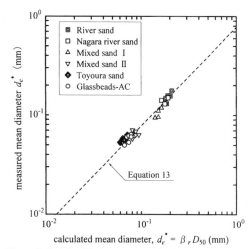

Figure 8. Relationship between calculated mean diameter of void and measured diameter by air intrusion method.

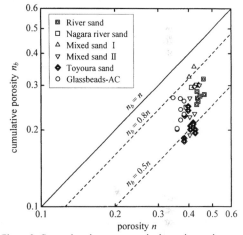

Figure 9. Comparison between porosity by moisture characteristic curve method and cumulative porosity by air intrusion method.

$$d_e^* = 0.78 \beta_r D_{50} \quad (13)$$

However, in contrast to Equation 12 for the moisture characteristic curve method, it is necessary to use a coefficient of 0.78 on the right hand side of Equation 13 for the air intrusion method.

The comparison between the porosity, n, as the void volume evaluated by the moisture characteristic curve method and the cumulative porosity, n_b, as the volume by the air intrusion method is shown in Figure 9. n_b corresponds to 50% to 80% of n, that is, the void volume by the air intrusion method is smaller than the one by the moisture characteristic curve method. Therefore, it is surmised that the difference between Equation 12 and 13 is due to the difference in the evaluation of the void ratio, e, in Equation 11 for the β_r-value.

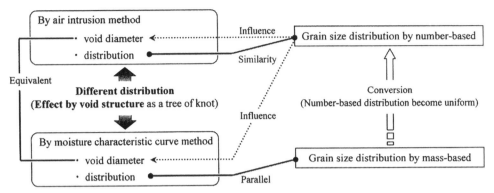

Figure 10. Correlation among void diameter distributions and grain size distributions.

3.5 *Consideration of void diameter distributions*

According to the results described above, the correlation among void diameter distributions and grain size distributions is shown in Figure 10. This correlation is recognized as a clue to understand the discrepancy between void diameter distributions by the air intrusion method and by the moisture characteristic curve method. Furthermore, the void structure in the three-dimensional void space is the cause of the discrepancy.

It is considered that the void space is mainly dependent on the numbers and sizes of particles as described by Kenney et al. (1985). It is, therefore, inferred that the number-based frequency of small size void space is higher than the large size void space. As mentioned above, only part of the air void flow channel is estimated by the air intrusion method. It is suggested that the flow depends on the number-based frequency of the void. Consequently, it is interpreted that the shape of the void diameter distribution by the air intrusion method becomes more uniform and is similar to that from the number-based grain size distribution.

4 CONCLUSIONS

In this paper, the following conclusions can be obtained.

1) The distribution of void diameter by the air intrusion method tends to be parallel to the particle number-based grain size distribution, and that by the moisture characteristic drying curve is parallel to the usual mass-based grain size distribution. The shape of the number-based grain size distribution becomes more uniform than the mass-based distribution.

2) The void diameter by both methods is estimated from the β_r-value and the particle size, in which the β_r-value is calculated using the void ratio, the shape factor of particles and the uniformity coefficient of the mass-based grain size distribution.

3) The relationship between void diameter distribution and grain size distribution is a clue to understanding the void structure in the three-dimensional void space and is the cause of the discrepancy between void diameter distributions by the air intrusion method and by the moisture characteristic curve method.

REFERENCES

Carman, P. C. 1937. Fluid flow through granular beds, *Transactions of the Institute of Chemical Engineers* 15: 150-166.

Fukuda, M. & Uno, T. 1997. Analysis of "method of classification of soils" based on proposed "diameter estimating grain-size distribution". *Journal of Getechnical Engineering on JSCE* (582/III-41): 125-136. (in Japanese)

Haverkamp, R. & Parlange, J.-Y. 1986. Predicting the water-retention curve from particle-size distribution: 1. Sandy soils without organic matter. *Soil Science* 142(6): 325-339.

Kamiya, K., Uno, T. & Matsushima, T. 1996. Measurement of the distribution of sandy soil void diameter by air intrusion method. *Journal of Geotechnical Engineering on JSCE* (541/III-35): 189-198. (in Japanese)

Kenney, T. C., Chahal, R., Chiu, D. I., Ofoegbu, G. N. and Ume, C. A. 1985. Controlling construction sizes of granular filters. *Canadian Geotechnical Journal* 22: 32-43.

Uno, T., Sugii, T. & Kamiya, K. 1993. Considerations of permeability related to physical properties of particles based on the measurement of specific surface area. *Journal of Geotechnical Engineering on JSCE* (469/III-23): 25-34. (in Japanese)

Uno, T. & Kamiya, K. 1995. Air intrusion method to measure sand void diameter. *Proceedings of the Tenth Asian Regional Conference on SMFE* 1: 99-102.

Uno, T., Kamiya, K. & Tanaka, K. 1998. The distribution of sand void diameter by air intrusion method and moisture characteristic curve method. *Journal of Geotechnical Engineering on JSCE* (603/III-44): 35-44. (in Japanese)

Uno, T. & Kamiya, K. 1999. Estimation of void diameter distribution of sands. In Hong, S.W. et al. (eds) *Soil Mechanics and Geotechnical Engineering – Eleventh Asian Regional Conference* 1: 129-132. Rotterdam: Balkema.

van-Genuchten, M. Th. 1980. A closed-form equation for predicting the hydraulic conductivity of unsaturated soils. *Soil Science Society of America Journal* 44: 892-898.

Zagar, L. 1956. Ermittlung der Größenverteilung von Poren in feuerfesten Baustoffen und Glasnutschen Teil II . *Archiv für das Eisenhüttenwesen* 27(10): 657-663. (in German)

Unsaturated Soils for Asia, Rahardjo, Toll & Leong (eds) © 2000 Taylor & Francis, ISBN 90 5809 139 2

Investigation of seepage behavior through unsaturated soil

R. Kitamura & M. Seyama
Department of Ocean Civil Engineering, Kagoshima University, Japan

H. Abe
Chubu-Chishitsu Company Limited, Japan

ABSTRACT: Kitamura et al.(1998) proposed a numerical model for seepage through unsaturated soil. In the model voids and soil particles in a small soil element are modeled by pipe and other impermeable parts respectively. The void ratio, water content, unsaturated-saturated permeability coefficient and suction can be derived from the model based on some probabilistic consideration. In order to prove the validity of the proposed model, the unsaturated permeability test and the water retention test were simultaneously carried out by a steady state type's permeability test apparatus which was newly manufactured in our laboratory. The numerical experiment and permeability test on unsaturated soil were carried out in parallel for Toyoura sand and Shirasu. The soils were sampled in Kagoshima Prefecture which is located in the southern part of Kyushu Island, Japan. From the comparison of the results obtained by the numerical experiment with those obtained by the laboratory test it was found out that the proposed model could simulate the seepage behavior of pore water through unsaturated soil.

1 INTRODUCTION

The soil mechanics has been established by Terzaghi (1943) and developed by many researchers and engineers. However, the emphasis has been put on the mechanical behaviors of saturated soil and the literature has been biased toward researches involving saturated soils (Fredlund and Rahardjo, 1993). On the other hand most soils above the ground water table are under unsaturated condition. Consequently the mechanical behavior of unsaturated soil should be investigated for the analyses of earth pressure, bearing capacity, slope stability etc.

In this paper the permeability and water retention tests on Toyoura sand and Shirasu were carried out to prove the validity of the numerical model for seepage of pore water through unsaturated soil as proposed by Kitamura et al. (1998).

2 KITAMURA'S MODEL FOR SEEPAGE BEHAVIOR IN UNSATURATED SOIL

Kitamura et al. (1998) proposed a numerical model for seepage of pore water through unsaturated soil. Figure 1(a) is a typical picture showing the arrangement of a few soil particles in a soil mass, which is commonly observed by the optical microscope. The voids and soil particle in Fig.1(a) can be modeled by a pipe and other parts as shown in Fig.1(b), where a few soil particles and voids in the element correspond to the impermeable part (shaded part in Fig.1(b)) and permeable part (part of pipe in Fig.1(b)) respectively. The diameter and inclination angle of pipe are defined as D and θ respectively.

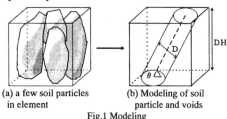

(a) a few soil particles in element   (b) Modeling of soil particle and voids

Fig.1 Modeling

The diameter D and inclination angle θ can be regarded as random variables because the shape and size of soil particles are random, and the structure of their assembly is irregular. Therefore the probability density functions of D and θ can be introduced to estimate the distribution of voids in soil mass. Consequently the void ratio, volumetric water content, suction, unsaturated-saturated permeability coefficient are derived by using the probability density functions of D and θ as follows.

$$e = \int_0^\infty \int_{-\frac{\pi}{2}}^{\frac{\pi}{2}} \frac{V_p}{V_e - V_p} P_d(D) P_c(\theta) d\theta dD \quad (1)$$

$$W_v = \frac{1}{1+e} \int_0^d \int_{-\frac{\pi}{2}}^{\frac{\pi}{2}} \frac{V_p}{V_e - V_p} P_d(D) P_c(\theta) d\theta dD \quad (2)$$

$$k_w = \int_0^d \int_{-\frac{\pi}{2}}^{\frac{\pi}{2}} \frac{\gamma_w \cdot D^3 \cdot \pi \cdot \sin\theta}{128 \cdot \mu_w \cdot \left[\frac{D}{\sin\theta} + \frac{DH}{\tan\theta} \right]} P_d(D) P_c(\theta) d\theta dD \quad (3)$$

$$h_c = \frac{4 \cdot T_s \cdot \cos\alpha}{\gamma_w \cdot D} \quad (4)$$

where e: void ratio; W_v: volumetric water content; k_w: permeability coefficient; h_c: water head due to suction; V_e: volume of element in Fig.1(b); V_p: volume of pipe in Fig.1(b); $P_d(D)$: probability density function of D; $P_c(\theta)$: probability density function of θ; DH: height of element in Fig.1(b); d: maximum diameter of pipe which can retain water; T_s: surface tension; α: contact angle between water and pipe; γ_w: unit weight of water; μ_w: viscous coefficient of water.

3 UNSATURATED PERMEABILITY TESTING APPARATUS

Figure 2 shows the whole view of unsaturated permeability testing apparatus. The apparatus is mainly composed of the pressure cell where the unsaturated specimen is set, the regulators for pressure control, the vessels where the volume of drained water is measured by the electronic balances and the personal computer by which the measuring data are acquired. In the apparatus the water circuit is clearly separated from the air circuit and then the water and air pressures can be independently controlled. By using this apparatus the suction and the unsaturated permeability coefficient are simultaneously obtained for an unsaturated specimen. Furthermore the hysterisis loop of moisture characteristic curves in the drying and wetting processes can be obtained from one specimen. Figure 3 is the plan view and cross section of pedestal and cap which are made of ceramic discs. The permeability coefficient and air entry value of ceramic disc is 10^{-7} cm/sec and about 200 kPa respectively. On the ceramic discs connected to air circuit a glass fiber sheet is pasted to block the water flow. The size of specimen is 5 cm in diameter and 2 cm in height.

The test material were Toyoura sand, in which fine grain (less than 75 μm) was removed, and Shirasu, which is defined as the non-welded part of pyroclastic flow deposits. Undisturbed Shirasu samples were obtained from the river embankment in Kagoshima Prefecture. Figure 4 shows the grain

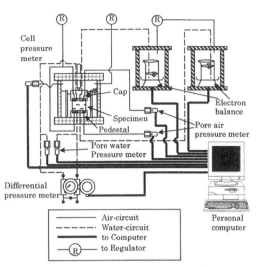

Fig.2 Unsaturated permeability testing apparatus (whole view)

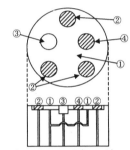

①Water-circuit for control (ceramic)
②Air-circuit for control (Glass filter)
③Water-circuit for measurement (ceramic)
④Air-circuit for measurement (Glass filter)

Fig.3 Plan view and cross section of pedestal and cap.

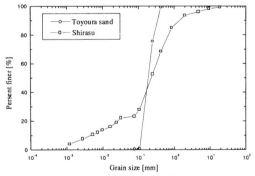

Fig.4 Grain size distribution curves of Toyoura sand and Shirasu

size distribution curves of Toyoura sand and Shirasu. The initial condition of specimen and applied pressure are listed in Table 1. The permeability test on these specimens was carried out in the wetting and drying processes. Figures 5 and 6 show the relations between suction, unsaturated permeability coefficient and volumetric water content.

Table.1 Physical quantities of sample and experimental condition

	Toyoura sand	Shirasu
Cell pressure [kPa]	50	
Pore air pressure [kPa]	30	
Net stress [kPa]	20	
Initial water content [%]	6.08	15.38
Dry density [g/cm³]	1.49	0.96
Void ratio	0.77	1.57
Height of specimen [cm]	1.96	1.87

4 NUMERICAL EXPERIMENT

The numerical experiment was carried out to obtain the relations between suction, unsaturated permeability coefficient and water content by using Kitamura's model. D_{10}, the diameter finer than 10 % is adopted as the height of element in Fig.1(b). The following logarithmic normal distribution is adopted as the probability density function of D in Fig.1(b).

$$P_d(D) = \frac{1}{\sqrt{2\pi}\zeta_v D} \exp\left\{ -\frac{(\log D - \lambda_v)^2}{2\zeta_v^2} \right\} \qquad (5)$$

where ζ_v: mean value, λ_v: standard deviation.

ζ_v in Eq.(5) is assumed to be the same as D_{10} which can be obtained by the grain size distribution curve. λ_v in Eq.(5) is obtained by assuming that the grain size distribution is regarded as the logarithmic normal distribution and its coefficient of variance is the same as that for $P_d(D)$. The pentagonal shape as shown in Fig.7 is assumed for the probability density function of $P_c(\theta)$, which can be prescribed by one parameter ζ_c. The values of these parameters are listed in Table 2. The results of numerical experiment are shown in Figs. 5 and 6. The relations between suction, unsaturated permeability coefficient and volumetric water content obtained by the numerical experiment are in good agreement with those obtained by the soil test for both Toyoura sand and Shirasu. The integration range, d, in Eq. (3) is 1/10 of that in Eq.(2), which means that the pore water related to water flow is much less than the retained pore water under unsaturated condition.

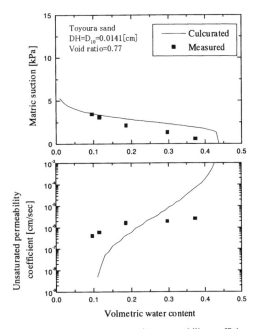

Fig.5 Relation between suction, permeability coefficient and volumetric water content for Toyoura sand

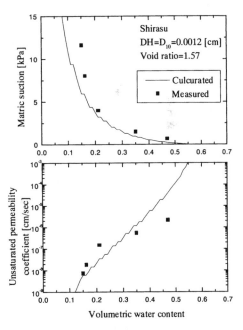

Fig.6 Relation between suction, permeability coefficient and volumetric water content for Shirasu

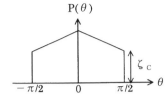

Fig.7 probability density function of θ

Table.2 Values of model parameters

	Toyoura sand	Shirasu
DH [cm]	0.0141	0.0012
Particle density [g/cm³]	2.64	2.45
Mean [cm]	0.0103	0.4322
Standard deviation [cm]	0.0036	147.942
ζ_c [cm]	0.159	

5 CONCLUSIONS

The numerical experiment and unsaturated permeability test on Toyoura sand and Shirasu were carried out to investigate the validity of Kitamura's model. As a result it is found out that the numerical simulation by Kitamura's model can describe the seepage behaviors of unsaturated soil. A great advantage of Kitamura's model is that the grain size distribution is the only data needed to simulate the seepage behaviors of unsaturated sandy soil. Additional unsaturated tests on various kinds of sandy soil should be carried out to prove Kitamura's model to be a useful model for the seepage model of unsaturated sandy soil in the near future.

ACKNOWLEDGEMENT

This research work is supported by the Grant-in-Aid for Scientific Researches (Project No.09555153 and No.10650490) from the Ministry of Education, Science and Culture of Japan.

REFERENCES

Fredlund,D.G. & Rahardjo,H.1993: Soil mechanics for unsaturated soils, John Wiley & Sons, INC.
Kitamura,R., Fukuhara,S., Uemura,K., Kisanuki,G. & Seyama,M. 1998: A numerical model for seepage through unsaturated soil, Soils and Foundations, Vol.38, No.4, pp.261-265.
Terzaghi,K. 1943: Theoretical soil mechanics, New York, Wiley.

Unsaturated Soils for Asia, Rahardjo, Toll & Leong (eds) © 2000 Taylor & Francis, ISBN 90 5809 139 2

Numerical simulation for pore fluid flow through unsaturated soil

R. Kitamura, Y. Miyamoto & Y. Shimizu
Department of Ocean Civil Engineering, Kagoshima University, Japan

ABSTRACT: In the southern part of Kyushu Island, Japan, slope failures often occur due to heavy rainfalls in the rainy season (from June to September). Most of these slope failures are classified as the surface slip failure whose depth is commonly less than 1m. The phase change of pore fluid due to temperature and pressure is deeply related to the occurrence of slope failure due to the change in effective stress in unsaturated soil. In this paper the phase change of pore fluid in unsaturated soil with the change of temperature, humidity and pressure in atmosphere is investigated to make clear the mechanism of surface slip. The numerical model is proposed to simulate the seepage behavior in unsaturated soil. Then the numerical experiment is carried out by the proposed numerical model. The results are compared with the data obtained by the field measuring of suction.

1 INTRODUCTION

Kagoshima Prefecture is located in the southern part of Kyushu Island, Japan, which belongs to both sub-tropical area and Aso-Kirishima volcanic zone. The pyroclastic flow deposits are widely distributed as the surface geo-material. The non-welded part of pyroclastic flow deposits is called Shirasu in Japanese which means white sand. In Kagoshima Prefecture failures often occur on Shirasu slopes due to heavy rainfalls in the rainy season (June-September). Most of these slope failures are classified as the surface slip failure whose depth is less than 1 m. The cyclic change in effective stress near the surface of slope is related to the surface slip failure because the seepage of rainwater and the evaporation of pore water in unsaturated soil causes the phase change of pore fluid followed by the change of pore pressure. These phenomena are qualitatively known as the cause of surface slip failure, but their qualitative estimation has not been made clear.

In this paper the numerical model is explained to simulate the seepage behaviour of rainwater into unsaturated soil, in which the void ratio, the volumetric water content, the suction and unsaturated-saturated permeability coefficient are derived. The field measuring system of suction, rainfall and temperature in soil is then explained. The results of numerical experiment is compared with those obtained by the field measurement to prove the validity of numerical model.

2 NUMERICAL MODEL

The water retention curve, unsaturated permeability coefficient and air permeability coefficient are necessary to estimate quantitatively the flow of pore water through unsaturated soil. Kitamura et al.(1998) proposed a numerical model for infiltration through unsaturated soil. Figure 1(a) is a typical picture showing the arrangement of a few soil particles in a soil mass, which is popularly observed by the optical microscope. The voids and soil particle in Fig.1(a) can be modeled by a pipe and other parts as shown in Fig.1(b), where a few soil particles and voids in the element correspond to the impermeable part (shaded part in Fig.1(b)) and permeable part (part of pipe in Fig.1(b)) respectively. The diameter and inclination angle of pipe are defined as D and θ.

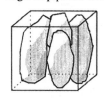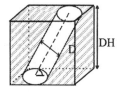

(a) A few soil particles in element   (b) Modeling of soil particles and voids

Figure 1 Modeling

The diameter D and inclination angle θ can be regarded as random variables because the shape and size of soil particles are random, and consequently the structure of their assembly is irregular. Therefore the probability density functions for D and θ can be introduced to estimate the distribution of voids in soil mass. The void ratio, volumetric water content, suction, unsaturated-saturated permeability coefficient and air permeability coefficient are derived by using the probability density functions of D and θ as follows.

$$e = \int_0^\infty \int_{-\pi/2}^{\pi/2} \frac{V_p}{V_e - V_p} P_d(D) P_c(\theta) d\theta dD \qquad (1)$$

$$W_v = \frac{1}{1+e} \int_0^d \int_{-\pi/2}^{\pi/2} \frac{V_p}{V_e - V_p} P_d(D) P_c(\theta) d\theta dD \qquad (2)$$

$$pF = \log_{10} h_c = \log_{10}\left(\frac{4 \cdot T_s \cdot \cos\alpha}{\gamma_w \cdot d}\right) \qquad (3)$$

$$k_w = \int_0^d \int_{-\pi/2}^{\pi/2} \frac{\pi \cdot \gamma_w \cdot D^3 \sin\theta}{128\mu(D/\sin\theta + DH/\tan\theta)^2} P_d(D) P_c(\theta) d\theta dD \qquad (4)$$

where e= void ratio;
V_e= volume of container in Fig.1(b);
V_p= volume of pipe in Fig.1(b);
$P_d(D)$= probability density function of D;
$P_c(\theta)$= probability density function of θ;
DH= height of container in Fig.1(b);
d= maximum diameter of pipe which can retain water;
T_s= surface tension;
α=contact angle between water and pipe;
ρ_w= density of water;
μ_w= viscous coefficient of water;
h_c= height of water due to surface tension (cm).

The basic equation for the heat conduction can generally be expressed as follows.

$$q = \lambda \cdot \Delta T \cdot \frac{A}{h} \qquad (5)$$

where q= heat flux;
λ =coefficient of heat conductivity;
ΔT= difference of temperature between two adjacent points;
A= cross section area;
h= distance between two adjacent points.

The unsaturated soil is composed of soil particles (solid phase), pore water (liquid phase) and pore air (gas phase) and then Eq. (5) can be modified as follows.

$$q_i = \lambda_i \cdot \Delta T \cdot \frac{A_i}{h}$$
$$q_t = \sum q_i (= q_{soil\ particle} + q_{water} + q_{air}) \qquad (6)$$
$(i = soil\ particle, pore\ water, pore\ air)$

where q_i = heat flux of phase 'i';
λ_i = coefficient of heat conductivity of phase 'i';
A_i = cross section area of phase 'i'.

The following logarithmic normal distribution equation is used as the probability density function of D.

$$Pd(D) = \frac{1}{\sqrt{2\pi}\zeta D} \exp\left\{-\frac{(\log D - \lambda)^2}{2\zeta^2}\right\} \qquad (7)$$

where λ ; mean value of log D,
ζ ; standard deviation of log D.

The pentagonal shape with the height of ζc as shown in Fig.2 is used as the probability density function of θ.

Figure 3 shows the relations between suction, permeability coefficient and volumetric water content which are obtained by the numerical experiment. The values of input parameters are listed in Table 1. These values are selected to correspond to Shirasu.

3 FIELD MEASURING SYSTEM OF SUCTION, RAINFALL AND TEMPERATURE

Figure 4 shows an arrangement of field measuring system of rainfall, suction and temperature with their data loggers. The suction in unsaturated soil is measured by the tensiometer. The tipping bucket type's rain gauge and thermister type's thermometer are used to measure the rainfall and temperature respectively. Figure 5 shows an example of measurement result.

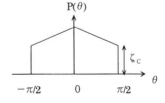

Figure 2 Probability density function of θ

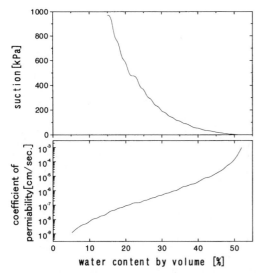

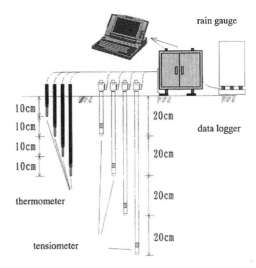

Figure 3 Relations between suction, permeability coefficient and water content by volume

Figure 4 Arrangement of field measuring system of rainfall

Suction and temperature with their data loggers

Table 1 Values of input parameters

Parameter	Value	
	Suction	permeability coefficient
λ_1 [cm]	$10^{-2.52298}$	$10^{-3.52298}$
ζ_1 [cm]	$10^{1.38352}$	$10^{1.38352}$
DH [cm]	$5.0*10^{-5}$	$1.0*10^{-5}$
ζ_c [cm]	0.159	0.159

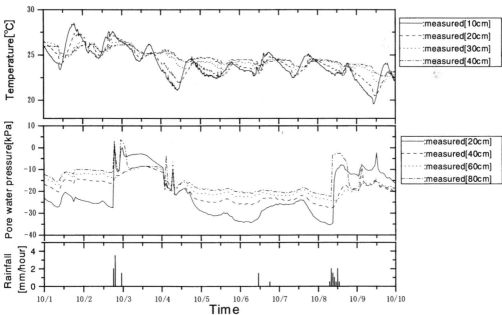

Figure 5 Field measuring results of suction, rainfall and temperature at Harihara in 1999

411

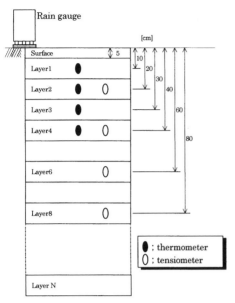

Figure 6 One-dimensional ground model

4 NUMERICAL EXPERIMENT

The numerical experiment was carried out to simulate one-dimensional heat conduction and seepage of rainwater in unsaturated soil. Figure 6 shows the ground model for one-dimensional heat conduction and seepage of rainwater in unsaturated soil. The bottom and side boundaries are assumed to be insulated and undrained condition.

Figure 7 shows the flow chart for the calculation procedure. The values of physical quantities in Table 2 are used to determine the initial and boundary conditions which represent the ground model in Figure 6. Figure 8 shows the results of numerical experiment.

5. CONCLUSION

The numerical model for the seepage behavior of unsaturated soil was introduced to simulate the seepage behavior of rainwater into unsaturated ground. The field measuring system was introduced and the data of suction and temperature in unsaturated ground, and rainfall measured by this system were shown. Then the change in suction due to rainfall with time

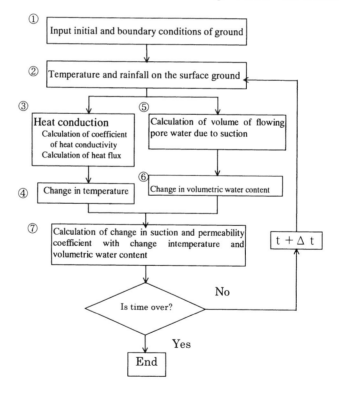

Figure 7 Calculation procedure

Table 2 Values of physical quantities used ground model

Number of layer	30
Cross sectional area	100 [cm^2]
Thickness of layer	10 [cm]
Density of soil particle	2.401[g/cm^3]
Specific heat(solid)	0.8 [J/(g·K)]
Coefficient of heat conductivity	3.0 [J/(m·K)]

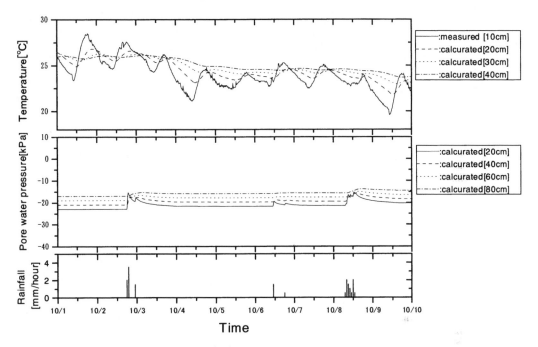

Figure 8 Results of numerical experiment

was simulated to compare it with that measured in the field. It was found out that the proposed numerical model can express the seepage behavior of rainwater into unsaturated ground behavior of pore water from soil, which reflect the change in suction due to the change in water content.

ACKNOWLEDGEMENT

This research work was supported by the Grant-in-Aid for Scientific Research (No.09555153, No.10650490) from the Ministry of Education, Science and Culture of Japan.

REFERENCES

Kitamura ,Uemura , Kisanuki and Seyama,(1998): A numerical model for seepage through unsaturated soil, Soils and Foundations, Vol.38,No.4, pp.261-265.

Unsaturated Soils for Asia, Rahardjo, Toll & Leong (eds) © 2000 Taylor & Francis, ISBN 90 5809 139 2

Impact of surface clogging on artificial recharge

D.J. McAlister & J. Arunakumaren
Water Assessment and Planning, Department of Natural Resources, Indooroopily, Qld, Australia

T. Grantham
Qld, Australia

ABSTRACT: This paper focuses on modelling fluid flow processes in the unsaturated zone beneath surface spreading artificial recharge facilities. A sixty layer model has been developed to simulate flow from a recharge basin through the unsaturated and saturated zones. Clogging of in-situ pore space is the major limiting factor in sustainable artificial recharge operations. Clogging rates have been simulated in the surface layers of the model by a power function. Several model scenarios have been generated with uniform and layered hydraulic conductivity soil media. Results of modelling indicate that the majority of recharge occurs in the initial 6 days after flooding of the basin. After this initial period recharge rates are reduced significantly by clogging. An understanding of these key parameters can help in generating and calibrating accurate numerical models. The data derived from these models can be used in management of artificial recharge operations.

1 INTRODUCTION

Artificial Recharge (AR) is a process where water is introduced into the sub-surface at an accelerated rate by anthropogenic means. AR can be carried out for aquifer storage and recovery, waste water filtration, permanent waste disposal, salt water intrusion remediation and subsidence control. Surface spreading and sub-surface injection are the two main techniques utilised for AR.

This paper focuses on modelling the fluid flow processes in the unsaturated zone beneath surface spreading AR facilities. Hypothetical examples have been generated to study controlling parameters associated with modelling fluid. An understanding of the parameters controlling flow in the unsaturated zone is vital for generating accurate predictive models. Factors such as variable vertical hydraulic conductivity, grain size of in-situ material and rates of clogging of pore space are investigated. Results from predictive modelling can be used in economic and management decision making.

2 BACKGROUND

Surface spreading AR facilities can be comprised of pits, basins, ponds, trenches, channels and weirs, this paper discusses the use of recharge basins. A recharge basin is a shallow excavation surrounded by an embankment, designed to be a water detention area. The AR process involves the introduction of waste water into the recharge basin. Infiltration occurs on the floor of the basin through pore space in the unsaturated zone.

Recharge basins can only be used over unconfined aquifers where the unsaturated zone contains highly permeable rock and soil. Most in-situ materials in the unsaturated zone are heterogeneous and thus vertical and horizontal hydraulic conductivity is anisotropic. Variations in the vertical hydraulic conductivity in lithological layers affect the rate and volume of water movement to the saturated zone and the aquifer.

The clogging of in-situ pore space reduces the rate of infiltration from the recharge basin. This is the significant retarding factor in AR operations. Clogging usually occurs from a combination of physical, chemical and biological processes. For the purpose of this paper the various forms of clogging have been grouped and demonstrated as a rate of reduction in permeability over time.

The types of waste water available for AR operations involve storm water run-off, municipal waste water, industrial waste, agricultural run-off or mine tailings. These water types may contain many suspended and dissolved organic and/or in-organic constituents. Treatment of waste water before recharge may reduce physical, health or environmental problems associated with AR.

A degree of filtration will occur as the waste water passes through the unsaturated zone. Aerobic and anaerobic bacterial conditions will reduce the organic load of the waste water, while physical clogging and chemical precipitation will reduce the in-organic load of the waste water.

3 UNSATURATED FLOW

AR by basin infiltration involves unsaturated and saturated groundwater flow processes. The hydraulic properties of the soil that control these processes are affected during recharge. When recharge commences, soil directly beneath the basin is unsaturated. This soil would normally have a lower hydraulic conductivity than the aquifer material. During recharge, clogging of the surface soil occurs, reducing hydraulic conductivity.

In unsaturated groundwater flow processes, hydraulic properties are dependent on the water retention characteristics of the soil. To study unsaturated flow processes, it is necessary to define the relationship between hydraulic conductivity versus water saturation and pressure head versus water saturation. Unsaturated hydraulic conductivity can be defined by Equation 1:

$$k = k_{rw} K \tag{1}$$

where:
k   = unsaturated hydraulic conductivity
k_{rw} = relative hydraulic conductivity
K   = saturated hydraulic conductivity

The relationship between relative hydraulic conductivity and water saturation is represented by the following equation (Huang et al, 1996):

$$k_{rw} = S_e^{1/2} \left[1 - \left(1 - S_e^{1/\gamma} \right)^\gamma \right]^2 \tag{2}$$

where:
S_e = effective saturation [cm^3/cm^3]

The relationship between pressure head and water saturation is represented by the following equation (van Genuchten, 1980):

$$S_e = \frac{\theta - \theta_r}{\theta_s - \theta_r} = \begin{cases} \left[1 + \left(\alpha |\psi| \right)^\beta \right]^{-\gamma}, & \psi < 0 \\ 1, & \psi \geq 0 \end{cases} \tag{3}$$

where:
θ = volumetric water content [cm^3/cm^3]
θ_r = residual water content [cm^3/cm^3]
θ_s = saturated water content [cm^3/cm^3]
S_e = effective saturation [cm^3/cm^3]
ψ = matric potential [cm] ($\psi \leq 0$)
α, β, γ = empirical parameters and $\gamma = 1 - 1/\beta$

Knowledge of hydraulic properties, as represented in the equations above, is essential in studying the movement of water and solutes through the sub-surface. Due to the spatial and vertical variability of soils, direct determination of these hydraulic properties in the field, is difficult and expensive. There are a number of indirect methods available to estimate hydraulic properties. These are based on soil texture and other readily available taxonomic information such as particle size distribution, exchangeable sodium percentage, bulk density and/or organic content. Figure 1 and 2 show a trend variation of the van Genuchten parameters α, β, θ_s and θ_r in relation to sand content. It should be noted that the variations in van Genuchten parameters cannot be directly attributed to sand content. These parameter variations depend on other factors such as the silt and clay content. Based on trend analysis, the variation of relative permeability in relation to matric potential and sand content was studied (Figure 3). Figure 3 shows that variations in relative permeability (curved lines) are not directly related to sand content, even if it is assumed that the van Genuchten parameters are directly correlated.

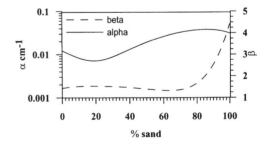

Figure 1. Variation of α and β van Genuchten parameters with sand content.

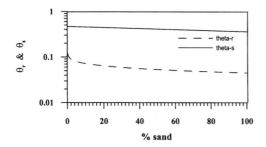

Figure 2. Variation of θ_s and θ_r van Genuchten parameters with sand content.

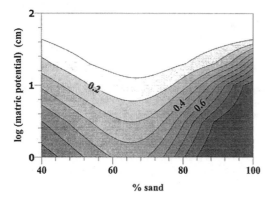

Figure 3. Variation of relative permeability with respect to matric potential and sand content.

4 MODELLING

A numerical model has been developed using MODFLOW-SURFACT (version 2.1) (MS) (HydroGeoLogic, 1998). The model has been used to study the unsaturated flow characteristics from AR basins, where hydraulic conductivity in the surface layers is reduced over time by clogging. MS is a comprehensive, three-dimensional, finite-difference flow and contaminant transport program. It is based on the United States Geological Survey modular groundwater flow model, MODFLOW (McDonald, 1988). The MS code is a more advanced code that allows simulation of groundwater flow under saturated and unsaturated conditions (based on Equations 1, 2 and 3).

The model simulates the flow from the surface (base of the flooded AR basin) to the water table (located at a depth of 3m from the surface). The water level in the recharge basin was kept at a constant head of 2m above the surface. The numerical model consists of 60 layers each of a uniform 5cm thickness. The clogging effect has been approximated in the first two layers of the model by a power function (Equation 4). Equation 4 was modified from an equation presented by Behnke (1969). A conceptualisation of the model is presented in Figure 4.

$$k = K t^{-n} \qquad (4)$$

where:

k	= hydraulic conductivity (m/d)
t	= time (days)
K and n	= constants

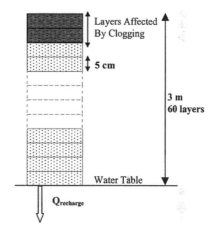

Figure 4. Conceptualisation of the model.

This study considers three different types of sub-surface media in five cases. The initial three cases shown in Table 1 are each of a uniform grain size media, representing low, medium and highly permeable sediment and soil. Case 4 and 5 were constructed using a layered combination of the different media of the first three cases to simulate more realistic hydrogeological conditions (Table 2).

Table 1. Input parameters used in model Cases 1, 2 and 3. Parameters were calculated from Figures 1 and 2.

	%sand	K (md⁻¹)	α (cm⁻¹)	β	γ	θ_s	θ_r	texture
Case 1	40	5	0.01298	1.46596	0.31785	0.4189	0.0557	Loam
Case 2	60	20	0.02605	1.33298	0.24980	0.3978	0.0509	Sandy Loam
Case 3	85	80	0.03780	2.00022	0.50006	0.3728	0.0469	Loamy Sand

Table 2. Percentage sand content in multi layer model-cases 4 and 5.

	Layer 1 (0 – 1 m) % sand	Layer 2 (0 – 2 m) % sand	Layer 3 (2 – 3 m) % sand
Case 4	40	60	85
Case 5	85	60	40

5 RESULTS

Results showed that in Cases 1 - 4 there is an initial period (0 – 6 days) of significant daily recharge. After this initial period daily recharge rates reduced significantly due to clogging in the surface layers of the model. Case 5 showed a linear decline in daily recharge rates. The higher conductivity materials (Case 3 and 2) allow a greater quantity of water to reach the water table over time. The combined layered media in Case 4 showed daily recharge figures similar to the low hydraulic conductivity sand content of Case 1.

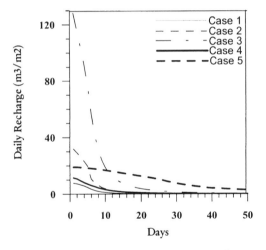

Figure 5. Model results showing daily recharge to the water table plotted against time.

Rate of reduction of recharge in Cases 1, 2 and 3 was then calculated by dividing the daily recharge results by the maximum daily recharge. These were then plotted against time and compared (Figure 6). Rate of reduction of recharge between cases was very similar. The greatest variation in results was gained by varying the n factor in Equation (4) that controls the rate of reduction in permeability (Figure 6). Varying the n factor (as in Figure 6) can be used in calibrating a model to actual observed data. There is a spike in the Case 2 curve where n = 1.5, which is probably attributed to numerical oscillation, generated by the MS program (Figure 6).

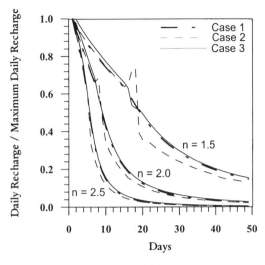

Figure 6. Daily recharge divided by the maximum daily recharge, plotted against time. Major differences in results are due to variance of the n factor in Equation 4.

The results of the different model runs can be presented by plotting water pressure on depth versus time graphs (Figures 7, 8 and 9). These graphs show the degree to which the soil is saturated or unsaturated (positive being saturated, negative being unsaturated).

Cases 1, 2 and 3 have similar saturation results from modelling (Figure 7). In the initial six days after flooding a large proportion of the profile is saturated. The profile becomes progressively less saturated with time as the surface layers clog (Figure 7).

In Case 4 the surficial low permeability zone (layer 1) impedes water transport to the underlying more permeable layers. Layer 3 tends to be less saturated than layer 1 and 2 at any given point in time. This is due to layer 3 having a greater permeability, allowing water to migrate to the water table more rapidly. As in Case 1, 2 and 3, the profile becomes progressively less saturated with time as the surface layers clog (Figure 8).

The results from Case 5, where the surface layers are highly permeable, show that a large amount of fluid enters the profile and is retained longer in the less permeable under lying layers 2 and 3 (Figure 9).

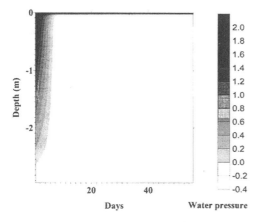

Figure 7. Water pressure plotted on a depth against time plot. Representative of Cases 1, 2 and 3.

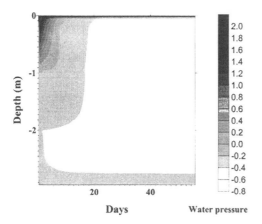

Figure 8. Water pressure plotted on a depth against time plot. Representative of Case 4.

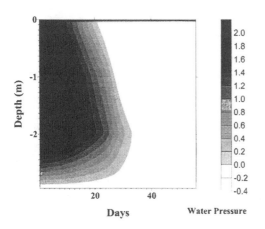

Figure 9. Water pressure plotted on a depth against time plot. Representative of Case 5.

6 ECONOMIC ANALYSIS

Results from interpretive modelling, show a distinct reduction in sub-surface permeability after a specific period of time. It is at this point that the economic integrity of the project is affected which can lead to unsustainability. It is important from an economic perspective that the monitoring and maintenance schedules allow appropriate cleaning and drying of the recharge basin at the point where reductions in permeability occur. These schedules are budget sensitive and will vary according to budget allocations related to water treatment and monitoring. Economically, this will be more efficient in allowing the maximum volume of recharge water to enter the aquifer.

7 CONCLUSION

1. Numerical modelling of the fluid flow processes in the unsaturated and saturated zones beneath artificial recharge basins has been carried out. This has been used to simulate the rate and amount of water entering the water table and aquifer. Results of modelling indicate that the majority of recharge occurs in the initial 6 days after flooding of the basin.
2. Physical, biological and chemical clogging are the major restrictions in surface spreading artificial recharge operations.
3. Clogging in surface layers of each model case has been simulated using a power function (Equation 4). Varying some of the parameters in the power function can be used to calibrate model results to observed field clogging rates.
4. Accurate approximation of the reduction in permeability in the surface layers below artificial recharge basins is the most important aspect of the modelling process.
5. Application of the knowledge gained can be applied to the design, maintenance and economic analysis of recharge schemes.

REFERENCES

Behnke, J.J. (1969). Clogging in surface spreading operations for artificial ground-water recharge. *Water Resources Research*. Vol. 5, NO 4:870-876.

Huang, K., B.P. Mohnaty, & M.Th. van Genuchten (1996). A new convergence criterion for the modified Picard iteration method to solve the variably saturated flow equation. *Journal of Hydrology*. 178:69-91.

HydroGeoLogic, Inc., (1998). MODFLOW-SURFACT): A comprehensive MODFLOW-based Flow and transport simulator. Code Documentation Report.

McDonald, M.G. & Harbaugh, A.W. (1988). A modular three-dimensional finite difference groundwater flow model. *US Geological Survey Techniques Water Resources Investigations Book 6*, Chapter A1.

van Genuchten, M. Th. (1980). A closed-form equation for predicting the hydraulic conductivity of unsaturated soils. *Soil Science Society of America Journal.* 4:892-898.

Unsaturated Soils for Asia, Rahardjo, Toll & Leong (eds) © 2000 Taylor & Francis, ISBN 90 5809 139 2

Determination of field-saturated hydraulic conductivity in unsaturated soil by the pressure infiltrometer method

T. Morii
Faculty of Agriculture, Niigata University, Japan

M. Inoue
Arid Land Research Center, Tottori University, Japan

Y. Takeshita
Faculty of Environmental Science and Technology, Okayama University, Japan

ABSTRACT: *In situ* measurements of field-saturated hydraulic conductivity, K_{fs}, are essential for accurate prediction of water movement in soil. The practical applicability of the pressure infiltrometer (PI) method to measure K_{fs} of soil is examined using field and laboratory tests and numerical calculations. Sand and loam soils were selected for the study. Theoretical features of the PI method are explained by the field permeability tests and the numerical calculations. It is shown that the PI method can be an excellent practical *in situ* permeability test. It is suggested that about 0.06 cm^{-1} or some smaller value of α^* may be more appropriate for sand and loam in calculating K_{fs} of soil, and that the measurement depth of the PI method can be controlled by combining the radius and insertion depth of the ring, and the constant head imposed on the soil surface within the ring.

1 INTRODUCTION

It is well established that *in situ* measurements of saturated hydraulic conductivity are essential for accurate determination of water movement in soil such as agricultural land, compacted soil and landfill. Because air bubbles are usually entrapped in porous media when they are saturated by infiltrating water, the saturated hydraulic conductivity measured in unsaturated soil is lower than the truly saturated hydraulic conductivity and is often referred to as a field-saturated hydraulic conductivity, K_{fs}. K_{fs} is considered appropriate to describe water movement in soil because many natural and man-made infiltration processes result in significant air entrapment within the soil.

The pressure infiltrometer (PI) method has been developed (Reynolds & Elrick 1990, Elrick & Reynolds 1992, Reynolds 1993) as a mean to measure K_{fs} based on a constant-head steady infiltration into soil from a single ring. It involves only a measurement of a steady-state infiltration rate required to maintain a steady head of water applied on the soil surface within the ring inserted a small depth into the soil. An apparatus to measure the steady-state infiltration as well as a procedure to calculate K_{fs} is quite simple. In this study, a theoretical feature of the PI method is examined by comparing the field experiment with the numerical calculation. The accuracy of measurement of K_{fs} is investigated through field and laboratory experiments, and the measurement depth of the PI method is estimated based on numerical calculations.

2 APPARATUS AND PROCEDURE

The PI apparatus consists of a single steel ring with a radius a inserted into the soil to a depth d, a water supply tube and a water reservoir tank as shown in Figure 1. The position of an air tube controls the constant head of water H applied on the soil surface within the ring. The steady-state infiltration rate Q_s is measured during the constant-head infiltration from the single ring into the soil.

Based on theoretical considerations and the

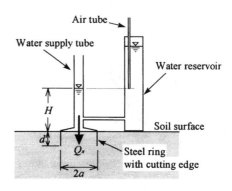

Figure 1. Schematic diagram of the PI apparatus.

numerical experiments of the three dimensional infiltration from the single ring into the soil, Reynolds & Elrick (1990) developed the equation to calculate K_{fs} of the soil:

$$K_{fs} = \frac{\alpha^* G Q_s}{\alpha^* a H + a + \alpha^* \pi a^2 G} \quad (1)$$

where G is a dimensionless shape factor which accounts for the geometry of the infiltration surface within the ring and is calculated by

$$G = 0.316 \frac{d}{a} + 0.184 \quad (2)$$

α^* in Equation 1 is a parameter which describes an exponential relationship of unsaturated hydraulic conductivity $K(h)$ and suction h of soil, and is interpreted as an index of texture/structure component of soil capillarity. The PI method requires that α^* be site-estimated by simple observation of soil. Suggested values of α^* for various soil textures and structures are given by Elrick & Reynolds (1992). If $K(h)$ is measured independently, the value of α^* can be calculated by

$$\alpha^* = \frac{K_{fs}}{\int_{-\infty}^{0} K(h)dh} = \frac{1}{\int_{-\infty}^{0} K_r(h)dh} \quad (3)$$

where $Kr = K_r(h)$ is a relative hydraulic conductivity of soil and is defined as a ratio of $K(h)$ divided by K_{fs} ($0 \le K_r \le 1$). α^* calculated by Equation 3 will be referred to as an integrally-equivalent α^* in the following.

3 FIELD EXPERIMENTS

3.1 Moisture movements in soil

The field permeability test was conducted at ALRC, Tottori, to examine the theoretical feature of the PI method. The soil tested is classified as sand with low content of fine soil particles. The maximum particle diameter is about 1 mm, as shown in Figure 2. The

steel ring with a=5.5 cm was inserted into the soil to a depth d=3.0 cm. H=9.7 cm was imposed on the soil surface within the ring for 10 minutes. The infiltration rate from the ring into the soil Q decreased soon after the beginning of the constant-head infiltration and approached the constant value. Q_s, which is calculated as an average of the infiltration rate measured during 5 to 10 minutes, was 12.89 cm³/s. h and the volumetric moisture content θ in the soil were monitored using four tensiometers and one moisture gauge, respectively, as shown in Figure 3a. The soil was dug, instrumented and uniformly back-filled before the permeability test. The average value of the dry density in the compacted soil was 1.48 g/cm³. The pressure head in the soil measured before the beginning of infiltration is given in Figure 3b.

Functional relationships between θ, h and K_r were previously determined (Inoue et al. 1982, Inoue 1987) and are shown in Figure 4. Numerically integrating the K_r-h relationship in Figure 4 and taking its reciprocal, the integrally-equivalent α^* is determined as 0.068 cm⁻¹ from Equation 3. Inserting values of the integrally-equivalent α^* and Q_s measured into Equation 1, then K_{fs}=2.74×10⁻² cm/s of the sand is obtained.

The axisymmetric water movement from the single ring into the soil is analyzed by numerical calculation using the finite element method (FEM) developed previously (Morii 1999). The entire region of the soil to be analyzed is 60 cm in depth

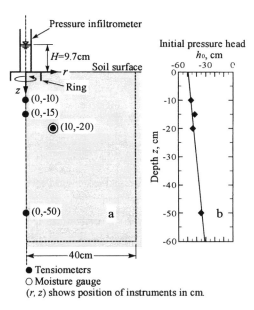

Figure 3. The monitoring instruments (a) and the initial pressure head measured in the soil (b) in the field permeability test.

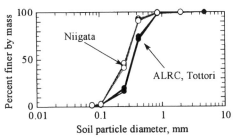

Figure 2. Grain size distribution of the soils tested.

and 40 cm in radius as shown by the dotted line in Figure 3a. The symmetric axis, the soil surface except for within the ring, and both sides and edge of the ring inserted into the soil are considered as impervious boundaries. 2.5 mm thickness of the steel ring is assumed in the numerical calculation.

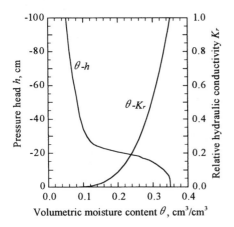

Figure 4. Unsaturated moisture properties of the soil.

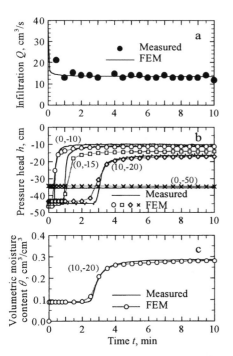

Figure 5. Comparison of (a) infiltration, (b) pressure head and (c) moisture content between the field measurement and the FEM calculation.

The pressure heads along the base and the vertical side of the entire region of the soil are fixed at the initial values. The initial pressure head in the soil at the beginning of the infiltration is assumed to be linearly distributed as shown by the regression line in Figure 3b. The allowable tolerance of the pressure head during the iterative computations in the FEM calculation is set at 0.01cm.

Q from the ring into the soil, h and θ in the soil measured with time are compared with calculated values in Figures 5a, 5b and 5c, respectively. Fairly good agreement between measurement and calculation confirm the theoretical effectiveness of the PI method. In Figure 6 the FEM calculation is plotted to show advances of the saturated water bulb and the wetted zone with time. The saturation at the wetting front shown in Figure 6 corresponds to 50 %. It is interesting to note that the size of the saturated water bulb remains almost constant during the constant-head infiltration, whereas the wetted zone expands with time. The size of the saturated water bulb depends on the constant pressure head imposed on the soil surface within the ring.

3.2 Measurement of K_{fs} in sand soils

The field permeability tests using the PI method were conducted at 142 plots in sand soils at ALRC, Tottori, and Niigata to examine the accuracy of measurement of K_{fs}. The sands with low content of fine soil particles as shown in Figure 2 were tested in the study. A steel ring with a=5.5 cm and d=3.0 cm was employed. H at about 5 to 15 cm was imposed on the soil surface within the ring, and Q_s

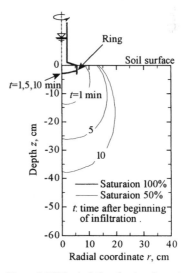

Figure 6. FEM calculation showing the moisture movement in soil with time during the constant-head infiltration from the ring.

from the ring into the soil was measured. Intact soil cores 100 cm³ in volume were bored from the test plot after the completion of the PI measurement, and were partially immersed into the water for one day. Then the hydraulic conductivity of the soil cores, $K_{fs(core)}$, was measured using the constant head permeability test in the laboratory.

Figure 7 compares K_{fs} determined by the PI method with $K_{fs(core)}$. The integrally-equivalent α^* of 0.06 cm⁻¹ obtained in Section 3.1 was used in Equation 1 to calculate K_{fs}. Note both K_{fs} and $K_{fs(core)}$ in Figure 7 are corrected to 15 °C of water temperature. It is found that the PI method yielded a factor of 1.2 to 1.3 higher K_{fs} than the soil cores. This level of accuracy may be sufficient for most practical applications of the PI method.

4 NUMERICAL EXPERIMENTS

4.1 Effect of α^* on the K_{fs} calculation

If an unsuitable value of α^* is estimated in the PI method, some error will be introduced in the K_{fs} calculations using Equation 1. This type of error is investigated based on the FEM numerical

calculations. The soil region to be analyzed and the boundary conditions of the soil region are the same as Figure 3a. At the beginning of the constant-head infiltration the pressure head is assumed to be distributed linearly from 0 cm at the groundwater level located at 100 cm in depth to -100 cm at the soil surface. Four soils published in the literature were assumed and listed in Table 1. The saturated hydraulic conductivity K_s used in the numerical calculations, the values of α^* suggested by Elrick & Reynolds (1992), and the integrally-equivalent α^* of the soils are given in Table 1. The unsaturated properties of the soils are given in the literature listed in the table. Three rings 2.5 mm thickness with $a=7.5$ cm and $d=3.0$ cm, $a=7.5$ cm and $d=5.0$ cm, and $a=5.0$ cm and $d=5.0$ cm were employed in the numerical calculations. For each ring, $H=10$ and 20 cm were assumed. Q_s was determined from numerical calculation, then two values of the field-saturated hydraulic conductivity, that is $K_{fs(suggest)}$ and $K_{fs(equivalent)}$, were calculated by Equation 1 using the suggested α^* and the integrally-equivalent α^*, respectively. The same tolerance of the pressure head as given in Section 3.1 was adopted in the calculations.

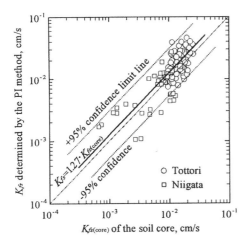

Figure 7. Comparison of K_{fs} determined by the PI method in the sand soils with the soil cores.

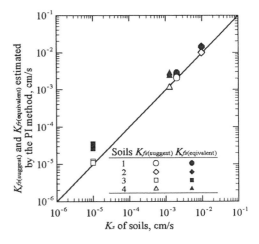

Figure 8. Difference of K_{fs} calculated using the suggested α^* and the integrally-equivalent α^*.

Table 1. Moisture properties of the soils employed in the FEM numerical calculations.

Soils	Saturated conductivity K_s cm/s	α^* suggested [++] cm⁻¹	α^* by Equation 3 cm⁻¹	Unsaturated moisture properties. See references.
1. Manawatu fine sandy loam	2.0×10^{-3}	0.12	0.065	Quadri et al. (1994)
2. Fine river sand	9.722×10^{-3}	0.12	0.047	Vauclin et al. (1979)
3. Weld silty clay loam	1.0×10^{-5} [+]	0.12	0.018	Yates et al. (1992)
4. Muir silt loam	1.278×10^{-3}	0.12	0.022	Sisson et al. (1992)

[+] Value of K_s is assumed.
[++] The value is suggested as the first choice for most soils (Elrick & Reynolds 1992).

Figure 8 shows the comparison between K_s and $K_{fs(suggest)}$ and between K_s and $K_{fs(equivalent)}$. It is noticed that $\alpha^* = 0.12$ cm^{-1} suggested as the first choice for most soils yields a factor of about 1.5-2 higher K_{fs} in sand soil, and a factor of about 2-3 higher K_{fs} in loam soil than the true value K_s. Of course this level of accuracy in determining K_{fs} should be judged based on the objectives of the field test. It may not be wrong to think that, from the results in Sections 3.1, 3.2 and 4.1, about 0.06 cm^{-1} or some smaller value of α^* is more appropriate for sand and loam soils than the suggested value of α^*.

4.2 Measurement depth of the PI method

FEM numerical calculations which simulate the constant-head infiltration from the single ring into the soil was conducted to investigate the measurement depth of the PI method. For this, it was assumed that a horizontal lower soil which was less permeable than the upper soil was present at a depth ζ in the soil region given in Figure 3a. Then ζ at which Q_s from the soil surface changes significantly can be regarded as the measurement depth of the PI method. Two 2.5mm-in-thickness rings with $a=7.5$ cm and $d=3.0$ cm, and $a=5.0$ cm and $d=5.0$ cm were employed, and $H=10$ and 20 cm were assumed for each ring in the numerical calculations. ζ was changed from 60 to 5 cm, where $\zeta=60$ cm means a homogeneous soil region. The field-saturated hydraulic conductivity of the upper soil as well as the homogeneous soil, $K_{fs(upper)}$, was assumed to be

the same value as the sand described in Section 3.1, and K_{fs} of the lower soil be one-tenth of $K_{fs(upper)}$. Both the upper and lower soils had the same unsaturated properties as shown in Figure 4. The boundary and initial conditions of the soil region are same as adopted in the preceding section.

Figure 9 shows the results of the numerical calculations, where K_{fs} estimated from Q_s using Equation 1 is divided by $K_{fs(upper)}$ that is a true value to be estimated. It is found that the ratio of $K_{fs}/K_{fs(upper)}$ begins to decrease from 1.0 when ζ approaches to 15-25 cm. This value of ζ may correspond to the measurement depth of the PI method, using the ring and H specified above, in the soil with K_{fs} of the order of 10^{-2} cm/s. A potentially important feature of the PI method induced from Figure 9 is that the measurement depth of the PI method can be controlled by combining a and d of the ring and H. This offers a practical advantage in conducting the *in situ* permeability test in compacted soil in earth works such as embankment dams and landfills.

A typical result of the FEM calculations is given in Figure 10 to show the water movement into the layered soil from the single ring, $a=7.5$ cm and $d=3.0$ cm, during the constant-head infiltration. The less permeable lower soil is located at $\zeta=10$ cm. Contrary to Figure 6 which shows the water movement in the homogeneous soil, the saturated water bulb is expanding with time. This is because the water flow from the ring is prevented from moving downward. Since the hydraulic gradient

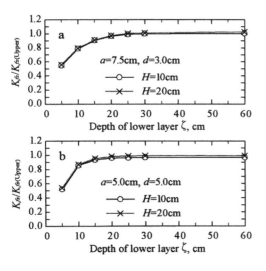

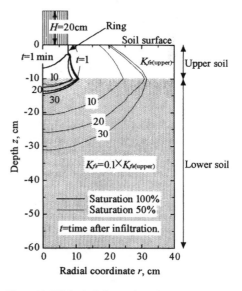

Figure 9. FEM calculation to estimate measurement depth of the PI method in sand soil. The rings with (a) $a=7.5$ cm and $d=3.0$ cm and with (b) $a=5.0$ cm and $d=5.0$ cm are employed in the calculations.

Figure 10. FEM calculation to show the water movement into the layered soil from the single ring during the constant-head infiltration.

beneath the soil surface within the ring decreases as the saturated water bulb expands, Q_s will also decrease, resulting into the lower estimation of K_{fs} by Equation 1. This is the reason that induced the decrease in $K_{fs}/K_{fs(\text{upper})}$ in Figure 9.

5 CONCLUSIONS

The practical applicability of the PI method to measure the field-saturated hydraulic conductivity of soil was examined by field and laboratory permeability tests and FEM numerical calculations. Sand and loam soils were selected for the study. The following are concluded:

1) Theoretical features of the PI method are well confirmed by the field permeability tests and the FEM numerical calculations.
2) The PI method can be a excellent practical *in situ* permeability test because of its simple and speedy procedure to determine the field-saturated hydraulic conductivity of soil.
3) About 0.06 cm^{-1} or some smaller value of α^* may be more appropriate for sand and loam soils in calculating the field-saturated hydraulic conductivity of soil.
4) The measurement depth of the PI method can be controlled by combining the radius and the insertion depth of the ring, and the constant head imposed on the soil surface within the ring.

ACKNOWLEDGMENTS

This study was supported in part by the Joint Research Grant made by the Arid Land Research Center, Tottori University, Japan. The authors are grateful to Misses E. Hata, S. Yoneguchi, Y. Akita and M. Sato for their help in conducting the field and laboratory permeability tests.

REFERENCES

Elrick, D. E. & Reynolds, W. D. 1992. Infiltration from constant-head well permeameters and infiltrometers. In G. C. Topp, W. D. Reynolds & R. E. Green (eds), *Advances in Measurement of Soil Physical Properties: Bringing Theory into Practice*. SSSA Special Publication 30: 1-24. Soil Science Society of America, Madison, WI.

Inoue, M. 1987. Evaluation of soil water properties during drainage periods in a sand field. *Bulletin of the Faculty of Agriculture, Tottori University, Japan*. 40: 119-129. (in Japanese with English summary)

Inoue, M., Yano, T., Yoshida, I., Yamamoto, T. & Chikushi, J. 1982. Determination of unsaturated hydraulic conductivity from soil water characteristic curve. *Research Association of Soil Physics, National Institute of Agricultural Sciences, Japan*. 46: 21-29. (in Japanese with English summary)

Morii, T. 1999. Prediction of water movement in soil by finite element method. *Bulletin of the Faculty of Agriculture, Niigata University, Japan*. 52(1): 41-54.

Quadri, M. B., Clothier, B. E., Angulo-Jaramillo, R., Vauclin, M. & Green, S. R. 1994. Axisymmetric transport of water and solute underneath a disk permeameter: experiments and numerical model. *Soil Science Society of America Journal*. 58: 696-703.

Reynolds, W. D. 1993. Saturated hydraulic conductivity: field measurement. In M. R. Carter (ed.), *Soil Sampling and Methods of Analysis*: 599-613. Boca Raton: Lewis Publishers.

Reynolds, W. D. & Elrick, D. E. 1990. Ponded infiltration from a single ring: I. Analysis of steady flow. *Soil Science Society of America Journal*. 54: 1233-1241.

Sisson, J. B. & van Genuchten, M. Th. 1992. Estimation of hydraulic conductivity without computing fluxes. In M. Th. van Genuchten, F. J. Leij & L. J. Lund (eds), *Indirect Methods for Estimating the Hydraulic Properties of Unsaturated Soils; Proceedings of the International Workshop, Riverside, California, 11-13 October 1989*: 665-674. University of California, Riverside, CA.

Vauclin, M., Khanji, D. & Vachaud, G. 1979. Experimental and numerical study of a transient, two-dimensional unsaturated-saturated water table recharge problem. *Water Resources Research*. 15(5): 1089-1101.

Yates, S. R., van Genuchten, M. Th., Warrick, A. W. & Leij, F. J. 1992. Analysis of measured, predicted, and estimated hydraulic conductivity using the RETC computer program. *Soil Science Society of America Journal*. 56: 347-354.

Unsaturated Soils for Asia, Rahardjo, Toll & Leong (eds) © 2000 Taylor & Francis, ISBN 90 5809 139 2

Hydraulic conductivity measurements in two unsaturated soils of Querétaro Valley, Mexico

A. Pérez-García & D. Hurtado-Maldonado
Engineering Faculty, Universidad A. de Querétaro, Mexico

ABSTRACT: In this paper the authors describe an experimental apparatus capable of measuring the hydraulic conductivity of unsaturated soils. The method uses the principle of instantaneous modified profile for the permeability measurements. The apparatus consist of a series of rings in the vertical and horizontal position with a soil sample inside. Hydraulic head is applied and kept constant during all the test. All rings have a hole with a psychrometer probe for measuring the soil suction in each ring of soil. With increase of water volume in the soil we determine the flow rate and the hydraulic conductivity at each point. We present the results obtained from two compacted soils, namely a silty sand and montmorillonite clay of Queretaro Valley. Hydraulic conductivity values presented in this paper seem to agree with those reported in the literature.

1 INTRODUCTION

1.1 *General*

The importance of knowing the phenomenon of water flow in unsaturated soils, for both geotechnical projects and environment protection, has generated the necessity to measure the hydraulic conductivity of the soil in the laboratory and to estimate it *in situ*.

This parameter is very important for the understanding and the solution of the phenomena present in:
a) expansive soils, b) the problem of environmental protection by means of compacted clay soils in sanitary fills, and c) polluting liquid deposit lagoons.

It is necessary that engineers have experimental tools to describe the behavior of these soils. Currently, several procedures are available to measure the rate of water flow in unsaturated soils. However, it is necessary to delve more deeply in this respect.

In the case of unsaturated soils, the flow is due to hydraulic gradients in the soil mass, making the water move from a point with a higher hydraulic head to a smaller one.

In the case of partially saturated soil, one generally has negative pore-water pressure and the flow is in fact generated by the gradients or differences of negative pressures in the soil.

The water moves from areas of smaller suction to areas of larger suction. The gas bubbles prevent an easy circulation of the water, and as a consequence a decrease of the hydraulic conductivity, $k(y)$.

The permeability that corresponds to the saturated soil has a constant value. When the suction increases the degree of saturation reduces and $k(y)$ decreases. However, hysteresis is present. This implies that the reverse path is not the same; that is, for the same suction, the permeability on the drying path is different from that on the wetting path. The permeability decreases when the volumetric water content also decreases.

The main objective of the paper is the description of an experimental laboratory test for measuring hydraulic conductivity, as well as to show the curves of hydraulic conductivity obtained and other parameters of the studied unsaturated soils. The method used corresponds to that of the instantaneous profile method for unsteady flow.

2 PREVIOUS STUDIES

The instantaneous profile method in the laboratory was first proposed by Richards in the fifties. Hamilton used the same idea and proposed a similar apparatus in 1981. Richards indicated that the method can be useful for *in situ* tests. Klute in 1972 considered it the best field test method.

Alimi-Ichola & Bentoumi (1995) showed several infiltration tests in unsaturated soils placed in a special permeameter that uses the instantaneous profile method. The apparatus provided results in the vertical as well as in the horizontal positions. These tests were compared with the hydraulic conductivity and the diffusivity of an unsaturated soil obtained in a theoretical way. Infiltration tests were made in a column 225 mm high, formed by soil rings. The suction and humidity were measured at different heights of the column, in the soil rings to give suction profiles and humidity versus depth.

Juca & Frydman (1996) reported that there were enough measurement data on the flow properties of unsaturated soils. These data indicate a trend towards direct measurement of the hydraulic conductivity. However, the results are still within a small range of soil suction.

Benson & Gribb (1997) analyzed a total of 40 methods to measure the hydraulic conductivity of unsaturated soils and discussed its advantages, disadvantages and costs. The methods analyzed include both laboratory and field methods. They recommended the instantaneous profile method for tests of fine grain soils and also when these have high suction values. Benson and Gribb reached the conclusion that it is necessary to continue with the development of methods for the conductivity measurements where the stress state can be controlled during the test.

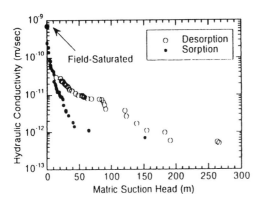

Figure 1. Tests of hydraulic conductivity obtained in an silty clay (Wenatchee) on the drying and wetting path (Meerdink et al. 1996, mentioned by Benson & Gribb 1997).

3 DARCY'S LAW FOR UNSATURATED SOILS

The water flow in a saturated soil can be described using Darcy's law (1856). In this law it is demonstrated that the speed of water flow through a soil mass is proportional to the hydraulic head gradient.

$$v_w = -k_w \frac{\Delta h_w}{\Delta y} \qquad (1)$$

Where:

v_w = Water flow speed; $\Delta h_w / \Delta y$ = Hydraulic head gradient in the y direction = i_{wy}; k_w is the permeability coefficient; for saturated soils k_w is considered to have a constant value.

Several authors have demonstrated that Darcy's law is also applicable for water flow through an unsaturated soil, that is to say, it has been demonstrated that the flow speed through a partially saturated soil is linearly proportional to the hydraulic head gradient.

However, the value k_w is not constant, thus it is more convenient to use the concept of hydraulic conductivity $k(y)$ instead of the coefficient of permeability.

During the wetting and drying processes the soils exhibit hysteresis. This arises mainly in the relationship between hydraulic conductivity, $k(y)$, and matric suction or in the relationship between volumetric water content and matric suction.

3.1 Instantaneous profile method.

This direct method is classified as unsteady flow. A cylindrical specimen of soil was subjected to water flow from one end.

Several procedures exist, which have variants in the form of flow entrance, in the measurement of the hydraulic head gradient and in the flow speed.

The hydraulic head gradient and the flow speed are determined at several points along the specimen, using some variants described by Fredlund (1993). In one of them, the water content and the distribution of the pore water pressure head were measured in an independent way.

The water content profile is used to calculate the flow speed. The pressure head of pore water can be calculated from the measurements of the soil suction by means of tensiometers and psychrometers.

4 MODIFIED APPARATUS TO MEASURE THE PERMEABILITY

Hamilton et al. (1981) had used the instantaneous profile method. The water flow is controlled at one end of the soil specimen, while the other end is opened to the atmosphere. Hamilton et al. used the final volumetric water content along the specimen, and the corresponding measured suction to produce the soil-water characteristic curve.

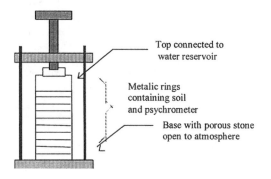

Figure 2. Apparatus used in the measurement of soil permeability.

The apparatus consists of ten rings of 5 cm inner diameter and 2 cm thick. It also has, in the ends, two porous stones of 0.5 cm thickness. The water flow is controlled at one end. The other one is opened to the atmosphere. The test is carried out with specimens of compacted soil or with natural specimens of expansive soil. The soil is placed inside each ring and later, all the rings are assembled to form a column of nine or ten rings, which are sealed with O-Rings.

Water is supplied at one end under a constant hydraulic head in all the tests. For this, a device based on the Mariotte bottle principle was built. Each ring is provided with a psychrometer. The set-up was placed in the horizontal position for maintaining a constant head. The apparatus was placed inside a constant temperature chamber maintained at 20°C +/- 1° C and under a relative humidity of 70%.

4.1 Preparation of the samples

The soil used in the tests was a fine sand with 22% of fine, classified as SM according to the Unified soil classification system.

The soil was previously sieved through mesh number 20 to remove large grains that could interfere with the compaction process. The soil was compacted such that it begins with a maximum suction. The soil was compacted inside rings of stainless steel using static compaction in order to achieve a soil, with the following initial characteristics:

Saturation S = 35%, dry unit weight γ_d = 17.7 kN/m^3, initial water content w_i =5.5%

The contact between the soil of two rings was made through two filter paper disks. We believe that, there is no influence of the paper in the measurements due to the high permeability of the material with which the filter paper is made of.

4.2 Calculations

To determine the permeability of the soil, it is necessary to know the flow speed and the hydraulic gradients that act on the specimen which can be calculated using the hydraulic head (h_w) against distance (x) -or depth - for each time. The hydraulic head that acts on the specimen consist of the gravity head and the pressure head:

$$h_w = y + \frac{U_w}{\rho_w g} \qquad (2)$$

Where:

y = Elevation of the point in study referred at a reference level; U_w = pore water pressure, in this case the measured suction; g = Gravity acceleration; ρ_w = Water density.

Once the hydraulic head was obtained, one proceeds to the gradient calculation:

$$i_w = \frac{dh_w}{dx} \qquad (3)$$

Where:

i_w = Hydraulic head gradient at a point for a specific time; dh_w/dx = slope of the profile of hydraulic head gradient at the point under consideration.

The flow speed " v_w ", at a point is similar to the water volume that flows through the area of the specimen traverse section, during an interval of time " dt ". The water flows from the permeameter left-end to the right end.

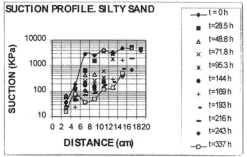

Figure 3. Suction profile measured at different times in the rings of the first studied soil (silty sand).

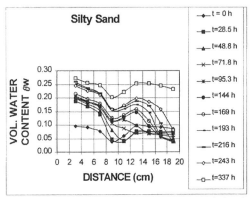

Figure 4. Evolution of the volumetric water content during the flow test in the studied soil (silty sand).

The total volume of water that passes through a point (for example the center of any ring) in the soil specimen, during a period of time, is similar to the change of water volume that passes between the point under consideration and the right end of the specimen during the period of specified time.

To find the changes of water volume in each ring it would be enough to divide weight increment in each ring by the water specific weight:

$$d v_{wi} = \frac{dw_{wi}}{\gamma_w} \qquad (4)$$

Where:

dv_{wi} = Water volume increase in a point "i", in an interval of specific time; dw_{wi} = water weight increase at a specific point "i", in an interval of specific time; γw = Water specific weight.

The total water volume "dV_{wj}" that passes through a point "j" during the time interval under consideration "dt" and is the sum of the volume changes that passes between the point under consideration and the last ring (m = 10):

$$d V_{wj} = \sum_{i=j}^{m} d v_i \qquad (5)$$

Where:

dV_{wj} = total water volume that passes through the point "j" in an interval of time "dt"; dv_i = water volume increment in each ring in an interval of specific time " dt".

The flow speed at a considered point "j", is calculated as follows:

$$v_w = \frac{dV_{wj}}{A\,dt} \qquad (6)$$

Where:

v_w = Flow speed of water, A = Area of the cross section of the specimen, dt = time interval.

The water flow speed corresponds to the average of the hydraulic head gradients "i_{prom}", obtained from two time series. Thus, the water permeability "k_w" is calculated dividing the flow speed "v_w" by the gradient of average hydraulic head "i_{prom}":

$$k_w = \frac{v_w}{i_{prom}} \qquad (7)$$

Where:

v_w = Flow speed; i_{prom} = Gradient of average hydraulic head.

The calculations can be repeated for different points and at different times. As a result, many permeability values can be calculated for several water contents or for suction values obtained in the test.

5 OBTAINED RESULTS

The graphs obtained are shown in Figures 4 and 5 for the first studied soil. In these graphs, both the suction evolution and the volumetric water content as a function of time is observed. With time, the suction profile decreases because of the advance of the wetting front. Clearly, the water content in the soil rings increases with time. These profiles allow the hydraulic conductivity in each of the soil rings to be calculated.

Figure 5 and Table 1 show some of the most important results obtained by this procedure for silty sand.

The points represent results of different rings, on the wetting path. The data shown in Figure 5 can be compared with the hydraulic conductivity data obtained by Meerdink et al. (1996) shown Figure 1. In Figure 1, the experimental points on the wetting path are very similar quantitatively in terms of hydraulic conductivity $k(y)$ for the suction values obtained in this study.

In the studied soil the lowest hydraulic conductivity, measured and calculated is in the order of 5.83 E-11 cm / s.

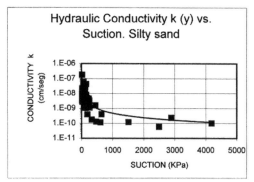

Figure 5. Hydraulic conductivity *k(y)* of the soil studied versus the soil suction.

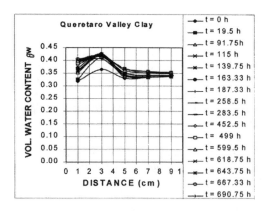

Figure 7. Volumetric water content measured at different times in the rings of the studied soil (clay CH Queretaro, Mexico).

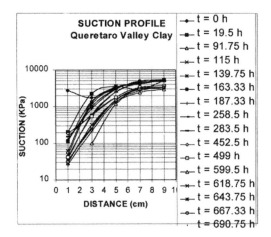

Figure 6. Suction profile measured at different times in the rings of the studied soil (clay CH Queretaro, Mexico).

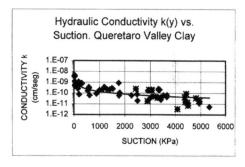

Figure 8. Hydraulic conductivity *k(y)* of the clay soil studied versus soil suction. The points represent results of diverse rings on the wetting path.

6 CONCLUSION

Hydraulic conductivity for a silty sand and a clay soil were obtained as a function of the suction using the instantaneous profile method.

The results are in agreement with those reported in the literature as can be observed in Figures 1 and 5

Table 1. Example of calculation of the hydraulic conductivity by using the data taken at different times in one of the rings of the apparatus (for silty sand).

RING Number 4

TIME (HOURS)	WATER VOLUME ACCUMULATED	CORRECTED SUCTION IN 4	PREVIOUS SUCTION IN 3	POSTERIOR SUCTION IN 5	HYDRAULIC HEAD	DISTANCE ΔX	HIDRAULIC GRADIENT	INTERVAL Δt	HYDRAULIC CONDUCTIVITY k
	(cm3)	(KPa)	(KPa)	(KPa)	(METERS)	(METERS)	(i)	(SECONDS)	(cm/seg)
0.0	0.00	2492	2891	3912	102.1	0.04	2552	102600	5.83E-11
28.5	0.30	1515	636	3887	325.1	0.04	8127	72900	1.20E-10
48.8	1.40	468	441	4192	375.0	0.04	9376	82800	1.30E-10
71.8	1.98	253	246	604	35.8	0.04	895	84888	1.10E-09
95.3	1.64	194	204	324	12.1	0.04	302	175500	2.38E-09
144.1	2.48	149	175	234	5.8	0.04	146	88200	7.03E-09
168.6	1.78	143	113	226	11.4	0.04	284	88200	7.98E-09
193.1	3.93	93	95	171	7.6	0.04	191	81612	8.57E-09
215.8	2.62	83	68	113	4.4	0.04	111	99000	1.43E-08
243.3	3.08	116	107	146	3.9	0.04	97	337500	2.75E-08
337.0	17.69	37	77	50	--	0.04	--		--

for the silty sand soil and in Figure 8 for the clay soil.

The method has the disadvantage of providing results after some four weeks. But the results demonstrate that it is feasible to obtain low values of hydraulic conductivity for soils with high suctions if these are within the measurement range of the psychrometers. The method was used to measure the conductivity following a wetting path in the soil. Very likely, the same procedure can be used in the drying process to complete the $k(y)$ graph and also for undisturbed soils.

A limitation of the apparatus is that it does not consider the control of the stress as recommended by Benson & Gribb (1997). It was demonstrated that the method works for two compacted soils that possess very low permeabilities.

ACKNOWLEDGEMENTS

The present work was financed entirely by the U.A. de Querétaro, Mexico. The authors would like to thank C. Lopez-Cajun for his help to prepare this paper.

* (e-mail: alfre@sunserver.uaq.mx)

REFERENCES

Alimi-Ichola I.& Bentoumi O. 1995. Hydraulic conductivity and difusivity in vertical and horizontal inflow. *1st. Int. Conf. on Unsaturated Soils*. Paris, France.

Benson C.& Gribb M. 1997. Measuring unsaturated hydraulic conductivity in the laboratory and the field. *Unsaturated Soil Eng. Practice*. ASCE GSP No. 68.

Fredlund D.G.& Rahardjo H. 1993. *Soil Mechanics for Unsaturated Soil*. Wiley inter science.

Hamilton J.M., Daniel D.E., Olson R.E. 1981. Measurement of hidraulic conductivity of partially saturated soils. *ASTM STP 746*, cited by Juca & Frydman 1995.

Juca J. & Frydman S. 1995. Experimental techniques. *1st. Int. Conf. on Unsaturated Soils*. Paris, France

Pérez-García A. 1997. Medición de succión con psicrómetro en suelo no saturado de Querétaro, Mex. Simp. Brasileiro de solos ñao saturados, Rio de J. Brazil 1997, in spanish.

Zepeda-Garrido A. 1989. Propiedades mecánicas e hidráulicas en suelos no saturados. Curso internacional de mecánica de suelos arcillosos. U.A. Querétaro-U. Laval, in spanish.

Unsaturated Soils for Asia, Rahardjo, Toll & Leong (eds) © 2000 Taylor & Francis, ISBN 90 5809 139 2

Temperature effects on water retention and water permeability of an unsaturated clay

E. Romero, A. Gens & A. Lloret
Geotechnical Engineering Department, Technical University of Catalunya, Barcelona, Spain

ABSTRACT: Laboratory tests were conducted on artificially prepared unsaturated fabrics obtained from natural clay to investigate hydraulic changes induced by heating (up to 80°C) in relation to water retention and water permeability characteristics. Retention curves at different temperatures show that total suction tends to reduce with increasing temperatures at constant water content in the high-suction or intra-aggregate zone. Temperature influence on water permeability is more relevant at low suctions corresponding to free water preponderance (inter-aggregate zone), whereas below a degree of saturation of 70% (intra-aggregate zone) no clear effect is detected. Experimental data show temperature dependence of permeability at constant degree of saturation smaller than could be expected from the thermal change in water viscosity, indicating that thermochemical effects altering clay fabric and pore fluid chemistry could be relevant.

1 INTRODUCTION

The influence of temperature on hydraulic properties of clays is of major concern in the design of engineered barriers in underground repositories for radioactive high-level waste disposal. There are a number of laboratory results concerning thermal effects on saturated water permeability (Towhata et al. 1993, Khemissa 1998), but in contrast, experimental information concerning unsaturated states is very limited and restricted to sandy and silty soils (Hopmans & Dane 1986). In addition, testing results of temperature effects on water retention have been usually limited to low suction values (sandy and silty soils) and/or low temperatures (Nimmo & Miller 1986, Constantz 1991, She & Sleep 1998).

To gain insight into these aspects of behaviour, a systematic research programme has been carried out on artificially prepared clayey samples to investigate changes in hydraulic properties induced by heating. In this study, maximum temperature is limited to 80°C, which is a little lower than the value of 100°C prescribed for unsaturated backfill in many repository reference designs. The paper presents water retention results of water content – temperature relationships at constant suction and isochoric retention curves at different temperatures. Water permeability dependence on degree of saturation, void ratio and temperature obtained from inflow/outflow test data is also presented. Finally, a phenomenological interpretation of temperature and suction effects on hydraulic properties is described.

2 MATERIAL, TESTING PROCEDURES AND EQUIPMENT

2.1 *Material*

Laboratory tests were conducted on artificially prepared (statically compacted on the dry side) powder obtained from natural Boom clay (Mol, Belgium). This moderately swelling clay (20%-30% kaolinite, 20%-30% illite and 10%-20% smectite) has a liquid limit of $w_L = 56\%$, a plastic limit of $w_P = 29\%$ and 50% of particles less than 2μm. In preparing specimens for vapor equilibrium control, powder was left in equilibrium with the laboratory atmosphere at an average relative humidity of 47% (total suction of $\psi \approx 100$ MPa) to achieve a hygroscopic water content of $(3.0\pm0.3)\%$. For samples tested using the air overpressure technique for matric suction control, the required quantity of demineralised water to achieve a predetermined gravimetric water content of $(15.0\pm0.3)\%$ was added to the powder, previously cured at a relative humidity of 90%. After equalization, an initial total suction of approximately $\psi \approx 2.3$ MPa is achieved. A static compaction procedure has been followed until a specified final volume is achieved under constant water content.

Experimental data is interpreted based on the existence of two main pore size regions, as observed from the analysis of mercury intrusion / extrusion porosimetry results (Romero et al. 1999). Firstly, an intra-aggregate porosity with quasi-immobile water at gravimetric water contents lower than 15% and pore sizes smaller than 150 nm. Secondly, an inter-aggregate and interconnected porosity containing free water. Experimental results presented by Romero et al. (1999) show that intra-aggregate water represents between 54 and 59% of the total volume of water in soil in a low-porosity packing compacted at a dry unit weight of 16.7 kN/m³, whereas it corresponds to around 28 and 38% in the case of a packing compacted at a dry unit weight of 13.7 kN/m³.

2.2 *Testing procedures for prescribing suction*

Results in the high suction range were obtained from isothermal main wetting paths performed with single-stage vapor equilibrium technique at different temperatures and under free swelling conditions (($\sigma_m - u_a$) = 0). Sodium chloride solutions with varying solute quantities were used in order to achieve the predetermined soil-water potential. The relation between the relative vapor pressure or activity a_l of the NaCl aqueous solution and the molality of the solute m (mol of NaCl /kg of pure water), as a function of temperature $T(°C)$, is given by the following empirical expressions (Horvath 1985, Romero 1999):

$$a_l = 1 - 0.035m - m(m-3)f(T) \qquad (1)$$

$$f(T) = 1.977 \times 10^{-3} - 1.193 \times 10^{-5}T \text{ for } m \geq 3 \text{ mol/kg} \quad (2)$$

$$f(T) = 1.142 \times 10^{-3} \text{ for } m < 3 \text{ mol/kg} \qquad (3)$$

A 3 MPa value was selected as the suction lower limit due to the difficulty of controlling with accuracy high relative humidity values that require small amounts of solute. The suction upper limit is controlled by the salt solubility that limits the relative humidity to a minimum of 0.75 at 22°C.

Compacted soil samples at seven different initial dry unit weights ranging from 14.7 to 20.6 kN/m³ were equilibrated for two weeks in hermetic jars at specified temperatures (22, 40, 60 and 80°C) and relative humidity of the air (32, 10, 8, 6 and 3 MPa). At the end of each single-stage equilibrium test the seven samples were carefully weighed (gravimetric water content resolution 0.02%) and measured (volumetric resolution less than 0.2%), and the water content of the different specimens determined. More details concerning the experimental procedure are presented in Romero (1999).

2.3 *Testing procedures and equipment for water permeability measurement*

A suction controlled oedometer cell surrounded by a thermostatically controlled heater was employed for the study in the low suction range (up to 0.5 MPa). A water volume change indicator connected to the high air-entry ceramic disc was used to obtain the values of inflow and outflow, as well as the soil water retention for each suction and temperature condi-

tion. Measurements of the water permeability were performed under controlled matric suction (both wetting and drying paths) and under constant net vertical stress states or constant volume (isochoric) conditions. A step of matric suction decrease / increase is applied to the soil and the transient inflow/outflow of water is carefully measured with time and interpreted with a simplified solution of Richards's equation corrected for non-negligible ceramic disc impedance (Kunze & Kirham 1962).

In order to determine accurately water volume changes it is necessary to account for water volume losses due to evaporation in the open air overpressure chamber (evaporative fluxes originated due to the difference in vapor pressure between soil voids and the overlying air). Evaporative fluxes are detected in the water volume change device under steady state conditions of volume change behavior. Evaporative fluxes reduce with increasing matric suction, with increasing relative humidity in the open air chamber and with lower soil porosity (Romero 1999). If evaporative fluxes measured at a reference 22°C are kept under a value of 1×10^{-6} (mm³/s)/mm², a relative humidity higher than 98.5% is ensured in the open air chamber, which limits the water vapor transfer at high temperatures and permits the attainment of matric suction equalization over the entire sample height.

It is also important to flush air bubbles periodically from below the high air-entry disc in order to avoid progressive cavitation of the system and the consequent loss of continuity between the pore water and the water in the measuring system, specially at high temperatures and high-applied matric suction values. An increment of the air diffusion coefficient through saturated ceramic discs at higher temperatures has been measured (Romero 1999). A diffused air volume indicator was therefore incorporated in the open water pressure system to allow periodic flushing and measurements.

The testing program has included two main soil packings of clay aggregates: a high-porosity fabric with collapsible tendency structure at a dry unit weight of 13.7 kN/m³ and a low-porosity structure with swelling tendency at 16.7 kN/m³.

3 TEST RESULTS AND INTERPRETATIONS

3.1 *Retention curve results*

Figure 1 summarizes the data corresponding to the variation of gravimetric water content with applied total suction observed at different temperatures. It can be seen that moisture retention capacity of the clay in the intra-aggregate zone is influenced by temperature. At a given total suction, water content reduces with increasing temperatures.

Lower values of moisture retention under constant suction and higher temperatures are in agreement with experimental results reported by Wan (1996) testing Boom clay up to 50°C and using vapor equilibrium technique. In addition, lower water content values for the same total suction and higher temperatures are also consistent with data reported in the low suction range (sandy and silty material) by Nimmo & Miller (1986), by Hopmans & Dane (1986), by Constantz (1991) and by She & Sleep (1998).

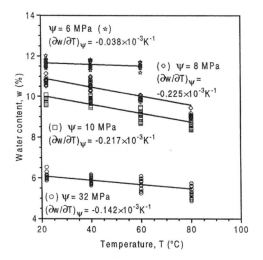

Figure 1. Water content-temperature plots at constant total suction.

Following the suggestion of Philip and de Vries (1957), numerous experimental results of temperature effects on matric suction s at constant water content w have been interpreted based on the temperature dependence of surface tension σ:

$$\left(\frac{\partial s(w)}{\partial T}\right)_w = \frac{s(w)}{\sigma}\frac{\partial \sigma}{\partial T} \tag{4}$$

Assuming a linear variation of surface tension σ with temperature T that vanishes at the critical point and integrating between a reference temperature T_r and an observational temperature T, temperature effects on energy status of soil water can be derived based on the following coefficients affecting a reference matric suction $s(w,T_r)$ (Grant & Salehzadeh 1996):

$$\frac{s(w,T)}{s(w,T_r)} = \left(\frac{a_1 + b_1 T}{a_1 + b_1 T_r}\right)^{b_1} \tag{5}$$

where, a_1 and b_1 are empirical coefficients reflecting the actual behavior compared to the surface tension mechanism. If values estimated from analysis of experimental data tend to $b_1 = 1$ and $a_1 = -767$ K, then

temperature induced changes may be described by temperature effects on surface tension.

Total suction – temperature plots at constant water content varying between 6% and 11% are indicated in Figure 2, where for a given water content in the intra-aggregate zone the total suction tends to decrease with increasing temperatures. The figure shows temperature influence that cannot be interpreted as representing only temperature dependence on surface tension. Additional thermal disturbances altering clay fabric and intra-aggregate fluid chemistry are to be postulated. Deviations at 80°C may be associated with some free water evaporation affecting water content determinations.

Comparisons between total suction – temperature plots at constant water content (ranging from 6% to 11%) and predicted in terms of surface tension mechanism of pure water are shown in Figure 2. At higher temperatures, the theory under-predicts the influence of temperature on total suction. Therefore, the capillary model cannot solely explain the effect of temperature on high suction values, as expected for relatively active clays and for electrolytic soil water. Similar results of greater temperature dependence have been reported by Hopmans & Dane (1986), Nimmo & Miller (1986) and Constantz (1991) in the low suction range for sandy and silty soils.

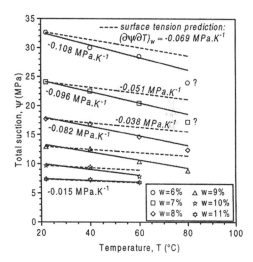

Figure 2. Total suction-temperature plots at constant water content.

In Figure 3, the relationships between total suction and water content (main wetting paths) at fixed dry densities (isochoric conditions) and two different temperatures (22 and 80°C) have been plotted. Data at high suction values (intra-aggregate zone) have been interpolated at constant dry unit weight from free swelling and shrinking data. Main wetting data at low suction values (under 0.50 MPa) and high temperature are obtained from constant volume swelling pressure tests. Both techniques can be overlapped in main wetting paths showing the overall retention curve. The end of the main adsorption curves at 80°C is somewhat lower compared to the ending point of the main wetting curves at ambient temperature because of the lower gravimetric water storage capacity for the same soil packing at higher temperatures, mainly caused by water dilatation and possibly due to thermal expansion of entrapped air.

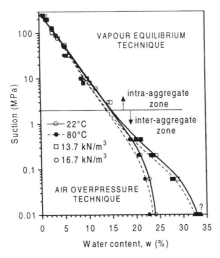

Figure 3. Retention curves for different temperatures at fixed dry densities.

Testing results have been fitted to a modified form of van Genuchten (1980) expression for water content w as a function of soil suction s at the observational temperature T:

$$\frac{w}{w_{sat}} = C(s)\left(\frac{1}{1+(\alpha_T s)^n}\right)^m \tag{6}$$

$$\alpha_T = \alpha_r \left(\frac{a_1 + b_1 T_r}{a_1 + b_1 T}\right)^{b_1} \tag{7}$$

where $C(s)$, n, m and α_r are a correction function (to force water content to be zero at suctions higher than 1000 MPa) and empirical parameters related to the slope of the inflection point, to the residual water content and to the air-entry value, respectively. w_{sat} represents the gravimetric water content at saturated conditions. $C(s)$, n and m are considered temperature independent. Temperature effects are introduced by incorporating Equation 5 into the parameter affecting

air-entry value, as suggested by Grant & Salehzadeh (1996). As it is expected that $a_1 \rightarrow -767$ K, $b_1 \rightarrow 1$ for high water content levels near the soil air-entry value (inter-aggregate zone in Figure 3 under the preponderance of free water), it is accurate enough to consider $b_1 = 1$ in the above expression. Retention curve data for both packings at high temperatures have been fitted to the proposed expression, maintaining parameters $n = 1.14$, $m = 0.20$ and $\alpha_r = (21.3\pm0.8)$ MPa^{-1} for the low-density packing and $n = 0.75$, $m = 0.35$ and $\alpha_r = (1.55\pm0.43)$MPa^{-1} for the high-density fabric. Fitted values of α_T are (25.0 ± 0.7) MPa^{-1} for the low-density packing and (1.98 ± 0.75) MPa^{-1} for the high-density fabric. Values of a_1 are $-(608\pm89)$ K, which are somewhat higher than the surface-tension prediction. As observed, Equation 6 is adequate to fit retention curve data at different temperatures over a wide suction range. However, the temperature parameter α_T cannot be interpreted as representing only temperature dependence on surface tension affecting soil air-entry value, but as an empirical fitting parameter.

3.2 Water permeability results

Observed water permeability values are represented in Figure 4 for different degrees of saturation, void ratios and two temperatures (the upper plot represents data at 22°C and the lower graph shows data at 80°C). The contour levels representing constant void ratio values were obtained from the data using polynomial regression. Certain temperature effect is observed in the lower graph, where solid contour lines obtained from water permeability results at 80°C are compared to dashed contour lines representing equivalent conditions in terms of void ratio at 22°C. In general, this temperature effect is more important at near-saturated conditions, whereas below a degree of saturation of 70%, corresponding to the intra-aggregate zone, no clear trend is detected. In addition to the dependence of water permeability on degree of saturation and temperature, a large void ratio dependence is observed, which appears to increase up to a certain limit with decreasing degree of saturation (refer to the upper plot in Figure 5). In order to observe this trend, the water permeability values have been conveniently grouped in approximately constant degree of saturation values and plotted versus void ratios, as shown in Figure 5 for both temperatures (the upper plot represents data at 22°C and the lower graph shows data at 80°C). The slight increase of permeability with temperature at $Sr \approx (95\pm5)\%$ of $k_w(80°C)/k_w(22°C) \approx 1.3$ cannot be solely explained in terms of the reduction of free water viscosity in the same interval of temperature. (solid lines at 80°C and dashed lines at 22°C in the lower graph of Figure 5). Water viscosity change is associated with a value of $k_w(80°C)/k_w(22°C) = 2.7$.

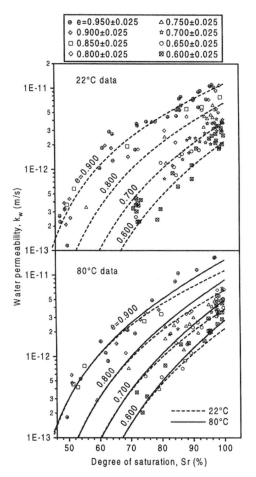

Figure 4. Water permeability vs. degree of saturation obtained in different suction steps at 22°C and 80°C.

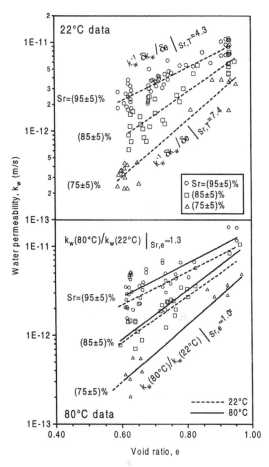

Figure 5. Water permeability vs. void ratio for constant degrees of saturation at 22°C and 80°C.

Temperature effects on saturated water permeability are usually interpreted based on viscosity changes under free water considerations, according to the following expression:

$$\left.\frac{k_w(T)}{k_w(T_r)}\right|_{e,w} = \frac{\rho_w(T)\mu(T_r)}{\rho_w(T_r)\mu(T)} \approx 1 + \beta_T(T - T_r) \qquad (8)$$

where ρ_w is the water density, μ_w the absolute viscosity and $\beta_T \approx 0.030$ K^{-1} an empirical coefficient that fits relative viscosity data over the temperature range of 22°C < T < 80°C. Experimental data, which show temperature dependence smaller than could be expected from the thermal change in water viscosity, are associated with $\beta_T \approx 0.005$ K^{-1} at nearly saturated conditions $Sr \approx (95\pm5)\%$ and tending to $\beta_T \to 0$ at lower free water availability in the intra-aggregate

zone $Sr \approx (75\pm5)\%$. Although no detailed study has been performed on unsaturated soils that corroborates the information presented in this research, it is possible that the nature of soil fabric, clay particles and pore fluid chemistry change when heated, even though the total porosity and degree of saturation appear to have changed only slightly. However, for all controllable factors remaining constant, it appears correct to state that temperature effect on the viscosity of the permeant represents the dominant influence.

Concerning saturated states, Towhata et al. (1993) studied the effects of temperature on kaolinite-clay up to 90°C using indirect methods, where an additional temperature effect over that predicted based on free water considerations was observed (an estimated value of $\alpha_T \approx 0.042$ K^{-1} is obtained from experimental data). Recently, Khemissa (1998) reported careful direct measurements of permeability for a saturated kaolinite-clay up to 130°C, where an estimated of $\alpha_T = 0.010$ K^{-1} is obtained from experimental data up to 80°C.

However, scarce experimental data of temperature effects on water permeability of unsaturated silty and sandy soils have been limited to low suction values and low target temperatures. Experimental results have been reported by Haridasan & Jensen (1972) on a silty soil up to 35°C using pressure plate outflow methods. In these results, a somewhat greater temperature dependence seems to be observed at higher volumetric water contents (although this fact has not been stressed by the authors). Hopmans & Dane (1986) tested a sandy soil using tensiometers up to 45°C.

In general, an increase in water permeability is detected at higher temperatures, which is not accounted for entirely by the temperature dependence of water viscosity. Higher values of permeability than predicted from viscosity changes can be interpreted at constant porosity if the amount and cross section of free water in macropores increases. This is sustained on the concept that quasi-immobile adsorbed water changes at around 70°C its nature to that of free inter-aggregate water (Derjaguin et al. 1986). However, the opposed trend could be presented at lower temperature increments, which would induce aggregate expansion upon heating leaving smaller inter-aggregate voids between them that affect free water flow. Flocculation / dispersion phenomena can also occur under constant porosity conditions when temperature is increased in clays, altering the distribution of pores and creating preferential pathways or obstructing macropores that affect free water flow.

4 CONCLUSIONS

Experimental results, which are presented for different artificially prepared unsaturated clay packings, offer a consistent pattern of temperature effects on water permeability and water retention in a wide range of suction and temperature. Moisture retention capacity of the clay in the intra-aggregate zone is influenced by temperature, where for a given water content the total suction tends to decrease with increasing temperatures. However, temperature influence cannot be interpreted as representing only temperature dependence on surface tension, where a somewhat greater dependence has been observed. Additional thermal disturbances altering clay fabric and intra-aggregate fluid chemistry are likely. Temperature influence on water permeability is more relevant at low suctions corresponding to free water preponderance (inter-aggregate zone), whereas below a degree of saturation of 70% (intra-aggregate zone) no clear effect is detected. Experimental data show that temperature dependence at constant degree of saturation and void ratio is smaller than could

be expected from the thermal change in water viscosity, indicating that thermo-chemical effects altering clay fabric (flocculation or dispersion), adsorbed / free water distribution, porosity distribution and pore fluid chemistry could be of importance.

Finally, expressions and material parameters for retention curves and water permeability functions at different temperatures, based on phenomenological evidence, have been presented.

REFERENCES

Constantz, J. 1991. Comparison of isothermal and isobaric water retention paths in nonswelling porous materials. *Water Resour. Res.* 12(12): 3165-3170.

Derjaguin, B.V., V.V. Karasev & E.N. Khromova 1986. Thermal expansion of water in fine pores. *J. Colloid and Interface Sci.* 9(11): 586-587.

Grant S. & A. Salehzadeh 1996. Calculation of temperature effects on wetting coefficients of porous solids and their capillary pressure functions. *Water Resour. Res.* 32(2):261-270.

Haridasan, M. & R.D. Jensen 1972. Effect of temperature on pressure head – water content relationship and conductivity of two soils. *Soil Sci. Soc. Am. Proc.* 36: 703-708.

Hopmans, J.W. & J.H. Dane 1986. Temperature dependence of soil hydraulic properties. *Soil Sci. Soc. Am. J.* 50: 4-9.

Horvath, A.L. 1985. *Handbook of aqueous electrolyte solutions: physical properties, estimation and correlation methods*. New York:.Ellis Horword Limited.

Khemissa, M. 1998. Mesure de la perméabilité des argiles sous contrainte et température. *Rev. Franç. Géotech.* 82: 11-22.

Kunze, R.J. & D. Kirkham 1962. Simplified accounting for membrane impedance in capillary conductivity determinations. *Soil Sci. Soc. Am. Proc.* 26: 421-426.

Nimmo, J.R. & E.E. Miller 1986. The temperature dependence of isothermal moisture vs. potential characteristics of soils. *Soil Sci. Soc. Am. J.* 50: 1105-1113.

Philip, J.R. & D.A. de Vries 1957. Moisture movement in porous materials under temperature gradients. *Eos. Trans. American Geophysical Union* 38(2): 222-237.

Romero, E. 1999. *Characterization and thermo-hydromechanical behavior of unsaturated Boom clay: an experimental study*. PhD Thesis, Universidad Politécnica de Cataluña.

Romero, E., A. Gens & A. Lloret 1999. Water permeability, water retention and microstructure of unsaturated Boom clay. *Engineering Geology* 54: 117-127.

She, H.Y. & B.E. Sleep 1998. The effect of temperature on capillary pressure-saturation relationships for air-water and perchloroethylene-water systems. *Water Resour. Res.* 34(10): 2587-2597.

Towhata, I., P. Kuntiwattanakul, I. Seko & K. Ohishi 1993. Volume change of clays induced by heating as observed in consolidation tests. *Soils and Foundations* 33(4): 170-183.

van Genuchten, M.Th. 1980. A closed-form equation for predicting the hydraulic conductivity of unsaturated soils. *Soil Sci. Soc. Am. J.* 44: 892-898.

Wan, A.W.L. 1996. The use of thermocouple psychrometers to measure in situ suctions and water contents in compacted clays. PhD Thesis, University of Manitoba.

Unsaturated Soils for Asia, Rahardjo, Toll & Leong (eds) © 2000 Taylor & Francis, ISBN 90 5809 139 2

Measuring hydraulic properties for unsaturated soils with unsteady method

T. Sugii & K. Yamada
Department of Civil Engineering, Chubu University, Kasugai, Japan

M. Uemura
Institute of Space Create, Japan

ABSTRACT: The instantaneous profile method is an effective method for estimating the permeability of unsaturated soils. However, it requires the use of many expensive measuring sensors and a large amount of soils, and in the case of fine soils, it is difficult to set up the specimen. Therefore, an experiment method has been developed in Chubu University to estimate the permeability of unsaturated soils. It is an improved type of the original instantaneous profile method, and is able to measure the permeability of many kinds of soils. The method is characterized by the unsteady condition of water profile and linear approximation of moisture distribution. In this paper, this improved instantaneous profile method (approximated profile method) is applied to sandy soils. Application of this experiment method is discussed in comparison with other steady methods, agreement between this method and others is demonstrated, and several notes concerning directions of use are given.

1 INTRODUCTION

Various methods of measuring the hydraulic properties of unsaturated soils have been proposed. These methods are classified as steady or unsteady methods. Although the former type can easily obtain the stable hydraulic properties of unsaturated soils, it is tedious and lengthy. The latter is often easier to perform and requires less time. Watson (1966) proposed an instantaneous profile method as an unsteady method for measuring the hydraulic conductivity of unsaturated soils. In almost every unsteady method, diffusivity is determined from instantaneous outflow at first, then, permeability is calculated from the diffusivity with the water retention curve. Therefore, the accuracy of water retention curve affects the hydraulic conductivity of unsaturated soils. On the other hand, the instantaneous profile method measures the transient flow and the hydraulic gradient during which conditions are approximately steady. As the hydraulic conductivity is determined directly, the water retention curve is not needed. We improved the original experiment apparatus of the instantaneous profile method of Watson (1966), which consisted of several tensiometers and water content sensors. In the improved instantaneous profile method, a water content sensor (ADR: Amplitude Domain Reflectometry) and one electronic balance are substituted for many measuring implements. As a result, the specimen and sensors are easier to set up in the specimen.

2 PRINCIPLES OF MEASUREMENT

2.1 Darcy's law

Darcy's law for one-dimensional vertical flow is given by

$$\int_0^z \frac{\partial \theta}{\partial t} dz = K(\theta)\left(\frac{\partial \varphi}{\partial z}+1\right) \qquad (1)$$

where θ =volumetric water content; t=time; $K(\theta)$=permeability; φ =pore water pressure; z=elevation of specimen.

Rearranging equation (1), permeability at time t, and elevation of specimen are given by

$$K(\theta)=\frac{\left(\int_0^z \frac{\partial \theta}{\partial t} dz\right)_{z,t}}{\left(\frac{\partial \varphi}{\partial z}+1\right)_{z,t}} \qquad (2)$$

where the numerator shows the velocity at elevation, and the denominator the hydraulic gradient.

The permeability of unsaturated soils against the volumetric water content corresponding to the point at the elevation can be obtained by measuring the velocity and hydraulic gradient at time t. The permeability of variable degree of saturation can be obtained as time proceeds.

2.2 Calculation of velocity

In the instantaneous profile method, volumetric water content is measured at several points. However, in our method, volumetric water content is measured at the top of the specimen only. The amount of drainage from the specimen is measured over time. After measurement is finished, the water content of the bottom of the specimen is measured using the oven-dried method. If the vertical distribution of moisture is assumed to be linear, the volumetric water content at the bottom of the specimen can be estimated as follows.

The amount of drainage through the specimen, q, between time, t_{n-1}, and time, t_n is given by the hatched area in Figure 1.

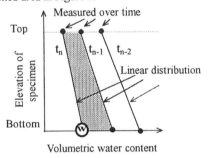

Figure 1 Distribution of volumetric water content in specimen over time.

The volumetric water content at point W is determined from the water content after the measurement is finished. Since the volumetric water content at the top of the specimen is measured over time, the volumetric water content in the previous time at the bottom can be calculated from the area of the trapezoid section (the amount of drainage per unit area of specimen).

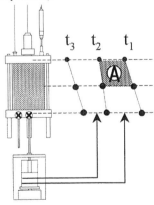

Figure 2. Amount of flow through the specimen.

The amount of flow from t_1 to t_2 through the middle of the specimen is calculated as shown in Figure 2.The numerator of equation (2) is calculated from following equation:

$$\left[\int_0^{z_m} \frac{\partial \Theta}{\partial t} dz \right]_{z_m, \frac{t_1+t_2}{2}} = \frac{A}{t_2 - t_1} \quad (3)$$

where z_m = elevation of middle of specimen, A= hatched area in Figure 2.

This experimental method is characterized by linear approximation of the vertical distribution of moisture in the specimen. In this paper, this method is called "Approximated Profile Method" (A.P.M.).

2.3 Calculation of hydraulic gradient

The change of pore water pressure over time is measured as in shown Figure 3.

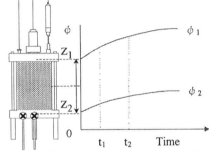

Figure 3 Change of pore water pressure over time.

The hydraulic gradient at the middle of the specimen is given by

$$i_{z_m} = \left(\frac{\partial \psi}{\partial z} + 1 \right)_{z_m, \frac{t_1+t_2}{2}} = \left(\frac{\psi_2 - \psi_1}{z_2 - z_1} \right)_{\frac{t_1+t_2}{2}} + 1 \quad (4)$$

The pore water pressure is controlled by elevation of the drainage tank or supplied air pressure. As there is filter on the bottom of the specimen, the pore water pressure is corrected using the equation:

$$p_w = p_{wf} + \frac{dq}{AK_f} \quad (5)$$

where p_w = pore water pressure at bottom of specimen, p_{wf} = calculated pressure from elevation of drainage tank or supplied air pressure, d = thickness of filter, q = amount of flow per unit time, A = area of specimen section, K_f = permeability of filter.

3 EXPERIMENT APPARATUS

In the original instantaneous profile method, many sensors are used to measure volumetric water content and water pore pressure. Watson (1966) used RI sensors (using gamma-ray), to measure

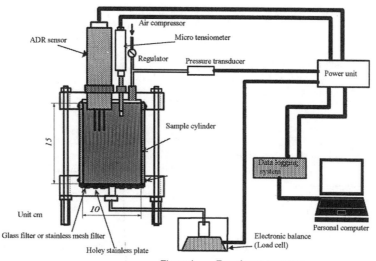

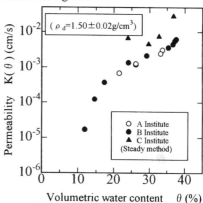

Figure 4.   Experiment apparatus

water content. Instead of RI sensor, recently, FDR (Frequency Domain Reflectometry (Topp et al. 1988)), ADR (Amplitude Domain Reflectometry (Gaskin et al. 1996)) are used for moisture measurement. Their sensors can accurately measure soil moisture easily and safely. Our method uses an ADR sensor to measure the moisture of soils, and a micro tensiometer to measure pore water pressure. To measure the amount of drainage over time, a load cell is used. In the case of sandy soils, pore water pressure is controlled by changing the elevation of the drainage tank (suction method). Figure 4 shows the experiment apparatus in this study, in which a sample cell unit that includes container filter is attached to the bottom of the sample specimen. The load cell can measure to an accuracy of 0.1g. When air pressure is necessary, air pressure is supplied from the top of the sample cell. The data logger and personal computer record the measured data of pore water pressure, volumetric water content, supplied pressure and amount of drainage over time. This study used a wire mesh filter and a glass filter.

4 DESCRIPTION OF SAMPLES AND TEST

4.1 Description of samples

In order to compare other methods, this method is applied to Toyoura sand as shown in Table 1.

Table 1. Toyoura sand

Property		Value
Grain size	D	105-300 μ m
Density of soil particles	ρ_s	2.65 g/cm^3
Dry density	ρ_d	1.50 g/cm^3
Permeability of saturated soils	k_{sat}	1.10E-02 cm/s

Some different permeability tests under the same conditions were performed by the Research Committee of JGS (Sugii et al. 1997). Their results are shown in Figure 5.

Figure 5 Permeability of unsaturated Toyoura sand.

4.2 Soil specific calibration of ADR sensor

The relationship between the output voltage of ADR and the square root of dielectric constant can be represented as a polynomial approximation:

$$\sqrt{\varepsilon} = 1.07 + 6.4V - 6.4V^2 + 4.7V^3 \quad (6a)$$
$$(0 \le \theta \le 100\%)$$

$$or \quad \sqrt{\varepsilon} = 1.1 + 4.44V \quad (\theta \le 50\%) \quad (6b)$$

where ε = dielectric constant, V= output voltage.

The linear relationship between the square root of dielectric constant and volumetric water content was discussed by Whalley (1993).

441

$$\theta/100 = \frac{\sqrt{\varepsilon} - a_0}{a_1} \qquad (7)$$

where a_0 and a_1 are constants for a specific soil type having typical values of 8.1 and 1.6 respectively.

As the moisture of Toyoura sand is less than 50%, from equations (6b) and (7) the calibration curve becomes:

$$\theta/100 = \frac{[1.1 + 4.44V] - a_0}{a_1} \qquad (8)$$

Values of a_0 and a_1 are 2.1 and 43.5 respectively, from Figure 6.

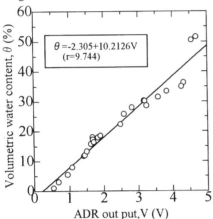

$\theta = -2.305 + 10.2126V$
$(r = 9.744)$

Figure 6 Calibration curve of ADR for Toyoura sand.

4.3 Kinds of test case

Two kinds of test are performed as follows.
(1) Stainless wire mesh filter and glass filter are used in order to investigate the difference of filter materials. The properties of filters are shown in Table 2.

Table 2. Properties of Filters

	Stainless wire mesh filter	Glass filter
Thickness	50 (μm)	4.72 (mm)
Opening diameter	20-30 (μm)	30-40 (μm)
Air entry value	4 (kPa)	9 (kPa)
Permeability	2.58E-5(cm/s)	1.45E-3 (cm/s)

(2) As the distribution of moisture is considered to vary according to the method used to control pore water pressure, differences between the suction method and pressure method are investigated.

5 DISCUSSION OF RESULTS

5.1 Influence of filter materials on permeability

This approximated profile method is based on a monotonous vertical distribution of moisture. It is considered that the vertical distribution of moisture

depends on the properties of filter. Figure 7 shows the results of tests using two kinds of filter.

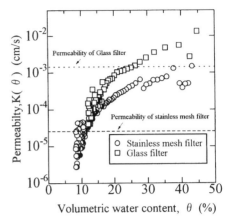

Figure 7 Influence of filter materials on permeability.

Permeability using the stainless wire mesh filter is less than that using the glass filter until the volumetric water content is less than 15%. From Figure 5 we considered that results of using the glass filter are more accurate than for the stainless wire mesh. A too-low permeability is measured due to the complicated moisture distribution. When the permeability of the filter is less than that of the soil, the filter restricts the velocity of flow, therefore, moisture stays at the lower part of the specimen. As soils are saturated in the lower part of the specimen, the moisture distribution cannot be approximated as a linear one. When the permeability of the filter is more than that of the soil, the velocity of flow is independent of the permeability of the filter, therefore, the moisture distribution can be approximated as a linear one. The higher the permeability of the filter, the lower the air entry value. When the air entry value of filter is less than the supplied air pressure, the filter is block, which restricts measurement of the drying side. In this approximated profile method, the choice of filter is significant.

5.2 Suction method and pressure method

The method of controlling pore water pressure influences the vertical distribution of moisture. In this study, the suction method is compared with the pressure method.

5.2.1 Suction method

The saturated soil sample is in hydraulic contact with bulk water through the glass filter. Atmospheric pressure is applied to the soil. The pressure in the bulk water is reduced to subatmospheric levels, thereby reducing its hydraulic head (-86 cmH$_2$0 =8.4kPa).

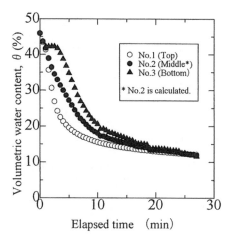

Figure 8.1 The change of volumetric water content with elapsed time(Suction method).

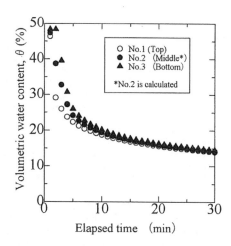

Figure 9.1 The change of volumetric water content with elapsed time (Pressure method).

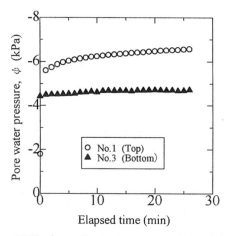

Figure 8.2 The change of pore water pressure with elapsed time (Suction method).

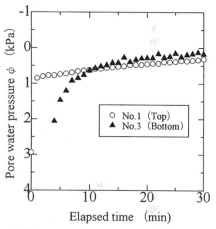

Figure 9.2 The change of pore water pressure with elapsed time (Pressure method).

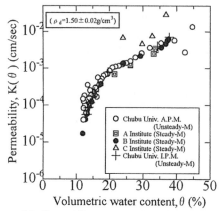

Figure 8.3. Permeability versus volumetric water content (Suction method).

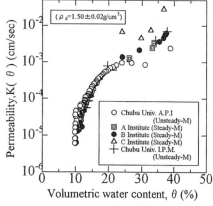

Figure 9.3. Permeability versus volumetric water content (Pressure method).

Measurement is continued until hydraulic equilibrium is reached.

5.2.2 Pressure method

Air pressure is supplied from the top of the soil sample at 7kPa. The bottom of the specimen is in contact with atmospheric pressure. Actual pressure is measured by the micro-tensiometer at the top of the specimen.

5.2.3 Contrast suction method with pressure method

Figures 8.1 and 9.1 show the measured volumetric water content (No.1 and 3) and calculated one (No.2). In the case of the pressure method, the vertical distribution of moisture quickly achieves a uniform condition, and the difference pressure in the top and bottom is reduced quickly (Figs. 8.2 and 9.2). Figures 8.3 and 9.3 give the relationship between permeability and volumetric water content. Measured data of the instantaneous profile method (I.P.M.) in Chubu University of the previous method (Sugii et al. 1999) is shown as a reference. The result of the suction method is roughly the same as other data over the full range of volumetric water content. The permeability of the pressure method is higher than that of the suction method, in the range of higher water content. It is considered that the moisture distribution of initial drain is complex. When a saturated soil is forced to drain with air pressure supplied, the moisture distribution tends to become an "S-curve" compared with drain by suction (Uno et al. 1991). The moisture distribution approaches a monotonous one with the reduction of volumetric water content. In this method, the previous volumetric water content is estimated from the last water content, therefore, permeability over the full range of volumetric water content is not affected by these errors.

6 CONCLUSIONS

The approximated profile method was developed to estimate the permeability of unsaturated soils. Though this method is an improved version of the instantaneous profile method (unsteady method), it is not necessary to use many expensive measuring sensors nor a large amount of soils.

As this method is based on a monotonous vertical distribution of moisture, the choice of filter affecting the moisture distribution is significant. It is advisable to use a filter having a permeability similar to that of the saturated soil.

Pressure method using supplied air pressure tends to contain errors of permeability in the range of high water content, but estimation of permeability is possible in the range of low water content.

The results of this method agreed with those of other methods. Partially, stable results over the full range of water content can be obtained by the suction method.

It takes about 20 minutes to measure sandy soils over the full range of volumetric water content. It was found that the proposed method is an effective for estimating the permeability of unsaturated soils directly.

ACKNOWLEDGMENTS

The authors wish to thank former graduate students of Chubu University Mr. T. Okumura and Mr. T. Saso for their support in carrying out the experiments. This study received support from a Grant-in-Aid for Scientific Research (C) titled "Evaluation of hydraulic properties of unsaturated soils considering bulk density" (No.11650515).

REFFERENCES

Gaskin,G.J. & Miller,J.D. 1996. Measurement of soil water content using a simplified impedance measuring technique, J. Agric. Engng Res.,Vol.63, 153-160.

Sugii,T. & Enomoto,M. 1997. Permeability test for unsaturated soils, Report of "Evaluation method of unsaturated ground", the Research Committee, Japanese Geotechnical Society, 82-90 (in Japanese).

Sugii,T. Yamada,K. & Uemura,M. 1999. Measuring unsaturated hydraulic conductivity and soil water characteristic curves in laboratory, Proc. symp. in Chubu-branch of JGS on Geotechnical Engng, 21-26 (in Japanese).

Topp,G.C, Yanuka,M., Zebchuk,W.D. & Zegelin,S. 1998. Determination of electrical conductivity using time domain reflectometry: soil and water experiments in coaxial lines, Water Resources Research, Vol.24, No.7, 945-952.

Uno,T. Sato,T. Sugii,T. & Tsuge,H. 1990. Method of for permeability of unsaturated sandy soil with controlled air pressure, J. Getechnical Engng., JSCE., No.400,Vol.III-13, 115-124 (in Japanese).

Uno,T., Sugii,T. & Tsuge,H. 1991. On the flow of pore water and air in sandy soil, Proc. national symp. on Multiphase Flow, 147-150 (in Japanese).

Uno,T. Sato,T. & Sugii,T. 1995. Laboratory permeability measurement of partially saturated soil: Alonso & Delage (ed.), Unsaturated Soils, Proc. intern. symp., Paris, 6-8 September, 573-578.

Watson,K.K. 1966. An instantaneous profile method for determining the hydraulic conductivity of unsaturated porous materials, Water Resources Research, Vol.2, 709-715.

Whalley,W.R. 1993. Considerations on the use of time-domain reflectometry (TDR) for measuring soil water content, J. Soil Science, Vol.44, 1-9.

Unsaturated Soils for Asia, Rahardjo, Toll & Leong (eds) © 2000 Taylor & Francis, ISBN 90 5809 139 2

Inverse analysis of in situ permeability test data for determining unsaturated soil hydraulic properties

Y. Takeshita – *Faculty of Environmental Science and Technology, Okayama University, Japan*

K. Yagi – *Graduate School of Engineering, Okayama University, Japan*

T. Morii – *Faculty of Agriculture, Niigata University, Japan*

M. Inoue – *Arid Land Research Center, Tottori University, Japan*

ABSTRACT: A new field experimental method of determining unsaturated soil hydraulic properties is proposed. In this method these properties are assumed to be represented by van Genuchten's closed-form expressions. Unknown parameters of this model are identified by Genetic algorithms (GA) incorporating finite element analysis of transient axisymmetric seepage flow. GA are search algorithms based on the mechanics of natural selection and natural genetics. They have become a popular global optimization method. A ponded single-ring infiltrometer technique, such as the Guelph Pressure Infiltrometer is performed. Simultaneously the measured change of soil water content with time and cumulative inflow data are used to identify unsaturated soil hydraulic function parameters. The advantage of the proposed method are that it allows estimation of the unsaturated soil hydraulic parameters and diminishes experimental time. The utility of our proposed method is demonstrated by using experimental data for Japanese dune sand.

1 INTRODUCTION

Knowledge of the unsaturated soil hydraulic properties is essential for prediction of seepage flow and contaminant transport through the vadose zone. These properties consist of the hydraulic conductivity as a function of pressure head and the soil water retention curve. Traditionally, steady state experiments have been used to estimate these properties. Recently, transient methods are becoming more popular. Transient experimental methods are inherently faster, because more powerful computers and seepage flow simulation technique are now available. The estimation of the hydraulic properties by using inverse methods can be performed easily.

The inverse methods for estimating unsaturated hydraulic parameter from transient experiments were reported in the past two decades (e.g. Kool et al. 1985, Parker et al. 1985, Eching & Hopmans 1993, van Dam et al. 1994, Eching et al. 1994, Inoue et al. 1998). These authors assumed the soil water retentions are defined by van Genechten (1980). Gradient-based optimization procedures were used to estimate the shape parameters α, n and saturated hydraulic conductivity Ks in van Genechten's equations. When some traditional gradient-based methods, however, are applied to identify the non-linear soil water retention and hydraulic conductivity experimentally, a number of problems related to ill-posedness as well as convergence, and parameter uniqueness, effects of initial value remain to be solved. The ill-posedness is generally characterized by the non-uniqueness of the identified parameters and instability of the identification procedures.

Most field experiments require restrictive initial and boundary conditions. Measurements are therefore time consuming, expensive and restricted to a certain range. One of the more recently developed field methods for measuring soil hydraulic properties of unsaturated soils is a single-ring infiltrometer known as the Guelph Pressure Infiltrometer (GPI) (Reynolds 1993, Elrick & Reynolds 1992). This method involves inserting a single ring at a certain distance into the soil, attaching a Mariotte reservoir, and calculating soil hydraulic properties (field-saturated hydraulic conductivity, matric flux potential, capillary length).

In this paper, a new field experimental method using the GPI is proposed. This method presents some advantages in the analysis of transient inflow data from GPI. Soil water content data is measured near the inserted single-ring of the GPI by using an amplitude domain reflectometry (ADR) sensor. Inverse analysis is applied to estimate the required parameters defined by van Genechten (1980). These parameters are identified by Genetic algorithms (GA) incorporating finite element analysis of axisymmetric transient seepage flow. The measured soil water content and cumulative inflow data as a function of time were used to evaluate the objective function in our GA-based method. Data from the experiment of the Japanese dune sand are used to verify our proposed method.

2 METHODS

2.1 Single-ring infiltrometer

Figure 1 illustrates the ponded single-ring infiltrometer, such as the Guelph Pressure Infiltrometer (GPI) and the location of soil water content measured by ADR sensor. The radius of the single-ring is 5.5 cm. The depth location is 3.0cm. The ADR sensor was installed at r=15cm and z=10.5cm. A constant head (H=14.9cm) condition was used to supply water to the soil surface using a Mariotte reservoir. The radius of the area considered is 100cm so that the flow out of the single-ring will not be affected significantly. For the distribution of initial pressure head, the hydrostatic pressure head is used. The depth considered is 300cm.

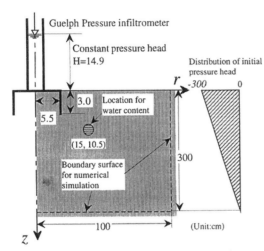

Figure 1. Schematic diagram of in situ experiment

2.2 Water flow

The proposed in situ experiment is based on the GPI. Although the water flow from a single-ring will occur spatially in three dimensions, we can assume axisymmetric transient seepage flow through the homogeneous, isotropic and rigid porous medium described by Richards equation.

$$C\frac{\partial h}{\partial t}=\frac{\partial}{\partial r}\left(K\frac{\partial h}{\partial r}\right)+\frac{K}{r}\left(\frac{\partial h}{\partial r}\right)+\frac{\partial}{\partial z}\left(K\frac{\partial h}{\partial z}+K\right) \quad (1)$$

where r is the radial coordinate, z is the vertical coordinate (positive upward) and h is pressure head. $C = d\theta/dh$ is the soil water capacity with the volumetric water content θ, K is the hydraulic conductivity and t denotes the time. The solution of Eq.(1) is obtained with a finite element model of transient saturated-un-

saturated seepage flow.

The soil hydraulic functions employed in numerical model are described by van Genuchten (1980). These functions will be referred as the VG model.

$$Se = \frac{\theta - \theta r}{\theta s - \theta r} = \left[\frac{1}{1+(\alpha h)^n}\right]^m \quad (2)$$

$$K(Se) = Ks\, Se^{0.5}\left[1-(1-Se^{1/m})^m\right]^2 \quad (3)$$

$$C= \alpha\,(n-1)(\theta s-\theta r)Se^{1/m}(1-Se^{1/m})^m \quad (4)$$

where m=1-1/n, Se is the effective saturation, θs is the saturated water content, θr is the residual water content, Ks is the saturated hydraulic conductivity, and α, n are the soil retention curve shape parameters (empirical parameters). In the parameter estimation problem Ks and θr have been measured independently. Ks can be determined by GPI theory. We assumed in our study that θr takes the value of 0.0 for sandy soil.

2.3 Parameter estimation using genetic algorithms

GA have been established as a reasonable and robust approach to inverse problems which requires an efficient search. The GA method is extremely simple to use, because it does not need to evaluate derivatives of objective functions. GA are search algorithms based on the mechanics of natural selection and natural genetics (Goldberg 1989). Compared to conventional optimization and search techniques, the advantages of GA are:

1. A coding of the parameter set is used.
2. Search begins from multiple locations.
3. Objective function is used directly.
4. Stochastic operators are used.

Various gradient-based inverse methods have been used to estimate unsaturated soil hydraulic functions. Generally the sums of weighted differences between the observed and predicted data are used as the objective function. In this study, the parameters α, n and θs in the VG model are estimated using genetic algorithms from change of water content and inflow data based on the same type of objective function.

A commonly used expression for the objective function $O(b)$ is the weighted least squares, as defined by Eq.(5).

$$O(b)=\omega_\theta\sum_{i=1}^{N}(\theta_m(t_i)-\theta_c(t_i,b))^2+\omega_Q\sum_{j=1}^{M}(Q_m(t_j)-Q_c(t_j,b))^2 \quad (5)$$

where ω_θ and ω_Q are weighting factor, N and M are the number of measurements of the water content $\theta_m(t_i)$ and cumulative inflow $Q_m(t_j)$, respectively. The weighting factors are used to account for differences in accuracy of measurements, and for correlation between residuals. We assumed that the weighting factors are determined by the following equations.

$$\omega_\theta = 1.0 \tag{6}$$

$$\omega_Q = \frac{\sum_{i=1}^{N} (\theta_m(t_i))^2 / N}{\sum_{j=1}^{M} (Q_m(t_j))^2 / M} \tag{7}$$

The water content $\theta_c(t, \boldsymbol{b})$ and the cumulative inflow $Q_c(t, \boldsymbol{b})$ are computed at prescribed times t_i and as a function of the set of optimized VG parameters in the vector $\boldsymbol{b}$.

Using GA, three unknown parameters are encoded as binary strings. The data structure of the encoded parameters are shown in Table 1. The range of α, n and θs are 0.005 to 0.068 cm^{-1}, 1.1 to 7.4, and 0.3 to 0.45, respectively. α and n are divided into 64 possibilities, which can be represented by a 6-bit binary code. θs is divided into 16 possibilities, which can be represented by a 4-bit binary code. There are 65,536 (= 64*64*16) different combinations of unknown parameters. The possible solutions can be searched from the immensity parameters by using GA. This results in a 16-bit long binary code for each individual string composed of all three unknown parameters. One evaluation of the string corresponds to each axisymmetric seepage flow simulation.

3 RESULTS AND DISCUSSION

GPI was carried out for the Japanese dune sand. The parameter Ks was determined by steady state inflow rate measured by GPI to be 2.97×10^{-2} cm/s. We employed the transient water content and inflow data to perform our GA-based parameter estimation procedure. In our GA operations the number of population of 30, the crossover probability of 0.6 and the mutation probability of 0.1 were used as the empirical parameters which may need frequent adjustment to ensure the successful and efficient application of

Table 1. Data structure of encoded parameters

String No.	α (cm^{-1})	n	6-bit binary code	String No.	θ_s	4-bit binary code
1	0.005	1.1	000000	1	0.30	0000
2	0.006	1.2	000001	2	0.31	0001
.	.	.	.	.	.	.
.	.	.	.	.	.	.
31	0.035	4.1	001111	7	0.36	0111
32	0.036	4.2	010000	8	0.37	1000
.	.	.	.	.	.	.
.	.	.	.	.	.	.
63	0.067	7.3	111110	15	0.44	1110
64	0.068	7.4	111111	16	0.45	1111

Table 2. Estimated VG model parameters from GA

GA run	Known parameters		Unknown parameters		
	K_{fs}(cm/s)	θ_r	α (cm^{-1})	n	θ_s
1			0.022	3.3	0.42
2	2.97×10^{-2}	0.0	0.022	3.4	0.41
3			0.022	3.4	0.41

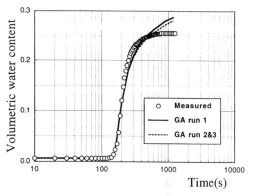

Figure 2. Change of estimated and measured water content with time.

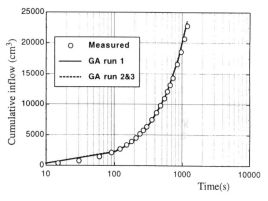

Figure 3. Estimated and measured cumulative inflow with time.

the GA method.

Table 2 shows the estimated VG model parameters from GA. Figure 2 and 3 show the measured data from the transient field experiment. The water content and cumulative inflow are presented in these figures, where the solid lines refer to the computed data using the estimated parameters of the VG model. The symbols denote the measured data. The behavior of computed transient water content and cumulative inflow rate agree with the measured data.

Figure 4 shows the evolution of the relationship between $O(\boldsymbol{b})$ and number of generations in the search process of the GA. Figures 5-7 also show the evolu-

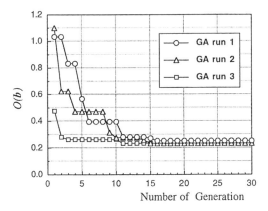

Figure 4. Evolution of $O(b)$ in the search process of the GA.

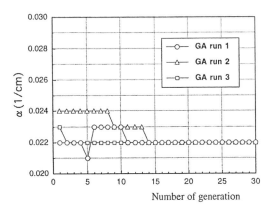

Figure 5. Evolution of unknown parameter α in the search process of the GA.

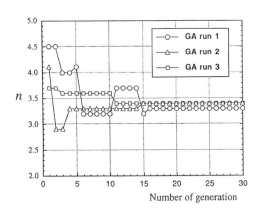

Figure 6. Evolution of unknown parameter n in the search process of the GA.

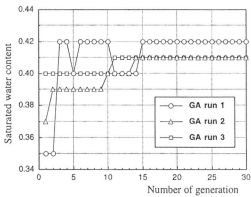

Figure 7. Evolution of unknown parameter θs in the search process of the GA.

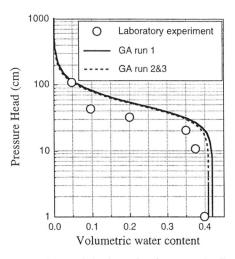

Figure 8. Estimated and measured soil water retention curves for dune sand.

tion of the relationship between estimated parameters and number of generations in the search process of the GA. The GA solution converged to a certain condition after about 15 generations runs. In our study, three parameters are unknown. It typically requires about 20 or 30 generations for the GA solution to converge. Each generation, in turn, requires many forward runs depending on the population size. The GA, however, can search for the reasonable solution which is difficult for conventional gradient-based procedure to estimate from immens solution sets.

The soil water retention curve and unsaturated hydraulic conductivity which were predicted by the VG model was compared with observed data from a laboratory experiment in Figure 8 and 9, respectively. As seen in these figure, a reasonable correspondence is obtained between predicted and measured data.

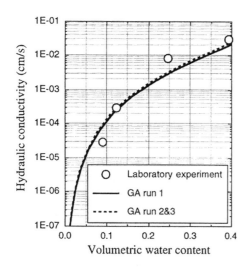

Figure 9. Estimated and measured unsaturated hydraulic conductivity for dune sand.

CONCLUSIONS

An in-situ permeability test using a Guelph Pressure Infiltrometer with transient water content measurement is proposed and its applicability is verified. The apparatus, as well as a procedure to measure the transient infiltration rate and water content is quite simple. We also developed a numerical genetic algorithm-based parameter estimation procedure for soil hydraulic parameters. The change of soil water content and cumulative inflow data were used to evaluate the objective function. The results presented in this study verify our proposed method. Unsaturated soil hydraulic parameters α, n and saturated water content in van Genuchten's equations were estimated simultaneously. Estimated parameters are accurate enough for practical use. The genetic algorithms are a very stable and robust approach compared to traditional gradient-based methods. This advantage will become more obvious when dealing with a lot of unknown parameters or a highly non-linear problem. On the other hand the limitation of genetic algorithm-based parameter estimation method is the large number of forward model simulation runs required.

REFERENCES

Eching, S. O., & J. W. Hopmans. 1993. Optimization of hydraulic functions from transient outflow and soil water pressure data. *Soil Sci. Soc. Am. J.* 57: 1167-1175.

Eching, S. O., J. W. Hopmans, & O. Wendroth. 1994. Unsaturated hydraulic conductivity measure ments from transient multistep outflow and soil water pressure data. *Soil Sci. Soc. Am. J.* 58: 687-695.

Goldberg, D.E. 1989. *Genetic Algorithms in Search, Optimization, and Machine Learning*, Addison-Wesley Pub. Co., 412p.

Inoue,M.,J. Simunek,J.W.Hopmans & V.Clausnitzer. 1998. In situ estimation of soil hydraulic functions using a multistep soil-water extraction technique. *Water Res. Res.* 34-5,1035-1050.

Kool, J. B., J. C. Parker, & M. T. van Genuchten. 1985. Determining soil hydraulic properties from one-step outflow experiments by parameter esti mation, I, Theory and numerical studies, *Soil Sci. Soc. Am. J.*49:1348-1354.

Parker, J. C., J. B. Kool, & M. T. van Genuchten. 1985. Determining soil hydraulic properties from one-step outflow experiments by parameter estimation, II, Experimental studies, *Soil Sci. Soc. Am. J.*49:1354-1360.

Reynolds, W. D., & D. E. Elrick. 1990. Ponded infil tration from a single ring: I. analysis of steady flow. *Soil Sci. Soc. Am. J.* 54: 1233-1241.

Reynolds, W. D. 1993. Saturated hydraulic conduc tivity: Field measurement. pp. 599-613. In: M. R. Carter (ed.), *Soil Sampling and Methods of Analy sis*, Canadian Society, of Soil Science, Lewis Pub lishers, Boca Raton, LA.

van Dam, J. C., J. N. M. Stricker & P. Droogers. 1994. Inverse method to determine soil hydraulic functions from multistep outflow experiments, *Soil Sci. Soc. Am. J.* 58: 647-652.

van Genuchten, M.Th.1980. A closed-form equation for predicting the hydraulic conductivity of unsat urated soils. *Soil Sci. Soc. Am. J.*44: 892-898.

Unsaturated Soils for Asia, Rahardjo, Toll & Leong (eds) © 2000 Taylor & Francis, ISBN 90 5809 139 2

The application of knowledge-based surface flux boundary modelling

G.W.Wilson & M.D.Fredlund

Department of Civil Engineering, University of Saskatchewan, Saskatoon, Sask., Canada

ABSTRACT: The moisture flux boundary condition at the ground surface is dynamic and continuously driven by atmospheric forcing events. The application of the SoilCover model to predict the flux boundary conditions for two soil cover systems is demonstrated in this paper. A knowledge-based system is used to predict the soil-water characteristic curves and associated soil property functions required for the SoilCover model. A case study for the Saskatoon landfill is reviewed. The analyses show that both soil properties and climate parameters are paramount. The final approach used for the design of any ground surface profile to control flux boundary conditions must include climatic parameters together with hydraulic properties of the soil profile.

1 INTRODUCTION

Geotechnical engineers are frequently called upon to design soil cover systems for the management of solid waste systems. The closure of municipal landfills is an important application for cover system design. In general terms, the primary function of the cover system is to control the net infiltration rate to the underlying waste that produces leachate at the base of the landfill. This paper illustrates the application of two computer models for the design of a soil cover system for a landfill situated in the City of Saskatoon, Canada. Surface flux boundary conditions for two potential cover profiles are evaluated using the SoilCover (1997) model. A newly developed knowledge based program system (Fredlund et al, 1998) is used to estimate the soil property functions necessary to run the SoilCover model.

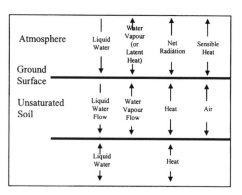

Figure 1. One dimensional view of fluxes in the soil-atmosphere profile.

2 THEORY

The moisture flux boundary condition is dynamic and continuously driven by atmospheric forcing conditions. Infiltration and evaporation across the soil/atmosphere boundary are a function of precipitation and potential evaporation as well as soil properties. Fluxes within the soil/atmosphere continuum are fully coupled in terms of heat and mass transfer. The flow of liquid water, water vapour, air and heat is shown in Figure 1.

The general equation for the flow of liquid water and water vapour are given by Wilson et al. (1997) as follows:

$$\frac{\partial h_w}{\partial t} = C_w^1 \frac{\partial}{\partial y}\left[k_w \frac{\partial h_w}{\partial y} \right] + C_w^2 \frac{\partial}{\partial y}\left[D_v \frac{\partial P_v}{dy} \right] \qquad (1)$$

where h_w = hydraulic head (m); C_w^1 = modulus of volume change with respect to the liquid water phase $(1/\rho_w g m_w^2)$; C_w^2 = modulus of volume change with respect to water vapour phase $[(P + P_v)/P(\rho_w)^2 g m_w^2]$; D_v = coefficient of diffusion for water vapour through soil (kg-m/kN-s); ρ_w = density of liquid water (kg/m^3); g = acceleration due to gravity (m/s^2); m_w^2 = slope of the soil-water characteristic curve (1/kPa); P = total atmosphereic pressure (kPa); and P_v = partial pressure in the soil due to water vapour (kPa).

Conductive heat flow is coupled with the interphase flux of liquid water and water vapour as follows:

$$C_h \frac{\partial T}{\partial y} = \frac{\partial}{\partial y}\left(\lambda \frac{\partial T}{\partial y}\right) - L_v \left(\frac{P + P_v}{P}\right)\frac{\partial}{\partial y}\left(D_v \frac{\partial P_v}{\partial y}\right) \quad (2)$$

where C_h = volumetric spcific heat (J/m³-°C); T = temperature (°C); λ = thermal conductivity (W/m-°C); and L_v = latent heat of vapouriztion (J/kg).

Infiltration events are simulated by applying a liquid flux equal to the rainfall intensity to the right side of equation 1. This procedure is straight forward. Alternatively the solution for evaporation events is more complex since the rate of actual evaporation is a function of both the rate of potential evaporation and the suction in the soil at the ground surface. The actual rate of evaporation is equal to the potential rate of evaporation until the value of suction at the soil surface exceeds approximately 3000 kPa. The value of actual evaporation progressively decreases during desiccation with increasing suction once the soil suction exceeds 3000 kPa. The decline in evaporation occurs due to depression of the vapour pressure within the voids of the soil with increasing suction. The relationship for relative humidity and suction given by Edlefsen and Anderson (1943) is shown in Figure 2 and written as:

$$RH = e^{\frac{\psi W_v}{RT}} \quad (3)$$

where RH = relative humidtiy of the soil surface as a function of total suction; ψ = total suction in the soil (kPa); W_v = Molecular weight of water (0.018 kg/mole); R = universal gas constant (8.314 J/mole/°K); and T = absolute temperature (°K).

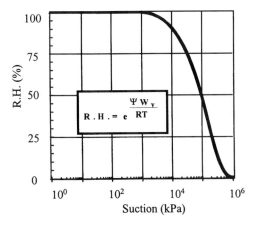

Figure 2. Relative humidity versus suction.

The relationship between actual evaporation (AE) and suction is universal for all soil types. Wilson et al. (1997) showed that the relationship is independent of soil texture for sand, silt and clay as shown in Figure 3. This is an important point to note since the term for suction can be defined as a stress state variable that controls the evaporative flux from a given soil surface.

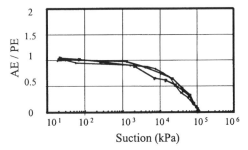

Figure 3. Ratio of actual evaporation and potential evaporation as a function of soil suction for a sand, silt and clay (after Wilson et al., 1997).

The principle outlined above may be used to modify the Penman (1948) method for potential evaporation (PE) to compute actual soil evaporation (AE) as follows (Wilson et al. 1994):

$$E = \frac{\Delta Q_n + \gamma E_a}{\Delta + A} \quad (4)$$

where E = evaporative flux (mm/day); Δ = slope of the saturation vapour pressure versus temperature curve at the mean temperature of the air (mmHg/°C); Q_n = net radian energy available at the surface (mm/day); γ = psychrometric constant; and

$$E_a = f(u)e_a(B-A) \quad (4a)$$

where $f(u)$ = 0.35 (1 + 0.146 Wa); Wa = wind speed (km/h); e_a = water vapour pressure of the air above the soil surface (mmHg); B = inverse of the relative humidity in the air; and A = inverse of the relative humidity at the soil surface.

The solution for the system of equations 1 through 4 requires that the temperature for the soil surface be defined. Surface temperature is computed as follows:

$$T_s = T_a + \frac{1}{\gamma (f(u))}(Q_n - E) \quad (5)$$

where T_s = temperature of the soil surface (°C); and T_a = air temperature above the soil surface (°C).

The solution for equations 1 through 5 outlined above is provided by the SoilCover (1997) computer model. Root-water uptake due to plant transpiration must also be included if the soil surface is vegetated. The method described by Tratch et al. (1995) and Ritchie (1972) is utilized in the SoilCover (1997) model.

3 APPLICATION OF THE KNOWLEDGE-BASE SYSTEM

The SoilCover (1997) model requires detailed input for the specification of soil properties and climatic parameters. Information for the climatic parameters can be obtained from routine weather station installations. The parameters include daily precipitation, net radiation, maximum and minimum temperature and relative humidity, wind speed and pan evaporation. In many cases this data can be easily obtained from government agencies (i.e., regional meteorological station and airports, etc.). The evaluation of soil properties is more difficult.

Soil property functions for the soil-water characteristic curve and the saturated/unsaturated hydraulic conductivity function are required as input parameters for the SoilCover model. The soil property functions may be measured in the laboratory using pressure plates and permeameters. These laboratory test procedures are expensive and often require several weeks to complete. In many cases, several soils must be evaluated. A knowledge-based system may be used to assist in the evaluation for a range of materials that are available for the design and construction of the cover. Application of the knowledge-based system described by Fredlund et al. (1998) for the selection of cover materials is demonstrated here.

Figure 4 shows the grain-size distribution for four potential material types that are available for construction of a cover system. The predicted soil-water characteristic curves given by the knowledge-base system (Fredlund et al. 1998) for each material can be seen in Figure 5. The clay material has an air entry value (AEV) greater than 100 kPa together with a low value of saturated hydraulic conductivity in the range of 1 x 10^{-10} cm/sec. In many cases this type of material is suitable for the construction of a sealing layer. However, a compacted clay cover system is not considered suitable for the semi-arid climate of Saskatoon. Experience has shown that this type of material is subject to shrinkage and cracking during dry periods. This will result in a loss of integrity with respect to restricting water infiltration. It can also be seen in Figure 5 that the AEV for the gravel is less than 1 kPa and that the suction at which the residual water content is reached is approximately 3 kPa. The high perme-

ability and poor water retention characteristics eliminate this material as a potential cover material.

The soil-water characteristic curves for the fine sand and silty loam suggest these soils may be suitable for the construction of a cover system. Figure 5 shows the AEV and the suction at residual water content for the fine sand to be 7 and 100 kPa respectively. The AEV for the silty loam is lightly higher. The gentle slope of the SWCC with increasing soil suction indicates this material will have excellent water retention characteristics for a store and release cover system. The fine sand and silty loam were therefore selected for laboratory testing based on the results of the predictions of the knowledge-based system. In summary, the number of pressure plate tests was reduced to two soil types from the original four samples.

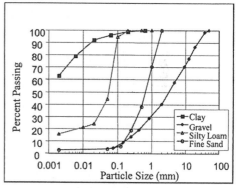

Figure 4. Grain-size distribution for four soil types available for cover construction.

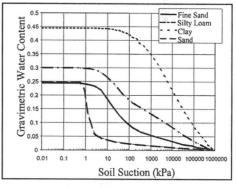

Figure 5. Predicted soil-water characteristic curves determined by the knowledge-based system.

4 SOILCOVER MODEL RESULTS

Figures 6 and 7 show the measured SWCC and estimated hydraulic conductivity function for the fine

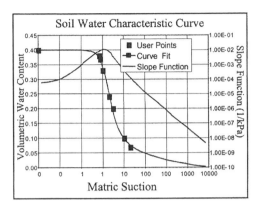

Figure 6. Soil-water characteristic curve for the fine sand specified in the SoilCover model.

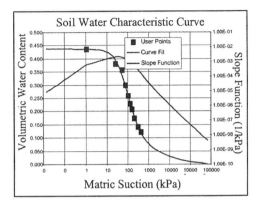

Figure 8. Soil-water characteristic curve for the fine sandy loam material specified in the SoilCover model.

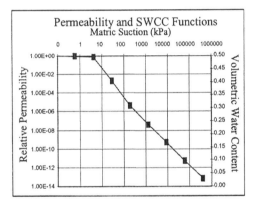

Figure 7. Hydraulic conductivity function for the cover material based on the Fredlund and Xing method.

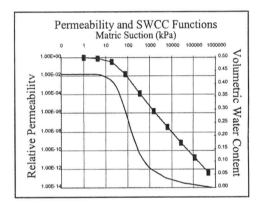

Figure 9. Hydraulic conductivity function for the fine silty loam material based on the Fredlund and Xing method.

sand. The hydraulic conductivity function was determined on the basis of the Fredlund and Xing (1994) method. The saturated hydraulic conductivity of the sand was measured to be 3×10^{-5} m/sec.

Figures 8 and 9 show the SWCC and hydraulic conductivity function for the fine sandy loam. The saturated hydraulic conductivity of the fine sandy loam was measured to be 1×10^{-7} m/sec. The soil property functions illustrated in Figures 6 through 9 were used to predict infiltration rates for two 1.5 m thick cover systems.

Figure 10 illustrates the cumulative ground surface fluxes for a 1.5 m sand cover at the Saskatoon landfill after approximately 150 days. The data for precipitation was obtained from historic meteorological data for the year 1993. It can be seen in Figure 10 that the total precipitation for the simulation period is 400 mm while potential evaporation is approximately double this value at 750 mm. Figure 10 shows the cumulative actual evaporation of 350 mm to be significantly less than the potential evaporation. This leaves the amount of water available

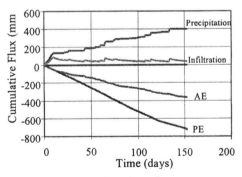

Figure 10. Cumulative fluxes for sand cover.

for infiltration equal to approximately 50 mm or about 10% of total precipitation.

Figure 11 shows the water content profiles for the 1.5 m thick sand cover on days 0, 10, 65 and 150. It can be seen that a wetting front advances through the profile on day 10 of the simulation followed by a second wetting period on day 65. This shows that

454

significant quantities of water readily infiltrate through the sand cover. In summary, the performance of the sand cover system is considered unacceptable.

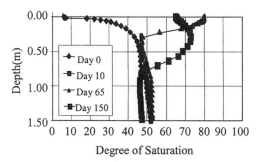

Figure 11. Water content profile for sand cover.

Figure 12 shows the cumulative ground surface fluxes for a cover system consisting of 1.0 m of silty loam over 0.5 m of fine sand. The climate data used for the simulation was exactly the same as that used for the previous analysis. It can be seen in Figure 12 that the net infiltration is reduced to zero for the 150 day period. This occurs as a result of the higher value of actual evaporation (AE) equal to approximately 400 mm. The performance of this cover system shows significant improvement compared to the homogeneous 1.5 m sand profile.

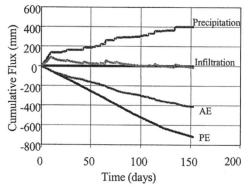

Figure 12. Cumulative fluxes for silt over sand cover.

Figure 13 presents the computed volumetric water content profiles for the 1.0 m silt over 0.5 m sand cover profile. It can be seen that the silty loam material maintains a relatively high value of water saturation. This can be attributed to the higher air entry value and water retention capacity of this material compared to the sand. The net result in retaining more soil water near the ground surface provides a higher value of actual evaporation. This translates to a reduction in net infiltration equal to a value approaching zero. The silty loam profile in the cover performs as a store and release cover system. Infiltration water that enters the soil profile early in the simulation period (actually a result of snow melt during the spring) is retained near the ground surface for a sufficient period of time such that it may be removed during the high evaporation period that occurs during the summer months. In summary, the analyses show that a material with a moderately low value of hydraulic conductivity (i.e., 1×10^{-7} m/sec) such as silt can provide a barrier to infiltration in semi-arid climates with wet and dry periods.

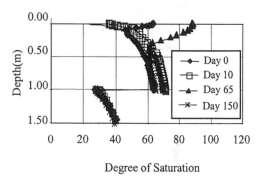

Figure 13. Water content profile for the silt over sand cover.

5 CONCLUSIONS

The application of the SoilCover model to the design of a soil cover system for a municipal landfill was demonstrated. The results of the simulation show that a silty loam material provides a barrier to infiltration in the semi-arid climate that prevails in Saskatoon. It is interesting to note that a silt rich material can provide better protection against infiltration than a highly plastic clay material since severe cracking associated with desiccation is avoided. The numerical modelling and design of the cover system was assisted with the use of a knowledge base system for the selection of suitable soils of testing and analysis. This approach reduced the amount of laboratory testing and time required to complete the analysis.

REFERENCES

Edlefsen, N.E. & Anderson, A.B.C. 1943. *Thermodynamics of Soil Moisture*, Hilgardia, 15(2), pp. 31-298.

Fredlund, D.G. & Xing Anqing, 1994. Predicting the permeability for Unsaturated Soils Using the Soil-Water Characteristic Curve. *Canadian Geotechnical Journal*, Vol.31, pp. 521-532.

Fredlund, M.D., Wilson, G.W., & Fredlund, D.G., 1998. Estimation of Hydraulic Properties of An Unsaturated Soil Using A Knowledge-Based System, *Second International Conference on Unsaturated Soils*, August 27-30.

Penman, H.L., 1948. Natural Evaporation From Open Water, Bare Soil and Grass. *Proceedings of the Royal Society of London, Series A,* 193; pp. 120-146.

Ritchie, J.T., 1972. Model for Predicting Evaporation From a Row Crop with Incomplete Cover. *Water Resources Research,* 8(5): 1204-1213.

SoilCover, 1997. *User's Manual,* Unsaturated Soils research Group, Department of Civil Engineering, University of Saskatchewan, Saskatoon, Canada.

Tratch, D.J., Wilson, G. W. & Fredlund, D.G., 1995. An Introduction to Analytical Modelling of Plant Transpiration For Geotechnical Engineers. *Proceedings of the 48th Annual Canadian Geotechnical Conference, Vancouver, BC,* September 25-27, pp. 771-780.

Wilson, G.W., Fredlund, D.G. & Barbour, S.L., 1997. The effect of soil suction of evaporative fluxes from soil surfaces. *Canadian Geotechnical Journal* (34), pp.145-155.

Unsaturated Soils for Asia, Rahardjo, Toll & Leong (eds) © 2000 Taylor & Francis, ISBN 90 5809 139 2

Influence of some hydraulic parameters on the pore-water pressure distribution in unsaturated homogeneous soils

L.T. Zhan, C.W.W. Ng & B. Wang
Department of Civil Engineering, Hong Kong University of Science and Technology, SAR, People's Republic of China

C.G. Bao
Yangtze River Scientific Research Institute, Wuhan, People's Republic of China

ABSTRACT: In this paper, a horizontal homogeneous soil layer subjected to rainfall infiltration is adopted to study the influence of some hydraulic parameters on the pore-water distributions systematically. The pore-water pressure profiles are calculated using a published analytical solution. The results from the study demonstrates that the influence of desaturation coefficient of the soil-water characteristic curve (SWCC) is the most significant, followed by the saturated permeability and the difference between saturated and residual volumetric water contents.

1 INTRODUCTION

Soil layers near the ground surface are normally unsaturated in arid and semi-arid areas. The existence and the variation of negative pore water pressure (i.e., matric suction in the case where the pore-air pressure is atmospheric) in unsaturated soils has great influences on soil properties (Fredlund and Rahardjo, 1993). Therefore, negative pore water pressure distribution and its influencing factors should be investigated firstly when confronting with engineering problems related to unsaturated soils. For instance, residual soils near the slope surface are mostly unsaturated, and negative pore water pressure is a major contributor to the shear strength of soil and therefore to slope stability during dry seasons. However, the value of negative pore water pressure can be reduced significantly or even destroyed completely when a certain amount of rain water infiltrates into the soil slope during wet seasons. Under such circumstances, the originally stable slope is likely to fail. Thus, to make reasonable evaluations on slope stability, it is necessary to know the distribution and variation of negative pore water pressure in the soil under various climatic conditions.

The distributions of negative pore-water pressure are affected by many intrinsic and external factors. The intrinsic factors here mainly refer to the hydraulic properties of soils, including SWCC and hydraulic conductivity. The external factors mainly refer to climate conditions, such as rainfall intensity and duration, rainfall pattern, evaporation rate. Ng and Shi (1998) and Ng et al. (2000) performed

parametric studies of the effects of rain infiltration on unsaturated slopes, in which the effects of various rainfall events and patterns, anisotropic soil permeability, presence of an impeding layer and conditions of surface cover were investigated using numerical simulation.

In this paper, three hydraulic parameters are firstly introduced to describe an idealized SWCC and hydraulic conductivity function. Following Srivastava and Yeh (1991), an analytical solution for 1-dimensional (1D) rainfall infiltration in homogeneous soil is derived and then applied to carry out a parametric study for investigating the influence of the three hydraulic parameters and the saturated water permeability of the soil on pore water pressure distributions.

2 HYDRAULIC PROPERTIES OF UNSATURATED SOILS

The hydraulic property of a soil affects the interaction between soil and water and water flow in the soil. It mainly includes the moisture retention characteristic and the hydraulic conductivity.

Moisture retention characteristic of a soil can be represented with a soil-water characteristic curve (i.e., SWCC), which is defined as the relationship between water content and matric suction. Figure 1 shows an idealized plot of SWCC with two characteristic points, A^* and B^*. Point A^* corresponds to the air-entry value (i.e., $(u_a-u_w)_b$), and B^* corresponds to the residual water content (i.e., θ_r). As shown in Figure 1, the SWCC between A^* and B^* is nearly a straight line on logarithmic scale. So the linear part of SWCC can be approximately represented with the saturated and residual volumetric water content, and the desaturation rate of SWCC.

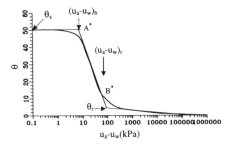

Figure 1 An idealized soil-water characteristic curve

In a saturated soil, the saturated hydraulic conductivity (permeability) is a function of void ratio. However, it is generally assumed be a constant. In an unsaturated soil, the hydraulic conductivity is significantly affected by both void ratio and degree of saturation or matric suction of the soil. Water flows through the pore space filled with water, so the percentage of the voids filled with water (degree of saturation) is an important factor. The relationship of degree of saturation to matric suction can be represented with SWCC. Thus the hydraulic conductivity of unsaturated soils with respect to matric suction bears a relationship to SWCC, and it can be estimated from the saturated permeability and SWCC (Marshall, 1958; Fredlund et al., 1994).

The hydraulic property of a soil can be approximately reproduced with the desaturation rate, the saturated and residual volumetric water content and the saturated permeability of the soil.

Figure 2 illustrates the SWCCs for Superstition sand and Guelph loam. As shown in the figure, the SWCC of the sand lies above that of the loam, which indicates that the volumetric water content of loam is larger than that of the sand with the same matric suction. In other words, the capacity of moisture retention for the loam is larger than that for the sand.

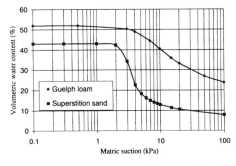

Figure 2. SWCCs for different types of soils (Data from Richards, 1952; and from Elrick and Bowman, 1964)

On the other hand, when the suction is beyond the air-entry value (about 1.5 kPa for the sand and 4 kPa for the loam), the desaturation rate of the sand is larger than that of loam. For the sand, a same increase in matric suction will lead to more water being expelled from the loam. In other words, more water content change is needed to reduce a same value of matric suction (e.g., 2-100 kPa).

Corresponding to Figure 2, Figure 3 shows the hydraulic conductivity curves together with measured data for Superstition sand and Guelph loam. As shown in the figure, there is an interesting phenomenon. For matric suction less than 2 kPa, the hydraulic conductivity of sand is larger than that of loam, but for matric suction greater than 2 kPa, it is on the contrary. This phenomenon may be attributed to the different desaturation rates of sand and loam (Figure 2). The former is larger, i.e., the moisture content in the sand decreases rapidly with an increase in matric suction. Therefore, the hydraulic conductivity of sand decreases sharply with the increasing matric suction due to a sharp reduction in the area of cross section for water flow.

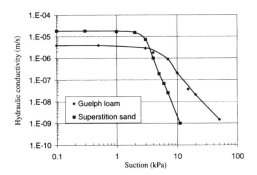

Figure 3. Hydraulic conductivity curves for different types of soils (Data from Richards, 1952; and from Elrick and Bowman, 1964)

The discussion above may be helpful to explain the influence of hydraulic parameters on the pore-water pressure distributions in the soil subjected to rainfall infiltration.

3 ANALYTICAL SOLUTION FOR 1D VERTICAL INFILTRATION IN AN UNSATURATED HOMOGENEOUS SOIL LAYER

The governing equation for 1D vertical infiltration in unsaturated soils is as follows, which is, namely, Richard's equation.

$$\frac{\partial}{\partial z}\left[k(\psi)\frac{\partial(\psi + z)}{\partial z}\right] = \frac{\partial \theta}{\partial t} \qquad (1)$$

458

where, ψ is the pressure head (negative for unsaturated flow)

θ is volumetric water content of a soil

The hydraulic conductivity k in unsaturated soils is a function of negative pressure head ψ, which results in the nonlinearity of this equation. To obtain a close form analytical solution, some assumptions have to be made to linearize the equation. In the past decades, many researchers have made great efforts to obtain the analytical solutions for the above equation (Philip, 1957a, Broadbridge and White, 1988, Pullan, 1990). Most of them assumed an exponential relationship between permeability and negative pressure head ψ. Srivastava and Yeh (1991) assumed the relationship of the hydraulic conductivity k and volumetric water content to ψ are both exponential,

$$\begin{cases} \theta = \theta_r + (\theta_s - \theta_r)e^{\alpha\psi} & (2a) \\ k = k_s e^{\alpha\psi} & (2b) \end{cases}$$

where,

θ_s and θ_r are the saturated and residual volumetric water content of soils;

k_s is the saturated hydraulic conductivity;

α is a coefficient representing the desaturation rate of the SWCC, defined as the desaturation coefficient. It is related to the pore size and its distribution.

The two functions can be used to approximate the hydraulic properties with the desaturation coefficient (α), the saturated and residual volumetric water content (θ_s and θ_r) and the saturated permeability (k_s) within most concerned matric suction range. Among them, α, θ_s and θ_r represent the moisture retention characteristic of soil (SWCC); α and k_s represent the hydraulic conductivity of unsaturated soils with respect to matric suction.

With these two functions, Richard's equation can be transformed to the following linear equation:

$$\frac{\partial^2 k}{\partial z^2} + \alpha \frac{\partial k}{\partial z} = \frac{\alpha(\theta_s - \theta_r)}{k_s} \frac{\partial k}{\partial t} \qquad (3)$$

Analytical solution to this equation requires one initial and two boundary conditions. In Srivastava and Yeh (1991) study, the one-dimensional infiltration problem in homogeneous soil is defined in Figure 4. $\psi_0 = 0$ is the prescribed pressure at the groundwater table that is assumed to remain stable during the rainfall duration. q_A is the antecedent flux at the soil surface which, along with ψ_0, determines the initial pressure distribution in the soil. q_B is the prescribed flux at the soil surface for times greater than 0. The thickness of the unsaturated soil layer is L. The solution for this equation can be obtained through Laplace's transformation. Details of the solution are given by Srivastava and Yeh (1991)

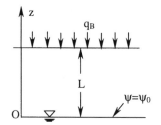

Figure 4. Homogeneous soil profile for calculation

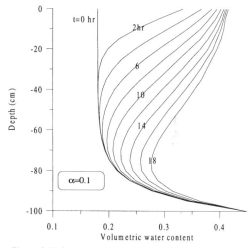

Figure 5. Volumetric water content profiles at various time intervals

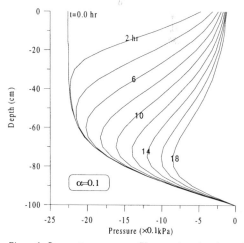

Figure 6. Pore-water pressure profiles at various time intervals

Provided with a certain geometry, soil hydraulic parameters (α, θ_s, θ_r, k_s), initial and boundary conditions, the moisture content distribution and the pore-water pressure distribution along depth at a particular instance can be calculated and the results are shown in Figure 5 & 6 respectively (the other parameters, $k_s=1$cm/hr, $\theta_s=0.45$, $\theta_r=0.15$, $q_A=0.1$ cm/hr, $q_B=0.9$ cm/hr). It can be seen that both the pressure profiles and the moisture content profiles are similar in shape, which is the result of the assumed exponential relations (Eq. 2a &2b)

4 PARAMETRIC STUDY OF HYDRAULIC PARAMETERS

A homogeneous soil layer (L =100cm) was chosen for this study. The cases shown in Table 1 are run to investigate influence of hydraulic parameters, namely, the desaturation coefficient α, the saturated permeability k_s, the saturated volumetric water content θ_s, residual water content θ_r, on the pore-water pressure distributions in homogeneous unsaturated soil.

Table 1. Analysis scheme for the parametric study

Cases	No. of run	α	k_s (cm/hr)	$(\theta_s-\theta_r)$	q_A (cm/hr)	q_B (cm/hr)
I	1	0.1	1.0	0.30	0.1	0.9
	2	0.01				
II	1	0.1	0.1	0.30	0.0	0.9
	2		1.0			
	3		10.0			
III	1	0.1	1.0	0.20	0.1	0.9
	2			0.30		
	3			0.40		

4.1 Effect of desaturation coefficient α

The desaturation coefficient α represents the decreasing rate of moisture content or hydraulic conductivity with respect to an increasing suction. It is related to the pore size distribution of a soil. Generally speaking, the bigger the particle size is, the larger the pore size will be and consequently the greater the value of α. In addition, the value tends to increase with the uniformity of the soil structure.

Figure 7 shows a series of pore-water pressure profiles with respect to two different α values (case I in Table 1). Firstly, it can be seen from Figure 7 that the two initial pore-water pressure profiles (t = 0 hr) differ greatly from each other. The greater the value of α is, on the whole, the lower the value of negative pore-water pressure will be. The maximum value is about 2.3 kPa for $\alpha=0.1$ and 8.2 kPa for $\alpha=0.01$. It is logical that the unsaturated soil with small particle sizes tends to possess high matric suction under identical condition.

As rainwater infiltrates into the soil, the wetting front advances downwards gradually. Comparing the two profiles corresponding to 4hr, it is clear that the smaller the value of α, the deeper the wetting front advances. For a given time (4hr), the wetting front reaches a depth of 40 cm for $\alpha=0.1$ and 100cm for $\alpha=0.01$. In addition, the two profiles corresponding to 18hr intersect with each other at the depth of 63 cm. In other words, the value of negative pore-water pressure for $\alpha=0.01$ is smaller above the depth but greater below the depth than that for $\alpha=0.1$. This could be explained as follows: Eq. (2b) indicates the hydraulic conductivity of soil with respect to a same ψ decreases with an increase in α provided with identical k_s. It should be noted that ψ is negative in Eq. (2a) & (2b). When the value of α is relative large, the hydraulic conductivity of the soil is lower than that of the soil with a smaller α. So more infiltrated water tends to retain in the soil near the ground surface, which leads to a lower negative pore-water pressure near the ground surface but a higher value within the lower part of the layer. On the contrary, when the value of α is small, the hydraulic conductivity is relatively high with respect to a same matric suction, the downward flow is easier and the moisture is more uniformly distributed. On the other hand, both Eq. (2a) and Fig. 2 suggest that for the soil with a higher α more water content change is needed to reduce a same value of matric suction (e.g., 2-100 kPa). Therefore, the reduction of matric suction in the soil near the ground surface needs more infiltrated water for the case with a higher α.

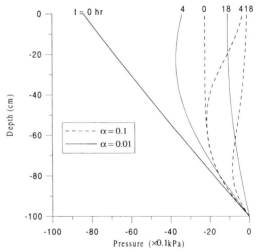

Figure 7. Pore-water pressure profiles at various time intervals

Based on the analysis above, it can be concluded that this parameter is of considerable influences on the initial pore-water pressure profile as well as those after the commence of rainfall.

460

4.2 Effect of the saturated permeability

It is known the permeability for saturated soils has a close relationship with the particle sizes and the dry density of soils. For homogeneous soil, the permeability generally increases with the particle sizes significantly and decreases with the dry density to some extent.

Figure 8 shows three pore-water pressure profiles for different k_s at t = 10hr, together with identical initial profiles (Case II in Table 1). As mentioned previously, the initial suction distribution is achieved from the steady state flow subjected to the antecedent infiltration rate q_A. If q_A is zero, which represents the initial infiltration rate equal to the evaporation rate, the hydrostatic condition is thus achieved. In Figure 8 the initial profiles are identical for different values of k_s.

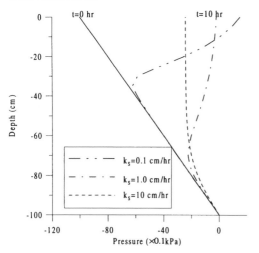

Figure 8. Pore-water pressure profiles with different k_s

As shown in Figure. 8 the saturated permeability has significant influences on the shape of pore-water pressure profiles. Provided the value of k_s is small, the negative pore-water pressure near the ground surface decreases rapidly. However it remains nearly at the initial value within the lower part of the layer. Provided the ratio of saturated permeability k_s to the infiltration rate q_B is smaller than 1 (k_s=0.1 cm/hr), a positive pore-water pressure will appear after a certain time of rainfall infiltration (corresponding to 10hr). It is attributed to the small k_s preventing all rainwater from infiltrating into the soil, which results in the development of ponding at the ground surface. On the other hand, if the value of k_s is relatively large, the depth affected by the infiltrated water will be larger. However, the magnitude of reduction in matric suction will be smaller because the infiltrated water is more uniformly distributed along the depth.

4.3 Effect of $(\theta_s\text{-}\theta_r)$

$(\theta_s\text{-}\theta_r)$ is the difference between the saturated volumetric water content and the residual volumetric water content, which is an index showing the capacity of water retention of the soil. Generally, the value of $(\theta_s\text{-}\theta_r)$ increases with particle sizes. To some extent, it is also related to the dry density of soils.

Figure 9 shows three pore-water pressure profiles for different θ_r at t = 10hr, together with identical initial profiles (Case III in Table 1). According to the solution (Srivastava and Yeh , 1991), the initial pore-water pressure profiles only depend on the parameters (α, k_s and q_A), so the initial profiles for different $(\theta_s\text{-}\theta_r)$ are identical. Of course, the parameters (θ_s and θ_r) will affect the initial moisture content profiles due to the different relationship between negative pore-water pressure and moisture content. In this paper, only the pore-water pressure is of concern.

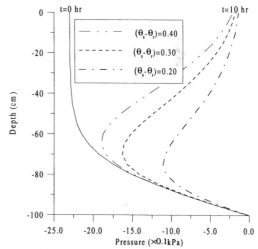

Figure 9. Pore-water pressure profiles with respect to different $(\theta_s\text{-}\theta_r)$

As shown in Figure 9, $(\theta_s\text{-}\theta_r)$ mainly affects the magnitude of the matric suction but not the shape of the profiles. A larger value of $(\theta_s\text{-}\theta_r)$ represents a higher water retention capacity of the soil, which results in a lower advancing rate of wetting front and a smaller reduction in negative pore-water pressure with respect to the initial value.

5 CONCLUSIONS

For investigating transient seepage in an unsaturated soil slope, it is necessary to know the SWCC and the saturated water permeability of the soil. For engineering applications, a SWCC may be approximated and represented by three hydraulic parameters : desaturation coefficient (α), saturated (θ_s) and residual (θ_r) volumetric water contents. These parameters are of great influence on the pore-water pressure distributions in soils. Based on the parametric study, α is found to be the most significant factor. It affects both the initial steady state and transient pore-water pressure distributions during rainfall. The greater the value, the smaller the initial matric suction will be. The effects of the saturated permeability of the soil are also considerable. However, difference of saturated volumetric water content and residual water contents has the least influence on the pore water pressure distributions.

ACKNOWLEDGEMENTS

The authors would like to acknowledge the finical support from research grants HKUST6046/98E, HKUST6108/99E and CA99/00.EG01 provided by the Research Grants Council of the Hong Kong Special Administrative Region. The authors are also grateful to Professor Wilson Tang for his continual support to their research.

REFERENCES

Broadbridge, P., and White I., (1988). Constant rate rainfall infiltration: A versatile nonlinear model, 1. Analytical solution, Water Resour. Res., 24(1), 145-154.

Fredlund, D. G. & Rahardjo, H., (1993). Soil Mechanical For Unsaturated soils. John Wiley & Sons, Inc., New York.

Fredlund, D. G., Xing, A. & Huang, S. Y., (1994). Predicting the permeability functionfor unsaturated soils using the soil-water characteristioc curve. Can. Geotech. J., 31, 533-546.

Marshall. T. J., (1958). A relation between permeability and size distribution of pores. Journal of Soil Science. 9, 1-8.

Ng, C. W. W., & Shi, Q, (1998). A numerical investigation of the stability of unsaturated soil slopes subjected to transient seepage. Computers and Geotechnics, Vol. 22, No. 1: 1-28.

Ng, C. W. W., Tung, Y. K., Wang, B. & Liu, J. K. (2000). 3D analysis of effect of rainfall patterns on pore water pressure in unsaturated slopes. Submitted to Geo Eng 2000 International Congress, Melbourne, Australia.

Philip, J. R., (1969). Theory of infiltration, Adv. Hydrosci., No. 5, 215-296.

Pullan, A. J., (1990). The Quasi Linear Approximation for Unsaturated Porous Media Flow. Water Resour. Res., 26(6), 1219-1234.

Srivastava, R. & Yeh, T. C. J. (1991). Analytical solutions for one-dimension, transient infiltration toward the water table in homogeneous and layered soils. Water Resour. Res., Vol. 27, No. 5, 753-762.

7 Mass transport

Unsaturated Soils for Asia, Rahardjo, Toll & Leong (eds) © 2000 Taylor & Francis, ISBN 90 5809 139 2

Experimental studies of salt movement through unsaturated soils

R. Bhargava & A. Gairola
Department of Civil Engineering, University of Roorkee, India

ABSTRACT: Ground water contamination is of increasing concern for environmentalists in developed and developing countries. To prevent the contamination of subsurface water, it is of paramount importance to understand the phenomenon which causes the transport of the salts and other contaminants. This information regarding the movement of dissolved substances in the subsoil is a vital requirement for successful subsurface management. The present work is a column study conducted to understand the parameters affecting flow through unsaturated media. Transport is also studied for saturated soil for comparison. This study is important to determine the time span or the time lag between the application of a solution or the waste on the land and the time for it to reach a certain depth, which is required to plan an effluent management system. This will also help in understanding the abrupt eruption of contaminant.

1 INTRODUCTION

Public awareness and concern about the leaching of nutrients and pesticides from agricultural lands has never been higher than it is today. This is not surprising as more and more evidence shows that diffused sources of pollution play a major role in the degradation of soil and ground water systems. Ground water pollution stemming from leachate generation in land fills, chemical-spillage, petroleum reservoirs, leaching off cesspools and septic tanks etc. and final disposal of sewage treatment plant effluents through sewage irrigation.

For understanding groundwater pollution, the soil between the ground surface and aquifer has to be studied and this is the unsaturated zone. The unsaturated zone plays an important role in preventing and delaying the contaminants from the surface in reaching the ground water but once the soil becomes saturated the movement is fast. The vadose zone also acts as a significant reservoir for the capture, storage and release of contaminants. The literature shows that large numbers of theoretical models have been developed in this field, but the experimental work is scarce (Porro and Wierengr 1993). The present work is an attempt towards experimental studies for understanding the transport phenomenon in the vadose zone.

2 EXPERIMENTAL SETUP

The experimental setup consists of a perspex column, 115 cm long and 10x10 cm square in cross

section. The column was fitted with the probes at 10-cm distance center to center. The column was filled with soil, gravel etc. as shown in Figure 1. Gypsum blocks and soil moisture meter were used to measure the moisture content of soil. The salt concentration was measured indirectly by measuring electric conductivity (EC). A reservoir of 4 hour capacity was installed to feed the column.

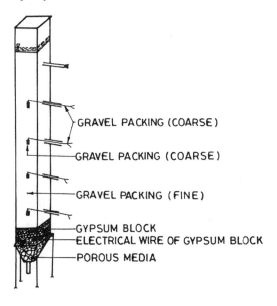

GRAVEL PACKING (COARSE)

GRAVEL PACKING (COARSE)

GRAVEL PACKING (FINE)

GYPSUM BLOCK
ELECTRICAL WIRE OF GYPSUM BLOCK
POROUS MEDIA

Figure 1. Experimental Column

3 EXPERIMENTAL PROCEDURE

Three types of soils were chosen for the tests. Initial test runs were conducted on all types of soil samples to decide the water column above the soil. The head was fixed to such a value that sufficient time was available for various measurements. Since the column was small in cross section the experiments were divided into two sets.

3.1 Soil Moisture and flood measurements

Gypsum blocks were placed at 10 cm, 30 cm, 50 cm, and 70cm depth. To maintain the porosity of the soil in all the runs a measured quantity of soil was used to fill the column. Clear water was then passed through the soil at a constant head of 1 cm and soil moisture meter reading were noted at 10 cm depth till the soil became saturated at that depth. Similar observations were recorded at 30, 50 and 70 cm depth.

The volume of water collected was recorded at the outlet point from the time the first drop appeared till the outflow rate became constant, which indicated the saturation of the complete soil column. The procedure was repeated three times for each of the soils.

3.2 The Salt Movement Observations

The soil was filled as above and the water containing 500mg /l salt was allowed to flow through the column for dry as well as for saturated soil. The EC was measured in the water passing through the outlet till saturation was achieved in the case of unsaturated soil, or till the effluent concentration is same as input concentration (which is also known as breakthrough) in the saturated column.

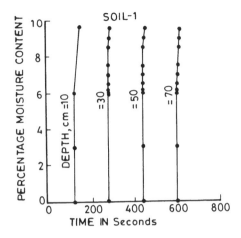

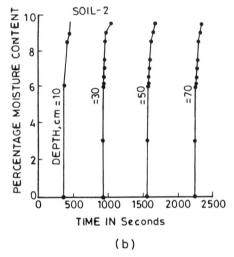

(b)

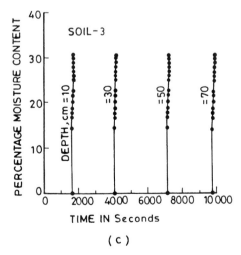

(c)

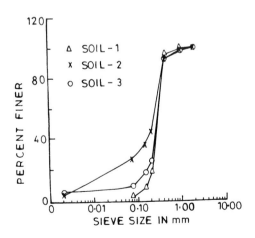

Figure 2. Particle size distribution for different soils

Figure 3a,b,c. Variation of percentage moisture content versus time

466

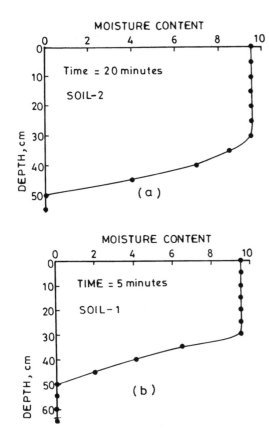

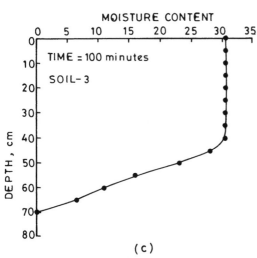

Figure 4a,b,c. Variation of percentage moisture versus depth

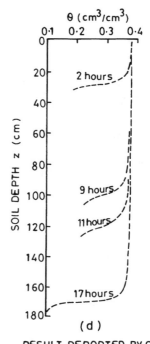

RESULT REPORTED BY GUPTA S.(1991)

Figure 4d. Variation of percentage moisture versus depth

4 RESULTS AND DISCUSSION

The physical properties of soils are listed in Table 1. Soil 1 is riverbed soil, soil 2 is topsoil and soil 3 is a mixture of top soil and Dhanouri clay. The particle size distribution is shown in Figure 2.

Table 1. Physical Properties of Soils

Soil Type	K (cm/s) Saturated	G	ρ_d Mg/m^3	e	n	S(m)
Soil 1	0.006	2.65	1.7	0.59	0.37	4.6
Soil 2	0.00016	2.67	1.373	0.97	0.5	2.79
Soil 3	0.000097	2.7	1.3	0.99	0.49	2.74

The movement of moisture with time is shown Figure 3(a, b, c) from which it is concluded that the variation of moisture content with time resembles plug flow (i.e. the complete front moves, also known as piston flow) in all the soils. Figure 4(a, b, c) shows the variation of moisture content with depth at specific time after water starts flowing.

The trend compares well with theoretical results (Gupta 1991) shown in Figure 4d. Suction head with percentage moisture is plotted in Figure 5 for which the trend compares with Abeliuk's (1981) work shown in Figure 6 i.e. initially the suction head decreases with increase in moisture contents faster than at later stage.

For study of salt movement, the results of EC vs time are shown in Figure 7. These are comparable with theoretical values plotted in Figure 8 (Gupta, 1991). The comparison of flow of salt through saturated and unsaturated column shows that in unsaturated soil the salt appears at a later time as compared to saturated soil and the eruption of salt is more or less abrupt, while in saturated column it slowly reaches the breakthrough. This explains the reasons for fast increase in the contaminant concentration in ground water after it's first appearance. Such study can be useful in planning solid waste/waste water disposal on the ground because the leachate shall be seeping down in the same manner.

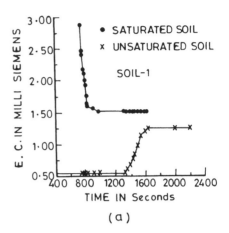

(a)

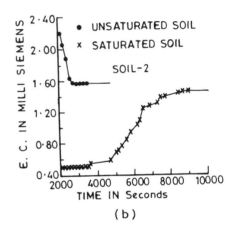

(b)

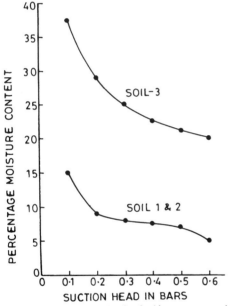

Figure 5. Variation of suction head with percentage moisture content in soil 1,2 and 3.

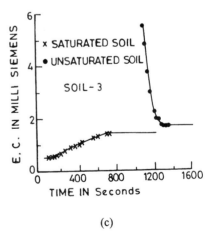

(c)

Figure 7a,b,c. Variation of electrical conductivity versus time

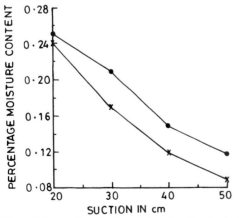

Figure 6. Reported results in variation of suction head with percentage moisture content by Abeliuk R. (1981)

468

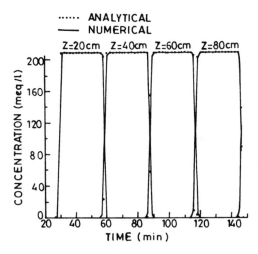

Figure 8. Concentration breakthrough curve (Gupta 1991)

CONCLUSIONS

From the figures and discussion above the following conclusions can be drawn:

(i) The experiment explains the abrupt eruption of contaminant into the ground or well.

(ii) Such study can be useful in planning solid waste/ waste water disposal on the ground.

REFERENCES

Abeliuk R. 1989. Simultaneous determination of unsaturated flow and solute transport parameters. *Contaminant Transport in Ground Water Conference Proceedings*, 413-420.

Gupta Sulekha, 1991. Modelling of solute transport through and unsaturated zone extending from ground surface to water table. *Ph.D. Thesis, Department of Hydrology*, University of Roorkee, Roorkee

Porro, I.and P.J. Wierengr, 1993. Transient and steady- state solute transport through a large unsaturated soil column. *Ground Water J.* 31(2): 193-200.

Role of unsaturated soil on aqueous diffusion

P.C. Lim
Geoconsult Asia Singapore Pte Limited, Singapore

ABSTRACT: The role of unsaturated soil on aqueous diffusion is examined in this paper. Transport processes governing the movement of inorganic contaminant in soil and the relative influence of each process is briefly outlined. Key mechanisms governing the molecular diffusion of inorganic contaminants in saturated and unsaturated soils are described. A review of existing functional relationship between the degree of saturation on the coefficient of diffusion is presented to provide an understanding of the role of water content on aqueous diffusion. The influence of the degree of saturation on the arrival time of a concentration front and the steady state mass fluxes were investigated. Results of numerical analyses show that the arrival time of the concentration front in general, increases with decreasing degrees of saturation. Decrease in degree of saturation is also accompanied by decreases in steady state mass fluxes, particularly, for cases near residual degree of saturation.

1 INTRODUCTION

The two key transport processes governing the movement of inorganic contaminants in the subsurface geologic material are primarily advection and molecular diffusion. Advection is related to seepage velocity, which is a function of the coefficient of permeability and the hydraulic potential gradient while diffusion is driven by differences in chemical potential.

In attempt to eliminate the component of advective transport, the last decade has seen the increasing application of fine-grained geologic materials as barrier materials. The use of compacted clay liner/barrier characterised by its low hydraulic conductivity is an example. This would have been an effective solution to provide a low or "zero" leakage impoundment for waste containment in the absence of molecular diffusion.

For cases in which the seepage flux or hydraulic conductivity is low, molecular diffusion could become a dominant transport process. Hence, in order to achieve an effective containment of the waste, positive means of reducing or even eliminating the diffusive mass flux would be desirable. As with the hydraulic conductivity, which is a function of the degree of saturation, the same would also hold true for the coefficient of diffusion. Reduction of diffusive mass flux is,

therefore, possible by decreasing the coefficient of diffusive via the degree of saturation. The functional relationship between the diffusion coefficient and the degree of saturation however, would need to be defined.

In this paper, the mechanism governing diffusive transport is briefly described. The relative importance of advection and molecular diffusion is illustrated by comparing the steady state mass fluxes through a saturated fine-grained soil barrier. A review of existing data on the influence of the degree o saturation on the coefficient of diffusion is also presented. The main objective is to present the practical implication of the role of the degree of saturation on diffusive transport in an unsaturated soil.

2 RELATIVE SIGNIFICANCE OF ADVECTION AND MOLECULAR DIFFUSION

To illustrate the relative influence of advectionand molecular diffusion, contaminant transport across a 0.3m thick saturated clay liner with volumetric water content of 0.4 was used as an example. A constant concentration source of 1000mg/L was assumed at the upper bound and zero concentration at the lower bound. The steady state advective mass fluxes across the 0.3m liner with hydraulic conductivity 1×10^{-10} m/s and unit hydraulic gradient is given as:

Advective mass flux $J_a = v\theta_w C$

$$= qC$$

$$= k\frac{\Delta h}{\Delta l}C$$

$$= 1 \times 10^{-10} * 1 * 1000 * \frac{10^3}{10^6}$$

$$= 3.1 * 10^{-3} \text{ kg/yr/m}^2$$

Mass flux from a 1000m by 1000m site = 3100 kg/yr.

The steady state diffusive mass flux across the same 0.3m thick liner with a diffusion coefficient of 5×10^{-10} m^2/s is as follows:

Diffusive mass flux $J_m = \theta_w D_s \frac{\Delta C}{\Delta l}$

$$= 0.4 * 0.016 * \frac{1000 - 0}{0.3} * \frac{10^3}{10^6}$$

$$= 0.021 \text{ kg/yr/m}^2$$

Mass flux from a 1000m by 1000m site = 21 000 kg/yr.

The calculations show that diffusive mass flux is about 7 times greater than that due to advection alone. Hence, in the case of a low permeability barrier, the diffusive mass flux constitutes a dominant fraction of the total mass flux.

3 BACKGROUND

Aqueous diffusion is a process whereby dissolved mass is transported from a higher potential towards a lower potential by random molecular motion (Robinson & Stokes 1959). Diffusion in free water can be described mathematically by Fick's law. A detailed treatment of the fundamental basis of Fick's First law for steady state diffusion has been presented by Shackleford & Daniel (1991a). The driving energy responsible for the diffusion of ions, atoms or molecules is attributed to a change in chemical potential, or concentration gradient for the case of a dilute solution.

The mechanism governing the diffusion of ions in free water is also operative in soil water. The diffusive mass flux in soil is, however, smaller than the flux in free water. The reduction in diffusive flux is attributed to two factors: first, the open area available for diffusion is reduced; and second, the diffusion pathways are more tortuous in soil. The one-dimensional steady state diffusive flux of ion i in soil can be written as

$$J_m = -\theta_w D_i^* \frac{dC_i}{dz} \tag{1}$$

where J_m = mass flux, dC_i = change in concentration of ion i in soil water over an elemental volume, dz = the macroscopic distance of an elemental volume, θ_w = volumetric water content of the soil, D_i^* = effective diffusion coefficient of ion i in soil.

For a saturated soil, the effective diffusion coefficient for a given soil type and stress states, is a constant and is defined by Porter et al. (1960) as

$$D_i^* = D_i^o (L/L_e)^2 \gamma \tag{2}$$

where D_i^o = diffusion coefficient of ion i in free water, $(L/L_e)^2$ = geometric factor which accounts for the effect of tortuous pathways, L = macroscopic distance across an elemental soil volume, L_e = length of actual diffusion pathway across the elemental soil volume, γ = a factor which includes both the effects of ionic interaction and the increased viscosity of water in the immediate vicinity of a soil particle.

For an unsaturated soil, the effective diffusion coefficient given in Equation 2 is no longer a constant but varies with water content:

$$D_i^*(\theta_w) = D_i^o (L/L_e)^2 \gamma \tag{3}$$

or with the degree of saturation:

$$D_i^*(S) = D_i^o (L/L_e)^2 \gamma$$

where S = degree of saturation.

3.1 Functional Relationship for Diffusion Coefficient

Studies have shown that the effective diffusion coefficient decreases with decreases in the water content of the soil (Porter et al. 1960, Romkens & Bruce 1964, Rowell et al. 1967, Warncke & Barber 1972, Barraclough & Tinker 1981). The reported data are given in Figure 1. The functional relationship between the coefficient of diffusion and water content for glass beads, sandy soils and gravel are nonlinear. For fine-grained soils such as loam, sandy loam, silty loam and clay, the functional relationships were either linear or slightly nonlinear.

Most researchers have attributed the decrease in the effective diffusion coefficient with a decrease in water content to an increase in the diffusion path length. Other factors that may have an important influence on the effective diffusion coefficient include increased viscosity of liquid next to the soil particle, increased ionic interaction along small pores and water films, and inter-connectivity between the water-filled pores and the water films (Porter et al. 1960, Romkens and Bruce 1964, Rowell et al. 1967, Warncke & Barber 1972, Barraclough & Tinker 1981).

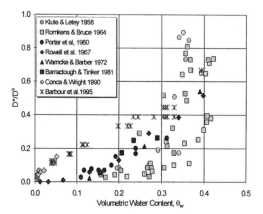

Figure 1. Relationship between the effective diffusion coefficient, normalized with respect to the diffusion coefficient in free water, and the volumetric water content for various geologic materials.

Functional relationship between the diffusion coefficient and water content for a uniform fine sand determined by a new technique described by Barbour et al. (1995) is also provided in Figure 1. The new test method uses the principle of axis-translation technique to control and maintain the stress-state of the soil (i.e., matric suction and total stress) throughout the test. This is deemed important particularly for test of long duration where redistribution of water content would affect the result interpretation. Potassium ion was used as the primary tracer. With potassium being a reactive chemical species, the influence of adsorption was taken into consideration in evaluating the diffusion coefficient.

The results show that a steeper decrease in the diffusion coefficient with initial decrease in water content, the trend is quite similar to that reported by earlier researchers. With further decrease in water content, the diffusion coefficient decreases almost linearly with decreasing water content. Values of the diffusion coefficient at lower water contents are generally higher than those determined using the method of the two half cells or one half cell with exchange membrane.

4 ROLE OF UNSATURATED SOIL ON DIFFUSIVE TRANSPORT

From the data presented in Figure 1, an understanding of the practical implications of the role unsaturated soil on diffusive transport in a containment system is, therefore, useful. Particularly, for cases near residual degree of saturation.

A composite barrier/liner in a waste containment system was chosen as an example in evaluating the practical significance of the role of the degree of saturation on diffusive transport. The main component of the composite barrier of interest is the secondary leachate collection layer. The leachate collection layer is generally constructed using free-draining granular material such as stone or gravel, which desaturates readily and exhibits steep functional relationship. For purpose of this study, the stone is replaced with sand whose diffusive characteristics over the range of degree of saturation are defined.

The role of drainage layer at varying degrees of saturation on diffusive transport was evaluated using numerical approach. One-dimensional diffusive transport of inorganic contaminant through the upper clay liner, granular drainage layer and lower clay layer into an aquifer was simulated using POLLUTE (Rowe & Booker, 1983). The role of the unsaturated soil in relation to the arrival time of the concentration front and the mass fluxes at the base of the lower clay liner was evaluated by varying the degree of saturation within the granular drainage layer. The effect of the thickness of the drainage layer on diffusive transport was also examined. The idealised model and the boundary conditions for the numerical simulations are shown in Figure 2. Key soil parameters for the clay liner and the granular layer are given in Table 1. For illustration purpose and simplicity, zero adsorption characteristics of the clay and granular layer were assumed.

Table 1. Soil parameters for diffusive transport simulations.

Soil layer	Degree of Saturation [%]	Volumetric water content	Effective diffusion coefficient [m²/a]
Clay liner	100	0.35	0.019
Granular layer	100	0.40	0.0377
	75	0.30	0.0226
	50	0.20	0.0160
	25	0.10	0.0109
	10	0.04	0.0075

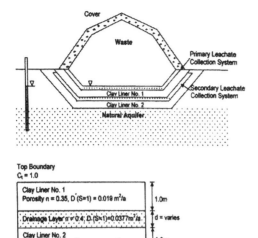

Figure 2. Idealised model of a typical composite barrier/liner.

4.1 Results and discussion

A summary of the results of the numerical simulations and a discussion of the results are outlined in the following.

4.1.1 Effect of degree of saturation on arrival time of concentration fronts

Breakthrough curves of the concentration front at the base of the granular layer for cases of different degrees of saturation for the 0.3m thick drainage layer are given in Figure 3. Due to differences in steady state concentration for the various cases of degrees of saturation, the concentrations were normalized with respect to that at steady state.

Results showed increases in arrival time with decreasing degrees of saturation. The influence of the degree of saturation on the arrival in general, was relatively insignificant except for cases near residual degree of saturation.

The effect of the degree of saturation on the arrival time of the $C/C_0 = 0.5$ and $C/C_0 = 0.9$ concentration fronts is further illustrated in Figure 4. The arrival times for the different cases were normalized with respect to the arrival of the respective concentration fronts of case with 100% degree of saturation.

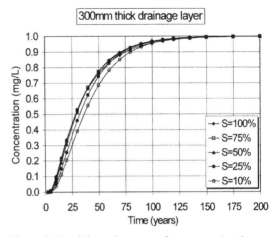

Figure 3. Breakthrough curves of concentration fronts at the base of the drainage layer for cases of different degrees of saturation.

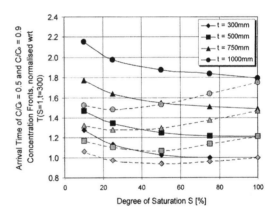

Figure 4. Arrival times of $C/C_0 = 0.5$ and $C/C_0 = 0.9$ concentration front versus the degree of saturation.

For the case of the $C/C_0 = 0.5$ concentration front, results show a gradual increase in arrival time with decreasing degrees of saturation initially followed by a steeper increase at near residual degree of saturation. With increasing thickness of the drainage layer, the arrival time also increases almost proportionately. At residual degree of saturation, the delay in arrival time increases from about 28% to 115% as the drainage layer increases from 300 to 1000 mm. In the case of the $C/C_0 = 0.9$ concentration front, the results show a reverse trend of earlier arrival as the degree of saturation decreases from 100% to 50% with a slight increase at near residual degree of saturation.

For a given drainage layer, the increase in arrival time of the concentration front with decreasing degrees of saturation in general is also quite insignificant. The increase in arrival time of the $C/C_0 = 0.5$ concentration front for cases at residual degree of saturation, is about 28% for the 300mm thick drainage layer and typically about 20% for thicker layer. As the concentration reaches that at steady state condition, the results show a reverse trend of earlier arrival with decreasing degrees of saturation.

4.1.2 Effect of degree of saturation on diffusive mass fluxes

The effect of the degree of saturation on diffusive mass fluxes was evaluated based on steady state condition. The results for cases of different drainage layer thickness are given in Figure 5. The steady state mass fluxes at various degrees of saturation were normalized with respect to that at 100% degree of saturation and 0.3m thick drainage layer.

For the case of 0.3m drainage layer, the mass fluxes decrease non-linearly with decreases in degree of saturation. The decrease in mass flux was gradual initially up to 50% degree of saturation and then more sharply at near residual degree of saturation. The percent decrease in mass flux from 100% degree of saturation to that at residual degree of saturation is about 75%. The decrease in mass fluxes can be attributed to decreases in both the diffusion coefficient and volumetric water content.

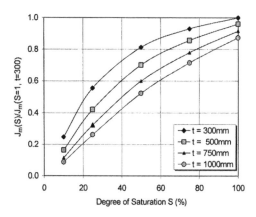

Figure 5. Effect of degree of saturation on steady state mass flux.

The effect of increasing thickness of drainage layer on steady state mass fluxes is replotted in Figure 6. Results show a proportionate decrease in mass flux with increasing thickness for cases at higher degrees of saturation. With further decrease in the degree of saturation, the rate of decrease in

mass flux increases with increasing thickness. For cases near residual degree of saturation the results however, show a smaller rate of decrease in mass flux with increases in thickness.

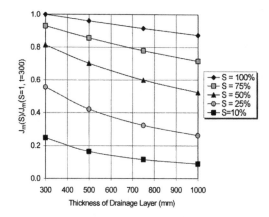

Figure 6. Effect of increasing thickness of drainage layer on steady state mass flux.

In summary, the analyses show that unsaturated soil has an influence on the arrival time of the concentration front. An increase in arrival time is noted with decreases in the degree of saturation. The delay in arrival time is relatively quite insignificant. As the concentration front approaches that at steady state condition, arrival time actually decrease with decreasing degrees of saturation. Decrease in the degree of saturation is also accompanied by a decrease in steady state mass flux. The rate of decrease increases at near residual degree of saturation. Increase in the thickness of the drainage layer in general result in a corresponding decrease in mass flux.

More importantly, diffusion is still evident at residual degree of saturation. Hence, molecular diffusion will always contribute to contaminant transport as long as the diffusive pathways are interconnected and continuous. To achieve "zero" diffusion, all pathways would need to be eliminated or by enhancing the particle surfaces with an infinite adsorption affinity for contaminants, or both.

Significant reduction of the diffusive pathways could be possible with gravel as a drainage material. A preliminary check based on the data reported by Conca & Wright (1990) shows a 200% decrease in mass flux for the case of a 0.3m drainage layer at a volumetric water content of 0.005. Nevertheless, for a given chemical potential gradient, molecular diffusion will remain active unless the liquid phase is completely hiatus.

5 CONCLUSIONS

Higher values of diffusion coefficient at near the residual degree of saturation were reported for fine sand determined using a new method for diffusion testing of unsaturated soil.

The effect of degree of saturation on the arrival time of concentration fronts in general show an increase in arrival time with decreasing degrees of saturation. The increase in arrival time is relatively quite insignificant. With increasing thickness of the drainage layer, the arrival time increases proportionately. As the concentration reaches that at steady state condition, the arrival time for any given drainage layer actually decreases with decreases in the degree of saturation.

In the case of the diffusive mass fluxes at steady state, the fluxes decrease non-linearly with decreases in degree of saturation. The decrease in mass flux was gradual initially and then increases more sharply at near residual degree of saturation. With increases in thickness of the drainage layer, a proportionate decrease in mass flux is observed. However, for cases near residual degree of saturation, the decrease in mass flux is less significant with increases in thickness. The decrease in mass fluxes can be attributed to decreases in both the diffusion coefficient and volumetric water content.

Gravel or stone at near residual degree of saturation can be an effective drainage material in providing significant reduction of the diffusive mass. Nevertheless, for a given chemical potential gradient, molecular diffusion will remain a valid process in contaminant transport unless the liquid phase is completely hiatus.

6 REFERENCES

Rowe, R.K. & J.R. Booker 1983. *Program POLLUTE – 1D pollutant migration analysis program*. Geotechnical Research Center, Faculty of Engineering Science, The University of Western Ontario, London, Ont. Canada.

Barbour, S. L., Lim, P. C. & D.G. Fredlund 1996. *A new technique for diffusion testing of unsaturated soil*. Geotechnical Testing Journal: ASTM. 19(3): 247-258.

Barraclough, P.B. & P.B. Tinker 1981. *The determination of ionic diffusion coefficients in field soils I. Diffusion coefficients in sieved soils in relation to water content and bulk density*. Journal of Soil Science. 32: 225-236.

Conca, J.L. & J. Wright 1990. *Diffusion Coefficients in Gravel under Unsaturated Conditions*. Water Resource Research. 26(5): 1055-1066

Klute, A. & J. Letey 1958. *The dependence of ionic diffusion on the moisture content of non-adsorbing porous media*. Soil Science Society of America Proceedings. 22: 213-215.

Porter, L.K., Kemper, W.D., Jackson, R.D., & B.A. Stewart 1960. *Chloride diffusion in soils as influenced by moisture content*. Soil Science Society of America Proceedings. 24: 460-463.

Romkens, M.J.M. & R.R. Bruce 1964. *Nitrate diffusivity in relation to moisture content of non-adsorbing porous media*. Soil Science. 98: 332-337.

Rowell, D.L., Martin, M.W. & P.H. Nye 1967. *The measurement and mechanism of ion diffusion in soils. III. The effect of moisture content and soil solution concentration on the self diffusion of ions in soils*. Journal of Soil Science. 18: 204-222.

Robinson, R.A. & R.H. Stokes 1959. *Electrolyte Solutions*. 2nd ed. Butterworths: London.

Shackleford, C.D. & D.E. Daniel 1991a. *Diffusion in saturated soils: I Background*. Journal of Geotechnical Engineering, ASCE. 117(3): 467-484.

Warncke, D.D., & S.A. Barber 1972. *Diffusion of zinc in soil: I. The influence of soil moisture*. Soil Science Society of America Proceedings. 36: 39-42.

Unsaturated Soils for Asia, Rahardjo, Toll & Leong (eds) © 2000 Taylor & Francis, ISBN 90 5809 139 2

States of saturation and flow as key inputs in modelling the biodegradation of waste refuse

J. R. McDougall & I. C. Pyrah
Civil Engineering, School of the Built Environment, Napier University, Edinburgh, UK

ABSTRACT: Moisture and moisture flow are now widely recognised as key factors in the biodegradation of waste refuse. This paper describes the use of coupled models of unsaturated moisture flow and biodegradation to simulate the decomposition of solid cellulolytic material in waste refuse, and illustrates the use of these models in an investigation of the biodegradation behaviour of an idealised waste column.

The numerical simulations indicate that under uniformly saturated conditions, a small advective flux of raw water produces a significant variation in the progress of biodegradation both over time and over the length of the column. In 20m depth of waste, biodegradation is most rapid at the surface and at lower elevations with slower decomposition occurring at approximately 1m to 10m below the surface. Under unsaturated conditions the same advective flux produced a more pronounced but generally slower pattern of biodegradation.

1 INTRODUCTION

In addition to self weight and other stress induced movements, the analysis of settlement in landfilled waste is complicated by a significant biodegradation related effect. Geotechnical techniques for calculating stress induced settlement are relatively well developed and have been used with some success in the case of landfilled waste. However, whilst these techniques can be extended to account for a time dependent settlement their usefulness is limited. Time dependent settlement is largely due to the decomposition of the solid organic phase and is governed by environmental conditions that are not accounted for in conventional settlement calculations.

One such environmental condition is moisture, and it is now widely acknowledged that both the presence and movement of moisture have a fundamental impact on biodegradation (Anon, 1995; Reinhart & Townsend. 1998). This paper first describes the use of two coupled numerical models - one, an unsaturated flow model, the other, an anaerobic digestion model - to provide a more realistic characterisation of waste decomposition. Specimen simulations are then presented to show the impact of a raw moisture flux on key biodegradation variables, particularly the progress of organic decomposition.

2. MODELLING FRAMEWORK

2.1 *Unsaturated flow model*

The unsaturated flow model is a finite element formulation of Richards' equation similar to that described by Thomas & Rees (1990) and implemented via linear quadrilateral elements. Fuller details of this model and its verification are contained in McDougall et al. (1996). The soil water relations are defined using van Genuchten's functions and depict a municipal solid waste. At the end of each time step during a simulation, moisture content and flow components are output directly to the biodegradation model for each element of waste.

2.2 *Biodegradation model*

The main objective of the biodegradation model is the determination of mass loss due to the conversion of an initial solid organic fraction.

Many studies of decomposition in waste refuse have been undertaken, e.g. Farquhar & Rovers, (1973); Barlaz et al., (1989), all of which refer to a complex set of biochemical and physicochemical processes. However, there is consensus on what can be regarded as the main biochemical conversion pathways for the organic fraction of waste refuse. Initially, microbial activity hydrolyses the solid organic fraction (SOF) to produce simple sugars. These are then fermented to volatile fatty acids (VFA), such as acetate, which in turn is a substrate for another, methane-producing, microbial population, or methanogenic biomass (MB).

Here, the fundamentals of a simplified biodegradation process are defined using two distinct but interdependent systems; they are, (a) an organic substrate, i.e. the concentration of VFA in the aqueous phase, and (b) a methanogenic biomass, also concentrated in the aqueous phase. Equations describing the transport, growth and decay of both systems then provide a framework for modelling the biodegradation process.

To capture the sensitivity of the biodegradation process to moisture, the hydrolysis term in the governing equations is controlled by an effective moisture content term. The application of such a term to this particular process is a simple yet physically plausible strategy as the process is essentially the only one which occurs at the solid/liquid interface. VFA depletion, and the growth and decay of the MB, all occur within the bulk aqueous phase and are defined on a per unit water volume basis by the governing equations. A fuller description of the model formulation and governing equations are given in McDougall & Pyrah (1999b).

The effective moisture content θ_E, is conventionally defined, i.e.,

$$\theta_E = \frac{\theta - \theta_R}{\theta_S - \theta_R} \tag{2}$$

where subscripts R & S refer to residual and saturated states respectively.

The transport, growth and decay of VFA and methanogenic biomass are modelled using,

$$D_C \frac{\partial^2 c}{\partial z^2} - \frac{q}{\theta} \frac{\partial c}{\partial z} + \theta_E r_g - r_h = \frac{\partial c}{\partial t} \tag{1a}$$

$$D_M \frac{\partial^2 m_w}{\partial z^2} - \frac{q}{\theta} \frac{\partial m_w}{\partial z} + r_j - r_k = \frac{\partial m_w}{\partial t} \tag{1b}$$

where D_C and D_M are the VFA and MB diffusion coefficients respectively, c is the VFA concentration and m is the methanogenic biomass per unit volume of the aqueous phase, θ is the volumetric moisture content, θ_E is the effective volumetric moisture content, q is the advective flux, r_g is the enzymatic hydrolysis term and r_h is the VFA decay term, r_j is

the methanogenic biomass growth term and r_k the decay rate.

At the end of each time step, the depletion of the SOF for each element of waste is determined using,

$$S^{t+\Delta t} = S^t - \theta_E r_g \Delta t \tag{4}$$

where S^t is the SOF remaining at the beginning of the current time step, t, and Δt is the time step.

The biodegradation model is a finite element model and like the hydraulic model is implemented using linear quadrilateral elements. Both models share the same mesh geometry.

2.3 *Parameter requirements and initial conditions*

Apart from numerical control parameters, the hydraulic and biodegradation models require the specification of five and nine physical parameters respectively. Initial conditions for the field variables in each of the models must also be provided. Values for the initial solid organic fraction define the amount of biodegradable material in any element. Table 1 summarises the hydraulic and biodegradation material parameters and initial values adopted for the simulations.

3 TRIAL SIMULATIONS

3.1 *Variation of VFA, MB and SOF in a column with zero advective flux*

By appropriate boundary moisture and flow control, the growth and decay behaviour of the governing equations can be observed without interference from transport processes. This is effectively an element test and is described in more detail in McDougall &

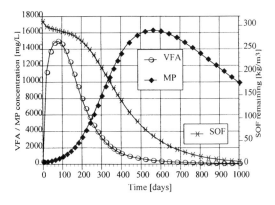

Figure 1. Landfill biodegradation field variables in zero flux element test.

478

Table 1. Summary of notation and assigned values for hydraulic & biodegradation model parameters and initial conditions.

Parameter Type	Parameter	Notation	Value
Hydraulic soil water	Residual moisture content	θ_R	0.14
	Saturated moisture content	θ_S	0.375
	van Genuchten	α	0.6
	van Genuchten	n	1.46
	Saturated conductivity	k_{sat}	5.0E-4 m/s
Biochemical transport	Diffusion coefficient, VFA	D_C	0.050 m²/day
	Diffusion coefficient, MB	D_M	0.050 m²/day
Biochemical growth/decay	Hydrolysis Rate	b	2500 mg/L/day
	Structural transformation	n	0.7
	Product inhibition	k_{VFA}	2E-4 L/mg
	Specific growth rate	k_0	0.02 day^{-1}
	Half-saturation constant	k_{MC}	4000 mg/L/day
	Methanogen death rate	k_2	0.002 day^{-1}
	Yield coefficient	Y	0.1
Biochemical moisture	Moisture content	θ	from hydraulic model
	Moisture flow	q_x , q_y	from hydraulic model
Initial conditions	Moisture content	θ	a) fully saturated
			b) hydrostatic equilibrium with base suction of 10kPa
	VFA	c	300 mg/L
	MB	m	250 mg/L
	SOF	S	310 kg/m³
Numerical	Time step	Δt	0.01 day

Pyrah, (1999a). Figure 1 shows the resulting variation in VFA, MB and SOF over time and compares well with the type of behaviour observed in actual landfill sites.

3.2 *Variation of VFA, MB and SOF in a saturated column with advective flux*

In this simulation, an advective flux condition is produced by Dirichlet hydraulic boundary conditions on both the upper (infiltration) surface and the base of the column. A constant downward flux velocity of 16.2mm/day is maintained with fully saturated conditions throughout the column. The magnitude of the prescribed flux is based on tests performed on waste refuse by Leckie et al (1979), in which one cell received a daily application of raw water. Accordingly, the biodegradation model prescribes zero concentrations for both VFA and MB at the upper surface. The waste column is 20m high with a finite element mesh based on a vertical discretisation of 0.5m.

Figure 2 shows the pattern of VFA concentration over time at five elevations, 0m, 5m, 10m, 15m and 19m from the column base. At the three lower elevations, the resulting pattern of VFA evolution is very much the same as that occurring under zero flux in the element test (Fig.1) but at 15m and 19m VFA concentrations fall at a slower rate.

The SOF depletion curves are shown in Fig.3 with, once again, the three lower elevations

revealing the closest agreement with the element trials. Except at 19m, the pattern of depletion is clearly non-linear with three or four identifiable phases. Phase I, occurring over the first 25 days, is characterised by a relatively rapid rate of depletion. This may be explained by the initial hydrolysis which results in a rapid accumulation of VFA. Phase II, occurring over days 25 to 200, shows a lower rate of depletion. This stage coincides with relatively high VFA concentrations, which will inhibit hydrolysis, and comes to an end when the methanogenic biomass is sufficiently well established to consume the VFA and reduce its

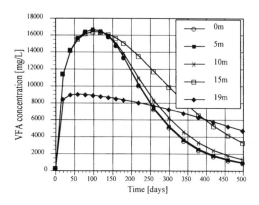

Figure 2. VFA concentrations at selected elevations in saturated waste column.

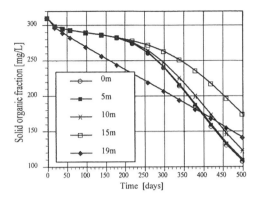

Figure 3. SOF depletion at selected elevations in saturated waste column.

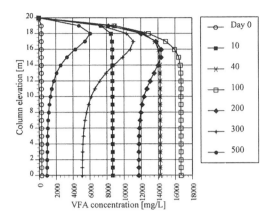

Figure 4. VFA concentration profiles in saturated waste column.

concentration. Phase III, occurring after day 200, shows an increased rate of depletion. The methanogenic biomass is well established so VFA is rapidly consumed and concentrations are held at relatively low levels. A fourth phase which is better distinguished in the longer run period of Fig.1 shows depletion rates gradually falling as the SOF tends to zero. Although the time scales are somewhat arbitrary, it is widely recognised that the sequence of biochemical processes described in these three or four SOF depletion phases are evident in the decomposition of waste refuse, e.g. Daniel, (1993); Anon, (1995).

As in the case of VFA concentrations, it is the pattern of SOF depletion at the upper two elevations which differ from the zero advection pattern shown in Fig.1. In the early stages of the simulation, SOF depletion at the 15m elevation is identical to the lower elevations but after about 175 days progresses at a slower rate. The 19m elevation has a distinctly different pattern of depletion which could be considered to reflect a real biochemical effect. For example, the continuous infiltration of raw water could be forcing a state of enzymatic hydrolysis, effectively holding this part of the column in phase I with the hydrolysis products removed by the advective flux. Alternatively, the pattern of depletion at 19m may be due to boundary condition interference. Further insight into biodegradation model performance and the nature of the interference from infiltrating raw water can be obtained from spatial distribution profiles.

For example, Fig. 4 shows the VFA concentration profiles over the length of the column at selected times during the simulation. In the lower part of the column, the rise and subsequent fall in VFA depicted in Fig.2 can be recognised. Since the

advective flux will only travel 8.1m in 500 days, there is little interference from the infiltrating moisture in this part of the column. However, in the region affected by infiltration, i.e. the upper half of the column, a combination of effects may be observed.

Firstly, although the infiltration of raw water means that VFA concentrations are zero at the upper surface, the percolating moisture encourages hydrolysis and accumulates VFA as it travels down through the column. Over the first 100 days, VFA concentration profiles in Fig.4 increase steadily from zero to the 'unaffected-by-infiltration' value lower down the column. Secondly, methanogenesis is simultaneously attempting to establish itself. Only after 200 days is any significant MB population evident (Fig.5) and even then it is mainly in the lower part of the column; in the upper part MB

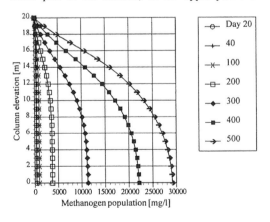

Figure 5. Methanogenic biomass profiles in saturated waste column.

accumulation is constrained by the infiltrating flux. The peak VFA concentrations in Fig.4, which appear in the column profiles after 100 days at around 15m to 18m, reflect an equilibrium condition between the flushing effect of the infiltrative flux and the inherent growth and decay functions. In this VFA-rich waste layer, the growth of MB struggles to offset its advective removal; methanogenesis is not well established so hydrolysis products accumulate and, as shown in Fig. 6, inhibit SOF depletion at this elevation. The SOF distribution profiles in Fig.6 not only reiterate the non-uniform variation in biodegradation over time, they also reveal a complex spatial variation.

3.3 Variation of VFA, MB and SOF in a partially saturated column with advective flux

In this simulation, the column is initially partially saturated in hydrostatic equilibrium with a suction of 10 kPa at the base. The infiltrative flux, again of 16.2mm/day, is defined using a Neumann boundary condition at the upper, infiltration surface. Figure 7 shows a series of moisture profiles at selected times during the simulation. It is evident that the column reaches its steady infiltration state moisture profile within 20 to 50 days. In fact, monitoring of the basal element shows that discharges first appear on day 3 and reach steady flux of 16.2mm/day by day 35.

Mass balance in the hydraulic model is good with a reported gross infiltration of 8092 litres and discharge of 7835 litres. For a metre square column the prescribed infiltration over a period of 500 days would result in a gross infiltration of 8100 litres. The discrepancy in the discharge figure is due to the amount of the moisture retained in the column at the end of the test.

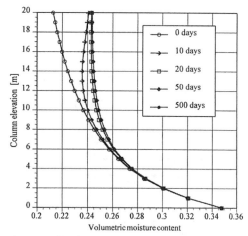

Figure 7. Infiltration moisture profiles at selected times

Whilst there are certainly differences in the detail of the time series plots for the saturated and partially saturated cases, the column profiles provide the more interesting comparisons.

Consider first the VFA concentration profiles shown in Fig. 8. The impact of non-uniform moisture content on the accumulation of VFA is clearly seen in the 10, 50 and 100 day profiles. Accumulation rates are generally slower but particularly so in the upper part of the column. Peak concentrations occur at about the same time as for the saturated case (100 days) but at lower magnitudes. Variations in post peak VFA concentration profiles are more pronounced with a more persistent and slightly deeper VFA-rich layer. All of these effects are entirely consistent with the much slower rate of methanogen growth evident in Fig. 9 compared with Fig. 5.

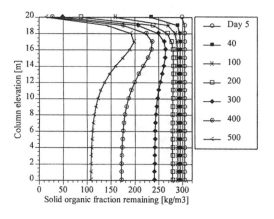

Figure 6. SOF depletion profiles in saturated waste column.

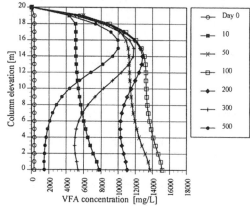

Figure 8. VFA concentration profiles in partially saturated waste column.

481

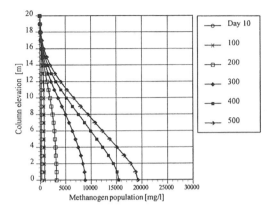

Figure 9. Methanogenic biomass profiles in partially saturated waste column.

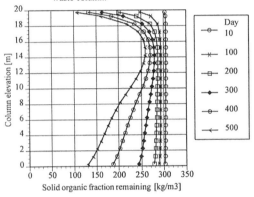

Figure 10. SOF depletion profiles in partially saturated waste column

From a mechanical point of view, it is decomposition of the organic fraction which is of most interest. Figure 10 shows SOF depletion profiles for the unsaturated case and also reveals a zone of inhibited decomposition. This zone is, however, larger, deeper and more pronounced than in the saturated case (Fig.6), particularly later in the simulation.

4. CONCLUSIONS

Through simulations of waste decomposition, this paper has revealed the operation of an unsaturated moisture flow model in conjunction with a modified landfill biodegradation model. The modifications in question account for moisture effects in the enzymatic hydrolysis of solid cellulolytic material within the framework of a two stage anaerobic digestion model. After showing how the fundamental characteristics of biodegradation, particularly VFA concentrations and SOF depletion, are interpreted by the biodegradation model, the coupled models have been used to simulate the process of biodegradation in a saturated waste column subjected to infiltration of raw water. SOF depletion is revealed to be non-linear over both time and space. Under conditions of partial saturation, a generally slower rate of decomposition occurred although the pattern of variation throughout the column was more pronounced.

ACKNOWLEDGEMENTS

This research has been supported by the Hanson Waste Management through the Environmental Services Association Research Trust.

REFERENCES

Anon (1995). Landfill design, construction and operational practice. *Waste Management Paper 26B*, Dept of Environment, H.M.S.O., London, 1995

Barlaz M.A., Ham R.K. & Schaefer D.M. (1989) Mass balance analysis of anaerobically decomposed refuse. *A.S.C.E., J. Env. Eng.*, 115, No.6, 1088-1102

Daniel, D.E (1993) *Geotechnical Practice for Waste Disposal*. Chapman & Hall.

Farquhar G.J. & Rovers F.A. (1973) Gas production during refuse decomposition. *Water, Air & Soil Pollution*, 2, 483-495.

Leckie J.O., Pacey J.G. & Halvadakis C. (1979) Landfill management with moisture control. *A.S.C.E., J. Env. Eng.*, 105, EE2, 337-355

McDougall J.R., Sarsby R.W. & Hill N.J. (1996) A numerical investigation of landfill hydraulics using variably saturated flow theory. *Geotechnique*, 46(2), 329-341.

McDougall J.R. & Pyrah I.C. (1999a) Modelling biodegradation and hydraulic effects in waste refuse. *Proc. Intl. Symp. Numerical Models in Geomechanics - NUMOG7*, eds. Pande, Pietruszczak & Schweiger, Balkema, Rotterdam, 615-620.

McDougall J.R. & Pyrah I.C. (1999b) Moisture effects in a biodegradation model for waste refuse. *Sardinia '99 - Proc. 7th Intl. Landfill Symp.*, CISA, S.Margherita di Pula, Sardinia, In press

Reinhart D.R. & Townsend T.G. (1998) *Landfill Bioreactor Design & Operation*, C.R.C.Press, Florida.

Thomas H.R. & Rees S.W. (1990) Modelling infiltration into unsaturated clays. *A.S.C.E., J.Geotech.Eng.*, 116, No10, 1483-1501.

Unsaturated Soils for Asia, Rahardjo, Toll & Leong (eds) © 2000 Taylor & Francis, ISBN 90 5809 139 2

Coupled transient heat and moisture flow in unsaturated swelling soil

A. M. O. Mohamed

Department of Civil Engineering, UAE University, United Arab Emirates

ABSTRACT: This study is designed to evaluate the transport coefficients involved in coupled heat and moisture flow equations in unsaturated swelling soil subjected to heat and moisture flow in opposite directions. The developed technique requires the use of both theoretical analysis and experimental data in conjunction with Powell's nonlinear optimization technique to calculate the transport coefficients governing the diffusion process during coupled heat and moisture flow. The calculated transport coefficients of a 50:50 sand-bentonite based barrier material are presented. Good agreement between calibrated values and experimental volumetric water content shows that the transport coefficients can be expressed in a linear function of the volumetric water content and temperature.

1 INTRODUCTION

Studies of moisture movement in unsaturated soils due to thermal gradients are important with relation to several practical engineering problems such as heat dissipation from buried power cables, moisture movement around soil warming system for plant growth, safe underground storage of hot materials like nuclear fuel waste and contaminant transport. In general, unsaturated coupled heat and moisture flow studies are aimed at controlling (a) heat, (b) moisture and (c) heat and moisture simultaneously. In all these cases, transport of contaminant can occur. An example of heat control may be found in a buried power cable wherein the goal is to dissipate the heat generated in order that the cable does not over heat and fail (Fig. 1(a)). Moisture control is a desired objective in a soil warming system, depicted in Fig. 1(b), where an attempt is made to transfer water from the ground water table towards the ground surface. A nuclear waste disposal system, Fig. 1(c), provides an example which requires both heat and moisture control. In the later example, heat generated from the canister directs the moisture outward, while the pore water pressure in the rock pushes the moisture in the opposite direction, i.e., toward the canister.

To properly reflect the sets of physics involved in the coupled heat and moisture transport process, the models developed to predict coupled heat and moisture flow in unsaturated clay-based materials must use transport coefficients that are representative of the coupled transport process. By and large, since it is difficult to properly uncouple the individual set of forces/fluxes responsible for the transport processes, determination of the transport coefficients associated with corresponding set of forces or processes become a problem. One observes, for example, that the difficulties encountered in obtaining successful prediction by transport models (Smith, 1943; Philip and de Vries, 1957; Taylor and Cray, 1960; Cassel et al., 1969; Nielsen et al., 1972) appear to be due to the manner in which the driving forces are expressed, and/or the validity of the transport coefficients used in the implementation of the analysis. From a mechanistic point of view, it is necessary to determine the transport coefficients in association with the corresponding responsible sets of driving forces. This procedure requires one to obtain a proper knowledge of the physical properties of the clay based barrier system which affect the transient heat and moisture flow of the system under consideration.

To overcome the difficulty, the following procedures have been developed to provide some capability in accommodating the results of heat and moisture transport, in a closed system with respect to water flux, as a coupled process.

1. Identification technique (Yong and Xu, 1988; Yong et al., 1990; Yong and Mohamed, 1996). The developed technique uses actual experimental data, i.e., moisture and temperature distributions in the time domain, for calculating transport coefficients. The underlying principle of the identification

technique used consisted of determination of the transport coefficients through a "matching" exercise using experimentally obtained values of volumetric water content and temperature at various times.

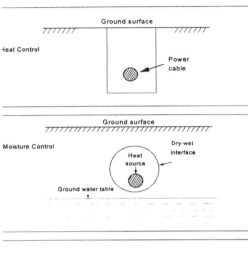

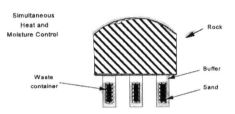

Figure 1: Schematic diagram of three different heat and moisture control applications.

2. Analytical technique (Yong et al., 1992; Yong and Mohamed, 1996). The developed technique is a phenomenological approach for evaluation of the transport coefficients based on an analytical (Fourier series) solution for the differential equations of coupled heat and moisture flow for specific initial and boundary conditions. This technique was developed to check the validity of the proposed "guess" functions used in the identification technique and to verify the general applicability of the identification technique.

3. Root-time procedure (Mohamed et al., 1996, Yong and Mohamed, 1996). The developed procedure is based on an analytical solution for the differential equations of coupled heat and moisture flow and a root-time method to identify the equilibrium times for temperature and moisture content distributions for calculation of the steady-state transport coefficients.

For all previously described solution techniques, it is necessary to identify the steady state time of the volumetric water content distribution which is difficult to obtain for swelling soils. In addition, the solution techniques were based on a compacted clay based barrier system which is subjected to heat flux at the boundary of one end of the specimen and the opposite end was closed with respect to water flux, i.e., no water flux is allowed to enter or leave the system (no volume change). For cases whereby the water flux is allowed to enter the system, the above solution techniques cannot be implemented. Hence, the present study focuses on developing a methodology for calculating the coupled transient heat and moisture transport coefficients in swelling soil subjected to heat and water fluxes in opposite directions.

2 MODEL DEVELOPMENT

In the absence of gravitational effects, the one-dimensional coupled heat and mass flow equation in unsaturated soil takes the following form:

$$\frac{\partial \theta}{\partial t} = \frac{\partial}{\partial x}\left(D_{\theta\theta}\frac{\partial \theta}{\partial x}\right) + \frac{\partial}{\partial x}\left(D_{\theta T}\frac{\partial T}{\partial x}\right) \quad [1]$$

$$\frac{\partial T}{\partial t} = \frac{\partial}{\partial x}\left(D_{T\theta}\frac{\partial \theta}{\partial x}\right) + \frac{\partial}{\partial x}\left(D_{TT}\frac{\partial T}{\partial x}\right) \quad [2]$$

where θ is the volumetric water content, x is the spatial coordinate, T is the temperature, t is the time, $D_{\theta\theta}$ is the moisture diffusivity, $D_{\theta T}$ is the moisture thermal diffusivity, $D_{T\theta}$ is the thermal moisture diffusivity and D_{TT} is the thermal diffusivity.

The boundary conditions corresponding to the experiments are given by:

$$T=T_1 \ at \ x=0 \ ; \ T=T_2 \ at \ x=L \quad [3]$$
$$\theta \geq 0.05 \ at \ x=0 \ ; \ \theta \leq 0.4 \ at \ x=L \quad [4]$$

The initial conditions corresponding to the experiments are given by:

$$T=T_1 \ at \ x=0 \ ; \ T=T_i \ for \ 0<x\leq L \quad [5]$$
$$\theta=\theta_0 \ for \ 0\leq x \leq L \quad [6]$$

where T_1 is the temperature at the hot end, T_2 is the temperature at the cooler end, T_i is the initial temperature, and L is the length of the sample.

In using the finite difference method to solve Eqs. [1] and [2] numerically, the volumetric water content and temperature at time $j+1$ is given by:

$$\theta_i^{j+1} = \theta_i^j + \frac{\Delta t}{2\Delta x^2} \ [\omega(AB-CD+EF-GH)^j$$
$$+(1-\omega)(AB-CD+EF-GH)^{j+1}] \quad [7]$$

$$T_i^{j+1} = T_i^j + \frac{\Delta t}{2\Delta x^2}\,[\omega(IF - KH + LB - MD)^j$$
$$+ (1-\omega)(IF - KH + LB - MD)^{j+1}] \qquad [8]$$

where:

$$A = D_{\theta\theta_{(i+1)}} + D_{\theta\theta_i}\ ;\quad B = \theta_{(i+1)} - \theta_i$$
$$C = D_{\theta\theta_i} + D_{\theta\theta_{(i-1)}}\ ;\quad D = \theta_i - \theta_{(i-1)}$$
$$E = D_{T\theta_{(i+1)}} + D_{T\theta_i}\ ;\quad F = T_{(i+1)} - T_i \qquad [9]$$
$$G = D_{T\theta_i} + D_{T\theta_{(i-1)}}\ ;\quad H = T_i - T_{(i-1)}$$

$$I = D_{TT_{(i+1)}} + D_{TT_i}\ ;\quad K = D_{TT_i} - D_{TT_{(i-1)}}$$
$$L = D_{\theta T_{(i+1)}} + D_{\theta T_i}\ ;\quad M = D_{\theta T_i} + D_{\theta T_{(i-1)}} \qquad [10]$$

For $\frac{1}{2} \le \omega \le 1$, the method is unconditionally stable. Here ω is considered $\frac{1}{2}$, yielding the Crank-Nicklson implicit method.

3 DETERMINATION OF TRANSPORT COEFFICIENTS

As indicated in Eqs. [7] and [8], if the transport coefficients and temperature and volumetric water content profiles at time j are known, then the temperature and volumetric water contents profiles at time $j+1$ can be found numerically. To express the diffusivity parameters as a function of volumetric water contents and temperature, the following linear function has been assumed:

$$D_i = a_i + b_i\,\theta + c_i\,T \qquad [11]$$

where a_i, b_i, and c_i are material parameters.

The optimum material parameters (i.e., a_i, b_i and c_i) in diffusion function are those which minimize the following function;

$$F = \sum_{i=1}^{N} \left| \theta_{Exp.} - \theta_{Calc} \right| \qquad [12]$$

where: N is the number of points at which the volumetric water contents are measured; θ_{Exp} is the experimentally obtained volumetric water contents; and θ_{Calc} is the calculated volumetric water contents. Whereas F can be calculated for each transport coefficient, it is not immediately clear how F will behave since the transport coefficients may vary. For the problem under consideration, since the derivative of F, with respect to a specific unknown material parameter, can not be determined in a simple manner, and since Powell's method (Powell, 1964) does not require derivatives of the objective function, Powell's conjugate directions method of non-linear optimization was used as a search technique to obtain a minimum value of F.

In order to calibrate the model, the finite difference solution of one dimensional heat and moisture flow equations was combined with Powell's optimization technique. The developed computer program is best illustrated in the flow chart shown in Fig. 2.

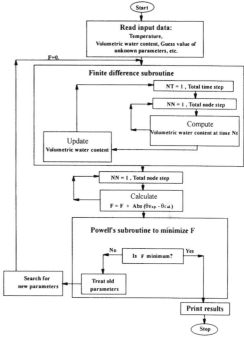

Figure 2: A flow chart illustrating computer program.

4 APPLICATION

To demonstrate the applicability of the present method, the following experimental results are used. The material used to represent the clay based barrier material was a laboratory mixture of sodium bentonite and graded silica sand in equal proportion by dry weight (Dixon and Gray, 1985). The "mixing" solution used was a "reference" synthetic Granitic Ground Water (GGW) formulated according to a recipe given by Abry et al., 1982 The geotechnical, mineralogical and physico-chemical properties of the materials was reported by Mohamed et al. (1996). One dimensional heat and moisture flow tests were performed for the buffer material compacted to a dry density of 1.67 Mg/m^3 and 17.7 percent of GGW. In all tests, water was allowed to infiltrate into a horizontal soil column from one end under a constant hydrostatic head of 276 kPa. Also the specimens were heated from the other end by the heater to a constant temperature of $100\,°C$. The basic arrangement of the compacted buffer, heater, water inlet, thermocouple and cooler system are shown schematically in Fig. 3.

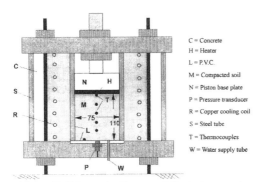

Figure 3: Schematic diagram of the test apparatus.

C = Concrete
H = Heater
L = P.V.C.
M = Compacted soil
N = Piston base plate
P = Pressure transducer
R = Copper cooling coil
S = Steel tube
T = Thermocouples
W = Water supply tube

The temperature distribution with space and time (Fig. 4) reveals that the temperature stabilizes in a short period. This provides a useful information about the thermal diffusivity function. That is, since the temperature essentially stabilizes in a short period, one can assume that the thermal field is not noticeably affected by moisture redistribution (Mohamed et al., 1993). Accordingly, in Eq. [2] the thermal moisture diffusivity, $D_{T\theta}$, can be assigned a zero value (Yong et al., 1992), and the thermal diffusivity, D_{TT}, can be considered solely as a function of temperature.

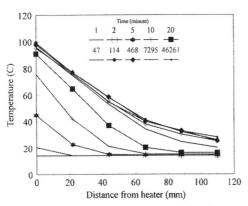

Figure 4: Variation of temperature with distance.

Figure 5 presents the volumetric water contents in relation to distance along the specimen for various time durations. The results indicate that there are two water fluxes. The first flux, which is due to the

externally applied hydraulic head, is in the direction away from the water intake, whereas the second, which results from the externally applied temperature gradient, is in the direction away from the heater. The superposition of these two water fluxes results in movement of water from both ends to the middle third of the soil specimen. The volumetric water content increases in the middle third as a function of time until the soil becomes saturated, whilst the volumetric water content of the soil column close to the heater decreases as a function of time because of the drying process.

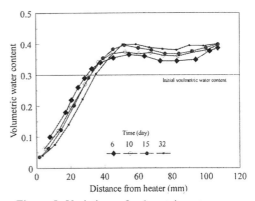

Figure 5: Variations of volumetric water content with distance.

Table 1: The calculated parameters of the diffusion functions.

Type of diffusivity and their dimension	Value of variables		
	a_i	b_i	c_i
Moisture, $D_{\theta\theta}(mm^2/day)$	0.100	0.12	0.398
Thermal moisture, $D_{\theta T}(mm^2/day/\,^\circ C)$	6.46E-3	1.54	3.27E-3
Thermal, $D_{TT}(mm^2/day)$	2504.95	-----	1.137

Using temperature profile at steady state and moisture profiles of 6, 10 and 15 days (Figs. 4 and 5) and aforementioned optimization technique, the unknown variables a_i, b_i, and c_i were determined for the thermal moisture, moisture and thermal diffusivity functions as shown in Table 1.

5 VALIDATION AND PREDICTION

To validate the calculated diffusivity functions, the volumetric water content profile of 32 days test was predicted using the calculated parameters given in Table 1 and compared with experimentally measured values as shown in Figs. 6. The comparisons between the predicted and experimentally measured moisture profiles indicate a good agreement, testifying to the validity of the calculated diffusivity functions. In addition, using the calculated diffusivity functions, the volumetric water contents were predicted for long time periods, as shown in Fig. 7. The results indicate that the material close to the heat source loses moisture at an early stage and as time progresses, the dry material starts to moisten. This phenomenon reveals that at an early stage the heat flux is high enough to push the moisture away from the heater. However, as the degree of saturation in the wet part increases, the established high moisture gradient at the interface of drying and wetting zone causes moisture movement toward the drying region. The moisture flow from the cold to the hot end continues until the thermal flux and the water flux induced by the moisture gradient are balanced. Since the trend of the predicted moisture profiles follows the experimental one, it indicates the correctness of calculated diffusivity coefficients.

6 CONCLUDING REMARKS

The numerical solution of the governing one-dimensional coupled heat and moisture flow equations were achieved via the use of an implicit finite difference method. Then it was combined with Powell's optimization technique to calculate the diffusivity coefficients governing the diffusion process. A computer program was developed to search for the optimum material parameters by comparing the predicted results with the experimentally measured data.

The calculated diffusivity coefficients showed that the moisture and thermal moisture diffusivities have increased as temperature and moisture content increase. The results also indicate that as temperature increases the thermal diffusivity increases in response to movement of water in both the liquid and vapor phases. Comparison between the measured and calculated moisture profiles shows a good agreement.

In addition, the following points can be noted:

1. Since the temperature profile reaches a steady state in a short period, thermal diffusivity can be taken to be a function of temperature.

2. Based on agreements between experimentally measured and calculated volumetric moisture content, it would appear that the diffusivity coefficients can be expressed as a linear function of volumetric moisture content and temperature.

3. Powell's optimization technique is an appropriate technique for calculating the material parameters that govern the diffusion process during coupled transient heat and moisture flow.

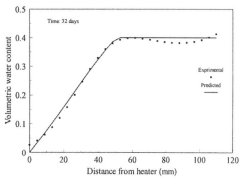

Figure 6: Comparison between predicted and measured volumetric water content.

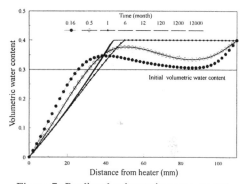

Figure 7: Predicted volumetric water content variations along the soil column.

REFERENCES

Abry, D. R. M., Abry, R.G.F., Ticknor, K.V. and Vandergraaf, T.T. 1982. Procedure to determine sorption coefficients of radionuclides on rock coupons under static conditions. *Atomic Energy of Canada Limited, Pinawa, Manitoba, Technical report TR-189.*

Cassel, D.K., Nielsen, D.R. and Biggar, J.W. 1969. Soil water movement in response to imposed temperature gradients. *Soil Science Society of American Proceeding, Vol. 33, pp. 493-500.*

Dixon, D.A. and Gray, M.N. 1985. The engineering properties of buffer material, research at Whiteshell nuclear research establishment. Proceeding, *17'th information meeting of nuclear* fuel waste management program, Toronto, Atomic Energy of Canada Limited, Technical report TR-350, Vol. 3, pp. 513-530.

Mohamed, A.M.O., Yong, R.N., Caporuscio, F., and Cheung, S.C.H. 1993. A coupled heat and water flow apparatus. *Geotechnical Testing Journal, GTJODJ, Vol. 16, No. 1, pp. 85-99.*

Mohamed, A.M.O., Y ong, R.N., Kjartanson, B.H. and Onofrei, C. 1996. Coupled heat and moisture flow in unsaturated expansive clay barriers. *Geotechnical Testing Journal, GTTODJ, Vol. 19, No. 2, pp. 155-163.*

Nielsen, D.R., Jackson R.D., Gray J.W. and Evans D.D., 1972. Soil water. *American Society of Agronomy and Soil Science Society of America, Madison, WI.*

Philip, J.R. and de Vries, D.A. 1957. Moisture movement in porous materials under temperature gradients. Trons. *American Geotechnical Union, Vol. 38, pp. 222-232.*

Powell, M.J.D. 1964. An efficient method for finding the minimum of a function of several variables without calculating derivatives. *Computer Journal Vol. 7, pp. 155-162.*

Smith, W.O., 1943. Thermal transfer of moisture in soils. *Transactions of the American Geophysics Union. Vol. 24, pp. 511-523.*

Taylor, S. A. and Cray, J. W., 1960. Analysis of the simultaneous flow of water and heat or electricity with the thermodynamics of irreversible processes. *Seventh International Congress of Soil Science Transactions, Vol. 1, pp. 80-90.*

Yong, R.N. and Xu D.M. 1988. An identification technique for evaluation of phenomenological coefficients in coupled flow in unsaturated soils. *International Journalof Numerical and Analytical Methods in Geomechanics. Vol. 12, pp. 283-299.*

Yong, R.N., Mohamed, A.M.O. and Xu, D.M., 1990. Coupled heat-mass transport effects on moisture redistribution prediction in clay barriers. *Engineering geology, Vol. 28, pp. 315-324.*

Yong, R.N., Xu, D.M., Mohamed, A.M.O. and Cheung, S.C.H. 1992. An analytical technique for evaluation of coupled heat and mass flow coefficient in unsaturated soil. *International Journal for Numerical and Analytical Methods in Geomechanics, Vol. 16, pp. 233-246.*

Yong, R.N. and Mohamed, A.M.O. 1996. Evaluation of coupled heat and moisture flow parameters in bentonite-sand buffer material. *Engineering Geology, Vol. 41, pp. 269-286.*

Unsaturated Soils for Asia, Rahardjo, Toll & Leong (eds) © 2000 Taylor & Francis, ISBN 90 5809 139 2

Dispersivity of partially saturated sand estimated by laboratory column test

T. Sato, M. Shibata & Y. Usui
Department of Civil Engineering, Gifu University, Japan

H. Tanahashi
Department of Civil Engineering, Shinshu University, Nagano, Japan

ABSTRACT: Dispersion coefficient was determined from the best fitting of one-dimensional advection-dispersion model to laboratory experiments done by a column with 5cm diameter and 30cm length. Suction is applied at the bottom to get rid of the end effect and to keep a constant water saturation throughout the column. The Toyoura sand was packed and sodium chloride solution is supplied with steady state flow condition. The experiments showed that dispersion coefficient increases dependently on decrease of water saturation and linear relationship exists between dispersion coefficient and average pore water velocity. The study highlights linear increase of dispersivity, which is defined by dispersion coefficient divided by average pore water velocity, as water saturation decreases. Discussions were placed on the comparisons with dispersivity of the glass beads and the applications to a numerical analysis for transient movement of chemical constituent through variably saturated sand.

1 INTRODUCTION

Advection and the molecular diffusion are the basic transport phenomena involved in solute transport through soil The complicated system of interconnected pores of different shapes and sizes causes a continuous subdivision of the initial solute mass within the flow domain to spread and occupy an ever-increasing volume of soil. This spreading phenomenon is termed as dispersion. Dispersion is the outcome of the flow and the presence of a pore system through which flow takes place. Besides advection and dispersion, various other mechanisms such as adsorption, various chemical reactions among the solute species, decay phenomena etc. affect the transport of the solute in soil. Various researches have been carried out during the last few decades to clarify the process of solute transport considering various aspects of the phenomenon, such as presence of mobile and immobile water and adsorption characteristics of the solute etc. There are, however, some aspects of the phenomena still not fully resolved, especially the nature of the dispersion.

The purpose of this study is to make clear the effects of water saturation on dispersion coefficient. Bresler(1973) may be the first researcher to describe dispersion coefficient as the function of pore water velocity and volumetric water content. His literature, however, intensively focussed on the dependency of molecular diffusion onto water content but does not address dispersion coefficient. There were no publications discussing the role of water saturation in solute dispersion phenomenon in soil. A special attention of this paper is put on the shape of breakthrough curve under steady state flow through partially saturated sand. Several literatures have discussed the behavior of aqueous constituents in unsaturated soil from the viewpoint of miscible displacement. Nielsen and Biggar(1961) reported the shape of breakthrough curve measured at bottom of partially saturated soil column. They revealed features of breakthrough curve for unsaturated soil and glass beads, such as curve with gentle slope, large amount of pore volume to be perfectly displaced by solution, etc. Krupp and Elrick(1968) compared breakthrough curve of partially saturated glass beads with that of saturated. They described that it continuously shifts to the left with decrease of water saturation. De Smedt and Wierenga(1984) discussed breakthrough curve by using the mobile-immobile model to estimate effects of stagnant zone within pore water on solute transport. They concluded that twenty times of dispersion coefficient in fully saturated soil are estimated as the unsaturated one when the one-dimensional advection-dispersion model considering mobile water(M-model) is applied, however, almost the same values are obtained by MIM-model. Sato et al.(1985) and Kodani and Yano(1988) showed that dispersion coefficient becomes larger in unsaturated soil than that in saturated one and it tends to increase with pore water velocity when we consider at an identical water saturation.

Transient behavior of chemical constituent is one of key issues in transport at shallower depth of ground. Contamination problem usually takes place near ground surface before its expanding to the underlying aquifer. In this region, degree of saturation is severely affected by precipitation or dehydration due to evaporation. Magnitude of dispersion coefficient is also expected to be changeable strongly influenced by degree of saturation as well as the other parameters characterizing solute transport.

2 COLUMN TEST WITH SUCTION

A device for laboratory column test is described in Fig.1. The Toyoura sand, of which particle size varies from 110 to 420 μm, was packed into a column at a saturation density of 1.55g/cm³, which was kept constant for all test cases. The column has 5cm of inside diameter and 30cm of length. The length of soil column was kept 12cm for all of the test cases. Drinking water was pumped up and supplied to the top of the soil sample at a constant flow rate. Pore water was drained from the bottom. A steady state condition was confirmed after the effluent volume equals to the influent. Solution of 0.0282N NaCl was supplied as a tracer instead of drinking water at the same flow rate after reaching steady state flow condition. Drained water flowed into fraction collector to measure concentration of chloride ion. The measurement was made on the basis of argentometric method. Concentration of background was also measured by drinking water before starting test. Air suction was applied to the sample at the bottom to eliminate the end effects and to keep a constant value in water content throughout the column. Magnitude of applied vacuum was

determined from water saturation and flow rate for a given test condition.

3 TEST RESULTS OF THE SAND

Water content was measured at each depth of the column after test operation. Distribution of water saturation was well controlled by air suction by keeping a constant level along the whole length as shown in Fig.2. The both cases in the figure were carried out at the same average pore water velocity. The figure shows that applied vacuum determines magnitude of water saturation when discharge rate is identical. Concentration measures were aimed to draw breakthrough curve at the exit of column. Breakthrough curves in Fig.3 show the transient behavior of tracer displacing with pore water in sand column. They were obtained at the exit of column with different values of water saturation under the identical pore water velocity. The vertical axis shows the relative concentration defined by ratio of Cl⁻ in effluent with respect to influent. The horizontal axis is pore volume, which is defined as vt/L, in which v: average pore water velocity, t: time since beginning of solution inflow and L: column length. It is convenient to describe the experiments by these two axes for characterization of solute transport affected by macroscopic flow condition.

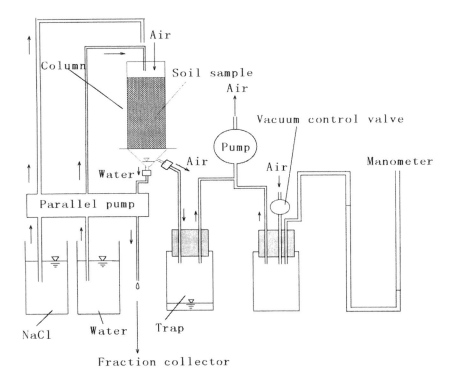

Figure 1. Experimental setup for laboratory column.

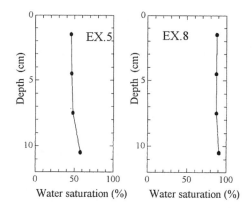

Figure 2. Water saturation in sand column.

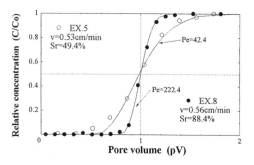

Figure 3. Breakthrough curves at different water saturation.

The two cases, of which test condition is the same in column length and velocity, show different shape of breakthrough curve. Continuous lines in this figure are mathematical solutions of one-dimensional advection-dispersion model being fitted to the experiments. The test results imply that the curve rotates clockwise with a decrease in water saturation. Breakthrough took place in Sr=88.4%. The amount of pore volume to reach 1.0 relative concentration becomes larger in Sr=49.4% than Sr=88.4%. These results indicate that preferential flow arises and displacement does not smoothly occur as water saturation decreasing.

4 DISPERSION CHARACTERISTICS

4.1 Dispersion coefficient of the sand

Dispersion coefficient was determined by fitting one-dimensional advection-dispersion model to the experiments. The test was made by keeping a constant value of flow rate and water saturation. The model is applicable to analysis of unsaturated soil as well as fully saturated one. Uniformity of volumetric water content and discharge velocity throughout the column is one of important conditions in column tests for the estimation of transport characteristics.

The best fitting was made around 0.9 of relative concentration to reduce errors due to lack of measured values around 0.5 of relative concentration and uncertainty of argentometric method at beginning of breakthrough.

Determined values are displayed with reference to average pore water velocity in Fig.4. Straight lines are regression of the almost same water saturation. Dispersion coefficient increases linearly with pore water velocity on every straight line. The slope, however, tends to be steeper with the decrease of water saturation except of more than Sr=70%. There were no measures less than 0.1 cm/min of pore water velocity. According to the literature describing relation in the Peclet number and the ratio of dispersion coefficient to molecular diffusion (Bear 1972), the region less than 0.1 cm/min becomes the so called diffusion controlled, which means that dispersion coefficient is given as a function of molecular diffusion and tortuosity. Dispersion coefficient is equal within this region, independent of pore water velocity. This is well understanding of the regression within the region of more than 0.1 cm/min in pore water velocity described by Fig.4. The relation is delicate at more than 70% of water saturation. The line of Sr=100% becomes inverse to that of Sr=70%. This results from difference of column length. The test was made by 24cm of column length for saturated soil. The test column was 12 cm for Sr=70%. The exact relation will be provided in future for this range of high water content by using the same column length as the test series.

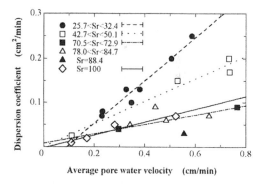

Figure 4. Relationship between dispersion coefficient and average pore water velocity.

4.2 Dispersivity of the sand

Dispersivity was estimated from definition of D/v in which D: dispersion coefficient and v: average pore water velocity. Spatial variance or distribution of location of chemical species at the front is given by $\sigma^2 = 2\alpha L$ in which σ: standard deviation of spatial distribution of chemical species, α: dispersivity and L: location. Dispersivity is one of important properties used to estimate the extent of contaminated soil by aqueous constituents. The experiments in Fig.5 show interesting prospects for pollutant progress. Dispersivity tends to increase linearly with the decrease of water saturation and does not strongly depend on pore water velocity. The experiments at $Sr=100\%$ were obtained from the tests with different column length. The preceding description may not be applied to the sand with more than 90% of water saturation. Independent behavior of dispersivity on pore water velocity implies that spatial distribution of chemical species can be uniquely determined by location and it does not depend on travel time from contamination source. Furthermore, the linear increase implies that the spatial distribution of the front linearly expands with the decrease of water saturation.

5. COMPARISON WITH GLASS BEADS

5.1 Dispersion coefficient

Glass beads, which particle size is 250μ m, were used in the column test instead of the sand. The length of the column and the density of the glass beads packed into the column are the same as the tests for the sand. Dispersion coefficient was compared in order to check the effects of grain size distribution and the shape of the particle. The comparisons are made by relationship between dispersion coefficient and average pore water velocity in Fig.6. The figure indicates that the amount of dispersion coefficient is not different so much and the dependency rate on pore water velocity is almost the same. The experiments show linear increases of the dispersion coefficient as average pore water velocity increases in both samples. The scatter in the glass beads becomes much smaller than that in the sand if comparisons were made at the same amount of pore water velocity. The tests were carried out using the samples with 30-90% of water saturation. The average particle diameter is almost the same while grain size distribution and shape of particle are different. The difference is expected to come from the spatial distribution of pore water velocity due to variety of pore size produced from the change of particle size and the shape.

5.2 Dispersivity

Dispersivity of the glass beads was determined and it was described with reference to the Toyoura sand in Fig.7. The figure shows that the dispersivity increases linearly with the decrease of water saturation. The amount does not differ from the sand above 60% of water saturation. The tendency meets with the identical linear relationship. Dispersivity does not increase in glass beads but keeps a constant value when the water saturation decreases below 60%. Dispersivity is defined as the spatial location of aqueous chemical at a travel length being equal to the column length. The test was done by 12cm length of the column. The results below 60% of water saturation have an effect of the column length, which may not be enough for travel

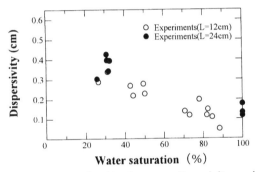

Figure 5. Relationship between dispersivity and water saturation.

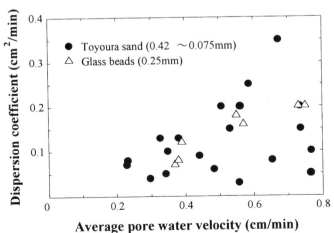

Figure 6. Comparison of dispersion coefficient of the sand with the glass beads.

492

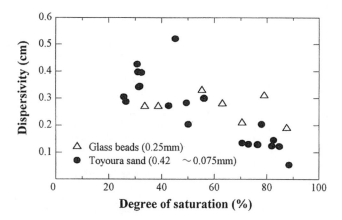

Figure 7. Comparison of dispersivity of the sand with the glass beads.

length to accomplish displacement with drinking water. There are still some unknowns in the scale effects on dispersion phenomenon. Further investigations will be continued to specify the effect of the column length on dispersion phenomenon.

6 NUMERICAL EXAMPLE

The relationship among dispersion coefficient, average pore water velocity and degree of saturation was applied to FE analysis for solute transport through a 1m length soil column with spatially variable saturation. Laboratory setup is described in Fig.8. The Toyoura sand was packed with 1.55 g/cm³ of dry density into the column in 3cm diameter. Gravitational drainage was carried out before starting the test for miscible displacement by 0.0282N NaCl solution. The seepage analysis was done to give the distribution of water saturation and of pore water velocity throughout the column at the beginning of transport analysis. The water retention characteristics and hydraulic conductivity of the sand was determined
by the laboratory tests(Uno et al. 1995). The laboratory results were mathematically formulated by the van Genuchten model(1980). The solution was supplied at 0.071 cm/min of discharge velocity to the top of the soil column. The breakthrough curve was obtained at the bottom.The transport analysis takes into account the dependency of dispersion coefficient on pore water velocity and water saturation. Comparisons are made between the experiments and FE results in Fig.9. The computation was done at the steady state flow condition given by degree of saturation in Fig.10. The FE model reflects the dependency of dispersion coefficient on pore water velocity and degree of saturation by specified four blocks with different values of dispersion coefficient depending on water content and pore water velocity. The computations described by continuous line show a good similarity with the experiments as shown in Fig.9. Some

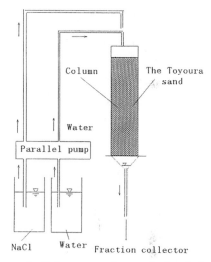

Figure 8. Column test for gravitational drainage flow in 1m length column.

discrepancies occur around 0.9 of relative concentration. The computations were based on the relationship estimated from the column test of 12cm length. There were no measures for more than 70% of water saturation in 12cm column length. The relationship at high water saturation is needed for the exact computation showing a good similarity to the experimental breakthrough curve.

7 CONCLUSIONS

Dispersion coefficient was determined from the best fitting of one-dimensional advection-dispersion model to the experiments of laboratory column test for the Toyoura sand by using chloride ion as a non-reactive tracer. The experimental results imply strong effects of water saturation on hydrodynamic dispersion as well as pore water velocity. The

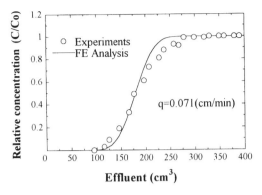

Figure 9. Comparison between computed and measured breakthrough curves.

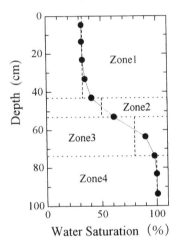

Figure 10. Degree of saturation induced by gravitational drainage in 1 day.

conclusions described in this paper are;
(1) Dispersion coefficient increases linearly with pore water velocity when water saturation is identical,
(2) Dispersivity increases with a decrease of water saturation but does not strongly depend on pore water velocity within the range of 0.1-0.6cm/min,
(3) The amount of dispersion coefficient is almost the same as that of the glass beads while the tendency of dispersivity is a little different below 60% of water saturation, and
(4) FE model, which reflects the dependency of dispersion coefficient on water saturation and pore water velocity in 12cm column length, shows a good similarity with the experimental breakthrough curve at the bottom of a 1m soil column with variably saturated sand.

REFERENCES

Bear J. (1972) Dynamics of fluids in porous media. American Elsevier: 607.
Bresler E.(1973). Simultaneous transport of solutes and water under transient unsaturated flow conditions, Water Resources Research, 9(4): 975-986.
De Smedt F. & Wierenga P.J.(1984) Solute transport through columns of glass beads, Water Resources Research 20(2): 225-232.
Kodani Y. & Yano T. (1988) Measurement of salt dispersion coefficients in unsaturated soils by the steady state method. Bull. Sand Dune Research Inst. Tottori University 27:1-7.
Krupp H.K. & Elrick D.E. (1968) Miscible displacement in an unsaturated glass bead medium. Water Resources Research 4(4): 809-815.
Nielsen, D.R. & Biggar, J.W. (1961). Miscible displcement in soil: 1. Experimental information. Soil Science Society of American Proceedings 25(1): 1-5.
Rose D.A.& Passioura J.B.(1971) The analysis of experiments on hydrodynamic dispersion. Soil Science 111(4): 252-257.
Sato K., Muraoka K.& Ito Y. (1985) Dispersion coefficient of solute in unsaturated flow. Tsuchi- to- Kiso 47(9): 45-50
Schwartz C.E.& Smith J.M. (1953) Flow in packed beds.Int. Eng. Chem. 45: 1209-1218.
Uno T., Sato T. & Sugii T.(1995) Laboratory permeability measurement of partially saturated soil. Proc. of the first Int. Conf. on Unsaturated Soil 2:573-578.
Van Genuchten, M. Th.(1980) A closed-form equation for predicting the hydraulic conductivity of unsaturated soils, Soil Sci. Soc. Am. J.:892-898.

Unsaturated Soils for Asia, Rahardjo, Toll & Leong (eds) © 2000 Taylor & Francis, ISBN 90 5809 139 2

Impact of the 'Alsacian vineyard ingrassed' on nitrate transfer

J.Tournebize & C.Gregoire-Himmler

ENGEES-CEREG, Ecole Nationale du Génie de l'Eau et de l'Environnement de Strasbourg, France

ABSTRACT: The Alsacian aquifer is the target of an action program from Nitrate Directive against groundwater pollution. The wine profession is aware of the role it plays in the reduction of nitrate leachate. This association recommends the use of ingrassed vineyards (growing grass between the wine rows). Within the scope of the interdisciplinary project EVA (Enherbement du Vignoble Alsacien), an analysis of the 98/99 allows to assess the impact of vineyard ingrassed on vine plot instrumented scale in Rouffach site (Infiltration and Run-Off). It is the first stage in the understanding process, the aim of which is to adapt a specific model of slope vineyard transfer. The interrow ingrassed method influences the vine roots, the water balance (-20%), the soil nitrate concentration (-90%) and thus the nitrate leachate flux (-91%) reloading aquifer.

1 INTRODUCTION

The presented work takes place in the interdisciplinary research program EVA (Enherbement du Vignoble Alsacien). Its focus is to quantify the effect which ingrassing (to put grass between vine rows) has on nitrate retention beneath vine during winter period. The final aim is to limit the measured increase of the nitrate concentration in Alsace aquifer. This project which has a regional predictive objective, takes into account the determination of ingrassing rate evolution for 4 years, the characterisation of the vineyard technical soil tillage, and the quantification of pollutant flux at determined different scales.

Through the description of the Rouffach (France) experimental site and the first results analysis, this presented work takes place in the initial phase to understand the dominant processes in the water and pollutant transfers phenomena in the continuum soil-plant-atmosphere system. We bring to attention the processes influenced by ingrassing on a small plot scale, from a pluri-annual monitoring (from June 98 to September 1999) of the dynamics of water and nitrate balances.

2 PRESENTATION OF EXPERIMENTAL SITE AND SOIL TILLAGE.

2.1.1 *Geographical and pedological context :*

The Lycée d'Enseignement Général et Technique de Rouffach (Alsace, France) put one site at our disposal. It is composed of two plots (a weeded witness and an experimental plot ingrassed every other row). They are situated on the same pedological bloc : a loamy soil (it represents 20% of the alsacian vineyard surface) on conglomerate.

2.1.2 *Hydrological characterisation of the site and*

of studied years :

The meteorological year is cut in two separate hydrological and vegetative seasons : SUMMER (from May to October) and WINTER (from November to April).

The 1998 hydrological year takes place in the inter annual average (583 mm). The year 1999 (until the October month) is superior in precipitation to the average for this period by approximately 37%. This allows us to assess ingrassment for two periods with contrasting amounts of precipitation.

2.1.3 *Practised fertilization on experimental plots :*

To compare the two monitored plots, the same Nitrate fertilizations were applied. For the year 1998, 22 units of nitrogen per hectare (Ammo-Nitrate) were spread across the width of the row on the 15th of May. In 1999, no fertilizer was used. It should be noted that vines effectively need about twenty units of nitrogen per hectare, year.

2.2 *Probes : presentation and validation.*

2.2.1 *Presentation of the experimental site non-destructive instrumentation :*

Every plot is studied on a system composed by a row and an inter-row, 100 meters long with a 3 meters separation. The rows are cut in 4 measured stations, at 4 monitored depths (30, 60, 90, 120 cm). The Run-Off monitoring system is composed of 3 micro-plots (25*3 m²) which are hydraulically insulated to intercept and collect the run-off water in 3 tanks. Infiltration (data weekly measured) is studied with 16 TDR profiles (water monitoring), 64 tensiometers (suction measurement), 64 ceramic cups (soil solution sampled to measure the soil nitrogen concentrations), 50 thermocouples (heat balance).

2.2.2 *Validation and use of installed captors :*

Two experimentations allowed us to better understand how the probes work. Firstly two couples of tensiometers (with hypodermic needle) among the 64 were monitored at the same depth of the same station to estimate the variations in the measured suction. The obtained results on the ingrassing plot at 30 cm deep are showed in Figure 1. We can see the two different seasons described in § 2.1.2. (weak suction for Winter and strong for Summer, except after storm). So we evaluate that the relative error between two probes, calculated from 70 weekly measurements, varies from 1 to 15%.

Secondly the TDR probe (Trime from IMKO) was studied. We compared the TDR probe value with the effective humidity (calculation of the infiltrated water quantity) in the same volume of Roufach soil. For the humidity measure domain $\theta_{vol} \in [5\%; 40\%]$ the calculated correction is:

$$\theta_{TDR}\% = 1,94 \cdot \theta_{vol} + 3.8 \text{ with } R^2 = 0,993.$$

3 IMPLICATION OF THE FIRST RESULTS ON THE KNOWLEDGE OF THE VINE ACTIVE ZONES IN THE FUTURE MODELISATION.

To approach this problem at the plot scale, we wanted to determine the main factors which must be taken into account for a modelisation (physical processes and geometrical mesh) using the field measures, the parameters, water and nitrogen monitoring.

3.1.1 *Description of the roots profiles :*

The description of the roots profiles provided evidence for the active zones in every plot. The parameters for describing the root profiles are taken from Curt & Al [1998]. Thus we establish the maximum depth of roots to be 120 cm in weeded plot and 170 cm in ingrassed plot. The depth corresponding to the highest density (we note PMAX), which indicates the most favourable horizon for prospecting, takes the value of 10 cm for the weeded plot and of 50 cm for the ingrassed plot. Lastly the depth corresponding to a density less than 10%, which indicates a strong decrease of roots prospecting, is evaluated at 20 cm for the weeded plot and at 60 cm for the ingrassing plot (See Figure 2).

When we compare the depths PMAX (10 cm in weeded plot without grass and 50 cm in ingrassed plot with grass), we can conclude that the vine adapts by pushing its roots downward, measured by its PMAX, to limit water concurrence. Moreover we conclude that vine spreads its roots in the soil until it meets a physical obstacle (hard rock in our case). This study reveals an axial dissymetry of the root exploration with the slope direction (the roots follow the slope). A lateral dissymetry shows the adaptation of the roots to deepend in the first 60 cm against soil packed down from wine vehicles circulation.

Whatever the soil tillage (weeded or ingrassed plots), the zones particularly dense with roots which are most active for water exchange are situated in the first meter as the literature proposes [Huglin & Schneider, 1998].

3.1.2 *Row influence :*

The probes stations are arranged all along the row and the inter-row. This disposition allows us to compare measurements according the station position in the plot. The analysis of the ceramic cup samples at different depths shows the row influence on nitrate concentration. This concentration influence is as easily observed in the weeded plot as in the ingrassed plot. The TDR humidity influence by the row is well observed only in the ingrassed plot. The nitrate concentrations are 30% higher under the row than under the inter-row. The inter-row humidity in the ingrassed plot is 20% less than beneath the rows. This result confirms the water stress which is generated by the grass against vine. This conclusion should be checked by taking into account the modelisation of the evapotranspiration under discontinued cover (ETP). In relation to suctions, there is no real row influence (the inter-row suction is 5 to 10% higher than this row one but we show before a 10% measurement error).

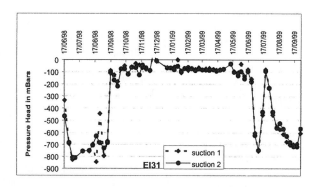

Figure 1: Example of comparison of weekly monitoring of 2 tensiometers on ingrassed plot (30 cm depth).

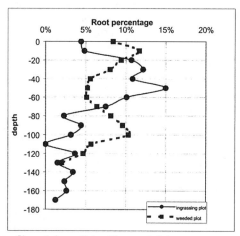

Figure 2: Comparison of roots profile made in weeded and ingrassed plots.

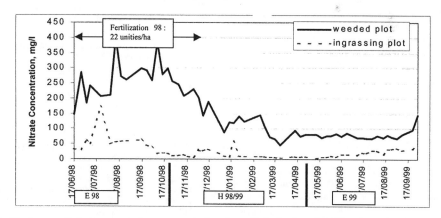

Figure 3: Weekly average of nitrate concentrations measured all stations, all depths, from June 98 to September 99 (E= summer, H= winter).

Thus the plots will be discretised in 30 cm deep layer (0 to 180 cm), 40 cm width band under vine and 100 cm in the inter-row.

4 RESULTS AND DISCUSSION : IMPACT OF INGRASSING ON THE NITRATE LEACHING TO THE AQUIFER.

The first obtained results since the beginning of the site Rouffach (France) monitoring show clearly the impact of ingrassing on the nitrate concentrations for seepage and run-off water : a decrease per a factor 10 in Winter and per 5 in Summer in average. Figure 3 illustrates this results for the three hydrological seasons (Summer 98, Winter 98/99, Summer 99).

4.1.1 *Impact on nitrate concentrations :*

The two measured summers allow us to appreciate the consequences of different fertilization policies (22 units of nitrogen in 1998 and 0 units in 1999). We show only the nitrate results because the others forms are really weak in soil solution. It is thus interesting to follow nitrate evolution brought by fertilization in layered soil studied over many years. Indeed we note that fertilization decreasing contributes to reduce the nitrate concentrations in seepage water from a factor 3. We advise that one 22 units nitrogen fertilization one year on two will be enough to an ingrassing vineyard.

4.1.2 *Impact on Winter nitrate and water flux :*

The aquifer reloading occurs generally during winter. The water balance of the winter stocks which are cumulated during 1998 and 1999 is carried out on each plot from the experimentally collected data on the site : hydraulical head and humidity variations. The results show a difference of 20% less for the water reserve in the ingrassing plot (ETP is considered as null in a first assumption in Winter). Note that the year 1998 was an average year for rainfall. Through the -120 cm horizon we assess the winter water flux about ten mm. The induced nitrate leaching reach about many ten kilograms of nitrate per hectare for the weeded plot. This flux is equivalent to about hundred mg/l nitrate concentration. For the ingrassed plot the flux is reduced to some kilograms of nitrate per hectare (it is equivalent to about ten mg/l of nitrate whose concentration is inferior to 50 mg/l, the drinking norm. Thus ingrassing technique allows to reduce the leaching by a factor 10, and decrease the seepage water volume too (See Figure 3).

4.1.3 *Impact on the Run-off :*

The run-off amounts to some millimeters (less than 10 mm) in weeded plot. It represents half or a quater infiltration. This result is the same observed by as Gaildraud [1996] in a small watershed. The run-off for the ingrassed plot is eight time less important than run-off in weeded plot.

5 CONCLUSION.

The in-situ instrumentation allows us to monitor and to quantify water and pollutant transfers by infiltration and run-off at the small scale for many years.

We show that ingrassing technique is efficient to limit nitrogen lost under vineyard during winter. The role of nitrate trap which is attributed to ingrassing cover is verified during winter. Indeed the nitrate leaching flux is reduced by a factor 10.

However hydraulical stress must also be taken into account. For an average hydrological year it percolates 20% water less in the 120 cm horizon under a ingrassing one to two row vine than under a weeded vine.

The voluntary suspension of the fertilization during the survey of the site led to a reduction of more than 60% of the nitrate content of the soil solution. This specific observation merits however to undertake a longer-term follow-up of the practices of fertilization.

The first achieved results will thus allow a modeling which is adapted to the parameters which are characteristic of the natural environment and will lead to making of a regional management tool to the service of the durable development.

This first experiment also opens on the paradox induced by the ingrassing practice grass: will it be necessary to choose between the protection of the quality of the water resource to the detriment of the quantity?

REFERENCES :

Curt T., Bouchaud M., Lucot E., Bardonnet C., Bouquet F., 1998. Influence des conditions géopédologiques sur le système racinaire et la croissance en hauteur du Douglas dans les monts du Beaujolais. Ingénieries – EAT, n°16 décembre 1998, p 29-46.

Gaildraud C., 1996, *Etude de l'impact du ruissellement dans le vignoble sur la qualité de la nappe phréatique d'Alsace.* DIREN, AERM, Région Alsace, 65 pages + annexes.

Huglin P., Schneider C., 1998, *Biologie et écologie de la vigne,* 2ième éditions; 370 p.

8 Strength and deformation

Unsaturated Soils for Asia, Rahardjo, Toll & Leong (eds) © 2000 Taylor & Francis, ISBN 90 5809 139 2

Triaxial testing of laterite soils

A.M. Deshmukh & N.N. Amonkar
Goa College of Engineering, Farmagudi-Ponda-Goa, India

ABSTRACT: Laterite soils which are obtained as weathering products are found commonly in many tropical countries such as India. In Goa, which lies on the west Coast of India, many constructions are undertaken in laterite soils, wherein they have exhibited unexpected behaviour. A laterite soil sample obtained from a typical mining area in Goa was subjected to a triaxial shear tests to investigate the shear strength properties. The parameters 'c' and 'φ' were determined under different drainage conditions and in partially saturated and fully saturated state. The effect of moisture content, diameter of sample and the rate of strain has been investigated. This paper presents the data so obtained and concludes that laterites exhibit different shear strength properties when partially or fully saturated and that the moisture content, diameter of sample and rate of strain influence the peak deviator stress.

1 INTRODUCTION

Laterite soils are most commonly found in the tropical Countries such as India. They are obtained as products of tropical and sub-tropical weathering of various crystalline igneous rocks, sediments and detrital deposits. The state of Goa which lies on the west coast of India has outcrops of laterite over 80% of the land area. In Goa, various major projects are undertaken which involve embankments and tunnels in laterite soils. At many of the sites, laterites have exhibited unexpected behaviour such as collapse, heaving etc. The Mine dumps are usually proposed to be used for such constructions, hence a laterite soil sample obtained from a typical mining area in the state of Goa has been considered for study. Rational design of any structure can be accomplished only after proper assessment of the shear strength behaviour of soil. Of all the tests, the triaxial test is the most reliable for assessing the shear strength behaviour. The factors thought to be relevant are:
1) Drainage conditions 2) Degree of saturation
3) Moisture content 4) Diameter of Sample 5) Rate of strain.

In the present paper, experimental data of the triaxial tests indicating the influence of these factors on the shear strength behaviour of a lateritic silty soil is briefly presented.

2 EXPERIMENTAL PROGRAMME

A laterite sample was obtained from a typical mining area in Goa. The soil was classified as sandy silt with clay. An experimental programme was carefully planned and executed. In all 27 triaxial tests were carried out. Cylindrical samples of 38 mm diameter and 76 mm height were prepared in a 3 - part split mould at a density at 1.6 g/cm^3. Conventional triaxial tests such as unconsolidated undrained test, consolidated undrained test and drained test were each conducted on three different samples subjected to different confining pressures. To determine the effect of moisture content on peak deviator stress, samples of 22%, 29% and 32% moisture content were tested. The effect of the diameter of the sample was determined with samples of diameters 38 mm, 50mm and 70mm. The effect of the rate of strain on the peak deviator stress was ascertained by varying the rate of strain as 6, 50, 75 and 100 mm/h.

The peak deviator stress criteria was assumed as the failure criteria. It is the maximum deviator stress beyond which there is an increase in strain with no increase in stress.

The tests were conducted using a GDS Triaxial testing system, which is a sophisticated system designed for automatic controlled testing using a software program.

The GDS System consists of a Bishop - Wesley hydraulic triaxial Cell, Digital Controller, Digital pressure Interface, Computer and its peripherals.

3   TEST RESULTS

Typical experimental data is presented so as to assess the influence of the above listed factors on the shear strength behaviour of the laterite soil sample.

3.1 *Shear strength parameters 'c' and 'ϕ'*

The partially saturated sample was subjected to three kinds of triaxial tests namely unconsolidated undrained test, consolidated-drained test   and drained test. Confining pressures of 50, 75 and 100 kPa were applied and the deviator stress was applied at a strain rate of 30 mm/h. The values of 'c' and 'ϕ' obtained in the above three tests are presented in Table 1.

Table 1 - 'c' and 'ϕ' values for partially saturated sample.

Sr. No	Test Condition	c kPa	$\phi°$
1	Unconsolidated undrained	48	24
2	Consolidated undrained	65	20
3	Drained	40	33

Similar tests were conducted on fully saturated samples and the values of c' and ϕ' obtained based on effective stresses are presented in Table 2 :

Table 2 - Effective c' and ϕ' values for fully saturated sample.

Sr. No	Test Condition	c' kPa	$\phi'°$
1	Unconsolidated undrained	75	35
2	Consolidated undrained	75	15
3	Drained	80	16

3.2 *Influence of moisture content of the Sample*

Samples were prepared with moisture contents of 22%, 26% ,29% and 32%, and tested at a strain rate of 30mm/h and a confining pressure of 75 kPa. The stress- strain curves obtained are shown in Figure 1(a). The variation of peak deviator stress with moisture content is shown in Figure 1(b).

It is observed from the graph, that as the moisture content increases, the peak deviator stress decreases.

3.3 *Influence of diameter of the Sample*

Samples of diameters 38 mm, 50 mm and 70 mm were tested at a confining pressure of 75 kPa and at a strain rate of 30 mm/hr. The stress - strain curves obtained are shown in Figure 2(a). The variation of peak deviator stress with diameter of the sample is shown if Figure 2(b).

It is observed that as the diameter of the sample increases, the peak deviator stress decreases.

3.4 *Influence of rate of strain*

To study the effect of the strain rate, samples were tested by varying the rates of strain as 6, 30, 50, 75 and 100 mm/h. The cell pressure was maintained at 75 kPa. The variation of peak deviator stress with

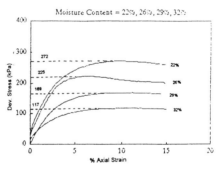

(a) Stress - strain curves

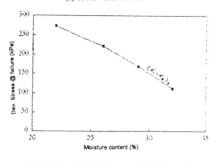

(b) Variation of peak deviator stress with moisture content.

FIG 1- INFLUENCE OF MOISTURE CONTENT

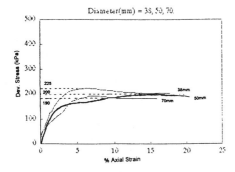

Diameter(mm) = 38, 50, 70.

(a) Stress - strain curves

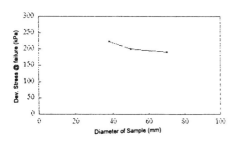

(b) Variation of peak deviator stress with diameter of Sample

FIG 2- INFLUENCE OF DIAMETER OF SAMPLE

Variation of peak deviator stress with rate of strain

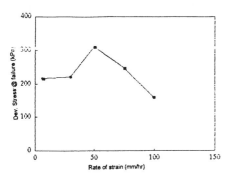

FIG 3- INFLUENCE OF RATE OF STRAIN

the rate of strain is shown in Fig 3.

It is observed that the peak deviator stress increases as the strain rate increases upto 50 mm/h and decreases thereafter with increase in strain rate beyond 50 mm/h.

4. CONCLUSIONS

(i) An increase in moisture content was found to decrease the peak deviator stress.

(ii) The peak deviator stress was found to decrease with increase in the diameter of the sample .

(iii) The peak deviator stress was found to increase as the rate of strain increases upto 50 mm/h and decreases beyond this value.

REFERENCES

Amonkar, N. N. (1995). *Computer controlled Triaxial testing of laterite soils.* M.E Dissertation submitted to Goa University.

Bishop, A. W. et al (1964). *The mearurement of soil properties in the Triaxial test,* (83 - 175), Edward Arnold Publishers., London

Deshmukh, A. M. (1988) *Geotechnical properties of laterites of Goa.* 143-145, 2nd International Conference on Geomechanics, Singapore.

Gidigasu, M. D. (1976) *Laterite Soil Engineering, Pedogenesis and Engineering principles,* Elsevior Scientific Publishing Co., New York.

Unsaturated Soils for Asia, Rahardjo, Toll & Leong (eds) © 2000 Taylor & Francis, ISBN 90 5809 139 2

Mechanism of rain-induced slope failure in residual soils

K. K. Han
Scott Wilson (Malaysia) Snd. Bhd., Kuala Lumpur, Malaysia

H. Rahardjo
NTU-PWD Geotechnical Research Centre, Nanyang Technological University, Singapore

ABSTRACT: Dissimilar water contents exists in soils under a same matric suction. The stress path in a slope during infiltration has been modelled using a modified direct shear equipment in order to investigate the influence of dissimilar water content on the mechanism of rain-induced landslides. The results indicate that failure due to infiltration is characterised by excessive reduction in matric suction and much lesser time to failure in soils at drying.

1 INTRODUCTION

Stability of many steep slopes in residual soils is due in part to the presence of matric suction in the soils. The high shear stress mobilised in a steep slope coupled with the deep groundwater table have elevated the role played by matric suction in maintaining the stability of the slope.

Unsaturated soils are characterised by hysteresis where the soils can exist under a similar matric suction yet at two dissimilar water contents. It will be demonstrated in this paper that the hysteresis can influence markedly the mechanism of rain-induced slope failures.

A modified direct shear equipment has been used to simulate the field stresses in a slope before and during infiltration in order to investigate the mechanism of rain-induced landslides.

2 HYPOTHESIS

Hysteresis, a common feature of unsaturated soils, is the ability of the soils under a similar matric suction to have two different water contents when the soils are being wetted or dried, as demonstrated in soil-water characteristic curves. The soils being wetted have less water content than the soils being dried. In short, there are no unique relationship between the matric suction and water content for unsaturated soils.

Within a group of soil grains or an aggregate, pores of varying sizes or diameters exist, which can be visualised as many interconnected bottlenecks. The ink-bottle effect is the main cause for the hys-

teresis in unsaturated soils (Han, 1997). The smallest pores at the outermost of an aggregate govern the maximum curvature of menisci to be formed around the pores. The maximum curvature in turn dictates the maximum difference between the pore-air and pore-water pressures and hence the maximum matric suction (air entry value) of a particular aggregate of particles. Accordingly, the air entry value of a soil is governed by the smallest pores at the outermost of the soil aggregate.

As the pore sizes are not uniform throughout an aggregate, larger pores can be found within the soil aggregate. These larger pores do not control or affect the air entry value of a group of soil particles. They have the tendency to retain water if they are surrounded by pores of smaller diameter when the soil is being dried under a constant matric suction. On the other hand, the larger pores do not contain water when the soil has been previously dried prior to being wetted under the similar matric suction. Hence the soils at drying always have greater water content than the soils at wetting.

During shearing of soils at drying, the particle rearrangement unleashes the water held in the larger pores. The sudden availability of more water to smaller pores will reduce the matric suction substantially. The opposite happens to soils at wetting where there is no 'excess' water to reduce the matric suction during shearing.

In unsaturated slopes, the decrease in matric suction brought about by infiltration induces strains and shearing in the slopes. Therefore for soils at drying, the induced shearing will indirectly decrease greatly the matric suction in addition to the reduction in matric suction due directly to infiltration. The sudden and significant reduction in matric suction explains

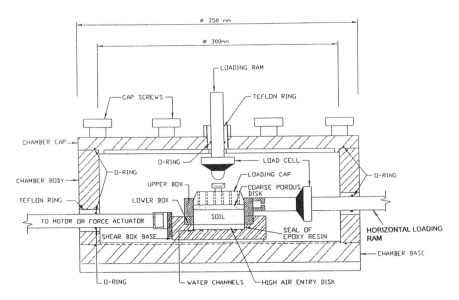

Figure 1. Cross-sectional view of shear box and pressure chamber.

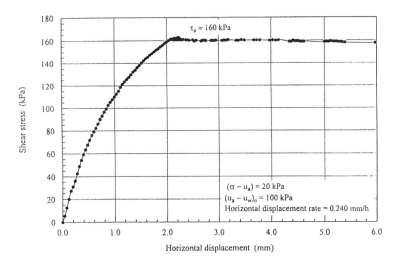

Figure 2. Shear stress versus horizontal displacement during shearing, creep monitoring and infiltration.

the many slope failures that took place without prior warning.

3 DIRECT SHEAR TESTS UNDER INFILTRATION

The above hypothesis was investigated using direct shear tests modified for infiltration and measurement of pore-air and pore-water pressures. The modified direct shear box is presented in Figure 1. Only brief description of the testing is explained here and a full description can be found in Han (1997).

A residual soil of granitic origin was statically compacted in a mold which was specially fabricated for preparing soil specimens for direct shear tests.

The only difference between the soil specimen tested at wetting and drying is the saturation stage for specimen at drying. The soil specimen at drying was saturated before equalised to a specified net normal stress and matric suction whereas the soil specimen at wetting was equalised without being saturated. After attaining equilibrium during equalisation, the specimen was sheared to a predetermined shear stress with a rate of 0.004 mm/minute under a constant matric suction. This predetermined shear stress had been evaluated to ensure that the soil would fail at unsaturated condition during the infiltrating process. This shear stress was maintained constant under the same matric suction for several hours until the creep in the soil decreased to a negligible value.

Infiltration was initiated with injecting water into the soil specimen through the high air entry disk at the base. The rate of water injection or infiltration was 0.04 mm³/s which had been determined to en-sure that the pore-water pressure in the specimen was the same as that measured in the water compartment. During the infiltrating process, the shear stress was kept constant while the matric suction, horizontal and vertical displacements of the specimen were measured at constant intervals of time. Failure was deemed to have occurred when the matric suction reached a minimum value and when the displacement rate increased considerably.

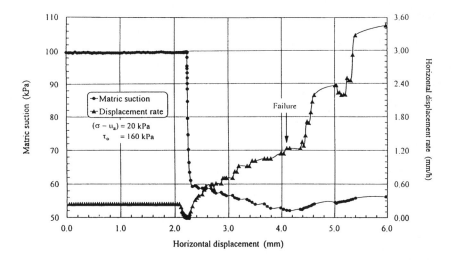

Figure 3. Matric suction and displacement rate versus horizontal displacement of specimen at drying during shearing, creep monitoring and infiltration.

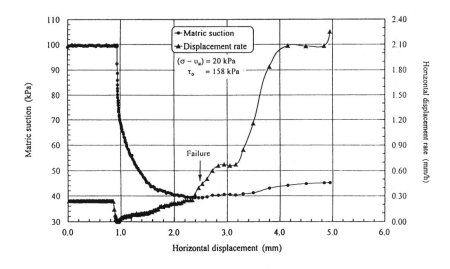

Figure 4. Matric suction and displacement rate versus horizontal displacement of specimen at wetting during shearing, creep monitoring and infiltration.

4 TEST RESULTS AND DISCUSSIONS

Figure 2 shows a typical plot of shear stress versus horizontal displacement during the shearing and infiltrating processes of a soil specimen at drying. The soil specimen was sheared to a shear stress of 160 kPa which was maintained constant thereafter. The constant matric suction and constant displacement rate during the shearing stage are depicted by the two horizontal lines on the left of Figure 3.

Upon the start of infiltrating process, the decrease in matric suction is significant and drastic for the soil at drying (Fig. 3) but the decrease in matric suction is gradual and much less rapidly for the soil at wetting (Fig. 4). The behaviour of the soil at drying during infiltration is therefore 'brittle' as compared to that of the soil at wetting. The infiltrating process initiates horizontal displacements in the specimens as shown in Figures 3 and 4. The displacement rates increase while the matric suction is further reduced.

It has been observed that much lesser time is required by the specimen at drying to reach failure in contrast to the much greater time required for the specimen at wetting. As the data were recorded at constant time intervals, the contrasting behaviour is proven from the discrete data points on Figure 3 and the overlapping and almost indistinguishable data points on Figure 4 before failure. The reason for the very short time taken by soils at drying to failure is the rapid reduction in matric suction made possible by the release of the trapped water volume (ink-bottle effect) in the soils from the shearing/strains induced by infiltration. The strains have been induced by the decrease of shear strength due to the reduction of matric suction.

The implication is that slopes with soils at drying can fail suddenly without warning, a phenomenon in rain-induced landslides. Whereas slopes with soils at wetting takes much longer time to fail only when the rainfall with a same intensity continues during this period of time.

5 CONCLUSIONS

Laboratory modelling of the stress path in a slope during infiltration has revealed the mechanism for the sudden rain-induced landslides. Large volume of trapped water, released in soils at drying due to induced strains from infiltration, causes the rapid reduction in matric suction and shear strength in the soils. The sudden decrease of shear strength takes relatively little time to occur and thus explains the sudden rain-induced landslides.

REFERENCES

Han, K.K. 1997. *Effect of hysteresis, infiltration and tensile stress on the strength of an unsaturated soil.* Ph.D. thesis. Nanyang Technological University, Singapore.

Unsaturated Soils for Asia, Rahardjo, Toll & Leong (eds) © 2000 Taylor & Francis, ISBN 90 5809 139 2

Estimation of the relation between suction and cohesion using unconfined compression tests

S. Kato & K. Kawai
Department of Civil Engineering, Kobe University, Japan

Y. Yoshimura
Department of Civil Engineering, Gifu National College of Technology, Japan

ABSTRACT: Unconfind compression tests, in which the suction in the specimens was measured, were carried out for statically compacted specimens of silty clay. The effect of degree of saturation on the unconfined compressive strength and deformation were examined. It was found that the suction stress affects the unconfined compressive strength, and that the relationship between the suction and the suction stress can be estimated by using the soil-water characteristic curve.

1 INTRODUCTION

The suction, that is defined as the pressure difference between the pore air pressure and the pore water pressure, affects on the behavior of unsaturated soil. At present, the triaxial compression test, in which constant suction is applied to a specimen, plays a central role for research on unsaturated soil. Based on the research results, some constitutive models have been proposed. In these constitutive models, the effect of the suction is taken as a cohesion or a stress component (the suction stress). Accordingly, it is necessary that the relationship between the cohesion and suction is known beforehand, when the behavior of the unsaturated ground is predicted by using these constitutive models. In order to decide the relationship between suction and cohesion, we must carry out triaxial compression tests under constant suction conditions. This will involve much time and money, and is not a suitable method for use in business practice.

The final purpose of this study is to propose a simple method for estimating the relationship between suction and the cohesion (or the suction stress) based on the results of the unconfined compression test in which the suction is measured. The unconfined compression test is a convenient test for undrained conditions. The estimation method proposed, when used with these proposed constitutive models, will contribute to the practical application of predictions for the behavior of unsaturated ground.

In this study, unconfined compression tests, in which the suction was measured, were carried out for statically compacted silty clay specimens. The relation between the suction stress and the unconfined

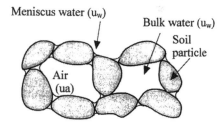

Fig.1 Bulk water and meniscus water in unsaturated soil

compressive strength and deformation characteristics, and the relationship between suction and the suction stress are examined.

Some research has been carried out using unconfined compression tests, in which the suction in nearly saturated soil specimen has been measured (Shimizu et al. (1993) and Mitachi et al. (1996)). However, in these studies, the suction worked as a negative pore water pressure under a high degree of saturation, and these researchers examined the effect of suction in a different saturation region from that in the present study.

2 SOIL WATER DISTRIBUTION IN SOIL MASS AND DEFINITION OF SUCTION STRESS

The relationship between soil water and suction under equilibrium is called the soil-water characteristic curve. The difference in the soil-water characteristic curve between wetting and drying processes is called hysteresis. Accordingly, there exist different soil moisture states under the same suction value. When soil moisture states are different, distribution patterns of pore water in the voids will consequently change.

Figure 1 shows a key sketch of the bulk water and the meniscus water. The tendency in the soil water distribution patterns is for the bulk water to exist more in the drying process, and the meniscus water to exist more in the wetting process. When we take the influence of suction on the behavior of unsaturated soil into account, we should know the different effects of the bulk water and the meniscus water on the skeleton of the soil mass.

The meniscus water increases the intergranular force that acts between soil particles, and it increases the stiffness of soil skeleton accordingly. The bulk water causes not only an increase of the stiffness of soil skeleton but also a decrease of volume of the soil mass with slippages at contact points.

Karube et al. (1996) named the stress component that arises from the effect of the meniscus water as the meniscus stress, p_m and the stress component that arises from the effect of the bulk water as bulk stress, p_b. They postulated that these two stress components consist of the suction stress (p_s) as shown in the following equation.

$$p_s = p_m + p_b \qquad (1)$$

They also proposed the following equation that defines the suction stress under some suction, s, the residual degree of saturation, S_{r0} of the specimen and degree of saturation, S_r.

$$p_s = p_m + p_b = \frac{S_r - S_{r0}}{100 - S_{r0}} \cdot s \qquad (2)$$

They proposed the following energy equation for unsaturated soil based on the energy equation used in the Cam clay model (Roscoe et al.,1963).

$$pdv + q \, d\varepsilon = M(p' + p_m)d\varepsilon + p' \, dv^e \qquad (3)$$

Here, M: the slope of the critical state line, $p' = p_t - u_a + p_b$; the mean net stress, $q = \sigma_1 - \sigma_3$;the shear stress, $d\varepsilon = 2/3 \ (d\varepsilon_1 - d\varepsilon_3)$;the increment of the shear stress, σ_1, σ_3 ; the maximum and the minimum total principal stress, $d\varepsilon_1, d\varepsilon_3$; the maximum and the minimum incremental strain, dv, dv^e ; the increment of the volumetric strain and elastic portion of it, $p = (\sigma_1 + 2\sigma_3)/3 - u_a$; mean principal total stress, u_a; the pore air pressure.

By dividing both sides of equation (3) with dv, the next equation is obtained.

$$q = M(p + p_s) + (p + p_b)\{(-dv/d\varepsilon) + (\, dv^e/\, d\varepsilon)\} \qquad (4)$$

They confirmed the validity of equation (4) for the results of triaxial compression tests carried out on compacted clay specimens. In practical applications, the next approximation based on equation (4) can be

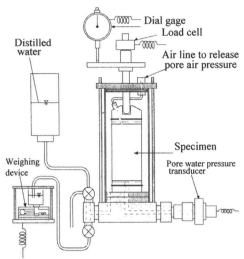

Figure 2 Sketch of the unconfined compression test cell used

applied to the results of unconfined compression test for unsaturated soil.

$$q = M(p + p_s) \qquad (5)$$

This equation means that the suction stress corresponds to the cohesion.

3 TEST PROCEDURE

3.1 Preparations for specimen

A silty clay, whose specific gravity of the soil particles, Gs is 2.68 and liquid limit is non-plastic, was used. After adding distilled water to the air-dried sample in order to adjust water content, it was stored over a whole day and night in a sealed container. The wetted sample was statically compacted in a mold of 50mm diameter and 300mm height. The inside of this mold was coated with Teflon. When the wetted sample was compacted in the mold, a spacer disk was used in order to adjust the specimen height to 120mm. The dry densities were set at three cases of 1.250, 1.330 and 1.405g/cm^3. The water contents adjusted were about 3~24% at which the preparation of the specimen was possible. After coating the sample with a plastic film and wrapping it with a wetted cloth to prevent the drying of the compacted sample, it was stored for a few days in a closed container, and was taken out before the test. By cutting from the top and bottom of the compacted sample, a specimen of 50mm diameter and 100mm height was made.

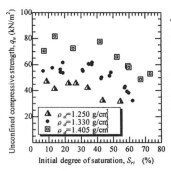

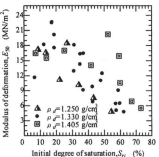

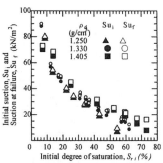

Figure 3 Relationship between unconfined compressive strength and the initial degree of saturation

Figure 4 Relationship between modulus of deformation and degree of saturation

Figure 5 Relationship between initial suction, suction at failure and initial degree of saturation

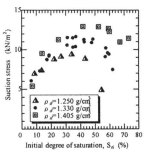

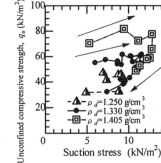

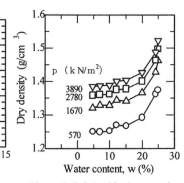

Figure 6 Relationship between suction stress and initial degree of saturation

Figure 7 Relationship between the suction stress and the unconfined compressive strength

Figure 8 Relationship between dry density and water content under constant static loads

3.2 Unconfined compression test measuring the suction

Figure 2 shows a sketch of the unconfined compression test cell used. After the specimen was set on the pedestal of the cell, the specimen was covered with a rubber membrane and a loading cap, and sealed with O-rings. Before the test was carried out, the cell was filled with water. During the test, the displacement of the cell water was measured with a weighing device that was connected to the inside of the cell. The volume of the specimen was obtained from adjusting the recorded displacement of the cell water for the piston penetration into the cell. A ceramic disk, whose air entry value was 500 kPa, was installed into the pedestal of the cell to measure the pore water pressure in the specimen. In the loading cap, a porous metal plate was installed. An air line was connected to it and released the pore air pressure in the specimen to atmosphere. The suction in the specimen was decided by value of the measured pore water pressure consequently. The rate of strain used was 0.1% per minute.

4. EXPERIMENTAL RESULTS AND DISCUSSIONS

4.1 Strength and deformation characteristics

Figure 3 shows the relationship between the unconfined compressive strength, q_u and the initial degree of saturation, S_{ri}, which was measured when the specimen was trimmed. The unconfined compressive strength increases as the dry density increases. For about 35, 40 and 45% of the degrees of saturation, the unconfined compressive strengths are constant in each case respectively. But the unconfined compressive strength decreases when the saturation increases from these degrees of saturation.

Figure 4 shows the relationship between the initial degree of saturation, S_{ri} and the modulus of deformation, E_{50}. From this figure, it is found that the initial degree of saturation increases when the dry density of the specimen increases, and the modulus of deformation decreases with an increase in the degrees of saturation. The rate of lowering in the initial degrees of saturation increases, when the initial degree of saturations, at which the unconfined compressive strength decreases in figure 3, are exceeded.

Figure 5 shows the relationship between the initial suction, su_i, the suction at failure and su_f with the initial degree of saturation, S_{ri}. The su_i and su_f values decrease with an increase in S_{ri}. But the increase of suction (su_f-su_i) is almost constant regardless of S_{ri}, and any effect of the density is hardly observed.

Figure 6 shows relationships between the suction stress, which is obtained from equation (2), and the initial degree of saturation S_{ri}. In each case, the suction stress increases to about 35, 40 and 45 (%) of S_{ri} respectively, and it begins to decrease, when S_{ri} increases over these degrees of saturation.

Accordingly, the relation between the suction and the unconfined compressive strength turns round as shown in figure 7. That is to say, the suction stress increases when the saturation is increased, and the unconfined compressive strength increases a little. At this time, the unconfined compressive strength, q_u is also dependent on the dry density. In addition, the unconfined compressive strength begins to rapidly lower, when the initial degree of saturation increases, and when the suction stress changes to a decrease. In this decreasing process, there is a unique relation between q_u and the suction stress in spite of the dry density.

Figure 8 shows the relationship between dry density, and water content, w during compactions of the sample under constant static loads, p. Under constant loads, p, the dry density increases rapidly at w=15~20%. This result means that, when specimens of some dry density are prepared under wetter condition, it will need less constant load. These water contents correspond to the initial degree of saturation of the lowering point of the unconfined compressive strength shown in figures 3,6 and 7.

4.2 Estimation of suction stress estimated from water characteristic curve

Various equations have been presented to express the relationship between suction and soil moisture. In the our study, the equation which was proposed by Van Genuchten (1980) (the VG model) is used, because it can be applied to various soil types from sand to clay. The VG model is given by the next equation.

$$Se(\theta) = \left\{1 + (\alpha \cdot s)^n\right\}^{-(1-\frac{1}{n})} \qquad (6)$$

Here, α and n are fitting parameters, and $Se(\theta)$ is the relative volumetric water content defined by the next equation.

$$Se(\theta) = \frac{\theta - \theta_r}{\theta_{sat} - \theta_r} \qquad (7)$$

Here, θ_r is the residual volumetric water content, θ_{sat} is the saturated volumetric water content, and θ

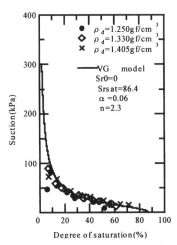

Figure 9 Relation between the suction at failure and the degree of saturation at failure

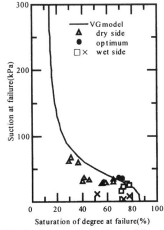

Figure 10 Relation between the suction at failure and the degree of saturation at failure in unconfined compression test for dynamically compacted specimen of the same silty clay.(Kato et al. 1999)

is the volumetric water content. In practice, the above parameters of α, n, θ_r and θ_{sat} are decided by the result of fitting soil-water characteristic curve.

When the volume change of a specimen is very small during the water retention test, the relative volumetric water content can be given by the next equation.

$$Se(\theta) = \frac{Sr - Sr_0}{100 - Sr_0} \qquad (8)$$

Figure 9 shows the relation between the suction at failure and the degree of saturation at failure. In this figure, the dots show the results of unconfined compression test, and the solid line shown in figure 9 is the approximation result by the VG model to the relation between suction at failure and degree of satu-

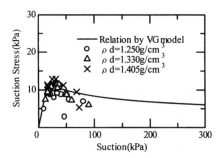

Figure 11 Comparison between prediction and test results of relationships between the suction stress and suction

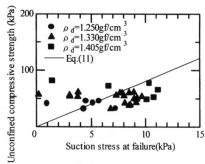

Figure 12 Relationship between the unconfined compressive strength and the suction stress

ration at failure. The used parameters in the VG model are $\alpha=0.06$, $n=2.3$, $Sr_{sat}=86.4(\%)$ and $Sr_0 =0\%$. Figure 10 shows the relation between the suction at failure and the degree of saturation at failure in unconfined compression tests for dynamically compacted specimens of the same silty clay (Kato et al. 1999). In this figure, the dots show the results of unconfined compression tests for specimens of 5 to 23% water contents, and the solid line is the approximation result by the VG model to the water retention curve for the dynamically compacted specimen of 17% water content.

From the figure 10, it is found that the test results are distributed along the solid line. The void diameter distributions in the specimens will change when the water contents at compaction are different, because the dry densities of the compacted specimens depend on the water contents at compaction. When the void diameter distributions show different tendencies, the water characteristic curves obtained from each specimen will be different. But the results showed in figure 10 mean that the relationships between suction and degree of saturation at failure show the same tendency with a water characteristic curve even if the dry densities of the specimens are different. Based on this result, it is supposed that the water retention curve for the statically compacted sample is almost the same as the solid line shown in figure 9.

Figure 11 shows a comparison between a prediction and test results of the relationships between the suction stress and the suction. In this figure, the solid line is a prediction obtained by the next equation that is derived from equations (2), (6) and (8).

$$p_s = \frac{Sr - Sr_0}{100 - Sr_0} \cdot s = \left\{ 1 + \left(\alpha \cdot s \right)^n \right\}^{-(1-\frac{1}{n})} \cdot s \qquad (9)$$

The parameters used are the same ones for the solid line shown in figure 9. The prediction shows a tendency for the suction stress to increase as suction increases to 30 kPa, above which it decreases and gradually approaches a fixed value. The dots shown in this figure are the calculated results obtained by

using eq.(2) for the unconfined compression test results. These calculated results show similar tendencies with the prediction. From this result, it can be stated that the suction stress at failure is known when using the results of the unconfined compression test.

4.3 Relation between unconfined compression strength and suction stress

Figure 12 shows the relationship between the unconfined compressive strength and the suction stress. In this figure, the dots show test results, and they are distributed around the solid line shown by eq.(11) mentioned later. From this result, it can be concluded that the suction stress affects the unconfined compressive strength.

The relationship between the unconfined compressive strength and the suction stress can be deduced as follows. Firstly, we postulate that the suction stress will act as a confining pressure for the specimen in an unconfined compression test, and that the Mohr's stress circle at failure is in contact with the failure criteria that corresponds to the suction at failure. Secondly, we postulate that the cohesion of the failure criteria is equal to the suction stress. Based on these assumptions, a geometric relationship between the failure criteria and the Mohr's stress circle at failure is given as shown in figure 13. In this figure, q_u means the unconfined compressive strength, and p_s means the suction stress at failure.

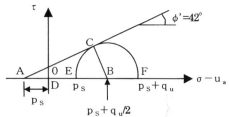

Figure 13 Geometric relationship between the failure criteria and the Mohr's stress circle at failure

From the geometric relationship concerning for the triangle ABC, an equation that connects the unconfined compression strength to the suction stress at failure is derived as follows.

$$q_u = \frac{4\sin\phi'}{1-\sin\phi'} \cdot p_S \qquad (10)$$

Here, ϕ' the internal frictional angle is based on the net stress that is defined by the difference between total stress and pore air pressure. The internal friction angle is slightly larger than that obtained from a saturated sample with an effective stress (Karube et al. 1986). Abe & Hatakeyama (1997) reported that the internal frictional angle based on effective stress is 42 degrees in saturated specimens that were made under the same condition as the specimens used in the unconfined compression tests. Substituting this value into the equation (10), the following equation is obtained.

$$q_u = 8.09 \cdot p_S \qquad (11)$$

The solid line shown in figure 12 represents the relationship obtained from the equation (11). It is found that the test results distribute around the solid line. This result demonstaretes the appropriateness of the assumptions mentioned above.

5. CONCLUSION

Unconfined compression tests, in which the suction was measured, were carried out on statically compacted silty clay specimens in order to establish a simple method for estimating the relationship between suction and cohesion. The relations between the suction stress, the unconfined compressive strength and deformation characteristics, and the relation between suction and the suction stress are examined. The results and discussions are summarized as follows.

1. Though the initial degrees of saturation and the dry densities of the specimens were different, the relationship between suction and the degree of saturation at failure was shown by one water retention curve. Also, the relation between the suction stress and suction showed a similar tendency with the prediction by equation (4) using the same parameters for the water retention curve mentioned above.

2. The unconfined compressive strength is affected by the suction stress, and the suction stress seems to work as a confining pressure for unsaturated specimens. The relation between the suction stress and the unconfined compressive strength showed the similar tendency as the relation shown in equation (11).

3. These results demonstrate the applicability of the unconfined compression test for estimating the relationship between the suction stress and suction.

REFERENCES

Abe, H. & Hatakeyama, M. 1997, Report of the committee for valuation of water permeability in unsaturated ground. Japanese Geotechnical Society, pp.82-88. (In Japanese)

Karube, D., Kato, S. & Katusyama, J. 1986, Effective stress and soil constants of unsaturated kaolin, *Journal of Japanese Society of Civil Engineering*, Vol.370, pp.179-188. (In Japanese)

Karube, D., Kato, S., Hamada, K. & Honda, M. 1996, The relationship between the mechanical behavior and the state of pore water in unsaturated soil, *Journal of Japanese Society of Civil Engineering*, Vol. 535, pp.83-92. (In Japanese)

Kato, S., Yoshimura, Y., Kawai, K. & Sunden, W. 1999, Effects of suction on strength and deformation of unsaturated soil in unconfined compression test, Poster session proceedings of 11th Asian Reginal Conference on Soil mechanics and Geotechnical engineering, Vol.1, pp.17-18.

Mitachi, T. & Kudo, Y. 1989, Method for predicting in-situ undrained strength of clays based on the suction value and unconfined compressive strength *Journal of Japanese Society of Civil Engineering*, No.541/III-35, pp.147-158.

Roscoe, K.H., Schofield, A.N. & Thurairajah, A. 1963, Yielding of Clay in State Wetter than Critical, Geotechnique, Vol.13, No.3, pp.211-240.

Shimizu, S. & Tabuchi, T. 1993, Effective Stress Behavior of Clays in Unconfined Compression Tests, *Soils and Foundations*, Vol.33, No.3, pp.28-39.

Van Genuchten, M. Th.,1980, A closed-form for predicting the hydraulic conductivity of unsaturated soils, *Soil. Sci. Soc. Am. J.*, Vol.44, pp.892-898.

Unsaturated Soils for Asia, Rahardjo, Toll & Leong (eds) © 2000 Taylor & Francis, ISBN 90 5809 139 2

Study on shear strength and swelling-shrinkage characteristic of compacted expansive soil

L.-W. Kong & L. R. Tan

LRSM Laboratory, Institute of Rock and Soil Mechanics, Chinese Academy of Sciences, Wuhan, People's Republic of China

ABSTRACT: The shear strength and swelling-shrinkage characteristic of a typical expansive soil from Yichang, Hubei province, China is systematically investigated. The correlation between the shear strength parameters, swelling pressure, expansion ratio under various loading pressures, shrinkage degree, shrinkage index and initial water content of unsaturated compacted expansive soil is established. The mechanism of some of the above mentioned parameters existing critical point is analyzed.

1 INTRODUCTION

Expansive soil is a very important type of regional soil. Expansive soil problems exist in most countries in the world, which causes serious hazards. Many researchers have studied the behavior of expansive soil. Indeed, the strength and swelling-shrinkage properties of expansive soil are more complex than ordinary clay soil, which results in significantly special characteristic in its deformation and strength (Tan 1987,1994). Up to date, it is reported that foundation damage and slope engineering damage due to unsaturated expansive soil behavior does happen. The fundamental behavior for expansive soil is not well understood, particularly, systemic research of expansive soil properties should be progressed further.

In this paper, the strength and swelling-shrinkage properties of a typical expansive soil is systemically investigated by using the methods of ordinary geotechnical testing so that its physico-mechanical properties and damage behavior can be understood.

2 SOIL PROPERTIES AND TEST METHOD

2.1 Properties of expansive soil

Soil samples were collected from Yichang, Hubei province, China. The soil is in shallow layer and has high swelling properties. Table 1 and Table 2 summarizes its property parameters and its physico-mechanical parameters for natural intact specimens, respectively. The corresponding compaction curve is plotted in Fig.1. It can be seen that the soil is a typical expansive soil.

Table 1 Basic property parameters of the expansive soil

Property	Value	Remark
Specific gravity	2.77	
Liquid limit	81.7 (%)	
Plastic limit	34.7(%)	
Plasticity index	47.0	
Dry water content	11.5~14.2	
Free swell	103.0(%)	
Shrinkage ratio	12.1(%)	Vertical
	11.7(%)	Horizontal
Grain size distribution	64.5(%)	<10μm
	50.0(%)	<5μm
	25.0(%)	<2μm

Table 2 Physico-mechanical parameters of natural unsaturated intact specimens

Item	Value	Remark
Natural water content (%)	35.3~40.9	
Natural density (g/cm^3)	1.76~1.85	
Void ratio	1.03~1.16	
Coefficient of shrinkage	0.511	Vertical
	0.448	Horizontal
Swelling pressure (kPa)	13.6	
Cohesion (kPa)	35.5	Slow direct
Internal friction angle (^{0})	11.7	shear test
Coefficient of compressibility (Mpa^{-1})	0.436	
Compression index	0.250	

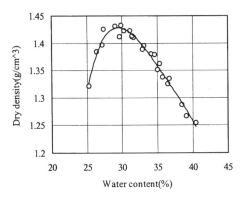

Fig.1 The compaction curve of expansive soil

Table 3 Relation between unsaturated shear strength parameters and initial specimen states

Water content (%)	Density (g/cm³)	Cohesion (kPa)	friction angle (⁰)
26.49	1.793	103.67	12.49
28.53	1.867	99.79	11.43
29.26	1.872	92.32	10.36
30.82	1.891	86.02	9.92
31.60	1.891	79.1	8.42
33.55	1.857	59.96	7.69
34.80	1.860	54.87	6.54
37.03	1.822	47.35	6.29
41.24	1.771	28.38	6.19

2.2 Test method

The compacted samples were prepared at various water contents and cured before compaction in the standard compaction device, in which standard effort was selected and used. The slow direct shear test method was adopted to measure the shear strength of the compacted specimens, in which the water content was maintained constant during the consolidation and shearing process. The "constant volume" test was used to measure the swelling pressure. The expansion ratio of compacted specimen was measured under loading of 50kPa .The load was maintained for some time (as the specimen tended to swell and continue to do so) until the specimen exhibited no further tendency to swell. Then the specimen was unloaded in a conventional manner. The shrinkage test was performed in the shrinkage limit apparatus. Besides the vertical shrinkage ratio was measured, the horizontal shrinkage ratio was also measured, in which the specimen slowly dried in air. The specimen was covered at various time intervals, as in this way, the specimen comes to equilibrium and shrinks uniformly.

3 RESULTS OF TESTS AND ANALYSIS

3.1 The shear strength properties of compacted expansive soil

Table 3 summarizes the results of slow direct shear tests on 9 sets of compacted specimens with various water contents in an unsaturated state. The corresponding shear strength versus vertical stress, cohesion c and friction angle φ versus water content are plotted in Fig. 2, Fig. 3 and Fig. 4, respectively.

From the above table and figures, the following trends in shear strength for compacted expansive soil can be observed

(1) The shear strength decreases as the initial water content increases along the compaction curve, whereas the effect of dry density on shear strength is

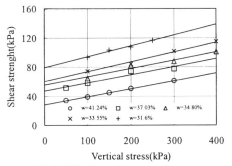

Fig 2(a) The direct shear strength versus water contant relation

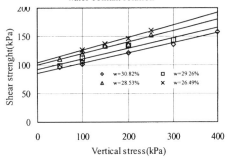

Fig.2(b) The direct shear strength versus water content relation

not very significant.

(2) The cohesion of the unsaturated compacted specimens reduces with the increase of water content, moreover, the trend is close to a line, which indicates matric suction varies inversely with water content.

(3) The trend of friction angle with water content is different from that of cohesion. The former decreases significantly in the range of water content less than the plastic limit, whereas as the water content is more than the plastic limit, the friction angle tends to a constant. That is to say, the friction angle shows a critical point at the water content close to

the plastic limit (w=35%). Furthermore, the slope of the relationship between cohesion and water content is more than that of the friction angle. It indicates that the friction angle does not vary with initial water content greatly once the degree of saturation reaches a certain value, which is determined by the intrinsic characteristics of a soil.

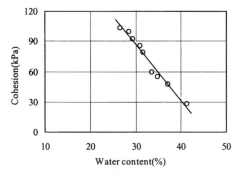

Fig.3 Cohesion versus water content relation

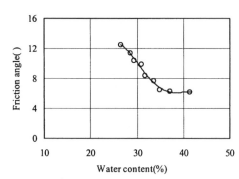

Fig.4 The friction angle versus water content relation

3.2 The swelling properties of compacted expansive soil

Table 4 and Table 5 summarize the results of swelling pressure tests on 17 sets of compacted specimens and expansion ratio tests on 8 sets of compacted specimens under various loading, respectively. The corresponding swelling pressure and expansion ratio versus initial water content is shown in Fig. 5 and Fig. 6, respectively.

From Fig. 5 and Fig. 6, it can be observed concerning the swelling properties of compacted expansive soil

(1) The swelling pressure decreases linearly with initial water content increase in the range of water content less than the plastic limit (w=35%), whereas once the water content is more than 35%, the swel-

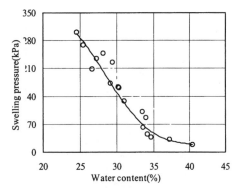

Fig.5 The swelling pressure versus water content relation

ling pressure is very small and tends to a constant value.

Table 4　Results of swelling pressure tests

Initial water content (%)	Final water content (%)	Swelling pressure (kPa)
24.55	29.99	301.10
25.47	30.32	269.04
26.58	31.01	208.40
27.17	30.53	234.70
28.10	31.11	247.50
29.12	31.80	173.40
29.39	31.45	225.60
30.17	32.10	163.00
30.31	32.12	162.60
30.93	32.89	128.80
33.46	35.54	102.40
33.58	37.10	62.13
33.97	34.96	86.40
34.16	35.69	46.00
34.58	35.79	38.10
37.22	38.35	31.70
40.35	42.64	18.10

Table 5 Results of expansion ratio test under various loading

Initial water content	Final water content	Expansion ratio under loading			
	(%)	50kPa	25kPa	12.5kPa	1.0kPa
24.98	39.69	6.15	7.02	7.88	10.45
25.17	36.18	4.94	6.36	7.33	10.02
28.15	38.24	4.14	5.15	6.04	9.06
30.57	38.70	2.48	3.42	4.70	8.35
32.00	38.57	1.70	2.57	3.65	6.81
32.91	38.00	1.07	1.75	2.58	5.85
34.96	38.05	0.51	1.05	1.58	2.58
37.32	39.97	-0.51	-0.20	0.16	1.49

517

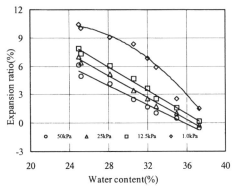

Fig.6 Expansion ratio versus water content
under different loading

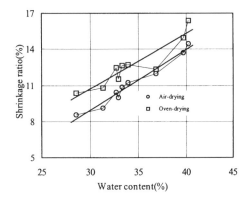

Fig.7(a) Vertical shrinkage ratio versus water
content

(2) The expansion ratio under various pressures increases with unloading for the same water content, whereas under the same unloading pressure, the expansion ratio decreases linearly with the initial water content, except for unloading pressures up to 1.0kPa. When the unloading pressure is equal to 1.0kPa, the expansion ratio versus initial water content relation is nonlinear.

3.3 The shrinkage properties of compacted expansive soil

Table 6 summarizes the results of shrinkage tests on 9 sets of compacted specimens. The corresponding shrinkage ratio and shrinkage degree versus water content are shown in Fig. 7 and Fig. 8, respectively. For the sake of contrast, the shrinkage ratio and shrinkage degree under the oven-dried condition were also measured. Its corresponding results are shown in Fig. 7 and Fig. 8. In order to reflect the shrinkage characteristics of specimens with high density, the horizontal shrinkage ratio of the specimens after swelling pressure tests under the oven-dried condition was also measured, the result was plotted in Fig. 9.

Table 6 Results of shrinkage tests

Initial	Final	Shrinkage ratio		Shrinkage index	
water content		Vertical	Horizontal	Vertical	Horizontal
(%)		(%)			
28.59	12.69	8.52	8.16	0.54	0.51
31.31	12.39	9.11	9.16	0.48	0.48
32.68	12.35	10.44	9.30	0.51	0.46
32.96	12.51	9.99	9.52	0.49	0.47
33.33	12.66	10.87	9.63	0.53	0.47
33.94	12.80	11.20	9.89	0.53	0.48
36.80	12.33	11.97	11.12	0.49	0.45
39.71	12.58	13.68	11.73	0.50	0.43
40.26	12.61	14.45	11.90	0.52	0.43

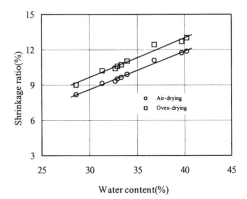

Fig.7(b) Horizontal shrinkage ratio versus
water content

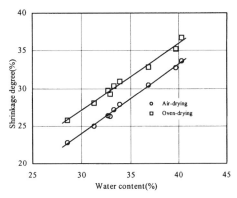

Fig.8 Shrinkage degree versus water content

From Table 6 and Fig. 7~Fig. 9, it can be observed concerning the shrinkage characteristic of compacted expansive soil:

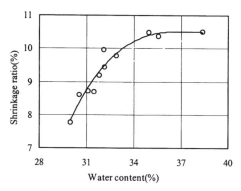

Fig.9 Horizontal shrinkage ratio of oven-dried condition versus water content for specimens after swelling pressure tests

(1) The effect of initial water content on dry water content and shrinkage index for compacted specimens is not noticeable.The above two parameters are determined by the intrinsic properties of a soil.

(2) The shrinkage ratio and shrinkage degree increase linearly with initial water content, moreover, the slope for air-drying is less than that of oven drying.

(3) For the specimens after swelling pressure tests, the horizontal shrinkage ratio increases linearly with water content in the range of water content less than 35%, whereas when the water content is more than 35%, the shrinkage ratio exhibits no further tendency to increase.

4 CONCLUSIONS

From laboratory tests on the shear strength and swelling-shrinkage characteristic of compacted expansive soil, the following conclusions can be obtained:

(1) When the initial water content of compacted specimens is less than the plastic limit, the shear strength parameters, swelling pressure, expansion ratio, shrinkage ratio and shrinkage degree change with water contents linearly. It indicates that a correlation between these parameters exists.

(2) The friction angle, swelling pressure and shrinkage ratio show a critical point when the water content is close to the plastic limit. When the water content is more than the plastic limit, the above parameters do not vary greatly with water content and tend to a constant value. It indicates that the behavior of unsaturated expansive soil is similar to that of saturated expansive soil as the degree of saturation reaches a certain value.

5 ACKNOWLEDGMENTS

The National Natural Science Foundation Commission of China supported the Project (Grant no: 19772068)

REFERENCES

Tan Luorong.1987.The Investigation of the Basic Properties of Expansive Soil. Chinese Journal of Geotechnical Engineering.9 (5): 31~42
Tan Luorong & Zhang Meiying et al.1994.Microstructure Characteristics and Engineering Properties of Expansive Soil. Chinese Journal of Geotechnical Engineering.16 (2): 48~58

Unsaturated Soils for Asia, Rahardjo, Toll & Leong (eds) © 2000 Taylor & Francis, ISBN 90 5809 139 2

Shear strength and swelling of compacted partly saturated bentonite

J. Kos
Department of Geotechnics, Czech Technical University of Prague, Czech Republic

J. Pacovský & R. Trávníček
Centre of Experimental Geotechnics, Czech Technical University of Prague, Czech Republic

ABSTRACT: Undrained shear strength and swelling pressure have been among the properties tested of bentonite seals for high-level radioactive waste repositories that are being prepared for use in the Czech Republic. The properties of Czech raw Ca/Mg bentonite and industrial Na bentonite have been tested. The specimens were compacted by pressures in the range of 0.5 to100 MPa. The influence of low water content on the pore water suction, shear strength and one-dimensional collapse or swelling has been measured and numerically modeled.

1 INTRODUCTION

Undrained shear strength and swelling pressure have been among the properties tested of bentonite seals to be used for high-level radioactive waste repositories that are being prepared for use in the Czech Republic. The properties of Czech raw Ca/Mg bentonite and industrial Na bentonite have been tested. There were some fears about the long-term chemical stability of Na bentonite. This is why Ca/Mg bentonite has been preferred in our research.

The tested specimens have been prepared from bentonite powder or crushed raw bentonite and compacted by pressures in the range of 0.5 to100 MPa. The water content of delivered bentonite (7-10%) has been modified and quartz sand has been added (up to 50% of total weight) before the compaction of some specimens.

2 SHEAR STRENGTH

Shear strength of all tested specimens has been measured under unsaturated conditions so that the influence of pore water suction on their strength could be observed in partly saturated bentonite. No pore water suction has been measured directly.

The shear strength of specimens compacted by low pressures (≤ 1500 kPa) was tested in a shearbox apparatus with internal dimensions of 84 mm × 84 mm and by uniaxial compression tests on specimens with a diameter of 38 mm. The results of shearbox tests on low pressure compacted bentonite are shown in Figures 1 and 2.

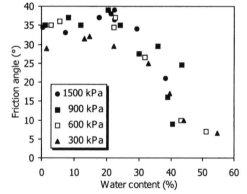

Figure 1. Friction angle-water content relationship for different compaction pressures.

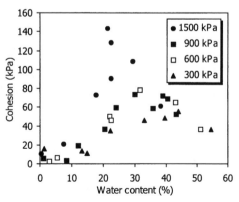

Figure 2. Cohesion-water content relationship for different compaction pressures.

The results of uniaxial compression tests are shown in Figure 3. Figure 4 shows estimated uniaxial compression strengths (calculated as the Mohr-Coulomb major principal stress for zero minor principal stress with φ_u, c_u from the shearbox test).

It is apparent that the estimated values of the uniaxial compression strength are lower than the measured ones. To explain these differences, the tests have been modeled numerically using the FLAC^{2D} computer program. Both test methods (shearbox and uniaxial compression) have been modeled in the same load steps as the real tests. Model results have been processed in the same way as test results. Differences in model input and output shear parameters have been observed. A constant error from the shearbox test has been detected. For example, the input soil parameters $\varphi = 29°$, $c = 90$ kPa have been lowered to output values $\varphi = 26°$, $c = 80$ kPa.

The specimens compacted by high pressures (≤ 100 MPa), with strength higher than 10 MPa, have been tested by the same methods as concrete and rock specimens, i.e. uniaxial compression, shear and flexure. These tests cannot be used for the direct evaluation of shear strength parameters. Therefore, the tests have been numerically modeled using the FLAC^{2D} program. The shear strength parameters have been found by taking the model input values that give the best coincidence between experimental and model results.

An example of the results of shear test modeling is shown in Figure 5. The specimens tested were cubes with an edge length of 100 mm. A sliding plane of loading jaws was inclined to the longitudinal axis of a press (30°, 45°, 60°). Model input parameters were higher than output (apparent) ones. It was similar to a shearbox test. This is the result of different inclinations of actual and predestined planes of failure.

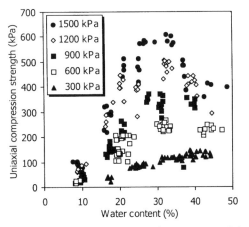

Figure 3. Uniaxial compression strength-water content relationship for different compaction pressures.

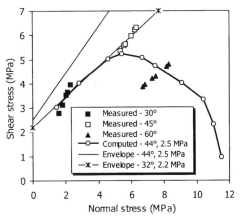

Figure 5. Measured and calculated results of shear test.

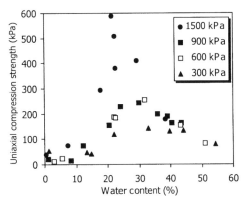

Figure 4. Estimated uniaxial compression strength-water content relationship based on shear box test results.

The tests have shown that there are certain limits of the water content for each compaction pressure within which the shear strength is maximal. The optimum water content decreases with the increasing compaction (consolidation) pressure. This is the result of two influences on the global effect of pore water suction: the force of adhesion between two soil particles and the relative amount of soil particles connected by capillary (contact) water. The first influence decreases with the increasing particle diameter, water content and water temperature. The second influence increases with the increasing water content (saturation ratio) and soil density. The increasing water content negatively influences the force of adhesion, but positively the total number of connected particles. It means that, as a compromise, there is an optimum water content. Dense soils need a smaller volume of contact water to connect more closely distributed particles than loose soils. Therefore, the optimum water content of dense soils is lower and the suction and shear strength is higher than those of loose soils. For example, bentonite powder with a water content of 7% compacted by 600 kPa had the cohesion of approx. 15 kPa (optimum water content ≅ 30%, Fig. 2), whereas the same powder compacted by 100 MPa had the cohesion of approx. 2.5 MPa (Fig. 5).

Low water content can be reached using high temperatures. However, without synchronous compression, shrinkage cracks may originate with negative influence on the soil strength. The effect of temperature (24 hours) on uniaxial and flexural strength of high pressure compacted bentonite is shown in Figure 6. (Each point represents the average of approx. five values.)

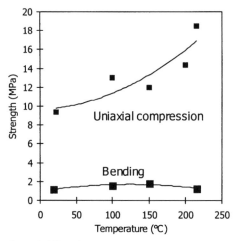

Figure 6. Effect of temperature on strength of high compacted bentonite.

3 SOIL COLLAPSE AND SWELLING

Bentonite powder compacted by low pressures (≤ 1.5 MPa) collapsed after inundating with water in the oedometer. After unloading, it swelled. Swelling pressures are too low for the bentonite sealing. Bentonite specimens compacted by high pressures (≤ 100 MPa) had always been unloaded before wetting because of the impossibility to compact them directly in the oedometer. Specimens were compacted in a special reinforced mould and then they were placed in the oedometer. The samples expanded after dismantling the mould. Therefore, the diameter of the mould was slightly smaller than the diameter of the oedometer and the specimen was abraded if necessary. High compacted specimens were disturbed from the geotechnical point of view. The state of

stresses changed after dismantling the mould. After wetting, the specimens swelled immediately with high swelling pressures, which are lower than in the case of compacting directly in the oedometer (Fig. 7). This method of preparation of samples, however, is near to the modelled production of bentonite buffer prefabricates.

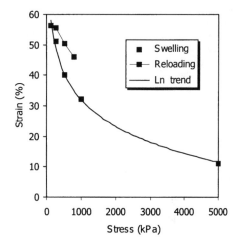

Figure 7. Swelling and reloading of dense bentonite powder.

Swelling is the result of the pore water suction decrease after wetting. If volume changes are restricted, the effective stress changes are restricted too and the decrease of the (internal) pore water suction has to be changed by the increase of the (external) normal stress, i.e. the swelling pressure:

$$\sigma' = \sigma_u - u_i = \sigma_u - u_a + \sigma_s \Rightarrow \sigma_s = u_a - u_i \qquad (1)$$

where σ_u (σ') = undrained (effective) normal stress; σ_s = swelling pressure; u_a (u_i) = actual (initial) average water pressure.

The swelling pressure is lower than in equation (1) if the volume changes are only partly restricted.

Equation (1) is not completely correct for soils with low water content. The contact water adhesion is compensated for by the increased normal stress near the particle contact due to action and reaction. At certain distance from the contact, the particle is practically not loaded by the effect of (internal) pore water suction. The whole particle, however, is overloaded after an increase in the external (swelling) pressure. Therefore, the swelling pressure should be somewhat lower than in equation (1). It also depends on the shape of soil particles.

Swelling and collapse are mostly observed and measured (oedometer) in an anisotropic state of stress. In this case, the most important factor is a reduction of cohesion induced by the pore water suction decrease. Collapse of the soil can be explained only by the loss of shear strength.

4 MODEL OF OEDOMETER TEST

The behaviour of wetted soils can be described using a numerical model created for the analysis of an oedometer test. This model can describe the influence of the pore water suction decrease on the reduction of the interparticle shear strength, which induces the soil collapse or swelling.

The model solves the relationship between a vertical stress and one-dimensional strain (oedometer test) using an incremental method. The non-linear oedometer modulus is a function of the actual soil strain. The oedometr modulus in the next step is calculated using the following method.

The corresponding stress is calculated from the actual strain:

$$\varepsilon_a = (\ln\sigma_a - \ln\sigma_i)/C \qquad (2)$$

$$\sigma_a = \exp(\ln\sigma_i + \varepsilon_a . C) \qquad (3)$$

where σ_a = stress corresponding to actual strain; σ_i = initial stress corresponding to zero strain; ε_a = actual strain; $C = \Delta\ln\sigma/\Delta\varepsilon$ (coefficient).

For all these calculations the one-dimensional stress-strain relationship of first loading is used only. In this way, the solution of a cyclic load is possible. During unloading and reloading, the soil is compacted by the precedent cycles. The soil density, the total amount of particle contacts and the average (global) soil water suction increase. Therefore, the soil is stiffer than during first compression. The actual effective stress (plus internal water suction) may be higher than the external (measured) value. That is why the oedometer modulus is calculated as dependent on the actual strain, not directly on the stress.

The oedometer modulus for the next step is calculated from the following equations:

$$\Delta\varepsilon = \Delta\sigma/E_{oed} = 1/C \times \ln[(\sigma_a + \Delta\sigma)/\sigma_a] \qquad (4)$$

$$E_{oed} = \Delta\sigma \times C/\ln[(\sigma_a + \Delta\sigma)/\sigma_a] \qquad (5)$$

where $\Delta\sigma$ = stress increment in the next loading step.

The oedometer modulus is then modified according to the following rule. If the soil is at rest, i.e. the strength of the soil structure (interparticle shear strength) is not overcome and the statistically decisive amount of particles is not in motion, the oedometer modulus is several times higher than the typical value. These high values of oedometer modulus can be derived from initial stages of one-dimensional unloading and reloading.

The state at rest of soils is understood as a state of stresses, in which the mobilized shear strength of the soil is so low that the strength of most (theoretically all) interparticle contacts is not overcome. It means that shear strength is only partly mobilized.

The limit values of shear parameters at rest can be derived from the equivalence of Jaky's coefficient and Rankine's coefficient of active stress for lower (partly mobilized) friction angle (Myslivec 1972):

$$K_0 = 1 - \sin \varphi = (1 - \sin \varphi_0)/(1 + \sin \varphi_0) \qquad (6)$$

$$\varphi_0 = \arcsin [\sin \varphi/(2 - \sin \varphi)] \cong 2/3 \ \varphi \qquad (7)$$

Assuming that the soil water suction u is constant:

$$u = c/\tan \varphi = c_0/\tan \varphi_0 \qquad (8)$$

it can be calculated:

$$c_0 = c \times \tan \varphi_0/ \tan \varphi \qquad (9)$$

where φ_0 (c_0) = partly mobilized friction angle (cohesion) at rest.

Lambe & Whitman (1969) published the equation similar to Eq. (6):

$$K_0 = (1 - \tan \beta)/(1 + \tan \beta) \qquad (10)$$

where β = slope of the stress path K_0.

The limit values of stress at rest can be computed as active and passive stresses from partly mobilized strength parameters. Generally, stress at rest can be computed using partly mobilized strength parameters from φ_0 (c_0) to zero (because after isotropic consolidation, the soil is also at rest).

The behaviour of the model and loose bentonite powder (w = 7%) is shown in Figure 8. The specimen was compacted during the first loading.

(It was slightly overconsolidated during its preparation - approx. 40 kPa. Back analyzed parameters of the bentonite powder are: $\varphi_u = 30°$; $c_u = 160$ kPa; $C (= \Delta\ln\sigma/\Delta\varepsilon) = 22$; C_0 (at rest) = 200.)

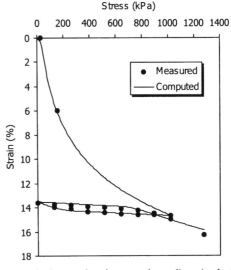

Figure 8. Computed and measured one-dimensional stress-strain relationship of bentonite powder.

At the end of the first loading, horizontal stresses reach their maximum values having positive influence on the extent of vertical stress interval at rest during unloading. On the contrary, reloading starts from lower vertical and horizontal stresses and the interval of one-dimensional stress at rest is shorter than that of unloading. This is the reason for soil hysteresis.

When the soil is wetted, suction and cohesion (at rest) are reduced. The new (computed) lower limit of stress at rest is higher than the actual stress. The soil becomes overloaded. The process of consolidation starts without an increment of external load, i.e. as the result of watering. The soil collapses. During this process, lateral stress increases from the actual value to a new value of stress at rest. It is not easy to solve this problem by a model based on the elastic continuum. The model described uses fictitious vertical stress increments to increase lateral stresses and also one-dimensional strain. The oedometer modulus is computed as dependent on the actual strain, not directly on stress.

Swelling is a similar process. It starts after wetting when the soil is overconsolidated. In this case, the lateral stress is higher than the vertical one. One-dimensional collapse and swelling are shown in Figures 9 and 10. The bentonite specimen (powder) was watered after unloading. The model computed "elastic" unloading (Fig. 9). Then, the soil computer model was wetted at the end of the first loading (it collapsed) and at the beginning of unloading (it swelled - Fig. 10).

Back analyzed parameters of the bentonite powder (Figs. 9,10) are: $\varphi_u = 30°$; $c_u = 160$ kPa; C ($= \Delta\ln\sigma/\Delta\varepsilon$) = 23; C_0 (at rest) = 180.

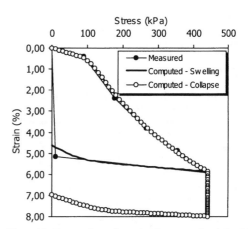

Figure 10. Computed one-dimensional stress-strain relationship of collapsing and swelling bentonite powder.

It is apparent that computed swelling values are lower than in reality, however, the computation of soil collapse is good. The material tested induces these differences. The macro-particles of powder (d ≅ 0.01 mm) are composed of micro-particles of bentonite (d ≅ 0.001 mm). The contact water suction is indirectly related to the particle diameter at the same water content. Therefore, the internal suction and cohesion of macro-particles should be at least 10 times higher than the suction and cohesion of the powder. That is why the maximum measured swelling pressure of bentonite powder is many times higher than the suction derived from shear strength.

5 CONCLUSIONS

There are certain limits of water content for each compaction pressure (dry density) of the soil within which the shear strength is maximal. The optimum water content decreases with the increasing compaction (consolidation) pressure. For example, bentonite powder with water content of 7% compacted by 600 kPa had a cohesion of approx. 15 kPa (optimum water content ≅ 30%) whereas the same powder compacted by 100 MPa had a cohesion of approx. 2.5 MPa. Dense soils need smaller volume of (contact) water to connect more closely distributed particles than loose soils. Therefore, the optimum water content of dense soils is lower and the suction and shear strength is higher than those of loose soils are. However, due to suction, soils with low water content have to be compacted using high pressures.

Low water content can be reached using high temperatures. However, without synchronous compression, shrinkage cracks may originate with negative influence on the soil strength.

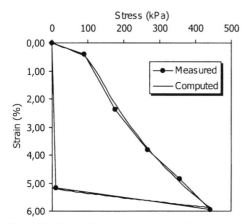

Figure 9. Computed and measured one-dimensional stress-strain relationship of swelling bentonite powder.

Swelling is the result of the pore water suction decrease after wetting. If volume changes are restricted, the effective stress changes are restricted too and the decrease of the (internal) pore water suction has to be changed by the increase of the (external) normal stress, i.e. the swelling pressure.

Swelling and collapse are mostly observed and measured (in the oedometer) in an anisotropic state of stresses. In this case, the most important is a reduction of cohesion induced by the pore water suction decrease. The soil collapse can be explained only by the loss of shear strength. After wetting, the suction and cohesion decrease and a new state of stress, at rest, has to be reached for the soil. Overconsolidated soils swell, normally consolidated soils collapse. Normally consolidated granular material with overconsolidated particles may collapse at first and then it may swell.

The state at rest of soils is understood as a state of stresses, in which the mobilized shear strength of the soil is so low that the strength of most (theoretically all) interparticle contacts is not overcome. It means that shear strength is only partly mobilized.

The limit values of shear parameters at rest can be derived from the equivalence of Jaky's K_0 coefficient and the coefficient of Rankine's active stress for lower (partly mobilized) friction angle.

The limit values of stress at rest can be computed as active and passive stresses from partly mobilized strength parameters. Generally, stress at rest can be computed using partly mobilized strength parameters from φ_0 (c_0) to zero (because after isotropic consolidation, the soil is also at rest).

ACKNOWLEDGMENTS

The Czech Grant Agency, through a grant No. 103/99/1288 has supported the research, on which this paper is based.

REFERENCES

Lambe, T.W. & R.V. Whitman 1969. *Soil mechanics*. New York: John Wiley & Sons.

Myslivec, A. 1972. Pressure at rest of cohesive soils. *Proc. 5th EC MSFE*. Madrid, Spain. Vol. 1: 63-67

Unsaturated Soils for Asia, Rahardjo, Toll & Leong (eds) © 2000 Taylor & Francis, ISBN 90 5809 139 2

Combined effect of fabric, bonding and partial saturation on yielding of soils

S. Leroueil
Université Laval, Ste-Foy, Qué., Canada

P.S.de A. Barbosa
Universidade Federal de Viçosa, M.G., Brazil

ABSTRACT: CRS oedometer tests performed at different suctions on a reconstituted residual soil in both non-cemented and cemented conditions show that the combined effect of partial saturation and bonding on yield stress is larger than the sum of the separated effects of the two factors. A model of yielding, which would apply to saturated or unsaturated, and structured or non-structured soils is then proposed with reference to the fabric of the soil which can be isotropic or anisotropic.

1 INTRODUCTION

Many soils, particularly in arid regions, are unsaturated and structured (bonded). This is often the case for residual soils, loess, weak rocks and volcanic agglomerates. Surprisingly, however, few studies have examined the combined effect of partial saturation and bonding. This paper presents the results of tests performed at different matric suctions on a reconstituted soil in non-cemented and cemented conditions. There is then a discussion on yielding of soils in relation with their fabric.

2 TESTED MATERIAL AND TESTING PROGRAMME

The soil tested is a saprolite from Brazilian gneiss that has been passed on through a 0.59 mm sieve. The grain size distribution indicates 50% sand, 45% silt and 5% clay. The specimens were statically compacted at water content of 14% to a void ratio of 1.4. In one of the two series of tests, 2% cement were added to the soil before compaction and the specimens were cured during 45 days before testing.

The soil specimens were subjected to 4 types of tests, in saturated conditions and at different matric suctions up to 500 kPa: drained triaxial compression (CID) tests; constant rate of strain (CRS) oedometer tests; constant $K = (\sigma_r - u_a)/(\sigma_a - u_a)$ triaxial compression tests and K_0 triaxial tests. In these tests, σ_r is the radial or horizontal stress and σ_a is the axial or vertical stress. With the exception of the oedometer tests, most of the tests were performed on the cemented material.

3 CRS OEDOMETER TEST RESULTS

The compression curves obtained at different matric suctions on the non-cemented and cemented soils are presented in Figs. 1 and 2 respectively. Yield stresses are well defined for all tests and generally increase with suction. In some of the tests performed in unsaturated conditions, suction was decreased to zero after reaching a volumetric strain larger than 10%. For the cemented soil, the compression curves then move on the saturated one (zero suction), indicating small variability of soil specimens; for the non-cemented soil, however, after saturation, the curves show larger strains than for the initially saturated specimen under the same net stress, indicating that, in this latter case, soil or compaction conditions could have been slightly different.

The yield stresses deduced from these tests are reported on Fig. 3 as a function of matric suction. In spite of some variability, the Loading-Collapse curves (Alonso et al., 1990), called here LC_a curve in reference to the fact that yielding is reached in the axial direction, are well defined. The LC_a curve for the cemented material, which has been drawn on the basis of all oedometer and triaxial test results, evidences that the yield stress obtained at a suction of 300 kPa is small in comparison with other results.

Figure 3 may be used to specify the influence of void ratio, bonding and partial saturation on yielding. The yield stress $\sigma_{y,e}$ for the non-cemented material at zero suction reflects the influence of void ratio; the difference between the yield stresses obtained in saturated conditions for the soil, cemented and non-cemented, $\Delta\sigma_{y,st}$, reflects the influence of bonding alone; and the influence of

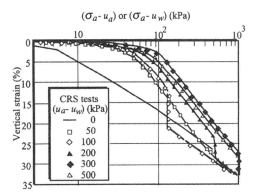

Fig. 1 - One-dimensional compression curves for non-cemented Viçosa residual soil.

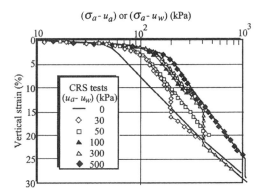

Fig. 2 - One-dimensional compression curves for cemented (2%) Viçosa residual soil.

matric suction alone is indicated by the Loading Collapse curve of the non-cemented material, $\Delta\sigma_{y,suc}$. The results show, however, that, as predicted by Alonso & Gens (1994), the yield stress resulting from the combined effect of void ratio, partial saturation and bonding is larger than the sum of the contributions of the three factors. This can be due to different distributions of water at the contacts between particles or, as mentioned by Alonso & Gens (1994), to bond strength that is influenced by water content, and indirectly suction, or to slightly different fabrics for the cemented and non-cemented soils. The yield stress of structured and unsaturated materials, σ_y, should thus be defined as the following sum, in which $\Delta\sigma_{y,st+suc}$ would be due to the interaction between bonding and suction:

$$\sigma_y = \sigma_{y,e} + \Delta\sigma_{y,st} + \Delta\sigma_{y,suc} + \Delta\sigma_{y,st+su} \qquad (1)$$

4 ISOTROPIC AND ANISOTROPIC TRIAXIAL COMPRESSION TESTS

The compression curves deduced from the isotropic and anisotropic triaxial compression tests performed at suctions varying between 0 and 300 kPa are similar to those obtained in CRS oedometer tests and presented in Fig. 2 (Barbosa, 2000; Barbosa and Leroueil, 1998). Only yield conditions will be considered here. They are shown in Fig. 4 in a (q vs (p – u_a)) diagram, with the yield stresses deduced from oedometer and K_o triaxial tests reported on the K = 0.47 line. This latter K value was observed in K_o tests, in the post-yield range.

It can be seen on Fig. 4 that, in spite of some scatter, yield stresses generally increase with suction On the same figure, yield curves represented by broken lines and corresponding to the yielding model that will be presented hereafter are also drawn.

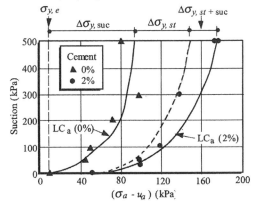

Fig. 3 - Loading-Collapse (LCa) curves for cemented and non-cemented Viçosa residual soil.

5 YIELDING OF SOILS

5.1 Yielding of soils in general

When considering soil behaviour, it is important to keep in mind the basic phenomena intervening at the level of particles or aggregates. One is that yielding mostly corresponds to inter-particles slippage, even when loading is isotropic; another one is that, when a soil is deposited, consolidated or compacted, the arrangement of particles (or aggregates), and the distribution of contacts between them, reflect the applied forces.

The first phenomenon can be examined by reference to Fig. 5 showing the contact between two particles. If the two particles are subjected to an external force with a component normal to the contact, N_n, the resistance to slippage, T_n, is equal to $N_n \tan\Phi'$ for a cohesionless material (Fig. 5a). If the same contact is bonded (Fig. 5b), the resistance to slippage is increased by the strength of the bond, T_{st}.

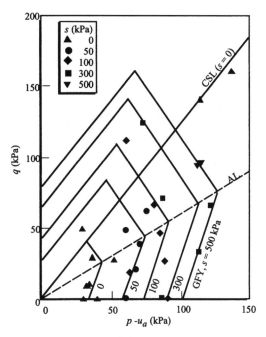

Fig. 4 - Yielding and GFY curves for cemented Viçosa residual soil.

If the soil is partially saturated, there is some water surrounding the contact, a capillary force, N_{suc}, normal to the contact and a resistance to slippage, $T_{suc} = N_{suc} \tan\Phi'$. The total resistance to slippage could simply be the sum of T_n, T_{st} and T_{suc}. However, as previously indicated, bonding and partial saturation may interact and result in a fourth component of resistance to slippage, T_{st+suc}.

$$T = T_n + T_{st} + T_{suc} + T_{st+suc} \qquad (2)$$

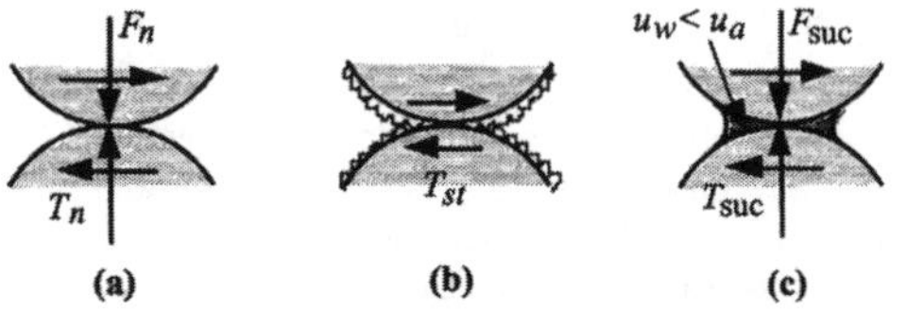

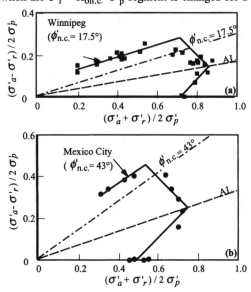

Fig. 5 - Resistance to slippage at a contact between particles due to: (a) external force; (b) bonding; and (c) partial saturation.

This is for a single contact. For a soil specimen or soil deposit, yielding results from slippage at many contacts, with characteristics that should reflect the distribution of these contacts. To examine this aspect, we will consider orthotropic soils only, with the same properties in all horizontal directions, and begin with saturated soils. When such a soil is normally consolidated under radial, σ'_{rm}, and axial,

σ'_{am}, effective stresses, there is a particle or aggregate arrangement, and a contact distribution so that the soil skeleton can resist to these effective stresses in the corresponding directions. In a stress diagram, the line reflecting this induced anisotropy is called here the anisotropy line (AL) and is characterised by $K_{AL} = \sigma'_{rm} / \sigma'_{am}$.

Stating that "a soft clay cannot be submitted to stresses in any direction higher than the previous stresses in that direction without yielding", Larsson (1977) proposed that yield curves of natural clays be schematised by four segments: two corresponding to the strength envelopes in compression and in extension; and two corresponding to $\sigma'_a = \sigma'_p$ and $\sigma'_r = K_{o.n.c.}\sigma'_p$. In these equations, σ'_p is the preconsolidation pressure of the clay and $K_{o.n.c.}$ is the coefficient of earth pressure at rest for the normally consolidated soil. When soft clays are considered, it is generally accepted that $K_{o.n.c.}$ is close to $(1 - \sin \Phi'_{n.c.})$, in which $\Phi'_{n.c.}$ is the friction angle of the normally consolidated soil.

Yield curves have been determined on natural clays from different origins and a compilation made by Diaz-Rodriguez et al. (1992) confirms that their shape reflects the anisotropy prevailing during deposition. This is evidenced on Fig. 6 showing the yield curves of Winnipeg and Mexico City clays, two clays presenting extreme $\Phi'_{n.c.}$ values of 17.5° and 43° respectively.

Analysis of the data gathered by Diaz-Rodriguez et al. (1992) shows that a good fit between observed yield curves and the schematic yield curves suggested by Larsson (1977) is obtained when the $\sigma'_r = K_{on.c.} \sigma'_p$ segment is changed for a

Fig. 6 - Yielding and GFY curves of natural saturated clays (after Diaz-Rodriguez et al., 1992).

$\sigma'_r = K^* \sigma'_p$ segment, with K^* being slightly larger than $K_{o.n.c.}$. The following expression is proposed by Leroueil (2000):

$$K^* = 0.85 - \sin(0.46 \, \Phi'_{n.c.}) \qquad (3)$$

The validity of this approach is confirmed by Fig. 6 where the schematised yield curves based on Eq. 3 are shown with broken lines for Winnipeg and Mexico City clays. It is not clear why K^* is higher than $K_{o.n.c.}$, but it is possibly due to an increase of K_o with secondary consolidation (Mesri & Castro, 1987). Yield curves of soils are certainly more rounded than those given by the model, but this latter represents a simple and useful approximation. K^*, as defined by Eq. 3, is K_{AL} for natural clays.

Diaz-Rodriguez (1992) and Leroueil (2000) show that the shape of yield curves of natural clayey soils depends on $\Phi'_{n.c.}$ and is essentially independent of the degree of structuring. This has to be associated to the fact that bonding applies at the contacts and that its global effect reflects the distribution of these contacts. An important consequence is that K_{AL} and the corresponding anisotropy line are not significantly influenced by structuring.

5.2 Yielding of unsaturated cohesionless soils

Considering yield curves of unsaturated soils, experience is rather limited (Zakaria et al., 1995; Maâtouk et al., 1995; Cui & Delage, 1996; Barbosa & Leroueil, 1998; Tang & Graham, 1998; Machado & Vilar, 1999). Figures 8 to 10 respectively show yield points obtained at different suctions by Zakaria et al. (1995), Machado & Vilar (1999) and by Cui & Delage (1996).

As previously indicated, matric suction generates a resistance to slippage at the contacts between particles (or aggregates). Its global effect should thus reflect the distribution of these contacts, and the anisotropy line (AL) should be the same as for the saturated soil. This is confirmed by the previously mentioned studies which show that yield curves obtained at different suctions essentially have the same shape (see Figs. 4, 8, 9 and 10).

If a soil in saturated conditions has a yield curve such as $OB_oA_oD_o$ in Fig. 7a, a given suction should extend the cap $B_oA_oD_o$ to $B_sA_sD_s$, with an increase in axial yield stress from σ^*_{ayo} to σ^*_{ays} and an increase in radial yield stress from σ^*_{ryo} to σ^*_{rys}. The variation of the axial and radial yield stresses with suction defines two Loading-Collapse curves, LC_a and LC_r respectively (Fig. 7b). However, the anisotropy line, AL, remaining the same independently of suction, the ratio $\sigma^*_{rys} / \sigma^*_{ays}$ at any suction is constant and equal to K_{AL}.

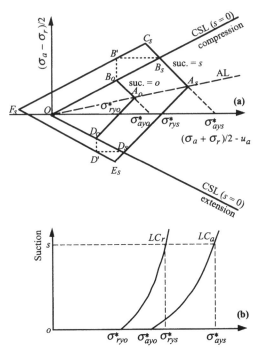

Fig. 7 - Description of the Given-Fabric-Yield (GFY) model

By increasing the axial yield stress from σ^*_{ayo} to σ^*_{ays}, matric suction increases the shear strength of the soil in compression from point B_o to point B_s, or from B_o to B', in Fig. 7a. It can be easily demonstrated that this compression strength increase is equivalent to a cohesion in compression due to suction, $c_{c,suc}$, defined as follows:

$$c_{c,suc} = \left(\sigma^*_{ays} - \sigma^*_{ayo}\right)\frac{\sin\Phi' \cos\Phi'}{1 + \sin\Phi'} \qquad (4)$$

This equivalent cohesion depends only on suction and on the distribution and number of contacts. So, as far as this distribution remains the same, which should be the case in non-expansive soils for net stresses smaller than yield stresses σ^*_{ays} and σ^*_{rys}, $c_{c,suc}$ should be a constant, as indicated on Fig. 7a. According to Eq. 4, this cohesion is typically 27% of the increase in axial yield stress due to suction.

In extension, the shear strength increase due to a given suction is associated to an increase in radial yield stress from σ^*_{ryo} to σ^*_{rys} (Fig. 7a). This corresponds to an equivalent cohesion in extension, $c_{e,suc}$, equal to:

$$c_{e,suc} = \left(\sigma^*_{rys} - \sigma^*_{ryo}\right)\frac{\sin\Phi' \cos\Phi'}{1 + \sin\Phi'} \qquad (5)$$

As in compression, if fabric remains the same for net stresses smaller than yield stresses, the cohesion in extension due to suction should be constant, as shown in Fig. 7a.

From Eqs. 4 and 5, it appears that cohesions in compression and extension due to suction are generally different. The ratio $c_{e,suc} / c_{c,suc}$ is equal to K_{AL}, and takes a value of 1.0 only for soils having an isotropic fabric.

The model previously described is called Given-Fabric-Yield (GFY) model to insist on the fact that it is closely associated to the hypothesis that fabric remains the same when there are changes in suction or net stresses. If there is a reduction in the number of contacts between particles when decreasing net stresses, the cohesion due to suction should decrease; also, if fabric is modified by loading the soil outside the yield curve or by shearing, the effect of suction should change accordingly. The GFY model has been described for cohesionless materials. It can, however, be easily extended to structured materials, the reference being in all cases the yield curve in saturated conditions.

The GFY model previously described has been applied to the sets of data provided by Zakaria et al. (1995) on Fig. 8, Machado & Vilar (1999) on Fig. 9, and Cui & Delage (1996) on Fig. 10. The experimental results and comparison with the GFY model are briefly described hereunder.

Zakaria et al. (1995) tested a kaolin, compacted and then subjected to an isotropic stress history. Figure 8 shows points obtained at yielding under suctions of 0, 100 and 200 kPa as well as the critical state lines obtained for the same suctions. It can be seen that the proposed model well fit the data points when assuming an isotropic AL reflecting the stress history of the material.

Machado & Vilar (1999) tested a natural residual soil from Brazil. The yield points are shown in Fig. 9. They are perfectly fitted by the GFY model, assuming an isotropic AL, which is in agreement with previous experience with residual soils (see Leroueil and Vaughan,1990).

Maâtouk et al. (1995) and Cui & Delage (1996) tested compacted silty soils and both observed yield curves reflecting some anisotropy. This is evidenced on Fig. 10 showing the results obtained by Cui & Delage (1996) on the compacted Jossigny silt at suctions from 200 to 1500 kPa. The GFY model has been applied to get a good "average fit". K_{AL} is in the order of 0.45 in this case. Comparison between the model and experimental results is not very good; in particular, the model overestimates the variations of the isotropic yield stress and the increase in strength with suction. Two remarks can, however, be made: the first one is that yield stresses were not easily defined and reflect some uncertainty; the second one is that, if the GFY model would have been based mostly on isotropic yields, the comparison between the strength envelopes defined from the model and measured strengths would have been much better.

Finally, the GFY curves shown in Fig. 4 were established on the basis of the test results obtained on the Viçosa residual soil (K_{AL} appears to be close to 0.58 in this case). They relatively well fit the data points below the critical state line associated to saturated conditions. On the other hand, and

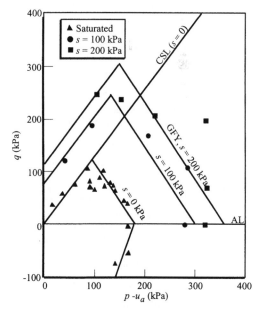

Fig. 9 - Yielding and GFY curves for a natural residual soil from Brazil (after Machado and Vilar, 1999).

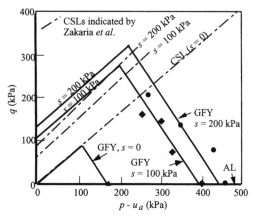

Fig. 8 - Yielding and GFY curves for compacted kaolin after isotropic consolidation (after Zakaria et al., 1995).

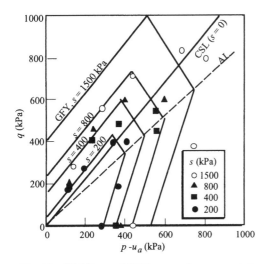

Fig. 10 - Yielding and GFY curves for compacted Jossigny silt (after Cui and Delage, 1996).

unfortunately, there is almost no strength measured in the pre-yield range.

6 DISCUSSION AND CONCLUSION

CRS oedometer tests have been performed at different matric suctions on a reconstituted residual soil in both non-cemented and slightly cemented conditions. The results show that, as predicted by Alonso & Gens (1994), the combined effect of partial saturation and bonding on the axial yield stress is larger than the sum of the separated effects of the two factors.

A model of yielding, which would apply to saturated or unsaturated, and structured or non-structured soils is proposed. This model (GFY model) emerges from considerations on fabric that can be isotropic or anisotropic, and assumes that the arrangement of contacts between particles is not influenced by bonding or partial saturation. This model is certainly idealised and approximate. In particular, yield curves of soils are certainly more rounded than those given by the model. It is thought, however, that it provides a rational physical framework for understanding yielding of soils.

The input parameters of the GFY model can be defined by oedometer tests performed at different suctions to determine the axial Loading-Collapse (LC_a) curve and a few isotropic triaxial compression tests, or oedometer tests on specimens trimmed vertically, at different suctions to define the fabric anisotropy characterised by K_{AL}.

The model gives representative results for unsaturated soils, at least below the critical state line of the saturated material. It still has to be confirmed for strength envelopes in the pre-yield stress range, mostly because of the lack of data in this domain. In fact, if several levels of strength have been defined in saturated soils (peak including dilatancy; peak in the pre-yield domain; critical state; residual state; etc.), it seems that we are far from this refinement with unsaturated soils. It is thought that the strength envelope provided by the GFY model is a peak strength envelope associated to the fabric of the soil within the yield curve. It could be different from the critical state strength envelope which, at large strains, certainly corresponds to a different fabric.

7 REFERENCES

Alonso, E.E. & Gens, A. 1994. Keynote Lecture: On the mechanical behaviour of arid soils. *Proc. Conf. on Engineering Characteristics of Arid Soils*, London, pp. 173-205.

Alonso, E.E., Gens, A. & Josa, A. 1990. A constitutive model for partially saturated soils. *Géotechnique*, 40(3): 405-430.

Barbosa, P.S. de A. 2000. Ph.D. Thesis in preparation.

Barbosa, P.S. de A. & Leroueil, S. 1998. Strength and compressibility of a reconstituted and slightly cemented saprolitic soil. *Proc. 2nd Int. Conf. on Unsaturated Soils*, Beijing, Vol. 1: 1-6.

Cui, Y.J. & Delage, P. 1996. Yielding and plastic behaviour of an unsaturated compacted silt. *Géotechnique*, 46(2): 291-311.

Diaz-Rodriguez, J.A., Leroueil, S. & Aleman, J.D. 1992. Yielding of Mexico City clay and other natural clays. *J. Geotech. Engrg. Div.*, ASCE, 118(7): 981-995.

Larsson, R. 1977. Basic behaviour of Scandinavian soft clays. *Swedish Geotech. Inst.*, Report No 4.

Leroueil, S. 2000. Paper in preparation.

Leroueil, S. & Vaughan, P.R. 1990. The general and congruent effects of structure in natural soils and weak rocks. *Géotechnique*, 40(3): 467-488.

Maâtouk, A., Leroueil, S. & La Rochelle, P. 1995. Yielding and critical state of a collapsible unsaturated silty soil. *Géotechnique*, 35: 465-477.

Machado, S. L. & Vilar, O.M. 1999. Personal communication.

Mesri, G. & Castro, A. 1987. C_α/C_c concept and K0 during secondary compression. *J. Geotech. Engrg. Div.*, ASCE, 113(3): 230-247.

Tang, G.X. and Graham J. 1998. On yielding behaviour of an unsaturated sand-bentonite mixture. *2nd Int. Conf. on Unsaturated Soils*, Beijing, Vol.1:149-154.

Zakaria, I., Wheeler, S.J. and Anderson, W.F. 1995. Yielding of unsaturated compacted kaolin. *Proc. 1st Int. Conf. on Unsaturated Soils*, Paris, Vol.1: 223-228.

Unsaturated Soils for Asia, Rahardjo, Toll & Leong (eds) © 2000 Taylor & Francis, ISBN 90 5809 139 2

Suction and strength behaviour of unsaturated calcrete

N.L.Ling
Mineral and Geoscience Department of Malaysia, Kuching, Malaysia

D.G.Toll
NTU-PWD Geotechnical Research Centre, Nanyang Technological University, Singapore

ABSTRACT: This paper describes a laboratory investigation of the behaviour of a construction material, calcrete when in an unsaturated state. The work has concentrated on a Powder Calcrete, which is essentially a very gravelly, fine to medium, calcified sand. Samples were compacted according to the British Standard heavy condition at a range of moisture contents below optimum moisture content. Suction measurement by the filter paper method, as well as strength measurement by the constant water content triaxial compression test were performed. The two sets of test results obtained support the concept that the strength is controlled by the inter-relationship between suction and degree of saturation.

1 INTRODUCTION

Conventionally, only clean graded aggregates, generally in the form of crushed rock are used in the construction of road bases. However, in many tropical countries, good quality rock for crushing is often scarce. Even if the rock is available, the cost of transportation and crushing may render many road construction projects unviable. In such cases, very often maximizing the use of on-site locally available materials would offer valuable options.

Calcretes, which are generally of low grade materials occurring naturally in huge deposits, have been demonstrated to perform successfully as road base construction materials in low-volume bituminous-surfaced roads in Botswana (Lionjanga et al, 1987). The ability to understand the material's behaviour can lead to the maximum use on-site, ensuring the roads are constructed to a satisfactory standard of design and longevity with the minimum cost (Toll, 1991).

This paper describes a laboratory investigation of the behaviour of compacted Botswana Powder Calcrete, BG7 when in unsaturated state. The testing programme included soil classification tests, and an attempt to measure the suction pressure quantitatively by the filter paper method on samples of varying moisture contents within the range of 4.5-10 %. A series of constant water content triaxial compression tests on samples within the same moisture content range was also carried out.

2 THE MATERIAL

The calcrete material studied in this investigation, designated as BG7 (Lionjanga et al, 1987) is a Powder Calcrete, essentially a calcified sand with weak cementation between the grains of the host Kalahari sand. The particle size analysis of Calcrete BG7 conducted in accordance with BS 1377 (1990): Part 2: 9.3 - the dry sieving method identifies the calcrete as a very gravelly, fine to medium SAND (Fig. 1). Each of the curves in Figure 1 shows a distinctive kink in the medium sand size range. Other properties are shown in Table 1.

Table 1. Some properties of Calcrete BG7.

Property	Value
Liquid limit	34 – 39 %
Plastic limit	19 – 23 %
Plastic index	14 – 17 %
Specific gravity	2.66

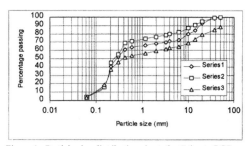

Figure 1. Particle size distribution charts for Calcrete BG7.

3 SAMPLE PREPARATION

A compaction test was first carried out in accordance with BS 1377 (1990): Part 4: 3.5, the BS heavy compaction test to obtain a dry density-moisture content curve that formed the targeted compaction curve (Fig. 2) for the subsequent sample preparation. The optimum moisture content was found to be 12.3 % with a maximum dry density of 1.87 Mg/m^3 (Ling, 1998).

Samples for suction measurement were formed by statically compacting a pre-calculated amount of the material and moisture mixture to make up the targeted moisture content and density. The material was compacted, as a single layer, into a cylindrical sample mould of internal diameter 106 mm, to a thickness of 20 mm to form a disc-shaped test sample of approximately 175 cm^3 in volume. Six sets of 12 samples each, totaling 72 samples, were produced and their dry densities and moisture contents are plotted in Figure 2.

The triaxial test samples were prepared in a similar manner to the samples for suction measurement tests. A cylindrical 3-segment split mould was used in this case. To ensure a uniform distribution of density throughout the volume, the samples were statically compacted in ten layers of equal thickness, using a compression machine. The resultant cylindrical samples formed were of 103 mm diameter and 200 mm height with an approximate volume of 1667 cm^3. A total number of 12 triaxial test samples were prepared and the dry densities and moisture contents of the samples are also shown in Figure 2.

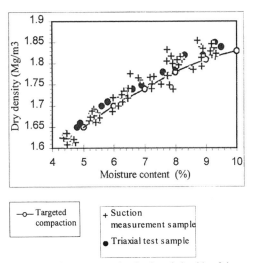

Figure 2. Moisture content-dry density relationship of the test samples prepared.

4 SUCTION MEASUREMENT

4.1 Filter paper method

The matric suction, which is closely related to the surrounding environment changes, is usually of interest in analyzing geotechnical engineering problems. The filter paper method, which is capable of measuring a wide range of suction, was chosen for this testing programme to measure the matric suction of the Calcrete BG7 material. Over the recent years, various attempts have been made, with a good degree of success; to use the filter paper method in geotechnical engineering especially in cases involving non-granular soils (Fredlund & Rahardjo, 1993).

The filter paper method of measuring suction is an indirect method in which the moisture content of the filter paper is measured and related to the suction through a calibration against known values of suction. It is based on the assumption that a filter paper, which is used as a sensor in this case, if allowed to absorb moisture from a soil specimen long enough to achieve a state of equilibrium, the suction in the filter paper will be equal to the suction in the soil.

For each of the test samples prepared, 3 air-dry, 90 mm-diameter, disc-sized, Whatman No. 42 filter papers were placed in direct contact with the top face (hereby named Filter Papers A, B and C from the top) and another 3 directly below the bottom face (hereby named Filter Papers D, E and F) to allow capillary flow of water to the filter papers as illustrated in Figure 3. In this type of arrangement, the middle filter papers in the stacks, that is, Filter Papers B and E, were used for the suction determination. The outer filter papers (Filter Papers A & C and Filter Papers D & F) were primarily used to protect the middle filter papers from soil contamination.

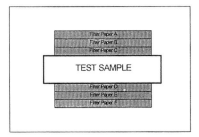

Figure 3. Contact filter paper method for measuring matrix suction.

Once the filter papers were put in place on the test sample, the set-up was wrapped with at least three layers of clear cellophane film and sealed in an airtight plastic bag to prevent moisture loss. To minimize the temperature variations during equilibration, all the sealed set-ups were stored in insulated plastic boxes which were kept in a closed cupboard in the laboratory. The details of the testing schedule for each set of the samples are shown in Table 2.

Table 2. Schedule of filter paper moisture content determination.

Sample No.	Equilibration time (days)					
	2/3	7	14	21	35	42
1	Y	Y	Y	Y	Y	N
2	Y	Y	Y	Y	Y	N
3	Y	Y	Y	Y	Y	N
4	Y	Y	Y	Y	Y	N
5	N	Y	Y	Y	Y	N
6	N	Y	Y	Y	Y	N
7	N	N	Y	Y	N	Y
8	N	N	Y	Y	N	Y
9	N	N	N	Y	N	Y
10	N	N	N	Y	N	Y
11	N	N	N	N	N	Y
12	N	N	N	N	N	Y

At the designated equilibrium times according to the schedule, the middle filter papers were weighed using an accurate balance with a readability of 0.0001 g. The weighings were carried out at a constant laboratory room temperature of between 22^0-23^0 C. At the end of the test, each of the middle filter papers was dried in an oven at a temperature of 105^0 C for a minimum of one hour in order to obtain the dry mass of the filter paper. The equilibration moisture contents of the filter papers for all the equilibration periods were computed from the differences between the dry masses of the filter papers and the wet masses of the respective filter papers that were obtained at the respective equilibration times.

The equilibration suction was obtained by correlating the measured equilibration moisture content of the filter paper with a widely accepted Whatman No. 42 filter paper calibration shown in Table 3.

Table 3 Whatman No. 42 filter paper calibration (after Chandler et al, 1992)

Moisture content of filter paper	$Log_{10}\left[suction(kPa)\right]$
< 47 %	4.842-0.0622.(w)
> 47 %	6.050-2.48.log(w)

(w) = Filter paper moisture content (%)

4.2 Test results

Filter Papers B and E, which were respectively the middle filter papers above and below the test sample, gave consistent, similar moisture content readings during each of the measurements (Ling, 1998).

For each set of the test samples prepared, the moisture contents of Filter Papers B and E were plotted against the equilibration time as shown in Figure 4a-f. It can be concluded from the graphs that it generally takes 14-21 days for the filter papers to achieve satisfactory equilibration with the test samples. However, equilibration time of 14 days would generally be sufficient as most of the graphs start to flatten off from the fourteenth day onwards.

The average of the moisture contents of Filter Papers B and E from the fourteenth day onwards was used to compute the suction of the test sample. The resulting suction values for all the 72 test samples were plotted against the moisture contents of the test samples in Figure 5. The plot suggests a bi-linearity with a change in sensitivity occurring at a sample moisture content of 7.6 % (or 62 % of optimum moisture content).

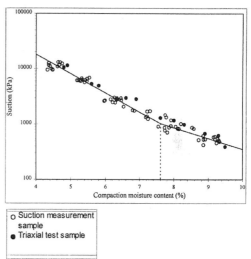

Figure 5. Plot of suction value obtained versus compaction moisture content of the Calcrete BG7 sample.

4.3 Discussion

Although the suction determined by the filter paper method in this test programme had not been compared with the suction values that would be obtained by using other devices, this test programme has inevitably established a reasonable degree of confidence in the quantitative estimation of suction in the Calcrete BG7 material. Moreover, the consistency in the test results obtained from the large number of tests in this test programme has also demonstrated quite convincely the possible applicability of the filter paper technique in deriving soil suction in granular material as a whole.

Nevertheless, this filter paper technique is highly user-dependent, and extreme care must be exercised when measuring the small masses associated with the filter papers. For example, a tiny speck of soil particles sticking to the filter paper can greatly affect the result. Also if the mass of the filter paper were not determined immediately in as short time as possible after it is taken out from the wrapping and es-

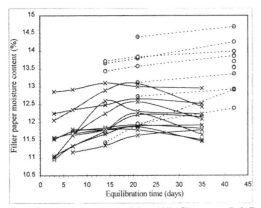

Figure 4a. Moisture contents of the middle filter papers B & E against equilibration time for Calcrete BG7 samples with moisture content 4.3 – 4.7 %.

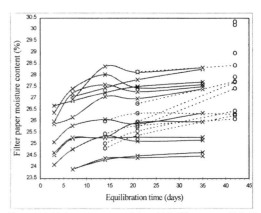

Figure 4d. Moisture contents of the middle filter papers B & E against equilibration time for Calcrete BG7 samples with moisture content 6.5 – 7.9 %.

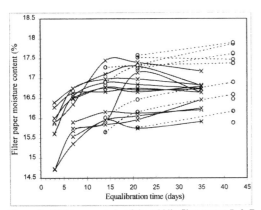

Figure 4b. Moisture contents of the middle filter papers B & E against equilibration time for Calcrete BG7 samples with moisture content 5.2 – 5.5 %.

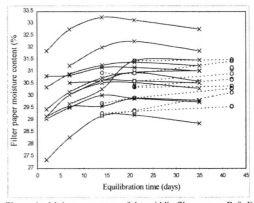

Figure 4e. Moisture contents of the middle filter papers B & E against equilibration time for Calcrete BG7 samples with moisture content 7.7 – 8.6 %.

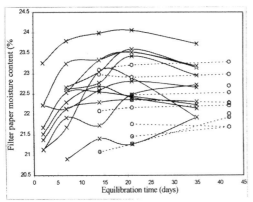

Figure 4c. Moisture contents of the middle filter papers B & E against equilibration time for Calcrete BG7 samples with moisture content 6.0 – 6.5 %.

(Note: The full line and dash line curves represent respectively samples 1 – 6 and 7 – 12 as shown in Table 2.)

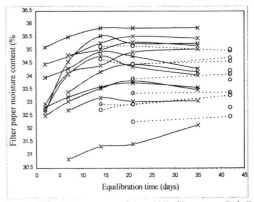

Figure 4f. Moisture contents of the middle filter papers B & E against equilibration time for Calcrete BG7 samples with moisture content 8.9 – 9.3 %.

pecially so if taken out from the oven, the filter paper would lose or gain moisture to or from the atmosphere, thereby affecting the final derivation of suction.

It is very interesting to note the bi-linearity of the suction graph with sensitivity changes occurring at about 7.6 % moisture content. The lower part of the curve represents the higher moisture content range of the calcrete samples. The upper part of the curve represents the calcrete samples of lower moisture content range where significant amount of suction increases had been observed.

During the course of the laboratory testing, some degree of adherence between the filter papers and also between the filter paper and soil sample were observed in the two wetter groups of the tests (i.e. the groups with test samples of moisture content 7.7 - 8.6 % and 8.9 - 9.3 %). The opposite was observed in the cases of the drier test samples; the filter papers were loose and did not adhere to each other nor to the test samples very much.

In the wetter test samples, the filter papers were held by the surface tension of the water in the sample and hence the moisture contents absorbed by the filter papers were predominantly through the liquid phase by capillary suction. Therefore, it could be assumed that it was the matric suction being measured in this case.

As for the drier samples, the surface tension was probably negligible as the water in the test sample might have retreated into the finer pores, thereby causing the menisci as well to recede completely inside the soil aggregates. Therefore, although the filter papers were kept in contact with the test samples, they were not necessarily in intimate contact with the soil water. The absorption in this case was likely to be predominantly through the vapour phase. Hence, increasingly it was the total suction instead of matric suction being measured; causing the upward deviation of the curve as the moisture content of the sample decreased.

5 STRENGTH MEASUREMENT

5.1 Constant water content test

The constant water content triaxial compression test, or CW triaxial test, had been selected to measure the shear strength of the Calcrete BG7 material. It was assumed that the strength measured in this way would have relevance to the drainage conditions being simulated in the actual road conditions. In carrying out the constant water content test, each of the samples was first consolidated to an equilibrium condition under an applied isotropic confining pressure of 50 kPa before the shearing, with the pore-air phase kept in a drained mode while the pore-water

phase was kept undrained, throughout both the stages. All the samples were sheared at a strain rate of 1.27×10^{-3} % per second until an axial strain of at least 20 % was achieved. With this strain rate, it generally took about 4.5 hours for the shearing of the sample to reach the 20 % strain.

5.2 Test results

The maximum deviator stress in each test, which is an indicator of the shear strength of the soil, has been plotted against the moisture content of the calcrete sample in Figure 6. A reasonably best-fit curve can be drawn through the points plotted as shown in the figure.

It can be seen that, the samples just dry of optimum moisture content gave increasingly higher shear strengths. The rate of increase in strength, however, decreased and below a moisture content of about 8.2 % (about 67 % of the optimum moisture content), instead of achieving higher shear strengths with decreasing moisture content, the samples registered drastic drops in their strengths.

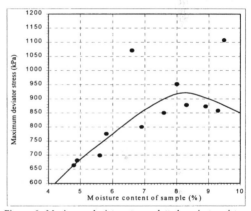

Figure 6. Maximum deviator stress plotted against moisture content of the Calcrete BG7 sample.

5.3 Discussion

Similar suction measurements were performed on the triaxial samples to measure the matric suction. The filter paper moisture contents registered after equilibration period of at least 15 days were translated into suction and are plotted together with those of the suction measurement samples in Figure 5. The measurement results from the triaxial samples were in fairly good agreement with that of the suction measurement samples except that the wetter triaxial samples seemed to give slightly lower filter paper moisture content which in turn were translated into slightly higher suction pressure. Nevertheless, the results from the triaxial samples still fall into the permissible limits and therefore the generalized bilinear curve for suction pressure can also be applied to the triaxial samples as well.

The strength measurement tests carried out had shown a maximum strength in samples compacted at a moisture content of about 8.2 %. This value is in fact very close to the moisture content value of the point where the suction pressure shows a change in sensitivity.

At moisture contents below optimum moisture content, the suction increases while the degree of saturation reduces. The suction is largely due to the matric suction in the form of the surface tension force at the air-water interface, being likened to the contractile skin by Fredlund & Rahardjo (1993). This air-water interface acts like a thin rubber membrane, pulling the particles together; thus increasing the shear strength of the soil.

Toll (1991) has pointed out the importance of fabric and soil suction in determining the unsaturated soil behaviour. The soil fabric, which is the geometrical arrangement of the particles, is held together by the suction in unsaturated soils. There are evidences from the previous studies (e.g. Zein, 1985) showing the development of an aggregated fabric in materials compacted dry of optimum moisture content. This type of aggregation was, however, not found to exist in the materials compacted on the wet side of optimum moisture content.

The aggregation will act like large individual particles if the suction is large enough to prevent breakdown (Toll, 1990). It is possible that this fabric would not be destroyed by shearing to large strains. Hence, as the suction increases while the degree of saturation decreases, the suction comes to play an increasingly important role in internally supporting the aggregations rather than pulling the aggregations together.

A maximum strength occurs around 67 % optimum moisture content where significant suction is present, and yet the degree of saturation is still sufficiently high for the suction to contribute to the strength of the sample.

Below 7.6 % which is about 62 % of optimum moisture content, the strength of the calcrete started to drop drastically despite the fact that the suction has started to deviate upward to even higher levels. It is possible that the suction contribution to strength has become less significant at this stage (Toll, 1990). The water phase in samples below this cut-off moisture content has started to recede into the finer pores within the soil aggregates and no longer affects the inter-aggregate contacts where shear is taking place. Hence, the shear behaviour at this stage has become increasingly dominated by the dry particle contacts only.

6 CONCLUSION

This test programme has shown, with a high degree of success, the use of filter paper method in determining the soil suction in a granular material such as a Powder Calcrete. The Calcrete is a highly gravelly, fine to medium sand and the test was carried out on the unsaturated samples being compacted to British heavy compaction standard.

In setting out to measure the matric suction of the unsaturated calcrete material however, it was found that in keeping the filter paper in direct contact with the test sample had not necessarily enabled the capillary equilibrium of the moisture content between the filter paper and the test sample. It is suggested that as the moisture content of the calcrete reduced to about 7.6 %, the filter paper started to equilibrate increasingly with the vapour pressure, resulting in measuring the total suction instead of matric suction. This change in the different types of suction is suggested by the bi-linearity of the suction curve obtained.

The strength measurement tests on the unsaturated samples have demonstrated that the strength of unsaturated Calcrete BG7 is controlled by the interrelationship between suction and degree of saturation of the material. The highest strength occurs when the degree of saturation is still sufficiently high to allow the increasing suction to contribute to the strength.

REFERENCES

British Standard 1377 1990. *Methods of Test for Soils for Civil Engineering Purposes.* British Standards Institution, London.

Chandler, R.J., Crilly, M.S. & Montgomery-Smith, G. 1992. A low cost method of assessing clay dessication for low-rise buildings. In *Proc. Inst. Civil Engineers: Civil Engineering* 92(2): 82-89.

Fredlund, D.G. & Rahardjo, H. 1993. *Soil Mechanics for Unsaturated Soils.* John Wiley and Sons Inc.

Ling, N.L. 1998. *Unsaturated calcrete behaviour.* MSc dissertation, Advanced Course in Engineering Geology, School of Engineering, University of Durham.

Lionjanga, A.V., Toole, T. & Greening, P.A.K. 1987. The use of calcrete in paved roads in Botswana. In *Proc. 9th African Regional Conf. Soil Mech. Found. Engg., Lagos,* vol. 1: 489-502. Rotterdam: Balkema.

Toll, D.G. 1990. A framework for unsaturated soil behaviour. *Geotechnique,* 40, No. 1: 31-44.

Toll, D.G. 1991. Towards understanding the behaviour of naturally-occurring road construction materials. *Geotechnical and Geological Engineering,* 9: 197-217.

Zein, A.K.M. 1985. *Swelling characteristics and microfabric of compacted Black Cotton Soil.* PhD thesis, University of Strathclyde.

Unsaturated Soils for Asia, Rahardjo, Toll & Leong (eds) © 2000 Taylor & Francis, ISBN 90 5809 139 2

Soil behaviour in suction controlled cyclic and dynamic torsional shear tests

C. Mancuso, R. Vassallo & A. d'Onofrio
Dipartimento di Ingegneria Geotecnica, Università di Napoli Federico II, Italy

ABSTRACT: The paper reports results of shear stiffness measurements performed with a resonant column-torsional shear device working in suction controlled conditions. Emphasis lays on the effects of compaction-induced fabric and suction on soil behaviour. Collected data indicate a significant variation of stiffness with moulding water content and show an S-shaped stiffness versus suction variations. This is explained with a progressive change of soil behaviour from a bulk- to a menisci-water regulated response. A model is proposed to account for the observed behaviour.

1 INTRODUCTION

Up to date research on unsaturated soils focused on the definition of constitutive relationships for deformation and failure problems of natural and reconstituted materials (Fredlund & Rahardjo, 1993) and on the development of critical-state models capable of fitting experimental observations (Alonso et al., 1990). These latter researches mainly concentrate on reconstituted clayey soils (Alonso et al. 1990; Wheeler, 1996), while the response of soils as construction materials has been rarely explored in suction controlled conditions (Rampino et al., 1999a; Vinale et al., 1999). Despite the clear need for analysing small strain behaviour of unsaturated soils (Wu et al., 1989; Quin et al., 1991), little attention has been devoted to the subject. Only in recent years have interesting contributions been given by Marinho et al. (1995), Picornell & Nazarian (1998) and Cabarkapa et al. (1999).

A step toward the analysis of this aspect of soil behaviour has been taken at the Dipartimento di Ingegneria Geotecnica of Naples, where a resonant column-torsional shear device working under suction controlled conditions has been developed (Vinale et al., 1999). The paper reports some of the results obtained with this new device, concentrating on the effects of moulding water content and suction on small strain stiffness of compacted soils.

2 WATER-SOIL INTERACTION

Compacted soils utilised in earth structures are in unsaturated state during construction and, in several

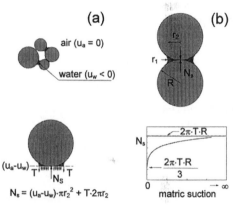

Figure 1. Effect of meniscus water.

cases, also during working stages. Their behaviour is thus affected by water and air in the pores, which make Terzaghi's principle no longer valid and require the use of two stress variables (Bishop & Blight, 1963; Fredlund & Morgernstern, 1977): net stress tensor $|\sigma-u_a|$ and matric suction $s=(u_a-u_w)$.

In order to explain such requirement, reference can be made to the simple model of Fig. 1a, taken from Fisher (1926). According to equilibrium consideration it is easy to demonstrate that the meniscus water at the spheres' contact induces a force (N_s) normal to the plane orthogonal to the line connecting their centres and passing through the contact. This force is the only one arising from menisci water, and increases towards a threshold value as the difference s between pore air and water pressures (i.e. matric suction) increases (Fig. 1b). Thus, the effect of matric suction results in an increase in the normal force tightening the spheres and, if the sphere contact behaves frictionally, in an increase of the limiting slippage strength between them. This is why suction affects the behaviour of unsaturated aggregates of spheres, inducing a stiffer and more resistant load response than dry or water submerged spheres. The above explains why unsaturated soil behaviour is not ruled by either the Terzaghi principle or its generalisations (Bishop, 1959). In fact, $|\sigma\text{-}u_a|$ represents normal and tangential forces induced by the external loads on inter-particle contacts, while suction affects normal forces acting at particle contacts and is associated to water content variations. The two variables are thus uncoupled.

Before leaving this section, it is worth noting that in unsaturated soils the particle-water arrangement may be more complex than described in Fig. 1. Both menisci- and bulk-water could be simultaneously present, bulk-water being that totally filling voids within clusters of grains surrounded by air. Bulk-water is subjected to negative pressure, as menisci-water. However, its pressure changes have the same effects as in traditional soil mechanics: in this portion of the soil, the Terzaghi equation works properly and changes in suction at constant net stress are equivalent to changes in p'.

3 THE TESTED MATERIAL

Classification - The tested soil comes from a decomposed granite extracted from a quarry located in Calabria (Italy) and used in the core of the Metramo River dam. It is a silty sand with a uniformity coefficient U_c of approximately 400 and a clay fraction of about 16%. The liquid limit w_L measured on the fraction having $D_{max}<0.4$mm is around 35% and the plasticity index PI is 14%. The finer fraction of the soil may be classified as lean clay according to the U.S.C.S. plasticity chart (ASTM D2487-93). The physical properties of the soil are reported in Table I.

Table I. Physical properties of Metramo silty sand.

Gs	D_{max} (mm)	sand (%)	silt (%)	clay (%)	Uc	w_L	wP	PI
2.64	2	63	21	16	400	35.4	21.7	13.7

Preparation procedure – The soil has been compacted with the modified Proctor procedure at the optimum (9.8%) and wet of optimum (12.3%) water contents. The under-compaction technique was adopted to improve homogeneity along the specimen height. Three specimens were obtained from the Proctor block using 36mm internal diameter moulds. The after-compaction suction was measured around 800 kPa and 60 kPa for optimum and wet specimens respectively, using the Imperial College tensiometer.

The tested soil has non-negligible fine-grained components. It is therefore expected that, depending on compaction water content, the soil may assume a different fabric due to water-soil particles interaction at a microscopic scale. This is supported by literature studies in which compaction induced microfabric was observed directly (Delage et al., 1996). Through these studies it was observed that optimum compaction results in a fabric made up of aggregates of different size and, to some extent, in a bimodal pore size distribution. On the other hand, wet compaction tends to produce matrix-dominated soil fabric and a single pore size distribution.

4 TESTING DEVICE

The suction controlled RCTS cell has been used to analyse the small strain behaviour of Metramo silty sand. Fig. 2 shows a general layout of the apparatus resulting from several modifications of the same device previously used for saturated soil testing.

Stress control systems - The new apparatus allows to control cell, pore air and pore water pressures separately, using three electro-pneumatic regulators (A, Fig. 2). Pressures are moved toward operator established values in feedback, taking advantage of the readings from transducers (B) connected to the pressure-lines. Suction is controlled using the axis translation technique. Both water- and air-pore pressures are applied through the base pedestal (Fig. 3). A high air entry value porous disk (A, Fig. 3) protects drainage line from short time de-saturation. The complex design of the pedestal rises from the choice of not varying the fixed-free torsional constraint in the RCTS-NA for unsaturated soils.

Strain measurements - Specific volume changes of the specimen are obtained measuring its axial and radial displacements. The former is measured using an LVDT coaxial to the specimen (C, Fig. 2) while the latter is obtained indirectly from the change in the level of a water bath surrounding the sample. The measuring chamber (D) consists of a 5mm thick annulus between an aluminium cylinder and the loading cap. It is connected to a differential pressure transducer (E) monitoring water level in comparison with that in a reference burette (F).

Water content measurements - The measurement of the water content changes is obtained through a system of two double-walled burettes coupled to a differential pressure transducer (G, Fig. 2). One burette (H) is connected to the specimen drainage line, while another (I) gives a reference value for the DPT readings. To protect the drainage line from possible air diffusion through the high air entry value disk, water is forced by a peristaltic pump (L) through a closed hydraulic circuit, thus transporting air bubbles toward the measurement burette and expelling them through the free air-water interface.

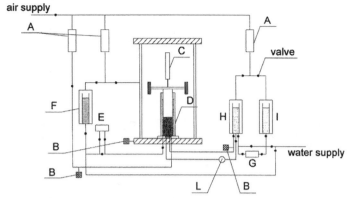

Figure 2. General layout of the resonant column–torsional shear device for unsaturated soils.

5 TESTING PROCEDURES AND PROGRAM

The testing procedure used with the new RCTS-NA device consists of three stages: equalisation, compression, and cyclic/dynamic torsional shear.

Equalisation – The equalisation stage pertains to unsaturated soil testing and is required to bring any specimen toward user established values of matric suction. It is achieved applying prescribed values of pore-air and pore-water pressures at the specimen boundaries and waiting for pore water pressure equalisation. The testing value reached after equalisation can be held constant during subsequent stages (as done in the present study) or varied following prescribed suction-paths.

Tests discussed in this paper refer to equalisation stages at net stress (σ-u_a) equal to 10 kPa, a sufficiently low value not to induce significant compression or collapse of the soil specimens. The constant water back-pressure set was equal to 50 kPa in all the tests reported below. Air pressure (u_a) and cell pressure (σ_c) were then moved to their target values. The end of the equalisation process was inferred by the stabilisation of the specific volume and of the specific water volume (v_w=1+S_R·e) or, according to Sivakumar (1993), by assuming a water content change rate limit of 0.04 %/day.

Compression - At the end of the equalisation stage each specimen was compressed by linearly increasing the net stress (σ_c-u_a) up to a final target value. A loading rate of 4 kPa/h was used to ensure fully drained and constant suction conditions for the Metramo silty sand (Rampino et al., 1999b).

Cyclic and dynamic torsional shear – Research on small strain behaviour of a single specimen at various net stresses required a multistage compression sequence, alternating compression with torsional shear stages. Several series of RC and TS tests were performed at each net stress level. To analyse strain rate effects the torsional series were performed at constant torque and variable loading frequencies.

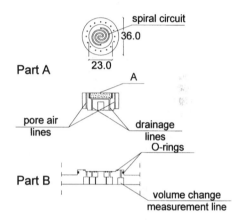

Figure 3. Base pedestal of the RCTS-NA.

Tests were initiated immediately after the end of compression in order to eliminate time effects. For the intermediate net stress values RC and TS tests were performed avoiding trespassing the elastic threshold strain of the soil. Once the highest scheduled (σ–u_a) was achieved, torque was increased up to the maximum value (0.43 N·m) to investigate medium strain soil behaviour.

The experimental program includes 5 tests on the optimum and 3 tests on the wet compacted soil. All the tests were performed at constant suction. Data are here completed by the tests on saturated soil specimens from d'Onofrio (1996). Investigated suction levels are 25, 50, 100, 200, 400 kPa for the optimum soil and 100, 200, 400 for the wet material. The (p-u_a) values chosen to analyse the material response are 100, 200 and 400 kPa. Only the properties pertaining to the small strain range were investigated with (p-u_a) equal to 100 and 200 kPa, while the widest allowable strain range was studied at (p-u_a) equal to 400 kPa. Due to space limitation, only initial shear stiffness is discussed in this paper.

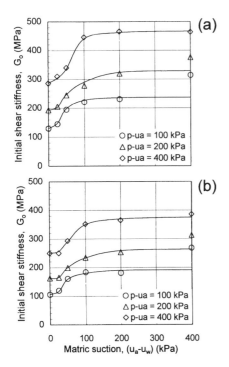

Figure 4. Initial shear stiffness versus suction for the optimum compacted soil: a) RC tests; b) TS tests.

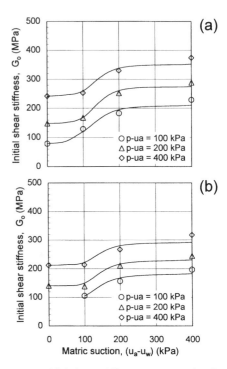

Figure 5. Initial shear stiffness versus suction for the wet compacted soil: a) RC tests; b) TS tests.

6 MEASURED SHEAR STIFFNESS'

Fig. 4 and Fig. 5 plot initial shear stiffness G_0 against matric suction obtained in the suction controlled resonant column and in torsional shear tests at a 0.5 Hz loading frequency.

According to previous observations at large strains (Rampino et al., 1999a), most of the suction effects are observed for s ranging from 0 to about 200 kPa. For suction higher than 200 kPa G_0 tends toward a threshold value that depends on the mean net stress level. The effect of suction is significant. Suction variation in the 0÷400 kPa range causes an increase in G_0 ranging from 85% to 50% for the optimum soil and from 165% to 40% for the wet compaction.

The stiffness values of both optimum and wet compacted materials are clearly distinct for constant stress levels and show an S-shaped increase with suction. This mechanical response complies with the typical S-shape of the soil-water characteristic curve, and can be explained dividing the G_0:s curves into three different zones:

Zone I – The first zone begins at null suction (i.e., saturated conditions) and is restricted to low suction values. Herein bulk-water effect dominates soil behaviour since the air entry value of the specimens is not trespassed. Variations in s are thus equivalent to

mean effective stress changes and the initial gradient of the G_0:s function is expected to be the same as that of the G_0:p' function at a mean effective stress equal to the (p-u_a) value where drying is initiated.

Zone II – In the second zone (intermediate suction) air starts to enter the pore voids. Increasing suction determinates a progressive shift of the soil response from bulk-water to menisci-water regulated behaviour, i.e soil response ranges between zones I and III.

Zone III – For suction values high enough to allow dominant menisci water effects soil response complies with the behaviour expected from the simple model presented in Fig. 1. The initial shear stiffness increases with suction at an initially fast rate and then tends toward a threshold value, near which further effects of suction are negligible.

Compaction water content seems to affect the ratio between the saturated value and the unsaturated threshold values of shear stiffness and the suction values [called s* from now on] characterising transition between bulk- and menisci-water regulated behaviour. This latter point could be justified on the basis of intuitive physical considerations. As a matter of fact, due to the increase in moulding water content, pore sizes of the wet material can be thought smaller than those of the optimum soil. This helps justify the higher s* value of the wet soil (around 100 kPa in Fig. 5) with respect to that of the optimum one, around a few tens of kPa.

Fig. 4 and Fig. 5 also reflect the idea that wet compaction induces a weaker soil fabric than optimum compaction. In fact, the increase in moulding water content has caused a strong reduction in the initial shear stiffness of Metramo silty sand, under both RC and TS conditions. For example, the RC data of the saturated samples indicate that the G_0 at $(p-u_a)$ equal to 100 kPa is about 80 MPa for the wet material and 130 MPa for the optimum compacted soil. Therefore, stiffness decreases by more than 60% with respect to the optimum by increasing moulding water content from 9.8% to 12.3%. Similar effects are observed for all the suction levels.

Besides the observed softening effect, the detected trend confirms that fabric is the key regulator of soil behaviour, since other factors such as ageing, stress and suction states or stress histories were not changed in the previous comparison. It must also be considered that void ratio variations are quite limited for Metramo silty sand (Rampino et al., 1999a). Thus, the observed trends would hold even if data correction were attempted with some function of the void ratio, as originally proposed by Hardin & Drnevick (1972).

7 MODELLING G_0–SUCTION BEHAVIOUR

The way in which a real soil moves from bulk- to menisci-water regulated behaviour depends on its response to de-saturation, i.e. on its characteristic curve. According to the previous discussion on the meaning of the S-shaped G_0:s relation, the null suction stiffness $(G_0)_{s=0}$ can be expressed as:

$$\frac{(G_0)_{s=0}}{p_a} = S \cdot \left[\frac{(p-u_a)_C}{p_a}\right]^n \cdot f(e) \qquad (1)$$

Eqn. (1) is similar to that proposed by Hardin (1978) for saturated soils, calculated at the mean net stress $(p-u_a)_C$ where drying starts. In the equation p_a is the atmospheric pressure, $f(e)$ a rational scaling function for void ratio-induced heterogeneity. S is the stiffness index and n the stiffness coefficient, which represent the stiffness of the material under the reference pressure and the sensitivity of the stiffness to the stress state, respectively.

The behaviour described by eqn. (1) should be valid not only when s equals zero, but should hold up to the air entry value (s_{EV}) of the soil. Therefore, in the zone I the G_0 versus suction relationship should follow the equation:

$$\frac{(G_0)_{s \leq s_{ev}}}{p_a} = S \cdot \left[\frac{(p-u_a)_C + s}{p_a}\right]^n \cdot f(e) \qquad (2)$$

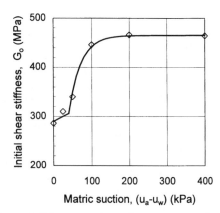

Figure 6. RC tests on the optimum compacted material (reference stress level 400 kPa).

Once the air entry value of the soil is reached, further increments of suction cause progressive de-saturation of the soil and move the G_0:s relationship from the bulk- to the menisci-water regulated zone. Here, it would be reasonable to assume that the effect of suction starts with a gradient equal to that defined by eqn. (2) at $(p-u_a)_C + s_{EV}$. The G_0:s curve in this zone is the link between the functions in zones I and III and must tend towards the latter for higher suctions. However, both the shape and the position in which menisci-water effects start prevailing (i.e. the end of zone II) depend on the details of the specific de-saturation process and are not straightforward.

In this paper a simplified shape for the G_0 versus s relationship is proposed, as derived from the experimental evidence. Fig. 6 displays an example of RC test data performed on the optimum compacted material at a reference stress level of 400 kPa. It shows that both zone II and III can be described modifying the equation proposed by Alonso et al. (1990) as a pattern for menisci-water regulated behaviour. Using the suction value s* defined above, the experimental data can be fitted by the relation:

$$G_0 = (G_0)_{s^*} \cdot \{[1-r] \cdot e^{-\beta \cdot (s-s^*)} + r\} \qquad (3)$$

where β is a parameter controlling the rate of increase of soil stiffness with suction and r defines the ratio between shear stiffness G_0 at s* and the threshold value for increasing suction.

On the basis of the former discussion it is possible to deduce that the s* value should fall into zone II. The assumptions made, however, imply a discontinuity between zones I and III, i.e. zone II has zero extension and s* is equal to s_{EV}.

The use of eqn. (3) does not follow directly from Alonso et al. (1990). As a matter of fact, they use a similar relationship to account for the variation with suction of soil parameters such as compressibility index λ. Therefore, if an analogy is drawn between G_0 and bulk modulus it would be expected that G_0 varied inversely to the Alonso et al. (1990) function. The shape of this function, however, is not different from that obtained by eqn. (3), and the latter is preferred since it is more manageable and yields to a clear meaning of the r and β coefficients. Furthermore, as this relation is assumed to describe the behaviour in the zones II and III, the variable s-s* is used, always being s greater than s*.

The example of Fig. 6 shows that the expression proposed can fit the experimental data properly, starting from stiffness values higher than those predicted by eqn. (2), extended to zone II.

8 CONCLUDING REMARKS

The experimental program here presented was conceived to analyse the effects of compaction-induced fabric and suction on small strain stiffness. Overall, results indicate a significant effect of these two variables on G_0 and confirm the functionality and the great potential of the new RCTS-NA.

It has been observed that soil fabric causes serious variations in stiffness and affects the ratio between the saturated and unsaturated threshold values of G_0.

Experimental results show a significant increase in G_0 with increasing suction. For both the optimum and wet compacted materials most of the suction effects are observed for s varying in the range 0÷200 kPa. Detected values are well grouped for constant stress levels and show an S-shape with increasing suction. This is explained by a progressive change of soil behaviour from a bulk-water to a menisci-water regulated response. If this remark is confirmed by further experiments, it will be concluded that menisci water effect dominates bulk water at relatively high suction, while low suction lead to the opposite.

This would also help reduce the number of experimental data required for the determination of the G_0:s function. As a matter of fact, G_0 at $(p-u_a)+s_{EV}$ could be inferred from the G_0:p′ function calculated at $(p-u_a)+s_{EV}$. The G_0 values required to fix r and β parameters should be measured at s higher than s_{EV}. One at very high suction would allow to fix the r coefficient and one or two at intermediate s values would be enough to reasonably set the β coefficient.

REFERENCES

Alonso, E.E., Gens, A., and Josa, A., 1990 - A constitutive model for partially saturated soils. Geotecnique, 40(3), 405-430.

Bishop, A.W., 1959 - The principle of effective stress. Proc. Tekenisk Ukebald, 39, 859-863.

Bishop, A.W., Blight., G.E., 1963 - Some aspects of effective stress in saturated and partially saturated soils, Geotechnique, 13(3), 177-197.

Burland, J.B., Ridley, A.M., 1996 - General report. The importance of suction in soil mechanics. Proc. UNSAT-ASIA'96.

Cabarkapa, Z., Cuccovillo, T., and Gunn, M., 1999 – Some aspects of the pre-failure behaviour of unsaturated soil. II Int. Conf. on pre-failure behaviour of geomaterials, Turin.

Delage., P., Audiguier, M., Cui, Y.D. & Howat, M.D., 1996 – Microstructure of a compacted silt. Can. Getotech. Juorn., 33, 150-158.

d'Onofrio, A., 1996 - Comportamento meccanico dell'argilla di Vallericca in condizioni lontane dalla rottura (in Italian). Doctoral Thesis in Geotechnical Engineering, University of Naples "Federico II".

Fisher, R.A., 1926 – On the capillary forces in an ideal soil, Journal Agr. Science, 16, 492-505.

Fredlund, D.G., Morgernstern, N.R., 1977 - Shear state variables for unsaturated soils, Journal Geotech. Eng. Div., ASCE, 103(5), 447-466.

Fredlund, D.G., Rahardjo, H., 1993 - An overview of unsaturated soil behaviour, ASCE. Geotech. Spec. Publ., 39.

Hardin, B.O., 1978 - The nature of stress-strain behaviour for soils. State of the Art, Proc. Geotech. Eng. Div. Specialty Conf. on Earthquake Eng. and Soil Dynamics, ASCE, Pasadena, California.

Hardin, B.O., Drnevich, V.O., 1972 - Shear modulus and damping in soils: measurements and parameter effects, Jou. Geotech. Eng. Div., ASCE, 98(6), 603-624.

Marinho, E.A.M., Chandler, R.J., Crilly, M.S., 1995 – Stiffness measurements on an high plasticity clay using bender elements. Proc. UNSAT'95, Paris, France, vol. 1, 535-539.

Picornell, M., Nazarian, S., 1998 – Effects of soil suction on the low-strain shear modulus of soils. Proc. UNSAT'98, Peking, China, 2, 102-107.

Quin, X., Gray, D.H., Woods, R.D., 1991- Resonant column tests on partially saturated sands. Geotech, Test. Journ., ASCE, 14(3), 266-275.

Rampino, C., Mancuso, C., and Vinale, F., 1999a – Mechanical behaviour of an unsaturated dynamically compacted silty sand. Italian Geotech. Journ., 33(2), 26-39.

Rampino, C., Mancuso, C., and Vinale, F., 1999b - Laboratory testing on a partially saturated soil: equipment, procedures and first experimental results. Can. Geotech. Journ. 36(1), 1-12.

Sivakumar, V., 1993 - A critical state framework for unsaturated soils. Ph.D. Thesis, Univ. of Sheffield.

Vinale, F., d'Onofrio, A., Mancuso, C., Santucci de Magistris, F., Tatsuoka, F., 1999 – The prefailure behaviour of soils as construction materials. II Int. Conf. on pre-failure behaviour of geomaterials, Turin.

Wheeler, S.J., 1996 - Inclusion of specific water volume within an elasto-plastc model for unsaturated soils, Canadian Geotech. Journ., 33(1), 42-57.

Wu, S., Gary, D.H., and Richart, F.E., 1989 – Capillary effects on dynamic modulus of sands and silts. Journ. Geotech. Eng. Div., ASCE, 110(9), 1188-1203.

Unsaturated Soils for Asia, Rahardjo, Toll & Leong (eds) © 2000 Taylor & Francis, ISBN 90 5809 139 2

Effects of moulding water content on the behaviour of an unsaturated silty sand

C. Mancuso, R. Vassallo & F. Vinale
Dipartimento de Ingegneria Geotecnica, Università di Napoli Federico II, Italy

ABSTRACT: This paper deals with the results of an experimental study carried out on an unsaturated silty sand to investigate the influence of compaction induced fabric and suction on soil behaviour. A series of triaxial tests was conducted on both the optimum and wet compacted material, under controlled suction conditions. The observed effects of fabric and suction are relevant in all the stages of the tests and highlight the need of carefully considering these two variables in handling engineering problems involving compacted soils as construction materials.

1 INTRODUCTION

Unsaturated soils can often be encountered in practical applications. In particular, even though index properties of a soil as a construction material are left unchanged, the compaction procedure (moulding water content, compaction energy, etc.) influences the mechanical response of the resulting material, its after compaction degree of saturation and, more generally, the earth constructions' behaviour (Vinale et al., 1999).

It is possible to identify two levels at which the selected preparation procedure influences soil response. First, it usually yields to an unsaturated material and determines its initial suction, i.e. its initial stress state. Second, it can profoundly affect soil fabric, thus causing a different response even under the same conditions of suction and after compaction testing path. At present, direct evidence of the separate effect of both variables (i.e. suction and fabric) is very limited (Gens et al., 1995). This lack is probably due, to some extent, to the need for more sophisticated experimental apparatuses designed for unsaturated soil testing, indispensable to isolate the above mentioned factors.

Regarding this point, it is worth recalling that experimentation on unsaturated soils requires the separate measure of water content and volume changes and the control of an additional stress state variable, which is suction. As a matter of fact, the presence of both air and water in the voids makes the pore fluid compressible, so volume changes can not be inferred from water content variations. Furthermore, it causes the stress-state of a soil element to be affected by air-water interfaces in the

soil mass, and as a consequence, Terzaghi's principle is no longer valid and a two stress variable approach is required. In this paper, reference is made to net stress $|\sigma-u_a|$, reducing to net mean stress $(p-u_a)$ and deviator q under triaxial conditions, and suction (u_a-u_w). The latter influences menisci curvature, essentially affecting normal stresses acting at the contacts between particles. As a consequence, increasing suction improves slippage strength between the contacts, i.e. stiffness and strength of the soil skeleton.

The discussed aspects of soil behaviour have been widely studied at the Dipartimento di Ingegneria Geotecnica of Naples through suction controlled devices (Mancuso et al., 1999; Vinale et al., 1999). This extensive work has included a series of triaxial tests performed on an optimum and wet of optimum compacted soil. Its main results are presented in this paper.

2 TESTED MATERIAL AND EXPERIMENTAL PROGRAM

The tested soil is a silty sand from the core of the Metramo dam (Italy). It is an inorganic clay of medium plasticity according to the Atterberg limits and Casagrande's classification. Table I summarises its main physical and after-compaction properties. The material was compacted at optimum $(w_{opt}=9.8\%)$ and wet of optimum $(w_{opt}+2.5\%)$ water contents according to the modified Proctor procedure. The under-compaction technique (Ladd, 1978) was used to improve soil homogeneity along the height of the sample. Using the Imperial College

Table I. Average properties of the tested soil.

	Index Properties			After Compaction at w=9.8%			After Compaction at w=12.3%		
G_S	w_L	w_P	PI	v	S_r	u_a-u_w	v	S_r	u_a-u_w
-	%	%	%	-	%	kPa	-	%	kPa
2.64	35.4	21.7	13.7	1.345	75	800	1.389	83	60

tensiometer the after compaction suction was measured to be as high as 800 kPa for the optimum material and as low as 60 kPa for the wet specimens – a first relevant effect of the moulding procedure.

This material has a non-negligible fine-grained component, hence, some dependence of the observed behaviour on the compaction procedure is foreseeable. As a matter of fact, compaction variables have a fundamental role in the structure of fine grained soils. In particular, soil fabric is strongly influenced by menisci-water and solid-water adsorption interaction occurring at a microscopic level. This statement is not supported by S.E.M. or mercury intrusion porosimetry measurements on the tested soil, but arises from literature studies devoted to direct observations of microfabric (Barden & Sides, 1970; Delage et al., 1996). The mentioned experimental works led to the conclusion that fine grained soils compacted dry of optimum generally exhibit a fabric made up of aggregates of varying size and tend to have a bimodal pore size distribution, whereas soils compacted wet of optimum tend to show a more homogeneous matrix-dominated fabric and a single pore size distribution.

The experimental program was conceived in order to analyse the effects of both suction and fabric on soil behaviour. Overall, it consisted of 23 triaxial tests, performed using a Bishop & Wesley cell. This device, suited for the control of suction through the axis translation technique and for the separate measurement of volume change and water content change, is described in detail by Rampino et al. (1999a). The investigated suction levels were 0, 100, 200, 300 kPa for the optimum compacted soil and 0, 100, 200 kPa for the wet compacted material. Isotropic compression stages, carried out at constant suction level, were preceded by an appropriate equalisation stage. The subsequent shear stages, i.e. both constant mean net stress (p-u_a) shear tests and "standard" shear tests, were all conducted under drained conditions and constant suction.

3 EQUALISATION STAGES

The equalisation stage is required to move suction from the value induced by compaction procedure to the level prescribed by the operator. It consists in imposing the appropriate value of pore air and pore water pressures, net mean stress (p-u_a) and deviator

q and then waiting for suction to become uniform in the whole specimen. In the present study (p-u_a) and q were chosen equal to 10 and 0 kPa respectively, so that the soil behaviour could be investigated starting from low net stress levels in the subsequent compression stages. The end of equalisation is deduced from the stabilisation of the monitored state variables, i.e. specific volume and water content or, equivalently, specific water volume (v_w=1+S_r·e).

The optimum and the wet compacted specimens exhibited very different behaviours. This is clearly shown by Figure 1a and Figure 1b, which plot the results in terms of water content versus logarithm of time. As a consequence of both the imposed suction level and the compaction induced suction, the optimum material follows a wetting path while the wet one follows a drying path. In other words, suction decreases for the former, causing a water flow directed towards the specimen, while it increases for the latter, making water flow out. For both materials, the change in water content is more pronounced as the absolute value of the imposed suction change increases. This observation is also valid for other state variables such as v and v_w. Under the same suction level, water content variations are greater for the optimum material, as a consequence of the greater difference between initial and final suction values.

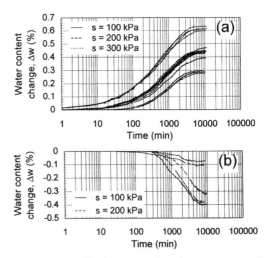

Figure 1. Equalisation stage: a) optimum compacted soil; b) wet compacted soil.

Likewise, the variations of specific volume are opposite for the two materials: as a matter of fact, the optimum compacted specimens swell while the wet ones shrink. The specific water volume follows the trend of specific volume but is always lower, under the same suction, as full saturation is never achieved.

4 ISOTROPIC COMPRESSION STAGES

Compression stages were carried out by linearly increasing $(p-u_a)$. A loading rate of 4 kPa/h proved to be sufficiently low to ensure water-drained conditions for the tested soil (Rampino et al., 1999b).

Figure 2a and Figure 2b represent the results in terms of specific volume versus net mean stress for the optimum and the wet materials. The effect of suction is significant: as (u_a-u_w) increases compressibility is reduced and the apparent over-consolidation pressure is increased. The stiffening of the soil due to suction also occurs with reference to unloading-reloading lines, although this is more evident for the wet compacted soil, while it is almost irrelevant for the optimum material. Compressibility index λ ranges from 0.022 to 0.015 for the optimum material and from 0.040 to 0.029 for the wet compacted soil. It can be said that λ follows an exponential trend: the greatest amount of suction effect occurs in a limited suction range $(0 \div 100$ kPa) while λ tends to a threshold value for (u_a-u_w) greater than 200 kPa. This agrees with the basic concepts of the modelling criteria for isotropic stress-strain behaviour from critical state models for unsaturated soil (Alonso et al., 1987; 1990). In particular, the dependence of λ on suction (s) is well interpreted by the equation proposed by Alonso et al. (1990):

$$\lambda(s) = \lambda(0) \cdot [(1-r) \cdot \exp(-\beta \cdot s) + r] \quad (1)$$

where β governs the rate of change of λ with suction and r defines the ratio between $\lambda(0)$ and the threshold value for increasing suction. The detected β and r parameters were 0.024 kPa^{-1} and 0.68 for the optimum material and 0.015 kPa^{-1} and 0.73 for the wet soil, respectively.

The compression curves in terms of specific water volume have the same trend as those in terms of specific volume and show smaller changes in v_w than in v. For both materials, water always flows out of the specimen: as the specific volume decreases, the soil expels water in order to hold suction at the imposed value. The lower compressibility of the optimum compacted material also implies a lower reduction of v_w upon compression. Again, the slopes of the normal compression lines at virgin state and those of the unloading-reloading cycles exhibit a sharp decrease when moving from saturated to unsaturated conditions.

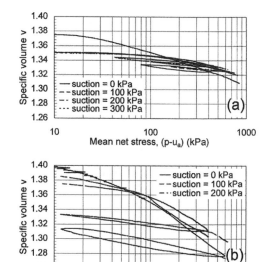

Figure 2. Isotropic compression stage: a) optimum compacted soil; b) wet compacted soil.

The degree of saturation always increases with confining pressure, as a consequence of changes higher in specific volume than in specific water volume.

Comparing experimental findings on the optimum and the wet specimens, it is observed that an increase in moulding water content induces a severe increase in compressibility. The indexes $\lambda(s)$, for instance, almost double when going from the lower to the higher compaction water content. This supports the previous opinion that different moulding water contents correspond to different soils, even if the physical and mineralogical characteristics of the grains and the grading curve of the material are left unchanged.

In addition, data indicate that both the shape and the position of the after compaction loading-collapse loci (LC) are largely affected by moulding water content. This is a further indication that soil susceptibility to collapse depends on that compaction variable. The apparent yield points were estimated using different procedures (Casagrande, 1936; Tavenas et al. 1979 and Graham et al., 1983). The obtained average values are well interpolated by a modified version of the Alonso et al. (1990) equation (Rampino et al., 1999b; Vinale et al., 1999), as displayed in Figure 3. At the applied suction levels the yield net stress is always greater for the optimum soil, except for the saturated one, in which these stresses have sensibly the same value for the two materials. In fact, the yield net stress ranges from 35 to 170 kPa for the optimum material

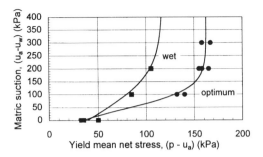

Figure 3. After compaction LC yield loci.

and from 39 to 105 kPa for the wet material. Furthermore, also the yield loci show that suction effect has a decreasing gradient.

The question whether the observed curves are effectively the "after-compaction" loci could arise. This would be true if the volume change of the specimen in the previous equalisation stage were reversible. Such a conclusion applies to the optimum compacted soil, expanding and reducing its suction during equalisation, while for the wet material it is valid under the hypothesis that LC and SI (i.e. suction increase locus) movements are independent from one another (Alonso et al., 1990). On the other hand, if the LC and SI movements were coupled by a hardening law (e.g., that proposed by Alonso et al., 1990), the conclusion that the elastic region found for the wet soil is larger than that corresponding to its after-compaction state would be reached, as this soil shrinks during equalisation. However, this hypothesis implies that the difference between wet and optimum materials is even more pronounced than displayed in Figure 3. Thus, regardless of the assumption made, it is possible to assert that, in its after compaction state, the wet material is more sensitive to potential collapse than the optimum.

It is noteworthy that Gens et al. (1995) observed a different behaviour through oedometer tests on a low plasticity compacted dry and wet of optimum silt. Their results, however, do not refer to the after compaction state, as do the data presented here, but to wet specimens brought to the same initial conditions as the dry ones, in terms of density, moisture content and suction. Working with this material and experimental procedure they found that the dry specimens have larger compression strains on saturation (i.e., larger sensitivity to collapse).

5 SHEAR STAGES

Figure 4 and Figure 5 plot the results of two shear stages, displaying deviator versus axial strain and volumetric strain versus axial strain. Observed trends can be considered representative of the

behaviour of the optimum and wet materials, respectively, in all the deviator stages.

The influence of the dynamic compaction procedure and water content becomes clear under shearing. The mechanical response to increasing deviator is always that of dense granular soils for the optimum compacted specimens (Rampino et al., 1999b) and that of loose materials in the wet of optimum case (Vinale et al., 1999).

The optimum soil is characterised by stress-strain curves which always exhibit softening, with a distinct peak in the deviator stress at $\varepsilon_a \cong 4\%$, final stable q value at $\varepsilon_a \cong 12\%$ and corresponding volume changes showing evident dilatancy. The variations of v_w follow the same trend as v, but are very small in absolute value and lower than the changes in specific volume. A final stable value is also achieved for these variables, at about the same axial strain level as that pertaining to deviator stress.

On the other hand, when tested under normally consolidated state, the wet soil shows no evident peak in the q:ε_a curve. This behaviour can be clearly classified as that of a loose granular soil, as mentioned above, because the deviator increases with axial strain moving towards an ultimate value and then becoming sensibly constant. The volume changes, however, do not confirm the latter statement. As a matter of fact, in all the shear stages it is observed that, after an initial shrinkage, dilatancy also occurs for the wet material. Furthermore, the ultimate state is not achieved in terms of specific volume: even for axial strains greater then 15% the gradient of the ε_a:ε_v curve is not negligible. Like in the case of the optimum soil, the diagrams ε_a:v_w have the same shape as the curves ε_a:v. This is probably due to the soil's tendency to keep a constant degree of saturation (because of the imposed constant suction) while its volume is changing.

A further difference between the two materials is the shape of the failed specimens. The optimum soil exhibits a clear bulging in saturated state, and with increasing suction evident shear bands appear. On the other hand, the wet material shows no shear bands when reaching the ultimate state, for all the applied suction levels.

The observed behaviour is strongly dependent on suction for the optimum and wet materials: increasing $(u_a - u_w)$ has a beneficial effect on both stiffness and shear strength. For instance, with reference to standard shear stages performed at (p-u_a) equal to 100 kPa and moving from 0 to 200 kPa of suction, the ultimate state deviator changes from about 345 to 720 kPa for the optimum compacted soil and from about 345 to 640 kPa for the wet material. A similar comparison can also be made by plotting secant Young modulus versus axial strain. However, it is worth highlighting that this calculations provide reliable values only in the large

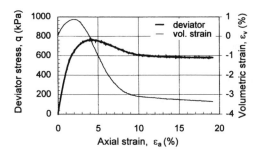

Figure 4. The shear stage performed on the optimum compacted soil at $(u_a-u_w)=100$ kPa, after isotropic compression up to $(p-u_a)=100$ kPa.

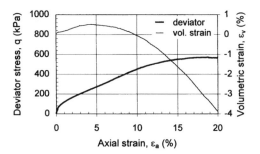

Figure 5. The shear stage performed on the wet compacted soil at $(u_a-u_w)=100$ kPa, after isotropic compression up to $(p-u_a)=100$ kPa.

strain range, as the used apparatus is not equipped with devices for local strain measurement. The small strain stiffness of Metramo silty sand was recently investigated through an RCTS cell, as described in a companion paper by the same authors.

If the final stable values of the stresses were plotted in a $(p-u_a)$:q plane, it would show that for both materials the data are well grouped for constant suction level and can be interpolated by the equation originally proposed by Wheeler & Sivakumar (1995):

$$q = M(p - u_a) + \mu(s) \qquad (2)$$

Thus, an ultimate state line can be drawn for each suction value. A constant value of 1.54 was detected for the M coefficient of the optimum material. The effect of growing suction on it is an increase in apparent cohesion, ranging from 0 to 240 kPa for suction varying in the entire investigated range. The wet material shows lower shear strength values for all the analysed suction levels. As fewer data on this soil are available, some uncertainties are found in interpolating ultimate strength values. The M coefficient seems to be different from the value

(1.54) of the optimum soil. However, if M is forced to 1.54, i.e. it is assumed to be dependent only on physical and mineralogical grain properties, the effect of changing moulding water content is a significant reduction in the apparent cohesion $\mu(s)$, which varies from 0 to 150 kPa in the investigated suction levels.

For both the optimum and the wet material the apparent cohesion shows an initially fast increase with suction and then tends towards a limiting value. Rampino et al. (1999a) observed that, for the optimum soil, $\mu(s)$ data can be fitted by the exponential law expressed by Equation (1) with r and β parameters different from those pertaining to compressibility indexes. This consideration can be extended to the wet soil, as it exhibits a similar trend.

By the extrapolation of the ultimate state lines in the $(p-u_a)$:q plane it is possible to draw another branch of the yield locus in the $(p-u_a)$:s plane. In fact, the yield stress pertaining to each suction is obtained from the intersection point of the relevant ultimate state line and the negative portion of $(p-u_a)$ axis (Alonso et al., 1990). Also on the basis of this observation, it is possible to affirm that one effect of moving from the optimum to the wet compaction water content is a reduction of the size of the after compaction yield surface.

6 CONCLUDING REMARKS

An experimental program has been carried out to investigate the effects of both compaction induced fabric and suction on the behaviour of an unsaturated silty sand.

The optimum and the wet compacted specimens exhibited very different behaviours in all the stages of the tests. For example, during the equalisation stage, optimum material swells and absorbs water, while wet soil shrinks and expels water. This response was explained as a consequence of the imposed suction levels (from 100 to 300 kPa) and of the after compaction suction, equal to about 800 kPa for the optimum material and to about 60 kPa for the wet specimens.

The results of the isotropic compression tests highlighted the strong effect of suction on soil compressibility. The greatest part of this effect occurs in a limited suction range (0÷100 kPa) while λ tends to a threshold value for (u_a-u_w) greater than 200 kPa. Comparing experimental findings on the optimum and the wet specimens, it was observed that an increase in moulding water content induces a severe increase in compressibility. In addition, data indicated that both the shape and the position of the after compaction loading-collapse loci (LC), expressing soil susceptibility to collapse, are largely affected by moulding water content.

Shear stages confirmed that moulding water content influences the soil stress-strain behaviour severely. For instance, a distinct peak followed by softening is observed in ε_a:q curves for the optimum material, while for the wet soil the deviator continuously increases with axial strain level, tending towards a limiting threshold value.

Suction significantly affects both stiffness and resistance. The ultimate state data points seem to be well grouped for suction levels and can be interpreted through straight fitting lines having the same M coefficient in the $(p-u_a)$:q plane. The detected effect of moving from the optimum to the wet compaction water content is a weakening of the material, as shown by the decrease in stiffness and in apparent cohesion.

For both isotropic and shear stages it was demonstrated that the experimental data can be satisfactorily modelled by the modern theories for unsaturated soils (Alonso 1990, Wheeler & Sivakumar 1995), if it is assumed that the different preparation procedures yield to different soils.

Altogether, the obtained results emphasise the relevant effect of fabric and suction on the material response and also highlight the need of appropriate testing devices and procedures to identify the separate influence of these variables on soil behaviour.

REFERENCES

Alonso, E.E., Gens, A., Hight, D.W., 1987 – Special problem soils. General report – IX ECSMFE, Dublin, 3, 1087-1146.

Alonso, E.E., Gens, A., and Josa, A., 1990 - A constitutive model for partially saturated soils. Geotechnique, 40(3), 405-430.

Barden, L., Sides, G.R., 1970 – Engineering behaviour and structure of compacted clay. Journ. Geotech. Eng. Div., ASCE, 96(4), 1171-1197.

Casagrande, A., 1936 – The determination of the pre-consolidation load and its practical significance. I ICSMFE, Harvard, 3, D-34, 60-64.

Delage., P., Audiguier, M., Cui, Y.D. & Howat, M.D., 1996 – Microstructure of a compacted silt. Can. Getotech. Juorn., 33, 150-158.

Gens, A., Alonso, E.E., Suriol, J. & Lloret A., 1995 – Effect of structure on the volumetric behaviour of a compacted soil. I Int. Conf. on unsaturated soils, Paris, 83-88.

Graham, J., NooNan, M.L. & Lew, K.W., 1983 – Yield state and stress-strain relationship in natural plastic clay, Can. Geotech. Journ., 20(3), 502-516.

Ladd, R.S., 1978 – Preparing testing specimen using under compaction. ASTM Geotech. Test. Journ., 1, 16-23.

Mancuso, C., Rampino, C., Vinale F., 1999 – Experimental behaviour and modelling of an unsaturated compacted soil. Submitted to Can. Geotech. Journ.

Rampino, C., Mancuso, C., and Vinale, F., 1999a – Mechanical behaviour of an unsaturated dynamically compacted silty sand. Italian Geotech. Journ., 33(2), 26-39.

Rampino, C., Mancuso, C., and Vinale, F., 1999b - Laboratory testing on a partially saturated soil: equipment, procedures and first experimental results. Can. Geotech. Journ. 36(1), 1-12.

Tavenas, F., des Roisiers, J.P., Leroueil, S., La Rochelle, P. & ROY, M., 1979 – The use of strain energy as a yield and creep criterion for lightly overconsolidated clays. Geotechnique, 29(3), 285-303.

Vinale, F., d'Onofrio, A., Mancuso, C., Santucci de Magistris, F., Tatsuoka, F., 1999 – The prefailure behaviour of soils as construction materials. II Int. Conf. on pre-failure behaviour of geomaterials, Turin.

Wheeler, S.J., Sivakumar, V., 1995 – An elasto-plastic critical state framework for unsaturated soils. Geotechnique, 45(1), 35-53.

Unsaturated Soils for Asia, Rahardjo, Toll & Leong (eds) © 2000 Taylor & Francis, ISBN 90 5809 139 2

Laboratory study of loosely compacted unsaturated volcanic fill in Hong Kong

C.W.W. Ng & C.F.Chiu
Department of Civil Engineering, Hong Kong University of Science and Technology, SAR, People's Republic of China

H. Rahardjo
NTU-PWD Geotechnical Research Centre, Nanyang Technological University, Singapore

ABSTRACT: A laboratory investigation was carried out to primarily study the unsaturated shear behaviour of loosely compacted fills. The soil used was a colluvial fill of volcanic nature and the soil samples were initially compacted to a dry density of 70 % of the maximum dry density obtained from the standard Proctor compaction test. Two series of triaxial tests were conducted. The first series were consolidated undrained shear tests on saturated samples and the second series were constant water content tests on unsaturated samples. From the test results, a nonlinear shear strength – matric suction relationship is observed for the loosely compacted unsaturated volcanic fill. The frictional characteristic of the fill is independent of matric suction while the extent of the increase in the shear strength with matric suction is influenced by the degree of saturation.

1 INTRODUCTION

Many loose fill slopes have been constructed in Hong Kong before the 1970s. The upper part of these slopes is generally unsaturated in the dry season. The existence of the matric suction helps to maintain their stability. Over the years, there have been a number of severe loose fill slope failures in Hong Kong during rainstorms, resulting in catastrophic flow slides and disasters (HKIE 1998). One of the major challenges for the geotechnical engineers in Hong Kong is to upgrade these marginal slopes safely, quickly and economically under congested working conditions. Recently, it has been proposed to use soil nails to reinforce the loose fill slopes because of its speedy and simple operation. Due to the lack of understanding of the fundamental behaviour of unsaturated loose fill, soil-nail interaction and long-term potential corrosion problems, no convincing conclusion can be reached after a large number of heated debates for sometime in Hong Kong.

In the modelling of unsaturated soils, it is now generally accepted to use two independent stress state variables to explain their mechanical behaviour (Fredlund & Morgenstern 1977). The most common stress state variables used are the net normal stress, $\sigma - u_a$ and the matric suction, $u_a - u_w$, where σ is the total stress, u_a is the pore-air pressure and u_w is the pore-water pressure. Since the early 1990s, several researchers have extended the critical state framework to unsaturated soils (Alonso et al. 1990, Toll

1990, Wheeler & Sivakumar 1993). Most of the theoretical work has been based on laboratory studies on well-compacted and intact soils. Relatively little research has been done on loosely compacted soils except Maatouk et al. (1995).

In this study, two series of triaxial tests have been conducted on a loosely compacted fill, namely (i) consolidated undrained tests on saturated samples and (ii) constant water content tests on unsaturated samples. The prime objective of the study is to investigate the influence of matric suction and degree of saturation on the shear behaviour of unsaturated loosely compacted fill.

2 SAMPLE PREPARATION AND TESTING PROCEDURES

The soil used in this study was a colluvium taken from a slope at the Victoria Peak, on the Hong Kong Island. The colluvium originated from decomposed volcanic. The soil can be described as slightly sandy SILT (GCO 1988). The particle size distribution is shown in Figure 1. Its liquid limit and plastic limit are 48 % and 35 %, respectively. Proctor compaction tests gave a maximum dry density of 1540 kg/m³ and optimum moisture content of 21 %. The testing programme is summarised in Table 1.

Triaxial samples were prepared by moist tamping. All soil specimens were compacted at a moisture content of 20 %, which is 1 % dry of the optimum obtained from Proctor compaction test. The samples

Table 1. Testing programme

Sample identity	Initial specific volume	Specific volume, v, before shear	Applied suction, s (kPa)	Applied net mean stress, p (kPa)
s1	2.422	2.065	0	25
s2	2.419	2.005	0	50
s3	2.43	1.938	0	100
s4	2.464	1.862	0	200
s5	2.435	1.797	0	400
u1	2.486	2.462	80	25
u2	2.503	2.416	80	50
u3	2.526	2.215	80	150
u4	2.504	2.068	80	310
u5	2.505	2.467	150	25
u6	2.493	2.408	150	50
u7	2.490	2.256	150	100
u8	2.489	2.178	150	150

were compacted in 10 layers and the under-compaction approach was employed to achieve a more uniform sample as suggested by Ladd (1978). All samples were compacted to a dry density of 70 % of the Proctor's maximum, which corresponds to a specific volume (v) of 2.47 and a degree of saturation (S_r) of 36 %.

Saturation of the samples was done by applying a vacuum of 15 kPa to evacuate air bubbles, followed by flushing them with carbon dioxide for at least 30 minutes, after which de-aired water was slowly introduced from the bottom of the samples. Back pressure was used until the B-value achieved a minimum value of 0.97. All the saturated samples were first isotropically compressed and then sheared under a strain-controlled loading environment.

For tests on unsaturated samples, a computer-controlled triaxial stress path apparatus equipped with three water pressure controllers and one air pressure controller was used. Matric suction (s) was applied to the samples through pore-water and pore-air pressures. Pore-water pressure was applied or measured at the base of the sample through a porous filter, which has an air entry value of 500 kPa. Positive pore-water pressure was maintained at the base by using the axis translation principle proposed by Hilf (1956). Pore-air pressure was applied at the top of the sample by a filter with a low air entry value. The initial suctions of the unsaturated samples were controlled to 80 kPa (u1 to u4) and 150 kPa (u5 to u8). Then they were isotropically consolidated to net mean stress (p = total mean stress - pore-air pressure), specified in Table 1. After that the samples were sheared under constant water content.

3 SHEAR BEHAVIOUR OF SATURATED SAMPLE

Figure 2 shows the effective stress paths in the deviator stress (q) – effective mean stress (p´) plane for five consolidated undrained tests on saturated samples of the loose volcanic fill. It can be seen from the figure that no strain softening is observed for all five tests. Bulging failure is observed for all samples at the end of the tests. Critical state is defined as the state in which shear produces no change in stress, or in volume of the soil. All the five tests approach the critical state line (CSL), with a gradient (M_{cs}) of 1.32 (Ng & Chiu 2000).

All the effective stress paths shown in Figure 2 have similar shape. In the beginning, each path moves towards the left-hand side until a turning point, after that it turns right and finally reaches the critical state at the end of the test. This turning point is termed the point of phase transformation (Ishihara, 1993), which is defined as a temporary state of transition from contractive to dilative behaviour. This observed behaviour resembles those of typical sand/clay on the dry side of the critical, even though the states before shear lie on the wet side of the critical. The dry side of the critical refers to the state lies below the CSL whereas the wet side of the critical refers to the state lies above the CSL. The relative states of the soil samples with respect to the critical state line will be discussed later.

4 SHEAR BEHAVIOUR OF UNSATURATED SAMPLE

Figures 3 and 4 show the stress-strain relationships for the constant water content tests on unsaturated samples of the loose volcanic fill at initial suctions

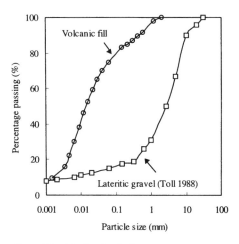

Figure 1 Particle size distributions

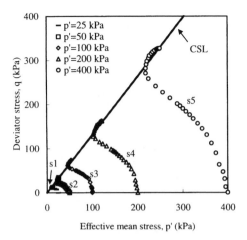

Figure 2. Effective stress paths of consolidated undrained tests on saturated samples

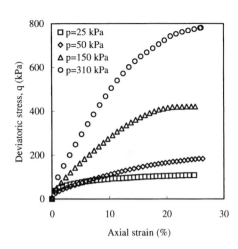

Figure 3. Stress-strain relationships for constant water content tests on unsaturated samples at initial suction of 80 kPa

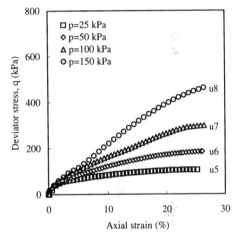

Figure 4. Stress-strain relationships for constant water content tests on unsaturated samples at initial suction of 150 kPa

of 80 kPa and 150 kPa, respectively. The shear strength increases steadily with axial strain and approaches almost a plateau at an axial strain over 25 %, with no evidence of any strain softening. It can be noted from both figures that the shear strength of the unsaturated specimens at constant water content increases with net mean stress. Similar to undrained shear tests on saturated samples, all unsaturated samples failed by barrelling. Most of the tests only approach the critical state but do not actually reach it even at strains in excess of 25 %. However, by inspecting the stress-strain relationships in both figures, it may be reasonable to use the stress state at the end of each test to approximate the critical state.

The suction during shear was measured by the difference between the positive pore-water pressure at the bottom of the sample and a constant applied pore-air pressure maintained at the top of the sample. The measured final suctions range from 72 to 79 kPa for samples u1 to u4 and from 140 to 144 kPa for samples u5 to u8. Hence, the change of suction at the end of the constant water content tests was at most 10 kPa for all unsaturated samples.

The approximated critical state of each unsaturated sample of the loose volcanic fill is plotted in v – log p plane in Figure 5. It can be seen from the figure that the critical state for all unsaturated samples may be represented by a single critical state line despite their final suctions which range from 72 to 144 kPa. It can be noted that all stress states before shear for saturated samples lie on the wet side of critical and all the samples change from contractive to dilative behaviour during undrained shear (refer to Fig.2). On the other hand, only contractive behaviour is observed for all unsaturated samples despite two of the unsaturated samples (u1 & u5) lie on the dry side of critical and the rest lie on the wet side of critical. The gradients of the critical state lines ($\psi(s)$)

for unsaturated and saturated samples in v - log p plane are 0.233 and 0.087, respectively.

Figure 6 shows the critical state of the loose volcanic fill in q – p plane. It can be noted that a single critical state line may be used to represent the critical state for all unsaturated samples at a range of final suctions from 72 to 144 kPa. For the loose fill, it appears that matric suction has a large contribution to its shear strength for suctions ranging from 0 to 79 kPa but does not have a significant effect on its shear strength for suctions ranging from 72 to 144 kPa. The critical state line for unsaturated samples has a gradient of 1.33 and an intercept of 33 kPa. In addition, this line is approximately parallel to the one for saturated samples. This may suggest that the gradient (i.e., the angle of internal friction, ϕ') of the critical state line in q – p plane is a constant for the

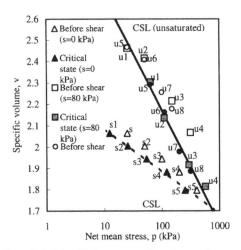

Figure 5. Relationship between specific volume and net mean stress for the loose volcanic fill

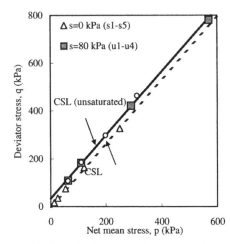

Figure 6. Relationship between deviator stress and net mean stress for the loose volcanic fill

limited range of applied suction. Gan & Fredlund (1996) reported that ϕ' is independent of the applied matric suction for two Hong Kong saprolites (Gan & Fredlund 1996). Hence the influence of matric suction on the shear strength of the unsaturated loose volcanic fill could be reflected by the intercept of the critical state line in the q – p plane and will be discussed in the next section.

5 COMPARISON BETWEEN EXPERIMENTAL RESULTS AND A THEORETICAL MODEL

5.1 Constitutive model (Toll 1990)

Under the extended critical state framework for unsaturated soils, Toll (1990) postulated the following shear strength equation for unsaturated soil,

$$q = M_a \cdot p + M_w \cdot s \qquad (1)$$

where M_a and M_w are two strength parameters related to net mean stress and matric suction, respectively.

Previously, Fredlund et al. (1978) suggested the following unsaturated shear strength equation:

$$\tau = c' + (\sigma - u_a) \cdot \tan\phi' + (u_a - u_w) \cdot \tan\phi^b \qquad (2)$$

where τ is the shear strength, c' is the effective cohesion, $\sigma - u_a$ is the net normal stress, $u_a - u_w$ is the matric suction, ϕ' is the angle of internal friction related to the net normal stress and ϕ^b is the angle indicating the rate of increase in shear strength with respect to the matric suction. In comparison to Equation 2, it appears that Equation 1 may be used

to model non-cemented soils only as no specific term can be used to describe effective cohesion. Besides, M_a and M_w are equivalent to $\tan\phi'$ and $\tan\phi^b$, respectively.

5.2 Comparison between volcanic fill and lateritic gravel

Based on test results of a lateritic gravel, Toll (1990) postulated that the strength parameters M_a and M_w depended on the degree of saturation. The above test results were conducted from soil samples compacted at different moisture contents (ranging from 17 to 31 %) and different compactive efforts (with specific volumes ranging from 1.766 to 2.219). As a result, different initial soil fabrics would be produced and this could contribute to the variation in the parameter M_a with respect to degree of saturation (Toll 1991).

In the present study, all samples were compacted at the same moisture content under the same compactive effort in order to produce the same initial soil fabric. Based on test results from samples with similar initial soil fabric, it may suggest that the parameter M_a is independent of matric suction ranging from 72 to 144 kPa for the volcanic fill. M_a is a strength parameter related to the frictional characteristic of the inter-particle contacts, which is an intrinsic property of the soil. As a result it may be independent of the stress state variable, such as matric suction.

Regarding the parameter M_w, Toll (1990) suggested that it could decrease with the degree of saturation and could become zero when the degree of saturation drops below 0.55. Figure 7 shows the variation in the parameter M_w with respect to degree of saturation for lateritic gravel and volcanic fill. Toll (1990) estimated the parameter M_w from Equa-

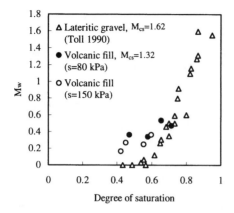

Figure 7. Relationship between M_w and the degree of saturation

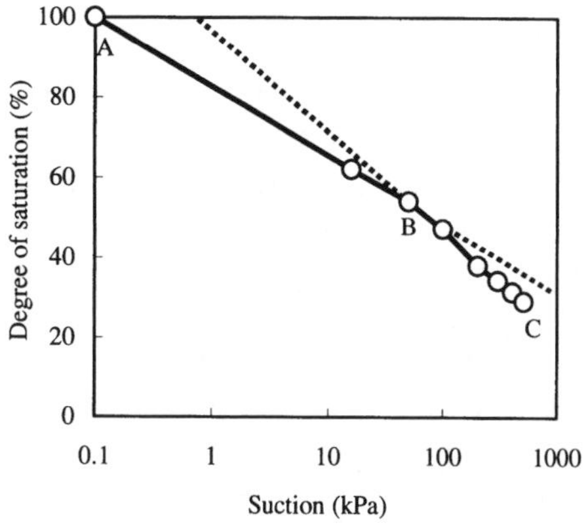

Figure 8. Soil water characteristic curve of loose volcanic fill

tion 1 by using multiple regression techniques and assuming both M_w and M_a are related to degree of saturation. In the present study, M_w for loose volcanic fill is estimated from Equation 1 by assuming a constant M_a. It can be seen from Figure 7 that the parameter M_w decreases with degree of saturation for both soils.

The desaturation of soils is caused by the matric suction exceeding their air entry values. When matric suction increases, the degree of saturation decreases and the parameter M_w also decreases. An increase in matric suction may not contribute to an increase in the second term of Equation 1 as the parameter M_w decreases at the same time. Hence the rate of increase in shear strength relative to matric suction may not be a constant and a nonlinear shear strength – matric suction relationship may be expected.

It can be seen from Figure 7 that most of the values of M_w of loose volcanic fill are greater than those of lateritic gravel over the range of degree of saturation used in this study. As lateritic gravel is much coarser than the loose volcanic fill (refer to Fig.1), the parameter M_w could be dependent on the grading of the soil. The contact area arising from the contractile skin (Fredlund and Morgenstern 1977) is related to the pore size of the soil, which is dependent on its particle size. Its influence on M_w will be discussed in the following section.

6 DISCUSSION

The above nonlinear shear strength - matric suction relationship has also been found for other soils (Gan & Fredlund 1996, Gan et al. 1988, Escario and Saez 1986). Gan & Fredlund (1996) postulated that this nonlinear relationship could be related to the soil water characteristic curve (SWCC). Figure 8 shows the measured SWCC for the loose volcanic fill used

in this study together with the extrapolations of the linear segments (AB & BC). It can be seen from the figure that the air entry value of the soil is not well defined for the loose volcanic fill. Nevertheless, it can be found from the figure that the loose volcanic fill exhibits two different rate of desaturation under suctions ranging from 0 to 500 kPa. The rate of desaturation at suction above 50 kPa is higher than that at a lower suction.

At low matric suction (below 50 kPa), the soil begins to desaturate. As enough air enters the pores, the contractile skin commences to form around the inter-particle contacts. The capillary effect arising from suction at the contractile skin may increase the normal forces at the inter-particle contacts (Wheeler and Karube 1995). These additional normal forces may enhance the friction at the inter-particle contacts. As a result, the unsaturated sample may exhibit a higher shear strength than the saturated sample. It agrees with the experimental results, where the shear strength of the unsaturated samples at matric suction around 72 to 79 kPa is higher than that of saturated samples.

When matric suction increases beyond 50 kPa, the degree of saturation drops to a lower value and the soil sample exhibits a higher rate of desaturation. The total contact area arising from the contractile skin also decreases significantly as a great amount of continuous air phase is formed throughout the soil. As a result, a further increase in matric suction may not be as effective as an increase in net normal stress for increasing the shear strength of the soil. Hence, M_w may decrease with a fall in the degree of saturation (refers to Fig. 7). For this range of suction, the effect of an increase in matric suction may be counter-balanced by a decrease in M_w. As a result, the shear strength of the unsaturated samples may not alter. This agrees with the experimental results observed for unsaturated samples at suctions ranging from 72 to 144 kPa.

555

7 CONCLUSIONS

Based on the test results of the loosely compacted volcanic fill, the following conclusions can be drawn:

Dilative behaviour is observed for saturated samples even though their states lie on the wet side of critical. However, contractive behaviour is observed for unsaturated samples despite the fact that some of their states lie on the dry side of critical.

A nonlinear shear strength – matric suction relationship is exhibited by unsaturated samples. For suctions ranging from 0 to 79 kPa, matric suction can have large contribution in their shear strength. However matric suction has relatively little effect on their shear strength for suctions ranging from 72 to 144 kPa.

The parameter M_a is related to the frictional characteristic at the inter-particle contacts of the soil and may be independent of matric suction. On the other hand, the parameter M_w, which indicates the rate of increase in shear strength with respect to matric suction, decreases with a decrease in the degree of saturation.

ACKNOWLEDGEMENTS

This research project is financially supported by the research grants HKUST6046/98E and CA99/00.EG 01 provided by the Research Grants Council of the Hong Kong Government of the Special Administrative Region (HKSAR) and the UK/HK Joint Research Scheme JRS96/39 provided by the British Council. The authors are grateful to the Geotechnical Engineering Office (GEO) of the HKSAR for providing soil samples to this project, in particular Drs Richard Pang and W.H. Sun and Messrs Y.C. Chan and W.K. Pun of GEO. The authors also wish to thank Prof. S. Wheeler for his invaluable discussion during second author's visit to University of Glasgow, UK.

REFERENCES

Alonso, E.E., Gens, A. & Josa, A. 1990. A constitutive model for partially saturated soils. *Geotechnique*, vol.40, no.3, 405-430.

Escario. V. & Saez, J. 1986. The shear strength of partly saturated soils. *Geotechnique*, vol.36, no.3, 453-456.

Fredlund, D.G. & Morgenstern, N.R. 1977. Stress state variables for unsaturated soils. *ASCE Journal of Geotechnical Engineering Division*, vol.103, 447-466.

Fredlund, D.G., Morgenstern, N.R. & Widger, R.A. 1978. The shear strength of unsaturated soils. *Canadian Geotechnical Journal*, vol.15, 313-321.

Gan, J.K.M. & Fredlund, D.G. 1988. Multistage direct shear testing of unsaturated soils. *Geotechnical Testing Journal, ASTM*, vol.11, 132-138.

Gan, J.K.M. & Fredlund, D.G. 1996. Shear strength characteristics of two saprolitic soils. *Canadian Geotechnical Journal*, vol.33, 595-609.

GCO 1988. *Geoguide 3: Guide to rock and soil descriptions*. Geotechnical Engineering Office, Civil Engineering Department, Hong Kong.

Hilf, J.W. 1956. *An investigation of pore water pressure in compacted cohesive soils*. Technical Memo 654, Denver, Bureau of Reclamation.

HKIE 1998. *Draft report : Soil nails in loose fill – A preliminary study*. The Geotechnical Division, Hong Kong Institution of Engineers (HKIE).

Ishihara, K. 1993. Liquefaction and flow failure during earthquakes. *Geotechnique*, vol.43, no.3, 349-416.

Ladd, R.S. 1978. Preparing test specimens using undercompaction. *Geotechnical Testing Journal*, Vol.1, 16-23.

Maatouk, A., Leroueil, S. & La Rochelle, P. 1995. Yielding and critical state of a collapsible unsaturated silty soil. *Geotechnique*, vol.45, no.3, 465-477.

Ng, C.W.W. & Chiu, C.F. 2000. Undrained behaviour of loosely compacted fill and its implications on slope stabilisation measures. Submitted to *8th International Symposium on Landslide*, Cardiff.

Toll, D.G. 1988. The behaviour of unsaturated compacted naturally occurring gravel. *PhD thesis, University of London*.

Toll, D.G. 1990. A framework for unsaturated soil behaviour. *Geotechnique*, vol.40, no.1, 31-44.

Toll, D.G. 1991. Discussion. A framework for unsaturated soil behaviour. *Geotechnique*, vol.41, no.1, 159-161.

Wheeler, S.J. & Karube, D. 1995. Constitutive modelling. *Proc. 1st International conference on Unsaturated Soils*, vol.3, 1323-1356, Paris.

Wheeler, S.J. & Sivakumar, V. 1993. Critical state framework for unsaturated soil. *Geotechnique*, vol.43, no.1, 35-53.

Unsaturated Soils for Asia, Rahardjo, Toll & Leong (eds) © 2000 Taylor & Francis, ISBN 90 5809 139 2

Direct shear properties of a compacted soil with known stress history

T. Nishimura
Ashikaga Institute of Technology, Japan

ABSTRACT: Stress history of soil profile is required in order to evaluate the shear strength of the compacted soil. Stress state variables of the *in situ* profile for an unsaturated soil consist of the confining pressure and matric suction. It is difficult to define overconsolidation ratio for unsaturated soil or compacted soils. This study used total stress ratio instead of overconsolidation ratio for compacted soils. Total stress ratio is defined as the ratio of compaction pressure to current confining pressure. In this study direct shear tests are performed on a compacted unsaturated soil subjected to known stress history. Shear stress and dilatancy of the unsaturated soil in direct shear were measured. The influence of the two stress variables on shear stress and dilatancy in compacted soils are noted.

1 INTRODUCTION

Soils in arid regions are generally unsaturated and, are characterized by high negative pore-water pressures. Regions related to arid climate are widely distributed in the world. Arid zone is climatically classified as extremely arid, arid and semiarid zones. Recently, the understanding of the unsaturated soil behavior extends to geotechnical engineers. Conventional equations for shear strength, volume change and seepage in unsaturated soil are established in the literature. In addition, technique of measurement soil suction or negative pore-water pressures is applied to unsaturated soil practice.

Geological deposition, erosion, drying, wetting and changes in environmental conditions determine the stress history of a soil profile. Such as unsaturated soils generally near the ground surface are commonly overconsolidated due to evaporation, permeation and environmental effects, the stress history is required in order to understand the behavior of unsaturated soils.

2 PURPOSE OF THIS STUDY

The stress state variables for an unsaturated, *in situ* profile consist of the net total stress and matric suction. Two stress variables in terms of effective stresses control the behavior of the unsaturated soil. The stress history of a soil is commonly determines using a one-dimensional oedometer test. Predicting of the stress history of an unsaturated soil is difficult.

Stress history is defined in terms of the preconsolidation pressure and current overburden effective stress in saturated soils. It is difficult to define overconsolidation ratio for natural unsaturated soils or artificially compacted unsaturated soils. This study used total stress ratio instead of overconsolidation ratio for a compacted unsaturated silty soil. Total stress ratio is defined as the ratio of compaction pressure to current confining pressure. As total stress ratio not included matric suction, total stress ratio is completely not the expression of present of stress states of unsaturated soils.

This study focuses on the direct shear properties of a compacted unsaturated silty soil with known stress history. The stress history of a compacted silty specimen is represented using the total stress ratio. Total stress ratio of compacted unsaturated silty soil varies from one to six. Direct shear tests are performed on a compacted unsaturated soil subjected to known stress history. Direct shear apparatus is modified to measure the unsaturated soil behavior. Modified direct shear apparatus in this test programs is possible to control both the pore-air pressure and the pore-water pressure in order to apply the matric suction and net normal pressure, respectively.

Shear stress and dilatancy of an unsaturated silty soil subjected to both matric suction and net normal pressure were measured. Relationship between total stress ratio and direct shear properties is discussed in this paper.

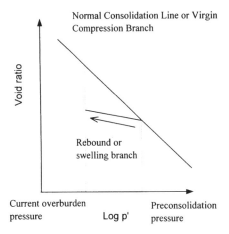

Figure 1. The relationship between effective stress and void ratio for a saturated soil.

3 LITERATURE

The overconsolidation ratio, OCR, had been used in geotechnical engineering as an expression of stress history for saturated soils (Fig. 1). The vertical stress versus void ratio relationship for a saturated soil is approximately linear on a logarithmic stress scale when the soil is normally consolidated. The stress states can not reach the region above and to the right of the virgin compression line. Any stress state inside the boundary surface gives rise to an overconsolidated soil. The overconsolidation ratio for a saturated soil is given by the following equation.

$$OCR = \frac{p_c}{p'_0} \qquad (1)$$

where: OCR = is overconsolidation ratio for a saturated soil, p'_0 = is the current overburden effective stress, and p_c = is the preconsolidation pressure.

The preconsolidation pressure, p_c, can be approximately as the stress at the intersection of the swelling line and the virgin compression line defined by one-dimensional loading. *In situ*, unsaturated soils and artificially compacted soils with negative pore-water pressure generally take on the character of being overconsolidated (Brooker and Ireland, 1965). The present *in situ* state of stress in a natural unsaturated soil and a compacted unsaturated soil is translated onto the total stress plane by performing oedometer tests with swelling processes. An overconsolidation ratio for unsaturated, compacted soils and unsaturated natural soils has not been clearly established.

Fredlund and Rahardjo (1993) suggested that the overconsolidation ratio, OCR, of an unsaturated soil. OCR suggested by Fredlund and Rahardjo (1993) is given by the following equation.

$$OCR = \frac{p_c}{p'_s} \qquad (2)$$

where: OCR = is the overconsolidation of unsaturated soils, p_c = is preconsolidation pressure, and p'_s = is corrected swelling pressure.

"Corrected" swelling pressure includes sampling disturbance, which is equal to the sum of the overburden pressure and the matric suction equivalent (Yoshida et al. 1983). The swelling pressure is not actually a characteristic or a property of the unsaturated soil, but is an indicator of the *in situ* stress states. The matric suction equivalent is not equal to the *in situ* matric suction but as the equivalent matric suction is reduced to zero and the soil is allowed to swell, the subsequent rebound to the overburden pressure corresponding to the rebound due to the reduction in the matric suction. The magnitude of the matric suction equivalent will be lower than the *in situ* matric suction. The difference between the *in situ* matric suction and the matric suction equivalent is related to degree of saturation of the soils (Fredlund and Rahardjo, 1993)

In this study, a total stress ratio, TSR, is defined to represent the loading history of a compacted unsaturated silty soil. A total stress ration, TSR, is given by the following equation.

$$TSR = \frac{p_{comp}}{(\sigma_v - u_a)} \qquad (3)$$

where: TSR = is total stress ratio, $(\sigma_v - u_a)$ = is current confining pressure, p_{comp} = is static compaction pressure, σ_v = is total stress, and u_a = is pore-air pressure.

Suggested the total stress ratio be not written in terms of the matric suction. This study involves the effect of matric suction on direct shear properties with test results. When the compaction pressure is equal to the current net vertical confining pressure, the total stress ratio is one. Highest overconsolidated soil specimen has a total stress ratio of six.

4 TEST PROCEDURES

4.1 *Soil material*

The physical properties of a silty soil are shown in Table 1. The silty soil was used in the test programs. The silty soil has a relatively uniform grain size distribution. The compacted soil specimens used in preparing the test specimens had a diameter of 50

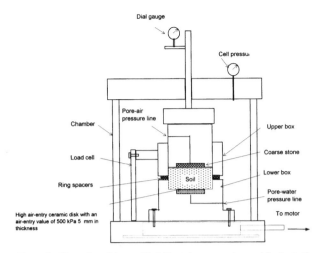

Figure 2. Illustration of the modified direct shear apparatus used for the testing program.

Table 1. Physical properties of the soil for the test program.

Specific gravity	2.65
Plasticity Index	non plastic
Maximum of soil particles	0.1mm
Sand (%)	0
Silt (%)	92
Clay (%)	8

mm and a height of 100 mm. The soil specimens were statically compacted at a water content of 13 %. Initially compacted silty soil had an initial matric suction of 43 kPa.

4.2 Direct shear apparatus

The direct shear apparatus used in the testing program is shown in Fig. 2. The apparatus is modified in order to investigate the unsaturated soil behaviors. Two stress state variables (matric suction and confining pressure) are possible to control to a compacted unsaturated soil in the direct shear apparatus. The apparatus was similar in design presented by Escario (1980), Gan and Fredlund (1988), Campus and Carrillo (1995); and Bastos et al. (1998). However, the soil specimens in this study were circular in shape. The cylindrical chamber was constructed of acrylic. The direct shear box was completely enclosed in a pressure chamber in order to elevate the ambient air pressure during the test. The shear box consisted of an upper shear box and a lower shear box. Five thin ring spacers were involved between the upper shear

box and lower shear box. In this study, preliminary pilot tests were performed to measure the friction between ring spacers. It was found that the friction due to the shear displacement of the box was a negligible value. Shear forces were induced to a soil specimen by displacing the lower shear box. The ceramic disk was installed on bottom portion of the shear box. The independent control of both the pore-air pressure and the pore-water pressure required the use of an air-entry ceramic disk. A controlled value of maximum matric suction depends on the air-entry value of the ceramic disk. The air-entry value of the ceramic disk used in this tests program was 500 kPa. The ceramic disk had a thick of 5 mm.

4.3 Test program

A one-dimensional static compaction pressure was applied to prepare the silty soil at a constant water content in the direct shear ring. Pore-air pressure, pore-water pressure and net normal stress were applied to the soil specimens for a period of twenty-four hours for the applied stress to reach equilibrium. The pore-air pressure was applied to the upper surface of the soil specimen in the chamber. The pore-water was allowed to drain. The pore-water pressure in the soil maintained at atmospheric conditions through the ceramic disk. The compaction pressure, total vertical pressure and matric suction associated with test program are shown in Table 2.

The compaction pressure for all specimens in test program was a constant of 600 kPa. Initial matric suction of soil specimen was 33.1 kPa when the soil was compacted at 600 kPa. Confining pressure applied to compacted soil specimen ranged from 100

Table 2. Stress states associated with test program.

Compaction pressure (kPa)	Confining pressure (kPa)	Total stress ratio	Matic suction at applied compaction pressure (kPa)	Applied matric suction (kPa)	Estimated matric suction after unloading (kPa), $B_w = 0$		Estimated matric suction after unloading (kPa), $B_w = 0.1$	
600	100	6	33.1	100	33.1	Drying	83.1	Drying
600	100	6	33.1	200	33.1	Drying	83.1	Drying
600	100	6	33.1	300	33.1	Drying	83.1	Drying
600	100	6	33.1	400	33.1	Drying	83.1	Drying
600	200	3	33.1	100	33.1	Drying	73.1	Drying
600	200	3	33.1	200	33.1	Drying	73.1	Drying
600	200	3	33.1	300	33.1	Drying	73.1	Drying
600	200	3	33.1	400	33.1	Drying	73.1	Drying
600	300	2	33.1	100	33.1	Drying	63.1	Drying
600	300	2	33.1	200	33.1	Drying	63.1	Drying
600	300	2	33.1	300	33.1	Drying	63.1	Drying
600	300	2	33.1	400	33.1	Drying	63.1	Drying
600	600	1	33.1	100	33.1	Drying	33.1	Drying
600	600	1	33.1	200	33.1	Drying	33.1	Drying
600	600	1	33.1	300	33.1	Drying	33.1	Drying
600	600	1	33.1	400	33.1	Drying	33.1	Drying

Drying indicates that water moved out of the soil specimen during testing.
Swelling indicates that water moved into the soil specimen during testing.

kPa to 600 kPa. Computed total stress ratio varied from one to six. Since the matric suction probably change when the vertical pressure was lowered from 600 kPa, the matric suction of each compacted specimen during unloading was computed using a pore pressure coefficient, B_w. The pore pressure coefficient, B_w, was estimated to be between 0 and 0.1 since the initial degree of saturation was approximately 30 % (Campbell (1973); and Skempton and Bishop (1954)).

5 TEST RESULTS AND DISCUSSIONS

Figure 3 shows the relationship between shear stress and horizontal displacement at a constant confining pressure of 100 kPa for compacted silty soil with a total stress ratio of six. Shear stress increased with increasing horizontal displacements and with increasing matric suctions. The shear stress versus horizontal displacement curve with total stress ratio of six shows a strain hardening type of behavior. The shape of stress versus horizontal displacement curve at a matric suction value of 100 kPa is different from the shape of curves at the matric suction value of 400 kPa. A peak shear stress could not be obtained for the compacted silty soils at matric suction of 400 kPa. The maximum or peak shear stress is usually used to establish the failure envelope in soil mechanics. If the shear stress versus horizontal displacement curves did not exhibit a peak stress, the shear stress at an arbitrary horizontal shear dis-

placement value (e.g., 12.0mm) was selected in the test results. It is found that the shape of the shear stress versus horizontal displacement curve is influenced by the matric suction.

Vertical displacements were induced by shear displacement. Relationship between vertical displacement and horizontal displacement at a constant confining pressure of 100 kPa is shown in Fig. 4. Compacted silty soil with a total stress ratio of six indicates the reduction of soil volume due to the increase in horizontal displacement. A negative vertical displacement increases with horizontal displacement. The negative vertical displacement is small at matric suction of 100 kPa. Negative vertical displacement value increases with matric suction.

Fredlund et al. (1978) presented the shear strength equation for unsaturated soil.

$$\tau = c' + (\sigma_n - u_a)\tan\phi' + (u_a - u_w)\tan\phi^b \qquad (4)$$

where: τ = shear strength, c' = effective cohesion intercept, σ_n = total normal stress, ϕ' = effective angle of internal friction with respect to confining pressure, u_a = pore-air pressure, u_w = pore-water pressure, ϕ^b = angle of internal friction with respect to matric suction, $(\sigma_n - u_a)$ = confining pressure, and $(u_a - u_w)$ = matric suction.

Confining pressure and matric suction are important stress state variables in shear strength of an unsaturated soils. Figure 5 shows relationship between shear strength and total stress ratio at a constant compaction pressure of 600 kPa and varying

560

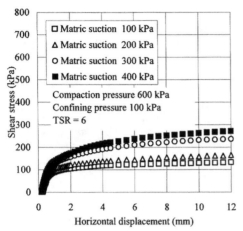

Figure 3. Stress versus horizontal displacement curves for compacted specimens.

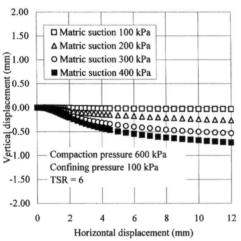

Figure 4. Change in vertical displacement for compacted specimens.

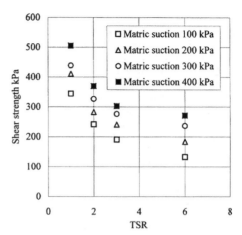

Figure 5. Relationship between TSR and shear strength (Compaction pressure 600 kPa).

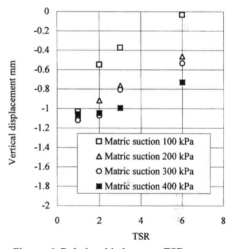

Figure 6. Relationship between TSR and vertical displacement (Compaction pressure 600 kPa)

matric suctions. Shear strength of compacted unsaturated silty soil decreases with total stress ratio. When total stress ratio ranges from one to three, shear strength reduces rapidly. After stress ratio reaches three, shear strength decreases slightly. Shear strength increases with increasing of matric suction at a constant total stress ratio.

Figure 6 shows relationship between total stress ratio and vertical displacement at a set arbitrary horizontal shear displacement value (e.g., 12.0mm). Compacted silty soil indicates the reduction of soil volume due to shear horizontal displacement. When total stress ratio is equal to one, vertical displacement shows the largest reduction. Vertical dis-

placement increases with increasing total stress ratios.

6 CONCLUSIONS

This study was performed on direct shear test for a compacted unsaturated silty soil using a modified direct shear apparatus. The modified direct shear apparatus can be used to control the two stress state variables (i.e., matric suction and confining pressure) in terms of effective stress for unsaturated soils. This paper report shear stress versus horizontal displacement and vertical displacement versus horizontal displacement of a compacted unsaturated silty soil. Stress paths of a compacted silty soil was evidenced using a total stress ratio in compaction process. A total stress ratio was defined as ratio of compaction pressure to current confining pressure for an unsaturated soil.

Shear stress of a compacted unsaturated silty soil applied matric suction increases in direct shear stress versus horizontal displacement. A reduction in volume of a compacted silty soil was due to direct shear forces. The shear stress versus horizontal displacement curve with a total stress ratio of six showed a strain hardening type of behavior. Shear strength of compacted unsaturated silty soil decreases with increasing total stress ratio at a constant compaction pressure. When total stress ratio is equal to one, the reduction of soil volume due to horizontal shear displacement indicates the largest value.

REFERENCES

Bastos, C.A., Gehling, W.Y.Y. & Bica, A.D. 1998. Some considerations about the shear strength and erodibility of unsaturated residual soils. *Proceedings of the Second International Conference on Unsaturated Soils, Beijing:* 19-24.

Brooker, E.W. and Ireland, H.O. 1965. Earth pressures at-rest related to stress history. *Canadian Geotechnical Journal:* Vol.2, No.1, 1-15.

Campbell, J.D. 1973. Porepressure and volume changes in unsaturated soils, Ph.D. dissertation. *University of Illinois at Urbana-Champaign.* 104.p.

Campos, T.M.P. and Carrillo, C.W. 1995. Direct shear testing on an unsaturated soil from Rio de Janeiro. *Proceedings of the First International Conference on Unsaturated Soils, Paris France:* Vol.1, 31-38.

Escario, V. 1980. Suction controlled penetration and shear tests. *Proceedings of the Fourth International Conference on Expansive Soils, ASCE:* Vol.2, 781-797.

Fredlund, D.G., Morgestern, N.R. and Widger, A. 1978. Shear strength of unsaturated soils. *Canadian Geotechnical Journal:* Vol.15, No.3, 313-321.

Fredlund, D.G. and Rahardjo, H. 1993. An overview of unsaturated soil behavior. *Proceeding of the American Society of Civil Engineering Annual Convention on Unsaturated Soil Behavior in Engineering Practice, Dallas. Texas. Oct:* 24-28, 1-31.

Fredlund, D.G. and Rahardjo, H. 1993. Soil mechanics for unsaturated soils. A Wiley-Interscience Publication, JOHN WILEY & SONS, INC. 517.

Gan, J.K.M. and Fredlund, D.G. 1988. Multistage direct shear testing of unsaturated soils. *Geotechnical Testing Journal, GTJOD, ASTM:* Vol.11, No.2, 132-38.

Skempton, A.W. and Bishop, A.W. 1954. Building materials. Their elasticity and inelasticity, In Soils, Amsterdam, North Holland. Chapter 10.

Yoshida, R.T., Fredlund, D.G, and Hamilton, J.J. 1983.The prediction of total heave a slab-on-grade floor on Regina clay. *Canadian Geotechnical Journal:* Vol.20, 69-81.

Unsaturated Soils for Asia, Rahardjo, Toll & Leong (eds) © 2000 Taylor & Francis, ISBN 90 5809 139 2

Relationship between shear strength and matric suction in an unsaturated silty soil

T. Nishimura
Ashikaga Institute of Technology, Japan

D.G. Fredlund
Department of Civil Engineering, University of Saskatchewan, Saskatoon, Sask., Canada

ABSTRACT: Unsaturated soil regions with high soil suctions are generally located near ground surface. This study reports the results of triaxial tests on a silty soil. The modified triaxial cell apparatus applied a constant net normal stress and varying soil suctions as well as having a constant high total suction with the net normal stress varying. Matric suctions were applied to soil specimens in the triaxial cell, through a ceramic base plate for suctions up to 450 kPa. Higher total suctions were applied to soil specimens in a glass desiccator where salt solutions were used to create a constant relative humidity chamber. Total suction values up to 292,400kPa were used in this test program. This study showed that the failure envelope with respect to suction was composed of two segments; namely, a curvilinear segment at low suctions and a horizontal segment at high suctions.

1 INTRODUCTION

An unsaturated soil profile can be classified as follows; namely, a dry soil zone, a two-phase zone and a capillary fringe zone. A soil is classified as dry when its water content is in the residual stage of unsaturation. This is often the situation near to the ground surface. Geotechnical engineers need to be able to assess the shear strength of unsaturated soils in the dry or high suction range. Dry soils have a behavior somewhat different in character from that of saturated soils.

A theoretical framework for volume change, shear strength and permeability for unsaturated soils has become generally accepted in geotechnical engineering. Unsaturated soil property functions for an unsaturated soil are required for numerical modeling. Examples of unsaturated soil property functions are the coefficient of permeability functions, the water storage functions, the shear strength functions and the volume change functions. One of several procedures can be used to obtain unsaturated soil property functions. These are direct measurements, the use of the soil-water characteristic curve and the use of classification tests such as the grain size test. The mathematical modeling of unsaturated soils through the use of soil-water characteristic curves is becoming more generally accepted. Several proposed mathematical procedures appear to provide a reasonable estimate of the non-linear character of unsaturated soil property functions.

A conventional triaxial test or a modified direct shear test can be used to measure the relationship between soil suction and shear strength. Its weakness is that it has proven to be costly and time-consuming. Unsaturated soil mechanics has a lack of experimental data on the shear strength of a soil in the residual stage of unsaturation. It is important to quantify the contribution to shear strength over the entire suction range (i.e., 0-1,000,000 kPa).

1.1 *Purpose of this study*

This study reports the results obtained from (i) triaxial tests at a constant net confining pressure (i.e., 25 kPa) and a varying matric suction (from the low suction range to the high suction range) and (ii) triaxial tests at a constant high suction (i.e., 217,000 kPa) and varying net confining pressure. The triaxial tests, using a modified, conventional cell, were performed on a statically compacted silty soil. Matric suctions were applied to the soil specimens placed on a ceramic plate for suctions up to 450 kPa. Soil specimens were brought to equilibrium with high matric suctions through the vapor pressure technique using a salt solution and the relative humidity control technique.

This study investigates the relationship between the shear strength and matric suction over the entire range of suction. This study determines a fitting parameter, κ, associated with the empirical equation proposed by Fredlund et al. (1995) in order to estimate the shear strength of an unsaturated soil. The shear strength parameter, ϕ', with respect to net confining pressure is assumed to be coincident with an effective angle of internal friction for the saturated soil.

1.2 Review of shear strength of an unsaturated soil

The shear strength of unsaturated soils been formulated in terms of two independent stress state variables (Fredlund et al. 1978). The linear form of the equation for the shear strength of an unsaturated soil is commonly expressed as follows:

$$\tau = c' + (\sigma_n - u_a) \cdot \tan\phi' + (u_a - u_w) \cdot \tan\phi^b \quad (1)$$

where: τ = shear stress on the failure plane at failure, c' = intercept of the "extended" Mohr-Coulomb failure envelope on the shear stress axis where the net normal stress and the matric suction are equal to zero, $(\sigma_n - u_a)$ = net normal stress on the failure plane, ϕ' = angle of internal friction with respect to the net normal stress, $(u_a - u_w)$ = matric suction, ϕ^b = angle indicating the rate of increase in shear strength relative to matric suction.

The shear strength of unsaturated soils can be determined in the laboratory using a modified, conventional triaxial cell and modified direct shear apparatus (Gan and Fredlund 1988). The shear strength versus matric suction envelope has been found to be non-linear (Gan et al. 1988). Donald (1956) shows that the slope of the shear strength versus matric suction envelope can even be negative.

Gan and Fredlund (1996) studied the saturated and unsaturated shear strength behavior of two saprolitic soils using direct shear tests and triaxial tests. Results show that the relationship between shear strength and matric suction takes on a non-linear form. The extent of the increase in the shear strength with matric suction is related to the soil-water characteristic curve for the soil and to the amount of dilation occurring during shear.

Mahalinga-Iyer and Williams (1985) conducted unconsolidated undrained triaxial tests on two lateritic soils. High suctions on the standard compacted soil specimens were measured using the filter paper method. The results show that the slope of the failure surface with respect to the matric suction decreases sharply in the suction range up to 1,000 kPa. When the matric suction exceeds 1000 kPa, the shear strength decreases slightly with an increase in matric suction.

Fredlund et al. (1995) proposed a model for predicting the shear strength of unsaturated soils using the soil-water characteristic curve and saturated shear strength parameters, c', ϕ'. The proposed equation is as follow:

$$\tau = c' + (\sigma_n - u_a)\tan\phi' + (u_a - u_w)\Theta^\kappa \tan\phi' \quad (2)$$

where: Θ = dimensionless volumetric water content, κ = is the fitting parameter.

The dimensionless water content is expressed as:

$$\Theta = \frac{\theta_w}{\theta_s} \quad (3)$$

where: θ_w = volumetric water content at any suc-

tion, and θ_s = volumetric water content at saturation. Fredlund et al. (1995) indicate that satisfactory predictions of the shear strength of an unsaturated soil with measured data. The fitting parameter, κ, depends on soil properties such as the plasticity of the soil; taken to be one or greater for highly plastic soils.

Vanapalli et al. (1996) emphasize that the soil-water characteristic curve is closely related to the shear strength of an unsaturated soil and suggested a second procedure for its estimation. The formulation for the estimation of shear strength was based on same concept as used in the empirical equation proposed by Fredlund et al. (1995).

$$\tau = c' + (\sigma_n - u_a)\tan\phi' + (u_a - u_w)\left[\tan\phi'\left(\frac{\theta_w - \theta_r}{\theta_s - \theta_r}\right)\right] \quad (4)$$

where: θ_w = volumetric water content at any matric suction, θ_s = saturated volumetric water content, θ_r = residual volumetric water content.

The relative water content term (i.e., $(\theta_w - \theta_r)/(\theta_s - \theta_r)$) is called the normalized water content. In recent years several investigators have proposed empirical procedures to predict the shear strength function for unsaturated soil by using the soil-water characteristic curve and the saturated shear strength parameters. The soil-water characteristic curve provides a conceptual understanding for the behavior of unsaturated soils. In 1999 Vanapalli and Fredlund summarized various mathematical prediction model for the shear strength of an unsaturated soil.

2 TEST PROCEDURE

A fine-grained silty soil with no cohesion was used for the shear strength tests. The statically compacted specimens had a height of 100 mm and a diameter of 50 mm. The initial water content was 9.6 %. The laboratory testing program consisted of Test Program 1 and Test Program 2. The two testing programs involved the use of a conventional triaxial cell modified for both air-pressure and pore-water pressure control. A five bar high air entry disk was sealed onto the bottom pedestal of the modified triaxial cell and was used to facilitate the separate control of the pore-air pressure and pore-water pressure.

2.1 Test program 1

Test Program 1 involved the consolidation stage and shearing stage under a constant net confining pressure but with varying matric suctions. The net confining pressure was 25 kPa. The matric suction applied to the soil specimen varied up to 450 kPa. The constant net confining pressure and matric suction were maintained during the shearing process.

Table 1. Salt solutions recommended for relative humidity control.

Salt Solution	Relative humidity %	Total suction kPa
Sodium Choride	75.5	38,600
Magnesium Nitrate	54.3	84,000
Magnesium Choride	33	152,400
Lithium Choride	11.5	292,400

High suction values were applied to the specimen using the vapor equilibrium technique and the relative humidity control technique. Salt solutions were used in the glass desiccator along with the soil specimens, in the vapor equilibrium technique. Four different salt solutions were used in these tests to obtain four relative humidities and soil suctions as summarized in Table 1 (Yong 1967). After the air voids in the soil were approaching equilibrium with each relative humidity applied by the salt solutions, the soil specimens was assumed to be desiccated. The desiccated soil specimens were consolidated under an isotropic confining pressure of 25 kPa in the triaxial cell. A low shearing rate of 0.014 mm per minute was adopted for strain. The deviator stress was applied until the axial strain exceeded 20%. In Test Program 1, the relationship between maximum deviator stress (i.e., shear strength) and matric suction was observed over a wide range of soil suctions.

2.2 Test program 2

All soil specimens in Test Program 2, were placed directly into a temperature and relative humidity controlled chamber in order to apply a high total suction. A relative humidity of 20% was selected for Test Program 2. A relative humidity of 20% corresponds to a total suction of 217,000 kPa. The soil specimens subjected to the high suctions was sheared under varying net confining pressures. The net confining pressures ranged from 24.5 kPa to 73.5 kPa. The rate of axial strain was the same as in Test Program 1. The shear strength parameter with respect to net confining pressure was measured.

The soil-water characteristic curve for the silty soil was also measured. It is generally accepted in geotechnical engineering that the estimation of the unsaturated soil property functions are closely related to the soil-water characteristic curve. The change in water content of the soil with increasing soil suction was measured in terms of gravimetric water content, w, using the pressure plate technique, the vapor pressure technique and the relative humidity control technique.

3 TEST RESULTS AND DISCUSSIONS

3.1 Soil-water characteristic curve

Figure 1 shows the soil-water characteristic curve for the silty soil. The soil-water characteristic curve shows the desaturation characteristics of the silty soil in the drying process. The gravimetric water content as the ordinate axis shows that the amount of water in the soil remains constant for suctions up to about 10 kPa. The key features of the soil-water characteristic curve are the air-entry value, the slope of the straight line portion of desorption branch, the residual water content and the residual suction. The air-entry value of the soil can be identified as the pressure at which air first enters the pores of the soil. The air-entry value for the silty soil was obtained by extending the constant slope portion of the soil-water characteristic curve to intersect the horizontal suction axis at saturated conditions. The air-entry value for the silty soil is about 10 kPa.

Figure 1 can be visualized as an integrated frequency distribution curve of the pore sizes in the soil. The curve has been defined using a numerical prediction model equation proposed by Fredlund and Xing (1994). All the soil pores are filled with water until the soil suction reaches the air-entry value. After the matric suction reaches the air-entry value, desaturation starts in the transition stage. The amount of water in the soil decreases significantly with increasing soil suction in the transition stage. A coarse-grain soil such as a gravel or sand shows a tendency to desaturate at a fast rate with increasing

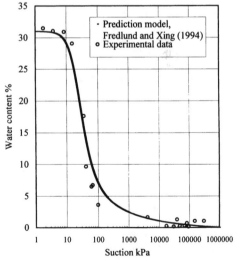

Figure 1. Soil-water characteristic curve for the silty soil.

Table 2. Summary of stress conditions during triaxial tests.

Net normal stress kPa	Soil suction kPa	Shear strength kPa
25	68	77
25	120	89
25	140	115
25	160	111
25	220	143
25	300	148
25	450	220
25	38600	204
25	84000	188
25	152400	166
25	292400	212
25	217000	212
49	217000	242
74	217000	272

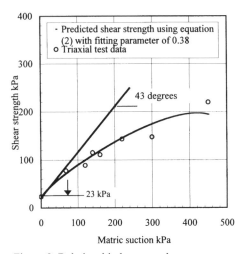

Figure 2. Relationship between shear strength and matric suction at constant net normal stress of 25 kPa.

suctions. The rate of desaturation, however, decreases with an increase in fine content. The slope of branch in the transition stage for the silty soil is steeper than that of kaolin.

Eventually the desaturation stage reaches the residual stage of unsaturation along the desorption portion of a soil-water characteristic curve. Large increases in soil suction lead to a relatively small change in water content in the residual stage of unsaturation. At suctions greater than the residual suction, flow occurs through the pores in the vapor phase. The wetted contact area has reduced significantly and the volume change in the unsaturated soil becomes negligible. The residual state of unsaturation can be defined using an empirical, graphical procedure. The residual suction was obtained as the point where the line extending from 1,000,000 kPa along the curve intersects the line tangent to the steeper portion of the soil-water characteristic curve. The residual water content and residual suction of the silty soil are 2.5 % and 200 kPa, respectively.

3.2 Relationship between shear strength and matric suction

The relevant stress conditions (i.e., shear strength) during the triaxial tests, either under a constant net normal stress and varying matric suction or under a constant suction and varying net normal stress, are summarized in Table 2. Figure 2 shows the relationship between shear strength and matric suction up to 450 kPa with the triaxial tests under a constant net normal stress. Shear strength increases with increasing matric suction. An "extended" form of the Mohr-Coulomb equation (i.e., Eq.1), is used to interpret the test results. When matric suction is zero (i.e., saturated conditions), the shear strength is calculated as 23 kPa with c' equal to 0 kPa, ϕ' equal to

43 degrees under a constant net normal stress of 25 kPa. That value shows the point where the failure envelope intersects the shear strength versus matric suction plane. The entire failure envelope is defined using the equation (2). The computed failure envelope represents the best-fit line with fitting parameter, κ, of 0.38. The failure envelope indicated a linear relationship until matric suction up to the air-entry value of 10 kPa. The initial slope of the shear strength versus matric suction envelope, ϕ^b, is equal to the effective friction angle, ϕ' at suctions below the air entry value. At matric suctions exceeding the air-entry value, the measured shear strength increases. The value of the friction angle, ϕ^b, however, decreases when the desaturation approaches residual conditions. The shear strength versus matric suction envelope was found to be non-linear. The peak strength envelope is composed of two segments, a linear segment and a curvilinear segment. The transition from the linear segment to the curvilinear segment occurs at approximately the air-entry value of soil. At or near a matric suction of 400 kPa, the slope of failure envelope, ϕ^b, approaches an angle of approximately zero. An increase in matric suction produces no further increase in shear strength. The effect of matric suction on the shear strength is clearly related to the soil-water characteristic curve of the silty soil.

3.3 Relationship between shear strength and total suction

The shear strength data from triaxial tests on a silty soil subjected to high soil suctions, using the salt solution desiccator technique, are presented in Fig. 3. A constant net normal stress of 25 kPa was applied and various total suctions were used. The soil

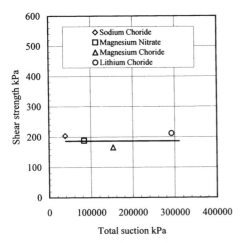

Figure 3. Relationship between shear
strength and total suction
at a net normal stress of 25 kPa.

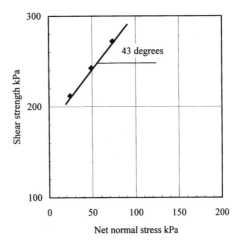

Figure 4. Relationship between shear strengrth
and net normal stress
at a constant suction of 217,000 kPa.

suctions applied were always larger than the residual
suction of 200 kPa. Figure 3 shows that a considerable increase in soil suction produces no increase in
shear strength. At suctions in excess of residual
suction, the shear strength envelope is essentially a
horizontal line. The suction in the residual stage
makes less of a contribution to increasing the shear
strength as compared to the suction in the transition
stage. Fredlund (1998) commented that the shear
strength of an unsaturated soil appears to remain essentially constant beyond residual state suction from
a theoretical and empirical model standpoint without
undertaking laboratory tests.

3.4 *Relationship between shear strength and net normal stress for unsaturated soil subjected to high total suction*

Figure 4 shows the relationship between shear
strength and net normal stress under a constant total
suction of 217,000 kPa. High total suctions were
applied to the silty soil equilibrated at a relative humidity of 20%. The failure envelope with respect to
the net normal stress was studied using the triaxial
test results. The slope of that failure envelope (i.e.,
the angle, ϕ', in Eq.1) was computed to be 43 degrees. The computed, ϕ', values are quite consistent
with the effective internal friction angle for the saturated silty soil. The effective angle of internal friction for the saturated soil appears to equally contribute to the shear strength of unsaturated soil.

4 CONCULUSIONS

This study conducted triaxial compression tests for
an unsaturated silty soil over a wide range of soil

suctions. Suctions were applied using the pressure
plate apparatus, salt solution desiccators and a relative humidity control chamber. The compacted, unsaturated silty soil was desaturated to the residual
state of unsaturation through the use of high suctions. The relationship between shear strength and
soil suction is non-linear as it goes into the transition
stage. After the suction exceeds the residual suction,
the shear strength remains a constant (i.e., essentially
horizontal failure envelope with respect to soil suction). The angle, ϕ^o, with respect to suction varied
from the angle of internal friction of the saturated
silty soil to zero at high suctions.

This study also used the triaxial test to study the
effect of net normal stress on shear strength of a soil
in the residual state of unsaturation. The shear
strength increases with the net normal stress for the
unsaturated silty soil with a suction of 217,000 kPa.
The ratio of shear strength to net normal stress is
coincident with the effective angle of internal friction of the saturated silty soil.

REFERENCES

Donald, I.B. (1956), Shear strength measurements in
  unsaturated non-cohesive soils with negative pore
  pressures, *Proceedings of 2nd Australia and New
  Zeland Conference on Soil Mechanics and Foundation Engineering, Christchurch, New Zealand.*
  pp. 200-205.
Fredlund, D.G. (1998), Bringing unsaturated soil
  mechanics into engineering practice. *Proceedings
  of the Second International Conference on UN-
  SATURATED SOILS, Beijing, China.* Vol. 2. pp.
  1-36.

Fredlund, D.G., Morgenstern, N.R. and Widger, R.A. (1978), The shear strength of unsaturated soils, *Canadian Geotechnical Journal*, Vol. 15, pp. 313-321.

Fredlund, D.G. and Xing, A. (1994), Equations for the soil-water characteristic curve, *Canadian Geotechnical Journal*, Vol. 31, pp. 521-532.

Fredlund, D.G., Xing, A., Fredlund, M.D. and Barbour, S.L. (1995), The relationship of the unsaturated soil shear strength to the soil-water characteristic curve, *Canadian Geotechnical Journal*, Vol. 32, pp. 440-448.

Gan, J.K-M. and Fredlund, D.G. (1988), Multistage direct shear testing of unsaturated soils, *Geotechnical Testing Journal, ASTM*, Vol. 11, pp. 132-138.

Gan, J.K-M. and Fredlund, D.G. (1996), Shear strength characteristics of two saprolitic soils, *Canadian Geotechnical Journal*, Vol. 33, pp. 595-609.

Gan, J.K-M., Fredlund, D.G. and Rahardjo, H. (1988), Determination of the shear strength parameters of an unsaturated soil using the direct shear test, *Canadian Geotechnical Journal*, Vol. 25, pp. 500-510.

Mahalinga-Iyer, U. and Williams, D.J. (1985), Unsaturated strength behaviour of compacted lateritic soils, *Geotechnique*, Vol. 45, No. 2, pp. 317-320.

Vanapalli, S.K. and Fredlund, D.G. (1999), Empirical procedures to predict the shear strength of unsaturated soils, *Proceedings of the Eleventh Asian Regional Conference on Soil Mechanics and Geotechnical Engineering, Seoul, Korea*, pp. 93-96.

Vanapalli, S.K., Fredlund, D.G., Pufahl, D.E. and Clifton, A.W. (1996), Model for the prediction of shear strength with respect to soil suction, *Canadian Geotechnical Journal*, Vol. 33, pp. 379-392

Yong, J.F. (1967), Humidity control in the laboratory using salt solutions-A review, *Journal of Applied Chemistry*, Vol. 17, pp. 241-245.

Unsaturated Soils for Asia, Rahardjo, Toll & Leong (eds) © 2000 Taylor & Francis, ISBN 90 5809 139 2

Analysis of a steep slope in unsaturated pyroclastic soils

A. Scotto di Santolo
Dipartimento di Ingegneria Geotecnica, Università di Napoli Federico II, Italy

ABSTRACT: This paper regards the properties of unsaturated pyroclastic soils involved in a shallow instability phenomenon in a cut slope. The slope is located along the road connecting the urban area of Naples to
the *Pozzuoli* suburb (10-km west from Naples, Italy). A complete analysis of stability in steep unsaturated soil
slopes requires the knowledge of the relationship between the shear strength, matric suction and rainfall.
Therefore mechanical and hydraulic behaviour of soil involved has been investigated through extensive laboratory activities. The main aim of this investigation is to define shear strength characteristics as well as to establish the relationship between shear strength and water content (or suction). Slope stability is then evaluated
with a simple approach that accounts for the contribution of suction to shear strength.

1 INTRODUCTION

In steep slopes with a deep groundwater table the
negative pore water pressures (or matric suctions)
play a positive role increasing the shear strength of
the soil and thus the stability of the slopes. The
safety conditions of slopes is therefore affected by
climatic conditions, such as rainfall and evapotranspiration, which control the changes of matric suction near the ground surface. As a matter of fact the
rainfall causes both an increase of pore water pressures (toward positive values), with a consequence
of a decrease in soil shear strength, and an increase
in water content, that results in the increase of soil
unit weight (or driving force). The combinations of
these effects triggered several slope failures
[Brand, 1985].

In order to predict the stability of a slope a value
of the shear strength of a soil is required.

Fredlund & Morgenstern [1977] showed that the
shear strength of unsaturated soil can be described
by two stress state variables net stress (defined as
the difference between total stress (σ) and air pressure (u_a)) and suction (defined as the difference between air and water pressure (u_w)). A linear shear
strength equation was proposed by Fredlund et al.
[1978]:

$$\tau = c + (\sigma - u_a) \cdot \tan\varphi + (u_a - u_w) \cdot \tan\varphi^b \qquad (1)$$

where τ = is the shear strength of an unsaturated
soil; c and φ = are soil strength parameters; φ^b = is
the angle of shearing resistant with respect to matric
suction; ($\sigma - u_a$) = is the net normal stress on the

plane of failure, and ($u_a - u_w$) is the matric suction of
the soil.

Other researchers demonstrated, by experimental
activities, that there is a nonlinear increase of shear
strength as a results of an increase in matric suction
[Escario & Saez, 1986; Fredlund et al., 1987; Gan et
al., 1988].

Unfortunately the analysis of stability requires,
according to equation (1), the prediction of both
shear strength parameters and of matric suction in
situ profile of unsaturated soil.

Tests on unsaturated soil require triaxial or direct
shear apparatus that have been modified to allow for
the control and (or) measurement of pore-air and
pore water pressures. These apparatuses are costly
and complex and the tests require time-consuming
procedures.

As shown the shear strength of unsaturated soil is
dependent on the matric suction, that is a function of
water content according to the soil-water characteristic curve. Therefore it may be assumed that the
shear strength of an unsaturated soil is also a function of water content [Fredlund et al., 1995; Vanapalli et al., 1996].

In this paper is proposed an approximate procedure to evaluate the influence of water content and
matric suction on the shear strength of an unsaturated soil. This procedure is based on the determination of the shear strength at different water content
and the soil-water characteristic curve. The main advantage of this approach is that require standard
laboratory shear apparatus and a pressure plate. This
procedure also require a relatively short time, and it

Figure 1. Geological sketch of *Campi Flegrei* and location of the study area (✦): 1) weathered pyroclastic deposits and alluvial deposit; 2) recent pyroclastic deposits and lava; 3) Neapolitan yellow tuff; 4) Grey tuff [from Di Vito et al., 1985].

seems to be effective method for the study of unsaturated shear strength in engineering practice.

2 SITE CHARACTERISTICS

The selected study area is located in the exterior edge of *Conca di Agnano* (Figure 1), an area northwest of Naples, along the road connecting the urban area of Naples to the Pozzuoli suburb (about 10 km west of Naples, Italy), where several instability phenomena occurred during our century [Pellegrino, 1994].

The geology of the investigated area consists of the pyroclastic deposits following eruptions of the volcanic district of *Campi Flegrei* (22,000 years Bp.). The studied zone is outcrop of pyroclastic deposits of eruptions of *Astroni* (3800 yr. Bp.) [Orsi et al., 1998]. These soils are usually unsaturated, structurally unstable and tend to collapse and flow due to heavy rainfall.

The cut studied has a slope angle of 60° and are 80-m long and 20-m high, and the surface is covered by bushes. The stratigraphy was identified by means of in situ testing and available geological data. Starting from the ground surface there are: weathered pyroclastic soils with roots and organic material (soil A), and intact pyroclastic soils (soil B).

3 LABORATORY ACTIVITY

In order to define the mechanical behaviour of the pyroclastic soils present in the studied area (A, B), direct shear tests were performed under natural, dry and saturated conditions for five different normal stress levels.

The hydraulic behaviour of the pyroclastic soils was investigated trough both Richards plate tests on soil samples under natural conditions and saturated conductivity tests on saturated samples in triaxial tests. The laboratory tests were carried out at the

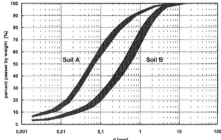

Figure 2. Grain size distributions

Table 1. Mean physical properties of natural soils.

SOIL	G_s	γ_d [kN/m^3]	γ_{sat} [kN/m^3]	e_0
A	2.503	9,838	15,908	1,566
B	2.524	13,763	18,310	0,837

Table 2. Physical properties of the soil samples, at the end of consolidation phases.

	γ_d [kN/m3]		e		w		S_r	
	min.	max	min.	max	min.	max	min.	max
B1	13.19	15.09	0.64	0.84	0.21	0.32	0.85	1.00
B2	13.32	14.02	0.77	0.86	0.05	0.13	0.15	0.39
B3	12.86	14.07	0.76	0.93	0.03	0.05	0.09	0.15

Department of Geotechnical Engineering (DIG) of the University of Naples *Federico II*.

In this paper are presented the results of direct shear test and of Richard's plate test relative to sample of soil B.

3.1 *Laboratory Tests*

The pyroclastic soils collected from pits of recent slope exposures consist of particles ranging from silty sand to sandy silt, Figure 2. The clay fraction was less than 10%. The mean physical property of natural soils is resumed in Table 1.

The tests were carried out on square intact samples (60 mm × 60 mm × 20 mm) of soil B. The physical properties of the soil samples, classified according to the degree of saturation at the end of consolidation phases, are given Table 2. Groups B1 get samples whose S_r are approximately equal to 1; group B3 consists of samples that are dried in pressure plate and groups B2 constitutes samples with a intermediate S_r.

The specimens with saturation degree (S_r) close to 1 were saturated in the shear box by immersion in distilled water for 24 h prior to testing, at the end of consolidation phase.

Considering that the initial stress state in the thin pyroclastic soil involved is very low (vertical stress not larger than 40 kPa), laboratory experimentation was carried out at low stress states with vertical stress values ranging from 7 kPa to 100 kPa.

In figure 3 diagrams are reported for three soil samples, at about the same initial density, which show the ratio τ/τ_u (where τ is the applied shear stress and τ_u is the shear stress at the end of the test) as a function of the shear box horizontal displacement dx. The tests on samples B1 and B2 have been done at the constant water content, instead for sample B3 the degree of saturation is increased by soaking while the soil is attained his peak strength.

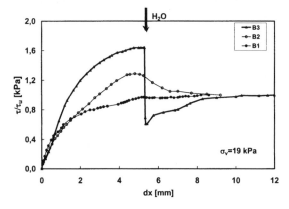

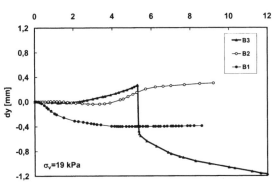

Figure 3. Example of results obtained from direct shear tests carried out at constant vertical stress and at different water content.

The curves show that the natural water content material (B2, B3) has a brittle behaviour, while the almost saturated one (B1) is always ductile and contractile. The B3 curve shows that, as already seen on other pyroclastic soils [Nicotera, 1998, Scotto di Santolo, 2000], an increase of the degree of saturation during a shear process leads to a sudden and sensible reduction of strength, and causes an instantaneous settlement. By keeping on shearing, strength increases up to a constant stationary value.

On the whole, a dilating behaviour was observed on low water content specimens. Specimens after saturation always showed a contracting behaviour.

Figure 4 shows the relationship between the ratio τ_p/τ_u (where τ_p is the peak shear stress) against de-

gree of saturation for all the sample analysed. It could be seen that the behaviour of soil changing from brittle to ductile as degree of saturation increase, independent of the applied vertical stress.

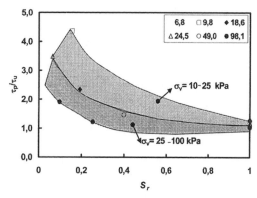

Figure 4. Ratio between peak shear stress, τ_p, and shear stress at the end of direct shear tests, τ_u, versus horizontal displacement, dx, from tests carried out at different normal stress and different degrees of saturation.

Peak strengths of the different specimens are represented on a (σ, τ) plane in figure 5. It may be observed that the peak shear strength of saturated soil (B1) is noticeably lower than that of the same material at lower S_r (B3). This is due to the positive contributions of the mechanical closing action exerted by capillary menisci of the unsatureted material at contact between particles [Burland, 1964].

These data are well interpolated on the space $\{\sigma, \tau, S_r\}$ with the function:

$$\tau = a + b \cdot \sigma + \beta \cdot \ln S_r \tag{2}$$

where a, b, and β = are fitting parameters.

It is obtained that the shear strength envelope increases linearly with vertical stress and logarithm of S_r. The friction term is independent of the degree of saturation. The envelope derived are shown also in figure 5. The fitting parameters are: a= 13.385 kPa; b=0.687; c=-24.902 kPa with R^2=0.909.

Fitting results show that the cohesion varies from 13 to 67 kPa for the degrees of saturation from 10 to 100%.

3.2 Soil-water characteristic curve

Soil–water characteristic curves were determined on undisturbed soil specimens (ϕ 56 mm; h 20 mm) cut from a block sample recovered from the field site. A 10 bar ceramic plate was used. The mean physical characteristics of tested specimen are presented in Table 3.

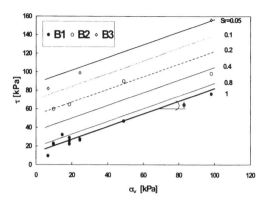

Figure 5. Failure envelopes of pyroclastic specimens obtained from fitting experimental data.

Table 3. Mean value of the physical properties of soil B tested in Richards's Plate.

SOIL	γ [kN/m^3]	γ_d [kN/m^3]	e_0	w_0	S_{r0}
B	15.96	13.55	0.87	0.18	0.52

Test results are presented in Figure 6 in terms of relative volumetric water content Θ, defined as the $(\theta_w - \theta_{wr} / \theta_{ws} - \theta_{wr})$ where θ_w is the volumetric water content (ratio of water volume to the total soil volume), θ_{wr} is the residual volumetric water content and θ_{ws} is the saturated volumetric water content [van Genuchten, 1980].

Experimental points (N=48) are interpolated using the correlation proposed by Brooks & Corey [1964]:

$$\Theta = \begin{cases} 1 & for \ \alpha \cdot (u_a - u_w) \leq 1 \\ [\alpha \cdot (u_a - u_w)]^{-\lambda} & for \ \alpha \cdot (u_a - u_w) > 1 \end{cases} \quad (3)$$

where α and λ are fitting parameters. The values of fitting parameters are summarised in Table 4.

Table 4. Fitting parameters for pyroclastic soil B.

N	R^2	θ_{ws}	θ_{wr}	α [kPa^{-1}]	λ
48	0,992	0,455	0	0,1065	0,4063

Experimental data and the fitting curve are shown in Figure 6. The air-entry pressure $(u_a-u_w)_e$, derived as α^{-1}, for this soil appears to be less than 10 kPa and the major variations are recorded for values less than 100 kPa.

The hydraulic conductivity function in terms of k_r (k_w / k_{sat}), plotted in Figure 6, has been obtained by indirect method as proposed by van Genuchten

[1980]. This method utilizes the soil-water characteristic curve and saturated conductivity (k_{sat}).

3.3 Relation between shear strength and soil suction

As said before the equation of shear strength of unsaturated soil can be written known the shear strength of saturated soil and soil-water characteristic curve. The expression utilized, as proposed by Vanapalli et al. [1996] follows:

$$\tau = [c + (\sigma - u_a) \cdot \tan\varphi] + (u_a - u_w) \cdot [(\Theta)^k \cdot \tan\varphi] \quad (4)$$

where τ = is the shear strength of an unsaturated soil; c and φ = are soil strength parameters for saturated soil; $(\sigma - u_a)$ = is the net normal stress on the plane of failure; $(u_a - u_w)$ = is the matric suction of the soil; Θ = relative volumetric water content and k is a fitting parameter.

The first part of the equation is the saturated shear strength, and it is a constant for particular net normal stress. The second part of the equation is the shear strength contribution due to the suction, which can be predicted using the soil water characteristic curve.

Substituting in (4) equation (3) proposed by Brooks and Corey [1964] in the case of the suction is greater than the air entry value give:

$$\tau = c + (\sigma - u_a) \cdot \tan\varphi + \left[\frac{(u_a - u_w)_e}{(u_a - u_w)} \right]^\lambda \cdot (u_a - u_w) \cdot \tan\varphi \quad (5)$$

This function is plotted in figure 7 on the plane $\{(u_a - u_w), \tau\}$ at different values of net stress by means of the results of direct shear test on saturated sample (B1) and soil-water characteristic curve reported in figure 6.

In figure 7 are also shown the peak shear strength performed on direct shear apparatus, with respect to suction. The matric suction of the specimens was obtained from measurements of water content by means of knowledge of the water-characteristic curve, or direct measurements of suction by means of Imperial College tensiometer [Ridley & Burland, 1993]. It is assumed that the matric suction suffers small changes during the tests. It is also assumed that at failure the pore-air pressure is atmospheric.

It can be seen that the predicted shear strength are generally greater than measured one at least for the range of stress tested.

Actually are in progress other experimental activities with the aim of consolidate these preliminary results.

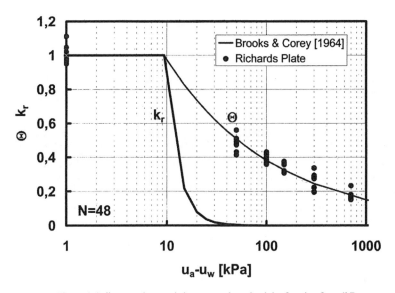

Figure 6. Soil-water characteristic curve and conductivity function for soil B.

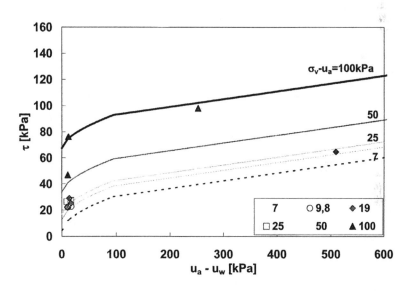

Figure 7. Comparison between measured and predicted values of shear strength against matric suction.

Table 5. Peak strength parameters derived from the envelope of experimental data.

GROUPS	φ	c [kPa]
B1		13.39
B2	34°48'	48.01
B3		69.99

Table 6 Safety factors determined by Bishop method

GROUPS	S_r	γ [kN/m³]	F
B1	1	18.42	0.94
B2	0.249	14.78	2.78
B3	0.103	13.87	3.97

4 STABILITY ANALYSES

The safety factor (F) variation as a function of S_r was investigated trough several analyses of stability.

Sixteen potential slip surfaces have been examined, according to the in situ observations.

These analyses were carried out in terms of total stress, using the Bishop's method, with reference to the physical and mechanical parameters of soils B in terms of friction angle and cohesion intercept, as reported in Table 5.

In all the examined cases the slip surface coincided with the one observed in situ. The results are shown in Table 6.

The results that emerged from these analyses were that the shallow instability phenomena could occur when the degree of saturation is greater than 80%.

5 CONCLUSION

The effect of water content on the shear characteristics of undisturbed samples of pyroclastic soils was investigated by means of several direct shear tests.

Test results proved that the most important factors affecting the strength of the soils tested are besides the initial void ratio the degree of saturation. Particularly the general stress-strain behaviour changes from brittle to ductile as the degree of saturation increases. This increase of the degree of saturation also produces a reduction in shear strength.

Moreover the comparison between the predicted and measured relationships of shear stresses and matric suction seem overestimate the contribution of suction to the soil strength.

The correlation expressed by equation (2) offers an easy and convenient alternative for take into account the contribution of suction to shear strength of an unsaturated soil.

6 REFERENCES

Brand E.W. (1985). *Predicting the performance of residual soil slopes*. Proc. 11th Int. Conf. On Soil Mechanics and Foundation Engineering, San Francisco, 5: 2541-2578.

Burland J.B. (1964) *Effective stress in partially saturated soils – Correspondence*. Géotechnique, 14:64-68.

Calcaterra D. & P.M. Guarino (1999a). *Morphodynamics and recent landslides in western hillslopes of Napoli*. Geologia Tecnica ed Ambientale, 2/99: 11-17, (in Italian).

Di Vito M., L. Lirer, G. Mastrolorenzo, G. Rolandi, R. Scandone (1985). *Volcanologic map of Campi Flegrei*. Dip. di Geofisica e Vulcanologia, largo San Marcellino 10, 80138, Naples, Italy.

Escario V. & J. Saez (1986). *The shear strength of partly saturated soils*. Géotecnique, 36: 453-456.

Fredlund D.G. & N.R. Morgenstern (1977). *Stress state variables for unsaturated soils*. J. Geotech. Eng. Div. ASCE, GT5 103: 447-466.

Fredlund D.G., Rahardjo H. & J.K.M. Gan (1987). *Nonlinearity of strength envelope for unsaturated soils*. Proc. 6th Int. Conf. Expansive Soils, New Delhi, 1: 49-54.

Fredlund D.G., A. Xing, M.D. Fredlund & S.L. Barbour (1995). *The relationship of the unsaturated soil shear strength to the soil-water characteristic curve*. Can. Geotech. J. 32: 440-448.

Gan J.K.M., D.G. Fredlund & H. Rahardjo (1988). *Determination of the shear parameters of unsaturated soil using the direct shear test*. Can. Geotech. J., 25: 500-510.

Nicotera M.V. (1998). *Effetti del grado di saturazione sul comportamento meccanico di una pozzolana del napoletano*. Tesi di dottorato consorzio tra le Università di Roma "La Sapienza" e Napoli "Federico II".

Oloo S.Y. & D.G. Fredlund (1996). *A method for determination of for statically compacted soils*. Can. Geotech. J., 33: 272-280.

Orsi G., M. Di Vito & R. Isaia (1998). *Volcanic hazards and risk in the Parthenopean megacity*. Field excursion guidebook, Int. Meet. on Cities on Volcanoes, 206 pp., Naples.

Pellegrino A. (1994). *I Fenomeni Franosi nell'area Metropolitana Napoletana*. Acta Nepolitana, 18, 237-256, Guida Ed., Naples, (in Italian).

Ridley A.M. & J.B. Burland (1993). *A new instrument for the measurement of soil moisture suction*. Géotecnique 43, No 2: 321-324.

Scotto di Santolo A. (2000). *Analisi geotecnica dei fenomeni franosi nelle coltri piroclastiche della Provincia di Napoli*. Ph.D. thesis, Universities of Naples & Rome, (in Italian).

Vanapalli S.K., D.G. Fredlund, D.E. Pufahl, A.W. Clifton (1996). *Model for the prediction of shear strength with respect to soil suction*. Can. Geotech. J. 33: 379-392.

van Genuchten M.Th. (1980). *A closed-form equation for predicting the hydraulic conductivity of unsturated soils*. Soil Science Society American Journal, 44: 892-898.

Unsaturated Soils for Asia, Rahardjo, Toll & Leong (eds) © 2000 Taylor & Francis, ISBN 90 5809 139 2

Shear strength behaviour of unsaturated granite residual soil

M. R. Taha, M. K. Hossain & S. A. Mofiz
Department of Civil and Structural Engineering, Universiti Kebangsaan, Malaysia

ABSTRACT: The shear strength behaviour of an unsaturated granite residual soil were investigated using consolidated drained (CD) and constant water content (CW) triaxial tests with special emphasis on the effect of the matric suction. Shear strength parameters (c', ϕ_b) were evaluated using two-dimensional graphical and numerical methods. It was found that a linear relationship can be correlated between shear strength and matric suction. The value of c' obtained using the graphical method is almost invariably greater than that obtained using the numerical method. However, both methods give similar values of ϕ_b. The c' values obtained in this study are significantly different from tests on saturated soil.

1 INTRODUCTION

Residual soils cover more than three-quarters of the land area of Peninsular Malaysia. A significant proportion of residual soils is granite residual soil. Many steep slopes in these residual soils often have a deep ground water table above which the soils possess high matric suction (Hossain 1999). It is well established that the stability of a natural or a cut slope in these soils depend on the shear strength which is affected by the matric suction. Thus, it is important to study the shear strength characteristics of the residual soils particularly with respect to changes in matric suction. This study attempts to find the shear strength parameters of the granite residual soil with respect to changes in soil suction utilising a modified triaxial test apparatus.

2 SHEAR STRENGTH OF UNSATURATED SOIL

Many researchers such as Croney (1952), Bishop (1959), Aitchison (1961), Jennings (1961) and Richards (1966), have developed relationships for effective stress equation for unsaturated soils. However, a more systematic approach was given by Fredlund et al. (1978) who developed a shear strength relationship for unsaturated soils in terms of two independent stress variables:

$$\tau = c' + (u_a - u_w) \tan\phi^b + (\sigma - u_a) \tan\phi' \qquad (1)$$

where c' = effective cohesion, σ = total stress, u_a = pore-air pressure, u_w = pore water pressure, ϕ' = effective angle of friction, ϕ^b = angle indicating the rate of increase in shear strength with respect to matric suction, $(u_a - u_w)$. The two-dimensional graphical and numerical methods used to evaluate of shear strength parameters (c', ϕ^b) from unsaturated triaxial tests were discussed in detail by Ho (1981).

3 SOIL PROPERTIES

The soil used in this study was obtained from a residual granite soil formation in Cheras, just 8 km south of Kuala Lumpur, the capital city of Malaysia. The block sampling technique was adopted for soil sampling in which thirty 300 mm cube samples were hand cut, wax around the sides, place in wooden boxes and stored in the laboratory until required for testing. The mean dry density obtained from field tests was 13.6 kN/m^3 with natural moisture content of about 31%. The specific gravity of this soil (2.55 to 2.61) shows a change from the specific gravity of fresh granites (2.65 to 2.68). Optimum moisture content and maximum dry density were found to be 23% and 14.7 kN/m^3, respectively from compaction test. Based on Unified Soil Classification System (USCS), the soil was grouped as "clay with high plasticity" (CH). It was also classified in the "A-7-6" group according to the AASHO classification system.

4 TESTING PROGRAM

The testing program consisted of consolidated drained (CD) and constant water content (CW) triaxial test on unsaturated specimens. The detailed testing program is shown in Table 1. These tests were performed under constant net confining pressure, $(\sigma_3 - u_a)$ and varying matric suction, $(u_a - u_w)$. The tests reported in this paper were carried out under initial effective confining pressure of 600 kPa and initial matric suction of 50 - 1000 kPa. The strain rates for CD and CW tests were used 0.00166 %/min and 0.00666 %/min, respectively.

Table 1. Test program for CD and CW triaxial tests on unsaturated specimens.

Test type	Test no.	$\sigma_3 - u_a$ (kPa)	$u_a - u_w$ (kPa)	Strain rate (%/min)
CD	1	600	50	0.00166
	2	600	100	
	3	600	200	
	4	600	400	
CW	5	600	100	0.00666
	6	600	200	
	7	600	400	
	8	600	800	
	9	600	1000	

5 TESTING PROCEDURE

The triaxial specimens (50 mm diameter and 100 mm high) were prepared by trimming the block samples. The specimens were set up between a saturated high air entry disk at its bottom and a porous disk at the top. Rubber membranes were placed over the sample using a membrane stretcher. O-rings were placed over the membranes on the bottom pedestal and upper cap. Saturation of test specimens was achieved by continuously increasing the cell pressure and back pressure. An unsaturated soil sample was prepared from a saturated sample by applying a desaturation and consolidation process. For CD tests, the specimens were sheared under a constant strain rate 0.00166 %/min and the air and water pressure valves were remained open. In CW test, specimens were sheared at a strain rate of 0.00666 %/min under drained condition for pore-air phase and undrained condition for the pore-water phase.

6 METHOD OF ANALYSIS

Test results were analysed using numerical and two-dimensional graphical methods. Ho (1981) used both methods as a means for evaluating the angle ϕ^b.

6.1 Two-dimensional Graphical Method

With the values of σ_1, σ_3, u_a and u_w at failure from triaxial tests on unsaturated soil samples, Mohr circles corresponding to various matric suction values can be drawn on τ vs. $(\sigma - u_a)$ plot. The ordinate intercepts of the various matric suction contours can be obtained by drawing tangent at all Mohr circles using the angle of friction ϕ' from triaxial test on saturated sample. Either by plotting or using linear regression analysis on the intercept (shear stress) vs. matric suction data, the angle ϕ^b of the unsaturated soil can be obtained. From the intercept of the best-fit straight line, c' could also be obtained.

6.2 Numerical Method

The numerical method is basically a geometrical interpretation of the three-dimensional plot between the shear strength (q) and the stress state variables (p, r). The shear strength parameters of saturated soil, c' and ϕ' can be obtained from conventional triaxial test data and the corresponding $d' = c' \cos \phi'$ and $\psi' = \tan^{-1} (\sin \phi')$ can be obtained. The values of σ_1, σ_3, u_a and u_w of an unsaturated soil sample at failure can also be found by carrying out triaxial test. Consequently, p_f, q_f and r_f at failure can be calculated. The projected shear strength $(d' + p_f \tan \psi')$ at the saturation plane can be obtained. Subtracting the projected shear strength from shear strength at failure (q_f) gives the increase in shear strength due to matric suction $(\Delta \tau_d)$. With several sets of triaxial test data, a plot between $\Delta \tau_d \cos \psi'$ vs. $(u_a - u_w)$ can be made. From the plot, the angle α_a can be obtained. From the obtained values of ψ' and α_a the values of $\psi^b = \tan^{-1} (\tan \alpha_a / \cos \psi')$ can be calculated. Finally, the shear strength parameter $\phi^b = \tan^{-1} (\tan \psi^b / \cos \phi')$ can be calculated using the values of ψ^b and ϕ'.

7 RESULTS AND DISCUSSION

The deviator stress vs. strain curves for various matric suctions obtained from the tests are shown in Fig. 1. A strain-hardening type of deviator stress-strain curves were obtained in most of these tests. The deviator stress vs. strain curve does not exhibit peak points, even at large strains. The limiting strain failure criterion (i.e., 18%) was selected to represent the failure condition (Fredlund & Rahardjo 1993). The figure illustrate that at a constant net confining pressure, shear strength increase as matric suction increases. Such phenomena were also found by Blight (1967), Mashhour et al. (1995) and Gan & Fredlund (1995).

The water volume and overall specimen volume (air and water) changes during shear are presented in Figs. 2 and 3. The volume of the specimens initially increased then started to decrease (at about 1%

- 2% axial strain) and continued to contract until failure. Although pore-air was forced into the specimen during compression, total volume of the specimen decreases. In other words, the specimen volume reduced during shear. The amount of contraction was low at higher matric suction (Fig. 3). Blight (1967) reported similar behaviour of consolidated drained tests on unsaturated silts.

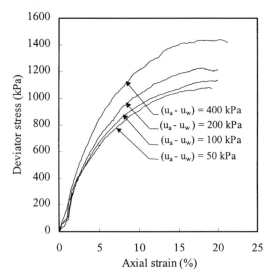

Figure 1. Deviator stress vs. strain curves.

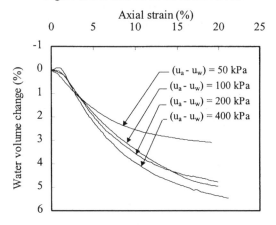

Figure 2. Water volume change vs. axial strain curves.

A typical two-dimensional projection of the failure envelope on the shear stress, τ vs. $(\sigma - u_a)$ plane is presented in Fig. 4(a). The shear strength parameters obtained were $c' = 40$ kPa and $\phi' = 26.5°$ from CD triaxial test on saturated specimens by Taha et al. (1999). They also mentioned that for a planar

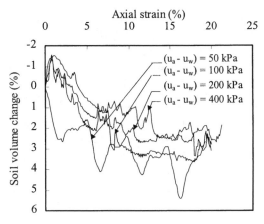

Figure 3. Specimen volume change vs. axial strain.

failure envelope, the internal friction angle, ϕ', remains essentially constant under saturated and unsaturated conditions. The value of intersections between the failure envelope and the ordinate are plotted with respect to matric suction in Fig. 4(b). For a constant net confining pressure, the shear strength increased with increasing matric suctions, as illustrated in Fig. 4(b). The effect of matric suction is clearly shown by the ϕ^b angle in Fig. 4(b). It shows a linear relationship between matric suction and shear strength. The value of cohesion intercept and ϕ^b angle obtained were 37.4 kPa and 17.8°, respectively from the two-dimensional graphical method. From the numerical method, the shear parameters obtained were $c' = 33.7$ kPa and $\phi^b = 17.2°$ (Hossain 1999). In two-dimensional graphical and numerical methods, the planar failure plane is assumed. However other authors, such as Rohm & Vilar (1995), obtained a hyperbolic relationship between shear strength and matric suction for lateratized sandy soil from Brazil. Ho & Fredlund (1982) reported a comparable $\phi^b = 15.3°$ for undisturbed decomposed granite from Hong Kong in multistage CD triaxial test. The results from the CD test conducted in this study show that the angle ϕ^b is less than the internal friction angle, ϕ', for the residual granite soil. This indicates that the shear strength of unsaturated soil is affected more by the change in effective confining pressure than by the change in matric suction. Fredlund (1985) measured the ϕ^b angle for various soils and the experimental results showed that the ϕ^b is always smaller than or equal to the internal friction angle, ϕ'. Gan & Fredlund (1995) showed that ϕ^b is equal to ϕ' at low matric suction when the soil is close to saturation.

Results of CW tests are shown in Figs. 5 to 7 respectively as plots of deviator stress vs. axial strain, air volume change vs. axial strain, and matric

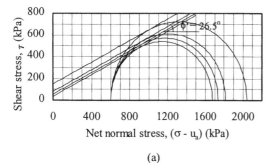

(a)

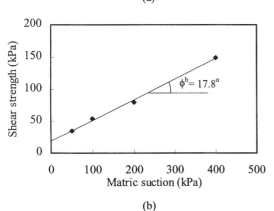

(b)

Figure 4. Two-dimensional presentation of failure envelope for CD test. (a) Failure envelope projected on the τ vs. ($\sigma - u_a$) plane; (b) intersection line between the failure envelope and the τ vs. ($u_a - u_w$) plane.

suction change vs. axial strain. The deviator stress vs. axial strain curves obtained from these tests also do not exhibit peak points even at large strains. Fig. 6 shows that the specimen undergoes contraction at suction less than 400 kPa and dilation at higher suction. Similar behaviour was also shown from CW tests on Dhanauri clay by Satija (1978). The suction of the test specimens decreased up to 7% axial strain and started to increase beyond that (Fig. 7). The suction decreases as the specimen volume decreases and the suction increases as the specimen volume increases. Similar behaviour was also shown by Salman (1995) for unsaturated granular soil. For the applied matric suction up to 400 kPa, change in matric suction during shear exceeded the applied matric suction, i.e., positive pore-water pressure development (Fig. 7). Significant effect of matric suction was not observed from deviator stress vs. axial strain curves for matric suctions 100 and 200 kPa. For higher matric suction ($u_a - u_w$) >400 kPa), net change in matric suction during shear was less than the applied matric suction.

The two-dimensional projections of the failure envelope on the shear stress, τ vs. ($\sigma - u_a$) plane are

presented in Fig. 8(a). The intersections between the failure envelope and the ordinate (shear stress) are plotted in Fig. 8(b). For a constant net confining pressure, the shear strength at failure increased with increasing matric suctions, as illustrated in Fig. 8(a). The values of c' and ϕ^b were found to have values of 58 kPa and 25.8°, respectively. From the numerical method, the respective values were 55 kPa and 26.1°. This indicates that the angle ϕ^b is also less than the internal friction angle, ϕ', for the residual granite soil.

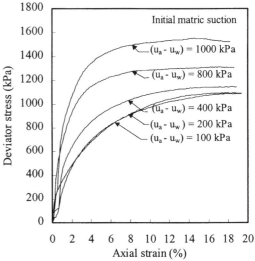

Figure 5. Deviator stress vs. axial strain curves for CW test on unsaturated specimen (net confining pressure 600 kPa at the beginning of shearing).

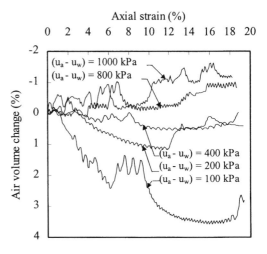

Figure 6. Air volume change vs. axial strain curves.

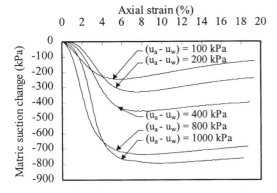

Figure 7. Matric suction change vs. axial strain curves.

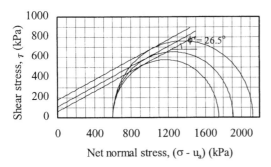

(a)

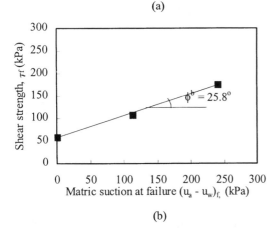

(b)

Figure 8. Two-dimensional presentation of failure envelope for CW test. (a) Failure envelope projected on the τ vs. (σ - u_a) plane; (b) intersection line between the failure envelope and the τ vs. (u_a - u_w) plane.

The shear strength parameters (c', ϕ^b) obtained from CD and CW tests using two-dimensional

graphical method and numerical method are tabulated in Table 2. The most significant difference between the numerical and two-dimensional graphical methods is in the values of c'. The value of c' obtained using the graphical method is almost invariably greater than that obtained using the numerical method.

Table 2 Shear strength parameters (c', ϕ^b) obtained from two-dimensional graphical and numerical method.

Test type	Shear strength parameters from analysis		Method of Analysis
	c' (kPa)	ϕ^b (degree)	
CD	37.4	17.8	Two-dimensional plot
	33.7	17.2	Numerical
CW	58	25.8	Two-dimensional plot
	55	26.1	Numerical

Similar behaviour was found by Wong (1984) from Bishop et al. (1960) test results. The difference between the values of c' obtained from these two methods is entirely due to the nature of the methods themselves. Considering the angle ϕ^b, the two-dimensional graphical and numerical methods gave very similar results.

In addition to the difference between the values of c' obtained from the numerical and two-dimensional graphical methods, either of these c' values differs significantly from the c' value obtained from tests on saturated soil samples. Similar phenomenon was reported by Ho & Fredlund (1982) from Satija's (1978) test data, and Wong (1984) from Ho's (1981) test data. The c' values obtained from unsaturated analysis may differ from c' from saturated CD test due to linear interpretation instead of a curved failure envelope (Fredlund & Rahardjo, 1993). The shear strength parameter (ϕ^b) obtained from CD and CW tests differs significantly and may have a relation to the degree of saturation varied during each type of test. During the shearing stage in CD tests, the degree of saturation continues to decrease, whereas in CW tests, the degree of saturation increased from its value at the end of the desaturation and consolidation stages. The shear strength parameter ϕ^b obtained from CW test should be used if the degree of saturation of a soil increase during loading. Conversely, if the degree of saturation of a soil continuously decreases after loading, the shear strength parameter obtained from CD tests should be used. Wong (1984) also considered the change of the degree of saturation of a soil during loading as the criteria to select test type which would simulated the field condition.

8 CONCLUSIONS

The following conclusions can be drawn from the test results:

1. The soil suction affects the shear strength, and an increase in soil suction increases the shear strength of the soil.

2. A linear relation between shear strength vs. matric suction was found. This finding indicates the suitability of using a planar failure envelope for natural residual granite soil.

3. The values of ϕ^b obtained were 17.8° and 25.8° respectively from CD and CW tests.

4. The value of c' obtained using the graphical method is almost invariably greater than that obtained using numerical method. Both methods gave similar values of ϕ^b.

5. The c' values obtained from numerical and two-dimensional graphical methods differ significantly from the c' value obtained from tests on saturated soil.

6. The shear strength parameter ϕ^b obtained from CW tests should be used if the degree of saturation of a soil increase during loading. Conversely, if the degree of saturation of a soil continuously decreases after loading, the shear strength parameter obtained from CD tests should be used.

REFERENCES

Aitchison, G.D. 1961. Relationship of moisture and effective stress functions in unsaturated soil. *Proceedings of Pore Pressure and Suction in Soils*: 47-52. Butterworths,

Bishop, A.W. 1959. The principle of effective stress. *Teknisk Ukeblad*, 106: 859-863.

Bishop, A. W., Alpan, I., Blight, G. E. & Donald, I. B. 1960. Factors controlling the shear strength of partly saturated cohesive soils. *ASCE Res. Conf. Shear Strength of Cohesive Soils*: 503-532. Univ. of Colorado, Boulder.

Blight, G. E. 1967. Effective stress evaluation for unsaturated soils. *ASCE J. Soil Mech. Found. Eng. Div.* 93(2): 125-148.

Croney, D. 1952. The movement and distribution of water in soils. *Geotechnique*. 3: 1-16.

Fredlund, D. G. 1985. Theory formulation and application for volume change and shear strength problems in unsaturated soils. *Proc. 11th Int. Conf. Soil Mech. Found. Eng.*, San Francisco.

Fredlund, D. G., Morgenstern, N. R. & Widger, R. A. 1978. The shear strength of unsaturated soils. *Can. Geotech J.* 15(3): 313-322.

Fredlund, D. G. & Rahardjo, H. 1993. *Soil mechanics for unsaturated soils*. John Wiley & Sons Inc., New York.

Gan, K. J. M. & Fredlund, D. G. 1995. Shear strength behavior of two saprolitic soils. *Proc. 1st Int. Conf. Unsaturated Soils*. 1: 71-76. Rotterdam: Balkema.

Hossain, M. K. 1999. *Shear strength characteristics of a decomposed granite residual soil in Malaysia*. Ph D thesis, Universiti Kebangsaan Malaysia.

Ho, D. Y. F. 1981. *The shear strength of unsaturated Hong Kong soils*. M. Sc. Thesis, University of Saskatchewan, Saskachewan, Canada.

Ho, Y. F. & Fredlund, D. G. 1982. A multi-stage triaxial test for unsaturated soils. *ASTM Geotech. Testing J.* 5: 18-25.

Jennings, J.E. 1961. A revised effective stress law for use in the prediction of the behaviour of unsaturated soils. *Proceedings of Pore Pressure and Suction in Soils*: 26-30. Butterworths.

Mashhour, M. M., Ibrahim, M. I. & El-Eemam, M. M. 1995. Variation of unsaturated soil shear strength parameters with suction. *Proc. 1st Int. Conf. Unsaturated Soils*. 3: 1487-1493.

Richards, B.G. 1966. The significance of moisture flow and equilibria in unsaturated soils in relation to the design of engineering structures built on shallow foundations in Australia. *Symp.on Permeability on Capillary*, ASTM.

Rohm, S.A., & Vilar, O.M. 1995. Shear strength of an unsaturated sandy soil. *Proc. 1st Int. Conf. Unsaturated Soils*, 1: 189-193. Rotterdam: Balkema.

Salman, T.H. 1995. *Triaxial behaviour of partially saturated granular soils*. PhD Thesis, University of Sheffield.

Satija, B. S. 1978. *Shear behaviour of partially saturated soils*. Ph. D. thesis, Indian Institute of Technology, Delhi.

Taha, M. R., Hossain, M. K., & Mofiz, S. A. 1999. Effect of Matric Suction on the Shear Strength of Unsaturated Granite Residual Soil. *Journal of the Institution of Engineers*, Malaysia (Accepted for June 2000 publication).

Wong, C. K. 1984. *The shear strength of unsaturated soils*. Ph. D. thesis, University of Glasgow.

Unsaturated Soils for Asia, Rahardjo, Toll & Leong (eds) © 2000 Taylor & Francis, ISBN 90 5809 139 2

Triaxial testing of unsaturated samples of undisturbed residual soil from Singapore

D.G.Toll, B.H.Ong & H.Rahardjo
NTU-PWD Geotechnical Research Centre, Nanyang Technological University, Singapore

ABSTRACT: A series of multi-stage controlled suction triaxial tests has been carried out on undisturbed samples of residual soil from the Bukit Timah and Jurong formations in Singapore. The Bukit Timah formation is granitic in origin while the Jurong formation is sedimentary. Tests were performed at three different constant net confining pressures and for various matric suctions. The angle of friction (ϕ') for the Bukit Timah soil was found to be 32°. The highest ϕ^b angle for the undisturbed residual soil from the Bukit Timah soil was determined to be 28° at matric suctions lower than 25 kPa. Beyond this matric suction value, the ϕ^b angle decreased and dropped to almost zero for matric suctions above 150 kPa. The results for unsaturated tests on the undisturbed residual soil from the Jurong Sedimentary Formation suggested an angle of friction (ϕ') of 51°. This very high friction angle could be attributed to the high density and cementation of the specimens. However, an alternative interpretation is also provided based on the results of saturated tests which showed an average ϕ' of 36°. This interpretation indicates high effective cohesion values between 91 and 173 kPa, consistent with saturated tests. In this case ϕ^b values of $10\text{-}22^\circ$ are indicated for the Jurong soil.

1 INTRODUCTION

Tropical residual soils cover almost two-thirds of the island of Singapore. They are derived mainly from the weathering of two geological formations: Bukit Timah and Jurong. Each of these two residual soil types covers approximately one third of the land area of Singapore (Fig. 1).

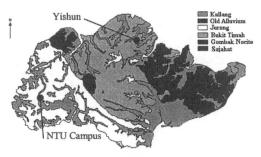

Figure 1. Geological map of Singapore (After Pitts, 1984) showing locations of sampling sites.

The Bukit Timah granite is a mainly acidic igneous rock formed in the Lower to Middle Triassic period (Pitts, 1984). Although known as 'granite', it actually varies from granite, adamellite to granodiorite. The Jurong formation is of late Triassic and Lower to Middle Jurassic age (Pitts, 1984). It is sedimentary in origin and the rock types include various types of conglomerate, sandstones and mudrocks.

The Bukit Timah residual soil varies from silty or clayey sands to silty or sandy clays (depending on the degree of weathering) but is commonly sandy clayey silt (Poh et al., 1985). The residual soils derived from the Jurong formation exist as interbedded layers of predominantly medium plasticity clayey silt, sandy clay and clayey to silty sand materials (Chang, 1988).

This paper describes a study of the shear strength (under unsaturated conditions) of the residual soils derived from the Bukit Timah and Jurong formations. The shear strength in unsaturated conditions was investigated by carrying out consolidated drained (CD) triaxial tests in which the matric suctions were controlled.

Samples for this study were obtained using a Mazier sampler from boreholes drilled at two sites. One site was at Yishun in the northern central area of Singapore (in the Bukit Timah formation). The other was adjacent to the School of Civil and Structural Engineering (CSE) on the Nanyang Technological University (NTU) Campus in the south-west of Singapore (in the Jurong formation). The sites are shown in Figure 1.

For the Bukit Timah material from Yishun, it was not possible to obtain three specimens from a single sample. In fact, the specimens had to be taken from different boreholes. Therefore, one specimen was taken from borehole Y3 (at the mid-height of the Yishun slope) at a depth of 2.5-3.5m. The other two specimens were taken from borehole Y4 (also located at the mid-height of the slope) at a depth range of 3-4m. In the case of the Jurong material (from the NTU-CSE site) three undisturbed specimens were taken from a single Mazier sample. The specimens were taken from Borehole N5 at a depth range of 4-5 m. Some classification parameters are shown in Table 1.

Table 1. Classification test results for residual soils from the Bukit Timah and Jurong formations.

Formation	Bukit Timah	Bukit Timah	Jurong
Borehole	Y3	Y4	N5
Depth (m)	2.5-3.5	3.0-4.0	4.0-5.0
Liquid limit (%)	47	42	36
Plastic limit (%)	28	29	21
Plasticity index	19	13	15
Sand (%)	9	53	13
Silt (%)	50	22	35
Clay (%)	31	25	52
Specific gravity	2.69	2.7	2.71
USCS classification	Clayey Silt, MH	Silty Sand, SM	Silty Clay, CL

It can be seen that a difference in grain size distribution exists between the specimens Y3 (2.5-3.5m) and Y4 (3-4m), particularly in the sand/silt ranges. These specimens were originally selected based on visual classification, which indicated similar materials. This heterogeneity, of course, is one of the difficulties faced when testing residual soils.

2 TEST PROCEDURES

The undisturbed specimens were prepared by trimming the samples to the required diameter (50mm) and length (100mm) after they had been extruded from the Mazier sampler. The soil specimens were saturated prior to each test.

Matric suctions were controlled using the axis-translation technique whereby the pore water pressure is elevated by applying a pore air pressure within the specimen. High air entry porous discs (with an air entry value of 500 kPa) were used to allow the control of pore water pressure.

After saturation, specimens were allowed to consolidate under an isotropic confining pressure, σ_3, a pore-air pressure, u_a, and a pore-water pressure, u_w. The magnitudes of the selected pressures for consolidation were based on the desired values of net confining stress, (σ_3-u_a), and matric suction, (u_a-u_w). Tests were performed at three different constant net confining pressures, (σ_3-u_a) of 50, 100 and 150 or 200 kPa. A single specimen was used for each net confining pressure. For each specimen a multi-stage

test was carried out in which matric suctions, (u_a-u_w) of 50, 100, 150, 200 kPa were imposed in each stage.

The specimens were sheared at a constant strain rate in triaxial compression. A strain rate had to be selected to ensure dissipation of excess pore-water and pore-air pressures during the drained shearing process. An initial assessment of an acceptable strain rate was made using the procedure described by Ho and Fredlund (1982) which considers the impeded flow through the high air-entry disc at the base of the specimen and the physical properties of the soil. This method suggested very low and impractical strain rates for the Bukit Timah and the Jurong residual soils ($8x10^{-5}$ %/min and $2x10^{-5}$ %/min, respectively). Such rates would have meant a single test would have taken more than one year.

Ong (1999) investigated the strain rate for constant water content (CW) tests on a compacted residual soil under the stress conditions of net total stress of 250kPa and matric suction of 400kPa. He found that changing the rate of shear from $9x10^{-3}$ %/min to $8.1x10^{-2}$ %/min did not produce a significant difference in the stress strain curve. However, strain rates above $1.8x10^{-2}$ %/min did show differences in matric suction values. Therefore, Ong (1999) adopted a strain rate of $9x10^{-3}$ %/min for CW tests. He found that with this rate the water content profiles after shear were uniform.

Satija and Gulhati (1979) suggested that 1/5th of the strain rate appropriate for CW tests was sufficient for dissipation of pore-water pressure in CD tests. In this study, an even lower rate of $9x10^{-4}$ %/min was adopted for the CD triaxial tests (i.e. 1/10th of the strain rate for CW tests).

The total and water volume changes were monitored throughout the shear process. Total volume change was measured by monitoring the volume of water flowing in or out of the triaxial cell surrounding the specimen. A correction was made to allow for the volume change induced by the movement of the loading ram. Water volume change was monitored by measuring the flow of water from the sample. Diffused air that accumulated underneath the porous ceramic disc was flushed regularly and measured using a diffused air volume indicator in order to minimize any error in the volume change measurement.

3 UNDISTURBED RESIDUAL SOIL FROM THE BUKIT TIMAH GRANITIC FORMATION

The results of the multi-stage CD triaxial test on undisturbed Bukit Timah soil at a constant net confining pressure of 50kPa and increasing matric suction are shown in Figures 2a, 2b and 2c respectively.

Water volume strain was calculated by expressing the water volume change as a percentage of the original specimen volume.

The deviator stress versus axial strain curve (Fig. 2a) shows an increase in shear strength with an increase in matric suction. The strain to failure decreased with each stage as the matric suction increased.

Figure 2b shows that the specimen compressed during shearing. By comparing Figures 2b and 2c it can be seen that the volume of water flowing out of the specimen was greater than the total volume change of the specimen. This means that the specimen was desaturating as the suction was increased at each stage, and also during shear. Therefore, the degree of saturation of the specimen decreased as the test progressed.

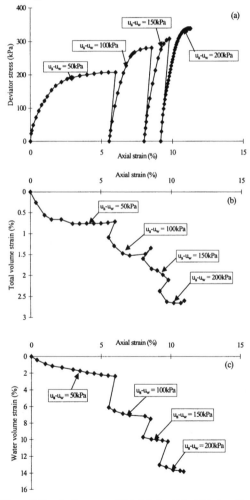

Figure 2. Results from consolidated drained triaxial test on Bukit Timah residual soil at 50kPa net confining pressure (a) deviator stress ($\sigma_1-\sigma_3$) (b) total volume strain and (c) water volume strain plotted against axial strain.

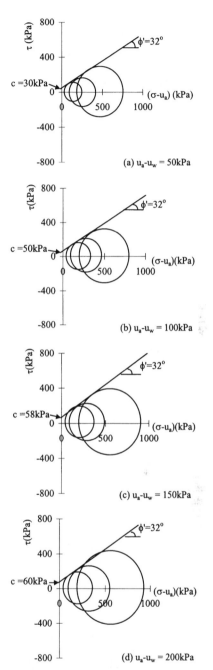

Figure 3. Extended Mohr-Coulomb failure envelopes for the Bukit Timah soil at matric suction values of (a) 50 kPa (b) 100 kPa (c) 150 kPa and (d) 200 kPa

Figure 3 shows Mohr circles for the three net normal stresses at each matric suction value. The same angle of friction in relation to net total stress (ϕ') of 32° can be seen for all four suction values (Figs. 3a-3d).

The results from a multi-stage consolidated undrained (CU) triaxial test on a saturated specimen of the Bukit Timah soil indicated shear strength paramaters c' = 10kPa and ϕ' = 33° (Ong, 1999). Therefore, the results from the unsaturated CD tests show a consistent ϕ' value to the saturated CU test.

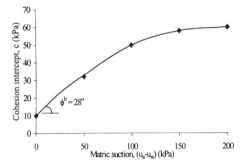

Figure 4. Total cohesion versus matric suction for the Bukit Timah soil

Figure 4 shows the values of total cohesion (c) obtained from the intercepts of the extended Mohr-Coulomb failure envelope, plotted versus matric suction. The highest ϕ^b angle for the undisturbed residual soil from the Bukit Timah soil can be determined to be 28° at matric suctions lower than 25kPa. This suggests that the undisturbed Bukit Timah soil desaturates near a matric suction of 25kPa since the ϕ^b angle obtained at a matric suction of 25kPa (28°) is already lower than the ϕ' angle of the soil (32°). Beyond the matric suction value of 25kPa, the ϕ^b angle decreases as shown in Figure 4, and the tangent value of ϕ^b falls to almost zero above 150 kPa.

4 UNDISTURBED RESIDUAL SOIL FROM THE JURONG SEDIMENTARY FORMATION

The results of the multi-stage CD triaxial test on undisturbed Jurong soil at a constant net confining pressure of 50kPa and increasing matric suction are shown in Figures 5a, 5b and 5c.

The graph of deviator stress versus axial strain curve in Figure 5a shows that the specimen exhibited very stiff and brittle behaviour. Due to this, the multi-stage triaxial test was very difficult to perform as excessive strains could have caused failure in a brittle manner during the early stages of shearing.

The strength of the Jurong soil at this particular location was much higher than that of the Bukit Timah soil under the same stress condition. This large difference in the strength can be seen by comparing Figures 2a and 5a. For example, under a net confining stress of 50kPa and a matric suction of 100kPa, the Jurong soil exhibits a much higher peak deviator stress of 600kPa as compared to the peak deviator of

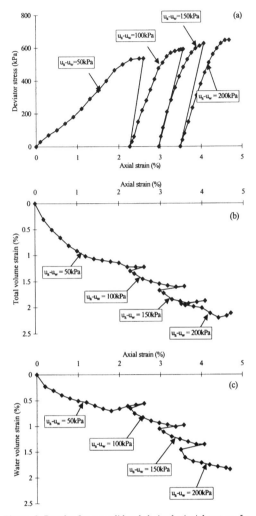

Figure 5. Results for consolidated drained triaxial test on Jurong residual soil at 50kPa net confining pressure (a) deviator stress $(\sigma_1-\sigma_3)$ (b) total volume strain and (c) water volume strain plotted against axial strain.

280kPa for the Bukit Timah soil. The failure mechanism for the Jurong soil was also more like that of rock behaviour as a longitudinal crack developed throughout the specimen rather than an inclined failure plane.

The Jurong soil compressed during shearing (Fig. 5b). The magnitude of compression was small and almost equal to the water volume change during consolidation and shearing as seen in both Figures 5b and 5c.

Figure 6 shows Mohr circles for the three net normal stresses at each matric suction value for the Jurong soil. A very high angle of friction in relation to net total stress (ϕ') of 51° appears to be indicated for all four suction values (Figs. 6a-d). Further consideration will be given to this high value later, but initially an interpretation will be considered based on this value.

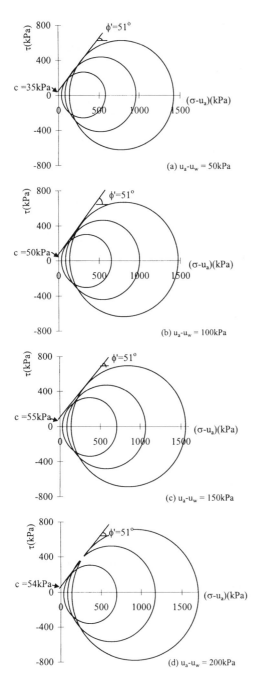

Figure 6. Extended Mohr-Coulomb failure envelopes for the undisturbed Jurong soil for matric suction values of (a) 50 kPa (b) 100 kPa (c) 150 kPa and (d) 200 kPa

Figure 7 shows the values of total cohesion (c) obtained from the intercepts of the extended Mohr-Coulomb failure envelope, plotted versus matric suction. A line has been sketched in such that the slope at low suctions approaches a ϕ^b value equal to ϕ'. Such a curve would suggest an effective cohesion, $c' = 10$ kPa. The tangent value of ϕ^b above 150 kPa approaches zero.

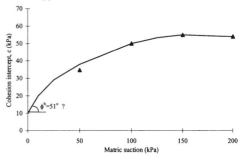

Figure 7: Total cohesion versus matric suction for the Jurong soil

It must be noted that the friction angle of 51° is exceptionally high for the fine-grained Jurong residual soil. The results of multi-stage CU triaxial tests carried out by the former Public Works Dept. of Singapore (now PWD Consultants Pte Ltd) and other results taken from the same site are tabulated in Table 2. High effective cohesion values have been observed (in excess of 100kPa) but ϕ' values are typically around 36°.

Table 2. Shear strength parameters measured in saturated tests on samples from the NTU-CSE site.

Source	Borehole	Depth (m)	c' (kPa)	ϕ' (°)
PWD	N1	8.5-9.5	139	34
PWD	N1	15.0-16.0	107	36
PWD	N1	18.0-19.0	167	37
Hritzuk (1997)	Trial pit	0.5	95	35

It is possible that the high friction angle observed in the unsaturated tests could be due to the high density and degree of cementation of the specimens. As noted earlier, the specimens did manifest brittle and rock-like behaviour during shearing indicative of a high degree of cementation. Even so, a ϕ' value of 51° does seem to be unrealistically high.

An alternative explanation could be that the three specimens tested, although coming from the same sample tube, did have differing degrees of cementation leading to different c' values. It must be remembered that the three Mohr circles shown for each matric suction value have been measured on different specimens.

Figure 8 shows an alternative interpretation of the data for a matric suction of 50kPa, assuming that $\phi' = 36°$ (the average value determined from saturated tests measured on single specimens). This results in a different cohesion value for each specimen.

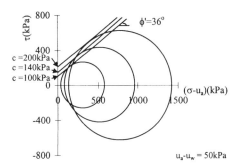

Fig. 8: An alternative interpretation of the data in Fig. 6(a) based on the assumption that $\phi'=36°$.

Figure 9 shows the values of total cohesion obtained from extending the approach used in Figure 8 to other suction values.

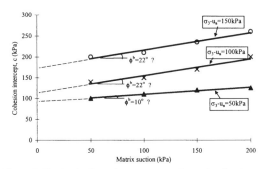

Figure 9. Total cohesion values versus matric suction for the Jurong soil based on the assumption that $\phi'=36°$.

This alternative interpretation suggests an almost linear relationship between total cohesion and matric suction over the suction range 50-200kPa. The ϕ^b values are 22° for two specimens and 10° for the third. It is interesting to note that if these linear trends are projected back to $u_a-u_w = 0$, the effective cohesion values predicted range from 91 to 173kPa, very similar to the range of values in Table 2.

Either of these interpretations could be correct, but clearly each interpretation suggests very different values for ϕ^b. This highlights the great difficulties in determining soil parameters for residual soils where the natural heterogeneity means that one is never sure that specimens used in testing are identical.

5 CONCLUSIONS

Consolidated drained triaxial tests with controlled matric suction have been used to determine the shear strength parameters of two types of residual soil from Singapore. The angle of friction (ϕ') for the residual soil formed from the weathering of the Bukit Timah granitic formation was found to be 32°. The highest ϕ^b angle for the soil was determined to be 28° at matric suctions lower than 25 kPa. Beyond this matric suction value, the ϕ^b angle decreased and dropped to almost zero for matric suctions above 150 kPa.

The results of unsaturated tests on the undisturbed residual soil from the Jurong sedimentary formation suggested an angle of friction (ϕ') of 51°. This very high friction angle could be attributed to the high density and cementation of the specimens. However, an alternative interpretation was also considered based on the results of saturated tests which showed an average ϕ' of 36°. This interpretation indicates high effective cohesion values between 91 and 173 kPa, consistent with the saturated test results. In this case ϕ^b values of 10-22° are indicated for the Jurong soil.

The possibility for alternative interpretations highlights the great difficulties in determining soil parameters for residual soils where the natural heterogeneity means that one is never sure that specimens used in testing are identical.

ACKNOWLEDGEMENTS

This work forms part of a project funded by the National Science and Technology Board of Singapore (NSTB 17/6/16: Rainfall-induced Slope failures). Thanks are also due to Mr Kevin Quan and Mr Tang Sek Kwan of PWD Consultants Pte Ltd for the triaxial test data shown in Table 2.

REFERENCES

Chang, M. F. 1988. In-situ testing of residual soil in Singapore, *Proc. 2nd Int. Conf. on Geomechanics in Tropical Soils, Singapore*, Vol. 1, 97-108. Rotterdam: Balkema.

Ho, D.Y.F. and Fredlund, D.G. 1982. A multi-stage triaxial test for unsaturated soils, *ASTM Geotechnical Testing Journal*, 5(1/2): 18-25.

Hritzuk, K. J. 1997. *Effectiveness of drainage systems in maintaining soil suctions*, MEng Thesis, School of Civil and Structural Engineering, Nanyang Technological University.

Ong, B.H. 1999. *Shear strength and volume change of unsaturated residual soil*, MEng Thesis, School of Civil and Structural Engineering, Nanyang Technological University.

Pitts, J. 1984. A review of geology and engineering geology in Singapore, *Quarterly Journal of Engineering Geology*, 17: 93-101.

Poh, K.B., Chuah, H.L. and Tan, S.B. 1985. Residual granite soils of Singapore, *Proc. 8th Southeast Asian Geotechnical Conf., Kuala Lumpur*, Vol. 1, 3-1 to 3-9.

Satija, B.S. and Gulhati, S.K. 1979. Strain rate for shear testing of unsaturated soil, *Proc. 6th Asian Regional Conference on Soil Mechanics and Foundation Engineering, Singapore*, 83-86.

Unsaturated Soils for Asia, Rahardjo, Toll & Leong (eds) © 2000 Taylor & Francis, ISBN 90 5809 139 2

Shear strength of undisturbed Bukit Timah granitic soils under infiltration conditions

J.C.Wong, H.Rahardjo, D.G.Toll & E.C.Leong
NTU-PWD Geotechnical Research Centre, Nanyang Technological University, Singapore

ABSTRACT: Shearing infiltration tests were conducted to simulate the stress path followed by unsaturated soils during infiltration due to rainfall. A modified direct shear and a modified triaxial apparatuses were used to carry out the tests on undisturbed Bukit Timah granitic residual soils from Singapore. The specimens were sheared under drained conditions to a pre-determined stress level about 90% of the peak shear strength of the specimen. After creep monitoring, water was injected through the base of the specimen at a constant rate until the specimen failed due to the decrease in matric suction. In triaxial testing, miniature pore-water pressure transducer was utilised to measure pore-water pressure changes at the mid-height of the specimen, especially during infiltration (before and after the failure). The results show that the specimens failed during infiltration due to the decrease of matric suction in both the direct shear and triaxial tests.

1 INTRODUCTION

Landslides occur frequently during heavy rainfall. Infiltration of rainwater into the soil slope results in the loss of negative pore-water pressures thus causing a reduction in shear strength. Understanding this reduction in shear strength can be quite critical to understanding the causes of slope instability. Therefore, shearing infiltration tests have been conducted to simulate the stress path as followed by an unsaturated soil element in a slope during rainfall. The normal stress, σ_n, and the shear stress, τ, essentially remain constant during rainwater infiltration into the slope. Consequently, the failure of the slope is primarily caused by a reduction in the matric suction.

Han (1997) and Melinda (1998) conducted a series of constant suction shearing and shearing infiltration tests on compacted residual soils from the Bukit Timah granitic Formation, Singapore, using a direct shear apparatus with suction control. Their results showed that the strength obtained from shearing infiltration tests was higher than that from constant suction shearing tests. The strength of the soils was affected by hysteresis (i.e. differences between wetting and drying). Lower strength was observed for the soils subject to wetting compared to soils subject to drying.

2 DESCRIPTION OF EQUIPMENT

Conventional direct shear and triaxial apparatuses have been modified to allow the shearing infiltration tests to be conducted on unsaturated soil with control of the matric suction using the axis translation technique. Data acquisition and control were carried out using the Triax4.0 program (Toll, 1999).

2.1 *Direct shear apparatus*

The modified direct shear apparatus was developed by Han (1997). A 500 kPa high air entry ceramic disk is secured into position on top of the water compartment and the pore-water pressure, u_w, in the water compartment is controlled using a digital pressure/volume controller (DPVC). A force actuator is used to maintain the desired shear stress during creep monitoring and infiltration.

2.2 *Triaxial apparatus*

For the purpose of measuring the pore-water pressure at the centre of the specimen, a miniature pore-water pressure transducer was installed at mid-height. Therefore, a larger triaxial cell was utilised to provide enough space for the installation of the miniature pore-water pressure transducer. Studies have been conducted using different miniature pore pressure transducers to measure the matric suction of unsaturated soils (Ridley and Burland 1993; Muraleetharan and Granger 1999). However, these previous studies have not installed the transducer directly onto a triaxial sample.

Figure 1 illustrates the set up of a triaxial soil specimen with the installation of a miniature pore-water pressure transducer. A 500 kPa high air entry disk was glued into the pedestal on top of the water compartment. The water compartment was connected to a DPVC to allow the measurement and the control of pore-water pressures at the base of the specimen. The water compartment was designed to allow for flushing of air bubbles on a periodic basis. The cell pressure, σ_3, was controlled by another DPVC. The pore-air pressure, u_a, was applied and controlled through a pore-air pressure line connected to a loading cap which was placed on top of the soil specimen.

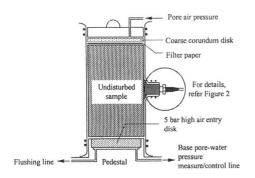

Figure 1. Set up of soil specimen with installation of miniature pore-water pressure transducer

The axial load, σ_1, was applied vertically to the soil specimen using a force actuator through a loading ram with a load cell attached and a loading cap. The force actuator was used to displace the soil specimen during shearing and to maintain the desired deviator stress during the creep monitoring and infiltration. A LVDT was attached to this vertical loading ram to measure the axial strain.

The miniature pore-water pressure transducer used was an Entran EPX-V03-250P (X-option) with a working range of up to 1500 kPa. After a few tests with the transducer in pressurized water, it was found that the transducer was not submersible. The water managed to seep through the connection between the wire and the transducer and subsequently short-circuited the transducer. Later, Araldite 2021 epoxy was used to seal the critical connection. Initially, the transducer worked well and read the pressure changes satisfactorily. However, after a few months, the pressure readings became unreasonable and it was suspected that the bonding of epoxy had broken down. The data of the pore-water pressures using the miniature transducer presented here were those when the transducer was functioning well.

Figure 2 shows the details of installation for the miniature pore-water pressure transducer. A silicone rubber grommet housed the miniature pore-water pressure transducer. The grommet was moulded from a mixture of Dow Corning 3120 RTV silicone

rubber and a fast cure catalyst F in 10:1 ratio. A stainless steel sheath was made to shroud the threaded miniature pore-water pressure transducer. A circular hole in front of the sheath was made to house a 5 bar ceramic disk. The sheath was designed and machined so that a small gap of about 250µm was created between the back of ceramic disk and the frontal sensor of the transducer. This gave a small volume of 3 mm^3 water in the gap so as to minimise the formation of the air bubbles in the gap (Ridley and Burland, 1993). Before the test started, the disk in the sheath was saturated by applying pressure cycles with de-aired distilled water for a couple of days. Increasing water pressures should dissolve the air trapped in the gap between the disk and the sensor and decreasing water pressures should assist dissolved air to flow out from the gap.

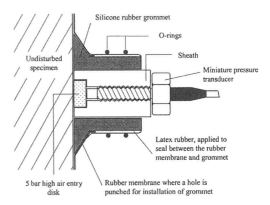

Figure 2. Details of the installation for the miniature pore-water pressure transducer

To install the grommet, a hole was cut in the rubber membrane at its mid-height by using a hole puncher. After inserting the grommet into the hole, latex rubber was applied between the rubber membrane and the grommet. Before pushing the transducer into the grommet, a thin pad of saturated kaolin was smeared on the disk in the sheath to ensure good pore-water phase contact between the specimen and the disk. After inserting the transducer into its housing, two O-rings were fastened onto the grommet, over the latex rubber coating to prevent leakage.

3 TEST PROCEDURES

The test procedure for conducting shearing infiltration tests for both direct shear and triaxial tests can be divided into three stages, i.e. saturation, equalisation and shearing-infiltration. Before saturation can start, the high air entry disk must be in a saturated condition to ensure correct measurement of pore-water pressures and to prevent 'blow-through' of air into the water compartment. The pore-air pressure, u_a, was maintained constant throughout the shearing infiltration test.

3.1 Direct shear test

Saturation of the specimen was achieved by applying a water pressure of 3-4 kPa at the bottom of the specimen. This small pressure difference causes water to flow from the bottom to the top of the specimen and force air out of the specimen. Meanwhile, 5 ml of de-aired distilled water was added to the top of the specimen to speed up the saturation process. During the equalisation stage, the specimen was brought to equilibrium under a net normal stress of 275 kPa and matric suction of 200 kPa.

The shearing-infiltration stage consisted of shearing, creep monitoring and infiltration processes. Having attained equilibrium at the end of the equalisation stage, the specimen was sheared to a stress level that was approximately 90% of the estimated shear strength of the soil. The rate of displacement used during shearing was 0.240 mm/h. This rate of displacement was found not to cause any excess pore-water pressure to build up during shearing (Han, 1997). The force actuator was then programmed to maintain the shear stress level during creep monitoring. Creep monitoring was considered to have finished when the rate of horizontal displacement approached zero. This was done to ensure that any increase in horizontal displacement during infiltration was not caused by creep. Once the creep monitoring was completed, the DPVC was programmed to inject water from the base of the specimen at a constant rate of 144 mm³/h. When there was a sharp increase in horizontal displacement during infiltration, failure was deemed to have occurred.

3.2 Triaxial test

The specimen was saturated until a value of the B_w parameter exceeding 0.9 was obtained (Head, 1986). During the equalisation stage, the specimen was brought to equilibrium under 50 kPa net confining pressure and 50 kPa matric suction. Upon attaining equilibrium, the specimen was sheared at a constant rate of 0.054 mm/h by a force actuator. This shearing rate was found to be suitable for CD tests for Bukit Timah granitic soils (Lim, 1995; Ong, 1999). The deviator stress was maintained constant by the force actuator once the pre-determined stress level was reached. The creep monitoring procedure was considered to have finished when the axial strain rate approached zero. During the infiltration procedure,

water was injected from the base of the specimen at a constant rate of 89 mm³/h. This water injection rate was calculated based on the same infiltration rate (height/time) that was used in the direct shear testing. Again, failure was defined as a point when there was a significant increase in axial strain.

4 SOIL PROPERTIES

Undisturbed samples of Bukit Timah Granitic residual soils were tested. They were taken out from a site in Yishun in north-central Singapore by Mazier drilling. The specimens used in direct shear and triaxial tests were from borehole Y1 at depths 5.5 to 6.5 m and 7 to 8 m respectively. The properties of the specimen are shown in Table 1.

Table 1. Soil properties for the specimens

Property	Direct shear test	Triaxial test
Bulk density (Mg/m³)	1.71	1.57
Initial water content (%)	48.4	56.1
Specific gravity	2.64	2.63
Liquid limit (%)	63	77
Plastic limit (%)	43	47
Sand (%)	6	3
Silt (%)	69	69
Clay (%)	25	28
USCS Classification	Clayey silt MH	Clayey silt MH

5 RESULTS AND DISCUSSIONS

A consistent sign convention is used for presenting both direct shear and triaxial test results. A negative water volume change means pore-water flows out from the specimen and vice versa. In direct shear test, a negative vertical displacement indicates a settlement of the specimen and vice versa. In the triaxial test, a negative axial strain shows a compression in the specimen in the axial direction and vice versa. A negative volumetric strain means compression in the total volume of the specimen and vice versa.

5.1 Direct shear test results

The results of the shearing-infiltration stage are shown in Figure 3. Figure 3(a) shows that the specimen was sheared to a stress level of 250 kPa and 3.65 mm of horizontal displacement was needed to mobilise this stress level. During shearing, the matric suction of 200 kPa was maintained constant as shown in Figure 3(b). The specimen compressed and the pore-water flowed from the specimen as depicted by the decrements in vertical displacement and water volume changes in Figure 3(c).

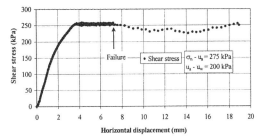

(a)

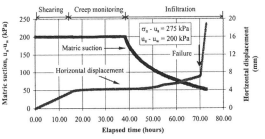

(b)

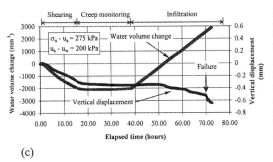

(c)

Figure 3. Results for shearing-infiltration direct shear test under 275 kPa net normal stress and 200 kPa matric suction (a) shear stress versus horizontal displacement (b) matric suction and horizontal displacement versus elapsed time (c) water volume change and vertical displacement versus elapsed time.

During creep monitoring, the horizontal displacement initially increased gradually but then levelled off with time as seen in Figure 3(b). Therefore, any horizontal displacements due to creep were completed before starting the infiltration procedure. No changes in water volume change suggests no movement of pore-water in or out of the specimen.

Once infiltration commenced, the matric suction decreased from 200 kPa and the rate of decrement became gradual with time. During infiltration before failure, the horizontal displacement increased gradually and the graph showed some changes in slope,

possibly indicating yielding points, as shown in Figure 3(b). Failure was deemed to have occurred at a matric suction of 56 kPa when there was a significant increase in horizontal displacement at 7.29 mm. Once failure occurred, the specimen compressed rapidly with a sudden decrease in vertical displacement as shown in Figure 3(c). This contraction at failure is probably due to the high net normal stress (275 kPa). Because of the contractant behaviour, the matric suction continued to decrease.

5.2 Triaxial test results

Figure 4 shows the results of the shearing stage. Figure 4(a) shows that the specimen was sheared to a pre-determined deviator stress of 187 kPa. After failure due to infiltration occurred, the deviator stress decreased to an ultimate strength which remained constant for large strains.

Figure 4(b) shows the behaviour in terms of axial strain, base and mid-height matric suctions with time. During shearing and creep monitoring, the mid-height matric suction remained constant. This proves that the drained shearing rate used was appropriate and it did not result in any build up of excess pore-water pressure. During creep monitoring, the axial strain decreased slowly (specimen compressed in the axial direction) but stabilised eventually. It shows that the effect of creep was negligible. Once infiltration started, the matric suctions at the base and mid-height decreased. However, the base matric suction reduced quicker than the mid-height matric suction. It clearly shows that the base and mid-height pore-water pressures were not the same during infiltration. Just before failure took place, the mid-height matric suction was 6 kPa higher than the base matric suction. This suggests that there was a water front moving vertically up from the base of the specimen. Once failure occurred, the axial strain decreased drastically (the specimen compressed in the axial direction).

The fluctuations of the volumetric strain during shearing and creep monitoring as shown in Figure 4(c) were due to expansion and contraction of the big triaxial cell due to temperature changes. This caused some difficulty in interpreting the small total volume change of the specimen. However, the volumetric strain increased greatly after failure. This is different from the behaviour in direct shear due to the lower net confining pressure used (50 kPa).

If the soil on a potential failure surface fails in a dilatant mode, the infiltration-dependent suction change and strain-dependent suction change counteract since for dilatant shear, a suction increase is induced by shear (Brand, 1981). The matric suction increased immediately after failure as seen in Figure 4(b) because the strain-induced suction change was greater than the infiltration-induced suction change. However, continued infiltration of water causes the infiltration-induced suction change to overcome the strain-induced suction change and this explains why the matric suction decreased eventually.

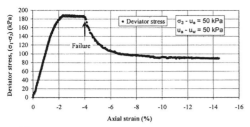

(a)

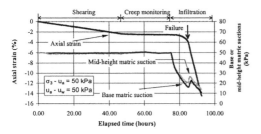

(b)

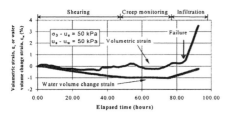

(c)

Figure 4. Results for shearing-infiltration triaxial test under 50 kPa net confining pressure and 50 kPa matric suction (a) deviator stress versus axial strain (b) axial strain and base or mid-height matric suctions versus elapsed time (c) volumetric or water volume change strains versus elapsed time.

During shearing, the decrease of water volume change strain (= percentage of water volume change relative to the initial total specimen volume) shows that the pore-water was squeezed out from the specimen. However, during creep monitoring, the pore-water volume was unchanged as shown by the water volume change strain. This was also observed in the direct shear test.

6 CONCLUSIONS

In both direct shear and triaxial tests, it is seen that failure of the specimens during infiltration was due to the decrease of matric suction.

Some of common behaviour observed in both direct shear and triaxial testing:

(a) During shearing, pore-water was expelled from the specimen as the specimen compressed.

(b) During creep monitoring, there were no changes in pore-water volume.

(c) During infiltration, the matric suction decreased. The rate of the suction decrement reduced with time.

(d) The specimen yielded before failure but showed a large increase in strain at failure.

The lower pore-water pressures at mid-height compared to that at the base during infiltration suggests that the pore-water pressure distributions were not uniform. Water takes time to travel from the base to the top of the specimen.

When failure occurs in dilatant mode, the strain-induced suction change counteracts the suction reduction due to infiltration, as the suction increases under shear. However, continued infiltration further reduces the suction.

ACKNOWLEDGEMENTS

This research was supported by a grant from the National Science and Technology Board (NSTB Grant 17/6/16: Rainfall-induced Slope Failures) in Singapore. Many thanks are due to Vincent Heng for his assistance and valuable comments to carry out the research.

REFERENCES

Brand, E.W. (1981). Some thoughts on Rain-Induced Slope Failures. *Proceedings of 10th International Conference on Soil Mechanics and Foundation Engineering.* Stockholm. Vol. 3: 373 – 376.

Han, K.K. 1997. Effect of hysteresis, infiltration and tensile stress on the strength of an unsaturated soil. *Ph.D. Thesis.* Nanyang Technological University, Singapore.

Head, K.H. (1986). Manual of Soil Laboratory Testing. Vol. 3. ELE International Limited.

Lim, T.T. 1995. Shear strength characteristics and rainfall-induced matric suction changes in a residual soil slope. *M.Eng Thesis.* Nanyang Technological University, Singapore.

Melinda, F. 1998. Shear strength of a compacted residual soil from unsaturated direct shear tests. *M.Eng Thesis.* Nanyang Technological University, Singapore.

Muraleetharan, K.K. and Granger, K.K. 1999. The use of miniature pore pressure transducers in measuring matric suction in unsaturated soils", *Geotechnical Testing Journal*, Vol. 22, No. 3, pp. 226 – 234.

Ong, B.H. 1998. Shear strength and volume change of unsaturated residual soil. *M.Eng Thesis*. Nanyang Technological University, Singapore.

Ridley, A.M. & Burland, J.B. 1993. A new instrument for the measurement of soil moisture suction. *Geotechnique 43.* No. 2: 321-324.

Toll, D.G. 1999. A Data Acquisition and Control System for Geotechnical Testing. *Computing Developments in Civil and Structural Engineering (edited by B. Kumar and B.H.V. Topping).* Edinburgh: Civil-Comp Press, p.p 237 - 242.

9   Volume change

9.1 Collapse

Unsaturated Soils for Asia, Rahardjo, Toll & Leong (eds) © 2000 Taylor & Francis, ISBN 90 5809 139 2

Collapsibility of Egyptian loess soil

F. M. Abdrabbo, R. M. El Hansy & K. M. Hamed
Structural Engineering Department, Faculty of Engineering, Alexandria University, Egypt

ABSTRACT: The work presented is dealing with experimental research project on the response of circular footing resting on collapsible soil. A special apparatus was prepared in laboratory to measure the collapse settlement of a footing model under the effect of several parameters. Most of the previous researches are dealing with the inundation of collapsible soil starting from the top surface, which represents the case of rainwater, or similar. In this research the inundation starts from bottom of the soil toward the surface which is representing the case of the breakage of sewage pipes and/or rise of ground water table. The used system for inundation enabled to study the effect of dry soil thickness beneath the footing on the value of collapse settlement. The effects of load ratio, initial dry unit weight and the molding moisture content are also studied. Sand cushion is placed beneath the footing as a stabilization method to improve the response of footing and to minimize the collapse potential of soil upon wetting. Comparison between the measured and the predicted collapse settlement using the collapse predictive model developed by Basma & Tuncer (1992) are made.

1 INTODUCTION

The collapsible soils are covering vast areas of several countries. Particularly in Egypt, they are widely spread throughout the Egyptian western desert, especially at Sidi-Baranee, New El-Ameria city, El-Boustan, Borg El-Arab city, the 6th of October city, etc.

Some collapsible soils are deposited in such a way that they exist in a loose honeycomb-type structure at a relatively low density. At their natural low moisture content they possess high apparent strength, but are susceptible to a large reduction in void ratio upon wetting. Some other collapsible soils are generally composed of uniform, silt-sized particles, which were loosely deposited, and are bonded together with relatively small fraction of clay forming the typical loess structure. Normally, loess has high shearing resistance and withstands high stresses without great settlement, when natural moisture content is low. However, upon wetting, the clay bond tends to soften and cause collapse of the loess structure inducing large settlement under low level of stresses. Therefore, it is dangerous to construct on these soils without improving their characteristics.

In Egypt, the formation is very dense; SPT values in the range of 250/300mm blows were recorded. Extensive work on Egyptian western desert collapsible soil was carried out (Abdrabbo &

Mahmoud 1988). The authors found that the angle of shearing resistance measured from shear box apparatus, decreases as the molding water content increases, meanwhile the cohesion increases with the increase of the molding water content up to the value of water content corresponding to optimum dry density, then decreases again. The authors also, found that the western desert soil exhibits collapse potential at initial void ratio bigger than 0.45.

Unstable collapsible soils have relative density ranging from 0.1 to 0.9, but for many stable soils they have relative density of 0.70 (Dudley 1970). But, it is misconception when thinking that all soils of low density will exhibit tendency to collapse (Jennings & Knight 1975). The U.S. Bureau of Reclamation (Gibbs & Bara 1967) demonstrated that if the soil exists in void ratio higher than that would exist at the liquid limit, the addition of water would result in collapse settlement. The post-soaked ($\Delta e/\Delta p$) values are independent of initial void ratio, but the pre-soaked values are dependent (Mahmoud & Abdrabbo 1992). The factors affecting the collapse potential of collapsible soil were reported by Abdrabbo & Mahmoud (1988), Mahmoud & Abdrabbo (1992), Houston et al. (1988) and Dudley (1970).

Many researchers reported serious damages that occurred to different types of structures due to collapse settlement (Capps & Hejj 1968) and (Rollins & Rogers 1994). Houston et al. (1988)

modified the procedure of double oedometer test developed by Jennings & Knight (1975) to estimate the magnitude of collapse settlement by conducting only one oedometer test. Many researchers made comparison between the output results of implementing the double and single oedometer tests in calculating the collapse settlement of a foundation; general agreement was found (Ismael 1988), (Lawton et al. 1992) and (Basma & Tuncer 1992). Houston et al. (1995) introduced an in-situ collapse test, which is called down-hole collapse test. The test results were used to develop stress-strain relationship of soil.

Basma & Tuncer (1992) developed the following equation to predict the collapse potential (S_c) as,

$$S_c = 48.496 + 0.102\ C_u - 0.457\ W_i - 3.53\ \gamma_d$$
$$+2.8 \ln P_w \qquad (1)$$

where C_u = coefficient of uniformity of the soil; W_i = initial moisture content, %; γ_d = dry unit weight in kN/m^3 ; and P_w= pressure at wetting in kN/m^2.

2 TESTING EQUIPMENT AND MATERIALS

A soil bin, used to contain the soil, was made of transparent plastic (perspex) plates; 345 mm side length, 400 mm height and 10 mm thickness connected together using a mixture of perspex powder and chloroform and tied together using steel ties at two levels as shown in Figure 1. The base of the bin was made of a square perspex plate; 365 mm side length and 20 mm thickness. The soil bin was placed on a square steel plate; 340 mm side length, 12mm thickness, which rests on two steel rings; 300 mm diameter, 450 mm height filled with compacted sand. The two rings were placed inside a cylindrical steel rigid container; 750 mm diameter and 600 mm height filled with compacted sand, in order to insure the stability of soil bin during running the test.

A calibrated proving ring, 49 N accuracy, recorded the load applied using a frictionless lever and guiding system. The displacement of the footing was recorded by two dial gauges; 0.01 mm accuracy fixed rigidly to reference beams using magnetic bases. During loading tests, the difference between readings of the two dial gauges, were kept to be within 1 % of the mean value, to accept the test results. The model footing 100 mm in diameter was machined from a steel plate 10 mm thickness. The loads transferred from the lever to the footing, were checked by the principles of static using the lever arm ratio. No appreciable difference what so ever were deserved between the calculated and the measured values. A circular water tank of 200 mm diameter was used to supply the soil bin with water through four plastic pipes. The water was controlled using four water valves, each of 10mm diameter, made from brass, and connected to the bottom side of soil bin symmetrically at equal distance.

The collapsible soil was obtained from New Borg El-Arab city, 60 km to the west of Alexandria city. The soil was dried in an electrical oven at 110 °C for 24 hours then, sieved on 0.425 mm B.S. sieve, the passing material was used. The main characteristics of the soil are shown in Table 1. The laboratory tests on soil are carried out in accordance with ASTM.

The soil bin, Figure 2 was placed and centered at its position with respect to the loading system. Fine gravel layer; 50 mm thickness acts as a filter, was placed and compacted to reach a dry unit weight of 17.56 kN/m^3 and then, covered with a sheet of felt to separate between collapsible soil deposit and the filter layer. The collapsible soil was prepared in six layers, each 50 mm thickness. The weight of dry soil was determined for each layer and compacted to achieve the desired predetermined dry unit weight.

After placing and forming the soil, the footing was placed accurately on the top of soil surface and loaded incrementally up to a predetermined percentage of the failure load. The load was kept constant, and the water allowed to flow to the soil via the four water valves. After water had reached the required level, the water valves were turned off. The water level in the water tank was held constant during the inundation process. The rise of water through the soil was recorded with time, if possible, before and after turning off the water valves. The collapse settlement due to wetting was also, recorded with time during this stage. The load was held constant for 24 hours until full collapse settlement takes place. The vertical load was then increased incrementally using the same procedure before soil inundation up to failure. For each load increment, the settlement of the footing was measured up to the rate of settlement becomes less than 0.005 mm / 20 min for three successive readings. A reference-loading test was conducted on similar soil having zero water content, but without inundation to determine the ultimate load (P_u) of footing-soil system. The ultimate load was defined as the load corresponding to abrupt change in the slope of the load-settlement relationship; presented in logarithmic drawing.

Table 1. Main characteristics of collapsible soil

% passing no 200 sieve	76 %
Effective diameter, $D_{10\%}$	0.008 mm
Uniformity coefficient, C_u	5.5
Coefficient of curvature, C_c	2.23
Liquid limit, W_l	30 %
Plastic limit, W_p	21 %
Plasticity index, I_p	9 %
Specific gravity, G_s	2.59
Maximum dry unit weight, γ_{dmax}	16.87 kN/m^3
Optimum moisture content, w_{opt}	21 %

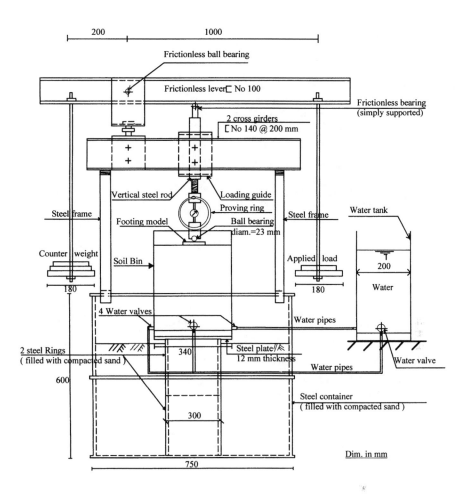

Figure 1. Testing equipment

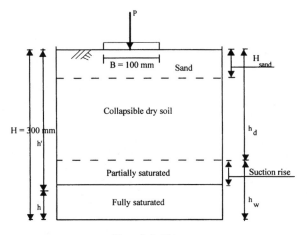

Figure 2. Soil bin

3 TESTING PROGRAMME

A testing program was designed to investigate the effect of the thickness of the inundated collapsible soil (h), the load ratio (P/P$_u$), the molding moisture content (W$_i$), the dry unit weight of soil (γ_d) and other many factors (Hamed 1998). The effect of the thickness of collapsible soil replaced by compacted sand, sand cushion, as a method of collapsible soil stabilization was also investigated.

A comparison between the measured values of collapse settlement from footing model, and the predicted values using the collapse predictive model developed by Basma & Tuncer, (Equation 1), was conducted.

4 TESTING RESULTS

4.1 *Effect of the inundation thickness*

Fourteen plate loading tests were performed to study the effect of the inundation thickness (h), Figure 2, on the collapse settlement. These tests were carried out at an initial dry unit weight (γ_d) of 11.87 kN/m^3 and at three load ratios of 0.25, 0.50 and 0.75. The load ratio was defined as the load (p) at which the soil is inundated divided by the ultimate load (P$_u$) obtained from the reference loading test. From these tests, it was found that, the collapse settlement (S$_c$) increases with the increase of the thickness of the inundated collapsible soil (h), that is to say, as the thickness of dry soil beneath the footing decreases.

Figure 3 shows the relationship between (S$_c$/B) and (h$_d$/B). This figure confirmed that, for load ratios 0.25 and 0.50 the collapse settlement of the footing is insignificant when the depth of the dry soil under the footing is nearly equal to 2.5 B. It is worthnoted from the figure that as the load ratio increased the collapse settlement increased, as long as, h$_d$/B is less than 1.0. Beyond this limit the collapse settlement becomes independent of load ratio (p/p$_u$). This can be attributed to the insignificance of the imposed stresses at depths beyond the limiting depth.

At load ratio of 0.75, high values of collapse settlement were developed which could not be measured, when inundation depth, h, was more than 100 mm which correspond to h$_d$/B less than 1.50.

The rise of water through the soil and the corresponding collapse settlement due to wetting were recorded. The relationship between (h$_w$ /B) and (S/B) during the inundation process was plotted as shown in Figure 4. These tests were carried out on dry soil having initial dry unit weight of 11.87 kN/m^3 and at three load ratios 0.25, 0.50 and 0.75. It is clear from the figure that the (h$_w$/B)-(S/B) relationship is similar to (P-S) relationship resulting from loading a footing. The failure pattern of the footing was recorded as punching failure.

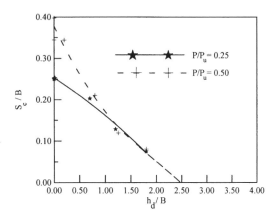

Figure 3. Variation of (S$_c$/B) with (h$_d$/B)

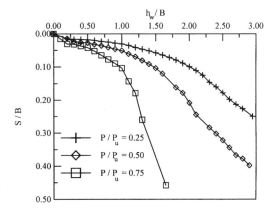

Figure 4. Variation of (S/B) with (h$_w$/B)

4.2 *Effect of dry unit weight*

The effect of initial dry unit weight (γ_d) on the values of ultimate load and collapse settlement was investigated. Obviously, as the initial dry unit weight of the soil increases, the ultimate load of the footing-soil system increases, Figure 5. Also the figure indicated that as the molding water content increases the ultimate load decreases, due to the decrease in shear strength. These finding agree with conclusions reported by Abdrabbo & Mahmoud (1988).

The effect of initial dry unit weight on the collapse settlement was studied by loading the soil until it reaches a load ratio, P/P$_u$ = 0.50, and then inundated, the height of the inundation, (h), was 300 mm.

It was found that the collapse settlement decreases with the increase of the dry unit weight as shown in Table 2.

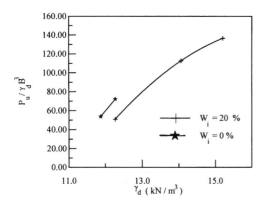

Figure 5. Dry unit weight of soil versus $P_u/\gamma_d B^3$

Table 2. Effect of dry unit weight

Wi, %	γ_d, kN/m³	P_u kN	$P_u/\gamma_d B^3$	S_c/B
0	11.87	0.64	53.719	0.39765
0	12.26	0.88	72.00	0.29715
20	12.26	0.62	50.912	Failure
20	14.06	1.58	112.649	0.01665
20	15.21	2.08	136.607	0.01375

4.3 Effect of molding moisture content

The effect of the molding moisture content, on both ultimate load and collapse settlement was studied. It was found that, the ultimate load increases with the increase in the molding moisture content up to a certain limit, and then decreases beyond this limit. The collapsible soil samples were completely inundated at load ratio $P/P_u = 0.50$ and the collapse settlement (S_c), was measured as shown in Table3. It was found that for low value of dry unit weight (γ_d =12.26 kN/m³) the footing exhibited high value of collapse settlement (S_c/B =0.297) where the molding moisture content was 0.0 %, while the footing failed when the molding moisture content (W_i =10% &20 %). For high value of dry unit weight =15.21 kN/m³ the collapse settlement decreases with the increase in molding moisture from 15% to 20%.

Table 3. Effect of molding moisture content

γ_d, kN/m³	Wi, %	P_u kN	$P_u/\gamma_d B^3$	S_c/B
12.26	0.0	0.88	72.0	0.29715
12.26	10	1.57	128.112	Failure
12.26	20	0.62	50.912	Failure
15.21	15	3.51	230.526	0.0634
15.21	20	2.08	136.607	0.01375

4.4 Effect of thickness of sand cushion

Sand cushion placed under the footing was used as a method of stabilization of the collapsible soil. Two tests were conducted. In the first test, the thickness of sand cushion was taken equal to the width of the footing, while in the second test, the thickness of sand cushion was taken equal to twice the footing width. The soil was prepared at dry unit weight of 12.26 kN/m³ and water content equal to zero. The collapse settlement S_c was measured at load ratio $P/P_u = 0.50$. The results of this study are shown in Figure 6 and Figure 7. From these figures, it was found that, the increase in thickness of sand cushion results in decrease in the collapse settlement. It was found that, the use of sand cushion of thickness (B) reduced the collapse settlement to nearly 50 % of its value without using sand cushion, while sand cushion of thickness 2B reduced the collapse settlement to nearly zero value.

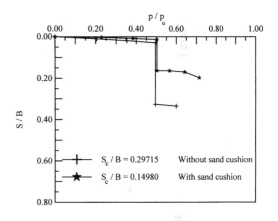

Figure 6. Effect of sand cushion, ($H_{sand} = B$)

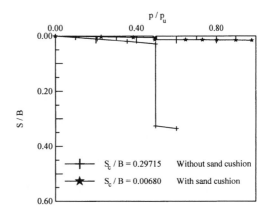

Figure 7. Effect of sand cushion, ($H_{sand} = 2B$)

5 COMPARISON OF RESULTS

A comparison between the measured values of collapse settlement from footing models, and the predicted values using the collapse predictive model, developed by Basma & Tuncer (1992) was conducted. The results of this comparison are shown in Table 4. The relationship between R (the ratio of measured collapse settlement/predicted collapse settlement using Equation 1) and (h_d/B) is shown in Figure 8. From this figure, it was found that, the measured collapse settlement is, in general, less than the predicted collapse settlement. But, this ratio increases with the decrease in the dry depth under the footing, and is closer to the measured values when h_d/B = 0, i.e. the collapse predictive model developed by Basma & Tuncer is more applicable when the inundation of collapsible soil starting from top surface of soil.

6 CONCLUSIONS

The following conclusions were drawn from the course of investigation:

1. The collapse settlement increases as the thickness of inundated collapsible soil increases; the load ratio increases and the initial dry unit weight decreases.

2. The collapse settlement at load ratios 0.25 and 0.50 is equal to inappreciable value when the depth of dry soil under the footing is equal nearly to two and half times the footing diameter.

3. Using a sand cushion with thickness equal to twice the footing diameter decreases the collapse settlement significantly.

4. Basma & Tuncer predictive model over estimates the collapse settlement of circular footing resting on collapsible soil and inundated far from foundation level.

Table 4. Measured and predicted values of collapse settlement

P/P_u	h_d/B	R = measured value/ predicted value
	1.80	0.44
	1.20	0.55
	0.70	0.68
0.25	0.20	0.69
	0.00	0.69
	0.00	0.71
	1.80	0.42
	1.25	0.46
	0.80	0.82
0.50	0.20	0.84
	0.00	0.85
	0.00	0.99

REFRENCES

Abdrabbo, F.M.& Mahmoud, M.A.1988.Shear strength and deformation of laboratory compacted collapsible soil. *Alexandria Engg. J.* 28(2): 553-564.
Basma, A.A. & Tuncer, E.R.1992. Evaluation and control of collapsible soils. *Geotechnical Engg.J.* 118(10) :1491-1504.
Capps, J.F. & Hejj, H.1968. Laboratory and field tests on a collapsing sand in northern Nigeria. *Geotechnique* 18(4) :506 - 508.
Dudley, J.H.1970. Review of collapsing soils. *J. Soil mech. and found.divison* 96(3) : 925 - 947.
Gibbs, H.J. & Bara, J.B.1967. Stability problems of collapsing soil. *J. Soil mech. and found. division* 93(4): 577 - 594.
Hamed, K.M.1998. Response of foundation on collapsible soil. *A thesis submitted to Alex. University in partial fulfillment of the requirements for the M.Sc. degree.*
Houston, S.L., Houston, W.N. & Spadola, D.J. 1988.Prediction of field collapse of soils due to wetting. *Geotechnical Engg.J.* 114(1): 40 - 58.
Houston, S.L., Mahmoud, H.H.& Houston, W.N.1995. Down-hole collapse test system. *Geotechnical Engg.J.* 121(4): 341 - 349.
Ismael, N.F.1988. Discussion of prediction of field collapse of soils due to wetting. *Geotechnical Engg.J.* 114(12): 1806 - 1808.
Jennings, J.E. & Knight, K.1975. A guide to construction on or with materials exhibiting additional settlement due to collapse of grain structure. *Proc. of sixth regional conf. for African soil mechanics and foundations engineering.* Durban 1: 99 - 105.
Lawton, E.C., Fragazy, R.J. & Hetherigton, M.D.1992. Review of wetting induced collapse in compacted soil. *Geotechnical Engg.J.* 118(9): 1376 - 1394.
Mahmoud, M.A.& Abdrabbo, F.M.1992. The collapse potential of collapsible soil. *Alexandria Engg.J.* 31(1):267-271.
Rollins, K.M. & Rogers, G.W.1994. Mitigation measures for small structures on collapsible alluvial soils. *Geotechnical Engg.J.* 120(9): 1533 - 1553.

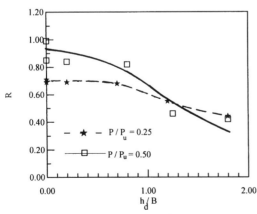

Figure 8. Variation of ratio (R) with (h_d/B)

Unsaturated Soils for Asia, Rahardjo, Toll & Leong (eds) © 2000 Taylor & Francis, ISBN 90 5809 139 2

Effect of dissolution of calcium sulfate on volume change characteristics of unsaturated soil sediments

S. Azam
Department of Civil and Environmental Engineering, University of Alberta, Edmonton, Alb., Canada

S. N. Abduljauwad
Department of Civil Engineering, King Fahd University of Petroleum and Minerals, Dhahran, Saudi Arabia

ABSTRACT: Calcium sulfate undergoes volume change due to its high solubility in water. Dissolution is accompanied by phase transformation, of which both cause serious engineering problems associated with, swelling, collapse, compressibility, hydraulic conductivity, and ion leaching. This paper describes the effect of dissolution of calcium sulfate on volume change behavior of unsaturated soil. Results of oedometer tests are studied in conjunction with those obtained from scanning electron microscopy, and chemical analyses.

1 INTRODUCTION

Unsaturated calcium sulfate also known as anhydrite $(CaSO_4)$, dissolves in flowing water that results in volume decrease (James & Lupton 1978). Such unrestricted access of water to the mineral converts it to gypsum $(CaSO_4.2H_2O)$. This process of phase transformation is accompanied by volume increase of 62% (Zanbak & Arthur 1986). Gypsification is followed by dissolution of gypsum, which is associated with decrease in volume. According to Azam et al. (1998), dissolution of anhydrite and gypsum results in formation of channels in poorly consolidated sediments.

Alternate volume change due to dissolution and phase transformation of anhydrite results in serious geotechnical problems. Gypsification of anhydrite is reported to cause excessive swell pressure in tunnels and floor heave in dams (Brune 1965). Similarly, the high solubility of calcium sulfate leads to loss in soil strength, increase in soil settlement and collapse (Azam 1997). Further, hydraulic conductivity may be multiplied manifold when the clay core in dams or clay liner in landfills contain even small amount of the mineral. Anhydrite dissolves in accelerating fashion as hairline cracks are enlarged within no time to produce unmanageable situations (James & Lupton 1978). Likewise, gypsum can be extremely dangerous when it cements the soil matrix because its dissolution leaves a mesh of soil (Abduljauwad & Al-Amoudi 1995).

In this paper, volume change characteristics of calcium sulfate are discussed. Based on laboratory investigation, results of conventional and modified oedometer tests, scanning electron microscopy (SEM), and chemical analyses are presented.

2 BACKGROUND

Calcium sulfate is a major component of evaporitic rocks, which underlay 25% of the continental areas (Blatt et al. 1980). These rocks are derived from seawater that on evaporation yields 3.6% by weight of the residue as calcium sulfate (Deer et al. 1972). Further, gypsum is the primary precipitate whereas anhydrite is its dehydration product (Murray 1964). Figure 1 gives a schematic diagram of gypsum-anhydrite cycle in nature. This cycle, which is controlled by the prevalent environmental conditions, is based on stability relations in calcium sulfate-water system and the sequential process of deposition-burial-uplift-erosion (Blatt et al. 1980).

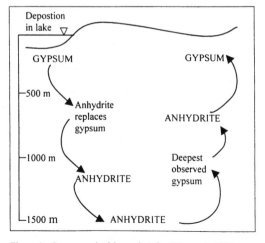

Figure 1. Gypsum-anhydrite cycle (after Blatt et al. 1980)

Gypsum and anhydrite dominate youthful soils, which are in their early weathering stages (Jackson & Sherman 1953). Such deposits are found in semi-arid and arid regions where limited water availability keeps chemical weathering to a minimum (Mitchell 1993). In these areas, periodic variations in temperature and relative humidity affect the behavior of indigenous sediments (Abduljauwad et al. 1999). Similarly, pressure as a measure of depth and brine chemistry has profound influence on engineering characteristics of these deposits. In such sedimentary basins, the ionic strength of formation waters increases with depth. At saturation, the predominant minerals are halite (NaCl), anhydrite and gypsum (He & Morse 1993). Figure 2 summarizes the intricate relationship between critical geothermal anhydrite-gypsum transformation temperature and solubility of anhydrite as a function of brine concentration and overburden pressure.

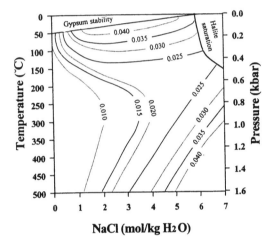

Figure 2. Solubility of anhydrite in sedimentary deposits (after Blount & Dickson, 1969)

The variability in the rate of solution formation of gypsum and anhydrite adds further complexity to the problem. Dissolution rates depend on flow velocity, hydraulic gradient, and salinity of water (Calcano & Alzura 1967). Gypsum dissolves according to first order kinetics (Liu & Nancollas 1971) whereas anhydrite goes into solution as per second order kinetics (Fabuss et al. 1969).

3 EXPERIMENTAL PROGRAM

Gypsum was retrieved from the coastal town of Dhahran in eastern Saudi Arabia. It was pulverized, to pass ASTM Sieve No. 40 (0.425 mm). Powdered gypsum was converted to anhydrite by heating it to 300 °C for 48 hours and thereafter keeping it at a

relative humidity not in excess of 70% (Azam et al. 1998). Identical anhydrite specimens were compacted to dry unit weight of 1.8 g/cm³. Samples were prepared according to ASTM Test Methods for One-Dimensional Consolidation Properties of Soils (D 2435) and ASTM Test Methods for Swell/Settlement Potential of Cohesive Soils (D 4546).

Volume change characteristics were determined using conventional and modified oedometer tests. In the latter test, Al-Amoudi & Abduljauwad (1994) modified the oedometer by making two holes beneath the specimen, an inlet for liquid supply and an outlet for overflow. For both tests, the specimen was placed in the oedometer and thereafter subjected to a seating pressure of 7 kPa (Abduljauwad & Al-Sulaimani 1993). When settlement of the sample stabilized, it was inundated and allowed to increase in volume vertically at the seating pressure. After completion of primary swell, the sample was loaded incrementally to cancel deformation due to flooding.

In the modified oedometer test, constant head was maintained by regulating flow of reservoir fluid. The amount of fluid flowing through the inlet valve was always kept more than that seeping through the consolidating specimen; the excess fluid drained out from the outlet valve. From the two holes below the consolidating specimen, the percolating fluid was collected using a graduated cylinder. The volume of this fluid was measured at regular intervals, thereby measuring the hydraulic conductivity with respect to both time and pressure during loading. Chemical analyses were conducted on this fluid to identify the various leached and dissolved ions. Concentrations of Ca^{2+} and SO_4^{2-} ions were determined using atomic absorption spectrophotometer (AAS), Perkin Elmer 4000 and calorimetric method, respectively (Sawyer et al. 1994).

To correlate morphology with volume change, a fabric and structure study was made on thin sections cut from the modified oedometer test samples. SEM analyses were conducted using JEOL (JSM-840), which performs morphological and micro-structural assessment. Each sample was held in an aluminum sample holder and sputters coated with a fine gold film. Micrographs were taken with enlargements of 500 times.

To elucidate the spontaneous effect of pore fluid chemistry on volume change, distilled water and brine were used in the modified oedometer test. Chemical analysis of brine indicated the presence of the following ions (g/l): Na^+ (78.80), K^+ (3.06), Ca^{2+} (1.45), Mg^{2+} (10.32), Cl^- (157.20), HCO_3^- (0.09), and SO_4^{2-} (5.48). These data suggest that the salinity of brine is five times that of typical seawater. According to Al-Amoudi & Abduljauwad (1995), the specific gravity and dynamic viscosity of brine at 20 °C were 1.207 and 1943 µNs/m², compared with 0.998 and 1009 µNs/m² for distilled water.

4 RESULTS AND DISCUSSION

4.1 *Void Ratio*

Table 1 gives oedometer test results at critical stages during the tests. The table indicates that in the conventional (un-leached) oedometer test, the void ratio increases from 0.31 to 0.45. This increase in volume upon flooding amounts to 45% and is due to gypsification of anhydrite. On the contrary, in the modified oedometer test, the void ratio increases by 25% for distilled water percolation. This is because of anhydrite dissolution, which does not allow the mineral to exhibit its volume increase capacity to mobilize completely. Similarly, for brine permeation, the void ratio increases by a mere 20% upon inundation. This means that the sample through which brine is allowed to pass through is more susceptible to solution formation as compared to the one through which water flows.

Table 1. Void ratio at critical stages in oedometer tests

Void ratio	Conventional oedometer	Modified oedometer	
		Water	Brine
Initial	0.31	0.31	0.31
On flooding	0.45	0.42	0.39
Final	0.37	0.34	0.31

Table 1 also suggests that percent decrease in void ratio on subsequent loading is about 20 for all samples. This means that the same pore space is left for the fluid to pass through the specimen. Figure 3, which compares results of un-leached, water-leached and brine-leached samples, confirms the above. The figure shows constant decrease in void ratio with successive load increments for all samples over the entire duration of the test.

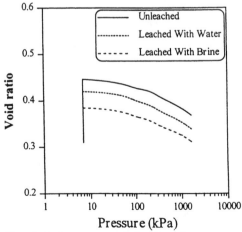

Figure 3. Oedometer test results for all samples

The collapse potential (*CP*) was estimated from the following equation:

$$CP\ (\%) = \Delta e\ /\ (1 + e_o) \qquad (1)$$

where Δe is the reduction in void ratio due to leaching at constant total stress, and e_o is the initial void ratio. The values of *CP* at a pressure of 7 kPa were 2.1% and 4.1% for water and brine infiltration, respectively. Thus, *CP* for brine is twice that for water. This is the result of the extremely high ionic strength of the fluid. Further, according to the classification of collapsible soils (Jennings & Knight 1975), the sample leached with brine offers moderate trouble whereas the sample leached with distilled water is not problematic.

Compressibility behavior is explained when final values of void ratio of leached samples are compared with the void ratio of the un-leached sample after the initial volume increase (i.e. 0.45). The identical samples are compressed by 0.11 and 0.14 for distilled water and brine percolation, respectively. These reductions in void ratio amount to 25% for distilled water permeation and 30% for brine percolation. It follows that dissolution of calcium sulfate results in appreciable increase in compressibility and the effect is more pronounced for brine infiltration. Further, for both percolating fluids, dissolution of anhydrite precedes gypsification, which is followed by dissolution of gypsum.

4.2 *Micro-Structure*

Examining its morphological details further highlights the volume change characteristics of calcium sulfate. Figure 4 (a) shows the SEM micrograph of calcium sulfate permeated with distilled water after the completion of the modified oedometer test. The micrograph shows lath-like anhedral gypsum crystals formed due to the hydration of calcium sulfate. Both long and short rods are visible, which are aggregated in separate but mutually conjoined colonies. Further, due to percolation of water and the resulting void spaces, gypsum crystals are oriented in all directions. This in turn has led to arching of gypsum crystals, which coupled by the low initial void ratio, is responsible for the low compressibility of the specimen. Further, due to chemical cementing, soil grains are held together causing decrease in compression (Clemence & Finbarr 1981).

Figure 4 (b) shows the SEM micrograph of identical sample after brine is allowed to pass through it. This micrograph shows larger and abundant cavities due to increased effusion of soil constituents, which is facilitated by Na^+ and Cl^- ions present in brine (Wigley 1973). These ions cause chemical leaching that precedes and creates favorable conditions for mechanical leaching in saline waters (Todd and McNulty 1976). Mechanical leaching is related to the removal of hard particles. Thus, the high concentration of halite forming ions not only increases the flow rate but also increases the amount of ions leached out of the sample. The micrograph also shows that dissolved halite coats gypsum crystals and occupies part of the pore space.

both flowing through identical samples. Further, the figure indicates high initial hydraulic conductivity values due to the high rate of dissolution of anhydrite and negligible loading, both of which lead to increase in pore area of the matrix. This is followed by sharp decrease in hydraulic conductivity due to gypsification. Finally, the curves flatten owing to subsequent loading and the low rate of dissolution of gypsum.

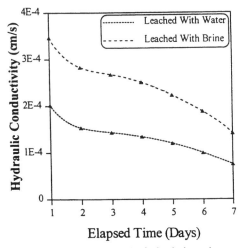

Figure 5. Hydraulic conductivity for leached samples

Figure 5 also shows that the hydraulic conductivity of brine percolated sample reduces from 3.5×10^{-4} cm/s on the first day, to a value of 1.4×10^{-4} cm/s on the seventh day which amounts to 60% reduction. This same percent reduction in hydraulic conductivity was recorded for distilled water permeated sample, which decreased from 2.0×10^{-4} cm/s to 7.5×10^{-5} cm/s in seven days. Further, hydraulic conductivity curves exhibit similar pattern for both fluids, confirming the phenomenon of almost equal reduction in void ratio with the application of loads (Figure 3). Therefore, hydraulic conductivity under all overburden pressures can be approximated as $e^{0.5}$ times that under the seating pressure of 7 kPa.

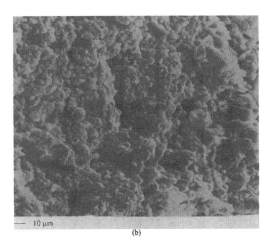

Figure 4. SEM micrographs of calcium sulfate: (a) leached with water and (b) leached with brine

4.3 Hydraulic Conductivity

Figure 5 presents results of hydraulic conductivity during the modified oedometer tests. Hydraulic conductivity was estimated from discharge measurements and knowledge of the cross sectional area of the sample. The figure shows high hydraulic conductivity for brine as compared to distilled water

4.4 Chemical Analyses

Figure 6 gives the results of the chemical analyses of water and brine percolating through calcium sulfate. The ion concentration given in the figure for samples leached with brine were corrected for original ion concentrations as determined prior to using the fluid in the oedometer test. The figure confirms the high concentration of ions leached from sample with brine infiltration as compared to that through which distilled water was allowed to pass. Figure 6 further indicates that for both the percolating fluids, the

604

concentration of Ca^{2+} ion was always higher than that of SO_4^{2-} ion. This high release of Ca^{2+} ions is responsible for the large and numerous interfacial vacancies left in the samples. Most of these vacancies are compressed by the subsequent addition of loads thereby leading to soil grain adjustment and hence denser structures. Figure 6 also shows that the ion concentration reduces with time. Initially high ionic strength is measured due to leaching of anhydrite. This is followed by low ion concentration owing to densification of the sample with loading.

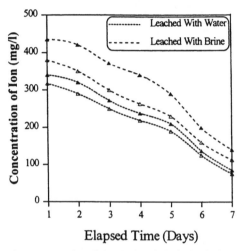

Figure 6. Results of chemical analysis for leached samples: Ca^{2+} ion (▲) and SO_4^{2-} ion (△)

4.5 Solubility

Figure 7 illustrates the variation in solubility index ($S_i = Q/K_{sp}$) with elapsed time for both distilled water and brine. For the dissociation reaction:

$$CaSO_4 \Leftrightarrow Ca^{2+} + SO_4^{2-} \qquad (2)$$

Q is the product of the activities of Ca^{2+} and SO_4^{2-} ions and $K_{sp} = 10^{-4.5}$ (Freeze & Cherry 1979). Based on concentration data presented in Figure 6, activities were calculated using appropriate activity coefficients (γ) as suggested by Berner (1971). The activity coefficients were estimated using Debye-Huckel equation (Klotz 1950). Values of various constants used in Debye-Huckel equation were taken from Manov et al. (1943).

Figure 7 confirms the dissolution of calcium sulfate phases with percolation of fluids. The figure suggests decrease in solubility index with time. This is because heavy surcharge loads leave smaller void area for fluid permeation, which leads to reduction in hydraulic conductivity. Also, due to such small area in contact with fluid means lower rate of ion release.

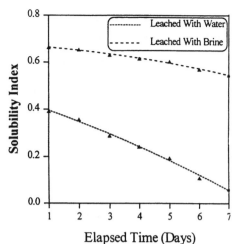

Figure 7. Solubility index for leached samples

Figure 7 also shows that the solubility index for brine-leached sample is higher than that for distilled water-leached sample. This should be attributed to common ion effect: Ca^{2+} and SO_4^{2-} ions are present in both brine and calcium sulfate. These common ions tend to precipitate and make stable compounds (McBride 1994). However, precipitation is not favored owing to the high concentration of Na^+ and Cl^- ions present in brine. The influence of Na^+ and Cl^- ions coupled by large sustained pressure leads to soil grain adjustment that in turn causes volume decrease. Conversely, for distilled water percolation, slower dissolution takes place in the absence of halite forming ions. Thus, dissolution of calcium sulfate depends on the nature of the incoming water and the natural rate of solution of the material.

5 CONCLUSIONS

1. Unsaturated calcium sulfate dissolves readily when it comes in contact with flowing water. This phenomenon tends to restrict the mineral from complete volume increase. Still, some water is imbibed that converts anhydrite to gypsum, which in turn dissolves although at a slower rate.

2. Dissolution of calcium sulfate results in a honeycombed structure, which undergoes volume decrease. Collapse caused by permeation of samples with alkaline water is twice that resulting from percolation with distilled water. Under the action of heavy surcharge load, compressibility is slightly affected by the chemical nature of the pore fluid.

3. Leaching of Ca^{2+} and SO_4^{2-} ions increases both with time and pore fluid chemistry. Brine. which is characterized by high concentration of halite forming ions (i.e. Na^+ and Cl^- ions). causes greater release of Ca^{2+} and SO_4^{2-} ions when compared to distilled water.

4. Irrespective of the percolating fluid, hydraulic conductivity varies at a constant rate and reduces by 60% in seven days. Further, hydraulic conductivity under all overburden pressures can be approximated as $e^{0.5}$ times that under the seating pressure of 7 kPa.

5. Solubility of calcium sulfate, which reduces with time under sustained loads, depends on the nature of the seeping water and the natural rate of solution of the material. Rate of dissolution in distilled water is slower than that in brine.

6 PRACTICAL IMPLICATIONS

Dissolution of calcium sulfate can lead to catastrophic volume change in arid regions, which are characterized by periodic variations in temperature and relative humidity. For any engineering structure in such regions, thorough site investigation is obligatory to establish phase (hydrated or dehydrated) and form (massive or particulate) of the mineral.

In aquifers with high flow velocities, the mineral can go into solution faster thereby causing runaway hazards. It is necessary to conduct hydro-geological studies to confirm whether the aquifer is confined or unconfined. Further, chemical composition of groundwater is pivotal to determine as it controls solubility of the mineral.

Since, both dissolution and phase transformation are affected by local geology, environment and climate, the two phenomena always occur simultaneously and therefore should be studied in conjunction in areas dominated by calcium sulfate.

ACKNOWLEDGEMENTS

The author is grateful to King Fahd University of Petroleum and Minerals for providing laboratory space and computer facilities. Thanks are due to Mrs. Shehla Farah and Ms. Ayesha Umme Kulsoom for providing invaluable help during the write up of the manuscript.

REFERENCES

Abduljauwad, S.N. & G.J. Al-Sulaimani 1993. Determination of swell potential of Al-Qatif clay. *Geotech. Test. J.* 16(4):469-484.

Abduljauwad, S.N. & O.S.B. Al-Amoudi 1995. Geotechnical behaviour of saline sabkha soils. *Geotechnique*, 45(3):425-445.

Abduljauwad, S.N., S. Azam, & N.A. Al-Shayea 1999. Effect of phase transformation of calcium sulfate on the volume change behavior of calcareous expansive soil. *Proc. 2nd Int. Conf. Eng. Calc. Sed.* Bahrain, 1.209-218.

Al-Amoudi O.S.B. & S.N. Abduljauwad 1995. Compressibility and collapse potential of an arid, saline sabkha soil. *Engineering Geology*, 39(3):185-202.

Al-Amoudi, O. S. B. & S. N. Abduljauwad 1994. Modified oedometer for arid, saline soils. *ASCE J. Geotech. Eng.* 120(10):1892-1897.

Azam S. 1997. *Effect of Gypsum-Anhydrite on the Behavior of Expansive Clays*. MS Thesis, King Fahd University of Petroleum and Minerals, Dhahran, Saudi Arabia.

Azam, S., S.N. Abduljauwad, N.A. Al-Shayea, & O.S.B. Al-Amoudi, 1998. Expansive characteristics of gypsiferous/anhydritic soil formations. *Eng. Geol.* 5:89-107.

Berner, R.A. 1971. *Principles of Chemical Sedimentology*. McGraw-Hill Inc. New York.

Blatt, H., G. Middleton & R. Murray 1980. *Origin of Sedimentary Rocks*. 2nd ed., Prentice-Hall Inc. New York.

Blount, C.W. & F.W. Dickson 1973. Gypsum-anhydrite equilibria in systems $CaSO_4-H_2O$ and $CaCO_3-NaCl-H_2O$. *Am. Mineralogist.* 58:323-331.

Brune, G. 1965. Anhydrite and gypsum problems in engineering geology. *Eng. Geol.* 2(1):26-38.

Calcano, C.E. & P.R. Alzura 1967. Problems of dissolution of gypsum in some dam sites. *Bul. Venezuelan Soc. Soil Mech. Found. Eng.* July-September:75-80.

Clemence, S.P. & A.O. Finbarr 1981. Design considerations for collapsible soils. *ASCE J. Geotech. Eng.* 107(3):305-317.

Deer, W.A., R.A. Howie & J. Zussmanet 1972. *Rock Forming Minerals*. Lowe & Brydon, London.

Fabuss, M., A Korozi, T.R. Middleton & J.P. DeMonico 1969. *Office of Saline Water Reproduction*, Contract No. 14-01-0001-1269.

Freeze, R.A & J.A. Cherry 1979. *Groundwater*. Prentice-Hall, Englewood Cliffs, NJ.

He, S. & J.W. Morse 1993. Prediction of halite, gypsum, and anhydrite solubility in natural brines under subsurface conditions. *Comp. & Geosci.* 19(1):1-22.

Jackson, M.L. & G.D. Sherman 1953. Chemical weathering of minerals in soils. *Advances in Agron.* 5:219.

James, A.N. & A.R.R. Lupton 1978. Gypsum and anhydrite in foundations of hydraulic structures. *Geotechnique*, 28(3):249-272.

Jennings, J.E. & Knight, K. 1975. A guide to construction on or with materials exhibiting additional settlement due to collapse of grain structure. *Proc. 6th Reg. Conf. Afr. Soil Mech.*, Durban, 1:99-105.

Klotz, I.M. 1950. *Chemical Thermodynamics*. Prentice-Hall, Englewood Cliffs, N.J.

Liu, S.T. & G.H. Nancollas 1971. The kinetics of dissolution of calcium sulfate dihydrate. *J. Inorg. Nuc. Chem.* 33:2311-2316.

Manov, G.G., R.G. Bates, W.J. Hamer, & S.F. Acree. 1943. Values of the constants in the Debye-Huckel equation for activity coefficients. *J.Amer. Cem. Soc.* 65:1765-1767.

McBride, M.B. 1994. *Environmental Chemistry of Soils*. Oxford University Press, New York.

Mitchell, J.K. 1993. *Fundamentals of soil behavior*, 2nd ed. New York: John Wiley & Sons, Inc.

Murray, R.C. 1964. Origin and diagenesis of gypsum and anhydrite. *J. Sedimentary Petrology*. 34:512-523.

Sawyer, C.N., L. McCarty & G.F. Parkin 1994. *Chemistry for Environmental Engineering*. 4th ed. McGraw-Hill Inc. New York.

Todd, D.K. & D.E.O. McNulty 1976. *Polluted Groundwater*. Water Information Center Inc. New York.

Wigley, T.M.L. 1973. Chemical evolution of the system calcite-gypsum-water. *Can. J. Earth Sci.* 10:306-315.

Zanbak, C. & R.C. Arthur 1986. Geochemical and engineering aspects of anhydrite/gypsum phase transitions. *Bull. Assoc. of Engg. Geologists* 23(4):419-433.

Unsaturated Soils for Asia, Rahardjo, Toll & Leong (eds) © 2000 Taylor & Francis, ISBN 90 5809 139 2

Boundary value problems associated with unsaturated soil collapse testing

R.A. Rohlf
Kentucky Department for Surface Mining Reclamation and Enforcement, Frankfort, Ky., USA

L.G. Wells
University of Kentucky, Department of Biosystems and Agricultural Engineering, Lexington, Ky., USA

ABSTRACT: Collapse behavior for hydrostatic and oedometer soil tests is simulated using a matric-stress critical state model. The model includes groundwater flow and elastoplastic soil constitutive relationships and is formulated in terms of fluid head and net stress plus matric stress, where matric stress is the stress generated by matric suction. An example simulation for a hydrostatic collapse test wetted from the top is presented. Model results show reasonable qualitative agreement with an oedometer collapse test wetted simultaneously from the top and bottom.

1 INTRODUCTION

One of the characteristics of unsaturated soil behavior that has proved difficult to address both experimentally and analytically is the phenomenon of collapse. Collapse occurs when a soil under a constant confining stress experiences a decrease in volume when moving from an unsaturated to a saturated state. The laboratory investigation of collapse is normally accomplished by wetting a sample at one or both ends during oedometer or hydrostatic testing. Wetting from the ends generates a matric-stress gradient and piezometric-head gradient or fluid drag force in a sample. The matric-stress and piezometric head gradients introduce shear and deviatoric stress and may be significant in initiating and controlling collapse behavior in a sample.

Collapse behavior for hydrostatic and oedometer tests is simulated using a matric-stress critical state finite element model. The model includes groundwater flow and elastoplastic soil constitutive relationships and is formulated in terms of fluid head and net stress plus matric stress, where matric stress is the stress generated by matric suction. Matric stress is included in both shear and volume relationships using a state function which expresses matric stress as a function of void ratio and degree of saturation. The soil-water characteristic curve is also expressed as a function of matric stress. Simulations for a hydrostatic test wetted from the top and an oedometer test wetted simultaneously from the top and bottom are presented.

2 STRESS STATE VARIABLES

The search for appropriate stress state variables to describe the behavior of unsaturated soil initially focused on the development of an effective stress equation similar to that used for saturated soils. Bishop (1959) proposed the relationship

$$\sigma' = \sigma - u_a + \chi(u_a - u_w) \tag{1}$$

where σ' is effective normal stress, σ is total normal stress, u_a is pore-air pressure, u_w is pore-water pressure, χ is a parameter initially assumed to be a function of degree of saturation, and $(u_a - u_w)$ is matric suction. Equation (1) predicts a decrease in effective stress under a reduction in matric suction and thus indicates soil expansion under wetting rather than contraction or collapse.

Primarily due to difficulties in defining the parameter χ for volume change behavior, the search for an effective stress relationship was gradually abandoned in favor of independent stress state variables. Fredlund and Morgenstern (1977) proposed the use of net stress $(\sigma_{ij} - u_a \delta_{ij})$ and matric suction $(u_a - u_w)\delta_{ij}$, where σ_{ij} is the total stress tensor and δ_{ij} is the Kronecker delta.

In contrast to the work of Fredlund and Morgenstern (1977), Karube (1988) described matric suction as an internal stress component which can be included in constitutive relationships without explicit addition as an independent stress state variable. For soils in which apparent cohesion was assumed to be a function of matric suction only, Karube proposed that the ultimate strength envelope be defined by

$$q_f = M[p_a + f(\mu)] \tag{2}$$

where q_f is the deviator stress at failure, $(\sigma_1 - \sigma_3)$; p_a is mean net stress, $(\sigma_1 + 2\sigma_3)/3 - u_a$; $f(\mu)$ is a function which converts matric suction, μ, into mean stress; M is the slope of the failure line; and σ_1 and σ_3 are the major and minor principle stresses, respectively. The stress produced by matric suction, $f(\mu)$, was determined from the intersection of the uncorrected ultimate strength line with the p_a-axis.

3 MATRIC-STRESS CRITICAL STATE MODEL

The University of Kentucky Department of Biosystems and Agricultural Engineering has been developing a model for predicting shear and volume behavior of unsaturated soil including collapse and a maximum collapse at a critical pressure. Essential elements of the model are described in the following sections with additional information available in Rohlf and Wells (1999).

3.1 Definition of matric stress

Rohlf et al. (1997) used the intergranular stress tensor developed by Matyas and Radhakrishna (1968) to define matric stress as the stress generated by matric suction. By applying equilibrium conditions along a curved surface normal to solid contact points in an unsaturated soil matrix, Matyas and Radhakrishna (1968) showed that the intergranular stress could be expressed as

$$\overline{\sigma}_{ij} = \left(p_a + A_w\mu + \int Tdx\right)\delta_{ij} + S_{ij} \tag{3}$$

where $\overline{\sigma}_{ij}$ is the intergranular stress or equivalent stress transmitted by solid contacts; A_w is a measure of pore space saturation; T is surface tension; x is the perimeter of the air-water meniscus; and S_{ij} is deviatoric stress. Since $\int Tdx$ is a function of water content and thus μ, Equation (3) can be written

$$\overline{\sigma}_{ij} = [p_a + f(\mu)]\delta_{ij} + S_{ij} \tag{4}$$

where

$$f(\mu) = A_w\mu + \int Tdx \tag{5}$$

can be considered as a matric stress term. Matric stress can be determined using ultimate strength and energy methods (Rohlf et al. 1997 & 2000).

The matric-stress critical state model is developed under the assumption that the state of a soil element can be described by the intergranular stress tensor, $[\sigma_{ij} - u_a\delta_{ij} + f(\mu)\delta_{ij}]$; void ratio, e; water content, w; and soil fabric. Matric stress is treated as an applied confining stress.

3.2 Groundwater governing equations

The flow of groundwater in a saturated or unsaturated porous medium can described by

$$\frac{\partial v_i}{\partial x_i} = (n\frac{\partial S}{\partial h} + nS\gamma_w\beta)\frac{\partial h}{\partial t} + S\frac{\partial \varepsilon_v}{\partial t} \tag{6}$$

where v_i is specific discharge; n is porosity; S is degree of saturation; γ_w is fluid unit weight; β is fluid compressibility; ε_v is volumetric strain; h is head, $h = z + u_w/\gamma_w = z + \psi$, z is elevation head, and ψ is pressure head.

The soil-water characteristic curve provides a relationship between matric suction and degree of saturation for the groundwater constitutive relationship. To provide a smooth transition in matric stress between unsaturated and saturated conditions, the water characteristic function can be formulated as a function of matric stress

$$(u_a - u_w) = [1 + g(1-S)]f(\mu) \tag{7}$$

where g is a constant. At saturation, pore-air pressure is zero and $f(\mu) = -u_w$.

Differentiation of Equation (7) gives the storage term in Equation (6)

$$\frac{\partial S}{\partial h} = \gamma_w \frac{1}{gf(\mu) - [1 + g(1-S)]\frac{\partial f(\mu)}{\partial S}} \tag{8}$$

3.3 Soil governing equations

The equilibrium equation for a differential soil structure element is given by

$$\frac{\partial \overline{\sigma}_{ij}}{\partial x_j} - F_i = 0 \tag{9}$$

where $\overline{\sigma}_{ij} = \sigma_{ij} - u_a\delta_{ij} + f(\mu)\delta_{ij}$, and F_i is body or phase interaction forces. Body and interaction forces incorporated in the formulation include unit weight of solids, unit weight of fluid, and fluid drag. The fluid drag force is given by

$$F_d = -\theta_w\gamma_w\nabla h \tag{10}$$

where F_d is drag force per unit volume, and θ_w is volumetric water content.

3.3.1 Ultimate strength and volume change

As suggested by Karube (1988), the ultimate strength line in the matric-stress critical state model is given by Equation (2).

Volume behavior considering mean net stress alone can be described by

$$e = e_o - \lambda_a \, Ln\!\left(p_a / p_{ao}\right) \tag{11}$$

where λ_a is the slope of the normal compression line for mean net stress and e_o is the void ratio at p_{ao}.

Volume behavior considering mean net stress and matric stress can be described by following Karube's (1988) suggestion and replacing p_a with $p_a + f(\mu)$

$$e = e_o - \lambda Ln\!\left[\frac{p_a + f(\mu)}{p_{ao} + f(\mu)_o}\right] \tag{12}$$

where λ is the slope of the normal compression line and e_o is the void ratio at $p_{ao} + f(\mu)_o$. Differentiating Equations (11) and (12) and equating differential void ratios gives

$$\lambda = \lambda_a \, \frac{p_a + f(\mu)}{p_a} \, \frac{1}{1 + \dfrac{\partial f(\mu)}{\partial p_a}} \tag{13}$$

which corrects λ_a for the effects of mean net stress and matric stress.

3.3.2 State functions for matric stress and slope of the normal compression line

Analysis of matric stress values determined from ten sets of triaxial tests conducted at low confining pressures of 0 - 7 kPa resulted in a matric stress state function of the form (Rohlf et al. 1997)

$$f(\mu) = ae^b \exp[c(1-S)^2], \quad S \prec S^* \tag{14a}$$

$$f(\mu) = f(e, S^*) \frac{(1-S)}{(1-S^*)}, \quad S \geq S^* \tag{14b}$$

where a, b, and c are coefficients and S^* a degree of saturation which defines the breakpoint between the two matric stress relationships. At saturation, Equation (14a) reduces to $f(\mu) = ae^b$ which is considered to be the air entry pressure.

Since Atterberg limits have been used for years to describe the consistency of remolded soils, slope of the normal compression line, λ_a, was assumed to be a function of water content. Analysis of hydrostatic compression tests showed that the slope of the normal compression line could be represented by the relationship

$$\lambda_a = \exp(d + fw) \tag{15}$$

where w is water content, and d and f are coefficients with d dependent on the formation void ratio. A relationship similar to Equation (15) can also be developed for slope of the recompression line κ_a.

3.3.3 Collapse mechanisms

Collapse is controlled by (1) the collapse yield curve, and (2) equilibrium in response to a critical stress distribution and the matric stress state function. The collapse yield curve is given by Equations (12) - (15) and is illustrated in Figure 1 for a range in water content and mean net stress. A soil elementary volume under hydrostatic loading at constant water content, w, follows a single normal compression line in $p_a + f(\mu)$ space with varying slope, λ. However, if water content increases at constant load, p_a, the soil elementary volume moves to a new normal compression line with an accompanying reduction in void ratio. Lines of constant mean net stress, p_a, are shown in Figure 1. The reduction in void ratio occurs because the new normal compression line has a greater slope than the previous normal compression line.

In the matric-stress critical state model, the normal compression line and collapse yield curve are the same. With an increase in water content at constant load, the normal compression line moves to the left producing an unstable stress state which lies outside the yield surface. Collapse occurs as the stress state follows the yield surface.

In addition to the collapse yield curve, collapse can be caused by equilibrium requirements in combination with the matric stress state function. The matric stress state function is presented in Figure 2 for lines of constant water content. Note that along lines of constant water content relatively large changes in void ratio can occur for small changes in matric stress. The matric-stress state function allows equilibrium to be satisfied at constant water content with a small change in matric stress and a large change in void ratio. Thus under a critical stress distribution, collapse can occur in a region of the sample where the advancing wetting front has not increased water content. In this case, the stress state is assumed to remain on the yield surface and plastic conditions are maintained.

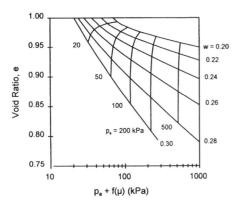

Figure 1. Collapse yield curve.

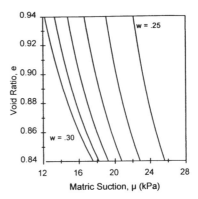

Figure 2. Matric stress at constant water content.

3.4 *Finite element implementation*

The previously described matric-stress critical state model is implemented in a two-dimensional saturated-unsaturated groundwater flow and elastoplastic stress-strain finite element program (Rohlf & Wells, 1999). The soil constitutive relationship uses the modified Cam clay yield function.

A Galerkin and finite difference solution is used for the groundwater equation, and a virtual displacement solution is used for the soil equation. The solution is uncoupled and loops over both the groundwater and soil equations until convergence is achieved. Since slope of the normal compression line, λ_a, is a function of water content, solving for unsaturated soil collapse requires integration of the soil constitutive relationship from initial conditions for each incremental change in water content.

4 MODEL SIMULATIONS

The matric-stress critical state model was used to simulate collapse for a hydrostatic test wetted from the top and an oedometer test wetted simultaneously from the top and bottom. Dimensions of the hydrostatic and oedometer samples are 50.8 mm diameter x 101.6 mm length and 62.0 mm diameter x 25.4 mm length, respectively. Input parameters for both analyses are presented in Table 1. A linear function was used to express the relationship between unsaturated water permeability and matric suction, $k_w = k_s - r\mu/\gamma_w$ with $k_w > k_r$, where k_w is unsaturated water permeability, k_s is saturated water permeability, k_r is residual unsaturated water permeability, and r is a constant. Pore-air pressure was assumed to be zero throughout the simulation.

The hydrostatic simulation was conducted for Maury silt loam (CL; LL = 34, PI = 10; 27% clay, 65% silt, 8% sand). Development of parameters for Maury silt loam is presented in Rohlf et al. (1997). Parameters for the oedometer test were based on the previous work with Maury silt loam with revision of the matric-stress state function parameters and water permeability to allow modeling of an oedometer test conducted by Tadepalli and Fredlund (1991).

4.1 *Hydrostatic test*

Simulation of collapse for a hydrostatic test wetted from the top is presented in Figures 3 and 4. Figure 3 shows several distinct segments of deformation at both the top and bottom of the sample: (1) A-B, normal consolidation in response to loading from 5 to 60 kPa at a constant matric suction boundary condition, μ, of 58.9 kPa; (2) B-C and B-D, application of the matric suction boundary condition from 58.9 to 0 kPa over 3 minutes; (3) C-F and D-F, continued deformation as the wetting front moves through the sample; (4) E, transition in the matric-stress relationship, Equation (14), at $S^* = 0.9$; and (5) F, zero matric stress and a final mean net stress equal to the hydrostatic loading of 60 kPa..

The top of the sample shows a general decrease in mean stress (B-C) due to the reduction in matric stress during wetting. The bottom of the sample shows an initial increase in mean stress (B-D) due to the fluid drag force incurred during application of the pore-water boundary condition and then a decrease in mean stress due to wetting. Plastic conditions are assumed to hold throughout collapse.

Table 1. Model Input Parameters.

Parameter	Hydrostatic	Oedometer
a	7.39 kPa	4.35 kPa
b	-4.86	-1.75
c	9.65	3.81
S^*	0.9	0.9
d_λ	-7.9	-7.9
f_λ	17.0	17.0
d_κ	-13.3	-13.3
f_κ	29.4	29.4
M	1.09	1.09
G_s	2.68	2.68
v	0.3	0.3
e_o	1.008	0.917
w_o	.258	.118
g	7.00	2.30
r	9.4×10^{-7} /min	1.8×10^{-7} /min
k_s	1.0×10^{-5} m/min	8.0×10^{-7} m/min
k_r	$0.04\ k_s$	$0.1\ k_s$
γ_w	9.81 kN/m^3	9.81 kN/m^3
β	1.7×10^{-8} /kPa	1.7×10^{-8} /kPa

Figure 4 shows that most of the collapse occurs during application of the matric suction boundary condition. The bottom of the sample deforms essentially simultaneously with the top of the sample even though water content remains nearly unchanged at the bottom of the sample. Concurrent deformation at both the top and bottom of the sample can be attributed to the requirement to satisfy equilibrium for external and body forces, including the matric stress and piezometric head gradients, in conjunction with the matric stress state function.

4.2 Oedometer test

Simulation of collapse for an oedometer test wetted simultaneously from the top and bottom is presented in Figure 5. For comparison, an oedometer collapse test conducted on Indian Head silt (LL = 22, PI = 6; 6% clay, 32 % silt, 62% sand) is presented in Figure 6 (Tadepalli & Fredlund, 1991). Void ratios in Figures 5 and 6 are sample averages while matric suction is at the center of the sample.

The collapse test was conducted by placing dry porous stones on the top and bottom of the sample, loading the sample to a vertical stress of 99 kPa, and then inundating the sample to initiate wetting. For the model simulation, it was assumed that the dry porous stones would attenuate the matric suction boundary condition at the top and bottom of the sample, and the boundary condition was decreased from 64 to 0 kPa over a time period of one minute.

Both the simulation and test show a sizeable and rapid initial collapse prior to any significant change in matric suction at the center of the sample. A similar rapid initial collapse has been documented by Booth (1975). In the simulation, a change in the rate of collapse occurs at the end of application of the matric suction boundary condition (1 min.). Although the simulation shows less total collapse than the oedometer test on Indian Head silt, there is reasonable qualitative agreement with the collapse test.

5 CONCLUSIONS

The matric-stress critical state model is formulated in terms of fluid head and net stress plus matric stress, where matric stress is the stress generated by matric suction. Matric stress is expressed by a state function

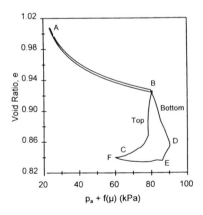

Figure 3. Hydrostatic collapse versus mean stress

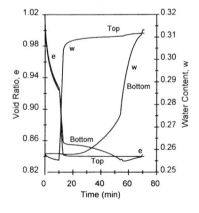

Figure 4. Hydrostatic collapse versus time.

611

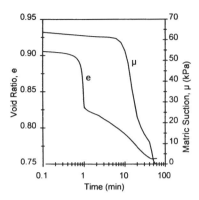

Figure 5. Simulated oedometer collapse.

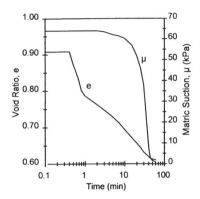

Figure 6. Oedometer collapse (after Tadepalli & Fredlund, 1991).

which includes void ratio and degree of saturation. Collapse is controlled by (1) the collapse yield curve, and (2) equilibrium constraints produced by a critical stress distribution and the matric stress state function.

The model provides reasonable qualitative agreement with an oedometer collapse test wetted simultaneously from the top and bottom. Results from the model suggest that boundary conditions imposed by the matric stress and piezometric head gradients may be important factors in initiating and controlling collapse.

REFERENCES

Bishop, A.W. 1959. The principle of effective stress. *Teknisk Ukeblad* 106(39): 859-863.

Booth, A.R. 1975. The factors influencing collapse settlement in compacted soils. In Sixth regional conference for Africa on soil mechanics & foundation engineering. Durban, South Africa, 57-63.

Fredlund, D.G.& Morgenstern, N.R. 1977. Stress state variables for unsaturated soils. *J. Geotech. Engrg.*, ASCE, 103 (GT5): 447-466.

Karube, D. 1988. New concept of effective stress in unsaturated soil and its proving test. *Advanced triaxial testing of soil and rock*, ASTM STP 977, Robert T. Donaghe, Ronald C. Chaney, and Marshall L. Silver, Eds., ASTM: 539-552.

Matyas, E.L. & Radhakrishna, H.S. 1968. Volume Change Characteristics of Partially Saturated Soils. *Geotechnique* 18(4): 432-448.

Rohlf, R. A., Barfield, B.J., & Felton, G.K. 1997. Ultimate strength matric stress relationship. *J. Geotech. and Geoenv. Engrg.*, ASCE, 123(10): 938-947.

Rohlf, R. A. & Wells L.G. 1999. Matric-stress critical state model for soils. In *Fourth international conference on constitutive laws for engineering materials*. R.C. Picu & E. Krempl (eds), Rensselaer Polytechnic Institute, Troy, NY: 363 - 366.

Rohlf, R. A., Xeng, X. & Wells, L.G. 2000. Simulating deformation of an excess spoil fill. ASCE/GI Geo-Denver Conference, Denver, CO.

Tadepalli, R. & Fredlund, D.G. 1991. The collapse behavior of a compacted soil during inundation. *Can. Geotech. J.* 28: 477-488.

Unsaturated Soils for Asia, Rahardjo, Toll & Leong (eds) © 2000 Taylor & Francis, ISBN 90 5809 139 2

Building on unsaturated fills: Compaction, suction and volume change

H. D. Skinner
Building Research Establishment, Watford, UK

ABSTRACT: An increasing proportion of new construction in the UK is founded on partially saturated fills. Where a foundation is to be constructed on either an engineered fill or a non-engineered fill which has undergone in-situ ground treatment, the engineering properties of the fill will be governed by the compacted state and subsequent history of the fill. An improved understanding of the compaction behaviour is required in order to more confidently predict the likely performance of the fill.

1 INTRODUCTION

Foundations for low-rise buildings are usually founded at shallow depths where the soil is partially saturated. The performance of the foundation will depend on the stiffness and compressibility characteristics of the foundation soil when subjected to the building load, but also on the volume changes which may occur in the soil subsequent to building due to changes in degree of saturation. The typical model for assessing foundation behaviour is concerned primarily with the application of a load to the ground and determining whether or not the allowable bearing capacity of the ground will be exceeded or unacceptable settlement will be caused. However, this is not the dominant type of problem for low-rise building foundations which apply relatively small stress increments that only affect the soil close to the ground surface. For lightly loaded shallow foundations, most problems are due to ground movements caused by factors not related to the weight of the building. In order to obtain an economic and safe design the designer must be able to predict with reasonable confidence the behaviour of the foundation and the soil.

An increasing proportion of new construction in the UK is founded on partially saturated fills. Where a foundation is to be constructed on either an engineered fill or a non-engineered fill which has undergone in-situ ground treatment, the engineering properties of the fill will be governed by the compacted state and subsequent history of the fill. Compaction of foundation fills should increase their bearing capacity and stiffness and reduce the susceptibility to collapse compression on inundation.

The volume change potential of a fill is very difficult to quantify and is the cause of many foundation problems. An improved understanding of the compaction behaviour is required in order to more confidently predict the likely performance of the fill.

It is important to understand the factors which control the effectiveness of the compaction process and the relationship between the compacted state of the fill and its subsequent performance as a foundation material. Suction has an important effect on the behaviour of partially saturated soils and its relationship to compaction characteristics and compacted behaviour has been explored by a programme of laboratory tests. The measurement of suction in compacted fills with a large proportion of particles larger than silt or sand sizes is difficult and the filter paper technique has been developed in order to measure suctions in fills compacted to a range of dry density and moisture content.

2 LABORATORY TEST PROGRAMME

A programme of laboratory testing has been carried out using three fill types; a medium plasticity boulder clay fill, a colliery spoil and uniformly graded sand of similar particle density. Grading curves for the fills are shown in Figure 1. The test programme included both routine compaction tests and some specially developed suction measurements.

3 SUCTION MEASUREMENTS

Suction is a significant parameter for understanding

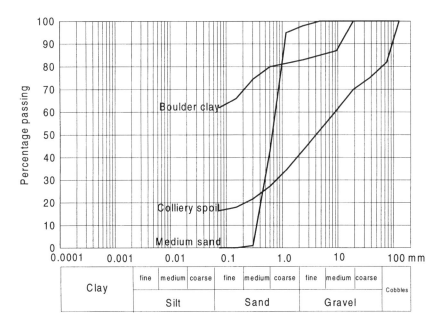

Figure 1: Grading curves for the fill types tested

the behaviour of partially saturated soils. Although many direct and indirect methods of suction measurement are available, in practice it can be difficult to measure suctions. Several methods for the measurement of suction in the laboratory have been used to characterise the variation of suction with moisture content and dry density.

A new variant of the filter paper technique (Chandler and Gutierez, 1986) has been developed in order to provide a simple and cost-effective means of measuring suction in re-compacted samples. Samples 60mm high were compacted into a 100mm diameter mould at the required density and moisture content. One or two pieces of filter paper were weighted onto the top of the sample using a steel plate and the whole mould was covered in cling film and wax. The samples were allowed 10 days to come to equilibrium before breaking open the moulds. The moisture content of the filter paper can be related to the suction present in the sample. Suctions have also been measured using a large pressure plate device (Fredlund and Rahardjo, 1993) capable of testing 150mm diameter samples and using measurements of the electrical resistance of gypsum blocks buried in large recompacted samples.

Similar values for suction have been obtained by each method. For example a variation in suction of 10-25kPa was found in tests on the colliery spoil using the three methods for moisture contents of 11-12%. Suctions were measured in sand fill samples using the filter paper method only. Some degree of

scatter was found in the results of all the suction tests on the fill materials.

4 SUCTION AND COMPACTION

Laboratory compaction tests and contours of suction interpreted from the test results for the three fill types are presented and discussed in the light of the properties which are likely to govern the soil response to the impact process and the subsequent performance of the fill. The contours are only plotted over the limited range of dry density and moisture content covered by the suction tests; tests at dry densities higher than those achieved by the 4.5kg rammer test were not carried out.

The measurements of suction for the boulder clay fill are plotted on a dry density-moisture content plot in Figure 2. Suction is principally a function of moisture content except at low air voids. This is consistent with testing on a compacted silt by Gens et al (1995). Load tests at constant moisture content and high air voids can therefore be assumed to be constant suction tests. It can be seen that the suction in the samples increased rapidly as the moisture content fell below 10%. Above 10% moisture content the suction decreased more slowly with increase in moisture content and had some dependence on dry density. Load tests at low air voids will need careful interpretation since changes in suction occur during compression.

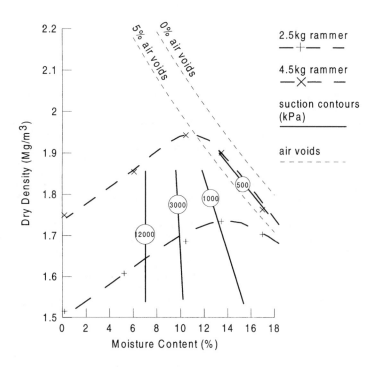

Figure 2: Suction contours and compaction results for boulder clay fill (particle density 2.66Mg/m³)

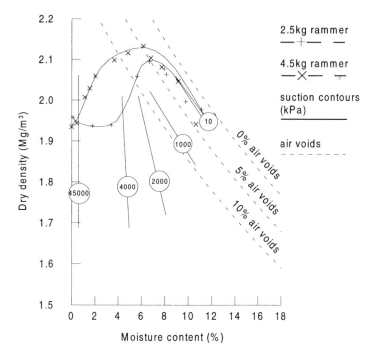

Figure 3: Suction contours and compaction tests on Bentinck colliery spoil (particle density 2.65Mg/m³)

A plot of suction-moisture content-dry density is shown in Figure 3 for the granular colliery spoil. Suctions of around 50kPa were present in fill samples close to the natural water content of 8%. There was little evidence of dependence of suction on dry density, except at low air voids. In this respect the colliery spoil was similar to the boulder clay although the boulder clay contained 40% more fines by weight.

Suction measurements on clay fills by Olsen and Langfelder (1965) indicated that there was little dependence of suction on dry density and that the magnitude of suction could be related to the specific surface area of the clay minerals. One of the clays had similar plasticity to the boulder clay and the suctions were similar. Tests by Acer and Nyeretse (1992) showed that the suction present in clay and sand mixtures compacted at optimum moisture content increased with increasing fines content and that this was particularly noticeable for samples with 0-20% clay fraction and for high activity clays. Dineen et al (1999) also found that suction was principally a function of moisture content for a bentonite enriched sand.

Suction tests were also carried out on a uniformly graded sand with no fines content. The measurements (Figure 4) indicated that very low suctions were present in samples at greater than 2% moisture content, but at moisture contents close to

zero the suction in the samples rose to very high values. Within the range of accuracy of the test there was no evidence of dependence of suction on dry density, in common with the other fill types.

The compaction results for the boulder clay plotted on Figure 2 show a significant decrease in the effectiveness of the compaction occurred as the moisture content fell below 10% and the suction in the boulder clay samples increased rapidly. With the moisture content above 10% the suction decreased slowly with increase in moisture content and the effectiveness of the compaction was little affected by change in suction. Compaction tests on the granular colliery spoil using the BS 1377 2.5kg rammer test indicated a maximum dry density at 7% and a minimum dry density at between 2 and 3% moisture content. At moisture contents below this the dry density achieved by the 2.5kg rammer test increased with decreasing moisture content. In this respect there was some similarity with the behaviour of the sand. Although there may be very high suctions at very low moisture contents in granular soils, the lack of fines means that the suctions have little effect on the compaction process. The increased energy input during compaction with the 4.5kg rammer did not produce a similar increase in density at very low moisture contents as gross disturbance took place.

In compaction tests on the sand fill there was a large range of moisture content over which there was

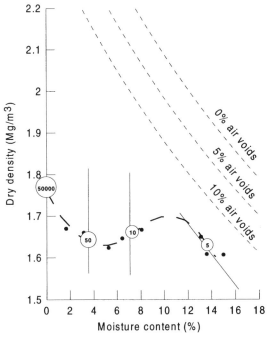

Figure 4: 2.5kg Proctor compaction and suction measurements for a uniformly graded sand (particle density 2.64Mg/m^3)

little variation in the dry density. Over the same range there was little variation in the suction present in the samples. Only at lower moisture contents was a higher dry density achieved which coincided with an increase in the measured suction.

Previous work on the factors which influence compaction of fills (Parsons, 1992) has highlighted the importance of shear strength on the maximum dry densities achievable by a given compactive effort, governed by the work required to rearrange a given volume of soil. For clay fills the energy (E) required to compact a soil of volume V varies with undrained shear strength (c_u) such that $E/V \approx 2.5c_u$ (Charles, 1993). Undrained shear strength is a function of moisture content in clay fills. It seems most likely that these factors are controlled by the suction present in the sample.

5 SUCTION AND COMPRESSIBILITY

Figure 5 shows the variation in one-dimensional compressibility for the colliery spoil expressed as a constrained modulus over a range of dry density and moisture content. The secant modulus for a load increment from 0-60kPa increased with increasing dry density and reducing moisture content. The stiffness appears to be a function of both the structure achieved by the compaction and the suction associated with a given moisture content.

The state of the fill is not, of course, fully defined by the dry density and moisture content. This is illustrated by two tests for which the same dry density and moisture content was achieved, in the first test by impact compaction alone and in the second by impact compaction followed by the temporary application of a static load. The stiffness of the compacted sample was only half that of the sample which had been preloaded. However, the compacted sample had a collapse potential which was half that of the preloaded sample. The structure built up during the compaction process was different from that developed during the static loading process and the resistance to other forms of stress increment was then different. Similar results for saturated compacted samples were found by Santucci et al (1998). The stress-strain behaviour of samples compacted dynamically was characterised by a lower initial stiffness than those whose density resulted from static overconsolidation.

6 SUCTION AND COLLAPSE

Collapse compression contours which have been found from previous tests are shown in Figure 6 for the colliery spoil. Collapse can be seen to be principally a function of dry density at low moisture contents and percentage air voids at moisture contents higher than optimum. The results, when compared with suction contours, indicate that collapse potential is not simply a function of suction. At moisture contents higher than optimum contours of collapse and suction appear aligned, but at low moisture content there is no such relationship. Gens et al (1995) show that the structure built up during compaction has an effect on the collapse. Two samples, one compacted below optimum water content and one above optimum, were brought to the same value of suction, moisture content and dry density before loading and wetting took place. The samples compacted wet of optimum had some

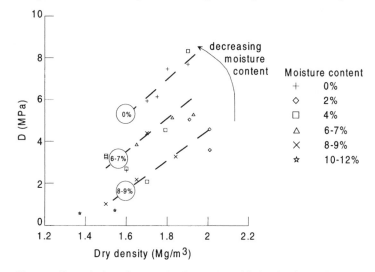

Figure 5: Bentinck colliery spoil - variation of constrained modulus with dry density and moisture content

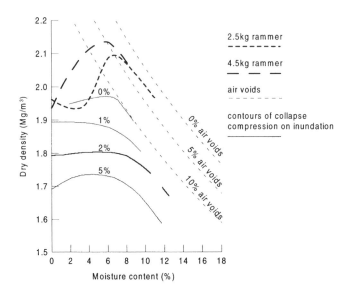

Figure 6: Collapse compression tests on Bentinck colliery spoil at 60kPa applied vertical stress

collapse potential but not as great as those compacted dry of optimum. Collapse potential can be seen to be both a function of structure and suction.

7 CONCLUSIONS

Compaction and suction tests have been carried out on three fill types in order to determine the relationship between suction and dry density and moisture content. The results provide the outline of a framework for understanding the response of fills to compaction. As the moisture content falls below optimum for fills with a significant fines content, a significant increase in suction is accompanied by a rapid decrease in effectiveness of the compaction. This indicates the major influence of suction on the compaction process for such fills. The constrained modulus of the colliery spoil increased with both the suction and dry density. Collapse compression behaviour could not be related solely to the suctions but was also a function of the structure built up during the compaction process. The structure built up during compaction and loading, and the suction which results from a given moisture content, have a controlling effect on the performance of a compacted fill, both in terms of the load response and the wetting response of the foundation fill.

8 REFERENCES

Acer Y.B. and Nyeretse P. (1992). Total suction of artificial mixtures of soil compacted at optimum water content. ASTM Geotechnical Testing Journal, vol 15(1), pp 65-73.

Charles J.A. (1993). Building on fill: geotechnical aspects. BRE Report BR 230, 163pp.

Chandler R.J. and Gutierez C.I. (1986). The filter paper method of suction measurement. Geotechnique, vol. 36, pp. 265-268.

Dineen K., Colmenares J.E., Ridley A.M., Burland J.B. (1999). Suction and volume changes of a bentonite-enriched sand. Geotechnical Engineering, Proceedings of the Institution of Civil Engineers, vol. 137, Oct., pp. 197-210

Fredlund D.G. and Rahardjo H. (1993). Soil mechanics for unsaturated soils. Wiley and Sons. 517pp.

Gens A., Alonso E.E. , Suriol J. and Lloret A. (1995). Effect of structure on the volumetric behaviour of a compacted soil. Unsaturated soils, Proceedings of the First International Conference on Unsaturated Soils, vol 1, pp83-88. Alonso E.E. and Delage P. (eds), Balkema, 1995.

Olsen R.E. and Langfelder L.J. (1965). Pore water pressures in unsaturated soils. Journal of the Soil Mechanics and Foundations Division, ASCE, SM4, July, pp 127-150

Parsons A.W. (1992). Compaction of soils and granular materials: A review of research performed at the Transport Research Laboratory. 323pp, HMSO.

Santucci F. de M., Silvestri F. and Vinale F. (1998). Influence of compaction on the mechanical behaviour of a silty sand. Soils and Foundations, vol 38, no 4, pp 41-56.

9   Volume change

9.2 Shrinkage and swelling

Unsaturated Soils for Asia, Rahardjo, Toll & Leong (eds) © 2000 Taylor & Francis, ISBN 90 5809 139 2

Soil structure influence on the heave reduction factor

W.S.Abdullah
Jordan University of Science and Technology, Irbid, Jordan

A.Basma
Sultan Qaboos University, Muscat, Oman

M.Ajlouni
University of Illinois at Urbana-Champaign, Ill., USA

ABSTRACT: Heave prediction by the direct method or the Jennings and Knight "double oedometer test" over-predicts footing's heave in a multi dimensional swell condition. Both mentioned methods are based on the one-dimensional swell condition, implying that swell potential is manifested in the vertical direction only. Accordingly, and in order to get more accurate results the predicted heave value must be multiplied by a reduction factor "R_f". Soil structure is one of the important factors influencing the heave reduction factor. Therefore and in order to simulate different types of soil structures compacted clay was used. Clay compacted dry of optimum simulates a flocculated structure, clay compacted wet of optimum simulates a dispersed structure, and clay compacted at optimum moisture content simulates an intermediate soil structure. The study of model footing resting on compacted expansive clayey soil showed that the heave reduction factor R_f ranges from 0.93 for a flocculated structure to 0.63 for a dispersed structure.

1 INTRODUCTION

Oedometer based heave prediction methods (such as the direct method, Jennings and Knight double oedometer method, and Sullivan and McClelland method) are recommended and widely used in engineering practice since early fifties up till now, (Jennings and Knight, 1957,1975), (Zeitlen 1969), (U. S. Army Corps of Engineers 1983), (Alonso et al. 1987), (Dalmatov 1991), and (Popescu 1992). The direct method of heave prediction is one of the simplest methods and widely used by practicing engineers. This method takes into account soil in situ conditions (soil structure, dry density, moisture content, etc.).

One-dimensional swell condition, practically, is seldom if ever satisfied. Multi-dimensional swell, however, is the common case. Thus, by assuming the volumetric strain due to soil swell is manifested in one-dimension represents the major disadvantage of these methods. In order to obtain a closer estimate to the actual heave, the predicted heave value should be multiplied by a reduction factor "Heave Reduction Factor R_f".

The discrepancy between actual (measured) and predicted heave using these methods is dependent on the percentage swell manifested laterally. Lateral strain is, in turn, dependent on the applied footing pressure, pre-wetting dry density and pre-wetting moisture content of the soil. Abdullah et al. 1999 studied the influence of applied pressure on the heave reduction factor R_f. It was found that the direct method is more accurate than the double-oedometer method. Also, the heave reduction factor R_f was found to decrease rapidly with increase in footing pressure. For instance, an increase in footing pressure from 25 kPa to 50 kPa the heave reduction factor was reduced from 0.92 to 0.62 for the direct method and from 0.88 t0 0.53 for the double-oedometer method.

The direct method is based on oedometer swell test of undisturbed soil specimens extracted from the sub-soil and from different elevations below foundation level. Each specimen is subjected to a swell test with the sum of field overburden pressure and the anticipated additional pressure due to foundation load is applied as surcharge load on the specimen. The soil specimen is, then, given access to water until it reaches saturation. The change in thickness of the specimen (Δh) or the change in void ratio (Δe) of the soil specimen is calculated. The total or the ultimate heave is determined as:

$$\delta = \sum_{i=1}^{n} \{\frac{\Delta e}{1+e_0}\}_i * \Delta H_i \tag{1}$$

$$\frac{\Delta h}{h} = \frac{\Delta e}{1+e_0} \tag{2}$$

where δ = total or the ultimate heave; e_0 = initial or pre-wetting void ratio; Δe = change in void ratio; h = specimen thickness; Δh = change in specimen thickness; ΔH = sub-layer thickness; and n = number of sub-layers.

2 SCOPE OF WORK

The present work studied the influence of soil structure on the heave reduction factor R_f. Test experiments of model footing on compacted expansive soil were used for the study. Compacted soil offers a good opportunity for obtaining a desirable soil structure, (Lambe 1958, 1962), and (Seed et al 1959). Various soil structures were produced by varying molding moisture content. Flocculated soil structure, dispersed structure, and an intermediate soil structure.

Model footing test experiments were conducted, footing's heave as well as heave of the soil surface at a number of points away from the model footing was measured and compared with the predicted values using the direct method.

3 MODEL FOOTING SETUP

The test experiments were conducted on a model footing 25 cm long, 5 cm wide, and 2 cm in thickness. The dimensions of the model footing were so chosen to simulate a plane strain condition. A metal box container with a square cross sectional area with inside dimensions of 50 cm by 50 cm and 60 cm in height was used to contain the soil.

The model footing was placed at the center of the soil surface. To prevent tilting of the model footing upon loading, a cylindrical solid steel rod (2 cm in diameter) was screwed 1 cm into the model footing at its center of gravity. This rod was greased and passed through a circular pivot (2.2 cm in diameter) in a steel plate that was tied to the sides of the metal box container. A schematic diagram of the test model setup is shown in Figure 1.

Heave movements of the model footing and the surrounding soil surface was measured by using number of dial gages. A total of 20 dial gages were used which were distributed as follows: four electronic gages on the four corners of the footing, eight standard gages on each side of the model footing placed at various distances from the footing centerline.

4 SOIL USED FOR THE INVESTIGATION

The soil used for this investigation was taken from a highly expansive soil deposit. The soil deposit is dark brown in color and about 9 m in thickness,

overlying basaltic bedrock formed during the Tertiary Era.

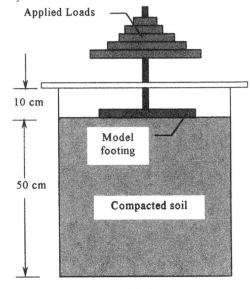

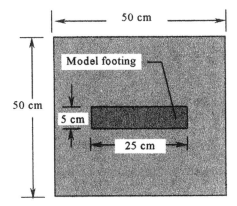

Figure 1 Model footing test set-up

The clay deposit is stiff and having prominent inclined slickensided fissures. In some areas, the clay deposit is found mixed with pebbles and cobbles of limestone, silicified limestone, and Chert. The surface of the clay deposit is characterized, in summer, by widespread shrinkage cracks of about 10 to 15 cm in width and extending to about 3 m in depth. The basic properties of the selected soil are given in Table 1 and the moisture-dry density relation is shown in Figure 2.

X-ray diffraction study was conducted on random sample, oriented, heated and glycolated soil samples to determine types of clay minerals present in the soil deposit. The study shows presence of mont-morillonite, mixed layer of illite smectite, and small amount of kaolinite.

Table (1) Properties of the tested soil

Soil property	Value
Depth of sample (m)	2.5
Specific gravity	2.712
Activity	0.67
USC classification	CH
Gravel size (%)	0.5
Sand size (%)	6.0
Silt size (%)	34.0
Clay size (%)	59.5
Liquid limit LL	71
Plastic limit PL	31
Shrinkage limit SL	18
Plasticity index PI	40
Maximum dry density (kN/m^3)	14.0
Optimum moisture content (%)	25.0

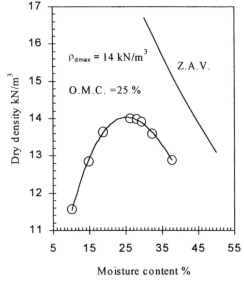

Figure 2 Moisture - dry density
relation, standard compaction test
ASTM D698-78

5 TESTING METHODOLOGY

Using a rubber tipped pestle the soil was broken down to small sized pieces in order to prepare the soil for placement and compaction in the test metal box container. The soil was, then, sieved through #40 sieve with the larger sizes being discarded. Then, the soil was placed into the metal box container and compacted to a prescribed moisture content and dry density (Table 2) in layers of 5 cm in thickness using a flat based square rammer

measuring 10 cm by 10 cm. The total thickness of the compacted soil was 50 cm leaving 10 cm free space at the top of the container. The box container walls were thoroughly greased to minimize friction between the compacted soil and the container's walls due to soil heave.

Three experiments were conducted on the model footing. Each experiment was conducted at a pre-scribed state of compaction (dry density and mois-ture content) in order to investigate the effect of soil structure on the heave reduction factor R_f. The first state of compaction was dry of optimum to simulate a flocculated structure, the second state of compaction was wet of optimum to simulate an oriented structure, and the third state of compaction was at near optimum moisture content, an interme-diate type of soil structure. The applied pressure on the footing during these experiments was 25kPa. The states of compaction and the applied loads are given in Table 2.

Table 2 States of compaction and applied load for
the test experiments

State of compaction	Dry density (kN/m^3)	Moisture content (%)	Applied pressure (kPa)
1	13.5	18	25
2	13.5	28	25
3	13.5	38	25

After compaction of soil, placement of footing and arranging dial gages, the load was applied on the ring-bearing load on the solid steel rod. Immediately after the load was applied, settlement was measured by the gages at lapsed times of 0.5, 1, 2, 4, 8, 15, 30, 60 minutes and every 60 minutes until deformations ceased. At this stage, the dial gages were reset to zero and water was added to the soil surface. As the soil begins to swell the soil surface begins to heave. The resulted heave of the soil surface and the footing was measured at lapsed time of 1, 2, 4, 8, 15, 30, 60 minutes, 2, 4, 16, 24 hours and every day thereafter until readings stabilized. Water was continuously added to the soil to ensure full saturation throughout the testing period.

At the end of each test, samples were extracted from various depths of the soil to determine the final moisture content of the soil. These values ranged from 42% to 45% for all experiments.

6 HEAVE PREDICTION FROM LABORATORY TESTING

Laboratory oedometer tests were conducted on soil specimens extracted from the compacted clayey soil in the test container box. The results of these tests were used to predict the amount of heave of the tested model footing. The adopted procedure takes into account pre-wetting dry density, pre-wetting moisture content, and applied footing pressure. The following steps explain the adopted method:

1. Oedometer undisturbed core samples (76 mm in diameter and 20 mm in thickness) were obtained from the compacted soil for the following molding moisture contents, 18%, 28%, and 38%. These are the molding moisture contents for the model footing experiments as given in Table 2.

2. Under each applied pressure, p, the dial gage was set to zero and the specimen was inundated permitting it to swell. The final amount of expansion (taken after dial gage readings are stabilized) divided by the initial height of the sample is the percentage swell of the soil under the prescribed conditions.

3. The process is repeated for various applied pressures using samples extracted from the compacted soil of the test experiments. The applied pressures used were 1, 5, 10, 25, 50, 100 kPa.

4. Swelling percentages versus corresponding pressure (logarithmic scale) yields a straight-line relationship (Figure 3). This plot is used to estimate soil expansion under any pressure between 1 kPa to 100 kPa.

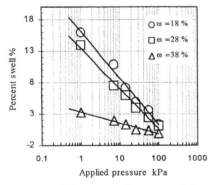

Figure 3 Percent swell versus applied pressure

7 HEAVE PREDICTION USING THE DIRECT METHOD

The following procedure was adopted to predict heave of the model footing and the soil surface upon soil wetting:

1. The soil profile beneath the model footing is divided into 10 equal subdivisions.

2. At the midpoint of each subdivision, determine the overburden pressure and the additional vertical pressure due to the model footing pressure using elasticity methods. The total vertical pressure acting at this level under the present footing pressure is then the summation of these two pressures.

3. Knowing the total vertical pressure at the midpoint of each subdivision and the pre-wetting moisture content, the percentage expansion for each subdivision is determined graphically using Figure 3.

4. The amount of expansion of each subdivision is determined as the percentage expansion of the midpoint of the subdivision under consideration times the subdivision thickness (5 cm).

5. Summing up the amount of expansion of all subdivisions yields the heave amount under the prescribed loading and soil condition.

8 TEST RESULTS AND ANALYSIS

Three test experiments were conducted on a model footing resting on compacted expansive soil at different compaction states; dry of optimum (ω_1=18%, ρ_{d1} =13.5 kN/m^3), at near optimum moisture content (ω_2=28%, ρ_{d1}=13.5 kN/m^3), and wet of optimum (ω_3=38%, ρ_{d1}=13.5 kN/m^3). The predicted and measured heave of the footing and the soil surface are presented in Figure 4. The predicted heave of the model footing, for the three states of compaction, is higher than the actual (measured) heave. The predicted heave of the soil surface, however, is lower than the actual heaves. It can, also, be said that the over-prediction of footing heave is not constant over the range of the tested moisture content (Figure 5). The heave reduction factor R_f ranges from 0.93 for the case of dry of optimum to 0.64 for the case of wet of optimum. Therefore, the dryer the soil the better the prediction by the direct method. The over-prediction of footing heave by the direct method can not be attributed to sample disturbance, since the direct method under-predicted heave values of the soil surface.

The prediction sensitivity to the pre-wetting moisture content is inherited in the approach of the direct method. The volumetric strain due to swelling (in the direct method or any other method based on the one dimensional heave condition) is manifested totally in the vertical direction, since it is based on the one-dimensional oedometer test. The loaded area in a footing problem is finite and the condition of no lateral strain (one-dimensional condition) is no longer valid. Thus, for a finite loaded area (as the case of plane strain condition) the volumetric strain due to swelling is manifested vertically as well as laterally. Consequently, the direct method over-predicts footing heave, since the volumetric strain due to swelling is partly vertical, while the direct method considers it to be totally vertical.

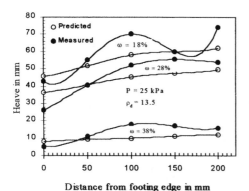

Figure 4 Predicted and measured heave of
the footing and soil surface

The variability of the heave reduction factor over the pre-wetting moisture content range can be attributed to the variability of the effective stress due to changes in the moisture content. The vertical and lateral proportion of the volumetric strain due to swelling, at a point within the soil mass, depends on the vertical and lateral pressures as well as the swell potential at that point. Vertical effective pressure (σ'_v) is composed of the effective overburden pressure ($\gamma'z$) plus the vertical pressure due to footing load (σ_{vf}). The lateral effective pressure (σ_{lf}) is composed of lateral component of the effective overburden ($K_0\gamma'z$) and a lateral pressure due to the footing load (σ_{lf}). When the soil is on the dry side the effective stress is high (due to high negative pore water pressure) in comparison to a soil on the wet side. Since the lateral pressure for a soil on the wet side is less than for a soil on the dry side; therefore, the lateral manifestation of the volumetric strain due to swelling increases as the pre-wetting moisture content increases. Consequently, as lateral swell increases heave over-prediction increases too.

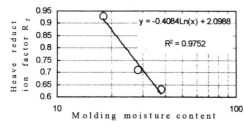

Figure 5 Heave reduction factor R_f versus
moisture content relationship

9 CONCLUSIONS

An experimental study was conducted on model footing resting on compacted expansive soil. The following conclusions were drawn from the study:

1. Swelling percentages versus applied pressure (logarithmic scale) yields a straight-line relationship.

2. The direct method of heave prediction is based on the one-dimensional oedometer; therefore, it overestimates heave value when applied for cases of multidimensional swell such as plane strain condition.

3. The heave reduction factor R_f is not constant and is highly influenced by the soil structure and the pre-wetting moisture content (Figure 5).

4. A value of 0.8 – 0.9 is recommended for the heave reduction factor R_f for soils on the dry side. And a value of 0.6 – 0.7 is recommended for the heave reduction factor R_f for soils on the wet side.

5. The upper limit of the heave reduction factor R_f is recommended for footings with partial saturation. The lower limit of the heave reduction factor R_f is recommended for footings with full saturation

REFERENCES

Abdullah, W. S., Jundi, M. and Nabulsi, F., 1999, "Influence of lateral swell on the accuracy of the methods of heave prediction" Submitted for publication.

Alonso, E. E., Gens, A. and Hight, D. W., 1987 "Special problem soils" General Report – Session 5, Proc. 9th European Conf. Soil Mechanics and Foundation Engineering, Dublin, vol. 2, pp. 1087 – 1146.

Dalmatov, B. I., 1991. Soil Mechanics, Foundations and Footings, A. A. Balkema Publ., Rotterdam.

Jennings, J. E., and Knight, K., 1957, "The prediction of Total Heave from the Double Oedometer Test", Transaction of the South African Institution of Civil Engineers, Vol. 7, No. 9, pp.13 -19.

Jennings, and Knight, K., 1975. "A guide to construction on or with materials exhibiting additional settlement due to "collapse" of grain structure, Proc. 6th Reg. Conf. for Africa on Soil Mechanics and Foundation Engineering, Durban, South Africa, vol. 1, pp. 99-105.

Jennings, J. E., 1969, "The prediction of Amount and Rate of Heave Likely to be Experienced in Engineering Construction on Expansive Soils", Proceeding of the Second International Research and Engineering Conference on Expansive Clay Soils, Texas A&M University, Texas, pp. 99 - 109.

Lambe, T. W., 1958, "The engineering behavior of compacted clays," Journal of the Soil Mechanics and Foundations Div., ASCE, Paper 1655.

Lambe, T. W., 1962, "Soil Stabilization," chapter 4, Foundation Engineering, G. A. Leonards (ed.), McGraw-Hill, New York.

Popescu, M. F., 1992 "Engineering Problems Associated with Expansive and Collapsible Soils Behavior", Proceedings of the 7th International Conference on Expansive Soils, Dallas, Texas (USA), pp. 25-46, August 3-5, 1992.

Seed, H. B., and Chan, C. K., 1959, "Structure and strength characteristics of compacted clays," Journal of Soil Mechanics and Foundations Div., ASCE, vol.85, no. SM5.

Sullivan, R. A., and McClelland, B., 1969, "Predicting Heave of Buildings on Unsaturated Clay", Second International Research and Engineering Conference on Expansive Clay Soils, Texas A & M University, College Station, Texas, pp. 404-420.

U. S. Army Corps of Engineers, Foundations in E - pansive Soils, Department of the Army, Technical Manual TM 5-818-7, Washington, DC, 1 September 1983.

Zeitlen, J. G., 1969, "Some approaches to foundation design for structures in an expansive soil area", Proceedings of the 2nd International Research and Engineering Conference on Expansive Clay Soils, Texas A&M University. pp 148 – 155.

Unsaturated Soils for Asia, Rahardjo, Toll & Leong (eds) © 2000 Taylor & Francis, ISBN 90 5809 139 2

Vertical swelling of expansive soils under fully and partially lateral restraint conditions

M.A.Al-Shamrani & A.I.Al-Mhaidib
Department of Civil Engineering, King Saud University, Riyadh, Saudi Arabia

ABSTRACT: This paper presents the results of an experimental investigation in which the feasibility of using a stress path triaxial cell for evaluating the vertical swell of expansive soils under multi-dimensional loading conditions was examined. Series of triaxial swell tests were conducted, and the influence of confinement and initial water content on the predicted vertical swell was evaluated. The results of these tests were compared with the volume changes observed for samples tested under identical initial conditions in the oedometer. The vertical swell linearly decreased with increasing confining pressure and initial water content. The percentages of ultimate vertical swell obtained from triaxial swell tests were considerably lower than the corresponding values measured in the oedometer tests. Besides, the rate of decrease in the vertical swell with increasing applied pressure is higher for the oedometer tests.

1 INTRODUCTION

Increased awareness of the structural damages associated with expansive soils has led to more attention being focused on different aspects of their characteristics and behavior. Among the various aspects associated with building on expansive soils, a reliable estimate of the anticipated heave represents the most important factor influencing the selection of treatment alternatives to minimize volume change, or preparation of a foundation design to accommodate the anticipated volume change. The majority of volume change testing of expansive soils has been performed under one-dimensional loading conditions in the oedometer. However, due to differences between laboratory test constraints and field conditions the amount of volume changes measured in various oedometer-testing methods may differ dramatically from heaves actually realized in the field.

This paper presents the results of an experimental investigation in which the feasibility of using a hydraulic triaxial stress path apparatus for evaluating the vertical swell of expansive soils under multi-dimensional loading conditions is examined. Series of triaxial swell tests are conducted, and the influence of confinement and initial water content on the predicted vertical swell is evaluated. The results of these tests are also compared with the volume changes observed for samples tested under identical loading and initial conditions in the oedometer.

2 EXPERIMENTAL PROGRAM

2.1 *Material and sample preparation*

The soil samples for this study were brought from the town of Al-Ghatt, 270 km northwest of Riyadh, the capital of Saudi Arabia. The region experiences some of the largest differential ground movements in Saudi Arabia. Soil samples were obtained from a test pit at a depth of about three meters. The expansive formation at the site is a gray-green to yellowish weathered clayey shale. The pit was excavated in an area where many residential buildings, sidewalks, and pavements had been severely damaged due to differential heave of the underlying soft rock shales.

Laboratory testing on extracted samples included routine classification and mechanical property tests. The liquid and plastic limits are 60 and 30, respectively, and the soil has a specific gravity of 2.8. The natural moisture content is about 22%, which is far below the plastic limit and hence relatively high values for swell parameters are expected. The soil contains 28% silt-size particles (0.075-0.002 mm) and 72% clay-size particles (< 0.002 mm). It is classified as a CH soil according to the Unified Soil Classification System (USCS).

The index properties of the tested shales put it into the "high' to "very high" categories of four of the most commonly methods of determining a soil's intrinsic expansiveness (Holtz and Gibbs, 1956; Seed et al., 1962; Van Der Merwe, 1964; Dakshanamurthy and Raman, 1973). The relatively high expansion in these deposits was primarily attributed to the destruction of capillary stresses resulting from extreme desiccation, coupled with the destruction in the stratified soil structure through water infiltration (Dhowian et al., 1990). In addition, the natural moisture contents were well below the plastic limit of the soil, resulting in a considerable water deficiency and high water intake potential.

In order to minimize variations in test data remolded samples were used in the testing program. To ensure uniformity of testing, the block samples were dried, pulverized into a powder in batches and screened through sieve No. 40. The batches were thoroughly mixed to obtain a uniform bulk sample. Individual samples were oven-dried for about 24 hours and then thoroughly mixed to the chosen molding moisture content. They were then stored in air-tight plastic bags and sealed in plastic wraps to avoid loss in moisture content, and were allowed to cure at room temperature for 24 hours to allow uniform distribution of moisture content.

The 38 mm-diameter, 76 mm-long specimens for the triaxial tests were statically compacted in a vertically-split compaction mold. The soil was compacted in three layers of about 25-mm thickness to maximize uniformity. Static compaction furnishes the most uniform and repeatable samples (Booth, 1976), hence it was chosen over other compaction methods such as impact compaction and kneading. The use of a split mold ensures that the density of the sample will not be altered, as would be the case if it extruded from the mold.

2.2 *Testing procedure*

The triaxial tests were carried out in a hydraulic triaxial stress path cell of the type reported by Bishop and Wesley (1975). A schematic diagram of the testing system is shown in Figure 1. Unlike the conventional triaxial cell, the stress path cell is self-contained, portable, and requires no loading frame. Instead, the axial load is applied to the sample by pressurizing the lower chamber at the bottom of the cell. The piston pushes up a loading ram, at the top end of which is the pedestal on which the soil sample is mounted. The sample is pushed upward against a stationary submersible load cell. This is a salient feature of the stress path cell that makes it possible to measure the vertical swell. In the conventional triaxial apparatus it has only been possible to measure the total volumetric swell of the tested soil sample.

Vertical displacement is measured with a dial gauge and/or a displacement transducer mounted on the top of the cell and deflected by two vertical extension rods which pass through clearance holes in the cell base and connected to the loading ram.

After the specimen was removed from the mold and enclosed into a single watertight rubber membrane, it was mounted on the base of the triaxial cell. The base platen was lightly coated with a film of thin grease prior to attaching the membrane. The specimen was seated against a saturated porous stone, and a second similar porous stone was placed on the top of the sample. The membrane was then sealed to the top loading cap and the bottom platen with O-ring seals. Then, valves 2 and 4, illustrated in Figure 1, were opened, and the perspex cell was filled with de-aired tap water flowing through valve 2. Then valve 1 was opened and the chamber pressure was incrementally increased until the loading ram, along with the sample, began to move upward. The chamber pressure required to move the ram upward is easily determined by knowing the combined weight of the loading ram and the specimen. A self-compensating mercury control (Bishop and Henkel, 1962) was used for applying both the lateral and the chamber pressure.

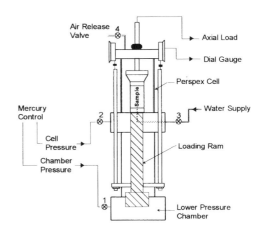

Figure 1. Schematic layout of the Bishop-Wesley triaxial stress path cell.

The upward movement of the loading ram slowly continued until the top loading cap touched the stationary load cell, which recorded the load. Utmost care was taken with the alignment of the top loading cap with the load cell. It was important that the loading cap gently touches the load cell so that the sample will not be subjected to axial compression. In the meantime, it was essential to ensure a full contact between the top loading cap and the load cell so that the entire vertical swell will be detected and measured by the dial gauges shown in Figure 1. If a space is left between the top loading cap and the load cell the specimen will expand upward, and this expansion will not be detected by the dial gauges. Therefore, to maintain contact between the top loading cap and the load cell a nominal axial pressure of about 2 kPa was applied to the sample.

After deciding the desired confining pressure for the given test, the chamber pressure was determined accordingly. With valves 1, 2 and 4 closed, the chamber and lateral pressures were increased to the desired values. Their values were monitored by pressure transducers connected to electronic readout devices. After reaching the specified values for the chamber and lateral pressures, initial dial gauge readings were taken and valves 1 and 2 were simultaneously opened.

The specimen was allowed to consolidate under the applied isotropic confining pressure for a period of about 24 hours before it was given free access to water. Initial dial gauge readings were taken, and water was introduced to the bottom of the sample by opening valve 3, which is connected to a 50-ml single-tube drainage burette. Readings were taken until the change in vertical swell under the applied confinement was negligible at which the test was terminated. The sample was then removed from the cell, weighed, and three slices were taken from the bottom, middle, and top of the specimen and used to find the final water contents.

3 RESULTS AND DISCUSSION

3.1 Effect of Confinement

The percentage of swell is dependent on the level of applied confining pressure. The first set of triaxial swell tests was conducted to examine the effect of confinement on the amount of vertical swell. A series of four tests was performed on samples compacted at an initial water content of 14% and dry density of 18 kN/m^3, under a range of confining pressure (viz. 25, 50, 100, and 200 kPa). The percentage of vertical swell versus time is shown in Figure 2. The percentage of vertical swell is defined as the ratio of the increase in height of the sample to its initial height before it was given free access to water.

As expected, the percentage of vertical swell decreases as the confining pressure increases. The increase in confining pressure from 25 to 200 kPa caused a reduction in the percentage of vertical swell from 8% to about 3.9%. Furthermore, following the commencement of the tests, the rate of expansion was relatively low especially for high confining pressure. This may be attributed to the slow rate of water sorption, and that some time was needed to pass before water starts to seep through the sample. The amount of vertical swell that occurs within a given period of time partially depends on the quantity of water that enters the soil. This in turns is influenced by the confinement. Under low confining pressures, water easily penetrates into the samples, whereas the entry of water is restricted by relatively high magnitude of confinement. Besides, saturation of the shale formations usually takes a very long time

because of their low permeability. Ruwaih (1987) found the permeability of various natural shale formations in Saudi Arabia, including the one considered in this study, to range from 10^{-9} to 10^{-12} cm/sec.

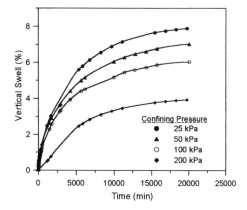

Figure 2. Vertical swell under different confining pressures.

It may also be noted from Figure 2 that during the early stage of the tests, confinement had less effect on the amount of vertical swell for relatively low confining pressures. Thus, up to a certain value of applied confining pressure, the magnitude of the pressure may have less effect on the vertical swell versus time relationship. This is in agreement with the results reported by Abduljauwad and Al-Sulaimani (1993).

3.2 Effect of Initial Water Content

In order to assess the effect of initial water content on the amount of vertical swell, a series of triaxial swell tests was conducted on soil specimens that were prepared at different initial water contents. Owing to the relatively long period of time required for a single test (about 15 days), the number of performed tests was limited to five specimens all having the same dry density of 18 kN/m^3. The first two tests were performed under a confining pressure of 25 kPa on two soil samples that were prepared at initial water content of 14% and 22%, respectively. Two other samples were also prepared at the same two values of initial water contents but subjected to a confining pressure of 100 kPa.

It can be seen from Figure 3 and Figure 4 that, irrespective of the magnitude of the confining pressure, the initial moisture content had a significant effect on the magnitude of the vertical swell of the tested soil. Moreover, the influence of initial water content is relatively higher for higher confinement. When the initial water content decreased from 22% to 14%, the ultimate vertical swell increased by about 300% and 400% for the confining pressures of 25 kPa and 100 kPa, respectively.

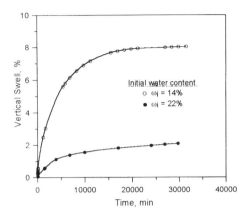

Figure 3. Vertical swell versus time for different initial water contents (confining pressure = 25 kPa)

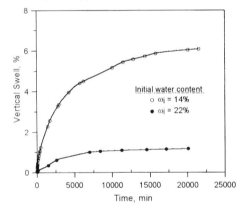

Figure 4. Vertical swell versus time for different initial water contents (confining pressure = 100 kPa)

To establish the relationship between swell and initial water content, the fifth specimen was prepared at an initial water content of 18% and tested under a confining pressure of 25 kPa. Figure 5 shows the expected trend of decrease in the ultimate vertical swell with the increase in the initial water content. The gradual reduction of ultimate swell with respect to initial water content implies that there should be an equilibrium position at which this value becomes almost stable. By extrapolating the linear relationship of Figure 5, the intersection with the abscissa indicates the initial moisture content at which the soil will experience no vertical swell. This value may be termed the *"null moisture content,"* can be found to be about 24% for the tested clayey shale. This value is quite below the plastic limit of 30%. Previous investigations (Dhowian et al., 1990; Edil and Alanazy, 1992) have shown that for soil samples

tested in the oedometer apparatus, swell stabilizes at a moisture content close to the plastic limit of the tested soils. The relationship between the vertical swell and the initial moisture content may not, therefore, be unique but rather dependent on the loading conditions of the swell tests.

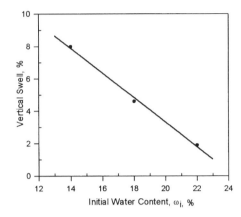

Figure 5. Relationship between initial water content and vertical swell (confining pressure = 25 kPa)

3.3 *Influence of Loading and Wetting Conditions*

To study the vertical swell under different loading conditions, two series of swell tests were conducted. Four confining pressures (viz. 35, 70, 140, and 210 kPa) were used in this set of swell tests. The first set of tests was performed in the oedometer, whereas the second series of tests was conducted in the stress path triaxial apparatus and subjected to the conventional triaxial state of stress. The samples for all tests of the two series were prepared at 22% initial water content and 18-kN/m^3 dry density. Testing and preparation of soil samples for the triaxial swell tests follow the same procedure described previously.

The swell overburden method was used and tests were conducted according to ASTM D4546-85. The 70 mm-diameter, 19 mm-thick oedometer specimens were statically compacted in two layers in the consolidation ring, which was enclosed in a special compaction mold. The sample was then transferred to a standard rear loading oedometer and subjected to a seating pressure of 7 kPa for a period of 5 minutes. Then, the deformation dial gauge was adjusted for the initial zero reading, and after 5 minutes the specified vertical pressure was applied to the specimen. Five minutes after applying the vertical pressure, water was introduced to the specimen and the swell test commenced. The expansion of the specimen was measured until the change was negligible, at which time the test was terminated and the final water content was measured.

Typical variation of the percentage of vertical swell with elapsed time is shown in Figure 6 for the applied pressure of 35. As expected, the vertical swells obtained from oedometer are considerably larger than the triaxial test measurements. Similar patterns have also been observed for the triaxial and oedometer tests of the other three values of applied pressure (i.e., 70, 140, and 210 kPa). The percentage of vertical swells measured in the oedometer and the stress path equipment are plotted against the logarithm of applied pressures in Figure 7. The data from both the triaxial and oedometer tests are fitted to a straight line. It is noticed, however, that the rate of decrease in the vertical swell with increasing applied pressure is higher for the oedometer tests.

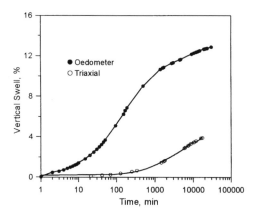

Figure 6. Percentages of vertical swell obtained from oedometer and triaxial swell tests (confining pressure = 35 kPa).

The values of the final water contents versus the applied pressures are shown in Figure 8. It is observed that, the final water content varies along the length of the 76 mm-long triaxial test samples. The variation, however, depends on the level of applied pressure. For low confinement (35 kPa) the difference between the values of the final water contents at the top and the bottom of the sample is about 4.5%, whereas it is only 2% for the confining pressure of 210 kPa. It can also be noted that, regardless of the value of the applied pressure, the final water contents for samples tested in the oedometer are higher than the corresponding values for samples tested in the stress path cell. In other words, the moisture deficiency or moisture intake potential, which is the difference between final and initial moisture contents, is higher for oedometer tests. This observation supported the suggestion made by Erol et al. (1987) that not only a lateral restraint factor but also a moisture factor should be considered in the analysis of predicted heave when using the results of oedometer tests. Another conclusion to be drawn from Figure 8 concerns a proposition made by some investigators (Weston, 1980; Hamberg and Nelson, 1985; Dhowian et al., 1990) that the ultimate swell condition occurs at water contents close to the plastic limit. However, the results of Figure8 indicate that this assumption may be erroneous especially for the case of high confinement under isotropic swelling condition.

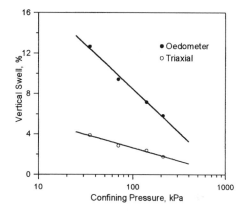

Figure 7. Relationship between applied pressure and ultimate vertical swell measured in oedometer and triaxial swell tests.

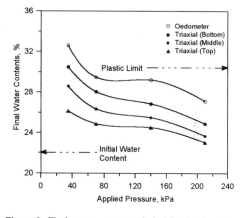

Figure 8. Final water contents of triaxial and oedometer samples at different confining pressures.

4 SUMMARY AND CONCLUSIONS

The experimental study presented in this paper included the performance of several series of swell tests under conventional isotropic state of triaxial stress. The influence of confinement and initial moisture content on the measured vertical swell was investigated. The results of the triaxial swell tests were compared with the volume change measured in specimens tested under one-dimensional loading conditions in the oedometer. The following conclusions are drawn:

1. The vertical swell linearly decreases with increasing confining pressure and initial water content. However, the influence of the change in initial water content on soil expansion slightly depends on the magnitude of the applied confining pressure.

2. The percentages of vertical swell obtained from triaxial swell tests were considerably lower than the corresponding values measured in the oedometer swell tests.

3. The final water contents for specimens tested in the triaxial apparatus are quite below the plastic limit. The discrepancy between the plastic limit and the final water content increased with increasing confining pressure. Moreover, the moisture deficiency or moisture intake potential, which is the difference between final and initial moisture contents, was higher for oedometer tests. Therefore, as has already been suggested by a number of researchers, not only a lateral restraint factor but also a moisture factor should be considered in the analyses of predicted heave when using the results of oedometer tests.

REFERENCES

Abduljauwad, S. N. & G. J. Al-Sulaimani 1993. Determination of swell potential of Al-Qatif clay. *Geotechnical Testing J.*, ASTM, 16(4): 469-484.

Bishop, A.W. & D. J. Henkel 1962. *The measurement of soil properties in the triaxial test.* 2nd Ed., Edward Arnold, London.

Bishop, A.W. & L. D. Wesley 1975. A hydraulic triaxial apparatus for controlled stress path testing. *Geotechnique.* 25: 657-670.

Booth, A. R. 1976. Compaction and preparation of soil specimens for oedometer testing. soil specimen preparation for laboratory testing. ASTM. STP599. 216-228.

Dakshanamurthy, V. & V. Raman 1973. A simple method of identifying expansive soil. *Soils and Foundations.* 13(1): 97-104.

Dhowian, A.W., A. O. Erol & A. Youssef 1990. Evaluation of expansive soils and foundation methodology in the Kingdom of Saudi Arabia. Final Report, KACST, AT-5-88.

Edil, T. B. & A. S. Alanazy 1992. Lateral swelling pressures. Proc. 7th Int. Conf. on expansive soils, Dallas, Texas, 227-232.

Erol, A. O., A. W. Dhowian & A. Youssef 1987. Assessment of oedometer methods for heave prediction. *Proc. 6th Int. Conf. on expansive soils.* New Delhi, India, 99-103.

Hamberg, D. J. & J. D. Nelson 1985. Prediction of floor slab heave. *Proc. of 5th Int. Conf. on expansive soils*, Adelaide, Australia, 137-140.

Holtz, W. G. & H. J. Gibbs 1956. Engineering properties of e - pansive clays. *Proc.*, ASCE, Vol. 80, Separate No. 516, 1-28.

Ruwiah, I. A. 1987. Experiences with expansive soils in Saudi Arabia. *Proc. 6th Int. Conf. on expansive soils*, New Delhi, India, 317-322.

Seed, H. B., R. J. Woodward & Lundgren, R. 1962. Prediction of swelling potential for compacted clays. *J. of the Soil Mechanics and Foundation Division*, ASCE, 88(3):53-87.

Van Der Merwe, D. H. 1964. The prediction of heave from the plasticity index and percentage clay fraction of soils. *Civil Engineers in South Africa*, 6(6):103-107.

Weston, D. J. 1980. Expansive roadbed treatment from Southern Africa. *Proc. 4th Int. Conf. on Expansive Soils*, Denver, CO, 1:339-360.

Unsaturated Soils for Asia, Rahardjo, Toll & Leong (eds) © 2000 Taylor & Francis, ISBN 90 5809 139 2

Influence of structure and stress history on the drying behaviour of clays

F.Cafaro, F.Cotecchia & C.Cherubini
Technical University of Bari, Italy

ABSTRACT: An investigation of the influence of stress history and structure on the state-path followed by clays during drying has been carried out. The paper discusses the results of tests concerning the drying behaviour of both a natural stiff illitic clay and the same clay reconstituted in the laboratory. The main differences found between the behaviour of reconstituted samples with different stress histories concern both the air entry value and the specific volume at the shrinkage limit. Moreover also the structure seems to influence the desaturation process, as observed on the natural clays.

1 INTRODUCTION

Several recent studies have been concerned with the influence of drying-wetting processes on the mechanical behaviour of clays, starting from the idea that the development of suction and capillary stresses can represent an important factor of the stress history of soils and influence their mechanical response (Fredlund & Rahardjo 1993). Following this approach, Cafaro (1998) and Cafaro et al. (1998, 2000) have shown that weathering resulting from drying processes can cause changes in the clay state and structure, which can be of similar size to the changes resulting from external loading. The authors show that drying-wetting cycles, occurred in recent Quaternary within a stiff blue Pleistocene clay (Montemesola Clay, Southern Italy), have caused permanent reduction in the clay specific volume and changes in its structure, being structure the combination of both fabric and bonding (Cotecchia & Chandler 1997). Viceversa, the clay stress history and structure influence the way the clay responds to drying, e.g. the drying state-path. The present paper discusses the influence of the stress history and structure of clays, either reconstituted or natural, on their drying behaviour.

The drying behaviour has been investigated for both the natural blue Montemesola clays and the same reconstituted in the laboratory. Specifically the natural clays, which are illitic and overconsolidated, occur in situ as grey and unweathered (G) at depths larger than 30m below ground level, and as yellow and weathered (Y) at shallower depths. The particle size distribution curves of both the grey and the yellow clays are shown in Figure 1.

The study of the drying state-paths experienced by these two clays, when reconstituted in the laboratory, has aimed at identifying the basic effects of both particle size distribution (sorting) and stress history (applied in the laboratory) on the desaturation and shrinkage behaviour of the clay. In particular, the drying tests were carried out on reconstituted normally consolidated samples of both the grey and the yellow clay (Gnc and Ync), which are of different sorting (Figure 1), and on an overconsolidated sample of the reconstituted grey clay (Goc), similar to Gnc in sorting. Moreover, drying tests were carried out also on four samples of natural clay (Ga, Gd, Yb, Yc), in order to examine the additional effects of the natural structure on the clay drying behaviour.

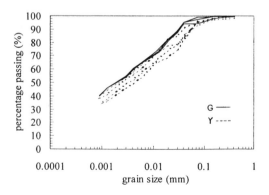

Figure 1. Particle size distribution curves of the Montemesola clays, G and Y.

2 TESTING PROCEDURE

The parameters of use to describe the state-path of a clay during drying generally are specific volume, v, degree of saturation, S(%), water content, w(%), and suction, p_k. (Toll 1988, Ridley 1993). In the following, the drying state-paths will be described, for the clay samples previously mentioned, by means of measurements of specific volume, degree of saturation and suction, taken at given time intervals during the drying tests.

These tests were carried out very slowly in the laboratory (in almost 10 months, starting from complete clay saturation), following the procedure suggested by Ridley (1993), which makes use of appropriately sealed perspex devices to place filter paper disks in-contact with both the top and bottom of the sample and to slow down the drying process. The volume changes were measured every 2 or 3 weeks by immersing the sample into mercury and by measuring the volume of the mercury being displaced. Suction was measured by means of the filter paper (in-contact) method (Chandler & Gutierrez 1986) and represents a matrix suction, at least until cavitation occurs. The filter paper calibration employed to deduce the suction values is that reported by Chandler et al. (1992) for filter paper Whatman n.42.

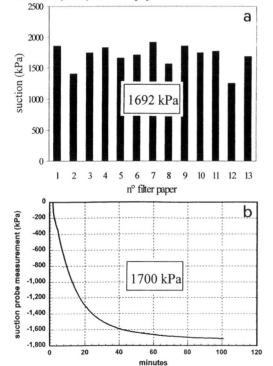

Figure 2. Suction measurements on the same clay sample by using two different methods (a: filter paper test; b: tensiometer I.C.).

The applicability of the filter paper method to measure the suction of the stiff clays being tested was checked by comparing 13 filter paper measurements, carried out on a clay block sample of 7000 cm³, with the suction measured on the same clay (block sample of 350 cm³) by means of the Imperial College tensiometer (Ridley 1993) (Figure 2). It appears that the average suction value resulting from the filter paper measurements (Figure 2a), 1692 kPa, is very close to the suction value measured with the tensiometer (Figure 2b). This value pertains to the in situ unweathered clay (G), whereas the suction in the weathered clay (Y) was measured to be more than 3000 kPa.

In the following the clay drying behaviour will be examined in the *degree of saturation* versus *suction* graph, where is possible to identify the suction value at which the first air enters the sample, and in the *specific volume* versus *suction* and the *specific volume* versus *degree of saturation* graphs, where both the minimum specific volume reached with drying and the corresponding suction value can be established. The minimum value of specific volume will be indicated as the "specific volume at the shrinkage limit", not in conformity with the standard definition of shrinkage limit.

3 RESULTS OF THE DRYING TESTS ON THE RECONSTITUTED CLAY

The normally consolidated reconstituted clay samples Gnc and Ync had been one-dimensionally normally consolidated from slurry up to 230 kPa vertical effective stress, σ'_v (K_0=0.55; p′=160kPa), before drying. The overconsolidated reconstituted sample Goc, instead, had been loaded to σ'_v=1100 kPa, before being unloaded to σ'_v=50 kPa (OCR=22).

Since the suction level sustainable by a pore before cavitation is a function of its radius (Klausner 1991), the specific volume of a soil is one of the main factors affecting the soil desaturation process. Specific volume of a reconstituted clay depends on both soil composition and stress history, so both of these factors affect indirectly the drying behaviour of a soil. In particular different sortings lead to different packings of the particles and so to different average pore sizes and pore-size distributions for a given consolidation stress state. In the following the effects of sorting and stress history on the drying behaviour of the reconstituted clay will be examined separately.

Figure 3 shows the three plots: v versus p_k (a), S versus p_k (b) and S versus v (c), which describe the drying behaviour of the normally consolidated samples Gnc and Ync, of similar stress history but different sorting. The suction values of the samples soon after one-dimensional consolidation were

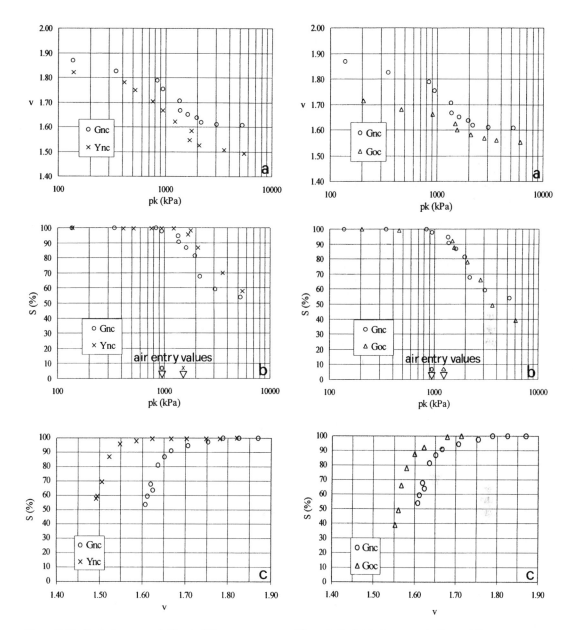

Figure 3. Drying tests on sample Gnc and Ync
(a: v-p_k, b: S-p_k, c: S-v planes).

Figure 4. Drying tests on sample Gnc and Goc
(a: v-p_k, b: S-p_k, c: S-v planes).

138kPa and 140kPa respectively. These values are close to the consolidation mean effective stress applied to both the samples, p'=160kPa. The drying curves in the v-p_k plane (Figure 3a) should be close to the Virgin Desaturation Lines, VDL (Toll 1988), of the two normally consolidated samples. It can be seen that sorting affects both the desaturation behaviour (Figures 3b and 3c) and the shrinkage behaviour of the soil (Figures 3a and 3c). Indeed the

data seem to indicate that, for the same stress history, the higher sorting of Ync makes desaturation start at lower specific volumes and higher suctions.

In Figure 4 the effects of stress history on the drying behaviour of the reconstituted grey clay are shown. For the overconsolidated sample (Goc), which has a gross yield vertical stress of 1100 kPa in one-dimensional compression, cavitation starts at suction values between 1100 kPa and 1300 kPa, when the specific volume is about 1.65, whereas for the sample normally consolidated up to 230 kPa vertical effective stress, cavitation starts at a suction value of 950 kPa, when the specific volume is 1.75. Differences are found also between the specific volumes at the shrinkage limit, that are 1.62 for sample Gnc and 1.55 for sample Goc. It follows that lower specific volumes due to overconsolidation at the start of the drying process involve larger suctions at cavitation and lower specific volumes at the shrinkage limit than for samples which start drying at higher specific volumes.

Unlike the normally consolidated samples, for the overconsolidated one a volumetric collapse occurs within a limited range of p_k values, which relates to a well defined segment of the drying curve in the v-p_k plane (Figure 4a). This has also been observed by Ridley (1993) with overconsolidated kaolin and by Cotecchia (1996) with other natural overconsolidated clay samples. This curve segment corresponds to the maximum gradient of specific volume decrease with increasing suction during drying and to the start of desaturation, i.e. degrees of saturation about 96-99% (Figure 4b). Moreover, for sample Goc the range of specific volumes within this curve segment is found to correspond to the range of specific volumes at which gross-yield develops in isotropic compression. We propose to define the v-p_k state in the centre of the above mentioned curve segment as a "pseudo-gross yield state", because it corresponds to the development of important permanent strains and structure degradation for the soil during drying (Cafaro et al. 2000). The specific volume decrease with desaturation beyond pseudo-gross yield is limited due to the mechanism of stress transmission for capillarity, as discussed by Burland & Ridley (1996).

It is possible to observe directly the correspondence between the gross yield state in isotropic compression and the pseudo-gross yield state of the same sample during drying by plotting its isotropic compression curve and v-p_k drying curve in a single plane, putting on the abscisse both the p' and the p_k values, although this comparison is rigorously valid only before cavitation occurs, when p_k may be assumed to equal p'. Indeed for a saturated material there is theoretical equivalence between the loading effects of a negative pore pressure and those of an isotropic effective pressure due to external loading. The joint observation, in the same v-p',p_k plane (Brady 1988, Esposito 1995), of the drying curve and of the Isotropic Normal Compression Line, INCL (Burland 1990) of sample Goc is possible in Figure 5. For the same sample the One-dimensional Normal Compression Line (K_0NCL) is also plotted in the figure. It appears that pseudo-gross yield occurs as the drying curve intersects the Isotropic Normal Compression Line of the clay, at the same

specific volume as that of the sample when was preconsolidated (on the K_0NCL) before drying.

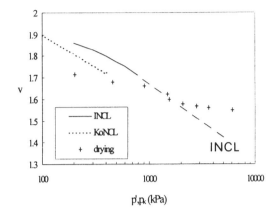

Figure 5. Drying, isotropic compression and one-dimensional compression curves of sample Goc.

In conclusion, isotropic gross yield appears to control both the strain behaviour and the desaturation of an overconsolidated clay during drying. For a one-dimensionally normally consolidated clay, although before drying starts the v-p'$\cong p_k$ state plots along the K_0NCL, soon after the v-p_k state shifts towards the INCL. In this respect, also the normally consolidated samples will exhibit, at the start of drying, a light apparent overconsolidation, as consequence of the shift from the K_0NCL to the INCL (Figure 3a).

4 RESULTS OF THE DRYING TESTS ON THE NATURAL CLAYS

A pseudo-gross yield state is recognized also along the state-paths in the v-p_k plane of the natural overconsolidated clay samples during drying. Figure 6 shows the state-path followed by sample Gd during drying as well as the one-dimensional compression curve of a similar sample. The two curves seem to confirm that pseudo-gross yield occurs about the same specific volume as that of the clay at gross yield in one-dimensional compression. Since gross yield for the natural clay is controlled by the natural structure, this is seen to influence the strain behaviour of the clay during drying and the start of desaturation, that starts just before pseudo-gross yield. For the clay here considered the current structure is not only the result of stress history, but also of structure modifications essentially due to diagenetic processes. The air entry occurs pre- pseudo gross yield, and so earlier for the natural clay than for the reconstituted clay during the drying process, since in all the four tests on natural clay the curve segment on the drying curve where the gradient in specific volume decrease is maximum corresponds to degrees of saturation between 85% and 90%.

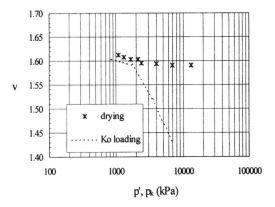

Figure 6. Drying and one-dimensional compression curves of sample Gd.

Figure 6 shows that the pseudo-gross yield state lies to the right of the K_0 gross yield state. Therefore, also for the natural clay pseudo-gross yield and cavitation appear to correspond to gross yield in isotropic compression, rather than gross yield in one-dimensional compression. This is probably due to the fact that the drying tests relate to free boundary conditions and the capillary stresses load the clay not one-dimensionally, but simil-isotropically.

Pre- pseudo gross yield, the drying curves have slopes $\Delta v/\Delta\ln(p_k)$ in the range $0.008 \div 0.013$, whereas the slopes $\kappa = \Delta v/\Delta\ln(p')$ of the swelling lines of the clay vary from 0.008 to 0.015 (Cafaro 1998), as measured during both K_0 and isotropic compression tests. Therefore, the drying curves of the clays studied appear to lie on the clay swelling lines pre-pseudo gross yield.

5 PERMANENT REDUCTION OF THE CLAY SPECIFIC VOLUME DURING DRYING

At Montemesola, the weathered clays (samples Y) have in situ specific volumes smaller than the unweathered clays (samples G), despite they lie above the latter of about 30m. Cafaro (1998) and Cafaro et al. (1998, 2000) discuss the reason why this difference in specific volume is the result of cyclic drying-wetting processes, which have involved the upper clays in the recent past.

Figure 7 shows the isotropic normal compression line (INCL) of the natural unweathered clay resulting from isotropic compression of grey clay samples post- gross yield. The same figure also shows the state-paths during drying of samples Gd and Yc,

in the v-p',p_k plane. The drying state-paths start, for each sample, from the in situ specific volume. The pseudo-gross yield states on these drying curves seem to fall on the INCL. The slope of the line which joins these pseudo-gross yield states, measured by means of the ratio $\Delta v/\Delta\ln(p_k)$, has value in the range $0.126 \div 0.143$. In particular the straight line regressing four points about pseudo-gross yield on each of the drying curves in the Figure has slope 0.126 (coefficient of correlation 0.999), which is very close to the slope of the INCL, $\lambda = \Delta v/\Delta\ln(p')$, that is 0.127 (Cafaro 1998).

Therefore it appears that drying-wetting cycles in situ have shifted the pseudo-gross yield state of the clay to lower specific volumes and higher suctions along the INCL, the same as an isotropic compression due to external loading would do with the clay gross yield state. Since during isotropic compression post- gross yield significant structural degradation occurs, also drying appears to cause a significant structural damage, as otherwise shown by Cafaro (1998) and Cafaro et al. (1998, 2000) by means of microstructural analyses.

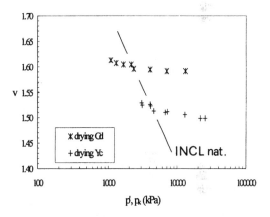

Figure 7. Drying state-paths and INCL of natural clay samples.

6 CONCLUSIONS

The research results indicate that the drying curves, of either natural or reconstituted clays, are characterized, in the v-p_k plane, by a curve segment within which major desaturation starts and major permanent changes in specific volume and structure occur. The state in the centre of this curve segment has been defined as a "pseudo-gross yield state". This has been found to correspond to the gross yield state of the same clay sample in isotropic compression. Therefore isotropic gross yield appears to control both the straining behaviour and the cavitation processes of clays during drying. Since gross yield is in turn controlled by the current clay structure, this appears to be the fundamental factor influencing the drying behaviour of clays, either natural or reconstituted.

ACKNOWLEDGEMENTS

The authors thank Dr. A.M. Ridley for the suction measurement carried out by means of the Imperial College tensiometer on an undisturbed clay sample.

REFERENCES

Brady, K.C. 1988. Soil suction and the critical state. *Géotechnique* 38 (1): 117-120.

Burland, J.B. 1990. On the compressibility and shear strength of natural clays. *Géotechnique* 40 (3): 329-378.

Burland, J.B. & A.M. Ridley 1996. The importance of suction in soil mechanics. *Twelfth Southeast Asian Geotechnical Conf.*, Kuala Lumpur.

Cafaro, F. 1998. *Influenza dell'alterazione di origine climatica sulle proprietà geotecniche di un'argilla grigio-azzurra pleistocenica.* Technical University of Bari, PhD thesis.

Cafaro, F., F. Cotecchia & C. Cherubini 1998. Weathering effects for stiff blue clays in Southern Italy. *The Geotechnics of Hard Soils – Soft Rocks.* 3: Rotterdam: Balkema.

Cafaro, F., F. Cotecchia & C. Cherubini 2000. Influence of weathering processes on the mechanical behaviour of an italian clay. Submitted to *GeoEng2000.*

Chandler, R.J. & C.I. Gutierrez 1986. The filter-paper method of suction measurement. Technical Notes. *Géotechnique* 36: 265-268.

Chandler, R.J., M.S. Crilly & G. Montgomery-Smith 1992. A low-cost method of assessing clay desiccation for low-rise buildings. *Proc. Instn. Civ. Engng.* 2: 82-89.

Cotecchia, F. 1996. *The effects of structure on the properties of an Italian Pleistocene clay.* University of London, PhD thesis.

Cotecchia, F. & R.J. Chandler 1997. The influence of structure on the pre-failure behaviour of a natural clay. *Géotechnique* 47 (3): 523-544.

Esposito, L. 1995. Comportamento dei terreni durante il percorso di primo essiccamento ed umidificazione ed applicazioni alla determinazione dei parametri di stato critico. *Rivista Italiana di Geotecnica* (4): 251-270.

Fredlund, D.G. & H. Rahardjo 1993. *Soil Mechanics for Unsaturated Soils.* John Wiley & Sons.

Klausner, Y. 1991. *Fundamentals of continuum mechanics of soils.* Springer-Verlag, London.

Ridley, A.M. 1993. *The measurement of soil moisture suction.* University of London, PhD thesis.

Toll, D.G. 1988. *The behaviour of unsaturated compacted naturally occurring gravel.* University of London, PhD thesis.

Unsaturated Soils for Asia, Rahardjo, Toll & Leong (eds) © 2000 Taylor & Francis, ISBN 90 5809 139 2

Study on the swell properties and soil improvement of compacted expansive soil

Y.J. Du & S. Hayashi
Institute of Lowland Technology, Saga University, Japan

ABSTRACT: In this paper, the swell behaviors of the compacted expansive soils in Ning-Lian Highway are introduced, and their swell mechanisms are discussed based on the change in soil water contents, dry densities, material compositions and microstructures. The improvement of the compacted expansive soils by lime is also discussed briefly. It is concluded that in terms of application practice careful attention should be paid to this type of compacted expansive soil.

1 INTRODUCTION

Under the undisturbed conditions, some clayey soils show certain swell-shrinkage, but their swell grades are low. When being disturbed, compacted and then used for the embankment and roadbed, their natural structures are destroyed and cementing bonds break, the water contents decrease, the dry densities become high and the swell indexes increase. By this they may become high swell grade expansive soils. Research has shown that most cracks of embankments and roadbeds are due to the ignorance of this problem. So in engineering works, enough attention should be paid to this type of expansive soil.

The purpose of this study is to investigate the swell behaviors, mechanisms and effect of lime improvement on this type of compacted expansive soil.

2 SOIL DESCRIPTION

In this study, investigated expansive soils are alluvial sedimented and overconsolidated ones. To accomplish the desired objectives, four typical types of undisturbed soils were selected in nearly the same locations of Ning-Lian Highway, which starts from Nanjing to Lian Yonggang City and allocates in Jiangsu Province, Southeast of China. Table 1 shows their physico-mechanical properties.

3 TEST METHODS

Compacted specimens were prepared according to the standard Proctor procedure (ASTM D 698 1991).

In this paper, swell indexes of the soils are defined as: swell pressure (kPa), swell percentage under 50 kPa vertical load, δ_{ep50} (%), and total swell percentage, δ_{ps} (%). To obtain the swell pressure and swell percentage under 50 kPa tests, the specimen of $\Phi60 \times H20$ mm disc size was placed in an oedometer consolidation cell, and two pieces of porous stone plate were put on the top and bottom of the laterally confined samples respectively. A vertical load was applied to the samples under 100% degree of saturation. In both of the tests, the samples were allowed to fully swell for at least 48 hours to complete swell. It should be noted that the samples were directly placed in the oedometer without curing because aging may result in the development of bonds which decreases the swell (Gizienski & Lee 1965, Day 1992, 1994).

Total swell percentage, δ_{ps} is calculated as

$$\delta_{ps} = \delta_{ep50} + \lambda_s \left(w - w_s \right) \qquad (1)$$

Table 1. Physico-mechanical properties of selected undisturbed soils.

Sample No.	Water content, w (%)	Liquid limit, w_L (%)	Plastic index, I_P	Organic matter (%)
1*	34.2	51	23	0.2
2**	43.9	58	32	1.5
3***	27.4	43	19	trace
4****	25.8	32	15	trace

*1-Grey soil **2-Black soil ***3-Greyish-yellow soil ****4-Yellow silty soil

in which λ_s is the soil shrinkage coefficient, w is the water content, and w_s is the shrinkage limit.

4 SWELL PROPERTIES

From Table 2 it can be seen that under undisturbed conditions, swell indexes of these soils are low, whereas they become considerably higher when remolded and compacted except for the Yellow silty soil. Swell pressure of a type of soil named Grey soil increases by 27 times varying from 22kPa to 587kPa, while that of so called Black soil increases by 39 times ranging from 10 kPa to 385 kPa. The swell pressure of Greyish-yellow soil increases by about 7 times from 40 kPa to 276 kPa. Only the Yellow silty soil remains non-expansive even after remolding and compaction. Another indication is that δ_{ep50} of undisturbed soil is below zero and it increases above zero (except for the Yellow silty soil) after being compacted. Total swell percentage (δ_{ps}) is a comprehensive index which well describes the swell properties of soil. After being compacted, δ_{ps} of the undisturbed soils become relatively higher than that of the undisturbed ones. δ_{ps} of the compacted Grey soil becomes the largest one. It increases by 4.3% to 11.8%, nearly 3 times, and that of Yellow silty soil is the lowest one. According to Li (1992), a soil with δ_{ps} greater than 0.7% should be considered as an expansive one. Therefore these compacted soils are expansive soils excluding Yellow silty one.

5 MECHANISMS

5.1 *Change in dry density and water content*

From Figures 1-2, it can be seen that dry densities of undisturbed soils are larger while water contents are lower compared with those of the compacted soils. The change in dry densities and water contents are the direct reasons for different swell properties between compacted soils and undisturbed soils.

The swell of soil is due to a moisture film forming around the particles as a result of reaction between the clay particles and water (Low 1987, 1992). Interaction of water with clay surfaces reduces the chemical potential of the water, thereby generating a gradient in the chemical potential that causes additional water to flow into the system. As the thickness of moisture film increases, the volume of the soil increases also. This phenomenon varies depending on the conditions of dry density and water content. For a specimen with low water content, formation of the moisture film is easy and it reaches maximum thickness in a short time. However, when the dry density is high, there are more clay particles per volume of soil, which is benefit to the swell, and thus the reactions between particles and water are more signifi-

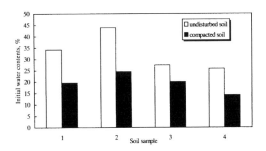

Figure 1. Comparison of initial water contents between undisturbed and compacted soils.

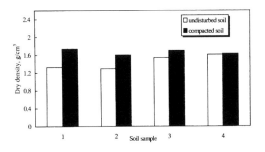

Figure 2. Comparison of dry density between undisturbed and compacted soils.

Table 2. Comparison of swell indexes between compacted and undisturbed soils.

Index	Sample No.	U*	C**	C / U
Swell pressure (kPa)	1	22	587	27
	2	10	385	39
	3	40	276	7
	4	6	7	1
δ_{ep50} (%)	1	-0.7	11.8	
	2	-1.9	7.3	
	3	-1.2	5.5	
	4	-3.7	-0.5	
δ_{ps} (%)	1	4.3	11.8	3
	2	7.3	7.5	1
	3	2.9	6.2	2
	4	-0.7	-0.4	1
Swell grade classification	1	moderate	high	
	2	high	high	
	3	low	high	
	4	non	non	

*U-Undisturbed soil **C-Compacted soil

cant. The dry densities of the compacted soils are higher while their water contents are lower than those of the undisturbed ones, which is in favor of swell. On the contrary, undisturbed soils have no such performances. Chen (1965), Osipov (1987), Dif and Bluemel (1991), and Day (1994) had studied how the initial water content affected the swell properties and their results are in accordance with that of the authors. Thus it is clear that degree of water saturation plays important roles in soil swell.

5.2 Material composition

However, it is not the case that all of soils may become "artificially expansive" after remolding and compaction. Only those which contain hydrophilic clay minerals and clay particle contents of expansive soils may change into "artificially expansive" after compaction. All the soils show in Table 3 have high illite contents, but illite-smectite mixed-layer mineral contents are high only in Grey soil and Greyish-yellow soil. Meanwhile, both smectite and illite are hydrophilic clay minerals, so the swell of compacted Grey soil and Greyish-yellow soil is high. Besides, clay particle content has also an important influence on compacted soil properties. Though Yellow silty soil composes mainly of illite, it contains only 2.8% of clay particles. Hence, its swell properties vary little after compaction. Tahir (1992) indicated clay content has influence not only on the change of the dry density but also on the feature of soil microstructure. Seed et al. (1962) put forward an empirical formula in which swell percentage increased as the clay content increased. These archived results show well how material composition affect on the swell properties of expansive soil.

5.3 Change in microstructure

Undisturbed soil properties result from the geological and stress/strain history. After long complicated natural events, undisturbed soil mass develops certain structural strength and strong connection form among soil particles. This can partly suppress the swell. On the contrary, in laboratory compacted soils have been ground, remolded and compacted, their original structure are destroyed, natural structure strength lows down and thus factors suppressing swell are removed, which make them easy to swell once reacting with water. Table 4, which well describe the microstructure differences between compacted and undisturbed soils can explain the phenomenon rather clearly. In order to get the Scanning Electronic Microscope (SEM) photographs, the freeze-vacuum-sublimation desiccator were employed to prepare the soil specimens. By doing so, specimens can well meet the microscopic test re-

Table 3. Material composition of investigated soils.

Sample No.	Clay mineral composition* %			Clay particle Content %
	I	I-Sm	Ch	
1	56	19	9	51.9
2	58	trace	19	56.1
3	39	35	20	50.3
4	32	trace	20	2.8

* I--illite, I-Sm--illite and smectite mixed-layer mineral, Ch--chlorite

Table 4. Comparison of microstructures between compacted and undisturbed soils.

Sample No.	Undisturbed soil	Compacted soil
1	turbulent	turbulent-orientation
2	aggregate	turbulent-orientation
3	matrix	turbulent
4	aggregate-skeletal	aggregate-skeletal

quirements. Additional details of the device are provided by Li (1992).

From Table 4 and Figures 3-8, it can be seen, except for Yellow silty soil, the microstructures of undisturbed soils have rearranged after compaction. The microstructures of compacted expansive soils are mainly of turbulent or turbulent-orientation aspects (Osipov, 1979) with higher structural particle orientation, and their pores distribute homogeneously and form rows of channels (in the case of Grey soil and Black soil). When soil immerses into water, water can penetrate easily along these channels which makes aggregated particles react with water sufficiently. As for the microstructures of undisturbed soils, they do not have the performance of high structural particle orientation, most of the pores are micro ones and they do not distribute homogeneously. So water penetrates into soils with difficulty. The comparison of the microstructures between those of compacted and undisturbed soils well explains this point due to "osmotic" swell. The main reason why the swell properties of Yellow silty soil do not differ much can be partly accounted to its little variation of microstructure prior and posterior to compact, therefore the size of structure elements does not change much either, and the swell properties of the undisturbed soil are close to that of the compacted one. Al-Homound & Basma (1995) studied the cyclic swell behaviors of clays and pointed out lower structural element orientation resulted in correspondingly lower water adsorption and swell ability. Seed et al. (1962) also showed the importance of the microstructure of compacted soil upon its swell characteristics.

Figure 3. The SEM photograph of Grey undisturbed soil (× 2000).

Figure 6. The SEM photograph of Black compacted soil (× 1300).

Figure 4. The SEM photograph of Grey compacted soil (× 2000).

Figure 7. The SEM photograph of Greyish-yellow silty undisturbed soil (× 500).

Figure 5. The SEM photograph of Black undisturbed soil (× 2000).

Figure 8. The SEM photograph of Greyish-yellow silty compacted soil (× 500).

6 LIME IMPROVEMENT

To reduce the swell, certain amount of lime was added into the compacted expansive soil specimens and homogeneously mixed. In this study, the percentage of lime in the mixture was 8%. Within laboratory environment under the conditions with 20°c

Table 5. Variation of swell index and unconfined compression strength of compacted soils before and after lime treatment.

Swell index	Sample No.	Pre-treatment	Post-treatment
	1	11.8	-0.4
δ_{ep50}	2	7.3	0.3
(%)	3	5.5	-0.3
	4	-0.5	-0.5
	1	collapse	893
q_u*	2	collapse	423
(kPa)	3	collapse	1191
	4	collapse	101
	1	high	non
Swell	2	high	low
grade classification	3	high	non
	4	non	non

*Unconfined compression strength after seven days soaking in water

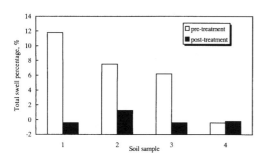

Figure 9. Comparison of total swell percentage between pre- and post-lime treatment of compacted expansive soil.

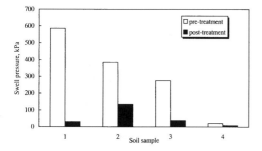

Figure 10. Comparison of swell pressure between pre-and post-lime treatment of compacted expansive soil.

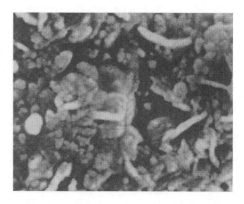

Figure 11. The SEM photograph of Grey soil after lime treatment (× 9000).

temperature and 22% humidity, twenty-four hours were allowed to pass for sufficient reaction between the lime and expansive soil specimens. To do the swell test, the water contents of the mixture were allowed to reach the optimum of the compacted soils respectively, and the dry densities of the specimens were also permitted to reach the maximum of the compacted soils respectively. For the unconfined compression test after seven days soaking in water, each specimen was formed into a Φ40×H50 mm cylinder.

From Table 5, Figures 9-10, it is clear that the improvement in the Grey soil is the best, with its swell pressure dropping to 33kPa, total swell percentage dropping to -0.4%, and the unconfined compression strength after seven days soaking in water increasing from 0 to 893 kPa. The improvement in the Black soil is relatively poorer, with its swell pressure dropping to 134kPa, and its total swell percent dropping to 1.3%. This soil therefore, still belongs to the category of expansive soil. The reason for this may be that the organic matter content of the Black soil is 1.5%, which is as much as 7 times that of the Grey soil. Among the soils, the improvement result in the Yellow silty soil is the poorest.

With sufficient reaction between the compacted expansive soils and the lime, a cation change between the adsorbed form on the soil particle surface and that existing in the dissolved lime causes the aggregation of the clay particles. The appearance of cemented fine needle lime among the clay particles, and a change in soil microstructure are also observed (Li et al. 1997). The microstructures become more dense compared with the untreated compacted expansive soils. This is also verified by SEM photographs. Figure 11, the SEM photograph of the Grey soil after the lime treatment well describes the phenomenon. The physical and mechanical properties of the compacted expansive soils are therefore improved.

7 CONCLUSIONS

Based on the analysis of the test data and photograph of SEM presented in this study, several conclusions can be obtained.

The breakdown of the cementing bonds is one of the reasons why compacted expansive soils show higher swell abilities and pressure compared with the undisturbed soils.

The obtained data certifies to the fact that the initial water contents play an important role in the swell abilities and pressure of expansive soils. The higher initial water content, the lower swell percentage and swell pressure.

Only those which contain hydrophilic clay minerals and clay particle contents of expansive soils may change into "artificially expansive" after compaction.

The SEM photographs of compacted expansive soils show higher structural element orientation as compared with those of undisturbed soils, and this rearrangement of the particles is another factor affecting the swell abilities and swell pressure.

In terms of practical applications, careful attention should be paid to such soils in engineering works. Improvement by lime treatment can ameliorate swell damage.

REFERENCES

Al-homoud, A. S. & Basma, A. A. (1995). Cyclic swelling behavior of clays. *J. Geotech. Engrg., ASCE.* 121(7):562-565.

Chen, F. H. (1965). The use of piers to prevent uplifting of lightly loaded structures founded on expansive soils. *Proc., Engrg. Effects of Moisture Changes in Soils, Int. Res. and Engrg. Conf. on Expansive Clay Soils.* Tex: Texas A & M press, College Station.

Day, R. W. (1994). Effective cohesion for compacted clay. *J. Geotech. Engrg., ASCE.* 118(4):611-619.

Day, R. W. (1994). Swell-shrink behavior of compacted clay. *J. Geotech. Engrg., ASCE.* 120(3):618-623.

Dif, A. E. & Bluemel, W. F. (1991). Expansive soils under cyclic drying and wetting. *Geotech. Testing J.* 14(1):96-102.

Gizienski, S. F. & Lee, L. J. (1965). Compaction of laboratory tests to small scale field tests. *Engineering Effects of Moisture Change in Soils; Concluding Proc., Int. Res. and Engrg. Conf. on Expansive Clay Soils.* Tex: Texas A & M press, College Station.

Li, S. L. et al. (1992). *Study on the engineering geology of expansive soils in china.* Nanjing: Jiangsu Science and Technology Publishing House.

Li, S. L. et al. (1997). Study on the engineering properties of expansive soils in China. *Ziran Zazhi.* 19(2):82-86.

Low, P. F. (1987). Structural component of the swelling pressure of clay. *Langmuir,* 3:18-25.

Low, P. F. (1992). Interparticle forces in clay suspensions: Flocculation, viscous flow and swelling. *Proceedings of the 1989 Clay Minerals Society Workshop on the Rheology of Clay/Water Systems.*

Osipov, V. I. (1979). *Nature of strength and deformation properties of clayey soils and rocks.* Moscow: Moscow University Publishing House.

Osipov, V.I.et al. (1987). Cycling swelling of clays. *Appl. Clay Sci.* l2(7):363-374.

American Society for Testing and Materials (1991). Test method for laboratory compaaction characteristics of soil using standard effort; Designation D698. *Annual book of ASTM standards.* 04. 08:165-172. Philadelphia, Pa..

Seed, H. B. et al. (1962). Prediction of swelling potential for compacted clays. *J. Soils and Foundations Division, ASCE.* 88(3):53-87.

Tahir, A. A. (1992). Collapse mechanisms of low cohesion compacted soils. *Bulletin of the Association Engineering Geologists.* 16(4):345-353.

Unsaturated Soils for Asia, Rahardjo, Toll & Leong (eds) © 2000 Taylor & Francis, ISBN 90 5809 139 2

Effect of mode of wetting on the deformational behavior of unsaturated soils

M.A. El-Sohby & M.M. Aboutaha
Department of Civil Engineering, AL-Azhar University, Cairo, Egypt

S.O. Mazen
Building Research Center, Egypt

ABSTRACT: Theoretically, there are two limiting conditions during wetting of expansive soils. These are: The completely unconfined condition in which the soil is allowed to swell freely along all directions; and the completely confined condition in which no overall volume change is allowed. In the present work, to simulate field conditions, partial loading was considered using two modes of wetting, the upward and downward water supply. Partial loading has been achieved experimentally using the oedometer by loading one plate of a fixed area on different specimen areas or by loading plates of different areas on a fixed specimen area. The upward water supply has been achieved by providing one filter paper and one porous plate under the specimen only. In case of downward water supply a steel plate was placed under the specimen and the free unloaded surface area surrounding the loading plate was provided with aluminum porous plate and filter paper. Test results indicated that, the mode of wetting has a significant effect on swelling characteristics, particularly swelling pressures and surface heave.

1 INTRODUCTION

Moisture is an integral part of clay soil. Their properties are affected by the total water content and by the energy with which the moisture is held. The force with which moisture is held has been defined by Bower (1956, p.227) as the force required to pull a unit mass of water away from a unit mass of soil.

Moisture migration may be directly responsible for changes in volumetric or linear dimensions of clay (Mitchel, 1976). On moisture uptake there is generally a volume increase and moisture loss is accompanied by shrinkage. This is very significant in case of expansive clay soils where the regime of water uptake affects the stress situation in the foundation soil. Moisture may migrate when the soil is either saturated or unsaturated. It moves downward under gravitational flow and moves upward under the influence of potential gradient against the force of gravity until equilibrium is established.

Since expansive clay soil deposits are not saturated, the water uptake resulting from access of water in the ground has the character of unsaturated flow. The regime of this uptake is of great practical significance since it affects the characteristics of the clay soil as well as the development of swelling presssures which affects the stress situation in the foundation soil. Therefore, in the present work some features of the unsaturated flow phase due to the effect of two modes of wetting are presented.

2 MATERIAL USED

Soil used was taken from an arid area called Nasr City at the geological boundaries of Cairo and to the north east of it. The site is adjacent to the Faculty of Engineering of Al-Azhar University and the soil represent desert clay deposit known for its expansive behavior.

The soil has a light grey colour. It has a high dry strength which is quickly lost when flooded in water. Its bulk unit weight was found to be 2.11 gm/cm3 and natural water content is 14.2%. Particle size analysis indicated that the clay content is 51%, silt and sand content are 43% and 6% respectively. The liquid limit is 78%, plasticity index is 47% and activity is 0.92. X-ray diffraction analysis indicated that the predominate clay mineral is montmorillonite.

According to the chart suggested by Van der Merwe (1975), the studied soil is classified as highly expansive.

3 PROGRAM OF INVESTIGATION AND TECHNIQUE

For an expansive soil there are two limiting conditions during wetting. These are the completely unconfined condition in which the soil is allowed to swell freely along all directions and the completely confined condition in which no overall volume change is allowed. Therefore, in the present work, to simulate field conditions, swelling pressures and surface heave were measured using partial loading and two different modes of wetting. The first mode of wetting, denoted Mode 1, is the upward water supply, and the second mode of wetting, denoted Mode 2, is the downward water supply (See Fig. 1).

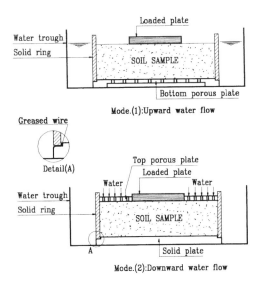

Mode.(1):Upward water flow

Mode.(2):Downward water flow

Fig. 1   Diagrammatic sketch showing the applied testing technique.

3.1 *Partial loading*

Partial loading has been achieved experimentally using the oedometer by loading one plate of a fixed area on different specimen areas or by loading plates of different areas on a fixed specimen area. In both cases, the specimens were laterally confined and partially loaded in the vertical direction.

3.2 *Mode of wetting*

The upward water supply (Mode 1) was achieved by providing one filter paper and one porous plate below the specimen only. In case of downward water

supply (Mode 2), a steel plate was placed below the specimen and the free unloaded surface area surrounding the loaded plate was provided with aluminium porous plate and filter paper which were of internal diameters equal to the loaded plate diameter while their external diameters were equal to specimen diameter.

3.3 *Specimen preparation*

Remolded soil specimens were used in the oedometer prepared as follows:

1. A predetermined weight of oven dried soil sample pulverized and passing sieve number 40 was used to get the required initial dry density.

2. Calculated amount of water was added to the predetermined weight of soil and mixed carefully to get the required initial water content. Then the mixed soil was kept in air tight container for 24 hours to allow for uniform distribution of moisture.

3. After lubricating the mould by high vacuum grease, the soil sample was then carefully powered into it at three layers and dynamically compacted until it reaches the height which gives the required density.

4. The final height of specimens was taken 3 cms and kept constant in all types of testing.

4 DETERMINATION OF SWELLING PRESSURE

In the present work the concept of constant volume method previously used (e.g. Porter and Nelson, 1980; Tisot and Aboushook, 1983) was adopted. However, the method was modified by loading the expansive soil specimen partially in its mold. Therefore, the swelling pressure can be defined as the pressure required to prevent the rise of a foundation model which is partially loaded on a soil specimen surface when the specimen is supplied with water.

4.1 Swelling pressure at constant specimen diameter

Specimens of diameter 17.7 cm with height of 3 cm were used. These soil specimens were loaded by plates of different diameters (dp) to get variable surface confinement percent (S.C%). The surface confinement (S.C.) is defined as the ratio between the loaded plate diameter to the specimen diameter. Results obtained when supplied with water using Mode 1 and Mode 2 are given in Table 1 and the relationship between swelling pressure and the surface confinement is shown in Fig. 2.

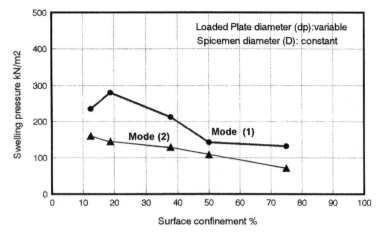

Fig. 2   Relationship between swelling pressure and surface confinement for different modes of wetting

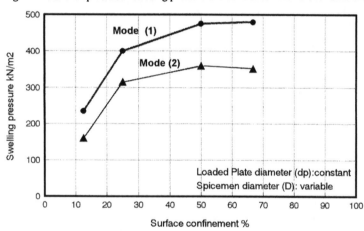

Fig. 3   Relationship between swelling pressure and surface confinement for different modes of wetting

Table 1. Swelling pressure at constant (D).

dp	S.C.	Swelling Pressure KN/m2	
(cm)	(%)	Mod 1	Mod 2
6.3	12.5	235	160
7.71	18.8	280	145
10.91	37.8	212	129
12.68	50.0	142	109
15.43	75.0	132	71

4.2 Swelling pressure at constant plate diameter

Another series of tests were done by using a plate of fixed diameter (dp=6.3 cm) and specimens of different diameters (D). The values of swelling pressures at different surface confinement (S.C.%) using the two modes of wetting are given in Table 2 and the relationship between swelling pressure and surface

confinement for the two different modes wetting are given in Fig. 3.

Table 2. Swelling pressure at constant (dp).

D	S.C.	Swelling Pressure KN/m2	
(cm)	(%)	Mod 1	Mod 2
17.74	12.50	235	160
12.68	25.00	400	315
8.88	50.00	475	359
7.71	66.70	480	351

4.3 Swelling pressure at constant (S.C. %)

A third series of test was done in which the swelling pressure was measured at constant surface confinement. This was verified using specimens as well as plates of different diameters as shown in Table 3. The relationship between swelling pressure, specimen d - ameter and loaded plate diameter for the two diffe - ent modes of wetting are given in Fig. 4.

Table 3. Swelling pressure at constant (S.C. %)

D (cm)	dp (cm)	Swelling Pressure KN/m2	
		Mod 1	Mod 2
8.88	6.30	475	359
12.68	8.88	360	281
17.74	12.68	142	109

5 DETERMINATION OF SURFACE HEAVE

The surface heaves surrounding the loaded plate were measured simultaneously with the swelling pressure. Three dial gauges were placed along the aluminum porous plate for their measurement and the average value was denoted as the average surface heave. As in the case of swelling pressure, the average surface heave was determined at different boundary conditions.

5.1 Surface heave at constant specimen diameter

Table 4 gives the values of average surface heave measured at a constant specimen diameter D=17.7 cm. The relationship between average surface heave and percentage of surface confinement for different modes of wetting at a constant specimen diameter is indicated in figure 5.

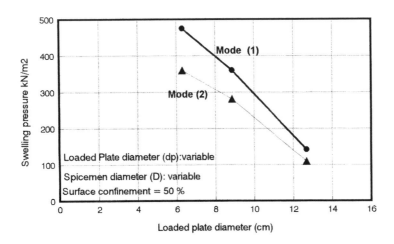

Fig. 4   Relationship between swelling pressure and loaded plate diameter for different modes of wetting

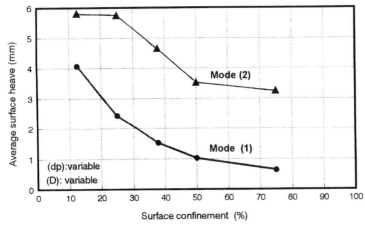

Fig. 5   Relationship between surface heave and surface confinement (%) at different modes of wetting

Table 4. Surface heave at constant (D)

dp	S.C.	Average Surface Heave(mm)	
(cm)	(%)	Mod 1	Mod 2
6.3	12.50	4.07	5.82
8.88	25.00	2.43	5.76
10.91	37.80	1.53	4.66
12.68	50.00	1.02	3.54
15.43	75.00	0.64	3.26

5.2 Surface heave at constant plate diameter

Table 5 gives the values of average surface heave measured at constant plate diameter dp=6.3cm and specimens of different diameter (D). Figure 6 indicate the relationship between average surface heave and percentage of surface confinement for different modes of wetting at a constant plate diameter dp=6.3cm.

Table 5. Surface heave at constant (dp)

D	S.C.	Average Surface Heave(mm)	
(cm)	(%)	Mod 1	Mod 2
17.74	12.50	4.07	5.82
12.68	25.00	2.81	5.37
8.88	50.00	1.46	4.56
7.71	67.00	1.26	3.76

5.3 Surface heave at constant surface confinement

The third series of tests are given in Table 6. Where the values of average surface heave were measured at

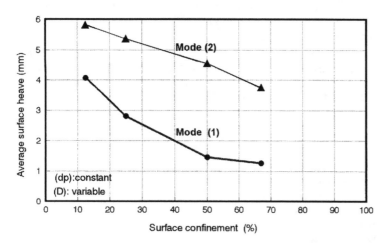

Fig. 6   Relationship between surface heave and surface confinement (%) at different modes of wetting

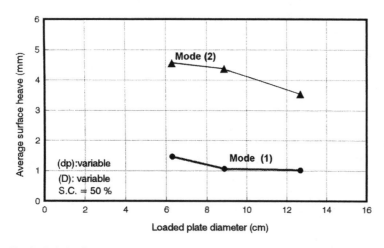

Fig. 7   Relationship between surface heave and loaded plate diameter at different modes of wetting

constant surface confinement. This was done by varying the specimen diameter and the plate diameter in the same time. Figure 7 indicate the relationship between average surface heave, loaded plate and specimen diameter for different modes of wetting at constant surface confinement.

Table 6. Surface heave at constant S.C %

D	S.C.	Average Surface Heave(mm)	
(cm)	(%)	Mod 1	Mod 2
8.88	6.30	1.46	4.56
12.68	8.88	1.06	4.37
17.74	12.68	1.02	3.54

6 ANALYSIS OF TEST RESULTS

From the previous tables and figures, we observe the following:

1. Swelling pressure values obtained when upward water supply (Mode 1) was used are higher than those when downward water supply (mode 2) was used. This was found for all different boundary conditions as indicated by figures 2,3 and 4. On the other hand, specimens wetted using Mode 2 gave the higher surface heave value than those wetted using Mode 1 (see figures 5,6,7).

2. When the loaded plate diameter was kept constant during testing, the measured swelling pressure was found to be directly proportional to the percentage of surface confinement (ratio between loaded plate diameter to specimen diameter x 100) as shown in figure 3. However for a constant percentage of surface confinement the measured swelling pressure was found to be inversely proportional to loaded plate diameter (Fig. 4). When the specimen diameter was kept constant, there were slight variations in the values of swelling pressure with the variations of loaded plate diameter (Fig 2). These relationships are non-linear and have similar trend for both modes of wetting.

3. At constant surface confinement (S.C.=50%) the measured swelling pressure was found to be inversely proportional to loaded plate diameter (Fig. 4). However the surface heave values were almost unchanged for different plate diameters (Fig. 7). This trend was observed for both modes of wetting.

7 CONCLUSION

The study indicated that the mode of wetting has a significant effect on the behavior of expensive clay soils. The swelling characteristics of the tested soils was found vary with the variation of the mode of wetting.

REFERENCES

Baver, L.D. 1956. Soil Physics, John Willley and Sons, Inc., New York.

El-Sohby, M.A. & M.M. Aboutaha 1995. Effect of surface confinement on swelling behavior of soils. 1 st int. Conf. on Unsaturated soils, Paris.

Mitchell, J.K. 1976 Fundamentals of Soil Behavior. John Wiley and Sons, New York.

Porter, A.A. & J.D. Nelson, 1980. Strain Controlled Testing of Expansive Soils. Proceedings, 4th International Conference on Expansive Soil. Denver, Colorado, Vol. 1, pp. 34-44.

Tisot J.P. & M.I. Abou Shook, 1983. Triaxial Study of Swelling characteristics. Proceedings, 7th Asian Regional Conference on S.M.F.E. Haifa, Vol. 1, p. 94-98.

Van der Merwe, D.H. 1975. Contribution to Speciality Session B" Current theory and Practice for Building on Expansive Clays, Proceedings, 6th Regional Conference for Africa on S.M.F.E. Duban, Vol.2, pp. 166-176.

Unsaturated Soils for Asia, Rahardjo, Toll & Leong (eds) © 2000 Taylor & Francis, ISBN 90 5809 139 2

An experimental study of elastic volume change in unsaturated soils

A.R. Estabragh, A.A. Javadi & J.C. Boot
Department of Civil and Environmental Engineering, University of Bradford, UK

ABSTRACT: In this paper the results of an experimental study on elastic volume change and shear strength behaviour of unsaturated compacted silty soil are presented. A comprehensive set of laboratory experiments has been carried out in a double-walled triaxial cell. In the experiments the soil samples were subjected to isotropic consolidation followed by unloading and subsequent isotropic or anisotropic reloading under constant suction. Furthermore, a number of experiments have been undertaken with constant net stress and varying suction. Volume change and shear strength data for the samples have been monitored continuously during the experiments. The test results have been used to study volume change and shear strength of unsaturated silty soils under various type of loading. The influence of suction on critical state shear strength has also been investigated. Comparison has been made between the experimental results and the theoretical estimates using relationships proposed in the literature.

1. INTRODUCTION

It is generally accepted that the mechanical behavior of unsaturated soils cannot be described in terms of a single effective stress (Jenning and Burland (1962) and Wheeler and Karube (1996)). The difficulty of using a single effective stress for unsaturated soils has led to the use of two stress state variables: net stress (the excess of total stress over pore air pressure) and matrix suction (the difference between the pore air pressure and pore water pressure). Many authors such as Matyas and Radhakrishna (1968), Fredlund and Morgenstern (1977), Llort and Alonso (1985) and Escario and Saze (1986) used the concept of two independent stress state variables to describe the behavior of unsaturated soils. In the three dimensional case four stress parameters are required to explain the stress state at any point within an unsaturated soil. For axisymmetric conditions the number of stress parameters reduces to a total of three, which are usually chosen as mean net stress (p), deviator stress (q) and suction ($u_a - u_w$) defined as:

$$p = \frac{(\sigma_1 + 2\sigma_3)}{3} - u_a \qquad (1)$$

$$q = \sigma_1 - \sigma_3 \qquad (2)$$
$$s = u_a - u_w \qquad (3)$$

where σ_1 and σ_3 are axial and radial total stresses and u_a and u_w are pore air pressure and pore water pressure respectively. Alonso, et al. (1987) presented an elasto-plastic critical state model in a qualitative form for unsaturated soil. The mathematical development of this model was presented by Alonso, et al. (1990). The model reduces to the modified Cam Clay critical state model for the fully saturated case. Wheeler and Sivakumar (1995) suggested a modification to this model based on the results of tests on compacted kaolin under different suctions. In the proposed method a loading-collapse curve is defined which is a part of the yield surface. It is assumed that only elastic strains occur if the state of stress in the soil remains inside a yield surface. Plastic volumetric and shear strains commence once the yield surface is reached. The elastic strain inside the yield surface is related to the changes of load (p) and suction (s). An increase in load (p) or decrease in suction (s) produces elastic deformation. Alonso, et al. (1990) suggested the following equation for the elastic variation of specific volume (v):

$$dv^e = \kappa \frac{dp}{p} - \kappa_s \frac{ds}{(s + p_{at})} \qquad (4)$$

where κ is the gradient of the unloading lines in $v:\ln p$ space, κ_s is the gradient of similar lines in the $v:\ln s$ and p_{at} is the atmospheric pressure. The two components of the elastic varation of v given by equation (4) are produced by elastic deformation of the soil particles. The elastic deformation is caused by varation of the inter-particle forces resulting from changes of either applied net stress or suction within water meniscus at the particle contacts. Wheeler (1997) highlighted the shortcomings of equation (4) in predicting the elastic volume change of unsaturated soils and concluded that this equation is not satisfactory in several aspects. Based on consideration of the physical mechanism responsible for elastic strains in soils, he proposed an equation for elastic deformation as follows:

$$dv^e = k\frac{d(p + f(s))}{(p + f(s))} \qquad (5)$$

where $f(s)$ is a function that shows the contribution of suction in increasing inter- particle forces and hence causing elastic deformation. The form of $f(s)$ is due to the gradual transition from bulk water at low suctions to a meniscus water at high suctions.

He suggested the following theoretical expression for the function $f(s)$:

$$f(s) = \alpha\tanh(\frac{s}{\alpha}) \qquad (6)$$

where α is a constant. Equation (5) overcomes the shortcomings of equation (4) and is consistent with the physical mechanism of elastic strains in unsaturated soils.

Considering the fact that the function $f(s)$ in equation (6) represents the effect of the inter-particle forces, it has also been suggested that $f(s)$ can be used to describe other aspects of the behavior of unsaturated soils, such as shear strength, which depend on the inter-particle normal forces. Karube (1988) suggested that the appropriate stress parameters for shear strength of unsaturated soil could be q (deviator stress), p (mean net stress) and $f(s)$ as follows:

$$q = M(p + f(s)) \qquad (7)$$

Wheeler (1997) states that the confirmation of the proposed equation requires a substantially large body of supporting experimental evidence.

In this paper, the application of equation (5) in prediction of elastic volume change and shear strength of unsaturated soils is examined in the light of a comprehensive experimental study. A series of tests under isotropic and anisotropic conditions with controlled suction were carried out on samples of silt. An analytical expression is presented for the function $f(s)$ in the form of equation (6) with α being a parameter of the soil.

Comparison has been made between experimental results and predictions using equation (5). It is shown that with the proposed modification in equation (6), satisfactory predictions can be obtained for the elastic deformations of unsaturated soils due to variation of load and suction.

2. EXPERIMENTAL STUDY

2.1 *Soil properties*

The soil used in the testing program was a low-plasticity silt. The soil consists of 5% sand, 90 % silt and 5 % clay. The soil has a liquid limit of 29 % and plasticity index of 19%

2.2 *Sample preparation*

Triaxial Samples 38 mm in diameter and 76 mm high were prepared by static compaction in a mould at water content of 10% (4% less than the optimum from the standard compaction test). All samples were compacted in nine layers, with each layer being compacted in an Instron machine at a fixed displacement rate of 1.5 mm/min, to a maximum vertical total stress of 1600 *kPa*. All samples were compacted in an identical fashion in order to produce the same initial soil fabric in each test.

2.3 *Experimental apparatus*

In order to measure the volume change of a sample a Bishop and Wesley triaxial cell was modified to a double-wall cell. Pore water pressure (u_w) was applied at the base of the sample through a high air entry disc. Pore air pressure (u_a) was applied at the top of the sample through a low air entry disc. The method of axis translation proposed by Hilf (1956) was used in order to create the desired suction in the sample. Four GDS controllers were connected to the apparatus and were controlled by a computer. The controllers were used to apply the pressures in the inner and outer cells, the pore water pressure (u_w) and the axial stress. All the experimental data were recorded continuously by the computer.

2.4 *Experimental procedure*

2.4.1 *Equalization*

After setting the apparatus, the first step of each test was equalization. At this stage, by applying the required air and water pressure, the sample was brought to the desired suction.

2.4.2 *Ramp consolidation and unloading*

After the equalization step each sample was consolidated isotropically to the pre-selected value of mean net stress (*p*) by ramp consolidation. The sample was then unloaded isotropically to a predefined lower value of mean net stress. Pore water and pore air pressures were maintained constant during the loading and unloading.

2.4.3 *Shearing*

On completion of ramp unloading, each sample was sheared by applying an increasing axial stress. Shearing was conducted at a constant rate of $2\mu m/\min$.

2.4.4 *Wetting and drying tests*

A number of tests were also carried out to study the behavior of soil due to variation of suction. For these tests, after the stages of isotropic consolidation and unloading, the soil suction was reduced continously to a pre-determined value. It was then increased to the initial value while the mean net stress was kept constant.

3. RESULTS AND DISCUSSION

Two main sets of experiments were carried out on the soil samples. In the first set, the mean net stress was increased and then decreased at constant suction. In the second set, the suction was increased and then decreased keeping the mean net stress constant. At the end of the unloading stage, the samples were sheared under drained conditions.

The results of the loading and unloading tests at constant suctions of 0, 50, 100, 200 and 300 *kPa* are shown in Figure 1. Analysis of the results shows that the value of κ is nearly constant for different values of suction ($\kappa = 0.013$).

Figure 2 shows the volume change of the soil resulting from the variation of suction at constant values mean net stress. In the first test (Figure 2a) the value of suction has been increased continuously from 50 to 300 *kPa* while in the second test (Figure 2b) the suction has been decreased from 300 to 100 *kPa*. The mean net stress was kept at a constant value of 20 *kPa* for both tests. The value of κ_s is obtained from the results ($\kappa_s = 0.0008$).

There is evidence that an increase in suction over conditions not previously reached by a partially

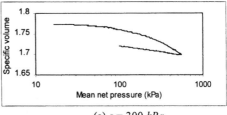

(a) $s = 300\ kPa$

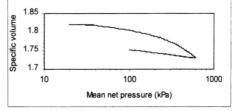

(b) $s = 200\ kPa$

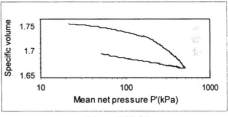

(c) $s = 100\ kPa$

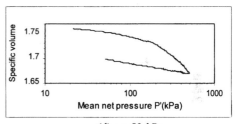

(d) $s = 50\ kPa$

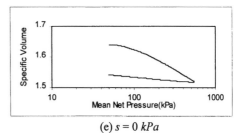

(e) $s = 0\ kPa$

Figure 1. Volume change of the soil in loading and unloading tests at constant values of suction.

saturated soil induces irrecoverable volumetric strain (Alonso, et al. 1987). This concept is very similar to that of pre-consolidation pressure for soils.

Therefore, following the same procedure as for determination of pre-consolidation pressure, the maximum suction to which the soil has been subjected during its stress history, (in this paper shown by the symbol s_c) can be obtained. For the soil tested, the value of s_c was obtained from Figure 2a as 190 kPa.

Figure 3 shows the critical state lines obtained from drained shear tests on the samples at three different suction values of 0, 200 and 300 kPa.

The average gradient of the critical state line (M) for the saturated case is 1.2. It is interesting to note that the critical state lines are very close for the suction values of 200 and 300 kPa. This indicates that as suction increases, the shear strength of unsaturated soils tends to a limiting value.

From the analysis of the experimental data a function $f(s)$ was proposed in the form:

$$f(s) = s_c \tanh\left(\frac{s}{s_c}\right) \qquad (8)$$

This function together with the equation (5) was used to estimate the elastic volumetric strains for the samples. The only parameter in equation (8) is s_c which can be obtained from triaxial tests by increasing the suction at constant mean net stress.

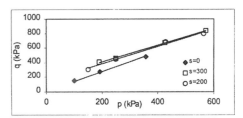

Figure 3. Critical state lines for different suctions

Figure 4 shows a comparison of the results for the estimated and measured elastic volumetric strains for different tests. The results show that equation (5) with the function $f(s)$ defined by equation (8) gives very good estimates of the actual elastic deformations.

Following the procedure proposed by Karube (1988), and using equation (7), the values of $f(s)$ corresponding to suctions of 0, 200 and 300 can be obtained from Figure 3. The best fit lines shown in Figure 3 correspond to equation (7), with $f(s)$ values of 119 and 123 kPa at suctions of 200 and 300 kPa respectively. The corresponding values of $f(s)$ predicted from equation (8) at the same suctions are 152 and 181 kPa respectively.

Comparison of the results shows that the analytical expression given by equation (8) does not seem to provide a satisfactory prediction for the critical state conditions of the soil. It is probable that one or more parameters should be introduced in equation (8) to account for the critical state of soils.

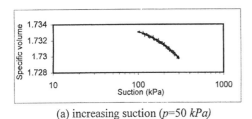

(a) increasing suction (p=50 kPa)

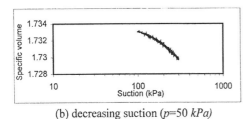

(b) decreasing suction (p=50 kPa)

Figure 2. Volume change of the soil due to variation of suction at constant net stress.

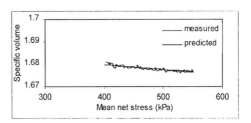

(a) s= 300 kPa

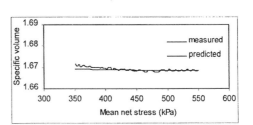

(b) s =200 kPa

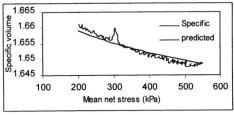

(c) $s = 100$ kPa

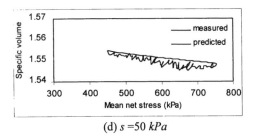

(d) $s = 50$ kPa

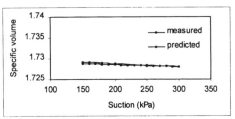

(f) Suction varied at constant p

Figure 4. Comparison of the predicted and measured elastic volumetric strains

4. CONCLUSION

The influence of suction on elastic volume change and critical state of unsaturated soils has been investigated by undertaking a number of triaxial tests.

A comprehensive set of triaxial tests has been carried out on a low-plasticity silt and the results have been used to examine the theoretical expressions given in the literature for the effect of suction on elastic volume change and shear strength of unsaturated soils. A modification to a theoretical relationship has been proposed which gives a very good prediction for elastic volume change.

Validation of a theoretical relationship requires a large amount of experimental data for various types of soil under different loading conditions. However, the proposed function $f(s)$ in equation (5) seems to predict elastic volumetric strains with good accuracy

at least for silty soils. Adaptation of this equation to extend its range of application up to the critical state requires further investigation.

REFERENCE

Alonso, E.E, Gens, A. and Hight, D.W. (1987). General Report: Problem soils. 9th. *ECMFE*, Dublin, Vol. 3, pp. 1087-1146.

Alonso, E.E., Gens, A. and Josa, A. (1990). A constitutive model for partly saturated soil. *Geotechnique* Vol. 40, No. 3, pp. 405-430.

Escario, V. and Saez, J. (1986). The shear strength of partly saturated soils. Technical Note, *Geotechnique* Vol. 36, No. 3. Pp. 453-456.

Fredlund, .G. and Morgenstern, N.R. (1977). Stress state variables for unsaturated soils. *Journal of the Geotechnical Engineering, Proc.ASCE*, Vol. 103, No. GT5, pp. 447-465.

Hilf, J.W. (1956). An investigation of pore pressure in compacted cohesive soils. *Technical Memorandum No. 654, Bureau of Reclamation, USDI*, Denver, Colorado, USA.

Jennings, J.E. and Burland, J.B. (1962). Limitation to the use of effective stresses in partly saturated soils. *Geotechnique*, Vol. 12, No. 2, pp. 125-144.

Karube, D. (1988). New concept of effective stress in unsaturated soil and its proving test. *Advanced Triaxial Testing of Soil and Rock*. (eds. R.T. Denaghe, R.C. chaney and M.L. Silver), ASTM STP977, pp. 539-552.

Lloret, A. and Alonso, E.E. (1985). State surface for partially saturated soils. 11th. *ICSMFE*, Sa Fransisco, pp. 557-562.

Matyas, E.L and Radhakrishna, H.S. (1968). Volume change characteristics of partially saturated soils. *Geotechnique*, Vol. 18, No.4, pp432-448.

Wheeler, S.J. and Sivakumar, V. (1995). An elastoplastic critical state framework for unsaturated soils. *Geotechnique*, Vol. 45, No.1.pp. 35-53.

Wheeler, S.J. and Karube, D. (1996). State of the art report on constitutive modelling. *Pro. 1st Conf. Unsaturated Soils*, Paris, Vol. 3.

Wheeler, S.J. (1997). Modelling elastic volume change of unsaturated soil, Proceedings of the 14th *Int. Conf. Soil Mechanics and Foundation Engineering*, Hamburg, 1997, pp. 427-430.

Unsaturated Soils for Asia, Rahardjo, Toll & Leong (eds) © 2000 Taylor & Francis, ISBN 90 5809 139 2

The influence of fabric and mineralogy on the expansivity of an Australian residual clay soil

S. Fityus & D.W. Smith
Department of Civil, Surveying and Environmental Engineering, The University of Newcastle, Callaghan, N.S.W., Australia

ABSTRACT: Preliminary results of a study of a residual clay profile developed in a temperate climate are presented. The study focuses on the mineralogy of the soil profile and its relationship to the expansive potential of the clay soils it contains. The soil mineralogy is quantitatively and semi-quantitatively assessed on the basis of two sets of xray diffraction results, which indicate a consistent dominance of randomly interstratified illite and smectite minerals throughout the profile. The soil mineralogy is compared with the expansive potential of the clay soils with respect to depth, assessed in terms of a measured shrink swell index, which is found to decrease consistently with depth. This reduction in expansive potential in a soil profile of consistent mineralogy is attributed to a consistent change in the fabric of the soil with increasing depth, which is due to an associated gradual decrease in the degree of weathering. The high proportion of smectite clay minerals in these soils explains why significant expansive soil movements are observed in an area where the influence of climate and the depth of clay soil only extend to the relatively shallow depth of 1.7m.

1. INTRODUCTION

Residual soil profiles develop from the in situ weathering of rock. They have particular characteristics which are specific to their origin, and which differ from those of soil deposits of other origins, such as alluvial, loessial and colluvial soils (Wesley, 1990). Rocks exposed at the earth's surface may weather to form residual soils of a clayey nature, although the mineralogy and distribution of clays within residual soil profiles will vary widely depending on parent rock type, climatic conditions and other factors (Velde, 1995; Blight, 1997)).

Residual soils usually occur in areas of elevated topography; that is, some height above sea level. As areas underlain by residual soils are generally less prone to serious flooding, they are often preferred as sites for human settlement. The properties of residual soils are thus of considerable interest, in that they affect the performance of the structures they support, and hence the everyday lives of countless numbers of people.

An important characteristic of residual clay soils is that they are usually only partially saturated, at least at shallow levels, and the degree of saturation can vary greatly under climatic influences. Where clays contain significant proportions of expansive clay minerals, such as sodium montmorillonite, they may respond to seasonal moisture changes by shrinking and swelling (Nelson and Miller 1992). The degree to which they

shrink and swell depends on many factors, including the depth of clay soil present and the severity of the climate For example, Dhowian (1990) reports that ground surface movements of up to 160mm are observed in a deep residual profile derived from shales in the arid climate of Saudi Arabia.

Efficient and effective foundation designs on expansive clay soils require accurate estimates of likely ground movements. An understanding of the relationship between expansive potential and soil fabric is thus crucial.

The aim of the work presented here was to examine the characteristics of a particular residual clay soil profile which is indicative of those underlying many of the populated areas of eastern Australia. The soil profile chosen for this study is considered important in that it realises substantial expansive movements in response to seasonal ground moisture changes, despite the existence of rock structure at shallow depths, and despite a relatively shallow depth of seasonal moisture influence.

2. SITE DETAILS.

The soil profile selected as the basis of this study is the Maryland soil profile from the Maryland Reactive Soils field site near Newcastle in Australia. Detailed descriptions of the Maryland site can be found in Allman. Smith and Sloan (1994) and Allman, Delaney, and

Smith (1998). Only details relevant to the present discussion will be presented here.

The expansive soils field monitoring site at Maryland, is on a gently sloping hillside, with a healthy cover of grass and occasional large trees. The region has a temperate, near coastal climate with an annual rainfall typically between 1000 and 1200mm per year.

The residual clays on the site are derived from carbonaceous siltstone parent rocks, with a small tuffaceous component. They lie between about 250mm of relatively non-expansive topsoil and weathered rock at about 1.7m. The Maryland expansive soil profile is described in Figure 1.

It is evident from Figure 1 that the fabric of the soil varies significantly with depth. Directly beneath the topsoil, the residual clay is of a uniform orange colour with no apparent structure, apart from desiccation cracking when in a dry state. With increasing depth, it becomes mottled, before becoming pale grey and exhibiting some relict rock structure, and then ultimately becoming weathered rock. The distinct colour changes from orange to grey with increasing depth indicate that more oxidising conditions prevail at shallow depths. This suggests that the degree of weathering might reflect a degree of chemical alteration throughout the profile.

The expansive potential of soils within the profile is not evident from soil profile observations.

Over the past 6 years, significant ground surface movements have been recorded at the field site. Ground movements of 47mm have been measured at the surface level control marker, with the average maximum movement for all site surface level markers being 58mm, and a range of movements of between 43 and 75mm (Fityus, 1999). The active depth is observed to be around 1.7m (Fityus and Delaney, 2000).

3. EXPANSIVE POTENTIAL.

The expansive potential of the clay soils with respect to depth was assessed in two ways. Firstly, it was measured directly by the shrink-swell testing of undisturbed cores (AS 1289.7.1.1-1992).

Secondly, it was assessed indirectly, by measurement of the relative surface area of the soil particles using methylene blue dye adsorption (Fityus, Smith and Jennar, 2000). The results are presented in Figure 2.

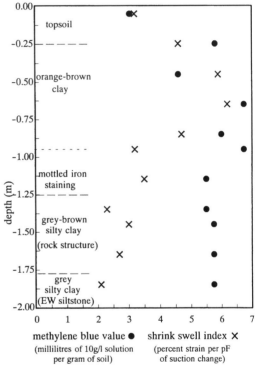

Figure 2. Assessment of expansiveness with depth.

From Figure 2, it is evident that the potential for physical shrinkage, as assessed by the shrink-swell index, decreases with depth, whereas the potential for dye adsorption does not. The decreasing trend in the physically measured shrink swell index is corroborated by a similar trend in the physically measured volume change index, as presented in (Fityus and Smith, 1998).

4. MINERALOGY.

The mineralogy of the Maryland residual soils has been assessed by both quantitative and semi-quantitative xray diffraction testing. The results are presented in Table 1.

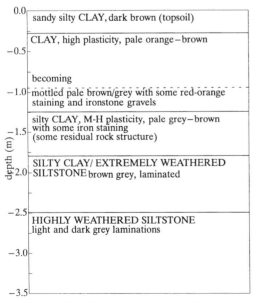

Figure 1 The Maryland expansive soil profile.

658

Table 1. Mineralogy of the Maryland clay soil profile using quantitative and semi-quantitative xray diffraction techniques.

Depth (m)	Whole soil sample							Sub 2 micron fraction		
	kaolinite	remaining clay (not differentiated)	quartz	muscovite and clay micas	feldspars (mostly plagioclase)	haematite and goethite	anatase	kaolinite	randomly interlayered smectite and illite	smectite + minor interlayered illite
0.15	Tr	D	SD	-	Tr	Tr–A	Tr	Tr–A	D	-
0.45	12%	42%	38%	4%	<1%	<1%	<1%	16%	51%smectite+7%Illite	
	Tr–A	D	SD	-	Tr	Tr	Tr	Tr–A	D	-
0.55	Tr–A	D	SD	_	Tr–A	-	Tr	Tr–A	D	-
0.75	Tr–A	D	SD	Tr	Tr–A	-	Tr	A	D	Tr
0.95	12%	47%	31%	5%	4%	<1%	<1%	13%	36%smectite+21%Illite	
1.05	A	D	SD	Tr–A	Tr–A	Tr	Tr	A	D	Tr
1.25	16%	42%	29%	5%	5%	<1%	<1%	15%	36%smectite+20%Illite	
1.55	A	D	SD	Tr–A	Tr–A	-	-	A	D	Tr–A
1.95	16%	45%	27%	6%	6%	<1%	<1%	19%	43%smectite+14%Illite	
	A	D	SD	Tr–A	Tr	-	-	A–SD	D	Tr–A
3.0	A	D	SD	Tr–A	Tr	-	-	A–SD	D	Tr–A

The semi-quantitative abbreviations used in Table 1 are:

    D = Dominant; used for the component which is most abundant, regardless of its proportion,

    SD = Sub-dominant; the next most abundant component(s), provided its proportion is greater than about 20%

    A = Accessory; present in a proportion between about 5% and 20 %,

    Tr = Trace; present, but in a proportion less than about 5%.

The analyses from which these results are derived were conducted by two different laboratories using different preparation and interpretation techniques, As such, the form of the results varies according to the methods employed.

It is apparent from Table 1 that the quantitative and semi-quantitative results are in good agreement. Their significance is discussed in the next section.

5. DISCUSSION.

An obvious yet important observation can be made in relation to the site data presented in section 2: that is, that movements of between 43 and 75mm are occurring within the 1.45m of residual clay between the topsoil and the active depth. This is a relatively substantial movement in a relatively small thickness of clay.

Reference to Table 1 confirms that the soils at all levels beneath the topsoil have a large component of smectite clay, which is amongst the most expansive of the clay minerals. This high proportion of potentially expansive clay is consistent with the large movements recorded in the Maryland soil profile.

Assessment of the data in Table 1 in conjunction with the soil profile details in Figure 1 reveals that the mineralogical variations within the soil profile are much less significant than a visual assessment of the soil profile might suggest. The textural and colour variations with depth are pronounced, as described in section 2. The mineralogical variations can be summarised as follows (topsoil excluded):

- The clay proportion increases slightly from about 54% in the structureless shallow clays, to about 61% in the extremely weathered rock.
- The proportion of free quartz increases from 27% to 38% from the weathered rock to the shallow clay.
- Randomly interlayered smectite and illite is the dominant mineral at all depths.
- In the clay fraction, smectite with minor interlayered illite increases from a trace mineral to an accessory mineral with depth, while kaolinite increases from a minor accessory mineral (13%) to an accessory which is becoming sub dominant (19%).

Most of the above trends can be explained in terms of the weathering processes which form residual soils from parent rocks, although a detailed discussion of soil chemistry is outside the scope of this paper. There is insufficient information in the data in Table 1 to enable implications of the apparent small increases in the illite content with depth to be assessed. The important outcome in-so-far as this work is concerned, is that the variations in mineralogy with depth, are generally small enough to support a conclusion that the clay soils are mineralogically similar at all depths. Further, it is apparent that in the Maryland soil profile, the soil forming processes involved in the in situ weathering of a siltstone rock have a greater effect on the soil fabric than they do on the soil mineralogy. This is may be due, in part, to the high detrital clay content which already exists in the parent rock material.

The relatively consistent mineralogy with depth in the Maryland soil profile would suggest that the expansiveness of the soil should be also similar with depth, since expansive potential is strongly influenced by clay mineralogy. Whether the potential for expansion is uniform with depth in the soil profile is difficult to assess from field observations alone. It is generally accepted that the expansive potential of clay soils in many parts of eastern Australia cannot be assessed visually (Fityus and Delaney, 1995). Differential ground movements taken at various levels beneath the ground surface at the Maryland field site indicate that more of the total movement was occurring at shallower levels in the soil profile (Fityus and Smith, 1998), but this is considered to be strongly influenced by a reduction in suction/moisture change with depth.

The expansiveness of Maryland clay soils from different depths has been measured by laboratory testing as described in section 3. The results given by the two methods employed appear to be conflicting. However, the conflict can be resolved by considering the differences between the two methods employed, and the quantities actually measured by each.

Indirect inference of expansivity using surface area measurements involves the titration of a small sample of powdered clay with a solution of adsorptive molecules. Methylene blue adsorption is a measure of the clay particle surface area, which is a function of clay type (Cocka and Birand, 1993), and an indicator of the potential for water adsorption by the clay, and hence, its potential to swell when wetted (Xidakis and Smalley, 1980). Surface area measurements are a direct reflection of clay mineralogy, but an indirect reflection of expansivity, which is influenced by other factors in an undisturbed soil.

It is thus not unexpected, that the surface area of a powdered clay sample should correlate well with its mineralogy. Examination of Figure 2 reveals that the methylene blue adsorption of Maryland clay is relatively consistent with depth, and consistent with the rela-

tively uniform mineralogy observed with depth in the soil profile.

The shrink swell index is a direct measure of physical expansiveness, involving a direct measurement of physical swelling and shrinkage of undisturbed clay soil samples subjected to imposed moisture changes.

Figure 2 indicates that the shrink swell index reduces by approximately one half over a depth of approximately 1m. If this decrease were in response to changes in the clay content of the soil, then a corresponding reduction in the smectite proportion of the whole clay soil might be expected. As already discussed, this does not seem occur.

The magnitude of the observed reduction in shrink swell potential is consistent, however, with the prominence of the textural changes observed in the Maryland profile. In fact, closer inspection of Figure 2 not only reveals a general tendency for shrink swell potential to decrease with increasing rock structure, but also, a correlation between particular values of expansive potential and particular soil layers. This can be summarised as follows:

- Shrink swell index values of 5-6% in the structureless orange-brown clays between about 0.25m and 0.9m.
- Shrink swell index values of 3-4% in the mottled and stained clays between about 0.9m and 1.25m.
- Shrink swell index values of 2 and 3% in the grey-brown clays with relict rock structure between about 1.25m and 1.7m.

These results provide strong evidence for the important influence that structure and fabric have on expansiveness in residual clay soils.

6. IMPLICATIONS FOR GEOTECHNICAL PRACTICE.

Effective foundation designs on expansive residual clay sites require good estimates of potential ground surface movements. In practice, this is often done by calculation using the results of laboratory testing and a simplified predictive model. Laboratory tests such as the shrink swell test are commonly used to provide quantitative data on the expansive potential of soils in such approaches.

Efficient foundation designs require that minimum testing be carried out in the estimation process. Superfluous testing adds unnecessary cost to foundation designs. The comprehensive data presented in Figure 2 involved a substantial amount of laboratory testing; considerably more than would be acceptable for the design of a small, lightly loaded structure in routine practice. Yet this data illustrates the variability in expansive potential which can exist over relatively short depths in residual expansive clay soil profiles.

660

In practice, the expansiveness of a soil profile may be estimated on the basis of one or two shrink swell tests, with values for the entire soil profile assumed on the basis of these. In the absence of other information, this may be insufficient data to accurately estimate the expansive potential of all soils throughout the depth in question, particularly if only one test is performed.

The observations presented here provide a starting point for a better understanding of the characteristics of residual expansive soil profiles. While no conclusive general rules can be stated from the limited results presented here, a number of possible site characterisation rules can be postulated, as a basis for further research.

The first of these is relatively simple. It is, that an accurate estimate of expansive potential with respect to depth can be achieved by firstly identifying the layers in the soil profile with similar textural characteristics, and secondly, testing a representative sample from each.

The second postulate is less straightforward, but potentially more useful. It is based on the observation that it is soil fabric, and not soil mineralogy, which most profoundly influences expansive potential. The postulate may be expressed as follows; that for a given parent rock type, there is a definable correlation between the degree of weathering and the relative expansiveness of the soil. In practice, proof of this hypothesis would enable the expansive potential of all soils in a profile to be reliably estimated on the basis of a single test, performed on, say, the most weathered/unstructured clay soil in the profile. The expansive potential of more heavily structured soils would simply be estimated as consistent proportions of the expansive potential of the least structured soil.

As noted above, this approach would have to be examined separately for specific parent rock types, as the structure of residual clays is is often remnant after the structure of the parent rocks, and this can vary greatly. Whilst this might at first appear to be a prohibitively involved task, it should be remembered that geotechnical practice in a particular geographical region is often confined to relatively limited geological conditions, and thus one or two locally derived correlations may serve well in the majority of cases.

7. CONCLUSIONS.

Residual clay soil profiles differ from other clay soil profiles in that they exhibit strong structural and textural trends with increasing depth, due to their unique formation through the in situ weathering of parent rock materials. The structure is generally characterised by a transition from structureless, desiccated clay near the ground surface, through soil with relict rock structure, to weathered rock at depth, with a decreasing degree of weathering with increasing depth. These textural changes are clearly evident from a visual assessment of the Maryland soil profile.

Despite the apparent severity of the textural changes which result from the weathering process, there is a relatively smaller corresponding effect on the soil mineralogy, with only small changes in the relative proportions of mineral species being evident from xray diffraction studies.

The changes in the soil profile due to weathering do significantly affect the expansive potential of the residual clays in the Maryland soil profile. While visual observation of the soil profile would suggest that this might be due to changes in the mineralogical assemblage with depth, the work presented here provides strong evidence that any reduction in expansive potential is almost entirely due to differences in soil fabric. It is concluded that the primary effect of the weathering of a siltstone rock in a temperate climate, is to destroy the compressed, laminated fabric of the rock, thereby enabling the clay fraction to realise its expansive potential to a greater extent.

This outcome has important implications for the assumptions which are made in regard to the expansivity of residual clay soils in many areas of Australian geotechnical practice.

8. ACKNOWLEDGMENTS

This research has been carried out with financial support from the Mine Subsidence Board of New South Wales, Robert Carr and Associates and the Australian Research Council.

9. REFERENCES

Allman, M. A., Delaney, M. G., and Smith, D. W., 1998. A field study of reactive soil movements in expansive soils. *Proceedings of the 2nd International Conference on Unsaturated Soils*, Beijing: 309–314.

Allman, M. A., Smith, D. W., and Sloan S. W., 1994. The establishment of a reactive soil test site near Newcastle. *Australian Geomechanics*, **26**: 46-56

AS 1289.7.1.1–1992 'Methods for Testing Soils for Engineering Purposes: Method 7.1.1: Determination of the shrinkage index of a soil; Shrink swell index.' Standards Australia.

Blight, G. E. 1997. *Mechanics of Residual Soils*. A.A. Balkema, Rotterdam.

Cocka, E. and Birand, A. 1993. Determination of cation exchange capacity of clayey soils by the methylene blue test. *Geotechnical Testing Journal - American Society for Testing and Materials*, **5**: 518-524.

Dhowian, A. W., 1990. Simplified Heave Prediction Model for Expansive Shale. *Geotechnical Testing Journal*, **13**: 323-333.

Fityus, S. G. (1999) *Transport processes in partially satu-rated soils.*, PhD. Thesis, The Department of Civil, Sur-veying and Environmental Engineering, The University of Newcastle, unpublished.

Fityus, S. G. and Delaney, M. G. 1995. The unique in-fluence of lower Hunter coal measures on reactive soil phenomena in the Newcastle area. *Proceedings of the 28th Symposium on Advances in The Study of the Sydney Basin*, University of Newcastle: 167-174.

Fityus S. G. and Delaney, M. G. 2000. Timber Pile Foundations for Expansive Soils. Submitted to *Geoeng 2000: An International Conference on Geo-technical and Geological Engineering.*, Melbourne.

Fityus, S. G., and Smith, D. W., 1998. A simple model for the prediction of free surface movements in swel-ling clay profiles. *Proceedings of Unsat '98: The 2nd International Conference on Unsaturated Soils*, Beijing, China: 473-478.

Fityus S. G., Smith, D. W. and Jennar, A. M. 2000 Sur-face Area using Methylene Blue Adsorption as a Measure of Soil Expansivity. Submitted to *Geoeng 2000: An International Conference on Geotechnical and Geological Engineering.*, Melbourne.

Nelson, J. D., and Miller D. J., 1992 *Expansive soils: problems and practice in foundation and pavement engineering.* John Wiley & Sons Inc., New York.

Velde, B. 1995. *Origin and Mineralogy of Clays.* Springer. Berlin.

Wesley, L. D. 1990. Influence of Structure and Compo-sition on Residual Soils. *Journal of Geotechnical Engineering.* **116:** 589-603.

Xidakis, G. and Smalley, I. 1980. Looking for expansive minerals in expansive soils: Experiments with dye ad-sorption using methylene blue". *7th Australia-New Zea-land Conference on Geomechanics*, Auckland, 203-206.

Unsaturated Soils for Asia, Rahardjo, Toll & Leong (eds) © 2000 Taylor & Francis, ISBN 90 5809 139 2

Estimation of volume change functions for unsaturated soils

M. D. Fredlund, D. G. Fredlund & G. W. Wilson
Department of Civil Engineering, University of Saskatchewan, Saskatoon, Sask., Canada

ABSTRACT: The estimation of unsaturated soil property functions has followed the pattern of using the saturated soil properties along with the soil-water characteristic curve to predict the behavior of the unsaturated soil. This has been the procedure for the permeability function and the shear strength functions for unsaturated soils. It has been more difficult to develop procedures for the volume change functions for an unsaturated soil; however, it is suggested that a similar procedure can be adopted. This paper suggests a series of postulates that can be used in estimating the volume change functions (during monotonic loading) for an unsaturated soil.

1 INTRODUCTION

Procedures to estimate the unsaturated soil property functions with respect to permeability and shear strength have been developed based upon the saturated soil properties and the soil-water characteristic curve. These procedures have proven to be of great value in the implementation of unsaturated soil mechanics into standard geotechnical engineering practice (Fredlund, 1999). Similar procedures are now required for the volume-mass constitutive relationships for an unsaturated soils.

2 ESTIMATION OF THE VOLUME-MASS FUNCTIONS

Over the past several decades, the volume change behavior of an unsaturated soil has been linked to two independent stress state variables; namely, $(\sigma - u_a)$ and $(u_a - u_w)$. This constitutive formulation forms the basis for modeling the volume change of an unsaturated soil. Modeling of such soil processes as stress/deformation, shrink/heave, and consolidation require an adequate description of the constitutive volume change behavior of a soil.

The overall volume change of an unsaturated soil can be defined as a change in void ratio in response to a change in the stress state (Fredlund and Morgenstern, 1976).

$$de = \frac{\partial e}{\partial(\sigma - u_a)} d(\sigma - u_a) + \frac{\partial e}{\partial(u_a - u_w)} d(u_a - u_w)$$

(1)

where: e = void ratio; and σ = total normal confining stress (e.g., isotropic confining pressure).

Equation (1) can be viewed as having two parts; namely, a part that is the designation of the stress state (i.e., $(\sigma - u_a)$ and $(u_a - u_w)$) and a part that is a designation of the soil properties (i.e., $(\partial e / \partial(\sigma - u_a))$ and $(\partial e / \partial(u_a - u_w))$). The soil properties can be viewed as the slope of the void ratio constitutive surfaces as shown in Figure 1. The soil properties are moduli that vary as a function of the stress state. The soil moduli associated with the net normal stress, $(\sigma - u_a)$, can be written in a general functional form.

$$\partial e / \partial(\sigma - u_a) = func[(\sigma - u_a),(u_a - u_w)] \quad (2)$$

The term *func* means that the soil property is a function of the stress state. At a particular stress state, the compressibility modulus for the void ratio constitutive surface with respect to $(\sigma - u_a)$, can be designated as a constant.

$$\partial e / \partial(\sigma - u_a) = m_1^s \quad (3)$$

Similarly, the soil moduli associated with soil suction, $(u_a - u_w)$, can be written in a general functional form.

$$\partial e / \partial(u_a - u_w) = func[(\sigma - u_a),(u_a - u_w)] \quad (4)$$

At a particular stress state, the compressibility modulus for the void ratio constitutive surface with respect to $(u_a - u_w)$, can be designated as a constant.

$$\partial e / \partial (u_a - u_w) = m \, {}^s_2 \qquad (5)$$

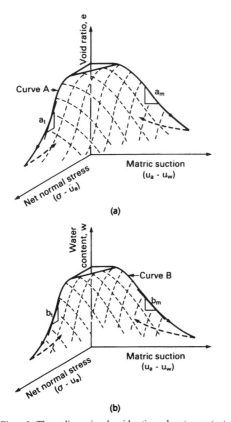

Figure1. Three-dimensional void ratio and water content constitutive surfaces for an unsaturated soil. (a) void ratio constitutive surface; (b) water content constitutive surface (Fredlund and Rahardjo, 1993).

Each of the soil moduli is function of both stress state variables. In order to define the magnitude of the soil moduli corresponding to any stress state, there needs to be a constitutive equation describing the entire void ratio constitutive surface. The equation then needs to be differentiated with respect to each of the stress state variables in order to obtain the compressibility moduli. At present, no equations have been published to represent the entire void ratio constitutive surface in terms of the stress state variables.

Two constitutive relationships are required to define the volume-mass variables in terms of the stress state variables. The need for two independent constitutive relations for an unsaturated soil can be demonstrated through the differentiation of the basic volume-mass relationship (i.e., $Se = wD_r$).

$$\int_{e_o}^{e_f} S \, de + \int_{S_o}^{S_f} e \, dS = D_r \int_{w_o}^{w_f} dw \qquad (6)$$

where: w = water content, with the subscript, o, and f, representing the initial and final states, respectively; S = degree of saturation; and D_r = relative density of the soil solids.

The water content constitutive surface can be used as a second relationship for defining the volume-mass behavior of an unsaturated soil (Fig. 1). The water content constitutive relationship can be written the following general form.

$$dw = \frac{\partial w}{\partial (\sigma - u_a)} d(\sigma - u_a) + \frac{\partial w}{\partial (u_a - u_w)} d(u_a - u_w)$$
$$(7)$$

Once again, Equation (7) has a part that designates the stress state and a part that designates an unsaturated soil property that is a function of the stress state. The soil moduli associated with the net normal stress variable, $(\sigma - u_a)$, can be written as a general function.

$$\partial w / \partial (\sigma - u_a) = func[(\sigma - u_a) , (u_a - u_w)] \qquad (8)$$

At a particular stress state, the compressibility modulus for the water content constitutive surface, with respect to $(\sigma - u_a)$, can be designated as a constant.

$$\partial w / \partial (u_a - u_w) = m_1^w \qquad (9)$$

Similarly, the soil moduli associated with the soil suction, $(u_a - u_w)$, can be written as a general function of the stress state.

$$\partial w / \partial (u_a - u_w) = func[(\sigma - u_a) , (u_a - u_w)] \qquad (10)$$

At a particular stress state, the compressibility modulus for the water content constitutive surface, with respect to $(u_a - u_w)$, can be designated as a constant.

$$\partial w / \partial (u_a - u_w) = m_2^w \qquad (11)$$

At present, there is no published equation to represent the entire water content constitutive surface. Once an appropriate equation is formulated, the derivatives will provide the soil moduli values corresponding to any stress state.

3 FORMULATION FOR THE ESTIMATION OF THE VOLUME-MASS CONSTITUTIVE SURFACES (LOADING)

The formulation of the constitutive surfaces is based on common laboratory compression, shrinkage and soil-water characteristic curve tests. Mathematical representation needs to be proposed for the compression, shrinkage, and soil-water characteristic curve soil property functions. Numerous equations have already been proposed for the soil-water characteristic curve.

A series of assumptions are needed to form a guide for combining the independent mathematical representations and the subsequent formulation of void ratio and water content constitutive surfaces. (Only the assumption or postulates relevant to the volume change constitutive relationship are presented in this paper). The desired end result is a mathematical representation of two independent constitutive surfaces. The mathematical representations can then be used to compute the compressibility (and subsequently the elastic moduli) of the soil corresponding to any stress state in the soil. These formulations provide the necessary information for the generation of elastic soil property functions that can be used for numerical modeling (i.e., finite element method) of soil behavior.

3.1 Volume Change Constitutive Surface (Loading)

The overall volume change can be defined in terms of void ratio, e, or specific volume, υ, (i.e., 1+ e). The void ratio is used herein to define the first constitutive surface.

A series of "postulates" are proposed for the prediction of the volume-mass constitutive relations (Fredlund, 2000). The postulates establish a series of priorities that must be adhered to when attempting to estimate the volume-mass relationships. Certain information has become well established in the research literature and this information forms a series of hierarchical priorities when predicting the constitutive surfaces.

The soil structure constitutive surface can be defined as the relationship between two independent stress state variables and a deformation state variable. The independent stress state variables are: $(\sigma - u_a)$ = net normal (isotropic) confining pressure; and ψ = soil suction.

The deformation state can be defined in terms of void ratio, e. The proposed "postulates" for the soil structure (i.e., void ratio) constitutive surface are given below for the case of an increase in both of the stress state variables (i.e., a monotonic decrease in volume). In addition, it is assumed that the testing of the soil starts with the specimen being in a saturated state. There are a number of loading stress paths as well as wetting and drying paths that could be analyzed; however, it is important to start by developing constitutive surfaces for the conditions on which the most information is available.

Postulate 1: The **primary reference condition** for the volume change (overall) constitutive relationship is determined by applying a net (isotropic) total stress loading of the soil with the pore-water and pore-air pressures maintained at zero, while measuring the change in void ratio.

This relationship is commonly referred to as the drained, effective stress loading path for a saturated specimen (Fig. 2). The term "isotropic" is placed in brackets to suggest that isotropic loading is the preferred form of loading. However, it is also possible for K_o or other forms of net total stress loading to be considered. Isotropic stress loading is preferable because: a) the stress path is the same as that used for critical state (or elasto-plastic) models, and, b) the matric suction stress state variable is also isotropic in character.

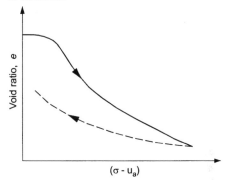

Figure 2. Typical loading and unloading curves of void ratio versus the applied load.

Postulate 2: The **secondary reference condition** for the volume change (overall) constitutive relationship is determined by applying various soil suctions to the soil with the net isotropic stress equal to zero, while measuring the change in void ratio (Fig. 3).

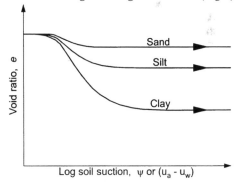

Figure 3. Typical void ratio versus soil suction plot for three soils (suction increasing).

There is a practical difficulty associated with directly measuring the volume change versus soil suction relationship. The difficulty is related to measuring volume change in three directions while changing soil suction. As a result of the above difficulty, "Postulate 2a" suggests an alternate means to indirectly provide the necessary secondary reference condition. An alternate procedure makes use of a combination of data from a shrinkage test and a soil-water characteristic curve test.

Postulate 2a: The void ratio versus soil suction relationship can also be computed using the soil-water characteristic curve for the soil along with the shrinkage curve, both sets of data are measured under condition of zero net isotropic stress.

Figure 4 shows three typical soil-water characteristic curves under drying conditions (or conditions of an increase in suction. Figure 4 shows a typical shrinkage curve associated with the drying of a clay soil.

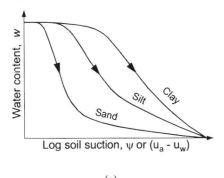

(a)

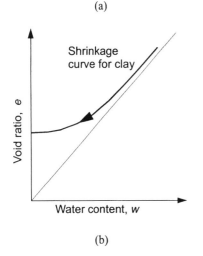

(b)

Figure 4. a) Typical soil-water characteristic curves for three soil types. b) Typical shrinkage curve for a clay soil.

It is possible to combine the results of a pressure plate test (i.e., soil-water characteristic curve data) and a shrinkage test to obtain a void ratio versus suction plot. The shrinkage test defines a curve that gives the ratio of change in volume of water to over-all volume, for a change in soil suction. Mathemati-

cally, the slope of the shrinkage curve can be written as follows.

$$\frac{de}{dw} = \frac{de/d\psi}{dw/d\psi} = \frac{a_m}{b_m} \qquad (12)$$

Combining the two sets of information makes it possible to compute the void ratio versus soil suction relationship. This forms the second reference (or limiting) condition for the soil structure constitutive surface.

Postulate 3: There is a unique volume change constitutive surface defined for conditions of monotonic deformation.

The surface for a decrease in volume under an increase in stresses is considered herein and shown in Figure 5.

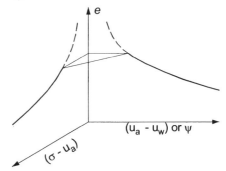

Figure 5. Three-dimensional plot showing the primary and secondary reference conditions for the void ratio constitutive surface.

The limiting (or reference) conditions associated with the void ratio constitutive surface have now been defined. The next step is to define the character of the constitutive surface between the limiting reference conditions. The remaining postulates pertain to establishing intermediate stress state conditions on the constitutive surface.

Postulate 4: The slope along any constant net total stress plane on the volume change constitutive surface is a function of the void ratio, as defined on the zero soil suction plane.

This postulate comes about as a result of Postulate 1 where it is stated that the void ratio versus net total stress is the primary and most fundamental relationship between void ratio and the stress state. As a consequence of Postulate 4, the slope of any line emanating from the soil suction versus void ratio curve, in a constant suction plane, must be equal to the compressibility defined on the primary reference curve at a corresponding void ratio. Appropriate

666

slopes for the constitutive surface can be determined by constructing a triangle in the horizontal plane, between the reference conditions.

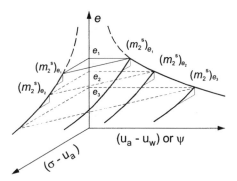

Figure 6. Illustration of the definition of the void ratio constitutive surface based on the slopes of the primary reference curve.

Postulate 5: There is a one-to-one relationship between the effects of changes in net total stress and a change in soil suction, when the soil suction is less than the air entry value of the soil (Fig. 7).

This mean that a 45 degree relationship will be defined between the two stress state variables when the void ratio constitutive surface is viewed along the void ratio axis.

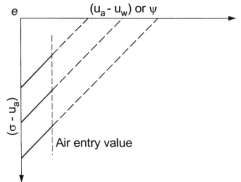

Figure 7. Variation of constant void ratio contours when the surface is viewed along the void ratio axis.

The straight line contour across the constitutive surface should be theoretically correct as long as the soil is saturated. This is in accordance with the effective stress concept for a saturated soil. It should be noted that the dashed lines drawn in Figure 7 may not intersect the secondary reference condition along the plane of net total stress equal to zero. It is necessary to comment further on the air entry value of the

soil before suggesting a further refinement on void ratio contours.

Postulate 5a: As a first approximation, the air entry value of the soil can be assumed to be a constant, but for greater refinement, the air entry value may need to be defined as a function of void ratio or the net isotropic stress (Fig. 8).

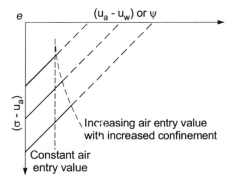

Figure 8. Effect of a variation in the air entry value on the void ratio contours.

The air entry value would be anticipated to increase with a decrease in void ratio. This means that the 45 degree contour would be adhered to for a greater distance from the net total stress reference plane. No attempt is made at this time to define the air entry value of the soil as a function of void ratio (or stress state).

Postulate 6: A gradual curve forms from the air entry value to the secondary reference condition, corresponding to a particular void ratio on the soil structure constitutive surface (Fig. 9).

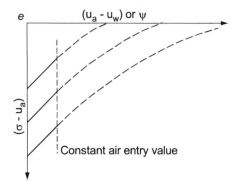

Figure 9. Variation in the constant void ratio contours as the soil becomes increasingly desaturated.

The curve must be tangent at the air entry value and increase in curvature as the secondary reference condition is approached. This means that it should always be possible to join the secondary reference curve provided it is positioned further from void ratio than is the primary reference curve. In other words, at a particular void ratio, the soil suction value should exceed the net total stress value.

The curves should always bend in the direction of the soil suction axis for a clayey, stable-structured soil. For a sandy soil, the curves will bend even more rapidly and may never reach the reference soil suction axis.

The loading portion of the void ratio constitutive surface can be approximated using the steps outlined above. The general character of the void ratio constitutive surface should apply for sands, silts and clays. The greatest difficulty should be observed in defining the constitutive surface near to initial conditions. This is due to the fact that not all of the tests are started from precisely the same stress state and volume-mass state. As well, different tests may follow different stress paths particularly near the start of the test. It is therefore necessary to take into consideration the initial state of the soil. For example, the soil could be initially slurried, compacted or be in an undisturbed state.

The above postulates do not cover all aspects of defining the void ratio constitutive surface. The postulates pertain to the loading (by net total stress or soil suction) constitutive surface of an initially saturated soil. Other postulates are required to define the unloading void ratio constitutive surface. Still other postulates are required for the case where one state variable is increased while the other one may be decreased. The scope of this paper is limited to monotonic loading of an initially saturated soil.

A series of similar postulates are required for the water content constitutive surface. The postulates for the water content surface are outside the scope of this research paper.

4. SUMMARY

A series of postulates have been provided for the estimation of the volume change constitutive surfaces for an unsaturated soil (for the case of monotonic loading). Once again, the estimation procedures make use of the saturated volume change properties of the soil along with the soil-water characteristic curve information. The suggested postulates need to be tested using experimental data on volume change.

REFERENCES

Fredlund, D.G. 1999. The implementation of unsaturated soil mechanics into geotechnical engineering. The 1999 R.M. Hardy Lecture, *52nd Canadian Geotechnical Conference, Regina, Saskatchewan, Canada.*

Fredlund, D.G. & H. Rahardjo. 1993. *Soil mechanics for unsaturated soils.* New York: John Wiley & Sons.

Fredlund, M.D. 2000. The role of unsaturated soil property functions in the practice of unsaturated soil mechanics. Ph.D. Thesis, University of Saskatchewan, Saskatoon, Canada (*In progress*).

Fredlund, D.G. & N.R. Morgenstern. 1976. Constitutive relations for volume change in unsaturated soils. *Canadian Geotechnical Journal* 13(3):261-276.

Unsaturated Soils for Asia, Rahardjo, Toll & Leong (eds) © 2000 Taylor & Francis, ISBN 90 5809 139 2

On the volume measurement in unsaturated triaxial test

F.Geiser, L.Laloui & L.Vulliet

Soil Mechanics Laboratory, EPFL, Lausanne, Switzerland

ABSTRACT: In unsaturated triaxial test, total volume change of the specimen is related to both water and air volume changes and cannot be measured with the same techniques as in conventional saturated experiments. Over the years several methods have been developed with various degrees of accuracy and complexity. This paper reviews and compares different approaches. Most of them have been tested by the authors on an unsaturated silt. Their advantages and disadvantages are discussed and their accuracy is compared. Finally recommendations and criteria are proposed concerning the choice of the best methods.

1 INTRODUCTION

Volume changes during triaxial tests are essential when analysing the mechanical behaviour of a soil. In the case of saturated triaxial test, specimen volume change is directly related to water exchange and can easily be measured with a glass burette or a pressure-volume controller. In the case of unsaturated soil, however, both air and water volume change and the total specimen volume change cannot be measured so readily.

Several methods have been described in the literature for the measurement of the unsaturated specimen volume change but their accuracy has not been compared. The choice of a method is mainly influenced by the type of tests, the stress and strain level, the cost, the existing equipments and the necessity for continuous data acquisition. In the particular case of unsaturated tests, many problems such as leakage and creep of the cells increase with test duration (from several days to several months, e.g. Geiser, 1999).

In this paper, after a presentation of existing measurement techniques, advantages and disadvantages are discussed. Several recommendations and criteria based on accuracy conclude the paper.

2 EXISTING METHODS

In this work, the focus is made on three categories of approaches:

(A) cell liquid measurement;

(B) direct air and water volume measurement;

(C) direct measurement on the sample.

2.1 Cell liquid measurement (A)

In this approach the sample volume changes are deduced from the volume changes of the confining *cell* liquid. Several problems are often encountered such as immediate expansion of the cell caused by the increase in pressure, Plexiglas creep under constant stress and possible water leakages. Consequently, the measurements have to be corrected at the end of each test for these effects (calibration process). Factors which affect the movement of water into or out of the cell and which must therefore be taken into account in volume change measurements have been listed by Head (1986).

The advantage of the cell liquid measurement method is simplicity. A standard triaxial cell can be used if it has been carefully calibrated. However the measurement of the sample volume remains indirect and the accuracy is related to the calibration procedure. Ideally, numerous calibrations are needed as the corrections depend on time, stress path and stress level.

As the error on the measurement increases with increasing surrounding confinement liquid volume, Bishop & Donald (1961) have proposed the addition of an *inner cylinder* sealed to the cell base to minimise the liquid volume. To enhance accuracy, mercury is used between inner cylinder and sample. The outer liquid is water. The rise and fall of the mercury level in the inner cylinder is measured by a cathetometer. The overall volume measurement can then be deduced. The accuracy of a similar device has been tested with air as outer cylinder liquid and water around the sample (Geiser 1999 after Cui 1993).

Finally, some authors have proposed to minimise the confinement liquid by using *double-walled cells* (e.g. Wheeler 1986, Sivakumar 1993). Here an inner cylinder is sealed to both top and base of the cell. Most of the possible errors discussed by Head (1986) can be reduced this way. Moreover, automatic readings are possible.

2.2 *Air-Water volume measurement (B)*

In this approach, the volumetric changes of the sample are deduced by simple addition of the air and water volume changes. Both volume changes are measured separately using pressure-volume controllers. The basic idea consists simply in an *air volume controller* filled with air instead of water. Very few experimental data exist with this method. The main advantage is that all kinds of test paths can be tested as the air volume and pressure can be controlled and imposed. However, undetectable air leakages through the tubes and connections occur and have to be estimated. Small temperature and atmospheric pressure changes also affect the volume measurement and have to be taken into account.

As the results are significantly influenced by those factors, an improved device has been proposed to minimise the errors: *a mixed air and water controller* is used, that allows to reduce the air volume to connection tubings only (Figure 1) (for details see Geiser 1999).

2.3 *Direct measurement on the sample (C)*

Three direct measurement approaches are considered.

1.   The most commonly used involves local *displacement sensors* attached directly onto the sample, measuring axial and radial deformations during the test (e.g. Clayton 1986). Radial displacements are measured at one to three discrete points and assumptions are made as to the shape of the samples to extract the volumetric strain. Placing the sensors is quite delicate: if not done correctly, it may lead to experimental errors. Furthermore, this approach is best suited for small deformation tests and requires an initial fairly rigid sample.

A radial Hall effect transducer was tested by the authors as shown on Figure 2. Note that when the sample tends to a barrel shape, a unique measure of the radial changes in the middle of the sample is not accurate enough.

2.   The second approach involves using the *laser* technique as proposed by Romero et al. (1997). A sweeping with the laser on the whole sample's height allows a more accurate determination of the sample volume. This approach also allows detection and identification of non-uniformities and localised deformations. This technique is powerful, but requires costly and sophisticated installation procedure.

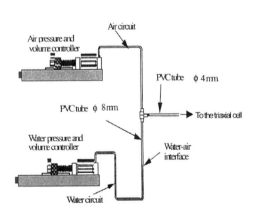

Figure 1: Sketch of the mixed air-water volume device

Figure 2: Radial Hall effect device sealed and pined on a sample

3. The third approach is an alternative direct measurement technique based on image processing (Gachet et al., in prep.). The approach consists of taking photographs through the Plexiglas cell during the test and then analysing the *images* to obtain the sample profile and volume. This approach is similar to the one proposed by Macari et al. (1997) except that in this case, the method used to correct the magnification effect due to the cell form and the different refraction indices (water, Plexiglas, air) is simpler. The images are taken using a camera, fixed at a constant distance from the triaxial cell as shown on Figure 3. Before each test, a calibration procedure is applied to a false rigid sample to optimise the accuracy. The soil sample is then placed into the triaxial cell and several photos are taken and analysed during the test. As for the laser method, no direct contact with the sample is required during the test and the profile is measured on the whole sample's height. With our actual device, volume measurement is only possible on symmetric samples, since only one fixed camera is installed on the cell.

3 COMPARISONS

3.1 *Advantages, disadvantages and accuracy of the different methods*

Table 1 summarizes the advantages and disadvantages of the previously described methods and compare them. An estimation of the absolute error is also given.

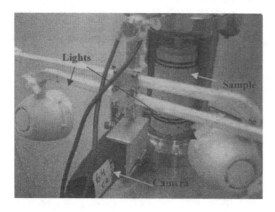

Figure 3: Image of the experimental device with the camera, the halogen lights and the triaxial cell.

The exact values of volume change ΔV, respectively volumetric strain ε_v, are related to the measured values (ΔV^*, ε_v^*) by:

$$\Delta V = \Delta V^* \pm \alpha$$

$$\varepsilon_v = \varepsilon_v^* \pm \beta$$

where α and β are absolute errors.

Except for the double-walled cell and the laser method, the devices have all been tested in the Soil Mechanics Laboratory at the EPF Lausanne at a quasi-constant temperature (19°C ± 0.4°C) for similar samples (volume varying between 200 and 250 cm^3 depending on the cell used). Comparison of accuracy values is rather indicative and should not be taken as absolute value since factors such as laboratory temperature, type and material of the cell, connections, data acquisition system and sample size can directly affect the measurements.

It should also be noticed that some methods could still be improved in order to obtain a better accuracy, either through more complete calibration procedures or technical modifications.

For most methods, the identification and quantification of the individual errors is quite difficult and time consuming. The accuracy has consequently been evaluated on the base of global calibration procedures. This can explain why the absolute error calculate by Sivakumar (1993, see Table1, A3) is much higher than what is observed with a standard cell by Geiser (1999), as it corresponds to the addition of all the possible individual errors. The value given for the standard cell A1 (Table 1) is on the contrary based on global comparisons of the water outflow and the cell volume change (with the calibrated corrections) during the consolidation of saturated samples. It is however obvious that the accuracy of the measurement should be improved with a double-walled cell.

For method using local Hall-effect transducers it was not possible yet to clearly evaluate the accuracy on the base of the test done on an unsaturated silt, as the results were not convincing. This was mainly du to the difficulty of setting the device on the samples.

Several methods (A2, B2, C2, C3) offer a precision of $\pm 10^{-3}$ in the volumetric strain determination. Such an accuracy is adequate for test involving strains of the order of magnitude of 10^{-1} to 10^{-2} (commonly observed in practical).

Table 1: Recapitulation and comparison of the different volume measurement methods

Method	Advantages	Disadvantages	Absolute error on the volume variation measurement and on the volumetric strain
A1- Standard triaxial cell	- use of a standard cell, without modifications	- indirect method, involving long calibration process	$\alpha = \pm 0.45$ cm^3 $\beta = \pm 2.2$ 10^{-3}
A2- Cell with inner open cylinder	- minimise or strongly decrease the undesired volumetric changes observed with A1, as the confinement pressure is imposed on both sides of the inner cell - minimise the inner volume cell and consequently increases the accuracy in comparison to A1	- indirect method, involving calibration process - no possible automatisation of the measurements	$\alpha = \pm 0.21$ cm^3 $\beta = \pm 0.8$ 10^{-3} Bishop & Donald (1961) found with a similar device for ≈ 100 cm^3 samples: $\alpha = \pm 0.1$ cm^3 $\beta = \pm 10^{-3}$
A3- Double walled cell*	- same as A2 - enables continuous measurement	- indirect method, involving calibration process	Sivakumar (1993): For ≈ 100 cm^3 samples: $\alpha = \pm 0.6$ to 1.02 cm^3 $\beta = \pm 6$ 10^{-3} to 10^{-2} depending on the cell The average global accuracy is believed to be better.
B1-Air filled controller	- direct measurement or imposition of the volume of air - enables continuous measurement	- air volume is strongly influenced by the temperature and the atmospheric pressure - undetectable air leakages	$\alpha = \pm 2.2$ cm^3 $\beta = \pm 1.1$ 10^{-2} + an additional continuous air leakage of 2-3 cm^3/ day
B2- Mixed air-water filled controller	- same as B1 - minimise the air volume and thus the possible temperature and atmospheric pressure influence and the leakages	- as B1, but less important	$\alpha = \pm 0.22$ cm^3 $\beta = \pm 1.1$ 10^{-3} + an additional continuous air leakage of 0.2 cm^3/ day
C1- Hall effect captor (radial strain measurement)	- direct measurement on the sample - enables continuous measurement	- planed for small strain measurement - problems of accuracy for barrel-shape sample equipped with only one radial strain gage. - Pining or sealing the transducer on the sample is quite delicate and requires an initial fairly rigid sample	Not defined yet
C2- Laser*	- direct measurement on the sample without contact - enables continuous measurement - measurement of the profile on the whole sample height - possible measurement all around the sample	- high costs and long calibration process to allow accurate measurements	Estimation based on Romero et al. (1997): $\beta = \pm 0.7$ 10^{-4}
C3- Image processing	- direct measurement on the sample without contact - measurement of the profile on the whole sample height - computer-controlled calibration process	- not valid for asymetric sample when using only one camera	$\alpha = \pm 0.25$ cm^3 $\beta = \pm 10^{-3}$

*not tested by the authors

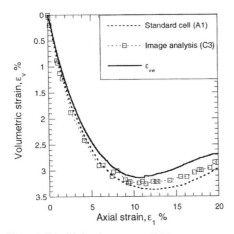

Figure 4: Triaxial shearing test at s=50 kPa

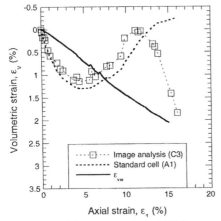

Figure 5: Triaxial shearing test at s=280 kPa

3.2 *Validation of the measurements*

In order to validate our measurements, several methods have been used simultaneously when possible. **Figures 4 & 5** show as examples the volumetric strain ε_v as a function of the axial strain ε_1 for two shearing tests at two different suction levels (s=50 kPa and s=280 kPa).

The measured volumetric strains obtained with the image processing (C3) and with a standard cell (A1) are close for the test at s=50 kPa. The maximum observed difference is around $1.5 \ 10^{-3}$ on the absolute volumetric strain. The average difference is of $5 \ 10^{-4}$. It can also be observed that the total volumetric strains are close to the water volume changes ε_{vw} (defined as the water volume changes over the total volume). This was expected as the sample is very close to saturation for this level of suction.

The results on **Figure 5** were obtained at a much

higher suction and the total volume change cannot be determined using the water volume changes. It can be observed that with the image processing the volumetric strain diverges at 11% of axial deformation. This corresponds to the point where a clear shear band was seen on the sample and where consequently the image processing was not valid anymore. Below this axial strain, both methods (A1 & C3) are comparable with a maximum observed difference of $3.5 \ 10^{-3}$ and an average of $1.5 \ 10^{-3}$.

4 CONCLUSIONS AND RECOMMENDATIONS

Several volume measurement techniques have been explained and tested. For most of them a long calibration procedures is necessary to obtain a satisfactory accuracy. The choice of the methods will be influenced for each laboratory by the available material and the required accuracy.

If one chooses to use the cell liquid approach (A), the most reliable method is the double-walled cell (A3).

The methods with direct air volume measurements are dependent of undetectable air leakages and are consequently quite delicate. However, with a limitation of the air volume (B2), a reasonable accuracy can be obtained.

The direct measurement approaches are interesting; here methods without contact should be preferred. Both the laser (C2) and image processing (C3) methods give good quantitative results.

ACKNOWLEDGEMENTS

The authors thank P. Gachet for working with them on the image processing technique. This work was supported by the Swiss NSF, grant 21-42360.94

REFERENCES

Bishop, A.W. &. I.B. Donald 1961. The experimental study of partly saturated soil in the triaxial apparatus. *Proc. 5th Int. Conf. Soil Mech. Found. Eng., Paris*:13-21.

Cui, Y. J. 1993. *Etude du comportement d'un limon compacté non saturé et de sa modelisation dans un cadre élasto-plastique.* PhD thesis, Ecole Nationale des Ponts et Chaussées, Paris

Clayton, C.R.I. & S.A. Khatrush 1986. A new device for measuring local axial strains on triaxial specimens. *Géotechnique* 36(4): 593-597.

Gachet, P., F. Geiser, L. Laloui & L. Vulliet, in preparation. An automatic method of digital image processing to measure volume changes in a triaxial cell.

Geiser, F. 1999. *Comportement mécanique d'un limon non saturé: étude expérimentale et modélisation constitutive.* PhD Thesis, Swiss Federal Institute of Technology, Lausanne.

Head, K.H. 1986. *Manual of soil laboratory testing.* Pentech Press.

Macari, E.J., J.K. Parker & N.C. Costes 1997. Measurement of volume changes in triaxial tests using digital imaging techniques. *Geotechnical Testing Journal* 20(1): 103-109.

Romero, E., J.A. Facio,A. Lloret, A. Gens & E.E. Alonso 1997. A new suction and temperature controlled triaxial apparatus. *Proc. 14th Int. Conf. on Soil Mech. and Found. Eng., Hambourg*: 185-188.

Sivakumar, V. 1993. *A critical state framework for unsaturated soils*, PhD thesis, University of Sheffield.

Wheeler, S.J. 1986. *The stress-strain behaviour of soils containing gaz bubbles.* PhD Thesis, Oxford University.

Unsaturated Soils for Asia, Rahardjo, Toll & Leong (eds) © 2000 Taylor & Francis, ISBN 90 5809 139 2

Shrinkage behaviour of some tropical clays

P. R. N. Hobbs, K. J. Northmore, L. D. Jones & D. C. Entwisle
British Geological Survey, Keyworth, UK

ABSTRACT: The results of a series of shrinkage limit and linear shrinkage tests are described. These were carried out on bonded, tropical, residual clay soils from Indonesia and Kenya, containing the clay minerals halloysite, allophane, and kaolinite. The results show the effect on shrinkage behaviour of the breakdown of the bonded fabric by both mixing energy input during preparation and by compaction. A new shrinkage limit test apparatus is described.

1 INTRODUCTION

A three part laboratory investigation is reported of the shrinkage behaviour of kaolinitic, halloysitic, and allophanic tropical residual clay soils from Java, Indonesia and from Kenya. The first part aimed to examine the influence of soil fabric on results of the shrinkage limit test carried out on undisturbed and compacted samples. The second examined the effect of mixing time on shrinkage limit, and the third the effect of cycles of shrinkage on the results of the linear shrinkage test (Anon, 1990, Part 2, Test 6.5), and hence inferred the influence of soil fabric. In the first part a new shrinkage limit test method developed at the British Geological Survey (BGS) was used rather than either the 'preferred' British Standard (BS) 1377 method (Anon, 1990, Part 2, Test 6.3), or the American Society of Testing & Materials (ASTM) method (Anon, 1995, Test D427-93) also referred to in BS1377 as the 'subsidiary' method (Anon, 1990, Part 2, Test 6.4).

The shrinkage limit test determines the water content below which an undisturbed (or compacted or remoulded) clay soil ceases to shrink with further drying. The shrinkage limit (SL) complements the Atterberg limits, liquid limit (LL) and plastic limit (PL), and may be considered more fundamental than them (Sridharan & Prakash, 1998). The linear shrinkage test is a rapid method of directly measuring the amount of shrinkage of a soil in a totally remoulded state, and is commonly carried out in tandem with the liquid limit test. The two soils from Java tested using the new shrinkage limit method were incapable of being tested by the 'preferred' BS1377 method due to their highly porous fabric.

However, comparative tests on British clay soils are to be published shortly and are described in Kadir (1997), Marchese (1998), and Hobbs (in preparation). The soils were chosen as being representative of common types of clay-rich, highly plastic, tropical residual soil. The linear shrinkage tests were carried out on samples from Kenya and Indonesia, representing various clay-rich residual soils formed on different rock types.

The work reported here formed part of a wide-ranging collaborative study of the geotechnical properties of tropical clay soils from Kenya, Indonesia, and elsewhere (Northmore et al., 1992a; Hobbs et al., 1992), and from more recent work using a shrinkage limit apparatus, newly-developed at the BGS, using soils from the earlier study. The weakly

Figure 1 New shrinkage limit apparatus

bonding characteristic of the halloysitic and allophanic tropical clays results in some interesting geotechnical properties (Vaughan, 1988).

Whilst allophane and halloysite are present together in varying proportions in many tropical residual clay soils the contribution of the minerals to the properties is different. The allophane-rich clays tested are derived from young volcanic ash, are typically a light-brown colour, and have a characteristically low density, bonded fabric, and a plasticity related to rework-energy input (Wesley, 1973; Vargas, 1988). The halloysite-rich clays are derived from older volcanic (and other) rocks, are dark red in colour, and are denser, less bonded and, whilst also having a plasticity which is sensitive to rework-energy input, the relationship is different from that of the allophane-rich clays.

2 NEW SHRINKAGE LIMIT TESTS

2.1 Method

The new shrinkage limit test apparatus makes use of a laser range-finder combined with a digital height gauge, and an electronic balance (Figure 1) to fulfil the essential requirements of the 'preferred' shrinkage limit test described in BS1377 (Anon; 1990: Part 2: Test 6.3). Unlike the 'preferred' BS1377 method, which makes use of Archimedes' principle and immersion in mercury to determine the volume of the sample, the new method determines a pseudo-volume by measuring the x, y co-ordinates of microscope targets fixed to the sample. Thus, immersion in a liquid is not necessary and, as the sample is mounted on the balance, it does not require handling during the test. The new test apparatus, at present in manually-operated prototype form, has produced shrinkage limit results extremely close to those from the 'preferred' BS1377 test for a variety of British clay soils (Kadir, 1997; Marchese, 1998; Hobbs, in preparation).

Most residual tropical clay soils, in common with many British clay soils, are incapable of measurement in the 'preferred' BS1377 test apparatus in an undisturbed state. This is due to the highly porous and highly fissured natures of tropical and temperate clay soils, respectively. In the 'preferred' BS1377 test droplets of mercury tend to enter the sample, and be retained within it on its removal from the mercury bath. This introduces errors into both volumetric displacement and weighing, and the risk of mercury contamination. Mercury droplets enter, and are retained in, fissures and pores, particularly where the surfaces are rough or silty. This may not be a problem with homogeneous, remoulded, or compacted samples, but frequently is a problem with undisturbed samples. A further problem is the loss of fragments of sample into the mercury during handling, particularly if the sample is fissured, low-strength, weakly-bonded, or sensitive.

The new shrinkage limit test uses a cylindrical sample of 100 x 100 mm (approx.), that is about nine times greater in volume than the 'preferred' BS1377 test sample (typically 38 x 76 mm) and about forty times the ASTM or 'subsidiary' BS1377 test sample. The sample is thus more representative of the soil's structure and fabric, and untrimmed U100 and Proctor compaction samples may be tested. The BS1377 and ASTM tests have the advantages that a true volume, rather than a pseudo-volume, is measured (mercury penetration and sample loss excepted), and the tests are relatively cheap. The new method has the advantages that safety hazards in the use of mercury are avoided, the sample is handled only at the beginning and end of the test, and lateral and vertical shrinkage strains are measured independently. Shrinkage strains of different layers within a sample can also be distinguished, depending on the disposition of targets. A fully computer-controlled version of the apparatus is under development at BGS.

Experience has shown that air drying in the new test takes place at a similar rate to that of the BS1377 tests for British soils, except that in the latter the test duration is shorter due to the much smaller size of the test specimen. The new shrinkage limit tests on samples A and B (see below) took between 17 and 42 days to complete at an ambient temperature of 20 ° C. The rate of air-drying may be reduced to minimise cracking. Between 23 and 45 determinations of water content and (pseudo) volume were made during each test, this being significantly greater than for the BS1377 tests.

2.2 Samples

The two soil types tested were as follows:
Sample A: a halloysite-rich, red-brown, latosol clay soil, derived from an older Quaternary volcanic breccia, lahar, or lava, from a depth of 4.0 m, 7km north of Bandung, W. Java, Indonesia.
Sample B: an allophane-rich, yellow-brown, andosol clay soil, derived from the youngest Quaternary volcanic eruption, from a depth of 3.0 m, 2 km west of Lembang, near Bandung, W. Java, Indonesia.
Undisturbed samples were taken in-situ using plastic tubes (Culshaw et al., 1992).

2.3 Results

The results of the new shrinkage limit tests, each pair consisting of an undisturbed and a compacted

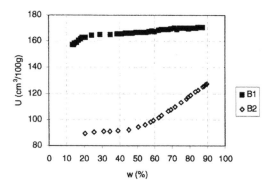

Figure 2. New shrinkage limit test: Volume per 100g dry soil, U vs. Water content, w. (A1 undisturbed, A2 compacted)

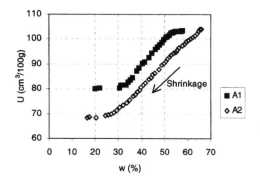

Figure 3. New shrinkage limit test: Volume per 100g dry soil, U vs. Water content, w. (B1 undisturbed, B2 compacted)

sample, are shown in Table 1, along with liquid limit data, and in Figures 2 and 3 (U = volume normalised by dry weight). It can be seen that dramatic differences in shrinkage behaviour exist both between soils A and B and between the undisturbed and, compacted samples from each type, particularly in the case of the allophane-rich soil (B). The differences are in terms not only of shrinkage limit (SL), but also of shrinkage amount (ΔV) and the shape of the shrinkage curves (Figures 2 and 3). Both soils A and B have shown an increase in shrinkage amount (ΔV) from undisturbed to compacted, but the difference is much greater (nearly four times) for the allophane-rich soil (B). However, the change in shrinkage limit from undisturbed to compacted is reversed for the two soils; soil A shows a small decrease while soil B shows a very large increase. This behaviour appears to be consistent with the opposite trends in liquid limit with mixing method, also shown in Table 1, and reported elsewhere for these soil types (Northmore et al, 1992a; Wesley, 1988).

Shrinkage amount, unlike shrinkage limit, is a function of initial water content. Whilst attempts were made to make initial water contents the same for undisturbed and compacted samples the comparison is not exact. Despite this, the contrast in behaviour is dramatic. It is believed that this is due to the fabric of the soils, the halloysite-rich fabric having less influence overall than the allophane-rich. The shrinkage curve for the undisturbed allophane-rich soil is very unusual in itself, in that it appears to be the reverse of a normal curve. Little shrinkage took place over a lengthy period (30 days) of air-drying from the start of the test, followed by a relatively sudden shrinkage. However, the overall shrinkage amount is small, compared with the compacted sample, and other soils. The soil fabric, and any bonding within it, appears to have the capability of reducing the amount of shrinkage, and particu-

larly in the case of the allophane-rich soil (B), the onset of a significant stage of shrinkage within which the shrinkage rate is higher. This may demonstrate a form of structural collapse due to yield at high suctions. The relative strengths and densities of the undisturbed and compacted fabrics, and their resistance to shrinkage stresses are key factors.

3 ASTM SHRINKAGE LIMIT TESTS

3.1 Method

A series of shrinkage limit tests using the ASTM (Anon, 1995, Test D427-93), or 'subsidiary' BS1377 (Anon, 1990, Part 2, Test 6.4) method, was carried out. The tests were done on remoulded samples and included an investigation into the effect of preparatory mixing time on the results. Initial volume of the test specimen was approximately 18 cm^3, and the initial water content at, or slightly above, the liquid limit.

3.2 Samples

Testing was carried out on samples of latosol and andosol collected from W. Java, Indonesia from eight hand-dug trial pits (IP1 to IP8). Index, triaxial, consolidation, and compaction tests were carried out (Hobbs et al., 1992; Northmore et al., 1992a). One of the test samples (IP1) was from the same location as the new shrinkage limit test sample A.

3.3 Results

The results of the tests are shown in Table 2. The effect of increasing the mixing time in the shrinkage limit test from 10 minutes to 60 minutes was generally inconclusive, although notable increases up or down were seen. Shrinkage limits were generally

Table 1 Summary of new shrinkage limit test results

Samp No.	Geol.	Soil type	LL* (%)	PI* (%)	LL+ (%)	PI+ (%)	Preparation		Time (dy)	ΔV (%)	SL (%)
A	oqv	Latosol	114	46	119	53	Undisturbed	A1	34	22.4	31.5
							Compacted	A2	41	34.4	27.7
B	yqv	Andosol	107	35	100	30	Undisturbed	B1	42	7.8	13.0
							Compacted	B2	17	28.7	49.0

Key:

oqv = Old Quaternary volcanics

yqv = Young Quaternary volcanics

* = Hand-mixed as per BS1377 (Anon, 1990), not pre-dried

+ = Grease-worked, not pre-dried (Northmore et al., 1992a)

Table 2 Summary of ASTM shrinkage limit test results

Samp No.	Soil type	Prep.	Depth (m)	LL (%)	PI (%)	SL (%) 10 min. mix	ΔV (%)	SL (%) 60 min. mix	ΔV (%)	SL diff (%)
IP1#	Latosol	Remld	1.5	94	45	27.8	50		51	
IP2	Andosol	Remld	0.5	56	14	33.7	28	32.4	29	+4
IP3	Latosol	Remld	0.5	92	42	32.6	51	29.5	56	-10
		Remld	1.5	105	45	26.9	54	29.2	56	+2
IP4	Latosol	Remld	0.5	79	35	23.4	50	23.3	48	0
		Remld	1.5	83	38	31.4	41	28.8	46	-8
IP5	Latosol	Remld	0.5	76	34	28.6	43	32	46	+3
		Remld	1.5	77	33	30.1	46	29.8	48	-1
IP6	Latosol	Remld	0.5	97	36	29.0	56	27.4	57	-6
		Remld	1.5	99	38	25.3	56	30.4	55	+20
IP7	Andosol	Remld	0.5	186	69	34.4	54	34.9	62	+1
		Remld	1.5	201	52	81.1	57	87.2	62	+8
IP8	Latosol	Remld	0.5	100	48	26.7	52	29.9	53	+12
		Remld	1.5	80	35	30.5	52	30.3	54	-1

Key:

= Location close to sample A (Table 1)

Remld = Remoulded

LL = Liquid limit (60 min. mix, not pre-dried)

PI = Plasticity index (60 min. mix, not pre-dried)

SL = Shrinkage limit (ASTM)

ΔV = Overall percentage reduction in volume during test

SL diff = Percentage change in SL from 10 min. to 60 min. mix time.

Table 3 Summary of linear shrinkage 'reversibility' results

Samp No.	Soil type	Prep.	Depth (m)	LL (%)	PI (%)	LS (%)	LS re-test (%)	LS re-test / LS (%)
K1	Nitisol	gw	3.0	84	27	21.5	10.7	50
		h	3.0	83	28	21.5	11.3	53
K2	Nitisol	gw	0.3	85	27	21.7	6.9	32
		h	0.3	85	24	21.2	7.1	33
K3	Nitisol	h	1.7	58	11	18.6	10.6	57
K4	Nitisol	h	3.5	64		16.8	16.5	98
I1	Andosol	h	2.0	220	77	30.8	0.9	3
I2	Latosol	h	2.0	100	33	28.8	11.7	41
I3	Andosol	gw	3.8	108	25	24.6	7.0	28
		h	3.8	110	26	22.7	7.5	33

Key:

gw = grease-worked (Jones, 1992)

h = hand-mixed (Jones, 1992)

LS = Linear shrinkage (BS1377)

high, and particularly so for the andosols. There appears to be a positive correlation between shrinkage amount (ΔV) and plasticity index (PI). The andosol sample IP7 (1.5 m) stands out from the remainder in terms of shrinkage limit (SL).

4 LINEAR SHRINKAGE TESTS

4.1 *Method*

A series of linear shrinkage tests was carried out according to BS1377 (Anon, 1990, Part 2, Test 6.5). The reversibility of the shrinkage process was also investigated (Jones, 1992) by carrying out a linear shrinkage test (dried at 40 °C), mixing the sample, and carrying out the linear shrinkage test again (dried at 105 °C). The purpose of this was to examine the role of bonding in fabric breakdown, and hence susceptibility to shrinkage. Elsewhere, investigations have been made of the effects of drying temperature, preparation method, and the relationship between linear shrinkage and plasticity index (Northmore et al., 1992; Newill, 1961).

4.2 *Samples*

Testing was carried out by BGS on samples of nitisol collected from Kenya (samples K1 to K4) and samples of latosol and andosol collected from W. Java, Indonesia (I1 to I3). The nitisols are kaolinitic and halloysitic red clays derived from Tertiary volcanic rocks (Northmore et al., 1992b).

4.3 *Results*

All samples showed reduced linear shrinkage on re-testing (Table 3). In the case of the Kenyan halloysite-rich nitisols (K1 to K4) and the Indonesian latosol (I2), the re-test linear shrinkage is between 32 and 98 % of the original amount. However, for the Indonesian allophane-rich andosols this proportion is between 3 and 33 %. Of these, sample I1 gave virtually no shrinkage on re-testing. One sample (K4) showed virtually no reduction in shrinkage on re-testing. The results suggest that pre-shrinkage reduces the ability of the soil to re-shrink, and that this is particularly the case for allophane-rich clays. The relative strengths and densities of the mixed and re-mixed fabrics, and hence their resistance to shrinkage stresses are key factors. The tests also showed that grease-worked (totally de-structured) samples gave higher linear shrinkages than hand-mixed samples.

5 CONCLUSIONS

The new shrinkage limit tests, on two different tropical clay soils, showed the dramatic influence of weakly bonded fabric on shrinkage behaviour. Compaction has broken these bonds, reducing the soils to de-structured, homogeneous materials with 'normal' shrinkage curves characteristic of temperate clay soils. The allophane-rich clay showed an unusual 'undisturbed' shrinkage curve with a distinctive delayed stage. This curve, and the amount of shrinkage, were totally changed by compaction. The amount of shrinkage increased by a factor of four with compaction, despite the sample's greater density. The relative strengths of the undisturbed and compacted fabrics, and their resistances to shrinkage stresses are key factors.

The duration, and hence energy-input, of preparatory mixing has been shown to have an influence on the results of the standard shrinkage limit test on remoulded samples at water contents close to the liquid limit. However, results were inconsistent for this test. Shrinkage limits were found to be generally very high, and extremely high for one andosol. The results suggest that pre-shrinkage reduces the ability of the soils to re-shrink, and that this is particularly the case for allophane-rich clays. This is probably related to a particular mineral's morphology and its associated fabric, structural resistance to breakdown of bonding on re-mixing, and hence reduction of the clay's susceptibility to shrinkage.

The types of shrinkage behaviour described suggest that shrinkage is primarily influenced by the nature and mode of breakdown of the clay fabric, rather than by the mineralogy. However, the fabric itself is strongly influenced by mineralogy, as has been shown by the differing response of halloysite-rich and allophane-rich clay soils.

ACKNOWLEDGEMENTS

The contribution of the Institute of Road Engineering (Bandung, Indonesia) and the Ministry of Public Works, Materials Testing & Research Dept. (Nairobi, Kenya) to the sampling and testing programme is gratefully acknowledged. The paper is published with the permission of the Director of the British Geological Survey (NERC).

REFERENCES

Anon. 1990. BS1377 British Standard Methods of test for civil engineering purposes. *British Standards Institution* London.

Anon. 1995. Annual Book of ASTM Standards, Vol. 04.08, Soil & Rock (1); Building Stones. *Philadelphia: American Society for Testing and Materials [ASTM]*

Culshaw, M.G., Hobbs, P.R.N., & Northmore, K.J. 1992. Manual sampling methods. Technical Report WN/93/14. *British Geological Survey.*

Hobbs, P.R.N., Entwisle, D.C., Northmore, K.J., & Culshaw, M.G. 1992. Engineering geology of tropical red clay soils: geotechnical characterisation: mechanical properties and testing procedures. Technical Report WN/93/13. *British Geological Survey.*

Hobbs, P.R.N. New method for determining the shrinkage limit of clay soils (in preparation).

Jones, L.D. 1992. Test specification to assess the 'reversibility' of shrinkage characteristics (linear shrinkage) on selected Kenyan and Indonesian samples. Technical Report WN/92/3. *British Geological Survey.*

Kadir, A.A. 1997. The determination of shrinkage limit for clay soils using a travelling microscope with comparison to the established definitive method (TRRL). *M.Sc. dissertation. University of Leeds* (unpublished).

Marchese, D. 1998. The determination of shrinkage limits of compacted clayey soils – Mercia Mudstone and Gault Clay using a travelling microscope with comparison to the definitive method (TRRL, BS1377). *M.Sc. dissertation. University of Leeds* (unpublished).

Newill, D. 1961. An investigation of the linear shrinkage test as applied to tropical soils in relation to plasticity. *British Road Research Laboratory*, Note 4106, 6p (unpublished).

Northmore, K.J., Entwisle, D.C., Hobbs, P.R.N., Jones, L.D., & Culshaw, M.G. 1992a. Engineering geology of tropical red clay soils: geotechnical characterisation: index properties and testing procedures. Technical Report WN/93/12. *British Geological Survey.*

Northmore, K..J., Culshaw, M.G., & Hobbs, P.R.N. 1992b. Engineering geology of tropical red clay soils: project background, study areas and sampling sites. Technical Report WN/93/11. *British Geological Survey.*

Sridharan, A. & Prakash, K. 1998. Characteristic water contents of a fine-grained soil-water system. *Géotechnique* 48, 3, 337-346.

Vargas, M. 1988. Characterisation, identification, and classification of residual soils. *Proc. of the 2nd International Conference on Geomechanics in Tropical Soils*, 12-14 Dec. 1988, Singapore, 1, 71-75. A.A. Balkema, Rotterdam.

Vaughan, P.R. 1988. Characterising the mechanical properties of in-situ residual soil. *Proc. of the 2nd International Conference on Geomechanics in Tropical Soils*, 12-14 Dec. 1988, Singapore, 2, 20-37. A.A. Balkema, Rotterdam.

Wesley, L.D. 1973. Some basic engineering properties of halloysite and allophane clays in Java, Indonesia. *Géotechnique*, 23, 4, 471-494.

Unsaturated Soils for Asia, Rahardjo, Toll & Leong (eds) © 2000 Taylor & Francis, ISBN 90 5809 139 2

Effects of compaction on the reduction of undrained strength and stiffness of Cikarang expansive clay due to wetting

H. Jitno & B.T. Rayadi
Department of Civil Engineering, Institute of Technology Bandung, Indonesia

ABSTRACT: Before the economic crisis in 1997, Cikarang was one of the most actively developed regions in the eastern of Jakarta. Residential and industrial infrastructures were built extensively in the region. However, the infrastructure development was challenged by the many problems associated with the expansive nature of the clay in most of the region. Study has been conducted to investigate the behavior of compacted expansive clay from this region due to wetting. This paper presents the results of laboratory study on the effects of compaction on the reduction of undrained strength and stiffness of Cikarang expansive clay due to wetting. The results show that the undrained strength and stiffness reduction is highly dependent on the initial water content during compaction. The samples compacted at dry of optimum exhibit the largest reduction in the strength and stiffness, whereas the samples compacted at wet of optimum and optimum water content show smaller reduction. Effects of compaction on the changes of other parameters such as CBR values, permeability and swelling pressure are also presented. Practical implications of the research findings are highlighted.

1 INTRODUCTION

Cikarang is located in the eastern Jakarta about 40 km from Sudirman central Business District of Jakarta. Cikarang is one the suburbs at the north coastal area of west Java that enjoys vast infrastructure development. This area is suitable for industrial zone since it is relatively close to Tanjung Priok International Harbour. In addition to the industrial zone, some of the area was also developed for residential area for the people who work in the area or those who commute to Jakarta. The high price of housing in the Jakarta city pushes people to buy houses in the outskirts of the city such as Cikarang.

The rapid development of the infrastructure, which was often not supported by proper design and sufficient understanding of the soil behaviour in the area, has resulted in many failures of the civil structures constructed. Numerous houses, shops, factories, canals, roads and bridges have been reported to experience significant damage due to soil heave or subsidence. In some area, slope failures were found at almost the whole length of the canal protected by masonry wall. In other areas, the canal slope was protected by reinforced earth structures for about 1 km long. The protected slope failed completely just after the first rainy season came.

To overcome the problems, the first author was retained by one of the developers to develop a construction procedure that is able to prevent similar problems to occur. This paper presents some of the study conducted in order to understand the soil characteristics at the site.

2 SUBSURFACE SOIL CONDITION

Cikarang is located in the Jakarta (Batavia) alluvial plain which extends from Sunda Strait to the Bay of Cirebon (Bammellen, 1949). In the south of the marshy coastal belt, the young Pleistocene sediments in the area mostly consist of pumiceous tuff-sandstones presumably deposited during the eruption of Salak and Gede mountains. The tuffs contain quartz and acid plagioclase and fragments of andesite and basalt. Some areas are covered by young alluvial deposits of Cikarang River.

The Cone Penetration Tests conducted at the sites typically show the cone resistance of 5 to 15 kg/cm2 at a depth to about 6 m and reach 150 kg/cm2 at depth varying from 7 to 15 m.

3 SOIL PROPERTIES

The samples were taken from one of the project area in Lippo Cikarang. Both disturbed and undisturbed samples were recovered from the ground. Several tests were carried out to provide a set of data for identifying the degree of expansiveness of the soil including Atterberg limit and chemical soil tests. The results are presented in Tables 1 to 3.

Table 1. General Soil properties

Natural water content, w_n	35.86 %
Unit weight, γ (t/m^3)	1.77
Specific Gravity, Gs	2.57
Plastic Limit, PL (%)	35.82
Liquid Limit, LL (%)	110.17
Plasticity Index, PI (%)	74.35
Shrinkage Limit, SL (%)	8.92
Clay content (%)	61.6
Classification	CH

Table 2. Chemical Composition of Soil

SiO_2	59.43 %
$Al2O3$	24.12 %
$Fe2O3$	2.20 %
CaO	1.87 %
MgO	0.53 %
Na2O	0.91 %
K2O	0.72 %
Organic content	10.17 %

Based on the quantitative analyses and diffraction tests, the mineral composition of the clay was estimated as follows:

Table 3. Mineral Composition of Soil

Montmorrilonite	15.72 %
Kaolinite	55.17 %
Alpha Quartz	27.09 %

The degree of expansiveness of the clay was identified following the procedures proposed by several investigators including Seed et al. (1962), and Chen (1988). The clay is classified as expansive soil with very high swelling potential.

Based on the Standard Proctor Compaction test, it was found that the optimum moisture content is 32% and the maximum dry density is 13.5 kN/m^3.

4 TESTING PROGRAMS

Most of the problems at the site usually occur during or at the beginning of the rainy season, particularly during the rain after a long period of dry season. Some of the river slopes experienced large displacements due to creep after being subjected to several dry and wet seasons. However, most of the slope failures usually happen in a relatively short period of time. Considering the low permeability values of the clay, the drainage condition during failure can be considered fully undrained. Thus, the testing program relevant to this case will be under undrained condition.

The first series of tests were intended to study the effects of wetting on the strength and stiffness of clay compacted at different moisture contents. The samples were prepared at three initial water contents: 27.2% (dry of optimum), 32% (optimum), and 35.46% (wet of optimum). At each initial water-content, six samples were compacted using Standard Proctor procedure. The relative compaction of these samples is respectively 98%, 100% and 98%. Each sample was respectively submerged under the water for 0,1,2,3,4, and 5 days, before subsequently subjected to Unconfined Compressive Strength (UCS) and California Bearing Ratio (CBR) tests.

The second series of tests were conducted to investigate the effects of compaction and initial water content on the swelling of clay. The samples were prepared at five different initial water contents: 21.67%, 23.69%, 28.98%, 32.44%, and 36.54%. At each water content, seven samples were compacted by seven different compactive energy: 10 blows, 15 blows, 20 blows, 25 blows, 35 blows, 45 blows and 55 blows per layer. The samples were submerged under water and the swelling was monitored.

The third series of tests were carried out to investigate the effects of compaction and initial water content on the swelling pressure of the clay. The samples were prepared under three different compactive energy and five different initial water contents. The results were also compared with the swelling test of the undisturbed clay.

5 TEST RESULTS

Figure 1 shows the undrained strength of compacted expansive clay as a function of soaking time, for different initial water contents. Two samples were compacted to relative compaction of about 98% (wet and dry of optimum) and one sample to 100%.

We can see from the figure that the undrained strength (unsoaked) is the greatest on the sample compacted at dry of optimum (w=27.2%) although the relative compaction is about the same. However, the strength drastically reduced to about a half of its original strength after submerged with water for 3 days. The sample compacted at wet of optimum and at optimum water content only experienced about 20 percent reduction of its original strength after 3 days of soaking. The rate of strength reduction for these two samples appears to be similar, being about 8 to 9 percent/day of soaking.

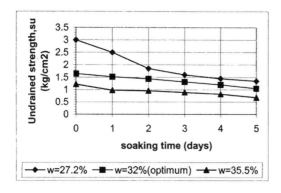

Figure 1. Strength reduction of Cikarang clay due to wetting for different initial water contents

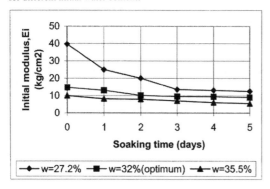

Figure 2. Stiffness reduction due to wetting for different initial water contents.

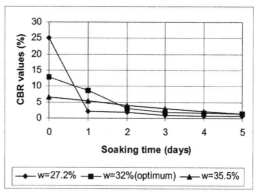

Figure 3. CBR reduction due to wetting for different initial water contents.

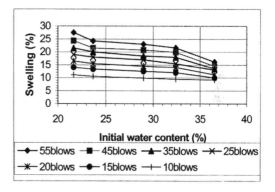

Figure 4. Effects of compactive energy and initial water content on the swelling potential of Cikarang expansive clay.

Figure 2 shows stiffness reduction with duration of soaking for the same samples as those in Figure 1. The trend is similar to that found for the undrained strength. Sample compacted at dry of optimum is much stiffer than the other samples. This sample has strain at failure of about 10 percent. The strain at failure tends to increase with increasing water content, or with duration of soaking. For example, the wet samples (water content higher than optimum) have the strain at failure of about 15%. The soil stiffness drastically reduced to about 1/3 of its original stiffness when the sample was soaked for three days.

The relationship between the CBR values and the duration of soaking is shown in Figure 3. Again, the sample compacted at dry of optimum shows the largest CBR unsoaked value, but the CBR decreases drastically from 25% to only about 2% after 1 day of soaking. The sample compacted at wet of optimum, on the other hand, only shows a slight reduction in CBR values, from 6% to 5%.

After 4 days of soaking, all samples practically have the same CBR values, i.e. about 1 to 2%.

The relationship between swelling potential and water content for different compactive effort is shown in Figure 4. Each curve represents the samples compacted using 55, 45, 35, 25, 20, 15, and 10 blows/layer. These represents the relative compaction ranging from 105% to 92% of Standard Proctor, with a maximum dry density of 13.5 kN/m^3.

It can be seen from Figure 4 that the swelling potential increases with the increase of relative compaction. Samples compacted with 55 blows/layer consistently show the greatest swelling potential at a given initial water content, confirming the findings by many previous investigators (e.g. Lambe, 1958). Moreover, for a given compactive effort, the swelling potential increases with decreasing initial water content. Thus, the sample compacted with the highest compactive energy at the driest water content shows the largest swelling potential. According to Lambe (1958), this is mostly due to the highly oriented soil structure of the sample.

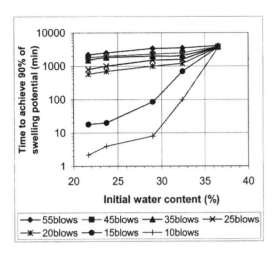

Figure 5. Time to achieve 90% of the swelling potential as a function of water content for different compactive efforts.

Although the swelling potential of the dense samples tends to be higher than those of the low density samples, the time to achieve most of the swelling is also longer, as shown in Figure 5. For a given compactive effort, the time required to achieve its maximum swelling increases with increasing water content. This is particularly true for the samples with low compactive energy. The time required to achieve its maximum swelling for dense samples is also affected by the increase in water content, but it is not as drastic as the low density samples. This is probably because dense samples have already had low permeability values and the increase in water content does not drastically reduce its permeability.

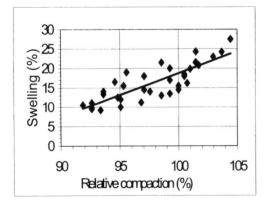

Figure 6. Effect of relative compaction on the swelling potential.

The swelling potential for different relative compaction is plotted in Figure 6. Although the data is rather scattered, there is a general trend that the

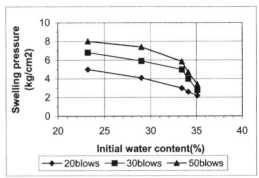

Figure 7. Effect of compactive energy and initial water content on the swelling pressure of Cikarang expansive clay.

higher the relative compaction, the higher the swelling potential.

Swelling pressure is closely interrelated with swelling potential. Therefore, it is expected that the trend will be similar to that obtained in swelling potential. The swelling pressure is found to be the highest on samples compacted at dry of optimum with high compactive effort (50 blows/layer), as shown in Figure 7. The swelling pressure also decreases with the increase in water content.

6 DISCUSSION

Several major problems encountered at the site were river slope failures, road maintenance, slab cracking, and bearing capacity failures.

Design of river slope must involve slope stability limit equilibrium analysis and the deformation analysis of the slope upon wetting, by incorporating appropriate strength and stiffness reduction in the soil model. This is important since a Factor of Safety (FOS) of 1.5 (or stress level of 0.67) in slope consisted of soft wet expansive clay will have larger deformation than that in slope of stiff clay. For example, in the clay tested, the stress level of 0.67 corresponds to axial strain of about 10%, which is large enough for slopes. Thus, the design criteria will not only acceptable FOS but also acceptable deformation. Deformation criteria will usually be the governing factor in designing slopes in soft clay.

Fifty percent strength reduction and 70 percent stiffness reduction was found to be adequate for the analyses. Using this approach, Jitno et al. (1996) presented a finite element analysis to estimate the deformation of Cilemahabang river slope protection due to wetting. The 1-km river slope designed using this approach has been performing well to date without any sign of distress.

In practice, expansive fill is often compacted at 4 to 5% wet of optimum water content. Although it is generally acceptable to obtain low permeability engineered fill, it may have lower strength and may be too difficult to compact (too wet). For slopes, as long as the surface is drained properly or protected with non-expansive clay layer, the Cikarang expansive clay has been compacted successfully at 2% dry of optimum to build up to 6-m high slopes without any failures.

7 CONCLUSION

Based on testing programs conducted on Cikarang expansive clay, the following conclusion can be derived:

For the expansive clay tested, it has been shown that the undrained strength and stiffness is highly dependent on the initial water content during compaction. The samples compacted at dry of optimum show the largest reduction in strength and stiffness which may reach 50% to 70% reduction, whereas the samples compacted at optimum and at wet of optimum only experience about 20% reduction after three days of soaking.

The samples compacted at dry of optimum show the largest swelling potential and swelling pressure. These two parameters decrease with increase in initial water content.

The time required for the samples to achieve its maximum swelling potential increases with increasing density and water content. Thus, as long as drainage is properly maintained, expansive soil compacted to achieve high relative compaction may not achieve its full swelling potential if water seepage is prevented.

Earth structures related to expansive soil are best analyzed using combination of both limit equilibrium slope stability analysis and deformation analysis. Both stability and deformation criteria must be satisfied. However, the deformation criterion is usually the governing factor in slope design involving soft plastic clay. Furthermore, the strength and stiffness of the soil must be chosen appropriately to represent the possible conditions in the field. This includes strength and stiffness reduction due to wetting during rainy seasons.

8 ACKNOWLEDGEMENT

The authors wish to express their gratitude to Bapak Teddy Hioe, General Manager of Housing Division, Bapak Djoko Susanto and Ibu Noormalasari of Lippo Cikarang for the opportunity to involve the authors in this interesting project.

REFERENCES

Bammelen, R.van (1949). *The Geology of Indonesia*. Martinus Niyhoff, The Haque. 732 pp.

Chen, F.H. (1988). *Foundations on Expansive Soils*. 1st ed. Amsterdam, Elsevier, 463pp.

Jitno, H., Hioe, T. and Susanto, D. (1996). Cilemahabang River Slope Protection, *Proceeding of National Geotechnical Seminar on Expansive Soil*, Sponsored by Lippo, Civil Engineering-ITB and Jayabaya, Lippo Cikarang.

Lambe, T.W. (1958)."The engineering behaviour of compacted clay," *Journal of Soil Mechanics and Foundation Engineering*. ASCE, Vol. 84, SM2, pp. 1-35.

Seed, H.B., Woodward, R.J. & Lundgren, R. (1962). Prediction of swelling potential for compacted clays. *Journal of Soil Mechanics and Foundation Engineering*. ASCE, Vol.88. SM4,pp.57-59.

Unsaturated Soils for Asia, Rahardjo, Toll & Leong (eds) © 2000 Taylor & Francis, ISBN 90 5809 139 2

Evaluation of ground movement using a simple soil suction technique in Western Australia

A. Khan & H. R. Nikraz

Curtin University of Technology, Perth, W.A., Australia

ABSTRACT: This paper reports the finding of a comprehensive investigation in which various methods of seasonal heave prediction were compared and estimates of ground movement made and reviewed in light of the seasonal soil movements observed at a site in Swan View in Perth, Western Australia. The methods used included empirical, semi empirical, mathematical and soil suction models. In addition, with a view to determine the vertical swell pressure and the accompanying volume changes under simulated field conditions, a modified oedometer ring was used. This instrument allowed the measurement of lateral pressure and lateral swelling pressure of a soil sample in the laboratory, without lateral strain. The paper contains a description of the equipment commissioned, test techniques, results, analysis and interpretation of the data obtained. The testing and evaluation techniques are general in nature and can be applied to field situations in locations where similar soils occur.

1 INTRODUCTION

The prediction of vertical movement of reactive clay soil profiles resulting from soil moisture changes is concern to many geotechnical engineers. The problem arises in those parts of the world where the climate is arid for all or part of the year. Such a climate exists over large regions of Australia, and considerable damage to road pavements and light structures is the result. In particular, wall cracking in masonry domestic houses is often severe, and has attracted wide public interest.

In Perth, Western Australia, several large scale housing developments in the foothills area of the Darling Range have recently proven to be bad investments for both individual block buyers and the developer because of widespread presence of active soils derived from dolerite dyke weathering. There is an urgent need for assessment of the seasonal movement of weathered dykes with different expansive properties in the foothills area. The determination of this movement may be determined by analytical methods and compared with observed seasonal heave. Some methods available to the design engineer for predicting heave are presented and assessed. To assess the validity of the methods, a series of 24 field surveying pegs were placed in four adjacent blocks of land in Swan View, Perth, Western Austra-

lia, as shown in Figure 1. Vertical seasonal ground movements were monitored. Soil suction, swelling pressure and moisture contents were assessed at strategic intervals to enable comparison of seasonal ground movements with predicted ground.

Concurrently, a comprehensive set of traditional engineering index and other soil tests were conducted on the soils in an attempt to compare empirical and semi-empirical methods of seasonal heave prediction with the observed ground movements.

2 THE SUCTION MODEL

Moisture-related movements of expansive clays can be predicted using a simple formula proposed by Aitchison (1973).

$$\delta = \sum_{n=1}^{m} I_{pt} \Delta u \Delta Z_n \qquad (1)$$

where

δ = the vertical shrinkage or heave,
Ipt = the instability index,
Δu = the soil suction change in pF units,
ΔZ_n = the thickness of the nth soil layer,
m = the total number of soil layers considered.

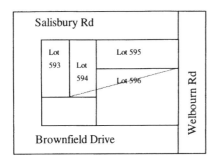

Figure 1. Plan view of project land. (Swan View)

Equation 1 expresses the suction model for calculating moisture-related ground movements. The model assumes that movement is dependent on three main factors:

* the instability index, Ipt, which is the rate of vertical strain of the clay with change in suction
* the depth of the clay, Z, and
* the soil suction changes that occur in the clay, Δu.

3 INSTABILITY INDEX

The instability index is a measure of the reactivity of a clay soil and is the percentage vertical strain per unit change in suction. It may be determined by a number of methods including; core shrinkage index, loaded shrinkage index and shrink-swell index. However, Cameron (1989) following a comprehensive investigation of the various forms of the reactivity tests, concluded that shrink-swell index is the most reliable method. The shrink-swell index (Iss) is given by

$$I_{ss} = \left[\frac{(\varepsilon_{sw}/2) + \varepsilon_{sh}}{1.8} \right] \qquad (2)$$

where
ε_{sw} = swelling strain
ε_{sh} = shrinkage strain to the oven-dry condition.

4 MEASUREMENT OF SOIL SUCTION

Total soil suction can be reliably measured in an exceptionally well-controlled environment using the conventional psychometric (Richards,1980) and the mechanistic approach using oedometer and pressure membrane devices (Johnson, 1974; Aitchison, 1973).

However, several calibration problems complicate analysis of results from pressure membrane tests. The test procedures are often laborious and time consuming. The use of thermocouple psychrometers to measure the soil suction is simple and reliable. But the range of measurements is limited. Low suction levels of less than 100 kpa cannot be measured by the thermocouple psychrometer, so that it does not cover the full range of values encountered in many environments.

There is a need for a single, effective and rapid routine method for measuring soil suction over the full range encountered in the field. Such a method has existed for a considerable period of time but it has not yet found much favour with the geotechnical community. Gardner (1930) first proposed the use of filter paper to measure soil suction.

A comprehensive correlation of the equilibrium water contents of Whatman's No. 42 filter paper at known suctions was made by Fawcett and Collis-George (1967). A variety of techniques such as pressure plate, pressure membrane and vacuum desiccator were used by Fawcett and Collis-George (1967) and Hamblin (1981) to obtain a range of suction from pF 1.0 (1 kPa) to pF 6.0 (100 Mpa).

Chandler and Gutierrez (1986), repeated the above correlation using Whatman's No. 42 paper up to a suction of pF 4.3 using a conventional oedometer. Their results were compared to those obtained by Hamblin (1981) and Fawcett and Collis-George (1967) to produce a line of best fit described by the relation

$$pF = 5.850 - 0.0622 \, w \qquad (3)$$

where
'w' = moisture content of filter paper.

The above relation was used to describe a range for pF of 2.9 to 4.8. The authors have adopted this relation for finding pF values of filter paper moisture contents in the range of 17 to 47%. In this study Equation 3 was extrapolated for finding suction values of filter paper moisture contents below 17%. For filter paper moisture contents of above 47% the following best-fit linear relation was adopted (Hamblin, 1981);

$$pF = 3.67 - 0.0159 \, w \qquad (4)$$

In this study the filter paper method has been preferred and all suction quoted are total suction. The test procedures are reported by Nikraz (1985). A typical graph of total suction with respect to soil moisture content for station 4 is given in Figure 3.

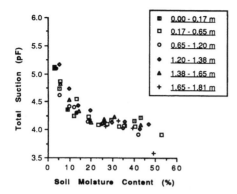

Figure 2. Total suction versus soil moisture content

5 CONSOLIDATION-SWELL TEST METHOD

Vertical swelling pressure is often determined by use of a consolidometer or other special apparatus but there has been little attention paid to the need for simultaneous measurement of lateral and vertical swelling pressure of expansive soils. Experience in Israel has shown that piles and other structures buried in a clayey soil are damaged by the action of horizontal swelling forces as well as vertical forces (Zeilten and Komorink, 1960; Kassiff and Zeitlen, 1962; Kassiff and Baker, 1969). In some of these cases the clay was free to expand in the vertical direction and yet appreciable horizontal forces appeared. For proper analysis of actual conditions, the values of both vertical and horizontal swelling forces need to be determined.

A correlation between the amount of swell and the lateral swelling pressure developed in a clay sample may be obtained from the lateral swelling pressure ring, which is a modified thin-walled consolidometer ring instrumented with strain gauges. The modified oedometer ring, the Lateral Soil Pressure Ring (LSP Ring) used for this study has been fully described by Nikraz (1985).

Swell testing was carried out on undisturbed samples. Approximately 12 tests were carried out with the lateral swelling pressure ring. The variation of the horizontal swelling pressure and vertical swell of the clay were determined for various values of vertical pressure, 7, 20, 47 and 77 kPa.

6 EMPIRICAL & SEMI APPROACHES

Seed et al (1962) proposed an empirical relationship between swell potential, soil activity (A) and clay content (C), which was later to be disputed by Ran-

ganathan and Satyanarayana (1965). They contended that swell potential could be more accurately assessed from the shrinkage index of the soil. Van der Merwe (1964) used a similar approach to Seed et al (1962). However, swell potential was ignored and the three basic soil properties of plasticity index (PI), clay content (C) and activity (A) were instead directly related to a general heave potential classification.

McDowell (1956) presented a method of predicting seasonal heave based upon generalised heave/swell pressure curves derived from a series of tests conducted on North American clay samples remoulded at various combinations of moisture and density. The percentage volumetric change can be either approximated from linear shrinkage (LS) or plasticity index.

7 MATHEMATICAL MOISTURE FLOW MODEL APPROACHES

A further estimate of movement was made with the moisture content data. The mathematical flow model of Richards (1967) expressed below was used:

$$\delta = \frac{1}{3} \sum_{n=1}^{m} \left[\frac{(\omega - \omega_o)G_s}{(100 + \omega_o G_s)} \right] \tag{5}$$

where

G_s = specific gravity of the soil
ω = final moisture content
ω_o = initial moisture content

Equation 5 assumes that the soil remain quasi-saturated throughout the shrinking and swelling phases, so that changes in water volume produce corresponding changes in total volume of the soil. If this assumption is valid, overestimates of movement will be produced. Researchers have previously reported good results using this expression with moisture contents converted from suction (Holland & Lawrance 1980). However in this study comparative movement estimates are based directly on water content data and do not require any suction input.

8 OBSERVED GROUND WATER MOVEMENT

Twenty-four surveying pegs approximately 2.5 metres apart were placed in the ground early January, as shown in Figure 1. The initial surveying peg levels were taken in April, when the soil was assumed to be at its driest state. The soil movement has been monitored every five weeks to determine the seasonal driest and wettest conditions. From the surveying peg levels, the ground movement profile of surveying pegs was outlined as shown in Figure 3.

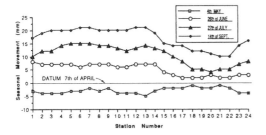

Figure 3. Observed soil movement

9 COMPARISON OF MOVEMENT PREDICTIONS AND ACTUAL GROUND MOVEMENTS

The comparison of the five estimates of surface movement, using different technique, for each of the stations and observed seasonal heave is given in Table 1. Typical input data required for these methods are given in Table 2, and Figures 4, 5 and 6.

From the summary of results in Table 1, Van der Merwe's method, which considers clay type and amount, depth of active clay and to some extent, overburden pressure, greatly overestimate for all the three stations. These overestimates suggest that Van der Merwe's proposed linear swelling strains corresponding to particular heave potential classifications must be varied to suit local climatic conditions and soil types, and that site drainage must be considered.

McDowell's method underestimated for all the three stations. This weakness is most likely related to the fact that McDowell does not consider either basic soil nature (ie. clay content, soil structures) or site drainage.

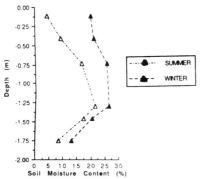

Figure 4. Soil moisture content vs depth.

In contrast, Table 1 illustrates that the Richard's method is generally conservative. It would be expected to produce overestimates when the assumption of quasi-saturated is invalid, as it is likely to be in desiccated soils. Furthermore, Equation 5 takes no account of soil history, mineralogy or composition, nor does it consider soil loading, and therefore it could not be expected to produce consistently accurate results.

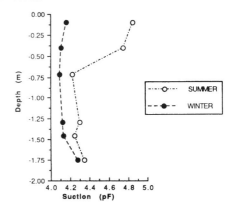

Figure 5 Suction vs depth

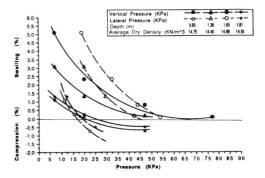

Figure 6 Typical vertical movement (station 4)

Typical consolidometer swell/vertical and lateral pressure curves for range of sampling depth are presented in Figures 6 for station 4. It can be seen from Table 1 that the swelling pressure model adequately reflected the trend of seasonal movement for the all three stations.

The ratio between the maximum lateral pressure after swelling and the overburden pressure was greater than 2, and decreased to less than 1 with an increase in the overburden pressure (see Table 2). The relative high lateral stresses measured are of great significance in explaining some of the damage observed to structures founded in swelling clay soil conditions.

The importance is obviously indicative of determining lateral pressures as a factor in design of structures exposed to forces from swelling clays.

The measured soil and total soil suction are used to develop total suction/depth profiles. Typical suction profiles for station 4 were plotted against depth in Figure 5. Surface movement (ys) was calculated over the depth of seasonal suction variation by using Equation 1. The value Iss were taken from Table 2 and corresponding values of soil suction were taken from suction profiles. From Table 1, it can be seen that the suction model approach of predicting heave from total suction profiles was able to provide better estimates than the other methods.

Although the suction model approach generates the smallest error, further studies of this type would be required before this method could be recommended with confidence.

10 CONCLUSIONS

The studies described in this paper lead to the following conclusions:

1. From the comparison of actual seasonal heave measurements to predicted values presented in this paper, it has been illustrated that methods of prediction based mainly on the type of clay, such as those developed by Van der Merwe and McDowell to a lesser extent, are of doubtful value. It is essential to consider at least soil profile and soil drainage in conjunction with the type and quantity of clay for accurate heave predictions.
2. The vertical swell pressure generated in expansive soil can be determined by employing the equipment described in this paper which is essentially a modified form of the thin walled consolidometer ring instrumented with strain gauges. The use of such equipment has the advantage of determining the lateral swell pressure at the same time.
3. Consolidation-swell tests method of predicting heave yielded excellent estimate of the seasonal heave for all three experimental stations. It was found that the lateral swelling pressure ring was also well adapted to measuring the at-rest pressure for the clay samples. The ratio of lateral to vertical stresses was higher for lower vertical stresses.
4. The suction model is a useful concept to employ in assessing ground movements. To enable the prediction, probable variations in soil suction over the depth of seasonal movement must be known as well as Instability Index of the soil.
5. The filter-paper method is a simple, inexpensive and reliable test that could prove to be of great value in measuring soil suction over a range of moisture contents that are commonly encountered in normal engineering situations.

Table 1. Comparison of Predicted and Observed Seasonal movement

Station No.	Predicted Method	Seasonal Heave (mm)	
		Predicted	Observed
4.	Suction Model	23.81	24.00
	Consolidometer Approach	27.6 (Nikraz 1985)	
	Empirical Approach	49.17 (Nikraz 1985)	
	Semi-Empirical Approach	21.64 (Nikraz 1985)	
	Mathematical Moisture Model	41.00 (Nikraz 1985)	
6.	Suction Model	26.11	24.00
	Consolidometer Approach	28.17 (Nikraz 1985)	
	Empirical Approach	41.32 (Nikraz 1985)	
	Semi-Empirical Approach	17.76 (Nikraz 1985)	
	Mathematical Moisture Model	50.37 (Nikraz 1985)	
13.	Suction Model	25.71	26.00
	Consolidometer Approach	30.76 (Nikraz 1985)	
	Empirical Approach	45.49 (Nikraz 1985)	
	Semi-Empirical Approach	19.76 (Nikraz 1985)	
	Mathematical Moisture Model	62.23 (Nikraz 1985)	

Table 2. Consolidometer Testing Results for Station 4

Depth (mm)	0 – 650	650 – 1380	1380 – 1650	1650--1810
Average Overburden Pressure (kpa)	4.79	14.83	22.11	25.11
$\Delta L/L$ % $\Delta L=27.60$	1.2	1.9	2.1	---
L (mm)	7.8	13.9	5.7	----
Lateral Pressure (kpa)	12	26	33	20
$K_{o(s)}$ *	2.50	1.75	1.50	0.80

$K_{o(s)}$ * = Coefficient of earth pressure at zero due to swelling Pressure

REFERENCES

Aitchison, G.C. (1973) *The quantitative description of the stress-deformation behaviour of expansive soil*, Proceedings of the 3rd International Conference on Expansive Soils, Haifa, Vol. 2, p. 79-82.

Alpan, I. (1957) *An apparatus for measuring the swelling pressure in expansive soils*, Proceedings of the 4th International Conference on Soil Mechanics and Foundation Engineering, London, Vol. 1, p. 3-5.

Cameron, D.A. (1989) *Tests for reactivity and prediction of ground movement*, I.E.Aust. Civil Engineering Transactions, VCE 31(3), p. 121-132.

Chandler, R.J. & Gutierrez, C.I., (1986) *The filter-paper method of suction measurement,* Technical Notes, Geotechnique, June, Vol. XXXVI, No. 2, pp.265-268.

Fawcett, R.G. & Collis-George, N. (1967) *A filter-paper method for determining the moisture characteristics of soil,* Australian Journal Experimental Agriculture and Animal Husbandry, Vol. 7, p. 162-167.

Gardner, R. (1937) *A method of measuring the capillary tension of the soil moisture over a wide moisture range.* Soil Sc., Vol. 43, p. 227-283.

Hamblin, A.P. (1981) *Filter paper method for routine measurement of field water potential.* Journal of Hydrology, Vol. 53, p. 355-360.

Holland, J.E. & Lawrance, C.E. (1980) *Seasonal heave of Australian clay soils*, Proceedings of the 4th International Conference on Expansive Soils, Vol. 1, p. 302-321.

Johnson, L.D. (1974) *An evaluation of the thermocouple psychometric technique for the measurement of suction in clay soils*, Technical Report S-74-1, U.S. Army Engineer Waterways Experiment Station, CE, Vicksburg, Miss., USA.

Kassiff, G. & Zeitlen, J.G. (1962) *Behaviour of pipes buried in expansive clays*, Proceedings ASCE, New York, Vol. 88(SM2), p. 133-148.

Kassiff, G. & Baker, R. (1969) *Swell pressure measured by uni and triaxial techniques*, Proceedings of the Seventh International Conference of Soil Mechanics and Foundation Engineering, Mexico City, Mexico, Vol. 1, p. 215-218.

Nikraz, H.R. (1985) *Instrumentation for laboratory and in-situ determination of lateral earth pressure, swelling pressure and heave prediction.* Msc Thesis, Curtin University of Technology, p. 158.

Ranganathan, B.V. & Satyanarayana, B. (1965) *A rational method of predicting swelling potential for compacted expansive clays*, Proceedings of the Sixth International Conference on Soil Mechanics and Foundation Engineering. Vol. 1, p. 92-96.

Richards, B.G. (1967) *Moisture flow and equilibria in unsaturated soils for shallow foundations. Permeability and capillary of soils*, ASTM Special Technical Publication, STP 417, p. 4-34.

Richards, B.G. (1980) *Measurement of soil suction in expansive clay.* Civil Engineering Transactions, I.E. Aust, Vol. CE22, No. 3, p. 252-261.

Seed, H.B., Woodward, R.J. Jr & Lundgren, R., (1962) *Prediction of swelling potential for compacted clays*, Journal of Soil Mechanics and foundations Division, ASCE, Vol. 88(SM3), p. 53-87.

Van der Merwe, D.H. (1964) *The prediction of heave from the plasticity index and percentage clay fraction of soils'*, Transactions, South African Institute of Civil Engineers, Vol. 6, p. 103-107.

Zeilten, J.G. & Komorink, A. *Concrete structures and foundation design as influenced by expansive clay in a semi-arid climate.* National Symposium on concrete and reinforced concrete in hot countries, Haifa, R.I.L.E.M. International Union of Testing and Research Lab. for Material and Structures.

Unsaturated Soils for Asia, Rahardjo, Toll & Leong (eds) © 2000 Taylor & Francis, ISBN 90 5809 139 2

Desiccation cracking of soil layers

J.K. Kodikara
School of the Built Environment, Victoria University of Technology, Melbourne, Vic., Australia

S.L. Barbour & D.G. Fredlund
Department of Civil Engineering, University of Saskatchewan, Saskatoon, Sask., Canada

ABSTRACT: The current paper presents a brief review of the available and emerging theories of desiccation cracking of clay soils, and an analysis of some laboratory tests undertaken on desiccation of relatively thin soil layers. The laboratory test data analysed included those reported by Corte and Higashi (1960) and Lau (1987). The experiments were conducted to study the effects of soil thickness, initial soil density, base adhesion and desiccation rate. The thicknesses of soil layers were in the range 3 mm to 60 mm. The results of the experiments indicated that cracking occurred predominantly in orthogonal patterns although non-orthogonal (hexagonal) patterns occurred in some instances. The results also indicated that the initial cracking water content decreases as the desiccation rate increases. The analysis of results indicated that cells created by cracking followed a log-normal distribution, and the mean cell area was a function of soil thickness, base adhesion, and initial soil density, but was not strongly related to the desiccation rate.

1 INTRODUCTION

It is common knowledge that clay soils can crack during desiccation. Cracks occur when soils are restrained while undergoing volume change produced as a result of the soil suction generated within the desiccating soil matrix. The desiccation cracking of clay soils can have a severe impact on the performance of clay soils in various geotechnical, agricultural and environmental applications. For example, desiccation cracking has the potential to render a low conductivity barrier constructed of clay soil virtually ineffective. Despite this significance, the essential mechanisms of desiccation cracking are not well understood, and consequently, predictive tools are inadequately developed.

Previous studies on desiccation cracking are spread across several disciplines and they date back to the early twentieth century. These studies have been predominantly qualitative and behavioural in nature. The field evidence reported is wide-ranging, generally incomplete in details and sometimes conflicting. For example, wide ranges in crack spacing, varying from about 75 mm to 76 m have been reported. Similarly, the depths of desiccation cracks ranging from several centimeters to over 10 m have been reported in the literature. The relationship between crack depth and spacing was also not very clear, although a general trend of larger spacing with deeper cracks seemed to exist. Several researchers

including Lau (1987), Morris et al. (1992) and Kodikara et al. (1998) have provided detailed summaries of the field evidence of desiccation cracking.

The apparent mysterious nature of desiccation cracking process can only be understood clearly by adopting a systematic research approach to analysis and experimentation. The current paper provides a brief review of the available and emerging theories and an analysis of some laboratory tests undertaken on desiccation of relatively thin soil layers. A fundamental understanding of desiccation of thin soil layers is an initial and useful step and is, specifically, relevant to the behaviour of soil liners used as covers and various material layers of road pavements.

2 THEORIES OF DESICCATION CRACKING

2.1 *General*

It is commonly considered that desiccating clay soils crack when the tensile stress developed in the soil due to the matric soil suction exceeds the tensile strength of the soil. Tensile stresses develop only when the soil is restrained in some way against shrinkage. The restraints can be external (e.g. rough layer interfaces) or internal (e.g. sections of soil undergoing non-uniform drying).

Observed field aerial cracking patterns can be divided into two broad categories, namely orthogonal patterns and non-orthogonal patterns. In orthogonal patterns, cracks tend to meet at right angles. In the evolution of these patterns, cracks usually occur sequentially. First, primary cracking develops dividing the clay surface into blocks and subsequent drying tends to further subdivide these blocks. In non-orthogonal cracking patterns, the cracks do not meet at right angles. Hexagonal patterns (cracks meeting at angles of 120^0) and their deviations fall into this category. The non-orthogonal cracks appear to originate simultaneously and connect up to form a blocky pattern. During, subsequent drying, however, (secondary and tertiary) cracking and opening of the cracks can occur over larger blocks encompassing a number of smaller blocks. Theoretically, this secondary and tertiary cracking behaviour can be considered to be a *bifurcation* from the primary cracking pattern (Bezant and Cedolin, 1991; Kodikara et al., 1999). As postulated by Bezant and Cedolin (1991) for cracks originated by uniform shrinkage (or cooling) in idealised media, Figure 1 shows the theoretically plausible crack patterns. The parallel cracks shown in Figure 1(a) can be considered as two-dimensional whereas other crack patterns are essentially three-dimensional.

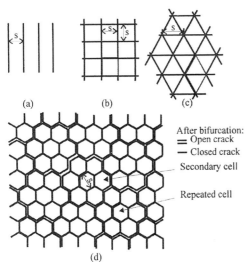

Figure 1. Theoretically plausible patterns of desiccation cracks in idealised media (after Bezant and Cedolin, 1991)

2.2 *Theoretical Models*

The theoretical research on desiccation cracking is not extensive with notable contributions made by Lachenbruch (1961), Fredlund and Morgenstern (1976), Lau (1991), Morris et al. (1992) and more recently, by Konrad and Ayad (1997). The original work of Lachenbruch (1961) involved the use of lin-

ear elastic theory and fracture mechanical principles to explain the magnitudes of depth and spacing of thermal contraction cracks in permafrost. Lau (1987) and Morris et al. (1992) made use of the consistent framework proposed by Fredlund and Morgenstern (1976) for unsaturated soils to develop theories for the prediction of the stable depths of desiccation cracking. Konrad and Ayad (1997) presented an idealised model for the analysis of clayey soils undergoing desiccation, utilising fracture mechanical concepts. Despite these advances, theoretical modelling approaches need further development, particularly explaining and predicting the relationship between crack spacing and depth and the evolution of crack patterns generally observed in the laboratory and in the field.

The crack spacing depends on the cracking pattern. Currently, there is no definite approach to the prediction of crack spacing for polygonal cracking patterns or the shape of the polygonal blocks. Instead, researchers have concentrated on prediction of the spacing of parallel cracks (in half-space) as a starting point (Lachenbruch, 1961, Konrad and Ayad, 1997). Lachenbruch (1961) and Konrad and Ayad (1997) considered the stress perturbation (or stress relief) caused by the formation of a single crack in a clay medium subjected to a tensile stress regime. According to this concept, the stresses are totally released at the face of the first crack, but will gradually rise away from the crack face. They considered that a second crack will occur when the tensile stress rises to a certain factor (α) of the tensile strength of the soil. In their analysis, the factor α is arbitrary and has no any physical meaning.

3 EXPERIMENTAL DATA

3.1 *General*

In the current paper, available laboratory experimental data on the desiccation of soil layers are analysed. The data include those reported by Corte and Higahsi (1960) and Lau (1987). The process of desiccation is strongly dependent on the local climatic conditions. The climatic conditions include temperature, relative humidity (RH), wind velocity and solar radiation. In these laboratory experiments, the temperature and RH are controlled. In the following sections, only a brief account of these tests is given.

3.2 *Data of Corte and Higashi (1960)*

Corte and Higashi (1960) carried out these experiments as part of early research on patterned ground by the US Corps of Engineers. These tests appear to be the most comprehensive series of laboratory tests undertaken on the desiccation of soil layers, but seemed to have not been published except in their original report. It is noted that about 60 tests have been conducted spanning a period of one year. The test parameters were divided into two categories, namely extrinsic and intrinsic. The extrinsic parameters controlled were temperature, relative humidity, thickness of the soil layer and the base material. The relevant intrinsic test parameters were initial moisture content and density or initial state of the soil.

The soil used for the experiments was Bloomington till obtained from a till deposit near Lily Lake, Illinois. The soil was processed by drying and crushing and passing through a 1mm screen. Table 1 shows the basic characteristics of the soil used.

Table 1.Basic physical characteristics of the soils used in the desiccation tests

Soil Description	Liquid Limit, Plastic Limit, Plasticity Index, Shrinkage Limit, Specific Gravity
Bloomington Till (Corte and Higashi, 1960)	31.4, 13.9, 17.5, 12.0, 2.63
Indian Head till (Glacial till) (Lau, 1987)	34.3, 14.2, 20.1, 14, -
Reginal clay (lake clay) (Lau, 1987)	89.5, 26.2, 54.3, 22.0, -

Most of the experiments were carried out in flat wooden containers of 600 mm x 840 mm plan area and 70 mm depth. The soil was used in two initial states: (1). as a slurry with initial water content of 60% and dry density of 1800 kg/m^3; and (2) as a loosely compacted soil with initial water content of about 45% and a dry density of about 1500 kg/m^3. Various materials were used at the base of the containers in order to provide differing base adhesion characteristics. The base materials used included plain wood, greased wood, and sheet of glass. A 20mm thick base layer of sand in the container having a wood base was also tested. For majority of the tests, the room temperature was kept about 22°C and the relative humidity was in the range 30 to 40%.

3.3 Data of Lau (1987)

These tests have been carried out in the soil laboratory at the University of Saskatchewan Canada. As detailed in Table 1, two types of local soils were used in the tests. The cracking tests were carried out in a flat wooden container of 610 mm x 610 mm plan area and 76 mm deep. The soils were prepared so that their initial moisture contents were close to their liquid limits and the soils were close to a slurry state at the beginning of the tests. An additional aspect of these tests was that the soil was instrumented with four embedded ceramic-cup tensiometers to

measure the soil suction during the tests. Furthermore, the vertical deformations of the top soil surface was also measured using dial gauges that were mounted at the four quadrants of the container. After the cracking commenced, the soil moisture content during the tests were measured by taking small samples of soil. No special attention was paid to the base condition of the wooden container. Photographs were taken during the tests to record the development of the crack pattern. Unfortunately, in most of these tests, the initiation of cracking was influenced by the presence of instruments in the soil.

4 ANALYSIS OF EXPERIMENTAL DATA

4.1 Desiccation rates and cracking water content

Figure 2 shows the reduction of moisture content with time for three typical tests (Experiments 15, 33 and 37) using soil at a slurry state, as conducted by Corte and Higashi (1960). The thicknesses of soil layer in these experiments were 9.6 mm, 9.5 mm and 28.1 mm respectively. Experiments 15 and 33 had wood bases and Experiment 37 had glass at the base. Experiments 15 and 33 had similar conditions except that Experiment 33 was conducted under a higher relative humidity of 90%. Also shown in Figure 2 are (desiccation) curves fitted to the experimental results. These curves are represented by $w = w_0 \exp(-kt)$, where k is referred to as the desiccation speed or rate, and w_o is the initial water content. The water contents at the crack initiation (cracking water contents) are also marked (by arrows) on the desiccation curves.

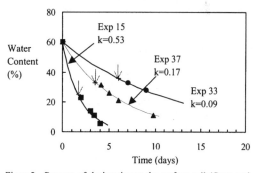

Figure2. Process of desiccation at the surface soil (Corte and Higashi, 1960).

The results of Experiments 15 and 33 show that desiccation rate, which is simplistically characterised by the parameter k in this instance, is strongly dependent on the local climatic conditions. The desiccation rate is also dependant on the thickness of the soil layer as evidenced from the results of Experiments 15 and 37, where the only relevant test variable is soil thickness. Similar observations can be made with Lau (1987) results. The test results of Corte and Higashi (1960) have generally indicated that the initial part of the desiccation curve becomes more linear as the initial soil density decreases (or is loosely packed). Approximate theoretical analyses of soil evaporation were carried out using SOIL-COVER model, a finite element program developed for modelling soil evaporation (Wilson et al., 1990). Owing to the limited space available, the results of these analyses are not presented here, but it was found that this model is capable of emulating test results and is suitable for modelling evaporation from soil layers.

Another important aspect evident from Corte and Higashi test data is the dependence of cracking water content on the desiccation rate. Their overall test data clearly indicate that the cracking water content decreases significantly as the desiccation rate increases. It is not possible to explain this phenomenon from the current desiccation theories, because they do not include a relationship of time (or desiccation rate) to the material behaviour. In the current understanding, the initial cracking is dependent on the soil matric suction, which, in turn, is uniquely related to the soil water content.

4.2 Observations on crack growth and pattern

Test results illustrate that crack growth occurs predominantly in orthogonal and sequential manner with primary cracks forming initially and secondary cracks subdividing the initial crack pattern (see Figure 3). However, there is evidence that at higher desiccation rates or at smaller thicknesses, non-orthogonal and predominantly 120^0 crack patterns evolve at the primary cracking stage, and they occur simultaneously and relatively fast over the soil surface. This pattern can be seen in Figure 4, which can be compared to Figure 3. The main test variable between these two tests is the thickness of the soil layer. In certain tests, mixtures of orthogonal and non-orthogonal cracks occur. In addition, hexagonal patterns were more common in loosely packed soil than in soil slurry, indicating an influence of the crack pattern on the soil density.

4.3 Influence of the base material

Test results indicate that base material has a significant influence on the resulting crack pattern and the block or cell sizes. This is illustrated in Figures 5 and 6, where the results of comparable crack patterns are shown for glass and wood bases respectively. It is clear that glass base produces smaller cell sizes in comparison to wood base. Generally, this effect is clearly evident as shown graphically in

Figure 7. The mean cell area ($\overline{A}$) is given by the total cell area divided by the number of cells when the soil was dry. It can be inferred from the adhesion tests undertaken separately by Corte and Higashi (1960) that the peak adhesion at the soil-base interface is approximately 40 kPa and 11 kPa respectively for glass and wood bases. The adhesion at the interface is, however, strongly dependant on the soil moisture content, where the value of adhesion rises gradually to a peak value as the water content decreases from the initial water content, and then decreases rapidly as the soil dries out further. The influence of shear characteristics of the base was further highlighted by a comparable test conducted on a sand base where no cracking was observed, owing to the lower shear restraint exerted on the soil by the base.

Figure 3. Experiment 7 of Corte and Higashi (1960) when the soil is dry (test details: soil slurry, soil thickness=33.5 mm)

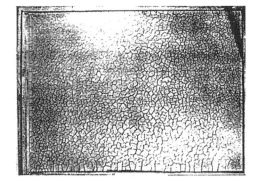

Figure 4. Experiment 1 of Corte and Higashi (1960) when the soil is dry (test details: soil slurry, soil thickness=3.4 mm)

Figure 5. Final crack pattern of Exp 19 of Corte and Higashi (1960) (test details: soil slurry, soil thickness=15.0 mm, glass base)

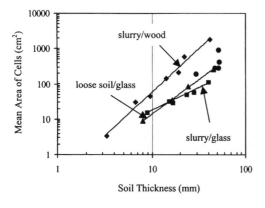

Figure 6. Final crack pattern of Exp 3 of Corte and Higashi (1960) (test details: soil slurry, soil thickness=14.7 mm, wood base)

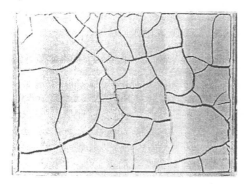

Figure 7. Relation between the mean cell area, $\overline{A}$, and thickness of the soil layer, d (Corte and Higashi, 1960 - ■, ▲, ♦ ; Lau, 1987 - ●)

4.4 Analysis of cell sizes, cracking patterns and shapes

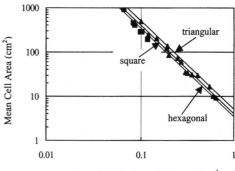

Figure 8. Relationship between the mean cell area, $\overline{A}$, and the length of the cracks per unit area, $\overline{L}$ (Corte and Higashi, 1960 - ▲; Lau, 1987 - ■)

As can be seen from Figure 7, the mean cell area, $\overline{A}$, increases as the layer thickness increases. The results also show that lower soil density or higher adhesion at the base tends to produce lower mean cell area for a given thickness. Total length of the cracks (L) seems to be inversely related to the layer thickness. Corte and Higashi (1960) have shown that cell areas generally follow a log-normal distribution. Interestingly, these relationships seem to be independent of the desiccation rate, despite apparent differences in the cracking patterns (i.e. development of orthogonal and non-orthogonal patterns).

As illustrated in Figure 1, theoretically plausible cracking patterns in idealised media undergoing uniform drying are considered as square, triangular and hexagonal patterns. Unique theoretical characteristics of these patterns can be established by considering the relationship between mean cell area ($\overline{A}$) and the specific crack length ($\overline{L}$), which is defined as the total crack length divided by the total plan area of the soil. These relationships for square, triangular and hexagonal shapes (considering one cell) can be respectively expressed as:

$$\overline{A} = 4/(\overline{L})^2, \overline{A} = 5.19/(\overline{L})^2 \text{ and } \overline{A} = 3.47/(\overline{L})^2 \qquad (1)$$

Figure 8 shows these relationships in graphical form, and they hold independent of the size of the cells. Also shown in this figure are the test results of Corte and Higashi (1960) and Lau (1987). An interesting feature of this presentation is that regardless of the differences in the test conditions, all the data tend to fall close to these lines. Specifically, the majority of the data fall close to the line representing square pattern. Virtually no data fall above the tri-

697

angular line, and this line represents the upper bound for the relationship between $\bar{A}$ and $\bar{L}$. Some data seem to fall close and below the line representing hexagonal pattern. Theoretically, the hexagonal pattern should provide the lower bound for the relationship between $\bar{A}$ and $\bar{L}$.

In an actual cracking pattern, there are mixtures of patterns containing cells that feature a distribution of number of sides in a cell. As the number of sides in cells of a pattern increases, the points representing the relationship between $\bar{A}$ and $\bar{L}$ will shift from triangular line towards hexagonal line. For instance, if a pattern contains mostly cells having five sides, then that result would plot between the square and hexagonal lines. As can be seen from Figure 8, this explanation agrees with the experimental evidence, which indicated that the majority of the cells had four and five sides. It is, however, possible to get values slightly under the hexagonal lower bound relationship when the pattern contains cells having more than six sides or when the sides of cracks are curved. However, the experimental evidence indicates that the occurrence of cells containing more than six sides is not common.

5 DISCUSSION AND CONCLUSIONS

Essential points of the preceding presentation can be considered as: (i) The dependence of cracking water content on the desiccation rate; (2) Apparent uniqueness of the relationship between $\bar{A}$ and soil thickness d for a given soil condition and base material; (3) The dependence of the relationship between $\bar{A}$ and d on base material and soil density; (4) The development of different cracking patterns predominantly depending on desiccation rate and soil thickness; and (5) The variation of A according to a lognormal distribution.

As far as the first point is considered, Corte and Higashi (1960) were able to explain this behaviour on the basis that the probability of cracking may be proportional to the tensile stress generated in the soil at a particular water content and to the time duration soil experiences at this water content. This explanation results in increase in tensile stress at cracking (or decrease in cracking water content) with the increase of desiccation rate. Alternatively, this phenomenon may be explained by arguing that soil material properties (e.g. tensile strength) may be dependent on the desiccation rate.

The Points (2), (3), (4) and (5) highlight the mechanics of desiccation cracking. The authors believe that explanations to these experimental findings may be found by considering the balance of energy components, which include energy used by cracks during their propagation and the strain energy released due to cracking. It can be seen from Equation 1 that the hexagonal pattern provides the most efficient release of energy because it has the lowest crack length per unit cell volume (The cell volume can be approximated by $\bar{A}\,d$). This may be a reason why soil containing high strain energy tends to crack in hexagonal pattern when subjected to high desiccation rates. It seems, however, that it is difficult to generate a perfect hexagonal cracking pattern in the laboratory. This difficulty may be associated with the dependency of this cracking pattern on material anisotropy and inhomogeneity, stress non-uniformity and the boundary shape.

The current paper elucidates some important aspects of the desiccation cracking process, specifically related to the desiccation of thin soil layers. Nevertheless, significantly more research is needed to further the understanding. In future, consistent explanations and quantitative relationships for these experimental findings need to be searched.

ACKNOWLEDGEMENT

The first author wishes to thank Victoria University of Technology, Melbourne for facilitating Outside Study Leave during which this study was undertaken. The support by Natural Sciences and Engineering Research Council of Canada is gratefully acknowledged.

REFERENCES

Bezant, P.Z. and Cedolin, L. (1991). *Stability of structures*, Oxford University Press.

Corte, A. and Higashi, A. (1960). Experimental research on desiccation cracks in soil. U.S. Army Snow Ice and Permafrost Research Establishment, Research Report No. 66, Corps of Engineers, Wilmette, Illinois, U.S.A.

Fredlund, D.G. and Morgernstern, N.R. (1976). Constitutive relations for volume change in unsaturated soils, Canadian Geotechnical Journal, 13, No. 3, pp. 261-276.

Kodikara, J., Barbour, S.L. and Fredlund, D.G. 1998. Clay structure development during wet/dry cycles, Research Report, University of Saskatchewan, Saskatoon, Canada.

Kodikara, J., Barbour, S.L. and Fredlund, D.G. 1999. Discussion: "An idealized framework for the analysis of cohesive soils undergoing desiccation" by J-M Konrad and R. Ayad, Canadian Geotechnical Journal, Vol. 34, pp. 1112-1114.

Konrad, J-M. and Ayad, R. (1997a). An idealized framework for the analysis of cohesive soils undergoing desiccation, Canadian Geotechnical Journal, 34, pp. 477-488.

Lachenbruch, A.H. 1961. Depth and spacing of tension cracks, Journal of Geophysical Research. 66.(12): 4273-4292.

Lau, J.T.K. 1987. Desiccation cracking of clay soils, MSc Thesis, Department of Civil Engineering. University of Saskatchewan, Canada.

Morris, P.H., Graham, J. and Williams, D.J. 1992. Cracking in drying soils, Canadian Geotechnical Journal. 29: 263-277.

Wilson, G.W., Fredlund, D.G. and Barbour, S.L. (1990). Coupled soil-atmosphere modelling for soil evaporation, Canadian Geotechnical Journal, 31-2, 151-161.

Unsaturated Soils for Asia, Rahardjo, Toll & Leong (eds) © 2000 Taylor & Francis, ISBN 90 5809 139 2

Analysis of swelling pressure for an expansive soil using the resistance concept

K. Mawire
Department of Civil Engineering, University of Zimbabwe, Zimbabwe

K. Senneset
Department of Geotechnical Engineering, Norwegian University of Science and Technology, Trondheim, Norway

ABSTRACT: The different stages of mobilisation of swelling pressure with time, for an instantaneously flooded soil sample, are analysed using the resistance concept. The laboratory data was obtained by flooding a partially saturated soil sample under completely confined conditions in a slit-ring oedometer. No external load was applied during the test, in order to study the soil response to suction reduction only. Both vertical and horizontal components of swelling pressure were simultaneously measured with the passage of time. The raw data was plotted to obtain dimensionless resistance numbers for the different stages. The dimensionless numbers are characteristic to the soil, for the given method of suction reduction. The simplicity and rationality inherent in the resistance concept makes it a very powerful and yet easy tool to use in characterising soil behaviour. This apparent success in characterising expansive soil behaviour makes the resistance concept a possible consistent framework that unifies saturated and unsaturated soil behaviour.

1 INTRODUCTION

The behaviour of unsaturated expansive soils has been the subject of much research activities in the last thirty-five years. A lot of effort was directed towards predicting the swelling pressure and heave, in a bid to solve practical problems related to engineering on expansive soils. At least forty different predictive methods are reported in the literature (Pellisier, 1991). These have met with different degrees of success, depending on their level of rationality. Sridharan et al (1986) and Snethen (1980) reviewed some of the different procedures used to determine swelling pressure. However, Mawire and Senneset (1998) pointed the shortcomings associated with all the methods. They noted that the test procedures hitherto, i.e. those that require application of an external load, introduced mechanical effects of the external load to the swelling process, a phenomenon that is essentially a result of physical-chemical interaction at the platelet level. In addition, it is not possible to separate the effects of suction reduction from those of the applied loads, on the soil behaviour. Thus, such procedures best simulate the soil-structure interaction effects of external loads. They then proposed a rational approach for measuring swelling pressure, without the interference of external loads. However, they did not present a method of analysis of the results. This paper deals with the analysis of such data.

The resistance concept, originally introduced into soil mechanics by Terzaghi (1925) and revisited by Janbu (1963,1998), is used here for the first time to analyse expansive soil behaviour. It is noted that the concept has been successfully used in consolidation settlement analysis (Janbu, 1963; 1967) and other brunches of engineering and applied science. Janbu (1967,1998) that, for engineering purposes, one can adequately cover the variations in compressibility of different types of soil from rock to soft clay, by means of one relatively simple resistance-based formula. In the 25th Rankine Lecture, Janbu (1985) presents resistance as a unifying concept in soil modelling. The apparent success of the resistance concept in characterising expansive soil behaviour makes it a possible consistent framework that unifies saturated and unsaturated soil behaviour. The concept is presented in terms of stress-strain relationship before it is applied in the analysis swelling pressure behaviour.

2 THE RESISTANCE CONCEPT

2.1 *Definition*

Any medium in a state of equilibrium tends to resist a forced change to its equilibrium condition. Such a resistance by a medium or part of it can be determined by measuring the incremental (internal) response to a given incremental (external) action.

$$\text{Resistance} = \frac{\text{external action}}{\text{internal response}}$$

Resistance is the first derivative or tangent to the action-response curve. It can be linear or non-linear without a change of the definition. For soils, the response is mostly non-linear. Usually effective stress change or time is the external action (given) while strain is the measured internal response. Figure 1 shows the graphical definition of the resistance concept, with time as the external action.

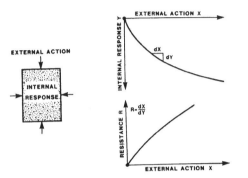

Figure 1. Definition of the resistance concept (after Janbu, 1998)

2.2 Time resistance, R

By taking time as the action and strain as the reaction, time resistance, R is defined by equation 1.

$$\frac{dt}{d\varepsilon} = R \qquad (1)$$

In this equation, t is the time (min.) and ε is the strain (dimensionless). Beyond a given time t_o, Janbu (1963) noted that the time resistance varies linearly with time indicating the creep stage, such that

$$\frac{dR}{dt} = r = \text{constant} \qquad (2)$$

This dimensionless constant, r is called the *resistance number*, and is the main parameter required to predict soil creep or secondary consolidation settlement. The resistance number is characteristic of the soil and its magnitude reflects the level of soil resistance mobilised. A small resistance number means high soil resistance, and big resistance number means low soil resistance.

2.3 Application to expansive soils swelling under completely confined conditions

In the case of confined unsaturated expansive soils with no externally applied load, suction reduction is the action and the developing swelling pressure is the response. Suction is a measure of the in situ 'effective' stress and its reduction is a form of 'effective' stress increase. Suction was reduced to zero by flooding the soil sample by inundation with water. The finite time associated with the initial flow of water during the flooding process is significantly small compared with the time taken to mobilise maximum swelling pressure. Thus flooding can be considered as 'instantaneous', making suction reduction a stepwise increase in effective stress.

The resistance number associated with swelling pressure as the internal response has the dimensions of kPa^{-1}. The resistance number is made dimensionless by normalising it with the atmospheric pressure.

3 RESULTS

3.1 Soil data

The data analysed in this paper is that presented by Mawire et al. (1998). Undisturbed samples were taken from a site within the University of Zimbabwe campus in Harare. The soil data is tabulated in table 1 below and the swelling pressure results are in Figure 2.

Table 1: Soil data

Soil type	Black cotton soil
Sampling method	Block sampling
Sampling depth	1.0 m
In situ unit weight	16.48 kN/m³
Grain density , G_s	2.58 kg/m³
Liquid Limit, LL	65 %
Plasticity Index, PI	45 %
Linear shrinkage, LS	19 %
Silt fraction	28 %
Clay fraction	40 %
Water content at test	12 %
Saturation water content	38 %

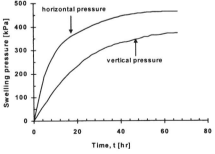

Figure 2. Swelling pressure-time plot (after Mawire et al, 1998)

3.2 *Resistance numbers*

Variation of time resistance with time for the vertical and horizontal swelling pressure are plotted in figures 3 and 4, respectively. The normalised resistance numbers, which are the slopes of the curves, respectively, are summarised in table 2.

Table 2   Summary of input data

Stage	Vertical pressure		Horizontal pressure	
	r	to (min)	r	to (min)
A-B	0	200	0	100
B-C	0.5	500	0.4	170
C-D	2.0	300	1.2	730

4 ANALYSIS

The general shapes of the two curves in figures 3 and 4 are the same. The orders of magnitude of the corresponding resistance numbers are the same. The analysis can therefore be made with reference to one graph (figure 3), without loss of generality. Figure 3 shows three distinct stages associated with development of swelling pressure with time. There is physical significance of the change in the soil resistance.

The early stage (A-B) of the swelling process is independent of time, having a constant time resistance (r = 0). The swelling pressure developed in the early stage is largely dependent on the increase in water content and its interaction with the surfaces of the clay minerals. A finite amount of time is required for water to penetrate the clay minerals, for the time-related swelling pressure to develop and dominate that of water increase. Increase in water concentration in the soil skeleton plays a more dominant role than that which is penetrating the clay minerals.

The early stage (A-B) of the swelling process is independent of time, having a constant time resistance (r = 0). The swelling pressure developed in the early stage is largely dependent on the increase in water content and its interaction with the surfaces of the clay minerals. A finite amount of time is required for water to penetrate the clay minerals, for the time-related swelling pressure to develop and dominate that of water increase. Increase in water concentration in the soil skeleton plays a more dominant role than that which is penetrating the clay minerals.

Thereafter, the time resistance increases with time at an increasing rate in two stages (r = 0.5) and (r = 2) respectively. The increase in time resistance is due to soil softening. The soil initially had a high resistance associated with the high suction value. With time, the rigid soil structure crumbles as the water penetrates the clay particles. The softening process can be said to take place in two stages. In the first

stage (A-B), the decrease in soil resistance is greatest when there is no swelling pressure to resist it. So the developing swelling pressure has to overcome the softening effect before it asserts its resistance. This explains the low rate of increase in resistance (r = 0.5). There is a marked increase in soil resistance at time t = 700 minutes, (r = 2). This is associated with the swelling pressure that is developing as a much faster rate than the soil is softening. Sufficient time has elapsed for the water to effectively penetrate and cause a release of the swelling pressure.

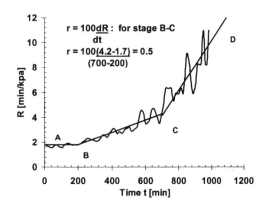

Figure 3. Time resistance, R for vertical swelling pressure

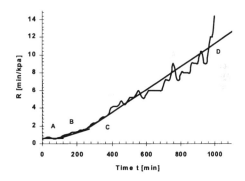

Figure 4. Time resistance, R for horizontal swelling pressure

5 CONCLUSIONS

The following conclusions can be drawn from the preceding analysis and similar analysis on results of 18 samples tested.
1. The resistance concept has been successfully used to study the effect of time on the development of swelling pressure.

2. The resistance number can be used to describe the swelling pressure-time relationship. It is characteristic of the soil type and is constant over specific ranges of initial water content
3. Swelling pressure can not be referred to in terms of 'primary' and 'secondary' values in the sense of consolidation settlement. The contribution by the time-related component far exceeds that of water content increase, which is a measure of suction or in situ 'effective stress'.
4. It is proposed that swelling pressure be referred to as having the water content-dependent and time-dependent components.

ACKNOWLEDGEMENTS

The authors acknowledge the support of the Department of Geotechnical Engineering, Norwegian University of Science and Technology, Trondheim in this research work. The research is part of a collaborative programme between the University of Zimbabwe and Norwegian University of Science and Technology (NTNU). The research is funded by the Norwegian Council of Universities' Committee for Development Research and Education (NUFU).

REFERENCES

Habib, S.A., Kato, T. and Karube, D., 1992. One Dimensional swell behaviour of unsaturated soil, *Proceedings of the 7th International Conference on Expansive Soils*, Dallas: 222-226.

Janbu,N. 1963. Soil compressibility as determined by oedometer and triaxial test. *Proceedings of the European Conference, vol. 1, Wiesbaden.*

Janbu,N. 1967. Settlement calculation based on the tangent modulus concept, *Three guest lectures at Moscow State University. Bulletin No. 2 of Soil Mechanics and Foundation Engineering*, Department of Geotechnical Engineering, NTNU, Trondheim, Norway.

Janbu, N. 1969. The resistance concept applied to deformations of soils. *Bulletin No. 3 of Soil Mechanics and Foundation Engineering*, Department of Geotechnical Engineering, NTNU, Trondheim, Norway.

Janbu, N. 1985. 25th Rankine Lecture, Geotechnique 35, No. 3, pp 241-281

Janbu, N. 1998. Sediment deformations, a classical approach to stress-strain-time behaviour of granular media as developed at NTH over a 50-year period. *Bulletin No. 35 3 of Soil Mechanics and Foundation Engineering*, Department of Geotechnical Engineering, NTNU, Trondheim, Norway.

Mawire K and Senneset, K., (1998); "Mobilised swelling pressures of an expansive soil", *12th African Regional Conference of the International Society for Soil Mechanics and Geotechnical Engineering*, Durban, South Africa, in October 1999.

Pellissier, J.P., 1991. Heave prediction: State of the art. *Division of Building Research for South African Road Research Board*, Interim report IR 88/039/6.

Sayed, A., Habib, S.A. and Karube, D., 1993. Swelling pressure behaviour under controlled suction. *Geotechnical Testing Journal*, Technical note.

Schreiner, H.D. & Burland, J.B., 1991. A comparison of the three swell-test procedures. *Geotechnics in the African Environment.*

Snethen, D.R., 1980. Characterisation of expansive soils using soils suction data, *Proceedings of the 4th International Conference on Expansive Soils*, Denver: 54-75.

Sridharan, A., Rao, A.S., and Sivapullaiah, P.V., 1986. Swelling pressure of clays. *Geotechnical Testing Journal*, 9(4): 23-33.

Terzaghi,K. 1925. "Erdbaumechanik auf bodenphysikalischer Grundlage". Deuticke, Leipzig und Wien, pp109

Yong, R.N., Sadana, M.L., and Gohl, W.B., 1984. A particle interaction model for assessment of swelling of an expansive soil. *Proceedings of the 5th International Conference on Expansive Soils*, Adelaide: 4-12.

Unsaturated Soils for Asia, Rahardjo, Toll & Leong (eds) © 2000 Taylor & Francis, ISBN 90 5809 139 2

Clay stiffness and reactivity derived from clay mineralogy

K.G. Mills
School of Engineering, University of South Australia, Adelaide, S.A., Australia

ABSTRACT: Volume change in reactive clay soils is usually related to soil moisture status using the soil suction as a key state variable. Such relationships as are used are experimental and empirical. However, clay physics may be used to identify the individual roles of solute and matric suction, and provide a background theory to better inform the study of the stiffness and swelling in these reactive soils. The paper applies osmotic swell pressure theory to help predict the behaviour of clay soils comprising mixtures of clay types and inert material. It explicitly identifies the role of solute suction.

1 INTRODUCTION

In foundation engineering on reactive soils, there has long been a need to estimate volume change as a response to moisture availability. The methods in use generally account for reactivity empirically. Nevertheless, by considering the physico-chemical properties of clays, the role of clay mineralogy and soil suction in determining soil stiffness and volume change can be clarified. This paper describes a semi-theoretic model for reactivity in soils that may find use in other mathematical soil models. It derives from the PhD dissertation of Mills (1998)

2 EXISTING APPROACHES

Commonly, soil swell is related to the field stress state and moisture content range, with various methods being used to quantify the swell. In Australian practice, *Australian Standard AS 2870* (1990), the change in wetness state is specified by an increment of soil moisture suction.

However, the use of "suction" in engineering is bedeviled by the complexity in measuring and even defining it. Conventionally, total suction, μ, is the sum of matric suction μ_m and solute suction, μ_s. Arguably, matric suction is akin to a physical pore pressure, and it is the effect of solute suction that is less predictable. Fredlund and Rahardjo (1993) *Sec 3.6 Role of Osmotic Suction,* note when solute suction needs explicit consideration, and when it may be neglected. Ridley and Wray (1995) commented that few authors in the First International Conference

Unsaturated Soils, addressed solute suction and distinguished between total suction and its components. The theoretic treatises by Mitchell (1976) and Yong et al (1992) show the role of suction. They point up the factors that may influence swell.

Much recent work seeking to understand the complete stress strain response of unsaturated soils, has shown the importance of pore space distribution. Generally the pore size distribution in a clay soil is seen to be bi-modal. (Delage et al 1995). For soils compacted dry, there are large voids between aggregates of clay (peds), and finer voids within the clay aggregates or domains themselves. When the soil is wetted, the clay domains themselves may swell, but with the application of stress they will be distorted and the larger pores will collapse in part. Alonso (1990) included pore collapse in a critical state model, with the state boundary surface being a function of suction. These writers presumed that the smaller voids are the seat of reactivity, causing clay domains to swell. When the domains swell, the larger inter-domain voids likewise swell. The resulting constitutive relationship includes suction explicitly, and will track the shrink/swell behaviour of soils, and also collapse and overconsolidation.

In such modelling, parameters have a physical relevance as far as possible rather than being just a curve-fitting exercise. The work on particle size distributions outlined above places this paper in context. It examines aspects of soil suction and may ultimately inform the formulation of these more encompassing models.

3 OSMOTIC PRESSURE SWELL MODEL

The theory below though simplified, quantifies the effect of solute suction, in terms of soil water chemistry and clay type.

3.1 Classical DVLO Theory for Clay Swelling

Interaction between clay particles was examined in the DVLO theory (Iwata et al 1988), so named from the physicists Derjaguin, Landau, Verwey and Overbeek. They investigated the forces between clay particles from two sources:- van der Waals attractions, and electrostatic fields. The repulsions between particles in a face-to-face configuration, are the predominate force of the latter type. The volume change behaviour of idealised clay can be predicted. The theory is certainly valid for sodium montmorillonites, but only trends in behaviour can be predicted with other clays (Mitchell 1976).

For a clay domain surrounded by water, cations are preferentially attracted into the space between the clay platelets. Thus the van't Hoff equation for osmotic pressure, Π, may be used to find the interparticle repulsion (Bolt 1956):-.

$$\Pi = R\,T\,\Sigma(c_i \text{ mid-plane} - c_i \text{ free solution})$$

where c_i are the molar concentration of the ion species. For homo-ionic dilute solutions, with Π replaced by the interparticle swelling pressure, P_s :-

$$P_s = R\,T\,c_o\,(\frac{c_c}{c_o} + \frac{c_o}{c_c} - 2\,) \qquad (1)$$

where c_c and c_o are the cation concentrations midway between platelets and in the soil water respectively. c_c may be found from the solution of

$$v\sqrt{\beta c_o}\left(x_o + \frac{e}{\rho_s\,A_s}\right) = 2\sqrt{\left(\frac{c_o}{c_c}\right)}\,E \qquad (2)$$

where v is the cation valence, $\beta \cong 10^{16}$ m/mole a constant related to temperature and dielectric constant, e the soil's void ratio, ρ_s the particle density, A_s the specific surface of the soil solids; x_o is $4/v\beta\Gamma$, (Γ is the surface charge in equivalents per unit area), $\cong 1/v$, $2/v$, and $4/v\,10^{-4}\mu$m for kaolinite, illite, and montmorillonite respectively. The fraction,

$\frac{e}{\rho_s\,A_s}$, is the interparticle spacing. E is the complete elliptic integral of the first kind with argument c_o/c_c.

3.2 Usage of Osmotic Pressure Theory

In most natural clay soils, other effects overwhelm the interactions prescribed by the DVLO theory. Probably as a result, engineers do not use clay mineralogy and soil water chemistry explicitly to find soil swell. However, osmotic theory leads to useful relationships between solute suction and swelling in reactive clays, provided the other forces are separately considered.

Swell pressure on inundation from Equation 4 relates directly to matric suction release. Pile (1980) in oedometer tests, found of swell pressure almost independent of any solute in the solution used for inundation. This implies the immediate relief of matric suction, with any solute effects being delayed.

4 FORMULATION OF A QUASI-THEORETIC MODEL FOR STIFFNESS AND SWELLING

4.1 Preamble

Given the concept of osmotic swell pressure, the swelling behaviour over a full range of solute conditions, can be calculated from conventional volume change test results on one soil specimen, by applying Bolt's theory to predict the osmotic pressure changes as a perturbation on the observed test results. At any given particle spacing, short range forces other than the osmotic ones are assumed constant. If several different pure clays are used in the base test, a compound hypothetical model results that can simulate a real soil. It is semi-quantitative, and partially theoretical rather than empirical. Its main features are:-

- Separate contributions are retained from the different clay minerals in a soil.
- Salt content is treated as a separate phase

In soil mechanics terms, a new "e-log P" curve is derived. The volume change response to solute suction is explicitly predicted. It is presupposed that the "source" e-log P curve is a unique swelling curve, and that structural hysteresis can be ignored.

4.2 Application of the model to Pure Clays

To set up the model, experimental data on compression testing by Olson and Mesri (1970) were used. It is convenient to separately identify P_S the component of total stress that arises from osmotic swelling.

For saturated soil with zero matric suction μ_m, $\sigma_{oedometer} = \sigma^* + P_s$, σ^* being the portion of effective stress excluding osmotic repulsion. For some new condition at the same void ratio,

$$\sigma_{oedometer\ new} = \sigma^* + P_{s\ new}$$
$$= \langle \sigma_{oedometer} - P_s \rangle_{source} + P_{s\ new} \qquad (3)$$

and for non-zero matric suction,

$$\sigma_{Total\ new} = \langle \sigma_{oedometer} - P_s \rangle_{source} + P_{s\ new} + |\mu_m| \qquad (4)$$

The data comprise the swelling or rebound coefficient for consolidation testing for pure clays with several salt types and concentrations. To extend the range of the "source" curve, an empirical function was adopted which would tend asymptotically to a small limiting e for very large P. The function used was

$$\log(e - e_{minimum}) = a \log(P)^m + b \qquad (5)$$

where a and b are fitted constants, m was taken as 3. P is the summed term, $\sigma_{effective} = \sigma + P_s$

The minimum e corresponds to some minimum particle separation within the clay domains plus an allowance for the inter-domain pores.

The equation gives an e that becomes increasing large at small pressures. Hence the tests chosen for source curves were for low concentration sodium salts, so the clay is continuously dispersing.

4.3 Adaptation to Soils of Mixed Composition

Soils may contain several clay minerals plus inert material. In calculations for such soils, this study assumes that the pressure exerted by the different clays, would by physical reasoning, have to be similar for each component.

At reasonable clay contents, as in a clayey soil with significant moisture reactivity, the soil moisture is mainly associated with the clay phase. Also any inert phase will exists in a matrix of moist clay. For these conditions the void ratios and moisture contents of the clay and the soil mixture are related by:-

$$e_{mixture} = e_{clay} \text{ by clay content, and}$$
$$w_{mixture} = w_{clay} \text{ by clay content} \qquad (6)$$

Provided that the densities of the clay and the silt/sand are nearly equal, then the "clay content" in Equation 6 will be the mass fraction of soil that is clay. It is also true that the compressibility will arise mainly from the clay component. If it is assumed, as is reasonable, that the osmotic swell pressure is the dominant stress component for the clays, then from

Paragraph 3.1, $\left(x_0 + \dfrac{e}{\rho_s A_s} \right)$ will be the same for each clay mineral.

An effective or average x_0 and A_s can be defined and thus the theory may be readily extended to handle mixtures of different clay types. Considering only the clay phase,

Aggregate clay void ratio, $\quad e = \Sigma\, p_i\, e_i \qquad (7)$

$$C_1 = \left(x_0 + \frac{e}{\rho_s A_s} \right) \text{ for each clay type I} \qquad (8)$$

where the p_i are the fractional proportions of each clay type, and C_1 is a constant for a particular stress state, applying to all the clay types. Substituting for e_i from the second of these equations in the first, (omitting the subscript i for brevity)

$$C_1 = \frac{\Sigma\, p\, x_0\, \rho_s\, A_s}{\Sigma\, p\, \rho_s\, A_s} + \frac{e}{\Sigma\, p\, \rho_s\, A_s} \qquad (9)$$

If ρ_s is constant or reasonably so, the effective parameters for the clay mixture are

$$\text{Mixture } x_0 = \frac{\Sigma\, p\, x_0\, A_s}{\Sigma\, p\, A_s} \text{, and}$$
$$\text{Mixture } A_s = \Sigma\, p\, A_s \qquad (11)$$

The base source curve is a simple weighted average curve for the constituent clays.

4.4 Adaptation to Soils with Packing Voids

The DVLO theory presumes a concentration of ions in the soil water. This is not the inter-layer water. It is rather the water surrounding the clay in void spaces in the soil fabric. Bolt employed the term dead voids for these packing voids. Dead voids may arise in pure clays due to terracing in the lattice.

Yong et al (1992) suggest that the "dead" volume may be taken as the voids pertaining at the shrinkage limit. This is used below, but with an allowance for some residual hydration of the clay. They also expected the dead voids to expand with swell in the clay domains. The dead voids can be included in the phase relationships. If V is the total volume of a specimen of saturated soil, then

$$V = V_c + V_{cv} + V_{dv} + V_i, \qquad (11)$$

where the subscripts refer to clay particles, clay voids, dead voids, and inert grains respectively. Consistent definitions of clay and dead void ratios are (taking density as constant again):-

$$e_c = \frac{V_{cv}}{V_c} \text{ and } e_d = \frac{V_{dv}}{V_c}. \qquad (12)$$

$$e = X(e_c + e_d)$$
$$= X(e_c + e_{do}(1 + e_c))$$

whence

$$e_c = \frac{e/X - e_{do}}{1 + e_{do}} \text{ and } e_d = e/X - e_c \qquad (13)$$

In this expression, X is the clay content, and the total dead void space is presumed to expand in proportion to the change in volume of the clay and the clay voids. A new parameter e_{do} is introduced, defined as the dead void volume at the limit when e_c goes (mathematically) to zero. Physically, the minimum e_c is non-zero due to the limit set to the minimum particle separation.

It is reasoned that the inert material and the "clay" and its voids transmit swelling stress. If the stress is assumed proportional to the area on a cross-section in the clay - clay voids - dead voids system,

$$P_s \cong P_{S-clay} \frac{(1 + e_c)^{2/3}}{(1 + e_c + e_d)^{2/3}} = P_{S-clay} \frac{(1 + e_c)^{2/3}}{(1 + e/X)^{2/3}} \qquad (14)$$

705

where $P_{s\text{-}clay}$ is the stress within the clay and its intrinsic voids, and P_s is the nominal stress over the whole external surface. Estimates of the dead void parameter e_{do} are given in Table 1. The shrinkage limits are taken from the ranges given in Whitlow(1995), other clay properties from Table 2.

Table 1. Estimates of Dead Voids Parameter

Clay type	Shrinkage Limit (SL) %	Void ratio at SL.	Inter-lamella voids at SL	Dead voids at SL	e_{do} (5)
Mont-morillo-nite	12.^{5}	0.34	0.22	0.12	0.1
Illite	16.^{5}	0.44	0.03	0.41	0.4
Kaolin	26.^{0}	0.70	0.005	0.70	0.7

Assumes the soil is still saturated as it approaches SL
For an inter lamellar spacing d of $1 \times 10^{-4} \mu m$. $e = \rho_s\, d\, A_s$

Concentration values for salts in the soil water require consistent interpretation, since mobile salt only occurs in the dead voids. When concentrations are defined as an average over the total water content, they must be corrected for use in computation both for the clay regime and solute suction. The correction factor is:-

$$\frac{e_c + e_d}{e_d} \text{ or alternatively } \frac{e}{e_d} X \qquad (15)$$

Minimum separations in all but the driest of soils still involve some hydration. (Handy and Turgut 1987). For Montmorillonite with perfect packing at the shrinkage limit, the minimum inter-lamellar spacing would be 1.6×10^{-4}. As dead voids are non-zero, a value of 1×10^{-4} μm was adopted. (There are instances where this assumption should be refined.)

The model may be applied in a stress analysis. Iteration is needed for some tasks, and Poisson's ratio must be assumed, 0.3 being used by the author. Young's modulus can be calculated from the void ratio change for a small change to confining stress.

5 TESTING THE MODEL

Two examples are given of usage of the model. These comprise a check for pure clays in different solutes, and a prediction of behaviour in a pressure membrane test on an Adelaide clay.

5.1 Prediction for pure clays

The data of Olsen and Mesri used for the base or source curves, also includes observations of clay compressibility in differing solutions. The model was applied for these altered solute conditions. Tests for illite and montmorillonite were calculated:-

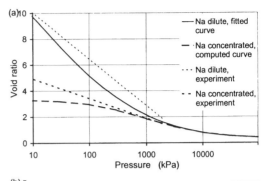

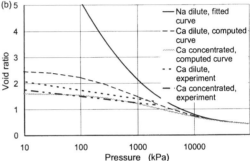

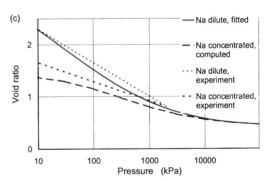

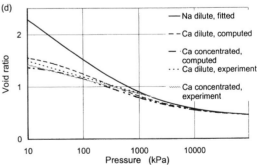

Figure 1 Void ratio prediction for pure clays.
a Sodium Montmorillonite,   b Calcium Montmorillonite
c Sodium Illite,   d Calcium Illite

- with a change in concentration in the soil water cation, and
- with a change in valence in the soil water cation,

The prediction was then compared with the experimental values. Figure 1 presents the results. Qualitatively at least, the prediction is encouraging, matching the swell reduction for increasing cation concentration, and, for the sodium/calcium systems tested, the swell suppression with increasing valence. The clays' properties are as per Table 2.

Table 2. Typical Clay Mineral Properties

Property	Kaolinite	Illite	Montmorillonite
Specific surface (m^2/gm)	5-20	80-120	700-800
Cation exchange capacity (meq/100gm)	3-15	15-40	80-100

5.2 Application to a natural soil

A comparison of the mixed soil model with experiment was made using data from Richards et al (1984) for volume change response in a suction controlled oedometer where the "flushing solution" is changed. The clay content (65%) and composition were estimated at the average figure suggested by Cox (1970) for similar soils.

Figure 2 is taken from Richards et al (1984). Figure 3 gives the predictions of the model. There is a good fit again for general response. However there are two marked limitations that appear. Firstly by comparing both tests with the experimental data, the soil shows too great a compression under increasing applied stress and too great a swell. This may be a matter of the assumed base curve, which is the same as was derived above. It is also difficult to be sure of the clay's composition. Evidently, changing these assumptions would alter the model predictions. The second difference is much more significant. The experimental work used potassium salts as well as sodiums. For the model proposed herein, both should give the same result since both are monovalent. However, the experimental swell with the potassium ions is nearly identical to that with calcium solutions. The reasons for this lie in the clay mineral's structure, but it is apparent that the simple theory is suited to calcium/sodium exchanges and not to those involving potassiums. If the influence of potassium is to be considered, it will have to be separately identified by more than its valence. It could perhaps be modelled by noting that the platelet charge density is effectively much reduced.

The remaining mismatches could probably be arbitrarily reduced by changes to the base curve shape, and by allowing that some of the clay sized material is inert, or rendered non-swelling by ce-

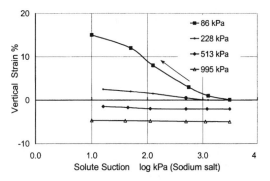

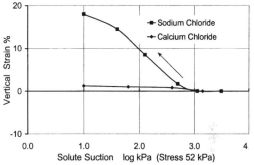

Figure 2 Consolidometer Tests at varying suction (after Richards 1984 Fig 9 and 10)

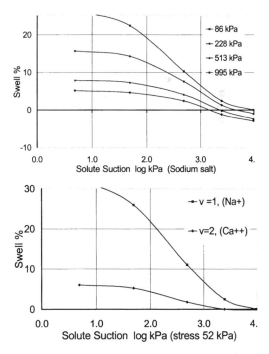

Figure 3 Consolidometer prediction matching Richards tests

707

mentation. There is simply not sufficient experimental work to justify this in this instance.

6 COMMENT ON THE QUASI-THEORETIC MODEL

6.1 Limitations to Osmotic Swelling Theory

Particle interaction as modelled has focused on one tractable process in a very complex system. A major simplification is that the effects of finite size for both cations and water molecules have been neglected. The theory for Stern layer formation allowing for cation size, which can be incorported into the model as an enhancement, may account for some discrepancies in the modelling.

The difficulty in diffusing salts into intact clay soils means that many non-equilibrium conditions persist for long periods. Cheney (1988) discusses an instance of on-going heave 25 years after tree removal from a site on London clay. It is an understatement to say this can complicate the interpretation or prediction of both field and laboratory situations.

Clay soils would generally have mixed cations solutions. To extend the theory for this, will involve tracking individual ion concentrations, noting that the cation concentrations in the soil water and in the clay exchange positions do not have the same relative values.

6.2 Conclusion

Osmotic pressure theory has been used to predict the swelling and stiffness characteristics of clay soils. Using generic properties for the mineralogy of the clays, it has produced a model that shows many of the major features observed in reactive soils. Thus the model is based on theory rather than experiment. It has two noteworthy characteristics.

- Fundamental properties of the clay minerals enter directly into the model,
- Effects from the soil water solutions are rationally separated out from the other components of suction.

It anticipates the nature of the relationship between volume change and solute suction and contributes to the better understanding of role of solutes in producing volume change. The model is not predictive in a fundamental sense, but it is rather intended to provide a theoretic framework within which experiment observation may be fitted to permit quantitative interpolation and extrapolation of test results to a wide range of external conditions.

Osmotic pressure theory provides a theoretical under-pinning for the modelling of reactive soils, establishing just what are the relevant variables. There is an obvious compatibility with the "double-structure" models and test specimens being used to study clay swell and collapse, (Alonso et al 1995). Such models could link the reactivity to solute conditions.

Actual experimental work designed specifically to check the theory is obviously desirable but is beyond this study.

REFERENCES

Alonso, F E, Gens, A, & Josa, A, (1990), A constitutive model for partially saturated soil, *Geotechnique*, 40, N0. 3, 405-430

AS 2870 - 1990, (1990), *Residential Slabs and Footings*, Standards Australia, Sydney.

Bolt, G H, (1956), Physico-Chemical Analysis of the Compressibility of Pure Clays, *Geotechnique*, Vol. 6 No. 1, p86-93

Cheney, J E, (1988), 25 years' heave of a building on clay due to tree removal, *Ground Engineering*, 1988, 21, No. 5, pp13-27.

Cox, J, (1970), A Review of the Geotechnical Characteristics of the Soils in the Adelaide Area, *Proc. Symp Soils and Earth Structures in Arid Climates*, Adelaide, Vol 1, p72-86.

Delage P. & Graham, J., (1995) Mechanical behaviour of unsaturated soils: Understanding the behaviour of unsaturated soils requires reliable conceptual models , *Proc 1st Int Conf on Unsaturated Soils*, Paris. Vol 3 p1223-1256, 1995.

Fredlund, D G, & Rahardjo, H, (1993), *Soil Mechanics for Unsaturated Soils*, John Wiley and Sons, New York.

Handy, R L, & Turgut, D, (1987), Swelling Pressure vs D-spacing of Montmorillonite, *6th Int Conf on Expansive Soils*, December 1987, New Delhi, Balkema, Rotterdam 1988. pp 161-166

Iwata, S, Tabuchi, T, & Warkentin, B P, (1988), *Soil-Water Interactions, Mechanics and Interactions*, Marcel Dekker, New York and Basel

Mills, K G, (1998), *The Response of Reactive Clay Soils to Wetting in the Presence of Active Vegetation*, Ph.D. Dissertation, Queensland University. of Technology.

Mitchell, J K, (1976) *Fundamentals of Soil Behaviour*, John Wiley and Sons, New York.

Olson, R E, & Mesri, G, (1970), Mechanisms Controlling the Compressibility of Clay, *Journal of the Soil Mechanics and Foundation Division*, A.S.C.E., Vol. 96, NoSM6, pp. 1863-1878

Pile, K C, (1980), The Relationship Between Matrix and Solute Suction, Swelling Pressure and Magnitude of Swelling in Reactive Clays, *Third Australian-New Zealand Conf on Geomechanics*, Wellington, Vol 1, 197-201, 1980

Richards, B G, Peter, P, & Martin, R, (1984), The Determination of Volume Change Properties in Expansive Soils, *Forth International Conference on Expansive Soils*, Adelaide, 21 23 May 1984.

Ridley, A.M., & Wray, W.K., (1995) Suction measurement: A review of current theory and practices, *Proc 1st Int Conf on Unsaturated Soils*, Paris 1995. Vol 3 p1293-1322.

Whitlow, R, (1995), *Basic Soil Mechanics*, Longman, UK, 1995

Yong R N, Mohamed A M O, & Warkentin B P, (1992), *Principles of Contaminant Transport in Soils*, Developments in Geotechnical Engineering, 73, Elsevier, Amsterdam.

Unsaturated Soils for Asia, Rahardjo, Toll & Leong (eds) © 2000 Taylor & Francis, ISBN 90 5809 139 2

The lateral swelling pressure on the volumetric behaviour of natural expansive soil deposits

A. Sabbagh
Ingeuras Limited, Shepperton, UK

ABSTRACT: For swelling soil volume changes are a function of state variable changes. Usually the most appropiate soil test used to simulate vertical displacements under a building constructed on expansive soil is the oedometer test, with vertical deformations allowed and deformations in the lateral direction prevented. In this paper, the volume change behaviour of the Madrid clay is reported. It was investigated by controlled wetting of undisturbed samples, subjected to constant vertical pressure and decreasing the sample's matrix suction, by means of a Cuellar's controlled suction oedometer. Lateral swelling pressure measurements were made using this apparatus. This null type device allows the modelling of changes in season or climate in the laboratory. The swelling test was performed considering that in situ, when water ingresses beneath a building, not only the soil suction changes, but the lateral pressure changes also. Therefore, this paper describes the importyance of predicting soil swell using the mean net stress, instead of the vertical net stress alone, as is usually the case.

1. INTRODUCTION

If elastic soil mechanics are applied to the expansive clays, there exist many problems of practical consequence because a linear form of stress-strain relationship is not preserved. Similarly, for Darcy's seepage law, the dependence of matric suction on the unsaturated coefficient of permeability gives non-linearity of the soil material properties in flow type situations.

While in linear problems the solution is always unique, this no longer is the case in many non-linear situations. Thus, if a 'solution' is achieved by some iterative technique, it may not necessarily be the solution sought. Physical insight into the nature of the problem and usually, small-step incremental approaches are essential to obtain significant answers. Such increments are indeed always required in expansive clays, where the constitutive laws relating stress, suction and strain changes are path dependent.

Furthermore, conditions in unsaturated expansive clays such as soil suction generally varies from a steady to a transient state and therefore, the time dimension has to be considered. For this, the finite difference approximation is widely applicable and provides the greatest possibilities.

This work aims to discuss the lateral pressure obtained from laboratory tests performed on unsaturated expansive clays.

The test program was conducted on undisturbed soil samples from the southern part of Madrid. A detailed description of the soil properties, equipment and testing procedures has been presented by Sabbagh, (1995), Oteo et al. (1996).

2. LATERAL PRESSURE

For a clay deposit formed by horizontal deposition, consolidation will have ocurred solely in the vertical direction with no lateral deformation across any vertical plane. The soil mass is in the rest condition, and the ratio of the horizontal and vertical principal effective stresses σ'_h and σ'_v, acting on an element is referred to as the at-rest earth pressure coefficient, denoted by K'_o. That is:

$$K'_o = \frac{\sigma'_h}{\sigma'_v} \tag{1}$$

This expression can be written in term of the constant elastic Poisson's ratio ν, as:

$$K'_o = \frac{\nu}{1-\nu} \tag{2}$$

If the soil mass is further consolidated vertically by a uniform loading $\Delta\sigma'_v$ applied over the whole of the surface area, the vertical effective stress increases to $\sigma'_v + \Delta\sigma'_v$ and the horizontal effective stress increases to $\sigma'_h + \Delta\sigma'_h$. The soil is still in the at-rest condition and normally consolidated, and therefore the ratio of horizontal to vertical effective stress remains unaltered.

If the surface loading is subsequently removed so that the soil becomes overconsolidated, the soil swells vertically and the vertical effective stress on the element reduces to its original value σ_v. But as no deformation occurs in the horizontal direction, there is only a partial relaxation of the horizontal effective stress, with some of the increase $\Delta\sigma_h$ remaining 'locked in'. In fact overconsolidation of a soil deposit results in an increase in the value of K'_o. However, the maximun value that K'_o can attain is that of K'_p, termed the effective passive earth pressure coefficient.

Let us consider the slow drying of a clay soil deposit, when the saturated soil becomes unsaturated. The soil will be unsaturated, for all depths, but the rate of reduction of degree of saturation is greater at shallow depths.

If now, the history of the soil deposit subjected to drying, is followed by wetting, the at-rest coefficient increases as the matric suction of the soil decreases. Therefore, the values of K_o^* can go from as low as zero, indicating a tendency for cracking, to as high as the coefficient of passive earth pressure. The at-rest coefficient under a constant matric suction $(u_a - u_w)_o$ can be defined (Fredlund & Rahardjo, 1993) by the ratio of the horizontal net normal stress $(\sigma_h - u_a)$ and the vertical net normal stress, as follows:

$$K_o^* = \frac{\sigma_h - u_a}{\sigma_v - u_a} \tag{3}$$

From considerations of elastic stress-strain equations, the at-rest coefficient can be written as:

$$K_o^* = \frac{\nu}{1-\nu} - \frac{E}{(1-\nu)H}\frac{(u_a - u_w)}{(\sigma_v - u_a)} \tag{4}$$

The above equation can be associated with the vertical net normal stress and matric suction changes to give:

$$K_o^* = \frac{\nu}{1-\nu} - \frac{E}{(1-\nu)H}\frac{\Delta(u_a - u_w)}{\Delta(\sigma_v - u_a)} \tag{5}$$

where:
E =modulus of elasticity or Young's modulus for the soil structure with respect to a change in the net normal stress.
H =modulus of elasticity or Young's modulus for the soil structure with respect to a change in matric suction.
ν =Poisson's ratio.

The elastic moduli, E and H, may vary in magnitud from one increment to another. Similarly ν will not remain a constant with the changes in stress and/or

suction (Wallace & Lytton, 1992, Lytton, 1996).

Inspection of the equations given above reveals that K_o^* in unsaturated soil depends basically upon the state variable changes and material parameters determined for every change, depending on the current state variables. Therefore, in estimating K_o^*, as has already been discussed in this section, due to the non-linearity introduced through the material properties of the soil, a convergent iterative process is required, following the physically correct path.

The equation (5), in an incremental form can be arranged to give the lateral pressure as follows:

$$d(\sigma_h - u_a) = (\frac{\nu}{1-\nu})d(\sigma_v - u_a) - \frac{E}{(1-\nu)H}(u_a - u_w) \tag{6}$$

The use of net normal stress and matric suction as state variables for describing volume change behaviour for an unsaturated soil have been demostrated by Fredlund and Morgenstern (1977). The soil structure constitutive relations associated with the normal strains, also written in an incremental form can be used to compute the volumetric strain increment for K_o^* loading conditions, as follows:

$$d\varepsilon_v = \frac{(1+\nu)(1-2\nu)}{E(1-\nu)}d(\sigma_v - u_a) + \frac{(1+\nu)}{H(1-\nu)}d(u_a - u_w) \tag{7}$$

But lateral earth pressures at rest are different for volumetrically inert soils to those active soils such as swelling and shrinking soils.

Furthermore, under field conditions when a building or a road is constructed on unsaturated expansive soil, the only components of stress state variables in Eqs. (6) and (7), that normally change are the matric suction $(u_a - u_w)$ and the lateral earth pressure $(\sigma_h - u_a)$ during wetting and drying due to infiltration and evaporation under the foundation. A steady state regime could take up to ten years to stabilise (Meintjes, 1992).

The material properties of the soil E, H and ν are dependent on the state stresses or strains developed, and under triaxial stress conditions soils exhibit a nonlinear stress-strain curve. In such cases, by using small increments of stress and strain, the nonlinear curve is assumed to be linear within each stress and strain increment and E or H can be computed. Now, therefore, the volume change corresponding to a specific point and particular time is computed according to Eq. (7), using a ν value obtained from an iterative procedure involving the Poisson's ratio and at-rest lateral earth pressure coefficient as in Equations (4) and (5).

Nevertheless, a simpler procedure is suggested for describing the volume change characteristics of an unsaturated soil, using the change in void ratio de, as the deformation state variable. This is the main focus of this investigation and will be discussed in the following sections.

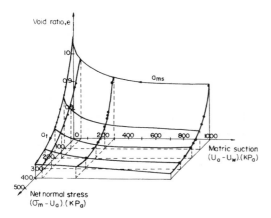

Figure 1. Constitutive surface for a set of undisturbed samples of Madrid grey clay, subjected to monotonic loading and wetting paths.

3 EXPERIMENTAL PROCEDURE

In the prediction of the change $d\varepsilon_v$, information on the present in situ stress state and the possible future stress state is required. To more closely simulate the in situ stress path, followed by the soil during wetting, a set of undisturbed specimens were subjected in the laboratory to vertical pressures under controlled matric suction, with measurement of the lateral swelling pressure. In the laboratory a modified oedometer apparatus, developed by Cuéllar (1978), was used to completely assess all of the above variables.

In this study, undisturbed unsaturated Madrid clay samples were subjected to constant surcharge loads of 10, 150 and 500 kPa and a controlled matric suction of 50, 300 and 1000 kPa.

Each specimen was subjected to different stress changes and its volume change monitored. Since in an expansive soil deposit as a result of the restricted lateral movement, during wetting paths the net horizontal pressure increases as the matric suction decreases, therefore the net horizontal pressure could be greater than the net vertical pressure and the net horizontal stress is the major principal stress and the net vertical stress is the minor principal stress. It is then necessary to measure the lateral pressure of the soil in order to obtain an indication of the complete stress state of the soil. These tests cannot be performed using conventional oedometers and soil mechanics testing procedures.

The effect of state variables on the volumetric strain in the expansive soil is illustrated in the form of a constitutive surface, representing the void ratio e against the state stress variables. Figure 1 illustrates the surface through various wetting paths, following monotonic deformation, for the series of experiments performed on Madrid grey clay samples.

4 VOLUME CHANGE INDICES

Volumetric strains, lateral pressures and swelling properties as a result of vertical surcharge and matric suction variations, were registered and represented in a three-dimensional plot of the void ratio e, the matric suction $(u_a - u_w)$, and the mean net normal stress $(\sigma_{mean} - u_a)$. This constitutive relationship can be formulated in incremental form as:

$$de = a_t d(\sigma_{mean} - u_a) + a_{ms} d(u_a - u_w) \qquad (8)$$

where:

$\sigma_{mean} = \frac{\sigma_v + 2\sigma_h}{3} =$ mean normal stress

$a_t =$ coefficient of compressibility with respect to a change in mean net normal stress, $d(\sigma_{mean} - u_a)$.

$a_{ms} =$ coefficient of compressibility with respect to a change in matric suction, $d(u_a - u_w)$

The a_t and a_{ms} coefficients become another form of the soil volumetric deformation coefficients. It is shown in Figure 1, generally they vary in a nonlinear manner, but can be considered as being constant for small increments of stress and void ratio.

These coefficients of compressibility represents the current elastic properties of the expansive soil E, H and v, and can be obtained as follows:

$$a_t = \frac{3}{(1+e_o)} \left(\frac{1-2v}{E} \right) \qquad (9)$$

$$a_{ms} = \frac{3}{(1+e_o)H} \qquad (10)$$

where:

$e_o =$ initial void ratio in the increment of stress

In this case, the coefficients of compressibility are associated with triaxial loading and include the variations of the elastic parameters due to the stress state changes. They are conceivable as two classes of general properties uniquely defining the state of strain by the state of stress or matric suction.

Therefore, the material properties of the soil can be computed as follows:

$$E = \frac{3}{(1+e_o)a_t}(1 - 2v) \qquad (11)$$

$$H = \frac{3}{(1+e_o)a_{ms}} \qquad (12)$$

In such case, E and v, are still functions of the stress state or strain components. However, it is now possible to proceed in an iterative way and to determine the values of E and v for any position in the soil profile, depending on the state of stress and matric suction (or of the strain).

5. F.E.M. NUMERICAL APPROACH

From the field and laboratory studies, the boundary conditions of the problem are established, including the presence of layers of different soils, and the material properties are characterized. Now, the field situation is always too complex to be tackled directly as it stands, so it must be idealized to a solvable problem. In this work the mathematical relations describing the idealization and the boundary conditions are formulated and solved using the well-known techniques of analysis, the finite difference and the finite element method.

From the analysis, stresses, matric suctions and displacements are obtained usually at a variety of locations in the idealized model. These have to be interpreted in terms of structural performance in problems of soil-structure interaction, that is to say, total and differential heave or settlement, bending moments, relation of stresses to yield values, and so on.

Much of soil engineering has to do with the interaction of structural elements such as beams, plates, or shells, generally straight or plane but occasionally curved, with soil. Pavements, individual and combined footings, mats, piles, walls, pipes are examples.

It could become common to apply the finite element method to three-dimensional analysis of state stress in expansive clays. However three-dimensional problems will naturally require a larger total number of elements to achieve a reasonable approximation and will, on occasion, tax the storage capacity and speed of most computers available in geotechnical engineering practice, at the present time.

Nevertheless, for some researchers, the two dimensional approximations give an adequate and more economical 'model'. Therefore, the method will be illustrated in terms of a plane strain problem, although there are no additional difficulties in considering a general three-dimensional situation.

Obviously space does not permit a treatment of the basic aspect of the finite element method. Although, some aspects of this process are described in the brief scope of this paper, many details of formulation had to be untouched, therefore for a full description the references should be consulted.

However, since the present study deals with the solution of stress and strain distributions due to the combined action of external loads and moisture flow in the soil, the elastic properties of the material, strains and strains induced by matric suction changes defining the state of stress throughout the element and on its boundaries will be discussed in the next sections.

5.1 Soil model

In general, an unsaturated soil within element boundaries will be subjected to strains such as heave or shrinkage due to matric suction changes. If these are denoted by $\{\varepsilon_o\}$ then the stresses will be caused by the difference between the actual strains and these strains.

The relationship between stresses and strains assuming nonlinear elastic behaviour, but using small increments of stress and strain, will be linear and of the form:

$$\sigma^* = \left\{ \begin{array}{c} \sigma_x - u_a \\ \sigma_y - u_a \\ \tau_{xy} \end{array} \right\} = \mathbf{D}^* \left(\left\{ \begin{array}{c} \varepsilon_x \\ \varepsilon_y \\ \gamma_{xy} \end{array} \right\} - \varepsilon_o \right) + \sigma_o^* \qquad (13)$$

where:
$\mathbf{D}^*$ = elasticity matrix
σ_o^* = initial stress matrix

The elasticity matrix $[D]$, containing the appropiate material properties E and ν, related to the problems discussed in the previous section, will be obtained from the usual isotropic stress-strain relationship, and is given for the case of plane strain by:

$$\mathbf{D}^* = \frac{E(1-\nu)}{(1+\nu)(1-2\nu)} \begin{bmatrix} 1 & \frac{\nu}{(1-\nu)} & 0 \\ \frac{\nu}{(1-\nu)} & 1 & 0 \\ 0 & 0 & \frac{(1-2\nu)}{2(1-\nu)} \end{bmatrix} \qquad (14)$$

In equation (13), the induced strain matrix $\{\varepsilon_o\}$, becomes:

$$\varepsilon_o = (1+\nu)\frac{\Delta(u_a - u_w)}{H} \left\{ \begin{array}{c} 1 \\ 1 \\ 0 \end{array} \right\} \qquad (15)$$

However, when moisture flows through an unsaturated clay soil, the matric suction $(u_a - u_w)$ varies with space and time. Therefore ε_o and σ^*, in equations (13) and (15) must take into account these matric suction changes.

5.2 Moisture flow model

The governing partial differential equation for unsteady state water flow in an anisotropic unsaturated soil can be written as follows (Fredlund & Rahardjo, 1993):

$$\frac{\partial}{\partial x}\left(k_{wxx}\frac{\partial \phi}{\partial x} + k_{wxy}\frac{\partial \phi}{\partial y} \right) + \frac{\partial}{\partial y}\left(k_{wyx}\frac{\partial \phi}{\partial x} + k_{wyy}\frac{\partial \phi}{\partial y} \right) + \bar{Q}$$

$$= m_2^w \rho_w g \frac{\partial \phi}{\partial t} \qquad (16)$$

where:

$k_{wxx} = k_{w1}\cos^2\alpha + k_{w2}\sin^2\alpha$

$k_{wxy} = (k_{w1} - k_{w2})\sin\alpha\cos\alpha$

$k_{wyx} = (k_{w1} - k_{w2})\sin\alpha\cos\alpha$

$k_{wyy} = k_{w1}\sin^2\alpha + k_{w2}\cos^2\alpha$

k_{w1} =major coefficient of permeability as a fuction of matric suction which varies with location in the s_1 direction which is inclined at an angle α, to the x-axis.

k_{w2} =minor coefficient of permeability as a fuction of matric suction which varies with location in the s_2 direction which is in a perpendicular direction to the major permeability.

$\phi = y + \frac{(u_a - u_w)}{\rho_w g}$ =hydraulic head

y =elevation

$\bar{Q} = f(x, y, t)$ =source or sink of moisture throughout the soil. A positive nodal flow $\bar{Q} > 0$, means that there is infiltration at the node or that the node acts as a source. A negative nodal flow $\bar{Q} < 0$, indicates evaporation, evapotranspiration at the node or that the node acts as a sink.

$m_2^w = \frac{\frac{\partial V_w}{V_o}}{\partial(u_a - u_w)}$ =coefficient of water volume change with respect to a change in matric suction

ρ_w =density of water

g =gravitational acceleration

t =time

The general governing equation has been obtained by considering the continuity for the phase water within a period of time. In this formulation, the net total stress is assumed to remain constant with time and the m_2^w coefficient are assumed to be constant for a particular time step.

The solution of the equation (16) defines the distribution of suction throughout the soil mass as a function of location and time. This solution can be uniquely determined by proper prescription of the boundary and initial conditions. The types of boundary and initial conditions associated with equation (16) in their most general forms are:

1. The essential boundary condition or Dirichlet type, and corresponds to the description of a known hydraulic potential or a matric suction over the boundary.

2. The natural boundary condition or Neumann type, corresponds to the description of the flow through the boundary. An impermeable part of the boundary represents a no-flow boundary.

3. The mixed boundary condition or Robinson type, describes the interaction between the soil body and the atmosphere over a part of the boundary, such as the evaporation process.

4. The initial conditions correspond to the description at the same initial instant, of a known

hydraulic potential or matric in the flow domain and of the conditions over its boundary.

All three types of boundary conditions may be specified for the same problem. However, in a well posed problem only one type of boundary condition is specified over a particular portion of the boundary.

Now, the governing differential equation (16) can be solved using a variational formulation. By minimizing the functional derived from the governing and the boundary condition equations. The variational approach and the weighted residual method or Galerkin's method are the most widely used (Fredlund & Rahardjo, 1993).

From the solutions of the flow equations for the system, the hydraulic heads is obtained at the nodal points $\{\phi_n\}$. However, equation (16) is nonlinear because the cofficients of permeability are a function of matric suction, which is related to the hydraulic head at the nodal points. Therefore, the hydraulic head must be obtained using a iterative procedure, involving the coefficient of permeability and the matric suction. Finally, until a tolerance specified is reached, the nodal matric suction of the system can be obtained.

The equation for the nodal suction is:

$$\{(u_a - u_w)_n\} = (\{\phi_n\} - \{y_n\})\rho_w g \qquad (17)$$

where:

$\{(u_a - u_w)_n\}$ =matrix of matric suction

$\{\phi_n\}$ =matrix of hydraulic heads at the nodal points

$\{y_n\}$ =matrix of elevation heads at the nodal points

Now the induced strain matrix $\{\varepsilon_o\}$ and the stress matrix $\{\sigma^*\}$, for one time step, can then be calculated using the converged nodal matric suction, and the soil problem can be solved following the standard computational procedures required in application discretized by the finite element method. Frequently some finite element computer programs fit different types of problem (Zienkiewicz & Taylor, 1991, Li & Cameron, 1995).

6. CONCLUSIONS

Expansive soils cause problems beneath structures because they change in volume when subject to variations in moisture content.

Steady state or equilibrium analyses are the most common use of the finite element method. Assuming elastic behaviour, an unsaturated soil body under equilibrium conditions can be analysed and its distorsion predicted. From the calculated values of displacement it is then possible to derive the strains and stresses experienced by the soil.

However, because in many situations in the field, the flow regime is unsteady and the coefficient of

713

permeability is a function of matric suction, matric suction varies with space and time. The governing equation for unsteady state flow of moisture is a partial differential equation that may be used together with some specified conditions to solve for soil suction.

For both soil problems, in finite element computer program a knowledge of the elastic moduli E, H and v, are needed as data input, to compute the volume change of the unsaturated soil.

However, test results indicate that the material properties in an unsaturated expansive soil are dependent of the variation of moisture and external loads. A plot of a constitutive surface for the Madrid grey clay, shows that material parameters are responsible for the non linearity in the surfaces representing the state stress variables and void ratio.

The tests were performed in a modified oedometer developed for measurement of lateral swelling pressure, under constant load and matric suction conditions. The experiment was conducted on undisturbed samples, under loading and wetting conditions.

A procedure is suggested to determine the material parameter. It considers the deformation state variable represented by the void ratio e. Therefore, the constitutive surface can be traced graphically by plotting e with respect to the mean net stress $(\sigma_{mean} - u_a)$ and the matric suction $(u_a - u_w)$.

Hence, the elastic parameters E, H and v can be calculated from the coefficient of compressibilities a_t and a_{ms}, obtained in the laboratory test.

This procedure consideres the effect on the material properties due to net lateral pressure $d(\sigma_h - u_a)$ increases, induced in the laterally confined expansive soil by reductions in matric suction $d(u_a - u_w)$.

7. ACKNOWLEDGMENT

The author would like to express his gratitude to Dr D G Toll of Nanyang Technological University, Singapore, for his valuable suggestions and corrections in preparing this paper.

8. REFERENCES

Cuellar, V. (1978). Analisis crítico de los métodos existentes para el empleo de arcillas expansivas. *Lab. Transp. Mec. Suelos Madrid.*

Fredlund, D.G. & Morgestern, N.R. 1977. Stress state variables for unsaturated soils. *J. Geotech. Eng. Div. ASCE. GT5.* Vol.103, 447-466.

Fredlund, D.G. & Rahardjo, H. 1993. *Soil Mechanics for Unsaturated Soils.* John Wiley & Sons, Inc., New York.

Li, J. & Cameron, D.A. 1995. Finite element analysis of deep beams in expansive clays. *Proc. 1st Int. Conf. on Unsaturated Soils,* Paris. Vol.2, 1109-1115.

Lytton, R. L. 1996. Foundation and pavement in unsaturated Soils. *Proc. 1st Int. Conf. on Unsaturated Soils,* Paris. Vol.3, 1201-1220.

Meintjes, H.A.C.1992. Suction load strain relations in expansive Soils. *Proc. 7th Int. Conf. on Expansive Soils,* Dallas. Vol.1, 51-54.

Oteo, C., Saez, J. & Esteban, F. 1996. Laboratory tests and equipment with suction control. *Proc. 1st Int. Conf. on Unsaturated Soils,* Paris. Vol.3, 1509-1515.

Sabbagh, A. 1995. Prediction of volume change in unsaturated clays *Proc. 1st Int. Conf. on Unsaturated Soils,* Paris. Vol.2, 791-796

Wallace, K. B. & Lytton, R. L. 1992. Lateral pressures and swelling in a cracked Expansive clay profile. *Proc. 7th Int. Conf. on Expansive Soils,* Dallas. Vol.1, 245-250.

Zienkiewicz, O. C. & Taylor, R.L. 1991. *The Finite Element Method.* McGraw-Hill Book Co., London.

Unsaturated Soils for Asia, Rahardjo, Toll & Leong (eds) © 2000 Taylor & Francis, ISBN 90 5809 139 2

Engineering characteristics of black cotton soils in Francistown

B. K. Sahu
Department of Civil Engineering, University of Botswana, Gaborone, Botswana

ABSTRACT: In the present investigation engineering properties of black cotton soils in Francistown, Botswana have been evaluated. Francistown is the second largest city of Botswana and is being given top priority for development. Six soil samples from four different locations were collected and tested in the laboratory. It was found that the laboratory results compare well with values predicted by Sahu, Vijayvergiya and Nayak. The expansiveness of almost all the soils vary from high to very high. For foundations, expansive soil stratum should be replaced by other inactive soils if its thickness is less than 2m, while under-reamed piles should be provided for strata deeper than 2m.

1 INTRODUCTION

Black cotton soil is a name believed to have first been used in India where areas with black or dark grey soils are the locations for growing cotton. Soils derived from the weathering of black trap rock, in particular, which are black, fine grained, heavy and climatologically suited to the growth of cotton are known as black cotton soils. These are expansive soils and can give rise to problems to various structures due to their tendency to undergo large volume changes with the change in moisture content.

The problems associated with expansive soils are not widely appreciated outside the areas of their occurrence. The amount of damage caused by expansive soils is alarming. In Botswana, statistics pertaining to the cost of damage caused by expansive soils do not exist due to lack of studies. However, from the statistics of other developed countries, it has been estimated that the damage to buildings, roads, and other structures founded on expansive soils exceeds two billion dollars annually.

Botswana is a small country in Southern Africa surrounded by South Africa, Zimbabwe, Zambia and Namibia. Almost two third land area of Botswana is covered either with Kalahari sand or sandy soils. Expansive soils are found only in patches. Quite a significant area of Francistown, the second largest city of Botswana, is covered with black cotton soil (fig.1). These areas covered with black cotton soils are posing serious problems in the planning and development of the township due to the problems associated with the expansive soils. The degree of volume change of an expansive soil causing upheaval and sinking of a structure founded on it depends upon the mineral composition , gradation, change in moisture content and the field density of the soil. Guidelines for the design of foundations and super structures are proposed by various investigators, but due to complexity of the problem it has not been possible for geotechnical engineers to come out with the definite solutions and guidelines for the design of foundations and superstructures which could be applied globally. Due to lack of any systematic work on local black cotton soils in Botswana, there are no definite guidelines available for the design and construction on these soils.

The present investigation on black cotton soils from various parts of Francistown consists of mineralogical analysis using an x-ray diffractometer, determination of index and swelling properties in the laboratory, prediction of swelling characteristics by various empirical correlations and its comparison with the values determined in the laboratory.

Various remedial measures have been critically examined and recommendations for the design of foundations have been made.

2 PHYSIOGRAPHY

Granitoid rocks underlie the northern part of Francistown. Topographically it is a flat pediment dipping gently south-east, broken by koppies and inselburgs of the more potassic of these rocks. Supercrustal strata in the southern half of the area form ridges and hills parallel to the general strikes of the rocks; thus an acruta series of ironstones form

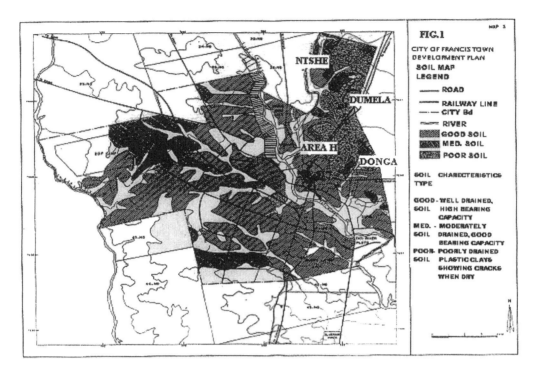

Figure 1. Soil type characteristics in the city of Francistown

the backbone of the Motasiloje Hills.

The rivers only flow infrequently during the rainy season (October to April). The drainage pattern is dendritic and controlled by the underlying geology. Exposure of the bedded strata is good, but most of the granitoid rocks are poorly exposed. Tree savanna vegetation is typical with C.mopane and acacia spp dominant.

3 GEOLOGY

Archean metasediments and metavolcanics of the TATI SCHIST GROUP, and a sequence of granitoid rocks, predominate the area. The oldest granitoid rocks (GI) are the elliptical tonalite plutons and their marginal munzonites; their emplacement post-dates the deposition of the Tati schist group supercrustal rocks. The youngest major suite of granitoid rocks includes the post tectonic (G4) tonalite stocks cutting the schist relic around Francistown. The Francistown diorite is a gently dipping sheet, which may be contemporaneous with the G4 tonalite along its southern contact.

Recent sediments mask the solid geology in large parts of the area. The granitoid rocks have a thin, grey, sandy soil cover generally less than 3 m thick. The Tati schist relic rocks are overlain by reddish, more fertile soil. Small floodplain deposits are lim-

ited to the larger watercourses where they cross the granitoid pediments.

4 SOILS USED

The soils used for the present investigation were collected from four different sites, namely Dumela, Ntshe, Donga and Area H of Francistown (fig.1). Francistown, the second largest city of Botswana, is situated about 450 km north of the capital city Gaborone and about 90 km from the southern border of Zimbabwe. During the dry season, wide and deep cracks occur in the soils, sometimes reaching a depth of 2 to 3m and a width of 75mm or more. These cracks do get filled during the wet season.

Historically, building construction, in these areas covered with black cotton soil, were avoided due to foundation problems. However, the rapid urban expansion and explosive industrial growth has resulted in an upsurge in construction in the areas covered with black cotton soils. In the areas where the black cotton soil was limited to a maximum depth of 2m, the soil was replaced by a better inactive foundation material but wherever the thickness of black cotton stratum was more and the buildings were founded on these soils, serious problems due to upheaval were encountered. Buildings of Bank of Botswana, Fire Department, Life Insurance Ltd and Ikhutseng

Table 1 Summary of test results.

	Dumela (1)	Dumela (2)	Ntshe (3)	Donga (4)	Area H (5)	Area H (6)
Clay fraction % (C)	43	41	72	73	62	50
Liquid limit W_l %	78	64	71	54	76	66
Plastic limit W_p %	40	31	39	29	48	28
Plasticity index I_p %	38	33	32	25	28	38
Activity Ac (I_p/ C-10)	1.15	1.06	0.52	0.40	0.54	0.95
Linear shrinkage LS %	21	22	20	21	22	24
Optimum moisture content %	28	23	31	23	32	20
Max. dry density kg/m^3	1457	1571	1451	1613	11479	1236

Table 2 Swelling Potentials

	Dumela (1)	Dumela (2)	Ntshe (3)	Donga (4)	Area H (5)	Area H (6)
Alawaji	-10.29	-8.1	1.19	5.34	0.77	-8.33
Chen	6.06	4.0	3.7	2.1	2.6	6.06
Seed et al (1)	15.46	11.0	10.16	5.6	7.3	15.46
Seed et al (2)	11.08	7.5	12.2	6.77	7.58	12.88
Sahu et al	10.02	6.43	4.77	2.77	6.87	4.85
Vijayvergiya et al	9.7	3.0	5.5	3.3	8.2	16.3
Nayak et al	8.3	8.3	8.8	9.0	8.2	9.5
Experimental	10.4	--	6.0	6.6	10.6	9.85

Swelling Potential =$[-3.39 + 0.364C + 0.001$ Wl $- 0.56$ Ip $- 0.00073C^2]$ Alawaji (1997)

Swelling Potential = $[0.2558 \ e^{0.0883 \ Ip}]$ Chen (1988)

Swelling Potential = $[3.6 \times 10^{-5} \ Ac \ C^{3.44}]$ Seed et al (1962) 1
    (Ac = Ip/C)

Swelling Potential = $[2.16 \times 10^{-3} \ (Ip)^{2.44}]$ Seed et al (1962) 2

Swelling Potential = $[1.76 \times 10^{-8}(\gamma_d) \ 2.75 \ (W_1 - W_0)^{3.27}]$ Sahu et al (1987)
    (γ_d in kN/m^3)

Log Swelling Potential = $1/12(0.44 \ W_1 - W_0 +5.5)$ Vijayvergiya et al (1973)

Swelling Potential =$[(0.0229Ip \times 1.45C)/W_0]+ 6.38$ Nayak et al (1974)

Primary School are some of the typical examples where the buildings have cracked due to upheaval of foundations and floors.

5 LABORATORY TESTS

Both disturbed and undisturbed soil samples were collected from all the four sites and laboratory tests were performed to determine soil gradation, Atterberg limits, linear shrinkage, natural moisture content, field density, optimum moisture content, and maximum dry density. All the tests were performed as per BS 1375: 1990 code of practice. The summary of the test results is shown in Table 1.

5.1 Swelling Potential

Swelling potential is an index that indicates the degree of volume change of the soils after saturation. From the swelling potential, it is possible to estimate the magnitude of heaving of floors and foundations. Empirical correlations have been suggested by various investigators to evaluate the swelling potential using the results of simple soil tests.

In the present investigation only a few correlations have been considered to compare with the experimental values. Alawaji (1997) regressed swelling potential against the clay content (C%), Wl%, Ip% and (C%)2..Chen (1988) has demonstrated that the plasticity index alone can be used as a preliminary indication of the swelling characteristics of most clays. Seed et al (1962) and Nayak (1974) have correlated the plasticity index and clay content of a soil with its swelling potential. Vijayvergiya (1973) has correlated liquid limit and initial water content with swelling potential. It has been observed that the factors which affect the amount of swelling of a given soil, also include initial moisture content and the density. Sahu et al (1987) correlated swelling potential and swelling pressure of a soil with its liquid limit, initial moisture content and initial dry unit weight.

To obtain experimental values of swelling potentials soil samples were remoulded in oedometer rings at its optimum moisture contents and maximum dry densites. These samples were then flooded and allowed to swell under a seating load of 6.9kPa. The percentage swells were recorded as the swelling potential of these soils. The experimentally obtained values were compared with the ones predicted by empirical correlations Table2.

5.2 Mineralogy of Soils

Mineralogical analysis of all the samples was done by X-ray diffractometer. Results are shown in Table 3.

Feldspar and amphiboles are the non clay minerals, which are more prevalent in soils formed from

Table 3 Mineralogical composition of all the soils

Site	Primary mineral	Secondary mineral
Dumela (1)	Feldspar, Montmorillonite	Amphibole, Chlorite
Dumela (2)	Feldspar, Montmorillonite	Amphibole, Chlorite
Ntshe (3)	Feldspar, Montmorillonite	Amphibole
Donga (4)	Amphibole, Feldspar	Montmorillonite
Area H (5)	Feldspar	Montmorillonite
Area H (6)	Feldspar	Montmorillonite

the igneous rocks. In all the soils quartz was also present in significant proportion. It is observed that all the soils contained mineral montmorillonite which is responsible for imparting swelling characteristics to these soils. It was rather difficult to do any quantitative analysis to get the proportion of mineral montmorillonite.

6 DISCUSSION

It is observed that the values of swelling potentials predicted by Chen and Alawaji are consistently low while those predicted by Seed et al are generally higher than the experimental values. The ones predicted by Sahu, Vijayvergiya, and Nayak, are quite comparable in most of the cases and these correlations can be applied to estimate the magnitude of swelling potential for the soils in Francistown.

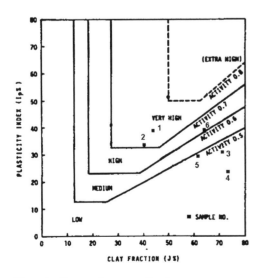

Fig. 2 Expansiveness Chart
(v.d. Merwe, 1975)

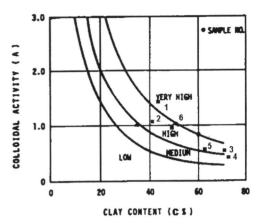

Fig. 3 Expansiveness Chart
(Seed et al, 1962)

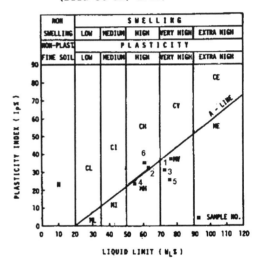

Fig. 4 Modified Plasticity Chart
(Dakshanamurthy and Raman, 1975)

Various investigators have developed charts and tables to assess the degree of expansiveness of expansive soils using simple index tests. Only a few are tested herein depending on simplicity and wide acceptance. Figures 2, 3 and 4 give the classification of the potential expansiveness of all the soils from Francistown.

It is observed that expansiveness of all the soils vary from high to very high according to Dakshanamurty (1975) and Seed et al (1962), while only three soils have high to very high expansiveness as per Van der Merve. The mineralogical analysis (Table 3) suggests that mineral montmorillonite, responsible for swelling characteristics, is present as a primary mineral only in three soils while in other three soils it is present as secondary clay mineral.

However, the quantitative proportion of the montmorillonite in various soils can not be taken as the only guiding factor to assess the degree of expansiveness as it also depends upon the type of exchangeable cation. The field observations of the distress caused by the upheaval of footing, floors and pavements in the all the areas indicate that the soils have high to very high degree of expansiveness in all the areas. It suggests that perhaps the charts developed by Dakshanamurty & Rahman (1975) and Seed et al. (1962) would make better predictions for the expansiveness of the soils in Francistown.

7 RECOMMENDATIONS

1. Removal of the black cotton soil if it is less than 2m deep and its replacement by an inactive soil duly compacted.
2. Under-reamed Pile: In deep deposits, the minimum length of under-reamed pile from ground level to toe should not be less than 3.5m so that the piles are not affected by seasonal ground movements. If the expansive soil deposits are shallow, overlying non-expansive soils of weak nature, the piles should pass the fill or weak soil and made to rest in good bearing stratum

These recommendations are based on the successful use of under-reamed piles in similar situations in India. In India, under-reamed piles were first introduced for expansive soil (black cotton soil) areas in mid fifties. But due to better performance and economy they have found a much wider use in almost all types of soil strata and for structures of varied description.

In Botswana there is no experience with under-reamed piles. It would, therefore, be wise to test a few prototype under-reamed piles before a proper guideline is recommended.

8 CONCLUSIONS

1. The mineral Montmorillonite is present in all the soils tested.
2. Swelling potential of all the soils vary from 6 to 10.6.
3. The degree of expansion of all the soils vary from low to very high.
4. Empirical correlations by Sahu, Nayak and Vijayvergiya make better predictions for soils tested.
5. Expansive soil should be replaced by an inactive soil if the thickness of expansive soil is less than 2 m.
6. If the expansive soil stratum is deep under-reamed pile foundations should by provided.

REFERENCES

Alawaji, H.A.1997. Swell and Strength Characteristics of Compacted sandy Clay Soils. *Proc.3rd Int. Geotechnical engineering Conf., Cairo, Egypt,* 303-314.

Baletsi, S.H. 1988. *Swelling Characteristics of Black Cotton Soils of Francistown.* B.Eng.IV Project - University of Botswana.

Chen, V.H. 1988. *Foundations on Expansive Soils,* Elsevier Science Publishers B.V.

Dakshanamurty, V. and Raman,V.1975 Review of Expansive Soils. Discussion, *J. of Geotechnical Engineering Div., ASCE,* Vol. 101, No. GT6.

Nayak, N.V & Christensen, R.W. 1970. Swelling Characteristics of Compacted Expansive Soils, *Clays and Clay Minerals,* Vol.19, No.4.

Sahu, B.K., Gichaga, F.J. & Viswesaraiya, T.G. 1987. Prediction of Swell of Black Cotton Soils in Nairobi, *Proc. Int. Symposium on Prediction and Performance in Geotechnical Engineering, Calgary, Canada,* 259-266

Seed, H.B., Woodward, R.J. & Lundgren, R. 1962. Prediction of Swelling Potential for Compacted clays, *J. of SMFE, Proc. of ASCE,* Vol. 88, No. SM3.

Sydney, M, 1997. *Francistown Black Cotton Soils,* B.Eng.IV Project, University of Botswana.

Van der Merwe, D.H. 1975. Contribution to Speciality Session B: Current Theory and practice for Building on Expansive Clays, *Proc. 6th Reg. Conf. for Africa on SMFE, Durban,* Vol.2

Vijayvergiya, V.N.& Ghazzaly, O.I. 1973. Prediction of Swelling Potential for Natural Clays, *Proc. 3rd Int. Conf. on Expansive Soils, Jerusalem,* Vol.1.

Unsaturated Soils for Asia, Rahardjo, Toll & Leong (eds) © 2000 Taylor & Francis, ISBN 90 5809 139 2

Behaviour of an unsaturated highly expansive clay during cycles of wetting and drying

R.S.Sharma
Department of Civil and Environmental Engineering, University of Bradford, UK

S.J.Wheeler
Department of Civil Engineering, University of Glasgow, UK

ABSTRACT: Cycles of wetting and drying were performed under isotropic stress conditions on compacted samples of an unsaturated highly expansive clay consisting of a natural sodium bentonite mixed with speswhite kaolin. The results showed that both irreversible shrinkage over wetting-drying cycles and irreversible swelling over wetting-drying cycles can occur on samples of the same soil, and this type of behaviour is not restricted to highly expansive clays. The precise forms of irreversible behaviour were not consistent with existing elasto-plastic constitutive models. An alternative physical explanation for irreversible volume changes is proposed, based on the occurrence of hydraulic hysteresis in the variation of degree of saturation.

1 INTRODUCTION

An unsaturated clay can either swell or exhibit collapse compression during wetting, depending upon the value of applied stress and the previous history of stress and suction variation. This applies even to highly expansive clays, containing active clay minerals (such as montmorillonites). The specific behavioural feature of highly expansive clays is that if swelling on wetting does occur, the magnitude of swelling can be very large.

Alonso, Gens and Hight (1987) presented an elasto-plastic framework to describe the mechanical behaviour of unsaturated soils, including the possibility of either swelling or collapse compression on wetting. This qualitative framework, which employed matric suction and net stress (the difference between total stress and pore air pressure) as stress state variables, was subsequently developed to a full elasto-plastic constitutive model by Alonso, Gens and Josa (1990). Subsequent refinements and variations to the model were suggested by authors such as Wheeler and Sivakumar (1995).

In the class of elasto-plastic models first presented by Alonso, Gens and Josa (1990), swelling on wetting and shrinkage on subsequent drying are both represented as elastic processes, whereas collapse compression on wetting is modelled as a plastic process (corresponding to expansion of a Loading Collapse (LC) yield curve). This type of model therefore predicts that, during cycles of wetting and drying (over a constant range of suction and at a constant value of net stress), the only possible source of irre-

versible volume change is if plastic collapse compression occurs during the first wetting path.

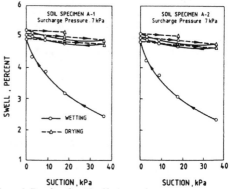

Figure 1: Results of wetting/drying cycles on a highly swelling soil (after Chu and Mou (1973))

Chu and Mou (1973) reported experimental test results from cycles of wetting and drying performed on a highly expansive clay, which showed a large irreversible component of swelling during the first wetting path (see Figure 1). Similar behaviour was reported by Pousada (1984). This form of behaviour cannot be represented by models of the type first developed by Alonso, Gens and Josa (1990). Subsequently, Alonso *et. al.* (1995) presented controlled-suction oedometer test results for another highly expansive clay, which showed significant irreversible components of shrinkage during the drying stages of wetting-drying cycles. An example of their results is shown in Figure 2. During the first wetting path C1, from a very high initial value of suction, initial swelling was subsequently followed by some collapse compression as the suction was progressively reduced. The subsequent cycles of drying and wetting showed significant components of irreversible shrinkage during drying paths C2 and C4, with a clear suggestion of a yield point in each drying path. The magnitude of the irreversible component of shrinkage was greater in the first drying path C2 than in the second drying path C4. Again, the form of behaviour observed by Alonso *et. al.* (1995), and illustrated in Figure 2, cannot be represented by the constitutive model of Alonso, Gens and Josa (1990).

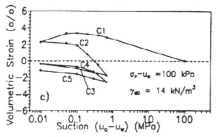

Figure 2: Wetting/drying cycles performed on Boom Clay (Alonso *et. al.* (1995))

In the light of experimental observations of irreversible volume changes in addition to those attributable to collapse compression on wetting, Gens and Alonso (1992) suggested a modified form of elasto-plastic model for unsaturated highly expansive clays. Further developments to the proposed model were presented by Alonso, Gens and Gehling (1994). In this new modelling framework for unsaturated highly expansive clays, irreversible swelling during wetting (such as reported by Chu and Mou (1973)) or irreversible shrinkage during drying (such as reported by Alonso *et. al.* (1995)) were attributed to elastic swelling or compression of the saturated microstructure of individual clay packets exceeding a maximum value that could be tolerated without plastic re-arrangement of the unsaturated macrostructure. In the proposed elasto-plastic framework, this was represented by the inclusion of Suction Decrease (*SD*) and Suction Increase (*SI*) yield curves.

The elasto-plastic framework for unsaturated highly expansive clays proposed by Gens and Alonso (1992) and Alonso, Gens and Gehling (1994) had not been satisfactorily validated by experimental tests. In particular, a number of important questions remained unanswered:

- could the two different types of behaviour illustrated in Figure 1 and Figure 2 both occur for the same soil (if cycles of wetting and drying were performed on samples with different stress histories);
- was the physical mechanism on which the proposed framework was based (plastic strains of an unsaturated microstructure caused by excessive

elastic strains of a saturated macrostructure) the true explanation for the observations of irreversible swelling on wetting or irreversible shrinkage on drying;

- could the proposed *SD* and *SI* yield curves be used to predict correctly the precise forms of the various different types of irreversible behaviour observed during cycles of wetting and drying;
- was the occurrence of irreversible swelling on wetting or irreversible shrinkage on drying specific to highly expansive clays or could it also occur during cycles of wetting and drying on unsaturated non-expansive clays?

A programme of experimental research was undertaken by Sharma (1998) to investigate these issues.

2 EXPERIMENTAL TESTS

Tests were performed on a highly expansive clay mix consisting of a natural sodium bentonite from Wyoming (10% by weight) combined with speswhite kaolin (90% by weight). Triaxial samples, 50 mm in diameter and 50 mm high were prepared by compaction in a mould at a water content of 25% (4% less than the optimum from standard Proctor compaction test). Three different static compaction pressures were employed: 400 kPa, 800 kPa and 3200 kPa.

Cycles of wetting and drying were performed under isotropic stress states in a controlled-suction triaxial cell employing the axis translation technique (Hilf (1956)). Pore water pressure u_w was controlled through a saturated porous filter with an air entry value of 500 kPa, whereas pore air pressure u_a was controlled through a dry filter with a low air entry value. During wetting or drying, matric suction $s = u_a - u_w$ was varied in a continuous fashion, using computerised control of stepper motors regulating u_a and u_w. Testing rates had to be very slow, because of the low permeability of the soil, and it was considered essential to limit the sample height to 50 mm. The resulting height-to-diameter ratio of 1 was considered acceptable, because the test conditions were limited to isotropic stress states.

Sample volume change was measured by monitoring the flow of water into the surrounding cell with a volume gauge. A double-walled cell was used, so as to avoid difficulties that would have resulted from creep or hysteresis of the inner acrylic cell wall, and the cell was calibrated for the effects of cell pressure, loading ram displacement and water absorption by the acrylic (see Wheeler (1988)). These measurements of sample volume change were used to monitor the variation of specific volume v. A second volume gauge was used to record the flow of water from the sample, and measurements from the two gauges were combined to monitor the variation of degree of saturation S_r.

3 EXPERIMENTAL RESULTS

A total of twenty tests were performed, and results from three selected tests, carried out on samples compacted to 400 kPa, 800 kPa and 3200 kPa, are presented here.

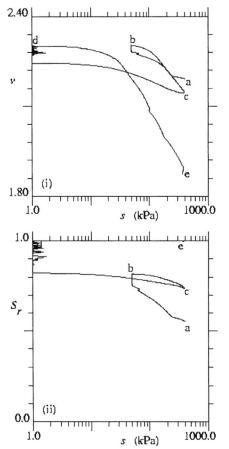

Figure 3: Results of wetting/drying cycles from a test on a sample compacted to 400 kPa

Figure 3 shows results from a test on a sample, which had been prepared by compacting to a vertical stress of 400 kPa. After equalisation at a matric suction s of 400 kPa and mean net stress $p - u_a$ 10 kPa the sample was subjected to two full cycles of wetting and drying, with $p - u_a$ maintained constant at 10 kPa throughout. During the first wetting path ab the suction was reduced from 400 kPa to 50 kPa. Suction was subsequently increased to 380 kPa (first drying stage bc), reduced to zero (second wetting stage cd) and finally increased to 370 kPa (second drying stage de). At the end of each wetting or drying stage there was an equilibration period, which was terminated when the measured rate of water inflow or outflow to the sample reduced to 0.05 cm³/day. The significant changes in v and S_r occurring during the equilibration periods at the end of the two wetting paths (apparent as the vertical steps in Figures 3(i) and 3(ii)) indicate that the rate of suction variation that was applied in this test was faster than desirable (despite a total test duration of almost 10 weeks).

Inspection of Figure 3(i) shows that the sample swelled monotonically during each of the wetting paths ab and cd, with no suggestion of wetting-induced collapse compression. Within the elasto-plastic model of Alonso, Gens and Josa (1990) this would be interpreted as indicating that the stress path remained inside the LC yield curve throughout the cycles of wetting and drying. During the first cycle abc there was, however, a small amount of net shrinkage, suggesting an irreversible component of shrinkage during first drying path bc. There was also a suggestion of a yield point during drying path bc. This pattern of behaviour was clearer during the second wetting-drying cycle cde, which was extended down to a lower value of suction. There was a large net shrinkage over this second cycle, and a strong suggestion of a yield point during drying path de. The form of behaviour shown in Figure 3(i) is qualitatively similar to that reported by Alonso et. al. (1995), as presented in Figure 2.

Figure 3(ii) shows the variation of degree of saturation S_r. The degree of saturation reached 1.0 during the equilibration period at the end of the second wetting path cd, when the suction was reduced to zero. The sample then remained fully saturated during the final drying path de (in fact the calculated values of S_r fluctuated between 1.00 and 1.01 during drying path de, indicating the level of measurement accuracy). Inspection of Figure 3(ii) shows that, at any given value of suction, the degree of saturation S_r was higher in a drying path than in the preceding wetting path. Over each cycle of wetting and drying there was a net increase of degree of saturation S_r, and this increase in S_r accumulated over the two cycles. This was the expected pattern of hydraulic hysteresis in the water-retention curve, reported by many previous authors (see, for example, Croney (1952)).

Results from a second test, in which the sample was prepared by compacting to a vertical stress of 800 kPa, are shown in Figure 4. After equalisation at $s = 400\,\text{kPa}$ and $p - u_a = 50\,\text{kPa}$, the sample was subjected to wetting and drying with $p - u_a$ held constant at 50 kPa. The suction was first reduced to 100 kPa (first wetting stage ab), then increased to 400 kPa once more (drying stage bc) and finally reduced to 10 kPa (second wetting stage cd).

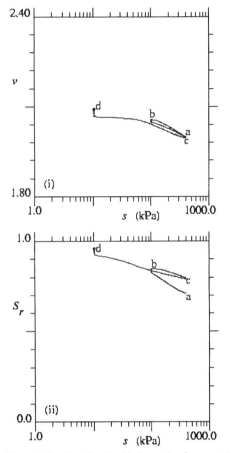

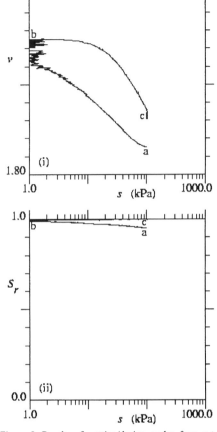

Figure 4: Results of wetting/drying cycles from a test on a sample compacted to 800 kPa

Figure 5: Results of wetting/drying cycles from a test on a sample compacted to 3200 kPa

Inspection of Figure 4(i) shows that no collapse compression occurred during wetting paths ab and cd, which would be interpreted as indicating that the stress path remained inside the LC yield curve. The volumetric response shown in Figure 4(i) appears almost reversible, in particular there was no net shrinkage or swelling over the wetting-drying cycle abc. The variation of S_r presented in Figure 4(ii) shows some evidence of hydraulic hysteresis, but the differences in the values of S_r during wetting and drying are relatively small.

Results from a third test, in which the sample was compacted to a vertical stress of 3200 kPa, are shown in Figure 5. After equalisation at $s = 100$ kPa $p - u_a$ constant at 10 kPa. Figure 5(i) shows that net swelling occurred over the cycle abc, but there was also some evidence of a yield point during drying path bc. Some hydraulic hysteresis is apparent in Figure 5(ii), the degree of saturation increasing from 0.95 to 1.00 during wetting path ab, whereas the

sample remained fully saturated during subsequent drying path bc.

4 DISCUSSION

The first point to emerge from the experimental results presented here is that net shrinkage over a wetting-drying cycle and net swelling over a wetting-drying cycle have both been observed on samples of the same soil (see Figures 3 and 5). The two different types of behaviour reported by Chu and Mou (1973) (see Figure 1) and Alonso *et. al.* (1995) (see Figure 2) can therefore occur in the same soil. Whether net swelling or net shrinkage will occur over a cycle of wetting and drying is clearly a function of the history of compaction and subsequent variation of net stress and suction, rather than being determined solely by soil type.

Sharma (1998) also performed tests on compacted samples of unsaturated pure speswhite kaolin (a non-expansive soil), and again observed either irreversible swelling or irreversible shrinkage over cycles of wetting and drying. This suggests that irreversible swelling or shrinkage is a feature of behaviour of all clays, rather than being restricted to highly expansive clays.

Clearly, the occurrence of irreversible shrinkage or swelling during cycles of wetting and drying cannot be represented by the conventional unsaturated elasto-plastic models of the type first proposed by Alonso, Gens and Josa (1990). The forms of irreversibility illustrated in Figures 1, 2, 3 and 5 are also not entirely consistent with the elasto-plastic models for unsaturated highly expansive clays proposed by Gens and Alonso (1992) and Alonso, Gens and Gehling (1994). For these models to predict irreversible shrinkage over a cycle of wetting and drying or irreversible swelling over more than the first cycle there would have to be yield points on both the wetting path (corresponding to the *SD* yield curve) and the drying path (corresponding to the *SI* yield curve). In practice, however, irreversible shrinkage seems to occur with evidence of yielding on drying paths but not on wetting paths (see Figures 1 and 3; similar behaviour has also been observed in many subsequent tests). In contrast, irreversible swelling over a cycle of wetting and drying appears to occur without evidence of yielding during either wetting or drying paths (see Figures 2 and 5; the possible yield point occurring during path *bc* in Figure 5 is not contributing to irreversible swelling, rather it tends to reduce the net swelling over the cycle by introducing a counteracting component of irreversible shrinkage).

Given that the observed patterns of behaviour were not entirely consistent with the predictions of their elasto-plastic model for unsaturated highly expansive clays, an alternative physical mechanism to that set out by Gens and Alonso (1992) and Alonso, Gens and Gehling (1994) is now proposed to explain the occurrence of irreversible shrinkage or swelling during cycles of wetting and drying. This alternative explanation for irreversible volume changes is closely linked to the observation of hydraulic hysteresis in the variation of degree of saturation.

The proposed explanation for irreversible volume changes is based on consideration of the differences in behaviour of water-filled voids and air-filled voids, and the response of a void on flooding or emptying of water (during wetting or drying respectively). The volume change behaviour of water-filled voids is likely to be largely controlled by the stress variable $p - u_w$. In contrast, the stress variable $p - u_a$ is relevant for the behaviour of air-filled voids, but in addition the volume changes of these air-filled voids are also affected by the presence of lenses of "meniscus water" at the surrounding inter-particle or inter-packet contacts (see Wheeler and Karube (1996)). Analysis of the contact between two idealised spherical particles, by, for example, Fisher (1926), indicates that the value of additional inter-particle normal force, caused by a lens of meniscus water, changes very little as suction increases (the limiting value as suction tends to infinity is only about 50% greater than the value at zero suction). To a first approximation, therefore, the additional inter-particle normal force due to meniscus water can be assumed to be constant, provided the surrounding voids remain air-filled and the meniscus water lens is present, but the additional force suddenly disappears if the voids on either side of the particle contact flood with water. Loss of this additional component of inter-particle normal force during flooding of the surrounding voids can have two effects: it will cause a component of swelling, due to elastic recovery of the particles under the reduced force, but it will also make the frictional contact more susceptible to slippage (giving rise to the possibility of wetting-induced collapse compression).

During wetting, by decrease of $u_a - u_w$ at constant $p - u_a$, two components of swelling can be expected. The first is an elastic increase in volume of water-filled voids due to the decrease in the value of the stress $p - u_w$. The second component of swelling is an increase in the volume of those voids that flood with water during the wetting process, due to the loss of the additional inter-particle force arising from meniscus water. Although this second component arises from an essentially elastic process, it is not necessarily reversible if the suction is increased again. It depends upon the relevant voids emptying of water once more, and would therefore require an appropriate decrease in the degree of saturation, rather than being dependent on the increase in suction. Clearly, if there is hydraulic hysteresis in the water-retention relationship, so that the reduction of S_r during a drying path is less than the corresponding increase of S_r during the preceding wetting path, the magnitude of shrinkage during drying will be less than the previous swelling during wetting. This would lead to net swelling over a cycle of wetting and drying, with no evidence of yielding during either wetting or drying. This is entirely consistent with the type of behaviour reported by, for example, Chu and Mou (1973) (see Figure 1).

Hydraulic hysteresis in the variation of degree of saturation S_r also explains the possibility of yielding during a drying path, which could then lead to net shrinkage over a cycle of wetting and drying. Hydraulic hysteresis means that some voids that were air-filled during a wetting path at a given value of suction will be water-filled at the same value of suction during the subsequent drying path. These voids will now be experiencing the highest value of $p - u_w$ that they have ever been subjected to while water-filled, and this may be sufficient to cause yielding (slippage of the surrounding inter-particle contacts); when the suction was of this magnitude during the previous wetting these voids were air-filled and therefore experiencing a lower stress $p - u_a$ and also the stabilising effect of meniscus water at the particle contacts. This mechanism for yielding during drying paths is entirely consistent with the behaviour illustrated in Figures 2 and 3, which resulted in net shrinkage over a cycle of wetting and drying.

5 CONCLUSIONS

Irreversible shrinkage over wetting-drying cycles and irreversible swelling over wetting-drying cycles have both been observed for samples of the same soil. Whether the former or the latter occurs appears to be dependent on the history of compaction and subsequent variation of net stress and suction. Both forms of behaviour have been observed on unsaturated non-expansive clays, and are therefore not restricted to highly expansive clays.

Net shrinkage over a wetting-drying cycle is accompanied by yielding during the drying path but no evidence of yielding during the wetting path. In contrast, net swelling over a wetting-drying cycle can occur with no evidence of yielding during either wetting or drying. This pattern of behaviour is not entirely consistent with the forms of irreversible behaviour predicted by the elasto-plastic model for unsaturated highly expansive clays presented by Gens and Alonso (1992) and Alonso, Gens and Gehling (1994). An alternative physical explanation for the two forms of irreversibility has therefore been proposed, linked to the occurrence of hydraulic hysteresis in the variation of degree of saturation. Work currently in progress is attempting to use this physical explanation in the development of an improved constitutive model.

ACKNOWLEDGEMENTS

The experimental tests reported in this paper formed part of a research project funded by the U.K. Engineering and Physical Sciences Research Council.

REFERENCES

Alonso, E.E., Gens, A. and Gehling, W.Y.Y. (1994). Elasto-plastic model for unsaturated expansive soils. *Proc. 3rd Eur. Conf. Num. Methods Geotech. Eng., Manchester,* 11-18.

Alonso, E.E., Gens, A. and Hight, D.W. (1987). Special problem soils. General report. *Proc. 9th Eur. Conf. Soil Mech., Dublin,* Vol.3, 1087-1146.

Alonso, E.E., Gens, A. and Josa, A. (1990). A constitutive model for partially saturated soils. *Géotechnique,* Vol.40(3), 405-430.

Alonso, E.E., Lloret, A., Gens, A. and Yang, D.Q. (1995). Experimental behaviour of highly expansive double-structure clay. *Proc. 1st Int. Conf. Unsaturated Soils, Paris,* Vol.1, 11-16

Chu, T.Y. and Mou, C.H. (1973). Volume change characteristics of expansive soils determined by controlled suction test. *Proc. 3rd Int. Conf. Expansive Soils, Haifa,* Vol.1, 177-185.

Croney, D. (1952). The movement and distribution of water in soils. *Géotechnique,* Vol.3(1), 1-16.

Fisher, R.A. (1926). On the capillary forces in an ideal soil; correction of formulae given by W.B. Haines. *J. Agr. Sci.,* Vol.16, 492-505.

Gens, A. and Alonso, E.E. (1992). A framework for the behaviour of unsaturated expansive clays. *Canadian Geotech. J.,* Vol.29, 1013-1032.

Hilf, J.N. (1956). *An investigation of pore water pressure in compacted cohesive soils.* Technical Memorandom 654. U.S. Department of Interior Bureau of Reclamation, Denver.

Pousada, E. (1984). *Deformabilidad de arcillas expansive bajo succión controlada. Tesis Doctoral,* Univ. Polit. Madrid.

Sharma, R.S. (1998). *Mechanical behaviour of unsaturated highly expansive clays. DPhil thesis,* University of Oxford, U.K.

Wheeler, S.J. (1988). The undrained shear strength of soils containing large gas bubbles. *Géotechnique,* Vol.38(1), 399-413.

Wheeler, S.J. and Karube, D. (1996). State of the art report: Constitutive modelling. *Proc. 1st Int. Conf. Unsaturated Soils, Paris,* Vol.3, 1323-1356.

Wheeler, S.J. and Sivakumar, V. (1995). An elasto-plastic critical state framework for unsaturated soils. *Géotechnique,* Vol.45(1), 35-53.

Unsaturated Soils for Asia, Rahardjo, Toll & Leong (eds) © 2000 Taylor & Francis, ISBN 90 5809 139 2

Swelling characteristics of compacted clayey soils related to their wetting curves

R.A.A.Soemitro & Indarto

Civil Engineering Department, Institut Teknologi Sepuluh Nopember, Surabaya, Indonesia

ABSTRACT: A laboratory investigation was conducted on two statically compacted clay soils to assess the influence of elevated temperature on the swelling characteristics (swelling deformation and swelling pressure). The tests were performed using an oedometer modified for unsaturated soils at temperatures of 30°, 50° and 70° C. It was observed that temperature influenced the swelling characteristics of the compacted clay soils. Wetting tests at 30° C were performed on each soil to determine the relationship between soil suction, water content and volume of the soil specimen. This paper illustrates how the wetting curve can be used to make preliminary assessment of the swelling characteristics.

1 INTRODUCTION

Soil, in nature, continuously undergoes wetting and drying cycles due to the alternance of dry and wet seasons. This plays an important role in activating the clay minerals in expansive soils.

Clays have several important influences on the soil properties and on the soil behavior. One of the most important factor is their water retention capacity (Chen 1980, Fleureau et. al. 1992). For soils having high clay content, the ability to absorb and to release water are main factors that affect the swelling shrinkage behavior. The swelling behavior of expansive soils often causes problems, such as differential settlements and ground heaves. But recently, expansive soils are attracting greater attention as back-filling (buffer) materials for repositories of waste materials. The function of the buffer material is to create an impermeable zone around the waste material dumping sites, as the waste must be kept separately from the surrounding environment. The buffer material must have swelling properties to fill up cracks, which may appear in the surrounding natural soil (Komine & Ogata 1994, 1999). Commonly, compacted bentonite is used as back-filling materials.

For this particular purpose, the swelling deformation and swelling pressure of standard proctor compacted bentonite were studied using an oedometer modified for unsaturated soils at temperatures of 30°, 50° and 70° C. Wetting tests were performed on each soil at 30° C (room temperature) to determine the relationship between soil suction, water content and volume of the soil sample. This paper illustrates how the wetting curve can be used to make preliminary assessment of the swelling characteristics.

2 SOIL PROPERTIES

The material used in this series of tests is a bentonite clay. Its characteristics are shown in Table 1.

Table 1. Characteristics of bentonite soil

	Bentonite A	Bentonite B
w_L (%)	170	140
w_P (%)	60	40
I_P (%)	110	100
Gs	2.70	2.65

3 TEST PROCEDURES

3.1 *Soil Specimen Preparation*

Statically compacted specimens were prepared for both wetting and swelling tests. Their initial conditions are shown in Table 2.

Table 2 Initial conditions of soil specimens in this study

	Bentonite A	Bentonite B
Water content-w_i (%)	41	32
Dry density -γ_d (kN/m3)	11.1	10.8
Suction --u_{wi} (kPa)	2000	1500
Degree of saturation - S_n (%)	70	50

The dimension of the soil specimen is governed by the requirements of both types of tests. For wetting test, the specimen should have a thickness as minimum as possible so that the test duration will not be unduly long. For the swelling test, the specimen should have the same diameter as of the oedometer. This implies that the minimum dimension of the soil specimen should be 40 mm in diameter and 10 mm in height.

Special attention was given to the procedure of preparing soil specimens at specified densities and water contents. The procedure adopted can be described as follows:

- Bentonite powder and water were thoroughly mixed in order to achieve predetermined water contents.
- Soil specimens was then placed in a hermetic plastic bag to avoid evaporation and a 2-days curing time was adopted to ensure homogeneous conditions in the sample.
- Predetermined quantities of soil specimens was then statically compacted in the oedometer. The swelling test should be carried out in this oedometer while the specimens for wetting test should be extruded and subjected to various negative pore water pressures.

3. 2. *Wetting Test*

The wetting curve was obtained by controlling the negative pore water pressure and by measuring the volume and water content of the specimens at the end of the tests. The osmotic solution technique was used in this experiment. The wetting test was performed at the room temperature which was 30° C.

The osmotic solution technique was conducted by placing a dialysis (semi-permeable) membrane to separate the soil specimen and a specific solution used to generate the osmotic pressure. The solution normally used is polyethylene glycol (PEG). The PEG has large molecules which cannot pass through the membrane. By varying the concentration, matric suctions up to 1500 kPa can be generated. The soil specimens were weighed and then placed on the dialysis membrane. The soil specimens in the chamber will slowly come to equilibrium with the applied matric suction. Periodically, the soil specimen was taken out of the chamber and weighed. The weight of the soil specimen with time was recorded and plotted to determine the equilibrium point that is the point at which there is no further change in water content of the soil specimen. The equilibrium time needed for the soil specimens to reach the suction generated by the osmotic solution was around 7 to 15 days, depending on the soil type and the initial suction of the soil specimen.

3. 3. *Swelling Test*

ASTM defines swelling pressure as the pressure which prevents the specimen from swelling or a pressure which is required to return the specimen back to its original state (void ratio, height) after swelling (Chen 1988).

The swelling tests were performed using two methods:

- Stress controlled test.
  The oedometer was placed in the compression machine and water absorption was allowed through the bottom of the apparatus. When swelling of the specimen has ceased, the vertical stress is increased in increments until it has compressed the specimen back to its original height. The stress required to compress the specimen back to its original height is commonly termed the zero volume change swelling pressure.
- Strain controlled test.
  The strain controlled test is based on the principle of controlling the strain that is developed as water is added. The oedometer was placed in the compression machine and a vertical adjustment bolt was fitted to control the strain. A load cell was mounted at the top to measure the load resulting from swelling during water absorption. Two types of test had been done: zero strain test and given strain test.

The above swelling test method was repeated at three different temperatures: 30°, 50° and 70° C. The heating conductor was placed outside the oedometer's mold in order to heat the soil sample. The calibration to adjust the heating conductor had been done to get the desired temperature.

4 TEST RESULTS AND DISCUSSIONS

The swelling tests and the final values of wetting test for both soils are shown in Figures 1 and 2, respectively. It can be clearly seen that the soil characteristics affect both the swelling deformation and swelling pressure. Soils of higher plasticity had higher swelling deformations and swelling pressures as well. The increase of temperature upon specimens

performed at the same temperature. For both soils, the wetting path can be used to predict the swelling pressure at a given volume deformation. It can be clearly stated that the swelling pressure curve is identical with the wetting path plotted as a function

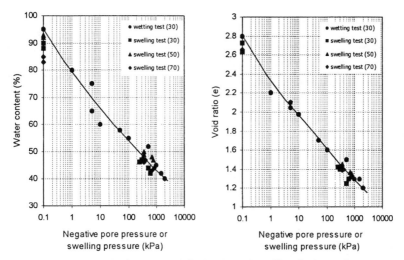

Figure 1. Wetting path and final values of swelling for bentonite A

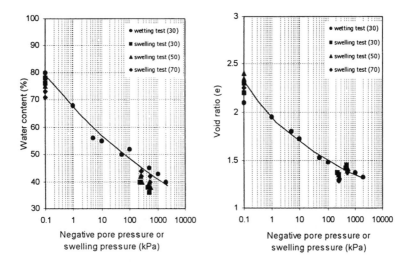

Figure 2. Wetting path and final values of swelling for bentonite B

at 50° C gave more significant effects compared to those at 70° C.

An interesting point is that for a given initial condition, the swelling pressure at a certain volume deformation are located on the wetting path having the same initial condition, if both tests were

of void ratio and the effective stress (Soemitro 1994).

Since the swelling clay minerals in bentonite used for this study are mainly montmorillonite, the swelling mechanism can be explained as follows: montmorillonite minerals in bentonite expand by absorbing water. Interlayer water and exchangeable cations exist between montmorillonite layers. The repulsive and attractive forces act between montmorillonite layers, so the swelling pressure and swelling deformation of montmorillonite minerals are caused by repulsive and attractive forces acting between two layers (Mitchell 1978, Komine & Ogata 1994, 1996, 1999).

Some previous investigators (Yong et al. 1969, Mitchell 1976, Baldi et. al. 1988, Romero et. al. 1995, Wiebe et. al. 1998) had tried to explain the temperature effect on the soil expansion with the clay-water interaction in soil systems. Yong et al. (1969) explained that in clay-water interaction in soil systems where solute effects are negligible is measured by the matric potential. The two forces contributing to the matric potential are the swelling and capillary forces. Theoretical expectation and available experimental evidence indicate that temperature changes affect the two components differently. Their experimental results in the swelling pressure test showed that where complete saturation exists, an increase in temperature resulted in a slight increase in water contents for the same swelling pressure. This is consistent with qualitative predictions based on the theory of swelling due to interaction of diffuse ion-layers. Where unsaturation occurs, an increase in temperature caused a decrease in water content for the same water potential. This is due to a decrease in surface tension caused by the increase in temperature, because water was considered to be held in soils primarily by surface tension forces at the air-water interfaces, i.e., capillary forces.

The large influence of temperature on water retention occasionally measured is thought to be due to temperature effects on the measuring instrument. Klute and Richard (1962) found that water content of a clay soil at a constant potential did not change with temperature, but the water content of a sodium montmorillonite increased with increase in temperature. Studies of the temperature effect on swelling pressures of high-swelling sodium montmorillonite have also shown that increased temperatures resulted in increased swelling pressures for the same volume. This is the temperature effect predicted from the diffuse ion-layer theory of swelling.

An important remark should be made upon findings of the previous references: there is no clear understanding of the condition under which the above mechanisms actually occur in clays. The theory of the double layer assumes a perfectly parallel arrangement of solid particles. A similar assumption is made in studies on hydration forces. Although this assumption helps in describing the mechanisms of the clay-water interaction in parallel structured pure clay pads, it can hardly furnish a quantitative assessment of the basic variables in clay soils, and in particular for natural soils.

5 CONCLUSIONS

The soil characteristics affect both the swelling deformation and swelling pressure. Soils of higher plasticity had higher swelling deformations and swelling pressures as well. The increase of temperature upon specimens at 50°C gave more significant effects comparing to those at 70°C.

For both soils the wetting path can be used to predict the swelling pressure at a given volume deformation, if both tests were performed at the same temperature. The wetting tests on different temperatures should be carried out to complete this study. It can be clearly stated that the swelling pressure curve is identical with the wetting path plotted as a function of void ratio and the effective stress at constant temperature.

REFERENCES

Baldi G., Hueckel T. & Pellegrini R. 1988. Thermal volume changes of the mineral-water system in low-porosity clay soils, *Canadian Geotechnical Journal*, Vol. 25: 807-825.

Chen F.H. 1988. *Foundations on Expansive Soils*, Elsevier Publisher B.V.

Fleureau J.M., Soemitro R. & Taibi S. 1992. Behavior of an expansive clay related to suction, *Proc. 7th Int. Conf. on Expansive Soils*, Dallas (USA), August 1992:173-178.

Hideo Komine & Nobuhide Ogata 1994. Experimental study on swelling characteristics of compacted bentonite, *Canadian Geotechnical Journal*, Vol. 31:478-490.

Hideo Komine & Nobuhide Ogata 1996. Prediction for swelling characteristics of compacted bentonite, *Canadian Geotechnical Journal*, Vol. 33:11-22.

Hideo Komine & Nobuhide Ogata 1999. Experimental study on swelling characteristics of sand-bentonite mixture for nuclear waste disposal, *Soils and Foundations*, Vol. 39, No. 2, April 1999: 83-97.

Mitchell J.K. 1976. *Fundamentals of soil behaviour*, New York, Univ. of California – Berkeley, John Wiley & Sons Inc.

Romero E., Lloret A. & Gens A. 1995. Development of a new suction and temperature controlled oedometer cell , *Proc. of the First Int. Conf. on Unsaturated Soils*, Paris (France), 6-8 September 1995:553-559.

Soemitro R.A.A. 1994. *Contribution à l'étude du rôle de la pression interstitielle négative dans le gonflement et d'autres aspects du comportement des sols non saturés*, Doctoral Thesis, Ecole Centrale de Paris, February 1994.

Wiebe B., Graham J., Tang G. X. and Dixon D. 1998. Influence of pressure, saturation, and temperature on the behaviour of unsaturated sand-bentonite, *Canadian Geotechnical Journal*, Vol. 35:194-205.

Yong R.N., Chang R.K. and B.P. Warkentin 1969. Temperature Effect on Water Retention and Swelling Pressure of Clay Soils, *Special Report 103*, Highway Research Board,:132-137.

10 Engineering applications

10.1 Foundations and roads

Unsaturated Soils for Asia, Rahardjo, Toll & Leong (eds) © 2000 Taylor & Francis, ISBN 90 5809 139 2

The blocking effect of piles on ground water level in slopes under rainfall

F.Cai & K.Ugai
Department of Civil Engineering, Gunma University, Kiryu, Japan

ABSTRACT: Piles are an effective measure to stabilize dangerous slopes, but they may also cause higher ground water level in the slope under rainfall due to the pile's blocking effect. The pile's blocking effect on the ground water level is evaluated using a three-dimensional finite element analysis of transient water flow through unsaturated-saturated soils. The influences of the spacing and positions of the piles on the ground water level in a typical slope under rainfall are numerically analyzed for three sets of hydraulic characteristics of soils. The numerical results show that the pile's blocking effect on the ground water level in the slope under rainfall is small, and the possible reason is explained.

1 INTRODUCTION

The use of piles to stabilize active landslides, and as a preventive measure in stable slopes, has been applied successfully in the past and proved to be an efficient solution since piles can be easily installed without disturbing the equilibrium of the slope. Some methods have been proposed to design slopes reinforced with piles, using the limit equilibrium methods (Ito et al. 1981, Hassiotis 1997), and more sophisticated numerical analyses (Poulos 1999, Chow 1996, Cai & Ugai 2000).

The pile's blocking effect on ground water level in slopes, especially under rainfall, affects the design of the slopes reinforced with piles. However, to the authors' knowledge, there is no research on it up to the present. One of the main reasons may be that three-dimensional numerical analysis is difficult for the computers in the past. In the present paper, the pile's blocking effect on the ground water level in the slopes under rainfall is evaluated by three-dimensional finite element analysis of transient water flow through unsaturated-saturated soils. The hydraulic characteristics of soils are modeled with van Genuchten model (van Genuchten 1980). Three sets of hydraulic characteristics parameters of the van Genuchten model are used to investigate their influences on the pile' blocking effect. The influences of the spacing and positions of the piles on the ground water level in the slope under rainfall are numerically analyzed for a typical slope. The numerical results show that the pile's blocking

effect on the ground water level in the slope under rainfall is small, and the possible reason is suggested.

2 MODELING WATER FLOW IN SOILS

2.1 *Fundamental flow equation*

Darcy's law has been shown to be valid for the water flow through unsaturated soils as well as the flow through saturated soils (Richards 1931). The main difference is that the hydraulic conductivity is assumed to be constant for saturated soils, while it depends on the pore volume occupied by water for unsaturated soils. Based on mass conservation and Darcy's law, the differential equation governing water flow through unsaturated-saturated soils is given by:

$$\nabla \cdot \left(K(\theta) \nabla (\Phi + z) \right) = c(\theta) \frac{\partial \Phi}{\partial t} \qquad (1)$$

where $K(\theta)$ is the hydraulic conductivity, θ is the volumetric moisture content, Φ is the pressure head, z is the elevation head, t is time, and $c(\theta)$ is the specific moisture capacity.

2.2 *Hydraulic characteristics*

Equation 1 includes two soil parameters that must be determined: the hydraulic conductivity and the specific moisture capacity. These parameters under unsaturated conditions are dependent on the volumetric moisture content, which is in turn related to the pressure head. A widely used representation of the hydraulic characteristics of unsaturated soils is the set of closed-form equations formulated by van Genuchten (1980), based on the capillary model of Mualem (1976). The soil-moisture retention, specific moisture capacity, and hydraulic conductivity are given by:

$$S_e = (\theta - \theta_r)/(\theta_s - \theta_r) = (1 + |\alpha \Phi|^n)^{-m} \qquad (2)$$

$$c(\theta) = \alpha(n-1)(\theta_s - \theta_r)S_e^{1/m}(1 - S_e^{1/m})^m \qquad (3)$$

$$K(\theta) = K_s K_r = K_s S_e^{1/2}\left[1 - (1 - S_e^{1/m})^m\right]^2 \qquad (4)$$

respectively, where

$$m = 1 - 1/n \qquad n > 1 \qquad (5)$$

and S_e is the relative degree of saturation, and θ_r and θ_s denote the residual and saturated volumetric moisture contents, respectively. K_s and K_r are the saturated and the relative hydraulic conductivity, respectively. α, n, and m are empirical parameters of the hydraulic characteristics. The hydraulic functions are determined by a set of five parameters, θ_r, θ_s, α, n, and K_s. The van Genuchten model is considered to provide a better match to the experimental data, although there are some alternative models whose parameters can be obtained more easily. Leong & Rahardjo (1997a, 1997b) had comprehensively evaluated the models for the hydraulic characteristics of soils.

2.3 Numerical approach

The finite element formulation for the transient water flow through unsaturated-saturated soils can be derived by the Galerkin principle of weighted residual. The numerical integration of the Galerkin solution to the governing equation is given by:

$$D\Phi + E\frac{\partial \Phi}{\partial t} = Q \qquad (6)$$

where, D is the seepage matrix, E is the capacitance matrix, and Q is the flux vector reflecting the gravitational flow and the flux boundary conditions.

The time derivative can be approximated with the following finite difference equation:

$$\left(\lambda D + \frac{E}{\Delta t}\right)\Phi_{t+\Delta t} = Q - \left[(1-\lambda)D - \frac{E}{\Delta t}\right]\Phi_t \qquad (7)$$

Different finite difference scheme is obtained if the parameter λ is selected differently. In the present analysis, the parameter $\lambda = 1/2$, i.e., the Crank-Nicolson scheme, is used as it gives a second order accuracy. Because the hydraulic conductivity and specific water capacity are functions of the volumetric moisture content, Equation 7 is highly nonlinear and is solved using an iterative method.

3 RESULTS AND DISCUSSIONS

3.1 Model slope

An idealized slope with a height of 10m and a gradient of 1V:1.5H is analyzed with a three-dimensional finite element mesh, as shown in Figure 1. Two symmetric boundaries are used, so that the problem analyzed consists of a row of piles with planes of symmetry through the pile centerline and through the soil midway between the piles. The initial ground water level is assumed to be horizontal and at the lower ground surface.

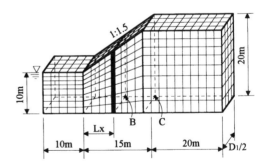

Figure 1. Model slope and FE mesh

The slope and the ground are assumed to consist of the same soil. Three sets of van Genuchten model parameters of the hydraulic characteristics, as shown in Table 1, are used to investigate their effects on the ground water level in slopes, stabilized with piles, under rainfall. The three types of soils are Glendale clayey loam (GCL) (van Genuchten 1980), Uplands silty sand (USS) (Russo & Bresler 1980), and Bet Degan loamy sand (BLS) (Staple 1969), and they can be considered to represent a wide range of soils. The saturated permeability of the pile material is 10^{-4} times of that of soils, so that the pile can considered as an impervious material, compared with the soils.

Table 1. Hydraulic characteristics

Soil type	GCL	USS	BLS
α (m^{-1})	1.060	7.087	2.761
n	1.395	1.810	3.022
θ_r	0.106	0.049	0.044
θ_s	0.469	0.304	0.375
K_s(10^{-4}cm/s)	1.516	18.29	63.83
i/K_s^*	1.83	0.152	0.0435

* Ratio of rainfall intensity to saturated permeability

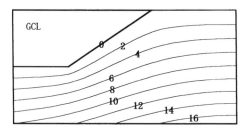

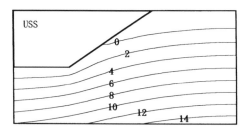

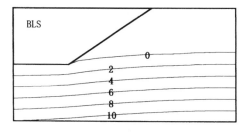

Figure 2. Contours of Pressure head for slopes without piles

The initial relative degree of saturation is assumed to be the same, i.e. S_e=0.617, at the slope crest, and linearly increases to unity at the height of the initial water level for each type of soil. The initial moisture content has no influence on the steady ground water level in the slopes. A rainfall of uniform intensity, i=10mm/h is assumed to last long enough to reach the steady ground water level, and the following results are at steady-state. The steady ground water level in the homogeneous slopes almost completely depend on the ratio of the rainfall intensity to the saturated hydraulic conductivity, although the other of the hydraulic characteristics of the soils have influences on the time-

histories of the ground water (Cai et al. 1998). Therefore, the following steady-state results are effective for the ratio of the rainfall intensity to the saturated hydraulic conductivity shown in Table 1. It is obvious that the steady ground water level is the highest and most dangerous for the slopes.

When there is no pile in the slope, the steady ground water level is shown in Figure 2 for the three soils, respectively. From this figure, it is indicated that the ratio of the rainfall intensity to the saturated permeability has significant influences on the steady ground water level in slopes. The larger the ratio of the rainfall intensity to the saturated permeability, less than unity, the higher the steady ground water level in the slopes. When the ratio is larger than or equal to unity, the steady ground water level must reach the surface of the slope.

3.2 *Effect of pile spacing*

When the piles are installed in the middle of the slopes, i.e., Lx=7.5m, the pile spacing is changed to evaluate its influence on the ground water level in the slopes under rainfall. The contours of the pressure head for the cross sections through the pile centerline and through the soil midway between the piles are shown in Figures 3 to 5 for three types of soils when D_1/D=1.5. Here, D_1 and D are the center-to center spacing between two piles and the side width of the piles, respectively. By comparing with Figure 2 and Figures 3 to 5, it is shown that the ground water level in the slopes rises slightly with the piles installed in the slopes, i.e., the pile's blocking effects is small for all three types of soils.

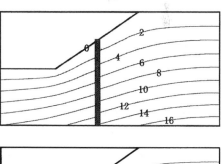

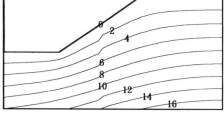

Figure 3. Contours of Pressure head for GCL slope

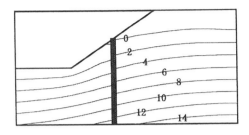

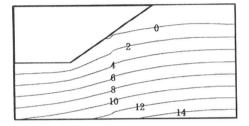

Figure 4. Contours of Pressure head for USS slope

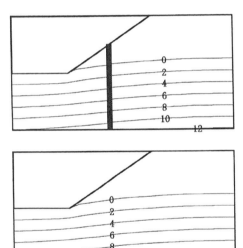

Figure 5. Contours of Pressure head for BLS slope

As maybe expected, the smaller the pile spacing, the more significant the pile's blocking effect. But even when the pile spacing D_1/D=1.5, the rise of the ground water level in GCL slope is only 0.22m. The safety factor of the slopes is almost linearly dependent on the pressure heads within the extent of the slip surface (Cai et al. 1998). Therefore, the pile's blocking effect on the ground water level in the slopes under rainfall can be indicated with the steady-state pressure heads at the typical points B and C, as shown in Figure 1.

Figure 6 shows the relationship between the steady-state pressure heads and the pile spacing at Points B and C, where the notation ∞ means that there is no pile in the slopes. The relationship between the steady-state pressure heads and the pile spacing shows that the pile's blocking effect on the ground water level in the slopes is small enough to be neglected for practical pile spacing of D_1/D=2 to 4 when the pile are installed at the middle of the slope.

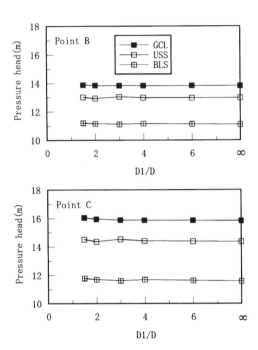

Figure 6. Pressure head versus pile spacing

The pile's blocking effect on the ground water level in the slopes can also be indicated with the difference of the steady-state pressure heads in front of and at the rear of the piles, dh. Figure 7 shows the relationship between the differences of the steady-state pressure head and the pile spacing for all three types of soils. The results once again indicate that the pile's blocking effect on the ground water level in the slopes can be neglected regardless of the hydraulic characteristics of the soils. One of the main reasons is that the hydraulic gradient is larger in the narrowest vertical section, where the piles are installed. This can be seen in Figures 3 to 5. Therefore, the discharge of the ground water through this section can be almost the same as that for the other sections.

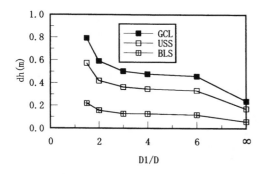

Figure 7. Difference of pressure head versus pile spacing

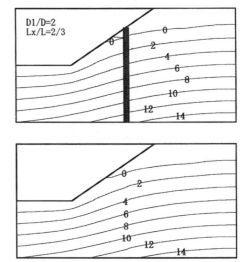

effect on the ground water level in the slopes can be neglected for the three types of soils.

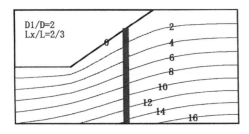

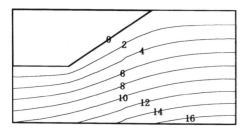

Figure 8. Contours of Pressure head for GCL slope

Figure 9. Contours of Pressure head for USS slope

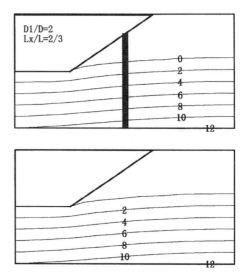

Figure 10. Contours of Pressure head for BLS slope

3.3 *Effect of pile position*

As indicated in the foregoing section, the smaller the pile spacing, the larger the pile's blocking effect on the ground water level in the slopes. Therefore, the smallest practical pile spacing $D_1/D=2$ is used to analyze the pile's blocking effect on the ground water level in the slopes for various pile positions. The pile positions in the slope are indicated with a dimensionless ratio of the horizontal distance between the slope toe and the pile position, Lx, to the horizontal distance between the slope toe and slope shoulder, L, as shown in Figure 1. When $Lx/L=2/3$, the contours of the pressure head for the cross sections through the pile centerline and through the soil midway between the piles are shown in Figures 8 to 10. By comparing Figure 3 and Figures 8 to 10, it is clear that the pile's blocking

Figure 11 shows the relationship between the steady pressure heads and the pile positions at Points B and C. Figure 11 shows that the pile's blocking effect on the ground water level in the slopes under rainfall is small enough to be neglected for the all pile positions.

Figure 12 shows the relationship between the steady-state pressure head differences and the pile position for all three types of soils. The results once again indicate that the pile's blocking effect on the ground water level in the slopes can be neglected regardless of the pile positions.

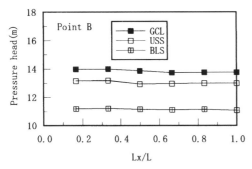

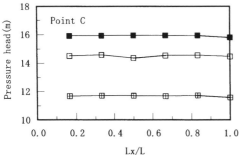

Figure 11. Pressure head versus pile position

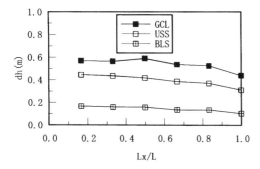

Figure 12. Difference of pressure head versus pile position

4 CONCLUSIONS

The three-dimensional finite element analysis of transient water flow through unsaturated-saturated soils is used to evaluate the pile's blocking effect on the ground water level in the slopes under rainfall. The influences of the spacing and positions of the piles on the ground water level are numerically analyzed for a typical slope with three sets of hydraulic characteristics. The numerical results show that the pile's blocking effect on the ground water level is small regardless of the pile positions and hydraulic characteristics of the soils, when the pile spacing is within the practical extent: $D_1/D=2$ to 4. One of the main reasons is that the hydraulic gradient is larger in the narrowest vertical section, where the piles are installed. Therefore, the discharge of the ground water, which is caused by the infiltration of the rainfall, through this section is almost the same as that for the other sections

REFERENCES

Cai, F., K. Ugai, A. Wakai & Q. Li 1998. Effects of horizontal drains on slope stability under rainfall by three-dimensional finite element analysis. *Computers and Geotechnics*, 23: 255-275.

Cai, F., & K. Ugai 2000. Numerical analysis of the stability of slope reinforced with piles. *Soils and Foundations*, 40(1).

Chow, Y.K. 1996. Analysis of piles used for slope stabilization. *Int. J. Numer. Anal. Meth. Geomech.*, 20: 635-646.

van Genuchten, M.T. 1980. A closed-form equation for predicting the hydraulic conductivity of unsaturated soils. *Soil Science Society of America Journal*, 44: 892-898.

Hassiotis, S., J.L. Chameau & M. Gunaratne 1997. Design method for stabilization of slopes with piles. *J. Geotech. and Geoenvir. Engrg.* 123(4): 314-323.

Ito, T., T. Matui & W.P. Hong 1981. Design method for stabilizing piles against landslide - one row of piles. *Soils and Foundations*, 21(1): 21-37.

Leong, E.C., & H. Rahardjo 1997a. Review of soil-water characteristic curve equations. *Journal of Geotechnical and Geoenvironmental Engineering*, ASCE, 123: 1106-1117.

Leong, E.C., & H. Rahardjo 1997b. Permeability functions for unsaturated soils." *Journal of Geotechnical and Geoenvironmental Engineering*, ASCE, 123: 1118-1126.

Mualem, Y. 1976. A new model for predicting the hydraulic conductivity of unsaturated porous media. *Water Resources Research*, 12: 513-522.

Poulos, H.G. 1999. Design of lope stabilizing piles. *Proc. Int. Symp. on Slope Stability Engrg., IS-SHIKOKU'99, Matsuyama, 8-11 November 1999: 83-102. A.A.Balkema.*

Richards, L.A. 1931. Capillary conduction of liquids through porous mediums. *Physics*, 1: 318-333.

Russo, D. & E. Bresler 1980. Field determinations of soil hydraulic properties. *Soil Science Society of America Journal*, 44: 697-702.

Staple, W.J. 1969. Moisture redistribution following infiltration. *Soil Science Society of America Proceedings*, 33: 840-847.

Unsaturated Soils for Asia, Rahardjo, Toll & Leong (eds) © 2000 Taylor & Francis, ISBN 90 5809 139 2

Total soil suction regimes near trees in a semi-arid environment

D.A.Cameron
Discipline of Civil Engineering, University of South Australia, Adelaide, S.A., Australia

ABSTRACT: Standards Australia (1996) provides guidance on classification of sites for movement and the subsequent design of shallow footings for residential buildings on expansive soils. It does not however provide any guidance on how the drying effects of vegetation might be incorporated in the footing design. Despite there not being a prescriptive Standard for designing for the effects of trees, engineers have been pressed by the community to allow for the extra soil drying on the ever diminishing size blocks which are used for housing. Some rules of thumb have been promoted and empirical design suction changes have been adopted. The design suction changes depend on the relative proximity of the tree to the building, as well as the site classification level and whether or not a single tree or group of trees is to be considered. Although it is understood that different species develop different root systems and so affect the soil differently, tree species is not a parameter that is required in the models.

The primary purpose of this paper is to present previously unpublished measured suction profiles pertaining to urban sites in a semi-arid climate (Adelaide, South Australia). All sites contained Australian native eucalypts. At least two suction profiles were determined for each site, close to and well away from the vegetation. A preliminary model is proposed to account for the drying settlement caused by trees.

1 CURRENT KNOWLEDGE

1.1 *General Effect of Trees on Clays*

It is generally accepted that vegetation is unable to draw moisture from the soil at high levels of soil desiccation or suction. The limit at which this occurs for a particular species is termed the wilting point and has been reported to reach values of total suction of 1.55 to 3.1MPa (McKeen 1992).

The way roots respond to soil water depletion depends on the hardiness of the plant. The roots may extend into new areas where there is greater water availability, or the vegetation may die, or the plant become dormant until water becomes available, all these possibilities being dependent on species.

Much of what civil engineers know about trees is based on indirect evidence. For example, the aggressiveness of different root systems of trees near water pipes was revealed in studies of root chokes by the Engineering and Water Supply Department (South Australia) in Adelaide. CSIRO (Australia) published information sheets on potential damage to houses by trees, which was based on this information. It was assumed that roots, which travelled long distances to get moisture from pipes, would potentially be the most treacherous plants to have on a suburban allotment. Most people would find little argument with the species that were revealed as potential threats to house footings. It was also assumed that climate differences around Australia would make little difference, although it is well known that the extent of a tree root system is greatly affected by water availability. For example, if water is plentiful (but not excessive), the tree root system is contained within a relatively small volume (Yeagher 1935).

Soil moisture data have been collected about trees in quite different environments, which generate different patterns of root development. It is contended that the most useful data for footing design will come from studies of desiccation in the urban environment. Investigations of damaged buildings can be fruitful, but require careful engineering judgement. The adequacy of the footings for the site classification and the structure must be incorporated into the assessment, as well as the possibility of other contributing factors such as plumbing leaks, overwatering of gardens, or previous use of the site. A number of researchers have simply equated a measure of building distress (crack width, differential movement) to the proximity of the vegetation to the building (Tucker & Poor 1978, Cutler & Richardson 1981, Wesseldine 1982, Cameron & Walsh 1981).

1.2 *Species and proximity*

Many authors have used the concept of proximity in their work. Proximity can be expressed by the ratio of the minimum horizontal distance, D, between the base of the tree and the building perimeter, to the height of the tree, H. The ratio, D:H, may be expressed as a range for a group of trees. Alternatively the proximity of the trees can be based on the average tree height within the group and the distance of a borehole from the nearest tree trunk, $D:H_{av}$.

Biddle (1983) conducted studies of soil moisture deficits around specimens of certain tree species in open grassland. The species included poplar, horse chestnut, silver birch, lime and a cypress. Five different clay soil profiles were investigated at locations in Cambridge, Milton-Keynes and London. Soil moisture was monitored with a down-hole neutron moisture meter to a maximum depth of 4m. Generally it was seen that the lateral extent of drying was contained within a radius equal to the height of the tree. However the extent of drying, both horizontally and vertically, appeared to be species dependent. Poplars caused drying to a radius of over 1.5 times the tree height and caused the deepest drying close to the trees, probably to a depth in excess of 4m. Other species dried the soil to a maximum depth of 2m. The cypress produced the smallest zone of drying within the soil.

Richards et al. (1983) studied total soil suction regimes near groups of three different species of trees in parkland in Adelaide, South Australia. The tree groups were described as eucalypts, casuarinas and pines, with average heights of 17, 13 and 10m and consisting of 13, 5 and 25 trees in each group, respectively. Boreholes for soil sampling were made progressively away from the tree groups. It can be deduced from this work that the eucalypts had the greatest drying effect while the pines had little effect on the surrounding clay soil. For the eucalypt site, the total soil suction reached 3.5MPa at depth below the group, and was not lower than 2MPa throughout the exploration depth of 8m. The radial extent of near-surface drying appeared to be $1.3H_{av}$. The depth of drying decreased with distance away from the group.

The differences in total suction across a site are more important than the absolute values. As pointed out by Richards et al (1983), soil salinity affects the suction values from one site to another. Where water tables are quite deep and significant vegetation is absent, it is generally accepted that soil suctions become relatively constant at depth. This suction value, u_{eq}, relates to the suction expected under the centre of a large paved area in the same environment (Richards & Chan 1971). Where practical, this value should be used to define suction differences due to trees across a site. As a soil's response to suction change (soil reactivity) is defined in terms of the change of the logarithm of suction, the suction change due to trees, Δu_{tree}, should be similarly expressed. For the eucalypt site the deep suction value or u_{eq} for $D:H_{av}$ greater than 1.5 averaged approximately 1.7MPa, giving a value of Δu_{tree} equal to 0.31.

1.3 *Species and proximity and building damage*

Probably the most substantial research into species was conducted by Cutler & Richardson (1981). They reported on the investigations of 2,600 cases of building damage in the UK. Trees implicated as a cause of the damage were recorded and their proximity to the building noted. A database of species was subsequently established and was reviewed to find the maximum distance for a tree to "cause" damage in 75% of the cases recorded for that species. The greater the distance relative to the height of the tree ($D:H_{75\%}$), the more treacherous the tree.

Unfortunately the information has had little impact on Australian practices owing to differences in climate, differences in range of tree species and different sensitivities to damage between homeowners in the UK and Australia. In fact, the degree of damage was not considered in Cutler & Richardson's report and may have often been insignificant in the Australian context. Furthermore, the usual construction in England is cavity brick walls on strip footings, which is much less tolerant to movement than masonry veneer walls on a stiffened raft, the dominant construction style in Australia.

In the USA in Texas, Tucker & Poor (1978) studied a housing estate, which was in the process of being demolished because of the extent of damage to the houses (masonry veneer walls on slabs). Tree species were mulberry, elm, cottonwood and willows. Differential movements were measured and compared with D:H ratios. The results of the study revealed a background movement due to site reactivity at D:H ratios above two. The data strongly indicated that tree effects were significant at D:H values greater than one.

In New Zealand, Wesseldine (1982) demonstrated the influence of the silver dollar gum (e. cinerea) on houses. A plot of the ratio D:H for damaged houses indicated a threshold value of 0.75 for single trees to cause damage and between 1 to 1.5 for groups of these trees.

In Melbourne, Victoria, studies of damaged masonry veneer houses on strip footings (Holland 1979 and Cameron & Walsh 1981) drew the conclusion that significant damage (cracking in walls greater than 3mm) was only likely if the proximity of a single tree was less than or equal to 0.5H.

From Adelaide in South Australia, Pile (1984) reported building distortions and suction profiles to a maximum depth of 4m for eight sites which were affected at least in part by trees. The approach taken by Pile is similar to the approach taken in this paper, however some deeper suction profiles are provided.

2 TOTAL SUCTION CHANGES IN THE BUILT ENVIRONMENT

2.1 New data

Suction profiles for three sites are given in Figures 1 to 3 and a summary of all the data is presented in Table 1. Generally these data have been obtained during investigations into building damage in Adelaide; one originates from Melbourne. Total suction profiles are provided at a location between the tree/s and the structure and a location remote from the tree/s. The trees were usually native Australian species, often eucalypts, but details on species of eucalypt was generally not available. Most of the information is for groups of trees.

For the purpose of design of footings in Adelaide, ground movements in the absence of vegetation are estimated on the basis of design suction changes representative of the range of wet to dry profiles expected for the site. The suction change is defined as a change of the logarithm of suction. Regional climate dictates the design suction change profile and for Adelaide, which enjoys a semi-arid climate, the surface suction change is 1.2, decreasing linearly to zero at a depth of 4m (Standards Australia 1996). The linearity is a convenient design simplification. All the Figures have a design envelope superimposed. The deep suction value was judged from the data to locate the base of the envelope and the surface suctions were postulated to give the wet and dry extreme suction profiles. Linearity of the profiles was assumed.

In Figure 1, the suction profiles relate to a row of trees consisting mainly of eucalypts. The dry suction profile at D:H of 0.4 lay outside the design envelope, below a depth of 2m. At greater depths, suction remained relatively constant at a value of 2.6MPa

down to a 5.7m. At a depth of 6.1m, the suction approximated u_{eq}, which was judged to be 1.1MPa. The logarithm of the suction difference, Δu_{tree}, is therefore 0.37.

It is contended that the almost constant value of suction at depth near vegetation represents the wilting point for the tree species, u_{wp}. A wilting point line has been superimposed on each of the Figures. When compared with the design envelope for the site, it can be seen that tree root systems are unlikely to be active close to the soil surface in dry periods. Moreover it indicates that trees cause problems by generating shrinkage of clay soils at depth in the soil profile, greater than the assumed depth of normal suction changes for a site.

Figure 2 provides another example for a row of eucalypts. The dry profile was located at a D:H of less than 0.4. The values for u_{eq}, u_{wp} and Δu_{tree} are 1.4MPa, 3.4MPa and 0.4, respectively. Below 0.9m, the suction values for the dry profile remained relatively constant down to a depth of 5.7m. and seemed to be heading towards u_{eq} by the exploration depth of 6m.

Figure 3 provides the data for another row of eucalypts, but in another climatic region, Melbourne in Victoria. Although the design envelope is quite different, being only 1.8m deep, the principles discovered at sites in Adelaide are evident at this site. The wilting point and equilibrium suction values are obvious. The values for u_{eq}, u_{wp} and Δu_{tree} are 0.42MPa, 0.93MPa and 0.35, respectively. Drying is deeper than the depth of exploration of 3m, which was limited by the equipment available on a tight residential site.

Table 1 includes data for a further six sites, two being provided by Jaksa (1998). The D:H value applies to the dry suction profile.

2.2 Observations

The value of u_{eq} was not always as evident as in the Figures, often because of limited depth of exploration. All the Adelaide sites were explored to the minimum depth of the design suction change envelope, with depths ranging between 4 and 6.1m. The Williamstown site was investigated to 160% of the normal design depth for the region. The Adelaide sites in Figures 1 and 2 represent the deepest investigations. Only at these two sites was there any indication of a lessening of the drying effect of the vegetation at depth.

For the Adelaide sites, the suction, u_{eq}, varied between 0.87 and 1.4MPa (average 1.1MPa), while

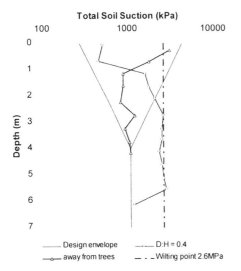

Figure 1. Total suction profiles near a row of trees of mixed species
(Ingle Farm, Adelaide, South Australia)

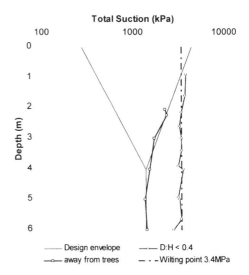

Figure 2. Total suction profiles near a row of large eucalypts
(Klemzig, Adelaide, South Australia)

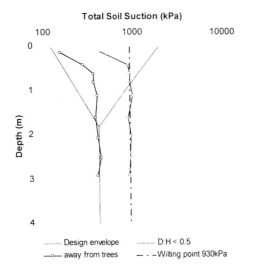

Figure 3. Total suction profiles near a row of large eucalypts
(Williamstown, Victoria)

Table 1. Summary of total suction data

Site	Trees (H)	D:H	$u_{eq and}$ u_{wp}	Δu_{tree}
			MPa	log(MPa)
Broadview SA	Eucalypt (9m)	< 0.5	0.98 2.2	0.35
Greenacres SA*	e. torquata (8.3m)	0.4 - 0.7	1.15 1.95	0.23
Greenacres SA*	e. mannifera (19m) and 2 smaller trees	0.6	1.15 1.95	0.23
Ingle Farm, SA	Row of eucalypts (8m)	1.5	1.1 1.95	0.25
Ingle Farm, SA	Row of eucalypts (10m)	0.4	1.1 2.6	0.37
Klemzig, SA	Row of mature eucalypts	< 0.5	1.4 3.4	0.39
The Levels, SA	Native plantation.	< 0.5	0.98 2.45	0.40
Hallett Cove, SA	Native plantation.	0.1	0.87 1.95	0.35
Williams-town, Victoria	Row of eucalypts	< 0.5	0.42 0.93	0.35

*From Jaksa 1998

the wilting point suction ranged between 1.95 and 3.4MPa, similar to the range suggested by McKeen (1992), with the upper limit close to the 3.5MPa reported by Richards et al (1983).

Although the Victorian site had remarkably lower values of both these suction signposts, the suction difference of 0.35 fell within the range of all the data, namely 0.23 to 0.40. The lower end of the range contained data for a single tree and a high D:H ratio. Ignoring these two sites would tighten the range of Δu_{eq} to 0.35 to 0.4.

3 DESIGN IMPLICATIONS

3.1 *Design philosophy*

Trees increase the dry side of the suction profile at depth and have little influence near the surface in a semi-arid environment. The effects extend past the normal design depth of soil movement. The dry suction profile will be dominated by the ability of the trees to extract moisture from the soil. It has been demonstrated in this paper that the wilting point of vegetation may be deduced from soil suction data gathered close to the trees. The suction, u_{wp}, forms a boundary, which can not be exceeded in the soil profile, beyond the bounds of the normal design suction envelope.

The value of u_{wp} will vary with species, but may also vary with soil salinity, as will the value of u_{eq}. Therefore estimates of the extra shrinkage settlement due to trees should be based on the suction difference, Δu_{tree}. This difference is moderate, being approximately a third of the current surface design suction change for Adelaide. However the suction changes persist to greater depths. To estimate movement reliably, site investigations need to be conducted to greater depths than is currently practised to evaluate the soil profile.

3.2 *Design guidelines*

In the absence of further data, for the purpose of estimating the extra settlement on semi-arid sites similar to Adelaide, containing groups or rows of eucalypts within D:H of 0.6, a design suction change of $0.4 \log_{10}$MPa, may be adopted to a depth of 6m.

Applying this proposal to a uniform soil profile and ignoring lateral confinement effects, the group of trees would potentially generate approximately 50% more movement. So a reactive site with an estimated design movement of 100mm would have to cope with 150mm if landscaping creates an Australian native

4 CONLUDING COMMENTS

The cases within this paper generally represent extreme conditions. Much more data are needed over a variety of conditions, which will enable some comprehension of the influences of species, climate and solute suction. There is mounting evidence that the conifer family has the least effect on subsoils. Eucalypts have often been blamed for shrinkage problems, however some eucalypts like *e. torquata* are known to be relatively benign.

The dissipation of movement with distance from vegetation is not well understood and comparative drying potential of single trees to tree groups needs to be better appreciated. The work of Biddle (1983), extended to other species and climates and including some suction data, would be most welcome. However, research funding has been limited by the complexity of the problem and the length of time over which monitoring of a site is required.

This paper suggests an alternative and supplementary approach, based on engineering assessment of damaged structures, which requires the accumulation of soil suction data at greater depths than previously recommended.

5 REFERENCES

Biddle, P.G. 1983. Patterns of soil drying and moisture deficit in the vicinity of trees on clay soils. *Geotechnique* 33(2):107-126.

Cameron, D.A. & Walsh, P.F. 1981. Inspection and treatment of residential foundation failures. *Proc., 1st National Local Government Eng. Conference*, Adelaide, pp 186-190.

Cutler, D.F. & Richardson, I.B.K. 1981. *Trees and Buildings*. Construction Press: London.

Holland, J.E. 1979. Trees - how they can affect footings. *POAV Journal.* pp. 11-14.

Jaksa, M.B. 1998. The influence of trees on expansive soils. Presentation to Footings Group, SA, IEAust. Adelaide.

McKeen, R.G. 1992. A model for predicting expansive soil behavior. *Proc., 7th Int. Conference on Expansive Soils*, Dallas, V1, pp. 1-6.

Pile, K.C. 1984. The deformation of structures on reactive clay soils. *Proc., 5th Int. Conference on Expansive Soils*, Adelaide, pp. 292-299.

Richards, B.G. & Chan, C.Y. 1971. Theoretical analyses of subgrade moisture under environmental conditions and their practical implications. *Aust.Road Research.* 4(6):32-49.

Richards, B.G., Peter, P. & Emerson, W.W. 1983. The effects of vegetation on the swelling and shrinking of soils in Australia. *Geotechnique*, 33(2):127-139.

Standards Australia. 1996. Residential slabs and footings – construction. AS2870-1996.

Tucker, R.L. & Poor, A.R. 1978. Field study of moisture effects on slab movements. *ASCE Journal of Geotechnical. Engineering.* 104(4):403-415.

Wesseldine, M.A., 1982. House foundation failures due to clay shrinkage caused by gum trees. *Transactions, Institution of Professional Engineers, N.Z..* March, CE9(1).

Yeagher, A.F., 1935. Root systems of certain trees and shrubs grown on prairie soils. *Journal of Agricultural Research* 51(12):1085-1092.

Unsaturated Soils for Asia, Rahardjo, Toll & Leong (eds) © 2000 Taylor & Francis, ISBN 90 5809 139 2

Distress caused by heaving of footing and floor slab – A case study in Gaborone

B. K. Sahu
Department of Civil Engineering, University of Botswana, Gaborone, Botswana

B. K. S. Sedumedi
Department of Architecture and Building Services, Government of Botswana

ABSTRACT: This case study investigates the cause of distress of the foundation of an industrial building. The investigations revealed that no soil investigations were done for the design of foundation. The foundation design was based on the satisfactory performance of the existing structure in the surrounding areas. Investigation consisted of digging four pits adjacent to the four walls. From the test results it was found that there was a great variation in the soil properties in lateral direction. Two adjacent walls were founded on black cotton soil with high degree of expansion, while the other two on the soil of low to medium degree of expansion. Flowerbeds around three walls added to the ingress of water to the foundation soil and caused upheaval of the walls and the floors. The relative movement of the foundation and floor is still taking place. It is recommended that remedial measures to minimise the ingress of water to the foundation soil would provide a satisfactory and economical solution to the problem.

1 INTRODUCTION

The phenomenon of swelling of soils is known for about as long as the field of geotechnical engineering has been practised. However, the practice of foundation engineering has its origins primarily in locations having deep deposits of soft clays such as the large coastal, North American, European and Mexican cities. As construction grew in arid regions, the problems associated with swelling and shrinking of soils began to receive more attention. However, most universities did not offer formal courses in foundation and pavement engineering on expansive soils. Thus, practising engineers usually learned about expansive soils the "hard way," after buildings and pavements began to experience distress. Expansive soil is a term generally applied to any soil or other rock material that has a potential for shrinking or swelling under changing moisture conditions. The primary problem that arises with regard to expansive soils is that deformations are significantly greater than elastic deformations and they cannot be predicted by classical elastic theory. Movement is usually in an uneven pattern and of such a magnitude as to cause extensive damage to the structures and pavements resting on them.

Expansive soils cause more damage to structures, particularly light buildings and pavements, than any other natural hazard, including earthquakes (Chen, 1988). Selected annual US losses from expansive soils were $798.1 million in 1970 and are expected to rise to $997.1 million by the year 2000 (Jones et al. 1973, Nayak et al. 1974). These values are for residence losses only. The cost of damage to other structures such as commercial/ industrial buildings and transportation facilities raises these total estimated values by a factor of 2 to 3. Despite of the fact that a huge amount of yearly damage losses have been attributed to these problem soils, the state of practice in design and construction is severely limited by continual lack of understanding of expansive soil behaviour and soil-structure interaction. Also a greater appreciation of risk associated with building on expansive soils must be developed on the part of various, lending institutions, regulatory officials, builders, architects and engineers.

Botswana is a small country with a total population of about 1.8 million. About two-third of the country is covered either with Kalahari sand or sandy soil. Black cotton soil is found only in smaller areas. There has not been much experience of foundation problems on expansive soils as these areas were avoided for any development.

In the present paper a case history of building damages due to expansive soil is presented. This case study is a typical example of distress caused by heaving of continuous footings and floor slabs having under-gone differential movements repeatedly

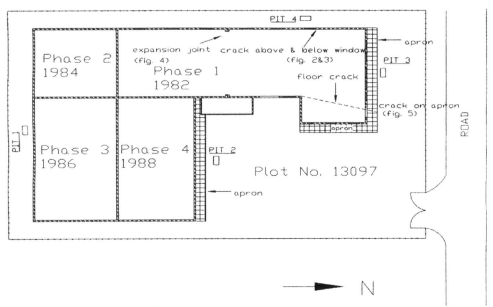

Figure 1. Floor plan showing construction phases

due the variation in the degree of saturation of the foundation soils. Because of cost, remedial measures were only partially carried out at regular intervals. Even after 25 years the building is functionally sound in spite of the cracks, which keep appearing regularly.

2 HISTORY

The building investigated is a furniture factory "DEE-ESS (Pty) Ltd in Broadhurst Industrial sites plot no.13097 in Gaborone, Botswana. The site is allocated for the development of business and industries. The building was completed in four phases. Phase 1 was completed in 1982, phase 2 in 1984, phase 3 in 1986 and phase 4 in 1988. Expansion joints were provided between various portions completed at various stages(Figure.1). Rough concrete floor was provided for phase 1 and 2 while power float concrete floor was provided in phase 3 and 4. A 1.2m wide concrete apron was provided on the front side (north) of the building, while the two complete sides and the third side partially of the building were left for flowerbeds. Due to scarcity of water in Gaborone, no proper flowerbeds were properly maintained and the area around the building was left for the growth of bushes and the wild plants, which needed only rain water. The building did experience a severe drought of 5 years between 1987 and 1992.

No soil exploration was done for foundation design. Perhaps, it was not deemed necessary as other buildings in that area had no history of any distress. As a result the building was founded on a standard

630mm wide and 870mm deep continuous concrete wall footing. Walls were constructed of 230mm thick concrete blocks with cement mortar. Similar foundations were used for other adjoining structures, which had shown no problem.

3 DISTRESS

Cracks first appeared in the building in 1984 and have continued steadily since. Observation revealed that the cracks were confined only to the portion completed in phase 1 as shown in Figures 1-5.

1. Cracks developed on the west wall of the building were of 'V' shape running from foundation to lower level of window and from upper level of window to the roof. Widths of the cracks on the walls varied from 10 to 20mm.

2. Spacing between the expansion joint on west wall kept varying from 0 to 40 mm depending on the degree of saturation of the soil.

3. The apron on the front (north) side of building kept on cracking at every 2-3 years. Every time the cracks were filled with cement mortar as a repair.

4. The cracks developed on the apron continued through the wall, room floor and the rear wall. These cracks were about 15 mm wide and divided the room into two parts.

Because of the cost, remedial measures were only partially carried out in the form of filling the cracks with cement mortar. This exercise has been quite regular as the cracks keep reappearing in a year or two after they were filled.

Figure 2. Wall crack above window

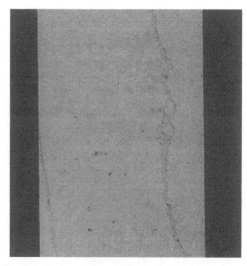

Figure 4. Crack on concrete apron

Figure 3. Wall crack below window

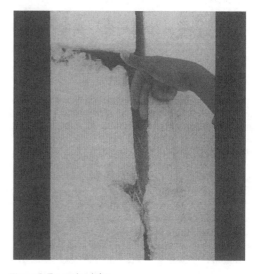

Figure 5. Expansion joint

4 INVESTIGATION

The investigation was carried out in Oct. 1998 just before the start of the wet season when the soil is likely to be most unsaturated. The degree of saturation of the soil is shown in Table 1. The following observations were made:

1. The cracks were first observed in 1984 i.e. two years after the completion of phase 1.
2. The cracks were confined only to the portion of the building completed in phase 1.
3. The cracks were of 'V' shape indicating the cause to be upheaval.

4. The cracks kept on reappearing after they were filled with the cement mortar as a repair.
5. There was no proper drainage for surface water, which was flooding the area around the building and was seeping into the foundation soil during wet seasons.
6. There was no soil investigation done for the foundation design.

From the distress history it was obvious that the cause of cracks was soil upheaval and hence soil exploration on all the four sides of the building was carried out. To achieve this, four 2.20 m deep pits were dug, with one on each side of the walls, as shown in Figure 1.

It was noted that the soil stratum was uniform below the top soil (0.50m). Both disturbed and undisturbed soil samples were collected from a depth just below the foundation. Various tests performed consisted of sieve analysis, hydrometer analysis, Atterberg limits, field density, natural moisture content, and linear shrinkage. All the laboratory tests were performed as per BS1377: 1990 code of practice. Summary of the test results is shown in Table 1.

Table 1: Test Results

| Tests | Pits | | | |
	1	2	3	4
Field m/c %	19.3	16.6	18.0	18.3
Degree of saturation %	37.8	33.2	36.0	36.0
Field Density Mg/m^3	1.30	1.32	1.32	1.30
Liquid limit (W_L)	40	38	54	65
Plastic limit (Wp)	18	18	20	20
Plasticity Index ,Ip	22	20	34	45
Linear Shrinkage,LS	11	14	19	18
Clay Content (C) %	34	28	17	22
Activity (A_c)	0.92	1.11	4.86	3.75
Classification	CI	CI	CH	CH

$A_c = I_p / (C-10)$

5 ANALYSIS

From the results of indicator tests it was noted that there was a great variation of soil properties in the lateral direction from northwest to southeast. The soils below the north and west walls were highly active while the soils below south and east walls were of low to medium activity. It was, therefore, decided to evaluate the swelling potential of the soils on each side of the building.

Swelling potential is an index that indicates the degree of volume change of the soil after saturation. From the swelling potential, it is possible to estimate the magnitude of heaving for floor and foundation. To determine swelling potential, the soils were remoulded at natural moisture content and field density in the oedometer rings and flooded. After flooding the samples were allowed to swell under a seating pressure of 6.9 kPa. The percentage expansion was recorded as the swelling potential of the soils. Swelling potentials were also estimated by various empirical correlations (Table 2).

Swelling potentials follow the same pattern as the activity. This shows that walls on the north and west side of the building were founded on soils of high to very high expansiveness while walls on south and east side of the building are founded on soils of low to medium expansiveness. It leads to the conclusion that the distress is primarily caused by the differential upheaval of the foundations.

Due to the absence of any reinforcement in the

Table 2. Swelling Potentials of the soil obtained in the laboratory and by various empirical correlations.

| Swelling Potential | Pits | | | |
	1	2	3	4
Seed et al	4.2	3.2	12.3	23.9
Vijayvergiya & Ghazzaly	2.11	2.94	8.67	20.9
Nayak & Christesen	7.68	7.5	7.5	8.19
Schneider & Poor	0.71	0.78	3.47	11
Chen	1.7	1.4	4.6	11.5
Sahu et al	0.4	0.5	2.7	6.3
Experimental	2.7	2.9	12	13

Seed et al. (1962)
  Swelling Potential $= 0.00216 I_p{}^{2.44}$
Vijayvergiya & Ghazzal (1973)
  Log Swelling Potential $= 1/12 (0.44LL\text{-}w_o + 5.5)$
  from w_o to saturation under 6.9 kPa.
Nayak & Christensen (1974)
  Swelling Potential $= (0.0229PI* 1.45C)/ w_o + 6.38$
  for compacted soil under 6.9 kPa.
Schneider & Poor (1974)
  Log Swelling Potential $= 0.9(I_p/w_o) - 1.19$
Chen (1988)
  Swelling Potential $= 0.2558e^{0.0838 (Ip)}$
Sahu et al,(1987)
  Swelling Potential $= 4.4 * 10^{-7} (\gamma_d)^{2.75}(w_1\text{-}w_o)^{3.27}/25$
  (γ_d in kN/m^3)

foundation, there was structural weakness resulting into severe cracking of the walls, floors and the pavements, due to foundation and floor movements.

6 TREATMENT

Since the building is being used for a furniture factory, the owner was mainly concerned with the stability of the structure and not with its aesthetic values. With the result that as long as it was insured that the building is not going to collapse, only patchworks were being done which were needed at regular intervals of every 2 to 3 years.

Since the process of cyclic drying and wetting causes reduction in the degree of expansion of an expansive soil, it is expected that in future the amount of upheaval of foundations and floors would not be as much as it was in the past. It can further be minimised by preventing the ingress of water beneath the building, which can be achieved by:

1. Providing 2.0-m wide downward sloping concrete aprons and a ditch drain all around. The drains would channel away the water and would prevent any ponding of the surface run-off.
2. A 10-mm plastic liner attached to the outside edge of the foundation, extending to a depth of about 2m and sloping away from the foundation. The liner should then be covered with a relatively inactive soil.

In our opinion, underpinning of the foundation might be too costly a proposition specially when the foundations are structurally weak due to the absence any reinforcement.

7 CONCLUSIONS

1. Botswana is a small country with a total population of about 1.8 million. The development of the country is rather recent and no past experience is available to the geotechnical engineers regarding the behaviour of structures on local expansive soils. The case history discussed appears to be the first of its kind in the country.
2. There was a significant variation in the soil properties in the direction of north-west to south-east.
3. There was a great variation in the values of swelling potentials for all the samples between the ones estimated by empirical correlations and the experimental values. However, the correlations by Seed et al. (1962) and Vijayvergiya & Ghazzaly (1975) seem to be closest and can be used for making predictions.
4. Foundations on north and west side of the building were founded on soils of high swelling potentials (12 to 13) while the walls on south and east side of the building were founded on soils of low swelling potential (2.7 to 2.9)
5. The distress was caused due to the differential upheaval of foundations due to variation in soil properties.
6. The development of cracks can be controlled by minimizing the ingress of water below the foundation by providing:
a. 2.0 m wide downward sloping concrete apron and a ditch drain all around. This is the cheapest solution which should suit the owner of the factory. Or
b. A strong plastic attached to the outside edge of the foundation, extending to a depth of about 2 m and sloping away from the foundation. The liner then should be covered with an inactive soil.
7. The case study presented should caution the local authorities of Botswana against the construction of buildings without any adequate geotechnical investigations.

REFERENCES

Chen F. H. 1988, *Foundations on expansive soils*. Development in Geotechnical Engineering Vol.54. Elsevier Science Publishing Company Inc.

Jones, D. E. & W. G. Holtz 1973. Expansive soils: the hidden disaster. *Civil Engineering: ASCE* 43 (8) 49-51.

Nayak N. V. & R. W. Christensen 1974. Swelling Characteristics of Compacted Expansive Soils. *Clay and Clay Minerals*, 19(4).

Nelson J. D. & D. J. Miller 1992. *Expansive soils: Problems and practice in foundation and pavement engineering*. John Wiley & Sons Inc.

Petak, W. J. 1978. *Natural hazard: A building loss mitigation assessment (Final Report)*. J. H. Wiggens Co. report under NSF Grant ERP-75-09998 (June).

Sahu B. K., F. J. Gichaga & T. G. Visweswaraiya 1987. Prediction of Swell of black cotton soils in Nairobi. *Proceedings of the International Symposium Prediction and Performance in Geotechnical Engineering University of Calgary, Alberta, Canada*. 259-266.

Schneider G. L. & A. R. Poor 1974. *The prediction of Soil heave and Swell Pressures Developed by an Expansive Clay*. Research Report TR-9-74, University of Texas Construction Research Center, Arlington Texas.

Sedumedi B. K S. & T. R. Keikitse 1994. *Investigation of cracks on a building*. Department of Civil Engineering, University of Botswana.

Seed H. B., R. J. Woodward & R. Lundgren 1962. Prediction of Swelling Potential for Compacted Clays. *Journal of the Soil Mechanics and Foundations Division; Proceedings of the ASCE*, 88 (SM3).

Vijayvergiya V. N. & O. I. Ghazzaly 1973. Prediction of Swelling Potential for Natural Clays. *Proceedings of the Third International Conference on Expansive Clay Soils*, Vol. 1, Jerusalem Academic Press, Jerusalem.

Wiggens, J. H 1978. Natural Hazards: *Earthquake, landslide, and expansive soil*. J. H. Wiggens Co. report for National Science Foundation under Grant ERP-75-09998 (Oct) and AEN-74-23993.

Unsaturated Soils for Asia, Rahardjo, Toll & Leong (eds) © 2000 Taylor & Francis, ISBN 90 5809 139 2

Properties of swelling soils in West Java

B. Wibawa & H. Rahardjo
NTU-PWD Geotechnical Research Centre, Nanyang Technological University, Singapore

ABSTRACT: Properties of swelling soils in Karawang, Subang and Cikampek in West Java, Indonesia are presented. A case study involving an industrial building founded on an unsaturated and compacted cohesive soil in Karawang is also discussed. Rainfall during an excavation resulted lateral deformation of building's foundation and horizontal cracks at the top of the pile foundations. Swelling pressure of the compacted soil is maximum at its optimum water content. In addition, properties of undisturbed swelling soils from Subang and Cikampek are also presented.

1 INTRODUCTION

1.1 Background

Expansive soils are found in many parts of the world, particularly in semi-arid areas. An expansive soil is generally unsaturated due to desiccation. Expansive soils also contain clay minerals that exhibit high volume change upon wetting. The large volume change upon wetting causes extensive damage to structures, in particular, light buildings and pavements (Fredlund & Rahardjo 1993). For example, according to a review from Chen (1988), expansive soils are detected in Australia, Canada, China, Israel, Jordan, Saudi Arabia, India, South Africa, Sudan, Ethiopia, Spain, United States. In United States alone, the damage caused by the shrinking and swelling soils amounts to about US$9 billion per year (Jones and Holtz 1973).

Fig 1 Map of Java Island (not to scale)

West Java is one of the provinces in Indonesia. (see Fig 1). West Java consists of mountaineous area in the central and southern part and low land areas in the northern part. It is situated around the capital (Jakarta). Therefore, this province has become an important region that supports the development in Jakarta.

The economic development in Jakarta has resulted in the growing need for infrastructures, such as industrial areas and housings in West Java. For example, industrial areas are found in Karawang, Cikampek and Subang. Many industrial areas have been developed on former rice fields or uncultivated lands. Cut and fill are commonly done in preparing the land for future industrial buildings. In general, most of the cut and fill works utilise local materials to save cost. The materials are usually cohesive soils cut from surrounding high level areas or certain quarries, and the materials are then used to fill the lower areas (Wibawa 1995).

Fill materials are usually compacted to a specified density and water content. In general, the dry density of the compacted cohesive soil must be greater than 90% or 95% of the maximum standard dry density. The maximum dry density relates to optimum water content of the compacted soil. Water content in the compacted cohesive soils can change because of the environment or man-made causes. Non uniform changes in water content can cause damage to the structure and foundation.

Cohesive swelling soils are found in several areas of West Java. Their locations, for example, are not only in the northern part but also in the central part of West Java (Indonesian Ministry of Mines & Energy 1996).

1.2. *Climate conditions*

As a tropical country, Indonesia including West Java, has two seasons annually. In general, dry season starts from April to October and rainy season begins from October to April.

Monthly rainfall in several sites of West varied considerably Java in the year 1997. The monthly rainfall data were obtained from three observation stations near the locations of swelling soils. The observation stations are located at Bekasi, Sukamandi, and Aljasari. Two of the sites of swelling soils are Karawang and Cikampek, which are located between observation stations Bekasi and Sukamandi. Another location of swelling soil is Subang, which is situated nearest to Aljasari observation station.

Measurements of monthly rainfall as monitored at three observation stations can be seen in Fig. 2 (Indonesian Meteorology & Geophysics 1997). In Bekasi observation station, the monthly rainfall reached its peak in January and it decreased significantly in the subsequent months. The monthly rainfall was very low in the period from July to October, even zero during several months. It was also quite low in November and December. The total rainfall in Bekasi was 1269 mm in the year 1997. In Sukamandi observation station, the monthly rainfall was quite high in the period from January to April. The monthly rainfall was very low or zero in the period from May to November. The rainfall was also quite low in December. The total rainfall in Sukamandi was 2401 mm in the year 1997. In Aljasari observation station, the monthly rainfall reached its peak in December and it fluctuated in the period from January to May. The monthly rainfall was very low or even zero in the period from June to November. The total rainfall in Aljasari was 3327 mm in the year 1997.

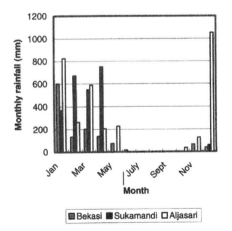

Fig. 2 Monthly rainfalls for three observation stations Bekasi, Sukamandi, and Aljasari in West Java in 1997

1.3 *Objective & scope*

The aim of this paper is to present properties of compacted swelling soils in Karawang area and of undisturbed swelling soils in Subang and Cikampek. A case study involving lateral deformation of the foundations and horizontal cracks of reinforced concrete piles of an industrial building in Karawang due to the swelling pressure induced by rainfall is also presented.

2. SWELLING SOIL IN KARAWANG

2.1 *Sub soil condition*

A case study involving damage of foundations due to the swelling pressure of a compacted swelling soil in Karawang is presented in this section. The foundations are located in a compacted cohesive fill layer that is underlain by the original cohesive soil layer. The compacted cohesive fill has an average thickness of 2 m. According to the Unified Soil Classification System, the cohesive soils are classified as a high plasticity and compressible silty clay (CH). Index properties of the cohesive soil are: Liquid limit w_L = 97.4%, Plasticity index PI = 78.24% and Specific gravity of solids G_s = 2.65. The soil consists of 52% silt and 48% clay particles. Activity of the cohesive soil is 1.63. In this case, Activity is the ratio of Plasticity index to percentage of clay fraction. According to Skempton (1951), there is a tendency of the presence of montmorillonite in the clay mineral if Activity is greater than 1.25. The soil has a very high swelling potential according to the classification given by Van de Merwe (1964) on the basis of the percentage of clay and Plasticity index. In addition, original soil layers underlain by the compacted cohesive fill were found to be a high plasticity and compressible silty clay (CH). Its thickness was 4 to 6 m. Index properties of the soil are: Liquid limit w_L = 92% - 96%, Plasticity index PI = 68.19% - 72.19%, Shrinkage limit = 9.7%, Specific gravity of solids G_s = 2.65, Void ratio e = 1.01 – 1.06, Water content w = 38.14% - 39.43%, Degree of saturation S = 96.5% – 99.4%. Composition of soil particles are 6% - 8% of sand, 55% - 57% of silt, and 37% of clay. Activity is in the range between 1.45 to 1.97.

2.2 *Compaction tests*

Compaction tests were carried out in order to simulate field conditions where compaction was carried out during the filling up process. The field specification required that the cohesive fill material be to a dry density greater than 90% or 95% of the laboratory maximum dry density (Wibawa 1995). Laboratory compaction tests were performed based on ASTM standard D698. Figure 3 shows the relationship between dry density and water content for the cohesive soil of Karawang. The range of water contents for the compaction tests were 16.3% - 38.9%. The soil has a maximum dry density of 1.52 g/cm³ at an optimum water content 26%.

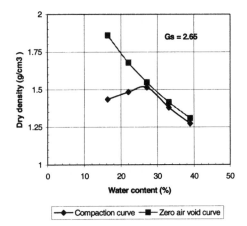

Fig. 3 Compaction and Zero air void curves of the cohesive fill material in the building area in Karawang

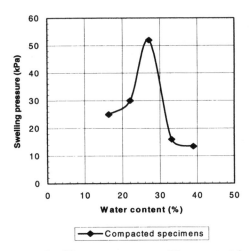

Fig. 4 Swelling pressures versus Water content for the compacted cohesive soil in Karawang

2.3 Oedometer tests

Free swell tests using oedometer (Smith 1973) were carried out on compacted specimens of the swelling soil from Karawang. For each condition of water content and dry density, one specimen was taken for the oedometer swelling test. Therefore, the soil specimens varied in their initial water content and dry density.

Figure 5 shows a relationship between void ratio and effective stress. The curve is for the soil specimen that has an initial water content corresponding to the optimum water content (w_{opt} = 26%). The curve shows the swelling of the specimen as indicated by an increase in void ratio to a value greater than the initial void ratio. This means that water was absorbed by the soil specimen so that volume of the soil became greater than the initial volume. Subsequently the void ratio decreases as water comes out from the soil specimen due to the increasing pressure beyond the swelling pressure. The rebound curve is found to reach a certain void ratio lower than the initial void ratio. Slope of the rebound curve is called swelling index.

2.4 Swelling pressures

Swelling pressure can be defined as the pressure which prevents the specimen from swelling or the pressure which is required to return the specimen back to its original state (void ratio, height) after swelling (Chen 1988). In the above case, the swelling pressure is found to be 52 kPa and its free swell index is 1.2%.

Water absorbed by the specimen results in the swelling of the soils. Figure 4 shows a relationship between the swelling pressures of the compacted cohesive soil from Karawang and their compaction water contents. Figure 4 indicates that the maximum swelling pressure corresponds to the optimum water content. In other words, it is also related to the max-

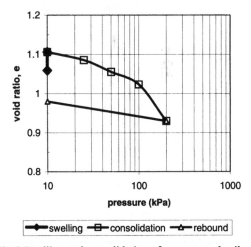

Fig 5 Swelling and consolidation of a compacted soil specimen from Karawang with an initial water content of w_{opt} = 26%

imum dry density of the compacted cohesive soil. This can be attributed to the specific surface of soil particles that has the greatest value at the maximum dry density condition (Hillel 1980). As a result, the soil absorbs the largest amount of water as compared conditions at other initial water contents. In addition, Activity of the soil is around 1.6 indicating the presence of monmorillonite mineral (Skempton 1951). This mineral can absorb a large amount of water compared to the other clay minerals, such as, kaolinite and illite.

Swelling pressure of the soil in the dry side of the optimum water content was found to be lower than that of the optimum water content. This can be explained by the lower dry density compared to the maximum dry density. Hence, the specific surface of the soil specimen is smaller than that of the maximum dry density. As a result, the amount of water absorbed by the specimen is also smaller.

The wet side of the optimum water content shows a decrease in the swelling pressure from that of the optimum water content. In this case, the swelling pressure drops sharply with increasing water contents. The decrease in the swelling pressure is not so significant when water content is greater than 32%. This is because the soil specimen has sufficient water in the pore and even it has nearly reached its saturated condition.

2.5 Structure and foundation of the building

An industrial building of 20 m in width and 90 m in length was designed using steel materials for upper structures. The span of the roof was 20 m and the distance between adjacent roofs was 6 m. The building had sheet walls, which would be built on the L-shaped reinforced concrete pile cap (see Fig 6). The reinforced concrete L-shaped pile cap had a height of 120 cm and a thickness of 25 cm. Equilateral triangular reinforced concrete piles with 28 cm in size were used as the foundations. Lengths of piles varied from 9 to 11 m and the longitudinal distance between adjacent piles was 3.0 m. The concrete strength of the pile is 40 Mpa and the steel grade is 400 MPa. Three reinforced bars, each 16 mm in diameter, were used for the pile. Floor slabs were made from reinforced concrete. The slab's dimension was 4 m by 2 m and 25 cm in thickness.

2.6 Damage of foundation

An excavation for a drain next to a row of pile foundations on the eastern side of the industrial building was dug. Other drains had existed on the northern and southern sides of the building before the excavation of the eastern side was carried out. . The excavation had not finished, so the new drain had no connection with other drains. A sudden rainfall occurred in May 1997 when the excavation was still under construction (see Fig 2).

Rainwater fell into all drains and the excavation of the new drain. As a result, water infiltrated and percolated into the surrounding soil along the excav-

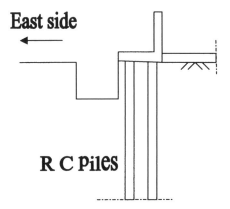

Fig 6 Cross section of the eastern foundation of the industrial building in Karawang

ation. As a result, the surrounding soil along the excavation swells especially towards the unconfined space along the excavation.

Lateral deformation at the L-shaped pile cap and horizontal cracks was observed at the top end of pile foundations on the eastern side of the industrial building. In addition, a gap between the edge of floor slabs and L- shaped pile cap was also detected. Rainwater infiltration along the excavation caused an increase in the water content of the surrounding soils. The absorbed water resulted in the soil swell and the development of swelling pressures. The lateral pressure due to the swelling of the wetted ground forced the L-shaped pile cap and piles to move outside toward the excavation.

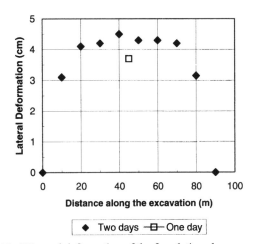

Fig 7 Lateral deformation of the foundation along the excavation of the building in Karawang

Figure 7 shows lateral deformations of the L-shaped pile cap including the pile foundations along the excavation as measured one day and two days after the rainfall. Figure 7 indicates that lateral deformations on both edges of the pile cap at northern and southern end points are zero. It can be understood because both edges are fixed by northern and southern sides of the base foundations as one solid body of reinforced concrete beam. . In other words, the pile cap can be assumed as a long beam fixed at both ends. Lateral deformations along the excavation increases with distance, and it reaches its peak around the middle of the length of the excavation.

3. SWELLING SOIL IN SUBANG

3.1 Sub soil condition

Five boreholes were drilled for investigating the swelling soil in Subang, (i.e., BH 5, BH 8, BH 10, BH 17 and BH 20). Undisturbed soil samples were taken from the boreholes. According to the Unified Soil Classification System, the cohesive soil was classified as brownish yellow Silty Clay (CH) with medium to stiff consistency. Degree of saturation are from 85.13% to 98.14%. Water contents of the undisturbed samples are from 34.3% to 42%. Liquid limits are 61% – 70%, Plasticity Indexes are 30.8% – 38.9%, and Shrinkage limits are 10.6% – 23.96%. Its Activities are 0.57 – 0.77. Ground water level are from – 1.54 m to - 3.45 m.

3.2 Swelling pressure

Figure 8 shows the swelling pressures of undisturbed samples taken from several depths. Figure 8 indicates that swelling pressures of BH 5 and BH 10 were very small (10.5 – 12 kPa). They were almost equal to the initial surcharge in the consolidation tests (10 kPa). However, the swelling pressures of BH 8 and BH 20 were from 18 to 20 kPa. The figure also shows larger swelling pressures at the depth

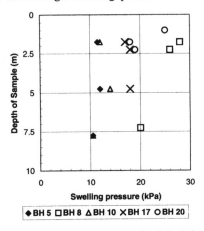

Fig. 8 Swelling pressures versus depth in Subang

around –2.0 m as compared to the swelling pressures at – 7.5 m depth. The swelling pressures of BH 17 were around 18 kPa. The soil has a low Activity, i.e., from 0.57 to 0.77. Hence, there is no indication of the presence of montmorillonite in the clay mineral because Activity is less than 1.25 (Skempton 1951). As a result, low swelling pressures were measured from the undisturbed cohesive soils.

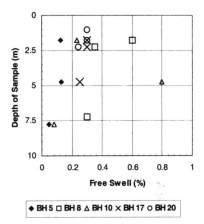

Fig 9 Free swell versus depth in Subang

3.3 Free swell and swell index

Figure 9 shows Free swell of undisturbed cohesive soil specimens in Subang for several boreholes. Free swell is the ratio of the difference between final volume and initial volume to the initial volume. Free swell are from 0.05% to 0.35%. Free swell of BH 5 were very small (0.05 – 0.1%). However, free swell of BH 8 were small at – 2.0 m and – 7.75 m, but quite high at – 5.0 m. Free swell of BH 8, BH 17 and BH 20 were between 0.25% - 0.6%. Most of them are around 0.3%. Swelling index is from 0.015 to 0.09 and all specimens were overconsolidated (Preconsolidation pressure is from 100 to 220 kPa).

4. SWELLING SOIL IN CIKAMPEK

4.1 Subsoil condition

Four boreholes were drilled for investigating the swelling soil in Cikampek (i.e., SB I, SB II, DB I and DB II). Undisturbed samples were taken from the boreholes. According to Unified Soil Classification System, the soil can be classified into silty clay (CH) with medium – stiff consistency. Index properties of the soils are: Degree of saturation are 86.97% to 98.4%. Water contents are from 31.96% to 49.6%. Liquid limit are 76.5% - 96.5%, Plasticity index are 53.88% - 74.83% and Shrinkage limit are 4.5% - 20.61%. The clay fraction are 45% - 58%. Activities of the soils are from 1.23 to 1.66

4.2 *Swelling pressure*

Figure 10 shows the swelling pressures of undisturbed samples taken from several depths in Cikampek. This figure indicates swelling pressures of SB I and SB II from 33 – 40 kPa. However, swelling pressures of DB I and DB II were around 50 kPa. Its Activity (1.23 – 1.6) shows that there is a tendency of the presence of montmorillonite in the clay mineral. The swelling pressures of the soil in Cikampek are greater than that of Subang.

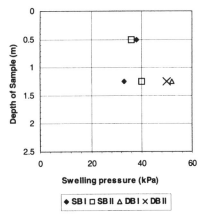

Fig 10 Swelling pressure versus depth in Cikampek

4.3 *Free swell and Swelling Index*

Figure 11 shows Free swell in several depths for undisturbed cohesive soils in Cikampek. Free swell of SB I and SB II were from 1.6% to 2.1% at - 0.50 m depth, and 0.5% to 1% at – 1.25 m depth. It indicates greater values at the depth near the surface. However, Free swell of DB I and DB II at – 1.25 m depth were around 0.9%. Swelling index was 0.095 – 0.17. Preconsolidation pressure was from 105 kPa to 140 kPa.

5. CONCLUSIONS

Conclusions for the particular sites are as follows:
1. In Karawang, the maximum swelling pressure of the compacted cohesive soil occurs at the optimum water content. Swelling pressure decreases on the dry and wet side of the optimum water content. Swelling pressure of the compacted cohesive soil, caused by rainfall infiltration, can exert unfavourable effect to the foundation. As a result, lateral deformation of the pile and L-shaped pile cap occurred.
2. In Subang, the swelling pressures of undisturbed cohesive soils are quite low values. Small Activity of the soils causes small swelling pressures.
3. In Cikampek, the swelling pressures of undisturbed cohesive soils are greater than that of Subang.

The tendency of present of monmorillonite in the clay mineral is detected by high values of Activity.

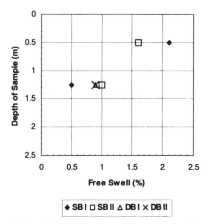

Fig. 11 Free Swell versus depth in Cikampek

REFERENCES

Chen, F.W. 1988. *Foundations on expansive soils, Developments in Geotechnical Engineering* 54, New York, Elsevier.

Fredlund D.G. & Rahardjo H. 1993. *Soil Mechanics for Unsaturated Soils*, New York, John Wiley & Sons, Inc.

Hillel D. 1982. *Introduction to Soil Physics*. New York. Academic Press.

Indonesian Meteorology & Geophysics. 1997. *Monthly rainfall for Bekasi, Sukamandi and Aljasari*, Jakarta.

Indonesian Ministry of Mines & Energy. 1996. *Engineering Geological Map*. Jakarta.

Jones D. E. & Holtz W. G. 1973. Expansive soils-The Hidden Disaster. *Civil Eng. ASCE*, New York, pp. 87-89.

Skempton W. 1951. The colloidal activity of clays. *Proc. 3rd ISSMFE*, Zurich, vol. 1, pp. 57-61

Smith, A. W. 1973. Method for determining the potential vertical rise, PVR. *Proc. Workshop Expansive Clays and shales in Highway design and Construction*, Laramie, vol. 1, pp. 189-205.

Van der Merwe, D. H. 1964 The prediction of heave from the plasticity index and percentage fraction of soils. *Civil Eng. In south Africa*, vol.6, no.6, pp 103 – 107.

Wibawa B. 1995. Relationship between optimum water content and index properties of cohesive soils. *Trisakti Science Magazine*, vol. 2, pp. 13-22. (in Indonesian)

Wibawa B. 1995. Cohesive soil at the ground surface as fill material. *Trisakti Science Magazine*, vol. 3, pp. 82-92. (in Indonesian)

Unsaturated Soils for Asia, Rahardjo, Toll & Leong (eds) © 2000 Taylor & Francis, ISBN 90 5809 139 2

Multistory apartment building on poznań clay – Case history

A.T.Wojtasik & J.Jeż
Technical University of Poznań, Poland

ABSTRACT: So called "poznań clay" is probably the most potentially expansive clay that is found in Poland. It is present in many parts of the country. However significant damage to many buildings have been encountered in the Wielkopolska Region. This paper presents a case study of a multistory apartment building founded on "poznań clay". The building has been damaged several times in a period of 30 years. The first major damage was caused by a cracked sewer pipe, and resulted in extensive heaving of clay under basement floors and foundations. After approximately ten years damage was once again initiated by a nearby group of trees. Tree roots have penetrated soil under foundations and dried out the soil causing its shrinkage and settlements of building foundations.

1 INTRODUCTION

In the last three decades geotechnical engineers in Poland encountered a significant number of buildings and other engineering structures that have been damaged due to differential movements of expansive clays. Over one third of the area of Poland has deposits of expansive clays. Most of the clays are Tertiary deposits with some present directly below ground surface, or in the range of foundation interaction. Among these deposits the most active are :
1. Pliocene clay known as "poznań clay";
2. Miocene clay (marine deposits);
3. Oligocene clay

Authors of this paper have analyzed over thirty cases of damaged buildings founded on "poznań clay" in Wielkopolska Region. In most cases the damage was initiated by nearby trees which caused shrinkage of clay under foundations. Only a few cases were related to clay swelling, which was usually induced by a broken sewer or water pipe. The case history, of a multistory apartment building, presented in this paper is an example of continuous destruction of the building that took place in a period of three decades (Jez & Wojtasik 1989).

2 SHORT DESCRIPTION OF THE SITE AND BUILDING

The building is located on a sloped terrace of Warta River. Inclination of the site is approximately 20%, in the South -East direction. The building, erected in the early sixties, is a 5-story apartment building with a partly underground basement. It is 32 m in length and 10 m in width. Its structure is traditional with brick walls and reinforced concrete floors and roof. Foundation slabs, located 1.3 - 2.0 m below ground surface, are made of reinforced concrete. Intensive damage caused by differential movements of soil occurred in the northern part of the building. About 1/3 of the building's length is founded directly on "poznań clay", the remaining part is founded on sandy boulder clay (Figure 1).

Deep geological investigations show that the site where the building is located, is generally made up of cohesive deposits mainly sandy boulder clay in the upper part and Pliocene clay directly beneath. In some locations the Pliocene clay was uplifted to the surface due to glacier action. This is why only the northern part of the building lies on "poznań clay", and therefore why the destruction process took place only there.

3 SOIL PROPERTIES

Soil investigations carried out several times during the 30 year period of continuous destruction of the building give detailed information on soil properties. Basic parameters of the sandy boulder clay and Pliocene "poznań clay" are given in Table 1. The given values are average values for the soil mass.
Natural moisture content of both clays varied depending on location, sampling depth, and year in which investigations were carried out.

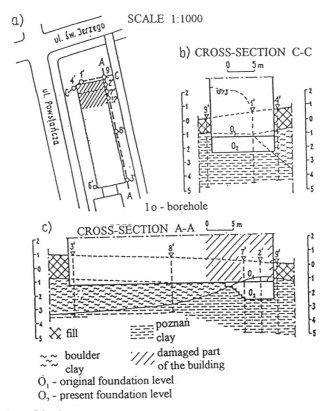

a) SCALE 1:1000

b) CROSS-SECTION C-C

1o - borehole

c) CROSS-SECTION A-A

⊠ fill

~~~ boulder
~~~ clay

‒‒‒ poznań
‒‒‒ clay

/// damaged part
/// of the building

O_1 - original foundation level
O_2 - present foundation level

Figure 1. Location and geology of the site.

Table 1. Parameters of sandy boulder clay and Pliocene "poznań clay"

	sandy boulder clay	poznań clay
Bulk density	2.22 g/cm³	2.05 g/cm³
Shrinkage limits	5.5 %	9.5 %
Plastic limit	14.0 %	27.0 %
Liquid limit	22.0 %	69.0 %
Plasticity index	8.0 %	42.0 %

Detailed soil moisture measurements were taken in the northern part of the building and in area between the building and trees. Figure 2 shows comparison of natural moisture content of "poznań clay" under damaged part of the building in years 1973 and 1986.

4 HISTORY OF BUILDING DESTRUCTION

Destruction of the northern part of the building, the part founded on "poznań clay", has a history of over 30 years. Phases of destruction are shown in Figure 3. Wide destruction took place twice, in 1972 and between 1982 -86. It resulted in closing part of the building for exploitation, and moving out residents. However, information on cracking walls, problems related to closing and opening doors and windows have been reported to the administration several times in that period of time.

Chronologically taking major problems were reported in the following order:

1. Construction phase - the building was erected in years 1961 to 1963. First problems related to clay heaving were noted during construction works. In January 1963 basement floors were lifted and intensively cracked soil swelling. All heave problems were related to the fact that rain water and surface water were gathering in excavations around the foundation walls. After installing a drainage system around the building all problems stopped for a period of 10 years

2. Year 1973 - northern gable wall and its foundation indicated extensive cracking. Basement floors were lifted and cracked. Detailed geotechnical and structural investigations were carried out. Results of soil investigations showed a large increase in moisture content of the "poznań clay" under basement floors and foundations in northern part of the building. This was caused by a cracked sewer pipe running under the building. Water, from the sewer pipe, gathered on top of the clay layer and initiated its swelling. After repairing the sewer pipe system, the cracked walls were repaired and reinforced. An ad-

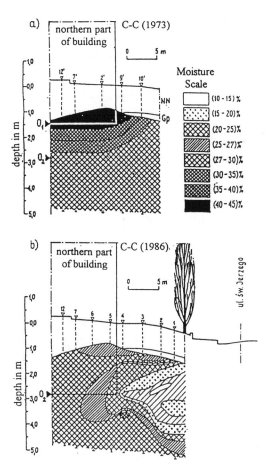

a) northern part of building   C-C (1973)

Moisture Scale

☐	(10 - 15)%
⫶	(15 - 20)%
▨	(20 - 25)%
▧	(25 - 27)%
▨	(27 - 30)%
▨	(30 - 35)%
▨	(35 - 40)%
■	(40 - 45)%

b) northern part of building   C-C (1986)

ul. św. Jerzego

Figure 2. Comparison of moisture conditions in "poznan clay" in 1973 and 1986.

ditional foundation wall was constructed. The new foundation was constru-cted 1.3 m below the previous foundation level, i.e. below the zone of swelled clay. This solution was found to be successful since for the next 10 years no further damaged to the building was noted.

3. Year 1982 - new cracks appeared on the northern gable wall and on basement floors located next to the wall. In the following 4 years the cracks widened and their number increased. No specialized investigations were carried out in that period of time.

4. Year 1986 - 26 gypsum seals were placed on the cracks in the northern wall. After approximately 2 months most of the seals were already broken. This indicated that the destruction process, that started in 1982, still continued. Detailed soil investigations and excavations of foundations under the northern gable wall indicated that the clay moisture conditions had dramatically chan-

ged compared to conditions reported in 1973. Soil samples taken from boreholes as well as uncovered clay in foundation excavations indicated a very large number of tree roots. The roots had penetrated under the foundation and were also present directly under the cracked concrete basement floor. The roots belonged to a group of trees growing nearby at a distance of around 11 m from north-eastern corner of the building (2 Italian poplars, 1 willow, 1 great maple). The trees were around 50 years old. Moisture content measurements of soil, from under and around the damaged part of the building, were taken every 20 cm in 6 boreholes to the depth of 5 m. Results of this measurements and their comparison with results taken in 1973 are shown in Figure 2. Significant settlements of road pavement and sidewalk of the neighboring św. Jerzy Street have also taken place.

5. Years 1989 and 1992 - progression of destruction process of northern part of the building. Soil borings made along the building indicated that expansion of three roots had progressed to approximately 1/3 of the building length. In 1992 a recommendation was made to construct a vertical moisture barrier between trees and the building or to remove all the trees. Building administration decided to cut down the trees in 1993. Since than no further damage to the building was registered.

5 RESULTS OF INVESTIGATIONS AND ANALYSIS

Analysis of available data i.e. results of all soil investigations carried out since 1963 and written descriptions of the continuing destruction of northern part of the apartment building, enabled identification of the mechanism responsible for the destruction process. The northern part of the building was founded on highly expansive clay, whereas the other part was founded on gray boulder clay - a stable clay not sensitive to limited moisture changes.

The first problems, that emerged during the construction period and the destruction that took place in 1973 resulted from a significant increase in moisture content of "poznań clay" under foundations and basement floors. The swelling of clay under basement floors caused its uplift and intensive cracking. Moisture increase in the clay, in the range of 40-45 %, directly under the northern gable wall foundation resulted in the increase in its plasticity and decrease in its bearing capacity. As a result of this process the plastic clay was pushed out from under foundation slabs horizontally. This induced settlements of the gable wall and caused intensive cracking of all outside and some inside walls in this part of the building. After ten years of stable exploitation of the building,

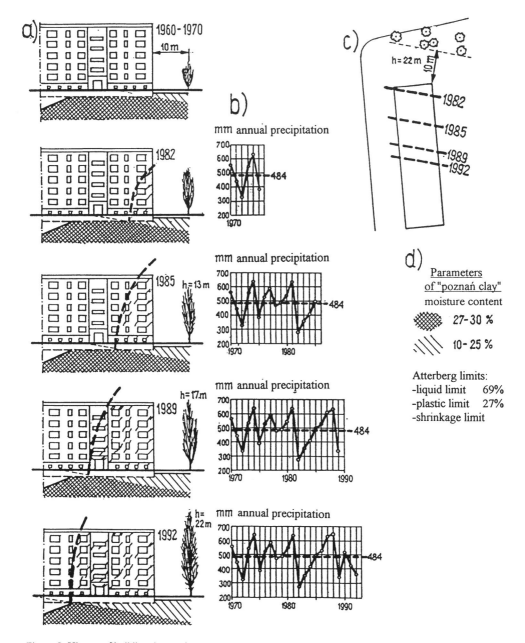

Figure 3. History of building destruction.

additional settlements of building foundations took place - between 1982 and 1986. Years 1989 and 1992 resulted in further settlements and destruction. This time settlements were related to shrinkage of "poznań clay". The shrinkage process of clay was caused by a radical drop in its moisture content to the level of approximately 10-20%. Low moisture content of soil in this location was due to relatively low annual amount of rainfall in these years and to the presence of expansively growing root systems belonging to the group of nearby trees. In 1992 tree roots were detected in boreholes 15 - 20 m away from the trees. After removing trees in 1993 the situation has been stable as far as the building is concerned and no further pavement deformations have been detected.

6 CONCLUSIONS

The case study described in this paper, one of very many that have taken place in the Wielkopolska Region in Poland, is an example of problems related to the presence of highly expansive "poznań clay" in subsoil. Both swelling and shrinkage of the expansive clay can be destructive to structures even in climates with relatively equal amount of precipitation throughout the year. It is evident that even minor changes in the moisture content of "poznań clay" can result in its heaving or shrinkage. The apartment building described above was unfortunate to be damaged continuously for three decades at first by clay swelling and later by shrinkage.

REFERENCES

Jez, J. & A.Wojtasik 1989. Wplyw drzew na awarie budynku posadowionego na gruncie peczniejacym. *Inzynieria i budownictwo 7/89*:264-266. Warszawa 1989.

Unsaturated Soils for Asia, Rahardjo, Toll & Leong (eds) © 2000 Taylor & Francis, ISBN 90 5809 139 2

On the bearing capacity of an unsaturated expansive soil basement

Y.F. Xu, D. Fu & G.J. Zhang
Shanghai Tunnel Engineering Company Limited, People's Republic of China

ABSTRACT: To determine the bearing capacity of a basement in expansive soil is very difficult and very important. By considering the expansive lateral pressure and the total cohesion induced by suction in Terzaghi's bearing capacity formula, two methods to determine the bearing capacity of unsaturated expansive soil basement are presented in this paper. The bearing capacity is verified by the results of plate loading tests on unsaturated expansive soil in Handan and Ningxia, P.R. China.

1. INTRODUCTION

Expansive soils are natural, highly dispersed and plastic soils, which contain mainly clay minerals and are very sensitive to either a dry or wet environment. Expansive soil is widely distributed in the world and found in more than 40 countries and regions. China is one of the countries with a large distribution of expansive soils, which have successively been discovered in more than 20 of its provinces and regions. In China expansive soils are mainly lacustrine, residual, slopewash, alluvial and diluvial in origin.

Studies of the mechanical and engineering character of expansive soil have been emphasized in all over the world because of their wide distribution and serious harm. Expansive soil is called the "hidden hazard" and it is reported that the economic losses caused by expansive soil amount to $ 2.3 billion per year, far more than from the total losses caused by floods, earthquakes and windstorms put together in U.S.A. In Japan expansive soil is called "difficult soil" and "problem soil", for it often brings about foundation deformation, frost soil and mud pumping of many roadbeds, the heaving of tunnel arches and landslides of embankments, etc. In China it is said that the "there is no cut that will not cause slides" and "there is no embankment that will not collapse " in regions of expansive soils.

Expansive soil is a typical kind of unsaturated soil. Many important results on the mechanics of expansive soils are obtained by using unsaturated soil mechanics. Fredlund & Rahardjo (1993) have studied the swelling and shrinking deformation using one dimensional consolidation theory of

unsaturated soils. Lu et al. (1992) have presented the linear relationship between swelling pressure and the cohesion of unsaturated expansive soils. Xu & Sun (1996), Xu & Liu (1999) and Xu & Deming (1999a) have studied the bearing capacity of basements in expansive soil by considering the influence of swelling pressure. In this paper, the character of bearing capacity of unsaturated soil basement is studied. Two methods to determine the bearing capacity are obtained for a basement in unsaturated soil and a comparison of the two methods is given in this paper. The theoretical methods of bearing capacity are verified using testing results of Handan and Ningxia expansive soil preformed on site.

2. BEARING CAPACITY CHARACTER

The character of bearing capacity of unsaturated expansive soil is shown as follows. (1) The bearing capacity is not a constant and it varies with water content (Figure 1). (2) The bearing capacity varies linearly with swelling pressure. If q_u is the ultimate bearing capacity of natural expansive soil and $(q_u)_{sat}$ is the ultimate bearing capacity of the expansive soil which is saturated by immersion in water, letting $\Delta q_u = q_u - (q_u)_{sat}$ then Δq_u linearly increases with swelling pressure (Figure 2).

3. BEARING CAPACITY DETERMINATION

Fredlund & Rahardjo (1993) have presented a method to determine the bearing capacity of an unsaturated soil basement by considering the

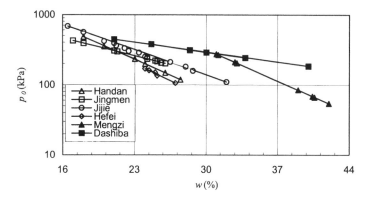

Figure 1. The relation between water content and bearing capacity.

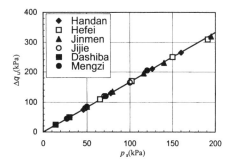

Figure 2. The relation of swelling pressure and bearing capacity.

influence of matrix suction. Xu & Sun (1996) and Xu & Deming (1999a) have studied the influence of swelling pressure on the bearing capacity of unsaturated expansive soil basement. The methods of the determination of the bearing capacity of unsaturated expansive soil basement are presented as follows.

3.1.*Method by using matrix suction* (METHOD I)

The shear strength of unsaturated soil is given as follows (Fredlund & Rahardjo 1993),

$$\tau_f = c' + (\sigma - u_a)tg\varphi' + (u_a - u_w)tg\varphi^b \quad (1)$$

where c' and φ' are the effective cohesion and effective internal frictional angle, respectively, φ^b is the frictional angle which describes the variation of shear strength with matrix suction and called suction frictional angle in short, u_a and u_w are pore air pressure and pore water pressure respectively, σ is total stress. By comparing equation (1) with the shear strength of saturated soil, a total cohesion of

unsaturated soil is obtained as follows:

$$c = c' + (u_a - u_w)tg\varphi^b \quad (2)$$

where c is total cohesion, $u_s = u_a - u_w$ is matrix suction. For unsaturated expansive soil, total cohesion is a power function of matrix suction and the function is given as follows (Xu & Deming, 1999b)

$$c = c' + K^\beta u_s^\alpha tg\varphi' \quad (3)$$

where $K = \dfrac{2T\cos\theta}{R}, \beta = 2 - \dfrac{2D}{3}, \alpha = \dfrac{2D}{3} - 1, T$ and θ are surface tension and contacting angle, R is maximum pore radius and D is fractal dimension.

The variation of strength parameters with water content is shown in Figure 3. It is seen that the internal friction angle varies slightly with water content and is nearly a constant. It is shown that the effective internal frictional angle is a constant for unsaturated soils with different matrix suction from test results (Chen, 1999).

The bearing capacity of an unsaturated soil basement can be obtained by using total cohesion instead of effective cohesion in the Terzaghi formula and the parameters in the bearing capacity formula of an unsaturated soil basement are equal to that for a saturated soil basement.

3.2. *Method using swelling pressure* (METHOD II)

Katti (1987) has studied the distribution of lateral pressure for sand, cohesive soil and expansive soil and he found that the lateral pressure of expansive soil is larger than that of the other two soils, and the swelling lateral pressure is in proportion to the swelling pressure of expansive soil. The swelling

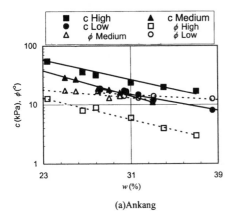

(a)Ankang

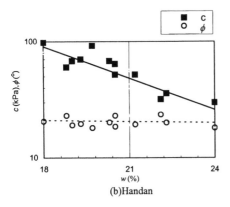

(b)Handan

Figure 3. The strength parameters for different water content

lateral pressure of expansive soil is given by:

$$p = \frac{p_s \dfrac{d}{d_0}}{a + 0.6 \dfrac{d}{d_0}} \quad (4)$$

where p_s is swelling pressure, p is swelling lateral pressure, a is cumulative amount of grains with radius less than 2μm, d is depth, d_0=1cm.

In the Terzaghi formula, it is assumed that the swelling lateral pressure is distributed along the side of an elastic wedge and its orientation is vertical to the orientation of an elastic wedge side. The total amount of swelling lateral pressure along the elastic wedge side is given as follows:

$$P_{ps} = \int_0^{\frac{B}{\cos\phi}} p_s\, d(\frac{d}{d_0})$$
$$= \frac{5 p_s B}{6 \cos\phi} + \frac{25}{9} p_s\, a \ln a + \frac{25}{9} p_s\, a \ln(a + \frac{B}{10\cos\phi}) \quad (5)$$

where ϕ is the angle between the elastic wedge side and horizontal plane, B is the width of the footing in meters, If $a \gg B$, the swelling pressure is written as follows:

$$P_{ps} = \frac{5 p_s B}{6 \cos\phi} \quad (6)$$

The ultimate load can be written as follows:

$$Q_u = 2 P_p \cos(\phi - \varphi) + cBtg\phi - \frac{1}{4}\gamma B^2 tg\phi + 2 P_{ps} \cos\phi \quad (7)$$

The ultimate bearing capacity is given by:

$$q_u = \frac{Q_u}{B} = cN_c + qN_q + \frac{1}{2}\gamma BN_\gamma + p_s N_s \quad (8)$$

where

$$N_c = tg\phi + \frac{\cos(\phi - \varphi)}{\cos\phi \sin\varphi}\left[e^{(2\pi/3 + \varphi - 2\phi)tg\varphi}(1 + \sin\varphi) - 1 \right]$$

$$N_q = \frac{\cos(\phi - \varphi)}{\cos\phi} e^{(2\pi/3 + \varphi - 2\phi)tg\varphi}\, tg(45° + \frac{\varphi}{2})$$

$$N_\gamma = \frac{1}{2} tg\phi\left[\frac{k_{p\gamma} \cos(\phi - \varphi)}{\cos\varphi \cos\phi} - 1 \right]$$

$$N_s = 5/3 \quad (9)$$

It can be seen from equation (8) that the ultimate bearing capacity of an unsaturated expansive soil basement contains two parts, one is the ultimate bearing capacity of a saturated soil basement, another is the ultimate bearing capacity induced by swelling pressure.

4. COMPARISON OF THE THEORETICAL RESULTS AND THE PRACTICAL RESULTS IN HABDAN AND NINGXIA

The region of Handan expansive soil is located between the foot of Taihang Mountain and the north China plain. The physical and mechanical properties of Handan and Ningxia expansive soil are shown in Table 1.

Table 1. Physical and mechanical indices

Index Type	Handan	Ningxia
Water content w(%)	18~25	16~24
Unit weight γ (kNm^{-3})	20.0	19.8
Relative density d_s	2.71	2.70
Void ratio e	0.54	.77
Degree of saturation S (%)	92	84
Liquid limit w_L(%)	44.0	50
Plastic limit w_p(%)	21.0	29
Content of ≤2μm a (%)	27	20
Free swell δ_{ef}(%)	77	64
Swelling pressure p_s(kPa)	159	90~101
Cohesion c (kPa)	15~92	19~70
Internal friction angle ϕ	6~13	14

The results of plate loading tests of the Handan expansive soil basement are shown in Figure 4. The ultimate bearing capacity of the unsaturated expansive soil basement is 400kPa in plate loading tests and it is 140kPa for the saturated soil basement immersed by water. For the Handan expansive soil, it is obtained that N_c=8.35 from ϕ =10. The calculated result of ultimate bearing capacity is $(q_u)_{sat}$=135kPa from c=16.2kPa and N_c=8.35. The swelling pressure is 158kPa and the calculated result of ultimate bearing capacity of the unsaturated soil basement is 398kPa by considering $(q_u)_{sat}$=135kPa. The calculated results by using the later two methods also are shown in Figure 4. It is found that the testing results and the calculated results using the method presented in this paper are similar, as seen in Figure 4.

For the Ningxia expansive soil, ϕ =14, c=25kPa, and N_c=10.4, $(q_u)_{sat}$=260kPa. Putting p_s=100kPa and $(q_u)_{sat}$=260kPa in Equation (8), the calculated result of ultimate bearing capacity is 427kPa. Letting R=100 m, T=375mN/m, D=2.63, the calculating bearing capacity by using method (I) is 393kPa by considering u_s=98kPa. The testing and calculating results of the bearing capacity of the Ningxia expansive soil basement are shown in Figure 5.

5. CONCLUSIONS

Two conclusions can be obtained as follows,

(1) The bearing capacity of an unsaturated expansive soil basement is related to the swelling pressure. The more the swelling pressure, the more the decrease of the bearing capacity when the basement is immersed to a saturated state by water.

(2) There are two methods to determine the bearing capacity of an unsaturated expansive soil basement. One method is that the bearing capacity is determined by using matrix suction, and another is by using swelling pressure. The calculated results of

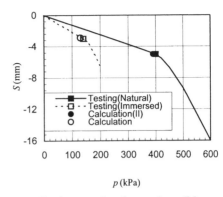

Figure 4. Bearing capacity of expansive soil basement in Handan

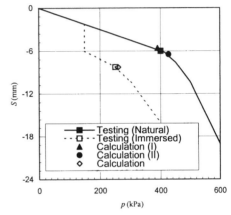

Figure 5. The bearing capacity of Ningxia expansive soil basement

the two methods are similar to each other, and very similar to the test results.

ACKNOWLEDGMENTS

The manuscript has been vastly improved by detailed and thoughtful reviews by Professor SUN Jun, the member of Chinese Academy of Science. We are also grateful to BAI Yun, the General Engineer of Shanghai Tunnel Engineering Co. Ltd. for his helpful discussion.

REFERENCES

Chen Zhenghang, Zhou Haiqing, 1998. A non-linear model for unsaturated soils, *Proceedings of 2nd International Conference on Unsaturated Soils, Beijing*: International Academic Publishers, 461~466.

Fredlund D.G., Rahardjo H., 1993. *Soil Mechanics for Unsaturated Soils*, John Wiley & Sons, Inc., New York.

Katti D.R. 1987. Role of CNS on passive resistance of saturated expansive soil, Katti R.K. eds, *6th Int Conf on*

Expansive Soils, Oxford & IBH Publishing, PVT, LTD, New Delhi, p87~89.

Lu Zhaojun, 1992. The relationship between shear strength and swelling pressure of unsaturated soils, *Symposium on the theory and practice of unsaturated soils, Beijing.*

Xu Yongfu, Sun Lianjing, 1996. Determination of the bearing capacity of unsaturated expansive soil basement, *J. of Harbor Engineering,* 5:28~32.

Xu Yongfu, Liu Songyu, 1999. *Theory on Shear Strength and Its Applications in Engineering Practice,* Southeast University Press, Nanjing, p264~270

Xu Yongfu, Fu Deming, 1999a. Study on the bearing capacity of unsaturated expansive soil basement, *J. of Rock Mechanics and Engineering,* 6:1~6.

Xu Yongfu, Fu Deming, 1999b. Study on the structural strength of unsaturated expansive soils, *J. of Engineering Mechanics,* 16(4): 72~77.

Unsaturated Soils for Asia, Rahardjo, Toll & Leong (eds) © 2000 Taylor & Francis, ISBN 90 5809 139 2

Measurement of soil pore pressures in the Danish Road Testing Machine (RTM)

W. Zhang
Institute for Planning, Technical University of Denmark, Denmark

R. A. Macdonald
Danish Road Institute, Road Directorate, Ministry of Transport, Roskilde, Denmark

ABSTRACT: The moisture contents and pore pressures in soils have a significant influence on the strength and structural performance of subgrade soils in road pavements. The road pavement deterioration rate is often dictated by the strength of the subgrade. Practical relationships between soil suction and structural behaviour are needed to predict pavement performance. In a recent study carried out in the Danish Road Testing Machine (RTM), the interrelationships between soil suction, layer moduli, pavement response and pavement performance have been investigated in two test pavements.

This paper briefly describes the structure of two RTM test pavements and their instrumentation. The subject of the paper is the soil pore pressure instrumentation that was installed in the test pavements for measuring pore pressures in the subgrade materials. The paper presents some of the results of the measurements made during the course of the studies.

1 INTRODUCTION

An international project sponsored by the Federal Highway Administration (FHWA) of the USA, to investigate the failure mechanisms of subgrade soils, commenced in 1994. A small, but significant component of this research project was the accelerated load testing of two test pavements undertaken in the Road Testing Machine (RTM) during 1995-97. This was carried out as a research partnership between the Danish Road Institute (DRI) and the Institute for Planning of the Technical University of Denmark (DTU/IFP). The objective of the Danish study was to evaluate different instruments, instrumentation and testing procedures, and to provide a preliminary pavement deterioration model.

Details of the materials used, the construction and instrumentation of the two test pavements have been reported [DRI, 1997a; Macdonald, 1997]. The data collected during the two test series has also been analysed and reported [Zhang, 1999 and 1998].

2 DANISH ROAD TESTING MACHINE (RTM)

The Danish RTM is a linear track, accelerated load testing facility (known as an ALT or APT) with a towed dual tyre loading cart which can apply up to 65 kN bi-directional constant loading to a test pave-ment in a temperature-controlled chamber [Zhang, 1999, 1998].

In the two studies, accelerated loading was applied to the two RTM test pavements, at a controlled surface temperature of approximately +25°C. Response instrumentation in the RTM-1 test pavement measured the stresses and strains of the unbound granular materials in three orthogonal axes. The instruments included: strain measuring sensors (Strain Deformation Transducers, SDTs, and Emu strain sensing induction coils developed and manufactured by researchers at Nottingham University, UK); Soil Pressure Cells (SPCs); Asphalt Strain Gauges (ASGs); Soil Moisture Probes and Pore Pressure Sensors (PPSs). In contrast, the RTM-2 test instrumentation focussed on measuring the vertical responses of the unbound granular materials (UGM) using SDTs and SPCs; ASGs were used to measure asphalt strains, while PPSs and thermometers recorded pore pressures and temperatures, respectively. Routine monitoring during both testing programs included Profilometer measurements and FWD testing.

The test section of the RTM has a width of 2.5 m and a length of 27 m. The central 9 m is the actual test section, which is approximately 1.8-2.0 m deep. A ramped approach to the central 9 m (from the Gate end) enables construction of the pavement using normal construction equipment. A plan and section of the RTM are shown in Figure 1. The test pavement and loading cart are enclosed within a climate controlled chamber equipped with temperature-control machinery, making it possible to maintain a temperature range of -10° C to + 40° C in the upper pavement layers. Ground water levels in the test pavement may also be raised and lowered by automatically regulating the water level in a well constructed alongside the test pit.

The wheel loading is hydro-pneumatically applied by a single or a dual wheel, using standard commercial vehicle tyres. For the two studies reported, the dual wheel load was applied by two 12R22.5 tyres, which are a common tyre type used in Denmark. The maximum bi-directional dual wheel load that can be applied is 65 kN at a maximum velocity of up to 25 km/h. During testing, however, the loading cart speed generally does not exceed 20 km/h. The dual tyres are individually controlled so that loading can be applied by each tyre separately actuated. Loading can be applied by the loading cart running on a fixed longitudinal path or with transverse movement (wander) at a constant

rate of 0.2% of the driving speed (roughly 11 mm/s), which is the normal operating function. The total width of the bituminous surfacing swept by the wheelpath under the applied dual wheel loading is about 0.90 m.

3 TEST PAVEMENT STRUCTURE, MATERIALS AND ACCELERATED LOADING TESTS

The structures of the RTM-1and RTM-2 test pavements are shown in Figure 2. The depths from the surface of the asphalt concrete to the PPSs in the three upper layers of the subgrade are also given in the Figure.

The unbound granular base is a graded crushed natural aggregate with an aggregate particle size ranging from 0-32 mm. It is non-plastic and has an optimum moisture content of 6.9 % and a maximum dry density of 2190 kg/m^3 (vibrating table, DRI).

A Danish "moraine clay", classified as a clayey silty sand (AASHTO classification A-4(0)), was used for the subgrade.

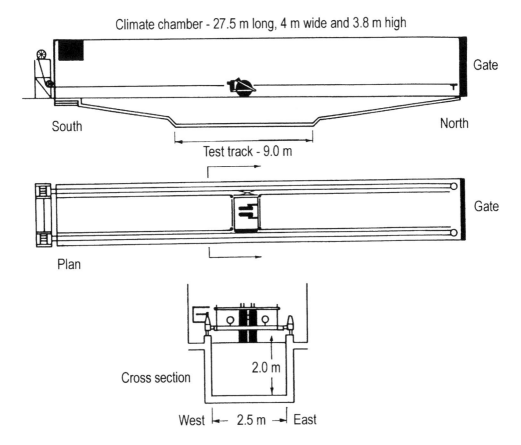

Figure 1. The Danish Road Testing Machine (RTM).

770

RTM-1	thickness
Asphalt Concrete (AC)	84 mm
Unbound Granular Base (BC)	172 mm
Clayey silty sandy gravel Subgrade (SG) Depth from AC surface to PPS installed in:	1308 mm
Subgrade layer 1	354 mm
Subgrade layer 2	460 mm
Subgrade layer 3	614 mm
Total of 9 layers of approx. thickness	145 mm
Filter Gravel (FG)	181 mm
Reinforced Concrete (CB) bottom slab	250 mm
Natural foundation soil	

Figure 2a. The structure of RTM-1.

RTM-2	thickness
Asphalt Concrete (AC)	84 mm
Unbound Granular Base (BC)	140 mm
Clayey silty sandy gravel Subgrade (SG) Depth from AC surface to PPS installed in:	1376 mm
Subgrade layer 1	300 mm
Subgrade layer 2	435 mm
Subgrade layer 3	568 mm
Total of 9 layers of approx. thickness	153 mm
Filter Gravel (FG)	181 mm
Reinforced Concrete (CB) bottom slab	250 mm
Natural foundation soil	

Figure 2b. The structure of RTM-2.

Table 1. Details of the accelerated loading applied to the RTM-1 and RTM-2 test pavements.

	RTM-1		
Dual wheel loads (kN)	20	40	
Tyre pressures (kPa)	500	600	
Load repetitions	50,000	50,000	

	RTM-2		
Dual wheel loads (kN)	40	50	60
Tyre pressures (kPa)	600	700	800
Load repetitions	50,000	50,000	50,000

This subgrade material has an optimum moisture content of 9.0 %, a maximum dry density of 2045 kg/m^3 (standard Proctor), and a liquid limit and plasticity index of 21% and 9%, respectively. The subgrade was constructed in nine layers, approximately 15 cm thick each, and is separated from the underlying 18 cm layer of porous granular aggregate by a geotextile. The porous aggregate layer enables regulation of the watertable level in the test pavements.

Details of the accelerated loading applied to the RTM-1 and RTM-2 test pavements are given in Table 1.

4 SOIL PORE PRESSURE MEASUREMENTS

Soil suctions, or negative pore water pressures, were measured by Soil Pore Pressure Sensors (PPSs). These were installed in the subgrades of both test pavements in two columns of five sensors each (one in each subgrade layer); one column was located in the negative section (from -4.50 m to 0 m) and the other in the positive section (from 0 m to + 4.50 m). Soil suctions were determined from the heights of the mercury columns in the manometers that were connected through high strength flexible tubing to porous ceramic cup tensiometers installed vertically in the five upper layers of the subgrade. The tensiometers of each PPS pair were separated by a horizontal distance of 3.0 to 3.4 m. The main response instrumentation did not extend below Subgrade Layer 3, hence the focus was on the upper subgrade layers.

The following formula has been used to convert the manometer readings in metre to kPa units of pressure:

$$\text{Suction (kPa)} = [(R_1 - R_2) \times 13600 - (R_1 + H_1 + Z_1) \times 1000] \times 9.81 / 1000 \qquad (1)$$

Where R_1 height of the mercury column (m)

R_2 reference level of mercury in the cup (m)

13600   unit weight of mercury (kg/m^3)

H_1 vertical distance from the test pavement surface to the mercury manometer (m)

Z_1 depth of each PPS (m) below the test pavement surface

1000   unit weight of water (kg/m^3)

9.81   gravitational conversion factor (m/second^2)

1000   conversion factor

In the routine procedure followed, each PPS system was flushed with de-ionised, de-aired water on day one, and then read on day two and/or day three. Soil suctions in the subgrade were monitored throughout the accelerated loading tests.

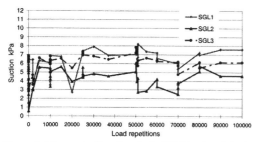

Figure 3. Measured soil suctions in the RTM-1 subgrade in the positive section 0 to +4.5 m.

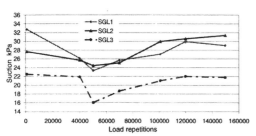

Figure 4. Measured soil suctions in the RTM-2 subgrade in the positive section 0 to +4.5 m.

5 SOIL PORE PRESSURES, MATERIALS RESPONSES AND PAVEMENT PERFORMANCE

Although the researchers attempted to construct two identical test pavements, the RTM-2 test pavement has consistently shown superior performance to the RTM-1 test pavement. The significant variation in soil suctions are a possible reason for this observed difference in performance. Figures 3 and 4 show the soil suctions measured in the subgrades of RTM-1 and RTM-2.

Figures 3 and 4 show that soil suctions in the subgrade remained almost constant throughout the accelerated loading tests, and that soil suctions in the RTM-1 subgrade were much lower than those in the RTM-2 subgrade.

During the course of accelerated load testing, which was carried out at regular, logarithmic intervals, the resilient and permanent responses of the subgrade and unbound granular materials were measured. Horizontal strains at the bottom of the asphalt concrete layer were measured with Asphalt Strain Gauges (ASGs); strains in the subgrade and unbound granular materials were measured with Soil Deformation Transducers (SDTs) and stresses were measured with Soil Pressure Cells (SPCs) [Zhang, 1999, 1998].

Pavement performance was monitored in terms of the pavement surface roughness and the depth of rutting, which were determined from longitudinal and transverse surface profiles with a dedicated Pro-

filometer. International Roughness Index (IRI) values were derived from the variation in the longitudinal profiles, and rut depths were computed from the variation in the transverse profiles. These procedures are described in the literature [DRI, 1997; Zhang, 1998].

Selected details of soil suctions measured in the subgrade, pavement responses and pavement performance of the RTM-1 and RTM-2 test pavements are given in Table 3.

Although RTM-2 was subjected to heavier loadings, the vertical strains measured in the subgrade layers are lower than those in the subgrade of RTM-1. The RTM-2 subgrade stresses are slightly higher, due to the higher levels of applied loads.

Soil suctions are inversely related to moisture contents, reaching high values at low moisture contents or in frozen conditions that may occur in roads through mountainous terrain at high altitudes and in other areas of the world subject to seasonal frost. It is believed that there is a strong relationship between soil suctions and the resilient moduli of unbound granular materials in road pavements. This is the subject of current research in the RTM.

Falling Weight Deflectometer (FWD) testing was used to measure surface deflections at specific (in logarithmic steps) stages during the accelerated loading period. Backcalculation is a complex procedure, and the outcome of the analysis often depends on how the pavement structure is treated, which analysis method is used and how the dynamic effects of FWD loading are considered. Because of the relative complexity of backcalculation methods, and since the effects of dynamic loading on the suction levels were not measured in this study, further research is recommended to investigate the relationships between soil suction and the resilient moduli of unbound granular materials.

6 CONCLUSIONS

The study found that the relatively simple, mercury manometer/tensiometer-type Soil Pore Pressure Sensors developed and installed in the Danish RTM test pavements are reliable tools for measuring negative soil pore pressures (suction) in subgrade materials.

Examples of the soil suctions measured in the subgrade, the responses measured in the pavement materials and the performance of the RTM-1 and RTM-2 test pavements are presented in graphs and tables. There is a clear relationship between soil suction and road pavement deterioration rate; more extensive research is, however, necessary to investigate and define this relationship.

Table 3. Soil suctions measured in the subgrade, compared against pavement responses and pavement performance of the RTM-1 and RTM-2 test pavements.

			Test pavement	
			RTM-1	RTM-2
Negative soil pore pressures	SGL 1		6	30
or suction (kPa)	SGL 2		6	27
	SGL 3		6	22
Total applied load repetitions:			100,000	150,000
Dual tyre load (kN):			40	60
Tyre pressure (kPa):			600	800
Tensile horizontal strains at	Longitudinal strain, ε_y ($\mu\varepsilon$)		-600	-600
underside of AC layer	Transverse strain, ε_x ($\mu\varepsilon$)		-400	-400
Compressive vertical strains	SGL 1	($\mu\varepsilon$)	4000	2500
in subgrade layers (ε_z)	SGL 2	($\mu\varepsilon$)	2300	1500
	SGL 3	($\mu\varepsilon$)	1200	2000
Compressive vertical	SGL 1	(kPa)	80	140
stresses in subgrade layers	SGL 2	(kPa)	60	100
(σ_z)	SGL 3	(kPa)	40	60
Pavement Performance	IRI	(m/km)	3.2	1.5
Indicators	Rut depth	(mm)	10	10

Key:
SGL 1, 2 , 3   Subgrade Layers 1, 2 and 3
IRI   International Roughness Index

7 ACKNOWLEDGEMENTS

In addition to both authors of this paper, the following persons participated in the construction and instrumentation works for the test pavements, assisted with managing the RTM, and also provided assistance, ideas and technical advice during the study, which was completed in 1997.
Dr. Per Ullidtz, Technical University of Denmark
Mr. H.J. Ertman Larsen, Danish Road Institute
Ms. Susanne Baltzer, Danish Road Institute
Mrs. Anna Marie Ørnstrup, Danish Road Institute
Mrs. Lise Bjulf, Danish Road Institute
Mr. Finn Sjølin, Technical University of Denmark
Mr. Finn Hansen, Danish Road Institute
Mr. Jan Sjølin, Technical University of Denmark.

8 REFERENCES

Danish Road Institute, 1997a. TRB Annual Meeting 97. DRI Paper Presentation at Session 13, Pavement Instrumentation, Part 1. DRI Report 84, Danish Road Institute, Roskilde.

Danish Road Institute, 1997b. Pavement Subgrade Performance Study, Preliminary Test Pavement Danish Road Testing Machine, Data Analysis Report, Danish Road Institute, December 1996, revised August 1997, Danish Road Institute, Roskilde.

Macdonald, RA & Baltzer, S, 1997. Sub-grade Performance Study, Part I: Materials, Construction and Instrumentation, DRI Report 85, Danish Road Institute, Roskilde.

Ullidtz, P, 1998. *Modelling Flexible Pavement Response and Performance*. ISBN-87-502-0805-5, Polyteknisk Forlag, Copenhagen.

Ullidtz, P, 1987. *Pavement Analysis, Developments in Civil Engineering, Vol. 19*. ISBN 0-444-42817-8, Elsevier, Amsterdam.

Zhang, W & Macdonald, RA, 1999. Pavement Subgrade Performance Study in the Danish Road Testing Machine, Paper No. GS2-2, *Proceedings of the 1999 International Conference on Accelerated Pavement Testing, October 18-20, 1999,* Reno.

Zhang, W, Ullidtz, P & Macdonald, RA, 1998. Pavement Subgrade Performance Study, Part II: Modelling Pavement Response and Predicting Pavement Performance, DRI Report 87, Danish Road Institute, Roskilde.

10   Engineering applications

10.2 Slopes

Unsaturated Soils for Asia, Rahardjo, Toll & Leong (eds) © 2000 Taylor & Francis, ISBN 90 5809 139 2

Field measurements of pore-water pressure profiles in residual soil slopes of the Bukit Timah Granite Formation, Singapore

M. S. Deutscher, J. M. Gasmo, H. Rahardjo & E. C. Leong
NTU-PWD Geotechnical Research Centre, Nanyang Technological University, Singapore

S. K. Tang
PWD Consultants Pte Limited, Singapore

ABSTRACT: This paper presents the results of field measurements of pore-water pressure profiles in the upper three metres of slopes consisting of residual soils derived from the Bukit Timah Granite Formation in Singapore. Two sites in north central Singapore were instrumented with a rain gauge, tensiometers and piezometers. The rain gauge provided site specific rainfall data (i.e., rainfall volume/duration/intensity), while the tensiometers and piezometers provided an indication of the corresponding pore-water pressure response of the soil.

The study showed that significant negative pore-water pressures (i.e., matric suctions) developed in the upper three metres of the slopes during prolonged dry periods. However, these suctions were readily dissipated during even a moderate wet period. Positive pore-water pressures were developed at all monitored depths; the pore-water pressure profiles measured at the sites approached that of a hydrostatic pore-water pressure distribution with the groundwater table at the surface (i.e., a fully saturated slope).

1 INTRODUCTION

Approximately one-third of Singapore's land area is covered with residual soils derived from the Bukit Timah Granite Formation. Slope failures frequently occur in the upper two to three metres of these residual soil deposits as a result of tropical rainfall events. These rainfall-induced slope failures can be a nuisance to property and infrastructure maintenance and are quite costly to repair, if not dangerous.

Design of slopes against the possibility of rainfall-induced slope failures is made difficult by a lack of understanding of the pore-water pressure distributions that can occur in the soil. An estimate of the worst-case pore-water pressure distribution that will occur during the design life of the slope is required to engineer an appropriate design. The worst-case pore-water pressure distribution is defined as the pore-water pressure distribution that results in the lowest factor of safety against a rainfall-induced slope failure. Generally speaking, higher pore-water pressures mean lower factors of safety. The worst-case pore-water pressure distribution that can occur is a function of both the soil (i.e., permeability, stratigraphy, secondary structure) and on the nature of the rainfall (i.e., duration, intensity, volume, antecedent rainfall).

Two slopes consisting of residual soils from the Bukit Timah Granite Formation were selected for the study. Site investigations were performed to determine the geometry (i.e., slope angle and height) and stratigraphy of the slopes. Field tests were performed to obtain an estimate of the permeability of the surficial soils. Laboratory tests were conducted to classify the soil comprising the slopes and determine its shear strength parameters. Each site was instrumented with a rain gauge, tensiometers and piezometers. The rain gauge provided site specific rainfall data (i.e., rainfall volume/duration/intensity), while the tensiometers and piezometers provided an indication of the corresponding pore-water pressure response of the soil. The instrumented slopes were part of a long term monitoring program to study the response of pore-water pressure changes in a slope due to climatic conditions. The results reported in this paper are for a period of two months, portions of which experienced above average amounts of rainfall.

2 CHARACTERISATION OF SELECTED SITES

The two sites selected for the study were located in the north central Singapore: one in Yishun and the other in Mandai. The two sites are hereafter referred to as the "Yishun" and "Mandai" sites.

2.1 Yishun Site

The Yishun site has a slope angle of approximately 23.5 degrees and a slope height of approximately 7 m. The slope face is well turfed and free of trees. The slope is a natural cut (i.e., as opposed to a man-placed fill).

The residual soil at the Yishun site extends to depths of about 20 to 25 m, at which point the intact parent rock of the Bukit Timah Granite Formation is encountered. The soil comprising the slope at the Yishun site is classified as silt with moderate to low plasticity. Properties of this soil are given in Table 1.

Table 1. Summary of soil properties for the soil comprising the slope at the Yishun site.

USCS classification	Test (units)	No. of tests	Average value	Max. value	Min. value
ML	w (%)	9	35	40	31
	LL	9	45	53	42
	PL	9	31	39	13
	PI	9	14	31	7
	fines (%)	9	58	69	47
	ρ_{total} (Mg/m^3)	7	1.88	1.96	1.73
	c' (kPa)	1	12	12	12
	ϕ'	1	33	33	33
	G_s	3	2.688	2.700	2.676
	C_c	1	0.30	0.30	0.30

Note: ϕ' and c' were determined through Consolidated-Undrained (CU) triaxial testing.

An open double-ring infiltrometer test was conducted to obtain an estimate of the surface permeability at the site. The permeability at the crest was measured to be 2.0×10^{-5} m/s while the permeability at the toe was measured to be 6.7×10^{-6} m/s.

2.2 Mandai Site

The Mandai site has a slope angle of approximately 31 degrees and a slope height of approximately 11 m. There are several trees located at the crest of the slope, as well as trees located a few meters from the toe. The slope face itself is well turfed and free of trees.

The residual soil at the Mandai site extends to a depth of 28 to 35 m beneath the surface, at which point the intact parent rock of the Bukit Timah

Granite Formation is encountered. The soil comprising the slope at the Mandai site is a man-placed fill consisting of residual soils from the Bukit Timah Granite Formation (classification and mineralogical tests were performed to confirm the origin of the fill). The soil is classified as silty sand. The fines have a moderate to low plasticity. Properties of this soil are given in Table 2

Table 2. Summary of soil properties for the soil comprising the slope at the Mandai site.

USCS classification	Test (units)	No. of tests	Average value	Max. value	Min. value
SM	w (%)	8	25	31	20
	LL	9	55	69	33
	PL	9	31	40	24
	PI	9	24	44	6
	fines (%)	8	41	63	28
	ρ_{total} (Mg/m^3)	2	2.02	2.09	1.95
	c' (kPa)	1	12	12	12
	ϕ'	1	30	30	30
	G_s	3	2.684	2.692	2.679
	C_c	2	0.19	0.19	0.18

Note: ϕ' and c' were determined through Consolidated-Undrained (CU) triaxial testing.

An open double-ring infiltrometer test was conducted to obtain an estimate of the surface permeability at the Mandai site. The permeability at the crest was measured to be 1.2×10^{-4} m/s while the permeability at the toe was measured to be 1.3×10^{-4} m/s.

3 INSTRUMENTATION

The instrumentation at both the Yishun and Mandai sites consisted of tensiometers, piezometers and a rainfall gauge.

Tensiometers were used for this study, as they are capable of measuring pore-water pressures in both the positive and negative range. The tensiometers were installed in six rows of five for a total of thirty tensiometers at each site. The tensiometer rows extended from crest to toe with more rows concentrated near the crest. The tensiometers in each row were installed to various depths to comprise a tensiometer "nest". The depths of the tensiometer tips (i.e., sensors) for each row at the Yishun and Mandai sites were 0.5, 1.1, 1.7, 2.3 and 2.9 m. Each tensiometer was fitted with a pressure transducer to facilitate automated data collection.

Cassagrande-type piezometers were located at the crest, mid-slope and toe of each slope to determine the location of the groundwater table. The piezometers were fitted with depth transmitters to facilitate automated data collection.

Each site was equipped with a tipping-bucket rain gauge to obtain site specific rainfall data. This was necessary as the rainfall that occurs at a given time in Singapore varies significantly across the island. The time of each tip was recorded by a data acquisition system (DAS) so that the intensity, duration and volume of rainfall events could be determined.

The data acquisition system chosen for the study was compatible with all the instrumentation at the site. This simplified the synchronization of the data collection process. In addition, the DAS was programmed to increase the frequency of monitoring in the event of a rainfall; readings were taken every 10 minutes during and immediately following rainfall events as opposed to every 4 hours during normal operation. The DAS used the rain gauge as a "trigger" to switch into the higher monitoring frequency mode.

4 CHARACTERISTICS OF THE STUDY PERIOD

Data were collected from the two sites for a two-month study period from 12-Oct-98 to 29-Dec-98. The study period consisted of a distinctive dry period, during which a below average amount of rainfall occurred, surrounded by two distinctive wet periods, during which an above average amount of rainfall occurred. A summary of the wet and dry periods for the Yishun and Mandai sites is given in Tables 3 and 4, respectively. Details of every significant rainfall event that occurred during the analysis period were recorded (i.e., rainfall volume, duration, and average intensity). However, these are not presented in this paper for lack of space.

Table 3. Summary of wet and dry periods at the Yishun site for the study period.

Period start	Period finish	Period description	Rainfall (mm)	10-year average rainfall (mm)[†]
12-Oct-98	10-Nov-98	Wet period 1	205.25[*]	215.96
10-Nov-98	4-Dec-98	Dry period	3	221.18
4-Dec-98	28-Dec-98	Wet period 2	269	257.65
		Total:	477.25	694.79

[*] Rainfall gauge experienced clogging problems prior to 06-Nov-98. Rainfall amounts for this period may be underestimated.

[†] 10-year average rainfall for the same period based on data collected from the Sembawang Met. Station (i.e., Station 80) by the Meteorological Service Singapore for 1989-98.

Table 4. Summary of wet and dry periods at the Mandai site for the study period.

Period start	Period finish	Period description	Rainfall (mm)	10-year average rainfall (mm)[*]
12-Oct-98	07-Nov-98	Wet period 1	307.25	187.42
07-Nov-98	04-Dec-98	Dry period	23.5	249.72
04-Dec-98	29-Dec-98	Wet period 2	323	267.96
		Total:	653.75	705.10

[*] 10-year average rainfall for the same period based on data collected from the Sembawang Met. Station (i.e., Station 80) by Meteorological Service Singapore for 1989-98.

5 FIELD MEASUREMENTS OF PORE-WATER PRESSURE PROFILES

Field measurements of pore-water pressure profiles that were recorded at the two sites during key times of the study period are presented here. Key times in the analysis period were considered to be:

- At the end of a prolonged dry period when pore-water pressures were at a minimum (i.e., slope stability at a maximum).
- Following one or a series of significant rainfall events when pore-water pressures were at a maximum (i.e., slope stability at a minimum).
- During a significant rainfall event when the pore-water pressure distribution was in transition.

5.1 Yishun site

One-dimensional plots of pore-water pressure profiles recorded at the crest of the Yishun site at key times in the study period are given in Figure 1. The piezometric levels recorded by piezometer at the crest of the Yishun site are included in Figure 1 for reference. Also included in the figure is a hypothetical hydrostatic pressure line for the case of a groundwater table located at the soil surface (i.e., a fully saturated slope).

5.2 Mandai site

One-dimensional plots of pore-water pressure profiles recorded at the crest of the Mandai site at key times of the study period are given in Figure 2. The pore-water pressure profiles recorded at the crest are typical for the Mandai site (i.e., similar to the toe). The piezometric levels recorded by the piezometer at the crest of the Mandai site are included in Figure 2 for reference. Also included in the figure is a hypothetical hydrostatic pressure line for the case of a groundwater table located at the soil surface (i.e., a fully saturated slope).

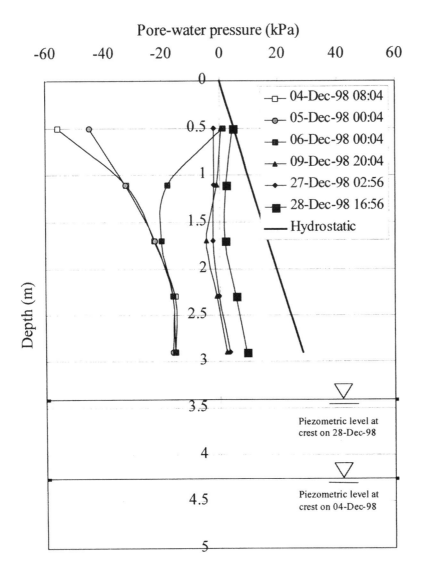

Figure 1. One-dimensional plot of pore-water pressure versus depth for a typical row at the crest of the Yishun site during key times of the study period. Piezometric levels and the hypothetical hydrostatic pressure line for the case of a groundwater table located at the surface (i.e., fully saturated slope) are included for reference.

6 DISCUSSION AND ANALYSIS

The following observations can be made with respect to the pore-water pressure profiles of the research sites:

- Significant negative pore-water pressures developed during prolonged dry periods (i.e., several days to several weeks).
- The negative pore-water pressures were dissipated in the upper 3 m with the occurrence of even a moderate rainfall event. Perched water tables readily developed.

Worst-case pore-water pressure conditions

The worst-case pore-water pressure conditions that can occur in a residual soil slope are of particular interest to the study of rainfall-induced slope failures. The worst pore-water pressure conditions were judged to be the pore-water pressure distribution with the highest magnitude of positive pore-water pressures for the purposes of this analysis. However, it is acknowledged that in non-homogeneous slopes, the slope stability may be dependent on the pore-water pressure along a weak layer or structural

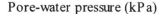

Pore-water pressure (kPa)

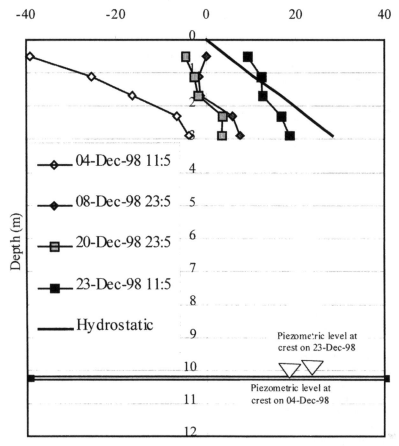

Figure 2. One-dimensional plot of pore-water pressure versus depth for a typical row at the crest of the Mandai site during key times of the study period. Piezometric levels and the hypothetical hydrostatic pressure line for the case of a groundwater table located at the surface (i.e., fully saturated slope) are included for reference.

discontinuity rather than the pore-water pressures in the slope as a whole.

The worst-case pore-water pressure profile that was measured in the upper 3 m at the Yishun site consisted of positive pressures at all depths monitored (i.e., 0.5, 1.1, 1.7, 2.2 and 2.9 m). The positive pore-water pressures did not exceed the hypothetical case of a hydrostatic pore-water pressure distribution with the ground water table at the surface (i.e., fully saturated slope).

The worst-case pore-water pressure profile that was measured in the upper 3 m at the Mandai site consisted of positive pressures at all depths monitored. The positive pore-water pressures that were measured at 0.5m and 1.1m slightly exceeded those of the hypothetical case of a hydrostatic pore-water pressure distribution with the groundwater table located at the surface (i.e., a fully saturated slope). The

positive pressures that were measured at depths of 1.7, 2.2 and 2.9 m fell below, but were near to, the hydrostatic pore-water pressure condition.

The pore-water pressures measured at the Mandai site were slightly higher than those measured at the Yishun site. This may be attributed either to the higher permeability of the soils at the Mandai site, or the fact that the Mandai site received slightly more rainfall than the Yishun site.

Further monitoring of the slopes will be required to capture rainfall events with a greater return period. Although the study period covered two distinctive wet periods, the climate of Singapore can produce significantly more precipitation than what was recorded during the study period. Pore-water pressures greater than those measured during the study period could likely occur. Monitoring at the research sites has been on-going since the study period in question.

Pore-water pressure measurements for this study were limited to the upper three metres of the soil profile. The pore-water pressure profile between three metres depth and the depth of the groundwater table is unknown. This information would be required in order to evaluate the factor of safety of slope failures with deeper slip surfaces (i.e., below three metres depth). However, it has been observed that most rainfall-induced slope failures in Singapore are limited to the upper two to three meters of the soil (Tan et al. 1987). Therefore, the information provided by this study is relevant to the design of slopes against rainfall-induced slope failures.

7 CONCLUSIONS

Pore-water pressure profiles were measured for the upper three metres of two slopes consisting of residual soils from the Bukit Timah Granite Formation. Although significant negative pore-water pressures (i.e., matric suctions) could develop during dry periods, these matric suctions could be readily dissipated by even a moderate wet period, and perched water tables developed in the soil.

The engineered design of residual soil slopes in Singapore should be based on the worst-case pore-water pressure distribution that can develop in the upper two to three metres of the soil. For a homogeneous slope (i.e., no structural discontinuities or obvious failure planes) consisting of residual soils from the Bukit Timah Granite Formation, the worst-case pore-water pressure profile consists of positive pore-water pressures at all depths above three metres. The pore-water pressure profile can at the very least approach that of a hydrostatic pore-water pressure distribution with the groundwater table at the surface (i.e., a fully saturated slope). Pore-water pressure profiles could likely become even more positive for rainfall events with a higher return period. Further monitoring of the two sites is required to assess this potential.

ACKNOWLEDGEMENTS

This research was funded by the National Science and Technology Board (NSTB), Research Project No. 17/6/16 entitled "Rainfall-Induced Slope Failures".

The authors would also like to extend their thanks to the Meteorological Service Singapore for providing the 10-year average rainfall data.

REFERENCES

Tan, S.B., Tan, S.L., Lim, T.L. & Yang, K.S. 1987. Landslide problems and their control in Singapore. Proc. 9th Southeast Asian Geotechnical Conf., Bangkok: 1:25-1:36.

Unsaturated Soils for Asia, Rahardjo, Toll & Leong (eds) © 2000 Taylor & Francis, ISBN 90 5809 139 2

Preliminary assessment of slope stability with respect to rainfall-induced slope failures

J. M. Gasmo
Clifton Associates Limited, Regina, Sask., Canada (Formerly: NTU-PWD Geotechnical Research Centre, Nanyang Technological University, Singapore)

H. Rahardjo, M. S. Deutscher & E. C. Leong
NTU-PWD Geotechnical Research Centre, Nanyang Technological University, Singapore

ABSTRACT: Rainfall-induced slope failures are difficult to analyse because they involve a complex interaction between saturated and unsaturated soil mechanics. To develop a better understanding of these failures, a parametric study was performed. Three slope heights and four slope angles were entered into numerical models of seepage and slope stability. The change in pore-water pressures throughout the slope during rainfall was modelled and used to calculate the change in the factor of safety over time. The results from the parametric study showed that similar pore-water pressure distributions developed in each of the slopes after the rainfall. The slope angle appeared to have the most significant effect on the stability of the slope. By compiling the results from the parametric study, useful relations for the preliminary assessment of slope stability were developed in the form of charts.

1 INTRODUCTION

Rainfall-induced slope failures pose a significant challenge when assessing the stability of a slope. Such failures are difficult to analyse because they involve a complex interaction between saturated and unsaturated soil mechanics. To develop a better understanding of rainfall-induced slope failures, a parametric study was performed using numerical models. The study focused on the infiltration characteristics of residual soil slopes to determine the change in the factor of safety for a slope during rainfall. The objectives were to identify parameters that had a significant effect on the stability of the slope and develop relations for the preliminary assessment of slope stability.

The parameters that are relevant for rainfall-induced slope failures can be summarised into three groups as shown in Table 1. The slope parameters include slope height, slope angle, and soil layer profile. The relevant soil parameters are the soil-water characteristic curve (SWCC), saturated permeability, permeability function (i.e. the variation of permeability with matric suction), and unsaturated shear strength parameters. The climatic and hydrologic parameters incorporate the ground-water level location, rainfall intensity, initial pore-water pressure distribution, elapsed time, and transient pore-water pressure distribution.

Table 1. Parameters relevant to rainfall-induced slope failures.

Slope Parameters	Soil Parameters	Climatic and Hydrologic Parameters
Slope Height	SWCC	Groundwater level location
Slope Angle	Saturated permeability	Rainfall intensity
Soil Layer Profile	Permeability function	Initial pore-water pressure distribution
	Unsaturated shear strength parameters	Elapsed time
		Transient pore-water pressure distribution

Of the parameters listed in Table 1, the transient pore-water pressure distribution (i.e., the changing pore-water pressure distribution) is especially important in the study of rainfall-induced slope failures. Changes in the pore-water pressure distribution cause changes in the stability of a slope during infiltration. The transient pore-water pressure distribution is dependent on many of the factors listed in Table 1. These included the SWCC, saturated permeability, and permeability function of the soil as well as the rainfall intensity, initial pore-water pressure distribution, and elapsed time.

2 PARAMETRIC STUDY

2.1 Description

For this parametric study, the slope height, slope angle and transient pore-water pressure distribution were varied. All other parameters remained constant. A finite element seepage model, SEEP/W (GEO-SLOPE, 1998a), was used to simulate rainwater infiltration into a slope and generate transient pore-water pressure distributions for each of the slope heights and slope angles. This provided information about the infiltration characteristics of each slope angle and slope height. A limit equilibrium slope stability model, SLOPE/W (GEO-SLOPE, 1998b), was used to determine the critical slip surface and the factor of safety for each of the slope angles and slope heights. The stability of the slope was determined at various times during the rainfall by using the respective transient pore-water pressure distributions. This provided information about the effect of infiltration on slope stability.

2.2 Parameter Values

Slope Geometry

Slope angles of 18°, 27°, 45° and 63° (3H:1V, 2H:1V, 1H:1V, and 0.5H:1V) were analysed in combination with slope heights of 10m, 20m and 40m. The soil layer profile selected for the parametric study consisted of a single homogeneous soil layer. The generalised geometry of the slope used in the parametric study is shown in Figure 1.

The slope height was defined as H, the slope length was defined as a variable, L, and the slope angle was defined as α. The variable L was set equal to 3H, 2H, 1H, and 0.5H to create the four different slope angles. The dimensions of the overall cross-section were also defined relative to the slope height. For the purposes of modelling, the distance to the left edge of the cross-section was 3H from the crest and the distance to the right edge of the cross-section was 3H from the toe. The cross-section was 3H in height on the crest side and 2H in height on the toe side.

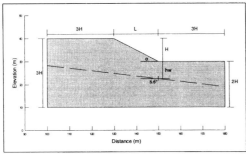

Figure 1. Generalised geometry of parametric study slope.

Location of Groundwater Level

The location of the groundwater level was defined with respect to the toe of slope. The depth to the groundwater level at the toe of the slope, h_w was set to a value of 0.75H. The groundwater table was arbitrarily defined by a straight line running at an angle of 6.6° (27°/4) with the horizontal, sloping downwards from the crest to the toe of the slope.

Soil Properties

The parameters for the single soil layer were based on laboratory tests of residual soil from the sedimentary Jurong Formation. The soil parameters were typical of a silty clay. The SWCC was determined from pressure plate tests. The saturated permeability specified for this soil was 1.0×10^{-4} m/s which was an approximate saturated permeability based on infiltrometer tests and was adjusted to account for cracks and root holes which increase the permeability at the ground surface. The permeability function for this soil was derived from the SWCC and the saturated permeability by using the estimate function provided in the numerical seepage model, SEEP/W.

Seepage Model Boundary Conditions

The bottom boundary was specified as a no flow boundary. The side boundaries were defined as head boundaries equal to the specific location of the ground water level as defined by the slope geometry. The top boundary was specified as a flux boundary which was used to apply the specified rainfall intensity to the slope.

Initial Pore-Water Pressure Conditions

The initial pore-water pressure distribution for each slope configuration was defined as a hydrostatic pore-water pressure profile with a limiting negative pore-water pressure of 75 kPa. This limit was selected based on field measurements of negative pore-water pressures. A limit was set to prevent the generation of unrealistic pore-water pressures and to provide a similar initial condition for each of the slope angles and slope heights.

Elapsed Time and Climatic Conditions

The elapsed time for the transient analysis of each slope configuration was 10.2 days. A rainfall intensity of 80 mm/hr (2.2×10^{-5} m/s) was applied from time equal to 0 hours until time equal to 4 hours using four time steps. A rainfall intensity of approximately zero (1.0×10^{-12} m/s) was then applied over a further period of 10 days using an additional five time steps which allowed the water that infiltrated the slope to redistribute.

Shear Strength

Shear strength parameters were defined using typical values measured for residual soils in Singapore. The effective cohesion, c′, equalled 10 kPa, the effective angle of internal friction, $\phi′$, equalled 26°, and the rate of increase in shear strength caused by matric suction, ϕ^b, equalled 26°. The unit weight of the soil, γ, equalled 20 kN/m² and was kept constant above and below the water table.

Stability Analysis

The factor of safety for each slope angle and slope height was calculated using the limit equilibrium slope stability model (SLOPE/W), the transient pore-water pressure distributions for the each of the time steps, and the shear strength parameters. A deep and a shallow depth range was used to find the critical slip surface before, during and after the rainfall. A minimum slip circle depth of 1.0 m was specified for both ranges. The factors of safety from each depth range were compared for each of the time steps and the lowest factor of safety was taken as the critical factor of safety.

3 PARAMETRIC STUDY RESULTS

3.1 *Infiltration*

The infiltration characteristics were found to be similar for each of the slope angles and slope heights. Initially the pore-water pressures at the ground surface were negative. Pore-water pressure contours from the 10 m, 27° slope in Figure 2 show that after 4.0 hours of rainfall, an area of positive pore-water pressures developed at the ground surface while an area of negative pore-water pressures still existed between the advancing wetting front and the groundwater table.

Pore-water pressure profiles in Figure 3 show that as rainwater infiltrated into the slope, the pore-water pressures at the ground surface increased first. As the rainfall continued, the pore-water pressures at deeper depths also increased over time. During the 4.0 hour rainfall, pore-water pressures increased significantly up to a depth of approximately 3.0 m.

After the rainfall stopped, the infiltration continued to redistribute itself through the slope and caused significant changes in pore-water pressures up to a depth of 6.0 m as shown in Figure 4.

3.2 *Slope Stability*

The slope stability results were found to be different depending on the specific slope height, slope angle and transient pore-water pressure distribution. The slopes standing at 18° and 27° did not have factors of safety less than one while the 45° and 63° slopes did. A plot of the factor of safety over time for the 45° slopes are shown in Figure 5. The three lines plotted on each graph represent the factors of safety for the three different slope heights.

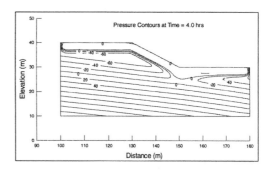

Figure 2. Pore-water pressure contours (kPa) at time equal to 4.0 hours.

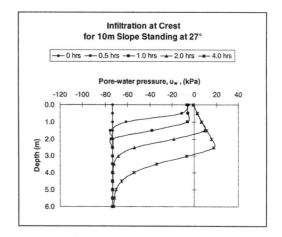

Figure 3. Pore-water pressure profiles at crest of 10 m slope standing at 27°.

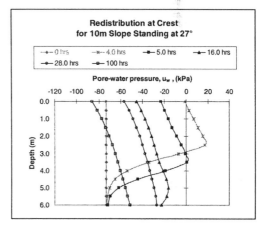

Figure 4. Redistribution of pore-water pressures for 10 m slope standing at 27°.

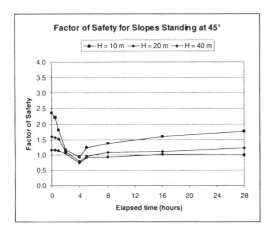

Figure 5. Factor of safety during rainfall and redistribution for 45° slopes.

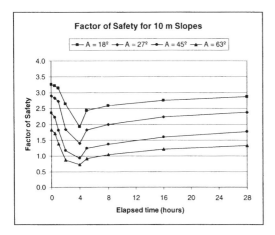

Figure 6. Factor of safety during rainfall and redistribution for 10 m slopes.

For each combination of slope height and slope angle, the stability of the slope decreased over time during infiltration and increased over time during redistribution. For the 45° slopes, all slope heights were found to be unstable after 4.0 hours of rainfall with a factor of safety less than 1.0. The critical slip surface was located at a depth of approximately 3.0 m. For the 63° slope, the 10 and 20 m slope heights were found to be unstable after 2.0 hours of rainfall with a factor of safety less than 1.0 and a critical slip located at approximately 3.0 m. The 40 m slope height at 63° was found to be completely unstable for all of the transient pore-water pressure distributions.

Some trends are apparent from these slope stability results. For the transient pore-water pressure distributions at time equal to 0 hours, the factor of safety was found to be higher for slopes with a lower height and lower angle. As well, the critical slip surface was found to be located at a greater depth as the slope height increased. These trends appear to be valid for each slope height and slope angle when positive pore-water pressures were not present. When positive pore-water pressures were present (between 1.0 and 4.0 hours), these trends changed.

After the positive pore-water pressure zone developed, it could be seen that slopes at the same slope angle had approximately the same factor of safety and the same depth of critical slip surface, regardless of the slope height (Figure 5). It appears that the stability of the slope is not significantly affected by the slope height. Instead, the transient pore-water pressure distribution appears to be a controlling factor.

A similar positive pore-water pressure zone developed for each slope angle and slope height. A maximum pore-water pressure of approximately 20 to 30 kPa was located at a depth of approximately 2.0 to 3.0 m.

By comparing slopes of the same height and different angles, as shown in Figure 6, it can be seen that the slope angle does have a significant effect on the factor of safety. Slopes standing at higher angles have a lower factor of safety for each transient pore-water pressure distribution. This is reasonable because the stability of a slope is directly related to the angle of the slope based on the shear strength of the soil.

3.3 Conclusions

The parametric study results show that for rainfall infiltrated slopes, the slope height and slope angle had little effect on the infiltration characteristics of the slope. Each of the slope heights and slope angles experienced the development of a wet zone near the ground surface as a result of rainfall. Negative pore-water pressures at the ground surface were lost and the pore-water pressures built up to form a perched water table. This is in agreement with the results that have been measured in the field and shows that the parameters input into the parametric study closely represented the conditions that occurred in the field.

The results of the parametric study also showed that the transient pore-water pressure distribution appears to be the main triggering mechanism behind rainfall-induced slope failures. Infiltration of rainfall into a slope causes positive pore-water pressures to develop after a period of time. As this positive pore-water pressure zone forms, the critical slip surface of the slope becomes a shallow slip surface through the positive pore-water pressure zone. Slopes with steeper slope angles are more susceptible to this type of failure in a shorter time.

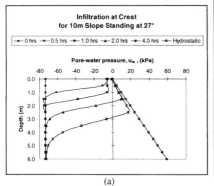

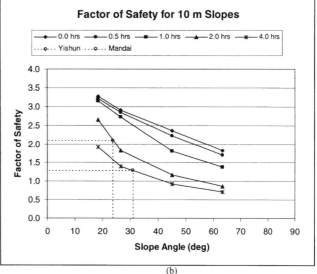

Figure 7. Preliminary Assessment of field sites using 10 m Preliminary Assessment Chart. Transient pore-water pressure distributions (a) used to calculate respective factors of safety (b).

4 PRELIMINARY ASSESSMENT CHARTS

The results from the parametric study can be used to perform a preliminary assessment of slope stability for field slopes that have similar parameter values as the ones used in the parametric study.

By combining the slope stability results with respect to slope height, summary charts can be generated as shown in Figure 7. This chart shows the pore-water pressure profiles and respective factors of safety for a 10 m slope. Similar charts were also developed for the 20 and 40 m slopes but are not shown here.

Because similar pore-water pressure profiles developed for each of the various slope angles, only the profiles from the 27° slopes were used in these charts. An additional line has been plotted on the pore-water pressure profiles to show the hydrostatic pore-water pressure profile. The factor of safety results from all four slope angles form a factor of safety line for one specific transient pore-water pressure distribution. The factor of safety line for the transient pore-water pressure distribution at 0 hours in Figure 7(a) is the uppermost line in Figure 7(b).

The preliminary assessment of a field slope can be carried out using the following four steps. Preliminary assessments of two research sites, Yishun and Mandai, are shown here to illustrate the use of the chart.

Step 1: Check field slope parameters.
Check to ensure that the field slope has similar parameters as outlined for the parametric study slope

(Section 2). The values used in the parametric study slope represent typical values for residual soils in Singapore.

Step 2: Select appropriate set of charts.
If the charts are deemed to be suitable for the field slope, select the appropriate set of preliminary assessment charts. For the assessment of the research sites, the 10 m assessment chart was selected.

Step 3: Determine the worst case pore-water pressure distribution.
The worst case pore-water pressure conditions for the Yishun and Mandai sites were determined from field measurements. The worst case field pore-water pressure profile for the Yishun site was a slight positive pore-water pressure build up. It is not as large a build up as the 4.0 hour profile in the 10 m assessment chart, but it is more than the 1.0 hour profile. It most closely matches the 2.0 hour assessment profile. The worst case pore-water pressure profile for the Mandai site most closely matches the 4.0 hour profile in the assessment charts.

Step 4: Estimate the factor of safety.
To estimate the factor of safety, take the angle of each slope and move vertically till you intersect the appropriate factor of safety line for the worst case pore-water pressures. Then move horizontally to estimate the factor of safety as shown in Figure 7. The factor of safety for the Yishun site is estimated at approximately 2.1. The factor of safety at the Mandai site is estimated to be approximately 1.3.

This parametric study was performed as a portion of an extensive research project on rainfall-induced slope failures (NSTB 17/6/16). Through the research project, a detailed slope stability analysis was performed to determined the factor of safety for each site. The factor of safety calculated for the worst case pore-water pressure profile at the Yishun site was 2.0. The preliminary assessment of 2.1 agrees satisfactorily with the detailed stability assessment. The factor of safety calculated for the worst case pore-water pressure profile at the Mandai site was 1.4, which also agrees well with the preliminary assessment chart result of 1.3.

5 CONCLUSIONS

These results show that the preliminary assessment charts can provide a satisfactory estimate of the stability of a slope with respect to rainfall-induced slope failures. If the results of the preliminary assessment indicate that the slope is at risk of failure (approx. equal to 1.0), then it would be advisable to perform a detailed stability assessment.

These charts are intended to be used only for the preliminary determination of the stability of a slope and are meant to provide the user with an estimate of slope stability. They should not be used independently to produce a final slope design. The charts are only applicable for field slopes with similar parameters to those defined in Section 2.

Additional preliminary assessment charts can be generated for slopes of different soil types by expanding the parametric study described in Section 2. By inputting different types of slope, soil, and climatic and hydrologic parameters, other slope conditions can be studied and summarised.

ACKNOWLEDGEMENTS

The authors would like to thank Dr. D.G. Toll for his contributions. Funding for this research was provided by the research grant, NSTB 17/6/16: Rainfall-Induced Slope Failures.

REFERENCES

GEO-SLOPE, 1998a. SEEP/W v4.0. GEO-SLOPE International Ltd., Calgary, Alberta, Canada.

GEO-SLOPE, 1998b. SLOPE/W v4.0. GEO-SLOPE International Ltd., Calgary, Alberta, Canada.

Unsaturated Soils for Asia, Rahardjo, Toll & Leong (eds) © 2000 Taylor & Francis, ISBN 90 5809 139 2

Three-dimensional back analysis of unsaturated soil strength parameters

J.C. Jiang, T. Yamagami & K. Ueno
Department of Civil Engineering, The University of Tokushima, Japan

ABSTRACT: A three-dimensional (3-D) analysis method is proposed to back-calculate unsaturated soil shear strength from a slope failure. The Fredlund strength criterion, with three parameters (c', ϕ', ϕ^b), is incorporated with a 3-D simplified Janbu method for formulation of the factor of safety. The 3-D back analysis procedure for determination of the three strength parameters is described by combining the 3-D Janbu factor of safety equation with two essential conditions that take full advantage of the information of a failure surface in a unsaturated slope. Application of the procedure to a hypothetical 3-D slope failure indicates that a unique and reliable solution of (c', ϕ', ϕ^b) can be obtained. Finally, a comparison of 2-D and 3-D back analyses is made to show the importance of performing a 3-D analysis in back calculating the unsaturated shear strength.

1 INTRODUCTION

Back analyses of slope failures have become a useful tool to evaluate soil shear strengths especially for remedial measures of landslide slopes. Existing back analyses for obtaining strength parameters have been performed for slope failures under saturated conditions (e.g. Nyugen, 1984; Yamagami & Ueta, 1992, 1994). As most natural and manmade slopes are in unsaturated zones, it is necessary to incorporate the effect of suction due to unsaturation on soil strength in analyses of slope failures. A 2-D analytic method was presented (Jiang, Yamagami & Ueta, 1999) to back-calculate unsaturated strength parameters from a circular slope failure. This method was also extended to accommodate noncircular failure surfaces in unsaturated slopes (Jiang, Yamagami & Ueta, 2000).

A 3-D analysis was recommended for back calculating strength parameters (e.g. Leshchinsky & Huang, 1992; Yamagami & Ueta, 1994), since the 2-D simplification of an actual 3-D slope failure may lead to significant overestimation of back calculated strengths (and thus unsafe design). The Yamagami and Ueta's (1994) 3-D back analysis for a saturated slope failure showed that ignoring 3-D effects resulted in an increase of calculated values of both c and ϕ by as much as 30%.

In the present paper, a 3-D analytic method is presented to determine unsaturated soil strength parameters from back calculation of a 3-D slope failure. The method is based on two essential conditions that take full advantage of the information provided by a failed surface. The 3-D simplified Janbu method (Ugai, 1988) is first incorporated with the Fredlund strength criterion for formulation. Then, a back analysis procedure for determination of three strength parameters (c', ϕ', ϕ^b) is described by combining the 3-D Janbu factor of safety equation with the above-mentioned two conditions. Finally, this paper provides a comparison of 2-D and 3-D analyses to show the importance of using a 3-D analysis in back calculating the unsaturated shear strengths.

2 BASIC CONCEPT OF BACK ANALYSIS

2.1 Two essential conditions for back analysis

For simplicity, we assume the slope in question to be homogeneous in strength. In other words, c', ϕ' and ϕ^b are back calculated as an average of the strength parameters along a 3-D failure surface. Two essential conditions suggested for back analysis of strength parameters (Yamagami & Ueta, 1992, 1994), as shown in the following, are also employed in the present study.

I) Strength parameters to be determined must satisfy a condition where the factor of safety, F_0, of the failure surface is equal to 1.0;

II) Strength parameters to be determined must satisfy a condition where the factor of safety of the failure surface is the smallest among trial slip surfaces in the vicinity of the failure surface.

2.2 3-D simplified Janbu method with the Fredlund strength criterion

The 3-D simplified Janbu method proposed by Ugai (1988) is employed in this study. Fig.1 illustrates the forces acting on a typical column in which Q is the resultant of all the intercolumn forces acting on the sides of the column. In Ugai's method, Q is decomposed to a component Q_1 parallel to the x-axis and a component Q_2 inclined by $\beta=\tan^{-1}(\eta\tan\alpha_{yz})$ with respect to the y-axis (η is an unknown constant), as shown in Fig.1. It has been shown that sufficiently accurate results can be obtained by using $\eta=0$ when the slope angle is not larger than 60° (Ugai, 1988; Yamagami & Jiang, 1997). For simplicity, the constant of $\eta=0$ is employed in the present work.

The Fredlund failure criterion (Fredlund, 1979) for unsaturated soil can be expressed by

$$\tau_f = c' + (\sigma_n - u_a)\tan\phi' + (u_a - u_w)\tan\phi^b \qquad [1]$$

where u_a - pore air pressure, u_w - pore water pressure, (σ_n-u_a) - net normal stress state variable on a failure surface, (u_a-u_w) - matric suction, c', ϕ', ϕ^b – strength parameters. The value of u_a is usually taken to be zero for comparatively shallow slip surfaces.

Using the strength criterion Eq.[1], instead of the Mohr-Coulomb model, the 3-D factor of safety can be derived from the horizontal force equilibrium equation. The derivation of this expression is quite similar to the procedure presented by Ugai (1988), and only the final equation is shown here (Eq.[2]).

$$F = \frac{\sum\left\{(c' - u_w\tan\phi^b)\Delta x\Delta y + W\tan\phi'\right\}/(m_\alpha\cos\alpha_{xz})}{\sum W\tan\alpha_{xz}}$$

$$[2]$$

where

$$m_\alpha = (1 + \tan^2\alpha_{xz} + \tan^2\alpha_{yz})^{-1/2} + \sin\alpha_{xz}\tan\phi'/F$$

Note that u_a is not included in Eq.[2] as it is taken to be zero.

The factor of safety, F_0, of a slip surface at failure is usually equal to 1.0. Substituting the known value of $F=F_0$ (=1.0) into Eq.[2], we have:

$$F_0 = \frac{\sum\left\{(c' - u_w\tan\phi^b)\Delta x\Delta y + W\tan\phi'\right\}/(m_\alpha\cos\alpha_{xz})}{\sum W\tan\alpha_{xz}}$$

$$[3]$$

in which W, u_w, α_{xz} and m_α are evaluated based on the information of the failure surface.

Eq.[3] indicates the relationship among c', $\tan\phi'$ and $\tan\phi^b$, representing a 3-D surface, *egf*, in c'-andϕ'-tanϕ^b space (see Fig.2). This surface will be called the "relation surface" hereafter. Note that the c'-tanϕ^b relationship is linear when $\tan\phi'$ in Eq.[3] remains a constant value. In Fig.2, maximum values of the strength parameters, c'_{max}, $\tan\phi'_{max}$ and $\tan\phi^b_{max}$, limit possible ranges of variation of c', $\tan\phi'$ and $\tan\phi^b$. The c'_{max}, $\tan\phi'_{max}$ and $\tan\phi^b_{max}$ values can be calculated from Eq.[3] respectively by assigning the other two parameters a value of zero.

It has been shown (Jiang, Yamagami & Ueta, 1999, 2000) that when the value of a parameter out of (c', ϕ', ϕ^b) is known through experimental or empirical means, the other two parameters can be obtained uniquely from a 2-D slip surface using Yamagami and Ueta's back analysis (1992). In the next section, we will first explain that this is also valid for a 3-D failure surface in an unsaturated slope. Then, a procedure, which is able to back calculate the three unsaturated strength parameters simultaneously, is presented.

2.3 Application of Yamagami and Ueta's method

In this section, a similar procedure to the 3-D back analysis (Yamagami & Ueta, 1994) is applied to back calculation problems in which one of (c', ϕ', ϕ^b) is given. Since one parameter is given, say $\phi'=\phi'_0$ is assumed to be known here for convenience. Substituting $\phi'= \phi'_0$ into Eq.[3] and re-arranging the terms of the equation, the following can be obtained:

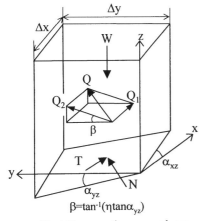

$\beta=\tan^{-1}(\eta\tan\alpha_{yz})$

Fig.1 Forces acting on a column

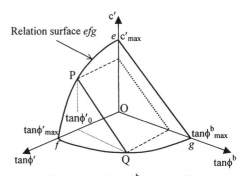

Fig.2 c'-tanϕ'-tan ϕ^b relationship

790

$$\sum \frac{(c' - u_w \tan\phi^b)\Delta x \Delta y}{m_\alpha \cos\alpha_{xz}} = \sum W \left(F_0 \tan\alpha_{xz} - \frac{\tan\phi'}{m_\alpha \cos\alpha_{xz}} \right)$$

[4]

This equation indicates the relationship between c' and $\tan\phi^b$, which can be represented by a straight line PQ on the relation surface, as shown in Fig.2. In the present case, condition I) means that strength parameters (c'_0, $\tan\phi^b_0$) to be identified must satisfy the c'-$\tan\phi^b$ relationship of Eq.[4]. In other words, a required solution of the parameters (c', $\tan\phi^b$) corresponds to a point on the line PQ in Fig.2.

In order to consider condition II), two trial slip surfaces are chosen above and below a general 3-D failure surface, respectively, as illustrated in Fig.3. The method (Yamagami & Ueta, 1994) for generating 3-D trial slip surfaces is also employed here. The factor of safety of a trial slip surface, F_t, may be expressed as:

$$F_t = \frac{\sum \left\{ (c' - \bar{u}_w \tan\phi^b)\Delta x \Delta y + \overline{W}\tan\phi'_0 \right\} / (\overline{m}_\alpha \cos\overline{\alpha}_{xz})}{\sum \overline{W}\tan\overline{\alpha}_{xz}}$$

[5]

where the overlined symbols are evaluated from the trial slip surface in question.

It has been shown that when (c', $\tan\phi^b$) vary along the line PQ in Fig.2, the F_t-c' and F_t-$\tan\phi^b$ relationships for a 3-D trial slip surface above the failure surface (such as $ao'd$ in Fig.3) can be represented as illustrated in Fig.4. Now, we consider the condition II) which requires that the factors of safety, F_t, of trial slip surfaces should be greater than F_0 (=1.0). This signifies that values of the required c'_0 and $\tan\phi^b_0$

cannot be beyond the ranges AB and DE, respectively. When the above procedure is performed with respect to a number of trial slip surfaces chosen above and below the failure surface separately, the range of variation of c' (or $\tan\phi^b$) can be restricted to an extremely narrow zone which can be regarded to be an approximate solution of the required c'_0 (or $\tan\phi^b_0$), as shown in Fig.5.

As done in Yamagami and Ueta's method (1994), the above back analysis procedure can be carried out more efficiently and systematically by using λ-c' and λ-$\tan\phi^b$ relationships (Fig.6). Note that λ is a speci-

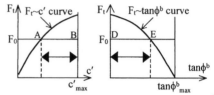

Fig.4 Restriction of ranges of variation of (c', $\tan\phi^b$)

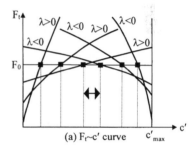

(a) F_t~c' curve

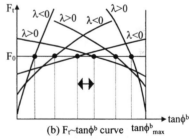

(b) F_t~$\tan\phi^b$ curve

Fig.5 Back analysis procedure based on F_t-c' and F_t-$\tan\phi^b$ relationships

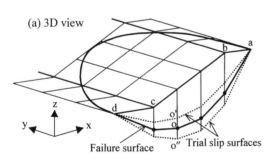

(a) 3D view

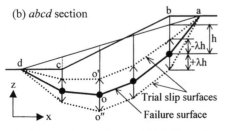

(b) $abcd$ section

Fig. 3 3D failure surface and trial slip surfaces

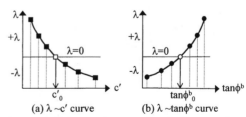

(a) λ ~c' curve

(b) λ ~$\tan\phi^b$ curve

Fig.6 An efficient and systematic solution procedure based on λ-c' and λ-$\tan\phi^b$ relationships

fied positive constant used in Fig.3 to generate trial slip surfaces. The abscissas of points ■ or ● in Fig.6 correspond to the abscissas of the points of intersection of F_r-c' or F_r-tanϕ^b curves with the line of $F_t = F_0$ shown in Fig.5 separately.

While the above procedure for back analysis of (c', tanϕ^b) is described by assigning a ϕ'_0 value to ϕ', when c' or tanϕ^b is given as a known value, a similar back calculation can also be carried out to determine (tanϕ', tanϕ^b) or (c', tanϕ'), respectively.

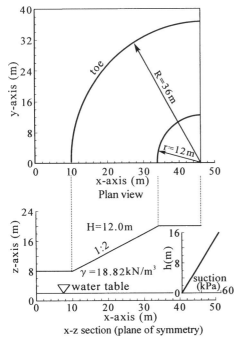

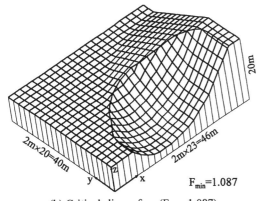

(a) Convex slope in plan view
and the distribution of suction

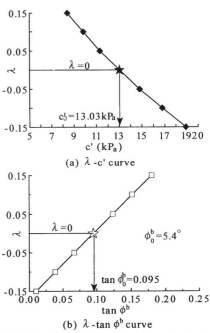

(b) Critical slip surface (F_{min}=1.087)

Fig.7 Example problem for 3D back analysis

2.4 Verification

An illustrative slope failure in a symmetrical and convex slope in plan view, as shown in Fig.7, is presented here to verify the effectiveness of the back analysis procedure described in the preceding section. By specifying c'=12.74kPa, ϕ'=10°, ϕ^b=6° and γ=18.82kN/m^3, the 3-D critical slip surface and the associated minimum factor of safety, as shown in Fig.7, were obtained from a search procedure combining Eq.[2] with dynamic programming (Yamagami & Jiang, 1997). Thus, the back analysis is to be conducted on condition that the critical slip surface, the associated minimum factor of safety (F_{min}= 1.087) and any one of (c', ϕ', ϕ^b) are known, but the others are not.

By assigning a known value to one of (c', ϕ', ϕ^b), computational procedures for back calculating the other two parameters were carried out using the efficient and systematic approach shown in Fig.6. When ϕ'=ϕ'_0=10° was given as a solution of ϕ', the values of (c'=13.03kPa, ϕ^b= 5.4°) were obtained (see Fig.8). In addition, (ϕ'=10.1°, ϕ^b=5.7°) and (c'=12.54kPa, ϕ'= 10.1°) were back calculated respectively by giving c'=c'$_0$=12.74kPa and ϕ^b= ϕ^b_0= 6.0°. It can be seen from these results that in each case the back calculated strength parameters agree with the correct (known) values quite well.

Stability analyses for the slope shown in Fig.7(a) were performed using the back-calculated strength parameters. As a result, the critical slip surface and

(a) λ -c' curve

(b) λ -tan ϕ^b curve

Fig.8 Back calculation results based on λ-c' and λ-tanϕ^b curves (when ϕ'= ϕ'_0=10°)

the associated minimum factor of safety obtained for each of the above three situations are almost the same as those shown in Fig.7(b).

3 BACK ANALYSIS PROCEDURE FOR IDENTIFYING (c', ϕ', ϕ^b) SIMULTANEOUSLY

It has been indicated that when one of the three parameters c', ϕ' and ϕ^b is given as a known value, a similar method to the Yamagami and Ueta 3-D back analysis (1994) can be applied to determine the other two strength parameters uniquely. Based on this fact, a computational procedure is presented here to back calculate the three unsaturated strength parameters (c', ϕ', ϕ^b) simultaneously where the following requirement is satisfied:

Required strength parameters (c'$_0$, ϕ'_0, ϕ^b_0) should satisfy a condition where the critical slip surface searched by c'$_0$, ϕ'_0 and ϕ^b_0 must be identical with the failure surface, and the associated minimum factor of safety must be equal to F_0 (=1.0).

By considering the above condition, the calculation procedure, which is able to determine (c', ϕ', ϕ^b) simultaneously, can be described as follows:
1) Calculate values of c'_{max}, $\tan\phi'_{max}$ and $\tan\phi^b_{max}$ based on the failure surface information.
2) Divide the range of 0~$\tan\phi'_{max}$ into an appropriate number of equal parts ($\tan\phi'_j$, j=1, 2, ...m) where $\tan\phi'_1$=0 and $\tan\phi'_{m+1}$=$\tan\phi'_{max}$.
3) Regard each value of $\tan\phi'_j$ to be a known solution of $\tan\phi'$, i.e. $\tan\phi'_0 = \tan\phi'_j$, then back calculate the associated values of (c'_j, $\tan\phi^b_j$) by the procedure described in the previous sections.
4) Search for the critical slip surfaces and the minimum factors of safety using each group of (c'_j, $\tan\phi'_j$, $\tan\phi^b_j$) obtained in 3).
5) Take such values of (c'_j, $\tan\phi'_j$, $\tan\phi^b_j$) as a required solution which give the critical slip surface that is most close to the failure surface.

An optimization approach is established to enhance the efficiency of the above calculation procedure. For a chosen $\tan\phi'$ between 0 and $\tan\phi'_{max}$, the associated values of c' and $\tan\phi^b$ can be back calculated by regarding the chosen $\tan\phi'$ to be $\tan\phi'_0$. As the parameters (c', $\tan\phi'$, $\tan\phi^b$) so obtained are usually not a required solution, the critical slip surface searched using these parameters differs from the original failure surface. It is obvious that the magnitude of differences between the locations of the critical slip surface and the failure surface depends upon the chosen value of $\tan\phi'$. In other words, differences between critical slip surfaces and the failure surface can be regarded to be a function of $\tan\phi'$. The difference between a critical slip surface and the failure surface can be defined by DIS shown in Fig.9. DIS varies with $\tan\phi'$, being a function of $\tan\phi'$. The

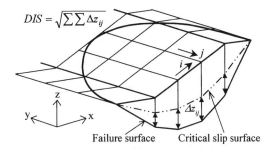

$$DIS = \sqrt{\sum\sum \Delta z_{ij}}$$

Failure surface   Critical slip surface

Fig.9 Optimization problem for back analysis

minimum value of DIS, DIS_{min} (=0), corresponds to a required solution of unsaturated strength parameters. This optimization problem, as done in the 2-D analyses (Jiang, Yamagami & Ueta, 1999, 2000), is solved by using *the golden section method*.

While the above procedure is described in terms of $\tan\phi'$, a similar back calculation can also be carried out in terms of c' or $\tan\phi^b$, respectively.

4 EXAMPLE PROBLEM

4.1 3-D back analysis results

Based on the critical slip surface, the minimum factor of safety and the distribution of suction shown in Fig.7, three unsaturated strength parameters were back calculated by the back analysis procedure described in the preceding section. The results obtained from the procedures in terms of each of (c', ϕ', ϕ^b) are summarized in Table 1. Their average values and the correct solution of (c', ϕ', ϕ^b) are also shown in this Table. It can be seen from Table 1 that the averages of the back calculated values of (c', ϕ', ϕ^b) were in good agreement with the correct values of the strength parameters, though small discrepancies are observed when the analysis is performed in each case. This indicates that the proposed method can provide sufficiently accurate results of the unsaturated strength parameters.

Table 1 Back calculated results from 3-D analysis

Cases	c' (kP$_a$)	ϕ'	ϕ^b
in terms of c'	11.56	10.4°	7.1°
in terms of ϕ'	13.92	9.8°	4.5°
in terms of ϕ^b	11.56	10.4°	7.1°
average value	12.35	10.2°	6.2°
correct solution	12.74	10.0°	6.0°

4.2 A comparison of 2-D and 3-D back analyses

In order to make a comparison between 2-D and 3-D analyses, a 2-D back calculation was also performed

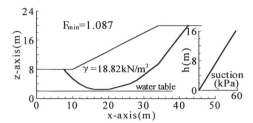

Fig.10 Trace of 3-D critical slip surface shown in
Fig.7 (b) on symmetry plane

Table 2 Back calculated results from 2-D analysis

Cases	c' (kPa)	ϕ'	ϕ^b
in terms of c'	22.46	8.5°	3.5°
in terms of ϕ'	21.65	8.7°	4.5°
in terms of ϕ^b	18.67	9.3°	7.6°
2-D average value	20.93	8.8°	5.2°
3-D average value	12.35	10.2°	6.2°
correct solution	12.74	10.0°	6.0°

to obtain values of (c', ϕ', ϕ^b). Fig.10 illustrates the
trace of the 3-D critical slip surface shown in Fig.7
on the symmetry plane. The value of F_0 was taken to
be equal to the 3-D minimum factor of safety (F_{min}=
1.087). The 2-D back analysis method (Jiang, Yamagami & Ueta, 2000) based on the simplified Janbu method was employed. The obtained values of
(c', ϕ', ϕ^b) are listed in Table 2 in which the averages
of the 3-D results and the correct solution of the
three parameters are also included.

It is seen from Table 2 that the 2-D back calculated value of c' is about 70 percentage larger than that
from the 3-D analysis which is almost the same as
the correct solution of c'. However, the 2-D results of
(ϕ', ϕ^b) are slightly smaller than their correct values.

5 CONCLUSION

A 3-D back analysis method has been presented to
determine the unsaturated strength parameters (c', ϕ',
ϕ^b) from a slope failure. This method is based on the
two conditions that take full advantage of the information provided by a slope failure. It is an extension
of the Yamagami and Ueta (1994) 3-D analysis for
saturated slope failures. The back calculation procedure for (c', ϕ', ϕ^b) was described by combining the
two conditions with the 3-D simplified Janbu factor
of safety equation for unsaturated situations.

The proposed method was applied to a 3-D hypothetical slope failure and the results showed that
the back calculated strength parameters agreed well
with the correct values of (c', ϕ', ϕ^b). This indicates

the potential of the proposed method to identify the
values of (c', ϕ', ϕ^b) from an actual slope failure as
long as the suction distribution along the slip surface
at failure is known.

A comparison between the 2-D and 3-D back
analyses has shown that the obtained value of c' was
greatly overestimated by the 2-D analysis. This signifies the importance of performing a 3-D analysis in
back calculating the unsaturated soil shear strengths.

REFERENCES

Fredlund, D. G. 1979. Second Canadian Geotechnical Col-
loquium: Appropriate concepts and technology for unsaturated soils. *Canadian Geotechnical Journal*. **16**. 121-
139.
Jiang J.-C., Yamagami T. & Ueta Y. 1999. Back analysis of
unsaturated shear strength from a circular slope failure,
*Proc. Int. Symp. on Slope Stability Engineering (IS-
Shikoku'99)*, Matsuyama, Japan, Nov. 8-11, Yagi N.,
Yamagami T. & Jiang J.-C. Eds, Rotterdam : Balkema, **1**,
305-310.
Jiang J.-C., Yamagami T. & Ueta Y. 2000. Back analysis of
unsaturated soil strength parameters from a noncircular
slope failure, *Proc. 8th Symp. on Landslides*, Cardiff,
UK, Jun. 26-30, (in print).
Leshchinsky, D. and Huang, C. C. 1992. Generalized three-
dimensional slope stability analysis, *Geotech. Engrg. Div.*,
ASCE, Vol. 118 (GT11), 1748-1764.
Nguyen, V. U. 1984. Back calculations of slope failures by
the secant method. *Geotechnique*. **34** (3). 423-427.
Ugai, K. 1988. 3-D slope stability analysis by slice methods,
*Proc. 6th Int. Conf. on Numerical Methods in Geome-
chanics*, Innsbruck, Austria, Rotterdam: Balkema, 1369-
1374.
Yamagami, T. & Jiang, J.-C. 1997. A search for the critical
slip surface in three-dimensional slope stability analysis.
Soils and Foundations, **37** (3), 1-16.
Yamagami, T. & Ueta, Y. 1992. Back analysis of strength
parameters for landslide control works. *Proc. 6th International Symp. on Landslides*, Christchurch, Feb. 10-14,
Bell D. H. Ed. Rotterdam : Balkema, **1**. 619-624.
Yamagami, T. & Ueta, Y. 1994. Three-dimensional back
analysis of strength parameters for landslide control
works. *Proc. Int. Conf. on Landslides, Slope Stability and
the Safety of Infra-Structures*. Malaysia, 393-400.

Unsaturated Soils for Asia, Rahardjo, Toll & Leong (eds) © 2000 Taylor & Francis, ISBN 90 5809 139 2

A new slope stability analysis for Shirasu slopes in Japan

R. Kitamura & K. Sakoh
Department of Ocean Civil Engineering, Kagoshima University, Japan

M. Yamada
Daiya Consultant Company Limited, Japan

ABSTRACT: The inter-particle force due to meniscus at a contact point of particles is derived and then the relationship between suction and apparent cohesion is numerically obtained based on some probabilistic consideration on the soil particle scale. Then a new slope stability analysis is proposed to analyze the stability of Shirasu slopes, in which Janbu method is applied to the slope failure due to heavy rainfall. In the proposed slope stability analysis the change in apparent cohesion with the change in water content due to rainfall is taken into account. The probability of slope failure can be calculated in the real time by a personal computer when the data on total amount and intensity of rainfalls are obtained in the real time.

1 INTRODUCTION

Kagoshima Prefecture is located in the southern part of Kyushu Island, Japan, and covered with various volcanic products. Shirasu is one of famous volcanic products in Kagoshima Prefecture and defined as a non-welded part of pyroclastic flow deposits. The density of Shirasu is smaller than usual sandy soil as particles are porous. Consequently the Shirasu ground is easy to be eroded by surface water flow. The slopes composed of Shirasu with other volcanic products often fail due to the heavy rain in the rainy season. Most of the slope failures are classified as the surface slip failure with less than 1 m in depth.

It is qualitatively well-known that one of the main causes of slope failure is the decrease in apparent cohesion due to the increase in water content which brings the decrease in inter-particle force between particles. But the quantitative estimation of change in apparent cohesion with the change in water content has not been established at present. Therefore the conventional slope stability analyses are not applicable to the Shirasu slope and a new method is expected to be able to predict the failures of Shirasu slope due to heavy rainfalls.

In this paper the inter-particle force due to meniscus at a contact point of particles is derived and then the relationship between suction and apparent cohesion is numerically obtained based on some probabilistic consideration in the soil particle size. Then a new slope stability analysis is proposed to analyze the stability of Shirasu slopes, in which the Janbu method is applied. In the proposed slope stability analysis the change in apparent cohesion with the change in water content due to rainfalls is taken into account.

2 DERIVATION OF COHESIONAL COMPONENT FROM INTER-PARTICLE FORCE

The inter-particle force between two adjacent particles is generated by the surface tension of pore water in soil mass as shown in Fig.1. The inter-particle force F_i can be expressed by the following equation.

$$F_i = 2\pi r' T_S + \pi r'^2 s_u \tag{1}$$

where F_i: inter-particle force,
T_S: surface tension ,
s_u: suction (= $u_a - u_w$),
r': radius of curvature of meniscus.

Eq. (1) includes the suction in the second term of right side and then the water content in soil mass must be estimated to obtain the inter-particle force. Kitamura et al.(1998) proposed the numerical model for seepage behavior of pore water in unsaturated soil. Figure 2(a) shows an element with a few soil particles. This situation is modeled by Fig.2(b), in which the voids and soil particles are modeled by the pipe with the diameter D and the inclination angle θ, and the other impermeable parts respectively. The void ratio e, volumetric water content Wv and suction s_u can be derived by some mechanical and probabilistic considerations in soil particle size as follows.

$$e = \int_0^\infty \int_{\frac{\pi}{2}}^{\frac{\pi}{2}} \frac{V_P}{V_e - V_P} \cdot P_d(D) \cdot P_c(\theta) d\theta dD \qquad (2)$$

$$W_V = \frac{e(d)}{1+e} = \frac{1}{1+e} \int_0^d \int_{\frac{\pi}{2}}^{\frac{\pi}{2}} \frac{V_P}{V_e - V_P} \cdot P_d(D) \cdot P_c(\theta) d\theta dD \qquad (3)$$

$$s_u = \gamma_w \cdot h_c = \frac{4 \cdot T_s \cdot \cos\alpha}{d} \qquad (4)$$

where e : void ratio,
γ_w: unit weight of water,
V_e: volume of element in Fig.2(b),
V_p: volume of pipe in Fig.2(b),
$P_d(D)$: probability density function of D,
$P_c(\theta)$: probability density function of θ,
W_v : volumetric water content,
s_u: suction,
h_c : water head,
α :contact angle between the pipe and water,
d: maximum diameter of pipe filled with water,
T_s :surface tension.

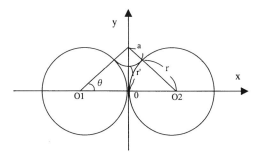

Fig.1 Inter-particle force between two particles due to surface tension

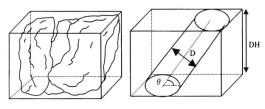

Fig.2(a) Soil particles in element

Fig.2(b) Pipe and other impermeable parts

Fig.2 Modeling of soil particles and voids

Substituting Eq.(4) into Eq.(1), the inter-particle force can be obtained.

Figure 3 shows the cross section of soil mass and the number of soil particles per unit area is obtained by the following equation.

$$Np = Nv \cdot 2 \cdot r = \frac{1}{1+e} \cdot \frac{3}{2 \cdot \pi \cdot r^2} \qquad (5)$$

$$Nv \cdot (1+e) = \frac{1}{\left(4 \cdot \pi \cdot r^2 / 3\right)}$$

where Np : average number of particles per unit area,
e: void ratio,
Nv : average number of particles per unit volume,
r : radius of soil particle.

In the slope the shearing force is generated and consequently the potential slip plane is supposed as shown in Figure 4. The number of contact point (coordination number) per unit area of potential slip plane can be assumed by the following equations(Kitamura,1980).

$$Nc = \frac{Ca}{2} \cdot Np$$

$$Ca = \frac{12}{1+e} \qquad (6)$$

where Nc: total number of contact points in the particulate material,
Ca: number of contact points per a particle.

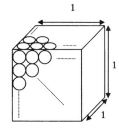

Fig.3 Cross section of soil mass

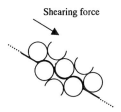

Fig.4 Potential slip plane in soil mass

Figure 5 shows the relation between void ratio and number of contact points per a particle obtain by using Eq.(6).

Figure 6 shows the semi-sphere particle point in Fig.4. The probability density function of contact points on the narrow belt designated by dark zone in Fig.6 is introduced to obtain the number of contact point at the narrow belt. Then the following equation is derived.

$$dN = Nc \cdot D(\eta) \cdot \cos(\eta) \cdot d\eta \qquad (7)$$

where $D(\eta)$: Probability density function of number of contact point on the narrow belt with angle η in Fig.6.

Supposing that the number of contact points is maximum at contact angle $\pi/2$ and minimum at contact angle zero, and changes linearly, the following equation can be derived as the probability density function of contact points.

$$D(\eta) = \frac{\eta}{\pi/2 - 1} \qquad (8)$$

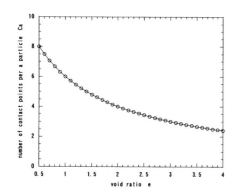

Fig.5 Relation between void ratio and number of contact points per a particle.

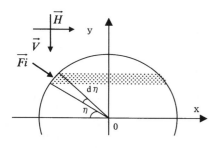

Fig.6 Equilibrium of force on the surface of a particle

Referring to Fig.6, the following equation can be derived from the equilibrium of force in the vertical and horizontal direction .

$$\frac{\vec{H} + \sum(\vec{F_i} \cdot \cos\eta)}{\vec{V} + \sum(\vec{F_i} \cdot \sin\eta)} = \tan\delta \qquad (9)$$

where $\vec{V}$: vertical force transmitted from adjacent particle,

   δ : frictional angle between particles,

   $\vec{H}$: horizontal force transmitted from adjacent particle,

   η : angle defined in Fig.6.

Changing the force to the force per unit area, the following equation is obtained.

$$\tau = \sum(\vec{F_i} \cdot \sin\eta) \cdot \tan\delta - \sum(\vec{F_i} \cdot \cos\eta) + \sigma \cdot \tan\delta \qquad (10)$$

where τ : shear stress,
         σ : normal stress.

Supposing that Eq.(10) corresponds to the equation of Mohr-Coulomb failure criteria, the following equations are obtained.

$$\phi = \delta \qquad (11)$$

$$c = \sum(\vec{F_i} \cdot \sin\eta) \cdot \tan\phi - \sum(\vec{F_i} \cdot \cos\eta) \qquad (12)$$

where ϕ : internal friction angle,
         c: apparent cohesion component.

Table.1   Values used for calculation

density of soil particle (g/cm³)	2.37
surface tension (N/m)	7.355E-02
radius of soil particle (cm)	0.012
internal friction angle (°)	35.0
height of container in Fig.1(b) DH(cm)	0.002
mean value diameter of pipe(cm)	0.0012
standard deviation diameter of pipe(cm)	0.0008
the most low height of probability density function of θ	0.159

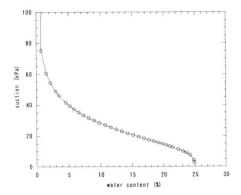

Fig.7 Relation between volumetric water content and suction

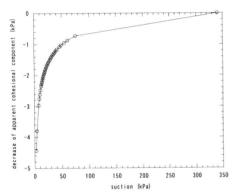

Fig.8 Relation between suction and decrease of apparent cohesional component

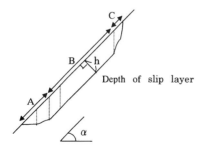

Fig.9 Steep slope with potential slip surface

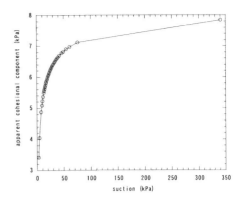

Fig.10 Relation between suction and apparent cohesion component

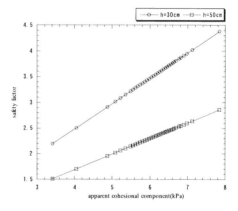

Fig.11 Relations between safety factor and apparent cohesional component

Table.2 Values used for calculation of safety factor

slope angle (°)	50.0
length of slip(m)	20.0
number of slice	20.0
width of slice(m)	1.0
length of A,C(m)	4.0
length of B(m)	12.0
depth of slip layer(cm)	30.0 , 50.0

3 SLOPE STABILITY ANAYLYSIS

Figure 9 shows the steep slope with the infinite and non-circle potential slip surface which is a typical pattern of failure at Shirasu slope. The slope is divided into three parts, i.e. a plain part and two curved parts. For each slip surface the Janbu method is applied as follows.

Substituting Eqs.(1), (7), and (8) into Eq.(12), the following equation is obtained.

$$c = \tan\phi \cdot \int_0^\pi (\overrightarrow{F_i} \cdot \sin\eta) dN - \int_0^\pi (\overrightarrow{F_i} \cdot \cos\eta) dN$$

$$= \frac{\pi}{\pi - 2} \cdot \overrightarrow{F_i} \cdot Nc \cdot \tan\phi \qquad (13)$$

798

$$F_s = \frac{\sum \left[(c \cdot b + (W - \Delta V - U \cdot \cos\alpha) \cdot \tan\phi)/m_j \right]}{\sum \left[\Delta E + (W - \Delta V) \cdot \tan\alpha \right]} \qquad (14)$$

$$m_j = \cos^2 \alpha + \sin\alpha \cdot \cos\alpha \cdot \tan\phi / F_s$$

where F_s: safety factor,
         W: weight of slice,
         b: width of slice,
         α: slope angle,
         ΔV: resultant vertical force in side of slice,
         U: resultant force of pore pressure at bottom of slice,
         ΔE: resultant force of pore pressure in side of slice.

Figure 11 shows an example of relations between safety factor and apparent cohesion component obtained from the relation in Fig 10. It is found out from Fig.11 that the safety factor decreases with the decrease in apparent cohesion component which relates to the decrease in suction due to the increase in water content.

4 CONCLUSIONS

In this paper the apparent cohesional component was related to the suction which changes with the water content. The slope stability analysis was carried out quantitatively to estimate the safety of Shirasu slope by using the relation between the apparent cohesional component and suction. This research aims to predict the shallow failure of Shirasu slope in the real time when the heavy rain falls. It is found out that the proposed approach may be promising as a new slope stability analysis for shallow slope failure.

ACKNOWLEDGEMENT

This research was supported by the Grant-in-Aid of Scientific Researches (Project No.09555153 and No.10650490) of the Ministry of Education.

REFERENCES

Kitamura,R,Fukuhara,S,Uemura,K,Kisanuki,G,and Seyama,M.(1998): A numerical model for seepage through unsaturated soil, Soils and Foundations, Vol.38, No.4, pp.261-265.
Field,W,G.(1963): Towards the Statistical Deformation of a Granular Mass, Proc. 4th A. and N.Z. Conf. SMFE, pp.143-148.

Unsaturated Soils for Asia, Rahardjo, Toll & Leong (eds) © 2000 Taylor & Francis, ISBN 90 5809 139 2

Rainfall-induced slope failures in Taiwan

H. D. Lin & J. H. S. Kung
Department of Construction Engineering, National Taiwan University of Science and Technology, Taipei, Taiwan

ABSTRACT: Slope failure in residual soils caused by rainfall infiltration is a major hazard in Taiwan. Rainfall-induced landslides are mainly attributed to the reduction of soil matric suction and its shear strength. In the past, slope stability analysis was often conducted using soil parameters corresponding to saturated conditions. The effects of soil suction, rainfall infiltration and soil conductivity were not considered. The objective of this research is to apply the principle of the unsaturated soil mechanics to study the effect of rainfall intensity on the soil suction profile and hence its effect on slope stability. Typical slopes of the lateritic terrace of residual soil origin in Taiwan are studied. Highlights from this study are summarized in this paper.

1 INTRODUCTION

Slope stability problems in residual soils are receiving increasing attention in recent years. Rainfall seems to be the most common cause of landslides in residual soil slopes in many countries such as Taiwan. Most of the rainfall-induced landslides in residual soils consist of relatively shallow slip surface above the groundwater table. Therefore, in situ pore water pressure in these residual soils is often negative with respect to the atmospheric conditions. This negative pore-water pressure is called matric suction. There has been increasing evidence that matric suction contributes towards the stability of nature slopes in residual soil (Fredlund & Rahardjo, 1985).

Lateritic soil terraces that are one kind of residual soil can be found in many regions of western Taiwan. The Linkou terrace, located in northwestern Taiwan, is the largest among them. The Linkou formation comprises of gravels and the lateratic mantle of a few meters to about ten meters. Lateritic soil has been weathered and leached from gravel and its fillings. The sediments are mostly loosely cohered and consolidated. Hence, the problems of erosion, slope instability, and landslides in this area have became more and more serious in the past decade. A number of landslides induced by rainfall have been reported.

In the past, the effects of soil suction, rainfall infiltration and conductivity of soil were not explicitly considered in the conventional slope stability analysis. However, slope instability in the lateratic terrace often involves shallow failures, deep groundwater tables, and climatic fluxes. Therefore, to better analyze this type of slope stability problem the principle of unsaturated soil mechanics should be applied. In this research, the interacting effects of soil suction, rainfall infiltration and soil conductivity are analyzed. In particular, emphasis is placed on the effect of rainfall intensity on soil suction profiles and hence its effect on slope stability. This paper will briefly describe the mechanism of rainfall-induced slope stability problem. Then, an analytical procedure that has employed the principle of unsaturated soil mechanics is presented. Subsequently, the results of the slope stability analyses on typical slopes of the lateritic terrace are discussed.

2 MECHANISM OF RAINFALL-INDUCED SLOPE FAILURE

Lumb (1975) and Brand (1984) provided some insights on the phenomenon of rain-induced landslide of Hong Kong residual soil. A rainfall intensity of about 70 mm/hour appears to be the threshold value above which landslide is likely to occur. These landslides take place when the peak hourly rainfall occurs. The failures always occur suddenly and rapidly without prior warning. The thickness of the failed zone is generally less than 3m. The scar is usually unsaturated and without any water piping out of it. The evidence available at present favors the rainfall infiltration from above as the principal cause of failure rather than the seepage from below.

Fredlund (1987) proposed a mechanism for this kind of slopes in residual soil. The slope maintains stable in dry season due to the matric suction that can increase the strength of soil. When heavy rain occurs, the percolating water would saturate the soil and then decrease the matric suction and shear strength at the same time. As a result, landslides may occur. This mechanism can explain the majority of the failures that take place during periods of heavy rainfall.

In light of the above discussions, to understand the failure mechanism of the residual soil slopes, two processes need to be further examined. One is the change of porewater pressure in the soil mass due to infiltration of rainwater. The other process is the change in shear strength of the soil due to the increase in pore-water pressure and hence its effect on the factor of safety against slope failure. Followings are brief discussion of the infiltration behavior and the stress paths of soil subject to rainfall infiltration.

2.1 Infiltration behavior

The process of infiltration is very complex. In this study, the process of rainwater infiltration is assumed as one-dimensional only. Mein & Larson (1973) used rainfall intensity (I), soil infiltration capacity (f_p), and saturated conductivity (k) to explain the infiltration behavior.

(1) $I < k$: For this condition, run off will not occur, since all the rainfall infiltrates. The percolating water will change the degree of saturation in the soil. But it cannot form the wetting front. Line A in Figure 1 shows this situation.

(2) $f_p > I > k$: During this stage, all the rainfall infiltrates into the soil, and the soil moisture level near the soil surface increases. Line B of curve BC in Figure 1 illustrates this case.

(3) $I > f_p > k$: The infiltration rate is at its capacity and decreases subsequently. Runoff is being generated. This is the case shown by curve C and also curve D in Figure 1.

When rainwater infiltrates into the soil, the redistribution of soil moisture will result in three different zones of soil mass: (1) infiltration zone, (2) transition zone, and (3) non-infiltration zone. Lumb (1975) simplified these three zones into a wetting front and a non-wetting front. The depth of wetting front can be expressed as (1):

$$h = (Dt)^{0.5} + \frac{kt}{n(s_f - s_i)} \qquad (1)$$

Where parameter t is the rain duration, D is the diffusivity parameter, n is the soil porosity, k is the saturated conductivity, s_f represents the degree of saturation in the wetting front, and s_i is the degree of saturation before rainfall.

If diffusion is assumed negligible at the end of an intensive rainfall, (1) can be simplified as (2):

$$h = \frac{kt}{n(s_f - s_i)} \qquad (2)$$

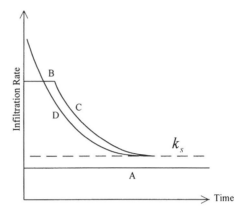

Figure 1.   Different cases of rainfall infiltration behavior (redrawn from Mein & Larson, 1973).

2.2 Stress paths of soil subjected to rainfall infiltration

Brand (1981) pointed out that rainfall-induced slope failure took place under conditions of almost constant total stress and increasing pore pressure. The discrepancy between the stress path normally followed in the triaxial test and that applied to the field conditions are illustrated in Figure 2. It can be readily shown that the stress path commonly followed in the triaxial test is not representative of the field conditions. The correct mechanism of failure can only be modeled in the laboratory by means of a constant load test in which the pore pressure is increased from an initial negative value until failure occurs.

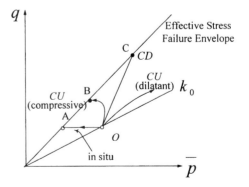

Figure 2. Field and laboratory stress paths (redrawn from Brand, 1981).

3 ANALYTICAL CONSIDERATIONS

Based on the characteristics of unsaturated soil and considering the influence of matric suction, slope stability analyses were carried out for the Linkou lateritic terrace. In the following, the engineering properties of soils obtained from laboratory and field tests previously reported in the literature are used.

3.1 Analytical method

Fredlund et al. (1981) proposed the General Limit Equilibrium method (i.e. GLE method) which could consider the matric suction effects on the soil shear strength. The conventional limit equilibrium slope stability method has been demonstrated to be special cases of the GLE method. This approach considers the matric suction term as part of the apparent cohesion (C). Therefore, the effective cohesion of the soil (c') used in the traditional analysis could be added with the matric suction term, as given in (3).

$$C = c' + (u_a - u_w)\tan\phi_b \qquad (3)$$

Where C = apparent cohesion of the soil, c'= effective cohesion, and $(u_a - u_w)$ = matric suction. ϕ_b is the angle indicating the rate of increasing shear strength relative to the matric suction. This approach has the advantage that the shear strength equation retains it conventional form. It is therefore possible to utilize a computer program written for saturated soils to solve the unsaturated soil problems. In this study, the well-known computer program "PC STAB5M" was used.

3.2 Assumptions and slope profiles

Slope stability analyses were conducted based on the following assumptions: (1) soil is homogenous and isotropic, (2) soil creep effect is neglected, (3) the influence on soil engineering property by surface vegetation is neglected, (4) the seepage force is not considered, and (5) ground water table is located in deep gravel stratum.

The assumed slope profile for the dry season is shown in Figure 3. The area of matric suction was divided into several zones, which were parallel with each other. Numbers in each zone represent the magnitude of matric suction in that area. The angle of these zones is assumed to be 45 degree with respect to the horizontal plane. The values of matric suction measured in the Linkou terrace by using the filter paper method ranged from about 10 to 200 kPa (Liu, 1990). The magnitude of matric suction is assumed to decrease from 300 kPa of the top zone to 50 kPa of the bottom zone. The differential matric suction value was 50 kPa between two neighboring zones. Each zone is of 1 meter thick.

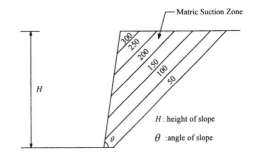

Figure 3. Slope profile in the dry season.

When the rainfall take place, a wetting front may be developed due to the infiltration of rainwater into the slope surface, as shown in Figure 4. The thickness of the wetting front will be discussed in the next section.

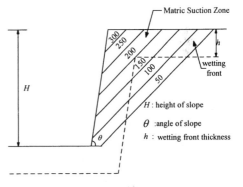

Figure 4. Slope profile in the rainy season.

3.3 Thickness of the wetting front

Recall the infiltration behavior described in section 2.1, only when the rainfall intensity is larger than the soil permeability the degree of soil saturation could increase. The infiltration front would then form near the ground surface and move gradually into the soil. The thickness of the wetting front can be calculated by (2).

The rain intensity must be greater than 36 mm/hr (the coefficient of soil permeability of the studied lateritic soil) in order to form the wetting front, if no run-off is considered. Assuming a return period of 50 years for the analysis and using the rain intensity-duration-frequency curve (Fig.5) from the Linkou rainfall data, the corresponding effective duration is determined to be 7.3 hours when the rain intensity is 36 mm/hr. Other parameters S_f, S_i, n, and k are determined as 1.0, 0.7, 0.495, 0.001 cm/sec (36 mm/hr), Respectively. Substituting these parameters into (2), the maximum thickness of the wetting front h is 1.77 m. However, according to the site condition, the run-off coefficient is 0.5. In such case, the rainfall intensity needs to be increased to 72mm/hr to form the wetting front. The duration corresponding to a rain fall intensity of 72 mm/hr is 2.5 hours. Then the thickness of wetting front became 1.2 m. In this study, the maximum thickness of wetting front is conservatively assumed to be 2 meters.

Table 1. Triaxial shear strength parameters of Linkou residual soil.

Test Type	Parameters
CID	$c'=0.2kg/cm^2, \phi' = 30°$
UUU	$c=1.3kg/cm^2, \phi = 24°$
SUU	$c=0.2kg/cm^2, \phi = 0°$

*CID= Isotropic Consolidated Drained Test
 UUU= Unsaturated Unconsolidated Undrained Test
 SUU= Saturated Unconsolidated Undrained Test

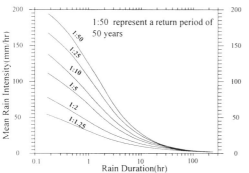

Figure 5. Rainfall data of the Linkou region (replotted from the data obtained from the Water Resource Planing Commission).

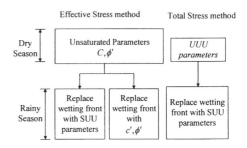

Figure 6. Soil parameters used in the slope stability analyses.

3.4 Soil parameters

Various triaxial shear strength parameters were summarized in Table 1 for the Linkou residual soil. Two sets of analyses were performed based on different combinations of these parameters. They are: (1) the effective stress method, and (2) the total stress method. Specific shear strength parameters used for each method are given in Figure 6.

For the effective stress method, apparent cohesion C is the summation of effective cohesion c' and the matric suction term ($(u_a - u_w)\tan\phi_b$) in the dry season. The effective strength parameters c' and ϕ' are determined from CID tests in this situation. Due to insufficient experimental data at present, ϕ_b was reasonably assumed to be 15°, 25° , 35° in the analyses.

When rainfall occurs, the wetting front will develop due to the infiltration of rainwater into the soil. As a result, the matric suction may decrease to zero in the wetting front. Thus the shear strength of the wetting front can be replaced by the saturated effective stress parameters (c' and ϕ'). Based on the proposed stress path by Brand (1981) as has been described in section 2.2 for the rainy season, SUU parameters can be used as an alternative.

For the total stress method, shear strength parameters determined from UUU tests were used for the dry season situation. For the rainy season, SUU parameters were used to simulate the reduced shear strength of the wetting front.

4 ANALYTICAL RESULTS

Both the analytical results using the effective stress method and the total stess method will be discussed below. Slope stability analyses using the effective stress method were conducted for various hypothetical cases by varying three variables. These variables include slope height (H), slope angle (θ), and friction angle of the matric suction term (ϕ_b). Values used in the analysis are shown in Table 2. For each analytical case, five scenarios were analyzed: (1) dry season, (2) rainy season with a wetting front of 1 meter thick and using c' and ϕ' for the wetting front, (3) rainy season with a wetting front of 1 meter thick and using SUU parameters for the wetting front, (4) rainy season with a wetting front thickness of 2 meter and using c' and ϕ' for the wetting front, and (5) rainy season with a wetting front thickness of 2 meter and using SUU parameters for the wetting front. The results are summarized as follows:

(1)   The results for constant slope height (H) of 20m but with different slope angle (θ) and ϕ_b are shown in Figure 7. The factor of safety apparently decreases with the increase of the slope angle regardless the value of ϕ_b for every scenario.

(2)   The results for constant slope angle (θ) of 75° but with different slope height (H) and ϕ_b are shown in Figure 8. It is obvious that the factor of safety decreases with the increase of the slope height. When the slope

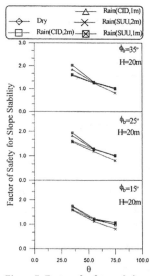 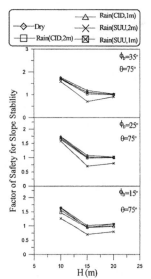 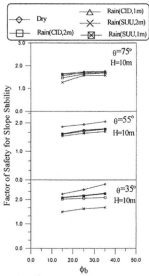

Figure 7. Factor of safety and slope angle (θ) for constant slope height H=20m.

Figure 8. Factor of safety and slope height (H) for constant slope angle $\theta = 75°$.

Figure 9. Factor of safety and matric suction term (ϕ_b) for constant slope height H=10m.

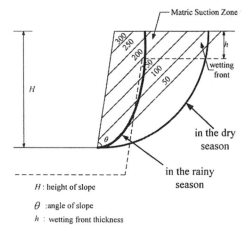

H : height of slope

θ : angle of slope

h : wetting front thickness

Figure 10. Potential failure surfaces.

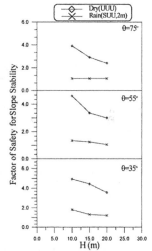

Figure 11. Analytical results of total stress method.

Table 2. Variables used in the effective stress method.

Variables	Values
slope height (H)	10m ,15m, 20m,
slope angle (θ)	35° , 55° , 75°
ϕ_b	15° ,25° , 35°

height increases from 10m to 15m, the factor of safety decreases sharply. The reason is that the matric suction zone occupies a larger portion of the slope for the 10m case.

(3) The results for slope height (H) of 10m but with different slope angle (θ) and ϕ_b are shown in Figure 9. The factor of safety increases with increasing ϕ_b for the dry season case. The increase is significantly less for the $\theta = 75°$ case, due to the fact that the slip surface passes through lower suction zone. Similarity, the effect of ϕ_b on the rainy season cases is also limited.

(4) Among the five scenarios the factor of safety is the highest in the dry season and the lowest in the rainy season if the SUU parameters were used for the wetting front of 2 meter thick. The analytical results support the observation that the unsaturated soil slopes maintain stable in a steep angle because of the matric suction in the dry season. In the rainy season, the rainwater would infiltrate into the soil, soften its shear strength and may eventually lead to slope failure.

(5) The potential failure surface in the dry season and the rainy season are shown in Figure 10. It is obvious that the potential failure surface is of deep sliding type in the dry season. For the rainy season, the slip surface is of shallow type and near the wetting front.

For the total stress method, the slope height (H), slope angle (θ), and ϕ_b were assumed as 20m, 75°, and 25°, respectively. The analytical results are shown in Figure 11. The factor of safety is very high in the dry season. The reason is that the shear strength determined from the UUU parameters used in the analysis is significantly higher than other types of soil parameters. The factor of safety becomes low enough to cause slope failure in the rainy season if the SUU condition is reached in the wetting front.

The above results agree very well with the case records reported by BSC (1989) in the Linkou terrace. It was also noted in the record that " the depth of slip surface is generally less than 2m." in short, slope failure in the Linkou terrace is mostly like to occur mainly in shallow depth due to rainwater infiltration. Therefore, deep drainage would not be very effective in preventing sliding in the rainy season. On the other hand, measures that can retard the rainfall infiltration, erosion, and forming of the wetting front are probably more effective in maintaining stable slopes. In this regard, slope surface protection and vegetation are most viable options.

5 CONCLUSIONS

Based on the analytical results of the slope stability of typical slopes of the Linkou lateritic terrace, the following conclusions can be derived.

(1) Matric suction increases the shear strength of unsaturated soils and thus increases the factor of safety of the slope in the dry season. Rain infiltration may result in the formation of a wetting front and reduction of the matric suction. Consequently, the apparent cohesion of the unsaturated soil would be decreased. Thus, a shallow-type of failure may take place within the wetting front.

(2) Most suitable preventive measures against the rainfall induced slope failure are slope surface covering and vegetation. These measures can prevent excessive rainfall infiltration and hinder the forming of the wetting front. Horizontal drains are ineffective in the unsaturated zone near the surface. Because the percolating water may saturate soil and decrease its apparent cohesion. Consequently, the slope may become unstable.

REFERENCE

Brand, E. W. 1981. Some thoughts on rain-induced slope failure. *Proc. 10th ICSMFE* (3): 373-376.
Brand, E. W. 1984. Landslides in southeast Asia, a state-of-the art report. *Proc. 4th int. symp. on landslides,* Toronto, Canada: 17-59.
Bureau of Soil Conservation 1989. *Investigation of landslides in Taiwan*: 33-46. (in Chinese)
Fredlund, D. G. et al. 1981. The relationship between limit equilibrium slope stability methods. *Proc. 10th ICSMFE* (3): 409-416.
Fredlund, D. G. & Rahardjo, H. 1985. Theoretical context for understanding unsaturated residual soil behavior. *Proc. 1st int. conf. geomech. in tropical lateritic and saprolitic soils,* Sao Paulo, Brazil: 295-306
Fredlund, D. G. 1987. *Slope stability.* John Wiely & Sons Ltd: 113-144.
Liu, C. N. 1990. *Study of slope failure in Linkou terrace.* National Taiwan University, Master Thesis, Taipei, Taiwan. (in chinese)
Lumb, P. 1975. Slope failures in Hong Kong. *Quarterly Journal Engineering Geoloy* (8): 31-65.
Mein, R. G. & Larson, C. L. 1973. Modelling infiltration during a steady rain. *Water Resource Research* 9(2): 384-394.

Unsaturated Soils for Asia, Rahardjo, Toll & Leong (eds) © 2000 Taylor & Francis, ISBN 90 5809 139 2

Suction and infiltration measurement on cut slope in highly heterogeneous residual soil

T.H.Low
Jurutera Perunding GEA(M) Sdn. Bhd., Malaysia

H.A.Faisal & M.Saravanan
Civil Engineering Department, University of Malaya, Kuala Lumpur, Malaysia

ABSTRACT: Instability is an extremely important consideration in the design and construction of man-made slopes and natural slopes. Slope failures and landslides are influenced by geologic topographic and climatic factors. In tropical region, most of the slope failure occurs during severe rainfall. Rain water infiltrates into the slope and reduce the soil matric suction the shear strength of the soil. In most condition residual soils are in unsaturated condition and therefore matric suction is an important factor to be considered in the design of slopes. Soil suction has an important bearing on water entry, structural stability, stiffness, shear strength and volume change. Soil matric suction, water content and the solute content variation with time are often the most important variables in unsaturated soil engineering design. The field instrumentation for automatic continuous measurement of soil matric suction, rainfall data and test results of field sprinkler system are presented in this paper.

1 INTRODUCTION

Construction activities in hilly terrain residual soil frequently confront geotechnical engineers with slope instability problems. Failures in both natural and cut slopes of Peninsular Malaysia are usually brought about by rainfall during the monsoon season. The upper layer of residual soil profile is always partially saturated , and has a relatively higher permeability to infiltrating rainwater. This causes the pore water regime to be governed largely by rainfall pattern.

The mechanism of slope failure is that the infiltration of rainwater that causes a reduction of matric suction in the unsaturated soil zone, resulting in a decrease of effective stress. This in turn reduces the shear strength to a point where equilibrium can no longer sustained in the slope. Good correlations between rainfall intensity and frequency of landslides were reported by some researchers from Hong Kong, Japan, United State and NewZealand.

The instrumentation in this study was attempted to study the change of soil matric suction with the rainfall on a cut slope along the link road of The Kuala Lumpur International Airport (KLIA) Malaysia. The cut slope mainly consists of two types of weathered sedimentary residual soil, i.e., weathered sandstone and shale. These residual soils come in alternate bedding which is almost vertical. The weathered sandstone beds were selected for the study. Figure 1 shows the geological formation of the study site. The soil consists of very fine sand and silt. The slope is covered by different types of synthetics (biodegradable) and non synthetic covers (poly-jute) after hydro-seeding to prevent erosion . Instrumentation was carried out on every berm with respect to different weathering grades of soil.

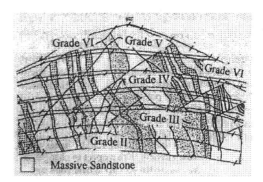

Figure 1. Geological formation of the study site

2 ROLE OF SUCTION IN SLOPE STABILITY

The principle of effective stress for unsaturated soil was first used by Terzaghi (1923) and proposed by him in the first International Conference on Soil mechanics in 1936. Numerous researchers have carried out work since then in order to confirm. Jennings and Burland (1962) questioned the validity of the principle for unsaturated soil mechanics. Following an extensive research program on unsaturated soil conducted in Imperial College the shear strength of partially saturated soil was hypothesised (Bishop, 1959) to be a function of an effective stress defined as:-

$$\sigma' = (\sigma - u_a) + \chi (u_a - u_w) \qquad [1]$$

Where σ' and σ are the effective and total stresses respectively, u_a is the pore air pressure and u_w is the pore water pressure. χ is a function that depends on the saturation with value 1 at 100% saturation and 0 for completely dry soil.

Fredlund and Morgentern (1987) showed from a stress analysis that any two combination of the three possible stress variables $(\sigma - u_a)$, $(\sigma - u_w)$ and $(u_a - u_w)$ can be used to define unsaturated soil. The equation for unsaturated shear strength τ is written in terms of the stress state variables for an unsaturated soil and is an extension of the form of equation used for saturated soils

$$\tau = c' + (\sigma - u_a) \tan\phi' + (u_a - u_w) \tan\phi^b \qquad [2]$$

where ,

c'   = effective cohesion
σ = total stress
u_a = pore air pressure
u_w = pore water pressure
ϕ' = effective angle of friction
$(u_a - u_w)$ = matric suction
ϕ^b = gradient with respect to changes in $(u_a - u_w)$ when $(\sigma - u_a)$ is held constant.

The factor of safety for slope stability analysis using method of slices can be derived using shear strength equation [2] above. The shear force mobilised at the base of slice can be written as:-

3 INSTRUMENTATION

20 numbers of tensiometer and 20 numbers of moisture block and a rain gauge were installed on the slope to monitor the changes of matric suction with respect to rainfall. The tensiometers and moisture blocks were installed at different depth of, 0.5m, 1m, 3m and 4m. An automatic data acquisition system was designed to record the output from tensiometers and moisture blocks. Automatic data acquisition system solves the problems of reliability, accessibility

$$S_m = \beta/F \{c' + (\sigma - u_a) \tan\phi + (u_a - u_w) \tan\phi^b \} \qquad [3]$$

Where
S_m = the shear force mobilised on the base of each slice.
F   = the safety factor
β = the sliding surface of the slice.

and safety, which were difficult with manual data recording. The data acquisition system was supported by a solar powered set and specifically designed for low power consumption with continuos logging.. The logging intervals were achieved by prescribing the appropriate interval during the set-up process. Two time intervals were adopted in this study, i.e., 30 minutes interval and 10 minutes. The system was set in such that when rapid changes in suction value the 10 minutes interval will be utilised, or else, the 30 minutes interval was used. The recorded data were downloaded from the data logger direct to a portable notebook computer through an RS232 interface.

An automatic logging tipping bucket rain gauge was installed at the study site. The rain gauge records rainfall events on a real time basis. The clock of the data logger for the tensiometers and moisture blocks and the rain gauge recorder are synchronised. Figure 2 shows the schematic arrangement of the instrumentation at the study site.

The infiltration characteristics of the soil were deduced by using a sprinkler system installed at the site. A V- Notch fixed with a flow meter was used to measure the surface run-off in a control section in the study (refer to Figure 3). In addition, an infiltrometer P-88 was used to obtain the infiltration capacity at the site with the sprinkler system test results. To investigate the water flow characteristics and suction variation in details, small tip tensiometers (refer to Figure 4.0) were also installed during the sprinkling process.

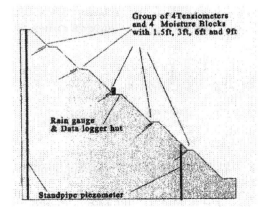

Figure 2.   Instrumentation Layout of the study Site.

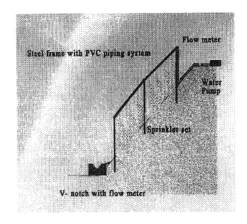

Figure 3. Field Infiltration Sprinkler System

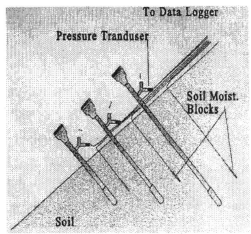

Figure 5. Sensors Layout For Each Berm.

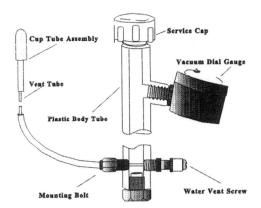

Figure 4. Small tip tensiometer

4 MATRIC SUCTION MEASUREMENT

In this study, matric suction of the soil was measured using jet-fill tensiometer and moisture block (Soil Moisture Corporation U.S.A). Moisture blocks were used because tensiometer can only measure matric suction below 1 bar of negative pressure while moisture block can measure more than 4 bar. The matric suction obtained from the tensiometer is the difference between the gauge reading and the head of the water in the stamp. The longer the tensiometer, the lower the suction it can measure.

In this study the tensiometer were installed perpendicular to the slope surface to reduce the head of water (refer to Figure 5).

Figure 5 shows the layout of the tensiometers for the sprinkler model at the study site. S1 to S7 in Figure 6 represent the position of the small tip tensiometers.

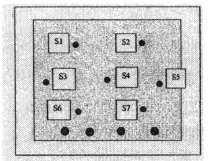

Figure 6. Typical Layout of Tensiometers For Sprinkler Model

The accuracy of the moisture block when the suction is below 1 bar is not as good as tensiometer. Both the tensiometer and moisture block were installed at the same depth for comparison and cross checking purposes.

5 INSTALLATION OF INSTRUMENT SENSORS

In this study, normal coring tools could not be used because the soil is brittle and hard. A special in-house designed motorised auger was fabricated for the installation purposes. The motorised auger consists of two motors with ½ and ¼ horse power. The quarter horse power motor was fitted at the top of the machine to push the auger into the slope as the half horse power motor was used to rotate the auger. Both the motors were controlled with a speed controller. During installation, four sand bags were placed at the base of the augering machine and the speed of the motors were properly controlled to prevent the auger machine from being lifted up when drilling through hard layers.

The auger used in this study was designed to suit the soil condition at the study site. The most suitable distance between the flights was determined to be 25mm in order to ease the process of augering. Due to the difficulty in fabricating longer auger, 2 or 3 lengths of short extension flights auger were used.

After the hole had been augered, a special tool was used to remove the residual from the hole. Precaution was taken to ensure good contact with the soil is necessary in order for the tensiometer and moisture block to function properly.

Moisture blocks were inserted into the bored hole using a fabricated tool in the form of a long rod with a modified tip. The soils were tamped firmly after placing the blocks and bentonite is placed at the surface of the hole in order to prevent the hole from becoming a water passage.

All the wire leads were inserted into a poly-pipe and are buried in a shallow trench in the ground. The wires were connected to the data logger situated at the midway between the berms. The adverse tropical climate and vandalism were major concerns in the installation. A lockable steel security cage grouted to the slope surface protects the installed tensiometers and moisture blocks. The data acquisition system was contained in a lockable steel hut. All the transducers for the tensiometers were protected from sunlight by wrapping them with a double layered foam rubber on the inside and aluminium foil on the out side.

6 DISCUSSION

Suction Variation With Rainfall

Figure 7 shows a typical suction variation with rainfall (one-month duration) for one of the berm. It clearly shows that as the depth increases, the matric suction reduces.

During the time interval of 14000 to 24000 minutes, there were no rainfall and and the four tensiometers experiences higher matric suction. When the rain starts to fall, matric suction does not reduce immediately. Due to the infiltration of rain water into the ground, after rainfall the matric suction reduce slowly for all depths.

From Figure 7 , In the time interval of 0 to 13000 minutes, suction for 3.0m depth tensiometer gives very low suction values. This is mainly due to the rainwater, which has infiltrated during the earlier rainfall periods.

From the field infiltration test, the water infiltration rate is 2.31 x10^{-6}m/s. The Rainfall intensity at the site can be as high as ., 1.13 x 10^{-4}m/s. From Figure 7 , it is clearly shown that during heavy rainstorm, the suction drops very fast and also recovers fast. One of the reasons is that the rainwater infiltrate into the slope through quartz veins in the soil.

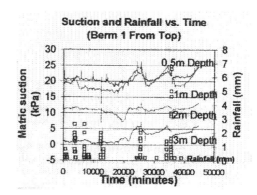

Figure 7. Typical suction variation with rainfall for one-month duration.

Rain Simulation Test Results

Figure 8 shows the typical suction variation at berm 4 from the top slope during the sprinkling process.

Four series of test were carried out with various types of surface cover and conditions. During the rainfall simulation process, continuous intense sprinkling was carried out for 120 minutes and concurrently suction changes were monitored. Monitoring of suction was continued even after the sprinkling to a time where the suction stabilize.

The suction reading for small tip tensiometers S2, S3 , S6 and S7 in all graphs in Figure 8 increases when the sprinkling process starts. This might be due to the effect of water flowing on slope that creates slight negative pressure at shallow depth of the slope. The increment is more significant for small tip tensiometers installed at the lower level of the slope because the water runoff velocity is higher at the slope toe.

For 120 minutes of sprinkling no significant changes in matric suction at depth i.e., beyond 2.0 meter from the slope surface. This is mainly due to very low infiltration rate of the soil ,6.07 x 10^{-3} mm/s.

From Figure 8 , shows all four condition of test, with the matric suction recovered slightly after the sprinkling process. The tests with grass as surface cover shows a greater and quicker increment in suction due to the effects of the grass roots

7 CONCLUSIONS

The automatic data logging system for monitoring tensiometers, moisture blocks and other devices has been detailed. The desired attributes of the system have been reasonably achieved and the advantages of fully electrical installation were credited to its flexibility and continuity of data obtained. Installation

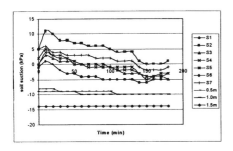

a) Slope covered with grass with erosion protection geotextile

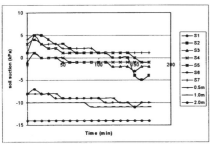

b) Slope covered with grass and erosion geotextile but was cut to 1 inch height

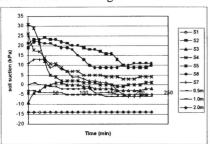

c)Slope covered with erosion protection geotextile

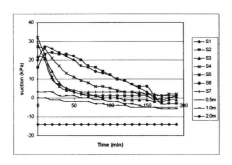

d) Bare slope surface

Figure 8. Results of suction variation for sprinkler test

method, minimise soil disturbance contribute to accurate data from the instruments.

Generally as the depth increases, matric suction reduces. But, in some circumstances, due to the flux condition, this can be the other way round.

Heavy rainfall intensity may not cause instability to slope but prolong rainfall can be significant.

Geological features like quartz vein could be significant in slope instability because it increases the rainwater infiltration.

Surface covers of slope also contribute to matric suction variation during rainfall. Bare surface causes higher reduction in matric suction during rainfall..

8 ACKNOWLEDGEMENT

Acknowledgement is due to Road Section of Public Work Department, Malaysia for providing part of the research grant. The authors also wish to acknowledge Mr. Lee Yik Seong and Mr. Theva Kumar s/o Kanan for assisting in the field sprinkling study.

REFERENCES

Abdullah, Affendi M.L. & Ali Faisal, "Field Instrumentation for Slope Stability in Residual Soil ". GEOTROPIKA 92' Johor Bahru, Malaysia.

Bishop, A.W. and Blight, G.E. (1967) "Effective stress evaluation for unsaturated soils" Jour. Soil Mech. & Found. Engg. Div., ASCE 93 (SM2), pp. 125 – 148.

Fredlund D.G. "Appropriate concepts and tecnology for unsaturated soils". Second Canadian Geotechnical Colloquim, Canadian Geotechnical Journal, No. 16 , 1979, pp 121-139.

Rahardjo.H, Loi, J., and Fredlund, D.G., "Typical matric suction measurements in the laboratory and in the field using thermal conductivity sensors", presented at Indian Geotechnical Conference (IGC-89), Vol.1, Visakhhapatnam, December 1989.

Low T.H., Faisal Hj. Ali. "Water infiltration on slopes" World Engineering. Congress, Kuala Lumpur,1999.

Low T.H., Faisal Hj. Ali, Saravanan.M."Field suction variation with rainfall on cut slope in weathered sedimentary residual soil", Slope Stability Engineering IS-Shikoku'99, Japan

Unsaturated Soils for Asia, Rahardjo, Toll & Leong (eds) © 2000 Taylor & Francis, ISBN 90 5809 139 2

Modelling variables to predict landslides in the south west flank of the Cameroon volcanic line, Cameroon, West Africa

E. E. Nama
Institute of Environmental Systems, Salford University, UK

ABSTRACT: The high rates of crustal movement and frequent seismic events mean that neotectonic terrains are subject to extensive development of natural hazards. Such terrains in tropical areas like the west flank of the volcanic province of Cameroon is subject to high annual precipitation levels (5000-9000 mm). The effect of precipitation on local soil texture increases the rate of landscape change.

Modelling the irregular shape of the volcanic region is not a simple problem, and the possibility of ever knowing the true elevation at every point is excluded. This paper examines surface elevation models based on samples of elevations. Two competing structures have been used to represent terrain in this study: Lattices and Triangular Irregular Networks (TINs). GIS, based on digital elevation models, offers the potential to be able to map lineament density, slope angles and facets and permit the modelling of a variety of stability criteria to predict surface processes such as slope failure.

1 INTRODUCTION

The occurrence of vertical crustal movements and frequent seismic events in the Mount Cameroon region mean that this region, as with many neotectonic terrains, is subject to the extensive development of natural hazards. This earthquake prone terrain is subject to high annual precipitation levels, greatly increasing the rate of landscape change and the development of natural hazards. Thus, there is an urgent need in this region to evaluate the geomorphology of the terrain to be able to delineate hazards with a view towards a predictive model. This paper describes the use of GIS derived data in collaboration with volcanic soil and precipitation data in detecting slope susceptibility to landslides on the slopes of Mount Cameroon (Fig. 1).

The Mount Cameroon region represents the site of an important population concentration in south west Cameroon as a result of the rich volcanic soils sustaining rubber, palm oil, banana and tea plantations. Hence, there is a need to derive a cost-effective, efficient and environmentally sensitive model to detect landslide susceptibility.

2 CLIMATE AND SOILS

The study area is affected by the equatorial climate regime with two distinct seasons. The Western flank

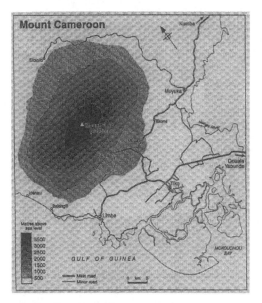

Fig. 1 The Mount Cameroon region

is the wettest part of the region with almost 9000 mm of total annual rainfall in almost 200 rainy days (Cameroon Development Corporation, 1999). The intensity of rainfall and rainy days reduce further north and east. On average the number of total annual sunshine hours per day is 1,270.2 in Debund-scha, 1200.8 in Idenau, 1521.5 in Limbe and 1120 in Buea (Cameroon Development Corporation, 1999). The mean monthly maximum and minimum temperatures for Buea are 23° C and 17.2° C and Limbe 27° C and 21° C.

The soils of the coastal lands of Cameroon west of Douala are the result of weathering, transportation and deposition from the steeper slopes of the region. Large scale volcanic activities, sedimentary coastal and fluvial deposition, differential rainfall patterns and temperatures, erosion and the process of gradual uplifting, have all contributed to the pedologic variety of the area. The area has a good drainage system and permeability of the soils range from moderate to rapid.

3 PHYSICAL SETTING

The tectonic and geological setting of Mount Cameroon is complex (Gèze, 1943). It is a member of an alignment of volcanoes stretching from islands of the Atlantic ocean (Pagalu, Sao Tomé, Principe and Bioko) to the main land (Mounts Cameroon, Manengouba, Bamboutos and Oku), this alignment is known as the Cameroon Volcanic Line (CVL) (Ateba and Ntepe, 1997). On its northern extension, the CVL is divided into two parts, one trending northwards to Bui plateau, the other westward to Chad through the Foumban shear zone and the Adamawa plateau. Mount Cameroon is a strato-volcano with its summit at 4100 m and it has erupted six times this century (1909, two in 1922, 1954, 1959, 1982, 1999) (Déruelle et al. 1987; Nama, 1999). It is made up of three series of basaltic lavas: the black, white and medium series (Gèze, 1953). Many faults have been recognised in this region especially near the vents of the 1999 volcanic eruptions (Nama, 1999). The south east and the summit areas are the most seismically active locations in the region with the south west flank of the mountain being the most volcanically active part.

4 METHOD

Digital elevation modelling (DEM) as a GIS technique in generating 3-D terrain models of the Mount Cameroon region has been important in the detection of lineaments. The underlying concept is that the model will reflect the nature of the terrain for tectonic identification and isolation. Essentially, the method relies on identifying morpho-tectonic variables and overlaying the data set for each of these variables to detect proximity of factors for landslide susceptibility analysis. The impact of precipitation

and soils in this region are evaluated with the model to detect areas most prone to landslide failure in the slopes of the south west flank of Mount Cameroon.

A Triangular Irregular Network (TIN) topology for the study area was created by manually digitising a 1 : 50 000 topographic map which approximated the terrain surface by a set of triangular facets. Most TIN models assume planar triangular facets for the purpose of simpler interpolation in applications such as contours (Lee 1991). The purpose of using a TIN method in this study is to convert the dense vector data to a TIN model in such a way that the surface defined is as close to the DEM surface as possible. The tolerance level of the data during transformation determines to a large extent the accuracy of the output model. Ideally, it is preferable to have as small a tolerance as possible and to maintain as small as possible a data set. However, it was noted that smaller tolerance resulted in larger sizes of extracted TIN and vice versa. To maximise the output representation of the data, a moderate tolerance was used. The lattice method was also used in representing the terrain in 3-D. The data input comprised contour lines in vector format which were dynamically rasterised into floating point topology using TIN-LATTICE (ESRI 1991) with a resolution of 60 m and a 20 m spacing. The TIN database from Arc/Info was converted to a raster structure, through TIN-to-grid linear algorithms (Fig. 2). Since such a structure transformation may degrade the quality of the elevation models, contour lines were first calculated from both the Arc/Info TIN data and the derived lattice. By overlaying the two contour sets, only very small discrepancies resulted. Hence, the TIN-to-grid conversion did not significantly alter the information content of the original TIN structure Ideally, a cell resolution of 10 m would have been perfect to adequately represent more accurately the terrain surface (DEM). However, the choice of a 51.082 m cell resolution was determined by the capacity of the computer workstation; hence the best possible digital terrain surface representation. Carrara et al. (1997) revealed that a similar method of interpolation was undertaken with the same results in Calabria, southern Italy while comparing techniques for generating digital terrain models from contour lines.

The TIN and Lattice models were used for morphological hydrologic modelling to examine the distribution of runoff in the catchment. The drainage network reflecting basement lineaments was digitised.

The surface DEM mapping has been applied in this study as an alternative derivative for the detection of brittle tectonic structures such as fractures, faults, large-scale fractures and fracture zones. The scale of the mapped fractures are commonly denoted as lineaments (O'Leary et al. 1976). This method as in remote sensing entailed a subjective visual identification of lineaments extracted manually on a transparent overlay of the Lattice model. Lineaments in a DEM are defined by a drop in elevation for a short distance (Wladis 1999) therefore, they have been described by a certain frequency in this region in terms of their representation in a digital elevation model.

The concept for transformation of a DEM into a shaded relief image first explained by Batson et al. (1975), was implemented in this study. Three dimensional renditions of the DEM were produced in which different visual effects were achieved by varying the values of surface/sun orientation (215°), sun-elevation angle (30°) and vertical exaggeration (800 m) to obtain the best brightness information of the shaded-relief to identify and map lineaments (Fig. 3).

5 RESULTS

By using the DEM to visually detect and manually digitise lineaments, a total of 529 lineaments were identified which did not reflect the entire lineament density of the region. However, using drainage network as a basis for lineament detection in this region proved more successful as a total of 573 lineaments were identified forming 51.9% of the total lineaments in the region. This result suggests that river network more readily reveals lineaments.

The density of detected lineaments in this region vary with total precipitation distribution of the region (Fig. 3). A total of 1102 lineaments were detected and mapped from the DEM.

Soil samples collected from four sites in this region show a variety of constituents with various grain size characteristics (Table 1).

Sample location	% Clay	% Silt	% Coarse silt	% Sand	% Gravel
Site 1	8.70	0.21	0.81	90.91	53.01
Site 2	4.40	2.73	0.27	92.57	61.26
Site 3	25.19	29.09	28.84	16.88	25.42
Site 4	15.72	19.42	0.65	64.21	75.19

Table 1 Soil samples and percentage of constituents from four sites in the study area

The effects of precipitation and infiltration on clay fine silt, coarse silt, sand and gravel are shown in the landslide susceptibility model in Figure 4.

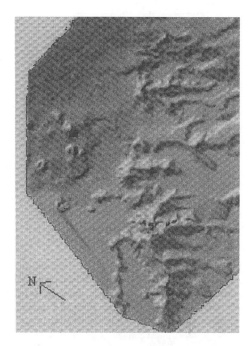

Fig. 2 Tin-Lattice terrain model of the study area

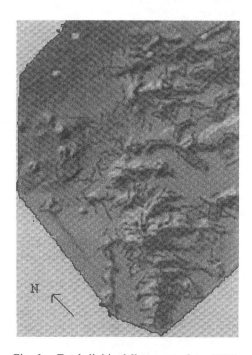

Fig. 3 Total digitised lineaments from DEM and drainage network of the study area

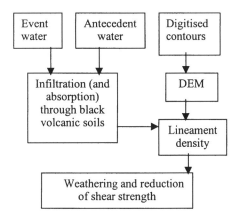

Event water	Antecedent water	Digitised contours

Infiltration (and absorption) through black volcanic soils

DEM

Lineament density

Weathering and reduction of shear strength

Fig. 4 Landslide susceptibility model

In this model the importance of soil texture and the excellent drainage conditions of the volcanic soils of Cameroon during rainfall increases the effects of erosion and weathering along the lines of discontinuities in the structure. This 'overloading effect' enhanced by the absorption capacity of clay in the soils increases the pore pressure and eventually reduces the strength of slopes especially where lineament density is high.

6 DISCUSSION

6.1 Systematic spatial variability of lineament density and volcanic soils

It is hypothesized that the location, direction and density of lineaments significantly influence the stability of slopes in the study area. Two forms of influence are postulated: (1) direct influence, which assumes that slope failure has been partially or wholly caused by one or more lineaments (2) indirect influences, which assumes lineaments affect the orientation and length of slopes and channels, and which in turn affect the relative susceptibility to slope failure.

Fracture traces or lineaments are likely to stimulate slope failures for several reasons. First, they form discontinuities of low strength, which interact with gravitational forces wherever local relief and dissection permit. Second, they tend to concentrate infiltration water, giving rise to localised increases in pore pressure. Third, faulting may create lithological junctions that bring permeable and impermeable materials into contact and increase the conditions for a rise in pore water pressure. Fourth, different vertical movements associated with faulting may create scarps and subsided blocks which

erosional forces may degrade. Fifth, fracture traces are likely to be attacked selectively and preferentially by the agents of erosion.

The distribution of lineaments show a strong concentration in the south west and south where the concentration of sand per 100 cm have ± 85% sand and ± 15% clay suggesting high infiltration capabilities of the volcanic soils leading to subsequent weathering of the basaltic rocks and the collapse of susceptible slopes in the region. In this neotectonic terrain, the slopes in the south west and south are steep and are exposed to very high annual total precipitation which means they form the most susceptible slopes in the region. This assertion was confirmed from morphometric findings during a field trip to the region which clearly showed that more than 48 % of these slopes are greater than 15°.

Although soil depth is a function of clay soil moisture retention capacity, the micro-variation that occurs in this region nonetheless make some slopes more prone to increased stress than others.

7 CONCLUSION

Although the slopes of Mount Cameroon have, in the past, received little attention it is clear that the recent volcanic activities are drawing more attention to specific earth science studies. This paper has introduced a GIS perspective by detecting landslide susceptibility, which is hoped will increase the awareness of the hazard in this region.

The spatial variability of lineaments and soil grain size percentage in the south west flank of the mountain have implications for environmental planning and risk assessment studies. Observed lineament density derived from digital elevation modelling and annual rainfall data of the region suggest that the south west and south slopes of Mount Cameroon are the most susceptible to landsliding.

REFERENCES

Ateba, B. & Ntepe, N. 1997. Post-eruptive seismic activity of Mount Cameroon (Cameroon), West Africa: a statistical analysis. Journal of Volcanology and Geothermal Research, 79, 25-45.

Batson, R.M. 1975. Computer generated shaded relief images. Journal of Research U.S. Geological Survey, 3, 4, 401-408.

Cameroon Development Corporation, 1999. Group oil palm, tea and banana estate rainfall data.

Carrara, A. Bitelli, G. & Carla, R. 1997. Comparison of techniques for generating digital terrain models from contour lines. International Journal of Geographical Information Science, 11, 5, 451-473.

Gèze, B. 1943 Géographie physique et géologique du Cameroun Occidental. Memoire du Muséum National d'Histoire, Paris, XV11, pp-12

Nama, E.E. 1999. The relationships between earthquakes, structure and landslides on the slopes of Mount Cameroon using GIS and remote sensing. Unpublished PhD thesis, University of Salford, UK.

O'Leary, D.W. Friedman, J.D. & Pohn, H.A. 1976. Lineaments, linear, lineations: Some proposed new standards for old terms. Geological Society of American Bulletin, 87, 1463-1469.

Wladis, D. 1999. Automatic lineament detection using digital elevation models with second derivative filters. Photogrammetric Engineering and Remote Sensing, 65, 4, 453-458.

Unsaturated Soils for Asia, Rahardjo, Toll & Leong (eds) © 2000 Taylor & Francis, ISBN 90 5809 139 2

Laboratory experiments on initiation of rainfall-induced slope failure

M. Nishigaki & A. Tohari

Department of Environmental Design and Civil Engineering, Okayama University, Japan

ABSTRACT: The results of a study of the process leading to the initiation of rainfall-induced slope failure are presented. The study included measurements of pore-water pressure at failure initiation on a series of soil slope models. The experimental results indicated that the main failure of soil slopes was initiated by development of localized unstable area at the toe of the slope models upon the formation of seepage face. The measurements of pore-water pressure showed that the soil at the slope toe was saturated at failure. This clearly demonstrates that rainfall-induced slope failure is a consequence of instability of the slope toe area of the slope induced by water seepage under drained condition. However, the major portion of sliding mass may still be in an unsaturated condition.

1 INTRODUCTION

Rainfall-induced slope failures are one of the most destructive natural disasters, which frequently occur in natural and man-made soil slopes in many areas around the world. Such slope failures have affected human lives and properties. In Japan, rainfall-induced slope failure has resulted in the loss of 30 lives in a recent occurrence in Hiroshima Prefecture.

Recently, some investigations have been carried out to study the condition leading to such slope failures (Iseda & Tanabashi, 1986; Kato & Sakajo, 1999; Sasaki et al. 1999; Sakajo & Kato, 1999). While all investigators agree that the slope failure during rainfall occurred due to the rise of groundwater level, and the increase of positive pore-water pressure is a major causative factor of these types of failures, the actual process leading to failure initiation has not been well understood and documented. The understanding on failure initiation is essential for performing analysis and taking subsequent measures to predict and to prevent the occurrence of rainfall-induced slope failure.

The objective of the work presented in this paper was to investigate the process leading to initiation of rainfall-induced slope failures and the critical influence of seepage on failure initiation.

2 EXPERIMENTAL PROGRAM

Failure was induced in a number of 1 m high soil slopes by raising water level within the slopes. Three different modes of raising the water level in the experimental slope were considered in the experiments. In the first experiment, a slow rise of water level was introduced to the slope model from a constant head tank at the rate of 5 cm/h. The second experiment considered a condition of rapid water level increase by raising the constant head tank at the rate of 0.63 cm/min. In the third experiment, rainfall at intensity 100 mm/h was simulated to induce the rise of water level in the slope model. Prior to the first two experiments, a water level of about 20 cm was introduced to produce initial water level across the base of each slope from a constant head tank.

3 EXPERIMENTAL APPARATUSES

3.1 *Soil Properties*

In all the experiments, sandy soil was used for construction of the slope models. Soil particles varied from fine sand to coarse sand. The properties of the sandy soil determined by laboratory experiments are summarized in Table 1.

Table 1. Properties of the sandy soil.

D_{10} (mm)	D_{60}/D_{10}	Density of particle, ρ_s (g/cm^3)	K_s (cm/s)	Cohesion (kPa)
0.175	7.14	2.69	0.029	0.0103

3.2 *Slope failure tank*

The slope failure tank as the basic apparatus and the profile of the slope models are shown schematically in Figure 1. This large metal tank was designed to accommodate 1 m high soil slope that could be brought to failure by seepage of water towards the toe. The dimensions of the slope model were chosen to reduce the effects of side friction limit (width and height ratio of 1.0), and not to confine the nature of slope failure. The slope angle was 45°, which is close to the angle of repose of 50°. The whole structure was conservatively designed to limit deflection. To generate an impervious boundary at the end of the slope model, a 1 cm acrylic board of 25 cm height was secured at the down slope end. The properties of the slope models for each experiment are given in Table 2.

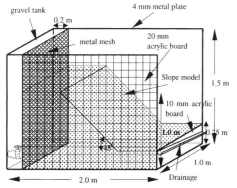

Figure 1. Slope failure tank.

An acrylic board of 15 mm thick was installed on the one side of the tank for allowing simple installation of instrument system and observation of the deformation process. Prior to each experiment, blue tracers were injected through the injection holes to visualize the flow lines within the slope profile.

Water was allowed to enter the slope from the constant head tank through a coarse gravel-filled tank with porous sheet outlets to generate lateral inflow in the first two experiments. To inhibit direct sliding of the base along the soil/floor interface, a thin layer of sand was created over the floor after glue was applied on the floor of the tank.

Table 2. Properties of experimental slope models.

Exp. No	Dry density, ρ_d (g/cm^3)	Void ratio	Initial moisture content
1	1.434	0.875	0.074
2	1.409	0.909	0.066
3	1.429	0.882	0.081

3.3 *Pore-water pressure transducers*

Eight pore-water pressure transducers were installed through the observation window with porous cup submerged about 5 mm into the soil to ensure effective measurements of the changes of pore-water pressure. Each of the transducers, monitored by microcomputer-based acquisition system, was logged at approximately 5 s and 60 s intervals during the first experiments and the last two experiments, respectively. Figure 2 shows the configuration of the pressure transducers.

3.4 *Rainfall simulator*

A rainfall simulator was set up above the slope model to introduce the change in pore-water pressure and the failure due to rainfall infiltration in experiment 3. The simulator was designed to produce an effective rainfall intensity of approximately 100 mm/hr. It mainly consisted of the sprayer arms, sprayers and flow meter.

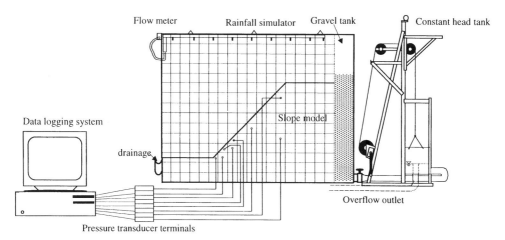

Figure 2. Overview of set up of instrumentation.

Due to fluctuation of water pressure in the water supply, the amount of water flowing into the sprayer arms was carefully controlled and monitored through a flow meter to maintain the rainfall at the prescribed intensity. As the simulated rainfall would introduce surface erosion, the entire surface of slope model was covered by a plastic net to reduce this effect.

3.5 Soil placement

Prior to placement, the sandy soil was mixed thoroughly with water to give an approximately 5% water content using turbo mixer for each experiment. The average moisture content was determined from ratio of total amount of water to total dry soil used to construct the slope profile. Subsequently, the soil was placed in a series of horizontal layers of 5 cm thick to the full width of the tank before being compacted using a light compactor to give a slight apparent cohesion throughout the slope profile. The average dry density of the profile calculated from the total density and average moisture content was used to determine void ratio for each slope profile (Table 2). Overall view of the tank and constructed slope profile is shown in Figure 2.

Table 3. Slope toe saturation.

Exp. No.	Time for toe saturation (min)	Water Level at the back of slope (cm)
1	73	29.7
2	10	51.95
3	54	24.0

3.6 Photography

A video camera was set up in front of the slope face to give a better view and complete record of the saturation process, failure initiation and movement. A 35-70 mm auto focus camera was used to record the flow lines, slope profile before and after failure, and displacement of sliding block.

4 FAILURE DEVELOPMENT

4.1 Overall failure process

Slope model under slow rise of water level (Exp.1) failed by multiple regressive non-circular slips. The whole process required 20 to 60 minutes for slope wetting and a few minutes for collapse. In contrast, slope models under quick rising (Exp.2) and rainfall infiltration (Exp.3) failed by non-circular shallow slide over a period of 16 minutes and 90 minutes from the start of the experiment, respectively. The failure process for each case is described in detail in this section. The detail history of failure varied with each experiment and needs to be interpreted in asso-ciation with the pore-water pressure records and the mode of rising water level.

The duration for toe saturation in each experiment is tabulated in Table 3. It is worth-noting that the wetting front advanced more rapidly under the quick rising condition compared with the other two conditions.

The three dimensional nature of the overall failure processes is illustrated schematically in Figure 3. Failures of each slope model commenced by toe saturation and development of tension crack at the toe of the slope. In experiments 2 and 3, tension cracks were also developed at the top of the slope before toe saturation. In experiment 1, upon the development of seepage face, retrogressive failure commenced by localized shallow sliding at the toe area. The next sliding did not initiate until the previous slab had virtually come to rest and the next rise of water level had increased the seepage face area further. The local failure proceeded with increasing water level, which finally resulted in a total failure of the slope model.

For comparison, development of tension crack at the toe and the top of the slope directly generated total instability to the whole portion of the slopes in experiments 2 and 3. The main failure took place a few seconds following the growth of the tension crack at the toe.

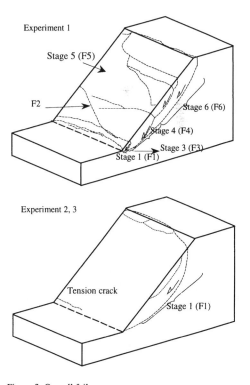

Figure 3. Overall failure sequences.

4.2 Failure modes

Failure modes observed in all of experiments are summarized in Figure 4. Most dominant mode of failure deduced from the experiments is non-circular shallow slide, frequently retrogressive with failure surface cut through the slope toe. This failure mode was very consistent with that which characterized failure of natural or man-made slope of residual soil.

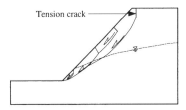

Retrogressive non-circular sliding (experiment 1)

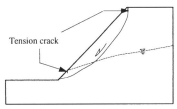

Shallow non-circular sliding (experiment 2)

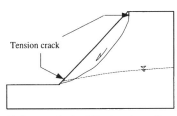

Shallow non-circular sliding (experiment 3)

Figure 4. Summary of failure modes for all experiments.

Although placement conditions for all experiments were as nearly identical as possible, the failure modes were obviously different. It is believed that the mode of raising water level in the slope controlled the failure modes. Experiment 1 failed by retrogressive non-circular sliding rather than non-circular slide apparent in experiments 2 and 3. However, it is interesting to note that all failures were preceded by toe instability. This clearly indicates that the initiation of rainfall-induced slope failures observed in all experiments was due to the loss of lateral support resulting from the toe failures.

5 PORE-WATER PRESSURE AT FAILURE INITIATION

Pore-water pressures during each experiment increased throughout the experiments in response to the advancing wetting fronts as shown in Figures 5-7. Saturation is indicated by initially negative pore-water pressures increasing to zero, and the increasing water level had introduced the development of a seepage face on the slope surface.

Detailed pore-water pressure records at failure for experiment 1 are shown in Figure 5. It should be noted that times representing initiation of the different stages of failure was determined from the observations and the video and camera records. The records showed that pore-water pressures increased slowly throughout the experiment during which most of failures took place. Transducers located near the toe of the slope, (P1 and P4) indicated the progress the development of the seepage face that contributed to the occurrences of localized shallow slips at the toe area (F1). However, the transducers displayed no pressure response to toe failures since the failures predominantly occurred in the middle section of the slope toe. The next stage of water level rise introduced further development of seepage face as shown by transducers P2 and P4 during which the next stages of sliding were initiated (F2-F6).

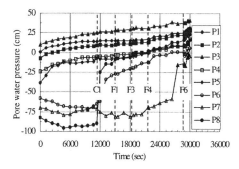

Figure 5. Overall pore-water pressure changes, experiment 1.

In contrast, the advancing wetting front resulted initially in significantly rapid increase in pore-water pressures throughout experiments 2 and 3 (Figs 6-7). The sharp increase in pore-water pressures near the slope surface was very significant at the beginning of experiment 3. Transducers located at the slope toe area (P1 and P2) showed increase in positive pore-pressures prior to failure initiation, indicating the progressive development of seepage face near the slope toe area at failure initiation. The pressure records indicated a quite rapid initiation of main failure (F1) after generation of tension crack (C1). Careful examination of pore pressure record of experiment 2 at P3 showed that a significant excess pressure of about 0.3 kPa was still observable shortly after the main failure. However, transducers located far above the toe (P4-P8) still recorded negative pore pressures decreasing to zero after the cessation of the main failure (F1). This suggests that the shear zone cut across the slope toe, and the failures involved mainly the unsaturated portion of the slope.

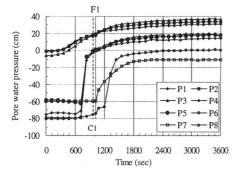

Figure 6. Overall pore-water pressure changes, experiment 2.

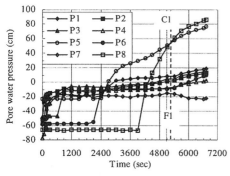

Figure 7. Overall pore-water pressure changes, experiment 3.

6 DISCUSSION

This discussion focuses mainly on some general inferences about the critical influence of water seepage on the initiation of rainfall-induced slope failures. The principal result of our experiments showed that the initiation process of rainfall-induced slope failures was preceded by the development of seepage face, and was apparently independent of the mode of introducing the rise of water level. The experimental results also indicated that non-circular shallow sliding was the most dominant failure mode, frequently retrogressive, with shear plane cutting across the toe of the slope.

The detailed observations of failure process and development demonstrated that rainfall-induced slope failures were essentially initiated by the loss of lateral support resulting from earlier seepage-induced failures of the toe area under drained condition. Stated another way the development of local instability in seepage face area is prerequisite for failure to take place. This contradicts some theoretical hypotheses that failure always starts with shear band mechanism.

Detail study on the process of failure initiation under the effect of the seepage of water indicated that the increase of failure potential predominantly occurred in the near-surface region. The increases were therefore particularly significant in the vicinity of the toe area due to the direction of seepage, which made an acute exit angle from the slope surface. This seepage direction is the most likely to cause slope instability (Nishigaki et al. 1996). Although the conventional stability analytical method has included pore-water pressure due to seepage, however, it has not considered the directional nature of seepage. In effect, the idea to analyze the stability of slope under the influence of water seepage should take into account the nature of the seepage. An example of analytical approach to assess the effect of groundwater seepage on the potential for slope instability can be found in Nishigaki et al (1996) among others.

Examination of pore-pressure records indicated that excess pore-water pressure was clearly generated in a saturated shear zone during the slope movement. It also should be emphasized that the shear zones inferred from the pore pressure records were mainly located in the unsaturated soil of the slope.

The overall failure initiation and mechanism owing to the action of seepage on the toe area clearly illustrates the critical influence of toe failure of soil slope on overall instability, and the importance of effective toe drainage in preventing premature, retrogressive sliding due to locally high seepage pressures. Therefore, the stability of the seepage area can be used to infer the overall stability of the slope.

As far as prediction of slope failure during rainfall is concerned, our experimental results clearly suggested that monitoring the stability of slopes should be focused in the area where the seepage face is likely to develop. The present-day unsaturated seepage numerical analysis can be used to determine the location and the extent of seepage face area during rainfall infiltration.

7 CONCLUSIONS

The results of this experimental study support the following conclusions.

1. Rainfall-induced slope failure is always preceded by the formation of seepage face, and is a consequence of instability of slope seepage face area. Therefore, such failure is initiated under drained condition. The main portion of soil slope involved in the movement, however, is still in unsaturated condition.

2. Excess pore-water pressures are generated in saturated shear zone during the slope movement.

3. Under the water seepage effect, the most dominant mode of rainfall-induced failure of slope is noncircular shallow sliding, often retrogressive.

4. The experiments suggest the prevention of premature seepage-induced toe failure by effective drainage on seepage area to maintain overall stability of the slope.

REFERENCES

Iseda, K & Tanabashi Y. 1986. Mechanism of slope failure during heavy rainfall in Nagasaki July 1982. *Natural Disaster Science*. 8 (1): 55-84

Kato, K. and Sakajo, S. 1999. Seepage analyses of embankments on Tokaido-Shinkasen in long term rainfalls. *Proc. International Symposium on Slope Stability Engineering, vol. 1, Matsuyama, 8-11 November 1999*: 521-526. Rotterdam: Balkema.

Nishigaki, M., Tohari, A. and Komatsu, M.: Stress and seepage vector system on stability of hillslope. *Proc. China/ Japan International Conference on Water Environment and Disaster Prevention, Zhengzhou, 5-7 November 1996*: 160-166.

Sasaki, S., Araki, S &Nishida, K. 1999. Seepage characteristics of decomposed granite soil during rainfall. *Proc. International Symposium on Slope Stability Engineering, vol. 1, Matsuyama, 8-11 November 1999*: 423-428. Rotterdam: Balkema

Sakajo, S &Kato, K. 1999. Failure analyses of embankments on Tokaide-Shinkansen in heavy rainfalls. *Proc. International Symposium on Slope Stability Engineering, vol. 1, Matsuyama, 8-11 November 1999*: 527-532. Rotterdam: Balkema.

Unsaturated Soils for Asia, Rahardjo, Toll & Leong (eds) © 2000 Taylor & Francis, ISBN 90 5809 139 2

On the stability of unsaturated soil slopes

L.T.Shao & Zh. P.Wang

Dalian University of Technology, People's Republic of China

ABSTRACT: The model for the prediction of shear strength with respect to soil suction, developed by Vanapalli and Fredlund, provides the possibility of applying shear strength parameters for saturated soils to practical problems in unsaturated soils. Based on this model, the method of slices is employed to analyse the stability of unsaturated soil slopes in this paper. Results of two numerical examples show the effects of the rise of the groundwater level on the stability of slopes. A technique to find the soil water content profile most dangerous to the stability of unsaturated soil slopes is also presented in this paper.

1 INTRODUCTION

Negative pore water pressure or soil suction (matric suction) has been found to play an important role in the stability of unsaturated soil slopes (Fredlund & Rahadarjo 1993). When the pore water is in hydrostatic equilibrium, matric suction at a point of the soil can be determined by the height of the point over the ground water table (GWT). In this case GWT is crucial to the stability of unsaturated soil slopes. When the pore water is in movement, which is most common in nature, matric suction may be determined by investigating water content (degree of saturation) and soil water characteristic of the soil. Once the soil suction is determined, its effect on the stability of unsaturated soil slopes can be evaluated using the model for predicting the shear strength of unsaturated soils, which was developed by Fredlund et al. (1978, 1993, 1996).

Soil embankments with 1:2 slope are taken as examples in this paper to show the influence of GWT on the factor of safety of the slopes using the method of slices. A technique to find the soil water content profile most dangerous to the stability of unsaturated soil slopes is also presented.

2 SHEAR STRENGTH OF SOILS

Mechanical behaviors of a soil are governed by stress state variable that controls the equilibrium of the soil structure. Equilibrium analysis for free bodies of total soil element and each phase indicates that the most fundamental components of the stress state variable are skeleton stress σ, pore water pressure u_w

and pore air or gas pressure u_a (Shao, unpubl.). For saturated soils, the expression of shear strength, which was presented by Skempton (1961), can be written as $\tau_f = c' + s \cdot f$ with stress state variable $s = \{\sigma_t - u_w, u_w\}$ and behaviour parameter vector $f = \{\tan\phi', a\tan\psi\}^T$, in which $\sigma_t - u_w$ is the skeleton stress, σ_t is the total stress. However the term related to pore water pressure in the expression above can be neglected because a is extremely small at pressures mostly encountered in engineering and geological problems. Thus the shear strength of a full saturated soil in geological engineering can be evaluated by taking into consideration only skeleton stress, or effective stress defined by Terzaghi (1936). For unsaturated soils, as it had been recognized that to describe the behaviour of shear strength with a single component of stress state variable is impossible, Fredlund et al. (1978) developed a modified form of the Mohr-Coulomb failure criterion with $\sigma_t - u_a$ and $u_a - u_w$ as components of stress state variable. In 1996 Vanapalli and Fredlund presented an empirical and analytical model for predicting the shear strength of unsaturated soils with respect to soil suction:

$$\tau_f = c' + \left[\sigma_t - u_a + S_e(u_a - u_w)\right]\tan\phi' \tag{1}$$

in which

$$S_e = \frac{S - S_r}{1 - S_r} \tag{2}$$

where S is the degree of saturation and S_r is the residual degree of saturation. This model makes it easy to assess the stability of unsaturated soil slopes using shear strength parameters for saturated soils.

Postulating that the adsorbed water film around soil particles is constituent of the soil skeleton and corresponding to the residual water content, the authors define the skeleton stress for unsaturated soils as

$$\sigma = \sigma_t - u_a + S_e(u_a - u_w) \qquad (3)$$

This expression presents the relationship of skeleton stress, total stress, pore water pressure and pore air pressure, which is derived from the equilibrium analysis for free bodies of every phase of a complete soil. For practical problems *in situ* pore air pressure u_a can be considered to be zero (relative atmosphere pressure or the gas pressure on the water in reference state) when it remains constant as atmosphere pressure everywhere in the soil. In this case, the skeleton stress can be calculated using

$$\sigma = \sigma_t - S_e u_w \qquad (4)$$

3 PROFILE OF THE PORE WATER PRESSURE AND SKELETON STRESS FOR UNSATURATED SOIL STRATUM

The energy status of pore water is described by making use of the concept of soil water potential in soil physics or soil thermodynamics. Precise definitions of total potential and its various component potentials were provided by the International Soil Science Society (ISSS) in 1963 and slightly modified in 1976. Total potential of soil water can be divided into three components:

$$\psi_t = \psi_p + \psi_g + \psi_o \qquad (5)$$

The pressure potential ψ_p is defined as the amount of useful work that must be done per unit quantity of pure water to transfer reversibly and isothermally to the soil water an infinitesimal quantity of water from a pool at standard atmospheric pressure that contains a solution identical in composition to the soil water and is at the elevation of the point under consideration. Similar definitions have been given for gravitational potential, ψ_g, and osmotic potential, ψ_o, which refer to the effects of elevation (i.e. the gravitational field) and of solutes on the energy status of soil water, respectively. Matric potential ψ_m is a sub-component of pressure potential and is defined as the value of ψ_p where there is no difference between the pressure of air or gas in the soil and the pressure of gas on the water in the reference state. The sum of matric and osmotic potential is referred to as the water potential ψ_w and is directly related to the relative humidity of vapour in equilibrium with the liquid phase in soils.

The ISSS definition of pressure potential already given includes (1) the positive hydrostatic pressure that exists below a water table, (2) the potential difference experienced by soil that is under a gas pressure different from that of water in the reference state, and (3) the negative pressure or suction experienced by soil water as a result of its affinity for the soil matrix. Some authors depart from this definition and use the term 'pressure potential' to refer only to sub-components (1) and (2). Fortunately, all authors agree on the equivalent definitions for matric potential, which is sub-component (3). Matric suction is defined as negative matrix potential. It refers to the same property but takes the opposite sign to matric potential.

Taking the free water in atmosphere as reference state, pressure potential is equal to pore water pressure when it is defined as energy per unit volume, and the magnitude of the pressure difference, $u_a - u_w$ is in general equal to the value of matric suction.

Soil water that is in equilibrium with free water is by definition at zero matric potential. As a soil dries, matric potential decreases and large pores are emptied of water. Progressive decreases in matric potential will continue to empty narrower pores until eventually water is held only in water film and in the finest pores. Therefore a decreasing matric potential is associated with a decreasing soil water content. Laboratory or field measurements of these two parameters can be made for the soil at a certain density and stress state and the relationship that is plotted as a curve is called soil water characteristic.

The soil water characteristic curve shows the relationship between the soil water content and the suction. If hysteresis can be neglected or the process (drying or wetting) is known for a practical problem, soil water potential can be determined directly from water content using the soil water characteristic curve. In the field, one can assign pore water pressure values of hydrostatic pressure at known depths below, or of the suction at known heights above the water table if the soil water is in hydrostatic equilibrium. Such a state of equilibrium is fortuitous in nature and no doubt rare, but provides a convenient starting point for subsequent discussion in situations involving soil water movement.

The water table is defined to be that level in the soil at which the hydrostatic pressure of soil water is zero. If a well is dug, the water table is the level at which water stands in it, for at that level the hydrostatic pressure of water in the well is clearly zero, and is at the same time in equilibrium with the pressure in the adjoining soil at that level. It follows that the pressure at a depth d below the water table, both in the well and in the soil when the two are in equilibrium, corresponds to a head d of water; i.e. the pressure is $\gamma_w \cdot d$ where γ_w is the unit weight of water. Similarly at a height z above the water table the pressure is less than that at the water table by a head

z, i.e. there is a suction of magnitude measured by the height z. So a measure of the height above the water table is thus also a measure of the prevailing suction; and a plot of soil water content against prevailing suction, i.e. the soil water characteristic curve is the same as a plot against the height. The latter is called the soil water profile.

Thus the pore water pressure profile in hydrostatic equilibrium can be expressed as

$$u_w = -\gamma_w z = -\gamma_w (H - h) \quad (6)$$

in which z is the coordinate in z-direction with origin at water table, h is the depth from ground surface, H is the depth from the ground surface to the water table.

As the density of a soil varies with depth because of the variation of water content for a homogenous soil layer, the total stress should be calculated by

$$\sigma_t = \int_0^h \gamma_{(h)} \cdot dh = \bar{\gamma} \cdot h \quad (7)$$

in which $\gamma_{(h)}$ is the unit weight of the soil which is a function of the water content and $\bar{\gamma}$ is the average of the unit weight over the height h.

Substituting Equations 6 and 7 into Equation 4 yields the expression of the skeleton stress profile in hydrostatic equilibrium:

$$\sigma = S_e \gamma_w H + (\bar{\gamma} - S_e \gamma_w) h \quad (8)$$

where S_e is the effective degree of saturation at the point under consideration. It should be noted that Equation 8 is valid everywhere along the depth in the soil layer no matter whether it is saturated or unsaturated.

4 STABILITY OF UNSATURATED SOIL SLOPES

Figure 1 illustrates a soil slope with a part above the GWT. When taking no soil suction into consideration the stability of the slope can be assessed with the method of slices by the factor of safety K:

$$K = \frac{\sum c_i' l_i + \left[b_i (\bar{\gamma} \ h_{1i} + \gamma' h_{2i}) \cos \alpha_i \right] \tan \phi_i'}{\sum b_i (\bar{\gamma} \ h_{1i} + \gamma_{sat} h_{2i}) \sin \alpha_i} \quad (9)$$

where h_{1i} is the length of the slice above the water table; h_{2i} is the length of slice below the water table; γ' is submerged unit weight of the soil; γ_{sat} is saturated unit weight of the soil; b_i is the width of the soil slice; $l_i = b_i / \cos \alpha_i$ is the length of the circle for the slice; α_i is the orientation angle for the circle of the slice.

According to the discussion on the skeleton stress and shear strength for unsaturated soils, the factor of safety of the slope, when taking into account the effect of soil suction, can be calculated by

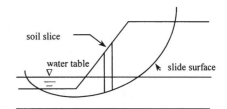

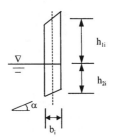

Figure 1. Partially submerged soil slope and soil slice

$$K = \frac{\sum c_i' l_i + \left[b_i (\bar{\gamma} \ h_{1i} + \gamma_{sat} h_{2i}) \cos \alpha_i - S_e u_w l_i \right] \tan \phi_i'}{\sum b_i (\bar{\gamma} \ h_{1i} + \gamma_{sat} h_{2i}) \sin \alpha_i} \quad (10)$$

in which S_e is the effective degree of saturation, u_w is the pore water pressure on the circle of the soil slice.

Provided that there is no movement of the water in an unsaturated soil layer in isothermal condition, the factor of safety K can be rewritten by substituting Equation 6 into Equation 10 as

$$K = \frac{\sum c_i' l_i + \left[b_i (\bar{\gamma} (h_{1i} + \gamma_{sat} h_{2i}) \cos \alpha_i + S_e \gamma_w (H - h_i) l_i \right] \tan \phi_i'}{\sum b_i (\bar{\gamma} \ h_{1i} + \gamma_{sat} h_{2i}) \sin \alpha_i}$$

$$(11)$$

in which H is the distance from the top of the slice to the ground water table and $h_i = h_{1i} + h_{2i}$, which is the total length of the slice.

Two numerical examples shown in Figure 2 are presented in this paper to show the influence of the variation of the GWT on the stability of unsaturated

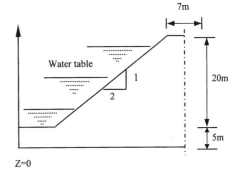

Figure 2. Soil slope as example for evaluation of stability

827

soil slopes. One is a sand slope of which the soil water characteristic is assumed to be that of F.F. sand, tested by Aitchison (1956). The other is assumed to be of glacial till, making use of the soil water characteristic curve by Vanapalli & Fredlund (1996). Soils in the slopes for these two examples are assumed to be homogenous. The parameters for shear strength are $c' = 0$ and $\phi' = 32°$ for the sand, and $c' = 0$ and $\phi' = 23°$ for the glacial till.

From Figure 3 one can see the influence of the rise of the ground water table on the factor of safety for the sand slope. When the water table is low, the factor of safety remains the same value as that for dry slope because the ground water has no effect on the factor of safety as the water table is far away from the potential slip surface. In this case, the water content of the sand in the zone above the potential slip surface is a constant, i.e. residual water content. After the water table reaches a level higher than 10m, the factor of safety decreases as the water table continues to go up without soil suction taken into consideration (curves with hollow dot in Figure 3 and Figure 4). It reaches its minimum value around 1.32 when the water table is about 12m to 20m in height and then shows a slight increase to 1.41 when the whole slope is submerged in water. On the other hand, when the soil suction is taken into account (curves with black dot in Figure 3 and Figure 4) the factor of safety increases with the rise of the GWT where matric suction can affect the stability of the potential slip surface. The value of the factor of safety reaches its maximum value of 1.48 when the height of the water table is around 11m. Thus it demonstrates the same increasing and decreasing tendency as the case with no soil suction considered.

When the potential slip surface is fixed to be that obtained for the dry sand slope, the factor of safety varies similarly but is greater than that for the potential slip circle surface.

Similar results for the slope in glacial till are

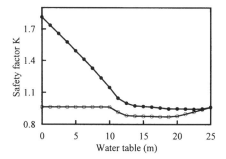

Figure 4. Variations of the factor of safety with the rise of the ground water table for glacial till slope

shown in Figure 4. The effect of the soil suction on the stability of the slope is apparently more significant comparing to the sand slope. Because the height of the soil layer that may be influenced by soil suction is up to about 100m, the deeper the ground water table the greater the factor of safety for the slope.

5 WATER CONTENT PROFILE MOST DANGEROUS TO SLOPE STABILITY

On the condition that no water movement is taken into account, soil suction can be determined directly from water content according to the soil water characteristic curve. For a slope with a defined GWT, there may exist a water content profile that is most dangerous to its stability. This most dangerous water content profile is abbreviated to DWCP in this paper. Provided that the water content is increasing from the hydrostatic equilibrium state, the DWCP can be determined by solving an optimization problem of minimizing the factor of safety in Equation 10 subjected to the condition that $S \in (S_{0(z)}, 1.0)$, i.e.

$$\begin{cases} K_{min} = K_{(S(z))} \\ S.t. \quad S \in (S_{0(z)}, 1.0) \end{cases} \qquad (12)$$

in which K_{min} is the minimum factor of safety that is corresponding to DWCP, $S_{(z)}$ is the profile of degree of saturation, $S_{0(z)}$ is the profile of degree of saturation in hydrostatic equilibrium.

With the degree of saturation changed from $S_{0(z)}$ to 1.0 at any height above GWT, the optimization problem can be solved using Hooke-Jeeves pattern search technique. The results of the DWCP for these two examples are shown in Figure 5 and Figure 6. Comparison of the factor of safety in hydrostatic equilibrium and in DWCP is given in Table 1.

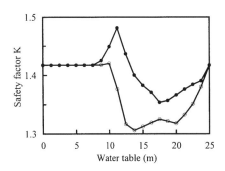

Figure 3. Variation of the factor of safety with the rise of the ground water table for the sand slope

Table 1. Comparison of the Factor of safety

Water table	Equilibrium profile		Dangerous profile	
m	Sand	Glacial till	Sand	Glacial till
0.0	1.42	1.81	1.42	0.96
10.0	1.42	1.50	1.42	0.96
15.0	1.39	0.97	1.30	0.88

For the slope in F.F. sand, the DWCP is the residual water content when GWT is lower than 10m. When the GWT is about 15m in height, the factor of the safety in DWCP is 5% lower than that in hydrostatic equilibrium. The profile of the degree of saturation (curves with hollow dot in Figure 5 and Figure 6) is significantly different from that in hydrostatic equilibrium (curves with black dot in Figure 5 and Figure 6). For the glacial till slope the DWCP is the same as that in full saturation because of high soil suction and the factor of the safety is much lower than that in hydrostatic equilibrium state.

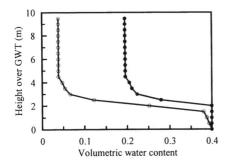

Figure 5. Profile of the water content in hydrostatic equilibrium and in DWCP (sand slope, GWT = 15m)

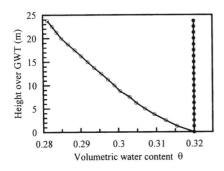

Figure 6. Profile of the water content in hydrostatic equilibrium and in DWCP (glacial till slope, GWT = 0m)

6 CONCLUSION

When using the model for the prediction of the shear strength of unsaturated soils with respect to soil suc-

tion, which was developed by Vanapalli and Fredlund, the formula for the factor of safety in the method of slices has a unified form for saturated and unsaturated soil slopes in open atmosphere.

As soil suction play an important role in the stability of unsaturated soil slopes, a stable slope will be unstable once the suction is decreasing when water content varies. The rise of GWT changes the water content in soils and affects the stability of the slope. For slopes in unsaturated soil with lower suction characteristic, the factor of safety may increase with GWT rising in some extent but decrease when GWT continues to rise. In this case the factor of the safety is not the smallest when the slope is fully submerged in water. On the other hand, for slopes in unsaturated soils with higher suction characteristic, the factor of safety decreases with the rise of GWT. The higher the GWT, the smaller the factor of safety. The factor of safety of the slope in glacial till in this paper decreases by 50% when the slope is fully saturated. However the rise of GWT would not affect the factor of safety significantly if no soil suction is taken into account.

As water movements require a change of the pressure gradients, so what has been discussed in this paper has no relevance to slopes in which water is moving. Similar results for slopes in which pore water is in movement will be presented in another paper.

REFERENCES

Aitchison, G.D. 1956. Some preliminary studies of unsaturated soils. Proceedings of the second Australia-New Zealand conf. on soil mechanics and foundation engineering.

Aslyng, H.C. 1963. Soil physics terminology. Int. soc. soil sci. bull., 23.

Bolt, G.H. 1976. Soil physics terminology. Int. soc. soil sci. bull., 49.

Fredlund, D.G., Morgenstern, N.R. & Widger, A., 1978. Shear strength of unsaturated soils. Can. getech. J., Vol. 15: 313-321.

Fredlund, D.G. & Rahardjo, H. 1993. Soil mechanics for unsaturated soils. John Wiley & Sons, Inc.

Fredlund, D.G., Anqing Xing, Fredlund, M.D. & Barbour S.L., 1996. The relationship of the unsaturated soil shear strength to the soil-water characteristic curve, Can. geotech. J., Vol. 33: 440-448.

Skempton, A.W., 1961. Effective stress in soils, concrete and rocks. Pore pressure and suction in soils: 4-25, conf. organized by the British national society of int. society of soil mech. and found. eng., Butterworths, London.

Shao, L.T. unpbl.. The axiomatic basis for the effective stress equation.

Terzaghi, K. 1936. The shearing resistance of saturated soils and the angle between the planes of shear. Proc. 1st conf. soil mech., Vol. 1: 54-56.

Vanapalli, S.K., Fredlund, D.G., Pufahl, D.E. & Clifton A.W., 1996. Model for the prediction of shear strength with respect to soil suction. Can. geo. J. Vol. 33.

Unsaturated Soils for Asia, Rahardjo, Toll & Leong (eds) © 2000 Taylor & Francis, ISBN 90 5809 139 2

Water retention in a steep moraine slope during periods of heavy rain

Ph. Teysseire, L. Cortona & S. M. Springman
Institute of Geotechnical Engineering, Federal Institute of Technology, Zürich, Switzerland

ABSTRACT: A 100 m^2 surface area on a 31° slope at around 2900 m ASL near to the Gruben glacier was selected as a field test site, to investigate slope stability in moraine as a function of degree of saturation and relative density of the soil. The moraine included 'particle' sizes varying from boulders down to fine silt. Evaluation of the behaviour of partially saturated moraine is necessary if the relative stability of such slopes are to be estimated with any degree of reality, and this is strongly dependent on the water retention characteristics. In order to study the water retention in response to several intensities of rain, a sprinkler system was constructed in the middle of this test area. The sustained rain intensity over a period of 7 days was between 10-30 mm/hour, depending on the local wind conditions, via a continuous water supply from the nearby glacier lake. A saturation of approximately 90 % could be reached after 7 days in the very top layer, whereas no significant changes at depths below 0.5 m were observed. In future, the field test data will be linked to more controlled analysis of the water retention response obtained from the laboratory tests, and the slope will be analysed under a range of different saturation regimes.

1 INTRODUCTION

Outbursts of proglacial lakes, caused by recession of glaciers and retained behind moraine dams, constitute a major natural hazard and often produce extreme discharge. Five major glacier floods/ debris flow were documented during 19th and 20th century, the latest one in 1970, for the village of Saas-Balen, Saas Valley, Valais/Switzerland (Haeberli ,1992)

Soil properties of these moraines as well as the triggering factors for stabilities like debris flow during rainfall are quite unknown. A series of laboratory tests and an artificial rain field test on 100 m^2 surface area with artificial rain were carried out.

2 LABORATORY INVESTIGATIONS

2.1 *Generalities*

The grain size distribution of the soil, sampled from adjacent to the test field, and the soil parameters are shown in Tables 1 and 2. Three fractions with different grain size were extracted from the "whole" sample and analysed to investigate the influence of the grain size distribution.

Table 1. Grain size distribution and fractions

[mm]	max.	<45	<16	<2
0.002	1.4	1.7	2.1	3.6
0.02	8.8	10.8	13.2	22.6
0.063	15.9	19.4	23.8	40.5
0.25	23.9	29.3	35.9	61.2
1	32.8	40.3	49.4	84.1
2	39.2	47.9	58.7	100
4	46.8	56.9	69.8	
8	56.7	69.2	84.8	
16	67.1	81.6	100	
31.5	77.4	93.8		
45	83.0	100		
63	87.8			
90	93.5			
138	100.0			

Table 2. Void ratios and densities

[mm]	Value	<45	<16	<2
e_{max}	[-]	0.57	0.71	1.19
e_{min}	[-]	0.2	0.24	0.38
γ_d (min)	[kN/m³]	17.8	16.4	12.6
γ_d (max)	[kN/m³]	23.0	22.5	20.2
ρ_S	[g/cm³]	2.84	2.84	2.84

The compaction characteristics of the investigated soil were determined by means of the Proctor Standard test. The resulting compaction curves are plotted in Figure 1. The values of maximum dry unit weight correspond to those indicated in Table 2.

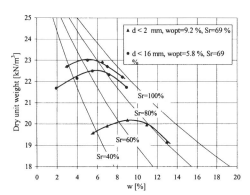

Figure 1. Proctor Standard tests

The degree of saturation at optimum condition of compaction is about 70 % for all fractions.

2.2 *Permeabilities*

The coefficient of saturated permeability is given in Figure 2. Permeabilities of the order of magnitude between 2E-06 and 8E-08 m/s can be expected for the range of densities measured in the field.

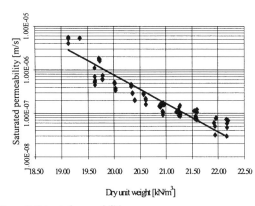

Figure 2. Saturated permeabilities

2.3 *Water Retention Curve (WRC)*

The evaluation of the WRC was done by performing tests using an adapted pressure plate cell (Figure 3) according to ASTM Standards (D2325-68).

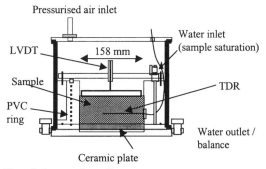

Figure 3. Pressure plate cell

The soil mass, with grain size smaller than 16 mm, was placed into the PVC ring on a saturated 1 bar high flow ceramic plate and compacted within the ring. Several types of compactions were performed (Table 3). Then the sample was saturated and the cell closed. Soil specimens were pressurised in the chamber and the water discharged against the influence of suction. Discharged water, deformation of the sample as well as a time domain reflectro-meter (TDR) were continously monitored. The steps applied were chosen between 1-10 kPa (Figure 4). Equilibrium was typically reached after 5-7 days. At the end of the test, water content was determined and back calculated for the water contents corresponding to the applied suction steps.

The initial conditions of the tests are given below in Table 3, and the results in Figure 4. They show typical behaviour for a silty/sandy soil with a steep rise in saturation within an interval of 5 – 15 kPa matric suction, between the air entry value and the residual water content. The air entry values vary between 4.5 and 5.5 kPa.

Table 3. Initial conditions of tests WRC 1,3,4 with grain size <16 mm

	Value	WRC1	WRC3	WRC4
e	[-]	0.59	0.40	0.36
γ_d	[kN/m³]	17.8	16.4	12.6
Sr	[%]	94.6	99.9	99.0
Compaction		Vibration	Static	Proctor
w at compaction		dry	w_{opt}	w_{opt}

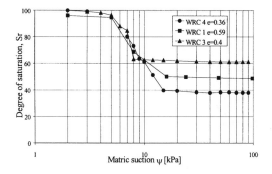

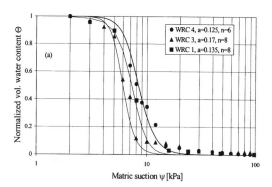

Figure 4. Water retention curves WRC1,3,4

An overview of equations for water retention characteristic curves are given by Fredlund and Xing (1994). The fit to the data (Figure 5a) was achieved using van Genuchten's (1980) equation.

$$\Theta = \left(\frac{1}{1 + (a\,\psi)^n} \right)^m \qquad [1]$$

Θ normalized volumetric water content;

ψ soil suction;

a, n, m material parameters, with $m = 1 - \frac{1}{n}$

and, Fredlund and Xing's equation (1994, Figure 5b)

$$\theta = C(\psi)\,\frac{\theta_s}{\left\{ \ln\left[e + \left(\frac{\psi}{a}\right)^n \right] \right\}^m} \qquad [2]$$

θ volumetric water content;

θ_s saturated volumetric water content

a suction related to the soil air-entry

n soil parameter related to the slope at the inflection point on the WRC

ψ total soil suction [kPa]

ψ_r suction value corresponding to residual water content, θ_r.

m soil parameter related to θ_r

e natural number, 2.71828...

$C(\psi)$ correction function

$$C(\psi) = \left[1 - \frac{\ln\left(1 + \frac{\psi}{\psi_r} \right)}{\ln\left(1 + \frac{1\,000\,000}{\psi_r} \right)} \right] \qquad [3]$$

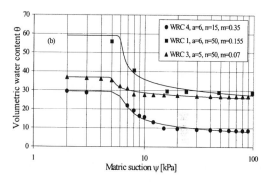

Figures 5a, b. Best-fitting curves to WRC1, 3, 4 (a) van Genuchten , (b) Fredlund/Xing

The selection of the associated material parameters is shown in Tables 4a and 4b.

Table 4a. Parameters for van Genuchten's model

	WRC1	WRC3	WRC4	Range
a	0.135	0.17	0.125	0.125-0.17
n	8	8	6	6-8

Table 4b. Parameters for Fredlund and Xing's model

	WRC1	WRC3	WRC4	Range
a	6	5	6	5-6
n	50	50	15	15-50
m	0.155	0.07	0.35	0.07-0.35

3 FIELD TEST

3.1 *Soil Properties*

The test field is located at 2800 m ASL below Lake 1 in the forefield of the Gruben glacier. Seismic refraction, D.C. resistivity together with gravimetry soundings were performed during a former geophysical campaign. The results indicated a maximum thickness of more than 100 m of moraine, with low seismic velocities in the range from 400 to 900 m/s, and resistivities from 2 to 15 kΩm within the uppermost 5 to 20 m.

The measurements of the unit weights up to 1 m depth was done with sand replacement and nuclear measurements. The soil properties are given in Table 5.

Table 5. Soil properties of test field (in situ)

Property	Value
γ_d	20.3 kN/m^3
Water content w	4.6 %
e	0.29
Dr (mean relative density)	77 %

3.2 Instrumentation layout

Field monitoring in slopes has been carried out by numerous researchers (e.g. Affendi and Faisal 1994, Lim et al. 1996). Tensiometers are often used to measure the matric suction in the soil, as long as the suctions are less than 100 kPa. The morainic soil covers a large range of the granulometric curve, varying from silt to gravel, so the instrumentation was mainly focused on indirect measurements with TDR. Two types of TDR were used.

A Moisture Point[1] system separates the probe into segments, allowing several values to be determined at different depths from one standard probe. It consists of two flat stainless side bars with black epoxy filler and electronic circuitry. It was inserted into the soil using a pilot rod. The conventional TDR probe consists of a 3 rod assembly. Both TDR systems are shown below in Figure 6.

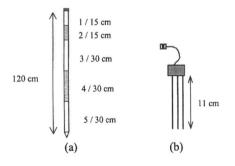

Figure 6a, b. Time domain reflectometry systems (a) Moisture Point for depths up to 1.2 m, (b) conventional TDR probe for shallow surface installation 0.3 m

The equipment used in this field test is listed in Table 6.

Table 6. Test equipment

Sensor	Measuring locations
TDR (3 Rod)	8
Moisture Point (4)	20
Extensometers	5
Fibre optic sensors	5
Rain gauges	2
Air temperature/humidity	2
Datalogger/transmission	1

3.3 Testing Procedure

Artificial rainfall was simulated on a hillslope over a test field area of approximately 100 m^2 with intensities varying between 10 and 30 mm/h, depending on the local wind conditions. This corresponds in order of magnitude to a 100 year rainstorm. The duration of the test was about 7 days with a total amount of rainfall of approximately 1900 mm. A continous water supply was supplied by pumping from a nearby glacier lake.

4 RESULTS AND DISCUSSION

After 7 days of intense rainfall, the soil reached 90 % of saturation. Figure 7 shows three TDR measurements (TDR 1, 2, 4) compared to rainfall intensity. A change in saturation degree from 60 – 75 % at the start, to over 90 % at the end of the test could be observed. A sharp trend can also be noticed for all curves in correspondence to pauses in the rain (first and second day). This wetting and changing of the degree of saturation is restricted to a depth up to 30 cm below surface.

Figure 8 shows the variation of the water penetration for depths up to 1.2 m. The top segment (No. 1) indicates the same behaviour as that measured by the TDRs. The changes in the degree of saturation with increasing depth develops much more slowly.

Figure 9 shows the behaviour after stopping the artificial rain. During this period (1 month after the test), the soil was only subject to the natural rainfall. Once partially saturated, the top layer is very sensitive to any changes to rainfall.

The results imply that to reach unstable conditions in such soils, large changes in the water content are required. The steep trend observed on the curves (see Figures 4,5a) suggested that small, but nevertheless significant, decreases in suction will be caused by increasing saturation by 50 %. This can

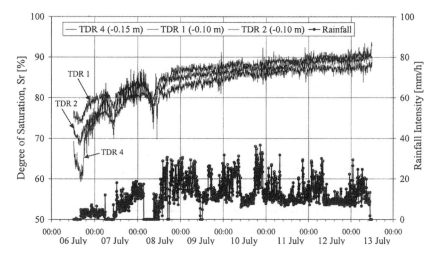

Figure 7. TDR measurements (artificial rain)

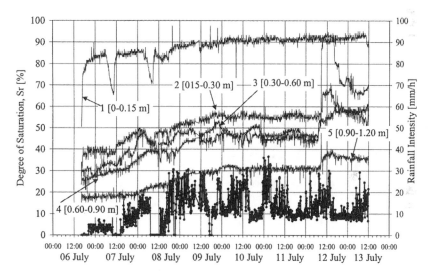

Figure 8. Moisture Points measurements (artificial rain)

induce relevant decreases of shear strength, so that as soon as critical conditions develop within the soil mass, sudden movements can occur and the unstable soil mass will collapse.

5 CONCLUSIONS

Saturation of the moraine slope (31°) of up to approximately 90 % was achieved after several days of raining (for the top soil layer depth up to 30

cm). No significant change in saturation was observed for depths below 0.5 –1.0 m.

Water retention curves show a behaviour typical of a silty sandy soil, with a sharp decrease in saturation within the range of matric suction between 5.5 -15 kPa.

Further long-term monitoring and observation on steeper slopes, based on this experience, will provide additional input for future stability analysis.

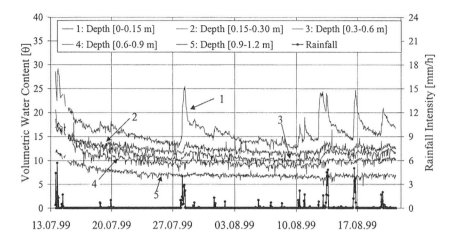

Figure 9. Moisture Point measurements (period following the test)

ACKNOWLEDGEMENT

The authors wish to extend grateful thanks to the Department for Construction, Environment and Traffic of the Canton of Valais and the Federal Office for Water and Geology for their technical and financial support for this project.

REFERENCES

ASTM D2325 – 68 (1994) Standard Test Method for Capillarity – Moisture Relationships for Coarse- and Medium- Textured Soils by Porous-Plate Apparatus. *In Annual book of ASTM standards,* Vol. 4.08, pp. 195-201. Philadelphia ASTM.

Affendi, A.A., and Faisal, A. 1994. Field measurement of soil suction, *13th Int. Conf. on Soil Mechanics and Foundation Engineering, New Delhi, India,* 1013-1016.

Fredlund, D.G., Xing, A., 1994, Equations for the soil water characteristic curve, *Can. Geotech. J.* 31:521-532.

Haeberli, W., 1992, Zur Stabilität von Moränenseen in hochalpinen Gletschergebieten, *Wasser-Energie-Luft* 84:361-364

Lim, T.T., Rahardjo, H., Chang, M.F., Fredlund, D.G., 1996, Effect of rainfall on matric suctions in a residual soil slope, Can. Geotech. J. 33:618-628.

Van Genuchten, M. T., 1980, A closed-form for predicting the hydraulic conductivity of unsaturated soils, *Soil Sci. Soc. Am. J.* 44, 892-898.

Unsaturated Soils for Asia, Rahardjo, Toll & Leong (eds) © 2000 Taylor & Francis, ISBN 90 5809 139 2

Influence of rainfall sequences on the seepage conditions within a slope: A parametric study

I. Tsaparas, D. G. Toll & H. Rahardjo
NTU-PWD Geotechnical Research Centre, Nanyang Technological University, Singapore

ABSTRACT: A parametric study has been carried out to investigate the influence of a slope's initial conditions on the infiltration from a major rainfall event. Several rainfall scenarios that combine minor antecedent rainfall with intensity of a few millimetres per day with a major rainfall event of many millimetres per hour are used for each simulation. The groundwater level and the initial pore-water pressure distribution in the soil quantify the initial conditions within the slope. The study was carried out using a commercial finite element software to simulate the seepage conditions within the slope under unsaturated conditions. The results of several simulations are used to study how infiltration develops under different conditions and how the redistribution of suctions takes place within the slope for each case.

1 INTRODUCTION

Rainfall induced slope failures are a very common problem, especially in the Tropics. Many slopes under unsaturated conditions have remained stable, but fail during or soon after a rainfall event. This negative effect of the infiltration on the pore-water pressures in the soil and subsequently on the stability of the slope is dependent on several parameters. The behaviour of the soil material under unsaturated conditions, the precipitation rate, the vegetation and the groundwater conditions within the slope at the time of the rainfall are the main factors that control the way in which infiltration takes place. There has been work in the past on the relation between rainfall events and slope stability (Ng et al. 1998) and also on the infiltration mechanism in unsaturated soils (McDougall et al. 1998) for simple rainfall events. In this paper consideration is also given to how minor antecedent rain events influence the seepage conditions within the slope before the major rainfall event occurs.

2 DESCRIPTION OF THE MODEL

2.1 *Infiltration and Unsaturated Flow*

The commercial finite element software Seep/W (GeoSlope Int. Ltd., 1998) was used for the seepage modeling in this parametric study. The analysis was two-dimensional and transient. Seep/W uses Richard's differential equation (Eq. 1) to model both the saturated and unsaturated flow:

$$\frac{\partial \theta_w}{\partial t} = \frac{\partial}{\partial x}\left(k_x \frac{\partial H}{\partial x}\right) + \frac{\partial}{\partial y}\left(k_y \frac{\partial H}{\partial y}\right) + q \tag{1}$$

where θ_w is the volumetric water content, k_x and k_y are the coefficients of permeability of the soil along the x- and y-coordinates respectively, H is the total head at a certain point and q is the flow rate at the specific time step t.

In a transient analysis the nodal values of head are time-dependent, as the pore-water pressures are changing with time. In the case of saturated conditions it can be safely assumed that the coefficient of permeability is constant and equal to the saturated value, k_s. However, when the soil is unsaturated the permeability depends on the degree of saturation. This non-linearity of the differential equation is handled numerically by the Galerkin's method.

For the purposes of the parametric study, the infiltration was assumed to be constant during the time that it was occurring and was modeled as a unit flux boundary (q) along the ground surface of the whole slope. All the nodes along the slope, to which the flux boundary was applied, were reviewed by the maximum pressure mode. That means that the head at the reviewed node is set equal to elevation for the first iteration.

2.2 *Dimensions of the slope and finite element mesh*

Transient analysis assuming that the flux is changing with time proved to be a numerically difficult problem to solve. There was some numerical instability in the results mainly at depths of 0.5 to 1.0m from the ground surface, although Seep/W produced reasonable predictions for the pore-water pressures at greater depths. The reason for this instability is related mainly to the complexity of the whole problem. The combination of the non-linearity of the flow equation and the use of a 'steep' permeability function has always been a reason for numerical instability. The development of the wetting front below the ground surface was another reason for such instability. Normally, by significantly reducing the size of the elements and by using a large number of nodes, such numerical difficulties can be overcome. On the other hand, to produce a very fine mesh in a slope requires many nodes and this can result in additional difficulties in convergence.

To produce a mesh that give reasonable results in the sensitive area near the wetting front requires many trials of alternative mesh configurations. Finally it was found that the mesh shown in Figure 1 produced reasonable results.

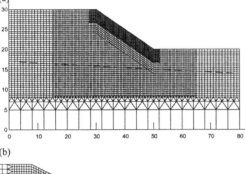

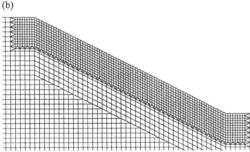

Figure 1. (a) Geometry of the slope and (b) detail of the design of the mesh in the area near the ground surface.

The inclination of the slope is 26.6° (i.e. inclination 2H:1V) and the height is 10m, which are typical dimensions of residual soil slopes in Singapore. The left and right boundaries of the mesh are at a distance of 30m from the crest and the toe of the slope respectively in order to avoid any possible effect of the boundary conditions on the way infiltration develops within the slope. The bottom of the mesh is at a depth of 20m below the toe of the slope. The mesh is a combination of very fine elements (0.25m x 0.25m with secondary nodes) near the ground surface of the slope to a depth of 3m and fine elements

(0.5m x 0.5m) from 3m depth and below. Head boundaries were applied along the edges of the slope in order to define the pore-water pressure profiles and the initial level of the groundwater table. The slope was assumed to overlie an impermeable rock so that no flow would take place at the bottom (zero flow boundaries at the base of the slope). The precipitation was modeled using flux boundaries along the ground surface. The inclination of the initial water table was assumed to be 12H:1V, while its level was varied as part of the parametric study.

2.3 Material Properties

The slope was assumed to be composed of a homogenous, isotropic, residual soil. The soil characteristics that were adopted were from soil samples of the Jurong formation (from the western part of Singapore) that were taken from one of the instrumented slopes on the Nanyang Technological University campus (Admin. Annex slope). This soil is an orange, residual, silty clay. The soil-water characteristic curve is presented in Figure 2.

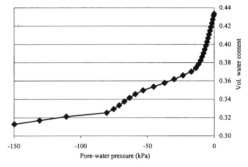

Figure 2. Soil-water characteristic curve for a residual soil from the NTU campus.

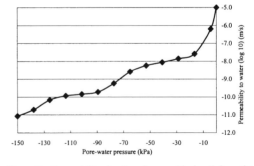

Figure 3. Permeability function for a residual soil from the NTU campus.

The function relating the coefficient of permeability to negative pore-water pressures, which is used for this study, was estimated by Seep/W (using Green and Corey's (1971) equation) from the soil-water characteristic curve. The permeability function that is presented in Figure 3 was estimated assuming a saturated coefficient of permeability of 10^{-5} m/s. The strength parameters that were used in the slope stability analyses were a low cohesion of 1kPa and an angle of friction ϕ' of $25°$ (relating to applied stress) and a ϕ^b angle of $24°$ (relating to suction).

2.4 *Plan of the study*

The objective of the study was to identify the influence of the initial conditions in the slope and also the influence of several rainfall sequences on the developments of the pore-water pressures within a slope. Observations of rainfall induced landslides in Singapore show that major slope failures occur if the daily rainfall exceeds 110mm. Minor failures can even take place when the total daily precipitation is below 50mm if there has been a significant amount of antecedent rainfall.

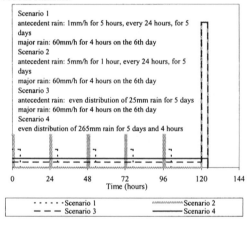

Figure 4. Rainfall scenarios used in the parametric study

In this study four rainfall scenarios have been considered. In all the four cases studied, the same total rainfall was applied but they differed in the intensities and the draining periods. The first three rainfall scenarios were divided into two periods, a period of antecedent rainfall followed by the major rainfall event. The total rainfall prior to the major event amounted to 25mm and was distributed over a period of 5 days. The major event had an intensity of 60mm/h, started on the 6th day and lasted for 4 hours. In the fourth scenario the same total amount of rainfall 265mm, was distributed over the whole 5 days and four hours for comparison. Figure 4 shows the characteristics of all four scenarios. The total duration of each simulation was 336 hours (i.e. 14 days) leaving the slope to drain for 9 additional days after the major rainfall event.

For the groundwater, two initial levels were used $h_w=0$ and 5m, where h_w is the depth of the groundwater table measured below the toe of the slope.

The maximum values of the initial suctions near the ground surface that were used were 25, 50 and 75kPa. The initial pore-water pressures in the unsaturated zone were assumed to be hydrostatic from the water level until they reached the appropriate limiting value (i.e. -25, -50 or -75kPa) and remained constant above that point.

This range of initial suctions has been taken to represent the typical range of suctions in Singapore. Field measurements from the instrumented slopes on the NTU campus, have shown that after a long dry period high suctions exceeding the 70kPa have developed near the ground surface (Lim et al. 1995).

3 PRESENTATION OF THE RESULTS

3.1 *Effect of different distributions of antecedent rain on the pore-water pressures*

The effect of the antecedent rain and the major rain is important mainly in the unsaturated zone from the ground surface until 7m depth. However, the changes in pore-water pressures are most significant in the first 3m and below that depth any changes are small.

Figure 5 presents pore-water pressure profiles at the crest of the slope for the case where $h_w=0$ and initial pore-water pressures were limited to -25kPa. At 0.5m from the surface the pore-water pressures before the major rainfall have increased from -25kPa (initial maximum pore-water pressure) to -22.5kPa in Scenario 1, to -17kPa in Scenario 2, to -13kPa for Scenario 3 and to -4.5kPa for Scenario 4 (Fig. 5a). Although the differences between these values are significant, after the major rainfall event these differences are largely eliminated for the first three scenarios (Fig. 5b). The pore-water pressures increased further to approximately -1kPa for Scenario 1, -0.5kPa for Scenario 2 and 0.5kPa for Scenario 3. In Scenario 4 the pore-water pressures within the first 5m are predicted to be -4kPa. However, in this last case the pore-water pressures reach this value almost two days after the beginning of the simulation.

Although at the end of the rain the pore-water pressures near the ground surface do not differ much in all four scenarios, the situation is different at deeper depths. From a depth of 1m and below in Scenarios 1 and 2 the water did not manage to infiltrate whereas pore-water pressure changes can be seen to 1.5m in Scenario 3 and to 5m in Scenario 4 (Fig. 5b).

Figure 6 presents the pore-water pressure profile, 2 days after the end of the rain. This shows the slower rate of wetting at depths greater than 1m. In Scenarios 1 and 2 pore-water pressures have managed to build up in the zone between 2m and 4m while in the other two cases pore-water pressures have built up from that depth and below, especially in the zone between 4 and 7m. For example, at the end of the major rain in Scenario 3 the pore-water pressures at 4m depth were -22kPa and two days later have built up to -13kPa. In Scenarios 1 and 2 the pore-water pressures at the same depth are respectively -24kPa and -22kPa.

(a)

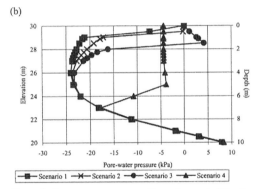

(b)

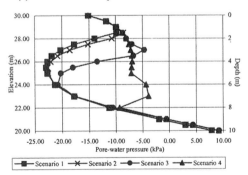

Figure 5. Pore-water pressure profile on the crest of the slope for the case of h_w=0 and initial pore-water pressures -25kPa. (a) at the start of the simulation and before the major rain event and (b) at the end of the major rain.

Figure 6. Pore-water pressure profile 2 days after the major rain on the crest of the slope for the case of h_w=0 and initial pore-water pressures -25kPa.

Scenario 3 also seems to have different effects on the development of the wetting front. Figure 7 shows the developments of the perched water table at the end of the major rain, as a result of the estab-

lished wetting front, for Scenario 1 (Fig. 7a) and Scenario 3 (Fig. 7b) for the case of initial pore-water pressures of-25kPa and for water table h_w=0m.

In Scenarios 1 and 2 the wetting front is about 0.25m deep from the ground surface at the time the major rainfall ends. In the case of Scenario 3 the wetting front moves down as deep as 1.8m.

(a)

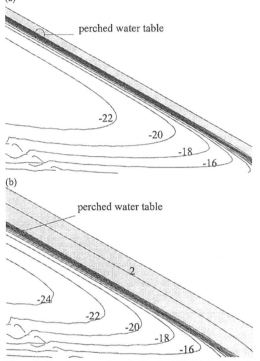

(b)

Figure 7. Development of perched water table (a) for Scenario 1 and (b) for Scenario 3

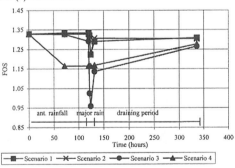

Figure 8. Development of the factor of safety (FOS) of the slope with time for the case of initial suction 25kPa and h_w=0.

Rulon et al. (1985) have described the effect of perched water tables within a slope due to soil layers with large differences in water permeability. Figures 5(b) and 7(b) show the sharp drop in suction across the wetting front, with a zone of positive pore-water pressures above it thus forming a perched water table. This condition leads to instability in the slope. Calculations of factor of safety presented in Figure 8 show that the worst case of the antecedent precipitation is that of Scenario 3 where the factor of safety drops below 1.0 at the end of the major rainfall. In the other three cases the factor of safety did not drop below 1.1. This lower factor of safety in Scenario 3 is consistent with the deeper zone of wetting front shown in Figure. 7b.

3.2 *Effect of the initial conditions within the slope in the development of the pore-water pressures*

3.2.1 *Crest of the slope*

Figure 9 shows the changes in pore-water pressures with time at the crest of the slope for Scenario 3. The effects of different initial suctions and different initial water table depths can be seen in Figure 9. The plots of pore-water pressures at a depth of 0.5m (Fig. 9a) for $h_w=0$ and $h_w=5m$ show that the initial groundwater table has no effect.

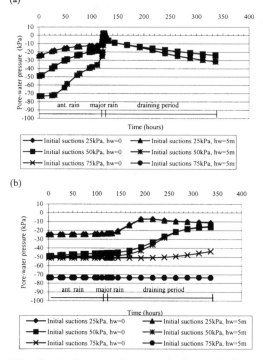

Figure 9. Developments of pore-water pressures with time at the crest of the slope (a) 0.5m deep and (b) 4.0m deep for Scenario 3.

However, at 4m (Fig. 9b) differences in pore-water pressures can be seen. This is particularly true for the case where the limiting suction is 75kPa as, with the higher water table ($h_w=0$), the suctions at 4m are less than the limiting value and start from a value of -51kPa.

The suctions that develop after the period of antecedent rainfall show a tendency to converge. At 0.5m depth (Fig.9a) the suctions at the end of the antecedent rain-period for the case of initial suction of 25kPa become 5.5kPa (80% drop), for the case of initial suction of 50kPa they become 10kPa (80% drop) and for the case of initial suction of 75kPa they become 25kPa (67% drop). At the end of the major event, pore-water pressures at this depth have become positive and have converged into a small range from 0.5kPa to 3kPa, showing that the initial suctions have little effect.

However, this kind of behaviour is only true for the first 2m. Below this, the large difference of pore-water pressures between each case is maintained throughout the entire simulation (Fig. 9b). The difference in pore-water pressures also affects the time for water to infiltrate the soil. The lower the pore-water pressures are the more impermeable the soil becomes (see Fig. 3). This is obvious at 4m depth (Fig. 9b) for initial pore-water pressures of -50kPa. In this case pore-water pressures start to change after the major rain event and continue to increase for 6 days due to the drainage from the upper layers. In the case of 75kPa initial suctions, the antecedent rainfall does not affect the pore-water pressures at this depth (4m), but small changes occur 6 days after the end of the major rain.

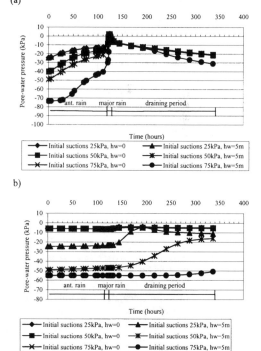

Figure 10: Development of pore-water pressures with time in the middle of the slope (a) 0.5m deep and (b) 4m deep

3.2.2 *Middle of the slope*

In the middle of the slope the influence of the initial suctions in the first 1m is similar to that described above for the crest of the slope. Figure 10 presents the development of pressures with time at depths of 0.5m and 4.0m. It shows that pore-water pressures near the ground surface convert into a small range of values after the end of the major rainfall (Fig. 10a). This range is between 0.5 and 2.5kPa, very similar to that at the crest.

Going deeper in the slope (Fig. 10b) profiles of pore-water pressures for high initial water level (h_w=0) do not change significantly. This is also the case for the lower water level (h_w=5m) with initial suctions of 75kPa. For h_w=5m and for initial suctions of 25kPa and 50kPa, pore-water pressures increase at this depth (4m), but again there is a delay of one day after the major rain event, before they start responding.

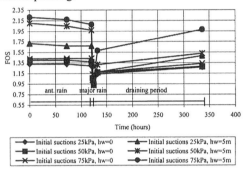

Figure 11. Development of factor of safety with time for Scenario 3.

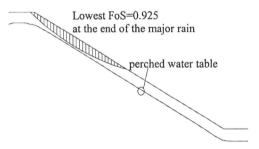

Figure 12. Slip surface with the lowest factor of safety (FoS) for Scenario 3. The entire slip does not go deeper than 1.5m from the ground surface.

Figure 11 shows the calculated minimum factors of safety for Scenario 3. The slope stability analysis shows that no matter how high or low the initial suctions are near the surface, the factor of safety after the major rain is practically the same for all of the cases and is below 1.0. The critical slip surfaces calculated at the end of the major rain are all shallow and do not go deeper than 1.0m below the surface.

Figure 12 presents the critical slip surface with the lowest factor of safety (0.925) as calculated at the end of the major rain for Scenario 3 with initial pore-water pressures of -50kPa near the surface and an initial water level h_w=5m. It can be seen that the failure occurs at shallow depth within the wetted zone.

4 CONCLUSIONS

This parametric study has investigated the effects that different distributions of antecedent rainfall have on the seepage conditions in a typical slope with an inclination of 2H:1V and a height of 10m. The influence of the initial conditions on the seepage has also been investigated. As far as the rainfall patterns are concerned, if 25mm of antecedent rainwater is distributed as a continuous event over 5 days then this has the most significant effect on the pore-water pressures during a major rainfall event (60mm/hr for 4 hours). This case shows a deeper zone of positive pore-water pressures near the surface (a perched water table). Slope stability analyses showed that the critical slip surfaces were only 1m or so deep. At this depth the initial conditions do not play any significant role. This is because, at the end of the major rain event, the pore-water pressures near the ground surface converge into a small range of positive values between 0 and 3kPa whatever the values of the initial suctions have been.

ACKNOWLEDGEMENTS

The National Science and Technology Board, Singapore kindly supported this work (NSTB Grant 17/6/16: Rainfall - Induced Slope Failures).

REFERENCES

Geo-Slope International Ltd. 1998. *SEEP/W for finite element seepage analysis (v.4),* Users Manual, Calgary, Alberta, Canada.

Lim, T. T., Rahardjo, H., Chang, M.F. & Fredlund, D.G. 1995. Effect of Rainfall on Matric Suctions in a Residual Soil Slope. *Canadian Geotechnical Journal* **33**(1996): 618-628.

McDougall, J.R., & Pyrah I.C. 1998. Simulating Transient Infiltration in Unsaturated Soils. *Canadian Geotechnical Journal* **35** (06, 1998): 1093-1100.

Ng, C. W. W., & Shi, Q. 1998. Influence of Rainfall Intensity and Duration on Slope Stability in Unsaturated Soils. *Quarterly Journal of Engineering Geology* 31(1998): 105-113.

Rulon, J.J., & Freeze, R.A. 1985: Multiple Seepage on Layered Slopes and their Implication for Slope - Stability Analysis. *Canadian Geotechnical Journal* 22 (1985): 347-356.

Green, R.E., & Corey, J.C., 1971. Calculation of Hydraulic Conductivity: A further evaluation of some predictive methods. *Soil Science Society of America Proceedings,* **35**: 3-8.

Unsaturated Soils for Asia, Rahardjo, Toll & Leong (eds) © 2000 Taylor & Francis, ISBN 90 5809 139 2

The measurement of matric suction in slopes of unsaturated soil

Z.Wang
Wuhan University of Hydraulic Electric Engineering and Tsinghua University, People's Republic of China

B.W.Gong & C.G.Bao
Yangtze River Scientific Research Institute, Wuhan, People's Republic of China

ABSTRACT: In order to study the variation of suction in natural and fill slopes and compare the efficiency of different measurement methods, the matric suction in situ at the Dagangpo Pump Station was measured with tensiometers, thermal conductivity sensors and filter papers for three months. The results of measurements and related meteorological information are presented.

1 INTRODUCTION

Based on the theory of unsaturated soils, seepage analysis, shear strength and volumetric deformation of unsaturated soil are related to matric suction (Fredlund and Rahardjo,1993). Prediction of the variation of suction with climatic conditions is key to the problem of stability of slopes in expansive soils.

The procedure for measurement of matric suctions was as follows. From 1997/3/17-3/21 soil investigations to determine the distribution of expansive soils around Zaoyang city in North-west of Hubei province were carried out and some representative samples were obtained. Based on the properties of samples, a slope with pipes and a slope forming part of bench canal at Dagangpo pump station were chosen as observation stations. The former is a grassed cut slope with a height of 19m and a slope angle of 24 degrees and the latter is a bare fill slope with a height of 7m and a slope angle of 22 degrees. Some more tests of physical and expansive properties were completed using undisturbed samples taken from both slopes. The two observation stations were established, in the cut slope on 1997/10/7 and in the fill slope on 1997/12/9, respectively. Because the plastic tubes of tensiometers were filled with deaired water and could not work under freezing conditions, both stations, after only 3 months and 1 month, were closed on 1998/1/7.

2 OBSERVATION STATION

The tasks at the observation stations were to measure

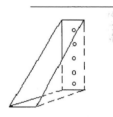

Figure.1 Observation gallery

in situ matric suction with time at various depths under the slope. Two observation galleries were constructed along the cut slope and fill slope, respectively. Fig. 1 illustrates the construction of the gallery. The vertical end face (width 600 mm) of the gallery functioned as an observation plane with access holes (depth 300mm) for installing sensors (tensiometers and thermal conductivities (TC)) at various depths . The observation plane were sealed with a plastic membrane to prevent evaoration and the top of the gallery was covered with asbestos lining . The directions of the two observation planes are shown in Fig.2.

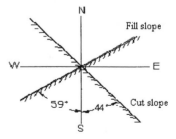

Figure.2 Direction of observation plane

Table 1 Calibration and fitted equations for thermal conductivity sensors .

No.	Calibration equation(kPa)	Fitted equation(kPa)	Location
705	s=2.5mv-7.375	s=5.000 mv²+12.918mv-29.47	cut slope/0.4m
b01	s=0.1413mv²-8.082mv+119.22	s=1.314 mv²+104.83mv-2024.1	cut slope/1.0m
c01	s=0.0134mv²−1.057mv+20.93	s=0.029 mv²-2.400mv+69.20	cut slope/2.0m*
d01		s=33.97 mv²+835.32mv-5041.3	fill slope/1.0m
d02		s=0.1538 mv²+0.209mv+11.709	fill slope/2.0m

* c01 was moved to fill slope/0.4m

The sensors were installed at various depths from the top of the observation plane, which are 0.2, 0.4, 0.7, 1.0, 1.3, 1.6, 2.0, 2.5, 3.0m in the cut slope and 0.4, 0.7, 1.0, 1.6, 2.0m in the fill slope. The TC sensors were installed at depths of 0.4, 1.0, 2.0m in both cut and fill slopes for parallel measurement.

The tensiometers were made by Wuhan University of Hydraulic and Electric Engineering and the TC sensors made by Tsinghua University. Table 1 gives the calibration equations and fitted equations of the TC sensors . The fitted equations were obtained by three points fitted with an observation curve based on matric suctions measured in situ by the tensiometers. When the observation station was constructed and closed, undisturbed samples were taken and matric suction was measured using the filter paper method in the laboratory with an equalization time of 10 days. Calibration equations for Shuang Quan ash free, quantitative filter paper, made in Hangzhou city of China, were obtained from calibrating against a pressure plate.

$Log_{10} s$ (kPa)=5.4628-0.07674w   when w<46

$Log_{10} s$ (kPa)=2.4701-0.01204w   when w>46

3 PROPERTIES OF EXPANSIVE SOILS

The soils around the observation stations mainly consist of brown-yellow, brown-red clays, occasionally mixed with yellow or grey-white thin layers. In places, the soils contain tumor-shaped

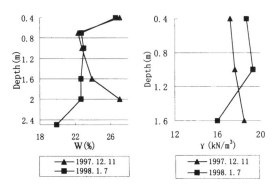

Figure.4. Profile of w and γ in filling slope

calciferous nodules with diameters of 30-150 mm.

The water content (w) and unit weight (γ) profiles at the cut and fill slopes are shown in Fig. 3 and Fig. 4. The w and γ slightly decreased with depth in the cut slope and kept almost constant with depth in the fill slope. The average properties of both slopes are given in Table 2.

Table 2   Average properties of expansive soil

Properties	Cut slope	Filling slope
w (%)	27.0	23.7
γ (kN/m3)	18.0	17.9
Ip *	42.6	24.4
δ ef ** (%)	60.7	40.7

 *   Plastic index

 **   Free swelling

4 RESULTS OF MATRIC SUCTION

4.1 Tensiometer

The time response curves for the tensiometers at both observation stations are shown in Fig. 5 and Fig. 6 respectively. From the curves some observation can be made.

1. The matric suctions near surface of slope were greater and decreased with depth in autumn and winter seasons.

2. When readings of suction exceeded 70-80 kPa,

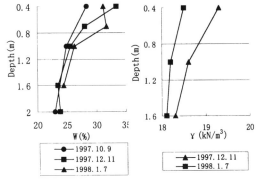

Figure.3.Profile of w and γ in cut slope

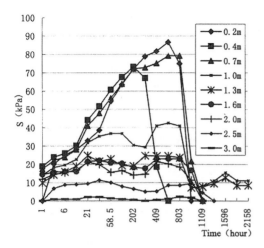

Figure.5.Time response curve (TS) in cut slope

air bubbles appeared in the tube connected the ceramic cup and mercury barometer. Even removal of air bubbles could not resume the readings, for example sensors of 0.2, 0.4, 0.7m.

3. Measurement of suction with tensiometer needs a balance time. From Fig. 5 it can be seen that readings from sensors at 1.3m and below gradually increased and reached their maximum after 20 hours.

4. The results of measurement can be considered reliable.

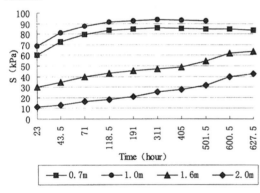

Figure.6.Time response curve(TS) in filling

4.2 Thermal conductivities (TC)

A comparison of the time response curves from the TC and tensiometers (TS) at depths of 0.4, 1.0 and 2.0m are given in Fig.7. In the legends, depths with " \ " and " / " are the results of calibration and fitted equations of TC, respectively. The other is the result from tensiometers (TS). From the comparison of

three depths, it can be seen that the results using the calibration equation for the TC have greater errors.

4.3 Filter paper methods

A comparison of three measuring methods in the cut slope are shown in Fig.8. In the legend TC2 and TC1 are the results of calibration and fitted equations for the TC, respectively. The others are the results of TS and filter paper (FP). From the comparison of the three methods, the results of FP, TS and TC with fitted equation are close to each other. However, TC with calibration equation has the greater deviation.

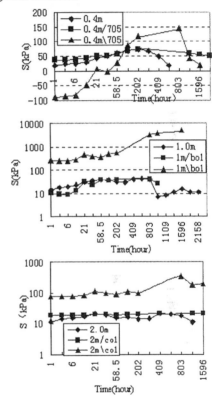

Figure.7. Time response curves of 0.4,1.0 and 2.0m by TS and TC in   cut slope

5 METEOROLOGICAL INFORMATION

Zaoyang has a typical dry climate. The average annual rainfall is less than 1000mm. Table 3 gives the monthly rainfall in 1997, which shows that the rainfall in three months (May, June and July) was more than half of the annual rainfall. The

845

average monthly temperatures are given in Table 3.

Table 4 gives rainfall's day and daily rainfall from September in 1997 (during observation station worked). Because of the observation period is after rain season, the suction in both slopes are not marked influenced by the rainfall.

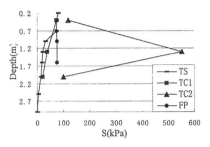

Figure.8. Comparison of three methods

Table 3　Monthly rainfall and average monthly temperature

Month	1	2	3	4	5	6
Rainf. (mm)	8.5	36.7	52.4	27.2	143.9	65.5
Temp. (C)	2.6	2.9	6.6	16.1	20.7	26.9
Month	7	8	9	10	11	12
Rainf. (mm)	202	74.7	44.7	52.7	75.3	11.3
Temp. (C)	29.3	26.2	21.5	15.9	11.6	4.6

Table 4　Rainfall　days and daily rainfall

Rain day	15/9	16/9	3/10	4/10	25/10	10/11	11/11
Rainf. (mm)	16.7	4.7	11.2	1.5	40	10	23

Rain day	12/11	13/11	16/11	24/11	25/11	27/11	28/11
Rainf. (mm)	8.1	13.9	5.0	1.5	3.0	3.4	5.6

Rain. day	29/11	30/11	15/12	22/12	28/12	31/12
Rainf. (mm)	0.5	1.3	0.5	1.2	8.9	0.7

6 CONCLUSIONS

1. Tensiometers gave direct and accurate results, However, the sensors were easy damaged. When suctions exceeded about 80 kPa, the readings gradually decreased, could not be resumed. They cannot work below zero degrees and the response time needed was about 20 hours.

2. Thermal conductivity has the advantage of rapid response, but when the condition of calibration equation is not consistent with in situ observations, the accuracy is low.

3. Filter paper method has the advantage of simplicity of equipment, but great care must be taken in weighing the filter paper, in order to get better accuracy.

4. The galleries could only be used to monitor suction in situ for a short period (no more than 3 months).

5. The matric suction as an independent variable in soil mechanics for unsaturated soils is very difficult to measure, especially in situ.

ACKNOWLEDGEMENTS

The authors wish to thank Mr. An Jiunyoun and Ms. Liu Yianhua et al for their taking part in the tests.

The paper was supported by the Natural Science Foundation of China and Foundation of Canal from South to North by YRSRI, China.

REFERENCES

Fredlund, D.G .& Rahardjo, H. 1993, Soil mecharics for unsaturated soils, New York:John Wiley &Sons,Inc

Lim, T.T., Rahardjo, H., Chang, M.F. & Fredlund, D.G. 1995, Effects of rainfall on matric suction in a residual soil slope. Can. Geotech. J.

Unsaturated Soils for Asia, Rahardjo, Toll & Leong (eds) © 2000 Taylor & Francis, ISBN 90 5809 139 2

Suction measurement for the prediction of slope failure due to rainfall

N. Yagi, R. Yatabe, K. Yokota & N. P. Bhandary
Department of Civil and Environmental Engineering, Ehime University, Matsuyama, Japan

ABSTRACT: Many prediction methods for rainfall induced slope failures have so far been proposed but their application in an effective way is yet to be practiced. Predicting any natural disaster has always been a challenging task to the engineers. Only 10 minutes before the occurrence of a soil disaster will be more than sufficient to save, if not the property, the human life. So as a prediction method for rainfall induced slope failures by critical rainfall criterion, measurement of suction due to unsaturated state of soil mass at a possible failure site was carried out. The data obtained were then utilized to establish initial conditions for the calculation of critical rainfall, which is a key to predict the rainfall-induced slope failure at a particular area.

1 INTRODUCTION

Not only in Japan, but also in many other countries where the number of mountains and man made slopes for engineering works is high, frequent slope failures in rainy seasons cause various problems like damage to civil structures and loss of lives and properties. So only a ten minutes' prediction of such slope failures with at least 50% reliability by some means will be more than sufficient to save, if not the structures, the human lives.

Slopes with unsaturated soils have a greater tendency to failure due to rainfall. As a general mechanism of rainfall-induced slope failures, unsaturated soil mass in a slope gets saturated during or after a continuous rainfall, and the saturation state causes development of a high pore water pressure. Increased pore pressure then results in the reduction of effective stress on the soil particles, which consequently causes a reduction in overall strength of the whole soil mass. This loss of strength by a certain soil layer below some depth in the slope then results in failure or slip of the soil mass above it.

In Soil Mechanics, saturation is a state when all the voids in a soil mass are 100% occupied by water. However, when some of the voids still contain air, the soil mass is referred to as an unsaturated one. In unsaturated state, the pores in soil mass behave as capillary tubes, which hold water by capillary action that creates suction (in other words, negative pore water pressure), which accelerates the entry of water into the soil mass. This fact implies that the

unsaturated soils when penetrated by water develop a sudden reduction in suction before getting saturated. Precise measurement of this suction, if made possible, can be a clue to the prediction of rainfall-induced slope failures.

Although various prediction methods, for example, methods depending on the amount of rainfall, the measurement of displacement, and the acoustic emission have already been tried, their effectiveness has yet to be confirmed. Therefore, efforts have been made to relate critical rainfall method of prediction for rainfall-induced slope failures with development of and drop in suction during seepage of water. For this, field measurements of suction before, during, and after the rainfall were carried out.

2 PREDICTION CRITERION

One of the major factors that help in predicting rainfall-induced slope failure is critical rainfall, which is the amount of rainfall causing collapse of a particular slope. Determination of critical rainfall is governed by many factors such as geology, geography, soil hydrology and hydraulics, soil characteristics, etc. However, at a particular site, all these factors except soil hydraulics remain same.

Soil hydraulics consists of those behaviors of soil that influence the seepage of water, development of suction or pore pressure, water table, etc in the soil mass. During or after a rainfall, all other factors that cause a slope failure remain unaltered except for some, for example, the coefficient of permeability, suction and pore pressure, bulk density, and in some extent the strength. However, the change in strength takes place only after a rapid increase in pore water pressure that occurs just before the failure. So, measurement of suction and establishment of initial conditions for the calculation of critical rainfall may effectively work for the prediction of a rainfall-induced slope failure at a particular site.

3 IN-FIELD SUCTION MEASUREMENT

3.1 *Site Description*

Two spots in the Ehime University agricultural forest area were chosen for in-site suction measurement during rainfall. Soils of both the points consist of Masado that is a remain of heavily weathered granite. The longitudinal sections of both the test spots and the suction measurement arrangements are shown in Figure 1 and Figure 2. The slope angle at spot A is 40^0 and that at spot B is 35°. The base rock at site-A is almost at constant slope with the ground surface, whereas that at site-B goes deepening towards downhill. Moreover, the depth of Masado soil at both the test points is about 1 m. To have an idea about soil condition, void ratios at both the places were determined, which came to be 1.0 and 1.5 at A and B respectively.

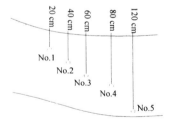

Figure 1. Suction measurement points at site-A.

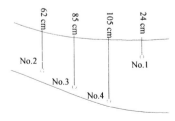

Figure 2. Suction measurement points at site-B.

3.2 *Suction Measurement*

For the measurement of suction in unsaturated soil mass, two arrangements were made. Moisture meters and suction-meters were set at test point A, whereas only the latter were set at B. Suction-meter can directly measure the negative pore water pressure developed in an unsaturated soil mass, whereas the data obtained by moisture meter needs calculation in order to determine the suction. The suction-meter used could measure pressures from 0.5 to 1.0 kgf/cm^2 with a least count of 1.0 gf/cm^2. As seen on Figure 1, points denoted by no. 1, 2, and 3 are the locations of moisture meter and the remaining 4 and 5 are those of suction-meter. The depths of these meters are as shown in the Figure 1 and Figure 2.

Suction measurement at point A was carried out for a total of 25 days in which the rainfall days were just five, while at point B, it was carried out twice, for 20 and 30 days with total of 8 and 13 days' of rainfalls respectively. The suction measurement data were then plotted against the time in days and hours along x-axis.

Figure 3 shows variation in suction with time at all the measurement points of site-A. As seen on the figure, the rainfall started on 13th day, the maximum rainfall of 67.5 mm is on 14th day, and the minimum of 10 mm is on 21st day from the beginning of the measurement. Looking at the variation of suction, it is seen that the maximum value has reached at point-1 (depth 20 cm) on 11th day when the rainfall has yet not started, and for other points, this values goes on decreasing with the increase in depth.

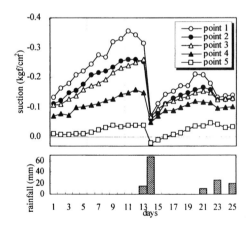

Figure 3. Suction data at site-A.

In about two weeks, the highest increase of about 0.23 kgf/cm^2 in suction has occurred at point-1 (20 cm), whereas the lowest of just 0.03 kgf/cm^2 has

848

occurred at point 5 (120 cm), and it is clear from the above figure that the rise in suction is decreasing with the increase in depth. A sudden drop in suction at all the measurement points can be seen on 14th day, where the suction at point 5 (120 cm) has dropped even to a negative value (pore pressure), and also there are slight drops in the values at all the points from 21st to 23rd day. Similarly, a slight drop in suction at point-1 can be seen on 13th day, whereas all other points show no drop; it is due to faster seepage of rainfall water up to 20 cm. These results clarify that the suction in an unsaturated soil considerably drops when the soil starts getting saturated, and the amount of drop depends on the suction developed and the depth from the ground surface.

non-rainy season. It is due to the reason that the moisture content of soil mass in rainy season is higher than that in non-rainy season, which causes a higher degree of saturation of soil in rainy season and a low suction development.

The values of suction developed at different depths of sites A and B, as plotted against rainfall days in above figures, are the lowest ones of the days. So to observe the hourly suction variation at all the points, suction data were also taken on the day of highest rainfall in one-hour interval. For site-A, the hourly drop in suction due to rainfall on 14th day is shown in Figure 6, and for site-B, that on 2nd day of rainy season is shown in Figure 7.

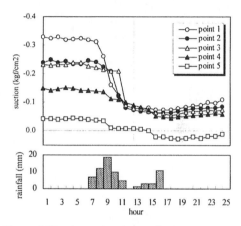

Figure 6. Hourly suction data at site-A.

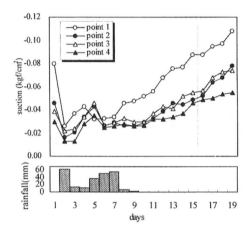

Figure 4. Suction data at site-B in rainy season.

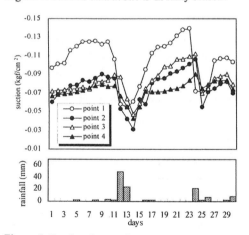

Figure 5. Suction data at site-B in non-rainy season.

It can also be seen that the suctions developed during rainy season are more than those during

As seen on Figure 6, the suctions at all the points have started dropping as soon as the rainfall started, and it is interesting to note that the drops at points 4 and 5 (depths 80 and 120 cm respectively) have started within three hours after the rainfall. Theoretically, the drops at points 1, 2, 3 should start little earlier than those at 4 and 5, so the earlier drop in suctions at latter points may be due to trapping of air in the voids. It is also seen that the suction at point-5 dropped to a negative value (pore water pressure) after about 7 hours of rainfall when the amount of rainfall is less compared to that at earlier hours. It may be due to impervious layer of base rock just beneath the measurement point.

Similarly in Figure 7, the hourly drop pattern of suction at site-B does not seem different from that at site-A. However, there is no point that showed negative suction (pore water pressure) in this site. It may be due to deepening rock bed that made an easy downhill flow of seepage water, which did not have much effect in developing positive pressure, as was in site-A.

Looking at these two figures (6 and 7), it can also be said that the lowest value of suction has, in either case, reached not immediately but some hours after the peak rainfall. It reveals that the pore water pressure in a soil mass starts increasing only after the saturation of the same, and the probability of slope failure is high at a time when the drop in suction is abrupt, and it has occurred only after rather than during the peak rainfall. However, if the duration of rainfall is longer, this statement may not be valid.

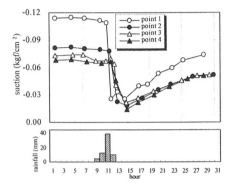

Figure 7. Hourly suction data at site-B.

4 ANALYSIS OF SUCTION DATA

The suction measurement data obtained were analyzed for the variation of suction and pore water pressure with the rainfall intensity and the number of no-rainfall days. Figure 8 and Figure 9 below show the drops in suction at different points of site-A and site-B, respectively, plotted against the amount of rainfall on the respective day.

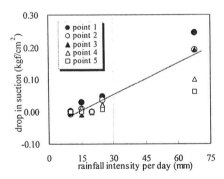

Figure 8. Drop in suction at site-A.

Although the drop in suction at a particular point may depend on various other factors, such as, number of no-rainfall days, moisture condition, soil type, etc., the results obtained as per the suction

measurement in this study have been interpreted as higher drop in suction with increased amount of rainfall. Both the figures (i.e., results of site-A and site-B) show a similar pattern of drop in suction; however, the amount of drop is not similar for the points of lower depths.

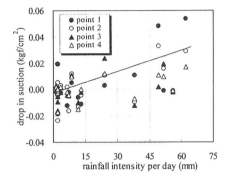

Figure 9. Drop in suction at site-B.

Likewise, the amounts of suction developed during no-rainfall days due to unsaturated state at both the sites are shown in Figures 10 and 11, plotted against the number of dry days.

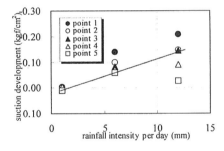

Figure 10. Suction development at site-A.

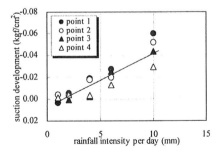

Figure 11. Suction development at site-B.

As seen on the figures, the development of suction rises with the increase in number of dry days (no rainfall). However, the rise in amount of suction developed is less at deeper points.

5 CONCLUSION

Determination of critical rainfall, which is the amount of rainfall that is just sufficient to cause a slope failure, for a particular slope is a key to have an idea about the possibility of failure. Failure of any slope can be predicted with a sufficient time margin if critical amount of rainfall for the slope soil is pre-estimated. However, estimation of the critical rainfall involves some calculations that require establishment of initial conditions such as suction value. So establishment of a relationship between suction and number of no-rainfall days may provide the required initial conditions for the determination of critical rainfall, which then can be used in predicting slope failure due to rainfall. For the purpose in-field suction measurements were conducted, whose results can be summarized at following points.

1. Use of two types of suction measuring devices resulted in a higher difference of suction values, especially at shallower points, although the ground as well as soil condition at both the sites were similar.

2. Development of suction takes place more quickly at shallower points than at deeper ones, and the value of suction developed is low at deeper and high at shallower points. It is due to quicker loss of moisture near the ground surface. Similarly, the drop in suction due to rainfall is more at the points near to the ground surface

3. Hourly suction measurement resulted in a time-lag between peak rainfall and drop in suction, which means the probability of failure is high not during but after the peak rainfall; however, if the rainfall continues for a time longer than the time-lag, this statement is no more valid.

4. Drop pattern in suction behaved somewhat linearly with the amount of rainfall on a particular day, and suction development did so with the number of no-rainfall days.

REFERENCES

Yagi, N., Yatabe, R. & Yamamoto, K. 1983. Slope failure mechanism due to seepage of rain water. *Proceedings of 7th ARCSMFE, Vol.1, Pp. 382-386.* Haifa.

Yatabe, R. 1986. Prediction and mechanism of slope failure in a *masado* soil mass due to rainfall. *Research work.*

Author index

Printed and bound by CPI Group (UK) Ltd, Croydon, CR0 4YY

28/10/2024

01780323-0001